Lecture Notes in Computer Science 7751

Commenced Publication in 1973
Founding and Former Series Editors:
Gerhard Goos, Juris Hartmanis, and Jan van Leeuwen

Jian Yang   Fang Fang   Changyin Sun (Eds.)

Intelligent Science and Intelligent Data Engineering

Third Sino-foreign-interchange Workshop, IScIDE 2012
Nanjing, China, October 15-17, 2012
Revised Selected Papers

 Springer

Volume Editors

Jian Yang
Nanjing University of Science and Technology
School of Computer Science and Technology
Nanjing 210094, Jiangsu Province, China
E-mail: csjyang@njust.edu.cn

Fang Fang
Peking University
Department of Psychology
Beijing 100871, China
E-mail: ffang@pku.edu.cn

Changyin Sun
Southeast University
School of Automation
Nanjing 210096, Jiangsu Province, China
E-mail: cysun@seu.edu.cn

ISSN 0302-9743   e-ISSN 1611-3349
ISBN 978-3-642-36668-0   e-ISBN 978-3-642-36669-7
DOI 10.1007/978-3-642-36669-7
Springer Heidelberg Dordrecht London New York

Library of Congress Control Number: 2013931226

CR Subject Classification (1998): I.4.8, I.4, I.5.4, I.5, I.3, I.2.6, I.2.7, I.2.10, H.5.1, H.2.8, F.2.1-2

LNCS Sublibrary: SL 6 – Image Processing, Computer Vision, Pattern Recognition, and Graphics

Typesetting: Camera-ready by author, data conversion by Scientific Publishing Services, Chennai, India

Printed on acid-free paper

Springer is part of Springer Science+Business Media (www.springer.com)

Preface

IScIDE 2012, the Sino-foreign-interchange workshop on Intelligence Science and Intelligent Data Engineering, took place in Nanjing, China, October 15–17, 2012. As one of the annual events organized by the Chinese Golden Triangle ISIS (Information Science and Intelligence Science) Forum, this workshop was scheduled as the third in a series of annual workshops that promote the academic exchange of research results on various areas of ISIS in China and abroad. In response to the call for papers, a total of 429 papers were submitted from 13 countries and regions, of which 105 were accepted (giving an acceptance rate of about 24.5%), including 19 oral papers and 30 spotlight papers. We would like to thank all the reviewers for spending their precious time on reviewing papers and for providing valuable comments that aided significantly in the paper selection process.

We would like to express special thanks to the conference General Chair, Lei Xu, for his leadership, advice, and help on crucial matters concerning the conference. We would like to thank all Steering Committee members, Program Committee members, Invited Speakers' Committee members, Organization Committee members, and Publication Committee members for their hard work. We would like to thank Jurgen Schmidhuber and Zong-Ben Xu for delivering the keynote speeches, and Baoquan Chen, James Kwok, Alessandro Giua, Zhouchen Lin, Vincent Tseng, Huafu Chen, Xiaochun Cao, Sayan Mukherjee, and Xinbo Gao for delivering the invited talks and sharing their insightful views on ISIS research issues. Finally, we would like to thank all the authors of the submitted papers, whether accepted or not, for their contribution to the high quality of this workshop. We count on your continued support of the ISIS community in the future.

October 2012

Jian Yang
Fang Fang
Changyin Sun

Organization

General Chair

Lei Xu   Chinese University of Hong Kong, Hong Kong, China

Program Committee Chairs

Jian Yang   Nanjing University of Science and Technology,
        Nanjing, China
Fang Fang   Peking University, Beijing, China
Changyin Sun   Southeast University, Nanjing, China

Program Committee Members

Akira Hirose	The University of Tokyo, Japan
Ikeda Kazushi	Nara Advanced Institute of Science and Technology, Japan
Kalviainen Heikki	Lappeenranta University of Technology, Finland
Fiori Simone	Università Politecnica delle Marche, Italy
Cheng Yuan Liou	National Taiwan University, Taiwan
Shu-Heng Chen	National Chengchi University, Taiwan
Dacheng Tao	University of Technology, Australia
Hujun Yin	The University of Manchester, UK
Karl Ricanek	UNC Wilmington, USA
Shuicheng Yan	National University of Singapore, Singapore
Yufei Tao	Chinese University of Hong Kong, Hong Kong, China
Wai-Kiang Yeap	Auckland University of Technology, New Zealand
Andrey S. Krylov	Lomonosov Moscow State University, Russia
Seungjin Choi	Pohang Univ. of Science and Tech., South Korea
Kenji Fukumizu	The Institute of Statistical Mathematics, Japan
Kazushi Ikeda	Nara Institute of Science and Technology, Japan
Minho Lee	Kyungpook National University, South Korea
Hava Siegelmann	University of Massachusetts, USA
Vikas Sindhwani	IBM T.J. Watson Research Center, USA
Cartic Ramakrishnan	USC/Information Science Institute, USA
Chengjun Liu	New Jersey Institute of Technology, USA
Jieping Ye	Arizona State University, USA
Jufu Feng	Peking University, Beijing, China
Kun Zhou	Zhejiang University, Hangzhou, China
Zhouchen Lin	Peking University, Beijing, China
Wenlian Lu	Fudan University, Shanghai, China

Xuegong Zhang	Tsinghua University, Beijing, China
Xuelong Li	Xi'an Optics and Fine Mechanics, Chinese Academy of Sciences, Xi'an, China
Yuanqing Li	South China University of Technology, Guangzhou, China
Lei Zhang	Hong Kong Polytechnic University, Hong Kong, China
Junzhou Huang	University of Texas at Arlington, USA
Yishi Wang	UNC Wilmington, USA
Dong Xu	Nanyang Technological University, Singapore
Liang Wang	Institute of Automation, Chinese Academy of Sciences, Beijing, China
Shiguang Shan	Institute of Computing Tech., Chinese Academy of Sciences, Beijing, China
Baojiang Zhong	Suzhou University, Suzhou, China
BaoLiang Lu	Shanghai Jiao Tong University, Shanghai, China
Changshui Zhang	Tsinghua University, Beijing, China
Changyin Sun	Southeast University, Nanjing, China
Daoqiang Zhang	Nanjing University of Aeronautics and Astronautics, Nanjing, China
Dewen Hu	National University of Defence Technology, Changsha, China
Fang Fang	Peking University, Beijing, China
Gang Pan	Zhejiang University, Hangzhou, China
Hua Huang	Beijing Institute of Technology, Beijing, China
Huchuan Lu	Dalian University of Technology, Dalian, China
Jian Yang	Nanjing University of Science and Technology, Nanjing, China
Jinhui Tang	Nanjing University of Science and Technology, Nanjing, China
Pingkun Yan	Xi'an Institute of Optics and Precision Mechanics, Xi'an, China
QingHua Hu	Har'bin Institute of Technology, Har'bin, China
Qingshan Liu	Nanjing University of Information Science and Technology, Nanjing, China
Sheng Li	Peking University, Beijing, China
Shengyong Chen	Zhejiang University of Technology, Hangzhou, China
Shutao Li	Hunan University, Changsha, China
Si Wu	Beijing Normal University, Beijing, China
Wenming Zheng	Southeast University, Nanjing, China
Xianglong Tang	Har'bin Institute of Technology, Har'bin, China
Xiang Li	Fudan University, Shanghai, China
Xiaofei He	Zhejiang University, Hangzhou, China
Xiaoyang Tan	Nanjing University of Aeronautics and Astronautics, Nanjing, China

Xiaoyuan Jing	Nanjing University of Posts and Telecommunications, Nanjing, China
Xihong Wu	Peking University, Beijing, China
Xinbo Gao	Xidian University, Xi'an, China
Xuesong Wang	China University of Mining and Technology, Xuzhou, China
Yanwei Pang	Tianjin University, Tianjin, China
Yiguang Liu	Sichuang University, Chengdu, China
Yizhou Wang	Peking University, Beijing, China
Yong Xu	Har'bin Institute of Technology Shenzhen Graduate School, Shenzheng, China
Yuan Yao	Peking University, Beijing, China
Yuan Yuan	Xi'an Optics and Fine Mechanics, Chinese Academy of Sciences, Xi'an, China
Yuhua Qian	Shanxi University, Taiyuan, China
Zhihua Zhou	Nanjing University, Nanjing, China
Zhong Jin	Nanjing University of Science and Technology, Nanjing, China
Liwei Wang	Peking University, Beijing, China
Qingshan Liu	Southeast University, Nanjing, China
Wangmeng Zuo	Har'bin Institute of Technology, Har'bin, China
WeiShi Zheng	Sun Yat-sen University, Guangzhou, China
Xin Geng	Southeast University, Nanjing, China
Jianjiang Feng	Tsinghua University, Beijing, China
Qi Wang	University of Technology and Science of China, Hefei, China
Xiaoqiang Lu	Dalian University of Technology, Dalian, China
Yi Tang	Hubei University, Wuhan, China

Invited Speakers' Committee Chairs

Xiaofei He	Zhejiang University, China
Xinbo Gao	Xidian University, China

Organizing Committee Chairs

Jianfeng Lu	Nanjing University of Science and Technology, China
Wankou Yang	Southeast University, China

Publication Committee Chairs

Yanning Zhang	Northwestern Polytechnical University, China
Xuelei Hu	Nanjing University of Science and Technology, China

Table of Contents

Nonnegative Matrix Factorization on Orthogonal Subspace with Smoothed L0 Norm Constrained

Jun Ye[1,2,*] and Zhong Jin[1]

[1] School of Computer Science & Technology, Nanjing University of Science and Technology,
Nanjing, 210094, China
[2] School of Natural Sciences, Nanjing University of Posts & Telecommunications,
Nanjing, 210003, China

Abstract. It is known that the sparseness of the factor matrices by Nonnegative Matrix Factorization can influence the clustering performance. In order to improve the ability of the sparse representations of the NMF, we proposed the new algorithm for Nonnegatie Matrix Factorization, coined nonnegative matrix factorization on orthogonal subspace with smoothed L0 norm constrained, in which the generation of orthogonal factor matrices with smoothed L0 norm constrained are the parts of objective function minimization. Also we develop simple multiplicative updates for our proposed method. Experiment on three real-world databases (Iris, UCI, ORL) show that our proposed method can achieve the best or close to the best in clustering and in the way of the sparse representation than other methods.

Keywords: NMF, Orthogonality, Clustering, Sparse representation, L0 norm.

1 Introduction

Nonnegative Matrix Factorization (NMF) is a recent method for finding a nonnegative decomposition of the original data matrix. Given an input data matrix V, each column of which represents a sample, NMF produces two factor matrices W and H using low-rank approximation such that $V \approx WH$. Each column of W represents a base vector, and each column of H describes how these base vectors are combined fractionally to form the corresponding sample in V. All entries in matrices are required to be nonnegative. Nonnegativity enables a non-subtractive combination of parts to form a whole, and make the encoding of data easier to interpret [1]. So, NMF is useful for learning parts-based representation and can be able to generate sparse representations of data. This caused the NMF method have been widely used in many applications, such as data mining, pattern recognition.

However, NMF cannot always guarantee an intuitive sparse representations of data. The parts-based representation of some facial images datasets reported in literature [2] was global rather than local. Later a multitude of variants have been proposed to improve NMF. A notable stream of efforts attaches various regularization

J. Yang, F. Fang, and C. Sun (Eds.): IScIDE 2012, LNCS 7751, pp. 1–7, 2013.

terms to the original NMF objective to enforce higher sparseness [3]. Recently, it has been shown that the orthogonality constraint on factor matrices can enhance the sparseness. Taking data clustering for example, it is conducted on the clustering of the columns of the input data matrix, and indicated by the matrix H, orthogonality on each row of H can improve clustering accuracy. It was proved that orthogonal NMF is equivalent to k-means clustering [4]. So how to enhance the ability of the sparseness representation of the data can be an important issue? Ding et al. [5] proposed Orthogonal Nonnegative Matrix Factorization (ONMF) firstly which orthogonality is achieved by solving an optimization problem with orthogonality constraints. However, their method requires an intensive computation, which is very expensive for clustering task. Zhao Li et al. [6] considered the deficiency of the computational complexity, they proposed a new method called NMF on Orthogonal Subspace (NMFOS), in which orthogonality constraint on one of the factor matrices is embedded as part of the objective function optimization. Thus, orthogonality is achieved through the process of factorization instead of using additional constraints.

To obtain the sparsest of the factor matrices, its essence is searching for a solution with minimal L0 norm for matrices, i.e., minimum number of nonzero components of W and H. It is stated that searching the minimum L0 norm is an intractable problem as the dimension increases (because it requires a combinatorial search), and it is too sensitive to noise (because any small amount of noise completely changes the L0 norm of a vector) [7]. Consequently, researchers consider other approaches. Zuyuan Yang et al.[8] introduced smoothed L0 norm constraints to the original NMF, denoted NMF-SL0, to enhance the ability of the sparseness.

In this paper, as the smoothed L0 norm of the factor matrices can reflect the sparseness intuitively and it is easy to be optimized, we consider NMF on orthogonal subspace with smoothed L0 norm constraints, called smoothed L0 norm constrained nonnegative matrix factorization on orthogonal subspace (NMFOS-SL0), and its application to the task of clustering, where an orthogonality constraint and the smoothed L0 norm constraints are imposed on the nonnegative decomposition of an inputing data matrix. We develop new multiplicative updates for NMFOS-SL0. Experiments on three different datasets show our method perform better in clustering task, and sparseness of the factor matrices, compared to other methods.

The rest of this paper is organized as follows. In Section 2, NMF and NMFOS are presented. Section 3 describes the NMFOS-SL0, and give new multiplicative updates for it. Simulations using three real databases are presented in Section 4. Finally, conclusions are summarized in Section 5.

2   Related Work

2.1   Standard NMF

Consider a data matrix $V = [v_1, v_2, \cdots, v_n] \in R^{m \times n}$, each column of which consists of m features, and represents a sample such as a text focument, or a face image. NMF aims to find two nonnegative matrices $W \in R^{m \times r}$, $H \in R^{r \times n}$, $r \ll \min\{m,n\}$, such that $V \approx WH$. So the objective optimization problem can be concluded:

$$\min_{W,H} \|V - WH\|_2^2 \quad s.t. \ W,H \geq 0 \tag{1}$$

The multiplicative update rules were investigated by Lee et al. [1], as follows:

$$a) \quad W \leftarrow W \otimes \frac{VH^T}{WHH^T}; \quad b) \quad H \leftarrow H \otimes \frac{W^T V}{W^T WH} \tag{2}$$

where $\otimes$ denote elementwise multiplication.

2.2   Nonnegative Matrix Factorization on Orthogonal Subspace (NMFOS)

The important application of NMF is the parts-based representation. But NMF cannot always guarantee the intuitive parts-based representation which is conducted on the clustering of the rows of the input data matrix, and is described by the factor matrix W. Local representation requires the base vectors of the factor matrix H, which represent the parts of data, to be distinct from each other. Local representation is related to orthogonality, the more orthogonal between the base vectors, the more distinct between the parts[6]. Ding et al [5] in order to enhance the orthogonality of the factor matrices, they introduced the orthogonality constraints to the original NMF. So the ONMF can be described as follows:

$$\min_{W,H \geq 0} \|V - WH\|_2^2 \ s.t. \ W^T W = I \ , \qquad \min_{W,H \geq 0} \|V - WH\|_2^2 \ s.t. \ H^T H = I \tag{3}$$

Considering the deficiency of the computational complexity, Zhao Li et al [6] proposed the method of NMFOS, in that orthogonality constraint on one of the factor matrices is embedded as part of the objective function optimization. And it was described as follows:

$$\min_{W,H \geq 0} \|V - WH\|_2^2 + \lambda \|W^T W - I\|_2^2 \ , \qquad \min_{W,H \geq 0} \|V - WH\|_2^2 + \lambda \|HH^T - I\|_2^2 \tag{4}$$

Also Zhao Li et al [6] developed the multiplicative update rules for problem (4) and could be find from the literature [6].

3   Related Work Smoothed L0 Norm Constrained Nonnegative Matrix Factorization on Orthogonal Subspace (NMFOS-SL0)

3.1   Smoothed L0 Norm

Let the function $f_\sigma(s) = \exp\left(-\dfrac{|s|^2}{2\sigma^2}\right)$, then $\lim\limits_{\sigma \to 0} f_\sigma(s) = \begin{cases} 1, if \ s = 0 \\ 0, if \ s \neq 0 \end{cases}$, where σ is a

positive constant and s is a variable. Let the function $J_W = m \times r - \sum\limits_{s=1}^{m} \sum\limits_{t=1}^{r} f_\sigma(w_{st})$ as measurement for the matrix W. It is obviously that when $\sigma \to 0$, $J_W \to \|W\|_0$. Therefore, J_W is called the smoothed L0 norm [9]. In a similar way, we can define the smoothed L0 norm for the matrix H as: $J_H = r \times n - \sum_{t=1}^{r} \sum_{u=1}^{n} f_\sigma (h_{tu})$.

3.2   NMFOS-SL0

As the smoothed L0 norm of the factorization matrices can reflect the sparseness intuitively and it is easy to be optimized, we consider NMF on orthogonal subspace with smoothed L0 norm constraints, called NMFOS-SL0. We introduce the measure functions J_W and J_H i.e. the smoothed L0 norm constraint with the factorization matrices to the NMFOS's objective function. And get the new problems as follows:

$$\min_{W, H \geq 0} : F_W = \|V - WH\|_2^2 + \lambda \|W^T W - I\|_2^2 + \alpha_W J_W \tag{5}$$

$$\min_{W, H \geq 0} : F_H = \|V - WH\|_2^2 + \lambda \|HH^T - I\|_2^2 + \alpha_H J_H \tag{6}$$

The partial derivatives of F_W in (5) with respect to W and H are as follows:

$$\begin{cases} \dfrac{\partial F_W}{\partial W} = -VH^T - 2\lambda W + WHH^T + 2\lambda WW^T W - \dfrac{\alpha_W}{\sigma^2} W \otimes exp\left(-\dfrac{W \otimes W}{2\sigma^2}\right) \\ \dfrac{\partial F_W}{\partial H} = -W^T V + W^T WH \end{cases} \tag{7}$$

where the parameter α_W is selected according to the following exponential rule: $\alpha_W = \beta_W \exp(-\tau_W k)$, k is the iteration number, β_W and τ_W are constants [10].

In order to constrain W and H to be nonnegative, let $\xi_H = \dfrac{H}{W^T WH}$,

$\xi_W = \dfrac{W}{WHH^T + 2\lambda WW^T W - \dfrac{\alpha_W}{\sigma^2} W \otimes exp\left(-\dfrac{W \otimes W}{2\sigma^2}\right)}$, Then, based on the widely

used alternate-least-squares multiplication updating rules, Substitute ξ_W and ξ_H to

$W = W - \xi_W \dfrac{\partial F_W}{\partial W}$ and $H = H - \xi_H \dfrac{\partial F_H}{\partial H}$ respectively. We can give the

multiplicative update rules of W and H for problem (5) as follows:

$$a)\, W \leftarrow W \otimes \frac{VH^T + 2\lambda W}{WHH^T + 2\lambda WW^TW - \frac{\alpha_w}{\sigma^2}W \otimes exp\left(-\frac{W \otimes W}{2\sigma^2}\right)}\, ;\, b)\, H \leftarrow H \otimes \frac{W^TV}{W^TWH} \qquad (8)$$

In similar way , we get the multiplicative update rules for problem (6) as follows:

$$c)\, W \leftarrow W \otimes \frac{VH^T}{WHH^T}\, ;\quad d)\, H \leftarrow H \otimes \frac{W^TV + 2\lambda H}{W^TWH + 2\lambda HH^TH - \frac{\alpha_H}{\sigma^2}H \otimes exp\left(-\frac{H \otimes H}{2\sigma^2}\right)} \qquad (9)$$

Based on the analysis above, the NMFOS-SL0 algorithm can be concluded as :

Step1: Initialization: input the nonnegative matrix V , and give the initial nonnegative matrices of W , H randomly . And set the parameters $\lambda, \sigma, \beta_i, \tau_i, i = W, H$, respectively;

Step2: Updating: update W , H using (a) and (b) (or (c) and (d)) respectively;

Step3: Stopping: if a stopping criterion is satisfied, the algorithm stops; otherwise, go to step 2.

4   Experiments

We tested our proposed method on three databases[11]. And we do the clustering and part-based representation experiment on these database. For comparisons, three other algorithms have been chosen: NMF [1], ONMF [5], and NMFOS [6].

Iris, a data set that contains 150 instances of four positive-valued attributes. The samples belong to three iris classes, each including 50 instances. This dataset is selected mainly for comparison with the following larger scale datasets.

• Digit, a subset containing "0," "2," "4," and "6" selected from UCI optical handwritten digit database. There are 2237 samples of 62 nonnegative integer attributes. This dataset is used to demonstrate the method behavior when samples are much more than attributes.

• ORL Database, a set of face images at different times, varying the lighting, facial expressions and facial details. There are 400 grayscale images from 40 distinct subjects and of size 92*112. This data set is used to study the case where the dimensionality is much higher than the number of samples.

A.   Clustering

Clustering is an important application of NMF and its variants. We have adopted two measurements, purity and entropy, which are widely used in nonnegative learning literature, for comparing clustering results. Suppose there is ground truth data that labels the samples by one of classes. Purity is given by $purity = \frac{1}{n}\sum_{k=1}^{r} \max_{1\le l \le q} n_k^l$, where n_k^l is the number of samples in the cluster k that belong to original class l .

A larger purity value indicates better clustering performance. Entropy measures how classes are distributed on various clusters. The entropy of the entire clustering solution is given by $entropy = -\dfrac{1}{n \log_2 q} \sum_{k=1}^{r} \sum_{l=1}^{q} n_k^l \log_2 \dfrac{n_k^l}{n_k}$, where $n_k = \sum_l n_k^l$.

Generally, a smaller entropy value corresponds to a better clustering quality. We set $r = q$ and repeated each algorithm on each data set 100 times with different random seeds for initialization. In our method, we set parameter $\lambda = 5$ and $\beta_i, \tau_i, i = W, H$ equal to 100 and 0.01, respectively. The parameter $\sigma \in [1, 0.5, 0.2, 0.1, 0.05, 0.02, 0.01]$ that we chosed following the literature [9] and then the best result was reported.The mean and standard deviation of the purities and entropies of each algorithm data set pair are shown in Table 1 and 2 , respectively. From the Table 1, 2, we can see that the NMFOS-SL0 algorithm improves clustering accuracy for Iris and ORL datasets, and on the Digit database our proposed method performs very closely to the best.

Table 1. Clustering performance of purity comparison on three database (mean± deviation)

Database	NMF[11]	ONMF[11]	NMFOS	NMFOS-SL0
Iris	0.78±0.05	0.85±0.03	0.86±0.02	**0.88±0.02**
Digit	**0.98±0.00**	**0.98±0.00**	0.95±0.01	0.96±0.01
ORL	0.47±0.03	0.72±0.02	0.74±0.03	**0.78±0.02**

Table 2. Clustering performance of entropy comparison on three database (mean± deviation)

Database	NMF[11]	ONMF[11]	NMFOS	NMFOS-SL0
Iris	0.42±0.08	0.30±0.05	0.26±0.04	**0.24±0.03**
Digit	**0.08±0.00**	**0.08±0.00**	0.16±0.01	0.14±0.01
ORL	0.34±0.02	0.17±0.01	0.16±0.01	**0.15±0.02**

B. Part-Based Representation

In order to study the sparseness ability of parts-based representation of proposed NMFOS-SL0 method. We introduce the sparseness according to the Hoyer[3], comparing the base sparseness ability with those learned by NMF, ONMF and the NMFOS on ORL database. And the sparseness was defined as:

$$sparseness(X) = \left(\sqrt{n} - \left(\sum_i |x_i| \right) \middle/ \sqrt{\sum x_i^2} \right) \middle/ \sqrt{n} - 1 \tag{10}$$

where n is the dimensionality of vector X. Table 3 shows the average sparseness of the columns in the learned basis by NMF, ONMF, NMFOS and NMFOS-SL0. It can be seen that the sparseness of the factorization matrices that used the method of NMFOS and NMFOS-SL0 are sparser than NMF and ONMF's.

Table 3. The sparseness of the factories matrices comparison

Hoyer's method	NMF	ONMF	NMFOS	NMFOS-SL0
W	0.38	0.58	0.69	**0.75**
H	0.34	0.49	0.60	**0.62**

5 Conclusion

In this work, a new NMF method based on the orthogonal subspace with smoothed L0 norm constrained is proposed, which can enhance the sparseness of the factor matrices. And we develop the multiplicative updates for the new method. This method introduce the additional parameters that balance the sparseness and reconstruction. However, the selection of the parameter value usually relies on exhaustive methods, which hinders their application. In future work, more efficient learning algorithms will be exploited.

References

1. Lee, D.D., et al.: Algorithms for non-negative matrix factorization. In: Advances in Neural Information Processing (Proc. NIPS), vol. 13, pp. 556–562 (2000)
2. Li, S., Hou, X., Zhang, H., Cheng, Q.: Learning spatially localized, parts-based representation. In: IEEE Comput. Soc. Conf. Comput. Vision Pattern Recognition, vol. 1, pp. 207–212 (2001)
3. Hoyer, P.O.: Nonnegative Matrix Factorization with Sparseness Constraints. J. Machine Learning Research 5, 1457–1469 (2004)
4. Yang, Z., Laaksonen, J.: Multiplicative updates for non-negative projections. Neurocomputing 71(1-3), 363–373 (2007)
5. Ding, C., Li, T., Peng, W., Park, H.: Orthogonal nonnegative matrix trifactorizations for clustering. In: KDD 2006: Proceedings of the 12th ACM SIGKDD International Conference on Knowledge Discovery and Data Mining, pp. 126–135. ACM, New York (2006)
6. Li, Z., Wu, X., Peng, H.: Nonnegative Matrix Factorization on Orthogonal Subspace. Pattern Recognition Letters 31, 905–911 (2010)
7. Donoho, D.L., Elad, M., Temlyakov, V.: Stable recovery of sparse overcomplete representations in the presence of noise. IEEE Trans. Info. Theory 52(1), 6–18 (2006)
8. Yang, Z., Chen, X., Zhou, G., Xie, S.: Spectral unmixing using nonnegative matrix factorization with smoothed L0 norm constraint. In: Proceedings of SPIE, vol. 7494 (2009)
9. Hosen Mohimani, G., Babaie-Zadeh, M., Jutten, C.: A fast approach for overcomplete sparse decomposition based on smoothed l0 norm. IEEE Transactions on Signal Processing 57(1), 289–301 (2009)
10. Zdunek, R., Cichocki, A.: Nonnegative matrix factorization with constrained second order optimization. Signal Processing 87, 1904–1916 (2007)
11. Yang, Z., Oja, E.: Linear and Nonlinear Projective Nonnegative Matrix Factorization. IEEE Trans. Neural. Networks 21(5), 734–747 (2010)

A Multiphase Entropy-Based Level Set Algorithm for MR Breast Image Segmentation Using Lattice Boltzmann Model

Souleymane Balla-Arabé and Xinbo Gao

School of Electronic Engineering, Xidian University, Xi'an 710071, China
balla_arabe_souleymane@ieee.org, xbgao@mail.xidian.edu.cn

Abstract. Recently, with the development of high dimensional large-scale medical imaging devices, the need of fast and accurate segmentation methods is increasing. In this paper, we propose a new variational multiphase level set approach to medical image segmentation. We first design an entropy-based energy functional, from which we derive the multiphase level set equations and a new entropic external forces for the lattice Boltzmann D2Q9 model. The method is accurate and highly parallelizable. The local nature of the LBM allows it to be suitable for fast segmentation methods implemented using some parallel devices such as the graphics processing unit. Experimental results on MR breast images demonstrate the effectiveness of the proposed method.

Keywords: Lattice Boltzmann method, entropy, multiphase level set method, image segmentation, partial differential equation.

1   Introduction

Segmentation [1] is one of the most important steps in image processing systems. It aims to extract objects boundaries in a given scene or to partition a given image into several distinct regions. The level set method (LSM) [2-3] is part of the whole family of image segmentation methods. Its original idea was the Hamilton Jacobi approach, i.e., a time-dependent equation for a moving surface in the seminal work of Osher and Sethian [4]. In two-dimensional (2D) space, the LSM represents a closed curve in the plane as the zero level set of a three-dimensional (3D) function. For instance, starting with a curve around the object to be detected, the curve moves toward its interior normal and has to stop on the boundary of the object [5].

The LSM has several advantages. For example, it can easily handle complex shapes and topological changes. Nevertheless, the method is computational expensive. The movement of the zero level set is driven by the level set equation (LSE), which is a partial differential equation (PDE). For solving the LSE, most classical methods such as the upwind scheme use some finite difference, finite element or finite volume based approximations and an explicit computation of the curvature of the evolving contour [6]. Unfortunately, these methods are computational intensive.

Lately, the LBM is used as an alternative approach for solving LSE [7-8]. It can better handle the problem of time consuming because the curvature is implicitly computed, the algorithm is simple and highly parallelizable.

J. Yang, F. Fang, and C. Sun (Eds.): IScIDE 2012, LNCS 7751, pp. 8–16, 2013.

In this paper, the LBM is used to solve a multiphase LSE. We firstly design an entropy-based energy functional from which we derive the multiphase level set equations, and then the new entropic external forces for the LBM-PDE solver [9]. The evolution of the active contour is effectively stopped when the entropy is at minimum. The proposed method combines both the advantages of the LSM and the LBM. It is effective when objects to be segmented are with weak or without edges like in MR breast image, and is highly parallelizable due the local nature of the LBM.

This paper is organized as follows. In section 2 an overview of the LSM and the LBM is presented. Section 3 explains the formulation of the proposed method. Section 4 demonstrates the performance of the proposed method through experimental results on MR breast images. The final is conclusion.

2   Background

The proposed method uses mainly two techniques belonging to different frameworks: the level set method and the lattice Boltzmann method.

2.1   Level Set Method

The level set method (LSM) is a numerical technique for tracking interfaces and shapes. Using an implicit representation of active contours, it has the advantage of handling automatically topological changes of the tracked shape. In 2D image segmentation, the LSM represents a closed curve as the zero level set of a given level set function (LSF) ϕ. The active curve evolution starts from an arbitrary contour, and is driven by the level set equation (LSE), which is a convection-diffusion equation expressed as

$$\frac{\partial \phi}{\partial t} + \vec{V}.\nabla \phi = \eta \Delta \phi \tag{1}$$

where $\nabla \phi$ and $\Delta \phi$ are respectively the gradient and the Laplacian of ϕ. The term $\eta \Delta \phi$ is called artificial viscosity. It is suggested to replace it with $\eta k |\nabla \phi|$ which can better handle the evolution of lower dimensional interfaces [10], with k the curvature of the LSF. The LSE can therefore be rewritten as

$$\frac{\partial \phi}{\partial t} + \vec{V}.\nabla \phi = \eta k |\nabla \phi| \tag{2}$$

As an alternative method for solving PDEs, the LBM has several advantages such as: parallelizability and simplicity.

2.2   Lattice Boltzmann Method

The LBM was firstly designed to solve macroscopic fluid dynamics problems [9]. It is second-order accurate both in time and space. In this paper, we use the D2Q9 (2D with 8 links with its neighbors and one link with the cell itself) LBM lattice structure. Each link has its velocity vector $e_i(\vec{r},t)$ and the particle distribution $f_i(\vec{r},t)$ that moves along this link. The evolution equation of LBM can be written as

$$f_i(\vec{r}+\vec{e}_i,t+1) = f_i(\vec{r},t) + \frac{1}{\tau}[f_i^{eq}(\vec{r},t) - f_i(\vec{r},t)] \tag{3}$$

where $\vec{r}$ is the position of the cell, t is the time, τ represents the relaxation time and f_i^{eq} the local equilibrium particle distribution. As shown in [9], LBM can be used to solve the following parabolic diffusion equation which can be recovered by the Chapman-Enskog expansion

$$\frac{\partial \rho}{\partial t} = \gamma \nabla . \nabla \rho \tag{4}$$

For performing the level set based image segmentation, an external force is incorporated in the LBM scheme, after the collision step, as the link with image data

$$f_i \leftarrow f_i + \frac{2\tau-1}{2\tau} B_i(\vec{F}.\vec{e}_i) \tag{5}$$

where B_i is a constant scalar coefficient specific to the chosen lattice geometry. And Eq. (4) becomes

$$\frac{\partial \rho}{\partial t} = \gamma \nabla . \nabla \rho + F \tag{6}$$

The level set equation can be recovered when replacing ρ by the signed distance function ϕ.

3   The Proposed Method

This section details the conception of the proposed multiphase entropy-based method. We design an energy function, from which we deduce the corresponding multiphase level set equations and the entropic external forces (EEF) for the LBM solver.

3.1   Energy Function Design

Let consider Ω the image domain, entropy-based gray level image segmentation consist to minimize a certain estimator. In this paper we use the Ahmed-Lin entropy estimator [11]

$$H(\Omega) = -\frac{1}{|\Omega|} \int_{\Omega} \log p(I(x),\Omega) dx \tag{7}$$

where $|\Omega|$ is the number of pixels in the domain Ω, and $p(I(x),\Omega))$ is the density of probability of the intensity estimated with the help of the Parzen's kernel estimator [12]

$$p(I(x),\Omega) = \frac{1}{|\Omega|} \int_\Omega K_h(I(x)-I(\hat{x}))d\hat{x} \tag{8}$$

where K_h is the Gaussian kernel of this estimation and h the band wide, $K_h(u) = K(u/h)/h$ and for a centered Gaussian kernel K is given by

$$K(x) = \frac{1}{\sqrt{2\pi}}\exp(-\frac{1}{2}x^2) \tag{9}$$

Note that all the logarithms used in this paper are binary logarithms. For the multiphase case the segmentation will divide the domain Ω into $N \geq 3$ regions $\Omega_i, i = 1, \cdots N$. The criterion to minimize can be rewritten as

$$H(\Omega) = H(\Omega_1) + \cdots + H(\Omega_N) \tag{10}$$

In this case, we need to use two or more level set functions $\phi_1, \cdots, \phi_k$. Let define N membership functions M_i

$$M_i(\phi_1(x), \cdots, \phi_k(x)) = \begin{cases} 1, x \in \Omega_i \\ 0, else. \end{cases} \tag{11}$$

The level set function can thus be introduced in the entropy criterion as follows

$$H(\Omega, \Phi) = \sum_{i=1}^{N} -\frac{1}{|\Omega_i|} \int_\Omega \log p(I(x), \Omega_i) * M_i(\Phi(x))dx \tag{12}$$

$$where \ \Phi = (\phi_1, \cdots, \phi_k) \ is \ a \ vector \ valued \ function.$$

The above term $H(\Omega, \phi)$ is used as the data link in our energy functional which is defined as

$$E(\Omega, \Phi) = H(\Omega, \Phi) + v\sum_{j=1}^{k}|C_j| \tag{13}$$

where $v\sum_{j=1}^{k}|C_j|$ is a regularization term which smooth the contour by penalizing its arc length, $v > 0$ a fixed parameter and C_j a given curve which is represented implicitly as the zero level of ϕ_j; $|C_j|$ is the length of C_j and can be expressed as in [13]

$$|C_j| = \int_\Omega |\nabla H(\phi_j)|dxdy \tag{14}$$

3.2   Level Set Equation

To obtain the LSE, we use the gradient descent method

$$\frac{\partial \phi_j}{\partial t} = -\frac{\partial E}{\partial \phi_j} \tag{15}$$

where $\partial E / \partial \phi_j$ is the Gâteaux derivative of E. We obtain the following level set equations

$$\frac{\partial \phi_1}{\partial t} = \sum_{i=1}^{N} \frac{1}{|\Omega_i|} \log p(I(x), \Omega_i) * \frac{\partial M_i(\Phi(x))}{\partial \phi_1} + v\delta(\phi_1)div\left(\frac{\nabla \phi_1}{|\nabla \phi_1|}\right),$$

$$\vdots \tag{16}$$

$$\frac{\partial \phi_k}{\partial t} = \sum_{i=1}^{N} \frac{1}{|\Omega_i|} \log p(I(x), \Omega_i) * \frac{\partial M_i(\Phi(x))}{\partial \phi_k} + v\delta(\phi_1)div\left(\frac{\nabla \phi_k}{|\nabla \phi_k|}\right).$$

3.3   Lattice Boltzmann Solver for LSE

Although the method can be easily extended to higher N values, let consider $N = 4$ and therefore using $k = 2$ level set functions. The membership functions M_i can be expressed as

$$\begin{aligned}
M_1(\phi_1, \phi_2) &= H(\phi_1)H(\phi_2), \\
M_2(\phi_1, \phi_2) &= H(\phi_1)(1 - H(\phi_2)), \\
M_3(\phi_1, \phi_2) &= (1 - H(\phi_1))H(\phi_2), \\
M_4(\phi_1, \phi_2) &= (1 - H(\phi_1))(1 - H(\phi_2)).
\end{aligned} \tag{17}$$

We then obtain the following level set equations

$$\begin{aligned}
\frac{\partial \phi_1}{\partial t} = \delta(\phi_1)\Bigg\{ &[\frac{1}{|\Omega_1|}\log p(I(x), \Omega_i) - \frac{1}{|\Omega_3|}\log p(I(x), \Omega_3)]H(\phi_2) \\
&+ [\frac{1}{|\Omega_2|}\log p(I(x), \Omega_2) - \frac{1}{|\Omega_4|}\log p(I(x), \Omega_4)](1 - H(\phi_2)) \\
&+ v div\left(\frac{\nabla \phi_1}{|\nabla \phi_1|}\right)\Bigg\},
\end{aligned} \tag{18}$$

$$\begin{aligned}
\frac{\partial \phi_2}{\partial t} = \delta(\phi_2)\Bigg\{ &[\frac{1}{|\Omega_1|}\log p(I(x), \Omega_i) - \frac{1}{|\Omega_2|}\log p(I(x), \Omega_2)]H(\phi_1) \\
&+ [\frac{1}{|\Omega_3|}\log p(I(x), \Omega_3) - \frac{1}{|\Omega_4|}\log p(I(x), \Omega_4)](1 - H(\phi_1)) \\
&+ v div\left(\frac{\nabla \phi_2}{|\nabla \phi_2|}\right)\Bigg\}.
\end{aligned} \tag{19}$$

By using the gradient projection method of Rosen [14], we can replace $\delta(\phi_k)$ by $|\nabla \phi_k|$ in the proposed level set equations. In this paper, we define ϕ as a signed distance function on the whole domain. We thus have $|\nabla \phi_k| = 1$, which simplify the proposed LSEs. Replacing ρ by the function ϕ, Eq. (6) becomes

$$\frac{\partial \phi}{\partial t} = \gamma div(\nabla \phi) + F \tag{20}$$

In this paper, we consider two kinds of particles, each kind has its own macroscopic fluid density ρ_k and is submitted to its own external force F_k. Therefore we have the following equations by replacing ρ_1 by ϕ_1 and ρ_2 by ϕ_2

$$\frac{\partial \phi_1}{\partial t} = \gamma \, div \, (\nabla \phi_1) + F_1, \tag{21}$$

$$\frac{\partial \phi_2}{\partial t} = \gamma \, div \, (\nabla \phi_2) + F_2. \tag{22}$$

By setting the external forces

$$F_1 = \lambda_1 ([\frac{1}{|\Omega_1|} \log p(I(x),\Omega_1) - \frac{1}{|\Omega_3|} \log p(I(x),\Omega_3)]H(\phi_2)$$
$$+ [\frac{1}{|\Omega_2|} \log p(I(x),\Omega_2) - \frac{1}{|\Omega_4|} \log p(I(x),\Omega_4)](1 - H(\phi_2))), \tag{23}$$

$$F_2 = \lambda_2 ([\frac{1}{|\Omega_1|} \log p(I(x),\Omega_1) - \frac{1}{|\Omega_2|} \log p(I(x),\Omega_2)]H(\phi_1)$$
$$+ [\frac{1}{|\Omega_3|} \log p(I(x),\Omega_3) - \frac{1}{|\Omega_4|} \log p(I(x),\Omega_4)](1 - H(\phi_1))), \tag{24}$$

where λ_1 and λ_2 are positive parameters, we can see that Eqs. (21) and (22) are only some variational formula of Eqs. (18) and (19), and thus can be solved by the LBM when inserting the above defined entropic external forces (EEF).

For each obtained LSE, the principal implementation steps of the proposed method are as follows:

Steps	Instructions
1	Initialize ϕ as a signed distance functions.
2	Compute the body force with Eqs. (23) and (24).
3	Resolve the LSE using the LBM evolution equation by incorporating the body force according to Eq. (6).
4	Update the LSF by accumulating the distributions function values at each grid point with Eq. (5).
5	Find the contours, if the segmentation is not done, go back to step 2 or change the value of λ.

4   Experiments and Analysis

In this section we demonstrate experimentally the performance of the proposed method in term of efficiency and accuracy when the object to be segmented is with weak or without edges. The method is compared with three level set image segmentation methods, the Chan and Vese's method (CV) [15], the Li's method [16] and the level set lattice Boltzmann based method proposed by Chen in [10].

Fig.1, Fig.2, Fig.3 and Fig.4 show the experimental results obtain when comparing the proposed method with the three methods stated above. It can be seen that the proposed method gives the best results, i.e., most of the object boundaries are detected; the contours are thin and present no discontinuities. Comparing with the Chan and Vese method, all the experiments demonstrates that the proposed method is effective when most of the time the CV method is trapped in local minima. The Li's method and the Chen's method are edge-based and are not effective because most of the boundaries present in the MR breast image are with weak or without edges. The experimental results show that, due to its multiphase nature, the proposed method can detect efficiently and effectively more useful boundaries.

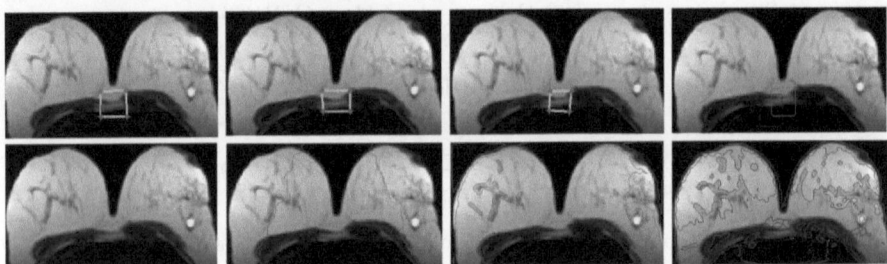

Fig. 1. First row: initial contours. Second row: the first image shows the result of the CV's method, the second image shows the result of the Li's method, the third image shows the result of Chen's method and the fourth image shows the result of the proposed method.

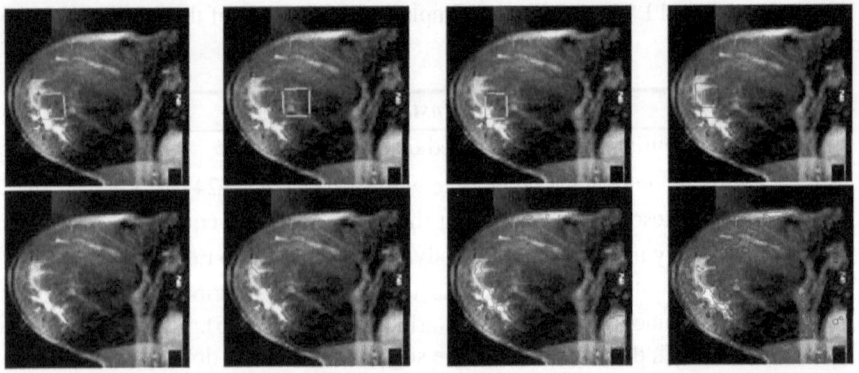

Fig. 2. First row: initial contours. Second row: the first image shows the result of the CV's method, the second image shows the result of the Li's method, the third image shows the result of Chen's method and the fourth image shows the result of the proposed method.

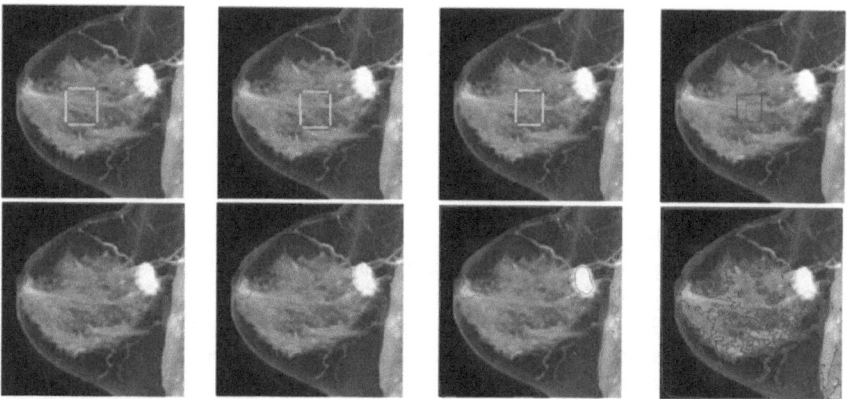

Fig. 3. First row: initial contours. Second row: the first image shows the result of the CV's method, the second image shows the result of the Li's method, the third image shows the result of Chen's method and the fourth image shows the result of the proposed method.

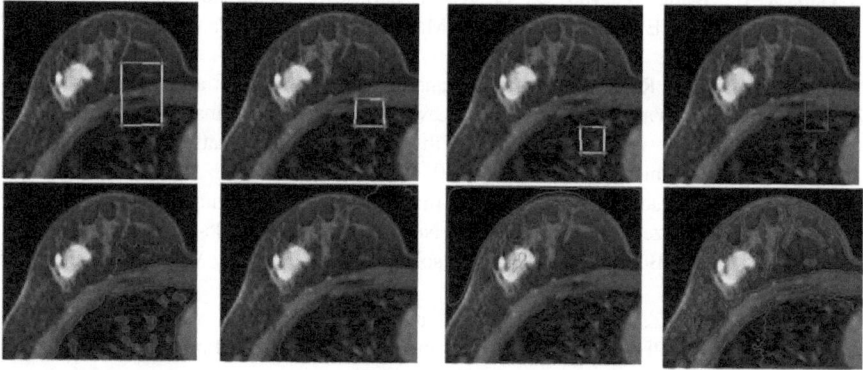

Fig. 4. First row: initial contours. Second row: the first image shows the result of the CV's method, the second image shows the result of the Li's method, the third image shows the result of Chen's method and the fourth image shows the result of the proposed method.

5   Conclusion

We have proposed a novel multiphase LSM based on LBM and the idea to stop the evolving contour when the entropy is at minimum. Comparing with the classical methods for solving the level set equation based on some finite difference or finite volume approximations, the proposed method costs less CPU time and is highly parallelizable. Experimental results on MR breast images have demonstrated the efficiency and the effectiveness of the proposed method. Future work will be its implementation using parallel devices such as the GPU for fast and accurate volume image segmentation.

Acknowledgements. This research was supported partially by the National Natural Science Foundation of China (Grant Nos. 61125204, 61172146), the Fundamental Research Funds for the Central Universities.

References

1. Heydy, C., Volodymyr, P., Luis, N., Victor, G.: Wavelet transform fuzzy algorithms for dermoscopic image segmentation. Computational and Mathematical Methods in Medicine 2012, Article ID 578721, 11 pages (2012)
2. Ben Salah, M., Ben Ayed, I., Mitiche, A.: Active curve recovery of region boundary patterns. IEEE Trans. on Pattern Analysis and Machine Intelligence 34(5), 834–849 (2012)
3. Balla-Arabé, S., Gao, X.-B., Wang, B.: A fast and robust level set method for image segmentation using fuzzy clustering and lattice Boltzmann method. IEEE Trans. on Systems, Man and Cybernetics Part B: Cybernetics PP(99), 1–11 (2012)
4. Osher, S., Sethian, J.: Fronts propagating and curvature dependent speed: algorithms based on Hamilton-Jacobi formulations. J. Comput. Phys. 79(1), 12–48 (1988)
5. Gao, X.-B., Wang, B., Tao, D., Li, X.: A relay level set method for automatic image segmentation. IEEE Trans. on Systems, Man and Cybernetics Part B: Cybernetics 41(2), 518–525 (2011)
6. Osher, S., Fedkiw, R.: Level set methods and dynamic implicit surfaces. Springer (2003)
7. Balla-Arabé, S., Wang, B., Gao, X.-B.: Level set region based image segmentation using lattice Boltzmann method. In: Proc. 7th Int. Conf. Computational Intelligence and Security, Sanya, China, pp. 1159–1163 (2011)
8. Balla-Arabé, S., Gao, X.-B.: Image multi-thresholding by combining the lattice Boltzmann model and a localized level set algorithm. Neurocomputing 93, 106–114 (2012)
9. Zhao, Y.: Lattice Boltzmann based PDE solver on the GPU. The Visual Computer 24(5), 323–333 (2007)
10. Chen, Y., Yan, Z., Chu, Y.: Cellular automata based level set method for image segmentation. In: IEEE/ICME Int. Conf. on Complex Medical Engineering, Beijing, pp. 23–27 (2007)
11. Boltz, S., Debreuve, E., Barlaud, M.: High-dimensional statistical distance for region-of-interest tracking: application to combining a soft geometric constraint with radiometry. In: IEEE Conference on Computer Vision and Pattern Recognition, pp. 1–8 (2007)
12. Parzen, E.: On estimation of a probability density function and mode. Annals of Mathematical Statistics 33(3), 1065–1076 (1962)
13. Evans, L.C., Gariepy, R.F.: Measure Theory and Fine Properties of Functions. CRC Press, Boca Raton (1992)
14. Rosen, J.G.: The gradient projection method for nonlinear programming, II, Non-linear constraints. J. SIAM 9, 514–532 (1961)
15. Chan, T., Vese, L.: Active contours without edges. IEEE Trans. on Image Processing 10(2), 266–277 (2001)
16. Li, C., Xu, C., Gui, C., Fox, M.: Distance regularized level set evolution and its application to image segmentation. IEEE Trans. on Image Processing 19(12),'3243–3254 (2010)

Understanding the Top Grass Roots in Sina-Weibo

Ze Huang[1], Bo Yuan[1], and Xuelei Hu[2]

[1] Intelligent Computing Lab, Division of Informatics,
Graduate School at Shenzhen, Tsinghua University, Shenzhen 518055, P.R. China
workthy@hotmail.com, yuanb@sz.tsinghua.edu.cn
[2] School of Computer Science and Technology,
Nanjing University of Science and Technology, Nanjing 210094, P.R. China
xlhu@njust.edu.cn

Abstract. Microblogging is now popular among everyday web users in China who have a common name called *grass roots* in Sina-Weibo, a major microblogging service similar to Twitter. In this paper, we investigate the properties of messages published by this group of users and classify the messages into various topic categories using text classification methods based on the Bag of Words (BOW) model. We find that, using Naïve Bayes, it is possible to achieve high accuracy in recognizing the topic of a message but the popularity of a message cannot be reliably predicated based on its contents. These findings are also further explored with visualization techniques.

Keywords: Microblogging, Text Classification, Bag of Words, Visualization.

1 Introduction

The popularity of social media is expected to be growing continuously world-wide. According to the recent report by Nielsen on American internet users, social networks and blogs account for 23% of time spent online, compared to 9.8% for online games and 7.6% for email [1]. In China, Sina-Weibo[1] is one of the most popular microblogging services. According to the quarterly reports of SINA Corporation, Sina-Weibo had more than 100 million registered users in March 2011, and this number doubled five months later with nearly 90 million messages published each day. Similar to Twitter[2], it allows users to post messages with a character limit, with optional links to other sources of information. There are also some differences between Sina-Weibo and Twitter due to local conventions.

Existing studies on applying text classification methods to English microblogging sites have been conducted in several aspects, such as sentiment analysis [2], topic detection [3], information filtering [4, 5] and performance comparison using different classifiers and feature selection methods [6, 7]. There are also some studies on understanding the information diffusion in Twitter and the structure of Twitter [8-12]. In Chinese context, an interesting case study was conducted in [13] to investigate how Chinese users used microblogging services in response to the 2010 YuShu Earthquake

---

[1] http://weibo.com
[2] http://twitter.com

J. Yang, F. Fang, and C. Sun (Eds.): IScIDE 2012, LNCS 7751, pp. 17–24, 2013.
© Springer-Verlag Berlin Heidelberg 2013

in China. The influence of Part-Of-Speech (POS) features on Chinese webpage classification was analyzed in [14]. A recent study based on 43,000 volunteer ratings on tweets shows that contents on information sharing, self-promotion and questions to followers were often valued highly [15].

Ordinary users in Sina-Weibo have a common name, grass roots, to be distinguished from famous users such as celebrities whose real identities are manually verified. Our study focused on analyzing the messages published by grass roots as they are representatives of the vast majority of microblog users. With a close observation of the message contents, we found that most of the non-private messages, especially those relatively long messages, may be mapped to a few topics. In this paper, messages were classified into five categories: Living Tips (LT), Design & Originality (DO), Fashion (F), Entertainment (E), Quotation & Sayings (QS). Certainly, for microblogging services where all messages have user specified tags, it is preferable to use the tags as class labels.

As a popular representation model used in text classification, the Bag of Words (BOW) model usually cannot achieve satisfactory performance on short text classification as the texts do not provide sufficient word occurrences, making the feature space quite sparse. To address this issue, one solution is to inflate the text by integrating meta-information and word-occurrence information from other sources, such as Wikipedia or search results returned by web search engines [16, 17].

However, we found that the messages published by top grass roots in Sina-Weibo had special properties different from ordinary short texts: they seemed to be well structured and contain good quality information for indicating a topic. By contrast, we found that the popularity of a message cannot be reliably predicated based on its contents and we concluded that there are some distinct characteristics of grass roots' messages in Sina-Weibo, which may reflect the special social phenomena behind the sociocultural system in China.

2   Data Preparation

Similar to other microblogging services, Sina-Weibo displays a list of the most influential grass roots based on their numbers of followers. This ranked list shows the top 300 grass roots and is updated on a daily basis since the number of followers may change continuously. We collected the messages of the top 300 grass roots and some randomly selected followers using Sina-Weibo API[3]. The API had an access limit and only returned up to 2,000 historical messages before the date that the API was invoked. As a result, the messages of some grass roots were collected completely while the others were not.

A closer look into the collected dataset revealed some special features in the messages of top grass roots:

- *Username.* Their usernames often directly implied a certain topic, such as "*The digest of cold joke*", "*Classic Quotations*" and "*Beauty and Health*". However, their messages were not always consistent with the topics revealed by the usernames. For example, many top grass roots publish messages related to quotations.

---

[3] http://open.weibo.com/development

- *Content.* They rarely published private messages, or original contents. Instead, they often shared information about fashion, constellation, jokes and classic quotations (in many occasions, they used software to automatically post messages). The contents of most of these messages were from the Web and had no connection with the social events at the time of publishing.

- *Hashtags and Personal Description.* Top grass roots widely used hashtags and personal signatures in their profiles for further explanation of the topics on which their microblogs were focused.

In addition to the text contents, messages obtained through Sina-Weibo API contained other types of information. For example, URL links to websites and videos were widely used in grass roots' messages. In this paper, the focus was on text classification and all non-text information was removed.

The main steps of data preprocessing for our experiments are as follows:

- Removed messages that contained "@*username*". Each user had a unique username, and "@*username*" was linked to the user's microblog. However, messages with multiple occurrences of "@*username*" often had only a few greeting or commentary words with little association with the message topic.

- Messages with string length less than 60 were removed to filter out those less meaningful messages and reduce the burden of manually labeling the messages.

- Messages with English words were discarded. Although there were some English messages in the dataset, we wanted to focus on Chinese text classification and also avoid the curse of dimensionality, as the feature space would expand a lot when considering English words.

- Removed URL links starting with "http:// ".

After this preprocessing procedure, the final dataset contained totally 40,636 messages (Table 1).

Table 1. The categorization of message topics

Category	Description	Count
Living Tips (LT)	Knowledge of daily life such as cooking and health	8989
Design & Originality (DO)	Novel design, new science & technology inventions	5665
Entertainment (E)	News and comments of movies	3884
Fashion (F)	Latest fashion trends and dressing advices	6029
Quotations & Sayings (QS)	Classic quotations, excerpts from literary works	16069

3   Classification and Prediction

Since in Chinese language there are no fixed separators between words, Chinese lexical analysis is required to segment the string of words into meaningful units, each of which is considered as a feature and then built into a vector space. We adopted a widely used Chinese lexical analysis system ICTCLAS[4] in our experiment, since ICTCLAS supports word segmentation, Part-Of-Speech (POS) tagging and unknown word entities recognition and has achieved satisfactory segmentation accuracy compared to other Chinese lexical analysis systems.

The experiments were conducted using the Multinomial Naïve Bayes classifier implemented in WEKA 3.6, using 10-fold cross validation. In Table 2, the first column shows the different groups of features adopted in classification (/n: nouns, /v: verbs and /a: adjectives). It can be seen that when only using nouns and verbs as the features, the size of the feature space reduced from 49,441 to 35,576 with little loss in accuracy. We also tested the effect of feature selection techniques such as information gain (IG) and χ^2-test (CHI). When only using nouns as the original features, both techniques achieved good accuracies close to 93.8% with around 15,000 features and the accuracy was already above 93% with only 4,000 features.

Table 2. Classification accuracy with different POS elements

Features	#Features	Accuracy (%)					
		DO	*E*	*LT*	*F*	*QS*	*Overall*
All	49441	91.2	92.7	96.5	97.8	95.6	**95.2**
/n	24236	88.4	89.9	96.2	95.3	94.6	**93.7**
/n/v	35576	90.2	91.6	97.2	97.0	95.1	**94.8**
/n/a	26885	89.2	89.4	96.2	95.7	95.1	**94.1**

Another interesting question is on the correlation between the contents of messages and their popularity, which can be measured by two figures: the number that a message was forwarded (*#relay*) and the number that a message was commented (*#cmt*). We focused on the top 50 grass roots as of 15/06/2011 in Sina-Weibo and collected 169,775 historical messages published in the original authorship. All URL links in messages were removed and messages with string length more than 80 were selected. There were 52,508 messages in the dataset, which were uniformly assigned to 3 levels or classes (low popularity, medium popularity and high popularity) according to their *#relay* and *#cmt* values.

Similarly, these messages were classified using Naïve Bayes and the performance was evaluated using 10-fold cross validation. According to the results in Table 3, the accuracies for both measures as well as for all of the levels were quite moderate (random guess: 33.3%). This evidence may indicate that using the pure text contents of messages to predicate their potential popularity is not reliable. We also observed that messages published in the early stage of the top 50 grass roots received little attention, regardless of their topics and their quality. By contrast, after these grass roots became influential (getting into the top list), their messages tended to have some good chance of being forwarded and commented.

---

[4] http://ictclas.org/

Table 3. Results of popularity prediction

Measures	Prediction Accuracy		
#relay	58.0% (Low)	41.5% (Medium)	56.0% (High)
#cmt	59.5% (Low)	52.4% (Medium)	60.2% (High)

4   Visualization

In order to provide deeper insights into the experiment results on topic classification and popularity prediction, we conducted some interesting text visualization with the help of an open-source software package called *Gephi*, which has previously been used in similar work [18]. Each message in the dataset was shown as a node in the graph. Edges were added only if two messages (nodes) were similar enough, which was measured by the cosine distance in our experiment (only nouns, adjectives and verbs were used as features). When the cosine similarity was above a threshold, an edge was drawn between the two nodes.

In Fig. 1 and Fig. 2, the size of a node was determined by its degree (the number of other nodes connected to it), and gray levels were used to distinguish messages in various categories. The Fruchterman-Reingold algorithm was used to create the layouts to bring together nodes with strong ties. Since a large number of nodes and edges would make the graph difficult to read, *Gephi* provides a filtering method to hide certain nodes and edges. For example, we can make a node or an edge invisible if the node's degree or the edge's weight is less than a threshold.

To create Fig. 1, 2,000 messages were randomly selected from each category. For the sake of clarity, messages belonging to QS, LT and E are shown in Fig. 1 (left) with threshold 0.6 while other messages are shown in Fig. 1 (right) with threshold 0.45. Nodes with degrees less than 4 were kept invisible (their edges were also kept invisible). Note that even a seemingly isolated node in Fig. 1 actually had at least 4 invisible edges linked to other invisible nodes.

The most important observation from Fig. 1 is that nodes within the same sub-graph (a set of connected nodes) often had the same gray level, which means that messages were generally similar to those in the same class and different from messages in other classes. We believe that this inherent similarity can largely explain the good classification performance observed in Section 3.

Fig. 2 was created in a similar manner, showing 391 nodes and 6640 edges generated from the original 52,508 messages in the dataset for popularity prediction (Section 3). There were three types of nodes in terms of *#relay*: white for low popularity, grey for medium popularity and black for high popularity. It is clear that each sub-graph typically contained messages with mixed popularity levels and it would be very challenging to accurately predict the popularity of messages.

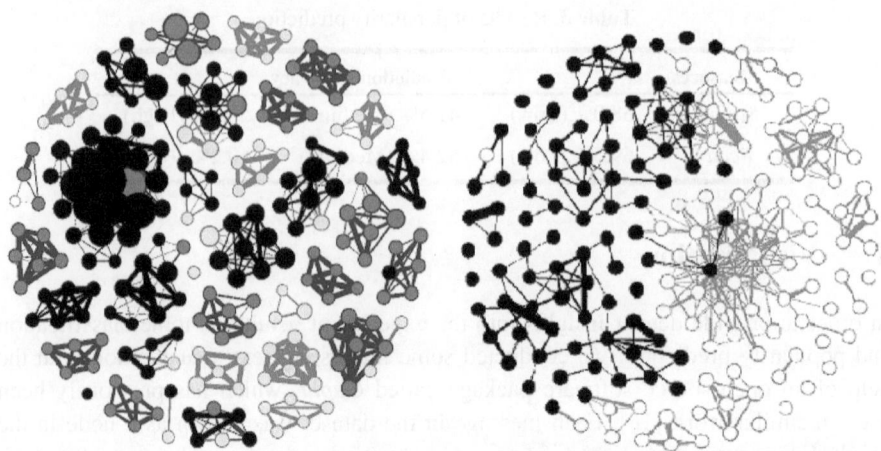

Fig. 1. Visualization of messages for topic classification: QS, LT & E (left) and DO & F (right). It shows that messages similar to each other often belong to the same topic category.

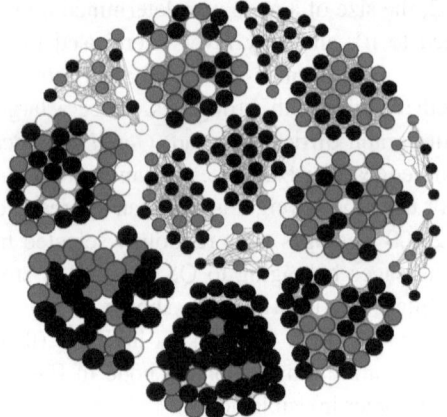

Fig. 2. Visualization of messages for popularity prediction: white (Low), gray (Medium) and black (High). It shows that similar messages often have different levels of popularity.

To better understand the forwarding relationship among users, Fig. 3 shows a directed graph describing the relationship among the top 100 grass roots as of 15/06/2011 and the authors from whom they have forwarded messages before this date. The size of a node was determined by the total times that his/her messages were forwarded by any of the top 100 grass roots. The white nodes are the top 100 grass roots and the grey ones are the non-top grass roots whose messages had been forwarded for more than 300 times and the black ones are other ordinary grass roots.

We tracked for 25 days the top 300 users with the most followers (Top-300) between August and September 2011 in Sina-Weibo. Interestingly, we found that, out of these 43 grey nodes, 16 entered into Top-300. So, it seems important for an ordinary grass root to become popular if his/her messages can be forwarded frequently by the top ones. We believe that this is the reason that many ordinary users actively contribute good quality messages to top grass roots.

Fig. 3. The forwarding relationship among users

5   Conclusion

The motivation of our work was to gain some appreciation of the characteristics of grass roots in Sina-Weibo, which represent the vast majority of microblog users. Experimental results showed that: (a) standard text classification methods performed well on topic classification with overall accuracy more than 95%; (b) using part of the POS features (e.g., nouns and verbs) can effectively decrease the feature dimension with little sacrifice of accuracy; (c) the pure text contents cannot provide sufficient information to accurately predicate a message's popularity.

These results suggest that the topics of non-private messages from top grass roots can be identified effectively with simple classification methods (e.g., Naïve Bayes). The reason behind this interesting phenomenon is, in our opinion, that the contents of such messages are often carefully organized to be concise (e.g., the frequent use of keywords) and of good quality compared to private messages to attract followers and improve their popularity. A possible incentive for publishing this type of messages may be the potential advertising value of these microblog accounts.

Acknowledgements. This work was supported by the National Natural Science Foundation of China (No. 60905030).

References

1. The Nielsen Company: State of the Media: The Social Media Report – Q3 (2011),
   http://www.nielsen.com/us/en/insights/reports-downloads.html
2. Bermingham, A., Smeaton, A.: Classifying Sentiment in Microblogs: Is Brevity an Advantage? In: 19th ACM Conference on Information and Knowledge Management, pp. 1833–1836 (2010)
3. Sankaranarayanan, J., Samet, H., Teitler, B., Lieberman, M., Sperling, J.: TwitterStand: News in Tweets. In: 17th ACM SIGSPATIAL International Symposium on Advances in Geographic Information Systems, pp. 42–51 (2009)
4. Laboreiro, G., Sarmento, L., Teixeira, J., Oliveira, E.: Tokenizing Micro-Blogging Messages Using a Text Classification Approach. In: Fourth Workshop on Analytics for Noisy Unstructured Text Data, pp. 81–88 (2010)
5. Sriram, B., Fuhry, D., Demir, E., Ferhatosmanoglu, H., Demirbas, M.: Short Text Classification in Twitter to Improve Information Filtering. In: 33rd ACM SIGIR Conference on Research and Development in Information Retrieval, pp. 841–842 (2010)
6. Ramage, D., Dumais, S., Liebling, D.: Characterizing Microblogs with Topic Models. In: Fourth International Conference on Weblogs and Social Media, pp. 130–137 (2010)
7. Rosa, K., Ellen, J.: Text Classification Methodologies Applied to Micro-Text in Military Chat. In: 2009 International Conference on Machine Learning and Applications, pp. 710–714 (2009)
8. Wu, S., Hofman, J., Mason, W., Watts, D.: Who Says What to Whom on Twitter. In: 20th International Conference on World Wide Web, pp. 705–714 (2011)
9. Naaman, M., Boase, J., Lai, C.: Is It Really About Me? Message Content in Social Awareness Streams. In: 2010 ACM Conference on Computer Supported Cooperative Work, pp. 189–192 (2010)
10. Java, A., Song, X., Finin, T., Tseng, B.: Why We Twitter: Understanding Microblogging Usage and Communities. In: 9th WebKDD and 1st SNA-KDD Workshop on Web Mining and Social Network Analysis, pp. 56–65 (2007)
11. Krishnamurthy, B., Gill, P., Arlitt, M.: A Few Chirps about Twitter. In: First Workshop on Online Social Networks, pp. 19–24 (2008)
12. Kwak, H., Lee, C., Park, H., Moon, S.: What is Twitter, a Social Network or a News Media? In: 19th International Conference on World Wide Web, pp. 591–600 (2010)
13. Qu, Y., Huang, C., Zhang, P., Zhang, J.: Microblogging after a Major Disaster in China: A Case Study of the 2010 Yushu Earthquake. In: 2011 ACM Conference on Computer Supported Cooperative Work, pp. 25–34 (2011)
14. Huang, W., Xu, L., Duan, J., Lu, Y.: Chinese Web-Page Classification Study. In: IEEE International Conference on Control and Automation, pp. 1553–1558 (2007)
15. Andre, P., Bernstein, M., Luther, K.: Who Gives a Tweet? Evaluating Microblog Content Value. In: 2012 ACM Conference on Computer Supported Cooperative Work, pp. 471–474 (2012)
16. Schonhofen, P.: Identifying Document Topics Using the Wikipedia Category Network. In: 2006 International Conference on Web Intelligence, pp. 456–462 (2006)
17. Broder, A., Fontoura, M., Gabrilovich, E., Joshi, A., Josifovski, V., Zhang, T.: Robust Classification of Rare Queries Using Web Knowledge. In: 30th ACM SIGIR Conference on Research and Development in Information Retrieval, pp. 231–238 (2007)
18. Stray, J.: A Full-text Visualization of the Iraq War Logs,
    http://jonathanstray.com/
    a-full-text-visualization-of-the-iraq-war-logs

Feature-Scoring-Based Multi-cue Infrared Object Tracking

Jiangtao Wang, Debao Chen, Suwen Li, and Yijun Yang

Huaibei Normal University, Huaibei, China
jiangtaoking@126.com, chendb_8@163.com,
suwen_li@yahoo.cn, yangyijun@sohu.com

Abstract. In this paper, we propose an effective tracker for infrared videos based on the multi-cue fusion. Under the particle filter tracking construction, a novel feature scoring scheme is introduced to evaluate different cue tracking ability, then the multi-cue fusion is executed in a weighted sum manner. In our tracking system, the score of each feature can be adaptively updated according to the current environment. Experimental results with various Infrared Video Database and different trackers are reported to demonstrate the accuracy and robustness of our algorithm.

Keywords: infrared video, object tracking, multi-cue fusion, particle filter.

1 Introduction

As a challenging problem within the field of computer vision, object tracking have attracted the attention of many researchers for several decades now[1]. Developing an accurate, efficient and robust visual tracker is always challenging, and the task becomes even more difficult when the target is expected to undergo significant and rapid variation in shape as well as appearance.

In this paper, we focus on the multi-cue based infrared object tacking problem under a particle filter framework. Object tracking in IR imagery is often more difficult than that in visible imagery. Unlike visible imagery containing color and texture cues, shape and intensity are arguably the only cue that can be exploited by tracking in IR imagery. To overcome the above difficulties, we propose a multi-cue fusion strategies that exploit both intensity and shape information adaptively at the same time. In our work, the discriminative scores of different cues are computed to measure the tracking ability of various cues. Each cue is weighted based on this score, and then all the cues are fused in a weighted sum manner.

The rest of the paper is organized as follows. In Section 2 we make a brief introduction of the particle filter framework and the state model. Then, in Section 3, after the feature representation model is introduced, we present our new feature scoring base multi-cue fusion scheme. Section 4 gives the experimental results; finally, conclusions are made in Section 5.

J. Yang, F. Fang, and C. Sun (Eds.): IScIDE 2012, LNCS 7751, pp. 25–31, 2013.

2   Particle Filter

Particle filters, also known as Sequential Monte Carlo methods (SMC) [2], are sophisticated model estimation techniques based on simulation within a Bayesian framework. Let $\mathbf{X}_t = \{\mathbf{x}_0,\mathbf{x}_1,..,\mathbf{x}_t\}$, $\mathbf{Y}_t = \{\mathbf{y}_0,\mathbf{y}_1,...,\mathbf{y}_t\}$ respectively be the state vector and the sensor measure up to time t. According to Bayesian estimation theory the optimal estimate of $\mathbf{x}_t$ is given by the posterior mean $E[\mathbf{x}_t|\mathbf{Y}_t]$. Given the posterior pdf $p(\mathbf{x}_{t-1}|\mathbf{Y}_{t-1})$ at time t-1, then current posterior pdf can be obtained by the following two steps:

Prediction Step:

$$p(\mathbf{x}_t \mid \mathbf{Y}_{t-1}) = \int_{t-1} p(\mathbf{x}_t \mid \mathbf{x}_{t-1}) p(\mathbf{x}_{t-1} \mid \mathbf{Y}_{t-1}) d\mathbf{x}_{t-1} \tag{1}$$

Update Step:

$$p(\mathbf{x}_t \mid \mathbf{Y}_t) = \frac{p(\mathbf{y}_t \mid \mathbf{x}_t) p(\mathbf{x}_t \mid \mathbf{Y}_{t-1})}{p(\mathbf{y}_t \mid \mathbf{Y}_{t-1})} \tag{2}$$

During prediction step, the prior $p(\mathbf{x}_{t-1}|\mathbf{Y}_{t-1})$ is propagated to $p(\mathbf{x}_t|\mathbf{Y}_{t-1})$ through a system dynamical model $p(\mathbf{x}_t|\mathbf{x}_{t-1})$. The predicted state is then modified by an observation likelihood function $L(\mathbf{Y}_t|\mathbf{x}_t)$ in the update step. In order to avoid the integral in Equation (1), the key idea of particle filtering is to approximate the posterior pdf by a weighted sample set $\mathbf{S} = \{(s^{(n)},b^{(n)}) \mid n = 1,...N\}$, Each sample s represents one hypothetical state of the object, with a corresponding discrete sampling probability b, where $\sum_{n=1}^{N} b^{(n)} = 1$. N is the number of particles. The mean state of an object is estimated at each time step by

$$E(\mathbf{S}) = \sum_{n=1}^{N} b^{(n)} s^{(n)} \tag{3}$$

Considering various types of motion, such as translations, rotations and changes of the object size, in this work we use two generic models to describe above mentioned motion. In those models, first order auto-regression (AR) dynamics is adopted for the translational motion and the random walk model is applied for the rotation and scaling.

3   Feature Scoring Based Multi-cue Fusion

3.1   Feature Model

For simple, in this work, we select two popular features: intensity and edge information as the representing cues. For these cues, we follow a conventional approach and use measurements based on histograms. The observation likelihood model for particles in each feature space can be gotten by the Bhattacharyya distance between candidate histograms and reference histograms. However, it is important to note out that any cues which can be represented by histogram are suitable for our method.

3.2   Scoring Feature

To emphasize the information derived from the reliable cues and ignore the information provided by the unreliable cues, we propose a novel approach for integrating different visual cues. The key idea of our approach is to give high score for the reliable cues and give low score for the unreliable cues, and the scores for all cues are gotten by the particle voting in current visual context. We grade each cues before tracking the target in every frame, and make it carry out time-adaptively according to the current context.

In order to illustrate the process of grading each cue by particle voting, as show in Fig.1, we generate N particles (in figure 1 N=10) around the target (white box in figure 1(a)). The center of the target (white dot in figure 1(b) and figure 1(c)) and the distribution of all particles (red dot in figure 1(b) and figure 1(c)) are known, thus the position of each particle is given. We then compute the observation model of each particle in all feature space. Fig.1 (b) and Fig.1(c) give observation likelihood map of edge cue and intensity cue respectively, and the Bhattacharyya distance between each particle and target in both feature space are also marked. The score of each feature's tracking ability can be gotten through the following way:

(a)

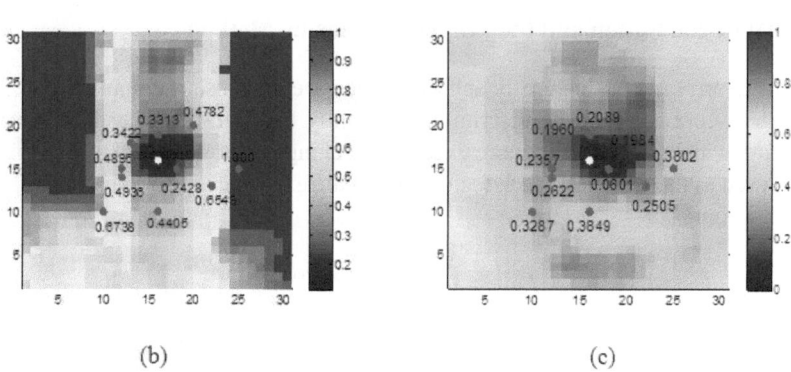

(b)   (c)

Fig. 1. Positions of particles and target and the Bhattacharyya distances between them

Algorithm 1 Scoring each feature based on the particle distribution

for i=1to N

for j=1to M

$$d_i^j = f_i^j - r^j$$

end for

$$\hat{z} = \underset{z \in \{1,2\cdots,m\}, \; d_i^z > th}{\mathrm{argmax}} \; d_i^z$$

$score(\hat{z})$++

end for

Here, N is the number of particles, M is number of cue to be fused, f_i^j is the Bhatta-charyya distance between particle i and the reference model for cue j, r^j is the Bhat-tacharyya distance between target and the reference model for cue j. In above process, we always vote for cue with maximum feature distance, and this distance should large than a threshold th to avoid distracters, Then the normalized score for each cue can be gotten as:

$$n\_score(j) = \frac{score(j)}{sum(score)} \tag{4}$$

From above analysis, we can see that the feature with more steep likelihood distriution, had been accessed with high score, on the other hand, the feature with flat likelihood map, had been accessed with low score. Feature score $n\_score(j)$ indicates the discriminate ability for cue j, so it can be used as the cue weight to strengthen the affact for cue with big score, and reduce the refluence for cues with small score.

3.3   Multi-cue Fusion and the Proposed Tracking Method

Once the score of each feature is achieved, then we can fuse the applied cues. The normalized score indicate the discriminative ability between object and background, and feature with high score has more excellent tracking performance, so the normalized score is used as the weight of each feature. We realize multi cue fusion by recomputing particles weight as:

$$B^{(n)} = \prod_{j=1}^{M} (b_f^{(n)})^{n\_score(j)} \tag{5}$$

Then the new weight is normalized to have a sum of 1:

$$\hat{B}^{(n)} = B^{(n)}\big/\text{sum}(B) \qquad\qquad (6)$$

Based on Equation (6) all cues are combined into one through fusing the particle weight in different feature spaces. This new weight is embedded in the particle filter tracking framework described in [3] to locate the infrared target. At the start frame, the reference models are initialized manually, and each cue weight is also computed by evaluating the performance of training samples. In the next coming frame, the initialized cue weights are used to estimate the target state in the particle filter framework, and then the reference models are updated by the currently estimated target models, the cue weights are also updated based on the new reference models. Thus we update reference models and cue weights at n-1 frame, then use it for the target tracking at $n(n>1)$ frames, by this operation, different cues can be weighted adaptively.

4   Tracking Experiments

In this section, we implemented the proposed approach in MATLAB and evaluated the performance on two video sequences. These experiments include a comparison of the proposed feature scoring tracker (FST) and other trackers: particle filter trackers using single cue of intensity and edge respectively, the standard Mean Shift tracker (MS) [4], the Incremental Visual Tracking (IVT) tracker [5] which using intensity cue. On a 2.8GHz Intel Core2 Duo CPU machine, our tracker runs at 3 frames per second. The number of particles used for our method, particle filter with single cues, IVT is 200 for all experiments. As for MS tracker, an intensity histogram with 256 bins is used. In all cases, the initial position of the target was selected manually in the first frame.

Fig. 2. Tracking results of test1 over representative frames with shape and scale changes

The first test sequence[6] shows a person walking from left to the right in the image with pose and scale changes. Fig. 2 shows some tracking results for both using our proposed method and particle filter based on single cue. The frame indices are 2041, 2129, 2217, 2387, 2463 from left to right. Where the first row images are tracking results for our proposed multiple cues method(FST), the second and the third row are results for singe gray cue(PF1) and single edge cue(PF2) each other. We can see that our multi cue method can track the target all the time despite of the pose and scale changes. However the tracker based on single cues loses some scale information and lead to partially drift. Fig.3 gives the normalized score of each cue, in the figure, the scores are changing adaptively, and in general the edge cue has a high score than the intensity cue, and this indicates that the edge feature is more robust in infrared videos.

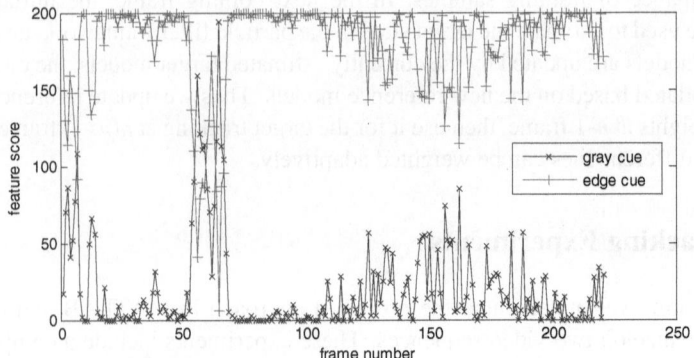

Fig. 3. Results for feature score in the tracking process

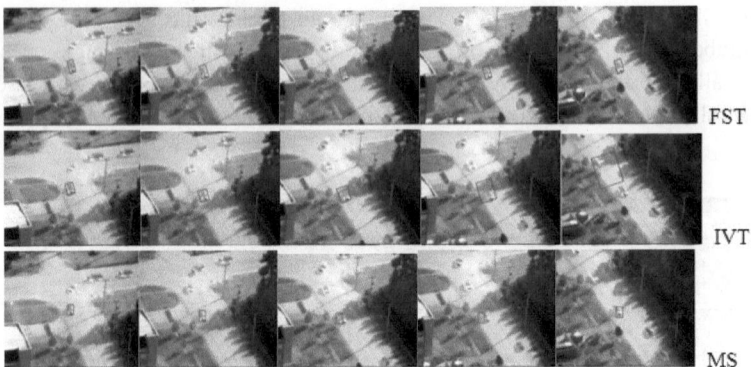

FST

IVT

MS

Fig. 4. Tracking results of test3 over representative frames with a car moves through shadows

The second test sequence[7] contains a car moving through shadows created by the trees on the roadside. In this test, we compare the tracking results of our proposed method (FST) with the MS and IVT tracker. Some samples of the tracking results are shown in Fig. 4. The frame indices are 781, 824, 847, 859, 900 from left to right. From the figure we see that our tracker is capable of tracking the car well even in the shadow. In this sequence, MS and IVT trackers can follow the car before it run into the shadow. However, the IVT tracker loses the scale information about the target as influence of the shadow, as the same the MS tracker can not rightly locate the car too. In comparison, our proposed method avoids this problem and is effective under shadow situations.

5   Conclusions

In this paper, we propose a multi-cue fusing scheme for robust visual tracking. Based on the particle filter, we compute each cue's score according to the particle distribution. Then an adaptive updating approach is designed to effectively combine different cues. The presented method is very flexible and can be applied on any histogram feature. In thorough experiments involving numerous challenging sequences and five other state-of-the-art trackers, our approach demonstrates very promising performance. The experimental results demonstrate clearly the effectiveness of our proposed method.

Acknowledgements. This work was partially supported by the Natural Science Foundation of China (61203272,41275027), Natural Science Foundation of Anhui Province (10040606Q56), Provincial Natural Science Research Project of Anhui Colleges (KJ2011A252), the Planned Science and Technology Project of Huaibei (2010211, 20110304), Research Project of Huaibei Normal University (jy10233).

References

1. Wang, R., Popovic, J.: Real-Time Hand-Tracking with a Color Glove. ACM Trans. Graph. 28(3) (2009)
2. Isard, M., Blake, A.: CONDENSATION-conditional density propagation for visual tracking. Int. J. Comput. Vis. 29(1), 5–28 (1998)
3. Nummiaro, K., Koller-Meier, E., Gool, L.: An adaptive color-based particle filter. Image Vision Comput. 21(1), 99–110 (2003)
4. Comaniciu, D., Ramesh, V., Meer, P.: Kernel-based object tracking. IEEE T. Pattern Anal. 25(5), 564–577 (2003)
5. Ross, D., Lim, J., Lin, R., Yang, M.: Incremental Learning for Robust Visual Tracking. Int. J. Comput. Vis. 77(1-3), 125–141 (2008)
6. OTCBVS benchmark dataset,
   http://www.cse.ohio-state.edu/otcbvs-bench/
7. VIVID database,
   https://www.sdms.afrl.af.mil/request/data-request.php#vivid

A Novel Unconstrained Correlation Filter and Its Application in Face Recognition

Yan Yan[1], Hanzi Wang[1,*], Cuihua Li[1], Chenhui Yang[1], and Bineng Zhong[1,2]

[1] School of Information Science and Technology, Xiamen University, China
[2] Department of Computer Science and Engineering, Huaqiao University, China
{yanyan,hanzi.wang,chli,chyang}@xmu.edu.cn,
bnzhong@gmail.com

Abstract. In this paper, a novel unconstrained correlation filter called Unconstrained Optimal Origin Tradeoff Filter (UOOTF) is presented and applied to face recognition. Compared with the conventional correlation filters in Class-dependence Feature Analysis (CFA), UOOTF increases the overall performance for unseen patterns by removing the hard constraints on the outputs during the filter design. Experimental results on the popular FERET, FRGC and CAS-PEAL R1 face databases show the effectiveness of the proposed unconstrained correlation filter.

Keywords: Unconstrained correlation filter, class-dependence feature analysis (CFA), face recognition.

1 Introduction

Face recognition has attracted much attention in computer vision and pattern recognition over the past few decades due to its important role in the areas of access control, security, and so on. However, it is still a very challenging task in practice due to great variations in facial appearances caused by pose, illumination, expression, etc. A variety of face recognition algorithms have been developed so far [1]. Among them, the appearance-based methods are one of the well-studied techniques where a face is usually represented as a high-dimensional vector. To overcome problems incurred by high dimensionality, subspace learning based methods, such as Principal Component Analysis (PCA) [2], Linear Discriminant Analysis (LDA) [3, 4], Locality Preserving Projections (LPP) [5], and Class-dependence Feature Analysis (CFA) [6, 7], have been proposed.

The projection axis obtained by traditional subspace learning methods, such as PCA, LDA and LPP, is used to preserve the dominant information or discriminate classes. One common problem of these methods is that they are not able to effectively discriminate classes close to each other since large class distances are often overemphasized during training. The resulting transformed subspace can preserve the distances of well-separated classes, while causing large overlaps between neighboring classes.

---

[*] Corresponding author.

J. Yang, F. Fang, and C. Sun (Eds.): IScIDE 2012, LNCS 7751, pp. 32–39, 2013.
© Springer-Verlag Berlin Heidelberg 2013

Compared with the traditional methods, the projection axis obtained by CFA can be used to distinguish one specific class from the other classes. The traditional methods [2–5] often employ features derived from the space domain, while CFA usually uses features derived from the frequency domain [6, 7]. The key step in CFA is the design of the correlation filters. Phase information which contains structural information for human perception is directly modeled by the correlation filters in CFA [7]. Furthermore, correlation filters offer some desirable properties, such as graceful degradation and closed-form solutions [6–9].

The traditional correlation filters, such as MACE [10], MVSDF [11], OTF [12], and OEOTF[8], assume that the distortion tolerance of a filter could be controlled by explicitly specifying desired correlation peak values for training images. As a matter of fact, the performance becomes worse if one enforces hard constraints on the correlation peak values. Relaxing the hard constraints by using the unconstrained form could increase the overall performance for unseen patterns[14]. However, experimental results show that the accuracy of UOTF [13, 14] cannot achieve as good performance as OTF [12]. The reason is that the design criterion of UOTF is not optimized for the feature vector extraction in CFA. Thus, it motivates us to design an unconstrained correlation filter in consistence with the feature vector extraction process.

In this paper, we propose a novel unconstrained correlation filter, called Unconstrained Optimal Origin Tradeoff Filter (UOOTF), to make the face feature extraction effective in CFA. As far as we know, very few work concerns the design of unconstrained correlation filters in the CFA framework.

2   Unconstrained Optimal Origin Tradeoff Filter

From Section 1, we can see that the traditional correlation filters [8, 10–12] are based on the assumption that the correlation peak amplitude should satisfy a specified value (i.e., the origin outputs are restricted to 1 for one specific class and 0 for the others). However, the overall performance of those filters may become worse for unseen patterns if the correlation peak values are constrained to some specified constant values. Although UOTF [14] tries to overcome this problem by removing the hard constraints, UOTF fails in 1D-CFA (see Section 3 for the experimental results). In this subsection, an effective correlation filter called UOOTF is proposed to overcome the limitations of the traditional unconstrained correlation filters.

For convenience, matrices and vectors are denoted by light face characters and bold characters, respectively. Upper case symbols refer to the frequency plane terms while lower case symbols represent quantities in the space domain.

1D-CFA designs a correlation filter for each class. Suppose a filter trained for the l-th class is $\vec{h}^l$. Let $\vec{o}_i{}^l$ be the output of $\vec{h}^l$ in response to $\vec{y}_i$. We obtain

$$\vec{o}_i{}^l(n) = \vec{y}_i(n) \odot \vec{h}^l(n), \tag{1}$$

where $\odot$ represents the correlation function; $\vec{y}_i$ is the low-dimensional PCA feature for the i-th training image; n represents the index in the spatial domain.

Equation (1) can be expressed in the frequency domain using the 1D Fourier transform as follows:

$$\vec{o}_i{}^l(n) = \sum_{k=0}^{p-1} \vec{\mathbf{Y}}_i(k)^* \cdot \vec{\mathbf{H}}^l(k)e^{\frac{j2\pi kn}{p}}. \tag{2}$$

Here, $\vec{\mathbf{Y}}_i$ and $\vec{\mathbf{H}}^l$ represent the 1D Fourier transform of $\vec{y}_i$ and $\vec{h}^l$, respectively; p is the reduced dimensionality of the PCA subspace; k represents the index in the frequency domain; '$*$' denotes the conjugate operator.

The design of UOOTF is to optimize the tradeoff between the origin correlation output energy and the origin correlation output noise variance for extra-class features given by

$$\min_{\vec{\mathbf{H}}^l}(\alpha\vec{\mathbf{H}}^{l+}\mathbf{R}_Y^l\vec{\mathbf{H}}^l + \sqrt{1-\alpha^2}\vec{\mathbf{H}}^{l+}\mathbf{C}\vec{\mathbf{H}}^l), \tag{3}$$

where $\mathbf{R}_Y^l = 1/(N - N_l) \sum_{i=1}^{N-N_l} \vec{\mathbf{Y}}_i^E\vec{\mathbf{Y}}_i^{E+}$; $\vec{\mathbf{Y}}_i^{E+}$ $(i = 1, \cdots, N - N_l)$ is the 1D Fourier transform of the extra-class feature for the l-th class. It can be easily derived by summing the origin energy $(\vec{o}_i{}^l(0)^2)$ of all the extra-class features according to (2). $\mathbf{C}$ is a diagonal matrix whose diagonal elements represent the noise power spectral density (i.e., usually the identity matrix); '$+$' denotes the conjugate transpose; α $(0 \le \alpha \le 1)$ is the tradeoff parameter; N is the number of training samples and N_l is the number of training samples for the l-th class.

For intra-class features, we try to maximize the average origin correlation output (called Average Correlation Height, i.e., ACH) which can be described as follows:

$$\max_{\vec{\mathbf{H}}^l}(\frac{1}{N_l}\sum_{i=1}^{N_l}\vec{\mathbf{Y}}_i^{I+}\vec{\mathbf{H}}^l) = \max_{\vec{\mathbf{H}}^l}(\vec{\mathbf{M}}^{l+}\vec{\mathbf{H}}^l), \tag{4}$$

where $\vec{\mathbf{M}}^l = 1/N_l \sum_{i=1}^{N_l} \vec{\mathbf{Y}}_i^I$ is the average value of all intra-class features for the l-th class.

By combining (3) and (4), similar to the ratio form of the Fisher criterion [3], we can have the following optimization criterion:

$$J(\vec{\mathbf{H}}^l) = \frac{|\vec{\mathbf{M}}^{l+}\vec{\mathbf{H}}^l|^2}{\alpha\vec{\mathbf{H}}^{l+}\mathbf{R}_Y^l\vec{\mathbf{H}}^l + \sqrt{1-\alpha^2}\vec{\mathbf{H}}^{l+}\mathbf{C}\vec{\mathbf{H}}^l}$$

$$= \frac{\vec{\mathbf{H}}^{l+}\vec{\mathbf{M}}^l\vec{\mathbf{M}}^{l+}\vec{\mathbf{H}}^l}{\vec{\mathbf{H}}^{l+}(\alpha\mathbf{R}_Y^l + \sqrt{1-\alpha^2}\mathbf{C})\vec{\mathbf{H}}^l}. \tag{5}$$

The unconstrained correlation filter UOOTF can then be derived by maximizing the criterion function $J(\vec{\mathbf{H}}^l)$, i.e.,

$$\vec{\mathbf{H}}^l = \arg\max_{\vec{\mathbf{H}}^l} J(\vec{\mathbf{H}}^l). \tag{6}$$

By using the Lagrange multiplier method [12], it is easy to prove that we have the following closed-form solution for UOOTF:

Algorithm 1. UOOTF based 1D-CFA for face recognition

Input: Query image $\vec{p}_q \in \Re^{m \times 1}$, training data matrix $D \in \Re^{m \times N}$ with L classes, where m is the dimensionality of the face vector.
Output: The class label of the query image.

Training Stage:
Step 1: Project the training data matrix $D \in \Re^{m \times N}$ to the PCA subspace to obtain the low dimensional PCA feature matrix $Y \in \Re^{p \times N}$ and the corresponding 1D Fourier transform matrix $\mathbf{Y} \in \Re^{p \times N}$.
Step 2: Do for $l = 1, \cdots, L$:
  2.1 Calculate the tradeoff $\mathbf{T}^l$ using the extra-class features of the l-th class;
  2.2 Calculate the average value $\vec{\mathbf{M}}^l$ of all the intra-class features of the l-th class;
  2.3 Design the correlation filter $\vec{\mathbf{H}}^l$ via (7).
Step 3: Construct the projection matrix $\mathbf{P} = [\vec{\mathbf{H}}^1, \cdots, \vec{\mathbf{H}}^L]$.
Step 4: Compute the feature matrix $X = \mathbf{P}^T \mathbf{Y}$.
Step 5: Normalize each column of the feature matrix X based on (8) to obtain the normalized feature matrix X_n .

Testing Stage:
Step 1: Project the query face $\vec{p}_q$ into the PCA subspace to obtain the low-dimensional feature vector $\vec{y}_q \in \Re^{p \times 1}$ and the corresponding 1D Fourier transform $\vec{\mathbf{Y}}_q \in \Re^{p \times 1}$.
Step 2: Compute the feature vector $\vec{x}_q = \mathbf{P}^T \vec{\mathbf{Y}}_q$.
Step 3: Normalize the feature vector $\vec{x}_q$ based on (8) to obtain $\vec{x}_{qn}$.
Step 4: Assign the class label to the query image $\vec{p}_q$ by using the nearest neighbor classifier based on $\vec{x}_{qn}$ and X_n.

$$\vec{\mathbf{H}}^l = (\mathbf{T}^l)^{-1} \vec{\mathbf{M}}^l, \tag{7}$$

where $\mathbf{T}^l = \alpha \mathbf{R}_Y^l + \sqrt{1 - \alpha^2} \mathbf{C}$.

Since the hard constraints are not used during the filter design in our method, the peak values at the origin vary for classes. To overcome the scale differences for different correlation filters, we normalize the feature vector by using the strategy as

$$\vec{x}_n = \frac{\vec{x}}{\max(\vec{x})}. \tag{8}$$

Here, $\max(\vec{x})$ returns the maximum value in vector $\vec{x}$; $\vec{x}_n$ is the normalized feature vector.

In Algorithm 1, we give the outline of our proposed UOOTF based 1D-CFA for face recognition.

2.1 Discussions

It is worth comparing the performance obtained by different types of unconstrained correlation filters based on various design criteria. The traditional unconstrained correlation filters are designed based on the overall correlation output plane. Nevertheless, such kind of filter design is not consistent with the feature vector extraction process where only the origin correlation output is used in 1D-CFA. In contrast, during the design of UOOTF, the optimization objective function only focuses on the origin correlation output which is in essence more appropriate for the feature vector extraction. Figure 1 shows the normalized origin correlation outputs (OCO) for UOOTF and UOTF on one test face on the PIE face database. We observe that UOOTF can produce only one large amplitude peak value (i.e., 1.0) for one class while it suppresses the peak values of the other classes. However, UOTF usually produces multiple large amplitude peak values (close to 1.0) for several classes due to overfitting.

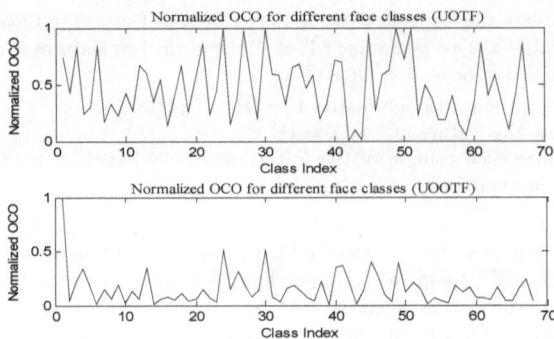

Fig. 1. Normalized Origin Correlation Outputs (OCO) for different face classes on the PIE face database. Top: Normalized OCO for UOTF; Bottom: Normalized OCO for UOOTF.

Similar to OEOTF, UOOTF only optimizes the origin correlation outputs. However, compared with OEOTF, the solution of UOOTF is simpler and the output distortion tolerance is enhanced by relaxing the constraints on the correlation peaks for intra-class samples.

3   Experiments

We evaluate the performance of the proposed UOOTF based 1D-CFA for face recognition on the FERET [15], FRGC [16] and CAS-PEAL face databases [17]. The algorithms chosen for the comparison are the Eigenface method (PCA) [2], the Fisherface method (PCA+LDA) [3], the Laplacianface method (PCA+LPP) [5], and 1D-CFA (including OTF [7], OEOTF [8] and UOOTF [14]).

All the images are cropped and normalized to the size of 64×64. Histogram equalization is applied to the face images for photometric normalization. The reduced dimensionality of the PCA subspace in 1D-CFA is set to $N - 1$. We have tried different settings of the tradeoff parameter and found that the performance is similar with different values. So, we set it to be a constant value ($\alpha = 0.6$) for all the correlation filters. All the computational time is reported on a workstation with 2 Intel Xeon E5620 (2.40GHz) CPUs (only one core is used) on the MATLAB platform.

3.1 Experimental Results on the FERET and FRGC Face Databases

A subset of the FERET [15] face database including 800 images and 200 individuals (there are 4 images for each person) is tested. Moreover, we select 6000 images for 300 individuals (20 images for each person) from the FRGC (Face Recognition Grand Challenge) face database [16].

A random subset with m images per individual is taken from the database to form the training set. The rest of the image database is used for testing. For each m, the experiments with randomly sampled subsets are implemented twenty times. We report the top average recognition rate and the corresponding dimensionality of the reduced subspace over the randomly sampled training sets as the final results. For all the databases, the value of m is set to either 2 or 3. The results are shown in Table 1. The highest value is shown in the bold font on each database.

From Table 1, we can see that the proposed method achieves the highest accuracy among the competing methods for all the experiments. This is due to the fact that the UOOTF based 1D-CFA enhances the discriminative ability of the feature vector by allowing the flexible distortion tolerance.

3.2 Experimental Results on the CAS-PEAL-R1 Face Database

To test the generalization ability of the proposed method, we use the CAS-PEAL-R1 database and the evaluation protocols in [17]. The CAS-PEAL-R1 database contains three types of data, i.e., the training set, gallery set and probe set. The training set consists of 1,200 images. The gallery set includes 1,040 images of the 1,040 subjects. The CAS-PEAL-R1 database contains six probe sets which correspond to six subsets under different conditions: accessory, age, background, distance, expression, and lighting. The results are shown in Table 2.

From Table 2, we can see that the UOOTF based 1D-CFA can achieve better recognition accuracy compared with the other competing methods. The overall recognition ability of the UOOTF based 1D-CFA is much better than the UOTF based 1D-CFA. Moreover, compared with OEOTF, UOOTF also obtains higher recognition rate. The traditional UOTF based 1D-CFA cannot achieve satisfying recognition accuracy in face recognition because the design criterion in UOTF is to optimize the whole correlation output plane which will lead to the overfitting problem in 1D-CFA (i.e., produce multiple large amplitude peak values). As a matter of fact, the feature vector extraction in 1D-CFA only considers the origin correlation output. Therefore, UOOTF is more effective for extracting the feature vector in the 1D-CFA framework. We also compare the CPU training time using different correlation filters in 1D-CFA on the training set of CAS-PEAL-R1 database, as shown in Fig. 2. From Fig. 2, we can observe that the training time of UOOTF is about 25% less than that of OEOTF. However, both OEOTF and UOOTF require more training time than OTF and UOTF. This is because the non-diagonal matrix inversion, which consumes the majority of the CPU time, is employed in both OEOTF and UOOTF during the filter design.

Table 1. The top average recognition rate (%) and the corresponding dimensionality

Algorithm	FERET ($m = 2$)	FERET ($m = 3$)	FRGC ($m = 2$)	FRGC ($m = 3$)
Eigenface	63.90 (397)	66.75 (595)	47.38 (127)	57.56 (136)
Fisherface	72.81 (82)	78.63 (190)	47.89 (148)	53.42 (199)
Laplacianface	74.69 (199)	82.15 (199)	53.31 (299)	61.21 (298)
1D-CFA (OTF)	75.07 (200)	82.25 (200)	54.32 (300)	62.03 (300)
1D-CFA (OEOTF)	83.65 (200)	92.40 (200)	62.98 (300)	74.95 (300)
1D-CFA (UOTF)	27.69 (200)	38.68 (200)	25.43 (300)	30.61 (300)
1D-CFA (UOOTF)	**84.95** (**200**)	**93.10** (**200**)	**64.89** (**300**)	**76.96** (**300**)

Table 2. The top average recognition rate (%) and the corresponding dimensionality

Algorithm	Accessory	Age	Background	Distance	Expression	Lighting	Average
Eigenface	59.39 (158)	57.58 (56)	95.84 (64)	93.09 (66)	73.69 (139)	10.16 (64)	51.00
Fisherface	45.95 (298)	33.33 (164)	87.70 (209)	77.45 (179)	61.34 (238)	4.95 (144)	40.67
Laplacianface	38.38 (319)	25.76 (317)	82.28 (320)	70.91 (294)	51.08 (315)	3.30 (265)	34.61
1D-CFA (OTF)	53.52 (300)	56.06 (300)	94.58 (300)	92.00 (300)	67.83 (300)	15.78 (300)	49.41
1D-CFA (OEOTF)	73.39 (300)	66.67 (300)	**98.19** (**300**)	98.18 (300)	**83.31** (**300**)	30.14 (300)	64.62
1D-CFA (UOTF)	13.74 (300)	6.06 (300)	57.14 (300)	44.00 (300)	41.34 (300)	0.62 (300)	13.74
1D-CFA (UOOTF)	**74.84** (**300**)	**71.21** (**300**)	**98.19** (**300**)	**98.55** (**300**)	83.12 (300)	**31.43** (**300**)	65.52

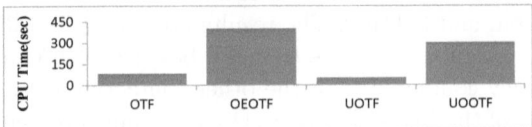

Fig. 2. Comparisons of the CPU time (in seconds) in the training step for different correlation filters in 1D-CFA on the CAS-PEAL-R1 database

4   Conclusions

In this paper, we present an effective unconstrained correlation filter and apply it to the task of face recognition. By removing the hard constraints during the filter design and focusing on the origin correlation outputs, UOOTF can extract discriminative face features more effectively. Experimental results show that the proposed method has achieved promising results in face recognition.

Acknowledgments. This work was supported by the National Natural Science Foundation of China under Grants 61170179 and 61201359, by the Natural Science Foundation of Fujian Province of China under Grant 2012J05126, by the National Defense Basic Scientific Research Program of China, by the Specialized Research Fund for the Doctoral Program of Higher Education of China under Grant 20110121110020, and by the China Postdoctoral Science Foundation under Grant 2011M501081.

References

1. Zhao, W., Chellappa, R., Phillips, P.J., Rosenfeld, A.: Face Recognition: a Literature Survey. ACM Comput. Surv. 35(4), 399–458 (2003)
2. Turk, M., Pentland, A.: Eigenfaces for Recognition. J. Cogn. Neurosci. 3(1), 71–86 (1991)
3. Belhumeur, P., Hespanha, J., Kriegman, D.: Eigenfaces vs. Fisherfaces: Recognition Using Class Specific Linear Projection. IEEE Trans. Pattern Anal. Mach. Intell. 19(7), 711–720 (1997)
4. Yang, J., Frangi, A.F., Yang, J.Y., Zhang, D., Jin, Z.: KPCA Plus LDA: A Complete Kernel Fisher Discriminant Framework for Feature Extraction and Recognition. IEEE Trans. Pattern Anal. Mach. Intell. 27(2), 230–244 (2005)
5. He, X., Yan, S., Hu, Y., Niyogi, P., Zhang, H.-J.: Face Recognition Using Laplacianfaces. IEEE Trans. Pattern Anal. Mach. Intell. 27(3), 328–340 (2005)
6. Xie, C., Savvides, M., Vijaya Kumar, B.V.K.: Redundant Class-dependence Feature Analysis Based on Correlation Filters using FRGC2.0 Data. In: Proc. IEEE Workshop on Face Recognition Grand Challenge Experiments in Conjunction with CVPR 2005, pp. 153–158. IEEE Press, New York (2005)
7. Vijaya Kumar, B.V.K., Savvides, M., Xie, C.: Correlation Pattern Recognition for Face Recognition. Proc. IEEE 94(11), 1963–1976 (2006)
8. Yan, Y., Zhang, Y.J.: 1D Correlation Filter Based Class-dependence Feature Analysis for Face Recognition. Pattern Recognit. 41(12), 3834–3841 (2008)
9. Bolme, D.S.: Theory and Applications of Optimized Correlation Output Filters. PhD Dissertation, Colorado State University (2011)
10. Mahalanobis, A., Vijaya Kumar, B.V.K., Casasent, D.: Minimum Average Correlation Energy Filters. Appl. Opt. 26, 3630–3633 (1987)
11. Vijaya Kumar, B.V.K.: Minimum Variance Synthetic Discriminant Functions. J. Opt. Soc. Amer. A. 3, 1579–1584 (1986)
12. Refregier, P.: Filter Design for Optical Pattern Recognition: Multi-Criteria Optimization Approach. Opt. Lett. 15, 854–856 (1990)
13. Vijaya Kumar, B.V.K., Mahalanobis, A.: Recent Advances in Composite Correlation Fitler Designs. Asian J. Phys. 8, 407–420 (1999)
14. Mahalanobis, A., Vijaya Kumar, B.V.K., Song, S., Sims, S.R.F., Epperson, J.: Unconstrained Correlation Filters. Appl. Opt. 33, 3751–3759 (1994)
15. Phillips, P.J., Wechsler, H., Huang, J., Rauss, P.J.: The FERET Database and Evaluation Procedure for Face-Recognition Algorithms. Image Vision Comput. 16(5), 295–306 (1998)
16. Phillips, P.J., Flynn, P.J., Scruggs, T., Bowyer, K.W., Chang, J., Hoffman, K., Marques, J., Min, J., Worek, W.: Overview of the Face Recognition Grand Challenge. In: Proc. IEEE Conference on Computer Vision and Pattern Recognition, pp. 947–954. IEEE Press, New York (2005)
17. Gao, W., Cao, B., Shan, S., Chen, X., Zhou, D., Zhang, X., Zhao, D.: The CAS-PEAL Large Scale Chinese Face Database and Baseline Evaluations. IEEE Trans. Syst. Man, Cybern. A, Syst., Humans 38(38), 149–161 (2008)

Salience-Based Prototype Selection for K-Nearest Neighbor Classification in Multiple-Instance Learning

Liming Yuan*, Qingcheng Huang, Jiafeng Liu, and Xianglong Tang

School of Computer Science and Technology, Harbin Institute of Technology,
Harbin 150001, China
yuanleeming@163.com, {huangqc,jefferyliu,tangxl}@hit.edu.cn

Abstract. The k-nearest neighbor algorithm (kNN) is one of the most well-known techniques in standard supervised learning. It could be adapted to the setting of multiple-instance learning (MIL) by using set-based distance metrics, such as Citation-kNN and Bayesian-kNN. However, kNN suffers from several drawbacks, including high storage requirements, low efficiency in classification response and low noise tolerance. These drawbacks would become particularly significant in MIL since every example here is a set of instances. One of the most promising solutions is dependent on prototype selection, and many prototype selection methods have been proposed in standard supervised learning. In this paper, we propose an efficient Salience-based Prototype Selection (MISPS) method to tackle the above problems in MIL. Then we present two variants of Citation-kNN and Bayesian-kNN based on MISPS, called MISPS-CkNN and MISPS-BkNN. Experimental results on five benchmark data-sets show that MISPS is effective and our MISPS-based algorithms are competitive to the state-of-the-art.

Keywords: Multiple-instance learning, Salience, Prototype selection, K-nearest neighbor classification.

1 Introduction

Multiple Instance Learning (MIL) is a variant of standard supervised learning, which was first introduced by Dietterich et al. when they were investigating the problem of drug activity prediction [1]. In this learning framework, every training example is not a single instance but a set of instances named bag. Labels are associated with bags rather than instances. A bag is labeled as positive if at least one instance in it is positive, and negative otherwise. The goal of a MIL algorithm is learning a classifier to predict the labels of unseen bags.

The notion of bag together with the labeling protocol often makes MIL more realistic than standard supervised learning for particular types of applications [2–6]. For instance, in content-based image retrieval (CBIR) [5, 6], each image

---

* Corresponding author.

J. Yang, F. Fang, and C. Sun (Eds.): IScIDE 2012, LNCS 7751, pp. 40–47, 2013.

contains many regions, but only a subset of them not all of them are of interest. Here each image can be considered as a bag and the image regions can be considered as instances in this bag, hence the CBIR problem can be better cast in a MIL setting than in a supervised learning context.

However, the hidden nature of instance labels makes MIL difficult to tackle and causes many MIL algorithms not working so well, including Citation-kNN and Bayesian-kNN [7]. As indicated in [7], the reason for the poor performance may be that positive bags contain *false* positive instances and one possible solution might remove them. Moreover, since the k-nearest neighbor (kNN) algorithm suffers from several drawbacks such as high storage requirements, low efficiency in classification response and low noise tolerance [8], kNN-based Citation-kNN and Bayesian-kNN algorithms also have these problems, especially Citation-kNN necessitating some additional consideration of citer examples [7]. These problems would become particularly significant in MIL since each example here is a bag composed of several instances rather than a single instance. Many solutions to these problems have been proposed in supervised learning and one of the most promising solutions depends on prototype selection. In this paper, we propose an efficient Salience-based Prototype Selection (MISPS) method to solve the above problems in MIL, aiming at keeping several representative instance prototypes (IPs) and eliminating other irrelevant or redundant instances from each bag. Then we present two variants of Citation-kNN and Bayesian-kNN based on MISPS, called MISPS-CkNN and MISPS-BkN, respectively.

The rest of this paper is organized as follows. In Sect. 2, we review some related work. In Sect. 3, we propose the MISPS method. In Sect. 4, we present our MISPS-CkNN and MISPS-kNN algorithms. In Sect. 5, we evaluate our methods on five benchmark data-sets. We conclude in Sect. 6.

2   Related Work

Many MIL algorithms have been proposed until now. They could be roughly divided into two main categories: generative and discriminative. Generative algorithms, including APR [1], DD [9] and EM-DD [10], use generative models to represent the target concept, aiming at finding a region in the instance feature space which includes all positive instances and excludes all negative instances. Alternatively, discriminative algorithms, such as Multi Instance Neural Networks (MI-NN) [11], Citation-kNN and Bayesian-kNN [7], MI-SVM and mi-SVM [12], use the discriminative learning paradigm to address MIL problems. Other variants are based on prototype selection and SVM, namely DD-SVM [13], MILES [14], MILD [15] and MILIS [16], which tackle MIL problems by converting MIL into single instance learning (SIL). The basic idea is mapping each bag into a new feature space constructed with some IPs selected from training bags, and then learning a SVM classifier in this new feature space.

Among all these algorithms, Citation-kNN and Bayesian-kNN [7] are two simple yet effective ones. They adapts the standard kNN algorithm to the MIL setting by using the set-based distance metrics, i.e. the minimal and maximal

Hausdorff distances [7]. In particular, Citation-kNN considers not only the nearest neighbors of an example but also examples which regard it as one of their nearest neighbors.

3   Salience-Based Prototype Selection for MIL

3.1   Notations

Let B_i^+ (B_i^-) denote a positive (negative) bag. Accordingly, B_{ij}^+ (B_{ij}^-) denotes an instance in B_i^+ (B_i^-). Let $\mathcal{B} = \{B_1^+, B_2^+, \ldots, B_{n^+}^+, B_1^-, B_2^-, \ldots, B_{n^-}^-\}$ denote a training set of n^+ positive bags and n^- negative bags. For the sake of simplicity, we will denote a bag as B_i with B_{ij}(s) when the label of it does not matter. Without ambiguity, B_{ij} also represents the feature vector of it depending on the context. $l(B_i)$ and $l(B_{ij})$ are the labels associated with B_i and B_{ij}, respectively. Note that $l(B_{ij})$ is not directly observable.

3.2   MISPS

Negative bags contain only negative instances while positive bags contain both positive and negative ones, so the ambiguity in labels of instances lies in positive bags. MISPS is aiming at identifying *true* positive instances in positive bags.

Assumption 1. *In general, any two positive or negative instances are close to each other while any positive and negative instance are far from each other.*

Definition 1. $\forall B_{ij} \in B_i$, *the salience of* B_{ij} *is defined as follows:*

$$Sal(B_{ij}) = \sum_{B_{ik} \in B_i \setminus \{B_{ij}\}} d(B_{ij}, B_{ik}) \ , \tag{1}$$

where $d(\cdot, \cdot)$ is the Euclidean distance between two instances.

Theorem 1. *Assume that $B_{ij} \in B_i^+$ is the unique positive or negative instance. Then*

$$\forall B_{ik} \in B_i^+ \setminus \{B_{ij}\}, \quad Sal(B_{ij}) > Sal(B_{ik}) \ . \tag{2}$$

Proof. By Definition 1, the salience of B_{ij} and B_{ik} is

$$Sal(B_{ij}) = \sum_{B_{it} \in B_i^+ \setminus \{B_{ij}\}} d(B_{ij}, B_{it}) = d(B_{ij}, B_{ik}) + \sum_{B_{it} \in B_i^+ \setminus \{B_{ij}, B_{ik}\}} d(B_{ij}, B_{it}) \ , \tag{3}$$

$$Sal(B_{ik}) = \sum_{B_{it} \in B_i^+ \setminus \{B_{ik}\}} d(B_{ik}, B_{it}) = d(B_{ik}, B_{ij}) + \sum_{B_{it} \in B_i^+ \setminus \{B_{ik}, B_{ij}\}} d(B_{ik}, B_{it}) \ . \tag{4}$$

Hence:

$$Sal(B_{ij}) - Sal(B_{ik}) = \sum_{B_{it} \in B_i^+ \setminus \{B_{ij}, B_{ik}\}} [d(B_{ij}, B_{it}) - d(B_{ik}, B_{it})] \ , \quad (5)$$

where $d(B_{ij}, B_{it})$ is the distance between a positive and negative instance, while $d(B_{ik}, B_{it})$ is the distance between two negative or positive instances. We would have $Sal(B_{ij}) > Sal(B_{ik})$ by Assumption 1.   □

Remark 1. We now consider the general case of Theorem 1, which is there are multiple positive and negative instances in a positive bag. Let m^+ and m^- denote the number of positive instances and the number of negative instances, respectively. Then the salience of any positive instance largely depends on m^- distances between the positive instance and other negative instances by Assumption 1 while the salience of any negative instance largely depends on m^+ distances between the negative instance and other positive instances. *Assuming that the distance between any positive instance and any negative instance fluctuates near a fixed value, then the above salience is largely dependent on m^- and m^+.* Therefore, the salience of any positive instance would be greater than that of any negative instance if m^- is greater than m^+; otherwise the former would be less than the latter. Strictly speaking, the above assumption does not always hold since there may exist outliers and noise in the data-set. However, this assumption might hold within the small scope of a single bag and could be largely satisfied according to the better experimental results in Sect. 5. We give the formal definition of the above description in Generalization 1.

Generalization 1. *Assume that $\{B_{i1}^+, B_{i2}^+, \ldots, B_{im^+}^+\} \subset B_i^+$ is the subset of positive instances and $\{B_{i1}^-, B_{i2}^-, \ldots, B_{im^-}^-\} \subset B_i^+$ is the subset of negative instances. Then, $\forall j \in \{1, 2, \ldots, m^+\}$, $k \in \{1, 2, \ldots, m^-\}$,*

$$Sal(B_{ij}^+) \begin{cases} > Sal(B_{ik}^-) & \text{if } m^+ < m^- \ , \\ < Sal(B_{ik}^-) & \text{if } m^+ > m^- \ . \end{cases} \quad (6)$$

If we can estimate which of m^+ and m^- is greater given a positive bag, we could depend on Generalization 1 to select *true* positive instances from the positive bag. When $m^+ < m^-$, the salience of any positive instance is greater than that of any negative instance by Generalization 1. Note that negative instances dominate the positive bag in this case. Thus, the higher its salience, the farther a candidate positive instance from all other negative instances by Definition 1 and Assumption 1, which means it is more likely to be positive. In this case, we can select instances with high salience as *true* positive ones. Similarly, when $m^+ > m^-$, we can select instances with low salience as *true* positive ones. The following Theorem 2 offers one possible solution to the above problem. We first give the definition of the probability that an instance is positive given a set of negative instances, then we formally introduce Theorem 2.

Definition 2. *Let* $\mathcal{B}^- = \{B_{rt} | B_{rt} \in B_r^-, r = 1, 2, \ldots, n^-\}$ *be the given set of negative instances. The probability that an instance B_{ij} is positive given $\mathcal{B}^-$ is:*

$$\Pr(l(B_{ij}) = 1 | \mathcal{B}^-) = 1 - \exp(-D(B_{ij}, \mathcal{B}^-)/\sigma^2) , \quad (7)$$

where σ is a scaling factor larger than 0, and

$$D(B_{ij}, \mathcal{B}^-) = \min_{B_{rt} \in \mathcal{B}^-} d(B_{ij}, B_{rt}) . \quad (8)$$

Remark 2. From Definition 2, we can easily deduce that $0 \le \Pr(l(B_{ij}) = 1 | \mathcal{B}^-) \le 1$, $\Pr(l(B_{ij}) = 1 | \mathcal{B}^-) = 1$ when $D(B_{ij}, \mathcal{B}^-) = +\infty$ and $\Pr(l(B_{ij}) = 1 | \mathcal{B}^-) = 0$ when $D(B_{ij}, \mathcal{B}^-) = 0$. This is well consistent with our intuition. If an instance is far from the set of negative instances, they would have a low similarity and hence the instance is likely to be positive; otherwise it is likely to be negative.

Theorem 2. $\forall B_i^+$, $\tilde{B}_i^+$ *is its corresponding re-sorted bag in descending order of salience of instances. Let m^+ and m^- be the number of positive instances and the number of negative instances in $\tilde{B}_i^+$, respectively, and $m = m^+ + m^-$. Assume that $\mathcal{B}^-$ is the given set of negative instances, then*

$$m^+ \begin{cases} < m^- & \text{if } \Pr(l(\tilde{B}_{i1}) = 1 | \mathcal{B}^-) > \Pr(l(\tilde{B}_{im}) = 1 | \mathcal{B}^-) , \\ > m^- & \text{if } \Pr(l(\tilde{B}_{i1}) = 1 | \mathcal{B}^-) < \Pr(l(\tilde{B}_{im}) = 1 | \mathcal{B}^-) . \end{cases} \quad (9)$$

Proof. Premise 1: $\Pr(l(\tilde{B}_{i1}) = 1 | \mathcal{B}^-) > \Pr(l(\tilde{B}_{im}) = 1 | \mathcal{B}^-)$. Assume $m^+ > m^-$, we would have, by Generalization 1:

$$Sal(\tilde{B}_{ij}^+) < Sal(\tilde{B}_{ik}^-) \quad \forall j \in \{1, 2, \ldots, m^+\}, k \in \{1, 2, \ldots, m^-\} . \quad (10)$$

Thus $\tilde{B}_{i1}$ is a negative instance and $\tilde{B}_{im}$ is a positive instance. Now by Assumption 1, $D(\tilde{B}_{i1}, \mathcal{B}^-) < D(\tilde{B}_{im}, \mathcal{B}^-)$. Then we will have, by (7):

$$\Pr(l(\tilde{B}_{i1}) = 1 | \mathcal{B}^-) < \Pr(l(\tilde{B}_{im}) = 1 | \mathcal{B}^-) . \quad (11)$$

This would contradict *Premise 1*, and thus cannot happen. So $m^+ < m^-$.
Premise 2: $\Pr(l(\tilde{B}_{i1}) = 1 | \mathcal{B}^-) < \Pr(l(\tilde{B}_{im}) = 1 | \mathcal{B}^-)$. Similarly, one can prove that $m^+ > m^-$ in *Premise 2*.   □

Given the above foundations, we can now present our MISPS method. First, we compute the salience of instances in each positive bag by Definition 1 and accordingly re-sort them in descending order. Next, we use **the instances in all negative bags** to compare the numbers of positive instances and negative instances based on Theorem 2. Finally, we select *true* positive instances from each positive bag by Generalization 1. Algorithm 1 summarizes the above process.

4   MISPS-CkNN and MISPS-BkNN

We use those selected IPs in Algorithm 1 to substitute the original positive bag. As for negative bags, we handle them in the same way since all the instances in

---

Algorithm 1. MISPS

Input: Training set $\mathcal{B}$, Number of salient instances per bag $SalNum$
Output: Set of IPs T

1: $\mathcal{B}^- = \{B_{rt}|B_{rt} \in B_r^-, r = 1, 2, \ldots, n^-\}$
2: **for** $i = 1$ to n^+ **do**
3:     Compute $Sal(B_{ij}^+)$ for each instance in B_i^+ by Definition 1
4:     Re-sort all instances in B_i^+ in descending order of salience
5:     Compute $D(B_{i1}^+, \mathcal{B}^-)$ and $D(B_{im^+}^+, \mathcal{B}^-)$ by (8) // m^+ is the number of instances

6:     **if** $D(B_{i1}^+, \mathcal{B}^-) > D(B_{im^+}^+, \mathcal{B}^-)$ **then**
7:        Add $B_{i1}^+, \ldots, B_{iSalNum}^+$ to T
8:     **else**
9:        Add $B_{i(m^+-SalNum+1)}^+, \ldots, B_{im^+}^+$ to T

---

negative bags are negative. Then we use the substitute bags to learn Citation-kNN or Bayesian-kNN [7]. Note that MISPS-CkNN has three parameters, i.e. the number of salient instances per bag $SalNum$ (refer to Algorithm 1), the number of the nearest neighbors of every bag $RefNum$ and the rank number between every bag and its citers $CiterRank$ [7], which can be tuned via the cross-validation method, while there are two parameters to be tuned in MISPS-BkNN, i.e. $SalNum$ and $RefNum$. Moreover, when the number of instances in a bag is less than $SalNum$, we select all instances from this bag and hence different substitute bags may have different numbers of instances.

5 Experiments

We use five MIL benchmark data-sets, i.e. Musk1, Musk2, Elephant, Fox and Tiger[1], to evaluate our MISPS-based algorithms. As introduced in Sect. 4, MISPS-CkNN and MISPS-BkNN have three parameters to tune, i.e. $SalNum$, $RefNum$ and $CiterRank$, where $CiterRank$ is 0 for MISPS-BkNN. In our experiments, $SalNum$ is restricted in $\{1, 2, \ldots, avgInstNum\}$, where $avgInstNum$ represents the average number of instances per bag. $RefNum$ is chosen from $\{1, 2, \ldots, 10\}$ and $CiterRank$ is set to $RefNum + 2$ based on the suggestion in [7]. Those giving the maximum 2-fold cross-validation accuracy on each training set are chosen and fixed in the subsequent experiments. Note that $SalNum$ in MISPS-BkNN is not the optimal one but empirically set to that in MISPS-CkNN and $CiterRank$ predefined as $RefNum + 2$ may not be the best choice.

We randomly run 10 times of 10-fold cross-validation on each data-set and report the average accuracies and corresponding 95% confidence intervals in Table 1 where the best performance is highlighted in boldface. We also list some other results for comparison, which are taken from [3], [15] and [17] . We can see that MISPS-CkNN and MISPS-BkNN are competitive to the state-of-the-art

---

[1] These data-sets are available at
http://www.uco.es/grupos/kdis/mil/fs/#experiments/

Table 1. Classification accuracies (%) of MIL algorithms on benchmark data-sets

Algorithm	Musk1	Musk2	Elephant	Fox	Tiger	Avg.
MISPS-CkNN	88.2[83.7 92.6]	82.3[77.3 87.3]	81.5[78.2 84.9]	**66.6[64.6 68.6]**	78.5[75.4 81.7]	79.4
MISPS-BkNN	87.1[83.6 90.6]	80.2[77.6 82.8]	79.7[77.5 81.8]	63.5[60.3 66.7]	78.0[75.2 80.7]	77.7
Citation-kNN	**88.8[84.1 93.5]**	82.6[78.5 86.6]	78.2[75.4 81.0]	58.9[54.6 63.1]	77.7[74.6 80.8]	77.2
Bayesian-kNN	87.8[83.4 92.2]	80.6[77.3 83.9]	75.8[73.3 78.4]	58.5[56.1 61.0]	76.4[73.9 78.9]	75.8
MI-SVM [12]	77.9	84.3	81.4	57.8	**84.0**	77.1
mi-SVM [12]	87.4	83.6	82.2	58.2	78.4	78.0
DD-SVM [13]	85.8	**91.3**	83.5	56.6	77.2	79.0
MILES [14]	86.3	87.7	**84.1**	63.0	80.7	**80.4**
MILD_B [15]	88.3	86.8	82.9	55.0	75.8	77.8
MILIS [16]	88.6	91.1	N/A	N/A	N/A	N/A

Table 2. Computation time (in seconds) of kNN-based algorithms on benchmark data-sets (time spent on prototype selection + time spent on classification)

Data-Set	MISPS-CkNN	MISPS-BkNN	Citation-kNN	Bayesian-kNN
Musk1	0.34 + 0.58	**0.34 + 0.37**	0 + 1.04	0 + 0.73
Musk2	20.40 + 5.60	**20.87 + 3.75**	0 + 154.41	0 + 105.48
Elephant	1.51 + 5.47	**1.53 + 3.59**	0 + 8.47	0 + 5.51
Fox	1.61 + 3.28	**1.59 + 2.08**	0 + 7.76	0 + 5.05
Tiger	1.59 + 4.75	**1.58 + 2.63**	0 + 6.71	0 + 4.40

MIL algorithms. However, it is more meaningful to note that they give similar or even higher classification accuracies as compared to Citation-kNN and Bayesian-kNN. In three out of five data-sets, MISPS-CkNN and MISPS-BkNN are more effective than Citation-kNN and Bayesian-kNN, respectively, and superior to them with respect to average accuracy.

Table 1 also shows that MISPS-BkNN is highly comparable to MISPS-CkNN, which indicates MISPS could keep the effectiveness of kNN without resorting to *Citation* [7]. This is beneficial to the learning process since not *Reference* but *Citation* leads to most of the time consumption in Citation-kNN. Therefore, we could substitute *Citation* with MISPS for these kNN-based algorithms. The quantitative empirical comparisons of 2-fold cross-validation on each data-set are given in Table 2. We can find that MISPS-BkNN is the most efficient one in Table 2. Overall, MISPS-BkNN is the best one.

6   Conclusions

Citation-kNN and Bayesian-kNN are two MIL variants of the standard kNN algorithms, so they also suffer several drawbacks like kNN, i.e. excessive storage requirements, low classification response and weak robustness to labeling noise. In this paper, we try to solve these problems with an explicit prototype selection method, and propose two variants of Citation-kNN and Bayesian-kNN based on our prototype selection method. The experimental results demonstrate the effectiveness of our methods.

Acknowledgments. This research has been supported by the National Natural Science Foundation of China under the Grant Nos. 61173087 and 61073128.

References

1. Dieterich, T.G., Lathrop, R.H., Lozano-Pérez, T.: Solving the Multiple Instance Problem with Axis-parallel Rectangles. Artif. Intell. 89, 31–71 (1997)
2. Viola, P., Platt, J., Zhang, C.: Multiple Instance Boosting for Object Detection. In: NIPS, pp. 1417–1424. MIT Press, Cambridge (2006)
3. Fung, G., Dundar, M., Krishnapuram, B., Rao, B.R.: Multiple Instance Learning for Computer Aided Diagnosis. In: NIPS, pp. 425–432. MIT Press, Cambridge (2007)
4. Babenko, B., Yang, M.H., Belongie, S.: Robust Object Tracking with Online Multiple Instance Learning. IEEE Trans. Pattern Anal. Mach. Intell. 33, 1619–1632 (2011)
5. Zhang, Q., Goldman, S.A., Yu, W., Fritts, J.E.: Content-Based Image Retrieval Using Multiple-Instance Learning. In: ICML, pp. 682–689. Morgan Kaufmann, San Francisco (2002)
6. Rahmani, R., Goldman, S.A., Zhang, H., Cholleti, S.R., Fritts, J.E.: Localized Content-Based Image Retrieval. IEEE Trans. Pattern Anal. Mach. Intell. 30, 1902–1912 (2008)
7. Wang, J., Zucker, J.D.: Solving the Multiple-Instance Problem: A Lazy Learning Approach. In: ICML, pp. 1119–1126. Morgan Kaufmann, San Francisco (2000)
8. Garcia, S., Derrac, J., Cano, J., Herrera, F.: Prototype Selection for Nearest Neighbor Classification: Taxonomy and Empirical Study. IEEE Trans. Pattern Anal. Mach. Intell. 34, 417–435 (2012)
9. Maron, O., Lozano-Pérez, T.: A Framework for Multiple-Instance Learning. In: NIPS, pp. 570–576. MIT Press, Cambridge (1998)
10. Zhang, Q., Goldman, S.A.: EM-DD: An Improved Multiple-Instance Learning Technique. In: NIPS, pp. 1073–1080. MIT Press, Cambridge (2001)
11. Ramon, J., De Raedt, L.: Multi Instance Neural Networks. In: ICML 2000 Workshop on Attribute-Value and Relational Learning (2000)
12. Andrews, S., Tsochantaridis, I., Hofmann, T.: Support Vector Machines for Multiple-Instance Learning. In: NIPS, pp. 561–568. MIT Press, Cambridge (2003)
13. Chen, Y., Wang, J.Z.: Image Categorization by Learning and Reasoning with Regions. J. Mach. Learn. Res. 5, 913–939 (2004)
14. Chen, Y., Bi, J., Wang, J.Z.: MILES: Multiple-Instance Learning via Embedded Instance Selection. IEEE Trans. Pattern Anal. Mach. Intell. 28, 1931–1947 (2006)
15. Li, W.J., Yeung, D.Y.: MILD: Multiple-Instance Learning via Disambiguation. IEEE Trans. on Knowl. and Data Eng. 22, 76–89 (2010)
16. Fu, Z., Robles-Kelly, A., Zhou, J.: MILIS: Multiple Instance Learning with Instance Selection. IEEE Trans. Pattern Anal. Mach. Intell. 33, 958–977 (2011)
17. Erdem, A., Erdem, E.: Multiple-Instance Learning with Instance Selection via Dominant Sets. In: Pelillo, M., Hancock, E.R. (eds.) SIMBAD 2011. LNCS, vol. 7005, pp. 177–191. Springer, Heidelberg (2011)

Calibrate a Moving Camera on a Linear Translating Stage Using Virtual Plane + Parallax

Xiaoqiang Zhang, Yanning Zhang, Tao Yang, and Zhengxi Song

ShaanXi Provincial Key Laboratory of Speech and Image Information Processing
School of Computer Science, Northwestern Polytechnical University, Xi'an, China
{vantasy,lirpa}@mail.nwpu.edu.cn, ynzhang@nwpu.edu.cn,
yangtaonwpu@163.com

Abstract. For a fixed camera array, the limited number of views restricts the quality of the imaging result. In this paper, we build a virtual dense camera array using different views from a single moving camera to increase the number of view conveniently. A virtual plane + parallax based calibration algorithm is proposed to estimate the relative positions between different views. The main characteristics of this system and calibration approach include: (1) Due to the mobility, one can vary views in the light of different scene configurations. (2) Via calibrating the moving camera under the plane + parallax framework, the computation cost of aligning images from all views is reduced. (3) A moving camera system is set up by mounting a camera on a translating stage which can slide on a linear track. Extensive experiment results demonstrate that our approach can enhance detail and quality of the occluded objects effectively compared to the fixed camera array synthetic aperture imaging approach.

Keywords: Camera Array Calibration, Synthetic Aperture Imaging.

1 Introduction

Recently, more and more researchers take interest handling computer vision or graphics issues using multiple sensors. Among all of the multi-view approaches, dense camera array, which aligns tens even hundreds of camera on a plane, has been frequently used when acquiring the light field of a scene [1]. According to light field theory, the light field or the radiance of all rays in a scene, can be sampled by a dense camera array and represented by images from different viewpoint. By sampling the rays of a scene, camera array can be used in many computer vision or graphics applications:

Synthetic aperture imaging. In this application, the dense camera array is utilized to simulate a virtual camera which has an extreme large virtual or synthetic aperture. By projecting images from different viewpoints onto some plane, the average of those projected images is called the synthetic aperture image (SAI) on the very focus plane. Due to the limited depth of field of the

J. Yang, F. Fang, and C. Sun (Eds.): IScIDE 2012, LNCS 7751, pp. 48–55, 2013.

(a) Moving camera system using Pointgrey Flea 3

(b) One view from the moving camera system

(c) Hidden object imaging result

Fig. 1. Moving camera system and synthetic aperture imaging result

virtual aperture, object on focus plane would stay sharp on SAI, while occlusion, which is off the focus plane, would become blur.The ability of seeing through occlusion is the main characteristic of SAI, and it makes camera array useful in video surveillance [2][3].

Natural Video Matting. Matting softly separates the foreground element from the background by setting per-pixel opacity on an image. Taking advantage of camera arrays, a high quality alpha matting result of an uncontrolled environment with textured background can be computed at near real-time rates [4]. The mean and variance of color value of corresponding pixels from different viewpoints is computed using projecting technique in SAI. Then those statistics are employed to construct a trimap and later upgrade to an alpha matte [4].

The basic intention to use multiple cameras in those applications is to observe a scene or an interest object from different viewpoints so that information could be aligned together. Therefore, the number of cameras in the array, or the amount of ray samples, is critical to the performance of the applications above. In camera array based natural video matting, since trimap and the final matting result are extracted from statistics of different viewpoints, adding more cameras could increase the accuracy when estimating the mean and variance, so that the final performance can be improved. However, in most of the camera array systems [1][2][3][4][5][6] cameras are fixed. Therefore, it is not convenient to add new cameras to the array. To overcome this difficulty, we place one camera on a linear translating stage (see Fig.1(a)). By sliding the camera on the track, light field in a static scene can be captured. Comparing with the fixed camera array, we could get arbitrary viewpoint along the baseline by shifting camera on the stage, making it easy to add new view or even vary view in the light of different scene configurations.

As same as many other computer vision applications, accurate calibration of the cameras is important to the implementations of the algorithms. Usually, a full metric calibration is employed on the camera array system. Since all cameras lie on the same plane or even on the same line, the co-planar or co-linear properties can be utilized to reduce the number of parameters in calibration [6]. In this paper, we also take advantage of co-linear property of all viewpoints along the stage, and intend to find a convenient way to calibration all views. First we use structure from motions techniques to estimate the pose of all viewpoints. Then a virtual Plane + Parallax procedure is employed to recover the relative positions between different views. When the calibration is done, we could simply shift images from different view to synthesize them together so that the computation cost is saved. In Fig.1(b), images from all 39 views are shown. In Fig.1(c) we give synthetic aperture imaging using 39 views with the parameters of the proposed calibration approach under the linear translating stage setup. From Fig.1 we can see that our moving camera system can be a powerful device for hidden object imaging.

The rest of the paper is structured as follows: A brief summary of the related work is given in Section 2. In Section 3, we introduce the proposed calibration algorithm, including viewpoint pose estimation, generating parallax and viewpoint position recovery. Section 4 shows the experimental results. Finally, we conclude the paper and discuss future directions in Section 5.

2   Related Work

In this section, we summarize a sample of work that covers the research we build upon in developing our algorithm, including light field and camera array calibration.

Light Field. The concept of light field was first defined by Arun Gershun in the year 1936, describing the amount of light traveling in every direction through every point in space[7]. Later in 1996, light field was introduced into computer graphic and vision field. Marc Levoy believe that light field is a 4D function that measures the radiance, which is the amount of light travelling along the ray, at a point in a given direction[7]. The scale of scene which is described by the light field range from city-scale to cell-scale and can be captured by totally different devices. For a static scene, a single moving camera can be used. Camera array is proper to capture the long baseline light field of a dynamic scene. In a short baseline scenario, multi-view can be replaced by a camera with array of lens. The recent light field camera Lytro[8] is an example. As for cell-scale scene, a light field microscopic is utilized. In our setup, we use a single camera moving on the linear translating stage to capture the light field of a static scene.

Camera Array Calibration. Generally speaking, there are two different ways to calibrate dense camera array, the metric calibration and calibration under Plane + Parallax framework [6]. The metric calibration estimates the focal length, distortion parameters and the pose of all cameras in the array. Zhangs method [9] is a commonly used algorithm when a metric calibration is needed. In [6],

Vaish found that the co-planar or co-linear properties can be utilized to reduce the number of parameters in calibration. The only parameter they estimate is the relative position of different cameras and it can be recovered by decomposing the rank-1 matrix of parallax from several parallel planes. Bundle adjustment is avoided so that there is no concern about the initialization. In this paper, we intend to use the Plane + Parallax framework to calibrate all views by creating control points on several virtual parallel planes.

3   Virtual Plane + Parallax Based Moving Camera Calibration

The main purpose of our algorithm is to calibrate all views of the moving camera under the Plane + Parallax framework so that the co-planar properties could be used. Inspired by the calibration procedure in [6], a set of 3D points on different parallel planes can be used to compute parallax between different views. In order to establish the projective relation between a 3D point and its corresponding pixel on each viewpoint, we need to know the orientation and position of each view with respect to the world coordinate system. After that, a reference plane and a reference view are defined and homography from each view to the reference view induced by the reference plane is estimated. The parallax on the reference plane can be computed using the view poses and the homographies. The final camera position is recovered by decompose the parallax matrix. A brief flow chart of the algorithm is shown in Fig.2.

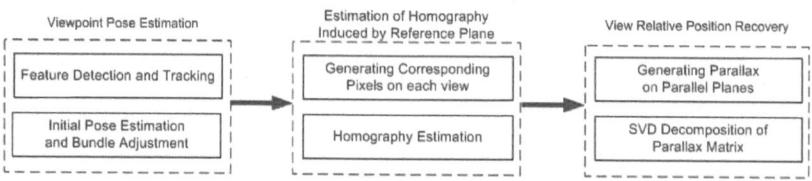

Fig. 2. Flow chart of the moving camera calibration algorithm

• Viewpoint Pose Estimation

Viewpoint pose estimation is essentially a structure from motion problem, that is, to simultaneity estimates the locations of 3D points, cameras pose and its internal parameters based on a set of sparse correspondence on 2D images. Because camera in our system is moving, feature points from each viewpoint are tracked. Then those potential correspondences are filtered by outlier elimination. An initial structure is finally refined using bundle adjustment to gain a global optimization of the entire image sequence.

As the first step of our calibration, we care more about the viewpoint pose than the 3D sparse structure. We use Voodoo Camera Tracker[10], which is a structure from motion software to estimate the pose. The camera pose is presented by the

CAHV camera model and parameterized by the CAHV vectors. Here we use matrix P to represent the entire projection matrix.

• Generating Parallax and Viewpoint Position Recovery

In Plane + Parallax framework, there are two main assumptions of the setup of the camera array system in order to simplify the computation when align images from different viewpoints together:

1. All cameras lie on the same plane called camera plane.

2. A reference plane that parallel to the camera plane is assumed. Images from all cameras are aligned on the reference plane first via homography projection. The focus plane is parallel to the reference plane.

The two parameters to estimate under this framework are the homography induced by the reference plane and viewpoint position which we use to compute parallax. The estimation steps are given as follows:

1. Homography induce by the reference plane.

Assume that there are k different views in the total sequence, namely $V_1, \ldots, V_R$, $\ldots, V_k$, where V_R is a reference view we pick to measure the parallax. Let Π_R represents the reference plane and H_i is the homography from V_i to V_R induced by Π_R. Since the pose of all views is known, H_i can be directly computed according to the camera matrix of the two views and the coordinate of Π_R. Here we use an alternative way to estimate this homography. Suppose that there are j control points, say $X_{R1}, \ldots, X_{Rj}$, on Π_R. By projecting those 3D points to the imaging plane of V_i and V_R , a set of corresponding pixels can be used to estimate the homography if $j \geq 4$.

2. Viewpoint position.

With those homograpies, images could be aligned on the reference plane, so that a synthetic aperture imaging on Π_R is computed. When another synthetic image on a new parallel focus plane is desired, we can shift those aligned images by different parallax with respect to views. Since there is similar triangles relation, parallax is proportional to the relative position from one viewpoint to the reference view. Suppose Π_i is a desired focus plane that parallel to Π_R. X_{ij} is a 3D point on Π_i. By projecting X_{ij} to the reference view V_R applying the projection relation, we get x_{ijR} on the imaging plane of V_R. Similarly, x_{ijk} is the projection of X_{ij} on view V_k. Applying homography projection to x_{ijk}, we get $\tilde{x}_{ijk}$.

$\tilde{x}_{ijk}$ is shown as the left red point on Π_R. Thus, the parallax p_{ijk} between V_k and V_R induced by 3D point X_{ij} is given by $p_{ijk} = \tilde{x}_{ijk} - x_{ijR}$.

Since p_{ijk} is proportional to the relative position d_k. Moreover, in [6], Vaish shows that we can take vantage of the proportion to recover the relative position. Suppose that there are I different parallel planes, say $\Pi_1, \ldots, \Pi_I$. On each planes, there are J control points. By computing parallax induced by these points and averaging them, we got the $p_{ik} = \frac{1}{J} \sum_{j=1}^{J} p_{ijk}$, which represents the averaging parallax from V_k to V_R on Π_i.

All the averaging parallax can be collected in a I-by-K matrix M. There exists a rank-1 constraint in M because of the proportion between parallax and relative position. Therefore, a SVD decomposition of M can be adopted to recover each view positions.

As mentioned before, by using the Plane + Parallax framework, the computation cost of the final synthetic imaging procedure is reduced. Here we give the comparison. When using the metric calibration parameters, the projective model is presented by a 3×4 P matrix. For a 3D point, computing the corresponding pixel in one view needs 12 times multiplication and 9 times addition. In Plane + Parallax framework, the projection process involves a homography projection and an image shift. The homography, which is induced by the reference plane, is fixed. Therefore, a matrix multiplication can be replaced by a simple table lookup. The entire projection process only needs a table lookup and 2 times addition.

4 Experimental Results

In order to evaluate the performance of the proposed approach, we have set up a virtual camera array with a camera moving a linear track. Experiments have been conducted using our own dataset in order to prove the calibration accuracy. Details of the result as well as the camera system will be given in the following subsections.

• System Setup
In our laboratory, we build a moving camera system (see Fig.1(a)). A Pointgrey Flea 3 camera was mounted on a linear translating stage and is connected to a single PC via PCI-E Firewire adaptors. The stage can slide on a 3-meter linear track, thus a 3-meter virtual aperture can be simulated. The main goal of this system is to increase the number of viewpoint by moving the camera so that the imaging performance can be improved. The input image sequence is captured with the size of 640×480 pixels. In order to achieve the imaging quality comparison between the moving camera system and the fixed camera array, we also setup a fixed camera array system using 7 Pointgrey Flea 3 cameras.

• Indoor Planar Imaging under Occlusion
As mentioned before, the main characteristic of synthetic aperture imaging based on camera array is hidden object imaging. In this scenario, the ability of seeing through occlusion of our system is verified. In our laboratory, an X-rite Color Checker (see Fig.3(b)) is placed behind the plants. Since the scope of this paper is limited, in Fig.3(a) we give 3 subimages of all 39 selected views from the moving camera system, from which we can see that, the color checker is partially occluded by the plants. The synthetic aperture imaging result on the plane of the color checker using all 39 views is given in Fig.3(c). The same scene is also captured by the fixed camera array system and the synthetic aperture imaging result in given in Fig.3(d). Please note that there are still some ghost pixels on Fig.3(d), while is clear on Fig.3(c).

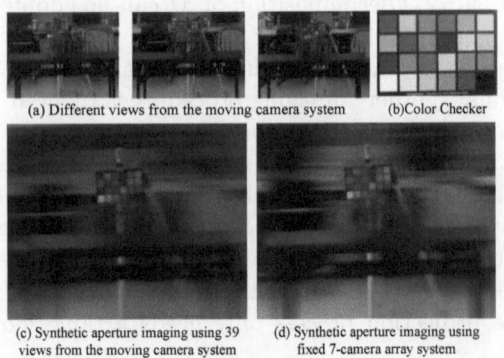

(a) Different views from the moving camera system   (b)Color Checker

(c) Synthetic aperture imaging using 39
views from the moving camera system

(d) Synthetic aperture imaging using
fixed 7-camera array system

Fig. 3. Comparison of the indoor hidden planar imaging result

(a) 4 of all 38 views from the moving camera system   (b) Hidden planar imaging result using all 38 select views

Fig. 4. Synthetic aperture imaging result of a hidden calibration board outdoor

• Outdoor Hidden Planar Imaging

Considering the scene scale is small and the light is easy to control in the laboratory, we present an outdoor experiment here to verify the applicability of our system outdoor. In this scene, a calibration board is placed behind a tree (see Fig.4(a)). The scene is captured by the moving camera system and we select 38 views. A sample of those views is shown in Fig.4(a), from which we can see that the board is partially occluded in each view. By calibrating all views and focus on the plane of the board, we get the hidden planar imaging result shown in Fig.4(b). The rays from the front tree are completely removed and a clear image of the board can be achieved.

5   Conclusion

In this work we intend to build a virtual camera array system by moving a camera and present a calibration algorithm under Plane + Parallax framework. A moving camera system is set up by mounting a camera on a translating stage that can slide on a linear track. By estimating the homography induced by the reference plane and the relative position between different views, one can align images from all viewpoints by a table lookup and 2 times addition. Experiment results demonstrate that the quality of synthetic aperture image is improved by increasing the number of views. In future work, we would like to break the co-linear constrain and to seek ways to align views from a freely moving camera. In the meanwhile, a visibility analysis of the scene is helpful to increase the quality of the synthetic aperture image, by which means we can filter rays so that ghost pixel of occlusion can be removed.

Acknowledgments. This work is supported by the National Natural Science Foundation of China (No.61272288,No.61231016,No.60903126), China Postdoctoral Special Science Foundation(No.201003685), China Postdoctoral Science Foundation(No.20090451397),the NPU Foundation for Fundamental Research(No.JC201120), and Plan of Soaring Star of Northwestern Polytechnical University(No.12GH0311).

References

1. Wilburn, B., Joshi, N., Vaish, V., Talvala, E., Antunez, E., Barth, A., Adams, A., Horowitz, M., Levoy, M.: High performance imaging using large camera arrays. ACM Transactions on Graphics 24(3), 765–776 (2005)
2. Yang, T., Zhang, Y., Tong, X., Zhang, X., Yu, R.: Continuously tracking and see-through occlusion based on a new hybrid synthetic aperture imaging model. In: 2011 IEEE Conference on Computer Vision and Pattern Recognition (CVPR), pp. 3409–3416. IEEE (2011)
3. Joshi, N., Avidan, S., Matusik, W., Kriegman, D.: Synthetic aperture tracking: tracking through occlusions. In: IEEE 11th International Conference on Computer Vision, ICCV 2007, pp. 1–8. IEEE (2007)
4. Joshi, N., Matusik, W., Avidan, S.: Natural video matting using camera arrays. ACM Transactions on Graphics (TOG) 25, 779–786 (2006)
5. Vaish, V., Levoy, M., Szeliski, R., Zitnick, C., Kang, S.: Reconstructing occluded surfaces using synthetic apertures: Stereo, focus and robust measures. In: 2006 IEEE Computer Society Conference on Computer Vision and Pattern Recognition, vol. 2, pp. 2331–2338. IEEE (2006)
6. Vaish, V., Wilburn, B., Joshi, N., Levoy, M.: Using plane+ parallax for calibrating dense camera arrays. In: Proceedings of the 2004 IEEE Computer Society Conference on Computer Vision and Pattern Recognition, CVPR 2004, vol. 1, pp. 1–2. IEEE (2004)
7. Levoy, M.: Light fields and computational imaging. Computer 39(8), 46–55 (2006)
8. Ng, R.: Lytro light field camera, http://www.lytro.com/
9. Zhang, Z.: A flexible new technique for camera calibration. IEEE Transactions on Pattern Analysis and Machine Intelligence 22(11), 1330–1334 (2000)
10. Thormahlen, T., Broszio, H.: Voodoo camera tracker, http://www.digilab.uni-hannover.de/docs/manual.html

Neural Population Decoding in Short-Time Windows

Wenhao Zhang[1,2] and Si Wu[2]

[1] Institute of Neuroscience, Shanghai Institutes for Biological Sciences,
Chinese Academy of Sciences, Shanghai 200031, China
[2] State Key Lab of Cognitive Neuroscience and Learning, Beijing Normal University, Beijing,
100875, China

Abstract. External information is encoded in spiking activities of neural population. The present study investigates the performance of population decoding in a
short-time window. Two decoding strategies, namely, maximum likelihood inference and template-matching, are explored. We find that in a short-time window, two
methods are not efficient and that their errors satisfy the Cauchy distributions. As
expected, maximum likelihood inference outperforms template-matching asymptotically. However, in a very short time window, template-matching has smaller
decoding errors than maximum likelihood inference. The implication of this result
is discussed.

1 Introduction

External information, no matter it is received through the visual, the auditory, the olfac-
tory, or the somatosensory system, is encoded in spiking activities of neurons. A neural
system typically uses the joint responses of a population of neurons to encode the stimulus information, which is called population code. Understanding the nature of neural
code is prerequisite for us to elucidate higher brain functions.

A large volume of theoretical work has investigated the performances of population decoding. These studies often analyze the Fisher information of the likelihood
function for a given encoding model (the latter specifies the probability of observing
a neural population response for a giving stimulus value) (e.g., [1,2]). According to the
Cramer-Rao, the inverse of Fisher information is the lower bound of decoding error
(measured by the variance) for any unbiased estimator can achieve [3]. It is also known
that maximum likelihood inference (MLI) can reach the Cramer-Rao bound asymptotically. Thus, Fisher information is a good indication of the performance of a population
code in many circumstances. This is not true, however, if a neural estimator has to
read-out the stimulus in a very short-time window: in this case MLI, and so do other
decoding methods, is no longer asymptotically efficient, implying that the Cramer-Rao
is not achievable in practice and the Fisher information is irrelevant to decoding performance.

With the pressure of surviving in a natural environment, animals need to respond to
external inputs as quickly as possible when dangers occur. Indeed, experimental data
has revealed that the brain has the capacity of extracting information in rapid speed. For
example, it takes only about 150ms for the brain to detect and recognize visual objects
from a complex natural scene [4], and during this process the stimulus information has

J. Yang, F. Fang, and C. Sun (Eds.): IScIDE 2012, LNCS 7751, pp. 56–63, 2013.

been processed through many layers of neurons. Thus, to understand brain functions, it is critical to analyze the performance of population decoding in short-time windows. To our knowledge, there is only one work in the literature formally studying the effect of short-time window on population decoding [5], which analyzed the optimal tuning function for decoding in a short-time window. It didn't explore the effect of short-time window on the performances of different decoding strategies.

In the present study, we analyze the performances of MLI and template-matching (TM), in short-time windows. TM is a simpler method compared with MLI, and it can be implemented by biologically plausible attractor networks [2,6]. As expected, MLI outperforms TM when the decoding time window is sufficiently large. Interestingly, we find that in a very short-time window, TM outperforms MLI. This gives us insight into understanding how neural decoding may work in practice.

2   The Encoding Model

Consider a continuous stimulus is encoded by an ensemble of N neurons. Without loss of generality, we assume the stimulus value x is in the range of $(-\infty, \infty)$. Denote c_i the preferred stimulus of the ith neuron, and c_i are uniformly distributed and cover the whole range of x.

The tuning function of a neuron, i.e., the averaged response over trials of a neuron to the stimulus, is given by $f_i(x) = Ae^{-(c_i-x)^2/2\sigma^2}$, where A is the maximal firing rate of the neuron and σ is the tuning width. Suppose in a time window t, n_i spikes are observed for the ith neuron. Denote $\mathbf{n} = \{n_i\}$, for $i = 1, \ldots, N$, the population response. Consider the spiking process of a neuron satisfies the Poisson statistics and neurons fire independently, the probability of observing the population activity $\mathbf{n}$ in a time window t is written as

$$P(\mathbf{n}|x) = \prod_i P_i(n_i|x) = \prod_i \frac{(f_i t)^{n_i}}{n_i!} e^{-f_i t}. \tag{1}$$

3   Population Decoding

3.1   Decoding Methods

Neural decoding is to estimate the stimulus value x from the observed neuronal activities $\mathbf{n}$. Two decoding methods are investigated in this work, which are MLI and TM.

Maximum Likelihood Inference. The result of MLI, denoted as $\hat{x}_{MLI}$, is obtained through maximizing the log likelihood function $\ln P(\mathbf{n}|x)$, i.e., by solving

$$\nabla \ln P(\mathbf{n}|\hat{x}_{MLI}) = 0, \tag{2}$$

where $\nabla g(x) = dg(x)/dx$.

Substituting Eq.(1) into (2), we obtain $\hat{x}_{MLI} = \sum_i n_i c_i / \sum_i n_i$. Note that for independent Poisson noises, the result of MLI agrees with that of Center-of-Mass (COM) [7], but in general MLI outperforms COM.

Template Matching. The result of TM, denoted as $\hat{x}_{TM}$, is obtained through maximizing the overlap between neuronal responses and a template given by the tuning function, that is, $\hat{x}_{TM} = \arg\max_{x} \sum_{i} n_i(x,t) f_i(x)$.

3.2   Decoding Performances

The decoding errors of MLI and TM are calculated.

Maximum Likelihood Inference. Suppose $\hat{x}_{MLI}$ is close enough to the true value x. We expand $\ln P(\mathbf{n}|x)$ at x to get

$$\nabla \ln P(\mathbf{n}|\hat{x}_{MLI}) \simeq \nabla \ln P(\mathbf{n}|x) + \nabla\nabla \ln P(\mathbf{n}|x)(\hat{x}_{MLI} - x) = 0. \tag{3}$$

Hence, the decoding error, $\xi_{MLI} \equiv \hat{x}_{MLI} - x$, is given by

$$\xi_{MLI} = \frac{\nabla \ln P(\mathbf{n}|x)}{-\nabla\nabla \ln P(\mathbf{n}|x)} = \frac{g_{MLI}^{(1)}}{g_{MLI}^{(2)}}, \tag{4}$$

where $g_{MLI}^{(1)}$ and $g_{MLI}^{(2)}$ are two stochastic variables, which are calculated to be

$$g_{MLI}^{(1)} = \nabla \ln \left[\prod_i \frac{(f_i t)^{n_i}}{n_i!} e^{-f_i t} \right] = \sum_i n_i \frac{c_i - x}{\sigma^2}, \quad g_{MLI}^{(2)} = \sum_i \frac{n_i}{\sigma^2}. \tag{5}$$

To get the above results, we have used the condition $\sum_i f_i' = (\sum_i f_i)' = 0$, due to the uniform distribution of neuronal preferred stimuli in the parameter space x.

In the large N limit, according to the central limit theorem, $g_{MLI}^{(1)}$ and $g_{MLI}^{(2)}$ satisfy the Gaussian distributions, whose means and variances are calculated to be (see Appendix 2)

$$E\left[g_{MLI}^{(1)}\right] = 0, E\left[g_{MLI}^{(2)}\right] = \frac{\sqrt{2\pi} A t \rho}{\sigma} = I_F,$$

$$V\left[g_{MLI}^{(1)}\right] = I_F, V\left[g_{MLI}^{(2)}\right] = \frac{\sqrt{2\pi} A t \rho}{\sigma^3} = \Omega^2, \tag{6}$$

where $E[\cdot]$ ($V[\cdot]$) denote the mean (variance) with respect to $P(\mathbf{n}|x)$, respectively. $I_F = E[-\nabla\nabla \ln P(\mathbf{n}|x)]$ is the Fisher information according to the definition. ρ denotes the neuron density.

Combining the above results, we can write the error of MLI as

$$\xi_{MLI} = \frac{\mu}{I_F + \nu}, \tag{7}$$

where μ and ν are independent Gaussian random variables having zero mean and the variances I_F and Ω^2, respectively (see Eq.(6)).

Template Matching. The decoding error of TM, $\xi_{TM} \equiv \hat{x}_{TM} - x$, can be calculated similarly. By expanding $\sum_i n_i f_i(\hat{x}_{TM})$ at the true stimulus value x, we get

$$\xi_{TM} = \frac{\nabla \left(\sum_i n_i f_i(x) \right)}{-\nabla\nabla \left(\sum_i n_i f_i(x) \right)} = \frac{g_{TM}^{(1)}}{g_{TM}^{(2)}}, \tag{8}$$

where

$$g_{TM}^{(1)} = \frac{A}{\sigma^2} \sum_i n_i(c_i - x)e^{-(c_i-x)^2/2\sigma^2}, \tag{9}$$

$$g_{TM}^{(2)} = \sum_i n_i A e^{-(c_i-x)^2/2\sigma^2} \left[\frac{1}{\sigma^2} - \left(\frac{c_i - x}{\sigma^2} \right)^2 \right].$$

In the large N limit, $g_{TM}^{(1)}$ and $g_{TM}^{(2)}$ can be regarded as Gaussian random variables with their means and variances given by (see Appendix 3):

$$E\left[g_{TM}^{(1)} \right] = 0, \, E\left[g_{TM}^{(2)} \right] = \frac{\sqrt{\pi}A^2 t\rho}{2\sigma} = M_2,$$

$$V\left[g_{TM}^{(1)} \right] = \frac{\sqrt{2\pi}A^3 t\rho}{3\sqrt{3}\sigma} = \Omega_1^2, \, V\left[g_{TM}^{(2)} \right] = \frac{2\sqrt{2\pi}A^3 t\rho}{3\sqrt{3}\sigma^3} = \Omega_2^2. \tag{10}$$

Thus, the error of TM is expressed as

$$\xi_{TM} = \frac{\phi}{M_2 + \psi}, \tag{11}$$

where ϕ and ψ are independent Gaussian random variables having zero mean and the variances Ω_1^2 and Ω_2^2, respectively.

4 Asymptotic and Finite Time Performances of Population Decoding

4.1 Asymptotic Performances

Let us study the asymptotic performances of two decoding methods first. It is easy to check that ν/I_F and ψ/M_2 are in the order of $1/(At\rho\sigma)^{1/2}$. As the decoding time window $t \to \infty$, ν/I_F and ψ/M_2 approach to zero, thus ν and ψ can be neglected compared to I_F and M_2 (see the denominators in Eq.(7) and (11)). With these conditions, ξ_{MLI} and ξ_{TM} are approximated to be $\xi_{MLI} = \nu/I_F$ and $\xi_{TM} = \phi/M_2$, respectively, and they satisfy the Gaussian distributions. Apparently, the variance of the decoding error of MLI, $V[\xi_{MLI}] = 1/I_F$, equals to the Cramer-Rao bound. This means that MLI is asymptotically efficient. On the other hand, the variance of the decoding error of TM is given by $V[\xi_{TM}] = \frac{4\sqrt{2}}{3\sqrt{3\pi}} \frac{\sigma}{At\rho} > \frac{1}{I_F}$. Thus, TM has a larger decoding error than MLI when the decoding time is sufficiently large.

4.2   Finite Time Performances

When the decoding time window t is short, ν and ψ are comparable to I_F and M_2, ξ_{MLI} and ξ_{TM} no longer satisfy the Gaussian distributions, but rather the Cauchy ones. In the below we analyze their properties.

Maximum Likelihood Inference. For a given value of the random variable ν, the conditional probability density function (p.d.f) of $\hat{x}_{MLI}$ is calculated to be

$$p(\xi_{MLI}|\nu) = p\left(\mu = \xi_{MLI}(I_F + \nu)|\nu\right)\left|\frac{d\mu}{d\xi_{MLI}}\right| \qquad (12)$$
$$= p\left(\mu = \xi_{MLI}(I_F + \nu)\right)|I_F + \nu|.$$

In the above the independency between μ and ν is used. Thus,

$$
\begin{aligned}
p(\xi_{MLI}) &= \int p\left(\xi_{MLI}|\nu\right)p(\nu)d\nu = \int |I_F + \nu|p\left(\mu = \xi_{MLI}(I_F + \nu)\right)p(\nu)d\nu, \\
&= \frac{1}{2\pi\sqrt{I_F}\Omega}\int |I_F + \nu|e^{-\xi_{MLI}^2(I_F+\nu)^2/2I_F}e^{-\nu^2/2\Omega^2}d\nu, \\
&\underset{t=I_F+\nu}{=} \frac{1}{2\pi\sqrt{I_F}\Omega}\left[\int_0^\infty te^{\xi_{MLI}^2 t^2/2I_F}e^{-(t-I_F)^2/2\Omega^2}dt\right. \\
&\qquad \left. + \int_0^\infty te^{\xi_{MLI}^2 t^2/2I_F}e^{-(t+I_F)^2/2\Omega^2}dt\right].
\end{aligned}
\qquad (13)
$$

After re-organization, we have $p(\xi_{MLI}) = \frac{e^{-I_F^2/2\Omega^2}\sqrt{I_F}\Omega}{\pi\left(I_F+\xi_{MLI}^2\Omega^2\right)}\left[1 + \sqrt{2\pi r}e^r erf\left(\sqrt{r}\right)\right]$, where $r = I_F^3/[2\Omega^2\left(I_F + \xi_{MLI}^2\Omega^2\right)]$ and $erf(x) = \frac{2}{\sqrt{\pi}}\int_0^\infty e^{-t^2}dt$ is the error function.

Template Matching. Similarly, the p.d.f of decoding error of TM is calculated to be
$P(\xi_{TM}) = \frac{\Omega_1\Omega_2 e^{-M_2^2/2\Omega_2^2}}{\pi\left(\Omega_1^2+\xi_{TM}^2\Omega_2^2\right)}\left[1 + \sqrt{\pi r}e^r erf\left(\sqrt{r}\right)\right]$, where $r = M_2^2\Omega_1^2/[2\Omega_2^2\left(\Omega_1^2 + \xi_{TM}^2\Omega_2^2\right)]$.

The p.d.f of the decoding errors of MLI and TM with different decoding time windows are shown in Fig.1. The narrower the distribution is (intuitively, this implies a higher chance of having a small decoding error), the better the decoding performance. For a large time window (Fig.1a), MLI outperforms TM as expected, since MLI is asymptotically efficient as proved above. Interestingly, for a very short-time window, TM outperforms MLI (Fig.1b).

Survival Function. For a short-time window, the errors of MLI and TM satisfy the Cauchy distributions, whose variances diverge. To measure the decoding accuracy, we introduce a survival function defined as

$$R(\theta) = 2\int_\theta^\infty P(r)dr, \qquad (14)$$

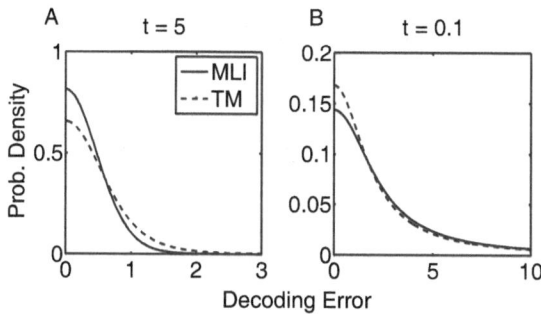

Fig. 1. The probability density functions of decoding errors of MLI and TM. The density function is symmetric and centralized at zero, therefore only the positive part is shown. a) For a large decoding time window. b) For a short decoding time window. Parameters: $A = \rho = 1, \sigma = 3$.

where θ is a positive number called the threshold. The survival function calculates the probability that decoding error is larger than the threshold θ, instead of counting the amplitude of error. This is different from the conventional variance measure, and is perhaps more relevant to animal behavior performances. The smaller the survival function is, the better the decoding performance.

Figure 2 shows how the survival functions of MLI and TM change with the decoding time window. It confirms that for short-time windows, TM outperforms MLI.

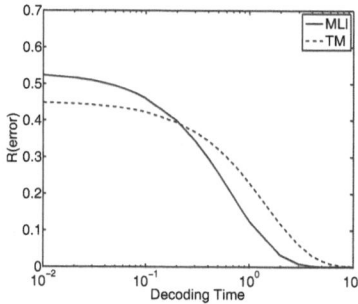

Fig. 2. Decoding performance of MLI and TM change with decoding time. Parameters: $A = \rho = 1, \sigma = 3, \theta = 1.5$.

5   Conclusions and Discussions

In the present study, we have investigated the performances of population decoding in short-time windows. Two methods, namely, MLI and TM, are explored. We consider neurons fire independently and satisfy the Poisson statistics. We find that in short-time windows, the decoding results of MLI and TM no longer satisfy the Gaussian distributions, but rather the Cauchy ones. The p.d.f of decoding errors of MLI and TM are derived. To measure the performances of MLI and TM in short-time windows (since the variance of Cauchy distribution diverges), we define a survival function counting the probability of decoding error larger than a threshold. As expected, MLI outperforms TM asymptotically. However, in very short-time windows, TM outperforms MLI.

It has been shown that TM can be implemented by continuous attractor neural networks (CANNs) [2,6], which hold a continuous family of stationary states having the same shape as the tuning function (i.e, the templates) [2]. Under the driving of an external input, the network state will move to a stable position having the maximum overlap with the external input as if implementing the operation of TM. CANNs have been successfully applied to describe the encoding of orientation [8], motion direction [7], head direction [9] and spatial location [10] of objects. Our results suggest that compared with other strategies, a CANN has an advantage of reading-out the stimulus information quickly in short-time windows.

Acknowledgement. We acknowledge the valuable discussions with Peter Dayan when this project was initiated. This work is supported by National Foundation of Natural Science of China (No. 91132702) and the 973 program of China (No. 2011CBA00406).

Appendix

1. Statistics of the Decoding Error of MLI

$$E\left[g_{MLI}^{(1)}\right] = \sum_i \frac{c_i - x}{\sigma^2} \sum_{n_i=0}^{\infty} n_i P_i(n_i|x) = \sum_i \frac{c_i - x}{\sigma^2} f_i t = 0, \tag{15}$$

$$E\left[g_{MLI}^{(2)}\right] = E\left[-\nabla\nabla \ln P(\mathbf{n}|x)\right] = I_F$$

$$= \sum_i \frac{1}{\sigma^2} \sum_{n_i} n_i P_i(n_i|x) = \frac{\sqrt{2\pi} At\rho}{\sigma}, \tag{16}$$

$$V\left[g_{MLI}^{(1)}\right] = E\left[\left(g_{MLI}^{(1)}\right)^2\right] = \sum_{i,j} \frac{(c_i - x)(c_j - x)}{\sigma^4} \sum_{\mathbf{n}} n_i n_j P(\mathbf{n}|x)$$

$$= \sum_{i,j} \frac{(c_i - x)(c_j - x)}{\sigma^4}(f_i f_j t^2 + f_i t) = \frac{\sqrt{2\pi} At\rho}{\sigma} = I_F, \tag{17}$$

$$V\left[g_{MLI}^{(2)}\right] = E\left[\left(g_{MLI}^{(2)}\right)^2\right] - E\left[g_{MLI}^{(2)}\right]^2$$

$$= \frac{1}{\sigma^4} \sum_{i,j} \sum_{\mathbf{n}} n_i n_j P(\mathbf{n}|x) - \left[\frac{1}{\sigma^2} \sum_i f_i t\right]^2$$

$$= \frac{1}{\sigma^4}\left(\sum_{i,j}(f_i f_j t^2 + f_i t) - \sum_{i,j} f_i f_j t^2\right) = \frac{\sqrt{2\pi} At\rho}{\sigma^3} = \Omega^2. \tag{18}$$

2. Statistics of the Decoding Error of TM

$$E\left[g_{TM}^{(1)}\right] = \frac{A}{\sigma^2}\sum_i (c_i - x)e^{-(c_i-x)^2/2\sigma^2} f_i t = 0 \tag{19}$$

$$E\left[g_{TM}^{(2)}\right] = \sum_n = At\sum_i e^{-(c_i-x)^2/2\sigma^2}\left[\frac{1}{\sigma^2} - \left(\frac{c_i - x}{\sigma^2}\right)^2\right] f_i t$$
$$= \frac{A^2 t\sqrt{\pi}\rho}{2\sigma}, \tag{20}$$

$$V\left[g_{TM}^{(1)}\right] = E\left[\left(g_{TM}^{(1)}\right)^2\right]$$
$$= \frac{A^2}{\sigma^4}\sum_{i,j}\left[(c_i - x)e^{-(c_i-x)^2/2\sigma^2}\right]\left[(c_j - x)e^{-(c_j-x)^2/2\sigma^2}\right] f_i f_j t^2$$
$$+ \frac{A^2}{\sigma^4}\sum_i (c_i - x)^2 e^{-(c_i-x)^2/\sigma^2} f_i t = \frac{\sqrt{2\pi}A^3 t\rho}{3\sqrt{3}\sigma}, \tag{21}$$

$$V\left[g_{TM}^{(2)}\right] = A^3 t\sum_i e^{-3(c_i-x)^2/2\sigma^2}\left[\frac{1}{\sigma^2} - \left(\frac{c_i - x}{\sigma^2}\right)^2\right]^2 = \frac{2\sqrt{2\pi}A^3 t\rho}{3\sqrt{3}\sigma^3} \tag{22}$$

References

1. Abbott, L., Dayan, P.: Neural Computation 11(1), 91–101 (1999)
2. Wu, S., Amari, S., Nakahara, H.: Neural Computation 14(5), 999–1026 (2002)
3. Cover, T., Thomas, J., Wiley, J.: Elements of information theory 6 (1991)
4. Thorpe, S., Fize, D., Marlot, C., et al.: Nature 381(6582), 520–522 (1996)
5. Bethge, M., Rotermund, D., Pawelzik, K.: Neural Computation 14(10), 2317–2351 (2002)
6. Deneve, S., Latham, P., Pouget, A.: Nature Neuroscience 2(8), 740–745 (1999)
7. Georgopoulos, A., Kalaska, J., et al.: The Journal of Neuroscience 2(11), 1527–1537 (1982)
8. Ben-Yishai, R., Bar-Or, R., Sompolinsky, H.: Proc. of the Natl. Acad. of Sci. 92(9), 3844 (1995)
9. Zhang, K.: The Journal of Neuroscience 16(6), 2112 (1996)
10. Samsonovich, A., McNaughton, B.: The Journal of Neuroscience 17(15), 5900 (1997)

Noise and Illumination Invariant Road Detection Based on Vanishing Point

Wei Luo, Heyou Chang, and Jian Yang

School of Computer Science, NJUST, Nanjing, P.R.C.
{lw860123,changheyoutd}@gmail.com, csjyang@mail.njust.edu.cn

Abstract. We propose a new method for robust road detection under noise and illumination varying conditions. Original input image is first divided into smooth and detailed component through structure-texture decomposition, where we verify the texture image is robust to various complicated road conditions. The texture image is then be used to compute each pixel's dominant orientation through Gabor wavelet analysis, followed by generating the vanishing point via grouping voters, which has an orientation confidence larger than a fixed threshold, in corresponding voting region through soft voting. Finally the road borders are constructed by feature inconsistency maximization criterion. Experiments on various road, weather, noise and lighting conditions are justified the accuracy and robust of our method. Furthermore, we analyze the applicability of texture based vanishing point method and conclude the main factors that degenerate the performance of this class method.

Keywords: road detection, vanishing point, noise and illumination invariant.

1   Introduction

Road detection as a subsystem of autonomous driving system, it should provide high performance for recognizing road region accurately, robust and real-time. Consequently, many researchers have focus on using integration method [1, 2, 3, 4], etc. While many researchers investigate this problem from the aspect of autonomous system design, some others pay more attention on purely computer vision techniques, which processing video stream frame by frame [5, 6, 7], and good surveys can be found in [5, 8, 9].

As a real environmental application system, noise and lighting impose serious challenge for vision based techniques. In some cases, it could completely change the road condition and cause the original robust features are invalid. Some examples are shown in figure 1(c). In order to relax this constraint, in this paper, we propose a method that first use a structure-texture decomposition to pre-filter the original input image, which has been verified very useful in recently developed optical flow computation [12, 13], and then adopt the texture image to compute candidate vanishing point, and finally we construct the borders of road based on the computed vanishing point and color/texture inconsistence maximization criterion. Figure 1 gives some examples of structured and unstructured road under various conditions.

J. Yang, F. Fang, and C. Sun (Eds.): IScIDE 2012, LNCS 7751, pp. 64–71, 2013.
© Springer-Verlag Berlin Heidelberg 2013

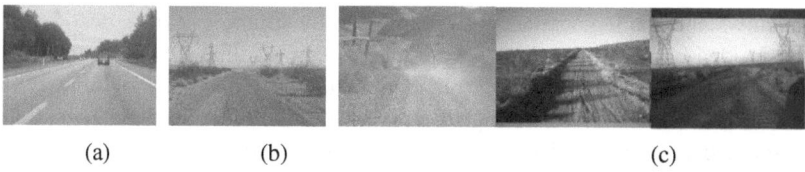

(a)   (b)   (c)

Fig. 1. Road images, (a) structured road, (b) unstructured road, (c) cluttered road images, from left to right each caused by dust, shadows and shading

The reminder structure of this paper is arranged as follows. Section 2 we introduce vanishing point related method for road detection. Structure-texture decomposition is presented in section 3. In section 4, we estimate vanishing point based on Gabor feature and construct the road borders based on color/texture inconsistency maximization criterion. Experiment results are detailed in section 5. In section 6, we conclude this paper.

2   Previous Work

A number of researchers use vanishing point as global constraints for road detection and tracking. In [14], concurrent lines, edges were used to find road markings for assisted outdoor navigation. Simond [15] compute the dominant vanishing point (DVP) using borders, and then the DVP and the borders of the road were used to constrain the region where the points of interest. Coughlan [16] propose using "Manhattan" three dimensional grid to describe visual scene, the hypothesis test verified it can give good estimates of vanishing points. Generally speaking, these methods first finding image lines, and then using voting method to find intersect point, e.g. Hough transformation.

While the above methods work well on well paved road surface, it failed in ill-conditioned roads. In [10], the author proposes using texture feature to compute eachpixel's dominate direction, and then group the local dominate direction to construct vanishing point. However, as argued in [11], the candidate vanishing point in the top part of image is incline to receive more votes than candidate vanishing point in the lower part, such that result in incorrect vanishing point. As an extended work, [11] introduce a soft voting method (SVVP) to relax this phenomenon, and make the voting processing more balance.

Nevertheless, Gabor wavelet texture feature on original image is sensitive to noise and lighting changing, moreover, the candidate vanishing point deterministic criterion is still under constructed. In this paper we propose a structure-texture decomposition method to relax the influence of noise and lighting condition. And we'll constrain our discussion to the first problem, while the second problem will leave for our future work.

3   Structure-Texture Decomposition

Recently development of optical flow computation has verified pre-filtering input image can improve optical flow accuracy efficiently [12, 13]. And draw a conclusion that shadows and shading artifacts can be relaxed by structure-texture decomposition. The basic idea behind this splitting technique is that an image can be regarded as a composition of a structure part, corresponding to the main body of objects in the image, and a texture part, reflecting fine scale-details. Figure 2 displays such an example for structure-texture decomposition, and we can find the shadows and shading mainly reflected in structure image, which is what we expected.

We take the idea of Rudin, etc. [17] to accomplish the structure-texture decomposition, which was first modeled to image denoise. For an input image $I(x)$, the structure part can be computed as the solution of

$$\min_{I_s} \int_{\Omega} \{|\nabla I_s| + \tfrac{1}{2\theta}(I_s - I)^2\}dx \tag{1}$$

where   is the structure image, θ is a small constant value, such that constrain I_s closely approximation to I, and the texture image can get as $I_t = I - \alpha I_s$. An efficient solver for Eq. (1) is based on gradient descent and subsequent re-projection using the dual-ROF model [18]. Due to the dual information in Eq. (1), it can iteratively converge to global minimum, and here we reproduce the formulation from [18], and the solution of Eq. (1) is given by $u = v + \theta div\mathrm{P}$, the dual variable $P = [p_1, p_2]$ is defined iteratively by

$$\widetilde{P}^{n+1} = P + \tfrac{\tau}{\theta}(\nabla(v + divP^n)), \quad P^{n+1} = \frac{\widetilde{P}^{n+1}}{max\{1, \widetilde{P}^{n+1}\}} \tag{2}$$

where $P^0 = 0$, the time step $\tau \le \tfrac{1}{4}$. For our model, u, v and θ corresponding to I_s, I and θ in Eq. (1). The original input image and target image are scaled to the range $[-1, 1]$ before computing the texture part. As the iteration process can converge very fast, we set $\alpha = 0.95$, $\theta = 0.125$ and $ite = 100$ in our experiments.

Fig. 2. Structure-texture decomposition results corresponding to fig. 1. (c). The first image in each pairs is structure image and the second is texture image.

4   Vanishing Point

Vanishing point is defined as the intersection of lines in prospective image space, which are parallel in physic world [19]. In this section, we first introduce pixel's dominant orientation, then detail voting scheme for our structure-texture decomposition vanishing point (STDVP), and followed by constructing road borders.

4.1   Pixel's Dominant Orientation

Our pixel's dominant orientation computed by texture analysis based on Gabor filter. For an orientation φ and a scale ω, the Gabor filters can be defined by [20]

$$\Psi_{\omega,\varphi} = \frac{\omega}{\sqrt{2\pi}c} e^{-\frac{\omega^2(4a^2+b^2)}{8c^2}} \left(e^{ia\omega} - e^{-\frac{c^2}{2}}\right) \tag{3}$$

where $a = x\cos\varphi + y\sin\varphi$, $b = -x\sin\varphi + y\cos\varphi$, and $c = 2.2$ in this paper. Given an input image $I(x,y)$, the filtered image obtained by convolution of image I with Gabor filter with orientation φ and scale ω defined by $F_{\omega,\varphi} = I \otimes \Psi_{\omega,\varphi}$. We then define each pixel's orientation response as weighted averaging Gabor response in all scales at different orientations $F_\varphi = \sum_\omega w_\omega F_{\omega,\varphi}$, where $\sum w_\omega = 1$. Then each pixel's dominant orientation (DO) and confidence value (CV) is computed by

$$\text{DO:} \quad O(x,y) = \max_\varphi F_\varphi \qquad \text{CV:} \ \text{conf}(x,y) = 1 - \frac{\text{mean}_{(O_5\cdots O_{15})}}{O} \tag{4}$$

where we reorder the sequence of orientation based on the Gabor response in decrease way, and O equivalent to O_1 represent the largest and dominant orientation. As we expected if O is obviously larger than other orientation value, we have strong confidence to believe O represents the true orientation, otherwise, it's not. From our experiment, we found the local maximum usually stays between O_5 and O_{15}, as we hypothesize O is the global maximum, so we average the value between O_5 and O_{15} to assign a confidence value to the global orientation. In our experiments, only those pixel's confidence larger than given threshold T are used to vote.

4.2   Voting Scheme

We follow the idea of [11] to construct vanishing point by soft-voting. The basic idea is voting for candidate vanishing point based on distance and angles, given a candidate vanishing point V and a voter $P(x,y)$, which inside the vote region R_V, the more shorten the distance between P and V, and the more smaller the angle between PV and O_P, $\gamma = \angle(PV, O_P)$, the contributions more significant. More details can refer to [11].

We then aggregate the contributions of all voters in the voting region for each pixel, and select the maximum as the true vanishing point. Not so complicated as method proposed in [11], we aren't proceeding to consider the vanishing point refinement process here, since the vanishing point computed based on our method is rather accuracy compare with original method. One point we should note here is we only set the lower 85% part of image pixels as voters, and the candidate vanishing point constrained to the region range from 20% to 85% of the image from the bottom. Figure 3 illustrate the voting scheme and a voting map computed by this method.

(a)   (b)

Fig. 3. (a) confidence map, (b) voting scheme

4.3   Constructing Road Borders

Since we only have vanishing point and no other prior knowledge about the image, we construct the road borders as follows: firstly, we evenly generate 29 lines from vanishing point to the lower part of image, which equally spaced 5 degree between each neighboring pair lines, comprising the angle range from 20 degree to 160 degree. Then we find the maximum orientation consistent ratio (OCR) line, l_{θ_1}, as the first road border. The OCR is defined as the ratio of number of points, which dominant direction are consistent with the line they lay on under a given threshold, divided by the total number of points that lie on the line. Finally, we construct the second road border by

$$\max_{l_{\theta2},l_{\theta2}\neq l_{\theta1}} \frac{mean(A_1)-mean(A_2)}{\sqrt{var(A_1)}+\sqrt{var(A_2)}+min\{mean(A_2)-mean(A_3),V\}} \tag{5}$$

s.t. $(\theta_1-90)(\theta_2-90)<0, |\theta_1-\theta_2|\geq 20$, where l_{θ_2} is the line with degree θ_2, A_1 is the area between l_{θ_1} and l_{θ_2}, A_2 and A_3 is the outside area of l_{θ_2} and l_{θ_1}, respectively. The object function imply the road region and non-road region should appear distinct as much as better based on the observed features, while the both sides of the road should present consistently as much as possible, which explicit modeled by the third term in the denominator. The parameter V is used to control the degree of how much we should believe our model, if the environment strong consistent with our model, we set large value for V, vice versa. And we formalize this criterion as feature inconsistency maximization criterion (FICMC). The constraints are all derived from our experiments, we find if the angle of the first border lies in the range of $[0, 90]$, then the other border always falls into range $(90, 180]$, vice versa. And the second constrain is based on our statistic of all detected images.

5   Experiments

We compare the performance of our method with ref. [11] which mainly contributed for soft voting. The test images are size of 180×240, confidence threshold $T = 0.3$ in both methods, the voting region is half the height of image below candidate vanishing point by the width of 100 neighbors in horizontal direction, and $\alpha = 0.95$, $\theta = 0.125$ and $ite = 100$ for structure-texture decomposition. We consider 5 scales and 36 orientations of Gabor filter, and the base frequency is set as $\{4, 8, 16, 32, 64\}$. For more thoughtful comparison, we classified original 1003 test images into 9 classes, corresponding to country road, desert, hill, hill-dark, nature, snow, structured, structured-dark and synthesis, and each class contain 47, 225, 102, 64, 273, 117, 109, 10 and 56 images, respectively. Figure 4 displays the result of the two methods in one instance of each class, and table 1 statistic the results in detail.

To verify our model is robust to noise and lighting challenge or even more complicated environment, the first group of our experiments, we focus on the accuracy of vanishing point. From comparison, we can find the STDVP can improve vanishing point accuracy very notably, esp. in desert and hill-dark class. The error of SVVP in desert class is mainly due to the dust and the edge between mountain and sky, while STDVP is more robust. In the class of hill, the error mostly caused by slope with curves, but STDVP can relax this case in some degree. Comparable improvement obtained in hill-dark class, while SVVP decrease performance dramatically due to dark lighting, STDVP keeps its performance as in normal lighting condition. The correctness of both methods in structured road is low mainly due to the strong edges of buildings, the rich texture of trees along with the road and dense cloud in the sky, while the accuracy increased in dark lighting condition, this mostly because many objects' edges and trees are become invisible, the algorithms can then efficiently extract road feature. What's interesting is the accuracy of SVVP is higher than STDVP in structured-dark class, we argue that this is mainly caused by small number of samples, as there only 10 samples in this class. In the class of synthesis, both methods perform equivalently with an average accuracy below other group definitely, except for structured group. And the same instance occurs in the nature class with a little high accuracy. This is mainly corresponding to rich of tress and structure edges in the two classes.

As a conclude, we find texture based vanishing point method especially applicable to extreme road, noise and lighting conditions. But for normal nature cases, its performance has a slightly decrease for the rich of trees, structure edges and clouds, but STDVP can efficiently improve the accuracy and robustness in all these conditions.

Table 1. Comparison of SVVP and STDVP in different classes

		conty.	desert	hill	Hill-dark	nature	snow	struc.	struc-dark	synthesis
SVVP	cor. V.	34	171	75	35	190	90	56	9	38
	cor. R.	0.735	0.760	0.735	0.546	0.696	0.769	0.513	0.90	0.678
STDVP	cor. V.	38	209	87	54	202	101	64	8	37
	cor. R.	0.808	0.928	0.852	0.843	0.739	0.863	0.587	0.80	0.660
SVVP	cor.Br.	27	143	58	25	136	68	48	7	31
	cor. R.	0.7941	0.836	0.773	0.714	0.715	0.755	0.875	0.777	0.815
STDVP	cor.Br.	29	179	68	42	158	85	58	8	33
	cor. R.	0.763	0.856	0.781	0.777	0.782	0.841	0.906	1.0	0.891
tol. img		1003	47	225	102	64	273	117	109	10	56

Note: *cor.* means correct, V., R. and Br. represents vanishing point and Ratio, and Borders, respectively.

In the second group of our experiment, we compare the accuracy of road segmentation between the two methods. Here we only use RGB feature for simplicity. As confirm to the FICMC, we find the second road border in R, G and B channels respectively and select the maximum as the final one. In SVVP, we refine the final vanishing point based on the refinement method given in [11], and then update the final road borders. In STDVP, we set $V = 10$ in all categories for simplicity. In this group, we statistic the number of images which road borders corrected construct under the precondition of correct vanishing point. For different road types, the correct segmentation evaluation criterion varied. In our experiment, we restrict the correctness to the road region maximize enough for driving while small enough to prevent intervene with other area for structured road, while we define safe driving area from the angle of the car as correct segmentation for unstructured road and wild environment.

From comparison, we can find both methods perform equivalently, except for structured-dark class, which we again argue the result by small number of samples in this class. Not as we expected, the ensemble correct ratio is slightly low. The reason for STDVP, we mainly account for two points except for features. One is the first constrain condition in our FICMC, which failed in case that the road cross the image in some like horizontal way but not in vertical way, thus the angle of borders violate the constrain condition. The other is the unified threshold value V we set in our criterion. And we believe the performance can be improved by adjust V according to different road, noise and lighting condition.

Fig. 4. Each column represents one road conditions of our test images. The first row is original image, the second and three row is vanishing point computed by SVVP and corresponding road borders, respectively. The fourth row is vanishing point computed by STDVP, and the last row is corresponding road borders. In the first four columns, we focus on the accuracy of vanishing point, while the others illustrate the accuracy of road borders of the two methods.

6   Conclusions

We have proposed a robust road detection method in this paper. We improve the vanishing point accuracy by structure-texture decomposition, which has been justified by experiments on various road, weather, noise and lighting condition. The proposed decomposition efficiently alleviated the vanishing point refinement process. We formulate the road region segmentation as feature inconsistency maximization criterion, and the robust term in denominate offered us flexibility to adjust the model to different road conditions. Furthermore, we analyze the applicability of texture based vanishing point method in various conditions, and conclude the main reasons which cause the failure of texture based vanishing point method.

References

1. Leonard, J., et al.: A Perception-Driven Autonomous Urban Vehicle. J. Field Robotics 25(10), 727–774 (2008)
2. Urmson, C., et al.: Autonomous Driving in Urban Environments: Boss and the Urban Challenge. J. Field Robotics 25(8), 425–466 (2008)
3. Montemerlo, M., et al.: Junior: The Stanford Entry in the Urban Challenge. J. Field Robotics 25(9), 569–597 (2008)
4. Miller, I., Campbell, M., et al.: Team Cornell's Skynet: Robust Perception and Planning in an Urban Environment. J. Field Robotics 25(8), 493–527 (2008)
5. McCall, J.C., Trivedi, M.M.: Video-Based Lane Estimation and Tracking for Driver Assistance: Survey, System, and Evaluation. IEEE TITS 7(1), 20–37 (2006)
6. Wang, Y., Teoh, E.K., Shen, D.: Lane detection and tracking using B-Snake. Image Vision Computing 22, 269–280 (2004)
7. Caraffi, C., Cattani, S., Grisleri, P.: Off-Road Path and Obstacle Detection Using Decision Networks and Stereo Vision. IEEE TITS 8(4), 607–618 (2007)
8. Bertozzi, M., et al.: Artificial Vision in Road Vehicles. Proceedings of the IEEE 90(7), 1258–1269 (2002)
9. Kastrinake, V., Zervakis, M., Kalaitzakis, K.: A survey of video processing techniques for traffic applications. Image and Vision Computing 21, 359–381 (2003)
10. Rasmussen, C.: Grouping Dominant Orientations for Ill-Structured Road Following. In: CVPR (2004)
11. Kong, H., Audibert, J.-Y., Ponce, J.: General Road Detection From a Single Image. IEEE TIP 19(8), 2211–2220 (2010)
12. Wedel, A., Pock, T., Zach, C., Bischof, H., Cremers, D.: An Improved Algorithm for TV-L^1 Optical Flow. In: Cremers, D., Rosenhahn, B., Yuille, A.L., Schmidt, F.R. (eds.) Statistical and Geometrical Approaches to Visual Motion Analysis. LNCS, vol. 5604, pp. 23–45. Springer, Heidelberg (2009)
13. Sun, D., Roth, S., Black, M.J.: Secrets of Optical Flow Estimation and Their Pricinples. In: CVPR (2010)
14. Se, S.: Zebra-crossing Detection for the Partially Sighted. In: CVPR, vol. 2, pp. 211–217 (2000)
15. Simond, N., Rives, P.: Homography from a Vanishing Point in Urban Scenes. In: International Conference on Intelligent Robots and Systems (IROS), vol. 1, pp. 1005–1010 (2003)
16. Coughlan, J.M., Yuille, A.L.: Manhattan World: Orientation and Outlier Detection by Bayesian Inference. Neural Computation 15(5), 1063–1088 (2003)
17. Rudin, L.I., Osher, S., Fatemi, E.: Nonlinear total variation based noise removal algorithms. Physica D 60, 259–268 (1992)
18. Chambolle, A.: Total Variation Minimization and a Class of Binary MRF Models. In: Rangarajan, A., Vemuri, B.C., Yuille, A.L. (eds.) EMMCVPR 2005. LNCS, vol. 3757, pp. 136–152. Springer, Heidelberg (2005)
19. Hartley, R., Zisserman, A.: Multiple View Geometry in Computer Vision, 2nd edn. Cambridge University Press (2003)
20. Lee, T.: Image representation using 2d gabor wavelets. IEEE TPAMI 18(10), 959–971 (1996)

Text-Like Motion Representation
for Human Motion Retrieval

Rongyi Lan, Huaijiang Sun, and Mingyang Zhu

School of Computer Science & Technology, Nanjing University of Science & Technology,
210094 Nanjing, Jiangsu, China
{brianlanbo,sunhuaijiang,zmy8475}@gmail.com

Abstract. Human motion capture (mo-cap) data has been increasingly applied in animation, movies and games in recent years due to its visual realism, and large amounts of them were accumulated. How to effectively search logically similar motions from large data repositories is a new challenge. The major limitation of existing methods is the semantic level of the features is not high enough to well distinguish different motion categories. In this paper, we propose a text-like motion representation based on key-pose extraction and hierarchical clustering (HC). This motion representation is easy to be extended or combined with topic models to obtain higher semantic-level features for motion retrieval. Our experiments demonstrate its scalability and performance in several applications, including motion retrieval and motion segmentation.

Keywords: motion capture, motion retrieval, motion representation, motion segmentation, hierarchical clustering, Latent Dirichlet Allocation, topic mining.

1 Introduction

Owing to the visual realism and the characteristics of detail-conservation, mo-cap data has been increasingly applied in animation, filmmaking and game industry. With the increasing needs of applications and researches, people accumulated a large amount of mo-cap data. If we can well manage and reuse the data, not only can the human animation be enriched, but also the time cost, money cost and human labor will drastically reduce. Although the problem of mo-cap data reusing has been researched for more than one decade, the related applications are still in the early stage. But it can't prevent motion capture becoming one of the mainstream media, like today's image, video and speech, in the future. By then, even the non-professional will be able to exploit their intelligence and creativity to make fantastic works.

One can hardly create animations without browsing and searching for materials. As a result, searching in large-scale motion databases requires efficient, accurate and reliable retrieval methods. However simple ones will fail because the mo-cap data is composed of high-dimensional, sequential vectors which are inherently complicated and sometimes logically dissimilar although they are numerically similar [1]. People did a lot of researches around this problem for years and produce many excellent

J. Yang, F. Fang, and C. Sun (Eds.): IScIDE 2012, LNCS 7751, pp. 72–81, 2013.

works [1-8], and content-based motion retrieval gradually developed into one of the most important branches in character animation.

Due to the complexity of human motion data, it is essential but difficult to precisely calculate the similarities among motions. One effective way is to use weighted or temporally aligned nature numerical features to calculate the similarities [1, 5, 7-8]. In these methods, the original low-level features are used to represent human motions and the details of the motion data are preserved. However, most of these methods need substantial computations in the searching phase, and the results are easy to be affected by disturbances in the data.

Transforming human motions into discrete data can effectively avoid the above problems, and actually it gets increasing popularities in recent researches [3-4,6]. Discrete representations, including strings [2,3], states [4] or text documents [6], are more meaningful in semantics and closer to people use. Some of the methods achieved quite good results, but the semantic level of the representations is still not high enough.

Lately, Zhu et al. [6] represent human motions as probabilistic distributions over motion topics (mo-topics) which are learned from training examples by Latent Dirichlet Allocation (LDA). This novel representation possesses more semantic information than most of the existing methods and achieves state-of-art retrieval results. However, firstly, there is no concept of motion documents (mo-docs) in their method, and LDA is applied on the approximated motion word (mo-word) frequencies, rather than text-like documents. So the combination of mo-cap data and topic models is strained and the way they model motions is hard to be extended or combined with other models and algorithms in text domain. Secondly, the chosen set of geometric features (geo-features) is only suitable for current dataset, which means they have to manually select appropriate geo-features as motion vocabulary (mo-vocabulary) for each potential dataset containing new motion categories. Thirdly, each pose in the mo-cap data is transformed into geo-features, so the mo-topics cannot be visualized, which is bad for browsing mo-cap datasets.

In this paper, we present a new motion representation which has a text-document-like structure, where the motion clips correspond to the documents and the motion frames correspond to the words. The most outstanding advantage of this representation is the conciseness and the scalability, which means it is very efficient and easy to be extended with any algorithms and models based on text documents. In addition, the transformation from mo-cap data to mo-docs can be automatically processed without human intervention. By naturally introducing Latent Dirichlet Allocation (LDA), we can visualize the mo-topics to give a quick overview of the mo-cap dataset and represent human motions in the perspective of mo-topics, which improves both the semantic level of the feature and the retrieval performance.

2   Related Works

Choosing an appropriate motion representation and good similarity metrics is the key point in motion retrieval, and it is also the most difficult part. In the early period, the mo-cap data was treated as continuous high dimensional vector-sequence, and the corresponding retrieval methods were based on numerical similarity metrics [1,5,7-8].

Unfortunately, similarity metrics based on low-level features often fail to get correct motions that users really want. Subsequently, geometric features (relational features) [2] and string-form representations [3] are proposed to overcome this problem. These methods transform the original continuous data into more concise discrete data. Although some details are lost, the semantic level of the data representation is heightened, and meanwhile, the retrieval results also get considerable improvements. Therefore, appropriately heightening the semantic level of motion representation is of help for finding the correct motions that are logically coherent.

Zhu et al. utilize LDA to find the latent topics of human motions, and represent motions as probabilistic distributions over these topics. In detail, they treat the set of geo-features as the mo-vocabulary and ignore the order of words, and use the changing times and moving ranges of each mo-word to approximate the mo-word frequencies of the motion. They provide a much more meaningful motion representation which achieves very good performance, but this method suffers from the lack of understandability and the difficulty when extending. We use a totally different way to construct the mo-vocabulary and mo-docs, and use LDA as well to find motion topics and represent motions.

3   Text-Like Motion Representation

The goal of this paper is to improve Zhu et al.'s model [6], in order to represent human motions as exactly the same form as text documents. Compared with their model, ours can easily be extended and works more naturally with topic models. Moreover, our method is automatic and can precisely visualize the data, which increases the understandability a lot.

In the field of text mining, vocabulary comes from human languages, and documents are composed of amounts of words that written under syntax rules. Unfortunately, there are no words and documents in human motions. In order to represent the mo-cap data in the text-like way, the corresponding concept should be properly created. In this paper, we serve the motion clips as the documents and regard the motion frames as the words. The first step is to use HC to merge similar human poses in the training data together to construct the mo-vocabulary, and then all the motions can be transformed into mo-docs by replacing each frame with its closest pose in the mo-vocabulary. It's noteworthy that the set of all the poses in the training data can be very huge and many of them are redundant. So, extracting the key-poses of each motion beforehand can speed up the clustering step and assure the mo-vocabulary are selected from relatively more meaningful poses.

We denote the training motion examples by $S_i, i = 1,...,M$, where M is the number of motion examples. The set of key-poses extracted from S_i is denoted by F_i, and the key-pose extraction method proposed in [9] is used here. Thus, the motion terms (mo-terms) $v_i, i = 1,..., K$ in the mo-vocabulary V are cluster centers selected from the fused key-pose set of $F_i, i = 1,...,M$ by HC, where K is the number of mo-terms. We make use of the distance metric between poses in [10]. Fig.1 demonstrates a mo-doc and the mo-vocabulary.

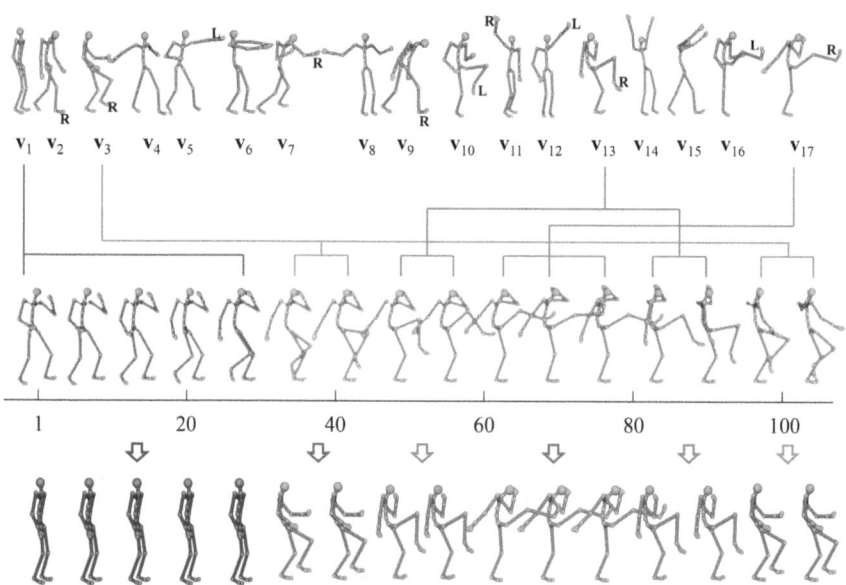

Fig. 1. The upper row is the mo-vocabulary, where the "R"/"L" markers identify right/left hands or feet. The middle row is the original "right kicking" motion (110 frames), and the lower row is the transformed mo-doc. The lines between the upper and middle row describe the nearest mo-term of each frame in the original motion.

Note that the modeling is slightly different from our earlier work [11].HC is used here rather than Affinity Propagation [12] (AP), because we can get logically more compact (poses are similar within a group) clusters by checking and tuning the number of clusters. In addition, the partitioning of the body and the term frequency adjustment are eliminated to keep this model more general and extensible. In essence, our method can be regarded as a discretization for the continuous mo-cap data. Although details loss can't be avoided in the process, the semantic level of data representation is heightened. Besides, identifying whether motions are logically similar is less dependent on details than motion editing or synthesis, and properly ignoring some details is even of great help to extract the discriminative information due to the elimination of disturbances in data. In summary, our text-like representation brings several important advantages:

- **Reduction of Dimensionality.** Original high-dimensional vectors are replaced by mo-words, namely one frame is described only by one number. It drastically improves the computational efficiency and raises the semantic level of the features.
- **Extensibility.** Mo-cap data can now be used or analyzed just like text documents. We apply this representation and its variations in motion retrieval and motion segmentation in the next section.
- **Automation.** The construction of mo-vocabulary and the transformation from mo-cap data to mo-docs requires little human intervention.

- **Visualization.** Each mo-term in the mo-vocabulary is a truly existed human pose, so we can easily visualize the mo-vocabulary, mo-docs and mo-topics, which is important for browsing a dataset.

4   Applications and Experiment Results

In this section, we introduce some classic approaches in text domain for our motion representation to cope with the motion retrieval and motion segmentation problems, and the corresponding experimental results demonstrate its effectiveness and extensibility.

4.1   Retrieval Performance Measure

To evaluate the performances of the retrieval methods, we calculate the query precision of each motion i according to the following equation:

$$precision(i) = \frac{N_C}{N_{t(i)}} \qquad (1)$$

, where $t(i)$ denotes the motion type of the query motion i, and $N_{t(i)}$ is the total number of motions in the dataset which is of type $t(i)$, also the number of return motions for motion i, and N_C is the number of motions whose types are identical with $t(i)$ in the return motions. It is note worthy that this measure also takes the recall into account.

4.2   Retrieval Using TF-IDF Framework

In the experiment of Section 4.1 and 4.2, 232 training examples and 66 querying examples (performed by 5 different actors) are involved, which are obtained from mo-cap database HDM05 of University of Bonn [13]. All the motion clips are of single category, and within the same type, the motions can have different numbers of cycles. We have 9 types of motions in total: "rotate both arms while walking / in-place", "rotate left / right arm", "left / right hand punch", "left / right kick" and "walk forward".

Since we have transformed the motion clips into the text-like mo-docs, the basic idea of using it may be follow the vector space model and term frequency – inverse document frequency (TF-IDF) framework. TF-IDF adjusts the weights of each term in the vocabulary and represents documents as term frequency vectors with the same length. By using the cosine distance metric, similar documents can be found for the query document.

4.3   Retrieval Using Topic Models

LDA is a generative probabilistic model that can discover hidden topics of huge stores of unstructured data, and it has been successfully used in text, music, images and even human motions [6,14-15]. As we know, LDA models the topic as a multinomial distribution over the vocabulary, which, generally speaking, captures the information of the co-occurrence of the words within the same document. The documents then can be represented by multinomial distributions over topics and the words in the document can be drawn from different topics.

We introduce LDA here to find mo-topics from the collection of mo-docs. Mo-docs are represented by mo-vocabulary, and the mo-terms in the mo-vocabulary are human poses. Therefore, the mo-topics depict several possible combinations of poses frequently occur together within a motion. So, mo-topics are some movement patterns or rules of human motions and possess more semantic information. The details of LDA are in [16] and how to combine LDA and human motions is in [6].

By applying LDA to our motion representation, we can obtain some meaningful and understandable mo-topics $\varphi_t, t = 1,...,T$, where T is the number of topics and φ_t is the multinomial parameter which is of length K. We ignore the mo-terms that have small proportions (less than 0.01) in each mo-topic and only visualize the frequently co-occurred poses. As can be seen from Fig.2, these mo-topics cover all the motion types in the dataset and can be labeled with short descriptions. This is very helpful when someone would like to have a quick view of a motion database. Most importantly, the mo-topics offer us a very high abstract-level representation for human motions. For example, a fighting motion should have larger proportions in hand punching topics and kicking topics, and a sprint kicking motion should primarily belong to the walking topic and kicking topics.

Fig. 2. The illustration and descriptions of the 12 mo-topics

Mathematically speaking, the ith mo-doc is a multinomial distribution governed by $\boldsymbol{\theta}_i = \{\theta_{i,t}, t = 1,...,T\}$, which is drawn from a Dirichlet distribution and describes its degrees of membership in each topic. We can make use of the symmetrical Kullback-Leibler (KL) divergence to measure the dissimilarity between two motions (multinomial distributions). Because multinomial distributions are discrete, so the formulation can be described as follows:

$$D(\boldsymbol{\theta}_q, \boldsymbol{\theta}_i) = D_{KL}(\boldsymbol{\theta}_q \parallel \boldsymbol{\theta}_i) + D_{KL}(\boldsymbol{\theta}_i \parallel \boldsymbol{\theta}_q) \quad , \tag{2}$$

$$D_{KL}(\boldsymbol{\theta}_q \parallel \boldsymbol{\theta}_i) = \sum_t \theta_{q,t} \log \frac{\theta_{q,t}}{\theta_{i,t}} \quad . \tag{3}$$

We compare the performances of Zhu et al.'s and Deng et al.'s methods and ours in this part of experiment (Fig.3). As the figure shows, Zhu et al.'s method performs the best on our experimental data. It proves the effectiveness of the combination of geo-features and topic models. Our main focus is on constructing a generic motion representation with understandabilities and extensibilities, so the performance is slightly behind Zhu et al.'s, but we think it can still be counted as a good method (precision > 0.9).Observing that LDA uses a bag-of-words assumption which drops the information of the word order, so Zhu et al.'s method might fail to distinguish the motions that are distinct in moving directions. In the next 2 sections, we further extend our representation to treat the motions which are direction-sensible, and use another simple text segmentation approach to segment long motion sequences into category-independent short clips. This is the most interesting characteristics of our representation.

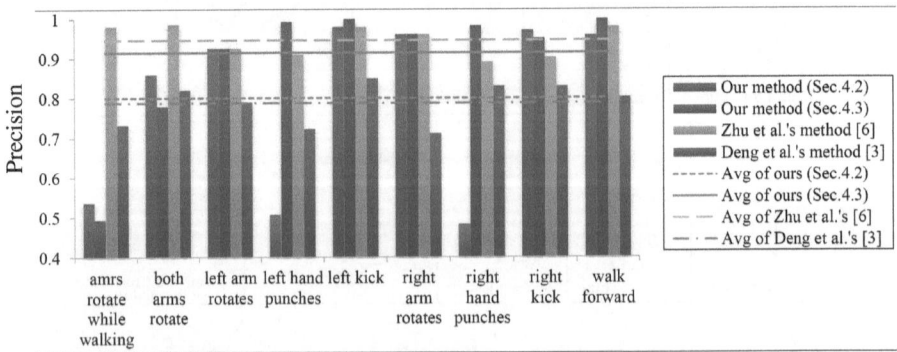

Fig. 3. The retrieval performances of each method on different 9 motion types. The horizontal lines (or dashed lines) indicate the average standards of these methods.

4.4   Retrieval Using Bigram Features

The experimental data used in this section is the same as the above, but the number of motion types is extended to 13, because the arm rotating motions are direction-sensible. We further separate them into forward and backward rotating ones.

In [6], they choose geo-features as the basic count of words in a mo-doc, and calculate the word switching frequencies and ranges of movement to adjust the mo-word occurrence of a mo-doc. This leads to the loss of word orders, directly affects the ability to represent direction-sensible motions. For example, walking forward and walking backwards can produce confusions to Zhu et al.'s model.

Taking the advantage of the extensibility of our motion representation, we can easily improve it by introducing bigram words to cope with the mentioned problem. Bigram motion terms (bi-mo-terms) are defined to be the ordered pairs (including self-pairs) of mo-terms in the mo-vocabulary. In specific, when the type of mo-words changes in the mo-doc, the previous mo-term and current mo-term forms a bi-mo-term. Take the mo-doc in Fig.1 for example, there're 5 switches in total: v_1 to v_3, v_3 to v_{13}, v_{13} to v_{17}, v_{17} to v_{13} and v_{13} to v_3. Therefore, there're $K_B = K \times K$ kinds of bi-mo-terms in total, and the self-pair bi-mo-terms appear when a mo-term keeps occurring for a long duration which exceeds a pre-defined threshold. The bigram vocabulary is denoted by $\mathbf{B} = \{\mathbf{b}_1, \mathbf{b}_2, ..., \mathbf{b}_{K_B}\}$, where the ith bi-mo-term $\mathbf{b}_i$ represents the ordered mo-term tuple (v_j, v_k), $j = 1, ..., K$ and $k = 1, ..., K$. Fig.4 demonstrates some of the bi-mo-terms of our dataset, and it can be seen from the figure that by adding the order information to the representation, the "forward rotate" motions and "backward rotate" motions can be well distinguished.

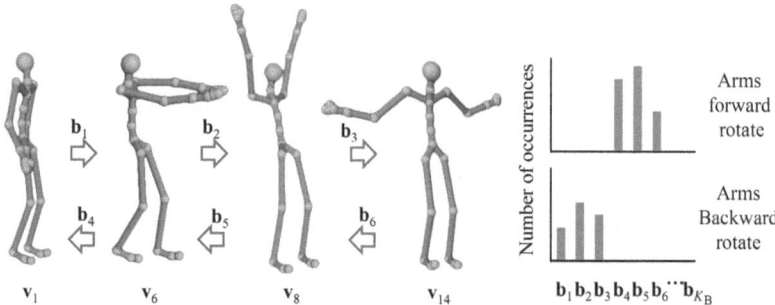

Fig. 4. Demonstration of 6 bi-mo-terms and the mo-doc distributions over bigram vocabulary. The arrows combine the connected 2 mo-terms into bi-mo-terms and indicate the order.

The retrieval precisions of these three methods are shown in Fig.5, and our method outperforms the others in average standard. Specifically, our bigram vocabulary representation successfully discriminates the directions of the rotating arms while Zhu et al.'s method is unable to handle these types of motions. However, motions of type "left hand punch" and "right hand punch" get low retrieval precision; the reason is that the number of frames of each punching motion is small comparing with other motion types, which causes small number of occurrences of the corresponding bi-mo-term. The motion vector turns out to be lack of statistical significance, so more work needs to be done to find a better bi-mo-term counting scheme. Anyway, the simple bigram modification exhibits the superiority of our motion representation from one aspect according to the average precision on 13 motion types.

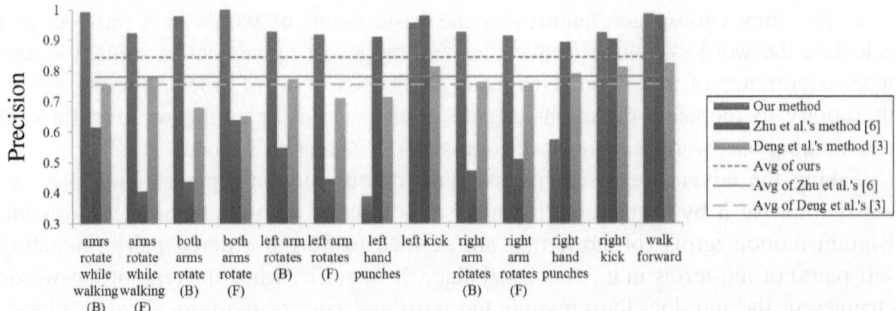

Fig. 5. The retrieval performances of each method on different 13 motion types. "B" / "F" is short for "backward" / "forward". The horizontal lines (or dashed lines) indicate the average standards of these methods.

4.5   Motion Segmentation

Due to the generality of our motion representation, a long motion sequence with multiple motion types can also be transformed into and treated as a mo-doc. Segmenting text documents into topic independent stories has been researched for years, and many of the approaches are quite effective. Since the motions are represented as text-like documents, these segmentation methods are available for mo-docs. We use a very simple statistics based segmentation method proposed in [17] to segment human motions, in which, the number of segments can be automatically determined using a dynamic programming strategy. The number of raw cuts generated by the algorithm is often more than the ones in the ground truth due to the repetitive cycles of motions. Therefore, we combine sufficiently similar neighboring segments together to obtain a finer result (Fig.6).

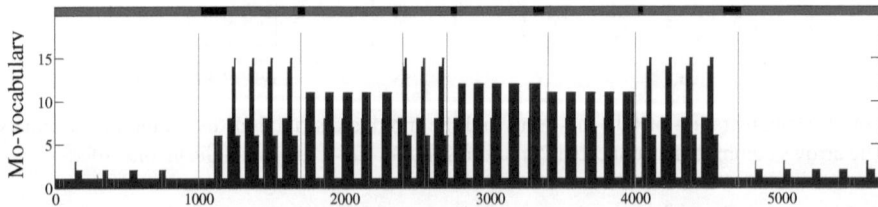

Fig. 6. The transformed mo-doc and segmentation results of the long motion sequence composed of multiple human behaviors. The stripe at the top of the figure exhibits the ground truth, and the vertical purple lines are algorithmic cuts.

5   Conclusion

In this paper, we propose a generic motion representation to overcome the shortages of Zhu et al.'s method [6]. Our method can automatically transform human motions into text-like documents and take advantage of any text algorithm to adapt kinds of requirements in motion analysis (in this paper, retrieval and segmentation). In conjunction with topic models (such as LDA), we are able to represent human motions at topic-level which possess much more semantics and get high-accuracy retrieval results. The experiments demonstrate the effectiveness and scalability of our motion representation.

References

1. Kovar, L., Gleicher, M.: Automated Extraction and Parameterization of Motions in Large Data Sets. ACM Transactions on Graphics 23, 559–568 (2004)
2. Müller, M., Röder, T., Clausen, M.: Efficient Content-Based Retrieval of Motion Capture Data. ACM Transactions on Graph 24, 667–685 (2005)
3. Deng, Z., Gu, Q., Li, Q.: Perceptually Consistent Example-Based Human Motion Retrieval. In: Proceedings of the 2009 Symposium on Interactive 3D Graphics and Games, pp. 191–198. ACM, New York (2009)
4. Wu, S., Wang, Z., Xia, S.: Indexing and Retrieval of Human Motion Data by a Hierarchical Tree. In: Proceedings of the 16th ACM Symposium on Virtual Reality Software and Technology, pp. 207–214. ACM, New York (2009)
5. Forbes, K., Fiume, E.: An Efficient Search Algorithm for Motion Data Using Weighted PCA. In: Proceedings of the 2005 ACM SIGGRAPH/Eurographics Symposium on Computer Animation, pp. 67–76. ACM, New York (2005)
6. Zhu, M., Sun, H., Lan, R., Li, B.: Human Motion Retrieval Using Topic Model. Computer Animation and Virtual Worlds 23(5), 469–476, doi: 10.1002/cav.432
7. Liu, F., Zhuang, Y., Wu, F.: 3D Motion Retrieval with Motion Index Tree. Computer Vision and Image Understanding 92, 265–284 (2003)
8. Chiu, C., Chao, S., Wu, M., Yang, S., Lin, H.: Content-Based Retrieval for Human Motion Data. Journal of Visual Communication and Image Representation 15, 446–466 (2004)
9. Assa, J., Caspi, Y., Cohen-or, D.: Action Synopsis: Pose Selection and Illustration. ACM Transactions on Graphics 24, 667–676 (2005)
10. Kovar, L., Gleicher, M., Pighin, F.: Motion Graphs. ACM Transactions on Graphics 21, 473–482 (2002)
11. Lan, R., Sun, H., Lian, H., Zhu, M., Li, B.: Vector Space Modeling and Retrieval of Human Motion Capture Data. Journal of Computer-Aided Design and Computer Graphics 23, 1357–1364 (2011)
12. Frey, B.J., Dueck, D.: Clustering by Passing Messages between Data Points. Science 315, 972–976 (2007)
13. Müller, M., Röder, T., Clausen, M., Eberhardt, B., Krüger, B., Weber, A.: Documentation Mocap Database HDM05. Computer Graphics Technical Report CG-2007-2, Universität Bonn (2007)
14. Wang, C., Blei, D., Li, F.: Simultaneous Image Classification and Annotation. In: Proceedings of the IEEE Conference on Computer Vision and Pattern Recognition, pp. 1903–1910. IEEE Press, Princeton (2009)
15. Hu, D.J., Saul, L.K.: A Probabilistic Topic Model for Unsupervised Learning of Musical Key-Profiles. In: International Conference on Music Information Retrieval, Kobe (2009)
16. Blei, D.M., Ng, A.Y., Jordan, M.I.: Latent Dirichlet Allocation. Journal of Machine Learning Research 3, 993–1022 (2003)
17. Utiyama, M., Isahara, H.: A Statistical Model for Domain-Independent Text Segmentation. In: ACL 2001 Proceedings of the 39th Annual Meeting on Association for Computational Linguistics, Stroudsburg, pp. 499–506 (2001)

A Novel Inductive Semi-supervised SVM with Graph-Based Self-training

ShengJun Cheng, QingCheng Huang, JiaFeng Liu, and XiangLong Tang

Department of Computer Science and Technology
Harbin Institute of Technology
Harbin, China
{chengsj,huangqc,jefferyliu,tangxl}@hit.edu.cn

Abstract. In this paper, a novel inductive support vector machine for semi-supervised learning, named IS3VM, is proposed, which aims to improve SVM by bootstrapping unlabeled data with self-training. The SVM classifier is iteratively refined through the augmentation of the training set. An improved self-training method is given by employing neighborhood graph for guarantying the reliability of newly added training examples. In detail, in each iteration of the self-training process, the local *cut edge weight* statistic is used to help estimate whether a newly labeled example is reliable or not, and only the reliable self-labeled examples are used to enlarge the labeled training set. Experiments show that, the improved self-training is beneficial and the proposed IS3VM algorithm can effectively exploit unlabeled data to achieve better performance, and is comparable to the-state-of-the-art semi-supervised SVM.

Keywords: Semi-supervised learning, Graph-based method, Self-training, Support vector machine.

1   Introduction

For many practical classification applications, labeled data can be difficult to obtain, but there may exist enormous amount of unlabeled data which is readily available. In recent years, semi-supervised learning has received considerable attention due to its potential for reducing the effort of labeling data. Existing semi-supervised learning can be categorized into several paradigms [1], including generative models (EM), Transductive SVMs, graph-based approaches and bootstrap learning.

Bootstrap learning [2] is referred to as a learner bootstraps from unlabeled data in order to augment the training data set. Self-training [3] is probably the most simple semi-supervised learning algorithm, which is characterized by the fact that the learning process uses its own predictions to teach itself. The assumption of self-training is that its own predictions, at least the high confidence ones, tend to be correct. This is likely to be the case when the classes form well-separated clusters. The major advantages of self-training are its simplicity and the fact that it is a *wrapper* method. Self-training improves the classification margin by selecting the unlabeled examples with the highest classification confidence, and assigns them the class labels that are predicted by simply the current classifier using its posteriori

J. Yang, F. Fang, and C. Sun (Eds.): IScIDE 2012, LNCS 7751, pp. 82–89, 2013.

probability outputs. The assigned labels are hereafter referred to as the *pseudo-labels*. The labeled data, along with the selected pseudo-labeled data are utilized in the next iteration for updating the classifiers parameters. This strategy is also well-applied in the famous co-training [4]. However, a problem with this strategy is that the introduction of examples with predicted class labels may only help to increase the classification margin, without actually providing any novel information to the classifier. Since the selected unlabeled examples are the ones that can be classified confidently, they often are far away from the decision boundary. As a consequence, adding these examples to the training set may not help improving the decision boundary; this is because by adjusting the decision boundary, the examples with high classification confidence will gain even higher confidence [5]. Besides, the pseudo-labels with high classification confidence may not be the ground-truth, since the confidence is estimated based on the current classifier and training set [6]. The estimated information may therefore be biased or distorted, thus, it is necessary to provide some means to escape from the distortion or bias from the current classifier. This implies that we may need alternative strategy other than only using classifiers' posteriori probability as the confidence measurement.

TSVM, also called as S3VM, is a popular method for employing SVM in the semi-supervised settings. This approach was introduced by Bennet&Demiriz [7]. S3VM reformulates the original definition by adding two constraints to the unlabeled examples. Considering a binary SVM, one constraint calculates the misclassification error as if the instance were in class 1 and the second constraint as if the instance were in class -1. S3VM tries to minimize these two possible misclassification errors [8]. The labeling with the smallest error is the final labeling. Moreover, TSVM is non-convex and finding its exact solution is NP-hard, several approximation algorithms have been established. However, when the size of test data set is big (e.g. larger than 1000), TSVM type algorithms are still time-consuming [3].

Besides TSVM, support vector machines are incorporated into semi-supervised settings in a different way. In several other studies, a multi-view co-training support vector machine and its variants were presented. For text classification, experiments have clearly shown that the co-training SVM outperforms the co-training Naive Bayes [9] Compared with TSVM algorithms, the computational burden of the co-training support vector machine is much lower.

In this paper, a novel inductive support vector machine for semi-supervised learning, named IS3VM, is proposed, which exploits the unlabeled data by leveraging SVM and a modified self-training. IS3VM not only does not require redundant views like co-training SVM, but also is more computational convenient than TSVM, since it does not need solve the optimization problem. By pseudo-labeling the unlabeled example with high classification confidence, IS3VM can be improved through the augmentation of the training set. This setting tackles the problem of determining how to label the unlabeled examples, which contributes much to the efficiency of the algorithm. Moreover, high reliability of the pseudo-label of the unlabeled example can be achieved through combining self-training with the local *cut edge weight* statistic [10] from the constructed neighborhood graph. Experiments on UCI data sets show that, the improved self-training is beneficial and the proposed IS3VM algorithm can effectively exploit unlabeled data to achieve better generalization performance, and is comparable to the-state-of-the-art semi-supervised SVM methods.

The remainder of this paper is organized as follows: Section 2 presents the graph-based self-training; Section 3 presents the proposed inductive S3VMs; Section 4 reports on the experiments on UCI data sets; Finally, Section 5 concludes and raises several issues for future work.

2   Graph-Based Self-training

Let X be the input space and $Y \in \{0,1\}$ be the output space. Suppose we have a small labeled training set $L = \{(x_i, y_i) \mid i = 1, 2, ..., l\}$ a large unlabeled set $U = \{x_j \mid j = l+1, ...u, u \gg l\}$.

First of all, a neighborhood graph is constructed from $L \cup U$, U is pseudo-labeled by the current classifier. The neighborhood graph which conveys the local information from all the examples in the training set, is conducted by the k-nearest neighbor criterion. In the graph, every example represent a vertex, and there exists an edge between two vertices a and b if either a or b is among the k nearest neighbor of the other. In this way, one example is *not only* related to its own neighbors, *but also* related to those ones which regard it as their neighbors. Furthermore, a weight $\omega_{ab} \in [0,1]$ is associated with the edge connecting a and b, which is computed as $(1 + dist(a,b))^{-1}$, where $dist(a,b)$ corresponds to the distance between a and b. In this paper distance is measured by the *EUCLIDEAN* distance.

After the graph is constructed, we evaluate the confidence of each pseudo-label being correct by employing the cutting edge technique. An edge in the graph is called a *cut edge* if the two vertices connected by it have different associated labels[11]. Intuitively, this coincides with the *manifold assumption* that examples with high similarity in the input space would also have high similarity in the output space. The basic assumption is that a correctly labeled example should possess the same label to most of its neighboring examples. Thus, the pseudo-label confidence of every x_i in U can be measured based on the following *cut edge weight statistic:*

$$J_i = \sum_{xj \in Ne_i} \omega_{ij} I_{ij} \tag{1}$$

where Ne_i is the neighborhood of x_i, ω_{ij} is the weight on the edge between x_i and x_j, I_{ij} are i.i.d random variables according to the Bernouilli law of parameter $P(y \neq \hat{y}_i)$, $\hat{y}_i$ is the pseudo-label of x_i produced by the current classifier. Let $H0$ be the null hypothesis that vertices of the graph are labeled independently according to distribution $D(Y) = \{Pr(y = 1), Pr(y = 0)\}$. Here, $Pr(y = 1)(Pr(y = 0))$ denotes the prior probability of an example being positive (negative), which is usually estimated as the fraction of positive (negative) examples. Hence, a good example will be *incompatible* with H_0. To test H_0 with J_i the distribution of J_i under H_0 is need. The distribution of J_i can be approximated by a normal distribution with mean and variance estimated by Eq.1 Recall the manifold assumption encoded in the neighborhood graph, correctly labeled examples tend to have few cut edges as its label should be consistent with most of its connected examples. Hence, the smaller the value of J_i, the higher the confidence of the pseudo-label $\hat{y}_i$ being correct. Therefore, we can select the candidate examples from U by the *cut edge weight statistic* J_i. The proposed method adds the examples whose neighbors have less cut edges to the training set. Instead of directing employing classifiers' predictions to teach itself, we accomplish this goal with an explicitly way, using the cut edge strategy to acquire more reliable candidates with high labeling confidence.

3   The Proposed Inductive IS3VM

In this section, we first present the steps of the proposed algorithm and conceive a method for choosing the appropriate set for parameter C in the SVM formula.

A standard SVM classifier for two-class problem can be defined as:

$$\min \frac{1}{2}\|w\|^2 + C\sum_{i=1}^{N}\xi_i \qquad (2)$$

s.t. $y_i(w^T x_i + b) \geq 1 - \xi_i$, $\xi_i \geq 0$, $i = 1,...,N$, where $x_i \in R^n$ is a feature vector of a training sample, $y_i \in \{-1,1\}$ is the label of x_i, $C > 0$ is a regularization constant. In the following paragraph, we will give the algorithm sketch of the proposed IS3VM.

IS3VM algorithm:

Initialize: binary classification problem: $D = L \cup U$

$L = \{(x_i, y_i) \mid i = 1,2,...,l\}$: Labeled training set,

$U = \{x_j \mid j = l+1,...u, u \gg l\}$: Unlabeled set

SVM model with parameters: w, ξ, b

$L' \leftarrow L, U' \leftarrow U$

$k = 0$

Step1   Train a SVM classifier f on L', perform classification on U', adding pseudo-labels to all the example in U'. The parameters of f are denoted as $w^0 \in R^n, \xi^0 \in R^l$ and b^0

Loop until $|f(w^k, \xi^k) - f(w^{k-1}, \xi^{k-1})| < \delta_0$

Step2   Construct a neighbor graph based on $L' \cup U'$, for each $x_i \in U'$, compute its cut edge statistic J_i, using Eq.0

Step3   Select 10% candidate examples from U' with the smallest J_i, associated by the corresponding pseudo-label, form a subset L'', $L' \leftarrow L' \cup L''$

Step4   Update f using L', the SVM parameters are denoted as $w^k \in R^n$, $\xi^k \in R^{l+u}, b^k$

Step5   Calculate the object function value in (1), $f(w^k, \xi^k) = \frac{1}{2}\|w^k\| + C\sum_{i=1}^{l+u}\xi_i^k$

Go to Loop

In [12], a standard EM algorithm with a naive Bayesian classifier was analyzed, which is a special case of self-training. Furthermore, the EM algorithm is convergent since the objective function of this algorithm monotonically increases during its iterations. The proposed algorithm has a similar working mechanism to the EM algorithm although the objective functions and classifiers of these two algorithms are different. Generally, if the distribution of the data is Gaussian (or close to Gaussian) and the dimension of the data is not very high, the EM algorithm may be used for classification; otherwise, the proposed IS3VM might achieve better results.

IS3VM can be regarded as a bridge between inductive models and transductive models. Our method takes advantage of both cluster assumption and manifold assumption, which can complement each other. We utilized a graph-based self-training method to give more reliable pseudo-labels to the unlabeled examples instead of the standard self-training which selects candidates only by the current classifiers' output posteriori probability. Especially, when the labeled examples are sparse, the current classifier is not strong enough to provide reliable predictions. In this case, the graph-based self-training method used in our algorithm can exhibit high advantage. Moreover, since IS3VM is more like a bootstrapping, the computational complexity is much lower than TSVM, which is a NP-hard problem, requiring a approximate optimization [14].

4   Experiments

In this section, we design experiments to verify the efficacy of the proposed IS3VM. 15 UCI data sets are used in the experiments. The characteristics of each dataset are shown in Table 1. Our experiments are configured as follows. For each data set, about 25% data are kept as test examples. 25% of the remaining data set is used as the labeled training set L ; and all the other examples are treated as the unlabeled set U .

Table 1. The characteristics of 15 UCI datasets

DataSet	Size	Attributes	Class	DataSet	Size	Attributes	Class
anneal	898	39	2	ionosphere	351	34	2
australian	690	15	2	kr-vs-kp	3916	36	2
breast-c	286	9	2	segement	2310	20	2
bupa	345	6	2	sick	3772	29	2
colic	368	22	2	vehicle	846	19	2
diabetes	768	8	2	vote	435	17	2
german	1000	20	2	wdbc	569	30	2
hypothyroid	3163	25	2				

The performance of IS3VM is compared with three algorithms, i.e. supervised SVM, self-SVM, TSVM. Supervised SVM is referred to as training a SVM classifier barely on the initial labeled training set, which is equivalent to other algorithms' initial. Self-SVM is based on the standard self-training algorithm, wherein SVM is used as the underlying classifier. TSVM is implemented as the SVM^{light}. *LIBSVM*[13] is used for all the other algorithms. Parameters are set as follows: The kernel trick is

RBF; for fair comparison, 50 examples are selected per round; maximal number of iterations is set to 50. The accuracy score is used to evaluate the performances of algorithms. In our experiments, the accuracy scores of each algorithm are obtained via 10runs of ten-fold cross-validation and evaluated on the same test set. Finally, we conduct two-tailed t-test with a 95% confidence level to compare the proposed algorithm to the others. The results are shown in Table 2.

Table 2. Average accuracy on 15 UCI data sets

Dataset	IS3VM	SVM	Self-SVM	TSVM
anneal	**85.36**	76.24	79.48	80.19
australian	78.83	65.91	78.5	79.87
breast-c	**72.25**	65.94	71.63	70.97
bupa	82.55	76.33	79.45	**85.87**
colic	68.42	**70.58**	66.23	64.38
diabetes	**71.66**	71.9	69.37	70.75
german	**84.64**	78.69	80.37	82.34
hypothyroid	68.54	75.44	71.31	**76.85**
ionosphere	**84.64**	80.34	80.6	82.73
kr-vs-kp	76.56	68.44	72.43	**81.35**
segement	**87.99**	85.77	85.43	78.44
sick	81.67	85.65	84.45	**86.23**
vehicle	81.45	77.4	83.47	**85.4**
vote	68.19	65.37	66.82	**69.38**
wdbc	**84.36**	79.43	81.65	80.12
w/t/l		13/0/2	11/4/0	7/4/4

The two-tailed t-test results are shown in the bottom row, where each entry has the format of *w/t/l*. This means that, comparing with IS3VM, the algorithm in the corresponding column wins *w* times, ties *t* times, and loses *l* times. Table 1 shows that IS3VM can effectively exploit unlabeled data to boost performance, and is superior to the other compared algorithms, where it wins 13 times and loses 2 times against SVM, wins 11 times and never loses against Self-SVM, wins 7 times and loses 4 times against TSVM.

Note that, although IS3VM achieves lower accuracy on 3 data sets (colic, hypothyroid, sick) than SVM, but on average, IS3VM actually achieves higher classification accuracy. One possible explanation of the degradation is that IS3VM suffers imbalance of the data set. In *hypothyroid* and *sick* data set, positive examples are much less than negative examples. Since the data set is unbalanced, a correctly labeled positive example could be easily mis-identified as mislabeled examples and rejected to be added to the labeled set for further training, due to the lack of neighbors possessing the same label, hence less chance for a correctly labeled positive example available for further training. The more the distribution of the training set is distorted, the easier for the learner to be mislabeled. Consequently, the performance degrades.

Furthermore, Table 2 also shows that IS3VM outperforms Self-SVM on 11 data sets, among which significance is evident in 8 data sets under a two-tailed pair-wise *t*-test with the significance level of 95%. This evidence supports our claim that the improved graph-based self-training is beneficial, and IS3VM is robust to noise in the self-labeled examples hence achieves better performance than Self-SVM.

Compared with TSVM, IS3VM achieve higher classification accuracy on 7 data sets under a two-tailed pair-wise *t*-test with the significance level of 95%. This suggests that our proposed method is comparable to the transductive SVM, sometimes even better than TSVM. While TSVM suffers from high computational complexity, our method is quiet computational convenient, especially when the data set has high dimensional attributes.

In summary, the experiments show that IS3VM can benefit from the unlabeled examples. The graph-based self-training used in IS3VM is robust to the noise introduced in self-labeling process and its learned hypothesis outperforms that learned via standard self-training. Moreover, IS3VM is more efficient than the famous TSVM, since it does not need solve the optimization problem.

5   Conclusions and Future Work

In this paper, a novel inductive support vector machine for semi-supervised learning, named IS3VM, is proposed, which aims to improve SVM by bootstrapping unlabeled data with self-training. In detail, during every iteration of self-training, the SVM classifier is refined through the augmentation of the training set. An improved self-training method is given by employing neighborhood graph for guarantying the reliability of newly added training examples. Specifically, the local cut edge weight statistic is used to help estimate whether a newly labeled example is reliable or not, and only the reliable self-labeled examples are used to enlarge the labeled training set. The experiment results on 15 UCI data sets show that IS3VM is able to benefit from the unlabeled examples, and the proposed graph-based self-training is able to provide more reliable pseudo-labels of the unlabeled examples. Since IS3VM is sensitive to imbalance data, exploring a way to solve this problem will be investigated in future.

In the future, we will combine our method with active learning for the purpose of obtaining better performance of the learned hypothesis. Theoretical verification of this method will be done, which might help to understand the functionality of this method. Moreover, it will also be interesting to apply IS3VM algorithm to real world applications, especially for the applications suitable for semi-supervised learning, such as natural language processing (NLP) and bioinformatics.

Acknowledgements. This paper is supported by the National Science Foundation of China (NSFC) under the Grant No. 61173087 and NO. 61073128.

References

1. Chapelle, O., Schölkopf, B., Zien, A.: Semi-Supervised Learning. MIT Press, Cambridge (2006)
2. Zhu, X.: Semi-supervised learning literature survey. Department of Computer Science, University of Wisconsin at Madison, Madison, WI, Tech. Rep. 1530 (2008)

3. Yarowsky, D.: Unsupervised word sense disambiguation rivaling supervised methods. In: Proceedings of the 33rd Annual Meeting of the Association for Computational Linguistics (ACL), pp. 189–196 (1995)

4. Blum, A., Mitchell, T.: Combining labeled and unlabeled data with co-training. In: Proceedings of the 11th Annual Conference on Computational Learning Theory, Wisconsin, MI, pp. 92–100 (1998)

5. Mallapragada, P.K., Jin, R., Jain, A.K., Liu, Y.: SemiBoost: Boosting for Semi-supervised Learning. IEEE Transaction on Pattern Analysis and Machine Intelligence 31(11), 2000–2014 (2009)

6. Li, M., Zhou, Z.-H.: SETRED: Self-training with Editing. In: Ho, T.-B., Cheung, D., Liu, H. (eds.) PAKDD 2005. LNCS (LNAI), vol. 3518, pp. 611–621. Springer, Heidelberg (2005)

7. Bennett, K., Demiriz, A.: Semi-supervised support vector machines. In: NIPS, vol. 11, pp. 368–374 (1999)

8. Joachims, T.: Transductive inference for text classification using support vector machines. In: Proceedings of International Conference on Machine Learning (ICML), Bled, Slovenia (1999)

9. Kockelkorn, M., Lüneburg, A., Scheffer, T.: Using Transduction and Multi-view Learning to Answer Emails. In: Lavrač, N., Gamberger, D., Todorovski, L., Blockeel, H. (eds.) PKDD 2003. LNCS (LNAI), vol. 2838, pp. 266–277. Springer, Heidelberg (2003)

10. Muhlenbach, F., Lallich, S., Zighed, D.A.: Identifying and handling mislabeled instances. Journal of Intelligent Information Systems 22, 89–100 (2004)

11. Zhang, M.-L., Zhou, Z.-H.: CoTrade: Confident co-training with data editing. IEEE Transactions on Systems, Man, and Cybernetics - Part B: Cybernetics 41(6), 1612–1626 (2011)

12. Xu, L., Jordan, M.I.: On convergence properties of the EM algorithm for gaussian mixtures. Neural Computing, 129–151 (1999)

13. Chang, C.C., Lin, C.J.: LIBSVM: A library for support vector machines. Optimization Techniques for Semi-Supervised Support Vector Machines

14. Chapelle, O., Sindhwani, V., Keerthi, S.S., Cristianini, N.: Optimization Techniques for Semi-Supervised Support Vector Machines. Journal of Machine Learning Research 9, 203–233 (2008)

15. Asuncion, A., Newman, D.J.: UCI machine learning repository (January 10, 2010), http://archive.ics.uci.edu/ml/datasets.html

Pseudo-Zernike Moment Invariants to Blur Degradation and Their Use in Image Recognition

Xiubin Dai[1,*], Tianliang Liu[2], Huazhong Shu[3], and Limin Luo[3]

[1] School of Geography and Biological Information,
Nanjing University of Posts and Telecommunications, 210046 Nanjing, China
[2] College of Telecommunications and Information Engineering,
Nanjing University of Posts and Telecommunications, 210003 Nanjing, China
{daixb,liutl}@njupt.edu.cn
[3] Laboratory of Image Science and Technology, School of Computer Science and Engineering,
Southeast University, 210096 Nanjing, China
{shu.list,luo.list}@seu.edu.cn

Abstract. The acquired images often provide a degraded version of the true scene due to the imperfect imaging devices or imaging conditions. Therefore recognition of blurred images has become a key task in pattern recognition and moment invariant-based methods play an important role in this field. In this paper, we construct a new set of invariants using Pseudo-Zernike moments which are invariant to convolution with circularly symmetric point spread function (PSF). The experimental results show that proposed invariants have better performance in terms of invariance and robustness to noise with the comparison to the blur invariants derived from Zernike moments whatever the PSF and noise.

Keywords. Blur invariants, pseudo-Zernike moments, circularly symmetric blur degradation.

1 Introduction

Since real imaging systems and imaging conditions are always imperfect, the acquired images are often distorted by the blur degradation. Therefore the recognition of blurred images becomes an important task in pattern recognition.

Numerous algorithms have been proposed to deal with the blurred images in which moment invariants-based methods play an important role. Generally, the existing moment invariants-based methods resort to find a set of invariants as feature descriptors which hold for blur degradation. The ground-breaking work for finding blur invariants of moment functions was performed by Flusser [1] who derived invariants to convolution with an arbitrary centrosymmetric PSF and reported their successful application in the registration of satellite images and image recognition [2]. Later, they extended the blur invariants to N-dimensions [3] and proposed a set of combined invariants to blur degradation and affine transformation based on geometric moments [4]. Besides, Zhang also construct a set of combined invariants by normalizations [5].

J. Yang, F. Fang, and C. Sun (Eds.): IScIDE 2012, LNCS 7751, pp. 90–97, 2013.
© Springer-Verlag Berlin Heidelberg 2013

However, the invariants mentioned above are mostly derived from non-orthogonal moments, such as geometric moments or complex moments which contained redundant information and become sensitive to noise especially when the high order moments are used.

Motivated by the work of [6] and [7] which reported that the orthogonal moments outperform non-orthogonal moments in terms of information redundancy and robustness to noise, Zhang [8] introduced blur invariants of orthogonal Legendre moments and prove their effectiveness in pattern recognition applications. And Zhu [7] proposed a set of combined blur-rotation invariants based on orthogonal Zernike moments. Unfortunately, their work only use Zernike moments whose order p is same as the repetition q to construct the invariants. To deal with the problems in [7], Chen [9] described a new set of combined invariants to blur and similarity transformations using orthogonal Zernike moments.

According to Zhang [10], Pseudo-Zernike moments have good properties of orthogonality and rotation invariance. Furthermore, it has been validated that Pseudo-Zernike moments are superior to Zernike moments in terms of feature representation [10]. Therefore, this paper proposes the use of the Pseudo-Zernike moments to construct a new set of blur invariants. A relationship is established between the Pseudo-Zernike moments of the blurred images and those of the original images which will be subsequently used to derive the blur invariants.

2   Some Basic Terms of Pseudo-Zernike Moments

This section briefly reviews the definition and properties of Pseudo-Zernike moments.

The Pseudo-Zernike moments of order p with repetition q is defined as [10]

$$PZ_{pq} = \frac{p+1}{\pi} \int_{-\pi}^{\pi} \int_0^1 \sum_{k=0}^{p-|q|} (-1)^k \frac{(2p+1-k)!}{k!(p+|q|+1-k)!(p-|q|-k)!} \quad p \ge |q| \ge 0 \quad (1)$$
$$\cdot r^{p-k} e^{-\hat{j}q\theta} f(r,\theta) r dr d\theta$$

Since the radial polynomial $R_{pq}(r)$ is symmetric with q, $R_{p,\,q}(r)$ is equal to $R_{p,\,-q}(r)$ for $q \ge 0$. Thus we only consider the case where $q \ge 0$.

Let $p = q + l$ in (2) with $l \ge 0$ and substitute it into (1), we can obtain

$$PZ_{q+l,q} = \sum_{k=0}^{l} c_{l,k}^q \cdot M_{q+k,q} \quad (2)$$

with

$$M_{q+k,q} = \int_{-\pi}^{\pi} \int_0^1 r^{q+k} e^{-\hat{j}q\theta} f(r,\theta) r dr d\theta \quad (3)$$

According to Zhang [10], we also have

$$M_{q+l,q} = \sum_{k=0}^{l} d_{lk}^{q} PZ_{q+k,q} \tag{4}$$

The parameters c_{ij}^{q} and d_{jk}^{q} can be referred to [10].

3   Blur Invariants of Pseudo-Zernike Moments

This section establishes the relation between the Pseudo-Zernike moments of the blurred images and those of the original images. Then we use this relation to construct the blur invariants of Pseudo-Zernike moments.

3.1   Pseudo-Zernike Moments of the Blurred Image

The blurred image $g(x, y)$ can be considered as the convolution of original images $f(x, y)$ with a point spread function $h(x, y)$. Based on the rotation invariance of Pseudo-Zernike moments [10], we can easily achieve that

$$PZ_{q+l,q}^{(h)} = 0, \quad \text{if } q \neq 0 \tag{5}$$

Using the similar deduction in Chen [9], we have

$$M_{q+k,q}^{(g)} = \sum_{m=0}^{q+\frac{k}{2}} \sum_{n=0}^{\frac{k}{2}} \binom{q+k/2}{m}\binom{k/2}{n} M_{m+n,m-n}^{(f)} M_{q+k-m-n,q+n-m}^{(h)} \tag{6}$$

Using (5), and Substituting (4) and (6) into (2) will yield

$$PZ_{q+l,q}^{(g)} = \sum_{i=0}^{l} PZ_{q+i,q}^{(f)} \cdot \sum_{j=0}^{l-i} PZ_{j,0}^{(h)} \cdot A(q,l,i,j) \tag{7}$$

where

$$A(q,l,i,j) = \sum_{k=i+j}^{l} \sum_{n=\frac{i}{2}}^{\frac{k-j}{2}} \binom{q+k/2}{q+n}\binom{k/2}{n} c_{l,k}^{q} d_{2n,i}^{q} d_{k-2n,j}^{0} \tag{8}$$

3.2   Blurred Invariants of Pseudo-Zernike Moments

Based on the subsection 3.1, we propose a new set of blurred invariants of Pseudo-Zernike moments in Theorem 1.

Theorem 1. Let $f(r, \theta)$ be an image function. Then the blur invariants of Pseudo-Zernike moment (PZMI) to circularly symmetric PSF is

$$I_{q+l,q}^{(f)} = PZ_{q+l,q}^{(f)} - \frac{1}{PZ_{00}^{(f)}\pi}\sum_{i=0}^{l-1}I_{q+i,q}^{(f)}\sum_{j=0}^{l-i}PZ_{j,0}^{(f)}A(q,l,i,j), \quad q\geq 0, l\geq 0 \quad (9)$$

The number $(q + l)$ is the order of the invariants. The proof of Theorem 1 is given in Appendix.

4   Experimental Results

In this section, we conduct some experiments to validate our method in terms of invariance and discriminative power, as well as its robustness to noise. The comparison with the blur invariants of Zernike moments [9] (ZMI) are also presented.

4.1   Test of Invariance and Robustness to Noise

In order to evaluate the performance of the invariants to blur and noises, the original image with size of 128×128 shown in Fig. 1 (a) will be respectively degraded by Gaussian blur (Fig. 1 (b)), averaging blur (Fig. 1 (c)), out-of-focus blur (Fig. 1 (d)) and motion blur (Fig. 1 (e)) with mask of sizes 5×5. Zero-mean Gaussian noises with different variances from 0.01Hz to 0.2Hz are added to the images distorted by Gaussian blur and averaging blur. And Salt-and-Pepper noises with densities from 0.01Hz to 0.2Hz are also added to the images distorted by Out-of-Focus blur and motion blur. The invariance between two images f and g can be measured by relative error given in [8]. The maximum order N of invariants to be used in this section was set to 5.

The relative errors between the original image (Fig.1 (a)) and the distorted images (Fig.1 (b)-(e)) using ZMI and blur invariants of Pseudo-Zernike moments (PZMI) are illustrated in Fig. 2 (a)-(d). From Fig.2, we can see that compared to ZMI, PZMI can obtain lower and more stable relative errors in all kinds of blurring modes with increasing densities of noises. Therefore, it can be concluded that PZMI have better invariance and robustness to noise than ZMI whatever the blurring mode and noise.

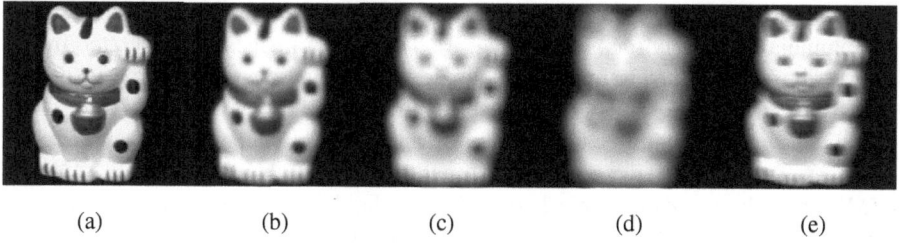

     (a)   (b)   (c)   (d)   (e)

Fig. 1. The original image and the distorted images

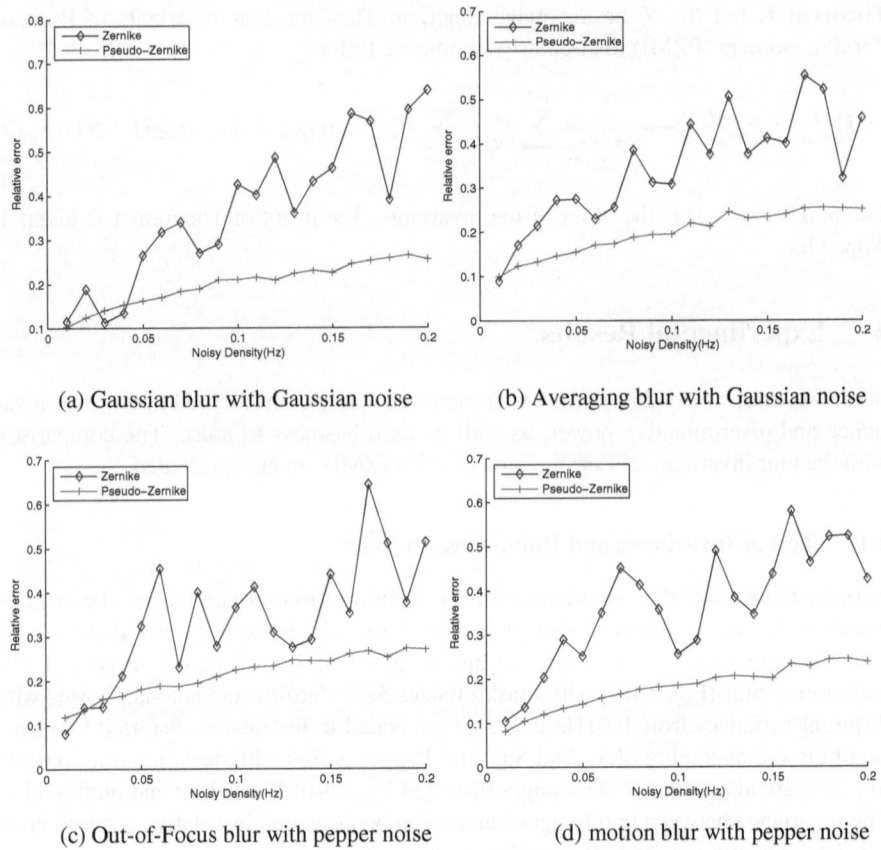

(a) Gaussian blur with Gaussian noise

(b) Averaging blur with Gaussian noise

(c) Out-of-Focus blur with pepper noise

(d) motion blur with pepper noise

Fig. 2. Relative error of ZMI and PZMI between original images in Fig. 1 (a) and distorted images in Fig. 1 (b)-(e) with different blur mode and noise

4.2   Classification Results

The next experiment was carried out to compare the discrimination power of the ZMI and PZMI. The Euclidean distance is used here as the classification measure. An original set of eight images of size 64×64 were selected from the Coil-100 image database of Columbia University (depicted in Fig. 3). The testing images are generated by using the above four blurring modes to degrade the original images, with masks of size 3×3, 5×5, 7×7, 9×9. To verify the robustness to noise, the testing images were subsequently corrupted by additive Gaussian noise and Salt-and-Pepper noise with different densities. Table 1 lists the recognition rates of the testing images using ZMI and PZMI. From the results in Table 1, it is obvious that both invariants can obtain pretty high recognition rates in the case of noise-free. However, the PZMI shows better discriminability in comparison with the ZMI when the density of noises goes up.

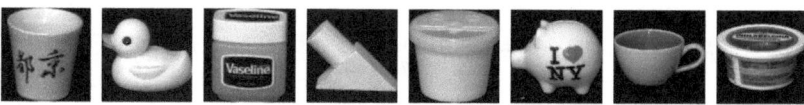

Fig. 3. Eight images chosen from Coil-100 image database

Table 1. The recognition rates of coil-100 images computed by ZMI and PZMI with different noised added

		ZMI	PZMI
Noise-free		99.85%	99.85%
Gaussian noise	va-riance=0.005	95.26%	98.22%
	variance=0.01	89.00%	92.08%
	variance=0.02	84.25%	87.97%
Salt-and-Pepper noise	density=5%	94.72%	97.03%
	density=9%	84.32%	88.67%
	density=11%	78.01%	81.55%

5   Conclusion

In this paper, we have proposed a new set of blur invariants of Pseudo-Zernike moments by establishing the relation between the Pseudo-Zernike moments of the blurred images and those of the original images. The experimental results demonstrated that compared to the blur invariants based on Zernike moment, those of Pseudo-Zernike moments have better discriminability in image recognition and are more robust to noise.

Appendix

Proof of Theorem 1: We prove this theorem by mathematical induction about l.
For $l = 0$, using (7) and (8) we have

$$I_{q,q}^{(g)} = PZ_{q,q}^{(g)} = PZ_{q,q}^{(f)} = I_{q,q}^{(f)} \tag{10}$$

Assume that Theorem 1 is true for $1, 2, ..., l-1$, then we obtain

$$I_{q+l,q}^{(g)} - I_{q+l,q}^{(f)} = PZ_{q+l,q}^{(g)} - PZ_{q+l,q}^{(f)}$$
$$- \frac{1}{PZ_{00}^{(f)}\pi} \sum_{i=0}^{l-1} I_{q+i,q}^{(f)} \sum_{j=0}^{l-i} [PZ_{j,0}^{(g)} - PZ_{j,0}^{(f)}] A(q,l,i,j) \tag{11}$$

If we can get that (11) is zero, Theorem 1 will be validated for the case of l. By using (7), we have

$$PZ_{q+l,q}^{(g)} - PZ_{q+l,q}^{(f)} = \sum_{i=0}^{l-1} PZ_{q+i,q}^{(f)} \sum_{j=0}^{l-i} PZ_{j,0}^{(h)} A(q,l,i,j) \tag{12}$$

Similarly,

$$PZ_{j,0}^{(g)} - PZ_{j,0}^{(f)} = \sum_{k=0}^{j-1} PZ_{k,0}^{(f)} \sum_{r=0}^{j-k} PZ_{r,0}^{(h)} A(0,j,k,r) \tag{13}$$

Introducing (12) and (13) into (11) leads to

$$I_{q+l,q}^{(g)} - I_{q+l,q}^{(f)} = \sum_{i=0}^{l-1} PZ_{q+i,q}^{(f)} \sum_{j=0}^{l-i} PZ_{j,0}^{(h)} A(q,l,i,j) - \frac{1}{PZ_{00}^{(f)}\pi} \sum_{i=0}^{l-1} I_{q+i,q}^{(f)}$$
$$\cdot \sum_{j=0}^{l-i} \sum_{k=0}^{j-1} PZ_{k,0}^{(f)} \sum_{r=0}^{j-k} PZ_{r,0}^{(h)} A(0,j,k,r) A(q,l,i,j) \tag{14}$$

Then we can obtain

$$PZ_{q+l,q}^{(f)} = \frac{1}{PZ_{00}^{(f)}\pi} \sum_{i=0}^{l} I_{q+i,q}^{(f)} \sum_{j=0}^{l-i} PZ_{j,0}^{(f)} A(q,l,i,j) \tag{15}$$

Replacing (14) with (15) can achieve

$$I_{q+l,q}^{(g)} - I_{q+l,q}^{(f)} = \sum_{i=0}^{l-1} \frac{1}{PZ_{00}^{(f)}\pi} \sum_{k=0}^{i} I_{q+k,q}^{(f)} \sum_{r=0}^{i-k} PZ_{r,0}^{(f)} A(q,i,k,r) \sum_{j=0}^{l-i} PZ_{j,0}^{(h)} A(q,l,i,j)$$
$$- \frac{1}{PZ_{00}^{(f)}\pi} \sum_{i=0}^{l-1} I_{q+i,q}^{(f)} \sum_{j=0}^{l-i} \sum_{k=0}^{j-1} PZ_{k,0}^{(f)} \sum_{r=0}^{j-k} PZ_{r,0}^{(h)} A(0,j,k,r) A(q,l,i,j) \tag{16}$$

By changing the order of summation and shifting the indices, (16) becomes

$$I_{q+l,q}^{(g)} - I_{q+l,q}^{(f)} = \frac{1}{PZ_{00}^{(f)}\pi} \sum_{i=0}^{l-1} I_{q+i,q}^{(f)} \sum_{k=0}^{l-i-1} PZ_{k,0}^{(f)} \sum_{r=0}^{l-k-i} PZ_{r,0}^{(h)} \cdot T_1 \tag{17}$$

where

$$T_1 = \sum_{j=i+k}^{l-r} A(q,j,i,k) A(q,l,j,r) - \sum_{j=r+k}^{l-i} A(0,j,k,r) A(q,l,i,j) \tag{18}$$

Substitution of (8) into (18) and use of the orthogonal property can produce

$$T_1 = \sum_{m=i+k}^{l-r} \sum_{n=\frac{i}{2}}^{\frac{m-k}{2}} \sum_{s=m+r}^{l} \binom{q+\frac{m}{2}}{q+n}\binom{\frac{m}{2}}{n}\binom{q+\frac{s}{2}}{q+t}\binom{\frac{s}{2}}{t} d_{2n,i}^{q} d_{m-2n,k}^{0} c_{l,s}^{q} d_{s-m,r}^{0}$$

(19)

$$- \sum_{s=r+k}^{l-i} \sum_{t=\frac{k}{2}}^{\frac{s-r}{2}} \sum_{m=s+i}^{l} \binom{\frac{s}{2}}{t}\binom{\frac{s}{2}}{t}\binom{q+\frac{m}{2}}{q+n}\binom{\frac{m}{2}}{n} d_{2t,k}^{0} d_{s-2t,r}^{0} d_{m-s,i}^{q} c_{l,m}^{q}$$

Changing the order of summation and shifting the indices again, we can obtain (19) as

$$T_1 = 0 \qquad (20)$$

Thus, that (11) is equal to 0 can be achieved. Finally, Theorem 1 has been proved.

Acknowledgement. This work was supported by the Starting Scientific Research Foundation of Nanjing University of Posts and Telecommunications for New Teachers (NY211030), by the National Natural Science Foundation of China under Grants 31200747 and 60911130370, by the Natural Science Foundation of Jiangsu Province under Grants BK2012437.

References

1. Flusser, J., Suk, T.: Degraded image analysis: an invariant approach. IEEE Trans. Pattern Anal. Mach. Intell. 20, 590–603 (1998)
2. Flusser, J., Zitova, B., Suk, T.: Invariant-based registration of rotated and blurred images. In: IEEE International Proceeding on Geoscience and Remote Sensing Symposium, pp. 1262–1264. IEEE Press, New York (1999)
3. Flusser, J., Boldyš, J., Zitova, B.: Moment forms invariant to rotation and blur in arbitrary number of dimensions. IEEE Trans. Pattern Anal. Mach. Intell. 13, 1123–1136 (2003)
4. Suk, T., Flusser, J.: Combined blur and affine moment invariants and their use in pattern recognition. Pattern Recogn. 36, 2895–2907 (2003)
5. Zhang, Y., Wen, C., Zhang, Y., Soh, Y.C.: Determination of blur and affine combined invariants by normalization. Pattern Recogn. 35, 211–221 (2002)
6. Coatrieux, J.L.: Moment-based approaches in imaging part 2: invariance. IEEE Engineering in Medicine and Biology Magazine 27, 81–83 (2008)
7. Zhu, H., Liu, M.: Combined invariants to blur and rotation using Zernike moment descriptors. Pattern Anal. Appl. 13, 309–319 (2010)
8. Zhang, H., Shu, H., Han, G.N., Coatrieux, G., Luo, L., Coatrieux, J.L.: Blurred image recognition by Legendre moment invariants. IEEE Trans. Image Process. 19, 596–611 (2010)
9. Chen, B.J., Shu, H.Z., Zhang, H., Chen, G., Toumoulin, C., Dillenseger, J.L., Luo, L.M.: Combined invariants to similarity transformation and to blur using orthogonal Zernike moments. IEEE Trans. Image Process. 20, 345–360 (2011)
10. Zhang, H., Dong, Z.F., Shu, H.: Object recognition by a complete set of pseudo-Zernike moment invariants. In: 35th IEEE International Conference on Acoustics Speech and Signal Processing, pp. 930–933. IEEE Press, New York (2010)

Gait Recognition Based on Partitioned Weighting Gait Energy Image

Xiaoxiang Li, Dimin Wang, and Youbin Chen

Graduate School at Shenzhen, Tsinghua University, China
{lixx06,insoliadi}@gmail.com, chenyb@sz.tsinghua.edu.cn

Abstract. Gait energy image (GEI) has been proved to be an effective gait feature representation method. But it's sensitive to the change of clothing and carrying conditions. We propose a novel gait recognition method called partitioned weighting gait energy image (PWGEI) to deal with these problems. A human body is divided into four parts and different weights are given to different parts to get the PWGEI from GEI. Two different weighting ways are conducted and a fusion of classifiers is adopted. We test our method on the USF database. Our average recognition rate is 48.87%, which is higher than GEI by 6% and higher than gait flow image (GFI) by 5.79%. The experimental results prove the effectiveness of our proposed PWGEI method.

Keywords: gait recognition, GEI, PWGEI.

1 Introduction

Unlike some other biometrics recognition tasks such as fingerprint recognition, face recognition and iris recognition, gait recognition does not need a close view and it's not necessary for human to hold still for a while for data capture. However, gait recognition is still a very challenging problem. A number of gait recognition methods have been proposed and can be divided into two categories: model-based methods [1–3] and model-free methods [4–15]. Model-based methods usually need to find several parts of the human body and model the relationship of the moving parts with one's gait information. The performance of these methods is usually limited because of the noise of the gait images and the difficulty in segmenting the parts. In model-free methods, the gait features are extracted from the silhouette images of the gait sequence using statistic methods. Among several ways of gait feature representation, gait energy image (GEI) [4] is the most widely used one. GEI can cope with the low quality of silhouettes and noise, but it doesn't contain the motion information and is sensitive to clothing and carrying conditions. Improvement on GEI has been made and some methods based on GEI have appeared such as gait flow image (GFI) [5], enhanced gait energy image (EGEI) [6], frame difference energy image (FDEI) [7], dynamic gait energy image (DGEI) [8], active energy image (AEI) [9], etc. These methods all concentrated on GEI without motion information. None of these methods dealt with the problem that static information contained by GEI

J. Yang, F. Fang, and C. Sun (Eds.): IScIDE 2012, LNCS 7751, pp. 98–106, 2013.

is sensitive to clothing and carrying condition changes. There are also some other methods of gait feature representation, like Shape Variation-Based (SVB) Frieze Pattern [10], motion silhouette contour templates and static silhouette templates (MSCT and SST) [11].

Aiming at solving the clothing and carrying condition sensitivity problem, we propose a method for gait recognition called partitioned weighting gait energy image (PWGEI). We try to give small weights to the parts of human body which may change easily with different clothing or carrying conditions. This will suppress the irrelative information and strengthen the useful information to identify a person. Principal component analysis and linear discriminant analysis are combined to reduce the dimension and separate the patterns [12] for gait recognition. Classifier fusion strategy is used to cope with different clothing and carrying conditions. The effectiveness has been proved on the USF database [13].

The rest of this paper is organized as follows. Section 2 describes our algorithm in detail. Section 3 presents the experiment results. Section 4 concludes the paper.

2   Algorithm

Our algorithm of gait recognition is described in this section. The algorithm is tested on the USF database. As all the samples in USF database have been segmented from the background and have the same image size, so our first step is to divide the sequence into several gait periods. Then all images among a gait period are accumulated to get a gait energy image (GEI) [4]. Different weights are given at different part of the body to generate the partitioned weighting gait energy image (PWGEI) from the GEI. We use the principal component analysis (PCA) and linear discriminant analysis (LDA) to reduce the dimension and separate the samples [12]. At the training phase, we use the PWGEIs from gallery set to get the PCA transform matrix and the LDA transform matrix. At the testing phase, we get the PWGEIs from probe set multiplying the PCA transform matrix and LDA transform matrix. We then use a nearest neighbor classifier to classify each sample from the probe set. Two classifiers are trained using different partitioned weighting method and a fusion is made to improve the recognition performance.

2.1   Gait Period Estimation

A gait image sequence may contain several gait periods. People may walk in different speeds and camera parameters are different in different scenarios. So we need to estimate the gait period and get the frame numbers in each gait period. We use the same feature as Toby et al. [5]: Counting the number of foreground pixels in the lower half part of the silhouette gait image. This feature works because the silhouette contains different pixels in the mid-stance position and double-support position. As shown in Fig. 1, the trough of the feature wave represents the mid-stance position while the peak point shows the double-support position. We then select the extreme values to divide the image sequence into several gait periods, each of which contains a whole gait cycle.

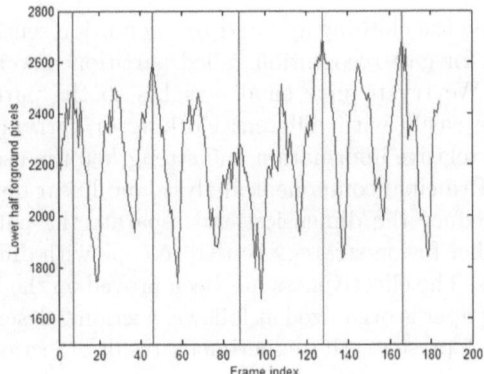

Fig. 1. Gait period estimation. The frames between each two adjacent red lines make up a gait period.

We find a robust method to detect the extreme points and estimate the gait period, which is implemented as follows: First we bring in the FFT transformation to get the main low frequency information. Then we use low frequency parameter to set up a sliding window and detect the extreme values including one peak point and one trough point in each window. At last the fake points are taken away if they are not far away from the neighborhood by a certain distance. One gait period starts from the first extreme point of the feature wave or the last gait period end point, and ends in the extreme points after two cycles, which is used to divide the image sequence into several gait periods. As shown in Fig. 1, the region between two red lines represents one detected period.

2.2   PWGEI Generation

The gait energy image is generated as formula (1), where $I_{k,i}(x, y)$ is the ith image in the kth period of a gait image sequence of one person. N_k is the number of images in the kth period of the same person.

$$GEI_k(x, y) = \frac{\sum_{i=1}^{N_k} I_{k,i}(x, y)}{N_k} \qquad (1)$$

We then use a human body partitioned model to partition the GEI into four parts from the top to the bottom: head and shoulder, upper half of the body, lower half of the body and noisy bottom. Each part of the body is given a different weight. So a GEI is then turned into a partitioned weighting gait energy image (PWGEI). The motivation to do this is that we think different parts of the body shows different human identification information. Besides, some parts of the body may be prone to noise and others do not. Some parts may change more frequently than others when people walking in different scenarios. In order to achieve high gait recognition rate, we should give big weights to the parts containing more human identification information, and to the parts that stay more stable when walking and camera environment are changed, and to the parts robust to noise. Our experiments have proved this thought is reasonable and we do get higher recognition performance than other methods published in recent years. Fig. 2 shows how we partition the body into four parts.

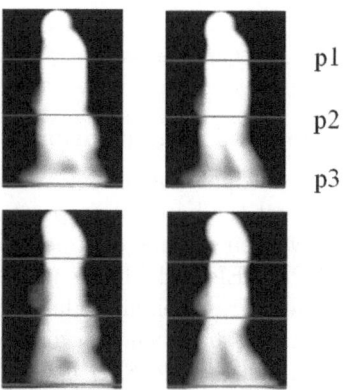

p1

p2

p3

Fig. 2. Partition the body into four parts. The three red lines are the boundaries of the four parts. The upper two figures correspond to a man with and without a bag, the lower two figures correspond to a woman with and without a bag.

We give a big weight value to the head and shoulder part, a middle weight value to the upper half of the body, a small weight value to the lower half of the body and a zero weight to the noisy bottom. This is because the head and shoulder and the upper half of the body shows significant human identification information and do not change so easily at different scenarios. However, the lower part of the body may change easily. For example, a person wears trousers or skirt or carries a bag or briefcase may change the lower half of the body in a gait energy image sharply. As to the noisy bottom, we give it a zero weight because there is a big difference whether a person walks on grass or on a concrete ground. The partition points of the four parts form a three dimension vector. For convenience, we expand it to a five dimension vector PWpoints and the weight values form a four dimension vector PW as shown in formula (2) and (3). We trained two classifiers using PWGEIs. Each is given a different weight vector. This is discussed in detail at the Classifier Fusion part. Formula (4) shows how PWGEI is generated, where x is the row index and y is the column index of images.

$$PWpoints = [0, p1, p2, p3, +\infty] \qquad (2)$$

$$PW = [pw1, pw2, pw3, pw4] \qquad (3)$$

$$if(PWpoints(m) < x \le PWpoints(m+1))$$
$$PWGEI_k(x,y) = GEI_k(x,y) * PW(m) \quad (m=1,2,3,4) \tag{4}$$

Note that we divide the sequence into several gait periods and get the GEIs as formula (1) in the training phase, but we need only one PWGEI to classify a person, so in the testing phase we don't divide the sequence into several gait periods, instead we use the total sequence GEIs as shown in formula (5), where N_{Total} is the number of images in the sequence and L is the number of periods in the sequence.

$$GEI_{TotalSequence}(x,y) = \frac{\sum_{k=1}^{L} \sum_{i=1}^{N_k} I_{k,i}(x,y)}{N_{Total}} \tag{5}$$

2.3 Dimension Reduction and Classification

Huang et al. [12] combined principal component analysis (PCA) and liner discriminant analysis (LDA) to get the best data representation and pattern separation. In our paper, we follow their method of dimension reduction. PCA is used to reduce the dimension of PWGEIs and LDA is used to separate the patterns.

Euclidean distance and the nearest neighbor classifier are used after the PCA and LDA procedure. The smaller the Euclidean distance between two samples, the more similar they are. A test sample S_{test} is converted into a $PWGEI_{test}$ as formula (1) and (4), and then transformed to Z_{test} using PCA and LDA. Then the distance between Z_{test} and those in the gallery are calculated, and the sample will be classified into the class the same as the nearest sample in the gallery according to (6), where $Z_{i,j}$ is the sample in the jth period of the ith class.

$$if((i',j') = \min_{i,j} ||Z_{test} - Z_{i,j}||) , \quad S_{test} \in \Omega_{i'} \tag{6}$$

2.4 Classifier Fusion

Two different classifiers are trained using two different weight vectors. The motivation is that the clothing condition of people has a big impact on the figures segmented from the gait image sequence. People may wear a close-fitting dress sometimes and wear cloaks and carry a briefcase or a bag at some other times. In order to cope with these problems, two classifiers are trained. One has big weight values in the upper half of the body and lower half of the body ($pw2$ and $pw3$), noted as Classifier1. The other has small weight values in these two parts, noted as Classifier2.

The way of the fusion of these two classifiers is like this: Firstly, a feature that indicates the cloth and bag condition of a person is extracted which is actually the average gray level from the waist to ham of the body of GEI, noted as Fcomb. Then a decision is made according to this feature. A threshold is used, noted as Fthr. If Fcomb is bigger than the threshold Fthr, Classifier2 will be used in later procedure, otherwise Classifier1 will be used. By doing this we can cope with the clothing and bag changing problem.

Fig. 3 shows the waist to ham region of the body used to extract the feature.

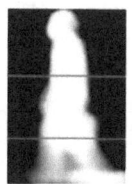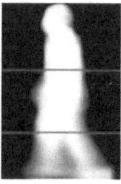

Fig. 3. The waist to ham region (between the two red lines)

3   Experiments

Our proposed method is tested on the USF database. The USF database has 1870 sequences from 122 subjects spanning 5 covariates. It was built by University of South Florida and was collected over four days, May 20-21, 2001 and Nov 15-16, 2001. The database is divided into 13 subsets: one gallery and 12 probes. The covariates are different between the 12 probes and the gallery and the number of samples in each probe is listed in Table 1.

Table 1. The USF database

Probe	Covariates Different Between Gallery and Probe	Number of samples
A	View	122
B	Shoe	54
C	View, Shoe	54
D	Surface	121
E	Surface, Shoe	60
F	Surface, View	121
G	Surface, Shoe, View	60
H	Briefcase	120
I	Shoe, Briefcase	60
J	View, Briefcase	120
K	Time (+Shoe, Clothing)	33
L	Surface, Time	33

The performance of our proposed method has been compared with other gait recognition methods in the USF database: gait energy image (GEI) [4], gait flow image (GFI) [5], and the baseline algorithm [13]. GFI is reported as a better gait representation than GEI. The recognition rate of GFI is from Toby et al.'s paper. The authors did two experiments on the GFI, one using direct matching and the other using PCA and LDA. Note that, Toby et al. [5] also did the GEI experiment, and because of different preprocessing procedures and different PCA and LDA parameters, the recognition rate of GEI is slightly different from both [4] and [5].

Table 2 shows the recognition rates of these methods. Compared with other methods, our method has a higher overall performance than others. The average recognition rate of PWGEI is 48.87%, which is 6% higher than GEI, 6.04% higher than GFI using direct matching, and 5.79% higher than GFI using PCA and LDA. It's reported that GFI has the best recognition rate of 44.83% when there are 300 principle components [5]. But our method is still 4.04% higher than that. It is obvious that our method has the highest recognition rate at the probe D, E, F, G and I. This shows our method copes well with the covariates difference of surface, shoes and briefcase.

Table 2. Comparison of recognition rate in USF database (%)

Probe	Baseline	GEI	GFI (direct matching)	GFI (LDA)	PWGEI
A	73	85.25	89	82	85.25
B	78	83.33	93	89	88.89
C	48	72.22	70	76	70.37
D	32	31.40	19	27	33.88
E	22	41.67	23	27	41.67
F	17	18.18	7	10	19.01
G	17	20.00	8	17	33.33
H	61	50.00	78	60	74.17
I	57	56.67	67	57	76.67
J	36	46.67	48	54	45.00
K	3	6.06	9	15	12.12
L	3	3.03	3	3	6.06
Average Rate	37.25	42.87	42.83	43.08	48.87

Fig. 4 shows that Classifier2 has higher recognition rates at the H, I, J probe, which has briefcases; however Classifier1 has higher recognition rate at other probes. This illustrates that by giving the lower half part of the body a small weight, our method can cope with the problem of briefcase varying condition. The average recognition rates of classifier1, classfier2 and the combined classifier are 45.35%, 44.70%, and 48.87%. So by combining two classifiers, we get a better performance than either one of the two classifiers, which shows the effectiveness of the feature Fcomb.

We did a series experiments to choose the optimal parameters. The PWpoints vector is set as $[0, 0.2813, 0.5938, 0.9766, +\infty]$*Rnum, where Rnum is the row numbers of the image. And the PW vector is set as [1,0.7,0.4,0] and [1,0.2,0.1,0] respectively in classifier1 and classifier2. The PCA and LDA dimension parameter Ethr and ldaDim are set as 61 and 53. As for the classifier fusion parameters, the start row and end row number of the waist to ham region of the body, we choose [0.3828,0.7344]*Rnum, where Rnum is the row numbers of the image. The average grey level threshold Fthr is chosen as 102 in all of our experiments.

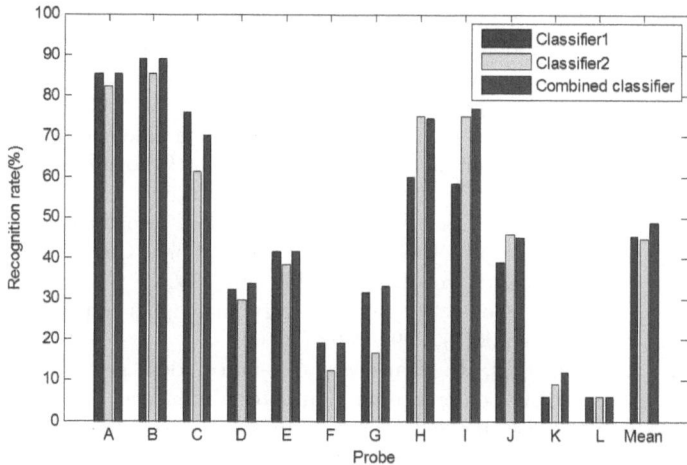

Fig. 4. Comparison of recognition rates between the combined classifier and two sub-classifiers

4   Conclusion

This paper describes a novel partitioned weighting gait energy image (PWGEI) method for gait recognition. We first get the GEIs and then use a partitioned weighting strategy to generate the PWGEIs. PCA and LDA are subsequently used to reduce the feature dimension and distinguish the patterns. Experiments on the USF database are conducted and the experimental results demonstrate the effectiveness of our proposed method. Our method outperforms GEI and GFI. Our future work will concentrate on the combination of partitioned weighting strategy and other gait feature representation methods such as EGEI, AEI, etc.

Acknowledgement. Portions of the research in this paper use the USF Gait Database collected by University of South Florida, Tampa.

References

1. Dockstader, S., Berg, M., Tekalp, A.: Stochastic kinematic modeling and feature extraction for gait analysis. IEEE Transactions on Image Processing 12(8), 962–976 (2003)
2. Singh, J., Jain, S.: Person identification based on gait using dynamic body parameters. In: Trendz in Information Sciences Computing (TISC), pp. 248–252 (December 2010)
3. Wu, J.: A novel approach for discrimination of human gait using kernel learning algorithm. In: 2010 Sixth International Conference on Natural Computation (ICNC), vol. 6, pp. 3253–3256 (August 2010)
4. Han, J., Bhanu, B.: Individual recognition using gait energy image. IEEE Transactions on Pattern Analysis and Machine Intelligence 28(2), 316–322 (2006)

5. Lam, T.H., Cheung, K., Liu, J.N.: Gait flow image: A silhouette-based gait representation for human identification. Pattern Recognition 44(4), 973–987 (2011)
6. Yang, X., Zhou, Y., Zhang, T., Shu, G., Yang, J.: Gait recognition based on dynamic region analysis. Signal Processing 88(9), 2350–2356 (2008)
7. Chen, C., Liang, J., Zhao, H., Hu, H., Tian, J.: Frame difference energy image for gait recognition with incomplete silhouettes. Pattern Recognition Letters 30(11), 977–984 (2009)
8. Zhang, E.H., Ma, H.B., Lu, J.W., Chen, Y.J.: Gait recognition using dynamic gait energy and pca+lpp method. In: 2009 International Conference on Machine Learning and Cybernetics, vol. 1, pp. 50–53 (July 2009)
9. Zhang, E., Zhao, Y., Xiong, W.: Active energy image plus 2dlpp for gait recognition. Signal Processing 90(7), 2295–2302 (2010)
10. Lee, S., Liu, Y., Collins, R.: Shape variation-based frieze pattern for robust gait recognition. In: IEEE Conference on Computer Vision and Pattern Recognition, CVPR 2007, pp. 1–8 (June 2007)
11. Lam, T.H.W., Lee, R.S.T., Zhang, D.: Human gait recognition by the fusion of motion and static spatio-temporal templates. Pattern Recognition, 2563–2573 (2007)
12. Huang, P.S., Harris, C.J., Nixon, M.S.: Recognising humans by gait via parametric canonical space. AI in Engineering, 359–366 (1999)
13. Sarkar, S., Phillips, P.J., Liu, Z., Vega, I.R., Grother, P., Bowyer, K.W.: The humanid gait challenge problem: Data sets, performance, and analysis. IEEE Trans. Pattern Anal. Mach. Intell., 162–177 (2005)
14. Yu, S., Tan, D., Tan, T.: A framework for evaluating the effect of view angle, clothing and carrying condition on gait recognition. In: ICPR (4) 2006, pp. 441–444 (2006)
15. Wang, L., Tan, T., Ning, H., Hu, W.: Silhouette analysis-based gait recognition for human identification. IEEE Trans. Pattern Anal. Mach. Intell., 1505–1518 (2003)

Facial Expressions Analysis Based on Cooperative Neuro-computing Interactions

Chao Xu, Guangquan Xu, Qinghua Hu, and Zhiyong Feng

School of Computer Science and Technology, Tianjin University, China
{xuchao,losin,huqinghua,zyfeng}@tju.edu.cn

Abstract. Facial expressions analysis is an important step in human-computer interaction and intelligence computing. Due to the complexity and uncertainty of real-time facial expressions, the performance of the existing algorithms is not satisfactory. In this paper, a novel approach is proposed to help to enhance the significance of analysis optimum. Based on the person-independent approach and the cooperative neuro-computing, multi-model interactions for facial expressions cluster structures are applied to improve the capacity of selection, distribution, and evaluation of the cluster centers. The resultant model is potentially capable of constructing high-quality clusters and achieving high efficiency of the convergence. It is suggested that the model with cooperative neuro-computing interaction has the characteristics to construct the cluster distribution rapidly, and can perform real-time analysis efficiently and accurately.

Keywords: cooperative neuro-computing interactions, cluster analysis, adaptive PCA.

1   Introduction

With the rapid development of pattern recognition, Human-Computer Interaction has come into many fields of human's life gradually, such as affective computing. If the interactive agents have the ability to recognize human facial expressions and emotions, the computational results can be used to understand and solve the problem of intelligent cognition and decision. It is even benefit to reduce the negative effects caused by frequent interruptions in HCI[1].

In spite of the challenges of facial expressions recognition for HCI and affective computing, there are many researches, such as the design and improvement of clustering algorithm for facial expressions. For example, Mok proposed the clustering analysis method to identify the desired cluster number and produces[2]; Cenk presented strategic group analysis of Turkish contractors to compare the performances of self-organizing maps (SOM) and fuzzy C-means method (FCM)[3]; Bidyut proposed the distance based clustering method (1-SL) to find arbitrary shaped clusters in large datasets[4];Marios presented the recognition algorithm using dynamic training in a multistage clustering scheme [5].

However, the approaches above classify facial expressions features in a static process. As we know that HCI and affective computing require facial expressions and

J. Yang, F. Fang, and C. Sun (Eds.): IScIDE 2012, LNCS 7751, pp. 107–114, 2013.

emotion recognition in real-time. Thus, static process should be improved to be benefit for the facial expressions analysis.

Since facial expressions analysis needs feedbacks and adjustments to recognize the affective status from coordinative perception and cognition, it is difficult to recognize each subject accurately by universal facial expressions modeling mechanism. Interactive neuro-computing is constructed to conduct the interactions of multi-models and deal with the traditional PCA algorithm for adaptive PCA improvement.

As person-independent models with interactions can be applied for facial expressions analysis, the context to be intra-changed is the core items and PCA can be used to extract facial features and reduce the computational complexity. Some works are concentrated in PCA, for instance, Wang proposed two dimensional principal components of natural images[6]; Yang proposed Laplacian BDPCA to non-Euclidean space[7]; and Diamantaras applied PCA for the blind separation of signals [8].

Neuro-computing can also be applied in researches of adaptive PCA and facial expressions analysis. For instance, Caridak is presented the approach using neural networks for emotion recognition [9]; Wang proposed the person-independent facial expressions space to analyze and synthesize facial expressions[10].

Our contribution, as well as the aim of this paper, is to present a mathematical framework of the person-independent models with neuro-computing interactions. Based on neural computing, adaptive estimation of principal components is proposed, and the adaptive PCA with neuro-computing interactions is feasible for real-time analysis of facial expressions.

Since the improved algorithm introduces interactions and neural networks to promote the learning algorithm of principal components, the person-independent models are applied to optimize the cluster structure and the estimation of principal components, respectively. It can improve the recognition precision and conduct intra-model cooperative interaction learning based on neuro-computing adaptive PCA. The essence of neuro-computing interaction is derived from affective resonance, including human-computer and computer-computer resonance, which are mutual affective feature signals, perceived and understood by interactive participants.

2   Person-Independent Analysis Model and Cooperative Interaction

In this paper, facial expressions are limited in a given context called social display rule which is used to determine the acceptable range of facial expressions in the given context. Suppose that different participants have similar perceptions of emotions, yet inner elements of emotions are different, such as personality factors, social status and expressions habit and soon.

Affective status composed of the same inner elements is approximately the same in facial expression patterns. For those elements, similarity generalization is conducted and facial expressions feature vectors are formulized in the process of model clustering. Then, facial expressions composed of different inner elements are expressed through interactions, and specific patterns are reflected in the information interaction.

Affective status can be expressed by facial expressions. However, participants in the similar affective status may give irrelevant facial expressions. Facial states and-features can be expressed by relevant and specific adjustments, and this experience reflects the inner needs of the participants exactly. If the inner affective status of the participants is understood, then facial expressions can be analyzed. We can promote the perception of the existent affective status, and even the physical and mental adjustments without specified facial expressions.

Person-independent models are designed for the process of cooperative interaction in the context. Person-independent models get facial data of inter subjective from perception. Facial expressions clusters are built and their distributions reflect inner affective patterns of participants. Person-independent models in the organizational context can conduct cooperative interaction, and the clustering centers can be mutually transferred between related models. Facial features of participant andparameters-tandards are evaluated based on the organizational constrains and the specifications. The selections of cluster centers of the models are eventually improved and the structure of the facial expressions database is optimized.

3   Improved Facial Expressions Clustering Algorithm Based on Neuro-computing Interaction

Facial expressions analysis of person-independent models is based on the optimized clustering algorithm. As we know that the algorithm, which simulates the adaptive learning process of human beings, is improved by introducing intra-model cooperative interaction and cooperative learning parameters. Cooperative learning parameters are introduced to improve the selections of cluster centers and the weight-values in the process of iteration and recognition analysis.

3.1   Facial Expressions Clustering Algorithm Framework with Cooperative Interactions

In the domain of facial expressions space, $cluster_j$ denotes the cluster j .For the sample x_i , the distance metrics is defined as $D(x_i, cluster_j)$, and $center_j$ denotes the cluster center. Since the choice of distance metrics function has effect on the cluster criterion function, Mahalanob is distance is applied to reflect the covariance distance of sample. As Mahalanob is distance represents the similarity between the unknown sample and the whole data set, it reflects the pattern distribution structure of clusters effectively.

$J_i = \min_j D^2(x_i, cluster_j) < \Sigma_m (w_j^{(m)} | \tilde{F}_j^{(m)} - center_j |)$ is defined as the criterion function, where $\tilde{F}_j^{(m)}$ denotes the features of class j acquired by the cooperative interaction with the m th person-independent model and $w_j^{(m)}$ denotes the weights of class j . The improved recognition algorithm is described as follows:

Improved recognition algorithm:

Step 1. Based on the cooperative interactions, $\tilde{F}_j^{(m)}$ and $w_j^{(m)}$ are constructed; the initialized cluster centers are $center_j^{(0)} = \Sigma_m (w_j^{(m)} \mid \tilde{F}_j^{(m)} \mid)$ for class j ;

Step 2. x_i is analyzed as belong to the class l after the k th iteration according to the criterion function $J_{il}^{(k)} = \min_j D^2(x_i, cluster_j^{(k)})$;then, $center_j^{(k+1)} = \Sigma_{x_i}(x_i / n_j^{(k+1)}) + \Sigma_m (w_j^{(k+1,m)} \mid \tilde{F}_j^{(k+1,m)} \mid)$ is generated and recalculated as cluster center at the next iteration, n_j is the sample number;

Step 3. If the cluster center satisfies the constrain as $\mid center_j^{(k+1)} - center_j^{(k)} \mid < threshold_j$, where $threshold_j$ is the threshold parameter of class j , finish the algorithm; otherwise go to step 2.

According to the analysis above, person-independent models are classified into the basic classes of facial expressions initially. The improved algorithm is used to adjust the cooperative interaction parameters and the cluster centers iteratively.

Mahalanob is distance and its independence of measurement scale are reflected in the interactions among various features in the sample space. For the square of the Mahalanob is distance $D^2(x_i, cluster_j)$, if the covariance matrix is a diagonal matrix, Mahalanob is distance can be reduced to the normalized Euclidean distance.

3.2   Algorithm Optimization Based on Adaptive PCA with Neuro-computing Interactions

From the section 3.1, it is clear that Mahalanob is distance can be reduced to the standard Euclidean distance if the covariance matrix is the element matrix. For example, PCA can be applied to achieve the element matrix from the covariance matrix.

However, cooperative interactions with other person-independent models cannot be implemented in this process. Neurons, the units of neuro-computing models, are biological neural cell's simulations that are fused into person-independent models, i.e. each person-independent model maintains a neuron. Adaptive PCA are extracted by neural computing estimation fused with multi-models interactions.

For the adaptive PCA form neuro-computing, normalized Hebbian learning rule of Oja for estimation of principal components is listed as follows

$$y_i^{(k)} = [\beta_i^{(k)}]^T \cdot x_i \tag{1}$$

$$\beta_i^{(k+1)} = \beta_i^{(k)} + \alpha \cdot y_i^{(k)} \cdot (x_i - y_i^{(k)} \cdot x_i) \tag{2}$$

where x_i is the data sample, $y_i^{(k)}$ is the first principal component of x_i in the k th iteration, $\beta_i^{(k)}$ is the weight vector of elements contained in x_i from the k th iteration, and $\alpha > 0$ is the learning rate parameter, which is considered to be time-varying, that is, $\alpha = \alpha(k)$.

Learning process of $\beta_i^{(k)}$ is converged if the following hypotheses are satisfied:

Hypotheses 1. Input vector x_i is taken from a stationary stochastic process, and its auto-correlation matrix R_{xx} has different Eigen-values;

Hypotheses 2. x_i and $\beta_i^{(k)}$ are independent for statistical significance;

Hypotheses 3. The weight learning is slow enough for the stationary process.

The hypotheses 1 and 2 are already true in the improved process. The collection of the new samples will cause neuro-computing interactions within person-independent models, and then weight vector will be updated through the iterations. The learning process of weight vector will not stop, so the hypothesis 3 is satisfied. Thus, the improved learning process is converged.

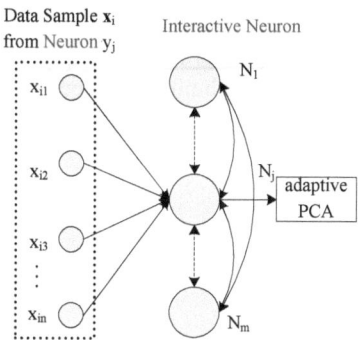

Fig. 1. Adaptive estimation of principal component fused with neuro-computing interaction

Fig.1presents the process of adaptive estimation of principal component fused with neuro-computing interactions within person-independent models. N_j is the neuron maintained in person-independent model j, x_i is the data sample vector collected by the neuron maintained in N_j. The solid lines mean the process of interactions described as: the transferring parameters of neuro-computing interaction including the data sample x_i and its weight vector elements β_i.

According to the analysis above, it is clearly seen that person-independent model's learning template does not change with traditional PCA, and is not directly related to the multi-model interactions, while adaptive estimation of principal component is much more in line with the facial expressions clustering algorithm based on affective information fused with neuro-computing interaction models.

4 Real-Time Facial Expressions Analysis Based on Cooperative Neuro-computing Interaction

Real-time facial expressions analysis model is designed based on the improved cooperative neuro-computing interaction algorithm proposed in this paper. Person-independent models in organizational context collect facial expressions data with camera, establish the personalized expressions clusters, and implement combining cooperative neuro-computing interaction with other person-independent models.

For the real-time requirement and experiments, 30 megapixel cameras are used. Person-independent models can be assigned to different cameras for the real-time analysis. Facial data acquisition and personal analysis on the multiple participants facilitate the achievement of cooperative neuro-computing interaction and adaptive clusters learning within different person-independent models at the same time.

Neuro-computing interacting with other person-independent models in the context, analysis models can adjust and feedback the distribution of facial expressions clusters. Since the improved model adjusts its own clusters from others and tends to be the same with most of the analysis models ultimately, it converges to the local optimum quickly and smoothly. The cluster distribution within person-independent model in the cooperative neuro- computing interaction comply with the same organizational context norms, thus the cluster distribution and recognition based on the improved algorithm is the global optimum.

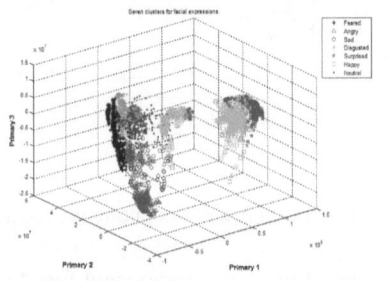 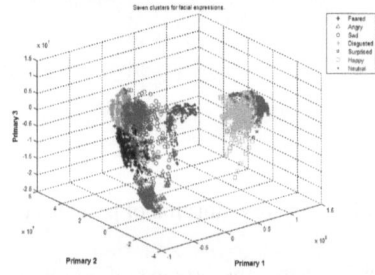

(a) Without neuro-computing interaction   (b) With neuro-computing interaction

Fig. 2. Contrast cluster distributions with cooperative neuro-computing interaction

Fig.2 presents the contrast cluster distributions with cooperative neuro-computing interaction in the three-dimensional space. Fig.2(a) shows the distribution result applied person-independent model interactions without neuro-computing; while Fig.2(b) shows the result with the improved algorithm. It is clearly seen that Fig.2(b) achieves clusters more correctly and the cluster boundaries are more obvious and demarcated. For example, Fig.2(a) indicates some of the neutral expressions are the happy ones, and divides some of the sad expressions into the neutral ones; while Fig.2(b) presents more realistic and accurate distribution.

Experiment results show that the improved algorithm gives more reasonable results on the cluster distributions of the facial expressions. Cooperative neuro-computing interactions avoid the selected cluster centers randomly, transfer the cluster structures, and construct the initial cluster centers with a certain practical significance. Calculated cluster distribution close to the realistic situation that is the global optimal result.

Tab.1 lists the convergence time and accurate rate of the contrast analysis algorithm between Person-independent interactions and cooperative neuro-computing interactions. Experiment results indicate that the improved algorithm based on cooperative neuro-computing interaction is more efficient with short convergence time.

With the iterations increasing, cooperation differences within person-independent models reduce gradually and approach equilibrium, and the analysis results are more accurate in real-time. With the number of samples increasing, facial expressions analysis with cooperative neuro-computing interaction improves the recognition rate.

Table 1. Convergence time (CT) and accurate rate (AR) of contrast algorithm models

Samples (frame)	Person-independent interactions		neuro-computing interactions	
	CT (s)	AR(%)	CT (s)	AR(%)
1000	0.66	82.4	0.65	82.7
2000	1.34	84.6	1.02	84.9
4000	2.49	87.2	2.21	88.3
8000	3.96	88.1	3.54	90.2

5   Conclusions

With the advances of the person-independent model with cooperative neuro-computing interaction, we propose and achieve an improved cluster algorithm model of facial expressions analysis with satisfactory results. Based on model interactions and cooperative neuro-computing, this paper presents the design and implementation of the analysis model in the organizational context, and shows facial expressions cluster distributions in three-dimensional space.

Additionally, experiments are conducted to evaluate the improved model. The results indicate that facial expressions analysis can be effective both in determining the cluster structures and in improving the rationality and collaboration of recognition rateby applying cooperative neuro-computing interaction.

Acknowledgment. This work was supported in part by a grant from the 985 Program of China.

References

1. Xu, C., Feng, Z.Y.: An Affective Modeling Approach to Interruptions in Proactive Computing Environments. In: The 12th International Conference on Human-Computer Interaction, pp. 628–632 (2007)
2. Mok, P.Y., Huang, H.Q., Kwok, Y.L., Au, J.S.: A Robust Adaptive Clustering Analysis Method for Automatic Identification of Clusters. Pattern Recognition 45(8), 3017–3033 (2012)
3. Budayan, C., Dikmen, I., Birgonul, M.T.: Comparing the Performance of Traditional Cluster Analysis, self-organizing maps and fuzzy C-means method for strategic grouping. Expert Systems with Applications 36(9), 11772–11781 (2009)
4. Patra, B.K., Nandi, S., Viswanath, P.: A Distance Based Clustering Methodfor Arbitrary Shaped Clustersin Large Datasets. Pattern Recognition 44(12), 2862–2870 (2011)
5. Kyperountas, M., Tefas, A., Pitas, I.: Dynamic Training Using Multistage Clustering for Face Recognition. Pattern Recognition 41(3), 894–905 (2008)

6. Wang, D., Lu, H.C., Li, X.L.: Two Dimensional Principal Components of Natural Images and its Application. Neurocomputing 74(17), 2745–2753 (2011)
7. Yang, W.K., Sun, C.Y., Zhang, L., Ricanek, K.: Laplacian Bidirectional PCA for Face Recognition. Neurocomputing 74(1-3), 487–493 (2010)
8. Diamantaras, K.I., Papadimitriou, T.: Applying PCA Neural Models for The Blind Separation of Signals. Neurocomputing 73(1-3), 3–9 (2009)
9. Caridakis, G., Karpouzis, K., Kollias, S.: Userand Context Adaptive Neural Networks for Emotion Recognition. Neurocomputing 71(13-15), 2553–2562 (2008)
10. Wang, H., Wang, K.Q.: Affective Interaction Based on Person-Independent Facial Expression Space. Neurocomputing 71(10-12), 1889–1901 (2008)

A Novel Moving Cast Shadow Detection of Vehicles in Traffic Scene

Yuhong Jia[1], Xiaoxi Yu[2], Jiangyan Dai[2], Wanjun Hu[2],
Jun Kong[2], and Miao Qi[2,*]

[1] Department of Information Engineering,
Jilin Business and Technology College, Changchun, 130062, China
[2] School of Computer Science and Information Technology, Northeast Normal University,
Key Laboratory of Intelligent Information Processing of Jilin Universities,
Changchun, 130117, China
{qim801,yuxx494}@nenu.edu.cn

Abstract. In the traffic video scene, the existence of shadows might generate negative effect on pattern analysis. This paper proposes a novel approach which adequately considers color space information to detect moving cast shadows of vehicles in traffic videos. Firstly, RGB component ratios between frame and background as well as blue and red colors ratio (B/R ratio) are taken into account to detect shadows respectively. Then we combine the two results for a refined shadow candidate. Finally, to improve the accuracy of shadow detection, post processing is adopted to correct the false detected pixels. Experimental results on several databases indicate that our approach not only achieves both high shadow detection and discrimination rates but takes on better performance than some state-of-the-art methods.

Keywords: Shadow detection, RGB component ratios, B/R ratio, Post processing.

1   Introduction

With the development of science and technology, intelligent transportation system (ITS) plays a key role in reducing traffic congestion and improving the efficiency of transportation. Due to the illuminations, the moving vehicles in the traffic video usually generate shadows which move as the vehicles moving. Shadows might lead to vehicle adhesion that affects the further research such as object tracking and object retrieval. Therefore, detecting and eliminating the dynamic cast shadows in videos are becoming a hot topic.

Generally speaking, the shadows in traffic videos are divided into two types. One is called self-shadow which generates via indirectly illuminating of sunlight. The other is cast shadow, which is the object projection in the direction of light source and forms darker areas in the background. Recently most research focus on the detection

---

* Corresponding author.

J. Yang, F. Fang, and C. Sun (Eds.): IScIDE 2012, LNCS 7751, pp. 115–124, 2013.

of cast shadow. In this paper, we dedicate the research of the cast shadow detection. Researchers have proposed a lot of shadow detection methods. Different color spaces (e.g. RGB [1, 17], HSV [2, 3, 16], c1c2c3 [2, 4, 17]) and image brightness were used to separate shadows from the objects. In Ref. [1], the ratios of each RGB space component between frame and background were utilized to detect shadow pixels in traffic imagery. The experimental results showed that the method was feasible. Color-space methods are easy to implement and have low computing complexity, but vulnerable to the influence of the noises. Physical property-based methods [5-7] supposed that there were two types of light source, one was white light from sun; the other was blue light that was the reflection of sky. The shadow areas were illuminated only by the diffuse light of sky. Ref. [5] introduced a new albedo test and dichromatic reflection model to predict the color change of shadow areas. The experimental results proved that it was effective and applicable in different background and foreground materials and illuminations. The restriction of physical property-based methods is only suitable for the situation that objects and background have similar chromacity. Geometric-based methods [8, 9] needed prior knowledge such as the light source, object shape and ground etc. al to predict the direction, size and shape of shadows. In Ref. [8], spectral and geometrical properties of shadows as well as wavelet transform were adopted to detect shadows in the premise of a flat background surface. These methods can deal with the input image directly and don't need estimating background, but the input image number is limited. Texture-based methods [10, 11] were based on the fact that the texture was the same before and after the region was covered by shadows. In Ref. [11], gradient magnitude and gradient direction for each pixel were calculated for every candidate region, and the gradient direction correlation between frame and background was estimated to determine the final shadow region. These methods don't rely on color and are not sensitive to the light, but computing complexity is high.

This paper proposes a shadow detection approach to detect shadow based on the color characteristics. Firstly, the RGB component ratios and B/R ratio are employed to detect shadow respectively. As a result, two rough shadow detection results are obtained. Then the two results are combined to acquire a relative refine shadow detection result. For improving the accuracy of shadow detection, post processing is carried out for final shadow. Experimental results show that our approach can detect shadows more accurately than some existing methods.

The rest of this paper is organized as follows. In Section 2, the proposed approach is described in detail. Experimental results and comparison are shown in Section 3. The Section 4 is the conclusion and future work.

2   The Proposed Approach

The approach in this paper includes moving foreground detection, analyzing the R, G, and B component ratios between frame and background, calculating the blue and red colors ratio and post processing. The flow chart is shown in Fig. 1.

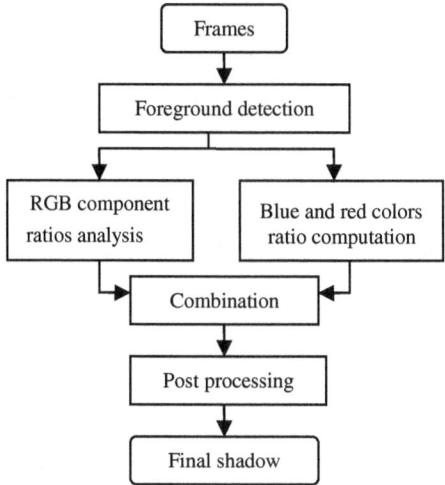

Fig. 1. The flow chart of the proposed approach

2.1 Foreground Extraction

Methods like background difference, single Gaussian Model and Gaussian Mixture model (GMM) [14], are usually adopted for foreground extraction. Comparing with other methods, GMM can update background in real time which is accord with the realistic situation. In this paper, we employ the GMM to detect foreground which is labeled Moving_pixels.

2.2 RGB Component Ratios

In the traffic video scenes, there are two types of light source. One is point light source (the sun); the other is the diffused light source (the sky). We suppose that the light from the point light source is white, and blue from the diffused light source. The moving objects have the Lambertian surface with a constant reflection coefficient. So the reflected intensity of the object Iv is calculated as Formula (1).

$$I_v = \int_{\Delta\lambda_1} K_1 L_1(\lambda_1) f(l,e,s) d\lambda + \int_{\Delta\lambda_2} K_2 L_2(\lambda) d\lambda, \tag{1}$$

where $\Delta\lambda_1$ and $\Delta\lambda_2$ are the wavelengths, K_1, K_2 and L_1, L_2 are the reflected coefficients and intensities of the sun and the sky respectively, $f(.)$ is geometric term, l, e and s are the incident angle, viewing angle and specular reflected angle. The first term in Formula (1) is generated due to the specular reflection and diffused reflection of point light source. The second is the influence of diffused reflection of the sky. It is not influenced by the object surface, thus there is no geometric term. As we known, the reason of shadow generation is that the light from the point light source is blocked by objects. Therefore, the shadow area is different from the background.

According to the Formula (1), the brightness of shadow pixel is less than the same pixel in the background. The color of each pixel can be represented by R, G, and B components and is affected independently by these three components. When shadow covers the background, Ref. [1] assumes the brightness decline proportion of each color component is independent, and the brightness ratio between shadow and background is constant as:

$$R_{shadow} = \alpha R_{bkg}, \qquad G_{shadow} = \beta G_{bkg}, \qquad B_{shadow} = \gamma B_{bkg}, \qquad (2)$$

where $R_{bkg}, G_{bkg}, B_{bkg}$ and $R_{shadow} G_{shadow}, B_{shadow}$ are the R, G, and B values of background and cast shadow pixels respectively. In most cases, α, β and γ are less than 1. However, the ratios change with pixel value. In order to establish a shadow model, the brightness ratios between shadow area and the same area in the background have been analyzed which is shown as Fig. 2.

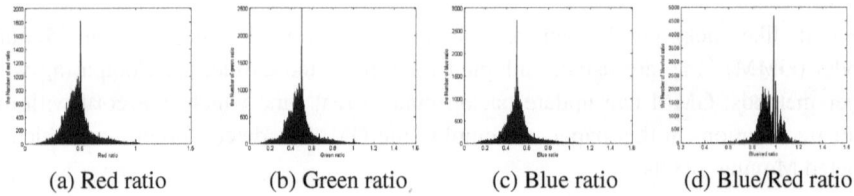

| (a) Red ratio | (b) Green ratio | (c) Blue ratio | (d) Blue/Red ratio |

Fig. 2. The R, G, B and B/R ratios

From the Fig. 2 (a)-(c), we find that the each brightness ratio of RGB components is subjected to Gaussian distribution. Therefore, threshold is decided using the Gaussian distribution inside $\pm 1.5\sigma$ (88.6%) and the pixels which all the three components meet the requirement retained as shadow. Formula (3) is used to determine whether pixel(x,y) is a shadow pixel or not.

$$Shadow1\ (x, y) = \begin{cases} 1 & \left| \dfrac{I_R(x, y)}{B_R(x, y)} - \mu_R \right| < 1.5\sigma_R\ and \left| \dfrac{I_G(x, y)}{B_G(x, y)} - \mu_G \right| < 1.5\sigma_G\ and \left| \dfrac{I_B(x, y)}{B_B(x, y)} - \mu_B \right| < 1.5\sigma_B \\ 0 & otherwise. \end{cases} \qquad (3)$$

where $I_R(x, y)$, $I_G(x, y)$, $I_B(x, y)$ and $B_R(x, y)$, $B_G(x, y)$, $B_B(x, y)$ are the RGB values at position(x,y) of foreground and background. μ_R, μ_G, μ_B and, σ_R, σ_G, σ_B are mean ratios and standard deviation ratios respectively. The detection results are illustrated in Fig. 3(a-b). The shadow is labeled by red, and the object is white. We can see that almost all the shadow pixels are successfully detected.

Fig. 3. The detected results based on RGB component ratios and B/R ratio

2.3   Blue and Red Ratio (BR Ratio)

As shown in Fig. 3(a)-(b), although the most shadow pixels have been detected successfully, a part of object pixels are detected as shadow incorrectly. In the shadow area, light from the point light source is blocked, so only diffused light illuminates the shadow area. According to Formula (1), a point covered by shadow will change the RGB components as blue component increases and red component decreases compared with its appearance when it is illuminated. Therefore, the B/R ratio of each pixel in shadow area is different from background/object. B/R ratio is analyzed in Fig. 2 (d) which is presented an approximation of Gaussian distribution. So we employ Formula (4) to judge shadow pixels:

$$Shadow2\ (x,y) = \begin{cases} 1 & |I_B(x,y)/I_R(x,y) - \mu_{B/R}| < 1.5\sigma_{B/R} \ . \\ 0 & otherwise\ . \end{cases} \tag{4}$$

where $\mu_{B/R}$, $\sigma_{B/R}$ denote the mean and standard deviation of B/R ratio respectively. The detection results are illustrated in Fig. 3(c)-(d).

Seen from Fig.3, some pixels which are detected as shadow incorrectly by using RGB ratios can be correct by B/R ratio. Therefore, after we get Shadow1 and Shadow2, "And" operator is used to combine the two shadow candidates to acquire a relatively refined shadow candidate as Formula (5):

$$Shadow=Shadow1\ \cap Shadow2. \tag{5}$$

From the Fig. 4(a)-(b), we can see that the false shadow pixels in the object area are reduced and shadow pixels are still successfully to be detected. After we get shadow candidate, object candidate is presented as Formula (6):

$$Object=Moving\_pixels-Shadow. \tag{6}$$

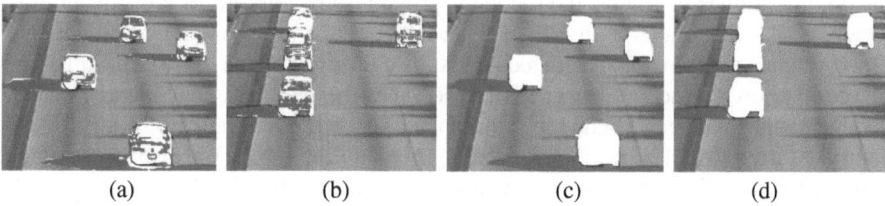

Fig. 4. The detected results of combined RGB with B/R ratios and results after post processing

2.4   Post Processing

Seen from Fig. 4(a)-(b), there are still errors in both shadow candidate and object candidate. In this section, we make good use of spatial adjusting to correct the errors.

(1)Area discrimination of moving object candidates and shadow candidates: In the process of shadow detection, the real shadow area contains small object blobs and the same to the object area. Taking the shadow candidate as an example, we use connected components labeling algorithm to label different areas, if a shadow area is smaller than a predefined threshold, it is corrected as object. The same operation is performed on object candidates.

(2)Border discrimination of moving shadow candidates: If a real object is detected as the shadow, this shadow is almost located inside of foreground. That is to say, if a shadow candidate area is real shadow, its most boundary pixels are adjacent to the boundary of the foreground. Firstly, the boundary of each foreground area is detected by Sobel edge detection algorithm. Then each shadow candidate is labeled and its boundary is also detected using Sobel edge detection algorithm. Finally, the number (N_s) of each shadow candidate boundary pixels and the number (N_f) of its corresponding boundary pixels that are adjacent to edge of the foreground are calculated. If N_f / N_s is greater than a, this shadow candidate can be regarded as real shadow, otherwise, it's object. It is worth noting that a is different for different videos and $a \in (0,1)$. Fig. 4(c)-(d) show the results after post processing. We can see that both true shadows and objects are classified correctly.

3   Experiment Results

3.1   Test Videos

The test videos in this paper are Highway which is usually selected as experiment data [1, 13], VideoI and VideoII collected ourselves. In Highway, we test the first 300 frames. The frame size is 320×240 Pixels. In VideoI, VideoII and VideoIII, the test frame numbers are 19, 92 and 95 respectively, and frame size is 320×180 Pixels. The proposed approach is evaluated both qualitatively and quantitatively.

3.2   Qualitative Results

In this paper, we have done a series of experiments to demonstrate the feasibility of our approach. Subjective comparison is given as Fig. 5 shown. The first row shows the original frames in videos, the second is ground truths. First RGB component ratios which is also the method in Ref. [1] and B/R ratio are applied singly, and post processing also employed. The detected results are shown as the third and fourth rows of Fig. 5 shown. And our approach is shown in the last row. It is clearly that the combined method gets better performance. Second, we compare our proposed approach with Ref. [2], Ref. [13], Ref. [16] and Ref. [17]. We can see that our approach successfully detects the shadow more accurately. Third we compare "Or" operator and "And" operator as ninth and tenth rows of Fig. 5 shown. Obviously, "And" operator is better than other combination methods.

Original
frames

Ground truths

Results in Ref.
[1]

Result based
on B/R ratio
singly

Results in Ref.
[2]

Results in Ref.
[13]

Results in Ref.
[16]

Results in Ref.
[17]

Results of
"Or"

The proposed
method

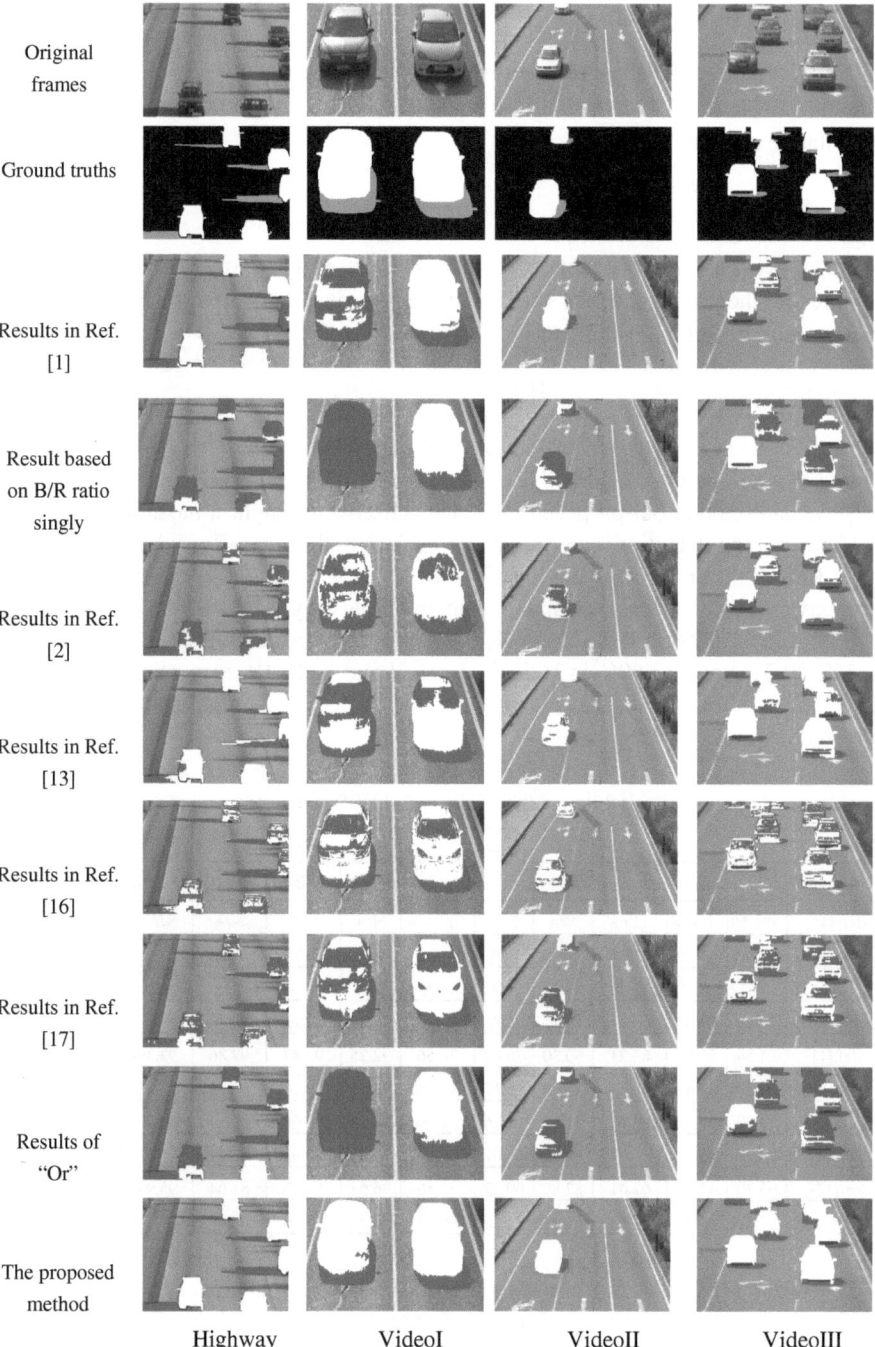

Highway   VideoI   VideoII   VideoIII

Fig. 5. Qualitative results

3.3   Quantitative Results

In experiments, all the ground truths are obtained manually. We adopt shadow detection rate η and shadow discrimination rate ξ presented in Ref. [12] as quantitative evaluation metrics.

$$\eta = \frac{TP_S}{TP_S + TN_S}, \qquad \xi = \frac{TP_O}{TP_O + TN_O}. \tag{7}$$

where subscripts S and O denote shadow and object, respectively. TP_S, TP_O and TN_S, TN_O denote the numbers of the shadow and object detected correctly and incorrectly, respectively. In Ref. [15], another metric is introduced:

$$Mean = \frac{\eta + \xi}{2}. \tag{8}$$

Table 1 lists the comparison of the proposed approach with some popular methods. Noting that the results of SP, SNP1, DNM1 and DNM2 are obtained directly from [12], we can see that our proposed approach achieves the highest mean detection rate. In Highway, the detection rate of our approach is higher than Ref. [1], Ref. [2], Ref. [13], Ref. [16] and Ref. [17] by 3.44%, 18.31%, 2.21%, 17.57% and 19.24% respectively. It enhances the mean detection rates by 12.31%, 16.31%, 19.08%, 22.89% and 19.31%; 1.05%, 26.96%, 3.77%, 17.44% and 12.27%; 4.13%, 21.90%, 4.21%, 29.21% and 13.60% in VideoI , VideoII and VideoIII, respectively. This table proves our approach habits excellent detection performance and is superior to other existing methods.

Table 1. Quantitative comparison results

Database	Highway			VideoI			VideoII			VideoIII		
Methods	η (%)	ξ (%)	Mean(%)	η (%)	ξ (%)	Mean(%)	η (%)	ξ (%)	Mean(%)	η (%)	ξ (%)	Mean(%)
SNP [12]	81.59	63.76	72.68	N/A	N/A	N/A	N/A	N/A	N/A	N/A	N/A	N/A
SP [12]	59.59	84.70	72.15	N/A	N/A	N/A	N/A	N/A	N/A	N/A	N/A	N/A
DNM1 [12]	69.72	76.93	73.33	N/A	N/A	N/A	N/A	N/A	N/A	N/A	N/A	N/A
DNM2 [12]	75.49	62.38	68.94	N/A	N/A	N/A	N/A	N/A	N/A	N/A	N/A	N/A
Ref.[1]	84.54	81.89	83.21	80.30	76.82	78.56	92.25	93.47	92.86	93.39	73.51	83.50
Ref.[2]	89.24	47.43	68.34	83.94	65.17	74.56	84.72	49.16	66.95	83.43	48.03	65.73
Ref.[13]	74.00	94.92	84.46	71.82	71.76	71.79	88.22	92.06	90.14	89.53	77.31	83.42
Ref[16]	88.22	47.93	68.08	69.95	66.02	67.98	75.02	75.93	75.47	56.55	60.30	58.42
Ref[17]	92.70	42.12	67.41	82.89	60.22	71.56	96.64	66.63	81.64	95.97	52.11	74.03
Proposed	87.37	85.37	86.65	89.17	92.56	90.87	89.57	98.25	93.91	93.73	81.54	87.63

In order to prove the stability of the proposed approach, the visualized comparison results of every frame in view of mean rate are illustrated in Fig. 6. For more clearly observation, the means of every 5 frames are shown in Fig. 6(a), the means of every 2 frames are shown in Fig. 6(c) and the means of every 5 frames are shown in Fig. 6(d). We can see that the curves of the proposed approach are almost on the top, which means that our approach is more stable and holds better detection results.

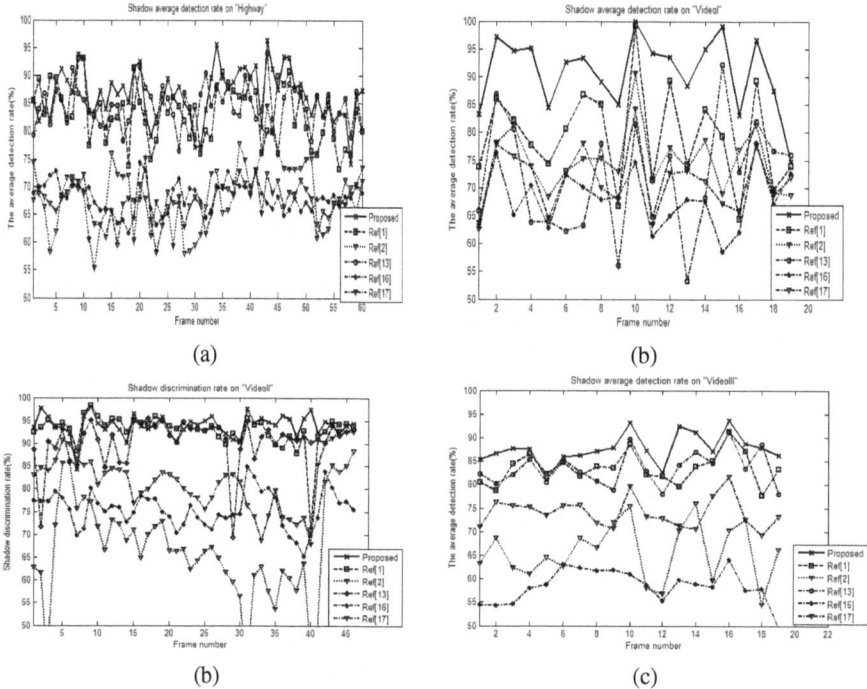

Fig. 6. Quantitative comparisons (a) Highway, (b) VideoI, (c) VideoII, (d) VideoIII

Seen the results of qualitative and quantitative evaluation, it indicates that our proposed approach is feasible and effective. Moreover the comparison results show the proposed approach outperforms the existing methods.

4   Conclusion and Future Work

This paper proposes a novel cast shadow detection approach in videos on the basis of analyzing the characteristics of shadow and the reason that shadow forms. Firstly, RGB component ratios between frame and background and the ratio of blue and red color are used to detected shadow preliminary and get two candidate shadows. Next, two candidate shadows are combined. Finally, spatial adjusting is adopted for further processing for the refined results. The experimental results show that the proposed approach can detect the shadows successfully. But the approach is not suitable for indoor scene where light are not from the sun and the sky. In the future work, we will explore other characteristics of shadow together to make it suitable for both indoor and outdoor scenes.

Acknowledgments. This work is supported bythe Young Scientific Research Foundation of Jilin Province Science and Technology Development Project (No. 201201070), the Jilin Provincial Natural Science Foundation (No. 201115003), the Fundamental Research Funds for the Central Universities (No. 10QNJJ004, No.

10JCXK008), the Fund of Jilin Provincial Science & Technology Department (No. 20111804), program for Changjiang Scholars and Innovative Research Team in University (PCSIRT).

References

1. Song, K.T., Tai, J.C.: Image-based traffic monitoring with shadow suppression. Proceedings of the IEEE 95, 413–426 (2007)
2. Sun, B., Li, S.: Moving cast shadow detection of vehicle using combined color models. In: Chinese Conference on Pattern Recognition, pp. 1–5 (2010)
3. Cucchiara, R., Grana, C., Piccardi, M., Prati, A.: Detecting moving objects, ghosts, and shadows in video steams. IEEE Transaction on Pattern Analysis and Machine Intelligence 25(10), 1337–1342 (2003)
4. Salvador, E., Cavallaro, A., Ebrahimi, T.: Cast shadow segmentation using invariant color features. Computer Vision and Image Understanding 95(2), 238–259 (2004)
5. Nadimi, S., Bhanu, B.: Physical models for moving shadow and object detection in video. IEEE Trans. Pattern Analysis and Machine Intelligence 26(8), 1079–1087 (2004)
6. Liu, Z., Huang, K., Tan, T., Wang, L.: Cast shadow removal combining local and global features. In: IEEE Conference on Computer Vision and Pattern Recognition, pp. 1–8 (2007)
7. Porikli, F., Thornton, J.: Shadow flow: a recursive method to learn moving cast shadows. In: Tenth IEEE International Conference on Computer Vision, pp. 891–898 (2005)
8. Fang, L.Z., Qiong, W.Y., Sheng, Y.Z.: A method to segment moving vehicle cast shadow based on wavelet transform. Pattern Recognition Letters 29(16), 2182–2188 (2008)
9. Nicolas, H., Pinel, J.M.: Joint moving cast shadows segmentation and light source detection in video sequences. Signal Processing: Image Communication 21(1), 22–43 (2006)
10. Leone, A., Distante, C.: Shadow detection for moving objects based on texture analysis. Pattern Recognition 40(4), 1222–1233 (2007)
11. Sanin, A., Sanderson, C., Lovell, B.C.: Improved shadow removal for robust person tracking in surveillance scenarios. In: International Conference on Pattern Recognition, pp. 141–144 (2010)
12. Prati, A., Mikic, I., Trivedi, M., Cucchiara, R.: Detecting moving shadows: algorithms and evaluation. IEEE Transactions on Pattern Analysis and Machine Intelligence 25(7), 918–923 (2003)
13. Choi, J.M., Yoo, Y.J., Choi, J.Y.: Adaptive shadow estimator for removing shadow of moving object. Computer Vision and Image Understanding 114(9), 1017–1029 (2010)
14. Stauffer, C., Grimson, W.E.L.: Adaptive background mixture models for real-time tracking. In: IEEE Conference on Computer Vision and Pattern Recognition, pp. 246–252 (1999)
15. Joshi, J., Papanikolopoulos, N.: Learning of moving cast shadows for dynamic environments. In: IEEE International Conference on Robotics and Automation, pp. 987–992 (2008)
16. Cucchiara, R., Grana, C., Piccardi, M., Prati, A., Sirotti, S.: Improving shadow suppression in moving object detection with HSV color information. In: Proceedings of IEEE Int'l Conference on Intelligent Transportation Systems, pp. 334–339 (2001)
17. Salvador, E., Cavallaro, A., Ebrahimi, T.: Cast shadow segmentation using invariant color features. Computer Vision and Image Understanding 95(2), 238–259 (2004)

Semi-supervised Learning with Local and Global Consistency by Geodesic Distance and Sparse Representation

Jie Gui[1,2], Zhongqiu Zhao[3], Rongxiang Hu[4], and Wei Jia[4]

[1] Intelligent Computing Lab,
Hefei Institute of Intelligent Machines,Chinese Academy of Sciences,
Hefei, Anhui 230031, China
[2] State Key Laboratory for Novel Software Technology,
Nanjing University, P.R. China
[3] College of Computer Science and Information Engineering,
Hefei University of Technology, China
[4] Institue of Energy Safety, Chinese Academy of Sciences, Hefei, Anhui 230031, China

Abstract. In many practical data mining applications, such as web categorization, key gene selection, etc., generally, unlabeled training examples are readily available, but labeled ones are fairly expensive to obtain. Therefore, in recent years, semi-supervised learning algorithms such as graph-based methods have attracted much attention. However, most of these traditional methods adopted a Gaussian function to calculate the edge weights of the graph. In this paper, a novel weight for semi-supervised graph-based methods is proposed. The new method adds the label information from problem into the target function, and adopts the geodesic distance rather than Euclidean distance as the measure of the difference between two data points when conducting the calculation. In addition, we also add class prior knowledge from problem into semi-supervised learning algorithm. Here we address the problem of learning with local and global consistency (LGC). It was found that the effect of class prior knowledge was probably different between under high label rate and low label rate. Furthermore, we integrate sparse representation (SR) in LGC algorithm. Experiments on a UCI data set show that our proposed method outperforms the original algorithms.

Keywords: Semi-supervised learning, Classification error, Label information, Geodesic distance, Class prior knowledge, sparse representation.

1 Introduction

In many applications of pattern classification and data mining, to collect lots of labeled samples would be costly and time-consuming, whereas getting unlabeled ones are far easier. For example, in web categorization, it is easy to get a lot of web pages, but to assign labels such as sports, news and economics to these pages requires laborious inspection, judgment, or even time-consuming reading

J. Yang, F. Fang, and C. Sun (Eds.): IScIDE 2012, LNCS 7751, pp. 125–132, 2013.

by human assessors, which is fairly expensive. Consequently, semi-supervised learning(SSL), which can solve these problems, has attracted many interests from researchers [1–3]. What is semi-supervised learning meant? Given a data set $X = \{x_1, x_2, ..., x_l, x_{l+1}, ..., x_n\}$, where the first l examples are labeled by $\{y_1, y_2, ..., y_l\}$, y_i are from the label set $L = \{1, 2, ..., c\}$, while the remained examples have no labels. The goal of semi-supervised learning is to predict the labels of unlabeled examples according to the knowledge hidden in the relation between the labeled and unlabeled data.

Essentially, there are many different classes of SSL algorithms reported in literatures. This paper concentrates on graph-based methods, which have been paid much attention to because of their solid mathematical background, close relationship with kernel methods, sparseness properties, model visualization, and good performances in many real applications. Generally speaking, graph methods are nonparametric, discriminative, and transductive in nature. Usually, the graph-based semi-supervised learning methods are to model the whole dataset as a graph. Although the graph is at the heart of these graph-based methods, its construction has not been studied extensively [3]. Among these methods, a promising family of the techniques are based on Gaussian fields, which assume that nearest neighbors (according to some similarity measure) in the high-dimensional input space will have similar 'outputs' or be close to each other in the low-dimensional manifold. More specifically, most of these methods [4–6] adopted a Gaussian function to calculate the edge weights of the graph. The edge link between data and is computed as:

$$w_{ij} = \exp\left(-\|x_i - x_j\|^2 \middle/ \left(2\sigma^2\right)\right) \tag{1}$$

where the range of the variance σ of the Gaussian function is from zero to infinity. So the work to find the best σ is very time-consuming. However, it is obvious that the label information and class prior knowledge, which can be more beneficial to classification, are ignored in Eq. (1). Though the labeled samples may be few, the label information and the class prior knowledge, as the priori information, are very important to improve the classification efficiency. In order to fully exploit the label information and class prior knowledge, we propose a novel adaptive weight to add the label information and class prior knowledge. In this paper, we address the algorithm named learning with local and global consistency (LGC) [4]. LGC designs a classifying function which is sufficiently smooth with respect to the intrinsic structure collectively revealed by known labeled and unlabeled points. A simple iteration algorithm is presented to obtain such a smooth solution. The keynote of LGC is to let every point iteratively spread its label information to its neighbors until a global stable state is achieved. In this paper, we propose a modified learning with local and global consistency algorithm to make full use of the label information and class prior knowledge. The experimental results show that the satisfying and better classification performance is obtained with respect to the original algorithms.

In the past few years, the theory of sparse representation (SR) and compressed sensing (CS) [7, 8] has received a great deal of attentions, which was initially proposed as an extension of traditional signal processing methods such as Fourier and wavelet. The problem solved by sparse representation is to search for the most compact representation of a signal in terms of linear combination of patterns in an over-complete dictionary. SR has been successfully used in image super-resolution [9] and signal recovery [10]. Recently, some new methods integrating the theory of SR, CS and semi-supervised learning have been proposed, and have been successfully applied in many practical applications [11, 8]. Here we propose LGC algorithm by sparse representation. In the proposed algorithm, the sparse representation is introduced to compute the weights between samples instead of directly setting them a simply value or a Gaussian kernel function.

In the following, we give two reasons why SR is more suitable than directly setting them a simply value or a Gaussian kernel function. (1) **Parameter-free**. SR does not have to encounter model parameters such as the Gaussian kernel parameter, which is generally difficult to set in practice. In contrast, SR does not need to deal with such parameters, which makes it very simple to use in practice. In fact, the data distribution probability may vary greatly at different areas of the data space, which results in distinctive neighborhood structure for each instance. However, the Gaussian kernel uses a predefined parameter to determine the neighborhoods for all the data. It seems to be unreasonable that all data points share the same parameter for Gaussian kernel, which may not characterize the manifold structure well, especially in under sampling case. Obviously, SR has the merit of being parameter-free. (2) **Robustness to data noise**. The data noises are inevitable especially for visual data, and the robustness is a desirable property for a satisfying graph construction method. The graph constructed by Gaussian kernel, is based on pair-wise Euclidean distance, which is very sensitive to data noise. It means that the graph structure is easy to change when unfavorable noises come in. However, SR has been shown to be robust to data noise in [7].

The remainder of this paper is organized as follows. Section 2 and Section 3 present the LGC algorithm by geodesic distance and sparse representation, respectively. Section 4 reports the experimental results on a UCI data set. Finally, some conclusive remarks are included in Section 5.

2   Learning with Local and Global Consistency by Geodesic Distance

Compared with the original learning algorithm LGC [4], our proposed algorithm, i.e. semi-supervised learning with local and global consistency (SLGC) has two characteristics. One is to consider not only the distances between the pairs of data points, but also their label information, though the labeled samples may be few. The basic idea for this change is to make two data points owing the same class label closer to each other, while two points owing different class label are far away from each other. The other characteristic is to use the geodesic distance rather than the straight-line Euclidean distance as the measure of the difference between two data points. The reason for doing so is that on an underlying manifold, the geodesic distance along the surface of the manifold can better preserve the intrinsic geometry of the data than the Euclidean distance. In addition, the second version named as SLGC-CMN adds the class prior knowledge based on SLGC.

Given a dataset $X = \{X_1, X_2, ..., X_l, X_{l+1}, ..., X_n\}$, the first l samples are labeled by $\{t_1, t_2, ..., t_l\}$, while the remained ones have no labels. The label set is assumed to be $L = \{1, \cdots, c\}$. Let F represent $n \times c$ matrices with nonnegative entries. A matrix $F = [F_1^T, \cdots, F_n^T]^T$ corresponds to a classification on the dataset X by labeling each point x_i according to a label $y_i = \arg\max_{j \leq c} F_{ij}$. We can regard F as a vector function $F : X \rightarrow R^c$ which assigns a vector F_i to each point x_i. Define an $n \times c$ matrix Y with $Y_{ij} = 1$ if x_i is labeled as $y_i = j$ and $Y_{ij} = 0$ otherwise. Clearly, Y is consistent with the initial labels according to the decision rule. Thus, our algorithm can be summarized as follows:

Step 1. Form the adjacent graph. Firstly set the parameter K. Then, define the graph G with all n data points, and construct an edge to link point X_i to X_j, with the length of

$$d(i,j) = \begin{cases} |X_i - X_j|, & if X_i \overset{K}{\sim} X_j; \\ +\infty, & otherwise \end{cases} \tag{2}$$

where $X_i \overset{K}{\sim} X_j$ denotes the points X_i and X_j are among the K-nearest neighbors of each other.

Step 2. Estimate the geodesic distances (In case of a neighboring point pair, the geodesic distance is approximated by the Euclidean distance. While in case of a faraway point pair, the geodesic distance is approximated by adding up a sequence of "short hops" between neighboring points.). Firstly initialize $d_G(i,j) = d(i,j)$; for each value of $k = 1, 2, ..., n$, in turn replace all entries $d_G(i,j)$ by $\min\{d(i,j), d(i,k) + d(k,j)\}$.

Step 3. Define the affinity matrix $W = (w_{ij})$ by the following rule:

$$\begin{aligned} &if \quad i \leq l \& j \leq l \\ &w_{ij} = \begin{cases} 1, & t_i = t_j \\ eps1, & t_i \neq t_j \end{cases} \\ &else \\ &w_{ij} = \begin{cases} \exp\left(-d_G(i,j)^2 \big/ \beta\right), & X_i \overset{K}{\sim} X_j \\ eps2, & otherwise \end{cases} \end{aligned} \tag{3}$$

The variables, $eps1$ and $eps2$, are two free parameters very close to zero. The variable, $eps1$, represents the similarity between two labeled samples which have different labels, while the variable, $eps2$, represents the similarity between two unlabeled samples which are not among K-nearest neighbors of each other. In our experiments, we set both $eps1$ and $eps2$ to 2^{-52} for simplification.

Step 4. Construct the matrix $S = D^{-\frac{1}{2}} W D^{-\frac{1}{2}}$, where $D = diag(d)$ is the diagonal matrix with entries $d_i = \sum_j w_{ij}$, and $W = (w_{ij})$ is the weight matrix.

Step 5. Iterate $F(t+1) = \alpha \times S \times F(t) + (1-\alpha) \times Y$ until convergence, where α is a free parameter between zero and one.

Step 6. Let F^* represent the limit of the sequence $\{F(t)\}$. Label each point x_i with a label $y_i = \arg\max_{j \leq c} F_{ij}^*$.

It can be proved that the sequence $\{F(t)\}$ will converge to $F^* = (I - \alpha S)^{-1} y$. The algorithm above is called as SLGC. If we incorporate the following step, the algorithm will become SLGC-CMN.

Step 7. Incorporating class prior knowledge: If there are c classes, let $F\_CMN(i, j)$ denote the probability of node i belonging to label j, $j = 1, 2, \cdots, n$. The obvious decision rule is to assign a label $y_j = \arg\max_{j \leq c} F\_CMN(i, j)$ to node i. Let $q(j)$ denote the proportion of class j in the labeled dataset. $F\_CMN(i, j)$ is defined as follows:

$$F\_CMN(i, j) = F^*(i, j) \times \frac{q(j)}{\sum_i F^*(i, j)} \qquad (4)$$

During the running of the algorithm, if there is a warning "Matrix is singular to working precision", it usually means the original graph is disconnected. The $eps1$ and $eps2$ in weight can be regulated tinily to a value very close to zero.

Note that though two free parameters, namely K and β, appear in our algorithm, the parameter β can be easily defined. Since it serves for preventing w_{ij} from falling too fast when $d_G(i, j)$ is relatively large, the value of β is set to be the average geodesic distance between all pairs of data points.

3   Learning with Local and Global Consistency by Sparse Representation

3.1   Graph Construction Based on Sparse Representation

Instead of considering k-nearest-neighbor and ε-ball based methods as in typical graph construction, we attempt to automatically construct a graph G and make it well preserve discriminative information based on SR.

SR has compact mathematical expression. Given a signal (or an image with vector pattern) $x \in R^D$, and a matrix $X = [x_1, x_2, \cdots, x_n] \in R^D$ containing the elements of an over-complete dictionary [7] in its columns, the goal of SR is to represent x using as few entries of X as possible. The objective function can be described as follows:

$$\min_{s_i} \|w_i\|_1 \qquad (5)$$
$$s.t. x_i = X w_i$$

Or,

$$\min_{s_i} \|w_i\|_1 \qquad (6)$$
$$s.t. \|x_i - X w_i\| \prec \varepsilon$$

where $w_i = \left[w_{i,1}, \cdots, w_{i,i-1}, 0, w_{i,i+1} \cdots, w_{in}\right]^T$ is an n-dimensional vector in which the i-th element is equal to zero (implying that the x_i is removed from X), and the elements $w_{i,j}, j \neq i$ denote the contribution of each x_j to reconstructing x_i. The l_1 minimization problem can be solved by LASSO or LARS. After repeating l_1 minimization problem to all the points, the sparse weight matrix can be expressed as $W = [w_1, \cdots, w_n]^T$. Then, the new constructed graph is $G = \{X, W\}$, where X is the training sample set and W is the edge weight matrix.

3.2 Learning with Local and Global Consistency by Sparse Representation

The detailed learning with local and global consistency by sparse representation algorithm is listed as follows:

Step 1. Symmetrize the graph similarity matrix by setting the matrix $W = (W + W^T)/2$.

Step 2. Construct the matrix $S = D^{-\frac{1}{2}} W D^{-\frac{1}{2}}$, where $D = diag(d)$ is the diagonal matrix with entries $d_i = \sum_j w_{ij}$, and $W = (w_{ij})$ is the weight matrix.

Step 3. Iterate $F(t + 1) = \alpha \times S \times F(t) + (1 - \alpha) \times Y$ until convergence, where α is a free parameter between zero and one.

Step 4. Let F^* represent the limit of the sequence $\{F(t)\}$. Label each point x_i with a label $y_i = \arg\max_{j \leq c} F_{ij}^*$.

We can prove that the sequence $\{F(t)\}$ will converge to $F^* = (I - \alpha S)^{-1} y$ in the similar way as in Section 2. The experiments of learning with local and global consistency by sparse representation algorithm are omitted due to space limitation.

4   Experiments and Discussions

In this section, we shall present an experiment on UCI dataset hypothyroid to verify our proposed algorithm. We select optimal parameters in all algorithms, except that the parameter α used in LGC was simply fixed at 0.99.

All of the data are used for learning, i.e., $L \cup U$. In each case, L and U are partitioned under different label rates including 10 percent, 50 percent and 90 percent. For instance, assuming the data sets contain 1,000 examples, when the label rate is 10 percent, 100 examples are put into L with their labels while the remaining 900 examples are put into U without their labels. The performance is measured by the error rate on these unlabeled points only.

The performance of SLGC and SLGC-CMN are compared with three representative semi-supervised learning algorithms, i.e., LGC, harmonic Gaussian field method (Harmonic) and harmonic Gaussian field method coupled with CMN (Harmonic-CMN) [6]. Ten independent runs are performed, and the average error rates and standard deviations are summarized in Table 1, which presents the performances of different algorithms under 10 percent, 50 percent and 90 percent.

Table 1. The average error rates and standard deviations of SLGC, SLGC-CMN, LGC, Harmonic and Harmonic-CMN on UCI data set under 10, 50 and 90 percent label rate

data set	SLGC	SLGC-CMN	LGC	Harmonic	Harmonic-CMN
hypothyroid 10%	0.116±0.012	0.091±0.011	0.128±0.009	0.115±0.018	0.118±0.020
hypothyroid 50%	0.049±0.003	0.055±0.007	0.123±0.010	0.076±0.013	0.072±0.013
hypothyroid 90%	0.058±0.023	0.050±0.027	0.101±0.026	0.061±0.036	0.058±0.030

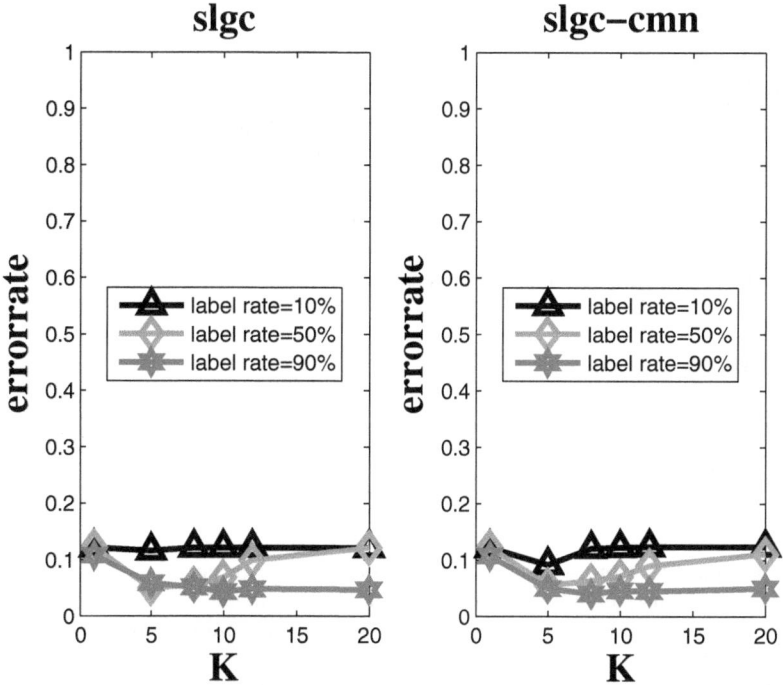

Fig. 1. Error rates under different K and label rates on hypothyroid

The average results in Table 1 show that SLGC and SLGC-CMN improve the performance of LGC obviously. Harmonic is a little better than SLGC under 10 percent label rate. Harmonic-CMN has the same error rate as SLGC under 90 percent label rate, but the standard deviation of SLGC is lower than Harmonic-CMN. These results suggest that our algorithm is more stable. In all other cases, SLGC and SLGC-CMN are better than Harmonic and Harmonic-CMN. The performances of SLGC-CMN are better than SLGC or at least nearly the same as SLGC in all cases.

From the Figure above, it can be easily seen that our algorithm is not very sensitive to K in these two cases. Moreover, there is not too much trouble in tuning the free parameter K. That is, our algorithm is more adaptive and robust.

5   Conclusions

In this paper, the semi-supervised algorithms SLGC and SLGC-CMN were proposed. By adding the label information and the class prior knowledge, and adopting the geodesic distance rather than Euclidean distance as the measure of the difference between two data points, SLGC and SLGC-CMN are facilitated with good efficiency and high generalization ability. Experiments on a UCI data set indicate that our approach could exploit unlabeled examples effectively. Furthermore, we derive the sparse version of LGC.

Acknowledgements. This work was supported by the grant of the National Science Foundation of China (grant nos. 61100161, 61272333, 61005007, 61175022, 61005010 and 30900321), the grant of the Knowledge Innovation Program of the Chinese Academy of Sciences (grant nos.Y023A61121, Y023A11292), Open Fund Project of State Key Laboratory of Software Novel TechnologyNanjing University, P.R. China (KFKT2012B26) and Postdoctoral Science Foundation of China(2012M520021). The author gratefully acknowledges the support of 2012 Endeavour Australia Cheung Kong research fellowship.

References

1. Chapelle, O., Schölkopf, B., Zien, A., et al.: Semi-supervised learning, vol. 2. MIT Press, Cambridge (2006)
2. Seeger, M.: Learning with labeled and unlabeled data, Tech. Rep., University of Edinburgh (2001)
3. Zhu, X.: Semi-supervised learning literature survey (2005), http://www.cs.wisc.edu/~jerryzhu/pub/ssl_survey.pdf
4. Zhou, D., Bousquet, O., Lal, T., Weston, J., Schölkopf, B.: Learning with local and global consistency. In: Advances in Neural Information Processing Systems, vol. 16, pp. 321–328 (2004)
5. Belkin, M., Niyogi, P.: Semi-supervised learning on riemannian manifolds. Machine Learning 56(1), 209–239 (2004)
6. Zhu, X., Ghahramani, Z., Lafferty, J.: Semi-supervised learning using gaussian fields and harmonic functions. In: Proceedings of the Twentieth International Conference on Machine Learning, vol. 20, p. 912 (2003)
7. Wright, J., Yang, A., Ganesh, A., Sastry, S., Ma, Y.: Robust face recognition via sparse representation. IEEE Transactions on Pattern Analysis and Machine Intelligence 31(2), 210–227 (2009)
8. Cheng, B., Yang, J., Yan, S., Fu, Y., Huang, T.: Learning with l_1-graph for image analysis. IEEE Transactions on Image Processing 19(4), 858–866 (2010)
9. Kim, K., Kwon, Y.: Single-image super-resolution using sparse regression and natural image prior. IEEE Transactions on Pattern Analysis and Machine Intelligence 32(6), 1127–1133 (2010)
10. Rish, I., Grabarnik, G.: Sparse signal recovery with exponential-family noise. In: Proceedings of the 47th Annual Allerton Conference on Communication, Control, and Computing, pp. 60–66. IEEE (2009)
11. Yan, S., Wang, H.: Semi-supervised learning by sparse representation. In: Proceedings of SIAM International Conference on Data Mining, pp. 792–801 (2009)

The Study of Rock Tunnel Stability Based on Extension Neural Network

Xihua Long[1], Bing Zhang[1], and Wanjun Ye[2]

[1] School of Computer Science and Technology,
Xi'an University of Science and Technology, Xi'an, China
mandy_zhangb@163.com
[2] School of Architecture and Civil Engineering, Xi'an University of Science and Technology,
The Western Mineral Resources and the Key Laboratory of Geological Engineering,
Chang'an University, Xi'an, China

Abstract. The study of surrounding rock stability is always an important subject in the field of tunnel engineering research. After comprehensively analyzed the factors of tunnel surrounding rock stability, combined the strong nonlinear mapping capability of artificial neural network with the compatibility and development ability of extension theory, putting forward a method of forecasting surrounding rock's stability based on the extension neural network.

Keywords: subway tunnel, stability, rock displacement, extension netural network.

1 Introduction

The forecast of surrounding rock stability in tunnel project is an essential part in surrounding rock tunnel construction. It is also a basic method on comprehensive evaluation of surrounding rock quality and the stability of underground projects. How to correctly determine the stability of surrounding rock becomes very important in practical significance. Now there are many methods used for surrounding rock's stability recognition. Basically there principle is simulation analysis based on the professional engineering' experience, or through proofreading the information in the tunneling process with the monitored information after the completion of the tunneling to distinguish its stability. Obviously, these two kinds of method in acquiring of the information have a strong human factor and uncertainty factors in the construction mode of operation. As far possible to eliminate the human and other factors in the process of distinguish, it is an urgent need for a new practical and reliable method for predicting the tunnel surrounding rock stability. As the nature of the tunnel wall rock showed a strong nonlinear characteristic, this paper introduces the extension neural network. Extension neural network is a new kind of neural network, it ably introduce the primitive model, the extension distance, the correlation function, the extension field, diamond thought such concepts of extension theory[1] to neural network technology, improved the training time of neural network, has more advantages than traditional neural network and extension theory method used lonely.

J. Yang, F. Fang, and C. Sun (Eds.): IScIDE 2012, LNCS 7751, pp. 133–139, 2013.
© Springer-Verlag Berlin Heidelberg 2013

2   Extension Neural Network

Though neural network[2] has a strong nonlinear mapping capability, its training standards to solve problems in pattern recognition is putting characteristic parameters region of a class to a point in the output space. This overly stringent requirement essentially determined the inconsistency of training guidelines and actual classification, leading to low sample identification and the slow speed of network training. However, extension is good at describing the nature of things in quantitative and qualitative, which is more similar to the human brain to learn to suit for expression of inaccurate and conflicting information. Besides, extension theory does a great expansion of traditional mathematics and logic. Therefore, extension neural network[3] combing with the advantages of the two methods is more superior than simple extension and the neural network.

2.1   The Extension Neural Network Design

Extension theory is the core of Extenics, namely turn no into yes, and turn can not into can[4]. In this paper, the scaling transformation of extension transformation method is introduced into BP neutral network. Extension neural network model[5] is the BP neural network with a momentum. The number of nodes M is determined by the number of characteristics of the problem itself. The output nodes L(L>=2) is determined by the classification of the problem. If the categories number of the problem is C, then $L = ceil(\log_2 C)$. Remember function $ceil(x)$ is x rounded up, and $floor(x)$ is rounded down. Activation function is $g(x) = 1.0 / (1 + \exp(-x))$, and particular teacher area is replaced with a teacher signal in output space. If the output falls within the region of this provision, the error is 0. From the perspective of the function nonlinear transform, the mapping from region to region is more reasonable and nature than the regional to a point, Which determines the extension neural network training speed is faster than BP network and ensure the consistency of the network training and classification criteria. Determine the area of teachers as follows:

Let the output of class c be the corresponding teachers regional v^c, and the goal of network training is that the output of various types of samples fall within their corresponding teacher's region. The selection rules of teachers areas are as follows:

a) $V^c \subset [0,1]^L, 且 \bigcup_{c=1}^{C} V^c \subset [0,1]^L$;

b) $d(V^c, V^{c'}) >= \sigma, c \ne c'$.

Where $d(V^c, V^{c'})$ is the square of the distance between V^c and $V^{c'}$, $\sigma > 0$. A hypersphere $V^c = \{(y_1, y_2, \cdots, y_l) | |y_i - \overline{y}_i| \le R^c, i = 1, 2, \cdots l\}$ is selected as the teacher's area, the center of which is ($\overline{y}_1^c, \overline{y}_2^c, \cdots, \overline{y}_l^c$)and the radius is R^c . The entire output space $[0\ 1]^L$ is divided into P^L super-sphere, just accommodating C classes. So we can get $P^L \ge C$ and $P = ceil(C^{\frac{1}{L}})$.

The center of the sphere is determined as following formula

$$\overline{y}_i^c = 1/P(q_i^c + 1/2), i = 1, 2, \cdots, L \tag{1}$$

Where,

$$\begin{cases} q_L^c = floor\left((c-1)/P^{L-1}\right), \\ q_l^c = floor\left(\left(c-1-\sum_{j=l+1}^{L} q_j^c P^j\right)\Big/P^{l-1}\right), l = L-1, L-2, \cdots, 1 \end{cases} \tag{2}$$

The radius

$$R^c = 0.8*(1/C) \tag{3}$$

When L=2, the teachers' area is a circle.

The k-th sample belongs to class c, and the actual network output is $(y_{k1}, y_{k2}, \cdots, y_{kL})$ .Its corresponding teachers area $V_k = V^c$, the $i(i = 1, 2, \cdots, L)$ output error of the k-th sample is

$$E_l(y_{kl}, V_k) = \begin{cases} 0 & \overline{y}_l^c - R^c \leq y_{kl} \leq \overline{y}_l^c + R^c \\ 1/2\left(y_{kl} - \overline{y}_l^c + R^c\right)^2 & y_{kl} \leq \overline{y}_l^c - R^c \qquad i = 1, 2, \cdots, L \\ 1/2\left(y_{kl} - \overline{y}_l^c - R^c\right)^2 & y_{kl} \geq \overline{y}_l^c + R^c \end{cases} \tag{4}$$

The error function of network is

$$E = \sum_{k=1}^{K} \sum_{l=1}^{L} E_l(y_{kl}, V_k) \tag{5}$$

2.2   Algorithm Steps of Extension Neural Network

With three layer BP neural network as an example, the number of nodes M is determined by the number of characteristics of the problem itself, and output layer node number is L. let hidden layers is G and K is the total number of training samples. w_{ij} is the weight of the i input layer to the j hidden layer. b_j is the threshold of the j hidden layer. h_{kj} is the j hidden layer output. V_{jl} is the weight of the j hidden layer to the l output layer. θ_l is the threshold of the l output layer. y_{kl} is the l output layer output. η is learning rate. α is momentum coefficient, Initial error Ez = 0.

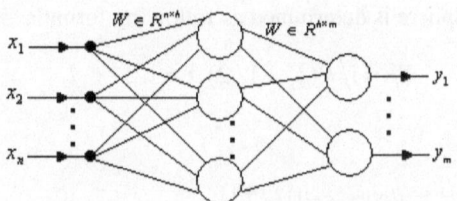

Fig. 1. BP neural network structure

So

$$h_{kj} = g\left(\sum_{i=1}^{M} w_{ij}x_{ki} - b_j\right) \tag{6}$$

$$y_{kl} = g\left(\sum_{j=1}^{G} v_{jl}h_{kj} - \theta_j\right) \tag{7}$$

Dynamic adjustment of learning rate:

$$\eta(t+1) = \eta(t) - \gamma * \left(\frac{E - Ez}{E}\right) \tag{8}$$

The formula of adjusting the weight from the input layer to hidden layer is as follows

$$w_{ij}(t+1) = w_{ij}(t) + \eta * \delta_{kj} * h_{kj} + \alpha\left(w_{ij}(t) - w_{ij}(t-1)\right) \tag{9}$$

$$\delta_{kj} = h_{kj}\left(1 - h_{kj}\right)\sum_{l=1}^{L}\delta_{kl}v_{jl} \tag{10}$$

Adjustment the weight from the hidden layer to output layer according to

$$v_{jl}(t+1) = v_{jl}(t) + \eta * \delta_{kl} * y_{kl} + \alpha\left(v_{jl}(t) - v_{jl}(t-1)\right) \tag{11}$$

Threshold adjustment formula is

$$b_i(t+1) = b_i(t) - \eta * \delta_{kj} \tag{12}$$

$$\theta_i(t+1) = \theta_i(t) - \eta * \delta_{kl} \tag{13}$$

$$\delta_{kl} = y_{kl}\left(1 - y_{kl}\right)E_{kl} \tag{14}$$

Where,

$$E_{kl} = \begin{cases} 0 & \bar{y}_l^c - R^c \leq y_{kl} \leq \bar{y}_l^c + R^c \\ \bar{y}_l^c - R^c - y_{kl} & y_{kl} \leq \bar{y}_l^c - R^c \\ \bar{y}_l^c + R^c - y_{kl} & y_{kl} \geq \bar{y}_l^c + R^c \end{cases} \quad i = 1, 2, \cdots, L \tag{15}$$

The specific steps:

1) Initialize w_{ij}, v_{jl}, b_j, θ_l, and determine the hidden layers G, η, α.

2) Identify the centers and radius of all class according to (1)-(3).

3) Calculate the actual output of each sample according to (6)-(7), and also calculate the error by (4)-(5). If the error equals 0, stop the training. Else k=1, Ez=E, and turn step *4)*.

4) Adjust the network weights, threshold, and learning rate by (8)-(15), then turn step *3)*.

3   Application and Analysis

Research shows the factors affected the surface subsidence caused by tunnel excavation include: the depth of the tunnel, underground water level, soil modulus of elasticity, soil shear strength, soil side pressure coefficient, soil weight, excavation clearance. The stability of surround-ding rock is divided into four types: here setting 1 said stability, 2 said more stable, 3 said relative stability, 4 said instability. The various parameters of the tunnel examples used in this paper come from different countries, with these data to train and test extension neural network.

3.1   Network Training

1) Data processing: In order to eliminate the influence of the dimension, we process the data. In this paper, we normalize the sample data into 0-1 interval.

$$\bar{x}_i = \frac{x_i - \min(x_i)}{\max(x_i) - \min(x_i)}$$

The training sample and test sample are used the same normalization method.

2) By the description of the problem, we can get M=8, C=4, $L = ceil(\log_2 C) = 2$, $p = ceil\left(C^{1/L}\right) = 2$, setting G=10. Initializing network parameters w_{ij}, v_{jl}, θ_l, b_j, the parameter for adjusting step $\gamma = 0.25$ .Setting $\alpha = 0.9$, $\eta = 0.1$ at the beginning, the upper and lower limits of learning rate are 1,0.1, maximum times of learning is 10000. Calculating the centers and radius of all classes by (1)-(3), the result is shown in Table 1.

Table 1. The centers and radius of each class

class	center	radius
1	(0.25,0.25)	0.2
2	(0.75,0.25)	0.2
3	(0.25, 0.75)	0.2
4	(0.75, 0.75)	0.2

3) Training the sample,when all the samples are trained one time,calculate the output and error of each sample by (6)-(7).

4) Dynamically adjust the learning rate η by equation(8), adjust w_{ij} , v_{jl} , θ_l , b_j by equation (9)-(15). If E=0, stop the training process; Or E=Ez,t=t+1,then repeat the process until E=0 or t=10000.

3.2   Network Text

Test sample is shown in Table 2.

Table 2. Test sample

Sample number	1	2	3	4	5	6
Tunnel depth	4.85	10.5	7.6	6.5	4.4	8
Underground water level	1.5	1.3	5	3.5	1.2	5
Tunnel diameter	2.74	2.47	2.25	3	2.74	3
Elastic modulus	4	13	20	6	4	20
Shear strength	10	35	35	12	10	12
pressure coefficient	0.7	1	0.5	0.7	0.7	0.7
weight	15	18	19	19	15	19
Excavation clearance	23	120	78	84	30	118
category	1	2	2	3	1	3

The data after normalized is generated into the extension neural network which has been trained, the output is shown in Table 3.

Table 3. The output of test sample

Sample number	Output	Category
1	(0.062231589,0.058837600)	1
2	(0.58069223,0.33582836)	2
3	(0.58067167,0.37007770)	2
4	(0.44999975,0.59478289)	3
5	(0.050001133,0.049999505)	1
6	(0.24891979,0.66919655)	3

From Table 3 we can see that with the trained extension neural network to classify the sample, the result and the actual are consistent. Which explain the accuracy of extension neural network in forecasting classification. The result of comparing with traditional BP neural network is shown in Table 4 below.

Table 4. The   Contrast between Extension Neural Network and the Traditional BP Neural Network

	Cost time (10000 times /s)	accuracy
Traditional network	41.172	66.7%
extension neural network	33.469	100%

From Table 4, we can see when the learning times is 10000, the time cost by the extension neural network is less than the BP neural network, and the classification accuracy is significantly higher than BP.

4   Conclusion

Extension neural network model combined the advantages of the Extenics and neural network which make it takes greater advantages compare to the traditional ones. Using the enlarge shrinks transform of extension theory, and replacing teachers signal with teachers regional, we reduced the possibility of local minima , improved the training accuracy, and shorted the training time.

Through the analysis of the application, we can get that: using the extension neural network to forecast the stability of the subway tunnel is completely feasible and effective. And compare to other neural network, extension neural network has the advantages in structure stability, speed and accuracy, making this method can forecast according to the data in any stage of the construction, with construction synchronously. It provides reliable scientific theory for the stability prediction of surrounding rock in engineering, and has the actual significance in the evaluation of underground surrounding rock.

References

1. Wen, C.: Matter Element Model and Its Application. Science and Technology Press, Beijing (1994)
2. Shi, Z.: Neural Network. Science Press, Beijing (2009)
3. Zhou, Y., Qian, X., Zhang, J.: Survey and Research of Extension Neural Network. Application Research of Computers 27, 1–5 (2010)
4. Zhao, Y., Su, N.: Extension Design. Higer Education Press, Beijing (2010)
5. Sun, B., Xing, A., Zhang, J.: The Extension Neural Network Model Design and Implementation. Journal of Harbin Institute of Technology 38, 1156–1159 (2006)
6. Kang, Z., Feng, X., Zhou, H.: The Application of Extenics Theory based on AHP in the Underground Carven Rock Quality Evaluation. Chinese Journal of Rock Mechanics and Engineering, 3687–3693 (2006)
7. Zhou, M., Yang, Y.: Implementation of Extension Neural Network of Dimamond thinking. Systems Engineering-Theory & Practice 6, 123–126 (2000)
8. Li, K., Xu, J., Li, S., Tao, Y.: Evaluation of Slope Stability based on Extension Theory. Journal of Chongqing University of Architecture 29, 75–78 (2007)

Head Pose Estimation via Background Removal

Bingpeng Ma[1], Xu Yang[2], and Shiguang Shan[3]

[1] GREYC – CNRS UMR 6072, University of Caen Basse-Normandie, Caen, France
[2] School of Computer Science and Technology,
Huazhong University of Science and Technology
[3] Visual Information Processing and Learning (VIPL) Group,
Institute of Computing Technology, Chinese Academy of Sciences (CAS)
bpma@hust.edu.cn, jasonyang.final@gmail.com,
shiguang.shan@vipl.ict.ac.cn

Abstract. In many applications, head pose estimation plays a very important role. Images with the large pose angle usually contain a proportion of the background which degenerate the performance of head pose estimation. In this paper, we propose a novel method to eliminate the influence of the background. An additional dataset with labeled background is introduced to benefit the head pose estimation. For the input image, several neighbors are first determined from the additional dataset, and then the background of this image can be estimated from the background of these neighbors. After determining the background of each image, the face region will be re-cropped. By this way, the background can be reduced greatly. The proposed method is evaluated on three datasets, MultiPIE, CAS-PEAL and our own database Multi-Pose. The promising results show that our method can improve the performance of head pose estimation significantly.

Keywords: Head Pose Estimation, Background Removal, Image Alignment.

1 Introduction

During the last decade, the research in face recognition developed a lot, but robust face recognition is still a problem, especially under pose variation. As precise pose estimation of input images is the prerequisite for solving above-mentioned problem, head pose estimation has attracted much attentions.

The existing methods for pose (yaw) estimation can be categorized into two main groups [11] [8]: model-based approach [4,12,8] and appearance-based approach [3,9,13]. The model-based methods exploit the 3-D structure of the human head. Typically, they build a 3-D model for the human faces and attempt to match the facial features, e.g., the face contour and the facial components of the 3-D face model with their 2-D projections. Appearance-based methods employ the whole face as the feature representation, which can avoid detecting the local face feature and modeling the faces.

The performance of pose estimation depends on the feature extraction algorithm and pose classification algorithm. For the automatic pose estimation

J. Yang, F. Fang, and C. Sun (Eds.): IScIDE 2012, LNCS 7751, pp. 140–147, 2013.

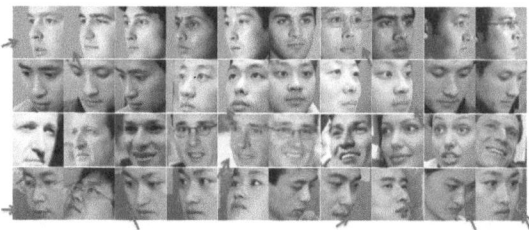

Fig. 1. Example of some detection results on several databases, including MultiPIE, CAS-PEAL, LFW, and our own Multi-Pose database. Detected by the existing face detection framework, there usually exists background in the detection result when samples are from a large pose. Red arrows point out the location.

system, the input image is the output of the face detection algorithm. Even if employing the most successful detection algorithm, these images may contain some background, i.e., the non-face area, which has no pose information. Fig. 1 displays some detection results produced by [15]. Generally, the background area will be increased with the increase of the head pose. The existed background will affect the representation ability of features that obtained from the feature extraction algorithm. Therefore, to improve the accuracy of head pose estimation, the background should be removed for images from multi-view.

In this paper, we proposed a template-based method for removing the background of images detected from multi-view, inspired by that images from similar pose will produce similar background under the same framework of face detection. Unlike the existing segmentation algorithms of distinguishing foreground and background, our method puts more attentions on the relationship between head poses and location of backgrounds, i.e., we only concern about the location of background, not the content of background.

2   Head Pose Estimation via Background Removal

In this section, we firstly introduce the motivation and the basic idea of our method, then describe algorithm in details. Fig. 2 shows the complete view of our method.

2.1   Motivation

Since face detection algorithm only focus on whether hitting faces or not, the output of an automatic face detection method may contain some background area especially if the image is with a large pose. However it is interesting that we find under the same face detection algorithm, samples in the near pose have similar background, location and shape. Fig. 3 give an example of this. In Fig. 3, the images in the first column are the face detection results of CAS-PEAL [6] and the images in the second column are the results of our private database Multi-Pose. We also show their relative ground-truth background in Fig. 3.

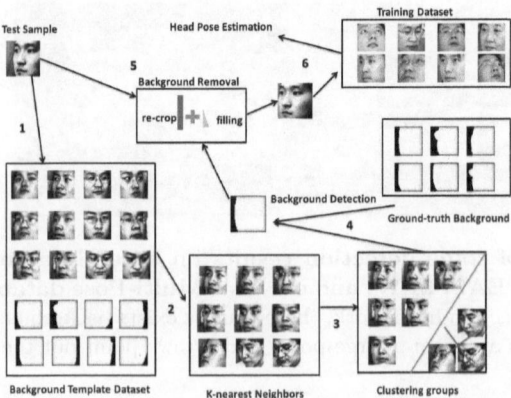

Fig. 2. Overview of our method. Firstly, N nearest neighbors of each sample are searched from a background template dataset and they are clustered into several groups. For the input image, the location and shape of the background is estimated by choosing the most believable cluster; finally, face region will be re-cropped by removing the background area according the detected background location.

Obviously, the samples that from different datasets with near pose have the similar background area along with the location and shape of the background. Based on this, a background template dataset that contains samples from extensive poses is introduced. Therefore, based on extern dataset our method can totally avoid analyzing the background itself, and needing no human interaction.

2.2 Background Template Dataset

The background template dataset is used to provide neighbors that has similar pose as the test image, so it needs to contain images from overall head poses. Therefore, a dataset collected by ourselves is employed as the background set.

Multi-Pose dataset is a private dataset. It consists of 3030 images of 102 subjects taken under normal indoor lighting conditions and fixed background with a Sony EVI-D31 camera. The yaw and pitch angles range within $[-50°, 50°]$ with intervals of $1°$ respectively. The number of the images for each subject is almost the same. For each pose, there are 30 samples.

For each image in the Multi-pose dataset, we mark manually the background region along the edge of the face using our own label tools. Therefore, each image has a binary mask that the region with value zero is the background. In Fig. 3, the images in the right column are the samples and the relative background region masks of Multi-Pose.

2.3 Background Detection

Given any image, we use K nearest neighbors searching method to find K neighbors from the background set. For the same image, using the different

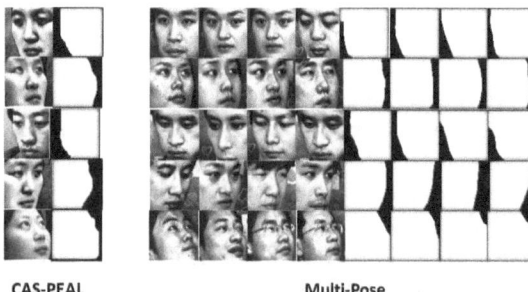

CAS-PEAL   **Multi-Pose**

Fig. 3. Samples and relative ground-truth backgrounds of different datasets.
The images in the left come from the CAS-PEAL database while the images in the
right come from the Multi-Pose database.

representation method of images, its neighbors could be quite different. Generally, if the representation contains more pose information, the K nearest neighbors are more similar in pose.

Considering the pose's continuous, the neighbors of the samples are not belong to the same poses, but the neighbors with the near pose can make contribution in fusing the background of test sample. Since the accuracy of pose estimation is very limited and the poses of some neighbors is far with the pose of the sample, we don't directly use all the neighbors. In our method, we first cluster the neighbors of the test sample into some groups based on their pose labels. Then, we choose the group with largest number of samples or minimum standard deviation as the candidate which is used in fusing the background. Then, for each pixel of the test image, a binary value representing whether it belongs to the background or not is determined by the relative pixels of the candidate neighbors. If the value of the pixels in more than half of the candidate neighbors is 0, the target pixel's value of the test image is set to 0, which means the pixel is in background; otherwise, its value is set to 1. At last, the fused background of test sample is formed.

2.4 Background Removal

Based on the fused background gained in Sec 2.3, face detection result will be updated using the following steps:

First, the rectangle region of background will be removed, actually, this is achieved by moving the face detection box until it does not contain any background area. The result is that one side of new face detection box will be next to the edge of face.

Then the irregular region such as the edge of face will be filled with middle grey level. The advantage of this operation is that using a fixed value instead of real background will make classifier ignore the region, since it does't change along with poses.

Finally, new faces will be cropped out for output.

2.5   Head Pose Estimation

For all evaluations on Multi-PIE, CAS-PEAL and Multi-pose datasets, we use the same training set constructed from the Multi-Pose dataset. As our target is to improve registration level for better yaw pose estimation, the result of using a less background training dataset is that the test sample will be re-cropped and filled for reducing background to reach a higher accuracy of yaw pose estimation.

We test our method with several works for feature extraction, i.e., Principal Component analysis (PCA) [14], Linear discriminant analysis (LDA) [1], histogram of oriented gradients (HOG) [5] and GaFour fisher feature (GFFF) [10]. Normally, PCA is used to decompose a multivariate dataset in a set of successive orthogonal components that explain a maximum amount of the variance. LDA is applied to after feature extraction to enhance the discriminative power and reduce the dimension. HOG feature is computed on a dense grid of uniformly spaced cells and uses overlapping local contrast normalization for the purpose of object detection. GFFF is a new feature proposed by Ma. et al., which is especially fit for yaw pose estimation with analysis in the asymmetry properties of head images.

For pose classification, Nearest centroid(NC) and Support Vector Machine (SVM) classifiers are employed. NC classifier assumes that the target classes correspond to individual (single) clusters and uses the cluster means (or centroid) to determine the class of a new sample point. SVM maps the data into a higher dimensional input space and constructs an optimal separating hyperplane in this space. We use the code of Libsvm [2] with radial basis function (RBF) kernel in our experiments. Generally, optimization of the gamma parameter is estimated by using the method of valid subset parameters searching.

3   Experiment

In this section, we first confirm the problem of improper registration level of pose images, and then evaluate the proposed method on three databases.

3.1   Improper Registration Level Problem

First, we design an experiment to show the performance of the pose estimation is effectively depended on the registration level of images. Tab. 1 shows the results when the gallery and probe images are cropped by different ways. D1 means the face images are cropped based on the output of the face detection method [15]. D2 means the face images are cropped manually for reducing background based on D1. All the images come from the Multi-Pose dataset. Head pose estimation is done under three different features, Nearest centroid(NC) classifier and 3-fold cross-validation method. From the table, we can know that when the grllery and probe images come from the same dataset(for example D1 VS D1, D2 VS D2), the mean average value of pose estimation is low, however, when they are improper alignment(D1 VS D2), the accuracy is decreased greatly.

Table 1. Mean average value of error of head pose estimation on two datasets (Unit: °). D1 **VS** D1 means using one part of D1 for test and the other two parts of D1 for training. D1 **VS** D2 means using one part of D1 for test and the other two relative parts of D2 for training.

Feature	Classifier	D1 **VS** D1	D2 **VS** D2	D1 **VS** D2
HOG+PCA	NC	4.71	4.48	6.4
PCA+LDA	NC	5.32	5.25	6.28
GFFF	NC	5.46	5.16	6.67

3.2   Experimental Results

MultiPIE [7] dataset contains 337 subjects, imaged under 15 viewpoints and 19 illumination conditions in up to four recording sessions. The database contains 13 yaw angles range from $-90°$ to $90°$ with intervals of $15°$. As focusing on head pose estimation, we use neutral images of all four sessions only different in head poses, the yaw angles range from $-45°$ to $45°$ which contains 8131 images.

CAS-PEAL database contains twenty-one poses combining seven yaw angles ($-45°$, $-30°$, $-15°$, $0°$, $15°$, $30°$ and $45°$) and three pitch angles ($30°$, $0°$ and $-30°$). We use a subset containing 4200 images of 200 subjects whose IDs range from 401 to 600.

Multi-pose is collected by ours. In experiment, each angle in yaw is taken as one separate class, which means that there are totally 101 classes, therefore, the dimension of LDA-based features will be not more than 100.

For all the images, the face detection method [15] is applied first to locate the face region from the input images, and then all the face regions are normalized to the same size of 64×64. Finally, histogram equalization is used to reduce the influence of lighting variations.

In the experiments, PCA is used before or after feature extraction to reduce the dimension of features and 95% of total energy of eigenvalues is kept. For the LDA, the dimension is the number of poses minus one. For the HOG descriptor, the cell window is 8×8 and the overlap is 0.5. For the GFFF, the GaFour features are first computed from the input samples, then LDA is applied to the GaFour features to improve the discriminative ability. For the NC classifier, the center number of each class is 4. The number K of the nearest neighbors in feature space is set by experiment experience, the value of the parameters should be set for avoiding the case that large K will slow down the speed of KNN algorithm and too many groups may can't tell the best group. For this paper K is set to 31 and the number of the cluster groups is 2.

To measure the performance of head pose estimation, we use *Mean Average Value of Error(MAVE)* on the Multi-Pose database, which is computed by the absolute value of the difference between the real poses of the samples and the ground-truth. For the CAS-PEAL and MultiPIE database, we use Accuracy Rate(AR) to measure the accuracy of estimation. For Multi-Pose, we use 3-fold cross-validation in order to avoid over-training and keep the test and training datasets with no overlap. And all the results of the experiments are the mean of

Table 2. Mean absolute value of error on Multi-Pose, accuracy on CAS-PEAL and MultiPIE, respectively. The unit of the mean absolute value of error is ° and the unit of accuracy is %.

Feature	Classifier	Multi-Pose original	modified	CAS-PEAL original	modified	MultiPIE original	modified
HOG+PCA	NC	6.40	5.80	39.5	48.9	34.5	37.8
PCA+LDA	NC	6.21	5.72	36.9	45.6	24.2	25.4
GFFF	NC	6.24	6.48	46.9	54.1	43.9	44.0
HOG+PCA	SVM	9.69	7.68	45.6	54.5	33.9	39.1
PCA+LDA	SVM	6.26	5.66	36.5	45.1	23.6	24.6
GFFF	SVM	6.97	6.82	45.5	53.9	41.2	42.2

Fig. 4. Original samples and modified samples. The first row is the original samples in CAS-PEAL, second row is the relative modified samples, third row is the original samples in MultiPIE, fourth row is the relative modified samples. Obviously, samples are re-cropped and filled while the background is reduced greatly.

results of all testing datasets. For CAS-PEAL and MultiPIE, we directly estimate head pose based on the training dataset (Multi-Pose), so the experiments on these two datasets are under cross-dataset, and we want to prove that our method can also improve the accuracy under the case of cross-dataset estimation.

Tab. 2 shows the original results and the results modified by our method on the different databases. Obviously, the accuracy of head pose estimation has been improved after the samples are recropped by our method. GFFF performs good in both accuracy and speed, the feature is more focus on the symmetry characteristic of image, which is especially good for yaw head pose estimation. Fig. 4 shows the original and modified samples. When sample has a large pose, our method can reduce the background greatly with cropping and filling, the result is to improve the face registration level.

4   Conclusions

Based on the fact that under the same face detection method, samples of similar pose have similar background with similar location and shape, we propose a novel method to improve the registration level between test dataset and training dataset by removing the background area of the images. And experiments show that our method can improve the accuracy of head pose estimation. In future, we will focus more on how to reduce the size of background template dataset.

Acknowledgment. This work is partially supported by National Natural Science Foundation of China under contract No. 61003103 and 61173065, Research Fund for the Doctoral Program of Higher Education under contract No. 20100142120029 and Hubei Provincial Natural Science Foundation under contract No. 2010CDB02304 and 2010CDB02305.

References

1. Belhumeur, P., Hespanha, J., Kriegman, D.: Eigenfaces vs. Fisherfaces: Recognition using class specific linear projection. IEEE Transactions on Pattern Analysis and Machine Intelligence 19(7), 711–720 (1997)
2. Chang, C., Lin, C.: LIBSVM: A library for support vector machines. ACM Transactions on Intelligent Systems and Technology 2, 27:1–27:27 (2011)
3. Chen, L., Zhang, L., Hu, Y., Li, M., Zhang, H.: Head pose estimation using fisher manifold learning. In: Proc. International Workshop on Analysis and Modeling of Faces and Gestures, pp. 203–207 (2003)
4. Cootes, T., Wheeler, G., Walker, K., Taylor, C.: View-based active appearance models. Image and Vision Computing 20(9-10), 657–664 (2002)
5. Dalal, N., Triggs, B.: Histograms of oriented gradients for human detection. In: Proc. Computer Vision and Pattern Recognition, vol. 1, pp. 886–893 (2005)
6. Gao, W., Cao, B., Shan, S., Chen, X., Zhou, D., Zhang, X., Zhao, D.: The caspeal large-scale chinese face database and baseline evaluations. IEEE Transactions on Systems, Man and Cybernetics–Part A: Systems and Humans 38(1), 149–161 (2008)
7. Gross, R., Matthews, I., Cohn, J., Kanade, T., Baker, S.: Multipie. Image and Vision Computing 28(5), 807–813 (2010)
8. Ji, Q., Hu, R.: 3d face pose estimation and tracking from a monocular camera. Image and Vision Computing 20(7), 499–511 (2002)
9. Krueger, V., Bruns, S., Sommer, G.: Efficient head pose estimation with gabor wavelet networks. In: Proc. British Machine Vision Conference, pp. 12–14 (2000)
10. Ma, B., Shan, S., Chen, X., Gao, W.: Head yaw estimation from asymmetry of facial appearance. IEEE Transactions on Systems, Man, and Cybernetics–Part B: Cybernetics 38(6), 1501–1512 (2008)
11. Murphy-Chutorian, E., Trivedi, M.: Head pose estimation in computer vision: A survey. IEEE Transactions on Pattern Analysis and Machine Intelligence 31(4), 607–626 (2009)
12. Nikolaidis, A., Pitas, I.: Facial feature extraction and determination of pose. In: Proc. NOBLESSE Workshop on Nonlinear Model Based Image Analysis, pp. 257–262 (1998)
13. Srinivasan, S., Boyer, K.: Head pose estimation using view based eigenspaces. In: Proc. International Conference on Pattern Recognition, vol. 4, pp. 302–305 (2002)
14. Turk, M., Pentland, A.: Eigenfaces for recognition. Journal of Cognitive Neuroscience 3(1), 71–86 (1991)
15. Yan, S., Shan, S., Chen, X., Gao, W., Chen, J.: Matrix-structural learning (MSL) of cascaded classifier from enormous training set. In: IEEE Conference on Computer Vision and Pattern Recognition, pp. 1–7 (2007)

Spatially Correlated Nonnegative Matrix Factorization for Image Analysis

Xinlei Chen, Cheng Li, and Deng Cai*

State Key Lab of CAD&CG, College of Computer Science,
Zhejiang University, China
{endernewton,licheng}@zju.edu.cn, dengcai@cad.zju.edu.cn

Abstract. Low rank approximation of matrices has been frequently applied in information processing tasks, and in recent years, Nonnegative Matrix Factorization (NMF) has received considerable attentions for its straightforward interpretability and superior performance. When applied to image processing, ordinary NMF merely views a $p_1 \times p_2$ image as a vector in $p_1 \times p_2$-dimensional space and the pixels of the image are considered as independent. It fails to consider the fact that an image displayed in the plane is intrinsically a matrix, and pixels spatially close to each other may probably be correlated. Even though we have $p_1 \times p_2$ pixels per image, this spatial correlation suggests the real number of freedom is far less. In this paper, we introduce a *Spatially Correlated Nonnegative Matrix Factorization* algorithm, which explicitly models the spatial correlation between neighboring pixels in the parts-based image representation. A multiplicative updating algorithm is also proposed to solve the corresponding optimization problem. Experimental results on benchmark image data sets demonstrate the effectiveness of the proposed method.

1 Introduction

Low rank approximation of matrices is one of the key problems in data analysis, and is widely used as a dimensionality reduction technique. In many problems of computer vision and information retrieval, we often need to deal with the data matrix $X \in \mathbb{R}^{p \times n}$, with each of the n columns representing a data sample in a p-dimensional space. The dimension p is usually very high, which makes learning from example infeasible [1]. Therefore, matrix factorization techniques have been developed to approximate X with a low-rank matrix, which can be expressed as the product of two matrices $U \in \mathbb{R}^{p \times r}$ and $V \in \mathbb{R}^{n \times r}$:

$$X \approx UV^T,$$

where $r \ll p$ is the rank of the approximation.

In the literature, there exist numerous variants of matrix factorization, emphasizing different objective functions measuring the quality of the approximation

---

* Corresponding author.

J. Yang, F. Fang, and C. Sun (Eds.): IScIDE 2012, LNCS 7751, pp. 148–157, 2013.
© Springer-Verlag Berlin Heidelberg 2013

and different constraints on the factors. Among them, Singular Value Decomposition (SVD) is the most renowned one, which underlies Principal Component Analysis (PCA) [34]. SVD uses the Frobenius norm of the approximation error as the objective function, and has been applied to many image analysis problems such as face recognition [32].

Recently, Nonnegative Matrix Factorization (NMF) [16] was introduced into the image analysis community. NMF focuses on matrices whose elements are all non-negative, which is often encountered in real world data such as intensity value of image pixels, document-term matrix, rating matrix, *etc.* Given such a matrix, NMF decomposes it into two non-negative matrices for low-rank approximation. In NMF, each data point can be explained as an additive-only linear combination of the nonnegative basis vectors, leading to a *parts-based representation* [16]. This straightforward interpretability makes it popular in various applications [27, 29, 35, 26], and a lot of research has been conducted [28, 7, 19, 2]. For image processing, since NMF enables the extraction of the different constitutive parts of a set of images, the resulted factorization is more robust to some unfavorable conditions like occlusion [10]. Therefore, NMF has been used for various image analysis tasks such as image clustering and image classification [17, 18]. However, all of the existing variants of NMF merely consider an image as a high-dimensional vector. As a result, the pixels are considered as independent pieces of information. Yet intrinsically, a $p_1 \times p_2$ image is a matrix and pixels are likely to have spatial correlation to their neighborhoods [33, 4, 6].

In this paper, we proposed a *Spatially Correlated Nonnegative Matrix Factorization* (SCNMF) trying to explicitly model the spatial correlation between neighboring pixels. Specifically, we introduce a discretized Laplacian smoothing term [25] to measure the derivatives of basis vectors along horizontal and vertical directions. This term is used as a regularizer in the objective function of SCNMF. This way, we can learn a set of non-negative spatially correlated basis vectors. We present a multiplicative updating algorithm to solve the optimization problem and the convergence proof of our algorithm is also provided.

2   A Brief Review of NMF

Suppose we have n images of size $p_1 \times p_2$. Let $X = [\mathbf{x}_1, \mathbf{x}_2, \dots, \mathbf{x}_n] \in \mathbb{R}_+^{p \times n}$ denote the corresponding data matrix, where $p = p_1 \times p_2$. Here $\mathbb{R}_+$ is the set of nonnegative real numbers. One of the standard NMF [17] objective function is based on the Frobenius norm:

$$\min_{U,V} \mathcal{O} = \|X - UV^T\|_F^2,$$

where $U = [\mathbf{u}_1, \mathbf{u}_2, \dots, \mathbf{u}_r] = [u_{ik}] \in \mathbb{R}_+^{p \times r}$ consists of r basis vectors, $V = [v_{jk}] \in \mathbb{R}_+^{n \times r}$ is the r-dimensional representation of the original inputs.

Although the objective function $\mathcal{O}$ is convex in terms of U or V, they are not convex in both variables together. In fact, it has been shown that NMF is $\mathcal{NP}$-hard [8] and we cannot expect to find a globally optimal solution in a reasonable computational time. Hence the state-of-the-art algorithms [12, 37, 20, 14] seek instead to obtain locally optimal solutions. For instance, the multiplicative update algorithm [16] minimizes it in an iterative manner:

$$u_{ik} \leftarrow u_{ik} \frac{(XV)_{ik}}{(UV^TV)_{ik}}, \qquad v_{jk} \leftarrow v_{jk} \frac{(X^TU)_{jk}}{(VU^TU)_{jk}}. \tag{1}$$

The non-negative constraints on U and V only allow additive combinations among different basis vectors. It not only induces a parts-based representation for images, which coincides with psychological and physiological evidences from the human brain [22], but also favors *sparse factors*, *i.e.* factors with relatively many zeros entries. This is because the stationary points (U, V) satisfying the Karush-Khun-Tucker conditions of NMF will typically be located at the boundary of the feasible domain $\mathbb{R}_+^{p \times r} \times \mathbb{R}_+^{n \times r}$, featuring zero components [8].

3   Spatially Correlated Nonnegative Matrix Factorization

In this section, we describe our *Spatially Correlated Nonnegative Matrix Factorization* (SCNMF) approach. Rather than considering the basis vector as a $p_1 \times p_2$-dimensional vector, we view it as a matrix, or a discrete function defined on a $p_1 \times p_2$ lattice. From this point of view, we can design constraints to explicitly model the spatial correlation between neighboring pixels.

3.1   Generalized Penalty Functionals

Let $u(\cdot)$ be a function defined on a region of interest, $\Omega \subset \mathbb{R}^d$. The function operator $\mathcal{L}^{(m)}(\cdot)$ can be defined as follows [13]:

$$\mathcal{L}^{(m)}u(\mathbf{t}) = \sum_{l=1}^{d} \frac{\partial^{(m)} u}{\partial (t^{(l)})^{(m)}},$$

where $m = 0, 1, 2, \ldots$ is the order of the derivative. The generalized penalty functional, denoted by $\mathcal{G}_q^m(\cdot)$, is defined by:

$$\mathcal{G}_q^m(u) = \int_\Omega |\mathcal{L}^{(m)}u|^q d\mathbf{t}.$$

Take $d = 1$ for instance, where $u(\cdot)$ can be viewed as a curve on a predefined range of variable t. Then $\mathcal{G}_q^0(u)$ penalizes the coefficients, $\mathcal{G}_q^1(u)$ penalizes *changes* of the coefficient curve, $\mathcal{G}_q^2(u)$ penalizes *curvature* in the coefficient curve [15]. Therefore, the later two functionals can be used to measure the spatial correlation of the function $u(\cdot)$. In fact, many notable regularization terms in regression can be viewed as the discrete analog of $\mathcal{G}_q^m(\cdot)$, *e.g.*, the Fused Lasso [30] algorithm corresponds to the combination of $\mathcal{G}_1^0(\cdot)$ and $\mathcal{G}_1^1(\cdot)$.

Since we primarily focus on images, which are essentially two-dimensional signals. Therefore, we take $d = 2$. We would also fix $q = 2$ for convenience of computation in the following derivation.

3.2   Discretized Spatial Correlation

As described previously, a $p_1 \times p_2$ image can be represented as a vector in $\mathbb{R}_+^p$ ($p = p_1 \times p_2$). And $\mathcal{G}_m^2(u)$ $(m = 1, 2)$ can be used to measure the spatial correlation of $u(\cdot)$ defined on the lattice. Let $D_m^{(l)}$ be a matrix that yields a discrete approximation to $\partial^{(m)} / \partial (t^{(l)})^{(m)}$ $(l = 1, 2)$. Then if $\mathbf{w} = (w(t_1), w(t_2), \ldots, w(t_{p_l}))$ is a p_l-dimensional vector which is discretized version of a continuous function $w(t^{(l)})$ on l-th dimension, then $D_m^{(l)}$ should have the property that:

$$[D_m^{(l)} \mathbf{w}]_\tau \approx \frac{\partial^{(m)} w(t_\tau)}{\partial (t_\tau^{(l)})^{(m)}}, \quad \tau = 1, 2, \ldots, p_l. \tag{2}$$

There are many options for $D_m^{(j)}$ [31, 4, 6], and in this paper, we define $D_1^{(l)}$ as:

$$D_1^{(l)} = p_l \begin{bmatrix} -1 & 1 & & & 0 \\ & -1 & 1 & & \\ & & \ddots & \ddots & \\ & & & -1 & 1 \\ 0 & & & & -1 & 1 \end{bmatrix} \in \mathbb{R}^{(p_l-1) \times p_l}.$$

And we define $D_2^{(l)}$ by modifying the Neumann discretization [25, 4, 6]:

$$D_2^{(l)} = p_l \begin{bmatrix} -1 & 1 & & & & 0 \\ 1 & -2 & 1 & & & \\ & 1 & -2 & 1 & & \\ & & \ddots & \ddots & \ddots & \\ & & & 1 & -2 & 1 \\ & & & & 1 & -2 & 1 \\ 0 & & & & & 1 & -1 \end{bmatrix} \in \mathbb{R}^{p_l \times p_l}.$$

It is easy to check that $\|D_m^{(l)} \mathbf{w}\|^2$ is proportional to the discrete approximation of generalized penalty of one-dimensional function $w(\cdot)$ defined over $t^{(l)}$. To give a more concrete idea of the nature of the discretized penalty, let us examine $\|D^{(j)} \mathbf{w}\|^2$ in the one-dimensional case where we set $m = 2$:

$$\begin{aligned} \|D_2^{(l)} \mathbf{w}\|^2 &= \frac{[w(t_2) - w(t_1)]^2}{h_l^2} + \frac{[w(t_{n_l}) - w(t_{n_l-1})]^2}{h_l^2} \\ &\quad + \frac{1}{h_l} \sum_{i=2}^{n_l-1} h_l \left[\frac{w(t_{i-1}) + w(t_{i+1}) - 2w(t_i)}{h_l^2} \right]^2 \\ &\approx [w'(t_1)]^2 + [w'(t_{n_l})]^2 \\ &\quad + \frac{1}{h_l} \int [w''(t^{(l)})]^2 dt^{(l)}, \quad l = 1 \text{ or } 2, \end{aligned} \tag{3}$$

where $h_l = 1/n_l$ is the step length of integration. Eq.(2) shows that $\|D_2^{(l)} \mathbf{w}\|^2$ is proportional to the discrete approximation of Laplacian penalty of function w defined over $t^{(l)}$.

For an image of size $p_1 \times p_2$, the region of interest Ω is essentially a two-dimensional rectangle. And in total there are p_2 p_1-dimensional rows and p_1 p_2-dimensional columns in it. Let $\mathbf{u} \in \mathbb{R}^p$ be one of the basis vectors obtained by NMF, and $\mathbf{u}^{(\tau,l)}, \tau = 1, 2, \ldots, p_l$ be a sub-vector of $\mathbf{u}$ corresponding to either the τ-th row or the τ-th column of the rectangle. Then according to Eq.(2), the spatial correlation on this lattice could be discretely approximated by the (weighted) sum of generalized penalty over all such sub-vectors:

$$\mathcal{G}_2^m(\mathbf{u}) = \sum_{l=1,2} \frac{1}{p_l} \sum_{\tau=1}^{p_l} \|D_m^{(l)} \mathbf{u}^{(\tau,l)}\|^2, \quad m = 1, 2.$$

This objective function can be rewritten in a more compact form of $\|\Delta_m \mathbf{u}\|^2$, with Δ_m defined as follows [25, 6]:

$$\Delta_m = \frac{1}{\sqrt{p_1}} D_m^{(1)} \otimes I^{(2)} + I^{(1)} \otimes \frac{1}{\sqrt{p_2}} D_m^{(2)}.$$

Here $I^{(l)} \in \mathbb{R}^{p_l \times p_l}$ is the identity matrix for $l = 1, 2$, and $\otimes$ is the kronecker product[1] [11].

From the above description and analysis, we hereby present the penalty term to measure the spatial correlation in the basis vectors U:

$$f_{\mathcal{G}}^{(m)}(U) = \|\Delta_m U\|_F^2 = \text{Tr}(U^T \Delta_m^T \Delta_m U).$$

Here $\text{Tr}(\cdot)$ returns the trace of the input matrix. The objective function of our SCNMF approach can be formally defined as the following:

$$\min_{U,V} \mathcal{O}^{(m)} = \|X - UV^T\|_F^2 + \lambda f_{\mathcal{G}}^{(m)}(U), \tag{4}$$

where U and V are nonnegative and $\lambda \in \mathbb{R}_+$ is the regularization parameter.

3.3   Multiplicative Update Rule

In this subsection, we present the multiplicative update rule for solving the optimization problem as follows:

$$\min_{U,V} \hat{\mathcal{O}} = \|X - UV^T\|_F^2$$
$$+ \lambda \text{Tr}(U^T P U) + \gamma \text{Tr}(V^T Q V), \tag{5}$$

where $P \in \mathbb{R}^{p \times p}$ and $Q \in \mathbb{R}^{n \times n}$ are positive semi-definite matrices, and λ and γ are nonnegative regularization parameters.

It is easy to check that our SCNMF in Eq. (4) is a special case of Eq. (5) where $P = \Delta_m^T \Delta_m$ and $\gamma = 0$. Other NMF variants like Graph Regularized NMF and

---

[1] Although $D_1^{(l)} \in \mathbb{R}^{(p_l-1) \times p_l}$ is not a square matrix, we can easily fix this problem by adding a dummy all-zero vector of size $p_l \times 1$ with no side effect on the approximation.

Table 1. Accuracy on CMU PIE (%)

k	Accuracy (%)				
	K-means	PCA	NMF	SCNMF-1	SCNMF-2
10	30.2±4.2	30.2±3.5	46.4±6.6	56.9±8.6	57.1±8.8
20	24.1±3.1	23.9±3.3	43.0±5.4	54.2±8.3	54.3±8.2
30	22.6±2.9	22.6±2.8	41.7±3.7	51.4±5.0	51.2±5.1
40	21.3±2.8	21.2±2.7	39.2±3.5	49.4±4.7	49.5±5.0
50	20.7±2.3	20.5±2.0	39.5±3.2	47.6±4.5	47.6±4.4
60	20.0±1.5	19.9±1.6	38.9±2.2	46.0±4.4	46.1±4.1
68	19.5	19.4	38.3	45.8	45.4
Avg.	22.6	22.5	41.0	50.2	50.2

its variants [5, 3, 9] can also be interpreted as Eq. (5) with different choices of P, Q, λ and γ.

Define operator $\lfloor A \rfloor_+$ to zero out all the negative entries of a matrix A, and $\lfloor A \rfloor_-$ to take the absolute values of the negative entries and zero out the others. Then A can be separated as $A = \lfloor A \rfloor_+ - \lfloor A \rfloor_-$. We have the following update rule:

$$u_{ik} \leftarrow u_{ik} \frac{(XV + \lambda \lfloor P \rfloor_- U)_{ik}}{(UV^T V + \lambda \lfloor P \rfloor_+ U)_{ik}}, \tag{6}$$

$$v_{jk} \leftarrow v_{jk} \frac{(X^T U + \gamma \lfloor Q \rfloor_- V)_{jk}}{(VU^T U + \gamma \lfloor Q \rfloor_+ V)_{jk}}. \tag{7}$$

Regarding these two update rules, we have the following theorem:

Theorem 1. *The objective function $\hat{O}$ in Eq.(5) is non-increasing under the update rule in Eq.(6) and (7). The objective function is invariant under these updates if and only if U and V are at a stationary point.*

This theorem can be proved in a similar way to [16], please refer to [3] for details.

4   Experiments

Image representation is a common test case for the NMF algorithms[17, 18, 21, 24, 36]. It has been shown that NMF can provide perceptually reasonable parts based representation [18], which is in turn beneficial for clustering [7]. In this section, we also evaluate our Spatially Correlated Nonnegative Matrix Factorization (SCNMF) on image clustering. Intuitively, a better clustering result indicates a better representation.

Two data sets are used in the experiment. The first one is the CMU PIE face database[2], which contains face images of 68 persons. We choose the near frontal pose (C05), in which each person has 49 images under different illuminations

---

[2] http://www.ri.cmu.edu/projects/project_418.html

Table 2. Accuracy on ORL (%)

k	Accuracy (%)				
	K-means	PCA	NMF	SCNMF-1	SCNMF-2
5	76.6±10.8	78.2±9.3	73.8±11.1	77.4±12.3	77.2±11.6
10	65.2±8.1	69.8±8.5	68.6±7.4	73.3±10.7	73.5±10.2
15	59.5±6.2	62.8±6.3	63.5±6.2	66.8±9.2	66.6±10.1
20	56.8±4.8	59.8±4.7	61.1±4.7	64.0±8.6	64.0±8.6
25	55.7±4.1	58.7±4.2	60.5±4.0	63.3±8.3	63.2±7.9
30	54.1±3.4	56.6±3.5	58.4±3.3	60.6±8.4	60.5±7.1
35	52.8±3.1	55.3±3.0	57.1±3.0	59.3±7.3	59.0±6.9
40	51.3	54.5	56.0	57.4	57.3
Avg.	59.0	62.0	62.4	65.3	65.2

and expressions. The second one is the ORL face database[3]. It contains 400 gray scale images of 40 individuals, each individual has 10 images. The images were captured at different times and have different variations including expressions (open or closed eyes, smiling or non-smiling) and facial details (glasses or no glasses). All the face images are manually aligned and cropped. The size of each cropped image is 32×32 pixels, with 256 gray levels per pixel. Thus, each face image is represented as a 1024 dimensional vector. The vectors are then normalized to have unit length.

4.1 Evaluation Metric

Following the convention of clustering studies, we use Accuracy (ACC) [3] as the evaluation metric. Denote q_i as the clustering result from the clustering algorithm and p_i as the ground truth label of $\mathbf{x}_i$, ACC is defined as:

$$ACC(p_i, q_i) = \frac{\sum_{i=1}^{n} \delta(p_i, map(q_i))}{n},$$

where n is the total number of samples and $\delta(x, y)$ is the delta function that equals 1 if $x = y$ and equals 0 otherwise, and $map(r_i)$ is the best mapping function that permutes clustering labels to match the ground truth labels using the Kuhn-Munkres algorithm [23]. A larger ACC indicates better performance.

4.2 Performance Evaluations and Comparisons

To demonstrate how the performance of clustering can be improved by our method, we implemented two versions of SCNMF corresponding to $m = 1$ and $m = 2$ (denoted as SCNMF-1 and SCNMF-2), and compared it with canonical K-means, K-means in the principal component subspace (PCA in short), and the NMF based clustering. The regularization parameter λ in SCNMF is fixed to be 0.1. The multiplicative update for both NMF and SCNMF can only generate local optimum. For fairness, we use the same initialization (U and V) for NMF

---
[3] http://cvc.yale.edu/projects/yalefaces/yalefaces.html

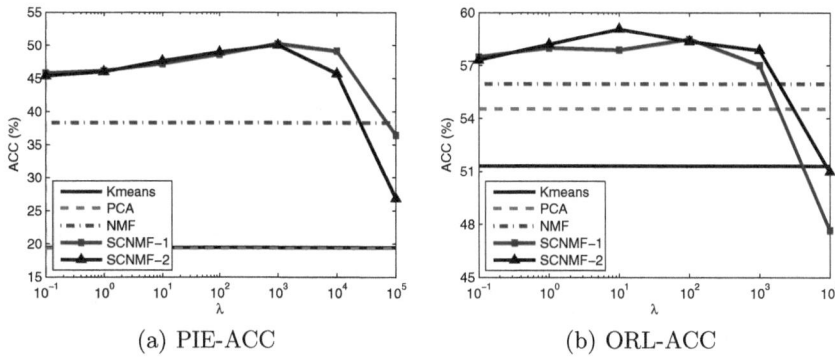

(a) PIE-ACC   (b) ORL-ACC

Fig. 1. Clustering performance with varying λ. The SCNMF is stable with respect to this parameter.

and SCNMF. For PCA, NMF and SCNMF, the reduced rank r is set to be the number of clusters.

Table 1, and 2 summarize the evaluation results on the PIE and the ORL data sets, respectively. For each cluster number K, 20 test runs were conducted on different randomly chosen clusters (except the last one which uses the whole data set) and the average performance as well as the standard deviation are reported in the tables. These results reveal a number of interesting points:

- The nonnegative matrix factorization based methods, both NMF and SC-NMF, outperform the PCA method, which suggests the superiority of parts-based sparse factors in discovering the appropriate representations.
- SCNMF-1 and SCNMF-2 get significantly better results than NMF. It demonstrates that by incorporating the prior information that neighboring pixels are spatially correlated in basis vectors, SCNMF can learn a better representation in image processing.
- SCNMF-1 and SCNMF-2 generate similar results in terms of clustering performance. Which version models the spatial correlation more robustly needs further investigation.

4.3   Model Selection

Our SCNMF model has a critical parameter λ, which controls the penalty of spatial correlation of the basis vectors. When $\lambda = 0$, the SCNMF model will reduce to the ordinary NMF. When $\lambda \to \infty$, the SCNMF model will choose the spatially smoothest basis vectors and completely ignore the information from the data. SCNMF with an appropriate λ is a trade-off between these two extreme cases. Thus a natural question would be how to choose the parameter λ, or how to select the model. We use a fixed λ ($\lambda = 0.1$) in our previous experiments. In this subsection, we study the impact of parameter λ on the clustering performance.

Figure 1 shows how the average performance of SCNMF varies with the parameter λ on the two datasets. As we can see, SCNMF is very robust with respect to λ. It achieves consistently better performance with λ varying in a wide range (from 10^{-1} to 10^3), though the optimal value is data dependent.

5   Conclusions

This paper proposes a new family of nonnegative matrix factorization methods termed Spatially Correlated Nonnegative Matrix Factorization (SCNMF) for image analysis. In contrast to the ordinary NMF, SCNMF explicitly considers the spatial relationship between the pixels in basis vectors. By introducing a penalized functional on derivatives of all directions, the number of degrees of freedom is significantly reduced, resulting in basis vectors more correlated than those obtained by the ordinary NMF. The promising empirical results show that SCNMF is able to provide a superior image representation for learning tasks.

Acknowledgments. This work was supported in part by National Natural Science Foundation of China under Grants 91120302 and 60905001, National Basic Research Program of China (973 Program) under Grant 2011CB302206.

References

1. Bishop, C.M.: Pattern recognition and machine learning, vol. 4. Springer, New York (2006)
2. Boutsidis, C., Gallopoulos, E.: Svd based initialization: A head start for nonnegative matrix factorization. Pattern Recognition 41(4), 1350–1362 (2008)
3. Cai, D., He, X., Han, J., Huang, T.: Graph regularized non-negative matrix factorization for data representation. IEEE TPAMI 33(8), 1548–1560 (2011)
4. Cai, D., He, X., Hu, Y., Han, J., Huang, T.: Learning a spatially smooth subspace for face recognition. In: CVPR. IEEE (2007)
5. Cai, D., He, X., Wu, X., Han, J.: Non-negative matrix factorization on manifold. In: ICDM, pp. 63–72. IEEE (2008)
6. Chen, X., Tong, Z., Liu, H., Cai, D.: Metric learning with two-dimensional smoothness for visual analysis. In: CVPR, pp. 2533–2538. IEEE (2012)
7. Ding, C., He, X., Simon, H.D.: On the equivalence of nonnegative matrix factorization and spectral clustering. In: SDM, vol. 4, pp. 606–610 (2005)
8. Gillis, N.: Nonnegative Matrix Factorization: Complexity, Algorithms and Applications. PhD thesis, University of Waterloo, Canada (2011)
9. Gu, Q., Ding, C.H.Q., Han, J.: On trivial solution and scale transfer problems in graph regularized nmf. In: IJCAI, pp. 1288–1293 (2011)
10. Guillamet, D., Vitrià, J.: Non-negative matrix factorization for face recognition. In: Topics in Artificial Intelligence, pp. 336–344 (2002)
11. Horn, R., Johnson, C.: Topics in Matrix Analysis. Cambridge University Press (1991)
12. Hoyer, P.O.: Non-negative matrix factorization with sparseness constraints. JMLR 5, 1457–1469 (2004)
13. Jost, J.: Riemannian Geometry and Geometric Analysis. Springer (2002)
14. Kim, J., Park, H.: Toward faster nonnegative matrix factorization: A new algorithm and comparisons. In: ICDM, pp. 353–362. IEEE (2008)

15. Land, S.R., Friedman, J.H.: Variable fusion: A new adaptive signal regression method. Technical report, Technical Report 656, Department of Statistics, Carnegie Mellon University Pittsburgh (1997)
16. Lee, D.D., Seung, H.S.: Algorithms for non-negative matrix factorization. In: NIPS, vol. 13. MIT Press (2001)
17. Lee, D.D., Seung, H.S., et al.: Learning the parts of objects by non-negative matrix factorization. Nature 401(6755), 788–791 (1999)
18. Li, S.Z., Hou, X.W., Zhang, H.J., Cheng, Q.S.: Learning spatially localized, parts-based representation. In: CVPR. IEEE (2001)
19. Li, T., Ding, C.: The relationships among various nonnegative matrix factorization methods for clustering. In: ICDM, pp. 362–371. IEEE (2006)
20. Lin, C.J.: Projected gradient methods for nonnegative matrix factorization. Neural Computation 19(10), 2756–2779 (2007)
21. Liu, W., Zheng, N.: Non-negative matrix factorization based methods for object recognition. Pattern Recognition Letters 25(8), 893–897 (2004)
22. Logothetis, N.K., Sheinberg, D.L.: Visual object recognition. Annual Review of Neuroscience 19(1), 577–621 (1996)
23. Lovasz, L., Plummer, M.: Matching Theory. Akadémiai Kiadó. North Holland, Budapest (1986)
24. Monga, V., Mihcak, M.K.: Robust image hashing via non-negative matrix factorizations. In: ICASSP, vol. 2. IEEE (2006)
25. O'Sullivan, F.: Discretized laplacian smoothing by fourier methods. Journal of the American Statistical Association, 634–642 (1991)
26. Pauca, V.P., Piper, J., Plemmons, R.J.: Nonnegative matrix factorization for spectral data analysis. Linear Algebra and its Applications 416(1), 29–47 (2006)
27. Sha, F., Lin, Y., Saul, L.K., Lee, D.D.: Multiplicative updates for nonnegative quadratic programming. Neural Computation 19(8), 2004–2031 (2007)
28. Sra, S., Dhillon, I.S.: Nonnegative matrix approximation: Algorithms and applications. University of Texas at Austin (2006)
29. Srebro, N., Rennie, J.D.M., Jaakkola, T.: Maximum-margin matrix factorization. In: NIPS, vol. 17, pp. 1329–1336. MIT Press (2005)
30. Tibshirani, R., Saunders, M., Rosset, S., Zhu, J., Knight, K.: Sparsity and smoothness via the fused lasso. Journal of the Royal Statistical Society: Series B (Statistical Methodology) 67(1), 91–108 (2005)
31. Trefethen, L.N., Bau, D.: Numerical linear algebra. Society for Industrial Mathematics, vol. 50 (1997)
32. Turk, M.A., Pentland, A.P.: Face recognition using eigenfaces. In: CVPR, pp. 586–591. IEEE (1991)
33. Vasilescu, M.A.O., Terzopoulos, D.: Multilinear subspace analysis of image ensembles. In: CVPR. IEEE (2003)
34. Wold, S., Esbensen, K., Geladi, P.: Principal component analysis. Chemometrics and Intelligent Laboratory Systems 2(1-3), 37–52 (1987)
35. Xu, W., Liu, X., Gong, Y.: Document clustering based on non-negative matrix factorization. In: SIGIR, pp. 267–273. ACM (2003)
36. Yang, J., Yang, S., Fu, Y., Li, X., Huang, T.: Non-negative graph embedding. In: CVPR, pp. 1–8. IEEE (2008)
37. Zdunek, R., Cichocki, A.: Non-negative Matrix Factorization with Quasi-Newton Optimization. In: Rutkowski, L., Tadeusiewicz, R., Zadeh, L.A., Żurada, J.M. (eds.) ICAISC 2006. LNCS (LNAI), vol. 4029, pp. 870–879. Springer, Heidelberg (2006)

A Competitive Sample Selection Method for Palmprint Recognition

Jiajun Wen[1,2], Jinrong Cui[1,2], Zhihui Lai[1,2], and Jianxun Mi[1,2]

[1] Bio-Computing Research Center, Shenzhen Graduate School,
Harbin Institute of Technology, Shenzhen, China
[2] Key Laboratory of Network Oriented Intelligent Computation, Shenzhen, China
{wenjiajun.hit,mijianxun}@gmail.com,
{tweety1028,lai_zhi_hui}@163.com

Abstract. In the field of palmprint recognition, traditional classification methods such as PCA and LDA only exploit training samples to build the model of classifiers while ignoring the feedback of test sample during the process of recognition. In this paper, both types of samples will be taken into account to make an optimal representation of the test sample. We first exploit nearest neighbor principle to construct the initial optimal training sample set for representation. This optimal set will be updated by adding a training sample which works well with current optimal set to obtain the least representation error to the test sample. The sample selection step does not stop until we acquire the optimal set with sufficient number of training samples for classification. Comparative experiments have been conducted on PolyU multispectral palmprint database which validates the proposed method. Moreover, a study on parameter points out that a small number of selected training samples is good for recognition in blue and green channels while much more selected training samples are suitable for improving recognition in red channel.

Keywords: Pattern recognition, Palmprint recognition, Sample selection.

1 Introduction

Palmprint has been a stable biometric for personal identification [1] and have drawn the attention of many security departments due to its high accuracy. It has been reported that competitive coding method [2] is one of best the feature-based methods owing to its effectiveness.

Besides, the subspace methods such as Eigenpalm [3] and Fisherpalm [4] have also obtained promising results. Eigenpalm aims to find a series of orthogonal axes in the subspace of palmprint so as to make a global representation of the test sample, while Fisherpalm tries to maximum between-class variance and minimum within-class variance to largely enhance discrimination of the classes. However, due to the inefficiency of PCA and LDA, novel methods such as 2DPCA [5, 6] and 2DLDA [7] directly map the whole palmprint into the subspace by using a linear transform which show an efficient computation and competitive effectiveness than PCA and LDA as

J. Yang, F. Fang, and C. Sun (Eds.): IScIDE 2012, LNCS 7751, pp. 158–164, 2013.

reported in [5-7]. Recently, Xu et al. [8] proposed a two-phase test sample representation (TPTSR) method and pointed out that sample selection is vital to representation-based methods. TPTSR shows a good performance with high accuracy. But it is not optimal in terms of representation error. The work in [9] selects m training samples for representation by iteratively excluding the one with least contribution to the test sample. Apparently, each selection is carried on in the subset of the former training sample set. Therefore, this method also cannot achieve global optimization.

The recent advance of representation-based method inclined that sample selection plays an important role in feature selection and even affects the performance of representation-based classifier [8, 9]. Hence, it is a valuable task to explore the way of sample selection. Unlike the sample selection scheme used in the literature mentioned above, in this paper, we devise a sample selection manner under the framework of representation-based method for palmprint recognition. We aim to find the training samples which can collaboratively work together to represent the test samples so as to construct optimal set. The sample selection step does not stop until we acquire the optimal set with sufficient number of training samples for classification.

The remainder of the paper is organized as follows. Section 2 presents the competitive sample selection method. Section 3 illustrates the experimental results. The conclusion is in Section 4.

2 The Proposed Classification Method

In this Section, we introduce a competitive way for sample selection so as to obtain an optimal training sample set for test sample representation. Under the representation framework, we use the representation deviation to construct the classifier.

2.1 The Competitive Sample Selection

Suppose that we have already obtained optimal set with n training samples which is highly correlated with the test sample. Then we search in the remaining training samples to find the next one that can collaboratively work with the training samples in optimal set to represent the test sample with minimum representation deviation. The one with larger representation deviation will not be considered in current selection. That is, the 'winner' in current round will be separated out from the remaining training sample set and be included in optimal set.

Define B_j as optimal set for representation in the j-th updates and m as the number of elements in B_j. Our method consists of two parts, namely optimal set initialization and optimal set updating.

(1) Optimal set initialization

The optimal set should contain the training samples which are the most correlated with the test sample. For the first time, there is no element in optimal set. It is not yet ready to execute the collaborative work to select training sample. To obtain the first element of optimal set, we use l_2 norm to measure the similarity between training sample and test sample. The training sample with minimum measure value is highly correlated with the test sample or even probably belongs to the same class as the test sample. Based on the nearest neighbor principle, we calculate the Euclidean distance between y and x_i using

$$d_i = \|x_i - y\|. \tag{1}$$

Find the nearest neighbor by using

$$j = \arg\min_i d_i. \tag{2}$$

Then x_j is taken as the first training sample in optimal set and denoted by b_1. Initially, we obtain the optimal set $B_1 = \{b_1\}$ and $m=1$.

(2) Optimal set updating
The task of optimal set updating is to find the training samples which are highly correlated with the test sample. There are two benefits of this procedure. For one thing, the training samples which are probably the same class as the test sample are gathered together to represent the test sample which facilitates the correctness of the final classification. For another, many irrelevant training samples which are quite different from the test sample will be excluded in advance.

Since the non-empty optimal set is obtained after the initialization step, we devise a competitive strategy to select training samples for representation. If the training samples that used for representation are from the same class as the test sample, its corresponding representation deviation will be closer to zero comparing with the training samples from other classes. According to this law, we have obtained the optimal set with the training samples which are highly correlated with the test sample. For any training sample in the remaining training sample set, if it is the most correlated one with the test sample, it should obtain the minimum representation deviation when it collaboratively works with all training samples in optimal set. In this way, any training sample, in the remaining training sample set, which is not the most competitive one will not be included in optimal set in current selection. After a certain number of competitive sample selection, we accomplish the construction of optimal set which is suitable for representation and classification.

Let R_m denotes the set which consists of remaining training samples in the m-th selection. For each sample selection, the most competitive training sample will be transferred from R_m to B_{m+1}. So we have

$$R_m = \{x_i, i = 1, 2, \cdots, N\} - B_m. \tag{3}$$

In order to find the most competitive x_i in R_m, we align all the samples in B_m and x_i as column vectors in matrix $X_m = \begin{bmatrix} b_1 & b_2 & \cdots & b_m & x_i \end{bmatrix}$. So we can compute the contribution of the corresponding training samples by (4).

$$\hat{c}_m = \left(X_m^T X_m + \gamma I \right)^{-1} X_m^T y, \tag{4}$$

where γ is a small positive constant and I is the identity matrix. Let $\| y - X_m \hat{c}_m \|$ be the representation deviation. Among all the x_i in R_m, the one that generates the minimum representation deviation is regarded as the competitive training sample and denoted as b_{m+1}. Therefore, the updating of B_m is as follow

$$B_{m+1} = B_m + \{b_{m+1}\}. \tag{5}$$

Set $m = m+1$. Then we have the optimal set $B_m = \{b_j\}$, ($j = 1, 2, \cdots, m$). Let s be a predefined parameter which controls the number of elements in B_m. The updating step does not stop until $m = s$.

2.2   Classification

For any test sample, the aim of classification is to classify the test sample into the existing classes so as to obtain the class label of the test sample. In this paper, we measure the distance of the test sample to the classes according to representation deviation between the test sample and selected training samples. Meanwhile, the training samples in T are from the same class and optimal set as well.

Since s training samples $b_1, b_2, \cdots, b_s$ in optimal set have been determined, let $B = [b_1 \ b_2 \ \cdots \ b_s]$ and the contribution of these training samples to test sample be $g = [g_1 \ g_2 \ \cdots \ g_s]^T$. The linear model of training samples to test sample is as follows:

$$y = g_1 b_1 + g_2 b_2 + \cdots + g_s b_s. \tag{6}$$

Then we can evaluate the corresponding contributions of the training samples by (7).

$$\hat{g} = \left(B^T B + \gamma I \right)^{-1} B^T y \tag{7}$$

Let T_l denote the set of the l-th class of training samples. The deviation between y and the contribution in representing y of the l-th class is calculated using

$$de_l = \left\| y - \sum_{b_i \in G_l} \hat{g}_i b_i \right\|, \tag{8}$$

where $G_l = B_s \cap T_l$. Finally, we determine the label of y by solving

$$k = \arg \min_l de_l \ (l = 1, 2, \cdots, L), \tag{9}$$

where L is the number of all classes.

3   Experiment Results

To validate the performance of the proposed method, experiments are conducted on PolyU multispectral palmprint database [11] which is collected from 250 people (195 males and 55 females) in two separate sessions. It includes the image data in three different channels, namely red, green and blue channels. For each channel, it contains 500 classes (every person provides his/her palmprints of both hands) with 12 images in each class, among which 6 images are from session 1; the other 6 ones are from session 2. All the images have been tailored by the method adopted in [1], leaving only the ROI of the palm. The resolution of the image is 128×128. In our experiments, three channels with the first 300 classes are used to make algorithm comparison, among which the first three images of session 1 will be used as training samples and the 6 images of session 2 will be used as test samples. Fig. 1 shows the ROI of a palm from three channels.

We compare the proposed method with the traditional appearance-based method including PCA, 2DPCA, LDA and 2DLDA.

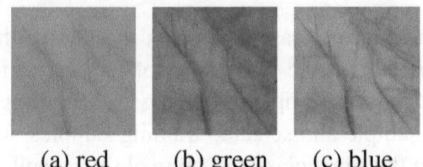

(a) red   (b) green   (c) blue

Fig. 1. Three ROI images of a palm

Table 1. Classification error rates in three channels

Method	Classification error rate		
	Red	Green	Blue
PCA(50)	6.50%	12.83%	7.61%
PCA(450)	4.11%	6.17%	3.94%
PCA(500)	4.06%	6.17%	3.89%
PCA(600)	4.06%	6.17%	3.89%
2DPCA(5)	20.61%	35.00%	22.78%
2DPCA(15)	14.56%	24.11%	14.67%
2DPCA(25)	14.22%	23.39%	14.61%
2DPCA(30)	14.22%	23.33%	14.44%
LDA	2.20%	3.70%	3.10%
2DLDA	5.70%	6.70%	5.47%
The global method in [9]	1.87%	3.53%	3.77%
The proposed method (s=100)	2.17%	2.06%	2.78%

Since the number of principle component have great influence on the performance of PCA and 2DPCA, we carry on the experiments of PCA and 2DPCA under different dimensionalities. Table 1 shows all the classification error rates of the mentioned methods in three channels of multispectral palmprint database. For PCA, as the increasing of dimensionality, the classification error rates in three channels reduce and reach a stable level since the dimensionality of 500. PCA gives better result in blue channel compared with other three channels. For 2DPCA, just like the situation of PCA, a suitable number of principle components are needed so that a stable recognition result is obtained. Unlike the report in [5], our experiment on PolyU multispectral palmprint database by using 2DPCA does not show a competitive result compared with PCA. On the contrary, 2DPCA does the worst among all the methods. The experiment results of LDA, 2DLDA and the proposed method can be seen in Table 1. LDA produces better results than that of 2DLDA, PCA and 2DPCA in three channels. Our method (parameter s is set to 100) surpasses LDA in all channels including the red, green and blue channels.

In addition, we also make a comparison between our method and the global method in [9]. The global method has no sample selection procedure. It classifies the test sample according to the sum of contribution of training samples from the same classes. See in Table 1, our method is better than global method in green and blue channel. We also obtain almost the same performance as global method in red channel with lower recognition rate of 0.3%. However, in the overall evaluation, our method is competitive as global method or even better in other two channels.

The only parameter of the proposed algorithm is s, it determines the number of selected training samples that are used for representing the test sample in the final round. We conduct the proposed method in three channels with s varying on 25 intervals from 25 to 150. Fig. 2 demonstrates the results that reflect the influence of on classification error rate. Different channels have different response to s. Obviously, with the increasing of s, the classification error rates of green and blue channels go up and then achieve a stable level. For red channel, the classification error rate will go down when more selected training samples are used for representation. Therefore, a better strategy for setting parameter s for each channel is as follow. A small value of s around 50 is a good choice for recognition in blue and green channel. A much bigger value of s around 150 is suitable for recognition in red channel.

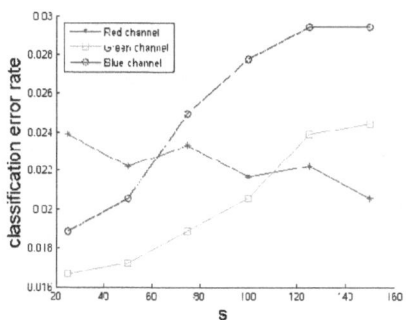

Fig. 2. Classification error rate with s varying on 25 intervals

4   Conclusion

In this paper we have proposed a competitive sample selection method for representation and classification. The method uses near neighbor principle to initially construct the optimal training sample set. Then it searches in the unselected training sample set to obtain the best one that can collaboratively work with the selected ones to represent the test sample based on the minimum representation deviation. On one hand, the proposed method exploits a competitive way to obtain optimal training sample set. On the other hand, it helps to reduce the influence of palmprint images with translation and rotation in the process of recognition. The experiment results show that the proposed method can perform very well in palmprint recognition and surpass the appearance-based method mentioned in the experiments. In addition, our study on parameter of the algorithm shows that a small number of selected training samples are good for recognition in blue and green channels while much more selected training samples are suitable for recognition in red channel.

References

1. Zhang, D., Kong, W.K., You, J., Wong, M.: Online palmprint identification. IEEE Trans. on Pattern Analysis and Machine Intelligence 25, 1041–1050 (2003)
2. Kong, A.W.K., Zhang, D.: Competitive coding scheme for palmprint verification. In: Proc. 17th Int. Conf. on Pattern Recognition, pp. 520–523. IEEE Press, Piscataway (2004)
3. Lu, G., Zhang, D., Wang, K.: Palmprint recognition using eigenpalms features. Pattern Recognition Letters 24, 1463–1467 (2003)
4. Wu, X., Wang, K., Zhang, D.: Palmprint recognition using Fisher's linear discriminant. In: Proc. Int. Conf. on Machine Learning and Cybernetics, pp. 3150–3154. IEEE Press, Piscataway (2003)
5. Sang, H., Yuan, W., Zhang, Z.: Research of Palmprint Recognition Based on 2DPCA. In: Yu, W., He, H., Zhang, N. (eds.) ISNN 2009, Part II. LNCS, vol. 5552, pp. 831–838. Springer, Heidelberg (2009)
6. Ma, Y., Sun, J.: Palmprint Recognition based on 2DPCA-moment invariant. In: Proc. 5th Int. Conf. on Image and Graphics, pp. 149–155. IEEE Press, Piscataway (2009)
7. Du, F., Yu, P., Li, H., Zhu, L.: Palmprint Recognition Using Gabor Feature-Based Bidirectional 2DLDA. In: Yu, Y., Yu, Z., Zhao, J. (eds.) CSEEE 2011, Part II. CCIS, vol. 159, pp. 230–235. Springer, Heidelberg (2011)
8. Xu, Y., Zhang, D., Yang, J., Yang, J.Y.: A two-phase test sample sparse representation method for use with face recognition. IEEE Trans. on Circuits and Systems for Video Technology 21, 1255–1262 (2011)
9. Xu, Y., Zuo, W., Fan, Z.: Supervised sparse representation method with a heuristic strategy and face recognition experiments. Neurocomputing 79, 125–131 (2012)
10. Golub, G.H., Van Loan, C.F.: Matrix computations. The Johns Hopkins University Press, Maltimore (1996)
11. PolyU multispectral palmprint Database,
    http://www.comp.polyu.edu.hk/~biometrics/
    MultispectralPalmprint/MSP.htm

A Local-Context-Based Fuzzy Algorithm
for Image Enhancement

Dansong Cheng, Daming Shi, Xianglong Tang, and Jiafen Liu

Department of Computer Science, Harbin Institute of Technology,
Harbin, China, 150001

Abstract. A grayscale images enhancement algorithm based on fuzzy Technique, with the ability to remove impulsive noise, while, simultaneously, enhancing contrast and preserving edges and image details efficiently, is proposed in this paper. To achieve these image enhancement goals, we first partition the pixels into smooth regions and boundary regions according to their neighborhood. Next we transform the image into several fuzzy sets corresponding to the smooth regions. The nonlinear enhancement is implemented in each fuzzy set. To demonstrate the capability of our filtering approach, it was tested on several different image enhancement problems. Comparing the classical methods, These experimental results demonstrate filtering quality, and image sharpening ability of the new filter.

Keywords: Fuzzy logic, Image enhancement, Edges detection, Medical image, ε -similar neighbor.

1   Introduction

Image enhancement is perhaps the most important low-level image processing task, which is among the simplest and most appealing areas of digital image processing. Enhancement is to bring out details that are obscured, or simply to highlight some features of interest in an image. A familiar example of enhancement is when we increase the contrast of an image because "it looks better". We have to take into account that enhancement is a very subjective area of image processing. These distortions cause an inadequate object-of interest presentation, which can result in inaccurate image analysis. The problem of image enhancement can be stated as that of filtering out impulse noise, smoothing out nonimpulse noise, and enhancing edges or certain other salient structures in the input image. Noise smoothing and edge enhancement are inherently conflicting processes, since smoothing a region might destroy an edge and sharpening edges might lead to unnecessary noise. As pointed out by many researchers in [1-4], enhancement procedures based on fuzzy theoretic considerations have a better favoring potential for fulfilling the above requirements. Their intrinsic natures facilitate the ambiguity and uncertainty of images analysis. Numerous methods have already been proposed for image enhancement [4-8]. For instance, Fan adopts the Canny edge detection principle based on the fuzzy

J. Yang, F. Fang, and C. Sun (Eds.): IScIDE 2012, LNCS 7751, pp. 165–171, 2013.

enhancement[7]. The algorithm enhances the pixel contrast of texture area and detects more texture details. Then it realizes image segmentation by using of inflation and region connection.

In this paper we propose an enhancement method, based on fuzzy-logic-control, that removes noise, while edges and image details are efficiently preserved. We will prepartition the image based on their gray level distribution and then map the image from the spatial domain to the fuzzy domain by a membership function. Finally, we will apply the novel adaptive fuzzy contrast enhancement algorithm to conduct contrast enhancement.

2   Nonlinear Fuzzfication

The most commonly used membership function of a gray level image is the π-function, To enlarge the contrast of boundaries between objective regions, we constructed a modified π-function, which is defined as following.

$$\mu_{G_i}(g_{ij},G_{i-1},y_i,G_i,y_{i+1},G_{i+1}) = \begin{cases} 0, & \text{if } g_{ij} \leq G_{i-1} \\ \dfrac{(g_{ij}-G_{i-1})^2}{(y_i-G_{i-1})(G_i-G_{i-1})}, & \text{if } G_{i-1} < g_{ij} \leq y_i \\ 1-\dfrac{(g_{ij}-G_i)^2}{(G_i-G_{i-1})(G_i-y_i)}, & \text{if } y_i < g_{ij} < G_i \\ 1, & \text{if } g_{ij} = G_i \\ 1-\dfrac{(g_{ij}-G_i)^2}{(G_{i+1}-G_i)(G_{i+1}-y_{i+1})}, & \text{if } G_i < g_{ij} \leq y_{i+1} \\ \dfrac{(g_{ij}-G_{i+1})^2}{(y_i-G_{i-1})(G_i-G_{i-1})}, & \text{if } y_{i+1} < g_{ij} < G_{i+1} \\ 0 & \text{if } g_{ij} \geq G_{i+1} \end{cases} \quad (1)$$

$$i = 1,2,\cdots,L$$

Membership $\mu_{G_i}(g_{ij},G_{i-1},y_i,G_i,y_{i+1},G_{i+1})$ can be abbreviated as $\mu_{G_i}(g_{ij})$. The shape of function is determined by the parameters. If $g_{ij} = G_i$, membership achieves the maximum value 1; the greater the gray value deviates from G_i, the lower Membership value is. When the gray values is between G_i and G_{i+1}, Membership value depends on the membership function of G_i.

A good enhancement algorithm should be able to automatically adjust the curve according to the statistical properties of the image. First, reasonable fuzzy partition is performed on the statistical properties of the histogram. Then, adaptive parameter selection is guided by the distribution, in order to achieve adaptive image enhancement.

3   Fuzzy Partition Based on Statistical Properties

An object can be easily detected in an image if the object has sufficient contrast from the background. Different objective regions often correspond to different gray level range, the gray value of pixels are similar in homogeneous region and is larger between objects. This results a series of peaks and troughs in histogram. The peak and troughs values of histogram are widely used in image segmentation and enhancement, which often used as a threshold. Histogram thresholding is one of the widely used techniques for monochrome image segmentation, but it is based on only gray level and does not take into account the spatial information of pixels with respect to each other. In this paper, we employ the concept of the ε-neighborhood to extract homogeneous regions in an image.

An image can be divided into two fuzzy regions: *smooth region,* and *boundary region.* For a pixel at location (*x, y*), if the gray-level characteristic is similar to its neighbor pixels, the pixel with high probability belongs to *smooth region.* Otherwise, the pixel is more likely belonged to *edge region.* $PH = \{ph_i, i = 0,1,\cdots\}$ consists all points corresponding to the peaks in the histogram of smooth region *HO.* For example, if there are three peaks in the histogram, the corresponding gray-scale are $k = 10$, $k = 120$ and $k = 220$, respectively, then $PH = \{10,120,220\}$. Number of elements in *PH* is number of fuzzy subset of the image, The entire image can be represented as a collection of fuzzy sets. Select one fuzzy subset *A* of them, whose corresponding gray level is ph_i, the membership function of *A* is defined in Eq.(1). the pixels satisfying $g_{ij} = ph_i$, gain the highest membership of set *A*, 1.

To increase the contrast of the edge region, the threshold value y_i in function (1) is determined adaptively referring to the histogram of the smooth regions either, which is the minimum value in the range of $[ph_{i-1} \quad ph_i]$。

If the magnitude differences between a pixel, *p*, with most of its surroundings are less than ε, the pixel is an inner pixel of an smooth region, that can be used as an initial seed to grow the region. Otherwise, if the magnitude of *p* is quite deferent from its neighbors, larger than ε, *p* is not an inner pixel of a region.

Referring to the definition of square neighborhood in reference [9], Suppose that the square neighborhood of each pixel is given, there are predicatively the pixels in the set

$$\Omega_p = \left\{q \in N_p : d(I(p), I(q)) \leq \varepsilon\right\} \tag{2}$$

and the pixels in the set

$$\Omega'_p = \left\{q \in N_p : d(I(p), I(q)) > \varepsilon\right\} \tag{3}$$

for an arbitrary threshold $\varepsilon \geq 0$, that should be set according to the image, where $I(p)$ is characteristic value of the pixel *p*. $d(I(p), I(q)) = abs(I(p) - I(q))$ is the magnitude difference between pixel *p* and pixel *q*. $\Omega_p \cup \Omega'_p = N_p$, N_p denotes the square neighborhood. The pixels in Ω_p are ε-*similar* with p. We defined *neighborhood coherence factor, NCF*, as:

$$NCF(p) = \frac{|\Omega_p|}{|N_p|} \qquad (4)$$

Where $|\bullet|$ refers to the cardinality of a set, i.e., the umber of elements in set Ω_p. NCF is defined as a ratio: the number of pixels with similar magnitude with p to that of the total number of the neighborhood pixels. Obviously, the ratio is different for different pixel. When $|\Omega_p| \geq |\Omega_p'|$, NCF($p$)≥0.5, p is similar with most of its neighbors, and it's likely within a smooth region. When $|\Omega_p| < |\Omega_p'|$, NCF(p)<0.5, p has similar intensity with the minority, so it is Likely in the boundary.

NCF is constructed referring to the value of ε, $\varepsilon = 0.5$ in this model. If NCF(p)<0.5 The pixel belongs to edge region, if NCF(p)≥0.5, the pixel belongs to smooth region.

4   Fuzzy Enhancement Algorithm

1) The definition of membership function transformed from image sets to fuzzy sets

The fuzzy membership function algorithm is shown as in Eq.(1):

Let $F = \{f_1; f_2; : : : ; f_L\}$ be a set of fuzzy membership values of the grayscales in G using the modified π-function and a suitable fuzzy region width. Let $G_1, G_2, \cdots, G_k$ be grayscales for different regions, i.e. the grayscales for different fuzzy set. Then the membership of g_{ij} is defined as:

$$\mu(g_{ij}) = \begin{cases} \mu_{G_{i-1}} \cup \mu_{G_i} = \max(\mu_{G_{i-1}}, \mu_{G_i}) \\ \qquad if \quad G_{i-1} < g_{ij} \leq G_i \quad i = 1, 2, \cdots, L \\ \mu_{G_i} \cup \mu_{G_{i+1}} = \max(\mu_{G_i}, \mu_{G_{i+1}}) \\ \qquad if \quad G_i < g_{ij} < G_{i+1} \quad i = 1, 2, \cdots, L \end{cases} \qquad (5)$$

Where $\cup$ denotes *OR* operator, μ_{G_i} is defined in Eq.(1). This is a piecewise modified π-function, to determine the membership of the grey level within the range. When $g_{ij} = G_i$, $\mu_{G_i}(g_{ij}) = 1$. The modified π-function enlarges the contrast near B_i. If $g_{ij} < B_i$, membership of g_{ij} will move closer to μ_{G_i}, otherwise, if $g_{ij} > B_i$, membership of g_{ij} will move closer to $\mu_{G_{i+1}}$.

2) Image fuzzy sets enhancement processing
The enhancement processing is completed by the following nonlinear transformation:

$$\mu'(g_{ij}) = \begin{cases} 2(\mu(g_{ij}))^{1/NCF(p)} & 0 \le \mu(g_{ij}) \le 0.5 \\ 1 - 2(1 - \mu(g_{ij}))^{1/NCF(p)} & 0.5 < \mu(g_{ij}) \le 1 \end{cases} \tag{6}$$

3) The inverse transformation from fuzzy sets to image sets

Through the inverse transformation of $T(\cdot)$, the gray level of the (i, j)th pixel in the enhanced image is obtained by $T^{-1}(\ )$in reference [10]

$$g'_{ij} = T^{-1}(\mu'(g_{ij})) \tag{7}$$

5   Experiments and Analysis

To investigate the performance, we experimentally carry out extensive comparisons with Fuzzy Entropy enhancement and Histogram Equalization enhancement on ultrasound images. Fig 1 shows the comparison of results.

Fig.1(a) shows a lobulated mass, whose shape and boundary are valuable to diagnose. According to the clinicians from a 3A hospital, because the of the mass is relatively low echo, low contrast and blurred boundaries, it is difficult to diagnose. We mark the region of interest in the image. Fig.1(b) shows result of the proposed method, which give the mass clearer appearance. Fig.1(c) shows the result of Histogram Equalization enhancement, the boundaries of tumor are sharpened as expected but subcutaneous tissue and muscle layer are over-enhanced, that is not conducive to diagnose. Fig.1(d) is the result of Fuzzy Entropy enhancement[11], in which the contrast has not been improved obviously.

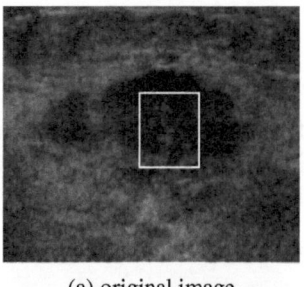

(a) original image

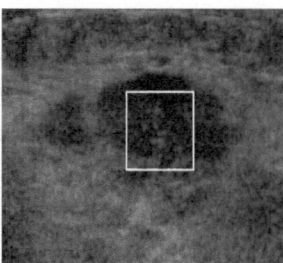

(b) Enhanced image of proposed method

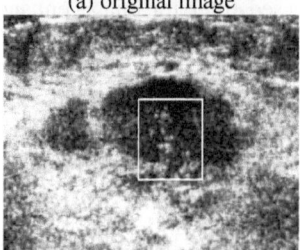

(c)   enhanced   image   with   Histogram Equalization

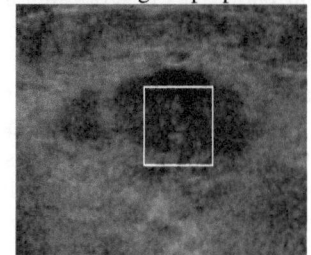

(d) enhanced image with Fuzzy Entropy

Fig. 1. Comparison of ultrasound image enhancement

170   D. Cheng et al.

For diagnosing, enhanced boundaries compared to the original image becomes clearer, fuzzy enhancement enhanced contrast between target area and the background area, reserving the details of the image features at the same time. Figure 2 is the image renderings of each processing stage and their corresponding three-dimensional surface. It can be seen from the enhanced image that, some pseudo-boundaries due to the presence of noise are removed while the heterogeneous areas are reserved and the contrast is stretched.

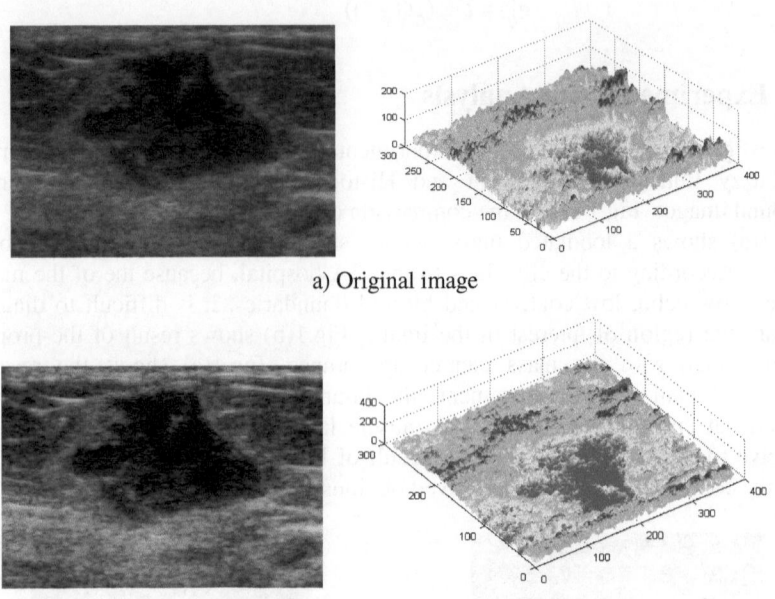

a) Original image

(b) Enhanced image with proposed method

Fig. 2. Enhancement of ultrasound breast image (the left are image, the right are corresponding three-dimensional surface)

6   Conclusion

Our contribution lies in proposing a fuzzy logic-based filter for image enhancement. Specifically, the fuzzification relies on an introduced ε-neighbor coherence segmentation criterion which is easy to interpret and implement. The membership function is defined with statistical properties of the smooth regions, which is extracted with ε-neighbor coherence. Nonlinear enhancement in fuzzy domain enlarges the contrast between the different grayscale regions, and thus protects the image texture features. The method is qualitatively effective to fulfill the task of semantic segmentation with an adaptively selected spatial scale and an appropriately determined intensity-difference scale, especially for medical images, significantly improve the effectiveness of the clinical diagnosis.

Acknowledgments. This research was supported by the National Natural Science Foundation of China (Grant No. 51077020).

References

1. Kunhua, Z., Xuan, Y., Li, Z.: Fractal feature enhancement based on fuzzy sets and its application in target detection. Computer Engineering and Applications 45(11), 172–174 (2009)
2. Ra, J., Jang, J., Bae, Y.: Contrast-Enhanced Fusion of Multi-Sensor Images Using Subband-Decomposed Multiscale Retinex. Image Processing 99(3), 1–12 (2012)
3. Li, L.: A Study of Primary Component Histogram Fuzzy Enhancement for Color Image Segmentation. Journal of Dalian University 32(3), 44–48 (2011)
4. Chen, Q., Huang, G., Sun, R., Shu, Y., Pu, Y., Zhou, J.: A Riemann-Liouville Fractional Differential Image Enhancement Algorithm Based on Human Visual Characteristics. Journal of Sichuan University 44(1), 99–105 (2012)
5. Wang, B., Liu, S., Fan, J., Xie, W.: An Adaptive Multi-level Image Fuzzy Enhancement Algorithm Based on Fuzzy Entropy. Acta Electronica Sinica 33(4), 730–734 (2005)
6. Fan, G., Su, H., Wang, C.: Image segmentation algorithm based on fuzzy enhancement. Computer Engineering and Design 33(4), 1463–1466 (2012)
7. Li, B., Guo, Z., Wen, C.: Multi-level Fuzzy Enhancement and Edge Extraction of Images. Fuzzy Systems and Mathematics 14(4), 77–83 (2000)
8. Hanmandlu, M., Verma, O.P., Kumar, N.K., Kulkarni, M.: A Novel Optimal Fuzzy System for Color Image Enhancement Using Bacterial Foraging. Instrumentation and Measurement 58(8), 2867–2879 (2009)
9. Cheng, D., Liu, X., Tang, X., Liu, J.: Image segmentation using neighborhood inspiring pulse coupled neural network. Journal of Huazhong University of Science and Technology 37(5), 33–37 (2009)
10. Cheng, D., Huang, J., Yu, Z., Tang, X., Yang, J.: Medical image enhancement based on fuzzy techniques. Journal of Harbin Institute of Technology 39(3), 435–437 (2007)
11. Li, H., Wang, T., Lin, J., Li, D.: Contrast enhancement of medical ultrasonic images based on exact histogram specification. Chinese Journal of Medical Imaging Technology 24(2), 278–281 (2008)

3D Partial Surface Matching Using Differential Geometry and Statistical Approaches

Kehua Guo[*] and Guihua Duan

School of Information Science & Engineering,
Central South University, Changsha, China
{guokehua,duangh}@csu.edu.cn

Abstract. 3D partial surface matching approach is universal to 3D object recognition. In this paper, a new solution utilizing Gaussian curvature and mean curvature to represent the inherent structure of surface is proposed, Point-Pair Set is constructed by means of filtrating points with similar inherent characteristic in partial surface, then Triangle-Pair Set is demonstrated after locating 3D surface by asymmetry triangle skeleton and searching similar triangles in Point-Pair Set, finally, optimal transformation is illustrated by scoring function to transformations in Triangle-Pair Set and optimal matching is determined. Experiments show that the algorithm is suitable for 3D partial surface matching, and an encouraging matching efficiency, speed and running time complexity to irregular surfaces is introduced.

Keywords: Curvature, Differential Geometry, Partial Surface, Pattern Recognition.

1 Introduction

Partially similar object matching approach in 3D space is universal to 3D object recognition. During the past decade, the problem of finding a partial match between 3D surfaces attracted considerable attention. In many cases, objects may be only partially visible because of occlusion or because the sensor usually cannot scan all sides of the objects, so the goal of 3D partial surface matching is to find a Euclidean geometry rigid transformation of the sample surface to overlap a large portion of the object surface. 3D partial surface matching has important applications in many cases such as the vision registration for automatic 3D models[1], content-based image retrieval from image database[2], assembly and restoration for fractured object[3], docking of proteins in molecular biology[4], and many additional industrial, military, and other applications.

There has been lots of research on 3D partial surface matching in recent decades. Literature[5,6] proposed a scoring schema for 3D partial surface matching, but this paper did not demonstrate a universal and high-efficiency algorithm for the constructing of scoring factor. Literature[7] proposed an approach based on Hausdorff and Frehet distance for geometry structure matching. This approach can be well

---

[*] Corresponding author.

J. Yang, F. Fang, and C. Sun (Eds.): IScIDE 2012, LNCS 7751, pp. 172–180, 2013.

applied in matching of points and line segments set, but it is unstable and cannot perform well in the presence of noise, occlusion and clutter. Literature [8] proposed a scoring schema for matching algorithm to 3D incomplete object. Literature [9-11] proposed some matching algorithm using differential geometry to 3D partial surface matching. In addition, statistical descriptors [12], probabilistic framework[13] and curve analysis[14] are also applied to the recognition of 3D partial surface, these approaches indicate encouraging matching results for 3D points, line segments set and mesh model, however, they will suffer from high computing complexity.

Our work is motivated by the technique of scoring schema and statistical approaches. This technique, which uses the so-called scoring-function method, was originally introduced for 3D partial surface and volumetric matching. This idea has been applied in some pattern recognition problem [15] and gets a good result. In this paper, we propose a new approach for the constructing of scoring factor based on differential geometry principle. In this algorithm, inherent characteristic of partial surface is fully employed for shape representation, point-pair with similar inherent characteristic is classified from sample and object surface.

2   Computing and Storing the Curvature

According to differential geometry principle, curvature is the inherent characteristic of a spatial surface. Therefore, Gaussian curvature and mean curvature are employed for the representation of spatial surface in this paper.

In 3D Euclidean space, given a parametric surface defined as:

$$S(x, y) = [x \ y \ f(x, y)]^T, (x, y) \in D \tag{1}$$

where X-Y is the reference plane in 3D space, D is projection region of the surface to X-Y plane, $f(x,y)$ represents the distance from the surface to point (x,y) in X-Y plane.

Gaussian curvature K and mean curvature H can be computed according to following formulas[16]:

$$K = \frac{f_{xx}f_{yy} - f_{xy}^2}{(1 + f_x^2 + f_y^2)^2}, H = \frac{(1 + f_x^2)f_{yy} + (1 + f_y^2)f_{xx} - 2f_x f_y f_{xy}}{2(1 + f_x^2 + f_y^2)^{3/2}} \tag{2}$$

For digital range image surface, approximations can be computing by local polynomial fitting approach, $n \times n$ operator is usually utilized to the convolution operation with original range image:

$$f_x = D_x * f, f_y = D_y * f, f_{xx} = D_{xx} * f, f_{xy} = D_{xy} * f, f_{yy} = D_{yy} * f \tag{3}$$

where D is $n \times n$ operator. For $n=7$, the parameters can be computed as follows:

$$\begin{cases} D_x = d_0 d_1^T, D_y = d_1 d_0^T, D_{xx} = d_0 d_2^T, D_{yy} = d_2 d_0^T, D_{xy} = d_1 d_1^T \\ d_0 = \frac{1}{7}[1\ 1\ 1\ 1\ 1\ 1\ 1]^T, d_1 = \frac{1}{28}[-3\ -2\ -1\ 0\ 1\ 2\ 3]^T, d_2 = \frac{1}{84}[5\ 0\ -3\ -4\ -3\ 0\ 5]^T \end{cases} \tag{4}$$

where d_0, d_1, d_2 are column vectors for window operator computing.

In order to measure the similarity between two points from sample and object surface, the definition of curvature distance is introduced as follows:

Definition 1. $\forall p_i^1 \in S_1, p_j^2 \in S_2$, curvature distance DC between point p_i^1, p_j^2 is:

$$DC(p_i^1, p_j^2) = \frac{1}{2} \times (\frac{|K(p_i^1) - K(p_j^2)|}{|K(p_i^1)| + |K(p_j^2)|} + \frac{|H(p_i^1) - H(p_j^2)|}{|H(p_i^1)| + |H(p_j^2)|}) \tag{5}$$

However, in many cases, a lot of plane points, whose Gaussian curvature and mean curvature are zero, can exist in the surface. They cannot contribute to the matching but will occupy large computing and reduce the matching efficiency, even lead to an error matching. So that these plane points will be discarded before matching.

Definition 2. A Point-Pair Set named PS is a set defined as:

$$\begin{cases} PS = \{(p_i^1, p_j^2) \mid p_i^1 \in S_1, p_j^2 \in S_2, DC(p_i^1, p_j^2) < \varepsilon c, |K(p_i^1)| > \sigma\} \\ \varepsilon c = (\max(DC(p_i^1, p_j^2) + \min(DC(p_i^1, p_j^2))/2, \sigma = (\sum_{i=1}^{n} |K(p_i^1)|)/N \end{cases} \tag{6}$$

where εc is an average threshold to guarantee the similarity of p_i^1 and p_j^2, σ is to discard plane points, N is the number of the points.

The cardinality of PS could be large in special case. For instance, cardinality of PS will be $|S_1| \times |S_2|$ when the sample and object surface are congruent sphere surface. A large Point-Pair Set will occupy large amount of storage space. Considering the quantity of curvature categories cannot exceed the quantity of pixel points in surface, list can be employed to store the set.

3   Matching Algorithm

3.1   Measuring Similarity for Spatial Triangles

We arbitrarily select three points from object surface to form an asymmetrical triangle, and then seek three corresponding points from sample surface to form another asymmetrical triangle, whose features of corresponding vertices and edges are similar to the previous triangle. The similarity of space triangles is measured by the definition as follows:

Definition 3. $\forall (p_1^1, p_1^2), (p_1^1, p_1^2), (p_1^1, p_1^2) \in PS$, $|p_1^m p_2^m| \triangleleft |p_1^m p_3^m| \triangleleft |p_2^m p_3^m|$, $m \in \{1,2\}$. The similar distance between $\Delta p_1^1 p_2^1 p_3^1$ and $\Delta p_1^2 p_2^2 p_3^2$ is:

$$DT = \frac{1}{3} \times \sum_{1 \le i < j \le 3} \frac{\|p_i^1 p_j^1| - |p_i^2 p_j^2\|}{|p_i^1 p_j^1| + |p_i^2 p_j^2|} \tag{7}$$

The absolute distance between $\Delta p_1^1 p_2^1 p_3^1$ and $\Delta p_1^2 p_2^2 p_3^2$ is:

$$DA = 2 \times \frac{\sum_{i=1}^{3} | p_i^1 p_i^2 |}{\sum_{1 \le i < j \le 3} (| p_i^1 p_j^1 | + | p_i^2 p_j^2 |)} \qquad (8)$$

3.2   Computing Geometry Rigid Transformation

Suppose that $\Delta p_1^1 p_2^1 p_3^1$ and $\Delta p_1^2 p_2^2 p_3^2$ are two similar asymmetrical triangles, $(p_i^1, p_i^2), i \in \{1,2,3\}$ are corresponding points, we divide the transformation T between $\Delta p_1^1 p_2^1 p_3^1$ and $\Delta p_1^2 p_2^2 p_3^2$ into a translation transformation N and a rotation transformation A:

$$T(\Delta p_1 p_2 p_3) = A[p_1 \ p_2 \ p_3] + N \qquad (9)$$

where $A = (n_1 \ n_2 \ n_3)^T$ is a rotation matrix, N is a translation matrix.

We firstly shift the triangle centers to the origin of coordinate, so that the transformation T satisfies:

$$\left[p_1^2 \ p_2^2 \ p_3^2 \right] - N_2 = A(\left[p_1^1 \ p_2^1 \ p_3^1 \right] - N_1) \qquad (10)$$

where N_1, N_2 is the center location matrix of $\Delta p_1^1 p_2^1 p_3^1$ and $\Delta p_1^2 p_2^2 p_3^2$.

Rotation matrix can be deduced as following formula:

$$A = (\left[p_1^2 \ p_2^2 \ p_3^2 \right] - N_2)(\left[p_1^1 \ p_2^1 \ p_3^1 \right] - N_1)^{-1} \qquad (11)$$

3.3   Generating Triangle-Pair Set and Determining Optimal Transformation

Similar triangle pairs from two surfaces will be stored in Triangle-Pair Set defined as follows:

Definition 4. A Triangle-Pair Set named TS is a set:

$$TS = \{ (\Delta p_1^1 p_2^1 p_3^1, \Delta p_1^2 p_2^2 p_3^2, T) | (p_i^1 p_i^2) \in PS, DT < \varepsilon \}, \varepsilon = (\max DT) + \min DT))/2 \qquad (12)$$

where ε is an average threshold to guarantee the similarity of $\Delta p_1^1 p_2^1 p_3^1$ and $\Delta p_1^2 p_2^2 p_3^2$. T is the transformation from $\Delta p_1^2 p_2^2 p_3^2$ to $\Delta p_1^1 p_2^1 p_3^1$.

The validity of every transformation will be test through applying it to every other triangle pair in TS. For a specified transformation T_0 in PS, we define the scoring function as follows:

$$Score(T_0) = \sum S(\Delta p_1^1 p_2^1 p_3^1, p_1^2 p_2^2 p_3^2, T_0) \qquad (13)$$

$(\Delta p_1^1 p_2^1 p_3^1, \Delta p_1^2 p_2^2 p_3^2)$ is an arbitrary similar triangle pair in PS, and function S is defined as follows:

$$\begin{cases} S(\Delta p_1^1 p_2^1 p_3^1, \Delta p_1^2 p_2^2 p_3^2, T_0) = \begin{cases} 1 & DA(T_0(\Delta p_1^1 p_2^1 p_3^1), \Delta p_1^2 p_2^2 p_3^2) < \varepsilon a \\ 0 & others \end{cases} \\ \varepsilon a = (\max(DA) + \min(DA))/2 \end{cases} \tag{14}$$

where εa is an average threshold to determine if $\Delta p_1^1 p_2^1 p_3^1$ can be transformed to $\Delta p_1^2 p_2^2 p_3^2$ by T_0. An optimal transformation T_P is the transformation with maximum score, satisfying:

$$Score(T_P) = \max(Score(T)) \tag{15}$$

where T is an arbitrary transformation in TS.

4   Experimental Results and Complexity Analysis

4.1   Detection of Partially Occluded Object

The goal of the first experimentation is to detect the partially occluded object from a range image using the algorithm proposed in this paper. The image data is established referred to [17]. In Fig. 1(a) and Fig. 1(b), two original objects (Duck and Venusm) in 3D space are demonstrated, the surface is formed by pixel points.

 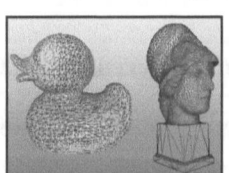  

(a) Original Object   (b) 3D Model of   (a)   Duck   is   (b) 3D Mode of
                        Original Object   Partially Occluded   the Scene

Fig. 1. Two 3D Objects   **Fig. 2.** Partially Occluded object

In Fig. 2(a), the duck is partially occluded by the venusm. In order to find the location of the occluded object(duck) in the 3D scene, we must deduce an optimal transformation from the scene to occluded object, then apply the transformation to occluded object and reconstruct it in the 3D scene in Fig. 2(b).

Optimal triangle in original object is demonstrated in Fig. 3(a), and the optimal triangle in occluded object is illustrated in Fig. 3(b). We apply the transformation to original object, the reconstruction result is shown in Fig. 3(c).

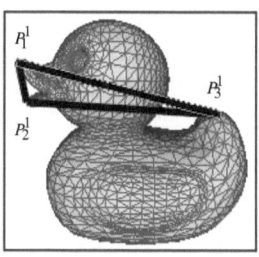
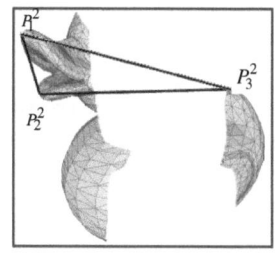
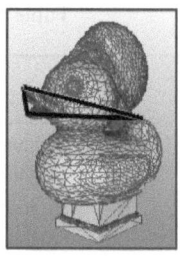

(a) Optimal Triangle in   (b) Optimal Triangle in   (c) Reconstruction Result
Original Object   Occluded Object   to Occluded

Fig. 3. Detection of Partially Occluded Objects

4.2   Complexity Analysis

Suppose that the two input sets have comparable sizes, we measure the computing complexity of the algorithm as a function of N, the cardinality of the input sets, k is the size of Point-Pair Set. Computing the curvature of every point takes $O(N)$ time. Preparing the Point-Pair Set can be executed with expected $O(N\log N)$ running time because we can sort the points based on curvature. In addition, seeking triangle pairs, computing and testing transformation based on Point-Pair Set will cost $O(k^3)$ running time. In total, the running time of the whole algorithm is: $O(N+N\log N+ k^3)$.

Obviously, the computing efficiency is mainly determined by the size of Point-Pair Set k. The quantity of point pairs will be smaller in case of the surface processing many different curvatures, our algorithm will cost less time at this time. Fig.4 shows the relationship between quantity of curvature categories and computing times for N=100(CPU:PIV2.0GHZ, RAM:1GB, Software:MATLAB7.0):

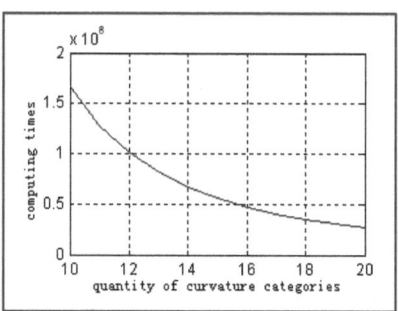

Fig. 4. Relationship Between Quantity of Curvature Categories and Computing Times (N=100)

In Fig.4, it is illustrated that computing times will be reduced when the quantity of curvature categories increases, so our algorithm is very suitable for the matching of irregular objects.

In order to explicitly describe the performance of our algorithm, we perform experimentations for objects with different shape. Running time of some algorithm for $N = 900$ is demonstrated in Table 1:

Table 1. Performance Comparison of Some Algorithms

Algorithm	Running time(ms)	
	Sphere Surface	Surface in Fig. 5
Algorithm in this paper	2635	758
Algorithm in [7]	1932	2575
Algorithm in [8]	2956	3548
Algorithm in [9]	1562	1652
Algorithm in [12]	1457	1238
Algorithm in [14]	3187	4572

We can see from Table 1 our algorithm cannot indicate an encouraging matching efficiency for regular objects such as sphere surface.

4.3   Noise Robustness

However, this work depends crucially on the Gaussian curvature and means curvature computation, which is notoriously sensitive to noise and local perturbation. We perform an experimentation to verify the noise robustness of this algorithm. In this experimentation, we generate 100 similar surfaces by matlab function, and Gaussian noise ($N(0,\sigma)$) is added to each image in the database and σ changes from 0 mm to 2.0 mm. We design the classification rate to represent the percentage of correct matching quantity in the database, the classification rates are shown in Fig. 5 for various σ values.

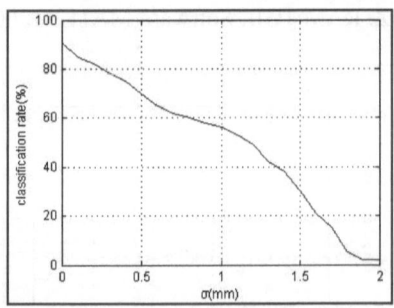

Fig. 5. Classification Rate when Gaussian Noise Increases

The result show the algorithm in this paper does not have good noisy robustness. At present, our approach can get encouraging matching efficiency and running time complexity in case of high signal to noise ratio. Therefore, the future work will concentrate to this problem.

5   Conclusion

In this paper, an algorithm for solving the 3D partial surface matching between two objects has been proposed. Firstly we utilized Gaussian curvature and mean curvature to represent the two object surfaces. Then curvature distance was defined to classify the point pair with similar curvature, and Point-Pair Set was employed to store point pairs. Secondly, we classified triangle pair with similar shape features from the two surfaces based on Point-Pair Set, similar distance was employed to measure the similarity of two triangles and then the Triangle-Pair Set was generated. Finally, absolute distance and score function were proposed for measuring the goodness of a given transformation in Triangle-Pair Set, a successful matching was determined based on the score of optimal transformation between the objects. Experimental result indicated our algorithm could be suitable for the matching of natural objects.

However, the efficiency of this approach would be reduced when the shape is regular and the robustness to noise is not very good. In future work, we will plan to explore several improvements of the algorithm, including the improvement of noise robustness, design of better distance definition, experimentation with other scoring functions and further study of statistical approaches, etc.

Acknowledgments. This work is supported by Research Fund for the Doctoral Program of Higher Education of China(20090162120069), Science and Technology Plan of Hunan (2009FJ3016), postdoctor fund of Central South University.

References

1. Ritter, L., Reiz, S.D., Rothamel, D., Dreiseidler, T., Karapetian, V., Scheer, M., Zöller, J.E.: Registration accuracy of three-dimensional surface and cone beam computed tomography data for virtual implant planning. Clinical Oral Implants Research 23(4), 447–452 (2012)
2. Buntea, K., Biehla, M., Jonkmanb, M.F., Petkova, N.: Learning effective color features for content based image retrieval in dermatology. Pattern Recognition 44(9), 1892–1902 (2011)
3. Tarte, S.M., Talib, H., Ballester, M., Langlotz, F.: Evaluating partial surface matching for fracture reduction assessment. In: 3rd IEEE International Symposium on Biomedical Imaging: Macro to Nano, vol. 4, pp. 514–517 (2006)
4. Shibberu, Y., Holder, A.: A spectral approach to protein structure alignment. ACM Transactions on Computational Biology and Bioinformatics 8(4), 867–875 (2011)
5. Barequet, G., Sharir, M.: Partial surface and volume matching in three dimensions. IEEE Transactions on Pattern Analysis and Machine Intelligence 19(9), 29–948 (1997)
6. Barequet, G., Sharir, M.: Partial surface matching by using directed footprints. Computational Geometry: Theory and Applications 12(122), 45–62 (1999)
7. Alt, H., Bras, P., Godau, M.: Computing the Hausdorff distance of geometric at terns and shapes. In: Discrete and Computational Geometry, Special Issue-The Goodman-Pollack-Festschrift 2003, pp. 65–76 (1999)
8. Mitra, N.J., Guibas, L.J., Pauly, M.: Partial and approximate symmetry detection for 3D geometry. ACM Transactions on Graphics 25(3), 560–568 (2006)

9. Gal, R., Cohen-Or, D.: Salient geometric features for partial shape matching and similarity. ACM Transactions on Graphics 25(1), 130–150 (2006)
10. Wang, S., Wang, Y., Jin, M., Gu, X.D., Samaras, D.: Confromal geometry and its applications on 3D matching, recognition, and stitching. IEEE Transactions on Pattern Analysis and Machine Intelligence 29(7), 1209–1220 (2007)
11. Tierny, J., Vandeborre, J.P., Daoudi, M.: Partial 3D shape retrieval by Reeb pattern unfolding. Computer Graphics Forum 28, 41–55 (2009)
12. Castellani, U., Cristani, M., Fantoni, S., Murino, V.: Sparse points matching by combining 3D mesh saliency with statistical descriptors. Computer Graphics Forum 27(2), 643–652 (2008)
13. Itskovich, A., Tal, A.: Surface partial matching and application to archaeology. Computers & Graphics 35(2), 334–341 (2011)
14. Tabia, H., Daoudi, M., Vandeborre, J.P., Colot, O.: A new 3D-matching method of nonrigid and partially similar models using curve analysis. IEEE Transactions on Pattern Analysis and Machine Intelligence 33(4), 852–858 (2011)
15. Guo, K.H., Liu, C.C., Yang, J.Y.: Differential geometry approach to 3D partially similar object matching. Pattern Recognition and Artificial Intelligence 21(5), 586–591 (2008)
16. Dubrovin, B.A., Fomenko, A.T., Novikov, S.P.: Modern geometry-methods and applications (I). In: GTM, 2nd edn., pp. 61–80. Springer (1999)
17. Mian, A.S., Bennamoun, M., Owens, R.: Three-dimensional model-based object recognition and segmentation in Cluttered scenes. IEEE Transactions on Pattern Analysis and Machine Intelligence 28(10), 1584–1601 (2006)

Dimensionality Reduction by Low-Rank Embedding

Chun-Guang Li, Xianbiao Qi, and Jun Guo

PRIS Lab., School of Information and Communication Engineering,
Beijing University of Posts and Telecommunications, Beijing 100876, P.R. China
{lichunguang,qixianbiao,guojun}@bupt.edu.cn

Abstract. We consider the dimensionality reduction task under the scenario that data vectors lie on (or near by) multiple independent linear subspaces. We propose a robust dimensionality reduction algorithm, named as *Low-Rank Embedding*(LRE). In LRE, the affinity weights are calculated via low-rank representation and the embedding is yielded by spectral method. Owing to the affinity weight induced from low-rank model, LRE can reveal the subtle multiple subspace structure robustly. In the virtual of spectral method, LRE transforms the subtle multiple subspaces structure into multiple clusters in the low dimensional Euclidean space in which most of the ordinary algorithms can perform well. To demonstrate the advantage of the proposed LRE, we conducted comparative experiments on toy data sets and benchmark data sets. Experimental results confirmed that LRE is superior to other algorithms.

1 Introduction

In the past decade the nonlinear dimensionality reduction by manifold learning has attracted a surge of interest in the field of data analysis, pattern recognition and machine learning, and a number of algorithms have been proposed, including Isometric feature mapping (ISOMAP)[1], Local Linear Embedding (LLE)[2, 3], Laplacian Eigenmap[4], Hessian LLE[5], Manifold Charting[6], Local Tangent Space Alignment (LTSA)[7], Semi-Definite Embedding (SDE)[8], LogMap[9, 10], Diffusion Map[11] and their variants[12–14].

Most of nonlinear dimensionality reduction techniques by manifold learning are designed based on pairwise Euclidean distance. ISOMAP[1] makes use of Euclidean distance between data points in local neighborhood to approximate geodesic distance. Laplacian Eigenmap[4] and Diffusion Map[11] use Euclidean distance to calculate weight function. SDE[8] preserves Euclidean distance in neighborhood and hence can be viewed as an implementation of ISOMAP in local manner. Euclidean distance can be regarded as a trustworthy guide to the local metric structure of manifold, however, it cannot capture the hidden nature structure when data points reside on multiple manifolds with subtle structures. In LLE[2] locally linear reconstruction coefficients are used to depict the linear relationship in neighborhood. However the local linear representation in LLE is still unable to discover the subtle natural structure.

J. Yang, F. Fang, and C. Sun (Eds.): IScIDE 2012, LNCS 7751, pp. 181–188, 2013.
© Springer-Verlag Berlin Heidelberg 2013

Besides, all the algorithms mentioned above are based on the basic assumption that *data points lie on a single subspace or manifold*. For the real world applications, the single subspace or single manifold assumption is hardly to be satisfied. In this paper we take into account the dimensionality reduction task under the scenario that data vectors lie on (or near by) multiple independent linear subspaces[1]. In some application scenario, it is more reasonable to imagine that data lie on *multiple independent subspaces* . For instance, the data vectors of frontal face images collection which consists of multiple individuals under different illuminations lie on multiple linear subspaces. However the aforementioned nonlinear methods by manifold learning cannot reveal the subtle multiple subspaces structure and hence fail in decoupling the subtle hiding structure into the target space. Due to the subtle structure, the performance of the ordinary algorithms which are designed for Euclidean space would be badly degenerated.

In this paper, we propose a novel robust dimensionality reduction algorithm, named as Low-Rank Embedding (LRE). Similar to the existing manifold learning techniques, the proposed LRE consists of two steps: a) calculating the affinity matrix and constructing an affinity relationship graph, and b) obtaining the embedding by spectral method. To reveal the latent subtle multiple subspace structure, we induce the affinity weights by low-rank model, which is proposed by Liu et al.[15].To demonstrate the validity of the proposed LRE, we conducted extensive experiments on toy data sets and two benchmark data sets. Experimental results show that dimensionality reduction by LRE can reveal the subtle multiple linear subspaces structure and yield superior results.

1.1   Previous Work

The common assumption of the previous dimensionality reduction by manifold learning is that data points lie on (or close to) a low dimensional manifold which is embedded in high dimensional Euclidean space. Most of manifold learning algorithms learn the low dimensional embedding by constructing a weighted graph to capture local structure of data set. Let $G(V, E, W)$ be the weighted graph, where the vertex set V corresponds to data points in the data set, the edge set E denotes neighborhood relationships between vertices, and W is a weight matrix. The options of the methods to define the weight matrix W (e.g. geodesic distance in ISOMAP, linearity in LLE, affinity in Laplacian Eigenmap and Euclidean distance in SDE) and the methods of optimization (i.e. scaling methods in Isomap, LLE, Laplacian Eigenmap and Hessian LLE, alignment methods in LSTA and manifold charting, semi-definite programming in SDE) will lead to different manifold learning algorithms [16].

Largely speaking, the measurements of neighboring relationship in the existing manifold learning algorithms can be divided into the following four ways: a) Euclidean distance, e.g. in ISOMAP [1, 12, 14], and SDE [8], b) Euclidean distance induced measurements, e.g. binary k nearest neighbors weights and heat

---

[1] The *independent linear subspaces* means $S = \bigoplus_i S_i$ and $\bigcap S_i = \{\mathbf{0}\}$ in which $\mathbf{0}$ is the zero vector, i.e. the origin of the space S.

kernel in [4] and Diffusion Map [11], c) local linear structure, i.e. in LLE [2, 3, 13], and d) sparse representation, i.e. in [17], [18].

The graph affinity matrix W encode all the subtle structure information in data set. When data vectors lie on *multiple independent subspaces*, it is crucial to construct an affinity matrix of the graph which can accurately reveal the subtle structure of multiple linear subspaces among data. However all exiting methods to define the graph mentioned above calculate the affinities between data vectors individually, there is no global constraint on the solution. As a result, they may fail to discover the subtle global structures (i.e. the multiple linear subspaces) of data and hence may not be competent for dimensionality reduction in the multiple subspace scenario. To this end, we propose to define the affinity weight by inducing from the Low-Rank Representation (LRR) [15]. There are two merits in LRE we should highlight:

- Owing to the affinity weight induced from LRR, LRE can reveal the subtle multiple subspace structure.
- In the virtual of spectral graph embedding, LRE transforms the data vectors which lie in each subtle structure (i.e. linear subspace) into each cluster in low dimensional target space.

2   Our Proposal: Low-Rank Embedding

2.1   Calculating Low-Rank Representation

Given a data set $X = [\mathbf{x}_1, \mathbf{x}_2, ..., \mathbf{x}_n]$, $\mathbf{x}_i \in R^p$ and a dictionary $A = [\mathbf{a}_1, \mathbf{a}_2, ..., \mathbf{a}_m]$, as a compressed sensing technique, LRR represents a data vector as a linear combination of the basis in a dictionary, i.e. seeks a representations Z of X by solving the following problem

$$\min_{Z} rank(Z) \quad s.t. \quad X = AZ \tag{1}$$

where $Z = [\mathbf{z}_1, \mathbf{z}_2, ..., \mathbf{z}_n]$ is the coefficient matrix with each $\mathbf{z}_i$ being the representation of $\mathbf{x}_i$. The optimal solutions Z^* of the above problem is referred as the 'lowest-rank representations' of data X with respect to the dictionary A.

Recent breakthroughs in convex optimization have shown that under fairly broad conditions, the objective function can be replaced by its convex surrogate [19], [20], [21], i.e. the $rank(Z)$ is replaced by the matrix nuclear norm $\|Z\|_*$. As a result, the problem (1) become a convex optimization problem:

$$\min_{Z} \|Z\|_* \quad s.t. \quad X = AZ \tag{2}$$

where $\|Z\|_*$ denotes the nuclear norm of matrix Z, i.e., the sum of all the singular values of matrix Z. In heavy noise corrupted case, when a fraction of data are grossly corrupted, the problem can be reformulated as follows:

$$\min_{Z} \|Z\|_* + \lambda \|E\|_l \quad s.t. \quad X = AZ + E \tag{3}$$

where the parameter $\lambda > 0$ and $\|E\|_l$ is a generic norm which depends on the noise model we assumed. For example, when the noise is 'sample specific', we may choose the $l_{2,1}$-norm, i.e. $\|E\|_{2,1} = \sum_{j=1}^{n} \sqrt{\sum_{i=1}^{p}(E_{ij})^2}$ is called as the $l_{2,1}$-norm. The problem (3) can be efficiently solved by exact or inexact Augmented Lagrange Multiplier (ALM) algorithm [22].

2.2   Inducing the Affinity Matrix and Computing the Embedding

To define the affinity matrix of the graph which encodes the pairwise affinities between data points, as in [23], [17], [18], we adopt the original data X as the initialization of the dictionary A. Let Z^* be the optimal solution of problem (3), we can induce the affinity weight for the graph. Notice that the value of $|Z_{ij}|$ indicates the membership strength between data vectors $\mathbf{x}_i$ and $\mathbf{x}_j$. If $\mathbf{x}_i$ and $\mathbf{x}_j$ lie in the same subspace, the $|Z_{ij}|$ will be of significant value; otherwise $|Z_{ij}|$ will be vanishing to zero. By taking into account the symmetry, it is reasonable to define the affinity matrix W as $W = |Z^*| + |Z^*|^T$.

The information of subtle global structure which is carried by the affinity weight matrix should be decoded into the embedding Y. Consequently, to find the low dimensional embedding $Y = (\mathbf{y}_1, ..., \mathbf{y}_n)^T$ which are consistent with the subtle latent structure, we should keep that: the data vectors which lie in the same linear subspace should be projected into a same cluster in target space and each of linear subspaces should be transformed into different cluster. In other words, if $\mathbf{x}_i$ and $\mathbf{x}_j$ lie in the same subspace, their corresponding low dimensional embeddings $\mathbf{y}_i$ and $\mathbf{y}_j$ should be near to each other. Therefore we define a cost function as follows:

$$\varepsilon(Y) = \tfrac{1}{2} \sum_{i,j=1}^{n} w_{ij} \|\mathbf{y}_i - \mathbf{y}_j\|_2^2 \tag{4}$$

where $Y = (\mathbf{y}_1, ..., \mathbf{y}_n)^T$ are the low dimensional embeddings and $\mathbf{y}_j \in R^d$. It is easy to show that the weight matrix W is symmetric. Notice that

$$\tfrac{1}{2} \sum_{i,j=1}^{n} w_{ij} \|\mathbf{y}_i - \mathbf{y}_j\|^2 = Y^T(D - W)Y = Y^T L Y \geq 0 \tag{5}$$

where D is a diagonal matrix whose diagonal elements are the row sums of the corresponding rows of W and $L = D - W$ is the graph Laplacian. Hence we can define the normalized graph Laplacian $\hat{L}$ as $\hat{L} = I - D^{-\frac{1}{2}} W D^{-\frac{1}{2}}$. To eliminate the arbitrary scaling factor in the embeddings, we added a constraint $Y^T Y = I$, in which I is the identity matrix. Consequently we obtain the following constrained minimization problem:

$$\min_Y \varepsilon(Y) = Y^T(I - D^{-\frac{1}{2}} W D^{-\frac{1}{2}})Y = Y^T \hat{L} Y, \quad s.\, t.\ Y^T Y = I. \tag{6}$$

The solution of the problem (6) can be transformed into an eigenvalue problem as $\hat{L}\mathbf{v}_j = \lambda_j \mathbf{v}_j$ where $j = 1, ..., d+1$ and d is the dimension of target Euclidean space. After sorting the eigenvalue in ascending order $0 = \lambda_1 \leq \lambda_2 \leq ... \leq \lambda_{d+1}$, we omit the minimal eigenvalue and keep the next d eigenvectors $\mathbf{v}_2, ..., \mathbf{v}_{d+1}$ which correspond to the next d minimal eigenvalues, denoted as $Y = (\mathbf{v}_2, ..., \mathbf{v}_{d+1})$.

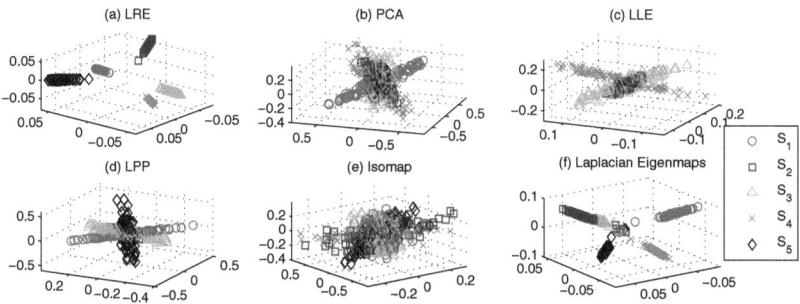

Fig. 1. Data visualization on the toy data set which consists of 5 subspaces in R^{600} without noise corruption. Each result is displayed in the best viewpoint.

Remarks

- Owing to the affinity weight induced from LRR [15] and the embedding computed by the spectral method, LRE can reveal the subtle multiple subspaces structure and transform it into multiple clusters in the low dimensional Euclidean space. In other words, the multiple linear subspaces are transformed into multiple clusters in low dimensional Euclidean space.
- LRE can be viewed as a special variant of Laplacian Eigenmaps [4] in which the affinity weight matrix is designed for revealing the subtle multiple linear subspaces structure. Laplacian Eigenmaps cannot reveal and decouple the subtle structure, whereas LRE is able to decouple themultiple linear subspaces structure into multiple clusters into low dimensional Euclidean space.

3   Experiments

To validate the effectiveness, we conduct experiments on toy data sets and the benchmark data sets, including Extended Yale B and CMU PIE.

3.1   Toy Data Set

Toy Data Set Preparation. We generate a toy data set X which consists of 5 independent linear subspaces $\{S_i\}_{i=1}^{5} \subset R^{600}$, whose bases $\{U_i\}_{i=1}^{5}$ are computed by $U_{i+1} = RU_i, i = 1, 2, 3, 4$ where R is a random rotation and U_1 is a random orthogonal matrix of dimension 600×5. Each subspace has a dimension of 5. We sample 100 data vectors from each subspace by $X_i = U_i Q_i, i = 1, 2, 3, 4, 5$ with Q_i being a 5×100 i.i.d. $\mathcal{N}(0; 1)$ matrix.

Data Visualization. At first, we compared the proposed LRE with other six well-known dimensionality reduction algorithms, including PCA, LPP, LLE, Isomap and Laplacian Eigenmaps. We set the noise level as 0% (i.e. noiseless), $\lambda = 0.10$ for LRR algorithm, $k = 30$ for the four manifold learning algorithms

LPP, LLE, Isomap and Laplacian Eigenmaps. Experimental results are given in Fig. 1. It is clear that LRE can perfectly decouple the subtle multiple subspace structures into multiple clusters, whereas other algorithms failed.

3.2   Experiments on Benchmark Data Sets

Data Set Description. The Extended Yale B data set contains 38 individuals and around 64 near frontal facial images under different illuminations per individual. The CMU pose, illumination, and expression (PIE) database contains more than 40,000 facial images of 68 people. In this experiment, we choose the images from the frontal pose (C27) and each subject has around 49 images from varying illuminations and facial expressions.

Experimental Protocol. To demonstrate the superiority of the proposed LRE, we choose the semi-supervised setting because it is a natural setting to compare different dimensionality reduction algorithms. Our basic assumptions are that: a) data vectors lie on (or near to) multiple independent linear subspaces, and b) data vectors in the same subspace are likely sharing the same label. At first we perform dimensionality reduction to the whole data matrix X to obtain a low dimensional data Y, and then conduct the Label Propagation (LP) framework [24] on Y.

Experimental Settings. In our experiments, all the faces are manually aligned, cropped and resized into 32×32 pixels, and each of them is stacked as a 1024-dimensional (column) vector. For both Extended Yale B and CMU PIE date sets, we randomly select $l(= 1, 2, 5)$ images per subject as the labeled data, and the remaining images are used as unlabeled data. For the dimensionality reduction algorithm, we compare the proposed LRE with other five well-known dimensionality reduction algorithms, including PCA, LPP, LLE, Isomap and Laplacian Eigenmaps.For the affinity weights in LP, we use the heat kernel with the self-tuning parameter [25], i.e. instead of the parameter t, we take the distance each data point to its k-th nearest neighbor as the local bandwidth parameter, in which $k = 20$. Each experiments are repeated 20 times and the averaged accuracy is reported under the optimal dimensionality of each algorithm.

Experimental Results. Experimental results on Extended Yale B and CMU PIE demonstrated the superiority of LRE over PCA, LPP, LLE, Isomap and Laplacian Eigenmaps for dimensionality reduction when data vectors lie on multiple independent linear subspaces, and are listed in Table 1. As can be seen, the proposed LRE yield overwhelmingly superior performance. This can be interpreted by the fact that both Extended Yale B and CMU PIE (C27) lie on multiple independent linear subspaces, and LRE can perfectly recover the multiple independent subspaces into multiple separate clusters such that the ordinary algorithms which used in Euclidean space can be competent. Unlike LRE, all the other algorithms, e.g. PCA, LPP, LLE, Isomap and Laplacian Eigenmaps, are based on the assumption that data vector lie on (or near to) a single subspace or

Table 1. Classification accuracy (%) on Extended Yale B with different numbers of labeled data in which $l(=1,2,5)$ indicates the number of labeled data. The integer number in bracket is the optimal dimensionality.

		Original (1024)	PCA (100)	LPP (300)	LLE (100)	Isomap (100)	Lap. Eigenmap (200)	LRE (100)
Extended Yale B	$l=1$	29.59	35.30	25.50	28.51	33.50	30.45	**62.31**
	$l=2$	41.31	45.59	39.35	42.28	40.37	47.32	**76.26**
	$l=5$	58.18	59.91	57.42	55.82	54.15	59.75	**88.06**
CMU PIE	$l=1$	30.26	33.50	34.42	35.71	34.20	39.48	**74.25**
	$l=2$	41.32	46.29	48.38	45.85	44.58	47.26	**83.62**
	$l=5$	54.15	57.18	59.30	60.27	58.94	61.34	**91.75**

manifold. There, however, in the two data sets is more subtle intrinsic geometry, *multiple independent subspaces*, which can not be recovered by PCA, LPP, LLE, Isomap and Laplacian Eigenmaps.

4   Conclusion

In this paper we proposed a nonlinear dimensionality reduction algorithm, Low-Rank Embedding, in which Low-rank Representation is used to define the affinity matrix and spectral method is adopted to compute the embedding. Experimental results demonstrated that: LRE can reveal the multiple subspace structure and transform it into multiple clusters in low dimensional Euclidean space. For the further direction, we will investigate the out-of-sample extension problem.

Acknowledgment. This work was partially supported by National Natural Science Foundation of China under Grant No.61005004 and 61175011, the 111 project under Grant No.B08004, and the Fundamental Research Funds for the Central Universities under Grant No.2012RC0108.

References

1. Tenenbaum, J.B., Silva, V., Langford, J.C.: A global geometric framework for nonlinear dimensionality reduction. Science 290, 2319–2323 (2000)
2. Roweis, S.T., Saul, L.K.: Nonlinear dimensionality reduction by locally linear embedding. Science 290, 2323–2326 (2000)
3. Saul, L.K., Roweis, S.T.: Think globally, fit locally: unsupervised learning of low dimensional manifolds. Journal of Machine Learning Research 4, 119–155 (2003)
4. Belkin, M., Niyogi, P.: Laplacian eigenmaps for dimensionality reduction and data representation. Neural Computation 15, 1373–1396 (2003)

5. Donoho, D.L., Grimes, C.: Hessian eigenmaps: Locally linear embedding techniques for high-dimensional data. Proc. Natl. Acad. Sci. USA 100, 5591–5596 (2003)
6. Brand, M.: Charting a manifold. In: NIPS, vol. 15. MIT Press, Cambridge (2003)
7. Zhang, Z., Zha, H.: Principal manifolds and nonlinear dimensionality reduction by local tangent space alignment. SIAM Journal of Scientific Computing 26, 313–338 (2004)
8. Weinberger, K., Packer, B., Saul, L.: Unsupervised learning of image manifolds by semidefinite programming. In: CVPR 2004, vol. 2, pp. 988–995 (2004)
9. Brun, A., Westin, C., Herberthson, M., Knutsson, H.: Fast manifold learning based on riemannian normal coordinates. In: Proc. 14th Scandinavian Conf. on Image Analysis, Joensuu, Finland (2005)
10. Lin, T., Zha, H.: Riemannian manifold learning. IEEE Trans. on PAMI 30, 796–809 (2008)
11. Coifman, R.R., Lafon, S., Lee, A.B., Maggioni, M., Nadler, B., Warner, F., Zucker, S.W.: Geometric diffusions as a tool for harmonic analysis and structure definition of data: Diffusion maps. Proc. of the Natl. Academy of Sciences 102, 7426–7431 (2005)
12. de Silva, V., Tenenbaum, J.B.: Global versus local methods in nonlinear dimensionality reduction. In: Advances in Neural Information Processing Systems, vol. 15, pp. 705–712 (2002)
13. Sha, F., Saul, L.K.: Analysis and extension of spectral methods for nonlinear dimensionality reduction. In: ICML 2005: Proceedings of the 22nd International Conference on Machine Learning, pp. 784–791. ACM, New York (2005)
14. Zha, H.Z., Zhang, Z.: Isometric embedding and continuum isomap. In: Proceedings of the Twentieth International Conference on Machine Learning, pp. 864–871 (2003)
15. Liu, G., Lin, Z., Yu, Y.: Robust subspace segmentation by low-rank representation. In: ICML (2010)
16. Cayton, L.: Algorithms for manifold learning. Technical report, UCSD (June 15, 2005)
17. Cheng, H., Liu, Z., Yang, J.: Sparsity induced similarity measure for label propagation. In: ICCV (2009)
18. Yan, S., Wang, H.: Semi-supervised learning by sparse representation. In: SIAM International Conference on Data Mining (2009)
19. Candès, E.J., Li, X., Ma, Y., Wright, J.: Robust principal component analysis (2009) (preprint)
20. Candès, E.J., Recht, B.: Exact matrix completion via convex optimization. Foundations of Computational Mathematics (2009)
21. Keshavan, R., Montanari, A., Oh, S.: Matrix completion from noisy entries. In: NIPS (2009)
22. Lin, Z., Chen, M., Wu, L., Ma, Y.: The augmented lagrange multiplier method for exact recovery of corrupted low-rank matrices. Technical Report UILU-ENG-09-2215, UIUC (2009)
23. Wright, J., Yang, A.Y., Ganesh, A., Sastry, S.S., Ma, Y.: Robust face recognition via sparse representation. IEEE Trans. Pattern Anal. Mach. Intell. 31, 210–227 (2009)
24. Zhu, X., Ghahramani, Z., Lafferty, J.D.: Semi-supervised learning using gaussian fields and harmonic functions. In: ICML, pp. 912–919 (2003)
25. Zelnik-Manor, L., Perona, P.: Self-tuning spectral clustering. In: NIPS (2004)

A Novel Cluster Combination Algorithm
for Document Clustering

Sen Xu, Zhenggang Wang, Xianfeng Li, and Rui Cao

Pattern Recognition and Data Mining Lab, Yancheng Institute of Technology,
Yancheng, China
xusen@ycit.cn

Abstract. Ensemble techniques have been successfully applied in the supervised machine learning area to increase the accuracy and stability of base learner. Recently, analogous techniques have been investigated in unsupervised
machine learning area. Research has showed that, by combining an ensemble of
multiple clusterings, a superior solution can be attained. In this paper, we solve
the cluster combination problem in term of finding a "best" subspace and formulate it as an optimization problem. Then, we get the solution according to
basic concept and theorem in linear algebra whereupon a novel cluster combination algorithm is proposed. We compare our algorithm with other common
cluster ensemble algorithms on real-world datasets. Experimental results demonstrate the effectiveness of our algorithm.

Keywords: clustering analysis, cluster combination, documents clustering.

1 Introduction

Clustering analysis is a useful tool and its task is to find the inherent structure in unlabeled data. The goal is to partition the data into several clusters such that the intra-
cluster similarity is maximized while the inter-cluster similarity is minimized. It has
been investigated for many years in pattern recognition and data mining areas for such
a technique is very useful. There have been many clustering algorithms proposed, but
no one is able to identify clusters with different sizes, shapes, and densities in practice
[1]. In the past, cluster analysis researchers often perform a clustering algorithm
for multiple times, and then manually select an individual solution that maximizes
(or minimizes) a user pre-defined criterion function. However, rather than merely
selecting a "best" solution, recent work has shown that combining an ensemble of
clusterings can often yield better results.

During the past decade, ensemble learning methods that train multiple learners and
then integrate their predictions to predict new instances have been a hot spot [2]. Recently, many attempts been made to apply analogous techniques to domains where
class information is unavailable. Researches indicate that the combination of multiple
clustering results could produce novel and robust clustering results [3-7]. The critical
problem in cluster ensemble is how to produce a final superior solution according to
the cluster ensemble members. There have been many approaches for solving cluster

J. Yang, F. Fang, and C. Sun (Eds.): IScIDE 2012, LNCS 7751, pp. 189–195, 2013.

ensemble problem, yet no one have both high efficiency and high quality (see next section for details).

In this paper, we solve the cluster combination problem from the view of finding a "best" subspace and formulate it as an optimization problem. Then, we get the solution according to basic concept and theorem in linear algebra whereupon a novel cluster combination algorithm is proposed. It needs only to solve a singular value decomposition problem to get the low dimensional coordinates of documents and then performs K-means algorithm to cluster documents according to their coordinates in the low dimensional space.

2   Related Research on Cluster Ensemble

Cluster ensemble usually consists of two stages. At the first, it stores the results of some independent runs of K-means algorithm or other clustering algorithms. Then, it uses the specific consensus function to find a final result from stored results.

The cluster ensemble (also called consensus clustering, clustering combination, etc.) problem is more difficult than classifier ensembles because in clustering analysis objects' labels are symbolic and thus we must also solve a label correspondence problem. Besides, the number of clusters provided by the cluster members may be different based on the clustering algorithm as well as on the particular view of the data available to that method. Furthermore, the true number of clusters is often not known in advance. In fact, the "right" number of clusters in a dataset often depends on the scale at which the data is viewed, and sometimes substantially different answers can be obtained from the same data. It has been denoted the clustering integration problem is NP-complete [3].

Many approaches have been proposed for solving the above problem. The most popular method is *pairwise* approach (also called *co-association* approach) which avoids the label correspondence problem by constructing a coincidence matrix between all pairs of instances. The matrices are then combined to form a new similarity matrix and a final clustering is achieved by applying agglomerative hierarchy clustering algorithms to produce a final clustering such as single-link (SL), complete-link (CL), average-link (AL) and ward linkage(WL) [4] or using graph partitioning algorithm such as CSPA (Cluster-based Similarity Partitioning Algorithm) [3]. Strehl and Ghosh [3] also proposed HGPA (Hypergraph Partitioning Algorithm) and MCLA (Meta-Clustering Algorithm). Recently, Nguyen & Caruana [5] presents an iterative EM-like method algorithm and its variations for finding clustering consensus. In contrast to classic, hard consensus functions that operate on labelings, Sevillano et al. [6] proposes BordaConsensus, a new consensus function for soft cluster ensembles based on the Borda voting scheme, which considers cluster membership information and is able to tackle multiclass clustering problems. Xu et al. [7] proposes a spectral clustering algorithm to solve document cluster ensemble problem.

The above methods more or less have some shortcomings, e.g. graph partitioning algorithms often impose strong constraint on the clusters' size so as to avoid trivial solution. Agglomerative hierarchy clustering algorithms explicitly or implicitly impose a structure to the final clusters, e.g. single linkage algorithm is prone to cause "chain effect", and complete linkage algorithm is prone to find clusters spherically like.

3   The Proposed Method

Let $X = \{x_1, ..., x_n\}$ be a dataset and let $P = \{P^{(1)}, ..., P^{(r)}\}$ be a set of clustering solutions on X. Let $H = H^{(1,...,r)} = (H^{(1)},...,H^{(r)})$ defines the adjacent matrix of a hypergraph with n vertices and $t = rk$ hyperedges, where r is the size of ensemble and k is the real number of classes in dataset (refer to [12] for details). Considering the hypergraph's adjacent matrix $H \in R^{n \times t}$, each row represents a data point and each column represents a hyperedge. Since K-means algorithm converges to local minima, each $H^{(i)}$ is not totally the same and thus each data point is described by t dimensionality (each dimensionality is a hyperedge). If we take these r $H^{(i)}$ as the r perturbed matrix corresponding to the r k-dimensional subspaces' orthogonal basis, then the problem comes down to how to get the "optimal" matrix which consists of k-dimensional subspaces' orthogonal basis according to these r $H^{(i)}$.

A natural idea is to find a rank-k matrix $H^* \in R^{n \times k}$ which makes the principal angle between the subspace corresponding to H^* and the r subspaces $(R^k)^{(i)}$, or the sum of cosine values of the principal angles. Note that, for unit vectors u and v, assume their starting points are origin and their finishing points are a and b, respectively, and the angle between them is Θ, then we get:

$$\cos(\Theta) = (\| u \|_2^2 + \| v \|_2^2 - d^2) / 2 \| u \| \| v \| = 1 - d^2 / 2$$

where d is the Euclidean distance between a and b.

Thus, the maximum problem of the above sum of cosine values can be formalized as minimizing SSE (Squared Sum of Euclidean distances) problem:

$$H^* = \arg \min_C \sum_{l=1}^{r} \sum_{j=1}^{k} \sum_{i=1}^{k} \left\| u_i - v_{lj} \right\|_2^2$$

where $u_1, ..., u_k$ are the standard orthogonal basis of C and $v_{l1}, ..., v_{lk}$ are the standard orthogonal basis of $(R^k)^{(i)}$.

According to linear algebra theory [8], the vector space R^{kn} is isometrically isomorphic with matrix space $R^{n \times k}$, thus the above minimizing SSE problem can be formalized as the following matrix low rank approximation problem:

$$H^* = \arg \min_{rank(C)=k} \left\| H_n - C \right\|_F^2 \tag{1}$$

where $H_n = HD^{-1/2}$, $D = diag(n_i)$, and n_is are the numbers of instances in each hyperedge.

Definition 1 Without loss of generality, suppose A is an $m \times n$ ($m \geq n$) matrix with rank(A) = p. The SVD (Singular Value Decomposition) of A can be defined as:

$$A = U \Sigma V^T$$

where $U^T U = V^T V = I_n$ and $\Sigma = diag(\sigma_1,..., \sigma_n)$, with $\sigma_i > 0$ for $1 \leq i \leq p$, and $\sigma_i = 0$ for $i \geq p+1$. The first p columns of the orthogonal matrices U and V, which is called left and right singular vectors of A, respectively, define the orthogonalized eigenvectors associated with the p nonzero eigenvalues of AA^T and A^TA, respectively. The singular values of A are defined as the diagonal elements of Σ which are the nonnegative square roots of the n eigenvalues of AA^T.

Let

$$H_k = \sum_{i=1}^{k} u_i \cdot \sigma_i \cdot v_i^T = U_k \Sigma_k V_k^T$$

where $k<p$, $U_k=[u_1...u_k]$, $\Sigma_k=\mathrm{diag}(\sigma_1, \quad ..., \quad \sigma_k)$, $Vk=[v_1...v_k]$, and $\sigma_1 \geq ... \geq \sigma_k \geq ... \geq \sigma_p > \sigma_{p+1}=...=\sigma_i=0$, according to the above definition 1 and theorem 1.2 in [9], we get:

$$\min_{\mathrm{rank}(C)=k} \left\| H_n - C \right\|_F^2 = \left\| H_n - H_k \right\|_F^2 = \sigma_{k+1}^2 + \cdots + \sigma_p^2 \tag{2}$$

Obviously, the solution of problem (1) is just H_k. We hereby get the k-dimensional embedding of both the objects and hyperedges. To obtain the final clustering results, we can simply use K-means algorithm to cluster instances according to their coordinates in the k-dimensional embedding space.

4   Experiments

In this section, we compare our proposed algorithm with other common cluster combination methods. The details are described in the following.

In our experiments, we used four different datasets, whose general characteristics are summarized in Table 1. We obtained them from different sources to ensure diversity in the datasets. For all datasets, we used a stop-list to remove common words. Moreover, any term that occurs in fewer than two documents is eliminated.

Table 1. Summary of experimental datasets

Dataset	# of documents	# of terms	# of classes
re0	1504	2886	13
re1	1657	3758	25
tr31	927	10128	7
tr41	878	7454	10

re0 and *re1*: These datasets are from Reuters-21578 text categorization test collection Distribution 1.0 [10]. We divided the labels into two sets and constructed data sets accordingly. For each data set, we selected documents having a single label. *tr31* and *tr41*: These datasets are derived from TREC-6, and TREC-7 collections [11]. The classes of these datasets correspond to the documents that were judged relevant to particular queries.

We use F1 measure -a popular measure in information retrieval communities to quantify the cluster quality. For a cluster C_k and a pre-defined class s, *recall* and *precision* are defined as:

$$recall(C_k, s) = n(C_k, s)/n_s$$

$$precision(C_k, s) = n(C_k, s)/n_k$$

where $n(C_k,s)$ is the number of documents both in C_k and s , n_k is the number of documents in C_k, and n_s is the number of documents in s.

The F1 measure of cluster C_k to class s is:

$$F1(C_k, s) = \frac{2 \times recall(C_k, s) \times precision(C_k, s)}{precision(C_k, s) + recall(C_k, s)}$$

The total F1 measure of clustering result is:

$$F1 = \sum_s \frac{n_s}{n} \max_k F1(C_k, s)$$

The greater F1 measure is the better cluster quality is.

For each dataset, we use *K*-means with cosine similarity function for fifteen times, each time with random initialization and generating a different clustering solution. We perform each cluster combination algorithm for ten times and each number reported below is obtained by averaging the corresponding F1 measures.

Fig. 1-Fig. 4 shows the F1 measures of the eight cluster ensemble algorithms over four datasets, respectively. In the figures, EASL, EACL, EAAL and EAWL represent the single linkage, complete linkage, average linkage and ward linkage algorithms proposed by Fred & Lourengo [4], respectively, and NCCA (Novel Cluster Combination Algorithm) is our proposed algorithm. A number of observations can be made by analyzing these results.

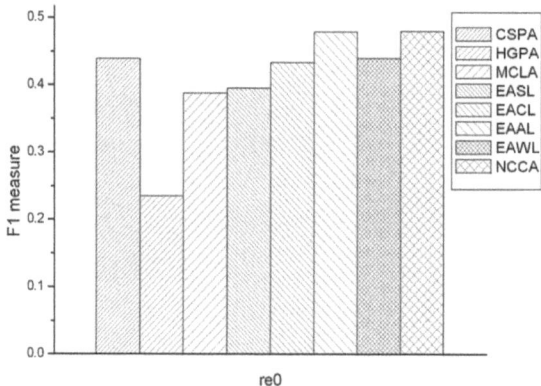

Fig. 1. F1 measures of different algorithms over *re0* dataset

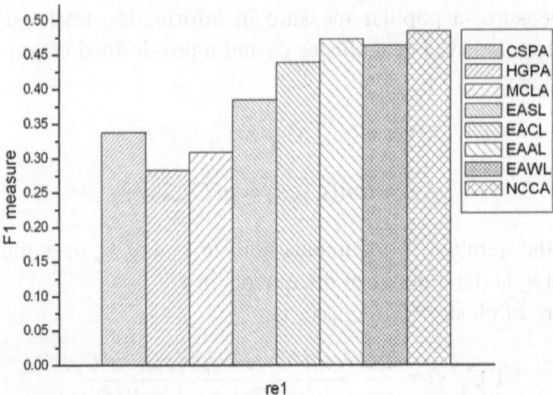

Fig. 2. F1 measures of different algorithms over *re1* dataset

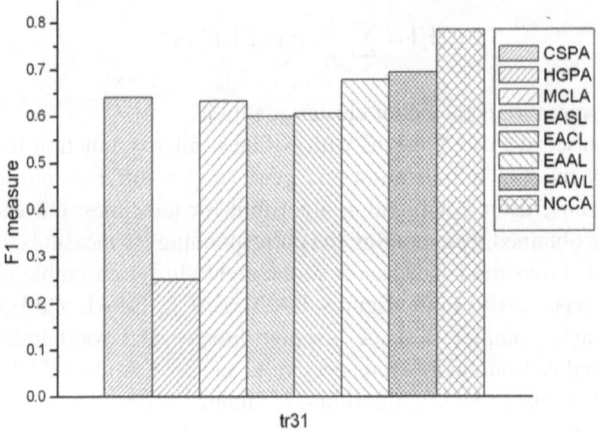

Fig. 3. F1 measures of different algorithms over *tr31* dataset

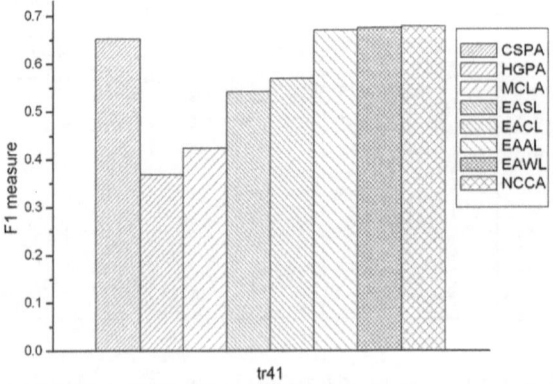

Fig. 4. F1 measures of different algorithms over *tr41* dataset

From the above figures, we can see: a) among graph partitioning algorithms, CSPA outperforms the other two algorithms, i.e., HGPA and MCLA, and this conforms to the result obtained in [3]; b) compare the four agglomerative hierarchy clustering algorithms, EAAL and EAWL is obviously superior to EASL and EACL and this accords with the result reported in [4]; c) our NCCA algorithm get the highest F1 measures over all the datasets.

From the above observations, we conclude that our NCCA outperforms the other cluster ensemble algorithms in terms of F1 measure.

5   Conclusions

In this paper, we proposed a novel cluster combination algorithm which only needs to solve a singular value decomposition problem and then performs K-means algorithm to cluster documents according to their coordinates in the low dimensional space. Experimental results on real-world document datasets indicate that our proposed algorithm outperforms other common cluster combination methods.

Acknowledgment. This research is financially supported by National Natural Science Foundation of China under Grant No. 60975042, 41006057, and 61105057, and the Talent Introduction Project of Yancheng Institute of Technology under Grant No. XKR2011019. All the persons involved in the research projects are thanked for their help.

References

1. Tan, P.N., Steinbach, M., Kumar, V.: Introduction to Data Mining. Addison-Wesley Longman Publishing, Boston (2010)
2. Maclin, R., Opitz, D.: Popular ensemble methods: an empirical study. Journal of Artificial Intelligence Research 11(8), 169–198 (1999)
3. Strehl, A., Ghosh, J.: Cluster ensembles - a knowledge reuse framework for combining partitionings. In: Eighteenth National Conference on Artificial Intelligence, pp. 93–98 (2003)
4. Fred, A., Lourengo, A.: Cluster Ensemble Methods: from Single Clusterings to Combined Solutions. In: Okun, O., Valentini, G. (eds.) Supervised and Unsupervised Ensemble Methods and their Applications. SCI, vol. 126, pp. 3–30. Springer, Heidelberg (2008)
5. Nguyen, N., Caruana, R.: Consensus clusterings. In: Proc. of the 7th IEEE ICDM, pp. 607–612 (2007)
6. Sevillano, X., Alfas, F., Socoró J, C.: BordaConsensus: a new consensus function for soft cluster ensembles. In: Proc. of the 30th Annual International, ACM SIGIR, pp. 743–744 (2007)
7. Xu, S., Lu, Z.M., Gu, G.C.: An efficient spectral method for document cluster ensemble. In: The 9th Intl. Conf. Young Computer Sci., pp. 808–813 (2008)
8. Dattorro, J.: Convex Optimization & Euclidean Distance Geometry. Meboo Publishing, USA (2005)
9. Berry, M.W.: Large-scale sparse singular value computations. The International Journal of Supercomputer Applications 6(1), 13–49 (1992)
10. http://www.research.att.com/~lewis
11. http://trec.nist.gov

Reliable Detection of Malignant Ventricular Arrhythmias Based on Complex Network Theory

Dong Yang and Xiang Li*

The Adaptive Networks and Control Laboratory, Electronic Engineering Department,
Fudan University, Shanghai 200433, China
{09210720043,lix}@fudan.edu.cn

Abstract. This paper presents a frequency-degree mapping algorithm which enables network analysis of human electrocardiogram (ECG) time series. Two important topological quantities, the average degree (AD) and the average shortest path length (APL) have been investigated in the associated networks of ECG time series. The results demonstrate that the quantity of AD can serve as an effective and reliable indicator in distinguishing malignant ventricular arrhythmias from normal sinus rhythm and other benignant arrhythmias. Meanwhile, the quantity of APL is shown to be capable of characterizing the heart rate, which may be helpful in detecting shockable rhythms.

Keywords: ventricular arrhythmias, detection, network analysis, electrocardiogram, time series.

1 Introduction

The extensive researches and fruitful achievements of complex network theory [1]-[6] over the last decade have provided us an effective tool for modeling various complex systems, which improve our understanding of the interplay between the structure properties and dynamics of complex networks. These dramatic advances of network theory recently motivate the attempts [7]-[15] to transform a time series to a network, which build bridges between time series analysis, nonlinear dynamics, and complex network theory. Although these methods are in their infancy, they have attracted considerable attentions and have been applied in several fields ranging from geophysics [16] to finance [17] and physiology [7]-[8], [13], [15], [18]-[20]. For example, the topological analysis of associated networks of human ECG time series, heart rate records and cardiovascular variability data have been applied to help identify arrythmia[7]-[8], [20], congestive heart failure [18]-[19] and to predict pre-eclampsia [13], respectively.

Ventricular fibrillation (VF) and ventricular tachycardia (VT) as the most serious cardiac arrhythmias deserve accurate and reliable detection to save lives.

---

* This work was partly supported by the NCET program (No. NCET-09-0317) of China.

J. Yang, F. Fang, and C. Sun (Eds.): IScIDE 2012, LNCS 7751, pp. 196–205, 2013.

The past decades have witnessed intensive efforts devoted to the detection of these arrhythmias [15], [21]-[25]. According to the involved types of cardiac rhythm, these techniques can be classified into two categories: One is designed to tackle certain kinds of cardiac rhythm, such as to separate VF from VT [21], or VF plus MVT from NSR [15], [22]. Though these algorithms perform well in the special case, they may fail to work in a large database. Because there are many other types of cardiac rhythm also present in real life problem. The other category is more practical for clinical applications, in which the techniques are designed to distinguish VF from non-VF [23] or shockable rhythms from other arrhythmias [24]-[25].

In previous work [15], we have proposed a novel method named as frequency-degree mapping algorithm which conserves both amplitude information and temporal information of the original time series. We have applied the method to a special case of arrythmia detection and found it is able to distinguish ventricular fibrillation (VF) and monomorphic ventricular tachycardia (MVT) from normal sinus rhythm (NSR) with high accuracy. In this paper, we aim to inquire the performance of the method in a whole electrocardiogram (ECG) database which contains various cardiac rhythms.

The basic principles of the frequency-degree mapping algorithm are introduced in Section 2. Then after mapping the ECG time series into networks, two important topological quantities, the average degree (AD) and the average shortest path length (APL), are investigated in details in Section 3. We find that the method can distinguish malignant ventricular arrhythmias, namely VF, VT and ventricular flutter (VFL), from normal sinus rhythm and other benignant arrhythmias using merely the quantity of AD. The impacts of threshold and algorithm parameters are also studied to verify the reliability of the method. Furthermore, the quantity of APL is found to be able to characterize the heart rate to some extent. Finally, the whole paper is concluded in Section 4.

2   Frequency-Degree Mapping Algorithm

To transform a time series into a network, we need to clearly define the nodes and the edges according to certain characteristics of original time series. As we know, temporal order and recurrence pattern are two crucial properties in the context of time series analysis. In the frequency-degree mapping algorithm, we aim to conserve the properties by encoding them into the edges of associated network. But before that, the nodes require to be defined at first. In the present algorithm, every node corresponds to one data point in the same time order, which means an one-to-one mapping from the data point to the associated node. Then we define the edges by two means, the first is temporal edge derived from temporal order, the other is recurrence edge obtained from recurrence pattern of original time series. In more details, temporal edge means every node is connected to its first left and right neighbors in temporal order.

The definition of recurrence edge deserves more explanation as the key point of the algorithm. Here the word "recurrence" means that two arbitrary values of the time series are approximately equal to each other. We preprocess the original time series by means of amplitude quantization, in which all values distributed in the same quantization interval would be assigned to the same integer and considered as recurrence. The more values distributed in a quantization interval, the high recurrence frequencies are obtained. The philosophy of the frequency-degree mapping algorithm is that the data with high recurrence frequencies would correspond to the nodes with large degree values. In other words, the recurrence nodes would be fully connected to form a clique in the associated network.

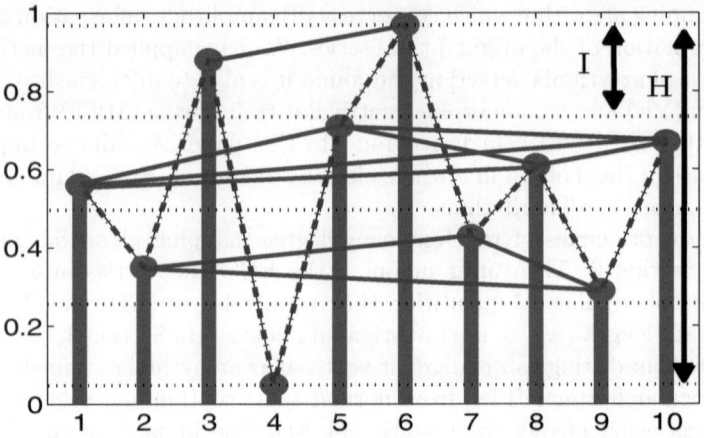

Fig. 1. An example of mapping a time series into a network with the frequency-degree mapping algorithm, wherein the quantization level $Q = 4$, the time series length $L = 10$. Every node corresponds to a data point in the same temporal order, which show as the ball on the top of vertical bar. The dash dot line represents temporal edge while the solid line represents recurrence edge.

For illustrative purposes, we present in Fig. 1 a scheme of the frequency-degree mapping algorithm. The single variable parameter in the present algorithm is the quantization level Q. The equation $I = H / Q$ always holds true in the process of amplitude quantization, in which H represents the amplitude range of original time series, I refers to the quantization interval. As shown in Fig. 1, the data values distribute in the same quantization interval are considered as recurrence, and these associated recurrence nodes are connected to each other. For instance, four recurrence nodes (node 1, node 5, node 8 and node 10) form a clique in the network, each node have three recurrence edges. Note that each node have two temporal edges, the degree value of a node (denoted as D) depend on its recurrence frequency (denoted as F), which can be expressed by the equation $D = (F - 1) + 2 = F + 1$. This is the reason that the method are named as the frequency-degree mapping algorithm.

3   Detection of Malignant Ventricular Arrhythmias

The database used in this paper is the MIT-BIH Malignant Ventricular Ectopy Database (VFDB) which is available online [26]. The VFDB contains 22 subjects, 2 channels per subject with each channel 2100 seconds long, and involves the following types of cardiac rhythm: atrial fibrillation (AFIB), ventricular bigeminy (B), first degree heart block (BI), high grade ventricular ectopic activity (HGEA), normal sinus rhythm (NSR), AV junctional rhythm (NOD), paced rhythm (PM), sinus bradycardia (SBR), supraventricular tachyarrhythmia (SVTA), ventricular escape rhythm (VER), ventricular fibrillation (VF), ventricular flutter (VFL) and ventricular tachycardia (VT). All signals in the VFDB are sampled at 250 Hz.

For a start, we export the ECG time series from the whole database as episodes of 10 seconds long. Note that noise signal and asystole signal are not included in these rhythms, we obtain 475 episodes of VT, 219 episodes of VF, 20 episodes of VFL, 1319 episodes of NSR, 304 episodes of AFIB, 141 episodes of SVTA, 340 episodes of BI, 43 episodes of B, 144 episodes of PM, 161 episodes of NOD and 71 episodes of HGEA. All these ECG episodes are preprocessed using a well-known filtering procedure [27] to suppress high frequency noise and baseline drift. After mapping each episode into an associated network with the frequency-degree mapping algorithm, we investigate the topological properties of these networks.

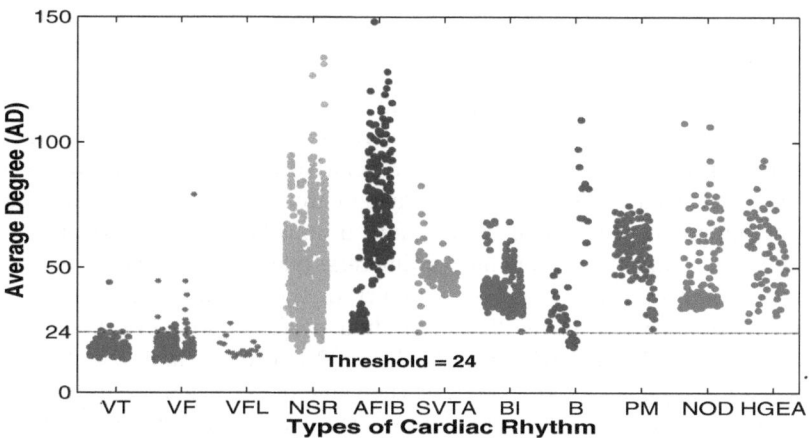

Fig. 2. The average degree values induced from various cardiac rhythms with $Q = 256$, $L = 10$ seconds

3.1   Average Degree

One of the basic measurements of a network is the average degree (AD). In an undirected and unweighted network, the degree of a node is the number of edges connected to the node [4]. The average degree values calculated from the associated networks of VFDB ECG episodes with $L = 10$ seconds and $Q = 256$ have been illustrated in Fig. 2, which manifests that the average degree values of malignant ventricular arrhythmias (VF, VT, VFL) are significantly smaller than those of other cardiac rhythms (NSR, AFIB, SVTA, BI, B, PM, NOD, HGEA). Thus we are able to detect malignant ventricular arrhythmias by selecting a proper threshold of the average degree, for instance, AD = 24 as plotted in Fig. 2. The results of detecting malignant ventricular arrhythmias can be seen in Table 1, in which the total accuracy (AC) and positive predictivities (PP) are all exceed 95 percentage.

Table 1. Accuracy (AC) and positive predictivity (PP) of malignant ventricular arrhythmias detection, $Q = 256$, $L = 10$ seconds, Threshold AD = 24

Type	VT	VF	VFL	NSR	AFIB	SVTA	BI	B	PM	NOD	HGEA
All	475	219	20	1319	304	141	340	43	144	161	71
True	452	214	19	1260	304	141	340	34	144	161	71
False	23	5	1	59	0	0	0	9	0	0	0
PP	95.2	97.7	95.0	95.5	100	100	100	79.1	100	100	100
and	95.9			97.3							
AC	97.0										

Note that the threshold choice is vital to detection effectiveness, we also evaluate the impact of threshold choice on the performance of the method. The quality parameters of sensitivity, specificity and accuracy, are calculated at different threshold, AD = 22, 23, 24, 25 and 26, respectively. As illustrated in Fig. 3, the specificity get better as the threshold increase, whereas the sensitivity decline as the cost. However, the total accuracy is relatively stable over the variance of threshold, which indicates that the average degree is an effective indicator for detecting malignant ventricular arrhythmias from other cardiac rhythms.

We further change the algorithm parameters, quantization level Q as well as time series length L, to verify the reliability of the frequency-degree mapping algorithm. As shown in Table 2, the method performs well in different cases of Q = 64, 128, 256, 512 with the fixed $L = 10$ seconds, and $L = 8, 6, 4, 2$ seconds with fixing Q as 256. Note that the value of detection accuracy exceeds 96 percentage in every case, which manifests that the frequency-degree mapping algorithm is a reliable method for detecting malignant ventricular arrhythmias.

The underlining reason for the reliability of the algorithm is that the associated network is capable of characterizing the amplitude distribution of ECG episodes. As we know, the width of the QRS complex is different for various cardiac rhythms which lead to different amplitude distributions. For normal sinus rhythm and other benignant arrhythmias, the majority data values are

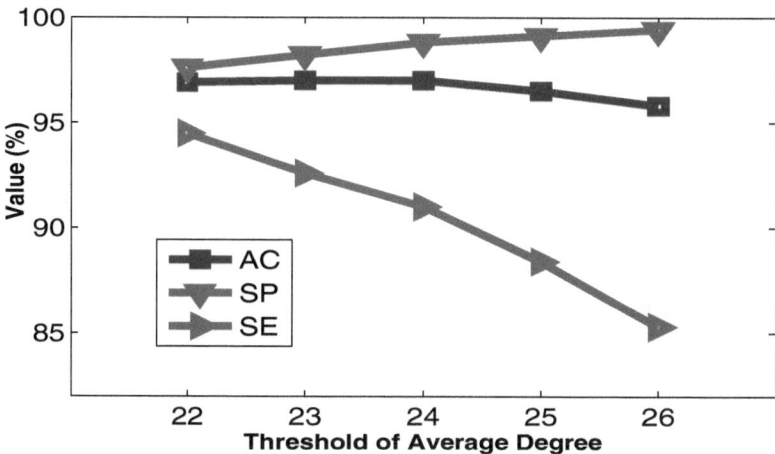

Fig. 3. The reliability measures of specificity (SP), sensitivity (SE) and accuracy (AC) of detecting malignant ventricular arrhythmias at different threshold, AD = 22, 23, 24, 25, 26

distributed around the baseline which contribute a high recurrence frequency. These recurrence nodes are fully connected to each other leading to a high value of the average degree. On the contrary, the amplitude distributions of malignant ventricular arrhythmias are more decentralized than those of other rhythms. For this reason, few recurrence nodes appear in the associated networks.

Table 2. The reliability measures of specificity, sensitivity and accuracy of detecting malignant ventricular arrhythmias with $Q = 64, 128, 256, 512$ and $L = 10, 8, 6, 4, 2$ seconds

Parameters	$L = 10$ seconds				$Q = 256$			
L and Q	$Q{=}64$	$Q{=}128$	$Q{=}256$	$Q{=}512$	$L{=}8$	$L{=}6$	$L{=}4$	$L{=}2$
Specificity	98.9	98.8	98.8	98.8	98.9	98.3	98.8	98.6
Sensitivity	90.8	91.1	91.0	90.7	89.4	92.2	88.6	88.3
Accuracy	97.0	97.0	97.0	96.9	96.8	97.0	96.4	96.3

3.2   Average Shortest Path Length

The average shortest path length (APL) is another important topological quantity of a network which measures the typical separation between two nodes in the network [4]. As illustrated in Fig. 4, the quantity of APL fails to distinguish malignant ventricular arrhythmias from some benignant arrhythmias such as BI and B. However, we find a remarkable feature of APL that it can characterize

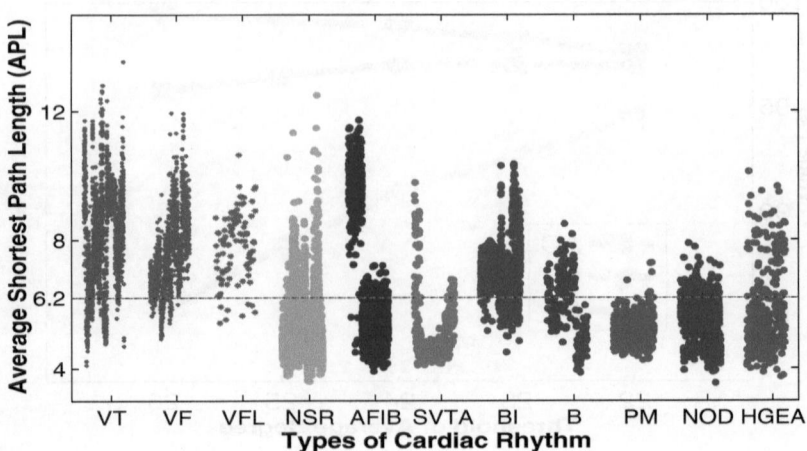

Fig. 4. The average shortest path length values induced from various cardiac rhythms with $Q =64$, $L = 2$ seconds

the heart rate in some way. Taking the case of ventricular tachycardia (VT) as an example, the waveforms of a rapid VT and a slow VT are plotted in Fig. 5. We applied the method to these two VT segments with $Q = 64$, $L = 2$ seconds. The values of AD, APL induced from associated networks has shown in Table 3. We clearly observe that the values of APL induced from a slow VT are significantly greater than those of a rapid VT while the values of AD show no evident difference.

Table 3. The values of AD and APL induced from the episodes of rapid VT and slow VT, wherein $L = 2$ seconds and $Q = 64$

ECG Episodes		0s-2s	2s-4s	4s-6s	6s-8s	8s-10s
Rapid	AD	12.02	14.00	12.12	12.29	13.54
VT	APL	5.39	5.33	5.31	5.19	5.16
Slow	AD	12.92	11.44	11.20	10.32	10.82
VT	APL	9.16	10.74	10.61	10.28	10.30

This phenomenon originates from the fact that the quantity of APL can conserve temporal information of original time series. As previously mentioned, the APL measures the separation among the nodes of a network, the more separated, the bigger value. Similarly, the transition in the data values also indicates the interaction of different data points. The VT with a high heart rate means the transition in values is more quick than the VT with a low heart rate. For this

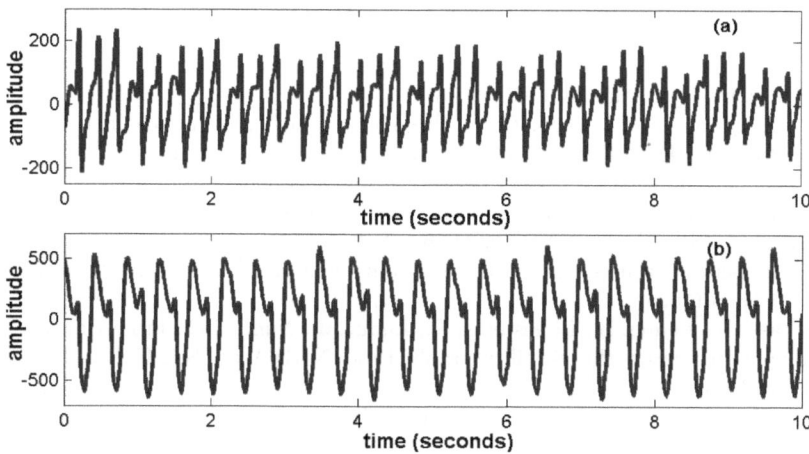

Fig. 5. Two typical waveforms of ventricular tachycardia (VT), (a) VT with a high heart rate, (b) VT with a low heart rate

reason, the quantity of APL is able to characterize the heart rate to some extent. Since VT with a rate exceed 180 bpm (beats per minute) are classified into shockable rhythms according to the AHA recommendations [28], we may use the quantity of APL to identify shockable rhythms within the scope of malignant ventricular rhythms in further research.

4   Conclusion

In summary, we have applied the frequency-degree mapping algorithm to electrocardiogram (ECG) time series in a large database. The associated networks have been in-depth studied with two important topological measurements, the average degree (AD) and the average shortest path length (APL). We found that the quantity of AD can serve as an effective and reliable indicator in distinguishing malignant ventricular arrhythmias from other cardiac rhythms. Meanwhile, the quantity of APL is capable of characterizing the heart rate to some extent, which can be used to identify shockable rhythms within the scope of malignant ventricular arrhythmias in future work.

References

1. Watts, D.J., Strogatz, S.H.: Collective dynamics of 'small world' networks. Nature 393, 440–442 (1998)
2. Barabási, A., Albert, R.: Emergence of Scaling in Random Networks. Science 286, 509–512 (1999)
3. Newman, M.E.J.: The structure and function of complex networks. SIAM Rev. 45, 167–256 (2003)

4. Boccaletti, S., Latora, V., Moreno, Y., Chavez, M., Hwang, D.U.: Complex networks: structure and dynamics. Phys. Rep. 426, 175–308 (2006)
5. Wang, X.F., Chen, G.Y.: Complex Networks: Small-World, Scale-Free and Beyond. IEEE Circ. Syst. Mag. 3, 6–20 (2003)
6. Li, X., Wang, X.F., Chen, G.Y.: Pinning a complex dynamical network to its equilibrium. IEEE Trans. Circuits Syst. I. 51, 2074–2087 (2004)
7. Zhang, J., Small, M.: Complex network from pseudoperiodic time series: topology versus dynamics. Phys. Rev. Lett. 96, 238701 (2006)
8. Zhang, J., Sun, J., Luo, X., Zhang, K., Nakamura, T., Small, M.: Characterizing pseudoperiodic time series through the complex network approach. Physica D 237, 2856–2865 (2008)
9. Lacasa, L., Luque, B., Ballesteros, F., Luque, J., Nuno, J.C.: From time series to complex networks: the visibility graph. Proc. Natl. Acad. Sci. U.S.A. 105, 4972–4975 (2008)
10. Xu, X.K., Zhang, J., Small, M.: Superfamily phenomena and motifs of networks induced from time series. Proc. Natl. Acad. Sci. U.S.A. 105, 19601–19605 (2008)
11. Marwan, N., Donges, J.F., Zou, Y., Donner, R.V., Kurths, J.: Complex network approach for recurrence analysis of time series. Phys. Lett. A. 373, 4246–4254 (2009)
12. Donner, R.V., Zou, Y., Donges, J.F., Marwan, N., Kurths, J.: Recurrence networks - a novel paradigm for nonlinear time series analysis. New J. Phys. 12, 33025 (2010)
13. Marwan, N., Wessel, N., Stepan, H., Kurths, J.: Recurrence based complex network analysis of cardiovascular variability data to predict pre-eclampsia. In: Proc. Biosignal, Berlin, Germany (July 2010)
14. Donner, R.V., Small, M., Donges, J.F., Marwan, M., Zou, Y., Xiang, R.X., Kurths, M.: Recurrence based time series analysis by means of complex network methods. Int. J. Bifurcat. Chaos 21, 1019–1046 (2011)
15. Yang, D., Li, X.: Bridge time series and complex networks with a frequency-degree mapping algorithm. In: IEEE International Symposium on Circuits and Systems, Seoul, Korea, pp. 910–913 (May 2012)
16. Elsner, J., Jagger, T.H., Fogarty, E.A.: Visibility network of United States hurricanes. Geophys. Res. Lett. 36, L16702 (2009)
17. Yang, Y., Wang, J., Yang, H., Ming, J.: Visibility graph approach to exchange rate series. Physica A 388, 44431 (2009)
18. Shao, Z.G.: Network analysis of human heartbeat dynamics. Appl. Phys. Lett. 96, 073703 (2010)
19. Dong, Z., Li, X.: Comment on "Network analysis of human heartbeat dynamics. Appl. Phys. Lett. 96, 266101 (2010)
20. Li, X., Dong, Z.: Detection and prediction of the onset of human ventricular fibrillation: An approach based on complex network theory. Phyc. Rev. E. 84, 062901 (2011)
21. Thakor, N.V., Zhu, Y.S., Pan, K.Y.: Ventricular Tachycaridia and Fibrillation Detection by a Sequential Hypothesis Testing Algorithm. IEEE Trans. BioMed. Eng. 37, 837–843 (1990)
22. Zhang, X.S., Zhu, Y.S., Thakor, N.V., Wang, Z.Z.: Detecting Ventricular Tachycardia and Fibrillation by Complexity Measure. IEEE Trans. BioMed. Eng. 46, 548–555 (1999)
23. Amann, A., Tratnig, R., Unterkofler, K.: Detecting Ventricular Fibrillation by Time-Delay Methods. IEEE Trans. Biomed. Eng. 54, 174–177 (2007)

24. Didon, J., Dotsinsky, I., Jekova, I., Krasteva, V.: Detection of Shockable and Non-Shockable Rhythms in Presence of CPR Artifacts by Time-Frequency ECG Analysis. Computers in Cardiology 36, 629–634 (2009)
25. Anas, E.M., Lee, S.Y., Hasan, M.K.: Sequential algorithm for life threatening cardiac pathologies detection based on mean signal strength and EMD functions. BioMedical Engineering (2010),
    http://www.biomedical-engineering-online.com/content/9/1/43
26. http://www.physionet.org/cgi-bin/atm/ATM.
27. http://homepages.fhv.at/ku/karl/VF/filtering.m.
28. Kerber, R., Becker, L., Bourland, J.: Automatic external defibrillators for public access defibrillation: Recommendations for specifying and reporting arrhythmia analysis, algorithm, performance, incorporating new waveforms and enhancing safety. Circulation 95, 1677–1682 (1997)

Face Registration: Evaluating Generative Models for Automatic Dense Landmarking of the Face

Karl Ricanek[1], Amrutha Sethuram[1], and Wankou Yang[2]

[1] ISIS Institute - Face Aging Group, UNC Wilmington, Wilmington, NC 28403, USA
{ricanekk,sethurama}@uncw.edu
[2] School of Automation, Southeast University, Nanjing 210096, China
wkyang@seu.edu.cn

Abstract. In this work we evaluate three generative techniques for automatic registration of more than 250 face landmarks (annotations). We compare/contrast these techniques based on developing general and a ethnic and gender specific models to detemine whether the specific, ethnic-gender, models can outperform the general model in accurately locating the dense landmarks. Further, we determine which of the three genrative tehcniques are more robust. The three techniques evaluted are the Active Shape Models (ASM), the Active Appearance Model (AAM), and the Constrained Local Model (CLM). In addition this work provides an understanding of the types of landmarks that each technique performs well on and the landmarks that the techniques perform poorly on. Further, it is shown that the performance of STASM and CLM are comparable and better than AAM and that specific models perform better than the general models.

1 Introduction

Automatic facial landmarking is a very important step that precedes any task involving face recognition or analysis. These landmarks, also referred to as iducial points or anchor points, are used for accurate registration of faces and have a significant effect on the impending analysis. While some applications, such as, face recognition and tracking use a few landmarks like the eye and eyebrow corners, centers of the iris, corners of the mouth, tip of the nose and chin for registration, other applications like age estimation, expression analysis, detection of intent or facial aging require a greater number of landmarks for analysis. Further, it is important that the detection of these landmarks be accurate and robust to environmental variables e.g. illumination, occlusion, expression, pose etc. It has been shown in [1] that precise landmarks are essential for face-recognition performance and in [2] that more landmarks results in higher recognition performance. However, under various conditions of image acquisition, automatic facial landmarking becomes a challenging task.

Performance on face landmarks are typically reported on a few well-defined internal to the face landmarks. An exhaustive analysis of the facial landmarking on face-boundary annotations and soft-tissue landmarks–driven by lines and folds from expression or age, has not been performed to date. However, in this paper, we consider an extremely dense landmarking scheme consisting of 252 points on the face that includes internal, boundary, and soft-tissue points. Such a dense scheme is helpful in

J. Yang, F. Fang, and C. Sun (Eds.): IScIDE 2012, LNCS 7751, pp. 206–215, 2013.
© Springer-Verlag Berlin Heidelberg 2013

applications in soft biometrics such as expression recognition, detection of micro gestures, age estimation and also for facial aging.

Although many algorithms exist that attempt to address the problem of automatic face registration via landmarking, there has not be a commonly adopted protocol for testing, evaluating and comparison of algorithms. The training and testing methods, the number of landmarks used, and strategies for reporting performance results vary widely in the literature. The contribution of this paper is two fold: 1) Introducing a dense landmarking scheme for the face and 2) Comparing the performance of three state-of-the art automatic landmarking methods using the dense scheme.

The remainder of this paper is organized as follows: A background on existing automatic landmarking methods are presented in the section 2. In section 3, we present the automatic landmarking methods that are considered for this work. Experiments and Results are presented in section 4. Conclusion and future work is discussed in section 5.

2 Background

Many algorithms have been proposed for automatic facial landmarking. These approaches can be broadly classified into two categories: image-based methods and structure-based methods. In image-based methods, faces are treated as vectors in high dimensional space, which are then modeled as a manifold. The variability in facial features is captured through popular transformations like the Principal Components Analysis, Independent Components Analysis, Gabor Wavelets, Discrete Cosine Transforms and Gaussian derivative filters. The appearance of each landmark is then learnt through the use of machine learning approaches like support vector machines, boosted cascade detectors and multilayer perceptrons. In [3], Viola and Jones describe a visual object detection framework to achieve high detection rates, Vukadinovic *et al.* [4] propose a method to automatically detect twenty facial feature points using Gabor feature based boosted classifiers. In [5], Valstar *et al.* use support vector regression and markov random fields to increase the efficiency and robustness of their algorithm. In [6], Efraty *et al.* present a system based on multi-resolution isotropic analysis and adaptive bag-of-words descriptors incorporated into a cascade of boosted classifiers. Recently, Dibekli-oglu *et al.* [7] proposed a generic statistical 2-D landmarking method and have assessed their methods under different performance criteria.

Structure-based methods use prior knowledge about facial landmark positions, and constrain the landmark search using heuristic rules that involve angles, distances and areas. Very popular methods in this category are Active Shape Models (ASM) [8], Active Appearance Models (AAM) [9] and Elastic Bunch Graphing Methods. ASMs model textures of small neighborhoods around landmarks and iteratively minimizes the differences between landmark points and their corresponding models. The AAM typically looks at the convex hull of landmarks, synthesizes a facial image from a joint appearance and shape model, and seeks to minimize similarity to the target face iteratively. Cristinacce *et al.* [10] proposed the Constrained Local Model (CLM) approach that uses a set of local feature templates for detection of landmarks.

3   Automatic Landmarking Methods

3.1   Landmarks and Dense Annotation Map

A landmark or an annotation can be defined as a point of correspondence on each object that matches between and within populations. Dryden et. al. [11] further discriminate landmarks into three subgroups: (1) Anatomical landmarks: Points assigned by an expert that correspond between organisms in some biologically meaningful way (2) Mathematical landmarks: Points located on an object according to some mathematical or geometrical property, i.e. high curvature or an extremum point (3) Pseudo landmarks: Constructed points on an object either around the outline or between landmarks.

Although the concept of landmarks is conceptually useful, the process of acquisition can be cumbersome. In addition to placing annotations manually, the process usually involves comparing annotations to ensure correspondence across the training set. Thus this work is extremely beneficial in comparing the capabilities of existing automatic landmarking methods. It is generally believed that increasing the number of landmarks in the model improves the mean fit of the model [12] [13]. As mentioned before, in this work, we consider an extremely dense annotation map to compare the landmarking methods as shown in Figure 1. Such a scheme was devised based on the literature found on the dynamics of craniofacial aging [14] and in consultation with medical professionals with a primary interest in facial morphology. The scheme includes anthropometric landmarks of the soft tissue and the skull with additional points selected based upon aging trends of the face–termed soft-tissue points above. This work represents the first attempt at developing a rich and diverse set of landmark registration points for the face. This can be attributed partially to the fact that obtaining ground truth information for such a dense scheme is challenging both in terms of time and resources.

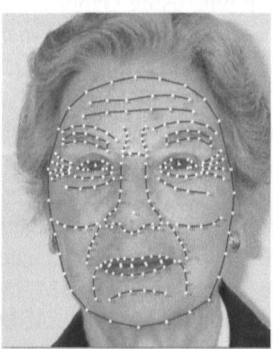   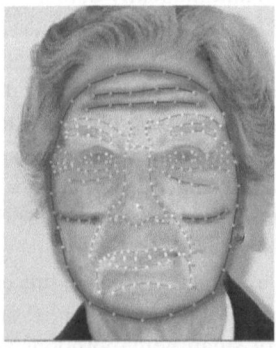

Fig. 1. Example of dense annotation map   **Fig. 2.** Component scheme adopted for landmarking. Each expert was trained on annotating a color coded feature.

The following three generative methods for registration of a dense set of craniofacial landmarks are considered for this work.

Active Appearance Models. Active appearance models (AAM), a group of flexible deformable models, have been widely used for automatic landmarking. First proposed by Cootes et. al [9], AAM decouples and models shape and pixel intensities of an object. The latter is usually referred to as texture. Accurate identification of a set of landmarks is a critical step in building an AAM model. As described in [9], the AAM model can be generated in three main steps: (1) A statistical shape model is constructed to model the shape variations of an object using a set of annotated training images. (2) A texture model is then built to model the texture variations, which is represented by intensities of the pixels. (3) A final appearance model is then built by combining the shape and the texture models. The AAM software used for this work was obtained from [15].

Constrained Local Models. Constrained Local Models are an off shoot of Active Shape Models (ASM). These are methods that make independent predictions regarding locations of the model's landmarks, which are combined by enforcing *a prior* over their joint motion. Most CLM variants implement a two step fitting strategy, where an exhaustive local search is first performed to obtain a response map for each landmark. Optimization is then performed using sophisticated strategies to maximize the responses of the landmarks. In this paper, one such strategy that uses a *Gaussian prior* over the model's PCA parameters is implemented. The algorithm is itself presented by Saragih *et al.* in [16].

STASM. STASM or Stacked Active Shape Model [13] is an extension of ASM proposed by Cootes [8]. As descibed in [13], STASM extends the active shape model by fitting more landmarks than actually needed, by selectively using two-instead of one-dimensional landmark templates and stacking two active shape models in series. The C++ software library to train and test the models for this work was obtained from [17].

4   Experiments and Results

4.1   Training

Databases Used. The automatic landmarking algorithms were trained using images obtained from MORPH [18] (a publicly available mugshot database), the PAL database [19] and the Pinellas County database (mugshot database with limited distribution). In addition, the ethno-gender groups: African American Male (AAM), African American Female (AAF), Caucasian American Male (CAM) and Caucasian American Females (CAF), were formulated with distribution of images over age ranges as shown in Table 1 and Table 2.

Obtaining Ground-Truth Data. Obtaining ground truth landmarks for face or any imagery is not only tedious–the points landmarks are manually selected one at a time–but, also a very time consuming task for all but the most trivial, in regards to the quantity of marks. The amount of time that is spent on training individuals to get the task done consistently, is non-trivial. In addition, one most account for interobserver variability in obtaining consistent marks. This is often achieved by requiring a minimum of

Table 1. Training Data: Distribution based on ethno-gender groups and age ranges

Age Range	AAM	AAF	CAM	CAF
18-30	50	50	50	50
31-40	50	50	50	50
41-50	50	50	50	50
51-60	50	50	49	50
61-70	50	50	48	50
71+	38	21	49	50

Table 2. Training Data: Distribution based on databases

Database	AAM	AAF	CAM	CAF
Pinellas	88	62	104	180
MORPH	200	199	149	67
PAL	0	10	43	53

three trained landmarkers to work on a single face image. The position of each mark is averaged over the trained annotators[20] [21]. Landmarks must be gathered for the training and testing imagery. Thus it has become customary for researchers to report the performance of their algorithms on databases that have ground truth as part of their distribution. However, since this work is based on a dense 252 landmark scheme for the face, which is not available in any database, it was necessary that we generate the ground truth in-house. Given the large number of training and testing data and the density of the annotation scheme itself (252 points on each image), it was not practical, both in terms of time and resources, to obtain repeat measurements of the ground truth data by 4 or 5 well-trained annotators. Instead, a component scheme was developed, in which, experts were trained to annotate a specific region or component of the face. Each expert annotated a specific feature/component of the face as color coded in Figure 2 across all the images in the training and testing data. This ensured that variations in annotating features of the face were kept at a minimum.

Training Methodology. For purposes of evaluation and comparison, each of the automatic landmarking methods were trained on the entire set of 1155 images, which will be henceforth termed as *general* model. Also, the methods were trained on images belonging to a specific ethno-gender group which will be henceforth termed as *ethno-gender* model or *hierarchical* model.

4.2 Testing

As already mentioned, two specific ethno-gender groups i.e. African American Females (AAF) and Caucasian American Males (CAM) were considered as test groups to evaluate the performance of the algorithms. In addition to the 1155 images that were manually annotated to train the models, ground truth for test images from the AAF and CAM ethno-gender groups were obtained using the same component scheme. The distribution of test images used in this work is as shown in Table 3 and Table 4.

For testing the performance of the landmarking algorithms, automatic landmarks were obtained on the two ethno-gender models separately using both the *general* and the *ethno-gender* trained models. These landmarks were then compared to the ground truth that was generated for the testing data. Thus for each of the three algorithms, there were two sets of annotations generated - using the model trained on general and a separate model trained on the specific ethno-gender model - on each of the two ethno-gender models used for testing.

Table 3. Testing Data: Distribution based on ethno-gender groups and age ranges

Age Range	AAF	CAM
18-30	50	50
31-40	44	49
41-50	41	50
51-60	49	50
61-70	11	50
71+	0	50

Table 4. Testing Data: Distribution based on databases

Database	AAF	CAM
Pinellas County	103	240
MORPH	80	44
PAL	12	15

4.3   Performance Measures

In our experiments, we compare the efficiency of the AAM, CLM and STASM landmarking methods on all of the 252 points. The interocular distance d_{io} is used as a normalization factor for computing error measures. Interocular distance is the distance between the centers of left and right eye and is often used in state-of-the art studies in 2-D facial landmarking. Since the distance error measure is scaled by the interocular distance, it is invariant to the variation in size of each individual face, which allows scaled comparison of point to point errors between images. The interocular distance varied between 56.8 and 129.2 pixels for the AAF test data and 49.35 and 115.7 pixels for the CAM test data. The detection error of a point i is defined as the Euclidean point to point distance between the ground-truth point T_i and detected point $\hat{T}_i$:

$$e_i = \frac{||T_i - \hat{T}_i||}{d_{io}} \tag{1}$$

For each of the two testing ethno-gender models, an *average error per annotation point* across all the images in the test set was computed using the formula

$$m_p = \frac{1}{m d_{io}} \sum_{i=1}^{m} d_i \tag{2}$$

where d_i are the Euclidean point t point errors for each individual annotation point and m is the number of images in the test data set. The average error for each of these individual points is as shown in Figure 3 for the AAF data and Figure 4 for CAM data.

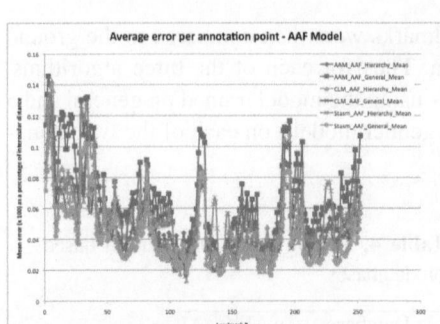 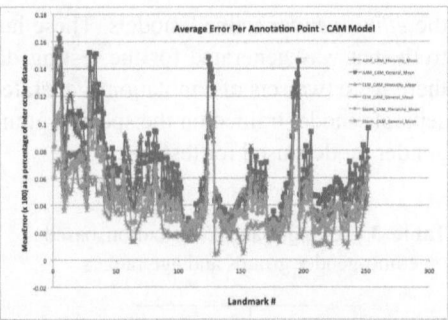

Fig. 3. AAF Test Data: Plot of average error per annotation point as a percentage of interocular distance vs. landmark

Fig. 4. CAM Test Data: Plot of average error per annotation point as a percentage of interocular distance vs. landmark

The classification rate C_i can be defined as:

$$C_i = \frac{\sum_{j=1}^{m} e_i^j < 0.1}{m} \tag{3}$$

where j is the image number and m is the total number of images in the dataset. The classification rate for the AAF data is as shown in Figure 5 and for the CAM test data is as shown in Figure 6.

The system error or the mean error for the entire system can be defined as the normalized mean distance of all facial feature points to their ground-truth locations. The system error, m_e can be defined as

$$m_e = \frac{1}{n d_{io}} \sum_{i=1}^{n} d_i \tag{4}$$

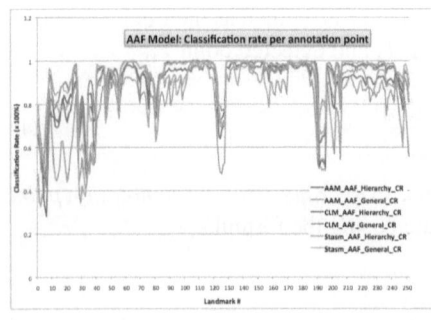 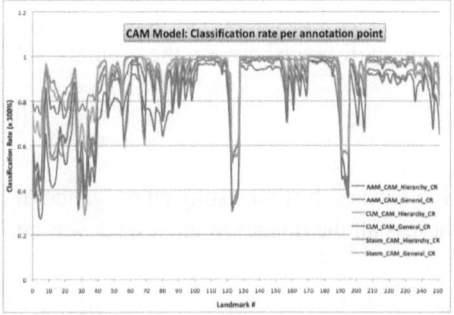

Fig. 5. AAF Test Data: Plot of average error per annotation point as a percentage of interocular distance vs. landmark

Fig. 6. CAM Test Data: Plot of average error per annotation point as a percentage of interocular distance vs. landmark

where n denotes the number of landmarks (252) and d_i values are the Euclidean point-to-point distances for each individual landmark location. This is then averaged over the entire set of images. The system errors for the AAF and CAM model on the landmarking methods using both the general and the ethno-gender trained models are as shown in Table 5. The term *Hierarchy* in the table and graphs mean that the ethno-gender trained model was used.

Table 5. System Error As A Percentage Of Inter Ocular Distance

Algorithm	AAF	CAM
AAM_Hierarchy	5.0	5.4
AAM_General	6.6	6.7
CLM_Hierarchy	4.8	5.6
CLM_General	4.6	5.6
Stasm_Hierarchy	5.3	3.1
Stasm_General	4.1	4.5

The cumulative error distribution of point to point error measured on the AAF and CAM test set are as shown in Figure 7.

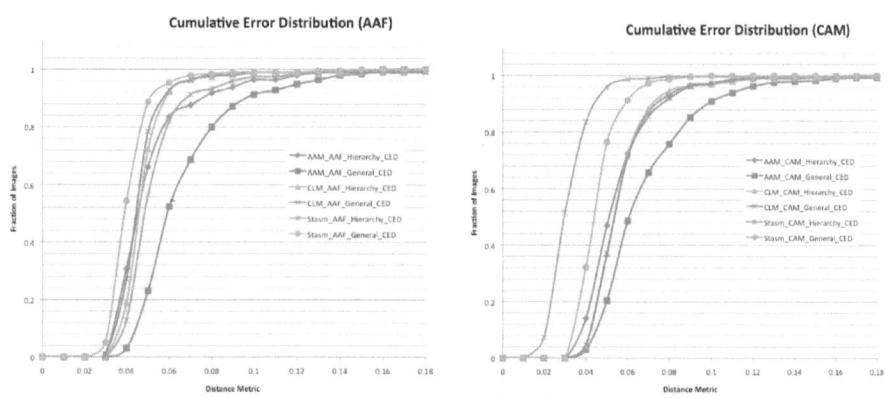

Fig. 7. Comparison of cumulative error distribution of point to point error

4.4   Analysis

It can be seen from Figures 3 and 4 that in general, annotation points on the soft tissue (Points 28-39, Points 123-127, Points 191-195) have higher mean error than those associated with an actual facial feature. STASM was more efficient in finding the points on the outline of the face (Points 0-27) from Figures 3 and 4. In general, the active appearance model has better performance when trained on the ethno-gender model as evidenced by better classification rates per annotation point as shown in the first row of Figures 6 and

5. For the CAM model, from Figures 6 and 4, it can be seen that algorithms when trained on ethno-gender models have better classification rates than general model. In general, from Figures 3 and 4, CLM and STASM have lower average error per annotation point. Also, Figures 7(a) and 7(b) show that STASM and CLM have a higher fraction of images on which the error was below 10% of the inter ocular distance. Also, CLM and STASM have a lower system error when compared to AAM. Finally, the time that it takes to train STASM and CLM is much lower than the time taken to train the AAM algorithm. From experiments performed, it was also found that the time required for landmark detection for the CLM and STASM are much lower than the AAM.

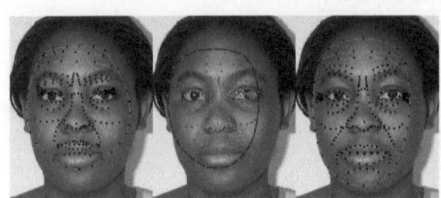

Fig. 8. AAF Test Data: An example on which the algorithms failed on a) AAM b) CLM c) STASM

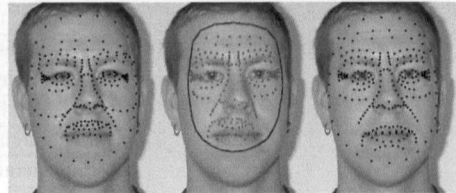

Fig. 9. CAM Test Data: An example on which the algorithms failed on a) AAM b) CLM c) STASM

5   Conclusion and Future Work

In this paper, three state-of-the art algorithms-AAM, CLM and STASM, are compared for automatic landmark detection on a set of dense landmarks on the face. In particular, the models are each trained on a general and specific group. African American Female (AAF) and Caucasian American Male (CAM) cohorts are used as test groups. The performance of each of the algorithms are evaluated in terms of their average error and classification rate per landmark. The cumulative error distribution of point to point error is also shown for all the three algorithms. The collection of ground truth data is in progress for the African American Male and Caucasian American Female ethno-gender groups and the evaluations will be extended to include these groups. It can be concluded from this paper that: 1) algorithms, generally, can generate more accurate landmarks when trained on ethno-gender models 2) CLM and STASM are more robust algorithms for automatic landmarking on dense scheme in terms of landmark accuracy, time taken to train the model and detection of landmarks. Although STASM performs better on the outline of the face, there are some images on which algorithms performed poorly for each of the ethno-gender models is as shown in Figure 8 and Figure 9. This is particularly problematic in applications which rely on auto-landmarking for facial synthesis.

References

1. Phillips, P.J., Flynn, P.J., Scruggs, T., Bowyer, K.W., Chang, J., Hoffman, K., Marques, J., Min, J., Worek, W.: Overview of the face recognition grand challenge. In: IEEE Computer Society Conference on Computer Vision and Pattern Recognition, vol. 1, pp. 947–954 (2005)

2. Beumer, G.M., Bazen, A.M., Veldhuis, R.N.J.: On the accuracy of eers in face recognition and the importance of reliable registration. In: 5th IEEE Benelux Signal Processing Symposium (SPS 2005), Antwerp, Belgium, Secretariaat in Delft, pp. 85–88 (April 2005)

3. Viola, P., Jones, M.: Robust real-time object detection. International Journal of Computer Vision (2001)

4. Vukadinovic, D., Pantic, M.: Fully automatic facial feature point detection using gabor feature based boosted classifiers. In: 2005 IEEE International Conference on Systems, Man and Cybernetics, vol. 2, pp. 1692–1698 (October 2005)

5. Valstar, M., Martinez, B., Binefa, X., Pantic, M.: Facial point detection using boosted regression and graph models, pp. 2729–2736 (June 2010)

6. Efraty, B.A., Papadakis, M., Profitt, A., Shah, S.K., Kakadiaris, I.A.: Facial component-landmark detection. In: Ninth IEEE International Conference on Automatic Face and Gesture Recognition (FG 2011), Santa Barbara, CA, USA, March 21-25, pp. 278–285. IEEE (2011)

7. Dibeklioglu, H., Salah, A., Gevers, T.: A statistical method for 2-d facial landmarking. IEEE Transactions on Image Processing 21(2), 844–858 (2012)

8. Cootes, T.F., Taylor, C.J., Cooper, D.H., Graham, J.: Active shape models-their training and application. Computer Vision and Image Understanding 61(1), 38–59 (1995)

9. Cootes, T.F., Edwards, G.J., Taylor, C.J.: Active Appearance Models. In: Burkhardt, H., Neumann, B. (eds.) ECCV 1998. LNCS, vol. 1407, pp. 484–498. Springer, Heidelberg (1998)

10. Cristinacce, D., Cootes, T.: Feature detection and tracking with constrained local models, pp. 929–938 (2006)

11. Dryden, I., Mardia, K.: Statistical shape analysis. John Wiley and Sons (1986)

12. Ramnath, K., Baker, S., Matthews, I., Ramanan, D.: Increasing the density of Active Appearance Models. In: IEEE Conference on Computer Vision and Pattern Recognition, CVPR 2008, pp. 1–8 (June 2008)

13. Milborrow, S., Nicolls, F.: Locating Facial Features with an Extended Active Shape Model. In: Forsyth, D., Torr, P., Zisserman, A. (eds.) ECCV 2008, Part IV. LNCS, vol. 5305, pp. 504–513. Springer, Heidelberg (2008)

14. Albert, A.M., Ricanek Jr., K., Patterson, E.: A review of the literature on the aging adult skull and face: Implications for forensic science research and applications. Forensic Science International 172(1), 1–9 (2007)

15. AAM, http://sourceforge.net/projects/asmlibrary/files/

16. Saragih, J., Lucey, S., Cohn, J.: Deformable model fitting by regularized landmark mean-shift. International Journal of Computer Vision 91(2), 200–215 (2011)

17. Stasm, http://www.milbo.users.sonic.net/stasm/download.html

18. Ricanek, K., Tesafaye, T.: Morph: A longitudinal image database of normal adult age-progression. In: 7th Int. Conf. on Automatic Face and Gesture Recognition, pp. 341–345 (April 2006)

19. Minear, M., Park, D.: A lifespan database of adult facial stimuli. Behavior Research Methods, Instruments and Computers: A Journal of the Psychonomic Society, Inc. 36, 630–633 (2004)

20. Vuini, P., Trpovski, E., Epan, I.: Automatic landmarking of cephalograms using active appearance models. European Journal of Orthodontics 32(3), 233–241 (2010)

21. Shi, J., Samal, A., Marx, D.: How effective are landmarks and their geometry for face recognition. Computer Vision and Image Understanding 102(2), 117–133 (2006) ISSN: 1077-3142

Face Recognition with Multi-scale Block Local Ternary Patterns

Lian Zhu[1], Yan Zhang[2], Changyin Sun[1], and Wankou Yang[1]

[1] School of Automation, Southeast University, Nanjing 210096, China
[2] The Third Research Institute of Ministry of Public Security, Shanghai 200031, China

Abstract. In this paper, we propose a novel approach to face recognition, called Multi-scale Block Local Ternary Patterns (MB-LTP), which considers both local and various scale texture information to represent face images. In MB-LTP, we compare average values of sub-regions and use a 3-valued codes method to get the MB-LTP value. The MB-LTP histograms are then extracted and concatenated into a single, spatially enhanced feature vector representing the face image in recognition. We use a nearest neighbor classifier in the computed feature space with Chi square as a dissimilarity measure. MB-LTP code presents several advantages: (1)It is more robust than LBP;(2)it is more discriminative and less sensitive to noise;(3)it encodes not only microstructures but also macrostructures of image patterns. Experiments on ORL and AR databases show that the proposed MB-LTP method significantly outperforms other LBP based face recognition algorithms.

Keywords: MB-LTP, LBP, Face recognition.

1   Introduction

Recently, face recognition has been a hot research topic in computer vision, since face recognition has potential application values as well as theoretical challenges. Recently, many appearance-based approaches have been proposed to deal with face recognition problems. They can capture small appearance details in the descriptors while remaining resistant to registration errors owing to local pooling. Another motivation is the observation that human visual perception is well-adapted to extracting and pooling local structural information ('micro-patterns') from images. Methods in this category include Gabor wavelets [4], local autocorrelation filters [2], and Local Binary Patterns [1].

Recently, one generalization called Local Ternary Patterns (LTP) of Local Binary Patterns (LBP) is introduced as a powerful local descriptor for noised images. As a computationally efficient local image texture descriptor, LTP has been used with considerable success in a number of visual recognition tasks including face recognition. However, the reliability of LTP decreases significantly under large illumination variations. Another limitation of LTP is its sensitivity to random and quantization noise in uniform and near-uniform image regions such as the forehead and cheeks. In this paper, we propose a novel approach for face recognition, called Multi-scale Block Local Ternary Patterns (MB-LTP), to overcome the limitations of LTP.

J. Yang, F. Fang, and C. Sun (Eds.): IScIDE 2012, LNCS 7751, pp. 216–222, 2013.

2   Related Work

2.1   Local Ternary Patterns(LTP)

In LTP, graylevels in a zone of width $\pm t$ around i_c are quantized to zero, ones above it are quantized to +1 and ones below it are quantized to -1. The indicator $s(u)$ is replaced by a 3-valued function:

$$s(u,i_c,t) = \begin{cases} +1 & u \geq i_c + t \\ 0 & |u - i_c| < t \\ -1 & u \leq i_c - t \end{cases} \tag{1}$$

Here, t is a user-specified threshold. The LTP encoding procedure is illustrated in Fig.1. Here the threshold t is set to 5, so the zone of width ± 5 is [49,59].

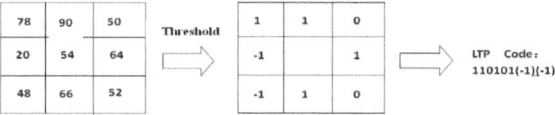

Fig. 1. The LTP operator

For simplicity, we use a coding scheme that splits each ternary pattern into its positive and negative parts as illustrated in Fig.2, subsequently treating these as two separate channels of LBP descriptors for which separate histograms and similarity metrics are computed, combining these only at the end of the computation.

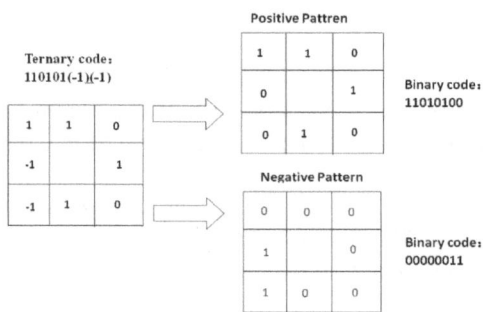

Fig. 2. Procedure to split an LTP code into positive and negative LBP codes

3   Multi-scale Block Local Ternary Patterns(MB-LTP)

3.1   MB-LTP

The original LBP operator labels the pixels of an image by thresholding the 3×3-neighborhood of each pixel with the center value. According to the research on

Multi-scale Block LBP [7], which replace the comparison operator between single pixels in LBP with comparison between average gray-values of sub-regions, the MB-LBP operator can be more robust to noise in uniform image regions and capture larger scale structure. In MB-LTP, the comparison operator between single pixels in LBP is simply replaced with comparison between average gray-values of sub-regions. To simplify the programme, we use the sum instead of the average gray-value of each sub-region. The graylevels' sum of each block in a zone of width $\pm t$ around the central sum graylevel s_c are quantized to zero, ones above that are quantized to +1 and ones below it to -1. The indicator $s(u)$ is replaced by a 3-valued function:

$$Bit(s_i, s_c, t) = \begin{cases} +1 & (s_i > s_c + t) \\ 0 & (|s_i - s_c| <= t) \\ -1 & (s_i < s_c - t) \end{cases} \qquad (2)$$

Here, s_i is the gray-values' sum of the i th block, s_c is the gray-value sum of the central block, t is a user-specified threshold. We illustrated the MB-LTP encoding procedure in the following Fig.3. Here the threshold t is set to 5.

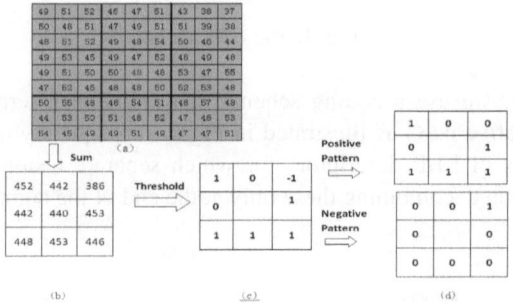

Fig. 3. The MB-LTP operator. (a) the 9×9 gray values of original; (b) compute the sum of each block; (c) encoding the block sum;(d) split the LTP code into positive and negative.

We take the size s of the filter as a parameter, and $s \times s$ denoting the scale of the MB-LTP operator. MB-LTP feature extraction can be very fast as the graylevel sums can be computed very efficiently. It only occupies a little more cost than the original 3×3 LTP operator.

Fig.4. gives example of MB-LTP filtered face images of filter scale 3×3, 9×9 and 15×15. Each image is filtered into two patterns. From this example we can see what influence parameter s would make.

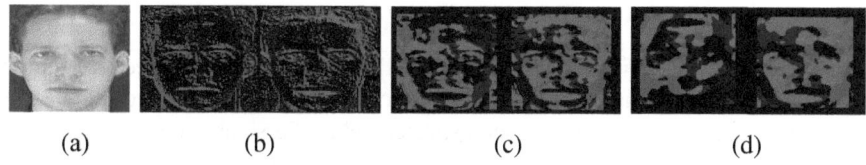

(a)   (b)   (c)   (d)

Fig. 4. MB-LTP filtered images in two patterns with different filter scales s×s.(a)original image;(b)filtered by 3×3 MB-LTP into positive and negative patterns; (c) filtered by 9×9 MB-LTP   (d) filtered by 15×15 MB-LTP .

3.2   Face Description with MB-LTP

In this part, the MB-LTP method presented in the previous section is used for face description. The procedure includes:(1) using the texture descriptor to build two feature images of the original image(positive and negative); (2) dividing the feature image into several small regions and compute the histogram of each region which is concatenated by the positive pattern histogram and the negative pattern histogram;(3) concatenating the histograms of all the regions into a single feature histogram efficiently representing the face image.

A histogram of the labeled image $f_{MB-LTP}(x, y)$ can be defined as

$$HP_i = \sum_{x,y} I\{f_{MB-LTP+}(x, y) = i\}, i = 0, \dots, n-1 \tag{3}$$

$$HN_i = \sum_{x,y} I\{f_{MB-LTP-}(x, y) = i\}, i = 0, \dots, n-1 \tag{4}$$

in which HP stands for the positive histogram, HN for the negative one and n is the number of different labels produced by the LBP operator and

$$I(A) = \begin{cases} 1, & A \quad is \quad true \\ 0, & A \quad is \quad false \end{cases} \tag{5}$$

In our paper, we use the uniform [6] patterns in LBP encoding to reduce the feature dimensions. A Local Binary Pattern is called uniform if it contains at most two bitwise transitions from 0 to 1 or vice versa when the binary string is considered circular.For example, 00000000, 00011110 and 10000011 are uniform patterns. Fig.5. show the histogram produced by the MB-LTP operator.

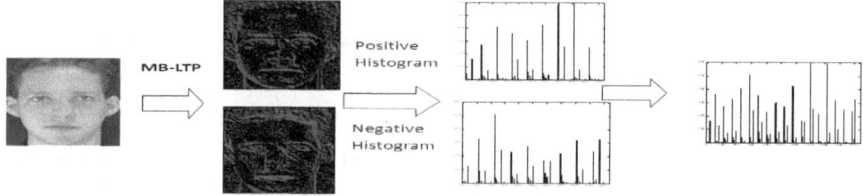

Fig. 5. MB-LTP histogram extraction steps of the image divided into 3×3regions

3.3   Distance Transform Based Similarity Metric

From the pattern classification point of view [8], a usual problem in face recognition is having a plethora of classes and only a few, possibly only one, training sample(s) per class. For this reason, more sophisticated classifiers are not needed but a nearest-neighbour classifier is used. The χ^2 measures perform slightly better than histogram intersection and loglikelihood, so we chose to use it in our experiments.

$$\chi^2(S,M) = \sum \frac{(S_i - M_i)^2}{S_i + M_i}$$ (6)

4   Experiments and Results

The proposed method is tested on the ORL face database (Olivetti Research Laboratory, Cambridge). All the images were taken against a dark homogenous background with the subjects in an upright, frontal position, with tolerance for some tilting and rotation of up to about 20 degrees. The images are grey scale with a resolution of 92×112.

4.1   Experiment Design

In the experiment, we select the first $l(l=3,4,5,6,7,8,9)$ images of an individual for the gallery set and the rest for the probe set. We divided the images into $m \times m$ ($m=3,4,5$)regions. To assess the performance of the proposed approach, we chose to use two different extensions of LBP, MB-LBP and LTP and LBP itself in windows of varying size. In both LTP and MB-LTP, we set the threshold t to $5(t=5)$. The recognition result is shown in the following Table 1.

Table 1. The performance of the LBP, MB-LBP, LTP and MB-LTP with different region numbers ($m \times m$)and different scales of gallery set

No Noise	$l=3$	$l=4$	$l=5$	$l=6$	$l=7$	$l=8$	$l=9$
LBP(3×3)	0.746429	0.850000	0.870000	0.937500	0.958333	0.950000	0.975000
MBLBP(3×3)	0.853571	0.933333	0.940000	0.968750	0.975000	0.975000	0.975000
LTP(3×3)	0.835714	0.895833	0.950000	0.975000	0.983333	0.962500	0.950000
MBLTP(3×3)	0.860714	0.941667	0.965000	0.987500	0.983333	0.987500	0.975000
LBP(5×5)	0.821429	0.887500	0.890000	0.937500	0.950000	0.962500	0.950000
MBLBP(5×5)	0.853571	0.933333	0.940000	0.962500	0.966667	0.962500	0.925000
LTP(5×5)	0.867857	0.925000	0.945000	0.968750	0.966667	0.975000	0.975000
MBLTP(5×5)	0.832143	0.933333	0.945000	0.968750	0.958333	0.962500	0.950000

To compare the robustness to noise of the MB-LTP operator against other operators, we add Gaussian white noise to the images (by the 'imnoise' function in Matlab with mean=0, std=0.05). Then we repeat the experiment above, and modify the threshold t to $18(t=18)$. The experimental results are shown in the following Table 2.

The data show that MB-LTP outperforms other operators in its robustness to noise.

Table 2. The performance of the LBP, MB-LBP, LTP and MB-LTP with Gaussian white noise added to the ORL database

+Noise	l=3	l=4	l=5	l=6	l=7	l=8	l=9
LBP(3×3)	0.339286	0.345833	0.390000	0.418750	0.425000	0.387500	0.550000
MBLBP(3×3)	0.846429	0.904167	0.915000	0.968750	0.975000	0.975000	1.000000
LTP(3×3)	0.539286	0.629167	0.630000	0.675000	0.700000	0.687500	0.700000
MBLTP(3×3)	0.878571	0.945833	0.935000	0.962500	0.966667	0.962500	1.000000
LBP(5×5)	0.525000	0.620833	0.640000	0.656250	0.683333	0.700000	0.700000
MBLBP(5×5)	0.814286	0.908333	0.910000	0.943750	0.958333	0.950000	0.950000
LTP(5×5)	0.653571	0.737500	0.760000	0.825000	0.858333	0.900000	0.925000
MBLTP(5×5)	0.846429	0.916667	0.930000	0.981250	0.975000	0.987500	0.975000

Additionally, to gain knowledge about the robustness of our method against slight variation of facial expression, illumination condition and occlusions, we tested our approach on the AR database. The database in our experiment contains 14 different images of 120 distinct subjects (individuals) and the images are grey scale with a resolution of 40×50.

In the experiment, we select the first l(l=1,2,3,4,5,6,7,8,9,10,11,12,13) images of an individual for the gallery set and the rest for the probe set. We divided the images into 3*3 regions. We set the threshold t to 5(t=5). The recognition result is shown in the following Fig. 6. From Fig.6, we can observe that recognition rate of MB-LTP based descriptor is apparently better than LBP, and a litter better than MB-LBP and LTP.

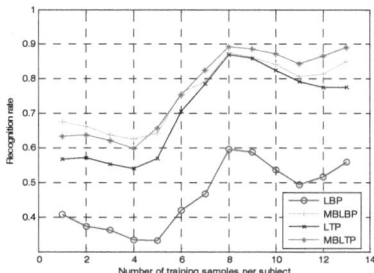

Fig. 6. Recognition rate of MB-LTP against LBP, MB-LBP, LTP on AR database

5   Conclusions

In this paper, we present a Multi-scale Block Local Ternary Patterns (MB-LTP) based operator for image representation. The Multi-scale Block Local Ternary Patterns (MB-LTP) use sub-region average gray-levels for comparison instead of single pixels and add threshold t to the operator. As a result, the coding method is changed to 3-valued. We simply use the distance between two histograms as dissimilarity measure. Experiments on ORL database show that MB-LTP significantly outperforms other LBP based methods.

Acknowledgment. This project is supported by NSF of China (61005008, 61273023).

References

1. Ahonen, T., Hadid, A., Pietikäinen, M.: Face Recognition with Local Binary Patterns. In: Pajdla, T., Matas, J(G.) (eds.) ECCV 2004. LNCS, vol. 3021, pp. 469–481. Springer, Heidelberg (2004)
2. Guodail, F., Lange, E., Iwamoto, T.: Face recognition system using local autocorrelations and multiscale integration. IEEE TPAMI 18(10), 1024–1028 (1996)
3. Lee, K., Ho, J., Kriegman, D.: Acquiring linear subspaces for face recognition under variable lighting. IEEE TPAMI 27(5), 684–698 (2005)
4. Liu, C.: Capitalize on dimensionality increasing techniques for improving face recognition grand challenge performance. IEEE TPAMI 28(5), 725–737 (2006)
5. Ojala, T., Pietikäinen, M., Harwood, D.: A comparative study of texture measures with classification based on feature distributions. Pattern Recognition 29, 51–59 (1996)
6. Ojala, T., Pietikäinen, M., Mäenpáá, T.: Multiresolution gray-scale and rotation invariant texture classification with local binary patterns. IEEE Transactions on Pattern Analysis and Machine Intelligence, 971–987 (2002)
7. Liao, S., Zhu, X., Lei, Z., Zhang, L., Li, S.Z.: Learning Multi-scale Block Local Binary Patterns for Face Recognition. In: Lee, S.-W., Li, S.Z. (eds.) ICB 2007. LNCS, vol. 4642, pp. 828–837. Springer, Heidelberg (2007)
8. Ahonen, T., Hadid, A., Pietikäinen, M.: Face Recognition with Local Binary Patterns. In: Pajdla, T., Matas, J(G.) (eds.) ECCV 2004. LNCS, vol. 3021, pp. 469–481. Springer, Heidelberg (2004)

Ensemble Haar and MB-LBP Features
for License Plate Detection

Qiuping Pan, Jifeng Shen,Wankou Yang, and Changyin Sun

School of Automation, Southeast University, Nanjing 210096, China
panqiuping914@126.com, shenjifeng1980@hotmail.com,
wankou_yang@yahoo.com.cn, cysun@seu.edu.cn

Abstract. This paper presents a new license plate detection algorithm, in which, the Haar and MB-LBP features are combined and the updated rules of the sample weights are revised. The cascade classifiers are used to detect digitals in the image, Non-Maximum Suppression and the license plate characteristics are applied to locate license plate area accurately. Experimental results show that the proposed method could effectively avoid the phenomenon of weights distortions and get higher detection rate while reducing false alarm rate.

Keywords: plate detection, AdaBoost, weights updated, Non-maximum Suppression.

1   Introduction

License Plate Recognition System (LPR) has been adopted widely in such as unattended parking, security control and stolen vehicle verification. LPR consists of three parts: the license plate detection, character segmentation, character recognition[1]. It is extremely difficult to detect license plate from cluttered background efficiently because of the affection of variant illumination, perspective distortion, interference characters, etc. Most of previous license plate detection algorithms are restricted in certain working conditions. Therefore, detecting license plate under various complex environments is still a challenging problem [2].

Much research has been done over the past decades, most methods are based on features of the license plates. Features commonly employed have been derived from the license plate format and the alphanumeric characters constituting license plate numbers [3]. The edge statistics and mathematical morphology [4-5] based techniques produced good results, while edges may be disturbed in complex scenes. Color model methods [6-7] were also proposed for detection, but they had a low degree of accuracy and were not robust enough in natural scenery, since color is sensitive to light changing.

Moreover, Learning-based methods have widely used in license plate recognition recent years, i.e. SVM(Support Vector Machine)[8], ANN(Artificial Neural Networks)[9], AdaBoost and so on. These methods need a large samples set in the training procedure. Recently, Haar-like features were widely used for object detection [10]. The classifiers based on Haar-like features can detect objects from cluttered background despite the variance of the illumination, the color, the position and the size of the objects.

J. Yang, F. Fang, and C. Sun (Eds.): IScIDE 2012, LNCS 7751, pp. 223–230, 2013.

Though AdaBoost is good at selecting efficient Haar-like features, the selected features set is still with a large dimension, which makes training process very complicated, unstable and extreme time consuming. What's more, in the training process, re-update rule makes the algorithm focus on relatively hard classification samples, but when the training samples have noise, these samples' weight will be presented exponential growth, eventually lead to serious distortions, reducing the detection performance of the algorithm. To overcome the training time-consuming and weights distortion of the traditional AdaBoost algorithm, Zuo Jinglong presented AdaBoost algorithm[11] based on the characteristic of the MB-LBP[12].

Since to calculate a MB-LBP characteristic need more time than to calculate a Haar characteristic, and some MB-LBP features have worse classification ability than some of Haar-like features, we present a new algorithm and apply it to the license plate detection in this paper. In this paper, we construct a cascade classifiers including both global statistical features and local Haar-like features and MB-LBP features; we present this new algorithm to detect the digital area, so digitals in license plates are used as training samples alone [13]; non-maximum suppression [14] and the number of transitions from white pixel to black pixel[15] are used to locate license plate area accurately.

2   AdaBoot Algorithm Based on Haar and the MB-LBP Features

2.1   The Structure of Weak Classifiers

Haar feature is a rectangular characteristic, the characteristic value is defined as pixels difference between the white rectangular area and the black rectangular area. These simple features consist in significant domain-knowledge information, they are sensitive to edge and line information. There are different types of rectangles with different sizes and location combinations, so the number of Haar characteristics of an image is far more than the number of pixels. Therefore, selecting appropriate Haar characteristics is crucial. In our work, five types of features were used to get a balance with training time and detection accuracy.

MB-LBP (Local Binary Patterns, Multi-Block) operator is a promotion of LBP operator. LBP encodes the relationship between the center pixel and its 8 neighbor pixels. MB-LBP calculates the characteristic value by comparing the average gray of the center sub-block and its eight average grays of surrounding neighborhood sub-block in a 3*3 box. The original LBP feature can only describe a small range of image information, and is susceptibility to noise, while MB-LBP could capture more details of the changes of the image.

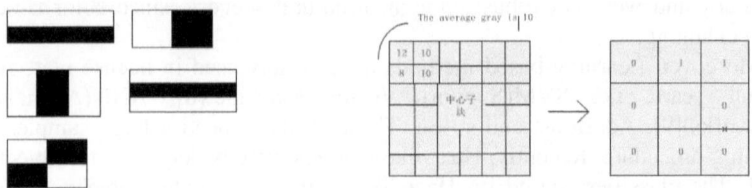

Fig. 1. Several Haar characteristics used in this article   **Fig. 2.** MB-LBP Eigen value

MB-LBP can be formalized described as:

$$f(g_c) = \sum_{i=1}^{8} s(g_i - g_c)2^i$$ (1)

$$s(x) = \begin{cases} 1 & if \quad x \geq 0 \\ 0 & if \quad x < 0 \end{cases}$$ (2)

Where g_c is the gray value of the center sub-block, $g_i(i=1, 1, \cdots 8)$ denotes the gray value of the eight surrounding neighborhood.

The value of the MB-LBP features is non-decimal, to get the effective integration of Haar and MB-LBP features, we need to modify the structure of Haar characteristics weak classifiers.

Multi-branching tree structure is used to construct the weak classifier of MB-LBP. The tree structure consists of 256 branches, the function value of each branch correspond to the value of the MB-LBP features. Final, the weak classifier based on MB-LBP is as follows:

$$f(x_i) = \begin{cases} a_0, & x_i = 0 \\ \quad \cdots \\ a_j & x_i = j \\ \quad \cdots \\ a_{255} & x_i = 255 \end{cases}$$ (3)

Where x_i denotes MB-LBP Eigen value of the *ith* sample ,which can be got by formula 1. a_j indicates the judgment coefficient, is defined as equation (4):

$$a_j = \frac{\sum_i w_i y_i \delta(x_i)}{\sum_i w_i \delta(x_i)}$$ (4)

Where $y_i \in \{+1, -1\}$, +1 for positive samples, -1 indicates a negative sample.

$$\delta(x_i) = \begin{cases} 1 & x_i = j \\ 0 & others \end{cases}$$ (5)

Know from equation (5), $a_j \in [-1, +1]$, when $a_j > 0$, it shows that MB-LBP Eigen values of j has a bigger possibility to be positive samples than negative samples.

Similarly, the Haar features values is divided into N regions, calculate $a_j(j = 1, ..N)$ value of each region as equation (6).

$$a_j = \frac{\sum_i w_i y_i \varphi(x_i)}{\sum_i w_i \varphi(x_i)},$$ (6)

The error of each region is:

$$e_j = \begin{cases} \dfrac{\sum_i w_i y_i + \sum_i w_i}{2}, a_j < 0 \\[2ex] \dfrac{\sum_i w_i - \sum_i w_i y_i}{2}, a_j > 0 \\[2ex] 0, a_j = 0 \end{cases} \qquad (7)$$

Each Haar feature's error is:

$$e_{Harr} = e_1 + e_2 + \ldots e_N \qquad (8)$$

Each MB-LBP feature's error is:

$$e_{MBLBP} = e_1 + e_2 + \ldots + e_{255} \qquad (9)$$

Weak classifier is designed to find out $f_{Harr}(x)$ or $f_{MBLBP}(x)$ that makes the error minimum.

2.2   The Structure of Strong Classifier

The proposed algorithm is described as follows:

(1) $(x_1, y_1), (x_2, y_2), \ldots, (x_N, y_N)$ represent the training samples, contains m positive samples, l negative samples, $y_i \in \{+1, -1\}$, +1 indicates positive sample, -1 denotes negative sample. x_i is on behalf of the *ith* sample.

(2) To initialize the weights: when $y_i = +1$, $w_{1i} = \dfrac{1}{2m}$, else $y_i = -1$, $w_{1i} = \dfrac{1}{2l}$.

(3) For t=1,...T, p=0,q=0(p, q are weak classifier number of MBLBP or Haar respectively)

   (1) Normalize weights: $w_{ti} = \dfrac{w_{ti}}{\sum\limits_{j=1}^{N} w_{tj}}$

   (2) Search for the best Haar characteristic, which make the error (quote 8) minimum, and construct weak classifier $f_{Harr}(x)$.
   Search for the best MBLBP feature, which make the error (quote 9) minimum, and construct weak classifier $f_{MBLBP}(x)$.

   The specific method to determine the optimal threshold for each weak classifier $f_{Harr}(x)$ or $f_{MBLBP}(x)$ is: a function value $f(x_i)$ is calculated for all the training samples, the larger the function value, the closer the license plate, thereby placing it in ascending order, set threshold equal to the value of sample x_k ,find the best threshold to make the formula(8)or(9) minimum.

$$\text{If } (e_{Harr} \le e_{MBLBP}) \quad \begin{cases} f_t(x) = f_{Harr}(x) \\ F_{Harr}(x) = F_{Harr}(x) + f_{Harr}(x), \\ q = q+1 \end{cases}$$

$$\text{else} \quad \begin{cases} f_t(x) = f_{MBLBP}(x) \\ F_{MBLBP}(x) = F_{MBLBP}(x) + f_{MBLBP}(x). \\ p = p+1 \end{cases}$$

Where $F_{MBLBP}(x)$ and $F_{Harr}(x)$ are strong classifiers consisting of MBLBP characteristic and Haar characteristic respectively.

(3) Update the weights

To avoid the weights distortions phenomenon, we just update the samples' weights when they are misclassified and their weights are less than current round of update weight threshold. The update weight threshold is defined as: $WT_t = \dfrac{1}{N}\sum_{i=1}^{N} w_i$

$$\text{When } h_t(x_i) \ne y_i, \quad w_{t+1i} = \begin{cases} w_{ti} e^{-y_i * f_t(x_i)} ; w_{ti} < WT_t \\ w_{ti} ; w_{ti} \ge WT_t \end{cases}$$

(4) Get strong classifier:

$$C(x) = \begin{cases} 1 & F_{MBLBP}(x) + F_{Harr}(x) \ge THRESH \\ 0 & F_{MBLBP}(x) + F_{Harr}(x) < THRESH \end{cases}$$

Where THRESH is the threshold of the strong classifier. THRESH is the sum of each threshold of weak classifiers.

3   Window Mergers and Screening

Our method detects digitals in the input image using the AdaBoost algorithm based on Haar and MB-LBP features. So the detection windows are the figure of a number of candidates, not the final plate region. In this paper, Non-maximum Suppression is adopted to merge these windows. What's more, Number of transitions has been applied to filter the candidate region, the proposed method is better to remove the interference.

An ideal fusion method would incorporate the following three characteristics:

(1) The higher the detection score, the higher the probability for the image region to be a true positive.

(2) The more overlapping detections there are in the neighborhood of an image region, the higher the probability for the image region to be a true positive.

(3) Nearby overlapping detections should be fused together, but overlaps occurring at very different scale or position should not be fused.

In order to achieve the above three requirements, detection score is used as the right of each region in the calculation, the three-dimensional input is the synthesis of location X, Y and scale, the main use of the thinking is Mean Shift algorithm.

By counting the number of transitions from white pixel to black pixel line by line, we can get a transition map of the entire image. Here we use gray-scale transition to test whether candidate window is a license plate. Progressively calculating the number of transitions of the gray level in the candidate region, setting a threshold value, if the number of transitions within the line is greater than the threshold, height will add 1. This threshold value is got from experience, we choose 14 as a threshold value in order to allow various types of license plate to meet this requirement. At last, we see if the height can meet the requirement of the license plate.

4   Experimental Results

Select 3500 pictures of number separated from license plates as positive samples, and 2391 pictures as negative samples. Training samples are resized into 18*30 pixels. The positive samples are digitals from 0 to 9, the negative samples are regions randomly selected from the background which do not contain any digitals.

The experiments are implemented on Intel 1.8G CPU, 2G RAM and programmed in the VC2008. In our experiments, we used a four-level strong classifier, the first three classifiers are based on the average gray-scale, variance of pixel values, edge information. The last classification is constructed of ten weak classifications constitute of two MBLBP characteristics and eight Haar characteristics. It needs about 48s to get a weak classifier. When calculating the detection time, we normalize the input picture to 352*288 sizes.

154 test pictures were selected randomly from the database, 146 license plates were detected. The detection rate is 94.8%. Some experimental results are shown in Table 1.

Table 1. Several methods of detection performance

Detection method	Detection rate	False detection rate	Detection time
Detection based on edge information	90.9%	4.5%	125ms
Ordinary AdaBoost	91.4%	1.9%	62ms
Our method	94.8%	1.2%	63ms

The table shows that, compared with the original method, the proposed method keeps a higher detection rate with lower false detection rate, and the detection speed can achieve real-time requirements.

License plate location results of some vehicle images, where the license plates are marked with red boxes.

Fig. 3. Detection of blue and white license plate in ideal condition

Fig. 4. Detection of license plate that has stains

5   Conclusions

We propose a new AdaBoost algorithm combined Haar and MB-LBP features to detect license plate in clutter image. The proposed method not only selects fewer features, but also overcomes the distortions of the weights with fast detection speed, a higher detection rate and lower false alarm rate.

Acknowledgment. This project is supported by NSF of China (61005008, 61273023).

References

1. Christos-Nikolaos, E., Anagnastopoulos, I.E.: Anagnastopoulos: License Plate Recognition From Still Image and Video Sequences: A Survey. IEEE Transactions on Intelligent Transportation Systems 9, 377–391 (2008)

2. Zhang, H., Jia, W., He, X., Wu, Q.: Learning-Based License Plate Detection Using Global and Local Features. In: The 18th International Conference on Pattern Recognition, Hong Kong, vol. 2, pp. 1102–1105 (2006)
3. Lin, B., Fang, B., Li, D.-H.: Character recognition of License Plate Image based on Multiple Classifiers. In: Proceedings of the 2009 International Conference on Wavelet Analysis and Pattern Recognition, Baoding, pp. 138–143 (2009)
4. Chen, B., Cao, W., Zhang, H.: An Efficient Algorithm on Vehicle License Plate Location. In: Pro. ICAL, pp. 1386–1389 (2008)
5. Saha, S., Basu, S., Nasipuri, M., Basu, D.K.: License Plate localization from vehicle images: An edge based multistage approach. International Journal on Recent Trends in Engineering (Computer Science) 1, 284–288 (2009)
6. Deng, H., Song, X.: A Novel Approach for License Plate Location in Natural Images, pp. 563–568 (2009)
7. Shi, X., Zhao, W., Shen, Y.: Automatic License Plate Recognition System Based on Color Image Processing. In: Gervasi, O., Gavrilova, M.L., Kumar, V., Laganá, A., Lee, H.P., Mun, Y., Taniar, D., Tan, C.J.K. (eds.) ICCSA 2005. LNCS, vol. 3483, pp. 1159–1168. Springer, Heidelberg (2005)
8. Ho, W.T., Lim, H.W., Tay, Y.H.: Two-stage License Plate Detection using Gentle Adaboost and SIFT-SVM. In: First Asian Conference on Intelligent Information and Database Systems, pp. 109–114 (2009)
9. Kim, K.K., Kim, K.I., Kim, J.B., Kim, H.J.: Learning-based approach for license plate recognition. In: Proceedings of the 2000 IEEE Signal Processing Society Workshop on NNSP, pp. 614–623 (2000)
10. Fang, Y., Lu, D., Ge, J.: License Plate Detection Classifier Cascade Fast Training Algorithm. Chongqing University of Posts and Telecommunications (Natural Science) 22, 104–107 (2010)
11. Zuo, J., Sun, C.-Y., Yang, W.: A face Detection Algorithm based on MB-LBP Features. Bulletin of Science and Technology 27, 652–656 (2011)
12. Zhang, L., Chu, R., Xiang, S., et al.: Face Detection Based on Multi-Block LBP Representation. In: IGB 2007, pp. 11–18 (2007)
13. Netzer, Y., Wang, T., Coates, A.: Reading Digits in Natural Images with Unsupervised Feature Learning. In: NIPS Workshop on Deep Learning and Unsupervised Feature Learning (2011)
14. Navneet DALAL: Finding People in Images and Videos. Institute National Polytechnnique de Grenoble, vol. 17 (2006)
15. Zhang, X., Shen, P., Bai, J., Lei, J.: License Plate Location Based on AdaBoost. In: International Conference on Information and Automation (2010)

Exploring Unknown Paths in Networks Based on Multiple Random Walks

Cunlai Pu[1], Jian Yang[1], Ruihua Miao[2], and Wenjiang Pei[2]

[1] School of Computer Science and Engineering,
Nanjing University of Science and Technology,
Nanjing 210094, People's Republic of China
[2] School of Information Science and Engineering, Southeast University,
Nanjing 210096, People's Republic of China
pucunlai@njust.edu.cn

Abstract. We study the problem of exploring unknown paths in networks through multiple random walks. It is assumed that a path is explored if it has been passed through by a random walker from the initial node to the terminal node continuously. We derive probability $\theta'(t)$ that a given path in a network is explored by one or more random walkers in t steps on condition that there are many random walkers traveling on the network. Results show that more random walkers are better for exploring the path. The larger length l of the path is, the smaller $\theta'(t)$ is. To explore paths with the same length in three kinds of networks, random walkers need least effort in SWW networks, most effort in BA networks and moderate effort in ER networks.

Keywords: random walks, path, networks.

1 Introduction

During the last decade, much attention has been focused on the interplay between topologies of complex networks and dynamical processes taking place on them [1,2,3,4]. Examples include information traffic [5,6,7,8,9,10], diffusive particle systems [11], epidemic spreading [12,13,14,15] etc. The random walk problem is one of the bases of these processes, which has been extensively studied for decades previously on regular lattices [16], fractal networks [17] and now on complex networks [18,19,20,21] characterized by power-law degree distribution and small-word property. Analysis on random walks has obtained fruitful results about hitting time, return time, cover time and so on. Expressions for the mean hitting time between two nodes in scale-free networks are obtained in Ref. 18. The authors also introduce the random walks centrality which describes the relative speed by which a node can receive and spread information using random walks method over networks. Properties of random walks in small-word networks are discussed in Ref. 22. The authors obtain that the average number of distinct nodes visited by the random walker, the mean-square displacement of the walker, and the distribution of first-return times all obey a characteristic

J. Yang, F. Fang, and C. Sun (Eds.): IScIDE 2012, LNCS 7751, pp. 231–237, 2013.

scaling form. Some other works have focused on the coverage problem, trying to find bounds for the expected number of hops taken by a random walk to visit all nodes in a network [23,24,25]. Random walks can also be used in a variety of network-related problems such as self-stabilization [26], dynamic routing [27] and so on.

Can random walkers carry out more delicate tasks, such as exploring unknown paths in networks? Wang [21] studied the problem of detecting unknown path in networks by a single random walker. This problem is appealing because it will bring about more precise information about the network on which the random walker travels. Similar phenomenons often emerge in complex systems such as an ecologist happens to find a new ecological chain, a citizen runs into an unknown path between two places in a city, a chemist accidentally invents a new chemical from an old chemical through a series of chemical reactions and so on. By means of iteration and generating function, Wang obtained the probability that a random walker from a source node successively passes through a given path from its initial node to its terminal node in t steps. Moreover he studied the factors including path structure and network topology affecting the properties of the path-detecting events. In this paper we further discuss the path-exploring problem in networks based on the theory of multiple random walks. We derive probability that a given path in a network is explored by one or more random walkers in t steps on condition that there are many random walkers traveling on the network. We investigate major factors affecting the path-exploring problem on complex networks by simulation.

2   Analytic Results

Let us describe the problem in detail. A network can be described as a graph $G(V, E)$ with node set V and edge set E. Here we consider only connected networks without loops and multiple edges. A path C of length l in a network is usually denoted by $\alpha_0\alpha_1 \cdots \alpha_l$, where α_i are distinct nodes in V and $\alpha_{i-1}\alpha_i$ are edges in E, $1 \leq i \leq l$. Suppose a random walker starting from node s travels on the network. Each time it jumps onto one of its neighboring nodes with equal probability. At time t its walking route $Q(t)$ can be denoted as $\delta_0\delta_1 \cdots \delta_t$, where $\delta_i \in V$, $0 \leq i \leq t$ (δ_i are not necessarily different from each other since a node in networks may be visited for many times by the random walker). Once the random walker continuously passes through path C from its initial node to its terminal node in turn, we assume that path C is explored by the walker. The probability $\theta(t)$ that path C is found in t steps can be expressed as:

$$\theta(t) = \text{Prob}\{C \subseteq Q(t), \delta_0 = s, \forall \delta_1, \ldots, \delta_t \in V\} \qquad (1)$$

Wang derived the generating function of $\theta(t)$ which is as follows [21]:

$$\Theta(x) = \frac{P_0 x^l \Phi_{s\alpha_0}(x)}{(1 - x)(1 + P_0 x^l \Phi_{\alpha_l\alpha_0}(x))} \qquad (2)$$

Where $P_0 = 1/K_{\alpha_0} K_{\alpha_1} \cdots K_{\alpha_{l-1}}$, K_i is the degree of node i. $\Phi_{ij}(x)$ is the generation function of $P_{ij}^{(t)}$ which is the t-step transition probability from node i to node j which can be calculated from the transition probability matrix $\mathbf{P}$ [18]. Then $\theta(t)$ can be calculated as follows:

$$\theta(t) = \frac{1}{t!} \frac{\partial^t}{\partial x^t} \Theta(x) \Big|_{x=0} \qquad (3)$$

If many random walkers travel on the network, what is probability $\theta'(t)$ that path C is explored in time t? It should be noticed that the path could be explored for many times by one or many walkers during that time. To calculate $\theta'(t)$, we should consider this problem negatively. Suppose there are n random walkers initially from the same source node s wandering in the network independently. At time t, the trails of the n walkers are $Q_1(t), Q_2(t), \cdots Q_n(t)$. We first calculate the probability $\bar{\theta}'(t)$ that the path is not explored in time t. It means none of the walkers has found C in time t, which can be expressed as follows:

$$\bar{\theta}'(t) = \text{Prob}\{C \nsubseteq Q_1(t) \cap C \nsubseteq Q_2(t) \cdots C \nsubseteq Q_n(t)\}$$
$$= (1 - \text{Prob}\{C \subseteq Q_1(t)\}) \cdots (1 - \text{Prob}\{C \subseteq Q_n(t)\}) \qquad (4)$$

Because the n random walks are the same independent random process, we get that:

$$\bar{\theta}'(t) = (1 - \theta(t))^n \qquad (5)$$

Then we obtain that:

$$\theta'(t) = 1 - (1 - \theta(t))^n \qquad (6)$$

From the above we get some analytic results of the $\theta'(t)$. Although we can not find out ultimate expressions of it, we obtain the main factors from the equations that affect the path-exploring problem are number n of random walkers, path length l, and source node s.

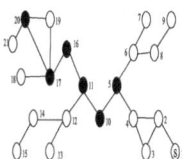

Fig. 1. A simple network. The given path is marked with black color. The starting node of the random walkers is node 1.

3   Simulation and Discussion

In this section we make numerical simulations in order to deepen our discussion as well as confirm analytic results. Fundamental factors that affect $\theta'(t)$ defined above are investigated, such as number n of random walkers, source node s, path length l, topological structure of the underlying network.

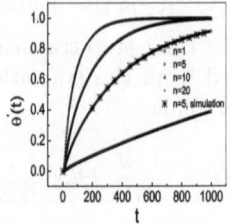

Fig. 2. $\theta'(t)$ is the probability that n random walkers have explored the give path in t steps. Analytic results and simulation results agree well in the figure.

3.1   A Simple Example

To test our analytic results, we first study the path exploring problem in a small network containing 21 nodes and 22 edges shown as Fig. 1. There is a given path in black in the network:5-10-11-16-17-20. At the beginning n walkers sets out from the source node 1 to explore the given path by walking randomly. For some walking time t, we conduct 100000 trials of random walks, and then compute the frequency of the successful exploring events in t which is the result of $\theta'(t)$. In Fig. 2, the simulation results for $n = 5$ are marked with asterisks (The other curves are analytic results obtained from the above equations). It can be seen from the picture that the analytic and simulation results are in good agreement. We can also see $\theta'(t)$ increases with step t and finally reaches 1. For example, as to the curve concerning $n = 20$, when $t > 500$, $\theta'(t) > 0.9934$. Convergence rate of $\theta'(t)$ increases as n increases. It can be inferred from the picture that although the random walkers are not sure where the path lies, as time grows, the chance for exploring the path grows. And they will surely explore the path as long as they travel long enough on the network. Furthermore, more random walkers are better for the efficiency of exploring the path.

3.2   Factors Affecting $\theta'(t)$

To demonstrate the influence of path length on probability $\theta'(t)$, we select 50 paths from the scale-free network [28] for each path length $l \in \{1, \ldots, 9\}$. The nodes in every path are different from each other. Each time 5 random walkers searching for a path set out from the initial node of the path. The total steps of each random walker is set as $t = 10000$. We calculate probability $\theta'(t)$ for each path, and then take average of the probability for each path length l. The results are depicted in Fig. 3. It shows that probability $\theta'(t)$ drops substantially when path length l increases. for example, the probability of exploring a path with length $l = 2$ is as high as 0.9999, but the probability of exploring a path with length $l = 7$ is as low as 0.0216. The topological structure of the underlying network has strong influence on probability $\theta'(t)$ which is reflected in $\Phi_{s\alpha_0}(x)$ and $\Phi_{\alpha_l\alpha_0}(x)$ in Eq. (2). Here we perform random walks on three kinds of networks, which are SWW network [29], BA network [28] and ER network [30]. The size of

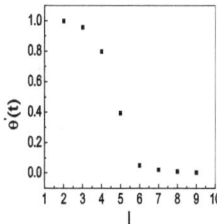

Fig. 3. Probability $\theta'(t)$ vs. path length l. The number of random walkers is $n = 5$. The size of the underlying scale-free network is $N = 100$, and the average degree of nodes is $< K >= 4$. All the data points are the results averaged over 50 paths.

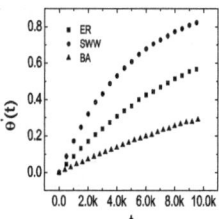

Fig. 4. Probability $\theta'(t)$ in three different network topologies. The number of random walkers is $n = 5$. All the data points are the results averaged over 50 paths with a fixed path length $l = 5$.

the three networks is 100 and the average degree is $\langle K \rangle = 4$. On each network, we randomly choose 50 paths with a fixed length $l = 5$. Every node is unique in a path. When searching for a path, 5 random walkers set out from the initial node of the path. Finally $\theta'(t)$ is averaged over 50 paths on each network, and results are shown in Fig. 4. $\theta'(t)$ converges to 1.0 quickly on the SWW network, but slowly on the BA networks, at a moderate pace on the ER networks. For example, when $t = 8000$, on the SWW network $\theta'(t) = 0.7750$, on the BA network $\theta'(t) = 0.2550$, on the ER network $\theta'(t) = 0.5146$, the probabilities are significantly different from each other. Therefore, we conclude that to explore paths with same path length l on the three kinds of networks, on average random walkers need least effort on SWW networks and most effort on BA networks, while on ER networks the walkers need more effort than on SWW networks and less effort than on BA networks, which indicates that the network homogeneity is a favorable factor for random walkers exploring a given path.

4   Conclusions

In this paper, we have studied the problem of exploring unknown paths by multiple random walks on both simple and complex networks, which is rarely studied before. we compute the probability $\theta'(t)$ that a given path in a network is explored by one or more random walkers in t steps when there are many random walkers wandering in the network. Simulations not only confirm our analytic results, but also demonstrate the factors affecting the path-exploring problem. Results show that more random walkers will enhance the exploring efficiency. Large length of the given path has a negative effect on the path-exploring problem. But the network homogeneity is a favorable factor for random walkers exploring the given path. In the future we will further the path-exploring problem on networks and study more properties related to it.

References

1. Barabási, A.L.: Scale-Free Networks: A Decade and Beyond. Science 325, 412–413 (2009)
2. Boccaletti, S., Latora, V., Moreno, Y., Chavez, M., Hwang, D.U.: Complex networks: Structure and dynamics. Phys. Rep. 424, 175–308 (2006)
3. Song, C.M., Qu, Z.H., Blumm, N., Barabási, A.L.: Limits of Predictability in Human Mobility. Science 327, 1018–1021 (2010)
4. Song, C.M., Havlin, S., Hernán, A.M.: Self-similarity of complex networks. Nature 433, 392–395 (2005)
5. Kleinberg, J.M.: Navigation in a small world. Nature 406, 845 (2000)
6. Wang, D.S., Wen, Z., Tong, H.H., Lin, C.Y., Song, C.M., Barabási, A.L.: Information Spreading in Context. In: Proceeding for the 20th International World Wide Web Conference, pp. 1–10. ACM, Hyderabad (2011)
7. Wang, W.X., Yin, C.Y., Yan, G., Wang, B.H.: Integrating local static and dynamic information for routing traffic. Phys. Rev. E 74, 016101–016105 (2006)
8. Yan, G., Zhou, T., Hu, B., Fu, Z.Q., Wang, B.H.: Efficient routing on complex networks. Phys. Rev. E 74, 046108–046112 (2006)
9. Wu, Z.X., Wang, W.X., Yeung, K.H.: Traffic dynamics in scale-free networks with limited buffers and decongestion strategy. New J. Phys. 10, 023025 (2008)
10. Yang, R., Wang, W.X., Lai, Y.C., Chen, G.R.: Optimal weighting scheme for suppressing cascades and traffic congestion in complex networks. Phys. Rev. E 79, 026112–026117 (2009)
11. Lee, S., Yook, S.H., Kim, Y.: Diffusive capture process on complex networks. Phys. Rev. E 74, 046118–046124 (2006)
12. Pastor-Satorras, R., Vespignani, A.: Epidemic spreading in scale-free networks. Phys. Rev. Lett. 86, 3200–3203 (2001)
13. Zhou, T., Liu, J.G., Bai, W.J., Chen, G., Wang, B.H.: Behaviors of susceptible-infected epidemics on scale-free networks with identical infectivity. Phys. Rev. E 74, 056109–056114 (2006)
14. Yang, R., Zhou, T., Xie, Y.B., Lai, Y.C., Wang, B.H.: Optimal contact process on complex networks. Phys. Rev. E 78, 066109–066113 (2008)
15. Yang, R., Huang, L., Lai, Y.C.: Selectivity-based spreading dynamics on complex networks. Phys. Rev. E 78, 026111–026115 (2008)
16. Weiss, G.H.: Aspects and Applications of the Random Walk. North-Holland, Amsterdam (1994)
17. Ben-Avraham, D., Havlin, S.: Diffusion and Reactions in Fractals and Disordered Systems. Cambridge University Press, Cambridge (2004)

18. Noh, J.D., Rieger, H.: Random Walks on Complex Networks. Phys. Rev. Lett. 92, 118701–118704 (2004)
19. Zhang, Z.Z., Qi, Y., Zhou, S.G., Gao, S.Y., Guan, J.H.: Explicit determination of mean first-passage time for random walks on deterministic uniform recursive trees. Phys. Rev. E 81, 016114–016121 (2010)
20. Zhang, Z.Z., Wu, B., Zhang, H.J., Zhou, S.G., Guan, J.H., Wang, Z.G.: Determining global mean-first-passage time of random walks on Vicsek fractals using eigenvalues of Laplacian matrices. Phys. Rev. E 81, 031118–031124 (2010)
21. Wang, S.P., Pei, W.J.: Detecting unknown paths on complex networks through random walks. Physica A 388, 514–522 (2009)
22. Almaas, E., Kulkarni, R.V., Stroud, D.: Scaling properties of random walks on small-world networks. Phys. Rev. E 68, 056105–056110 (2003)
23. Feige, U.: A Tight Lower Bound on the Cover Time for Random Walks on Graphs. Random Structures and Algorithms 6(4), 433–438 (1995)
24. Kahn, J.D., Linial, N., Nisan, N., Saks, M.E.: On the cover time of random walks on graphs. Journal of Theoretical Probability 2(1), 121–128 (1989)
25. Aldous, D.J.: Lower bounds for covering times for reversible Markov chains and random walks on graphs. Journal of Theoretical Probability 2(1), 91–100 (1989)
26. Dolev, S., Schiller, E., Welch, J.L.: Random Walk for Self-Stabilizing Group Communication in Ad Hoc Networks. IEEE Transactions on Mobile Computing 5(7), 893–905 (2006)
27. Tian, H., Shen, H., Matsuzawa, R.: Maximizing Networking Lifetime in Wireless Sensor Networks with Regular Topologies. In: Proceedings of the Sixth International Conference on Parallel and Distributed Computing, Applications and Technologies, Dalian, China, pp. 211–217 (2008)
28. Barabási, A.L., Albert, R.: Emergence of Scaling in Random Networks. Science 286, 509–512 (1999)
29. Watts, D.J., Strogatz, S.H.: Collective dynamics of 'small-world' networks. Nature 393, 440–442 (1998)
30. Erdös, P., Rényi, A.: On random graphs. Publ. Math. 6, 290–297 (1959)

Face Recognition by Using Overlapping Block Discriminative Common Vectors

Xiaohui Wang

Department of Computer Application and Technology
HanShan Normal University, Chaozhou, Guangdong Province, China
xiaohui_w@yahoo.cn

Abstract. Discriminative Common Vectors (DCV) has been widely used in face recognition. Previous literatures show that DCV can outperform PCA or LDA in classification accuracy of face images. In this paper, the author proposes a novel block DCV method, i.e. overlapping block DCV. This method first partitions every image into a number of blocks and views each block as a sample. Calculating the covariance matrix and solving its eigen values and eigenvectors are similar to PCA. Then the method chooses any sample from each class and projects it onto the null space to obtain the Common Vectors. DCV takes the Common Vectors as transform axes and exploits the transform axes to perform feature extraction. Compared with conventional DCV, overlapping block DCV seems to be more robust to the variation of facial details such as facial expression and can obtain a higher classification accuracy for face recognition.

Keywords: principal component analysis, common vectors, discriminative common vector, face recognition.

1 Introduction

Face recognition has been used to a large number of real-world applications such as the borderline control, the attendance system and the vision system of the robot [1-8].

The application of principal component analysis (PCA) can be viewed as milestone of face recognition [9,10]. But PCA often loses more useful information for classification in null space [11,12]. DCV performs feature extraction by taking the eigenvectors corresponding to a number of smallest eigenvalues (i.e. all zero eigenvalues) of the covariance matrix [11-16]. The common vector of DCV is obtained by eliminating all the components of a feature vector that are along the eigenvectors corresponding to the nonzero eigenvalues (i.e. a number of largest eigenvalues) of the covariance matrix [12]. Further improving the performance of face recognition, people also extended the methodology of DCV [13-16].

Block PCA is one of the most noticeable methods for face recognition. It can produce a better recognition performance than conventional PCA [17-20]. Block PCA has the following advantage: the sample's dimension of block PCA is lower than conventional PCA; it also has a lower-dimensional covariance matrix than PCA. For

J. Yang, F. Fang, and C. Sun (Eds.): IScIDE 2012, LNCS 7751, pp. 238–245, 2013.
© Springer-Verlag Berlin Heidelberg 2013

example, if the original image is a 100×100 matrix, then the covariance matrix of conventional PCA is a 10000×10000 matrix. As a result, to solve the eigen-equation of this covariance matrix needs a very high computational cost. However, if a block PCA partitions a face image into four blocks, the corresponding covariance matrix will be only a 2500×2500 matrix, which requires a much less memory space and lower computational cost.

We note that block PCA always partitions the original image into non-overlapping blocks. Recently, we see that Xu et al. proposed to first partition the original image into a number of overlapping blocks and to perform feature extraction for the obtained overlapping blocks [17]. We identify that the improvement to the interest operator proposed by Xu et al. is very effective and can achieve very high classification accuracy [17]. Motivated by this, we propose a novel block DCV method, i.e. overlapping block DCV (OBDCV). It uses the within-class scatter matrix of all the classes to obtain the common vectors. The major characteristic of the DCV method is that its projection matrix P_{DCV} resides in the null space of the within-class scatter matrix. OBDCV first partitions every image into a number of blocks and views each block as a sample. After calculating the within-class covariance matrix and solving its eigenvalues and eigenvectors, OBDCV takes the eigenvectors corresponding to all the nonzero eigenvalues of the within-class covariance matrix as transform axes and exploits the transform axes to perform feature extraction.

Let the training set be composed of C classes, with the ith class containing $N_i (> 1)$ d-dimensional samples. Suppose that the training samples are linearly independent, which can be generally satisfied in such applications as face classification. Then there will be a total of $N = N_1 + N_2 + \cdots + N_C$ linearly independent training samples. Note that in such high dimension data classification as face recognition, the SSS problem exists, namely $d \gg N$ generally holds.

Description $\{x_j^i\}$: Let x_j^i be a d-dimensional column vector which denotes the jth sample from the ith class, and then $X = [x_1^1, x_2^1, \cdots, x_{N_1}^1, x_1^2, \cdots x_{N_C}^C]$ contains all the training samples.

The remainder of this work is organized as follows: In Section 2 we simply describe Discriminant Common Vectors (DCV). In Section 3 we present our method, overlapping block DCV (OBDCV). In Section 4 we show the experimental results. In Section 5 we present the conclusion.

2   Discriminant Common Vectors (DCV)

Before describing DCV, we first introduce the idea of common vectors from which DCV is originated. The idea of common vectors is originally introduced for isolated word recognition problems[11,12], in the case where the number of samples in each class is less than or equal to the dimensionality of sample space. These approaches extract the common properties of classes in the training set by eliminating the differences of the samples in each class. A common vector for each individual class is obtained by removing all the features that are in the range space of the scatter matrix of its own class and then the obtained common vectors are used for recognition.

To solve the small sample size problem, the DCV method utilizes the idea of common vector. However, instead of using a given class's own scatter matrix, it uses the within-class scatter matrix of all the classes to obtain the common vectors. The major characteristic of the DCV method is that its projection matrix P_{DCV} resides in the null space of the within-class scatter matrix. Consequently P_{DCV} concentrates the samples from the same class to a unique discriminant common vector and the Fisher's Linear Discriminant criterion achieves a maximum (infinite in fact). In [11,12], the authors gave two theoretically identical ways for implementing the DCV method, i.e., one by eigen-decomposition and the other by difference subspace and the Gram-Schmidt Orthogonalization Procedure. Due to the latter's efficiency over the former, we introduce DCV implemented by the latter procedure as follows[16]:

Step 1: Calculate the range space of the within-class matrix, which is identical to the range space of the difference subspace H_w. Here, H_w is defined as

$$H_w = [b_1^1, \cdots, b_{N_1-1}^1, b_1^2, \cdots, b_{N_C-1}^1],$$
$$b_j^i = x_j^i - x_{N_i}^i, i = 1,2,...,C, j = 1,...,N_i - 1 \tag{1}$$

b_j^i is the jth difference vector of the ith class. Apply the Gram-Schmidt orthogonalization procedure to H_w, and get

$$H_w = UV_1 \tag{2}$$

Then U is an orthonormal matrix whose column vectors span the range space of the within-class matrix.

Step 2: Choose any sample from each class (typically, the last sample of the ith class $x_{N_i}^i$) and project it to the null space of the within-class matrix through the following equation:

$$x_{com}^i = x_j^i - UU^T x_j^i = x_{N_i}^i - UU^T x_{N_i}^i \tag{3}$$

Where x_{com}^i is a common vector of the ith class and is independent of index j.

Step 3: Form the matrix B_{com}, where

$$B_{com} = [b_{com}^1, b_{com}^2, \cdots, b_{com}^{C-1}],$$
$$b_{com}^i = x_{com}^i - x_{com}^C, i = 1,2,\cdots,C-1 \tag{4}$$

Apply the Gram-Schmidt orthogonalization procedure to B_{com}, and get

$$B_{com} = P_{DCV}V_2 \tag{5}$$

Then P_{DCV} is the projection matrix calculated by DCV.

3   Overlapping Block DCV (OBDCV)

Block DCV (BDCV) works as follows: BDCV first partitions each face image into M blocks. Let X_k^{ij}, $j=1,...,N_C, i=1,...,C, k=1,\cdots,M$ stand for the kth block of the jth face image for the ith person. BDCV treats each X_k^{ij} as a training sample and converts it into a one-dimensional column vector Y_k^{ij}. The covariance matrix of OBDCV is defined as

$$E = \frac{1}{MN}\sum_{i=1}^{C}\sum_{j=1}^{N_i}\sum_{k=1}^{M}(Y_k^{ij}-\overline{Y})(Y_k^{ij}-\overline{Y})^T \tag{6}$$

where $k=1,...,M, i=1,...,C, j=1,\cdots,N_i$, $\overline{Y}$ stands for the average value of all of Y_k^{ij}. BDCV computes all the eigenvalues of $\lambda_1 \geq \lambda_2 ... \geq \lambda_D$ and the corresponding eigenvectors $z_1, z_2,..., z_D$, respectively. D is the dimensionality of Y_k^{ij}.

If every sample should be transformed into a d-dimensional vector, then BDCV uses $z_1, z_2,..., z_d$ as transform axes and performs the transform using

$$R_k^{ij} = Z^T Y_k^{ij}, Z=[z_1 z_2 ... z_d] \tag{7}$$

We refer to R_k^{ij} as the feature of sample Y_k^{ij}. The feature of the jth face image for ith from the training set is

$$RR^{ij} = [R_1^{ij} ... R_M^{ij}], i=1,...,C, j=1,\cdots,N_i \tag{8}$$

Let t be one original face image from the testing set. It is clear that t consists of M test samples $t_1, t_2,..., t_M$. The features of these M test samples are

$$R_1 = Z^T t_1, R_2 = Z^T t_2, \cdots, R_M = Z^T t_M. \tag{9}$$

We combine all the features to form a vector

$$TR = [R_1 \quad R_2 \quad R_3, \cdots, \quad R_M] \tag{10}$$

In previous literature [14], an overlapping block partition scheme enables the following feature extraction method to be more robust for the variation of facial details such as facial expression. Motivated by this, we propose a novel block DCV method, i.e. overlapping block DCV(OBDCV). Firstly, OBDCV converts the original image into a one-dimensional vector and partitions the obtained vector into M overlapping sub-vectors. Then it views each sub-vector as a sample and applies block DCV to these samples.

OBDCV partitions $X^{ij}, i=1,...,C, j=1,\cdots,N_i$ (W is the dimensionality of X^{ij})into M overlapping sub-vectors

$$p_1^{ij} = X^{ij}(1:W/M + overl),\cdots,$$
$$p_k^{ij} = X^{ij}((W*(k-1)/M - overl):W*k/M,\cdots, \qquad (11)$$
$$p_M^{ij} = X^{ij}((W*(M-1)/M - overl):W)$$

OBDCV treats each $p_k^{ij}, k=1,...,M; i=1,...,C; \quad j=1,\cdots,N_C$ as one training sample and calculates the covariance matrix using

$$C' = \frac{1}{MN} \sum\nolimits_{i=1}^{N} \sum\nolimits_{j=1}^{N_C} \sum\nolimits_{k=1}^{M} (p_k^{ij} - \overline{p})(p_k^{ij} - \overline{p})^T \qquad (12)$$

where $\overline{p}$ denotes the average value of all of p_k^{ij} .If $z_1', z_2',..., z_d'$ are the d eigenvectors corresponding to the first d largest eigenvalues of C', then the features of one original image X^{ij} is

$$Q = Z'^T X^{ij},$$
$$Z' = [z_1' z_2'...z_d'],$$
$$X^{ij} = [p_1^{ij} p_2^{ij}...p_M^{ij}] \qquad (13)$$

$p_1^{ij}, p_2^{ij},..., p_M^{ij}$ are the M overlapping sub-vectors generated from X^{ij}.

OBDCV classifies t in the following way: let $dist_i$ stand for the Euclidean distance between t and RR^{ij}, i.e. $dist_i = \| RR^{ij} - TR \|, i=1,...,C, j=1,\cdots,N_i$. Suppose $k = \arg \min dist_i$, and then OBDCV assumes that t is from the same class as the ith face image from the training set.

4   Experiments

We used the ORL face database [21] and AR face database [22] to compare overlapping block DCV with conventional DCV and conventional block DCV. The ORL database contains a set of face images taken between April 1992 and April 1994 at the lab. It was used in the context of a face recognition project. There are ten different images of each of 40 distinct subjects. For some subjects, the images were taken at different times, varying the lighting, facial expressions (open/closed eyes, smiling / not smiling) and facial details (glasses/no glasses). All the images were taken against a dark homogeneous background with the subjects in an upright, frontal position (with tolerance for some side movement). The size of each image is 92×112 pixels, with

256 grey levels per pixel. Fig. 1 shows the face images of two persons in the ORL database. The AR face database contains more than 4000 face images. There are 26 different images of each of 126 distinct subjects(70 Male, 56 Female). Images feature frontal view faces with different facial expressions, illumination conditions, and occlusions (sun glasses and scarf). The pictures were taken at the CVC under strictly controlled conditions. No restrictions on wear (clothes, glasses, etc.), make-up, hair style, etc. were imposed to participants. Each person participated in two sessions, separated by two weeks (14 days) time [22]. Fig. 2 shows the eight images(cropped image 80×100) of one persons in AR database.

Fig. 1. Five face images of one person in the ORL database

Fig. 2. Some face images of one person in the AR database

In the experiments, the first two to five face images of each person in ORL database were used for training and the other face images were used for testing, and the first five to eight images of each person(the first 119 persons) in AR database . M is set to $M = 4$ and *overl* is set to *overl* = 100. Table 1 and Table 2 show the best classification accuracies and the right number for their test samples on ORL and AR database seperately. We see that overlapping block DCV produces a higher accuracy than both conventional DCV and conventional Block DCV.

Table 1. Classification result with the first two to five face images of each person being used for training on ORL database. The number in parenthesis denotes the total number of the test samples (shown in the first and second columns) or correctly classified test samples (shown in the third to fifth columns).

Train sample number	Test sample number	DCV(right number total)	BDCV(right number total)	OBDCV(right number total)
2(80)	8 (320)	89.125%(285)	89.688%(287)	89.375%(286)
3(120)	7 (280)	91.786%(257)	92.143%(258)	92.143%(258)
4(160)	6 (240)	95.833%(230)	96.25%(231)	96.25%(231)
5(200)	5 (200)	96.5%(193)	96.25%(193)	97%(194)

Table 2. Classification result with the first two to five face images of each person being used for training on AR database. The number in parenthesis denotes the total number of the test samples (shown in the first and second columns) or correctly classified test samples (shown in the third to fifth columns).

Train sample number per subject	Test sample number per subject	DCV(right number total)	BDCV(right number total)	OBDCV(right number total)
5(595)	21 (2499)	78.591%(1964)	78.431%(1960)	78.631%(1965)
6(714)	20 (2380)	83.782%(1994)	83.866%(1996)	83.950%(1998)
7(833)	19 (2261)	86.599%(1958)	86.732%(1961)	86.687%(1960)
8(952)	18 (2142)	87.535%(1875)	87.768%(1880)	87.815%(1881)

5   Conclusions

Previous literatures show that DCV can outperform PCA or LDA in classification accuracy of face images[13-15]. In this paper, the author proposes a novel block DCV method, i.e. overlapping block DCV. This method first partitions every image into a number of blocks and views each block as a sample. Calculating the covariance matrix and solving its eigenvalues and eigenvectors are similar to PCA. Then the method chooses any sample from each class and projects it onto the null space to obtain the Common Vectors. DCV takes the Common Vectors as transform axes and exploits the transform axes to perform feature extraction. Compared with conventional DCV, overlapping block DCV seems to be more robust to the variation of facial details such as facial expression and can obtain a higher classification accuracy for face recognition.

References

1. Kim, M., Kim, D., Lee, S.: Face recognition using the embedded HMM with second-order block-specific observations. Pattern Recognition 36, 2723–2735 (2003)
2. Torres, L.: Is there any hope for face recognition? In: Proc. of the 5th International Workshop on Image Analysis for Multimedia Interactive Services, WIAMIS 2004, Lisboa, Portugal, April 21-23 (2004)
3. Xu, Y., Feng, G., Zhao, Y.: One improvement to two-dimensional locality preserving projection method for use with face recognition. Neurocomputing 73, 245–249 (2009)
4. Gross, R., Shi, J., Cohn, J.: Quo vadis Face Recognition? - The current state of the art in Face Recognition. Technical Report, Robotics Institute, Carnegie Mellon University, Pittsburgh, PA, USAL
5. Xu, Y., Song, F., Feng, G., Zhao, Y.: A novel local preserving projection scheme for use with face recognition. Expert System With Applications 37(9), 6718–6721 (2010)
6. Sirovich, L., Meytlis, M.: Symmetry, Probability, and Recognition in Face Space. PNAS - Proceedings of the National Academy of Sciences 106(17), 6895–6899 (2009)
7. Xu, Y., Zhong, A., Yang, J., Zhang, D.: LPP solution schemes for use with face recognition. Pattern Recognition 43(12), 4165–4176 (2010)

8. Ekenel, H.K., Fischer, M., Gao, H., Toth, L., Stiefelhagen, R.: Face recognition for smart interactions. In: FG 2008, pp. 1–2 (2008)
9. Xu, Y., Zhang, D., Yang, J.Y.: A feature extraction method for use with bimodal biometrics. Pattern Recognition 43, 1106–1115 (2010)
10. Kirby, M., Sirovich, L.: Application of the KL procedure for the characterization of human faces. IEEE Trans. Pattern Analysis and Machine Intelligence 12(1), 103–108 (1990)
11. Gülmezoğlu, M.B., Dzhafarov, V., Keskin, M., Barkana, A.: A Novel Approach to Isolated Word Recognition. IEEE Transactions on Speech and Audio Processing 7(6), 620–628 (1999)
12. Gülmezoğlu, M.B., Dzhafarov, V., Barkana, A.: The Common Vector Ap-proach and Its Relation to Principal Component Analysis. IEEE Transactions on Speech and Audio Processing 9(6), 655–662 (2001)
13. Cevikalp, H., Wilkes, M.: Face Recognition by Using Discriminative Common Vectors. In: Proceedings of the 17th International Conference on Pattern Recognition, Cambridge,UK, vol. 1, pp. 326–329 (2004)
14. Cevikalp, H., Neamtu, M., Wilkes, M., Barkana, A.: A Novel Method for Face Recognition. In: Proceedings of the IEEE 12th Signal Processing and Communications Applications Conference, Kuşadası, Turkey, pp. 579–582 (2004)
15. Cevikalp, H., Neamtu, M., Wilkes, M., Barkana, A.: Discriminative Common Vectors for Face Recognition. IEEE Transactions on Pattern Analysis and Machine Intelligence 27(1), 4–13 (2005)
16. Liu, J., Chen, S.C.: Discriminant Common Vectors versus Neighbourhood Com-ponents Analysis and Laplacianfaces: A Comparative Study in Small Sample Size Problem. Image and Vision Computing 24, 249–262 (2006)
17. Xu, Y., Yao, L., Zhang, D.: Improving the interest operator for face recognition. Expert System with Applications 36(6), 9719–9728 (2009)
18. Wang, L.W., Wang, X., Zhang, X.R., Feng, J.F.: The equivalence of two-dimensional PCA to line-based PCA. Pattern Recognition Letters 26(1), 57–60 (2005)
19. Sun, X., Liu, B., Liu, B.Y.: Face Recognition Based on Block-PCA. Computer Engineering and Applications 27, 80–82 (2005)
20. Wang, D.R., Liu, H.L., Wang, A.Z., Li, M.D.: Performance analysis of PCA, 2DPCA and block-based PCA in face recognition. Science and Technology of West China 08(27), 14–16 (2009)
21. http://www.cl.cam.ac.uk/research/dtg/attarchive/
   facedatabase.html
22. Martinez, A.M., Benavente, R.: The AR face database. CVC Tech. Report #24 (1998)

Supervised Kernel Self-Organizing Map

Dongjun Yu[1,2], Jun Hu[1], Xiaoning Song[3], Yong Qi[1,2], and Zhenmin Tang[1]

[1] School of Computer Science and Engineering,
Nanjing University of Science and Technology, Nanjing 210094, China
[2] Changshu Institute, Nanjing University of Science and Technology,
Changshu 215500, China
[3] School of Computer Science and Engineering,
Jiangsu University of Science and Technology, Zhenjiang 212003, China
njyudj@njust.edu.cn

Abstract. We generalize the traditional supervised self-organizing map to supervised kernel self-organizing map by incorporating the kernel function to further improve its capability of solving non-linear problems. The kernel function maps the low-dimensional input space to high-dimensional feature space thus potentially makes the complex non-linear structure in the input space much easier in the mapped feature space. Qualitative and quantitative analysis of the experimental results on the two benchmark datasets illustrate the effectiveness of the proposed method.

Keywords: Self-organizing Map, Kernel Function, Supervised Learning, Non-linear System.

1   Introduction

The Self-Organizing Map (SOM) [1] proposed by Kohonen has been widely used in many fields such as pattern recognition [2, 3], data mining [4] and clustering analysis [5]. The traditional SOM is a typical unsupervised learning model and much effort has been made to facilitate the SOM to suit for supervised learning tasks. One of the most straightforward methods to construct supervised SOM (SSOM) was proposed by Barreto et al. in [6].

In SSOM [6], each training sample consists of two parts, which correspond to the input and the desired output of a input-output mapping, respectively. Accordingly, the weight of each output node of SSOM is divided into two parts. During the training stage, the first parts of training samples and weights are used to locate the best matching unit (BMU), while the second parts of training samples are used as teachers to guide the learning of the second parts of weights.

However, in SSOM [6], the Euclidean distance metric is taken for locating the BMU, thus the effectiveness will be significantly affected by the nonlinearity of the underlying data. In this paper, we generalize the SSOM to supervised kernel self-organizing map (SKSOM) by incorporating the kernel function into the SSOM. The essence of SKSOM is to map the low-dimensional input space to a high-dimensional feature space and potentially convert the complex nonlinear problem much easier in

J. Yang, F. Fang, and C. Sun (Eds.): IScIDE 2012, LNCS 7751, pp. 246–253, 2013.

the mapped space. Experimental results on two benchmark datasets demonstrate the effectiveness of the proposed method.

2   Supervised Kernel Self-Organizing Map (SKSOM)

Let $\Phi : \mathbf{a} \in L \rightarrow \Phi(\mathbf{a}) \in F$ be a non-linear mapping, L be the original input space, F be the mapped high-dimensional space. We defined the distance metric in the mapped space as:

$$KD(\mathbf{a},\mathbf{b}) = \|\Phi(\mathbf{a}) - \Phi(\mathbf{b})\|^2 \tag{1}$$

The kernel function satisfies the Mercer condition is defined as:

$$K(\mathbf{a},\mathbf{b}) = \Phi(\mathbf{a})^T \Phi(\mathbf{b}) \tag{2}$$

Thus, Eq. (1) can be further formularized as:

$$KD(\mathbf{a},\mathbf{b}) = \|\Phi(\mathbf{a}) - \Phi(\mathbf{b})\|^2 = K(\mathbf{a},\mathbf{a}) + K(\mathbf{b},\mathbf{b}) - 2K(\mathbf{a},\mathbf{b}) \tag{3}$$

Let $X = \{\mathbf{x}(i)\}_{i=1}^{N}$ be the training dataset, and each training sample $\mathbf{x}(i)$ consists of two parts as follows:

$$\mathbf{x}(i) = \begin{pmatrix} \mathbf{x}^{in}(i) \\ \mathbf{x}^{out}(i) \end{pmatrix} \tag{4}$$

where $\mathbf{x}^{in}(i)$ be the input part, $\mathbf{x}^{out}(i)$ be the desired output.

Accordingly, the weight of the output node j $(1 \le j \le K)$ also consists of two parts:

$$\mathbf{w}_j = \begin{pmatrix} \mathbf{w}_j^{in} \\ \mathbf{w}_j^{out} \end{pmatrix} \tag{5}$$

where K is the total number of output nodes.

Let

$$\mathbf{x}_t = \begin{pmatrix} \mathbf{x}_t^{in} \\ \mathbf{x}_t^{out} \end{pmatrix} \in X \tag{6}$$

be the chosen training sample at the learning step t.

Then, the BMU is located by the following formula:

$$j^* = \arg\min_{1 \le j \le K}\{KD(\mathbf{x}_t^{in}, \mathbf{w}_j^{in}(t))\} \tag{7}$$

where $KD(\mathbf{x}_t^{in}, \mathbf{w}_j^{in}(t))$ denotes the kernel distance between $\mathbf{x}_t^{in}$ and $\mathbf{w}_j^{in}(t)$ as defined in Eq.(3).

After locating the BMU, the input and output parts of the weights of the BMU and its neighboring nodes are updated as follows:

$$\mathbf{w}_i^{in}(t+1) = \mathbf{w}_i^{in}(t) + \alpha(t)h(j^*,i;t)\frac{\partial KD(\mathbf{x}_t^{in}, \mathbf{w}_i^{in}(t))}{\partial \mathbf{w}_i^{in}(t)} \tag{8}$$

$$\mathbf{w}_i^{out}(t+1) = \mathbf{w}_i^{out}(t) + \alpha(t)h(j^*,i;t)\frac{\partial KD(\mathbf{x}_t^{out}, \mathbf{w}_i^{out}(t))}{\partial \mathbf{w}_i^{out}(t)} \tag{9}$$

where $\alpha(t)$ is the learning rate, and $h(j^*,i;t)$ is the neighborhood function. Different neighborhood function can be applied. For example, a Gaussian neighborhood function is formularized as:

$$h(j^*,i;t) = \exp\left(-\frac{\| r_i(t) - r_{j^*}(t) \|^2}{2\sigma(t)^2}\right) \tag{10}$$

where $r_i(t)$ and $r_{j^*}(t)$ are the coordinates of the output node i and j^* on the output layer, respectively.

Obviously, the updating equations (8) and (9) will be different when applying different kernel function $K(\cdot,\cdot)$. Taking radial basis kernel function as an example:

$$K(\mathbf{x},\mathbf{y}) = e^{-\|\mathbf{x}-\mathbf{y}\|^2/2\sigma^2} \tag{11}$$

Then, Eq. (8) can be further formularized as follows :

$$\begin{aligned}
\mathbf{w}_i^{in}(t+1) &= \mathbf{w}_i^{in}(t) + \alpha(t)h(j^*,i;t)\frac{\partial KD(\mathbf{x}_t^{in}, \mathbf{w}_i^{in}(t))}{\partial \mathbf{w}_i^{in}(t)} \\
&= \mathbf{w}_i^{in}(t) + \alpha(t)h(j^*,i;t)\frac{\partial\left(K(\mathbf{x}_t^{in},\mathbf{x}_t^{in}) + K(\mathbf{w}_i^{in}(t),\mathbf{w}_i^{in}(t)) - 2K(\mathbf{x}_t^{in},\mathbf{w}_i^{in}(t))\right)}{\partial \mathbf{w}_i^{in}(t)} \\
&= \mathbf{w}_i^{in}(t) + \alpha(t)h(j^*,i;t)\frac{\partial\left(-2K(\mathbf{x}_t^{in},\mathbf{w}_i^{in}(t))\right)}{\partial \mathbf{w}_i^{in}(t)} \\
&= \mathbf{w}_i^{in}(t) - 2\alpha(t)h(j^*,i;t)\frac{\partial e^{-\|\mathbf{x}_t^{in} - \mathbf{w}_i^{in}(t)\|^2/2\sigma^2}}{\partial \mathbf{w}_i^{in}(t)} \\
&= \mathbf{w}_i^{in}(t) - \frac{2\alpha(t)h(j^*,i;t)e^{-\|\mathbf{x}_t^{in}-\mathbf{w}_i^{in}(t)\|^2/2\sigma^2}}{\sigma^2}\left(\mathbf{x}_t^{in} - \mathbf{w}_i^{in}(t)\right)
\end{aligned} \tag{12}$$

Similarly, Eq. (9) can be further formularized as:

$$\mathbf{w}_i^{out}(t+1) = \mathbf{w}_i^{out}(t) - \frac{2\alpha(t)h(j^*,i;t)e^{-\|\mathbf{x}_t^{out}-\mathbf{w}_i^{out}(t)\|^2/2\sigma^2}}{\sigma^2}\left(\mathbf{x}_t^{out} - \mathbf{w}_i^{out}(t)\right) \tag{13}$$

3   Experimental Results and Analysis

3.1   Sunspot Dataset

Sunspots are temporary phenomena on the photosphere of the Sun that appear visibly as dark spots compared to surrounding regions. The frequency and intensity will significantly affect the earth environment. Thus, accurately predicting the trend of sunspot is of great significance for human life and activities. In this study, we obtained the sunspot dataset from SIDC-Solar Influences Data Analysis Center (http://sidc.oma.be/sunspot-data/) [7]. Note that the yearly sunspot dataset (yearssn.dat) was taken as the benchmark dataset.

3.2   Nonlinear System Identification

The nonlinear system [8] defined as follows was taken to generate the second benchmark dataset:

$$y(t+1) = \frac{y(t)y(t-1)(y(t)+2.5)}{1+y^2(t)+y^2(t-1)} + u(t) \tag{14}$$

where $u(t) = \sin(2\pi t / 25)$ is the activation function, and the initial values of the nonlinear system are as follows:

$$y(-1) = 0.5, \; y(0) = 0.9 \tag{15}$$

3.3   Results and Analysis

Experimental Parameters Configuration

The maximal iteration is set to be $T_{max} = 20000$, the width of Gaussian kernel function $\sigma = \sqrt{2}$. The initial and end values of learning rate are $\alpha_0 = 0.8$ and $\alpha_T = 0.03$, respectively. In addition, the learning rate is monotonically decreasing with t according to Eq. (16):

$$\alpha(t) = \alpha_0 (\alpha_T / \alpha_0)^{t/T_{max}} \tag{16}$$

Square neighborhood function is applied. Let *outputwidth* be the width of the output layer, then $h(t)$, the width of square neighborhood at learning step t, is defined as:

$$h(t) = \left\lfloor \frac{1}{2} \times outputwidth \times (1 - t/T_{max}) \right\rfloor \tag{17}$$

The weights of the output nodes are randomly initialized with values among interval (-1, 1).

Data Pre-processing and Sample Generation

As the values of sunspot dataset are much bigger and there may exist noises, we pre-processed the sunspot dataset as follows:

For a sunspot time series $x_1, x_2, \cdots$, and x_n, we first normalize the time series using the following equation:

$$y_i = \frac{x_i - \min_{1 \le j \le n}\{x_j\}}{\max_{1 \le j \le n}\{x_j\} - \min_{1 \le j \le n}\{x_j\}} \tag{18}$$

Then, we perform the median filter on the normalized time series $\{y_i\}_{i=1}^n$ and the obtained time series are used.

On both the two benchmark datasets, 100 time series values are taken to generate training and testing samples. The first 50 values are used to generate training samples and the remaining 50 values are used to generate testing samples. Note that, we use 3 previous values to predict the next value, i.e., the dimensionality of the input is 3 while the dimensionality of the output is 1.

To comprehensively investigate the influence of output size on the performance of SKSOM, we also carried out experiments on both benchmark datasets with different output sizes, i.e., 8×8 and 16×16. Fig. 1 and 2 illustrate the performance of SKSOM on sunspot dataset with output sizes 8×8 and 16×16, respectively; while Fig. 3 and 4 show the performance of SKSOM on the nonlinear system as described in (14) with output sizes 8×8 and 16×16, respectively.

Fig.1 ~ 4 qualitatively illustrates the good performance of SKSOM. From Fig.1~4, we can find that the proposed SKSOM not only performs well on the training dataset, but also has good generalization ability on testing dataset. In addition, we also quantitatively investigate the performance of the proposed SKSOM by comparing it with the SSOM with different output sizes. Each experiment was independently performed 10 times and the averaged results were then reported in Table 1 and 2.

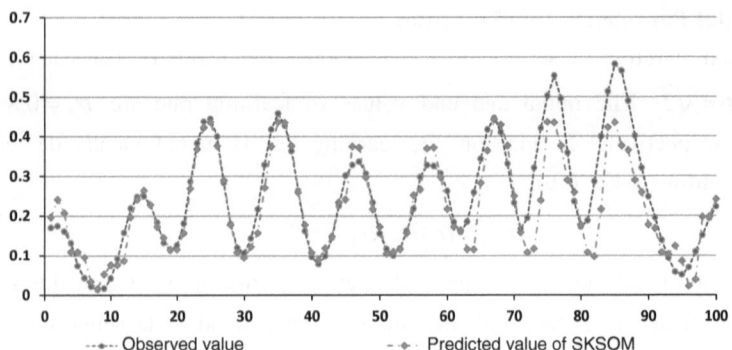

Fig. 1. Performance of SKSOM on sunspot dataset with output size 8×8

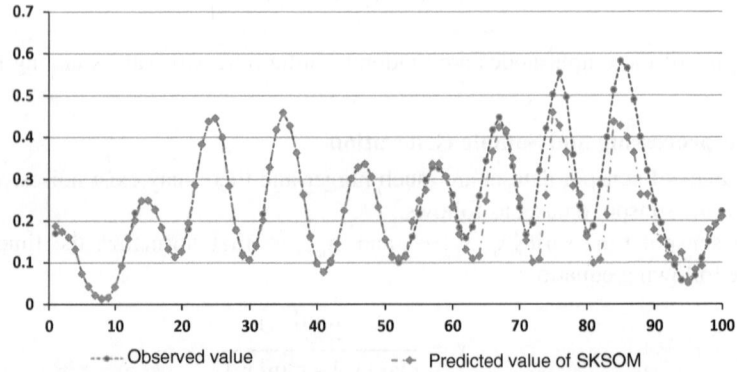

Fig. 2. Performance of SKSOM on sunspot dataset with output size 16×16

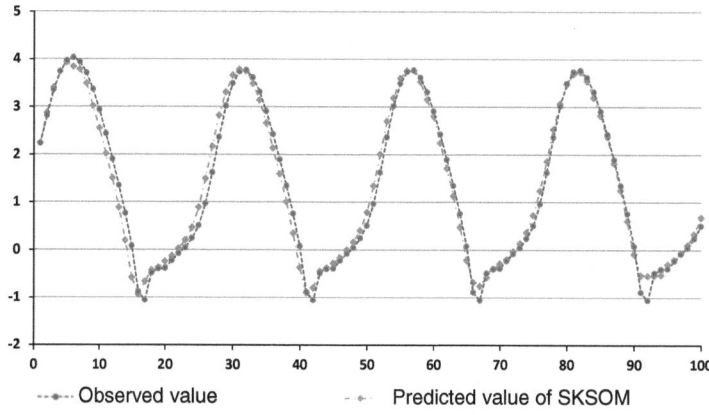

Fig. 3. Performance of SKSOM on nonlinear system with output size 8×8

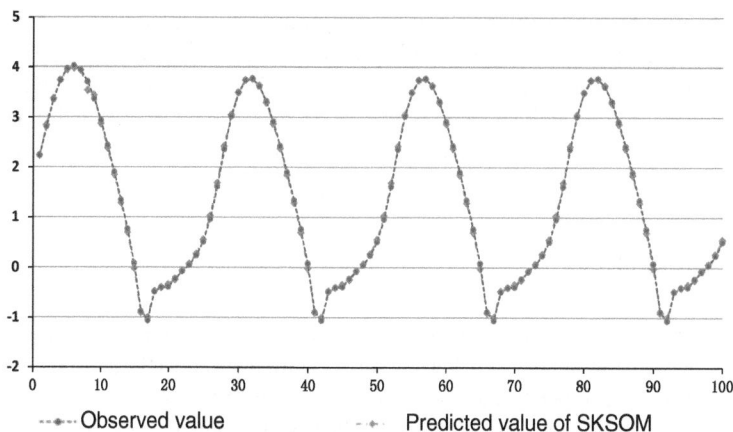

Fig. 4. Performance of SKSOM on nonlinear system with output size (16×16)

Table 1. Performance comparison between SSOM and SKSOM on sunspot dataset

		RMSE	
	Output size	8×8	16×16
SSOM	Error of training samples	0.001619	0.000057
	Error of testing samples	0.007872	0.007685
	Averaged error	0.004745	0.003871
SKSOM	Error of training samples	0.001318	0.000033
	Error of testing samples	0.007681	0.007309
	Average error	0.004500	0.003671

Table 2. Performance comparison between SSOM and SKSOM on non-linear system dataset

		RMSE	
	Output size	8×8	16×16
SSOM	Error of training samples	0.011752	0.000198
	Error of testing samples	0.017578	0.000203
	Averaged error	0.014665	0.000201
SKSOM	Error of training samples	0.011324	0.000070
	Error of testing samples	0.016563	0.000067
	Average error	0.013944	0.000069

By carefully observing Table 1 and 2, the performances of both SSOM and SKSOM increase when the output size is enlarged from 8×8 to 16×16. We explain this phenomenon as follows: when the output size increases, the number of weights connecting the input layer and the output layer will increase accordingly thus can more effectively learn the data distribution of the training samples. While the less number of weights will results in a much rough insight of the data distribution of the training samples thus leading to a poor prediction performance. In addition, we also found that the SKSOM consistently outperforms SSOM on both benchmark datasets. We speculate that the major reason is that in SKSOM the kernel function effectively maps the non-linear structure in the low-dimensional input space to an easier solving structure in the high-dimensional feature space.

4   Conclusion

By incorporating kernel function, we have generalized the supervised self-organizing map (SSOM) to the supervised kernel self-organizing map (SKSOM). On the one hand, the SKSOM inherits the good advantages of SSOM; on the other hand, its capability of solving non-linear problems has been improved by introducing kernel function into SSOM. Experimental results on sunspot and non-linear system datasets demonstrated the effectiveness of the proposed method.

Acknowledgments. This work was supported by the Natural Science Foundation of Jiangsu (Grant No. BK2011371), the Jiangsu Postdoctoral Science Foundation (Grant No. 1201027C), and the Industry-Academia Cooperation Innovation Fund Projects of Jiangsu Province (Grant No. BY2012022).

References

1. Kohonen, T.: Self-organizing map. Proc. IEEE 78, 1464–1480 (1990)
2. Kohler, A., Ohrnberger, M., Scherbaum, F.: Unsupervised pattern recognition in continuous seismic wavefield records using Self-Organizing Maps. Geophysical Journal International 182(3), 1619–1630 (2010)

3. Yu, D.J., Shen, H.B., Yang, J.Y.: SOMRuler: A Novel Interpretable Transmembrane Helices Predictor. IEEE Transactions on Nanobiosci. 10(2), 121–129 (2011)
4. Fan, C.Y., Fan, P.S., Chan, T.Y., Chang, S.H.: Using hybrid data mining and machine learning clustering analysis to predict the turnover rate for technology professionals. Expert Systems with Applications 39(10), 8844–8851 (2012)
5. Liu, Y.C., Wu, C., Liu, M.: Research of fast SOM clustering for text information. Expert Systems with Application 38(8), 9325–9333 (2011)
6. Barreto, G.A., Aluizio, A.F.R.: Identification and control of dynamical systems using the self-organizing map. IEEE Transactions on Neural Networks 15(5), 1244–1259 (2004)
7. Podladchikova, T., Van der Linden, R.: A Kalman Filter Technique for Improving Medium-Term Predictions of the Sunspot Number. Solar Physics 277(2), 397–416 (2012)
8. Lin, C.T. (ed.): Neural Fuzzy Systems. Prentice-Hall Press, New York (1997)

Automated Classification of Protein Subcellular Location Patterns on Images of Human Reproductive Tissues

Fan Yang[1,2,3], Ying-Ying Xu[1,2], and Hong-Bin Shen[1,2,*]

[1] Department of Automation, Shanghai Jiao Tong University, Shanghai, China
[2] Key Laboratory of System Control and Information Processing, Shanghai, China
[3] Key Laboratory of Optic-Electronic and Communication, Jiangxi Science & Technology
Normal University, Nanchang, China
hbshen@sjtu.edu.cn

Abstract. As one of the most significant characteristics of human cell, subcellular localization plays a critical role for understanding specific functions of mammalian proteins. In this study, we developed a novel computational protocol for predicting protein subcellular locations from microscope cell images in human reproductive tissues. Experiments are performed on a benchmark dataset consisting of 7 major subcellular classes in human reproductive tissues collected from Human Protein Atlas database. We first separated protein and DNA staining in the images with both linear and nonnegative matrix factorization separation methods; then we extracted protein multi-view texture features including wavelet haralick and local binary patterns; finally based on the selected important feature subset achieved by feature selection technique, we constructed ensemble classifier based on support vector machines for predictions. Our experimental results show that 84% accuracy can be achieved through current system, and when only considering the most confident classifications, the accuracy can rise to 98%.

Keywords: Subcellular location, Wavelet, Local binary pattern, Feature selection, Ensemble classifier, Support vector machine.

1 Introduction

To understand the specific functions of a protein, it is generally acknowledge that its subcellular location is one of the most crucial characteristic [1, 2]. For example, the subcellular localization can describe the behaviors of a protein under various cell conditions, more importantly, this knowledge can be helpful in early diagnosis, prevention, control of diseases and drug discovery [3]. Although the protein subcellular localizations can be identified by various experiments, in consideration of the premises of time consumption and artificially prone error, simply relying on manpower resources to annotate more than tens of thousands of proteins is considered impossible. It is highly desired to develop automated protein subcellular localization classification systems. Current efforts on automatic predictions of protein subcellular localizations can be generally grouped into two groups depending on how to represent

J. Yang, F. Fang, and C. Sun (Eds.): IScIDE 2012, LNCS 7751, pp. 254–262, 2013.

the protein targets: (1) 1-D sequence based approaches [4], and (2) 2-D image based classification systems [5].

In 2D image based protein subcellular location predictors, subcellular location features (SLFs) sets that can well describe protein subcellular distributions and robust to cell rotations and translation have been proposed in [6, 7]. Owing to the encouraging results obtained from the classifications of images from Chinese hamster ovary (CHO) cells database by SLFs, Boland et al. generated a new image dataset composed of 10 subcellular patterns in HeLa cells, then trained a back propagation neural network (BPNN) to confirm the performance of SLFs [8]. In order to solve the problem of collecting enough images for training classification system, Murphy and Kou developed a system called subcellular location image finder (SLIF) for collecting images from on-line publications based on genome-wide determination [9]. Furthermore, they have also trained BPNN to validate the improvements through SLFs features, and also demonstrated the consistency between subcellular location trees created by SLFs and biological knowledge at that moment [10]. Chen et al. proposed a new algorithm inferred classes from a graphical model combined with support vector machine (SVM) to improve the accuracy of classification of subcellular patterns in multi-cell fluorescence microscope images[11]. Newberg and Murphy have proposed a framework of automated SVM classification system of protein subcellular patterns in human protein atlas images using feature sets consisting of different features of SLFs [7]. It has also been noticed that many different prediction systems have been reported for various organisms because of their specific features and organization components, such as human cell [7], yeast cell [12] etc.

Although much progress has been achieved in this literature, no system has been constructed for predicting proteins' subcellular localizations in human reproductive tissue cells. Considering the importance to understand the protein functions in the reproductive systems, it is also critical to know its cell localizations. Because of this, current study aims to develop a specific prediction protocol in the human reproductive tissues. Beyond the image subcellular location features investigated in the literature such as the wavelet haralick texture feature, current system has also incorporated the local binary pattern (LBP) features. Both the feature selection outputs as well as the classification accuracies demonstrate that the LBP features are positive for improving the prediction performance.

2   Methods

2.1   Datasets

Images from the human protein atlas (HPA) (http://www.proteinatlas.org/) were employed in this study, which is a bountiful source of location proteomics data. Current release of HPA (version 9.0) contains 12238 genes with protein expression profiles based on 15598 antibodies corresponding to 46 different normal human tissues and 20 different cancer types, where the majority tissues images are generated using immunohistochemistry (IH) based on confocal microscopy. We have generated a collection of human reproductive tissue images from the HPA as our benchmark dataset, and a set of 14 proteins of 353 images in different reproductive tissues belonging to 7 subcellular locations have been collected in this study.

The information of the protein chosen, with their antibody identification numbers in HPA and the corresponding subcellular class, are described as follows: Golgin subfamily A member 5 (992, Golgi apparatus), ATP synthase subunit beta (1528, mitochondrion), Eukaryotic translation initiation factor 4E transporter (2078, vesicles), High mobility group protein B1 (3506, nucleolus), Endoplasmin (3901, endoplasmic reticulum/ER), Acetyl-CoA acetyltransferase, mitochondrial(4428, mitochondrion), Nucleolar transcription factor 1(6385, nucleolus), Bcl-2-associated transcription factor 1 (6669, nucleus), Ribosome-binding protein 1 (9026, ER), Cell division cycle 5-like protein(11361, nucleus), Golgin subfamily B member 1 (11555, Golgi apparatus), Phosphatidylinositol-binding clathrin assembly protein (19053, vesicles), Intraflagellar transport protein 20 homolog(21376, cytoskeleton), Cdc42 effector protein 4(23335, cytoskeleton).

2.2   Object Identification

Image samples in HPA are stored in RGB model and are the mixtures of two major components, i.e., DNA and protein, which are labeled with different colors. Since our task is to extract features from protein objects and then perform predictions, it is critical to separate protein parts from the DNA background, which means the first problem we faced is object separation or color separation because DNA and proteins are often labeled with different colors.

Linear Separation

Linear separation works under two assumptions below: first, the signals or images are linearly separable (Eq. 1); second, there is a certain color basis for each image. The performance of linear separation is shown in Figure 1. Suppose the images are linearly separable, we could then achieve the separation of different colors as long as determining the color basis of W in Eq. 1.

$$I_{orig} = W \times I_{separ} \tag{1}$$

where I_{orig} is the original image, W is the color basis, and I_{separ} contains 2 columns which indicate protein and DNA, respectively.

We first invert the background of each image from white to black, and the second step is to transform each image from RGB space into HSV space, and then a histogram in hue subspace is built. All of the bins of the histogram are used to define hues of brown and purple, where the bins above or equal to the threshold are considered as the spectrum of brown, otherwise the spectrum of purple. In this study, the threshold of hue is set to 0.3 according to our experiments and other studies [7]. Following above steps, we can get the color basis W of all the images on our benchmark dataset as:

$$W = \begin{bmatrix} coef_R^{DNA} & coef_R^{prot} \\ coef_G^{DNA} & coef_G^{prot} \\ coef_B^{DNA} & coef_B^{prot} \end{bmatrix} = \begin{bmatrix} 12.5728 & 10.3871 \\ 12.0870 & 12.4426 \\ 9.5335 & 14.0513 \end{bmatrix} \tag{2}$$

With W defined in Eq. 2, we then can separate the original images by:

$$I_{separ} = I_{orig} \times pinv(W)'$$ (3)

where $pinv(W)$ denotes the pseudo-inverse of matrix W.

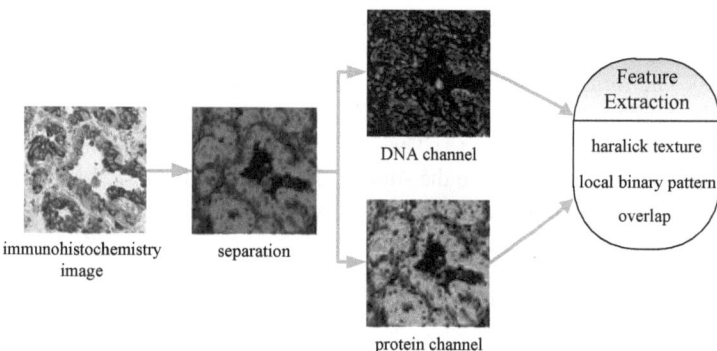

DNA channel

Feature Extraction

haralick texture

local binary pattern

overlap

immunohistochemistry image

separation

protein channel

Fig. 1. The performance of linear separation on the image of seminal vesicle tissue of HPA009026 antibody in Human Protein Atlas

Blind Source Separation by Non-Negative Matrix Factorization

An alternate approach of color separation is the blind source separation. In this paper, we employ non-negative matrix factorization (NMF) algorithm to separate the protein and DNA channels. NMF works under the physical constraint that each color will have a non-negative response in its own frequency band. Different from the linear separation that has a uniform color basis W for all the images, each image in NMF is corresponding to its unique color basis. Except for this, other steps of NMF-based color separation are similar to those of linear color separation. For more detailed mathematical derivations about NMF, readers are suggested to read ref. [13]

2.3   Feature Extraction

Haralick Texture Features

Due to the arbitrary location and orientation of cells in images, features sought were expected to be invariant to the translation and rotation. Here we employed Haralick texture features, which were extracted from the spatial grey level dependence matrices (SGLD) obtained by N gray levels at a fixed angle θ and a determinate d denoting the measurement of pixel distance [14]. In 2D space, the Haralick descriptor can be extracted from four directions of SGLD, namely horizontal, vertical, and two diagonal directions. In this work, we concatenate the features extracted considering horizontal and vertical direction, and similarly with another two diagonal directions,.

Next, discrete wavelet transform (DWT) was employed to decompose the protein image to one low and three high frequencies sub-bands. The energy of each three high frequency sub-band at each decomposition level was employed as another feature. In this work we extract the Haralick texture features by Daubechies wavelet with the vanishing moment from 1 to 10.

DNA Distribution Features

The significant difference between the eukaryotic and prokaryotic cells is the nucleus, a membrane-enclosed organelle. Because human belongs to the eukaryotic species, thus each image will contain the DNA parts. It has been pointed out in [7] that DNA spatial distributions of human cell is quite valuable for improving the prediction accuracy. Hence, we also extract 4 DNA overlap features on the resulting binary images: (1) The ratio of the sum of pixel value in protein area to DNA area; (2) The ratio of the number of pixel of protein area that co-localize with DNA area to the number of pixel of protein area; (3) The ratio of the sum of pixel value in protein area that co-localize with DNA area to the sum of pixel in protein area; and (4) The average distance between protein (overlap DNA area) and nearest nuclear pixel.

Local Binary Patterns

The local binary pattern (LBP) is a simple yet efficient operator proposed by Ojala et al. [15], and the descriptor describes the center pixel by the combination of a given threshold and the relative gray value of a neighboring pixel. The value of the pixel in a local neighborhood is set to 1 where the gray value of neighboring pixel is greater or equal than the threshold, otherwise is set to 0:

$$LBP_{K,R} = \sum_{k=0}^{K-1} s(g_k - g_c)2^k, \quad s(\cdot) = \begin{cases} 1, & x \geq 0 \\ 0, & otherwise \end{cases} \tag{4}$$

where K and R represent the number of neighboring pixels and the radius of the neighborhood, respectively, g_c corresponds to the gray value of the center pixel, and g_k corresponds to the gray value of neighboring pixel.

The histogram of the binary patterns obtained over the neighborhood is used for describing the local texture features [16]. We extracted the LBP features based on the configurations of $R = 2$ and $K = 16$ in uniform and rotation invariant pattern.

2.4   Feature Selections

Tremendous previous studies have demonstrated that not all the features that can be calculated are informative to the following classification. Considering of this, it is also important to perform feature selection step to get an important features subset from the original high dimensional features. To reach this goal, we applied the so called stepwise discriminant analysis (SDA) approach, which use Wilks' λ statistic to iteratively determine which features are the most discriminating in the feature space to separate the classes from one another [8].

2.5   Classifier Design

In this work, the LIBSVM-3.11 toolbox (from http://www.csie.ntu.edu.tw/~cjlin/libsvm) is employ to implement the 7-class classification of protein subcellular patterns, where the parameters, slack penalty and gamma of kernel function were determined by grid searching with 10-fold cross validations.

Both feature fusion and decision fusion mechanisms are implemented in our system. As shown in previous section, three kinds of selected features are combined as the input of single SVM for training process. The system of voting SVM classifier was also constructed based on 10 separate single classifiers, where features of the sample in these 10 classifiers are different from each other. Namely, Daubechies wavelet was employed to decompose the image in order to extract multi-resolution features from different vanishing moment, and then 10 separate classifiers can be constructed. Finally, we utilize these 10 different classifiers to decide a final prediction. In this work, we calculated the average posterior probability of 10 separate classifiers.

3   Results and Discussions

3.1   Results Affected by Color Separation Approaches

As discussed above, two approaches of color separation have been discussed in this study, i.e., linear separation and NMF approaches. Table 1 summarizes the classification accuracies on 10 SVMs on 10 different vanishing moment of Daubechies wavelet by taking linear and NMF separations respectively. As shown in Table 1, the designed linear color separation approach (Eqs. 2 and 3) outperforms those from NMF on all the 10 tests in current study by 5% to 18%. These results indicate that the classification accuracies are significantly dependent on the separation outputs. This is reasonable because good color unmixing methods will make us capable of clearly identify the protein targets. We also argue that although the performances of NMF are not as good as the linear approach in this study, the results can be still enhanced if we combine NMF and the linear approach together as demonstrated in the following experiments.

Table 1. Performance comparisons between linear and NMF separation approaches, where the input features to SVM are SLFs including Haralick texture features and DNA overlap features

Approach	db1	db2	db3	db4	db5	db6	db7	db8	db9	db10
Linear	0.790	0.670	0.722	0.659	0.688	0.653	0.710	0.761	0.744	0.636
NMF	0.664	0.551	0.659	0.620	0.625	0.506	0.602	0.580	0.614	0.557

3.2   Results Enhanced by Incorporating LBP Features

Besides the wide-used of image descriptors SLFs of Haralick texture features and DNA overlap features, we also investigated another image descriptor called LBP (Eq. 4). We systematically analyzed the outputs from SDA feature selection methods to demonstrate the importance of the LBP features in current study. On the list of top 14 features outputted from SDA selection approach, eight are haralick texture features, two are DNA overlap features, and the other two are LBP features, and LBP features are ranked high, see Table 2. In order to further demonstrate the roles of LBP, we compare the classification accuracies on cases of before and after adding LBP respectively. Figure 2 illustrates the results on the 10 different vanishing moment of Daubechies wavelet. Results shown in Figure 2 clearly demonstrate that by adding LBP, the performances are enhanced, and the improvements are observed from 2% to 14% and 3% to 17% in linear and NMF separation respectively. These results demonstrate the effectiveness and practicality of LBP features.

Table 2. Features ranked by SDA for haralick (H), DNA overlap (D), and LBP features (L)

Rank	1	2	3	4	5	6	7	8	9	10	11	12	13	14
Features	H	H	D	H	H	H	H	H	L	D	H	L	H	H

Table 3. Comparison between independent and ensemble classifications, where the input features are Haralick features, DNA overlap features, and LBP

Color Separation	Independent prediction	Ensemble prediction
Linear separation	0.7784	0.8295
NMF	0.7272	0.7500
Linear + NMF	0.8352	0.8352

Table 4. The Confusion Matrix for Classification of 7 classes under linear and NMF color separation, where the input features are Haralick features, DNA overlap features, and LBP

Classes	ER %	Cyto %	Golgi %	Mito. %	Nucleolus %	Nucleus %	Vesicles %	Number of images
ER	**83.3**	0	12.5	0	0	0	4.2	24
Cyto	0	**92**	0	4	0	0	4	25
Golgi	7.7	0	**84.6**	3.9	0	3.9	0	26
Mito	0	0	4.2	**91.7**	0	4.2	0	24
Nucleolus	0	0	0	0	**61.5**	38.5	0	26
Nucleus	0	0	0	4	0	**96**	0	25
Vesicles	7.7	0	0	7.7	3.9	3.9	**76.9**	26

3.3   Results Improved by Fusion Strategy

Our previous experiments have shown that results from the 10 different vanishing moment of Daubechies wavelet are different (Table 1), indicating that various information can be buried into the 10 channels. This also makes us capable of investigating the fusion strategy among 10 classifiers in order to make them being complementary to each other. Therefore, we have realized an ensemble classifier for this task, where we summed all the prediction probabilities over the 10 classifiers and then chose the highest likelihood to assign the sample to its correspond class. The results are shown in Table 3. From Table 3, we can find two things: (1) Ensemble strategy can improve the prediction accuracies compared with those independent predictors, the performances are enhanced, and the improvements are observed from 2% to 5%. (2) Results can be further enhanced if we combine the two different color separation approaches achieved by fusing the outputs of probabilities by SVM trained under the two conditions. Table 4 shows the confusion matrix of ensemble all classifiers under combination of linear and NMF separations. If we only consider the images with prediction probabilities higher than 0.5 of Table 4 (confident prediction), then the accuracy can reach 98.82%. All the above results demonstrate that constructing a proper ensemble classifier by fusing multi-view knowledge is very promising for improving the classifier performance.

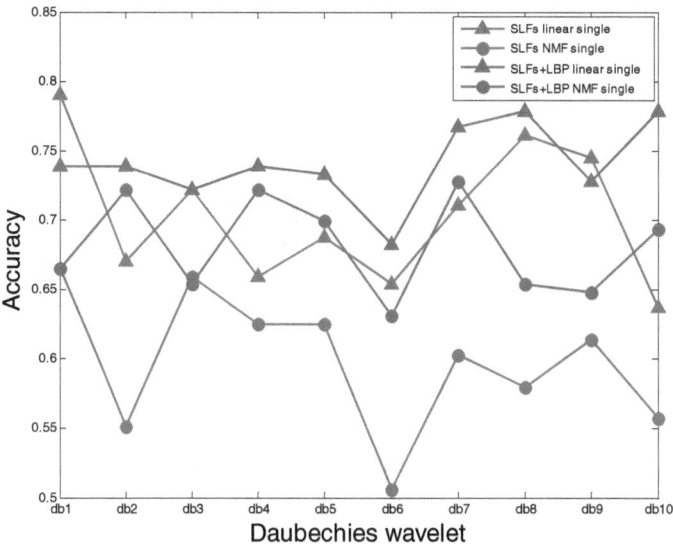

Fig. 2. Comparisons of classification accuracies before and after adding LBP on both linear and NMF separation results

4   Conclusions

We have presented a novel prediction model for classifying protein subcellular location images of human reproductive tissues from HPA database. This model features by multi-view information fusion and high prediction accuracy. Our results show that the newly introduced LBP features, as well as multi-resolution texture features and DNA overlap features are all informative in recognizing protein subcellular location patterns. Our results also show that different color separation methods have significant effects on the following feature extraction and classification processes. Although our method is promising, as shown in Tables 4, current predictor often confuses nucleolus with nucleus classes. The reason can be nucleolus itself also located in the nucleus; hence the pattern similarity between nucleolus and nucleus is high. In the future, we will design specific features for discriminating these two classes and design a kind of hierarchical system to solve this problem.

Acknowledgements. This work was supported by the National Natural Science Foundation of China (No. 61222306, 91130033, 61175024), Shanghai Science and Technology Commission (No. 11JC1404800), A Foundation for the Author of National Excellent Doctoral Dissertation of PR China (No. 201048).

References

1. Chou, K.C., Shen, H.B.: Cell-PLoc: a package of Web servers for predicting subcellular localization of proteins in various organisms. Nature Protocols 3(2), 153–162 (2008)
2. Shen, H.B., Chou, K.C.: Gpos-mPLoc: A top-down approach to improve the quality of predicting subcellular localization of Gram-positive bacterial proteins. Protein and Peptide Letters 16(12), 1478 (2009)
3. Chou, K.C., Shen, H.B.: Recent progress in protein subcellular location prediction. Analytical Biochemistry 370(1), 1–16 (2007)
4. Chou, K.C., Shen, H.B.: Euk-mPLoc: a fusion classifier for large-scale eukaryotic protein subcellular location prediction by incorporating multiple sites. Journal of Proteome Research 6(5), 1728–1734 (2007)
5. Peng, T., Bonamy, G.M.C., Glory-Afshar, E., Rines, D.R., Chanda, S.K., Murphy, R.F.: Determining the distribution of probes between different subcellular locations through automated unmixing of subcellular patterns. Proceedings of the National Academy of Sciences 107(7), 2944–2949 (2010)
6. Newberg, J., Hua, J., Murphy, R.F.: Location proteomics: systematic determination of protein subcellular location. Methods in Molecular Biology, Systems Biology 500, 1–20 (2009)
7. Newberg, J., Murphy, R.F.: A framework for the automated analysis of subcellular patterns in human protein atlas images. Journal of Proteome Research 7(6), 2300–2308 (2008)
8. Boland, M.V., Markey, M.K., Murphy, R.F.: Automated recognition of patterns characteristic of subcellular structures in fluorescence microscopy images. Cytometry 33(3), 366–375 (1998)
9. Murphy, R.F., Velliste, M., Yao, J., Porreca, G.: Searching online journals for fluorescence microscope images depicting protein subcellular location patterns, pp. 119–128. IEEE (2001)
10. Murphy, R.F., Velliste, M., Porreca, G.: Robust numerical features for description and classification of subcellular location patterns in fluorescence microscope images. The Journal of VLSI Signal Processing 35(3), 311–321 (2003)
11. Chen, S.C., Murphy, R.: A graphical model approach to automated classification of protein subcellular location patterns in multi-cell images. BMC Bioinformatics 7(1), 90 (2006)
12. Huh, S., Lee, D., Murphy, R.F.: Efficient framework for automated classification of subcellular patterns in budding yeast. Cytometry A 75(11), 934–940 (2009)
13. Lee, D.D., Seung, H.S.: Learning the parts of objects by non-negative matrix factorization. Nature 401(6755), 788–791 (1999)
14. Haralick, R.M.: Statistical and structural approaches to texture. Proceedings of the IEEE 67(5), 786–804 (1979)
15. Ojala, T., Pietikainen, M., Maenpaa, T.: Multiresolution gray-scale and rotation invariant texture classification with local binary patterns. IEEE Transactions on Pattern Analysis and Machine Intelligence 24(7), 971–987 (2002)
16. Heikkilä, M., Pietikäinen, M., Schmid, C.: Description of interest regions with local binary patterns. Pattern Recognition 42(3), 425–436 (2009)

Formation Flight of Multi-agent
Based on Formation Feedback Control

Lei Zhang, Yangwang Fang, and Xiang Gao

College of Aeronautics and Astronautics,
Air Force Engineering University, Xi'an 710038, China
szl1985@163.com

Abstract. In traditional leader-follower approach, there is no formation feedback from the followers to the leader. If the leader moves too fast, due to the control saturation, the followers will not be able to track the desired position to keep formation. In order to overcome the disadvantage of the leader-follower approach, based on the idea of formation feedback, a leader controller with formation feedback from the followers is designed to guarantee both formation maintenance and formation speed. In general flight, the angular acceleration of the leader can't be obtained by the follower. Based on the sliding mode control theory, the follower controller without the angular acceleration is designed to overcome the problem of information absence. Finally, a simulation of four agents flight in the three-dimension space is given to prove its effectiveness.

Keywords: Formation Flight, Multi-agent, leader-follower, Formation feedback, Sliding mode.

1 Introduction

The Coordination and control of formation of Multi-agent system has received significant attention because of the high performance, lower cost and better fault tolerance in recent years. Many different control approaches have been contributed to the formation control of multiple robots, autonomous underwater vehicles, unmanned air vehicles, spacecraft, aircraft and missiles [1-8]. All these approaches can be categorized as leader-follower approach, behavior based approach, and virtual structure approach. Each approach has its advantages and disadvantages. Due to the simplicity and easier implementation, the leader-follower approach has been applied extensively [1-3,7,8]. In the leader-follower approach, an agent is designated as the leader, moves along a predefined trajectory while the other agents, designated as the followers, are to maintain a desired distance and orientation to the leader. The disadvantage of the leader-follower approach is that there is no feedback from the follower to the leader. Thus, if the speed of the leader is too high, the follower may not be able to track the desired position and the formation can't be maintained.

In order to overcome the disadvantage of the leader-follower approach, based on the idea of formation feedback brought in [5, 6, 9], a leader controller with formation

J. Yang, F. Fang, and C. Sun (Eds.): IScIDE 2012, LNCS 7751, pp. 263–270, 2013.

feedback from the followers is designed to guarantee both formation maintenance and formation speed. As the sliding mode control system can overcome the uncertainty of the disturbance and has a strong robustness, it is introduced in this paper to solve the track problem while the angular acceleration of the leader can't be obtained by the follower. The sliding mode surface functions are designed to make agents asymptotically stabilize.

This paper is organized as follows. In Section 2, the dynamic model of a single agent is described and formation control problem based on the leader-follower approach is formulated. In Section 3, Control strategies are proposed for both the leader and the followers. In Section 4, numerical simulations are also provided to show the validity of the theories.

2   Problem Statement

In this section, we will first introduce inertial position dynamics of agents. Then, a typical leader-follower control problem is formulated and the desired position for each agent will be derived.

2.1   Inertial Position Dynamics

Consider a multi-agent system composed of N agents. We denote the state of agent i by $r_i =[x_i\ y_i\ z_i]$ $(i = 0,1,. . . ,N\text{-}1)$ and assume that the agents are moving in an 3 dimensional space with the motion of each agent being described by the integrators in the inertial frame:

$$\begin{aligned} \dot{r}_i(t) &= v_i(t) \\ \dot{v}_i(t) &= u_i(t) \end{aligned} \qquad i = 0,1,\cdots N-1 \qquad (1)$$

where r_i, v_i and u_i are respectively the position, velocity and control input of agent i. The kinematic equations of agent i are:

$$\begin{aligned} \dot{x}_i &= V_i \cos\theta_i \cos\psi_{Vi} \\ \dot{y}_i &= V_i \sin\theta_i \\ \dot{z}_i &= -V_i \cos\theta_i \sin\psi_{Vi} \end{aligned} \qquad (2)$$

where θ_i, ψ_{Vi}, V_i are respectively the path angle, heading angle and speed of agent i.

2.2   Leader-Follower Formation Formulation

In leader-follower formation control, the leader has to track the desired trajectory and the follower tries to maintain a desired distance and angles relative to the leader. When all agents are in expected positions, the desired formation is established. In all the N agents of the multi-agent system, an agent is designated as the leader (simply labeled 0), the others, designated as the followers.

For given waypoints $r_{0,j}^d = [x_0^d \ y_0^d \ z_0^d]_j^T$, $j = 1,2,\cdots m$, m is the number of waypoints, and j is the waypoint index. The leader is assigned to track these waypoints in inertial space. Given the position $r_0(t)$ of the leader, the reference trajectory for follower i ($i =1,\ldots,N$-1)is set in such a way that its position is shifted by a distance d_i, path angle θ_{0i}, heading angle ψ_{0i} relative to the leader. The desired position for follower i is given by:

$$r_{0i}^d = [d_i \cos\theta_{0i} \cos\psi_{0i} \quad d_i \sin\theta_{0i} \quad -d_i \cos\theta_{0i} \sin\psi_{0i}]^T$$
$$r_i^d(t) = r_0(t) + L[\theta_0(t),\psi_{v0}(t)]r_{0i}^d \tag{3}$$

where r_{0i}^d is the position of agent i in the reference frame C_0, C_0 is embedded in agent i as a velocity frame. $L[\theta_0(t),\psi_{v0}(t)]$ is the rotation matrix of the frame C_0 with respect to the inertial frame.

3   Controllers Design for Formation Flight

In this section we design controllers for both leader and followers. Formation feedback based on the relative distance between c and leader is introduced from the followers to the leader. The sliding mode controllers for the followers are designed to solve the problem of information absence.

3.1   Controllers Design

The main objective of the formation controllers is to drive $r_0(t)$ to the given waypoints $r_{0,j}^d$ in succession and guarantee $r_i(t)$ tracks $r_i^d(t)$ .Furthermore, while the speed of the leader is too high and the relative distance between the leader and followers is longer than desired distance, the speed of the leader need to slow down. Contrarily, the speed needs to increase. Aim to track waypoints and maintain formation flight, the controller is designed as follows:

$$u_0 = K_r(r_0^d - r_0) + \sum_{i=1}^{N-1} K_l(l_{0i}^d - l_{0i})\bar{n}_{0i} - K_v v_0 \tag{4}$$

where K_r , K_l and K_v are symmetric positive definite matrices, l_{0i} and l_{0i}^d are respectively the actual and desired distance between the leader and follower i, $l_{0i} = \|r_0 - r_i\|$, $\bar{n}_{0i} = \dfrac{r_0 - r_i}{\|r_0 - r_i\|}$ is unit vector.

The track error of agent i in the inertial coordinate frame is:

$$E_i(t) = r_i^d(t) - r_i(t) = r_0(t) + L[\theta_0(t),\psi_{v0}(t)]r_{0i}^d - r_i(t) \tag{5}$$

Differentiating (5) with respect to time yields:

$$\dot{E}_i(t) = \dot{r}_0(t) + L'[\theta_0(t), \psi_{v0}(t)]r_{i0}^d - \dot{r}_i(t) \tag{6}$$

Thus,

$$\ddot{E}_i(t) = \ddot{r}_0(t) + L''[\theta_0(t), \psi_{v0}(t)]r_{i0}^d - \ddot{r}_i(t) \tag{7}$$

In (7), $L''[\theta_0(t), \psi_{v0}(t)]$ is a function matrix about $\theta_0, \psi_{v0}, \dot{\theta}_0, \dot{\psi}_{v0}, \ddot{\theta}_0, \ddot{\psi}_{v0}$. If θ_0, $\psi_{v0}, \dot{\theta}_0, \dot{\psi}_{v0}, \ddot{\theta}_0, \ddot{\psi}_{v0}$ are all available for the follower i, the controller u_i can be designed as:

$$u_i = \ddot{r}_0(t) + L''[\theta_0(t), \psi_{v0}(t)]r_{i0}^d + k_1\dot{E}_i(t) + k_2 E_i(t) \tag{8}$$

where k_1, k_2 are symmetric positive definite matrices.

In general, $\ddot{\theta}_0, \ddot{\psi}_{v0}$ are not easy to be obtained and $L''[\theta_0(t), \psi_{v0}(t)]$ is a bounded matrix, i.e. there exist f_j satisfy:

$$|L_j^r| \le f_j, j = 1,2,3 \tag{9}$$

where L_j^r is element in $L''[\theta_0(t), \psi_{v0}(t)]r_{i0}^d$.

Define sliding surfaces $s = [s_1 \quad s_2 \quad s_3]^T = \dot{E}_i(t) + K_1 E_i(t)$, where K_1 is symmetric positive definite matrix, then differentiating s with respect to time yields:

$$\dot{s} = \ddot{E}_i(t) + K_1\dot{E}_i(t) = \ddot{r}_0(t) + L''[\theta_0(t), \psi_{v0}(t)]r_{i0}^d - \ddot{r}_i(t) + K_1\dot{E}_i(t) \tag{10}$$

The control law can be defined as follows:

$$u_i = u_0 + K_1\dot{E}_i(t) + \varepsilon \operatorname{sgn}(s) + K_2 s \tag{11}$$

where $\varepsilon = diag[\varepsilon_1 \quad \varepsilon_2 \quad \varepsilon_3]$ and K_2 are symmetric positive definite matrices.

3.2   Stability Analysis

Consider the following Lyapunov function candidate:

$$V_0 = \frac{1}{2}(r_0^d - r_0)^T K_r(r_0^d - r_0) + \frac{1}{2}\sum_{i=1}^{N-1} K_l(l_{0i}^d - l_{0i})^2 - \frac{1}{2}v_0^T v_0 \tag{12}$$

$$V_i = \frac{1}{2}s^T s \tag{13}$$

Differentiate V_0 along the system trajectories:

$$\dot{V}_0 = -\dot{r}_0^T K_r (r_0^d - r_0) - \sum_{i=1}^{N-1} \dot{r}_0^T K_l (l_{0i}^d - l_{0i}) \frac{\partial l_{0i}}{\partial r_0} - v_0^T u_0 \tag{14}$$

Substituting (4) and $\dfrac{\partial l_{0j}}{\partial r_0} = \dfrac{r_0 - r_j}{\|r_0 - r_j\|}$ into the expression of $\dot{V}_0$ yields:

$$\dot{V}_0 = -v_0^T K_v v_0 \le 0 \tag{15}$$

Differentiate V_i with respect to time and substituting (10),(11) into $\dot{V}_i$ yields:

$$
\begin{aligned}
\dot{V}_i &= s^T \dot{s} \\
&= s^T (-\varepsilon \operatorname{sgn}(s) - K_2 s + L''[\theta_0(t), \psi_{v0}(t)] r_{i0}^d) \\
&= -\varepsilon_1 |s_1| - \varepsilon_2 |s_2| - \varepsilon_3 |s_3| + L_1^r s_1 + L_2^r s_2 + L_3^r s_3 - s^T K_2 s \\
&\le -(\varepsilon_1 - f_1)|s_1| - (\varepsilon_2 - f_2)|s_2| - (\varepsilon_3 - f_3)|s_3| - s^T K_2 s
\end{aligned} \tag{16}
$$

For any $\varepsilon_j \ge f_j, j = 1,2,3$, $\dot{V}_i \le 0$ .Therefore, it can be ensured that the formation maneuver will be asymptotically achieved.

4   Simulation Result

In this section, the effectiveness of the formation controller with formation feedback is demonstrated by a simulation of four agents' flight in the three-dimension space. The simulation conditions are initialized as following:

(1)Desired position of followers
①follower 1: $d_1 = 1200m$, $\theta_{01} = -2°$, $\psi_{01} = 120°$;
②follower 2: $d_2 = 1200m$, $\theta_{02} = -2°$, $\psi_{02} = -120°$;
③follower 3: $d_3 = 1200m$, $\theta_{03} = -2°$, $\psi_{03} = 180°$;
(2) Initial state
①leader: $x_0 = 1300m$, $y_0 = 200m$, $z_0 = 0m$, $V_0 = 200m/s$, $\theta_0 = 0°$, $\psi_{v0} = 0°$;
②follower1: $x_1 = 0m$, $y_1 = 150m$, $z_1 = -500m$, $V_1 = 200m/s$, $\theta_1 = 0°$, $\psi_{v1} = 0°$;
③follower2: $x_2 = 0m$, $y_2 = 150m$, $z_2 = 300m$, $V_2 = 200m/s$, $\theta_2 = 0°$, $\psi_{v2} = 0°$;
④follower3: $x_3 = 0m$, $y_3 = 150m$, $z_3 = 0m$, $V_3 = 200m/s$, $\theta_3 = 0°$, $\psi_{v3} = 0°$;
(3) Limits: $|\dot{V}| \le 7g$, $150m/s \le V \le 300m/s$.

Three-dimension formation routes, track errors and flight speed of four agents without and with formation feedback are shown in Figs. 1 to 6 respectively. In the first case, the formation keeps moving to its final goal even without formation feedback. But the followers cannot track its desired position and the formation cannot maintain the desired configuration. Compared with the first case, leader of formation in the second case with formation feedback tracks the waypoints well and the formation can maintain the desired configuration.

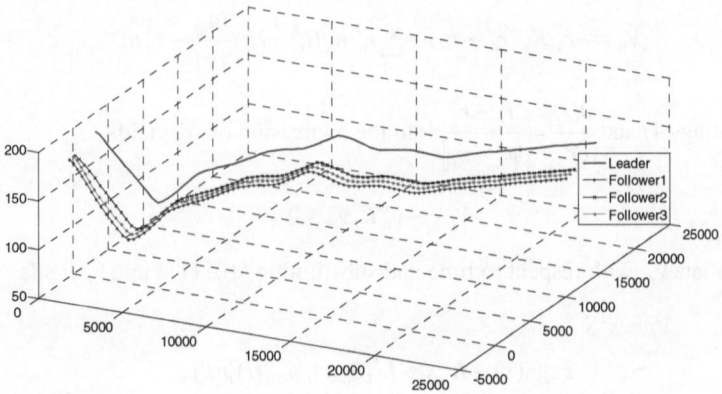

Fig. 1. Three-dimension formation route without formation feedback

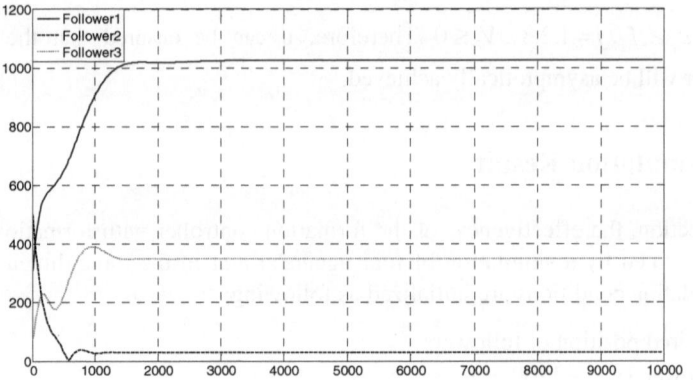

Fig. 2. Track error without formation feedback

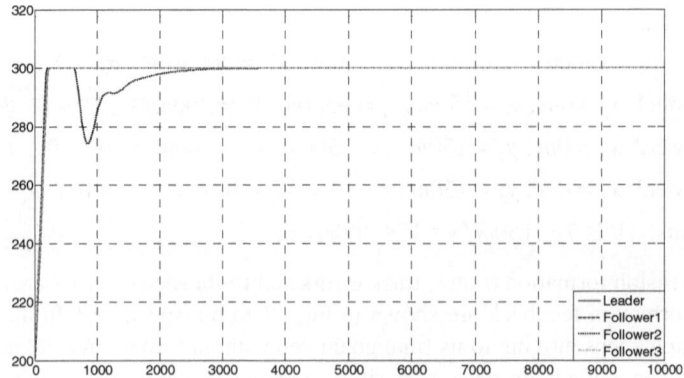

Fig. 3. Flight speed without formation feedback

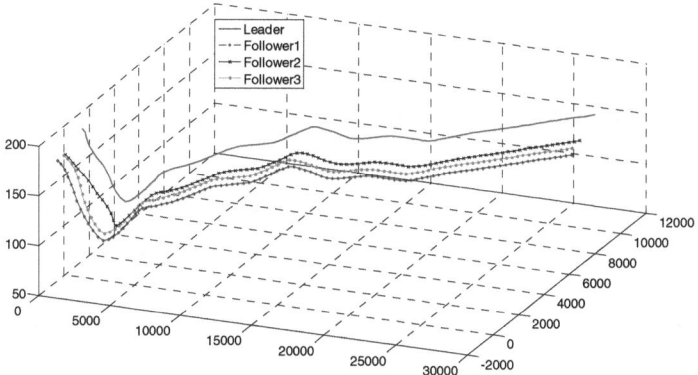

Fig. 4. Three-dimension formation route with formation feedback

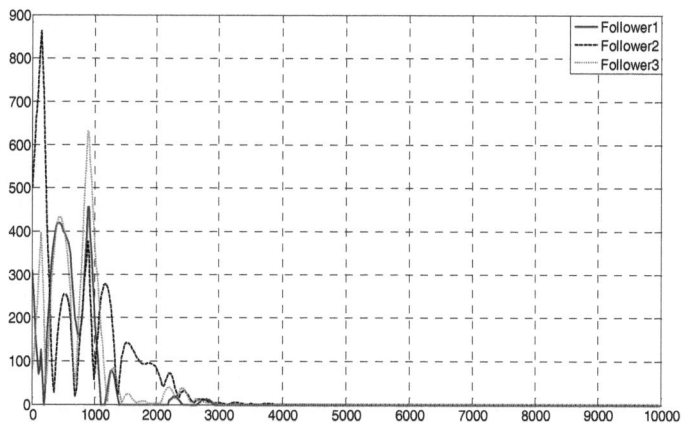

Fig. 5. Track error with formation feedback

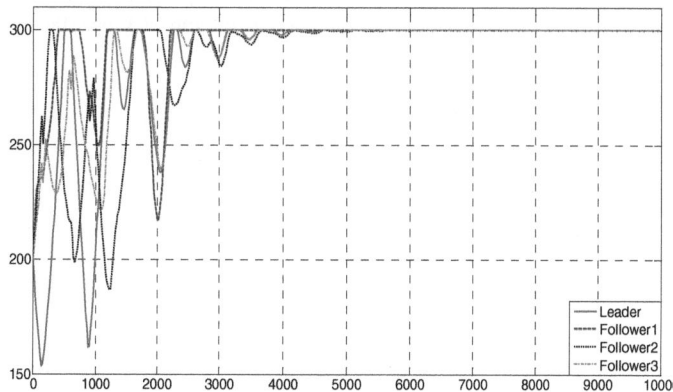

Fig. 6. Flight speed with formation feedback

5   Conclusion

In this paper, a new leader-follower approach is studied. A leader controller with formation feedback from the followers is designed to guarantee both formation maintenance and formation speed. A follower controller without the angular acceleration of the leader based on sliding-mode theory is designed to maintain desired formation. Simulation results indicated the robustness and effectiveness of the controller of each agent.

References

1. Desai, J.P., Ostrowski, J.P., Kumar, V.: Modeling and Control of Formations of Nonholonomic Mobile Robots. IEEE Transactions on Robotics and Automation 17(6), 905–908 (2001)
2. Consolini, L., Morbidi, F., Prattichizzo, D., Tosques, M.: Leader–follower formation control of nonholonomic mobile robots. Automatica 44, 1343–1349 (2008)
3. Cui, R., Ge, S.S., How, B.V.E., Choo, Y.S.: Leader–follower formation control of underactuated autonomous underwater vehicles. Ocean Engineering 37(7), 1491–1502 (2010)
4. Paul, T., Krogstad, T.R., Gravdahl, J.T.: Modelling of UAV formation flight using 3D potential field. Simulation Modelling Practice and Theory 16, 1453–1462 (2008)
5. Ren, W., Beard, R.W.: Formation feedback control for multiple spacecraft via virtual structures. IEE Proc. Control Theory Appl. 151(3), 357–368 (2004)
6. Ren, W., Beard, R.W.: A Decentralized Scheme for Spacecraft Formation Flying via the Virtual Structure Approach. In: Proceedings of the American Control Conference, pp. 1746–1751 (2003)
7. Wei, C., Guo, J., Cui, N.: Research on the Missile Formation Keeping Optimal Control for Cooperative Engagement. Journal of Astronautic 31(4), 1043–1050 (2010)
8. Ma, P., Ji, J.: Three-dimensional Multi-missile Formation Control. Acta Aeronautica et Astronautica Sinica 31(8), 1660–1666 (2010)
9. Young, B., Beard, R., Kelsey, J.: A control scheme for improving multi-vehicle formation maneuvers. In: Proc. American Control Conference, Arlington, VA, pp. 704–709 (2001)

Predicting the Age of Healthy Adults from Structural MRI by Sparse Representation

Longfei Su, Lubin Wang, and Dewen Hu[*]

College of Mechatronics and Automation
National University of Defense Technology
Changsha, Hunan 410073, China
dwhu@nudt.edu.cn

Abstract. It is generally accepted that degenerative brain diseases lead to abnormal aging process of the human brain. Thus, healthy brain aging model has great potential in clinical diagnosis and intervention. The aim of this work is to construct a regression model which is efficient for age prediction of healthy brain. Two groups of T1-weighted MRI images were involved. The first group was used for voxel selection then corresponding voxels in the second group were used for age prediction. Then mean absolute error (MAE) between the predicted age and the true age is obtained. The age prediction accuracy can reach as high as 4.67 years (MAE). In conclusion, the framework in current study can be a healthy aging model for abnormality detection of human brain. The brain regions identified by this model is sensitive to aging process which can be viewed as biomarker of brain age.

Keywords: healthy brain aging, age prediction, MRI; voxel selection, MAE, biomarker of brain age.

1   Introduction

Population aging-the older individuals become larger percentage of the total population - is a worldwide problem which will be faced by all the countries. Try our best to keep the elderly healthy is of ultimate importance. Aging research is now gaining momentum. Cognitive decline or dementia is one of the greatest health threats to the elders. Old age itself is the biggest risk factor for these neurodegenerative disease [1]. Hypothesis that some neurodegenerative diseases are accelerated aging process is supported by some previous studies [2-5]. Through comparing the chronological age and the predicted brain age, early diagnose of the pathologic brain abnormality before the onset of clinical symptoms is enabled. Efficient, reliable and robust methods are urgently called for age prediction of healthy brain.

Based on GM of structural MRI images, researchers took use of the framework of hidden Markov models for age prediction and got a high accuracy (MAE: 2.16 years for

---

[*] Corresponding author.

J. Yang, F. Fang, and C. Sun (Eds.): IScIDE 2012, LNCS 7751, pp. 271–279, 2013.

8 healthy subjects, age: 50-76 years; MAE: 2.41 years for 20 healthy subjects, age 50-86 years), but the sample size is too small [6]. Even higher age prediction accuracy was achieved (MAE: about 1 year), but the subject age range is 3 to 20 years [7]. Principle component analysis (PCA) and relevance vector regression (RVR) were both used for age prediction based on T1-weighted MRI images [4]. High prediction accuracy was achieved - mean absolute error (MAE) is 4.98 years. This confirms the predictive ability of T1-weighted MRI images. But, components obtained from PCA method consist of information from all the features (voxels).

It is universally accepted that human brain shrinks with age. For main reason of whole brain age-related volume decline is the decrease of GM volume [8], study about age effects on GM has a unique position in this field [9,10]. Most white matter (WM) changes occur during advanced aging process, but GM seems to be a constant, linear function of age [11-14]. VBM studies estimate that loss of GM is 0.18% per year [15]. Age-related changes are too subtle to be detected and lead to different even controversial conclusion. Thus, we need new technique to mine the implicit age information from MRI.

Currently, in the face of the problem named overfitting – that is too complex relationship will be established when too many features were involved which will lead to poor generalization. Many studies have confirmed that only part of the features were useful for classification or prediction [16-18]. And, sufficient evidence have shown that sparsity is promising in MRI studies [19]. Here, a feature selection method based on sparse representation was employed to select the most discriminative voxels for age prediction.

Combining of the sparse representation voxel selection and RVR, we predict the brain age of the subjects based on their T1-weighted MRI images.

2   Materials and Methods

2.1   Participants

This study's participants were selected from two databases. One group included 290 healthy subject's images which were available at the open access series of imaging studies (OASIS) website (http://www.oasis-brains.org). The initial data set in OASIS consists of a cross-sectional collection of 416 subjects with detailed description in [20]. One hundred subjects suffered with Alzheimer disease (AD) and 26 with excessive head motion were removed.

The second group consisted of eighty four healthy volunteers participated in this study. All of the subjects were left-handed except two were ambidextrous and four were left-handed. Characteristics of the two groups of subjects are displayed in Table 1.

Table 1. Characterization of subjects in this study

Variable	Group 1	Group 2
Sample size	290	84
Age (mean ± SD years)	43.25 ± 23.04	43.98 ± 17.38
Gender (male/female)	107/183	41/43
Age range (years)	18-91	19-79

2.2 Imaging Protocol

For group 1, T1-weighted structural magnetization prepared rapid gradient echo (MP-RAGE) images were obtained with the following parameters: TR = 9.7 ms, TE = 4 ms, slice thickness = 1.25 mm, flip angle = 10°, and in-plane resolution = 256 × 256 (1 mm × 1 mm). For each subject, 3-4 T1-weighted structural images were obtained on a 1.5 T Vision scanner (Siemens, Erlangen, Germany) during a single image session. In this study, one T1-weighted MRI image was selected randomly for each subject. For group 2, the T1-weighted structural MRI images were acquired with the following parameters: TR = 22 ms, TE = 9.2 ms, slice thickness = 1 mm, flip angle = 30° and in-plane resolution = 256 × 256 (1 mm × 1 mm). All scans were performed on the same Siemens Sonata 1.5 T MRI scanner.

2.3 Data Preprocessing

Data preprocessing was performed using SPM8 toolbox (http://www.fil.ion.ucl.ac.uk/spm/). First, the new segment procedure was used to segment the MRI images into GM, WM, cerebrospinal fluid (CSF) and three other background partitions. Next, one template was generated from each group of dataset using the diffeomorphic anatomical registration by exponentiated Lie algebra (DARTEL) technique [21] which matched the GM and WM to each other. Finally, GM images were spatially normalized to the template that was created in the second step and then smoothed by an isotropic Gaussian filter with an 8 mm full-width half-maximum kernel.

2.4 Voxel Selection

The proposed voxel selection method includes two steps: the t-test filter and the sparse representation algorithm. In order to achieve cluster effect and fix the computational problem faced by sparse representation algorithm, we filtered the original data by t-test and retained 20000 voxels in the first step. Then in the second step, the sparse representation algorithm was performed on the retained voxels. The purpose of the first step is to select voxels by considering the relationship between single voxel and age, while the second step aimed at selecting bundle of voxels based on the accumulating information contained in covarying relationship of voxels in different location [22]. Details of the sparse representation algorithm are shown below.

1. For $t = 1, 2, ..., T$, (in this study, we choose $T = 200$ empirically) perform steps 2 to 4.

2. Using matrix A', $(A^1 = A)$, and the label y (age), perform steps 2.1-2.2 for $k = 1, 2, ..., R$ times (in this study, we choose $R = 300$ empirically).

> 2.1: Randomly choose $q = 0.1n$ (n = rows of A) rows from A' to a construct submatrix A_k , corresponding q entries of y forms y_k .
>
> 2.2: Solve optimization problem

$$min\|w\|_1 , \ subject \ to \ A_k w = y_k \tag{1}$$

> We denote the solution of equation S.1 as w^k .

3. Let

$$w = \left| \frac{1}{300} \sum_{k=1}^{300} w^k \right| \tag{2}$$

4. According the weight vector w , the 100 voxels with highest elements are selected. Column index of these 100 voxels are defined as ind' . After these 100 columns removed from A' , the remaining columns form A^{t+1} .

5. $[ind^1, ind^2, ..., ind^T]$ is the new rearranged index of the 20000 voxels.

2.5   Age Prediction and Validation

RVR was used for regression. In this study, 10-fold cross validation for group 1, leave one out cross validation for group 2 was used to confirm the prediction accuracy.

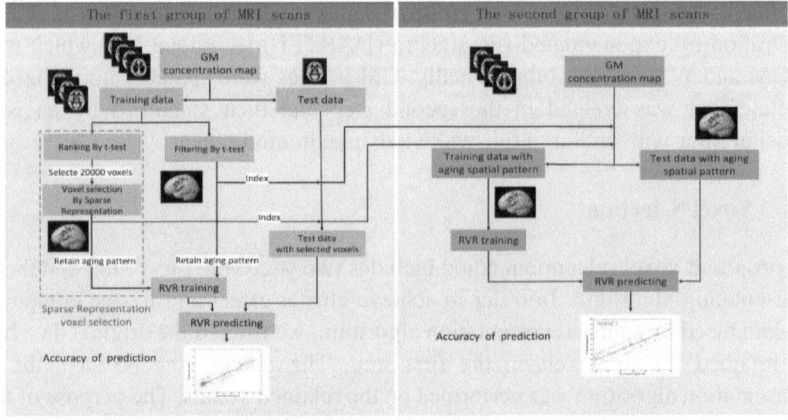

Fig. 1. Procedure of voxels selection and regression for age prediction

The first 1200 voxels in the intersection among the ten groups of rearranged voxels were chosen as the final aging spatial patterns. To confirm the robustness of the sparse representation method, we applied the spatial patterns of aging that were selected from the first group of MRI images to the second group of MRI images for age prediction. The voxels selection and the regression procedure are illustrated in Fig. 1.

3   Results

3.1   Results of the First Group of MRI Scans

For the first group of MRI scans (down load from website: http://www.oasis-brains. org) contain more young subjects especially the subjects aged 20 to 30 than other age steps, so in our results of Fig 2 more points concentrated in the 20-30 range. Voxel selection and the age prediction using RVR were all go through the ten-fold cross-validation. The voxel selection step consists of t-test filter and sparse representation based voxel selection. As a comparison, the 1200 voxels selected by our voxel selection method and the 1200 voxels selected by only the t-test filter were separately used for age prediction. The results were both shown in Fig 2. The left panel is the age prediction results using our voxel selection method; the right is the results using t-test filter. We take use of mean absolute error (MAE) and standard deviation (SD) to evaluate the performance of the two age prediction models. The age prediction results of the two models are both shown in Table 2.

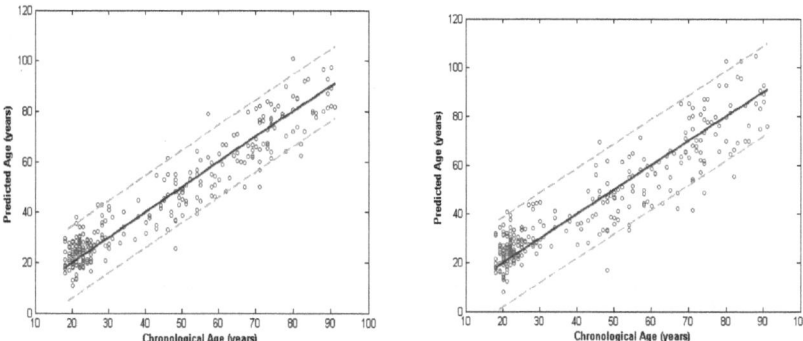

Fig. 2. Age prediction results of the first group of MRI scans. The left panel is the results corresponding to our method; the right panel is the results corresponding to t-test + RVR. Chronological age is shown on the x-axis and predicted age on the y-axis. Correctly prediction is shown by solid blue line, 95% predictions limits are shown in dash green lines.

Table 2. Age prediction accuracy of the presented method and the t-test + RVR in group 1

Group 1	MAE (years)	SD(years)
t-test + Sparse representation	5.69	7.15
t-test + RVR	6.93	9.23

From Fig 2 and Table 2 we can see that our voxel selection method was more efficient for age prediction than the t-test method. Not only the MAE but also the SD of our method is lower than the t-test method, which means our method can explore more age information and the selected biomarker is more stable and robust. But, due to the ten-fold cross-validation, the 1200 voxels in one fold were not completely identical to the 1200 voxels of the other fold. So, it cannot extract a general, stable and accurate biomarker for human brain aging. To fix this problem, we reevaluated the selected voxels in more wide range and identified the common voxels shared by all the ten folds. Then we get the final aging biomarker. For the final biomarker contain aging information of all the scans in the first group, we had to test it on another group of MRI scans.

3.2   Results of the Second Group of MRI Scans

According the index of the final biomarker obtained from the first group of MRI images, we select 1200 voxels as the input of the RVR. Using only these voxels, we successfully predicted the age of the subjects. The prediction results were more accurate than the results of first group. The details were all displayed in Table 3.

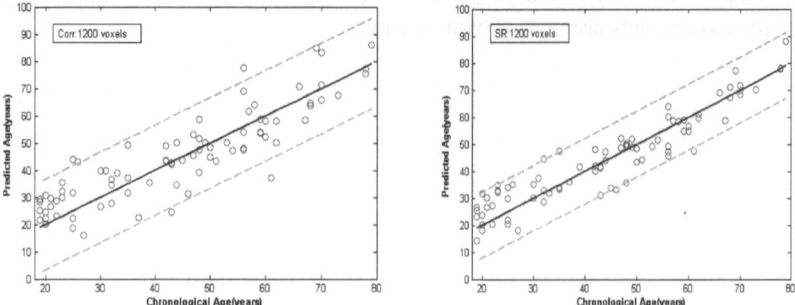

Fig. 3. Age prediction results of the second group of MRI scans. Different from Fig. 2, this results used the final biomarker extracting from the first group of MRI scans.

Table 3. Age prediction accuracy of the presented method and the t-test + RVR in group 2

Group 2	MAE(years)	SD(years)
t-test + Sparse representation +	4.67	6.10
t-test + RVR	6.54	8.33

Compare Table 3 with Table 2 we can see that the final biomarker of the human brain aging significantly increased the accuracy of the age prediction. Especially our new voxel selection method, the MAE and SD both declined.

3.3   Biomarker of Human Brain Aging

The 1200 voxels were selected by our method as the biomarker of human brain aging. The results were all shown in Fig 4.

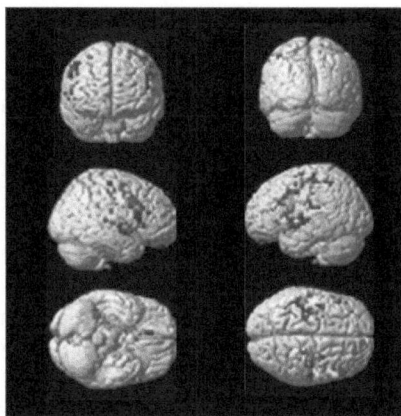

Fig. 4. Biomarker identified by our age prediction model

4   Discussion

In this paper, we presented a sparse representation based model to predict the brain age of the subject using T1-weighted MRI scans. Compare with previous study, our method increased the prediction accuracy and can further identify the most significant brain regions for age prediction.

There is a problem that both using all voxels and dimension reduction by PCA took use of the whole brain aging information, so the discriminative ability of different brain regions cannot be identified. In our model for age prediction, through voxel selection by sparse representation we used only the most discriminative voxels which only a small part of the whole brain voxels. By using only these few voxels, we got an even high accuracy of age prediction (MAE: 4.67 years, SD: 6.10 years). This confirmed the superiority of our age prediction model. From the aspect of physiological interpretation, identification of the aging brain regions is promising in diagnosis of neurodegenerative disease. Related work has also been reported [2-5,23,24].

The brain regions used for age prediction were also identified in this study. The most discriminative voxels were focus on the precentral and postcentral sulci, the medial temporal lobe, the caudate, the right inferior frontal regions, and the cerebellum. This result is highly consistent with our previous study which used sparse representation based voxel selection for age-related classification [18,25]. The difference between voxel selection for age prediction and for age-related classification is the label used in sparse representation algorithm. For classification the label is class information (1 or -1), but for age prediction, the label is the age information (age of subject). However, the results were almost the same. There were also many other studies reported similar results [26], which confirmed our voxel selection results. So, the regions identified by our method are reliable which can be recognized as biomarker of brain aging.

In conclusion, our new sparse representation based age prediction model can identify the biomarker of brain aging with highly age prediction accuracy. This healthy age prediction model has potential clinical use in early diagnosis of neurodegenerative disease such as AD before onset of clinical symptom.

Acknowledgments. This work was supported by the National Basic Research Program of China (2011CB707802), and the National High-Tech Program of China (2012AA011601).

References

1. Heemels, M.-T.: Ageing. Nature 464, 503 (2010)
2. Frisoni, G.B., Fox, N.C., Jack Jr., C.R., Scheltens, P., Thompson, P.M.: The clinical use of structural MRI in Alzheimer disease. Nature Reviews Neurology 6, 67–77 (2010)
3. Fan, Y., Batmanghelich, N., Clark, C.M., Davatzikos, C.: Spatial patterns of brain atrophy in MCI patients, identified via high-dimensional pattern classification, predict subsequent cognitive decline. Neuroimage 39, 1731–1743 (2008)
4. Franke, K., Ziegler, G., Kloppel, S., Gaser, C.: Estimating the age of healthy subjects from T1-weighted MRI scans using kernel methods: exploring the influence of various parameters. Neuroimage 50, 883–892 (2010)
5. Brickman, A.M., Habeck, C., Zarahn, E., Flynn, J., Stern, Y.: Structural MRI covariance patterns associated with normal aging and neuropsychological functioning. Neurobiol. Aging. 28, 284–295 (2007)
6. Wang, B., Pham, T.D.: MRI-based age prediction using hidden Markov models. J. Neurosci Methods 199, 140–145 (2011)
7. Brown, T.T., Kuperman, J.M., Chung, Y., Erhart, M., McCabe, C., et al.: Neuroanatomical Assessment of Biology Maturity. Current Biology 22, 1–6 (2012)
8. Taki, Y., Kinomura, S., Sato, K., Goto, R., Kawashima, R., et al.: A longitudinal study of gray matter volume decline with age and modifying factors. Neurobiol. Aging. 32, 907–915 (2009)
9. Salat, D.H., Lee, S.Y., van der Kouwe, A.J., Greve, D.N., Fischl, B., et al.: Age-associated alterations in cortical gray and white matter signal intensity and gray to white matter contrast. NeuroImage 48, 21–28 (2009)
10. Tisserand, D.J., van Boxtel, M.P.J., Pruessner, J.C., Hofman, P., Evans, A.C., et al.: A Voxel-based morphometric study to determine individual differences in gray matter density associated with age and cognitive change over time. Cereb. Cortex 14, 966–973 (2004)
11. Galluzzi, S., Beltramello, A., Filippi, M., Frisoni, G.B.: Aging. Neurol Sci. 29, s296–s300 (2008)
12. Ge, Y., Grossman, R.I., Babb, J.S., Rabin, M.L., Mannon, L.J., et al.: Age-related total gray matter and white matter changes in normal adult brain. Part I: volumetric MR Imaging analysis. Am J. Neuroradiol. 23, 1327–1333 (2002)
13. Giorgio, A., Santelli, L., Tomassini, V., Bosnell, R., Smith, S., et al.: Age-related changes in grey and white matter structure throughout adulthood. NeuroImage 51, 943–951 (2010)

14. Good, C.D., Johnsrude, I.S., Ashburner, J., Henson, R.N.A., Friston, K.J., et al.: A voxel-based morphometric study of ageing in 465 normal adult human brains. NeuroImage 14, 21–36 (2001)
15. Smith, C.D., Chebrolu, H., Wekstein, D.R., Schmitt, F.A., Markesbery, W.R.: Age and gender effects on human brain anatomy: A voxel-based morphometric study in healthy elderly. Neurobiol. Aging. 28, 1075–1087 (2007)
16. Shen, H., Wang, L., Liu, Y., Hu, D.: Discriminative analysis of resting-state functional connectivity patterns of schizophrenia using low dimensional embedding of fMRI. Neuroimage 49, 3110–3121 (2010)
17. Robinson, E.C., Hammers, A., Ericsson, A., Edwards, A.D., Rueckert, D.: Identifying population differences in whole-brain structural networks: a machine learning approach. Neuroimage 50, 910–919 (2010)
18. Su, L., Wang, L., Chen, F., Shen, H., Li, B., et al.: Sparse representation of brain aging: extracting covariance patterns from structural MRI. PLoS One 7, e36147 (2012)
19. Daubechies, I., Roussos, E., Takerkart, S., Benharrosh, M., Golden, C., et al.: Independent component analysis for brain fMRI does not select for independence. Proc. Natl. Acad. Sci. USA 106, 10415–10422 (2009)
20. Marcus, D.S., Wang, T.H., Parker, J., Csernansky, J.G., Morris, J.C., et al.: Open access series of imaging studies (OASIS): cross-sectional MRI data in young, middle aged, nondemented, and demented older adults. J. Cognitive NeuroSci. 19, 1498–1507 (2007)
21. Ashburner, J.: A fast diffeomorphic image registration algorithm. NeuroImage 38, 95–113 (2007)
22. Li, Y., Namburi, P., Yu, Z., Guan, C., Feng, J., et al.: Voxel selection in fMRI data analysis based on sparse representation. IEEE T. Bio.-Med. Eng. 56, 2439–2451 (2009)
23. Ecker, C., Rocha-Rego, V., Johnston, P., Mourao-Miranda, J., Marquand, A., et al.: Investigating the predictive value of whole-brain structural MR scans in autism: a pattern classification approach. Neuroimage 49, 44–56 (2010)
24. Fan, Y., Shen, D., Gur, R.C., Gur, R.E., Davatzikos, C.: COMPARE: classification of morphological patterns using adaptive regional elements. IEEE Trans. Med. Imaging 26, 93–105 (2007)
25. Duara, R., Loewenstein, D.A., Potter, E., Appel, J., Greig, M.T., et al.: Medial temporal lobe atrophy on MRI scans and the diagnosis of Alzheimer disease. Neurology 71, 1986–1992 (2008)
26. Li, S., Xia, M., Pu, F., Li, D., Fan, Y., et al.: Age-related changes in the surface morphology of the central sulcus. Neuroimage 58, 381–390 (2011)

A Multi-dimensional Data Association Algorithm for Multi-sensor Fusion

Wei Zou[1] and Wei Sun[2]

[1] Naval Academy of Armament, Beijing, China
[2] Nanjing Researcher Institute of Electronics Technology,
Nanjing, China

Abstract. Data Association for multi-sensor fusion is one of difficult points in target tracking. A multidimensional data association algorithm for Multi-sensor fusion is present. Firstly, we judge the mode of multi-target tracking, then choose varied multidimensional data association algorithm according to the mode, analyze and improve the existing problem of the assignment algorithm. We put forward a brief outline of the algorithm in pseudo-code form. Simulation and practical experimental results show that the algorithm is suitable for the ballistic target tracking and multi-target tracking in complex environment.

Keywords: Multi-sensor fusion, Data association, Multidimensional assignment, Ballistic target.

1   Introduction

The problem of data association for multi-sensor, namely, partitioning measurements across lists (e.g., sensor scans) into tracks and false alarms so that accurate estimates of true tracks can be recovered, has been extensively studied[1]. For multitarget tracking problems, a literature survey shows numerous well-known approaches proposed over the years, e.g.,(in order of decreasing complexity) multiple hypothesis tracking(MHT), multi-dimensional(S-D) assignment, joint probabilistic data assignment(JPDA), and two-dimensional(2-D) assignment(single scan processing)algorithms[2~7].

Data association using a multidimensional assignment algorithm such as S-D assignment has been shown to be a practical and feasible alternative to MHT. S-D assignment is a discrete mathematical optimization formulation of the data association problem that systematically resembles an MHT within a window of length(S-1). However, the main challenge to overcome in the S-D assignment problem is that of solving the ensuing NP-hard multidimensional assignment problem. In particular, an algorithm that determines the optimal solution is not only arduous, but also impractical for even fairly small sized problems, e.g., unsatisfactory results were reported for a problem as small as 10 targets and 3 scans in d dense scenario. However, satisfactory tracking and computational performance can be realized utilizing existing S-D assignment algorithm that provides good suboptimal solutions, of quantifiable accuracy, and in pseudo-polynomial time.

J. Yang, F. Fang, and C. Sun (Eds.): IScIDE 2012, LNCS 7751, pp. 280–288, 2013.

The focus of this paper is to study the data association of multi-sensor fusion and propose the self-adaptive multidimensional data association algorithm. Firstly, the multi-sensor detective zone is divided and the mode of multitarget tracking is judged; then the varied multidimensional data association algorithm is used to correlate according to the tracking mode. The existing problem about multidimensional assignment is analyzed and the pseudo-code of the algorithm is presented. The innovation point of this paper is that varied data association algorithm is self-adaptive choosed according to the tracking mode; Lagrangian technology is improved for S-D assignment algorithm and the pseudo-code of this algorithm is presented; the algorithm performed well in the simulation experiment on ballistic target tracking. Now the algorithm is used in practical system equipment, and performed well in complex environment.

The outline of this paper is as follows. Firstly the multi-dimensional assignment is simply reviewed in section 2. Self-adaptive multi-dimensional assignment association algorithm and the pseudo-code about it is presented in section 3. The result of the simulation and practical experiment on this algorithm is showed in section 4. The next studying emphasis is presented in section 5.

2   S-D Data Association Algorithm

The S-D data association is formulated as a constrained global optimization problem, where the objective is to minimize the total "cost" of associating (or not associating) a particular sequence of measurement in frames to the tracks.

To specify the constrained optimization problem more formally, define the binary assignment variable

$$
a\left(k,\{m_s\}_{s=k-S+2}^{k},n\right)=\begin{cases}1 & \begin{array}{l}measurements\ \mathbf{z}_{lj,mj}\left(t_{mj}\right),j=k-S+2,\dots,k\\ are\ assigned\ to\ track\ \tau_{n}(k-S+1)\end{array}\\ 0 & otherwise\end{cases} \tag{1}
$$

a cost associated with (1) given by

$$
c\left(k,\{m_s\}_{s=k-S+2}^{k},n\right)=-\ln\left(\frac{\phi\left(k,\{m_s\}_{s=k-S+2}^{k},n\right)}{\phi\left(k,\{m_s\}_{s=k-S+2}^{k},0\right)}\right) \tag{2}
$$

$$
m_s=0,\dots,M\ (s);s=k-S+2,\dots,k;n>0
$$

$\phi\left(k,\{m_s\}_{s=k-S+2}^{k},n\right)$ is the joint likelihood, given by (3)

$$
\phi\left(k,\{m_s\}_{s=k-S+2}^{k},n\right)=\begin{cases}\displaystyle\prod_{s=k-S+2}^{k}[1-P_D]^{1-u(m_s)}\left[P_D\Lambda(s,m_s,n)\right]^{u(m_s)} & for\ n>0\\ \displaystyle\prod_{s=k-S+2}^{k}V_f(s)^{-u(m_s)} & for\ n=0\end{cases} \tag{3}
$$

Where $V_f(s)^{-1}, j = k - S + 1,...,k$ is the spatial false alarm density in revisit j, P_D is the detection probability of target-originated measurements, $u(m)$ is a binary function such that

$$u(m) = \begin{cases} 1 & m > 0 \\ 0 & m = 0 \end{cases} \tag{4}$$

And $\Lambda(j,m,n), j = k - S + 1,...k$ is the filter-calculated likelihood when measurement is associated with track.

A complete set of valid associations satisfying all the constraints for each track and for each measurement is denoted by $a(k)$, i.e., $\Lambda(j,m,n), j = k - S + 1,...,k$

$$\mathbf{a}(k) = \left\{ a\left(k, \{m_s\}_{s=k-S+2}^{k}, n\right); m_s = 0,1,..., M_{ls}(s); s = k - S + 2,...,k; \right. \tag{5}$$
$$\left. n = 0,1,..., N(k - S + 1) \right\}$$

Where $a(k)$ satisfy the constraints discussed above. The cost of the complete set of associations $a(k)$ is given by

$$C(k|\mathbf{a}(k)) = \sum_{n=1}^{N(k-S+1)} \sum_{m_{k-S+2}=0}^{M_{lk-S+2}(k-S+2)} \sum_{m_{k-S+3}=0}^{M_{lk-S+3}(k-S+3)} \cdots \sum_{m_k=0}^{M_{lk}(k)} a\left(k, \{m_s\}_{s=k-S+2}^{k}, n\right) c\left(k, \{m_s\}_{s=k-S+2}^{k}, n\right) \tag{6}$$

Then the objective of the multi-dimension assignment is to find the best set of associations, denoted by $a^*(k)$, with the lowest cost, i.e.,

$$\mathbf{a}^*(k) = \arg \min_{\mathbf{a}(k)} C(k|\mathbf{a}(k)) \tag{7}$$

3   Self-adaptive Multi-dimensional Data Association Algorithm

3.1   Multisensor Detecive Zone Divisional Process

Suppose in radar net system , the location of radar R_i is $[L_i, \lambda_i, H_i]$, $i = 1,2,...,S$, S is the number of the radar in detective zone; L_i, λ_i, H_i are the longitude, latitude and altitude of radar R_i. The maximum detective distance of radar R_i is ρ_{max_i} . So the detective scope of radar R_i is $[L_{min\_i}, L_{max\_i}] \times [\lambda_{min\_i}, \lambda_{max\_i}]$ (Suppose the detective altitude is zero).

As the maximum detective zone of one radar is $\left[L_{\min}, L_{\max}\right] \times \left[\lambda_{\min}, \lambda_{\max}\right]$, then

$$L_{\min} = \arg \min_{i} L_{\min\_i}, L_{\max} = \arg \max_{i} L_{\max\_i} \tag{8}$$

$$\lambda_{\min} = \arg \min_{i} \lambda_{\min\_i}, \lambda_{\max} = \arg \max_{i} \lambda_{\max\_i} \tag{9}$$

Choose less size grid (such as $10Km \times 10Km, 30Km \times 30Km$ and so on) to split the multi-radar detective zone $\left[L_{\min}, L_{\max}\right] \times \left[\lambda_{\min}, \lambda_{\max}\right]$, and slide window process.

3.2   Self-adaptive Multi-dimension Data Association Algorithm

Data association includes coarse data association and fine data association. Coarse data association uses the general association gate to get all of the possible association. Coarse data association can not drop any possible correlated measurement on the one hand, and must reduce the false association as much as possible on the other hand.

The candidate correlated set is defined as follows at k th time:

$$\varsigma(k) \triangleq \left\{(n, m_k) : (n, m_k) \in N(k-1) \times M(k)\right\} \tag{10}$$

Where $N(k-1) \times M(k)$ denotes the cross multiplication between the index of tracks and correlated measurements in the same grid. The distance gate and the velocity gate are adopted in coarse data association. The result of coarse data association may be that many measurements are correlated to one track or one measurement is correlated to many tracks.

We get tracks $T_n(t_k - 1), n = 1,2,..., N(k-1)$ at $k-1$ scan and measurements $Z_m(t_k), m = 1,2,..., M(k)$ at k scan in the candidate set $\varsigma(k)$ gotten by coarse data association in the same grid, and judge whether there are shared measurements between tracks $T_n(t_k - 1), n = 1,2,..., N(k-1)$ at $k-1$ scan. If there are no shared measurements, the track is judged in the single target tracking mode; otherwise, the tracks which have same measurements are judged in the multarget tracking mode(track group).

In single target tracking mode, we adopt 2-D data association algorithm in fine data association; and in multitarget tracking mode, we adopt multi-dimensional data association algorithm in fine data association.

Deb and Pattipati developed an algorithm to associate measurements from multiple sensors to identify the real targets in surveillance region, and presented the pseudo code[6] about the algorithm. The S-D assignment algorithms are analyzed and some problems come up with as follows.

Lagrangian multiplier can be computed by many methods of selection of step size, including accelerated sub-gradient Update and Heuristic price update. We find that the actual update of Lagrangian multipliers u_{r+1} requires the solution of $r+1$ sub-problem. Due to J_{r+1} is not yet computed at this time, u_{r+1} cannot be computed to the next step. Deb explained the problem, but there exists no good improvement. Besides, $u(r+1)i_{r+1}$ is only suitable for the next iteration, and this point was not mentioned in the Deb's algorithm, which may bring the confusion.

Based on Deb's assignment algorithm, we improve it and present more explicit S-D assignment algorithm to avoid the problem pointed out above. Main idea is to compute respectively in current iteration. The algorithm in pseudo-code is put forward below. In the assignment algorithm by Deb, the calculation sequence is $J_2^* \rightarrow J_3 \rightarrow g_{3i_3} \rightarrow J_4 \rightarrow g_{4i_4}$, while in our algorithm, the calculation sequence is $J_2^* \rightarrow J_3 \rightarrow g_{3i_3} \rightarrow J_4 \rightarrow g_{4i_4}$. In our algorithm, we combine step 3(enforce constraint set), step 4(update Lagrangian multipliers), step 5(recursion) as new step 3(enforce restriction set and update Lagrangian multipliers) to ensure that J_{r+1} has been computed before computation of $u(r+1)i_{r+1}$. In order to be more convenient to calculate adaptive step-size parameter, f_{dual} is computed in step 3 in advance.

The S-D Assignment Algorithm

begin

Step 1 Initialize.

$$N = n_1 , \quad \gamma_k \leftarrow \{k\} \quad k = 0, 1, \dots, N$$

$$f_{dual} = -\infty ; \quad f_{primal} = \infty ; \quad \min gap = \varepsilon \quad (typically\ 0.01\ to\ 0.05) ;$$

$\max iter = 200$

$iter = 0 , u_{rir} = 0; \quad i_r = 1, 2, \dots, n_r; r = 3, 4, \dots, S$

Step 2 Compute reduced costs for r= S-1,S-2,...,2

$$d_{i1\dots ir}^r = \min_{i_{r+1}} \left(d_{i1\dots iri_{r+1}}^{r+1} - u_{(r+1)ir+1} \right)$$

Step 3 Enforce constraint set R and Update Lagrangian Multipliers.

$$r = 2 \quad , \left\{ J_r, \omega_{\gamma_k i_r}^* \right\} = \min_{\omega_{\gamma_k i_r}} \sum_{\gamma_k = 0}^{N} \sum_{i_r = 0}^{n_r} d_{\gamma_k i_r}^r \omega_{\gamma_k i_r} + \sum_{i_{r+1} = 0}^{n_{r+1}} u_{(r+1)ir+1} \cdots + \sum_{is = 0}^{ns} u_{Sis}$$

$$\sum_{i_r = 0}^{n_r} \omega_{\gamma_k i_r} = 1 \quad \gamma_k = 1, 2, \dots, N \quad , \sum_{\gamma_k = 0}^{N} \omega_{\gamma_k i_r} = 1 \quad i_r = 1, 2, \dots, n_r$$

Do $r < S$

$\quad N = 0$

If $\left(\omega_{\gamma_k i_r}^* = 1 \right) \left\{ N \leftarrow N + 1; \gamma_N \leftarrow \{\gamma_k i_r\}; \right\}$

$$r = r + 1$$

$$\left\{ J_r, \omega^*_{\gamma_k i_r} \right\} = \min_{\omega_{\gamma_k i_r}} \sum_{\gamma_k=0}^{N} \sum_{i_r=0}^{n_r} d^r_{\gamma_k i_r} \omega_{\gamma_k i_r} + \sum_{i_{r+1}=0}^{n_{r+1}} u_{(r+1)i_{r+1}} \cdots + \sum_{i_s=0}^{n_s} u_{Sis}$$

$$\sum_{i_r=0}^{n_r} \omega_{\gamma_k i_r} = 1 \quad \gamma_k = 1, 2, \ldots, N , \quad \sum_{\gamma_k=0}^{N} \omega_{\gamma_k i_r} = 1 \quad i_r = 1, 2, \ldots, n_r$$

$$g_{r i_r} = 1 \quad i_r = 1, 2, \ldots, n_r$$

If γ_N

$$\left\{ j = \arg \min_{i_r} c_{\gamma_N i_r \ldots is} - u_{r i_r} \cdots - u_{Sis} \quad g_{r i_r} = g_{r i_r} - 1 \right\}$$

$$u_{r i_r} = u_{r i_r} + adapt * g_{r i_r} \quad i_r = 1, 2, \ldots, n_r$$

End Do

Step 4 Iteration: Improve solution quality

$$f_{dual} = \max \left(f_{dual}, J^*_2 \right), \quad f_{primal} = \min \left(f_{primal}, J_s \right), \quad gap = \left(f_{primal} - f_{dual} \right) / \left| f_{primal} \right|$$

$$iter = iter + 1$$

If $\left(gap < \min gap \text{ OR } iter > \max iter \right)$

go to **Step 5**

$$N = n_1, \quad \gamma_k \leftarrow \{k\} \quad k = 0, 1, \ldots, N \quad \text{go to } \textbf{step 2}$$

Step 5 Final Results

Number of targets $\leftarrow N$, $S - tuples = \gamma = \left\{ \gamma_k; k = 1, 2, \ldots, N \right\}$

end

3.3 Milti-dimensional Assignment Algorithm Application

The optimal solution of 2-D assignment problem may be not corresponding to the optimization problem, especially, when the optimal solution of 2-D problem is not smaller than other solution at all. In this case, using the former M optimal solutions to 2-D assignment problem to joint the association relation between decision-making measurements and tracks can improve correct association probability effectively. On other hand, some problem had to be considered in application.

1) when the number of tracks is bigger than the gate,   all of the track are no longer update(may be in clutter region);
2) The algorithm is suitable for null logic. When the track grade is different distinctly with gate, track will no longer be updated.
3) The track correlated by batch and is more continuous is set higher believable degree.

4   Simulated and Practical Experiment Results

The result of simulated and practical experiment about self-adaptive assignment algorithm is presented.

In the experiment, all of simulate parameters are corresponding to practical scenarios; sensor's scan circle is the same as practical sensor's scan circle. All of the clutters are in possion distribution, the number of the dummy is alterable, practical clutters are added in some experiment. The curve in figure is target's horizontal projection on geographical coordinates. In this experiment, there are four three-coordinate radars deployed in different zones. Radar simulated parameter is shown in table 1.

Table 1. Radar Simulated parameter

Radar No	Location(longitude, latitude, altitude)	Radar cycle	Range error	Azimuth error
1	(118.870, 31.88, 100m)	5s	100m	0.30
2	(118.165, 32.35, 50m)	5s	80m	0.25
3	(118.859, 32.47, 17m)	5s	50m	0.25
4	(116.594, 30.43, 34m)	5s	100m	0.20

Fig.1 shows the result of tracking multi-warhead formation of ballistic target. There are satellite orbit targets and ballistic targets formation in figure. In the figure, the break line denotes predicting trajectory, the yellow line denotes the part zoom out picture of multi-warhead target. The distance between target formation is about one kilometer. The interacting multiple model filter with two models (CA/CV) is adopted. Divisional process with $30Km \times 30Km$ region is used. Due to the smaller interval between targets, it is judged multitarget tracking mode $S = 4$. From this part zoom out picture, we can see that this algorithm can track multi-warhead formation effectively.

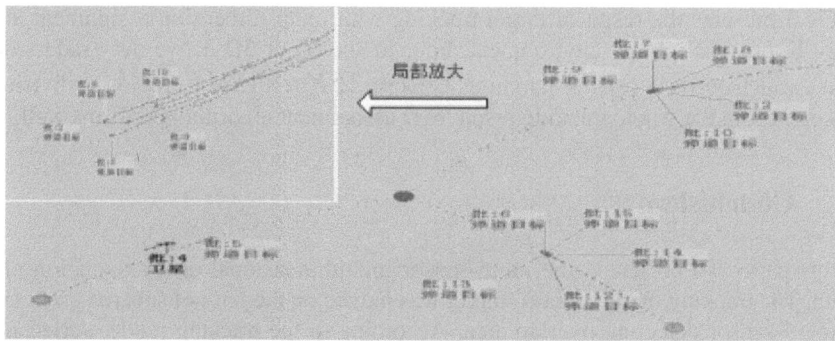

Fig. 1. Multi-warhead Target Track Effect

The result of tracking cross-cutting/evasive target with multi-dimensional assignment algorithm is shown in figure 2, two batch targets above do cross flying, and two batch targets below do evasive flying. We adopt interacting multiple model filter with two models (CA/CV) and choose $30Km \times 30Km$ region to divisional process. In non-cross-cutting/evasive region, the target is judged in the single mode, and is correlated with 2-D assignment algorithm; in non-cross-cutting/evasive region, targets are judged in multitarget mode, and are correlated with S-D assignment algorithm, $S = 4$. From the result, we can see that this algorithm has good effect on cross-cutting/evasive target tracking.

Fig. 2. Cross-cutting/evasive Target Tracking Effect

Fig. 3. Target Tracking Effect in Jamming environment

Fig.3 presents the result of target tracking with multi-dimension assignment algorithm in practical clutter environment. Four to six 2D or 3D active duty radars deployed, which one is in barrage jamming. Choose $15Km \times 15Km$ region to divisional process. From the whole tracking result, S-D assignment algorithm performs well.

5   Conclusion

In this paper we presented the multi-sensor multidimensional data association algorithm. The tracking mode of multi-target was judged on the base of subarea processing of multi-sensor detecting overlap area. According to the tracking mode, varied multi-dimensional assignment algorithm was used for correlation processing. This algorithm does well in ballistic target formation tracking or complex air scenario. The next emphasis in work is studying tracking grade technology based on track quality and parallel process of multi-dimensional data assignment algorithm.

References

1. Pulford, G.W.: Taxonomy of Multiple Target Tracking Methods. IEE Proc.-Radar Sonar Navig. 152(5), 291–304 (2005)
2. Blackman, S.S.: Multiple Hypothesis Tracking for Multiple Target Tracking. IEEE A&E Systems Magazine 19(1), 5–18 (2004)
3. Bar-Shalom, Y., Blackman, S.S.: Dimensionless Score Function for Multiple Hypothesis Tracking. IEEE Transactions on Aerospace and Electronics Systems 43(1), 392–400 (2007)
4. Popp, R.L., Pattipati, K.R., Bar-Shalom, Y.: m-Best S-D Assignment Algorithm with Application to Multitarget Tracking. IEEE Transactions on Aerospace and Electronics Systems 37(1), 22–39 (2001)
5. Deb, S., Yeddanapudi, M., Pattipati, K.R., Bar-shalom, Y.: A generalized S-D assignment algorithm for multisensor-multitarget state estimation. IEEE Trans. Aerospace and Electronic Systems 33(2), 523–538 (1997)
6. Kirubarajan, T., Bar-Shalom, Y., Pattipati, K.R.: Efficient Multisensor Fusion Using Multidimensional Data Association. IEEE Transactions on Aerospace and Electronics Systems 37(2), 386–398 (2001)
7. Wang, W., Luh, P.B., Yan, J.H.: An Improved Lagrangian Relaxation Method for Discrete Optimization Applications. In: 4th IEEE Conference on Automation Science and Engineering, Key Bridge Marriott, Washington DC, USA, pp. 359–364 (2008)

An Image Quality Assessment Algorithm Based on Feature Selection

Ting Lu, Yanning Zhang, and Haisen Li

Shaanxi Provincial Key Laboratory of Speech & Image Information Processing,
School of Computer Science, Northwestern Polytechnical University

Abstract. Image Quality Assessment(IQA) is of fundamental importance to numerous imaging and video processing applications. For most of the applications, the perceptual meaningful measure is the one which can automatically assess the quality of images or videos in a perceptually consistent manner. However, most commonly used IQA metrics are not consistent well with the human judgments of image quality. Recently, the SSIM metric which takes people's visual characteristics into consideration performs much better than the traditional PSNR/MSE. But the defects of it still exit on some specific kinds of distortions. A new algorithm of IQA based on feature selection is proposed in this paper. Local gradient entropy and phase congruency are added to the SSIM framework. Through in-depth feature selection and definition plus better pooling strategy, this algorithm performs much better in LIVE datasets.

Keywords: Image Quality Assessment, feature selection, phase congruency, local gradient entropy.

1 Introduction

With the rapid tremendous development of digital imaging and video processing, such as acquisition, compression, transmission, reproduction, enhancement, restoration and so on, the increasing importance of Image Quality Assessment(IQA) comes out, so does the need for the research in IQA. IQA aims to evaluate the fidelity or the intelligibility of an image. It can be used in many applications[1], for instance, in image processing field, it is important to evaluate the performance of the amounts of algorithms according to the IQA measure, besides, the result may assist the researcher to obtain the optimal design of the mathematical model and the optimal parameter.

According to the availability of the reference image in computing a perceptually relevant score, IQA indices can be classified into 3 categories[3], full reference(FR), no reference(NR)[4-6], reduced-reference quality assessment(RR). In this paper, we focused on the FR methods.

In the signal and image processing literature, the most common and conventional IQA indices are straightforward methods, such as the peak-signal-to-noise ratio(PSNR), the root-mean-squared error(RMSE). Although they are simple to get, it

J. Yang, F. Fang, and C. Sun (Eds.): IScIDE 2012, LNCS 7751, pp. 289–297, 2013.

has been widely acknowledged that that are not always in agreement with the subjective fidelity ratings because they operate directly on the intensity of the image[7]. Thus great efforts have been made to predict the human visual quality that take advantage of known characteristic of the human visual system(HVS)[8]. Modeling of the HVS has been regarded as the most suitable paradigm for achieving better results. Through computing the different sensitivity of the HVS, the error between the test image and reference image will be obtained. NQM[9] and VSNR[10] are two representative algorithm following this kind of paradigm. At the same time, Wang and Bovik[1] proposed a structural similarity index(SSIM) based on the hypothesis that the HVS is highly developed for extracting structural information from the visual scene.By measuring the loss of structure in the image, we can obtain the quality rating. This research is said to be a milestone in the research of the FR IQA models, changing the focus from error measurement to structural similarity. From then a new point of view is elicited into this field. A lot of research has been done to improve the performance of it[11, 12]. The multi-scale extension of SSIM produces better results than its single-scale[13].

It is true that the SSIM contributes a lot in the develop of the IQA, but further analysis shows out the SSIM failed to evaluate the quality of blurred image or the Gaussian noise image[14] and it is not sensitive enough to the edge of change in the image. In this paper, a new way to address the problem is introduced here. Under the SSIM framework, we introduce two new features to it to improve its performance. One is the local gradient entropy, it is computed from the gradient map which represents the edge of the image, and it can indicate the region of the interest at some point. The other is the phase congruency, it is a contrast invariant and can represent the structure of the image properly, combined with the contrast information, a quality map can be reached. At the pooling stage, the gradient entropy is used again as the weight function to improve the pooling strategy.

The rest of this paper is organized as follows: Section 2 discusses the feature selection procedures, and the detail implementation of the proposed method, section 3 shows the experimental results and the associated discussion, section 4 concludes the paper.

2   Feature Based Image Quality Assessment

Image quality assessment aims at evaluate the fidelity of the image. And it is indeed difficult to compare with the reference image directly by computer. Through feature selection, we can obtain the good description of an image. and then through the comparison of the features, we get the final evaluation.

2.1   Feature Selection

In the design of the algorithm of IQA, one of the key issue that should be handled properly must be what kinds of features can be used. So the feature selection problem comes out.

- Local Entropy of the Gradient Image

The gradient of image is a traditional topic in image processing. Image is always been seen as a 2-dimensional dispersed function, so to image $I(x, y)$, the gradient of it can be expressed as equation(1):

$$dx(i, j) = I(i+1, j) - I(i, j)$$
$$dy(i, j) = I(i, j+1) - I(i, j)$$

$$(1)$$

In application, the gradient operators can be expressed by convolution masks. There are many kinds of gradient operators at present[15]. we choose the commonly used Sobel Operator as the convolution mask due to its simplicity and efficiency, as a dispersed difference operator, it is always been used to compute the gradient of the luminance of image efficiently.

Suppose $M_i \times N_i$ is a part of the whole gradient map naming as Ψ_i, then the local entropy of the gradient image can be expressed as equation(2).

$$GE(\Psi_i) = -\sum_{j=0}^{L-1} p_j \log p_j \qquad (2)$$

Where $p_j = \dfrac{n_j}{M_j \times N_j}$ means the probability of the gray scale j in Ψ_i, L is the amount of the total gray scale, n_j indicates the total amount of the gray scale j in Ψ_i, $M_i \times N_i$ indicates the size of the image, and $GE(\Psi_i)$ is the local gradient entropy of this part of image.

The gradient image and the local gradient entropy image can be seen in Fig.1. The value of the gradient entropy will be larger if the area located in the place where there are more details or the change of the contrast is sharper. On the contrary, the area has less details, or smoother edges will have smaller gradient entropy.

a)Original Image   b)Gradient Image   c)Gradient Entropy Image

Fig. 1. The Entropy Image

- Phase Congruency

It has been proved that the phase information can represent the feature of image quite well[16, 17]. When an image is decomposed into Fourier Field, the magnitude part of the image characterizes the amount of the frequency of the image, it contains the contrast information of the image. For a specific image, it will turn to be much softer if there are many dark points on the magnitude map, while it will be much more sharp if the points on the magnitude map turn to be bright. The phase part of the image indicates the position of the specific frequency component, it contains much of the texture information. In order to show the typicality of the two features, we reconstruct the image using each of them. It can be seen clearly that the reconstruction image(Fig.2(b)) from the phase has much more details than it from the magnitude.

The phase congruency postulates that features can be detected at points where the Fourier components are maximal in phase[18]. The phase congruency refers to the phase similarity of each frequency component in every position. As a dimensionless quantity, the value of it has nothing to do with the change of the luminance and contrast. Physiology experiments have confirmed that the human visual system is very sensitive to the place where phase information consistently with each other[19]. So we can use it to represent the structure of the images.

To obtain the quantity, first execute the image filter convolution with a 2-dimensional Log Gabor filter. Then compute it using the 1-dimensional signal phase consistency model based on local energy. This model greatly simplifies the computation of the phase congruency. The 2D log-Gabor function has the transfer function like equation (3).

(a) The Reconstruction Image using the Magnitude Map

(b) The Reconstruction Image using the Phase Map

(c) Phase Congruency Map

Fig. 2. The Phase Image

$$G_2(\omega,\theta) = exp(-\frac{(log(\omega/\omega_0))^2}{2\sigma_r^2}) \cdot exp(-\frac{-(\theta-\theta_j)^2}{2\sigma_\theta^2}) \qquad (3)$$

Convolving the filer with the image, the responses at each point can be divided into a pair of even-symmetric and odd-symmetric.

$$[e_n(x), o_n(x)] = I(x) * G \qquad (4)$$

Then the local amplitude on scale n is :

$$A_n(x) = \sqrt{e_n(x)^2 + o_n(x)^2}$$ (5)

The local energy will be:

$$E(x) = \sqrt{(\sum_n F_n(x))^2 + (\sum_n H_n(x))^2}$$ (6)

Where $F(x)$ is the sum of the even-symmetric part of the response. And $H(x)$ is the sum of the odd-symmetric of all the response. Then the phase congruency is defined as:

$$PC(x) = E(x) \Big/ (\varepsilon + \sum_n A_n(x))$$ (7)

Fig.2(c)shows the phase congruency map of the original image, the most of the details in the original image can be seen in it, which validly prove the efficiency of the phase congruency as a feature .

2.2   Feature Based Image Quality Assessment

Getting these four features, then, come to the problem of how to use it in IQA. An traditional way of this problem is to compare the features in the test image with the reference image, and then combine them into one evaluation image, and the last step is to gather the map into one index. The index is the last result that can be used to assess the quality of the reference image.

According to the SSIM, the commonly used similarity measure function has the following form.

$$s = \frac{2xy + C}{x^2 + y^2 + C}$$ (8)

So the similarity measure for the reference image and the test image is defined as:

$$S_{GE}(x, y) = \frac{2GE_x GE_y + C}{GE_x^2 + GE_y^2 + C}$$ (9)

$$S_{PC}(x, y) = \frac{2PC_x PC_y + C}{PC_x^2 + PC_y^2 + C}$$ (10)

$$S_l(x, y) = \frac{2\mu_x\mu_y + C}{\mu_x^2 + \mu_y^2 + C} \tag{11}$$

$$S_c(x, y) = \frac{2\sigma_x\sigma_y + C}{\sigma_x^2 + \sigma_x^2 + C} \tag{12}$$

Then the feature based image quality assessment can be obtained as equation :

$$S(x, y) = [S_{GE}]^{\alpha} \bullet [S_{PC}]^{\beta} \bullet [S_l]^{\gamma} \bullet [S_c]^{\tau} \tag{13}$$

The $S(x, y)$ at each point is obtained so that the overall similarity between the two image has been obtained. Considering that the aggregation strategy will influence the result of the index very much and that different locations have different contribution to the quality of the image, we use the gradient entropy information to weight the importance of the $S(x, y)$ in the overall similarity between the two image. It can be comprehended easily that the edge in the image convey much more crucial visual information than the locations in the smooth area and people pay much more attention to it. So the feature based image quality assessment index is defined as:

$$index = \frac{\sum_{(x,y)\in I} S(x, y) \bullet GE(x, y)}{\sum_{(x,y)\in I} GE(x, y)} \tag{14}$$

Where the I is the whole image, $S(x, y)$ .is the quality map just obtained , and GE is the local gradient entropy as the weight function.

3   Experimental Results

In order to evaluate the performance of the algorithm we proposed, some experiments have been done on the LIVE(Laboratory for Image & Video Engineering) dataset. we go for the LIVE Image Quality dataset for in-depth discussion[20]. The LIVE dataset has 779 images in total, 29 reference images, each containing about 30 distortion images. These distortions cover a broad range of the image impairments and a broad range of quality, by containing 5 types of distortion, JPEG2000, JPEG, white noise, Gaussian blur, and simulated fast fading Rayleigh channel and different levels of distortion from imperceptible levels to high levels of impairment. The perceptual score of these images is obtained by 25000 individual human quality judgments removing outlier subjects and scores and compensating for the bias across reference images and subjects.

We design the algorithm to approximate the visual effect of the human eye, so the closer the better. Thus, we use the Spearman rank-order correlation coefficient(SROCC), the Kendall rank-order correlation coefficient(KROCC) and the

Pearson product-moment correlation coefficient(PLCC) between the objective and subjective scores to measure the correlation between them. The Pearson correlation coefficient is defined as the covariance of the two product of their standard deviations. The Spearman correlation is the Pearson correlation coefficient between the ranked variables while the Kendall correlation is a non-parametric hypothesis test. All of them can be used to test for statistical dependence or the association between two measured quantities. The higher the value is, the closer the quantities get, and the better the algorithm is.

First, we test the algorithm on each of the different types of distortions(as shown in Fig.3), and compare the performance of it with PSNR and MSSIM, the related results are listed in Table 1. From the results, we can see that our algorithm can perform well when the PSNR cannot distinguish between the distortions.

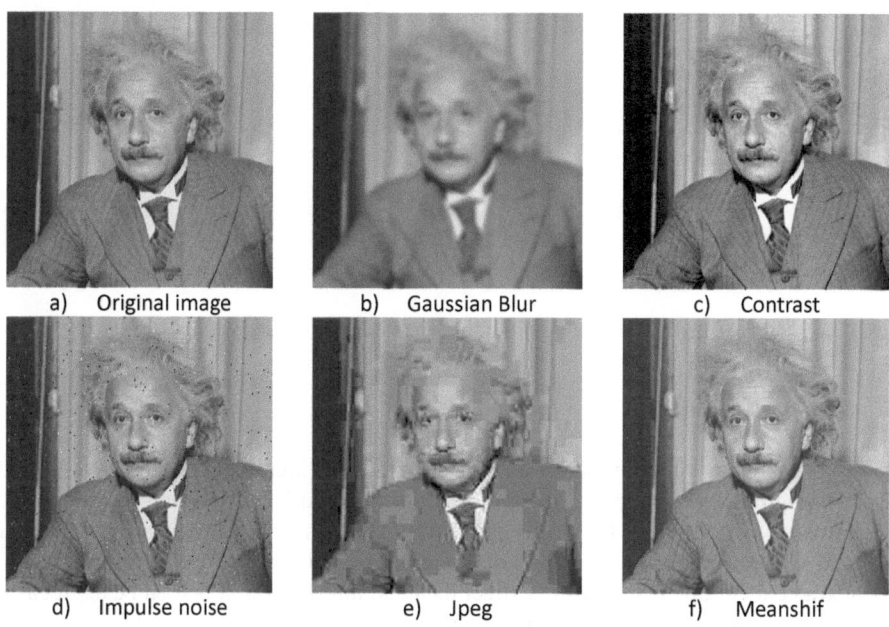

a)   Original image   b)   Gaussian Blur   c)   Contrast

d)   Impulse noise   e)   Jpeg   f)   Meanshif

Fig. 3. The Images for testing

Table 1. Performance comparison on specific distortion

Distortion type	PSNR	MSSIM	Our algorithm
Blur	26.5499	0.6940	0.6046
Contrast	26.5406	0.9133	0.8789
Impulse	26.5490	0.8396	0.8610
JPG	26.6094	0.6624	0.5439
Meanbshift	26.5473	0.9884	0.9859

Second, we test the algorithm on the entire datasets. Table 2 lists the SROCC and KROCC results of the algorithm we proposed, the PSNR and the MSSIM index on the entire datasets. As shown in the table, our algorithm has the highest coefficient, and is most relevant to the subject scores of the three. Fig.4 shows the scatter plot of each of the algorithm, it is obvious too that our algorithm is more convergent and linear than the rest of the algorithm. Both of them prove that the algorithm we proposed is competitive in state-of –art indices.

Table 2. Performance comparision of IQA metrics

Correlation coefficient	PSNR	MSSIM	Our algorithm
SROCC	0.8730	0.9226	0.9468
KROCC	0.6801	0.7474	0.7959
PLCC	0.8435	0.8072	0.8628

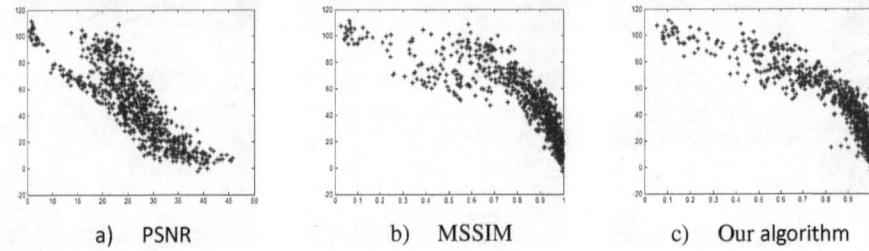

a)   PSNR   b)   MSSIM   c)   Our algorithm

Fig. 4. Scatter plots of the IQA method over the subject ones

4   Conclusion

In this paper, a new algorithm of evaluating the quality of image is proposed. We showed that through more in-depth feature selection plus complex pooling strategy, the performance of SSIM can be greatly improved. Three kind of features are used in our algorithm, local gradient entropy, phase congruency and the contrast information. The local gradient entropy is utilized to extract the edge information. It is then be aggregated to the final metric. Considering the fact that people pay more attention to the structure of the image, the phase congruency is utilized to extract the structural information instead of the original method of calculating. Taking into account that the visual sensitivity is paid more attention to the local edge information, we use the local gradient entropy again as the weight function to improve the pooling strategy. The experimental results on the LIVE datasets show that our algorithm does improve the performance a lot.

References

[1]   Wang, Z., et al.: Image quality assessment: From error visibility to structural similarity. IEEE Transactions on Image Processing 13(4), 600–612 (2004)
[2]   Tang, H., Joshi, N., Kapoor, A.: Learning a blind measure of perceptual image quality. In: Computer Vision and Pattern Recognition. IEEE Computer Society Press, Colorado Springs (2011)

[3] Sheikh, H.R., Sabir, M.F., Bovik, A.C.: A statistical evaluation of recent full reference image quality assessment algorithms. IEEE Transactions on Image Processing 15(11), 3440–3451 (2006)

[4] Gabarda, S., Cristóbal, G.: Blind image quality assessment through anisotropy. Virtual Journal for Biomedical Optics 24(12), B42–B51 (2007)

[5] Sheikh, H.R., Bovik, A.C., Cormack, L.: Blind quality assessment of JPEG 2000 compressed images using natural scene statistics. In: Conference Record of the Thirty-Seventh Asilomar Conference on Signals, Systems and Computers (2003)

[6] Saad, M.A., Bovik, A.C., Charrier, C.: A DCT statistics-based blind image quality index. IEEE Signal Processing Letters 17(6), 583–586 (2010)

[7] Lee, C., et al.: Objective video quality assessment. Optical Engineering 45(1), 17004-11 (2006)

[8] Al-Hinai, N., et al.: Optimum wavelet thresholding based on structural similarity quality assessment for FFT-OFDM. In: International Conference on Advanced Technologies for Communications, ATC 2008 (2008)

[9] Damera-Venkata, N., et al.: Image quality assessment based on a degradation model. IEEE Transactions on Image Processing 9(4), 636–650 (2000)

[10] Chandler, D.M., Hemami, S.S.: VSNR: A Wavelet-Based Visual Signal-to-Noise Ratio for Natural Images. IEEE Transactions on Image Processing 16(9), 2284–2298 (2007)

[11] Chun-Ling, Y., Hua-Xing, W., Lai-Man, P.: Improved Inter Prediction based on Structural Similarity in H.264. In: IEEE International Conference on Signal Processing and Communications, ICSPC 2007 (2007)

[12] Ho-Sung, H., Dong, O.K., Rae-Hong, P.: Structural information-based image quality assessment using LU factorization. IEEE Transactions on Consumer Electronics 55(1), 165–171 (2009)

[13] Xinbo, G., et al.: Image Quality Assessment Based on Multiscale Geometric Analysis. IEEE Transactions on Image Processing 18(7), 1409–1423 (2009)

[14] Bin, L., Yan, C.: An Image Quality Assessment Algorithm Based on Dual-scale Edge Structure Similarity. In: Second International Conference on Innovative Computing, Information and Control, ICICIC 2007 (2007)

[15] Chen, G.-H., Yang, C.-L., Xie, S.-L.: Gradient-based structural similarity for image quality assessment. IEEE Computer Society, Atlanta (2006)

[16] Huang, S., Burnett, T.J.W., Deczky, A.G.: The Importance of Phase in Image Processing Filters. IEEE Transactions on Acoustics, Speech, and Signal Processing (1975)

[17] Oppenheim, A.V., Lim, J.S.: The Importance of Phase in Signals. Proceedings of the IEEE 69, 529–541 (1981)

[18] Liu, Z., Laganire, R.: Phase congruence measurement for image similarity assessment. Pattern Recognition Letters 28(1), 166–172 (2007)

[19] Szilagyi, T., Brady, S.M.: Feature extraction from cancer images using local phase congruency: A reliable source of image descriptors. IEEE Computer Society, Boston (2009)

[20] Sheikh, H.R., et al.: LIVE Image Quality Assessment Database Release 2, http://live.ece.utexas.edu/research/quality

Force Work Induced Metric for Face Verification

Jianjun Qian[1], Jian Yang[1], Zhangjing Yang[1], and Weilan Wang[2]

[1] School of Computer Science, Nanjing University of Science and Technology,
Nanjing, 210094, China
[2] School of Mathematics and Computer Science, Northwest University for Nationalities,
Lanzhou, 730030, China

Abstract. This paper presents a robust and simple metric approach named Force Work Induced Metric (FWIM) according to a Physical model. A novel image local descriptor based on FWIM (FWIM-LD) is then introduced for face verification. FWIM-LD captures the local structure information between central pixel and its neighbors effectively. PCA thus is used to obtain the low-dimensional and significant features. Subsequently, we employ the binary-like face representation method to further improve the face verification rate. Experimental results on the challenging benchmark "Labeled Faces in the Wild" (LFW) dataset demonstrate that the proposed method achieves better performance than the state-of-the-art algorithms.

1   Introduction

Recently, the "Labeled Faces in the Wild" (LFW) image set and benchmarks have been published to promote the development of face recognition under challenge conditions (e.g. large variations arising from pose, lighting condition, facial expression, occlusion, misalignment, etc.). The faces of LFW are detected in images "in the wild", taken from the Yahoo! News.

Since LFW is published, plenty of approaches have been proposed to improve the performance on the benchmarks associated with the LFW database [1-4]. A texture image representation method Local Binary Pattern (LBP) has been demonstrated to be extremely effective for face recognition [5, 6]. The representation of a face image is obtained by dividing the image into a series of sub-windows and computing their histograms of LBP values. Lior Wolf et.al developed a family of novel face image descriptors (e.g. three-patch LBP (TPLBP) and four-patch LBP (FPLBP)) to capture statistics of local patch similarities [7]. Apart from these methods, there are some other approaches derived from LBP and successfully used in face recognition [8, 9].

Recently, H. J. Seo et al. proposed a novel nonparametric object detection framework without training process by using locally adaptive regression kernels descriptors (LARK) [10]. Furthermore, this object detection framework is extended to face verification [11]. However, why LARK descriptors can achieve better performance in the large real-world face image dataset LFW? The main reason is that LARK effectively captures the local structure information by using the geodesic distance to measure the similarities between the central pixel and its neighborhoods.

J. Yang, F. Fang, and C. Sun (Eds.): IScIDE 2012, LNCS 7751, pp. 298–305, 2013.
© Springer-Verlag Berlin Heidelberg 2013

The geodesic distance represents shortest path on the signal manifold between arbitrary points and is computed via the local gradient covariance matrix and spatial information. In other words, how to measure the structure information plays an important role in face representation.

Motivated by this idea, we propose a novel metric method named Force Work Induced Metric (FWIM) to describe the metric relation between central pixel and its neighbors in the local patch. Compared with geodesic distance, FWIM is computed easily and effectively. For this reason, a novel image Local Descriptor based on FWIM (FWIM-LD) is introduced for face representation. We also use a logistic function on FWIM-LD descriptors to further improve the face verification rate.

2   FWIM: A Novel Metric Approach

2.1   The Physical Model

Given an object M, and two points A and B (as shown in Fig. 1). S is the fixed-length, which represents the distance between A and B. The task is that moving the object M from place B to place A by using the tension F (leaving out other factors). Here, we assume that the size of F is changing and the direction remains constant. However, the tension value of F is different under different conditions. To complete the task, the force work on the object M is computed as follows:

$$W = FS \tag{1}$$

where, $F = \sum_{i=1}^{n} f_i$ and n is the number of element. In fact, S can be ignored since it is the fixed-length (unit length). So, the Eq. (1) also can be formulated as follows:

$$W = FS = \sum_{i=1}^{n} f_i S = \sum_{i=1}^{n} f_i \tag{2}$$

Suppose that there is a regular 2-D distribution of tension f. Under this distribution, the connection between the arbitrary two points includes different tensions, which are regarded as the elements of F. So the force work is evaluated by using F. Since the works can measure the differences of two groups of tension, which are used to move the object M from the place B to the place A, it actually reveals the metric relation of the arbitrary two points on the 2-D distribution of tensions. Further, we will introduce the novel metric approach in detail in the following section according to this model.

2.2   Force Work Induced Metric (FWIM)

First of all, one 2-D image is considered as 3-D curve surface. The data of z-axis is intensity value of an image. Intuitively, it's believed that the values of z-axis (intensity value) represent the energy of the corresponding points to a certain degree. Moreover, the intensity values of an image also follow the regular 2-D distribution. So, the intensity values of an image can be regarded as the distribution of tensions.

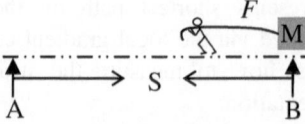

Fig. 1. The physical model: Giving a tension F to move the object M from place B to place A

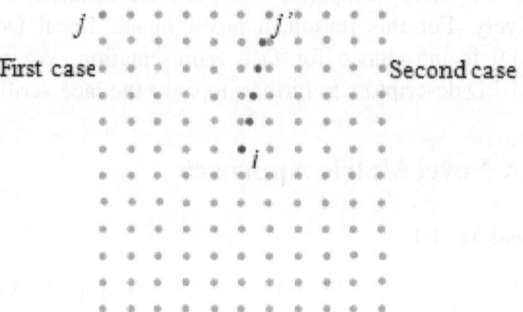

Fig. 2. One Local patch in an image, the values of red, orange, blue and black are known. In the first case, the relationship between i and j is computed directly. In the Second case, the relationship between i and j' is achieved using the points partial of which are achieved by interpolating.

For this reason, the above mentioned model is used to represent the metric relation between any of two points. In this way, we introduce a novel metric approach FWIM and give the definition as follows.

Definition 1. Let $i\ (x_i,\ y_i)$ and $j\ (x_j,\ y_j)$ (or $j'\ (x_{j'},\ y_{j'})$) represent the arbitrary two pixels in an image. The relationship between i and $j\ (j')$ is evaluated by accumulating the intensities of the pixels from i to j (or j'). Specifically, there are two cases are shown in Fig. 2. In the first case, the relationship can be directly obtained. In another case, we should interpolate some pixels on the line between i and j' using the two nearest neighbors at first. The relationship of them is then computed by collecting all the intensities. So, this procedure is called Force Work Induced Metric (FWIM).

$$R(i, j) = \sum_{\alpha \in \Psi} f(\alpha) \tag{3}$$

where $R(i, j)$ represents the relationship between i and j, Ψ is the set of pixels among i and j. $f(\alpha)$ is the corresponding intensity values of the α-th pixel.

So, the structure information between arbitrary two pixels in an image is achieved by using FWIM.

3   Local Descriptor Based on FWIM (FWIM-LD)

In this section, a novel image local descriptor based on FWIM is introduced. As described in the previous section. FWIM effectively and efficiently reveals the metric relation between any of two points. The main idea of FWIM-LD is to robustly capture the local structure by using FWIM to measure the relationship between central pixel and its neighborhoods in the local patch, in conjunction with spatial information. The image descriptor FWIM-LD is modeled as follows (how to compute the image descriptor is also shown in Fig. 3):

$$L(c,i) = \exp\{-w_{c,i}R(c,i)\} \quad i \in \Omega \tag{4}$$

where, $R(c,i)$ represents the metric relationship between the central pixel c and its i-th neighbor. Ω is a pixel set in the local patch. $w_{c,i} = \|x_c - x_i\|^2$ is the spatial relationship between the central pixel c and its i-th neighbor. x_c and x_i are the spatial coordinates. Subsequently, the local image structure features are essentially represented by using the FWIM based local descriptor $L(\cdot,\cdot)$. Moreover, the FWIM based local descriptor of the j-th patch is computed and normalized as follows:

$$D^j(c,i) = \frac{L^j(c,i)}{\sum_{i=1}^{P} L^j(c,i)}, \quad i \in [1,\cdots,P], j \in [1,\cdots,N] \tag{5}$$

where, P is the size of the local patch. N is the number of patches in an image. The dense FWIM-LD descriptors of an image D are composed of a serious of stacked version of $D^j = [D^j(c,1),\cdots D^j(c,P)]^T$.

$$D = [D^1,\cdots D^j,\cdots D^N] \tag{6}$$

Generally, it's believed that dense FWIM-LD descriptors contain rich information. However, it also encompasses redundant information. Therefore, PCA is used to obtain the low-dimensional and significant features of an image. The projection matrix V consists of the first t principal components. Finally, the significant features of dense FWIM-LD descriptors are evaluated by projecting V on to D as follows:

$$H = V^T D \in \mathbb{R}^{t \times N} \tag{7}$$

To further improve the face verification performance, the logistic function is employed to features H as follows [12]:

$$G = \frac{1}{1+\exp(-cH)} - 0.5 \tag{8}$$

This function is to make the features becomes more or less binary-like and enhance the discriminative power of any distance measure.

4   Experiments

The LFW database contains 13,233 target face images. There are 5,749 different individuals in the LFW. 1,680 people have two or more face images. The remainder 4,069 persons have just only one image. All of the images are the result of detections by using the Viola-Jones face detector, and have been rescaled and cropped to a fixed

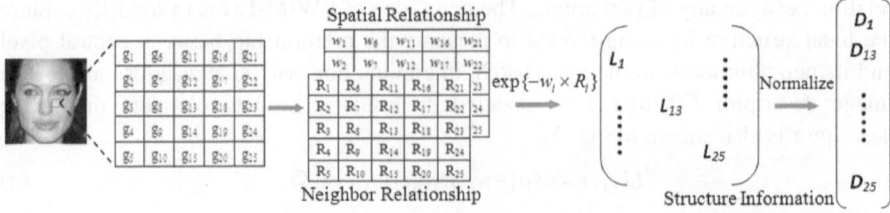

Fig. 3. The procedure of computing the local structure information (5x5) centered at g_{13} via FWIM

Fig. 4. Example faces from LFW

size (250x250). These images have a large degree of facial expression, occlusions, pose and illuminations (as shown in Fig. 4) since all of them are taken from the real world images. In this paper, we pay close attention to the challenge settings: the image restricted setting. Here, we use the aligned version of images (http://www.openu.ac.il/home/hassner/data/lfwa/) and simply crop the face image to remove the background, leaving a 128x128 face image as shown in Fig. 5.

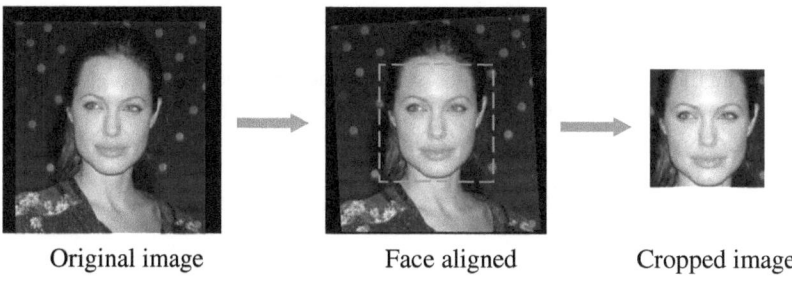

| Original image | Face aligned | Cropped image |

Fig. 5. Cropping the face image

Table 1. The average verification rates of LBP, TPLBP, FPLBP, LARK and FWIM-LD on the LFW dataset (View2)

Methods	L2	MCS
LBP	67.10	67.86
TPLBP	68.20	68.33
FPLBP	67.25	67.68
LARK	66.25	68.95
FWIM-LD	67.33	69.13

Table 2. The average verification rates (Mirror version) on the LFW dataset (View2)

Methods	L2	MCS
LBP(Mirror)	68.48	69.63
TPLBP(Mirror)	69.68	69.95
FPLBP(Mirror)	69.46	69.75
LARK(Mirror)	68.87	70.06
FWIM-LD(Mirror)	69.12	71.00

In this experiment, we examine the performances of various methods on the "image-restrict setting" benchmark. The benchmark test requires carrying out the method 10 times, each time one subset for testing, and other nine subsets for training. Moreover, the following state-of-the-art image descriptors LBP, TPLBP, FPLBP, LARK and the proposed method FWIM-LD are used for face verification. For LBP features, the image is divided into 8x8 LBP blocks, each with 59 dimension histogram. The parameters of TPLBP and FPLBP are copied from [7]. The parameters of LARK are the same with the paper [11]. FWIM-LD is computed by using FWIM in the local patches (13x13) of the face image. PCA is used to reduce the feature dimension from 169 to 12. Further, the final feature vector is achieved by employing logistic function (c=80). The MCS (matrix cosine similarity) score, is proposed in the paper [10], is evaluated from each of 6,000 pairs.

We use the descriptor vectors of LBP, TPLBP, FPLBP, LARK and FWIM-LD in the experiment. L2 and MCS are employed to report the performances. Further, we also perform the experiment with mirror version of these methods [11], which can improve the verification rate. The average verification rates of LBP, TPLBP, FPLBP, LARK and FWIM-LD with linear SVM (Support Vector Machine) classifier are listed in Table 1. Table 2 illustrated the average verification rates of mirror version of these methods. From Table 1 and Table 2, we can see that the mirror version of the proposed method is better than other state-of-the-art algorithms. Fig. 6 shows the comparative results between FWIM-LD, LARK and mirror version of them with different local patch size.

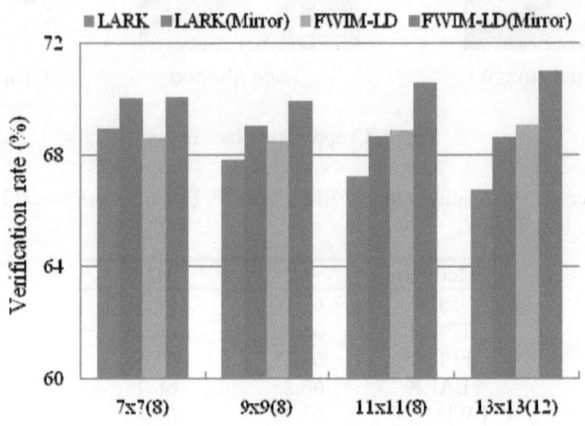

Fig. 6. The comparisons between LARK and FWIM-LD with different local patch size

5   Conclusions

In this paper, we present a simple and effective metric algorithm called FWIM for image feature representation. It is inspired by a physical model. Subsequently, the image descriptor is induced based on FWIM. FWIM-LD significantly captures the local structure information by describing the relationship between central pixel and its neighbors. PCA is then used to reduce the dimension and obtain the significant feature. Further, the logistic function is employed to encode the image feature. Experimental results demonstrate that FWIM-LD illustrates a favorable performance on the LFW benchmark compared with the state-of-the-art descriptors. Although there are many similarities between FWIM-LD and LARK, FWIM-LD is much more simple and effective than LARK.

In the future, we will still focus on how to explore the local structure information accurately. Meanwhile, how to exploit the proposed method for face detection and object recognition is another future work.

References

1. Guillaumin, M., Verbeek, J., Schmid, C.: Is that you? Metric Learning Approaches for Face Identification. In: 12th IEEE International Conference on Computer Vision, Kyoto, Japan, pp. 498–505 (2009)
2. Hua, G., Akbarzadeh, A.: A Robust Elastic and Partial Matching Metric for Face Recognition. In: 12th IEEE International Conference on Computer Vision, Kyoto, Japan, pp. 2082–2089 (2009)

3. Pinto, N., DiCarlo, J.J., Cox, D.D.: How far can you get with a modern face recognition test set using only simple features? In: IEEE-Computer-Society Conference on Computer Vision and Pattern Recognition Workshops, Miami Beach, FL, pp. 2583–2590 (2009)
4. Cao, Z., Yin, Q., Tang, X., Sun, J.: Face Recognition with Learning-based Descriptor. In: 23rd IEEE Conference on Computer Vision and Pattern Recognition (CVPR), San Francisco, CA, pp. 2707–2714 (2010)
5. Ahonen, T., Hadid, A., Pietikainen, M.: Face description with local binary patterns: Application to face recognition. IEEE Transactions on Pattern Analysis and Machine Intelligence 28, 2037–2041 (2006)
6. Ojala, T., Pietikainen, M., Maenpaa, T.: Multiresolution gray-scale and rotation invariant texture classification with local binary patterns. IEEE Transactions on Pattern Analysis and Machine Intelligence 24, 971–987 (2002)
7. Wolf, L., Hassner, T., Taigman, Y.: Effective Unconstrained Face Recognition by Combining Multiple Descriptors and Learned Background Statistics. IEEE Transactions on Pattern Analysis and Machine Intelligence 33, 1978–1990 (2011)
8. Tan, X.Y., Triggs, B.: Enhanced Local Texture Feature Sets for Face Recognition Under Difficult Lighting Conditions. IEEE Transactions on Image Processing 19, 1635–1650 (2010)
9. Lei, Z., Liao, S.C., Pietikainen, M., Li, S.Z.: Face Recognition by Exploring Information Jointly in Space, Scale and Orientation. IEEE Transactions on Image Processing 20, 247–256 (2011)
10. Seo, H.J., Milanfar, P.: Training-Free, Generic Object Detection Using Locally Adaptive Regression Kernels. IEEE Transactions on Pattern Analysis and Machine Intelligence 32, 1688–1704 (2010)
11. Seo, H.J., Milanfar, P.: Face Verification Using the LARK Representation. IEEE Transactions on Information Forensics and Security 6, 1275–1286 (2011)
12. Kavukcuoglu, K., Ranzato, M.A., Fergus, R., Le Cun, Y.: Learning Invariant Features through Topographic Filter Maps. In: IEEE-Computer-Society Conference on Computer Vision and Pattern Recognition Workshops, Miami Beach, FL, pp. 1605–1612 (2009)

Functional Connectivity-Based Parcellation of Human Medial Frontal Cortex via Maximum Margin Clustering

Lubin Wang[1], Qiang Liu[2,3], Hong Li[2,3], and Dewen Hu[1,*]

[1] College of Mechatronics and Automation, National University of Defense Technology,
Changsha, Hunan, China
[2] Key Laboratory of Cognition and Personality (SWU), Ministry of Education, China
[3] School of Psychology, Southwest University, Chongqing, China
dwhu@nudt.edu.cn

Abstract. Recently, the utilization of resting-state functional connectivity MRI (rs-fcMRI) for defining specialized functional regions has been explored. In this study, we employed a novel clustering algorithm, maximum margin clustering (MMC), for group parcellation of human medial frontal cortex using rs-fcMRI data. Experimental results showed that the medial frontal cortex can be divided into anterior and posterior clusters. The border between the two clusters located close to the vertical commissure anterior line, suggesting the homologues of the pre-supplementary motor area (pre-SMA) and supplementary motor area (SMA). The pre-SMA cluster exhibited stronger connectivity with cognitive-processing regions, whereas the SMA cluster exhibited stronger connectivity with the motor cortex. Our findings demonstrate that rs-fcMRI represents a useful and effective tool for parcellation of human cortex into functional regions.

Keywords: resting-state functional connectivity, medial frontal cortex, parcellation, maximum margin clustering.

1 Introduction

In human, the medial frontal cortex plays an important role for linking cognition to action. It has been found that the medial frontal cortex is likely made up of two anatomically and functionally distinct parts, and can be divided into the pre-supplementary motor area (pre-SMA) and supplementary motor area (SMA). While there is no precise boundary to distinguish pre-SMA and SMA, the vertical commissure anterior (VCA) line provides the best approximation [1]. Although previous neuroimaging studies provide insight into the medial frontal cortex, the complex functional attributes of this brain region remain unclear. Exploring the interrelation between medial frontal cortex and other brain regions will probably enhance our understanding of the specific contributions of its subregions.

Resting-state functional connectivity magnetic resonance imaging (rs-fcMRI) detects temporal correlations in spontaneous low-frequency (< 0.1 Hz) blood oxygen

---

* Corresponding author.

J. Yang, F. Fang, and C. Sun (Eds.): IScIDE 2012, LNCS 7751, pp. 306–312, 2013.

level-dependent (BOLD) signal fluctuations. During the "resting state", participants are typically instructed to keep their eyes closed, to keep as motionless as possible, not think of anything and not to fall asleep. This method demonstrates that spatially distinct but functionally related brain regions have strong functional connectivity with each other, and has been used to study normal and abnormal functional brain organization [2-4]. Based on previous findings that the pre-SMA and SMA have different functional roles, we hypothesized that these two regions may have different resting-state functional connectivity profiles.

Recently, the utilization of rs-fMRI data for defining functional regions has been explored [5, 6]. Most of such studies employed unsupervised clustering methods to divide a predefined brain region into spatially coherent subregions of homogeneous functional connectivity. Inspired by the maximum margin classification criterion in support vector machine (SVM), maximum margin clustering (MMC) is a recent approach which simultaneously learns the optimal hyperplane and cluster labels [7-9]. It has been showed that MMC often obtains more accurate results than conventional clustering methods. In this study, we utilized a two-level MMC method to divide the medial frontal cortex into the putative pre-SMA and SMA clusters. Using these two clusters as regions of interest, we further examined the possible functional connectivity differences between pre-SMA and SMA.

2   Materials and Methods

2.1   Participants and Data Acquisition

Participants included 40 healthy native Chinese speakers (3 females, age range 28-54). No one had suffered major head trauma, had a history of alcohol or drug dependence, or had neurological disorders. Written informed consents were obtained from all the participants who took part in this study. This study was approved by the Institutional Review Board of the Southwest University.

During the resting state, participants were simply instructed to keep still with their eyes closed, remain awake, and not to think of anything in particular. All the participants reported that they were awake in the experiments. Functional imaging was performed using a SIEMENS TRIO 3-T MRI scanner in the Key Laboratory of Cognition and Personality (Southwest University, SWU), Ministry of Education, China. The imaging parameters were as follows: 32 axial slices, TR = 2000 ms, TE = 30 ms, slice thickness = 3.0 mm, flip angle = 90°, FOV = 200×200 mm^2, in-plane resolution = 64×64. Each resting-state scanning lasted 8 minutes, and 240 volumes were obtained.

2.2   Data Preprocessing

Prior to preprocessing, the first 10 volumes of each scan were discarded to remove possible T1 stabilization effects. All resting-state images were preprocessed using the statistical parametric mapping software package (SPM5, Wellcome Department of Cognitive Neurology, Institute of Neurology, London, UK). The data were corrected for within-scan acquisition time differences between slices, and realigned to the first volume to correct for inter-scan head motions. All subjects in this study had less than 2 mm translation and 2° of rotation in any of the x, y, and z axes. Then, the volumes were normalized to the standard EPI template in the Montreal Neurological Institute (MNI) space and resliced to 3×3×3 mm^3. The resulting images were spatially smoothed with a Gaussian filter of 8 mm full-width half-maximum kernel. Subsequently, the data were temporally filtered with a brand-pass filter (0.01-0.1 Hz), followed by linear detrending to remove any residual drift. Nine nuisance signals were further removed from the data via multiple regression, including signals averaged from white matter, cerebrospinal fluid, and the whole brain, and six parameters obtained by head motion correction. This regression procedure was utilized to reduce spurious variance unlikely to reflect neuronal activity. Functional connectivity between a pair of resting-state time series was evaluated by using Pearson's correlation coefficient. Fisher's z-transform was applied to the correlation values to ensure normality.

2.3   Functional Parcellation of the Medial Frontal Cortex

A mask of the medial frontal cortex was created in the MNI space using the software WFU_PickAtlas (http://www.ansir.wfubmc.edu) (Fig. 1). The number of voxels in medial frontal cortex was about 1300. In this study, a two-level MMC method was employed to divide medial frontal cortex into subregions. We first performed individual-level clustering to categorize voxels according to their functional connectivity profiles, and then performed group-level clustering based on the consistency across the individual cluster maps.

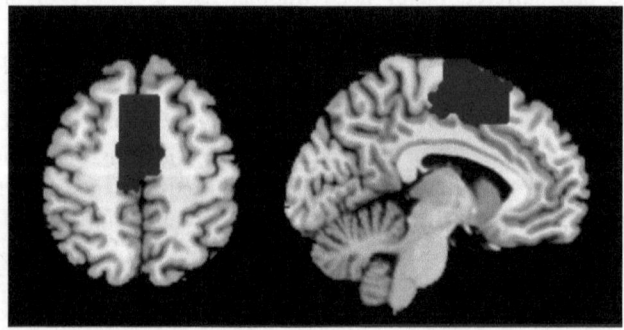

Fig. 1. Medial frontal cortex mask shown in axial and sagittal view

In the individual-level clustering stage, the connectivity profiles of voxels in the medial frontal cortex should be calculated first. To reduce computational burden, we randomly selected 7000 voxels from the rest of the brain. Each voxel in the medial frontal cortex was correlated with these 7000 voxels to represent its connectivity profile. The connectivity profiles were stored in a set of vectors $\mathbf{x}_i$ ($i = 1,2,...,n$), where n is the number of voxels in the medial frontal cortex. MMC was then used to assign $\mathbf{x}_i$ into two clusters $y_i \in \{-1, +1\}$. The principle of MMC is to find a labeling so that if one were to subsequently run an SVM, the margin obtained would be maximal over all possible labeling cases. To prevent a useless solution with the infinite margin by assigning all data points to a single cluster, a class balance constraint is introduced:

$$B = \left\{ \hat{\mathbf{y}} \big| \hat{y}_i \in \{\pm 1\}, -\beta \leq \mathbf{1}' \hat{\mathbf{y}} \leq \beta \right\} \tag{1}$$

Straightforwardly, MMC requires the solution of the following optimization problem:

$$\min_{\hat{\mathbf{y}} \in B} \min_{\omega, \xi} \quad \frac{1}{2} \|\omega\|_2^2 - \rho + \frac{C}{2} \sum_{i=1}^{n} \xi_i^2$$

$$\text{s.t.} \quad \hat{y}_i \, \omega' \varphi(\mathbf{x}_i) \geq \rho - \xi_i, \quad i = 1, ..., n \tag{2}$$

where ω is the optimal separating hyperplane, C is a regularization parameter that trades off the empirical risk and the model complexity, and φ is the feature map by some kernel function (linear kernel was used in this study). This is usually solved in its dual:

$$\min_{\hat{\mathbf{y}} \in B} \max_{\alpha \in A} \quad -\frac{1}{2} \alpha' \left(\mathbf{K} \cdot \hat{\mathbf{y}} \hat{\mathbf{y}}' + \frac{1}{C} \mathbf{I} \right) \alpha \tag{3}$$

where α is the vector of dual variables for the inequality constraints in (2), and $\mathbf{K}$ is the kernel matrix.

However, (3) is not a convex function, and this formulation does not lead to an effective algorithmic approach. In this study, we performed MMC using a novel convex optimization algorithm, referred to as Label Generating MMC (LG-MMC) [8]. LG-MMC maximizes the margin of opposite clusters by generating the most violated label vectors iteratively, and then combines them via multiple kernel learning (MKL). Because of the exponential number of possible labeling, the set of base kernels is also exponential in size and direct MKL is computationally intractable. The cutting-plane algorithm is applied to handle this problem, which is described as follows:

Step 1: Initialize $\alpha = 1 / n$. Find the most violated $\hat{\mathbf{y}}$ and set $\Gamma = \{ \hat{\mathbf{y}}, -\hat{\mathbf{y}} \}$

Step 2: Run MKL for the subset of kernel matrices selected in Γ and obtain α

Step 3: Find the most violated $\hat{\mathbf{y}}$ and set $\Gamma = \hat{\mathbf{y}} \cup \Gamma$

Step 4: Repeat steps 2 and 3 until convergence

For each subject, MMC simultaneously learned the cluster labels and an SVM classifier $f(\mathbf{x}) = \omega' \mathbf{x}$. Let s_{ij} denote the SVM score for the i-th voxel in the medial frontal cortex of the j-th subject, which indicated the margin between the connectivity profile $\mathbf{x}_i$ and the separating hyperplane ω_j. After performing individual-level clustering on all the subjects, the vector of SVM scores $\mathbf{s}_i = [s_{i1}, s_{i2}, ..., s_{im}]$ was obtained, where m is the number of subjects. In the group-level clustering stage, MMC was used to assign $\mathbf{s}_i$ ($i = 1, 2, ..., n$) to two clusters. Therefore, the resulting group cluster map expressed the consistency of parcellations across subjects.

Finally, we examined the possible functional connectivity differences of the two clusters. For each subject and for each cluster, the correlation map was created by calculating functional connectivity between the mean time series of the cluster and time series of all voxels in the brain. Voxel-wise two-sample t-test was performed between the correlation maps of the two clusters across all the subjects. Significance level was set at $p < 0.05$, corrected (false discovery rate).

3   Experimental Results and Analysis

Fig. 2 shows the clustering results of 4 sample subjects. Voxels with positive and negative clustering scores are displayed in hot and cold, respectively. After mapping each voxel in the medial frontal cortex back onto the brain, we can see that parcellation of medial frontal cortex results in anterior and posterior clusters (top row of Fig. 2). The reordered functional similarity matrices of medial frontal cortex for different subjects are also shown in the bottom row of Fig. 2, each element of which is quantified by the inverse of the Euclidean distance between connectivity profiles of two voxels. We can see that voxels within a cluster were functionally proximate, whereas voxels between clusters generally showed dissimilar connectivity patterns.

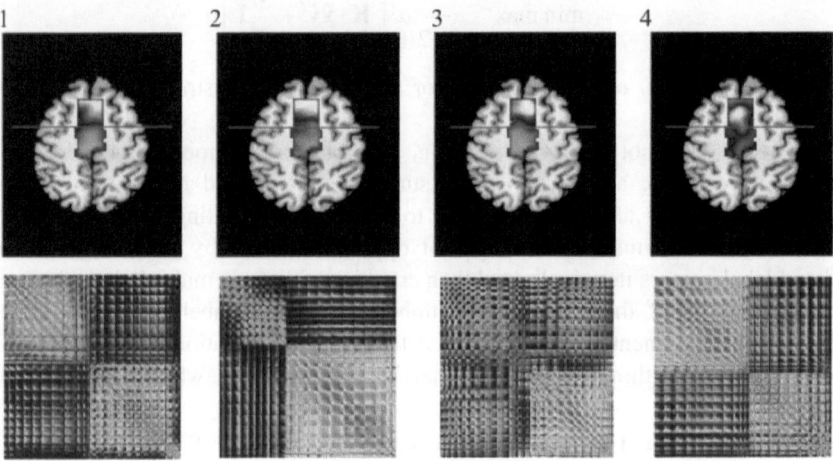

Fig. 2. Functional connectivity-based parcellation of the medial frontal cortex for 4 sample subjects. The top row shows the two clusters mapped on the brain. The green line indicates the VCA line (Y = 0). The bottom row shows the reordered functional similarity matrices of medial frontal cortex for different subjects, which are colored from blue to red.

Because of individual variation, the individual-level clustering results can not be applied directly for group analysis. To test the general reproducibility of the functional connectivity-defined clusters, we further performed MMC across the cluster maps of all the subjects. Our results showed that the border between the anterior and posterior clusters located close to the VCA line (Y = 0), suggesting that they corresponded to the putative pre-SMA and SMA, respectively (Fig. 3)

Surface rendering of functional connectivity differences between the pre-SMA and SMA is shown in Fig.4. Brain regions showing stronger connectivity strength with the pre-SMA are displayed in hot, whereas brain regions showing stronger connectivity strength with the SMA are displayed in cold. We found that the pre-SMA exhibited stronger connectivity with the prefrontal cortex, inferior parietal lobule, and basal ganglia, whereas the SMA exhibited stronger connectivity with the premotor cortex, primary motor cortex, and somatosensory cortex. Our results are consistent with previous structural connectivity study [10], suggesting a general correspondence between anatomical and functional landmark-based parcellation. Moreover, our results are also consistent with previous findings that the pre-SMA is more involved in complex/cognitive situations, while the SMA is more related to actions [1].

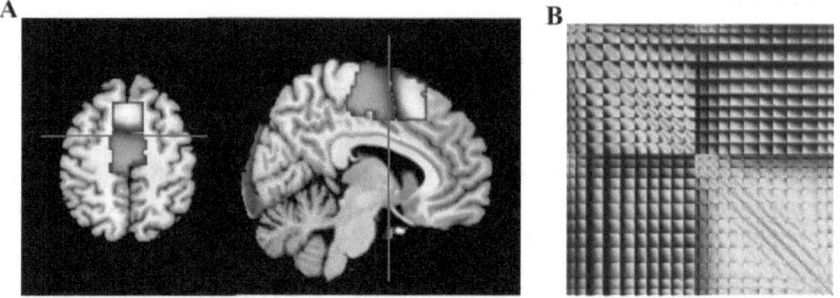

Fig. 3. Group-level parcellation of the medial frontal cortex. (A) Clustering maps for the putative pre-SMA and SMA. Voxels with positive and negative clustering scores are displayed in hot and cold, respectively. The green line indicates the VCA line (Y = 0). (B) The mean functional similarity matrix of medial frontal cortex across all the subjects.

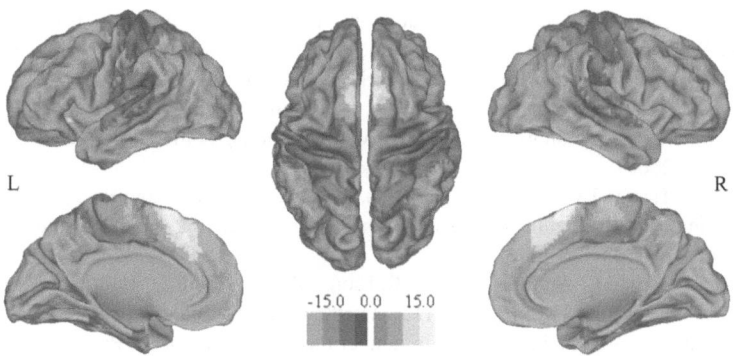

Fig. 4. Surface rendering of functional connectivity differences between the pre-SMA and SMA. L and R indicate the left and right side of the brain.

4   Conclusions

In this paper, we employed MMC for functional connectivity-based parcellation of human medial frontal cortex. We found that the medial frontal cortex can be divided into anterior and posterior clusters, which correspond to the putative pre-SMA and SMA, respectively. Our findings may shed light on the specific functional roles of the pre-SMA and SMA. Moreover, this study demonstrates the feasibility of defining specialized functional regions using rs-fcMRI.

Acknowledgments. This work was supported by the National Basic Research Program of China (2011CB707802), and the National High-Tech Program of China (2012AA011601).

References

1. Nachev, P., Kennard, C., Husain, M.: Functional role of the supplementary and pre-supplementary motor areas. Nat. Rev. Neurosci. 9, 856–869 (2008)
2. Fox, M.D., Raichle, M.E.: Spontaneous fluctuations in brain activity observed with functional magnetic resonance imaging. Nat. Rev. Neurosci. 8, 700–711 (2007)
3. Shen, H., Wang, L., Liu, Y., Hu, D.: Discriminative analysis of resting-state functional connectivity patterns of schizophrenia using low dimensional embedding of fMRI. NeuroImage 49, 3110–3121 (2010)
4. Zeng, L., Shen, H., Liu, L., Wang, L., Li, B., Fang, P., Zhou, Z., Li, Y., Hu, D.: Identifying major depression using whole-brain functional connectivity: a multivariate pattern analysis. Brain 135, 1498–1507 (2012)
5. van den Heuvel, M., Mandl, R., Hulshoff Pol, H.: Normalized cut group clustering of resting-state fMRI Data. PLoS ONE 3, e2001 (2008)
6. Nelson, S.M., Cohen, A.L., Power, J.D., Wig, G.S., Miezin, F.M., Wheeler, M.E., Velanova, K., Donaldson, D.I., Phillips, J.S., Schlaggar, B.L., Petersen, S.E.: A parcellation scheme for human left lateral parietal cortex. Neuron 67, 156–170 (2010)
7. Xu, L., Neufeld, J., Larson, B., Schuurmans, D.: Maximum margin clustering. In: Saul, L.K., Weiss, Y., Bottou, L. (eds.) Advances in Neural Information Processing Systems, vol. 17, pp. 1537–1544. MIT Press, Cambridge, MA (2005)
8. Zhao, B., Wang, F., Zhang, C.: Efficient maximum margin clustering via cutting plane algorithm. In: Proceedings of the 8th SIAM International Conference on Data Mining, pp. 751–762 (2008)
9. Li, Y.F., Tsang, I.W., Kwok, J.T., Zhou, Z.H.: Tighter and convex maximum margin clustering. In: Proceedings of International Conference on Artificial Intelligence and Statistics, pp. 344–351 (2009)
10. Johansen-Berg, H., Behrens, T.E.J., Robson, M.D., Drobnjak, I., Rushworth, M.F.S., Brady, J.M., Smith, S.M., Higham, D.J., Matthews, P.M.: Changes in connectivity profiles define functionally distinct regions in human medial frontal cortex. Proc. Natl. Acad. Sci. U S A 101, 13335–13340 (2004)

3D Facial Landmark Localization via a Local Surface Descriptor *HoSNI*

Xiaobo Zhang[1], Yueming Wang[2], and Gang Pan[1,*]

[1] College of Computer Science and Technology, Zhejiang Univ.
[2] Qiushi Academy for Advanced Studies, Zhejiang Univ.
{xb_zhang,ymingwang,gpan}@zju.edu.cn

Abstract. Facial landmarks of a 3D face model, such as nose-tip, inner-eyes, and mouth-corners, play an important role in many applications of 3D face models. This paper presents an effective approach to automatically detect landmarks of a 3D face. A novel discriminative surface descriptor, named *HoSNI*(Histogram of Shape Normal Information), is presented to characterize the local shape around a point on the facial surface. The *HoSNI* is applied to localize facial landmarks. The experiments are carried out to detect 19 facial landmarks on FRGC v2.0. The results demonstrate that our approach has high accuracy and is insensitive to expression variation.

1 Introduction

Three-dimensional capture technology has made considerable progress in the past decade. Acquisition of 3D face data is becoming cheaper and easier. Since 3D faces convey geometric shape information while images do not, applications of 3D faces emerge in large numbers in recent years, such as 3D face recognition [1][2][3], 3D face expression recognition, face animation, face synthesis, removal of 3D facial expressions [4], gender classification with 3D face models, and super-resolution of 3D faces [5].

Facial landmarks, such as nose-tip, mouth-corners, and inner-eyes, play an important role in many applications of 3D face models. They usually serve as anchor points during some key steps, for example, face alignment, face registration, face normalization, and feature extraction. Most of 3D face recognition methods need nose-tip for face alignment and facial region extraction.

Although a couple of algorithms have been proposed to detect facial landmarks in recent years, automatic accurate localization of 3D facial landmarks is still a big challenge in the 3D face processing community. Difficulties of landmark detection mainly come from three aspects: 1) data noise: the face models captured by 3D digitizer usually contain noise and outliers due to technical limitations; 2) facial expression: expression introduces non-rigid deformation into a facial surface, especially in mouth region, which makes localization more difficult; 3) inter-person variation: faces vary in size and shape from person to person.

---

* Corresponding author.

J. Yang, F. Fang, and C. Sun (Eds.): IScIDE 2012, LNCS 7751, pp. 313–321, 2013.
© Springer-Verlag Berlin Heidelberg 2013

In this paper, we propose an automatic landmark localization approach for a 3D face, which does not need any face image and texture information. Our approach employs a novel local descriptor, named *HoSNI* (Histogram of Shape Normal Information) to characterize local shape around a point on the 3D face.

The paper is organized as follows: Section 2 summarizes related work on facial landmark localization and detection; Section 3 describes the novel local descriptor *HoSNI* and how it is applied to landmark detection; experimental results are presented in Section 4; finally we conclude this paper in Section 5.

2   Related Work

There are lots of efforts on detection of landmarks or facial feature regions for a 3D face. The previous approaches can be roughly categorized into five groups.

1) **Relative-Position-Based Methods.** In early literature, the assumption that "nose-tip is the nearest point to 3D digitizer" was widely used to find the landmark of nose-tip from range data. For example, Lee et al. [6] employed the assumption to detect nose-tip. Hesher et al. [7] deemed the region with high density of points in point clouds as nose-tip region. Although the idea is simple and easy to implement, it does not work for other landmarks.

2) **Profile-Based Methods.** It uses the curve where a facial surface and its symmetry plane intersect at, named *facial profile*, to detect the landmarks located at the profile such as nose-tip and nose-bridge, for example, the work by Beumier et al. [8], Wu et al. [9], and Wang et al. [10]. The feature proposed by Timothy et al. [11] used characteristic of *facial profile* to find nose-tip.

3) **Curvature-Based Methods.** Curvature is one of the most important attributes of 3D surface, which is invariant to Euclidean transformation. This kind of methods employed curvature as the main feature for localization of landmarks or facial regions.Gordon et al. [12] proposed the seminal work that employed curvature for localization of 3D facial landmarks and regions. Colbry et al. [13] located the inner-eyes, nose-tip, mouth and chin point by shape index, which are calculated from principal curvature. Moreno et al. [14] employed mean and gaussian curvature to localize 7 facial regions and 2 lines. Nair et al. [15] combined point distribution model with shape index for 5 feature points. Akagunduz et al. [16] proposed a transform and scale invariant feature based on curvature information for nose and eyes regions detection. Chang et al. [17] utilized mean curvature and Gaussian curvature to detect facial regions of nose-tip, nose-bridge, and eye cavities. The curvature-based methods usually require 3D data of good quality due to their sensitiveness to noise.

4) **Feature-Descriptor-Based Methods.** A lot of work adopted feature descriptors to characterize the landmark-specific information. Wang et al [18] employed a local surface descriptor, named Point Signature, to localize 4 landmarks from 3D data. Xu et al. [19] proposed 2 features to describe local shape of 3D point cloud and combined them with *SVM* (Support Vector Machine) to localize nose-tip and nose-bridge. Breitenstein et al. [20] utilized depth information to find nose-tip using *GPU*. Timothy et al. [11] proposed a feature, named

Rotated Profile Signatures, to detect nose-tip. Fanelli et al. [21] used depth difference and random regression forests to localize nose-tip. Most of them utilized nose-tip detection to estimate head pose and achieve real-time performance.

5) **Image-Assisted-Based Methods.** 3D landmark detection with face image-assisted becomes a promising way when the corresponding face/texture image of a face model is available. A lot of work combine the results based on 3D face shapes with those based on face image, for example, the work by Boehnen et al. [22] and Wang et al [18].

3   3D Facial Landmark Localization via *HoSNI*

On 3D facial surface, the local shape is different from point to point. It can be used as the basic evidence to judge whether a point is a facial landmark. Inspired by this idea, we propose an automatic 3D facial landmark localization approach, which employs a novel local surface descriptor to characterize local shape around a point on the 3D face surface. Our method includes local shape information (*HoSNI*) extraction part and the most similar point searching part.

3.1   Local Shape Descriptor: *HoSNI*

The *HoSNI* combines basic attributes of surface and structure information together. The attributes we use are the normal and the area of triangles. Partitioning the local shape into several patches and extracting characteristics in each patch adds structure information into our descriptor. In detail, calculating *HoSNI* includes two steps, The first is transforming 3D data into *ELB* (*EGI* with Low-density Bins Image, where *EGI* is Extended Gaussian Image) Image, and the second is extracting *HoSNI* from *ELB* Image.

The first step is mainly "re-sampling" and characteristics extraction. It transforms discrete points of a 3D facial surface into a continuous data format. We partition a 3D facial surface into several patches. And in each patch, we extract the *ELB* histogram of the local shape. The *ELB* histogram is similar to *EGI* (Extended Gaussian Image) but calculated on a hemisphere with low-density bins. Each patch is regarded as a "pixel", then they can be deemed as an "image", while the "value" of each pixel is a histogram.

The second step is *HoSNI* extraction. Given a pixel P on *ELB* Image, we partition the nearby pixels into several cells. The histograms of the pixels in a cell are summed up to form the "value" of the cell. We concatenate all histograms of the cells to build the *HoSNI* of the pixel P. The general view is shown in Fig 1.

3.1.1   *ELB* Image

A facial model could be represented as a triangles mesh T. Let t_i be the ith triangle of it, and $\mathbf{c}_i = (x_i, y_i, z_i)$ be the mass center of t_i. We separate the triangles into $M \times N$ patches according to the locations of their mass centers on XY plane. The range of the (mn)th patch is made up of four coordinate in XY plane $\{x_{mn}^L, x_{mn}^U, y_{mn}^L, y_{mn}^U\}$, where x_{mn}^L and x_{mn}^U are the left and right

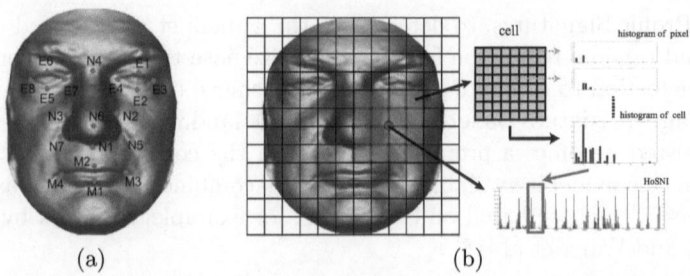

(a)   (b)

Fig. 1. (a) The 19 typical facial landmarks on the face. (b) *HoSNI* Extraction. The red block contains several cells. Each cell has several pixels. All histograms of the cells are concatenate to build the histogram of the block.

coordinate limits respectively, while y_{mn}^L and y_{mn}^U are the bottom and upper coordinate limits. The triangles belonged to the (mn)th patch are

$$T_{mn} = \{t_i | x_i \in [x_{mn}^L, x_{mn}^U), y_i \in [y_{mn}^L, y_{mn}^U)\}, m = 1, 2, .., M, n = 1, 2, .., N. \quad (1)$$

The normal vectors of triangles lie on a unit sphere. Considering that our approach focuses on frontal facial models, none of normal vectors would point to negative direction of Z axis. Therefore the normal vectors of triangles lie on a hemisphere on positive side of Z axis. We partition this hemisphere into several bins according to latitude and longitude on it, represented as $B = \{b_1, b_2, ..., b_k\}$. Given a patch, for example the (mn)th patch, let n_i's be the normal vectors of triangles within T_{mn}, and a_i's be the area of them. The *ELB* histogram $\mathbf{h}_{(mn)} = (s_1, s_2, ..., s_k)$ of the (mn)th patch is calculated through

$$s_j = \sum a_i, \ i \in \{r | n_r \text{ locates on } b_j\}, j = 1, 2, ..., k. \quad (2)$$

here, each patch is deemed as a "pixel" in an image, named *ELB* Image.

3.1.2 *HoSNI*

Given a pixel P on *ELB* Image, we partition its nearby pixels into several cells. The ith cell contains several pixels. Let $\mathbf{q}_i$ denote the histogram of the ith cell, then $\mathbf{q}_i$ is formed by summing up the histograms of the pixels in the cell. We normalize it by the area of the surface in this cell

$$\mathbf{q}_i = \sum \mathbf{h}_{(mn)}/A_i \ , \ (m, n) \in \{(p, q) | \text{ the } (pq)\text{th pixel} \in i\text{th block}\}, \quad (3)$$

where A_i is the surface area in this cell . All histograms of the cells are concatenated to the histogram of the pixel P, i.e., *HoSNI*.

In this paper, let the resolution of *ELB* Image be $M \times N = 100 \times 100$, the "re-sampling" interval is 2, which means $|x_{mn}^L - x_{mn}^U| = 2$ and $|y_{mn}^L - y_{mn}^U| = 2$. In *ELB* histogram extraction, the number of bins on hemisphere is $k = 3 \times 12$, where 3 is on latitude and 12 is on longitude. In *HoSNI* extraction, we set 4×4 as the number of cells, and each cell contains 8×8 pixels. Then the dimension of *HoSNI* is $k \times 4 \times 4 = 576$. Fig 1 shows the details of *HoSNI* .

3.2   Facial Landmark Localization

The deformations of the local shape of different points are different when facial expression and inter-person variation exist. To automatically localize 3D facial landmarks, we employ the following strategies in training and testing procedure.

1) **The Training Procedure.** Given 3D facial models and a facial landmark L, we transform facial models into *ELB* Images firstly. Then calculate the *HoSNI* of landmark L in each image, represented as $H_1, H_2, ..., H_n$. This paper uses K-Means Clustering to classify them into k clusters. Centroids of these clusters are used as templates of the landmark L, represented as $T_{L_i}, i = 1, 2, ..., k$.

2) **The Testing Procedure.** Given a testing facial model G, we transform it into *ELB* Image, and extract the *HoSNI* of each pixel in *ELB* Image. Then we compare each of them with templates of landmark L to find the most similar pixels by chi-square measure.

Given two histograms X and Y, the chi-square measure χ^2_{XY} is defined as

$$\chi^2_{XY} = \sum_i \frac{(x_i - y_i)^2}{(x_i + y_i)}, \tag{4}$$

where x_i is the ith bin of X and y_i is the ith bin of Y. In our testing procedure, the X is the template T_{L_i} and the Y is the *HoSNI* of the candidate pixel.

Let $\{I_1, I_2, ..., I_m\}$ denote the m most matching pixels of landmark L, $\chi^2_1, \chi^2_2, ..., \chi^2_m$ denote the chi-square distances of them. Then according to these matching pixels' positions on *ELB* Image, we find their nearest points $\{\mathbf{p}_1, \mathbf{p}_2, ..., \mathbf{p}_m\}$ in testing facial model G. The location of the landmark L on the model G is the weighted sum of them

$$\overline{\mathbf{P}} = \frac{1}{\frac{1}{\chi^2_1} + \frac{1}{\chi^2_2} + ... + \frac{1}{\chi^2_m}} (\frac{1}{\chi^2_1}\mathbf{p}_1 + \frac{1}{\chi^2_2}\mathbf{p}_2 + ... + \frac{1}{\chi^2_m}\mathbf{p}_m), \tag{5}$$

where reciprocal of χ^2_i is the weight of $\mathbf{p}_i$. Here, the smaller the chi-square distance, the greater the weight. $\overline{\mathbf{P}}$ is regarded as the landmark L on the facial model G.

4   Experimental Results

4.1   Database and Preprocessing

We use public 3D face database FRGC v2.0[23] to evaluate our approach. The FRGC v2.0 database contains 4950 face models of 577 individuals with several expressions in addition to neutral face. We define 19 typical facial landmarks for localization experiments, shown in Fig 1, and manually label the landmarks as ground truth during evaluation. In this paper, the pre-processing consists of hole-filling, noise-removing, smoothing, and extraction of facial region with the radius of 100 (mm). However we do not perform any alignment for them. We use 5-fold cross-validation and the accuracy of localization is defined with error tolerance of 10.0 (mm), the same as [11].

4.2   Performance against Data Resolution

To evaluate the localization performance against data resolution, we randomly pick out 500 facial models from the database, and simplify them into 5 different resolution levels in term of triangle number: $\{2000, 4000, 6000, 8000, 10000\}$ by the method of Qslim. Including original models, there are 6 levels of resolution in this experiment. The results are shown in Fig 2, where the horizontal axis is triangle number and the vertical axis is average error distance in Fig 2a, average accuracy rate in Fig 2b. We find that the average accuracy is 96.76%, and the average error distance is 4.12mm, when the resolution of data is 2000. The performance increases with the resolution of data. It becomes stable when the resolution is larger than 6000. the results are 3.87mm and 97.69%.

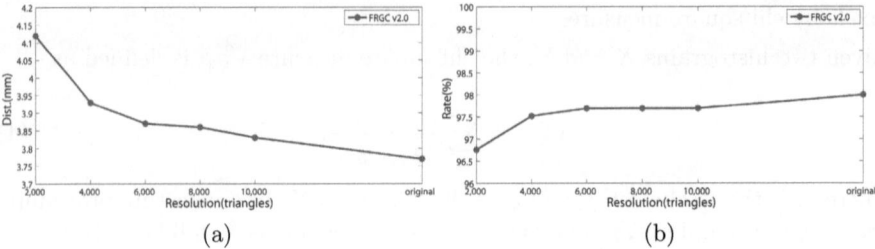

(a)   (b)

Fig. 2. Localization performance against data resolution. (a) The average error distance. (b) The average accuracy.

4.3   Precision of Localization

To evaluate the precision of our approach, we set different error distance ranges: $\{[0, 2], (2, 4], (4, 6], (6, 8], (8, 10]\}$ (mm). The result is shown in Fig 3, where the horizontal axis is different landmarks, and the vertical axis is the percentage of testing results in a certain distance range. From Fig 3, we find that most of the results are distributed in the range $[2, 4]$(mm). The precision of the inner-eyes (E4, E7) and the middle of upper-lip (M2) are better than the others, nearly 75% of their localization results distribute in the range $[0, 4]$(mm).

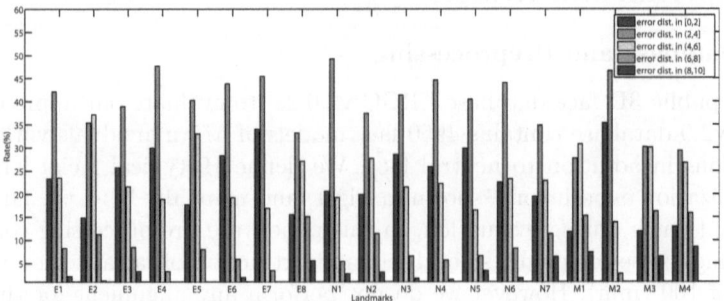

Fig. 3. The precision of our approach on FRGC v2.0

Some examples of facial landmark localization results by our approach are shown in Fig 4. The first row is localization results on neutral faces. The second row is visualization results on facial models with expression. The third row contains outlier or distortion and even contain shoulder and hair. These results prove effectiveness of our approach against outlier, noise and facial expression.

The experiments are carried out on the system of CPU X5670 2.93GHz and 32GB RAM. it only needs 13.15 seconds to localize 19 facial landmarks on a face containing 200000 triangles.

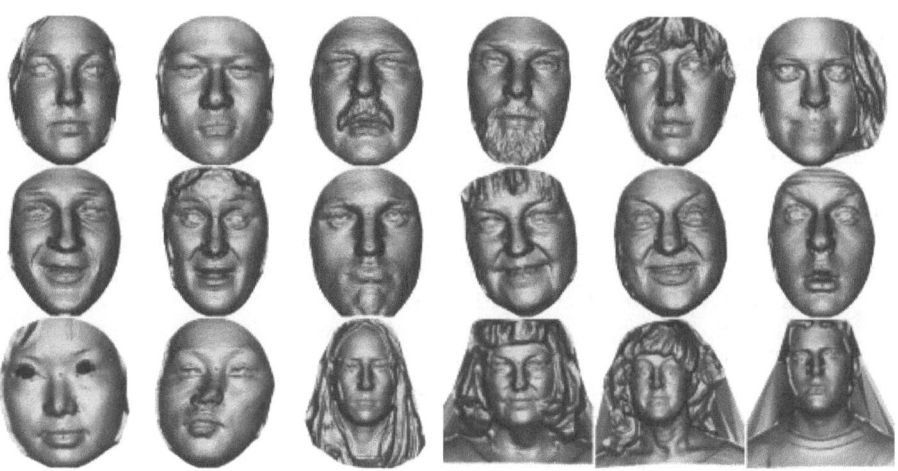

Fig. 4. The visualization results

4.4 Comparison with Other Descriptors

A lot of facial localization approaches only aim at nose-tip and inner-eyes, and do not work very well on the other facial landmarks. The accuracy rate of our approach on nose-tip and inner-eyes gets more than 99.00%. Therefore, we implement and optimize several typical surface descriptors for comparison. They are *EGI* (Extended Gaussian Image)[24] and *PS* (Point Signature)[25]. The results are shown in Table 1. From Table 1, we find that our approach significantly outperforms other existing methods on every facial landmark; the average accuracy rate of our method is 97.9%, and that of the other are less than 85.5%.

Table 1. Comparison with other two surface descriptors (%)

Method	E1	E2	E3	E4	E5	E6	E7	E8	N1	N2	N3	N4	N5	N6	N7	M1	M2	M3	M4	Avg.
Our	99.2	99.8	97.9	99.9	99.9	99.1	99.9	96.4	97.2	99.4	99.6	99.9	96.7	99.8	94.9	92.9	99.4	94.8	93.5	**97.9**
EGI	79.6	84.8	77.1	88.6	85.3	79.5	88.7	77.2	95.6	95.0	96.6	84.6	83.3	78.3	83.5	84.8	95.5	79.9	81.2	**85.2**
PS	88.6	93.8	23.8	94.0	94.1	88.4	94.0	24.2	92.8	98.8	99.2	90.0	95.9	85.8	96.8	29.1	92.5	33.9	34.6	**76.3**

5   Conclusion and Future Work

This paper attempts to tackle the problem of automatic localization of facial landmark for 3D facial shapes. We present a novel local surface descriptor, named *HoSNI*, to describe the local geometric characteristics of landmarks. We evaluate the proposed approach using a public 3D face database: FRGC v2.0. For localization of 19 typical facial landmarks, our approach achieve the average accuracy of 97.9%, which demonstrated the effectiveness to deal with facial expression variation. Our future work will mainly focus on practical applications.

Acknowledgments. This work was supported by grants from National Natural Science Foundation of China (No.61103107 and No.61070067).

References

1. Bowyer, K., Chang, K., Flynn, P.: A survey of approaches and challenges in 3D and multi-modal 3D + 2D face recognition. CVIU 101(1), 1–15 (2006)
2. Pan, G., Han, S., Wu, Z., Wang, Y.: 3D face recognition using mapped depth images. In: CVPR Workshops on FRGC (2005)
3. Pan, G., Wu, Z., Pan, Y.: Automatic 3D face verification from range data. In: ICASSP (2003)
4. Pan, G., Han, S., Wu, Z., Zhang, Y.: Removal of 3D facial expressions: A learning-based approach. In: CVPR (2010)
5. Pan, G., Han, S., Wu, Z.: Hallucinating 3D facial shapes. In: CVPR (2008)
6. Lee, Y., Park, K., Shim, J., Yi, T.: 3D face recognition using statistical multiple features for the local depth information. In: Int'l Conf. on Vision Interface (2003)
7. Hesher, C., Srivastava, A., Erlebacher, G.: A novel technique for face recognition using range imaging. In: ISSPA (2003)
8. Beumier, C., Acheroy, M.: Automatic 3D face authentication. Image and Vision Computing 18(4), 315–321 (2000)
9. Wu, Y., Pan, G., Wu, Z.: Face Authentication Based on Multiple Profiles Extracted from Range Data. In: Kittler, J., Nixon, M.S. (eds.) AVBPA 2003. LNCS, vol. 2688, pp. 515–522. Springer, Heidelberg (2003)
10. Wang, Y., Tang, X., Liu, J., Pan, G., Xiao, R.: 3D Face Recognition by Local Shape Difference Boosting. In: Forsyth, D., Torr, P., Zisserman, A. (eds.) ECCV 2008, Part I. LNCS, vol. 5302, pp. 603–616. Springer, Heidelberg (2008)
11. Faltemier, T.C., Bowyer, K.W., Flynn, P.J.: Rotated profile signatures for robust 3D feature Detection. In: FGR (2008)
12. Gordon, G.: Face recognition based on depth maps and surface curvature. In: SPIE Proceedings (1991)
13. Colbry, D., Stockman, G., Jain, A.: Detection of anchor points for 3D face verification. In: CVPR (2005)
14. Moreno, A., Sanchez, A., Vélez, J., Díaz, F.: Face recognition using 3D surface-extracted descriptors. In: Irish Machine Vision and Image Processing (2003)
15. Nair, P., Cavallaro, A.: Region segmentation and feature point extraction on 3D faces using a point distribution model. In: ICIP (2007)
16. Akagunduz, E., Ulusoy, I.: 3D object representation using transform and scale invariant 3D features. In: ICCV (2007)

17. Chang, K., Bowyer, K., Flynn, P.: Multiple nose region matching for 3D face recognition under varying facial expression. PAMI 28(10), 1695–1700 (2006)
18. Wang, Y., Chua, C., Ho, Y.: Facial feature detection and face recognition from 2D and 3D images. Pattern Recognition Letters 23(10), 1191–1202 (2002)
19. Xu, C., Tan, T., Wang, Y., Quan, L.: Combining local features for robust nose location in 3D facial data. Pattern Recognition Letters 27(13), 1487–1494 (2006)
20. Breitenstein, M., Kuettel, D., Weise, T., Van Gool, L., Pfister, H.: Real-time face pose estimation from single range images. In: CVPR (2008)
21. Fanelli, G., Gall, J., Van Gool, L.: Real time head pose estimation with random regression forests. In: CVPR (2011)
22. Boehnen, C., Russ, T.: A fast multi-modal approach to facial feature detection. In: WACV (2005)
23. Phillips, P., Flynn, P., Scruggs, T., Bowyer, K., Chang, J., Hoffman, K., et al.: Overview of the face recognition grand challenge. In: CVPR (2005)
24. Horn, B.: Extended gaussian images. Proceedings of the IEEE 72(12), 1671–1686 (1984)
25. Chua, C., Han, F., Ho, Y.: 3D human face recognition using point signature. In: FGR (2000)

Feature Reduction for Efficient Object Detection via L1-norm Latent SVM

Min Tan[1], Yueming Wang[2,*], and Gang Pan[1]

[1] College of Computer Science and Technology, Zhejiang Univ.
[2] Qiushi Academy for Advanced Studies, Zhejiang Univ.
{tanmin,ymingwang,gpan}@zju.edu.cn

Abstract. The deformable part model is one of the most effective methods for object detection. However, it simultaneously computes the scores for a holistic filter and several part filters in a relatively high-dimensional feature space, which leads to low computational efficiency. This paper proposes an approach to select compact and effective features by learning a sparse deformable part model using L1-norm latent SVM. A stochastic truncated sub-gradient descent method is presented to solve the L1-norm latent SVM problem. Extensive experiments are conducted on the INRIA and PASCAL VOC 2007 datasets. Compared with the feature used in L2-norm latent SVM, a highly compact feature in our method can reach the state-of-the-art performance. The feature dimensionality is reduced to 12% in the INRIA dataset and less than 30% in most PASCAL VOC 2007 datasets. At the same time, the average precisions (AP) have almost no drop using the reduced feature. With our method, the speed of the detection score computation is faster than that of the L2-norm latent SVM method by 3 times. When the cascade strategy is applied, it can be further speeded up by about an order of magnitude.

Keywords: object detection, feature reduction, sparse, L1 optimization, stochastic gradient descent, cascade.

1 Introduction

Object detection serves as one of the key problems in the computer vision research, which plays an important role in image retrieval, video surveillance and smart cars etc [1]. The target is to decide whether or not an object is present in an image and give the bounding box if present. It is a challenging task as most objects in images have large intra-class variations due to illumination variations, different views, occlusion, and different intrinsic appearances [2]. Many efforts have been made to design discriminative features [3, 4] and powerful model structures [5–8] to improve the detection performance. Recently, deformable part models have achieved strong performance on the difficult PASCAL datasets [6, 7] and become one kind of the most effective models for object detection. However, the detection has to be evaluated on a large amount of windows when the sliding

---

* Corresponding author.

J. Yang, F. Fang, and C. Sun (Eds.): IScIDE 2012, LNCS 7751, pp. 322–329, 2013.

sub-window method is used. The holistic detector (analogous to the Dalal-Triggs filter) and multiple part detectors in the deformable part models can lead to relatively low computational efficiency.

There are at least three schemes to improve the computational efficiency. The first one is to reduce the number of evaluating windows in the search space [9]. The second one is to apply a cascade of part/holistic detectors to reject most simple non-object windows quickly [7]. Besides the above schemes, one can reduce the feature dimension so as to simplify part/holistic detectors [10]. This paper focus on the third scheme to help improve the computational efficiency.

To reduce the feature dimension, Felzenszwalb et al. [8] proposed to project the Histograms of Oriented Gradients (HOG) features and the weight vectors in the filters to a low dimensional space by Principal Component Analysis (PCA), which needs additional projecting computation. Directly selecting the effective elements from the feature vectors is another way of feature reduction [11]. There are at least two feature selection schemes. One is to discard the ineffective feature elements based on the user-defined criteria, including the adaboost algorithm where the criteria is based on the classification error [12]. But it is heuristic and unable to generalize well [13]. Another scheme is to construct a sparse model. Recently, a cascaded L1-SVM has been proposed to learn a sparse human detector, which was solved by interior point and integer optimization [10]. High consumption of time and memory is the main drawback of this method.

In this paper, we propose a method to use L1-norm latent SVM (ℓ_1-LSVM) in the deformable part model [6] to construct compact part/holistic filters for efficient detection. Also, we have interest in the answer to the question: how sparse the features can be, in order for the sparse deformable part model to perform comparable performance to that of the state-of-the-art methods.

2   Review of Deformable Part Model

A deformable part model (DPM) [6] is a star-structured model including one root filter and several part filters. The score of a test sample x is computed by $f_\beta(x) = \max_{z \in Z(x)} \beta \cdot \Phi(x, z)$, where β is a weight vector concatenating the root filter, the part filters, and deformation cost weights, z is the latent value specifying the object configuration, $Z(x)$ is the configuration set for x, $\Phi(\mathbf{x}, z)$ is a concatenation of holistic features, part features, and part deformation cost determined by z. A mixture model may contain several components, each being a model $f_\beta(x)$. Given training data $(\mathbf{x}_i, y_i)$, where $y_i = \pm 1$ denotes the class label, the DPM is learned by solving the following latent SVM problem:

$$\beta^* = \arg\min_{\beta} \left(\frac{1}{2} \|\beta\|_2^2 + C \sum_{i=1}^{N} \max(0, 1 - y_i f_\beta(x_i))\right). \tag{1}$$

As stated in [6], the above problem is convex if the latent information of the positive samples is specified. It can be solved by the gradient descent method. We call the problem L2-norm latent SVM (ℓ_2-LSVM). The training framework of the DPM is shown in Figure 1. Please refer to [6] for details.

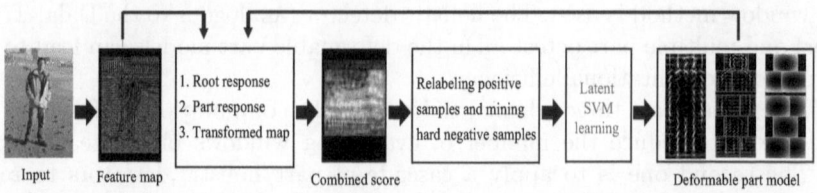

Input   Feature map   Combined score   Deformable part model

Fig. 1. The learning framework of the deformable part model

The DPM is a complex model, thus the dimension of the feature vector can be high. If the HOG feature is used and 8 parts are specified in a DPM, the dimension of $\Phi(x, z)$ is more than 20 thousand. This leads to expensive computation. However, the high dimensional feature vectors may be redundant [14], thus it is useful to suppress the small elements which may represent noise.

3   Our Method: ℓ_1-LSVM

3.1 ℓ_1-LSVM

To reduce the feature dimension, we design a sparse version of [6]. The common way of sparse modeling is either adding sparse constraints or constructing L1 regularized optimization directly. This paper focus on the latter as it does not need additional projection onto a convex set determined by the constraints [15].

Our ℓ_1-LSVM is constructed via L1 regularized optimization by formulating the ℓ_1-regularization term into (1). Replacing the regularization term in (1), we learn a sparse DPM by solving the following objective function,

$$\beta^* = \arg\min_{\beta}(\|\beta\|_1 + C\sum_{i=1}^{N}\max(0, 1 - y_i f_\beta(x_i))). \tag{2}$$

Let $\varphi_i(\beta) = \Phi(x_i, \arg\max_{z \in Z(x_i)}(\beta \cdot \Phi(x_i, z)))$ be the feature vector of the sample x_i having maximal score among all z and $L(\beta, \zeta_i(\beta)) = \max(0, 1 - y_i f_\beta(x_i)), \zeta_i(\beta) = (\varphi_i(\beta), y_i)$. Then (2) can be rewritten as

$$\beta^* = \arg\min_{\beta} R(\beta, C), R(\beta, C) = \|\beta\|_1 + C\sum_{i=1}^{N}L(\beta, \zeta_i(\beta)). \tag{3}$$

We would like to learn a sparse DPM by (3) using the framework of [6] (see Figure 1). This problem is semi-convex as it involves latent variables.

To solve (3), we use the sub-gradient descent method since (3) is not differentiable everywhere. Since the training data is very large, the stochastic scheme may be the best choice as it can reduce the memory requirements. However, common stochastic sub-gradient descent methods can not achieve sparsity in β^* due to the numerical problem. Recently, there is a growing research on algorithm making results sparse for L1 regularized models [16–18], including the L1 regularized SVM [16]. However, none of these models deal with latent variables. Thus, we extend [16] to our model and present an efficient stochastic truncated sub-gradient descent (STSGD) method to solve our problem.

3.2   STSGD Algorithm: Solving ℓ_1-LSVM

STSGD is a modification of the stochastic sub-gradient descent (SSGD) method. SSGD works in an iterative manner. In each step, a subset of training set are selected to estimate the sub-gradient and variables are updated by the negative sub-gradient direction. The sub-gradient of the objective function in (3) is

$$\nabla_\beta R(\beta, C) = \text{sign}(\beta) + C \sum_{i=1}^{N} \nabla L_\beta(\beta, \zeta_i(\beta)), \qquad (4)$$

where $\nabla L_\beta(\beta, \zeta_i(\beta)) = \begin{cases} 0, & if \ y_i f_\beta(x_i) \geqslant 1 \\ -y_i \varphi_i(\beta), & otherwise \end{cases}$.

We select one sample to estimate (4), and separate the update to two stages:

$$\begin{cases} \hat{\beta}_t = \beta_t - \eta_t CN \nabla L_\beta(\beta_t, \zeta_{it}(\beta_t)) \\ \beta_{t+1} = \hat{\beta}_t - \eta_t \text{sign}(\hat{\beta}_t) \end{cases}. \qquad (5)$$

Along with the iteration step, a part of elements in β approach zero progressively, but they can not be purely zero. The simple rounding is too aggressive and lacks of theoretical guarantee [16]. An alternate is to update the entries of β_{t+1} in the second stage of (5) by a truncation function. For the jth entry β_{t+1}^j

$$\beta_{t+1}^j = H(\hat{\beta}_t^j, \eta_t, \theta) = \begin{cases} max(0, \hat{\beta}_t^j - \eta_t), \ if \ \hat{\beta}_t^j \in [0, \theta] \\ min(0, \hat{\beta}_t^j + \eta_t), \ if \ \hat{\beta}_t^j \in [-\theta, 0) \ . \\ \hat{\beta}_t^j, \qquad\qquad otherwise \end{cases} \qquad (6)$$

This is less aggressive than the simple rounding method. Practically, the truncation is performed once every K steps to avoid local optimum. When the algorithm terminates, a simple rounding on the entries of β is applied because, at this time, rounding small values to zero is relatively safe:

$$f(\beta^j) = \begin{cases} \beta^j, \ if \ |\beta^j| > \alpha M \\ 0, \quad otherwise \end{cases}, \qquad (7)$$

where $M = \max_j(|\beta^j|)$ and α is a weight tuning the degree of the sparsity.

This STSGD method works well in both achieving sparsity and outputting good classifiers. We apply it to solve the ℓ_1-LSVM and learn a sparse DPM, shown in algorithm (1). The algorithm is convergent [16].

4   Experiment

In this section, we evaluate the detection accuracy and efficiency of our ℓ_1-LSVM based DPM. Firstly, we compare the proposed approach with that in

---

Algorithm 1. Learning of ℓ_1-LSVM by STSGD

Input:

$P = \{(x_i, z_i^p)|1 \leq i \leq n\}$: positive training samples with determined latent variables z_i^p; $N = \{(x_i, z_{i1}^n, ..., z_{ik_i}^n)|1 \leq i \leq m\}$: negative training samples with undetermined latent variables z_{ij}^n; k_i: number of latent variables for sample i. K: the truncation stride; θ: the parameter in the truncation function; T: the number of the iteration; η_0: the initial learning rate.

1 **Initialization:**
2 $\beta := \beta_0$.
3 **Learning:**
4 **for** $t = 1, 2, ..., T$ **do**
5   Randomly select a sample x_i; Set learning rate $\eta := \eta_0/\sqrt{t}$.
6   **if** x_i is positive,
7          $\varphi := \Phi(x_i, z_i^p)$, y $:= +1$.
8   **else**
9          $\varphi := \Phi(x_i, \arg\max_{1 \leq j \leq k_i}(\beta \cdot \Phi(x_i, z_{ij}^n)))$, y $:= -1$.
10   **end**
11   **if** $y \cdot \beta\varphi < 1$
12         $\beta := \beta + CN\eta y\varphi$.
13   **end**
14   **if** $\mod(t, K) = 0$
15         $\beta := H(\beta, K\eta, \theta)$.
16   **end**
17 **end**
18 Use (7) to round small entries of β to zero.
   Output: the sparse deformable model β

---

[6] on INRIA person dataset and all the 20 categories of PASCAL dataset [19]. Then, we extend our method to a cascade version and evaluate it on INRIA dataset. Though our method also can be deployed to be multi-thread, similar to ℓ_2-LSVM. We conduct all the experiments in a single thread setup on a PC with one Intel Core i3 530 2.93Hz CPU and 3G RAM to make a fair comparison. The baseline codes are downloaded from [20]. For the parameters, it reaches satisfied results to set $\theta = 1$, $\eta_0 = 10^{-4}$, $\alpha = 0.001$ and $K = 20$. Besides, we choose appropriate values of C on different datasets by cross validation.

4.1 Comparisons with ℓ_2-LSVM

We compare the average precision (AP) and efficiency of DPMs using l2- and l1- LSVM on both INRIA person and PASCAL dataset.

Figure 2 illustrates the DPMs trained on INRIA dataset using ℓ_2- and ℓ_1-LSVM respectively. Unlike ℓ_2-LSVM, the filter coefficients of ℓ_1-LSVM are very sparse, i.e., only a small portion (about 12%) of the HOG feature elements are selected. As shown in Table 1, the AP of the two DPMs are also similar. Due to the sparsity of our model, it is 4 times as computationally efficient as ℓ_2-LSVM.

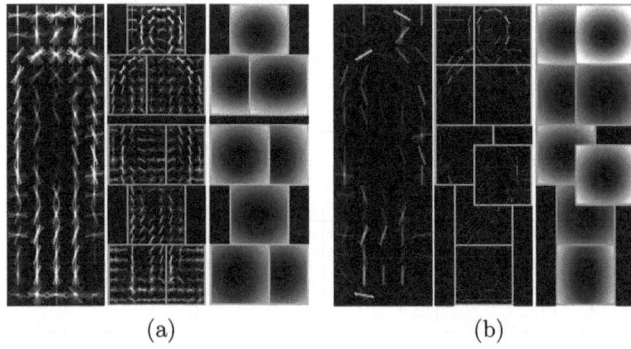

(a)   (b)

Fig. 2. Visualization of the learned DPMs by: (a) ℓ_2-LSVM; (b) ℓ_1-LSVM. From left to right, they are the root filter, part filters, and deformation cost weights in a sub-figure. The number of non-zero feature elements is 23232 and 2882 respectively.

Table 1. Comparisons with ℓ_2-LSVM based DPM on INRIA person dataset

Method	AP	Time*(s)	Sparsity
ℓ_2-LSVM	0.884	5.015	0
ℓ_1-LSVM (Ours)	0.873	1.251	0.878

* *Time* is the detection time consumption per image (for all the tables)

Also, we test ℓ_1- and ℓ_2-LSVM on all the 20 categories of PASCAL 2007 datasets. All the DPMs are trained with six components. For ℓ_2-LSVMs, we directly use the model available in [20]. Table 2 summarizes AP and model sparsity (the ratio of zero entries) of both ℓ_1- and ℓ_2-LSVM as well as the speedup of ℓ_1-LSVM. For most categories, ℓ_1-LSVM models have the sparsity rates as high as 0.7 and improve the computation by 2-3.5 times.

Besides, although the STSGD converges slower than the gradient descent method, the former needs less time for relabeling due to the sparsity property. Therefore, the training time of our method was close to that of ℓ_2-LSVM.

4.2   Performance of Cascade DPMs

In [8], Felzenszwalb et al. proposed a cascade technique to improve the detection efficiency of DPM while keeping its high accuracy. With concern about feature redundancy, they further used PCA to reduce the dimension of the HOG feature in each cell. These techniques can be naturally integrated into our method.

We compare the performance of different DPMs using the cascade technique with and without PCA. Table 3 shows the results. To ensure the consistency among different methods, the final classifier always uses the precision-equals-recall threshold, which leads to an experimental result different from Subsection 4.1. Note that this strategy is also used in [8]. From the table, we can see that AP of different approaches are at the same level, and the cascade technique

Table 2. Comparisons between the DPMs using ℓ_1- and ℓ_2-LSVM on PASCAL VOC 2007 dataset: γ is the sparsity, ℓ_p denotes ℓ_p-LSVM, Δ is the AP gap of ℓ_1-LSVM relative to ℓ_2-LSVM, and *speedup* is the speedup factor of ℓ_1-LSVM with regard to ℓ_2-LSVM. The classifier uses a fixed threshold -1.1.

Method		Aero	Bike	Bird	Boat	Bottle	Bus	Car	Cat	Chair	Cow	Table	Dog	Horse	Mbike	Person	Plant	Sheep	Sofa	Train	TV	Avg.
γ	ℓ_2	.002	.000	.000	.001	.000	.001	.000	.001	.000	.001	.001	.001	.001	.002	.000	.001	.002	.000	.000	.000	.001
	ℓ_1	.889	.898	.728	.707	.821	.906	.772	.505	.803	.815	.839	.731	.859	.820	.496	.682	.810	.830	.846	.846	.780
AP	ℓ_2	.289	.595	.100	.152	.255	.496	.579	.193	.224	.252	.233	.111	.568	.487	.419	.121	.172	.336	.448	.416	.322
	ℓ_1	.291	.593	.102	.136	.237	.491	.573	.171	.224	.192	.225	.115	.554	.441	.383	.121	.195	.280	.429	.393	.307
	Δ	-.002	.002	-.002	.017	.019	.005	.006	.022	.000	.060	.008	-.004	.014	.046	.036	.001	-.023	.056	.019	.023	.015
Speedup		3.71	4.32	2.39	2.37	3.17	4.12	2.88	1.70	2.89	3.30	3.54	2.50	3.93	3.06	1.68	2.15	2.88	3.23	3.19	3.27	3.01

Table 3. Comparisons with ℓ_2-LSVM with cascade DPMs on INRIA person dataset

Method		ℓ_2-LSVM	ℓ_1-LSVM(Ours)	Speedup*
Non-cascade	AP	0.804	0.809	
	Time (s)	**4.638**	1.229	3.774
Cascade (no PCA)	AP	0.805	0.809	
	Time (s)	0.462	**0.093**	4.966
Cascade (PCA)	AP	0.805	0.809	
	Time (s)	0.203	0.177	1.149

* *Speedup* is the speed-up factor of ℓ_1-LSVM with regard to ℓ_2-SVM

can boost the efficiency for both the ℓ_2 and ℓ_1 cases. Unlike ℓ_2-LSVM, cascade model without PCA is more efficient than that with PCA for ℓ_1-LSVM. One possible reason is that transforming the filter coefficients with PCA may break the sparseness. Thus, ℓ_1-LSVM detector should be only integrated with the non-PCA cascade strategy. The cascade ℓ_1-LSVM detector without PCA is faster than the non-cascade ℓ_2-LSVM by about 50 times, and about 5 times as computationally efficient as the cascade ℓ_2-LSVM detector without PCA, and still more than 2 times as efficient as the cascade ℓ_2-LSVM detector with PCA.

5   Conclusion

This paper proposed a method for feature reduction in ℓ_2-LSVM and learned a sparse deformable part model for efficient object detection. The experimental results showed that our model is highly sparse and a small amount of feature elements can reach comparable detection performance to that of ℓ_2-LSVM. Our method is 3 times faster than the state-of-the-art methods. We believe our method is valuable for both the detection efficiency and insightful understanding of feature capability. Besides, our work can be extended to many other fields, e.g., 3D face detection/recognition [21], object tracking, and human parsing etc.

Acknowledgments. This work was supported by grants from National Natural Science Foundation of China (No. 61103107 and No.61070067).

References

1. Sun, J., Wu, Z., Pan, G.: Context-aware smart car: from model to prototype. Journal of Zhejiang University-Science A 10(7), 1049–1059 (2009)
2. Geronimo, D., Lopez, A., Sappa, A., Graf, T.: Survey of pedestrian detection for advanced driver assistance systems. T-PAMI 32(7), 1239–1258 (2010)
3. Dalal, N., Triggs, B.: Histograms of oriented gradients for human detection. In: CVPR (2005)
4. Zhang, J., Huang, K., Yu, Y., Tan, T.: Boosted Local Structured HOG-LBP for Object Localization. In: CVPR (2011)
5. Bourdev, L., Malik, J.: Poselets: Body Part Detectors Trained Using 3D Human Pose Annotations. In: ICCV (2009)
6. Felzenszwalb, P., Girshick, R., McAllester, D., Ramanan, D.: Object detection with discriminatively trained part-based models. T-PAMI, 1627–1645 (2009)
7. Zhu, L., Chen, Y., Yuille, A., Freeman, W.: Latent hierarchical structural learning for object detection. In: CVPR (2010)
8. Felzenszwalb, P., Girshick, R., McAllester, D.: Cascade object detection with deformable part models. In: CVPR (2010)
9. Lampert, C., Blaschko, M., Hofmann, T.: Beyond Sliding Windows: Object Localization by Efficient Subwindow Search. In: CVPR (2008)
10. Xu, R., Zhang, B., Ye, Q., Jiao, J.: Cascaded L1-norm Minimization Learning (CLML) classifier for human detection. In: CVPR (2010)
11. Sun, Z., Bebis, G., Miller, R.: Object detection using feature subset selection. Pattern Recognition 37(11), 2165–2176 (2004)
12. Wang, Y., Tang, X., Liu, J., Pan, G., Xiao, R.: 3D Face Recognition by Local Shape Difference Boosting. In: Forsyth, D., Torr, P., Zisserman, A. (eds.) ECCV 2008, Part I. LNCS, vol. 5302, pp. 603–616. Springer, Heidelberg (2008)
13. Chen, Y., Lin, C.: Combining SVMs with various feature selection strategies. In: Guyon, I. (ed.) Feature Extraction, Foundations and Applications. Springer (2006)
14. Hussain, S., Triggs, W., et al.: Feature sets and dimensionality reduction for visual object detection. In: BMVC (2010)
15. Ratliff, N., Bagnell, J., Zinkevich, M.: Subgradient methods for maximum margin structured learning. In: ICML Workshop on Learning in Structured Output Spaces (2006)
16. Langford, J., Li, L., Zhang, T.: Sparse online learning via truncated gradient. Journal of Machine Learning Research 10, 777–801 (2009)
17. Yuan, G., Chang, K., Hsieh, C., Lin, C.: A comparison of optimization methods and software for large-scale L1-regularized linear classification. Journal of Machine Learning Research 11, 3183–3234 (2010)
18. Zhu, J., Rosset, S., Hastie, T., Tibshirani, R.: 1-norm support vector machines. In: NIPS (2003)
19. Everingham, M., Van Gool, L., Williams, C., Winn, J., Zisserman, A.: The PASCAL Visual Object Classes Challenge. VOC 2007 Results (2007)
20. http://www.cs.brown.edu/~pff/latent-release4/ (2008)
21. Pan, G., Wu, Z., Pan, Y.: Automatic 3D face verification from range data. In: ICASSP (2003)

Iterative Phase Unwrapping in Color Doppler Flow Mapping

Artem Yatchenko and Andrey Krylov

Laboratory of Mathematical Methods of Image Processing,
Faculty of Computational Mathematics and Cybernetics,
Lomonosov Moscow State University, Russia
artem@yatchenko.com.ua, kryl@cs.msu.ru
http://imaging.cmc.msu.ru

Abstract. Iterative regularization method to process captured data by color Doppler flow mapping has been proposed. It takes into account features of the ultrasound investigation and allows to solve the problem even in the case of noisy data and in the case of existing areas with unknown flow values. The influence of the values of regularization parameters on the obtained solution was analyzed. Test results for the left ventricle color Doppler flow mapping investigation showed effectiveness of the algorithm.

Keywords: Ultrasound color Doppler flow mapping, phase unwrapping, iterative regularization method.

1 Introduction

Color Doppler flow mapping (CDFM) is a method for noninvasively imaging blood flow through the heart by displaying flow data on the two-dimensional echocardiographic image. This capability has generated great excitement about the use of the technique for identifying different forms of heart disease, as the color flow image adds spatial information to the Doppler data. The colors displayed on the flow map image contain useful information. By convention, Doppler color flow systems assign a given color to the direction of flow; red is flow toward, and blue is flow away from the transducer. Progressively increasing velocities are encoded in varying hues of either red or blue. The more dull the hue, the slower the velocity. The brighter the hue, the faster is the relative velocity.

When the velocity range is set to high, low-flow states might not be displayed. Conversely, when the velocity scale is set to low flow aliasing occurs. This phenomenon is related to the fact that Doppler technique is based on a measuring of the phase shift of emitted pulsed signals. If the interval between pulses is small and the measured flow velocity is high, phase shift can exceed 2π. However, the transducer can estimate the phase shift only in the interval $[-\pi, \pi]$. So the so-called phase wrapping (or "aliasing") occurs.

The phase unwrapping problem is crucial moment in CDFM image processing. Formally, we have:

$$\varphi(x,y) = \psi(x,y) + 2\pi k(x,y) , \qquad (1)$$

J. Yang, F. Fang, and C. Sun (Eds.): IScIDE 2012, LNCS 7751, pp. 330–338, 2013.

where $\varphi(x, y)$ is true phase in (x, y), $\psi(x, y) \in [-\pi, \pi]$ is measured wrapped modulo-2π phase, $k(x, y)$ is an integer valued function.

In general, the problem of phase unwrapping is ill-posed. However, we know that the observed field is a blood flow and should be continuous and smooth. Using this additional information, one can build a regularizing algorithm, which minimizes the functional that takes into account the conditions of smoothness of the solution.

There are many different approaches of phase unwrapping problem solution: using graph-cut algorithm [1], [2]; the minimization of the L_2 norm [3] and L_p [4] of partial derivatives of the field; using complex-valued random Markov field [5]; using regularizing conditions [6]. We illustrate features of some of these methods in Fig.1.

Methods that use graph cuts for phase unwrapping have the advantage that the obtained field differs from the initial exactly by an integer number of 2π periods in each point. In addition, there exist a very fast solutions for these methods. However, these methods do not works well when a lot of data is missed or determined incorrectly (Fig. 1(c)). Also, these methods are not suitable for noise suppression.

The methods based on L_2 norm and L_p norm minimization of partial derivatives can suppress noise and are less sensitive to the miss of data. However, the result may be different from the original by a non-integer number of periods (Fig. 1(d)).

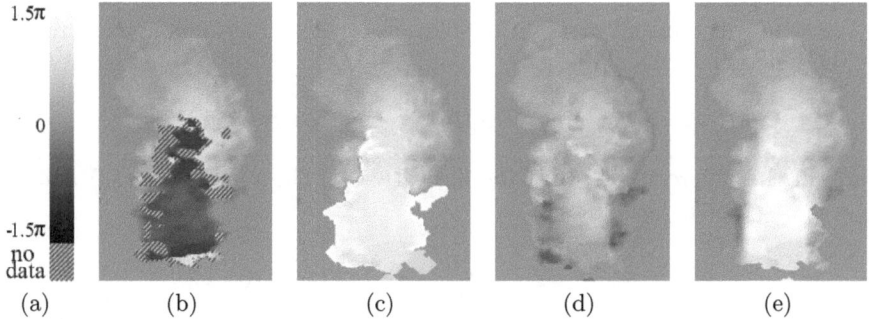

Fig. 1. Examples of different phase unwrapping methods results: (a) scale of the phase, (b) measures wrapped flow, (c) graph-cut based method [2], (d) minimization of L_p norm of derivatives based method [4] ($p = 1$), (e) proposed method

Regularization methods are also based on the minimization of the norms of the partial derivatives, but can be better adapted to CDFM task than [3] and [4]. The method proposed in this paper is similar to [6], but allows us to solve the problem in the case when there are many areas with unknown flow values. This is achieved by inclusion of additional stabilizing term with the first derivative of the solution. The direction of motion relative to the sensor is also taken into account in the minimizing functional.

2   Phase Unwrapping Using Regularization

We consider a rectangular lattice O with $M \times N$ sites: $\{(x, y) : x \in \{1, 2, \ldots, M\},$ $y \in \{1, 2, \ldots, N\}\}$. We also consider subsets of O: the set S of sites for which valid observations are available; the set D that corresponds to those sites across which one wishes to propagate the solution ϕ; the set E of sites where exact phase values are known. Initially E is empty and D is equal to O.

Scanning transducer makes several impulses and averages obtained velocities for greater stability [1]. If scan results in some area differ much then it usually means that this area is significantly turbulent, and it is impossible to detect direction and speed of the flow velocities. These areas $(D \setminus S)$ are marked in the image with a special color.

Let g is the measured signal (1). We are looking for a smooth solution $\phi(x, y)$, connected with the initial signal g by the following relationship

$$\phi(x, y) \approx g(x, y) + 2\pi k_g(x, y) , \tag{2}$$

where $k_g(x, y)$ is an integer-valued function. We define functions

$$g_x(x, y) = g(x, y) - g(x - 1, y) + k_x(x, y) ,$$
$$g_y(x, y) = g(x, y) - g(x, y - 1) + k_y(x, y) ,$$

where integer-valued functions $k_x(x, y)$ and $k_y(x, y)$ give $|g_x(x, y)| < \pi$ and $|g_y(x, y)| < \pi$.

We change the initial problem of solution of (2) to the minimizing problem for the functional

$$M_{\lambda, \beta, \gamma} = \|L\phi - h\|^2 + \lambda \|P_1 \phi\|^2 + \beta \|P_2 \phi\|^2 + \gamma \|P_3 \phi\|^2 . \tag{3}$$

Norm of the difference between the derivatives of the solution $\phi(x, y)$ and the values g_x, g_y looks as follows:

$$\|L\phi - h\|^2 = a_x \sum_{(x,y) \in S_x \cap D_x} [\phi_x(x, y) - g_x(x, y)]^2$$
$$+ a_y \sum_{(x,y) \in S_y \cap D_y} [\phi_y(x, y) - g_y(x, y)]^2 ,$$

$$h = (g_x, g_y) ,$$
$$S_x = \{(x, y) \in S : (x - 1, y) \in S\}, \quad S_y = \{(x, y) \in S : (x, y - 1) \in S\} ,$$
$$D_x = \{(x, y) \in D : (x - 1, y) \in D\}, \quad D_y = \{(x, y) \in D : (x, y - 1) \in D\} ,$$

where a_x, a_y are parameters of the method, $L\phi = (\phi_x, \phi_y)$ is the differential operator, $\phi_x = \phi(x, y) - \phi(x - 1, y), \phi_y = \phi(x, y) - \phi(x, y - 1)$.

$$\|P_1\phi\|^2 = \sum_{(x,y)\in D_{xx}} [\phi(x-1,y) - 2\phi(x,y) + \phi(x+1,y)]^2$$

$$+ \sum_{(x,y)\in D_{yy}} [\phi(x,y-1) - 2\phi(x,y) + \phi(x,y+1)]^2$$

$$+ \sum_{(x,y)\in D_{xy}} [\phi(x,y) - \phi(x-1,y) - \phi(x,y-1) + \phi(x-1,y-1)]^2 ,$$

$$D_{xx} = \{(x,y) \in D : (x-1,y) \in D, \ (x+1,y) \in D\} ,$$
$$D_{yy} = \{(x,y) \in D : (x,y-1) \in D, \ (x,y+1) \in D\} ,$$
$$D_{xy} = \{(x,y) \in D : (x-1,y) \in D, \ (x,y-1) \in D\}, \ (x-1,y-1) \in D\} .$$

$$\|P_2\phi\|^2 = \sum_{(x,y)\in E} [\phi(x,y) - c(x,y)]^2 .$$

$$\|P_3\phi\|^2 = \sum_{(x,y)\in D_x\backslash S_x} [\phi(x,y) - \phi(x-1,y)]^2 + \sum_{(x,y)\in D_y\backslash S_y} [\phi(x,y) - \phi(x,y-1)]^2 .$$

The term $\|P_1\phi\|^2$ smoothes the resulting derivatives of the field ϕ. We include term $\|P_2\phi\|^2$ because we know exact values of flow $c(x,y)$ in the points of the set E. The term $\|P_3\phi\|^2$ in (3) is included to make the solution smooth in the data-missed area $(D \setminus S)$.

3   Minimization Algorithm

To solve the variational problem for the functional (3) we find ϕ as a solution of

$$L^T L\phi - L^T h + \lambda P_1^T P_1\phi + \beta(\phi - c) + \gamma P_3^T P_3\phi = 0 ,$$

or

$$Q\phi = \rho , \qquad (4)$$
$$Q = L^T L + \lambda P_1^T P_1 + \beta I + \gamma P_3^T P_3, \quad \rho = L^T h + \beta c .$$

In the discrete case, these formulae look as follows:

$$(L^T L\phi)(x,y) =$$

$$= \sum_{k=0}^{1} \{\tilde{a}_x(x,y,k)[\phi(x+k,y) - \phi(x+k-1,y)]$$

$$+ \tilde{a}_y(x,y,k)[\phi(x+k,y) - \phi(x+k-1,y)]\} ,$$

$$(\lambda P_1^T P_1 \phi)(x,y) =$$

$$= \lambda \sum_{k=0}^{2} \{ \tilde{b_x}(x,y,k)[\phi(x-k,y) - 2\phi(x-k+1,y) + \phi(x-k+2,y)] \}$$

$$+ \sum_{k=0}^{2} \{ \tilde{b_y}(x,y,k)[\phi(x,y-k) - 2\phi(x,y-k+1) + \phi(x,y-k+2)] \}$$

$$+ \sum_{n=0}^{1} \sum_{k=0}^{1} \{ b_{xy}(x,y,k,n)[\phi(x-n,y-k) - \phi(x-n,y-k+1)$$

$$- \phi(x-n+1,y-k) + \phi(x-n+1,y-k+1)] \} ,$$

$$(\beta I \phi)(x,y) = \beta \tilde{c}(x,y)\phi(x,y) ,$$

$$(\gamma P_3^T P_3 \phi)(x,y) = \gamma \sum_{k=0}^{1} \{ \tilde{d_x}(x,y,k)[\phi(x+k,y) - \phi(x+k-1,y)]$$

$$+ \tilde{d_y}(x,y,k)[\phi(x+k,y) - \phi(x+k-1,y)] \} ,$$

$$\rho(x,y) = \sum_{k=0}^{1} \{ \tilde{a_x}(x,y,k) g_x(x+k,y) + \tilde{a_y}(x,y,k) g_y(x,y+k) \} + \beta \tilde{c}(x,y)c(x,y) ,$$

where

$$\tilde{a_x}(x,y,k) = \begin{cases} (-1)^k a_x, & \text{for } (x+k,y) \in D_x \cap S_x , \\ 0, & \text{otherwise} , \end{cases}$$

$$\tilde{a_y}(x,y,k) = \begin{cases} (-1)^k a_y, & \text{for } (x,y+k) \in D_y \cap S_y , \\ 0, & \text{otherwise} , \end{cases}$$

$$\tilde{b_x}(x,y,k) = \begin{cases} 1, & \text{for } k \in \{1,0\}, \ (x-k+1,y) \in D_{xx} , \\ -2, & \text{for } k = 2, \ (x-k+1,y) \in D_{xx} , \\ 0, & \text{otherwise} , \end{cases}$$

$$\tilde{b_y}(x,y,k) = \begin{cases} 1, & \text{for } k \in \{1,0\}, \ (x,y-k+1) \in D_{yy} , \\ -2, & \text{for } k = 2, \ (x,y-k+1) \in D_{yy} , \\ 0, & \text{otherwise} , \end{cases}$$

$$\tilde{b_{xy}}(x,y,k,n) = \begin{cases} (-1)^{k+n}, & \text{for } (x-k,y-n) \in D_{xy} , \\ 0, & \text{otherwise} , \end{cases}$$

$$\tilde{c}(x,y) = \begin{cases} 1, & \text{for } (x,y) \in E , \\ 0, & \text{otherwise} , \end{cases}$$

$$\tilde{d_x}(x,y,k) = \begin{cases} (-1)^k, & \text{for } (x+k,y) \in D_x \setminus S_x , \\ 0, & \text{otherwise} , \end{cases}$$

$$\tilde{d}_y(x,y,k) = \begin{cases} (-1)^k, & \text{for } (x,y+k) \in D_y \setminus S_y , \\ 0, & \text{otherwise} . \end{cases}$$

Equation (4) can be solved using conjugate gradient method:

1. Set $r(x,y) = \rho(x,y) - Q\phi_0(x,y)$ and $\phi(x,y) = \phi_0(x,y)$, $k = 0$.
2. Set $W = \sum_{(x,y)\in D} r^2(x,y)$.
3. If $k = 0$ set $p(x,y) = r(x,y)$, else set $p(x,y) = r(x,y) + (W/W_{\text{old}})p(x,y)$.
4. Set $u(x,y) = (Qp)(x,y)$.
5. Set $\alpha = W/\sum_{(x,y)\in D} p(x,y)u(x,y)$.
6. Set $\phi(x,y) = \phi(x,y) + \alpha p(x,y)$ and $r(x,y) = r(x,y) - \alpha u(x,y)$.
7. If $W < \epsilon$ stop, else set $W_{\text{old}} = W$, $k = k+1$ and go to step 2.

4   Iterative Construction of the Set E

P_2 in (3) is a sum of squared distances to the known values of velocities in the set E. Initially, we do not know exact values of the velocities, so the set E is empty.

The set E and values $c(x,y)$ are determined iteratively. Initially E is empty and we solve the equation $Q\phi = \rho$ (4). $\phi_0(x,y) = g(x,y)$, where $g(x,y)$ is the measured signal.

Then we update the set E and values $c(x,y)$:

$$E(x,y) = \{(x,y) : |g(x,y) - \phi(x,y) - 2\pi k(x,y)| < \tau\} , \tag{5}$$

where integer valued function $k(x,y)$ gives $|g(x,y) - \phi(x,y) - 2\pi k(x,y)| < \pi$,

$$c(x,y) = g(x,y) + 2\pi k(x,y) . \tag{6}$$

At the next iteration equation $Q\phi = \rho$ is solved for new calculated E and $c(x,y)$. Then new values of E and $c(x,y)$ are calculated again using (5) and (6). The process stops when change of E becomes small.

In our calculations we use $\tau = 0.3\pi$. Results at different iteration steps are illustrated in Fig. 2.

5   Setting Method Parameters

The proposed method has some parameters, which reflect both physics of the signal receiving process and the resulting solution constrains. Further, we describe their effects on result.

Parameters a_x, a_y are the weights of the derivatives of g with respect of x and y. Since the observed signal is the fluid flow measured in the direction to the transducer, the derivatives in the direction to the transducer should be smoother than in the orthogonal direction. Typically, the transducer is placed at the top of the image, i.e. direction to the transducer corresponds to the y axis, while the

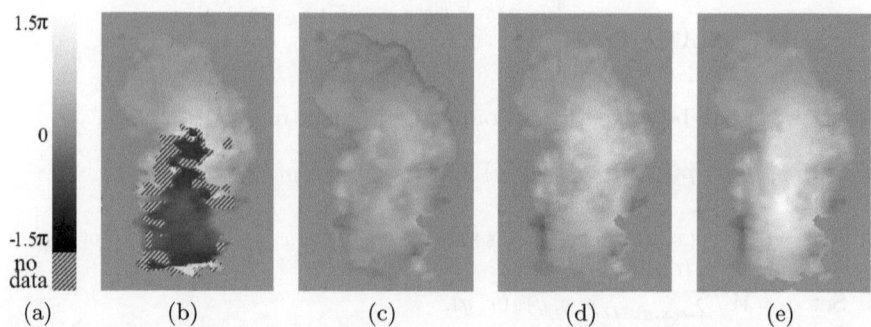

Fig. 2. Example of phase unwrapping after different number of iterations: (a) scale of phases, (b) input signal, (c) unwrapping result after the first iteration, (d) unwrapping result after the third iteration, (e) unwrapping result after the tenth iteration

orthogonal direction corresponds to the x axis. Therefore a_y should be greater than a_x.

Examples of results obtained for different parameters a_x and a_y are shown in Fig. 3. This picture shows the systole moment i.e. a powerful injection of blood into the left ventricle. The flow rate exceeds the measurement limit, and the aliasing occurs (the dark areas in Fig. 3(b)). If a_x is high, the algorithm will smooth the edges of the stream, thus understating the actual speed (Fig. 3(c), 3(d)). We selected $a_x = 0.1$, $a_y = 1$ as the most optimal parameters.

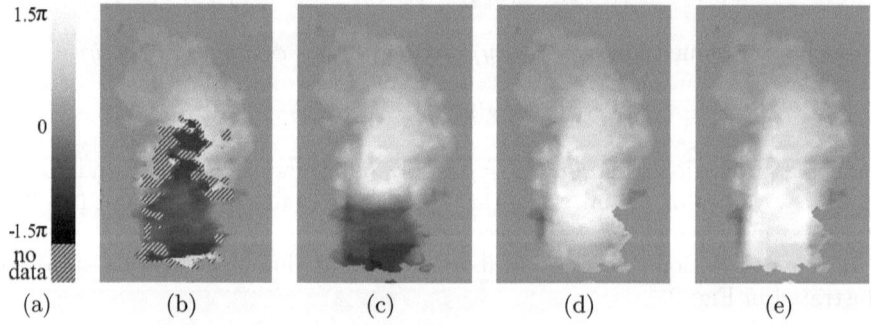

Fig. 3. Example of phase unwrapping for different a_x and a_y: (a) scale of phases, (b) input signal, (c) unwrapping result with $a_x = 1$, $a_y = 1$, (d) unwrapping result with $a_x = 0.5$, $a_y = 1$, (e) unwrapping result with $a_x = 0.1$, $a_y = 1$

The smoothness of the result depends on the used parameter λ. Large λ value leads to a smooth result. It helps to suppress the noise, but if λ is too high small details may be lost. If the parameter is too small the aliasing in some areas may remains. In our implementation we use the value $\lambda = 0.1$.

To give big weights for effective use of known values set E we used value $\beta = 1000$.

The derivatives ϕ_x and ϕ_y must be continuous in the set D. In the set S where g_x and g_y are known this condition is achieved by the choice of the discrepancy in the functional (3). In the set $D \setminus S$ derivatives ϕ_x and ϕ_y became continuous due to the term P_3. Therefore, the regularization parameter γ is responsible for the smoothness of the flow in areas where the data is missed. The phase unwrapping results, obtained with different parameter values γ are shown in the Fig. 4. This picture shows the systole moment. If γ is low then due to the large number of missed data areas algorithm cannot put in correspondence aliased areas with the neighbor areas (Fig. 4(c)).

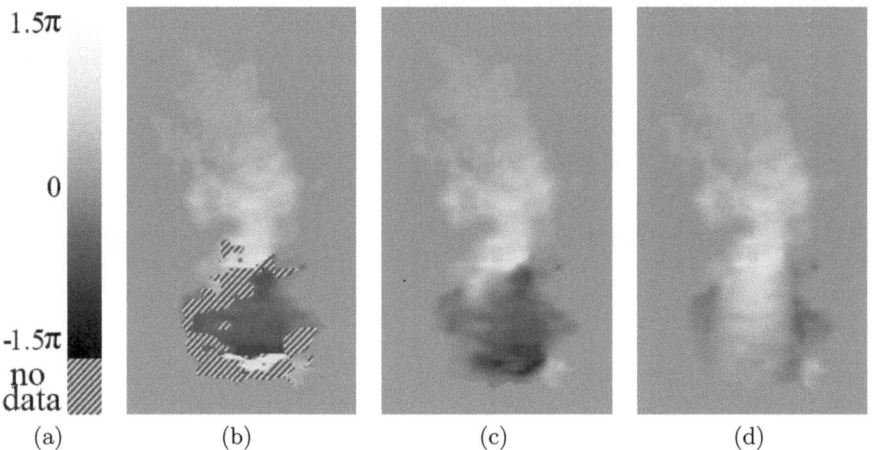

(a)   (b)   (c)   (d)

Fig. 4. Example of phase unwrapping for different values of γ: (b) scale of phases, (b) input signal, (c) $\gamma = 0$, (d) $\gamma = 1$

6   Conclusion

An effective method for iterative phase unwrapping in color Doppler flow mapping has been suggested. Selection of method parameters and the analysis of the results were performed on the medical images of patients with healthy hearts and on the images of patients with different diseases. The studies were conducted in Petrovsky National Research Center of Surgery of Russian Academy of Medical Sciences. The clinical studies showed that the proposed method better increases the CDFM images quality comparing with the existing methods.

Acknowledgment. The work was supported by Federal Targeted Programme "R&D in Priority Fields of the S&T Complex of Russia 2007–2013".

References

1. Bioucas-Dias, J.M., Valadao, G.: Phase Unwrapping via Graph Cuts. IEEE Transactions on Image Processing 16(3), 698–709 (2007)
2. Yatchenko, A.M., Krylov, A.S., Gavrilov, A.V., Arkhipov, I.V.: Graph-cut based Antialiasing for Doppler Ultrasound Color Flow Medical Imaging. In: International Conference Visual Communications and Image Processing (VCIP 2011), Taiwan, pp. 1–4 (2011), doi:10.1109/VCIP.2011.6115923
3. Pritt, M.D., Shipman, J.S.: Least-squares two-dimensional phase unwrapping using FFTs. IEEE Transactions on Geoscience and Remote Sensing 32(3), 706–708 (1994)
4. Ghiglia, D.C., Romero, L.A.: Minimum Lp-norm two-dimensional phase unwrapping. J. Opt. Soc. Am. A 13(10), 1999–2013 (1996)
5. Yamaki, R., Hirose, A.: Singular Unit Restoration in Interferograms Based on Complex-Valued Markov Random Field Model for Phase Unwrapping. IEEE Geoscience and Remote Sensing Letters 6(1), 18–22 (2009)
6. Marroquin, J.L., Rivera, M.: Quadratic regularization functionals for phase unwrapping. J. Opt. Soc. Am. A 12(11), 2393–2400 (1995)

TRUS Image Segmentation Driven by Narrow Band Contrast Pattern Using Shape Space Embedded Level Sets

Pengfei Wu[1], Yiguang Liu[1,*], Yongzhong Li[2], and Liping Cao[3]

[1] School of Computer Science, Sichuan University, Chengdu, China
[2] Ultrasound Department of West China Hospital,
Sichuan University, Chengdu, China
[3] Library, Sichuan University, Chengdu, China
`lygpapers@yahoo.com.cn`

Abstract. Prostate segmentation in transrectal ultrasound (TRUS) images is highly desired in many clinical applications. However, manual segmentation is difficult, time consuming and irreproducible. In this paper, we present a novel automatic approach using narrow band contrast pattern to segment prostates in TRUS images. Implicit representation of the segmenting level sets curve is firstly trained via principal component analysis, which also constraints the shape of prostate into a linear subspace. Then the model evolves to segment the prostate by maximizing the contrast in a narrow band near the segmenting curve. Many experimental results demonstrate the performance of the proposed algorithm, whose favorableness is validated by comparing to the state-of-the-art algorithms. Especially, the shape of prostate segmented by our algorithm is close to the one manually obtained by expert, and the mean absolute distance is only 1.07 ± 0.77mm, which is quite promising.

Keywords: Prostate segmentation, transrectal ultrasound images, active contours, level sets, shape prior, narrow band contrast pattern.

1 Introduction

Prostate cancer is one of the major public health problems. It is the second leading cause of cancer death among men, only surpassed by lung cancer [1]. Transrectal ultrasound (TRUS) has been the main imaging modality for prostate related applications for various reasons: It is real-time realization, low cost, simplicity and is not inferior to MRI or CT in terms of diagnostic value. Accurate prostate segmentation is significant in many clinical applications. For example, both biopsy needle placement [2] and the measurement of the prostate gland volume [3] require the segmentation information.

However, accurate computer aided prostate segmentation from TRUS images encounters considerable challenges due to low signal-to-noise ratio (SNR)

---

* Corresponding author.

J. Yang, F. Fang, and C. Sun (Eds.): IScIDE 2012, LNCS 7751, pp. 339–346, 2013.

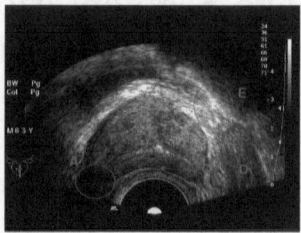

Fig. 1. Challenges for segmentation of prostate boundaries from TRUS images. (A) Shadow artifact and blurred boundary. (B) Area of calcification inside the prostate. (C) Low SNR and contrast. (D) Homogeneity in both prostate and non-prostate tissues. (E) Large speckles. (F) Heterogeneous intensity distribution inside the prostate.

and contrast, speckles, shadow artifact and heterogeneous intensity distribution inside the prostate etc.. Fig. 1 shows an illustration of these challenges. Accordingly, a few prostate segmentation methods have been proposed in literature [4]. Since the inherent difficulties of prostate segmentation in TRUS images, we should take as much prior knowledge, such as prostate shape, as possible into consideration to facilitate the segmentation procedure. As a consequence, statistical shape models gain the most interests and demonstrate a better results of segmentation [5,6,7,8,9]. Zhan et al. [7] used Gabor filter banks to extract texture features in TRUS images, then support vector machines (SVMs) were employed to classify them. After that, the classification result was used to drive the shape model to the prostate boundary. Although this method gains nice experimental results, it is too computational expensive. Yan et al. [6] proposed an automatic method to segment prostate using partial active shape model. Like may other active shape models (ASM) [10] based methods, the evolution of model relies on discrete control points on the model and is prone to be trapped by interferences such as speckles and small areas of calcification. More over, gradient based evolution can't handle the fuzzy edges well, which is often the case in TRUS images due to low SNR and homogeneity of both prostate tissue and non-prostate one.

In this paper, we attempt to create a novel automatic method to address the aforementioned problems. Our method is a level sets approach under the constraint of shape prior. During the segmentation, the model is driven by the proposed narrow band contrast pattern (NBCP) to the prostate boundary. Since the homogeneity of both side of prostate and the majority of useful information is near the prostate boundary, which makes Chan-Vese model [11] impractical, the NBCP only takes information in a narrow band near the model into consideration. At the same time, the NBCP shares part of advantages of Chan-Vese model and don't locally rely on the edge-function, which depends on the gradient of the image, to stop the curve evolution. The proposed model don't require large gradient value near the target boundary and has a large detection range, which is quite important to model initialization. Since our method is a level sets approach, no control points are used and the problem caused by ASM won't rise here. Besides, multi-resolution fashion is utilized to relieve the computational load and enlarge the detection range of the model.

The rest of this paper is organized as follows. How to embed shape prior into level sets is discussed briefly in Section 2. The details of narrow band contrast pattern and the use of it to segment prostate are given in Section 3. Section 4 shows the model initialization method, followed by Section 5 which demonstrates the performance of the proposed method by many experiments. Finally, concluding remarks are drawn in Section 6.

2 Level Sets Embedded with Shape Space

In this section, we briefly discuss the method in [8], which is utilized in our approach for its nice performance and easy to implement feature.

Firstly, a set of training images is segmented manually by expert. Then the segmentation results are aligned to eliminate the influence of different volumes as well as different places of prostate in the TRUS images. The aligned segmentation results are treated as the zero level sets and a set of signed distance functions (SDF) $\{\Psi_i\}$ are constructed from them. After that, mean shape $\bar{\Phi}$ is subtracted from them and they are converted to large column vectors $\{\psi_i\}$. Next, define the shape-variability matrix $\mathcal{S}$ as

$$\mathcal{S} = (\psi_1 \; \psi_2 \; \cdots \; \psi_n) \,. \tag{1}$$

We extract the first k eigenvectors of $\mathcal{S}$ and reshape them to the size the same as the training images. They are represented by $\{\Phi_i\}_{i=1,2,\ldots,k}$. With scale, translation and rotation of the shape taken into account, the implicit description of shape is finally given by the zero level set of the following function:

$$\Phi[\mathbf{w}, \mathbf{p}](x, y) = \bar{\Phi}(\tilde{x}, \tilde{y}) + \sum_{i=1}^{k} \omega_i \Phi_i(\tilde{x}, \tilde{y}) \,, \tag{2}$$

where ω_i is the weight of each eigen-shape Φ_i,

$$\begin{bmatrix} \tilde{x} \\ \tilde{y} \\ 1 \end{bmatrix} = T[\mathbf{p}] \begin{bmatrix} x \\ y \\ 1 \end{bmatrix} \,, \tag{3}$$

$$T[\mathbf{p}] = \begin{bmatrix} h\cos(\theta) & -h\sin(\theta) & a \\ h\sin(\theta) & h\cos(\theta) & b \\ 0 & 0 & 1 \end{bmatrix} \,, \tag{4}$$

and $\mathbf{p} = (a, b, h, \theta)^T$ is the pose parameter.

3 Narrow Band Contrast Pattern

In TRUS images, The prostate boundary has the property of dark-to-light transition of intensities from the inside of the prostate to the outside. However, this property is more notable in upper boundary in most cases. While the left,

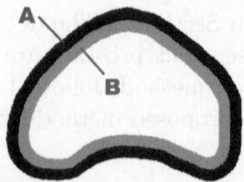

Fig. 2. Narrow band near the model boundary. Black area with label 'A' is outer band satisfying $0 < \Phi \leq h$, while gray area with label 'B' is the inner band satisfying $-h \leq \Phi < 0$.

right and bottom boundaries are significantly blurred, which might lead gradient based segmenting method to failure. Fortunately, the proposed NBCP not only can detect the blurred boundary well, but also is not sensitive to noises, speckles and areas of calcification etc., for its area based property. In this section, we first propose the energy function of NBCP, then the minimal cost approach of our method is discussed in details.

3.1   Energy Function

Because of the property of dark-to-light transition of intensities from the inside of the prostate to the outside, we hope to maximize the intensities in the narrow band outside the prostate boundary while minimize them in the narrow band inside it. So our NBCP energy function is defined as

$$E = - \iint_{\Omega} \tilde{H}(\Phi)I - \tilde{H}(-\Phi)I \, \mathrm{d}A \, , \qquad (5)$$

where Ω is the TRUS image, I is the intensity value and Φ is defined previously in (2). The function $\tilde{H}(\cdot)$ makes sure only the pixels in the narrow band of model boundary are taken into account

$$\tilde{H}(\Phi) = \begin{cases} 1, & \text{if } 0 < \Phi \leq h \\ 0, & \text{if } \Phi > h \text{ or } \Phi \leq 0 \end{cases}, \qquad (6)$$

where h is a predefined bandwidth value. It's worthy to point out that Φ got in (2) may not be a SDF and the narrow band is not strict h-pixel width. However, during our experiments, this fact doesn't cause any disastrous results. Fig. 2 shows an example of narrow band generated by (6), which is quite acceptable although Φ may not be a strict SDF.

3.2   Minimal Cost Approach

Since Φ is parametrical represented by $\mathbf{w}$ and $\mathbf{p}$, instead of using the Euler-Lagrange equations to do the minimal cost approach of (5), the gradient descent approach is proper here. The gradient of E is given by

$$\nabla E = -\iint_\Omega \nabla \tilde{H}(\Phi)I - \nabla \tilde{H}(-\Phi)I \, \mathrm{d}A$$

$$= -\iint_\Omega \delta(\Phi)\nabla \Phi I - \delta(\Phi - h)\nabla \Phi I$$

$$-\delta(-\Phi)\nabla(-\Phi)I + \delta(-\Phi - h)\nabla(-\Phi)I \, \mathrm{d}A$$

$$= -2 \oint_{\Phi=0} \nabla \Phi I \, \mathrm{d}s + \oint_{\Phi=h} \nabla \Phi I \, \mathrm{d}s + \oint_{\Phi=-h} \nabla \Phi I \, \mathrm{d}s \,, \qquad (7)$$

where $\delta(\cdot)$ is Dirac delta function. The gradient of Φ taken with respect to ω_i is

$$\nabla_{\omega_i}\Phi = \Phi_i \,, \qquad (8)$$

while it taken with respect to $\mathbf{p}^i$ (the i-th element of $\mathbf{p}$) by chain rule is

$$\nabla_{\mathbf{p}^i}\Phi = \begin{bmatrix} \frac{\partial \Phi}{\partial \tilde{x}} & \frac{\partial \Phi}{\partial \tilde{y}} & 0 \end{bmatrix} \frac{\partial T[\mathbf{p}]}{\partial \mathbf{p}^i} \begin{bmatrix} x \\ y \\ 1 \end{bmatrix} \,, \qquad (9)$$

where $T[\mathbf{p}]$ is perviously defined in (4).

The segmentation process is implemented by updating the parameter $\mathbf{w}$ and $\mathbf{p}$ based on gradient descent.

4   Initialization

The initialization of the model is a search procedure in the image Ω, by setting $\mathbf{p}$ with different values while $\mathbf{w} = \mathbf{0}$, which implies the mean model $\bar{\Phi}$ is scaled, rotated and translated to different places to find the minimal E using (5). Since the resolution of original TRUS image can be high, the search initialization procedure is performed in the coarsest resolution in our multi-resolution framework to relieve the computational load into a feasible scale. However, because $\mathbf{w}$ is set fixed during the search procedure, the initial model can't fit the target contour exactly. Fortunately, our energy function (5) is quite robust to this coarse fit, for the information in a narrow band near it is taken into account. What's more, for each tentatively set $\mathbf{p}$, a small number of evolution step is performed. Then the energy $E(\mathbf{p})$ is set to the minimal one during this tentative evolution. This procedure further enhance the accuracy of initialization and the problem caused by the fixed model shape is further reduced.

5   Experimental Results and Discussion

Ultrasound images used in the experiments were obtained by using a Philips HDI 5000 sonographic imaging system. Each image has 576×768 pixels. The pixel sizes of the images are 0.1384×0.1384mm. The settings of the device to acquire different images were the same. In our experiments, totally 132 TRUS images were segmented by one expert, 47 of them were used to train the model, while

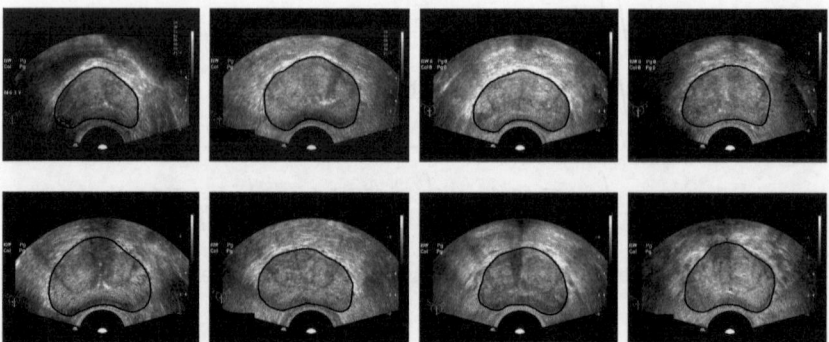

Fig. 3. Some segmentation results. Solid lines are the prostates segmented by the proposed algorithm, while the dash lines represent the contours segmented by expert.

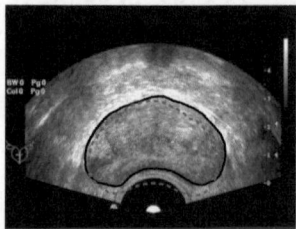

Fig. 4. Example of initialization, with initial model shown in dashes line, while the final segmentation result shown in solid line

the rest 85 ones were used to validate the algorithm. What's more, each slice was processed for individual patient and the consent of all patients was obtained for this study. Our method was implemented in Matlab on a notebook computer with Intel 2.4 Ghz processor. The mean processing time during our experiments of the entire method is about 17s with Matlab scripts. We are hoping to segment one TRUS image in seconds by fully optimizing the code and implementing it in C++ in the near future.

In order to evaluate the efficiency and robustness of our algorithm, distance and area criteria were used by comparing the automatic segmentation result with the segmentation made by the expert. For distance-based metric, the mean absolute distance (MAD) error [6] was utilized. For area-based metric, the coverage (Cov) [12] was employed.

In our experiments, the predefined bandwidth h is set to 20 pixels in the original image and the multi-resolution framework has 4 levels. First, the experiments were done to validate the effectiveness of our initialization algorithm. During our experiments, all the initial model were located in acceptable places. It's worthy to point out that the initial model is by no means to be accurate enough, for the variance of prostate shape while the shape parameter $\mathbf{w}$ is set to be fixed. However, the multi-resolution framework as long as the property of our energy function to consider the information around the boundary of model

Table 1. Prostate Segmentation Evaluation

Method	MAD	Cov
S. Ghose [5]	1.50±0.41mm	-
Yan [6]	2.01±1.02mm	-
N. Betrouni [12]	2.5±0.9mm	93±9%
Cosío [13]	1.65±0.67mm	-
Our Method	1.07±0.77mm	90±7%

enable our algorithm to have a large detection range to the target contour. Fig. 4 shows an example of initial model in dashes line and the final result in solid line. Because of the difference between the mean shape and the target prostate, the initial model didn't fit the target well. But our algorithm still got a good result.

Then experiments were done to segment the set of real TRUS images mentioned above. Some example segmentation results with many challenges mentioned previously to show the robustness of our algorithm is illustrated in Fig. 3, which are compared with the 'golden truth' given by the expert. Table 1 shows the quantitative evaluation results of the our automatic segmentation method, with a compare with some other state-of-the-art algorithms. The symbol '-' in Table 1 means not given. From the table we can see that our experimental results are quite inspiring, because of the small MAD error we got. Note that our method doesn't need any preprocess or postprocess procedure. The performance of our method can be further improved by adopting some postprocess methods, such as active contour model [14] used in [6]. Compared with the small mean value of MAD we got, the standard deviation seems large. The reason is that each slice used to validated our algorithm was obtained from individual patient. As a consequence, the TRUS images varied a lot and so many tough situations rose. While many other algorithms were only validated by limited number of data sets (e.g. [5,6]), though the number of images may be large.

6 Conclusions

In this paper, we have introduced a novel automatic approach using narrow band contrast pattern to segment prostates in TRUS images. Implicit representation of the segmenting level sets curve is firstly trained via principal component analysis, which also constraints the shape of prostate into a linear subspace. Then the model tries to maximize the contrast in a narrow band near the segmenting curve and drives the curve to prostate contour. The algorithm is robust to blurred boundary, low SNR, speckles and calcification, and the performance of it is validated by many experimental results. For example, the shape of prostate segmented by our algorithm is close to the one manually obtained by expert; the mean absolute distance is only 1.07 ± 0.77mm. All the results imply that our algorithm is quite promising.

Acknowledgement. This work is supported by NSFC under grants 61173182 and 61179071, as well as by Applied Basic Research Project (2011JY0124), International Cooperation and Exchange Project (2012HH0004) of Sichuan Province.

References

1. American Cancer Society: The prostate cancer quandary (2011),
   http://www.cancer.org/
2. Shen, D., Lao, Z., Zeng, J., Zhang, W., Sesterhenn, I.A., Sun, L., Moul, J.W., Herskovits, E.H., Fichtinger, G., Davatzikos, C.: Optimized prostate biopsy via a statistical atlas of cancer spatial distribution. Med. Image Anal. 8(2), 139–150 (2004)
3. Terris, M., Stamey, T.: Determination of prostate volume by transrectal ultrasound. J. Urol. 145(5), 984 (1991)
4. Noble, J.A., Boukerroui, D.: Ultrasound image segmentation: a survey. IEEE Trans. Med. Imag. 25(8), 987–1010 (2006)
5. Ghose, S., Oliver, A., Martí, R., Lladó, X., Freixenet, J., Villanova, J., Meriaudeau, F.: Prostate segmentation with local binary patterns guided active appearance models. Medical Imaging: Image Processing, France (2011)
6. Yan, P., Xu, S., Turkbey, B., Kruecker, J.: Discrete deformable model guided by partial active shape model for trus image segmentation. IEEE Trans. Biomed. Eng. 57(5), 1158–1166 (2010)
7. Zhan, Y.Q., Shen, D.G.: Deformable segmentation of 3-d ultrasound prostate images using statistical texture matching method. IEEE Trans. Med. Imag. 25(3), 256–272 (2006)
8. Tsai, A., Yezzi Jr., A., Wells, W., Tempany, C., Tucker, D., Fan, A., Grimson, W.E., Willsky, A.: A shape-based approach to the segmentation of medical imagery using level sets. IEEE Trans. Med. Imag. 22(2), 137–154 (2003)
9. Shen, D.G., Zhan, Y.Q., Davatzikos, C.: Segmentation of prostate boundaries from ultrasound images using statistical shape model. IEEE Trans. Med. Imag. 22(4), 539–551 (2003)
10. Cootes, T.F., Taylor, C.J., Cooper, D.H., Graham, J.: Active shape models-their training and application. Comput. Vis. Image Understanding 61(1), 38–59 (1995)
11. Chan, T., Vese, L.A.: An Active Contour Model without Edges. In: Nielsen, M., Johansen, P., Fogh Olsen, O., Weickert, J. (eds.) Scale-Space 1999. LNCS, vol. 1682, pp. 141–151. Springer, Heidelberg (1999)
12. Betrouni, N., Vermandel, M., Pasquier, D., Maouche, S., Rousseau, J.: Segmentation of abdominal ultrasound images of the prostate using a priori information and an adapted noise filter. Compt. Med. Imag. and Graph. 29, 43–51 (2005)
13. Cosío, F.A.: Automatic initialization of an active shape model of the prostate. Med. Image Anal. 12(4), 469–483 (2008)
14. Kass, M., Witkin, A., Terzopoulos, D.: Snakes: Active contour models. Int. J. Comput. Vis. 1(4), 321–331 (1988)

Multi-modal Based Violent Movies Detection in Video Sharing Sites

Xingyu Zou[1], Ou Wu[2], Qishen Wang[2], Weiming Hu[2], and Jinfeng Yang[1]

[1] College of Aviation Automation,
Civil Aviation University of China, Tianjin, China
{xyzou_yjs10,jfyang}@cauc.edu.cn
[2] National Laboratory of Pattern Recognition,
Chinese Academy of Sciences, Beijing, China
{wuou,qswang,wmhu}@nlpr.ia.ac.cn

Abstract. In this paper we present a method for the detection of violent movies in video sharing sites. The proposed method operates on three modalities: text, video and audio, the former being collected from the accompanying synopsis and user comments. Towards our goal, a multi-step approach is followed: initially, the text information is utilized to build a pre-classifier which selects the potential violent movie segments. At a second stage, a classifier is adopted, which combines the visual and audio information, in order to classify the potential violent movie segments as "violent" or "non-violent". The experimental results on 220 movie segments from YouKu and TuDou show the effectiveness of our method.

Keywords: Violent movies Detection, Multi-modal, Text, Audio, Video.

1   Introduction

Movies are an important part of the daily entertainment. With the rapid development of the Internet technology, people increasingly prefer to watch movies in video sharing sites instead of the cinema. A huge amount of movies can be easily uploaded and accessed by online users of all ages including children, teenagers, etc. However, violence in movies has harmful influence on children. So it is necessary to protect children from accessing violent content. But the manually labeling huge volumes of video data is a hard work.

In this paper, we present an automatic method to detect violent movies in video sharing sites by combining text, video and audio information. Our method is divided into two stages. At the first stage, a text classifier is built as a pre-classifier, using the synopsis and the user comments on each movie online. The pre-classifier is used to select potential violent movie segments. Furthermore, the potential violent movie segments are segmented to a set of shots, using a shot detection method. Then four video features and two audio features are extracted from each shot. The final binary problem (i.e., violent vs non-violent movies) on a 6-dimensional feature space is solved by using Support Vector Machine (SVM).

J. Yang, F. Fang, and C. Sun (Eds.): IScIDE 2012, LNCS 7751, pp. 347–355, 2013.
© Springer-Verlag Berlin Heidelberg 2013

The rest of the paper is organized as follows. Section 2 provides a brief overview of the related work. In Section 3, we detail the proposed method. Experiment results and discussions are presented in Section 4. Section 5 concludes this paper and gives the future work.

2   The Related Work

A few approaches have been proposed to violence detection in videos, owing to ambiguity in the definition of violence. It is difficult to give a precise definition about violence, Each related work has its own definition of violence. Current techniques for violence detection can be basically classified into three categories. The first one is based on visual cues. Using an accelerate motion vector, Datta et al. [1] addressed the problem of detecting human violence in videos such as fist fighting and kicking. The second category is based on audio cues. Cheng et al. [2] presented a hierarchical approach to locate gunplay and car racing. Giannakopoulos et al. [3] used eight audio features, both from the time and frequency domain, as input to a binary classifier which decides the video content with respect to violence. They proposed a multi-class classification algorithm for audio segments from movies in [4]. The third category is based on the fusion of visual and audio features. Nam et al. [5] combined multiple audio-visual features to identify violent scenes, in which flames and blood were detected by matching predefined color tables, and various representative audio effects (gunshots, explosions, etc.) were also exploited. Smeaton et al. [6] combined audio-visual features to select representative shots in an action movie to produce a trailer. Giannakopoulos et al. [7] presented a method for violence detection in movies based on audio-visual information that used a statistics of audio features and average motion and motion orientation variance features in video combined in a k-Nearest Neighbor classifier to decide whether the given sequence is violent. In this work, we research three typical events generally associated with violence, such as explosions, gunshots and fighting.

In this work, we use three modalities (text, video and audio) to detect violent movies. The use of text information can quickly filter out a large number of non-violent movies, reducing the cost of time. Moreover, the fusion of three modalities improves the detection accuracy compared with existing methods.

3   Our Method

Our method is composed of two stages, as shown in Fig. 1. In the first stage, a pre-classifier based text information is built to select the potential violent movie segments. In the second stage, we segment the video sequence into a set of shots. For each shot, video and audio features are extracted. Then a SVM classifier is trained to identify violent shots and movie segments are classified. The following subsections detail the two stages.

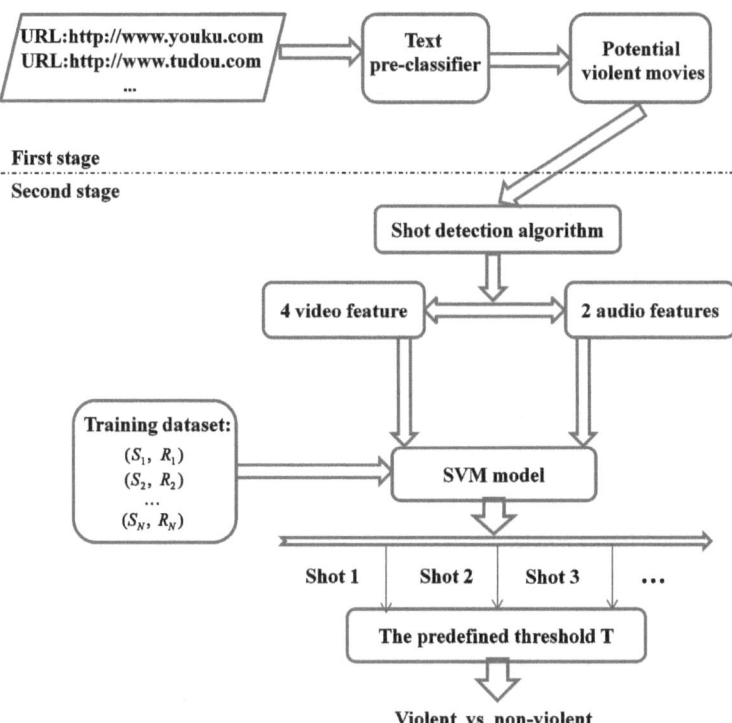

Fig. 1. The framework of our method

3.1   Detection of Potential Violent Movies

To a movie, the video sharing sites generally provide its synopsis and the users also give their own comments. In this paper, we consider the synopsis and comments as the text information associated with the movie. Compared with other types of movies, the movie types related to violence are usually action, war, crime and adventure, etc. We collect a dictionary of 250 keywords, which contains words frequently related to murder, crime, drug abuse, violence, war and weapon, etc. We have chosen Document Frequency (DF) in many text features, because it has low computational complexity and high computing speed. The processing steps of the text information are as follows.

The original text information is first preprocessed. Chinese word segmentation is carried out by ICTCLAS from ICT of Chinese Academy of Sciences [8]. Then the stop words are removed, such as preposition, conjunction and director, protagonist, etc, which frequently occur in the synopsis and comments and are not useful for the classification.

In order to computing convenience, we adopt Vector Space Model (VSM) [9] to represent the text information because of its good performance. Each text D_i is represented by a feature vector $(T_1, \omega_1; T_2, \omega_2; \cdots; T_n, \omega_n)$ based on the dictionary of

250 keywords, where T_i is the key word, ω_i reflects the degree of correlation between keywords and texts. The weight ω_i is computed by TF-IDF.

After VSM is built, the text vectors are normalized. Finally, a SVM classifier is trained to classify the movie segments as non-violent or potential violent ones. The process of pre-classification is presented in Fig. 2.

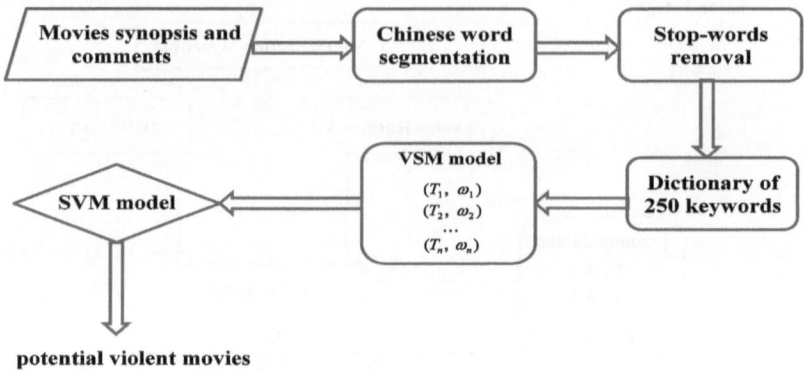

Fig. 2. The process of pre-classification

3.2   Video-Audio Classification

For each potential violent movie segment, we further check whether or not it is violent. First each video sequence is segmented into shots by the twin-comparison approach [10]. The shots whose duration is less than a threshold (e.g., 10 frames) are merged with their neighbor shots to avoid meaningless shots. Then each shot is identified by the characteristics of violent movies. Film-making is a creative process, some universal rules should be followed. Generally, violent movies have the atmosphere of fast-pace and it is created by high-speed visual movement and fast-paced sound. Thus, we identify violent shot through the detection of fast camera movement, frequent shot transitions, sudden sound and typical events generally associated with violence, such as explosions, gunshots and fighting.

Video Features. The visual signal is split into shots. Each shot includes several video frames, and the visual features are calculated on every frame. Based on the visual characteristics of violent movies, the following features are extracted.

Motion Intensity: Motion is an important visual feature which could describe the sustaining temporal variation of video streams. Also, it reveals the correlations between frame sequences within a video shot. Generally violent movie shots contain high activity and abrupt motion. To characterize the degree of motion within a shot, the average motion intensity is computed based on the motion vectors. First we compute the motion intensity of each frame. The kth frame of a shot is split into blocks of which we calculate the motion vectors using the previous frame as reference. The motion intensity of the ith block is defined as

$$M_k(i) = \sqrt{u^2(i) + v^2(i)} \qquad (1)$$

where (u_i, v_i) is the motion vector of the ith block of the kth frame. The average motion intensity of the kth frame is defined as

$$\overline{M_k} = \frac{1}{n}\sum_{i=0}^{n-1} M_k(i) \qquad (2)$$

where n is the number of   blocks in the kth frame. Assuming there are m frames in the shot, the average motion intensity of the shot is defined as

$$\overline{M} = \frac{1}{m}\sum_{k=0}^{m-1} \overline{M_k} \qquad (3)$$

Finally, this value is normalized to be in the interval [0, 1].

Flame: Gunshot and explosion are typical and distinct violent events. One obvious visual feature is flame which suddenly generates in gunshot and explosion events. The flames have dominant yellow, orange and red color components. Thus, a predefined color template which includes a range of colors is employed to match flame colors. The flame of violent events is from scratch, and rapidly changes in a short period of time. This characteristic reflected on the image is that flame pixels rapidly increase or reduce within a few continuous frames. We are interested in the variation of frames with time, which reveal an obvious change in the number of flame pixels, rather than in the number of flame pixels owing to the existence of possible flame-like colors. The variation of flame pixels in a shot is defined as

$$\overline{V_f} = \frac{1}{M_f}\sum_{i=0}^{M_f-1} |F_{i+1} - F_i| \qquad (4)$$

where F_i is the number of flame pixels in the ith frame, M_f is the number of continuous frames whose flame pixels continue to change, and the value of F_0 is zero representing the neighbor frame without flame pixels before the flame frame. Finally the value $\overline{V_f}$ is normalized to be in the interval [0, 1].

Bleeding: Three typical violent events (fighting, explosion, gunshot) often lead to bleeding. Thus, bleeding could be a useful visual feature associated with violence. A predefined color template is employed to match bloody colors and bloody color pixels in video frames are identified. Similar to flame, the bloody element of violent events is from scratch, and rapidly increases in a short time. This characteristic reflected on the image is that bloody pixels rapidly increase within a few consequent frames. We focus on the variation of frames with time, which reveals an obvious increase in the number of bloody pixels, rather than in the number of bloody pixels owing to the existence of possible blood-like colors. The variation of bloody pixels in a shot is defined as

$$\overline{V_b} = \frac{1}{M_b}\sum_{i=0}^{M_b-1} |B_{i+1} - B_i| \qquad (5)$$

where B_i is the number of bloody pixels in the ith frame, M_b is the number of sustained frames whose bloody pixels continue to increase, and the value of B_0 is zero representing the neighbor frame without bloody pixels before the bloody frame. Finally the value $\overline{V_b}$ is normalized to be in the interval [0, 1].

Shot Length: In order to attract attention, violent movies generally create a tense atmosphere. One of the most common methods is frequent conversion of the lens. As a result, the length of shots containing violence is commonly shorter than normal ones. That means violent shots have less frames than normal ones. Thus the length of shots also becomes a feature related to violence. The number of frames in a shot L represents the length of the shot. Likewise, the value of L is normalized to be in the interval [0, 1].

Audio Features. To exhibit exciting scenes we create the atmosphere of fast-tempo through not only visual sense but also auditory sense in violent movies. Generally, their audio characteristics include less speech, fast-paced music and typical audio events associated with explosions, gunshots and fighting. The common characteristics are the sound of intense fast-paced and short-term severe variations. Based on the visual characteristics of violent movies, the following features are extracted.

Audio Energy: In violent shots, the sounds of fast-paced music, fierce fighting, explosion and gunshot provide additional energy to the shots. Compared with normal movies, the sounds in violent shots have often higher energy. Thus, the audio energy also becomes a useful feature related to violence. The energy of the ith audio frame is defined as

$$E(i) = \sum_{n=1}^{N} x_i^2(n) \tag{6}$$

where N is the number of sampling points in the ith frame, $x_i(n)$ is the value of the nth sampling points in the ith frame. Assuming there are m audio frames in the shot, then the average audio energy of the shot is defined as

$$\bar{E} = \frac{1}{m} \sum_{i=1}^{m} E(i) \tag{7}$$

Finally, the value of $\bar{E}$ is normalized to be in the interval [0, 1].

Energy Entropy: In violent shots there are unique sound effects of beating, gunshot, explosion, crashing of objects and etc. They are mostly accompanied with a sudden burst in the audio level. We consider the abrupt change in the audio signal energy as another feature associated with violence. The energy entropy is employed to describe the abrupt change and defined as [11]

$$I_n = -\sum_{i=1}^{J} \sigma_i^2 \log_2 \sigma_i^2 \tag{8}$$

where J is the total number of segments in the nth audio frame (each frame is divided into smaller J segments) and σ_i^2 is the normalized energy of the ith short segment in a frame, J is chosen to be 10. The value of the energy entropy measure falls down in frames with sudden energy transitions while the energy entropy is largest in a frame with nearly constant energy. Then, the minimum of the energy entropy in a shot is defined as

$$I = \min_{n=1,\dots k} I_n \tag{9}$$

where k is the number of audio frames in a shot. Finally, I is normalized to be in the interval $[0, 1]$.

Binary Classification. The SVM classification model is adopted to solve this binary classification problem (violent vs non-violent content), where S_i is the feature vector of the ith video and R_i is the corresponding label , N shot objects make up of the SVM training data, i.e., N pairs of the form (S_i, R_i), $i = 1, \dots, N$. After all shots of a movie segment are detected, the movie segment is classified by a predefined threshold T, experimental results reach the best performance when the threshold is 0.1, i.e., when $P > T$, the movie segment is classified as "violent", where P is the percentage of violent shots in a movie segment. The false alarms will become less through the use of the threshold T.

4   Experimental Results

4.1   Dataset

220 movie segments from YouKu and TuDou are used in our experiment, which have been manually labeled and segmented into labeled shots. The normal movie segments include a wide range of categories consisting of comedies, dramas, documentaries, romances and dance films, etc. The total video duration is 634 minutes, the average video duration is 2.88 minutes and 40% of the videos are less than 1 minute long. 121 movie segments are annotated as "violent". The distribution of the film genres is presented in Fig. 3.

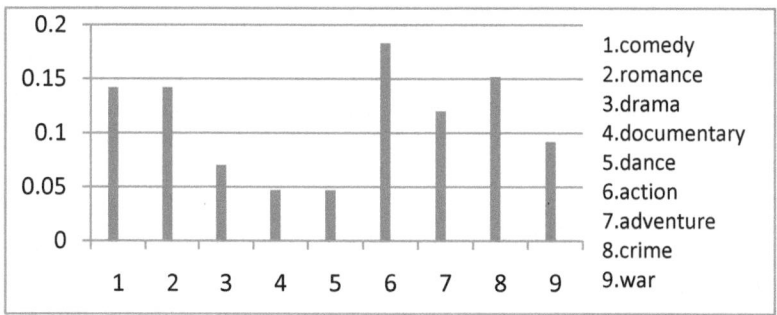

Fig. 3. The percentage of movie genres

4.2   The Result of Experiments

In our method, SVM with RBF kernel is used as the pre-classifier and video-audio classifier. In the SVM training process, 5-fold cross-validation is executed. We measure the performance of the proposed method by precision, recall and F_1-measure.

Table 1 lists the experimental results of the pre-classifier. In the process of pre-classification a large number of non-violent movie segments are filtered out. Because of the speed of text processing, the cost of time is greatly reduced. Meanwhile, most of violent movie segments are retained to enter the second step.

Table 1. The performance of the pre-classifier

Recall	Precision	F_1
88.89%	75.47%	81.63%

When only the video and audio information are used, shots are split from movie segments, and then they are detected by video-audio classifier as non-violent or violent. Table 2 shows the experimental results of shots classification.

Table 2. The performance of shots classification

Recall	Precision	F_1
79.74%	78.71%	79.22%

After the threshold $T=0.1$ is used, the result of movie segments classification is improved. Table 3 shows the result of movie segments classification without the pre-classifier.

Table 3. The result of movie segments classification

Recall	Precision	F_1
79.74%	80.64%	80.19%

Finally, all the three modalities are used. The result is shown in Table 4. Compared with the above results, the use of three modalities results in the precision improvement of 2.77%.

Table 4. The final result of our method

Recall	Precision	F_1
84.22%	83.41%	83.81%

5   Conclusion

In this work, we present an effective approach for automatically detecting violent movies in video sharing sites. While most of previous approaches addressed on one or two modalities only, our approach works on three modalities: text, video, and audio.

Hence, more useful information of movies is utilized for detection. Text information is used to design a pre-classifier to identify potential violent movie segments and they are classified by video-audio classifier as violent or non-violent. Though the extracted features are simple, experimental results show that the proposed method is efficient and effective. In the future, we plan to combine other useful information (e.g., other video information, other audio information and so on) to further improve the performance.

Acknowledgement. This work is partly supported by the National Nature Science Foundation of China (No.61005030, 60935002, and 60825204) and Chinese National Programs for High Technology Research and Development (863 Program) (No. 2012AA012503 and No. 2012AA012504).

References

1. Datta, A., Shah, M., Lobo, N.D.V.: Person-on person violence detection in video data. In: Proc. ICPR, pp. 433–438 (2002)
2. Cheng, W., Chu, W., Wu, J.: Semantic Context Detection Based on Hierarchical Audio models. In: Proceedings of the 5th ACM SIGMM International Workshop on Multimedia Information Retrieval, pp. 109–115 (2003)
3. Giannakopoulos, T., Dimitrios, K., Andreas, A., Sergios, T.: Violence content classification using audio features. In: Proc. HCAI, pp. 502–507 (2006)
4. Giannakopoulos, T., Pikrakis, A., Theodoridis, S.: A multi-class audio classification method with respect to violent content in movies, using Bayesian networks. In: IEEE International Workshop on Multimedia Signal Processing, pp. 90–93 (2007)
5. Nam, J., Alghoniemy, M., Tewfik, A.H.: Audio-Visual Content-Based Violent Scene Characterization. In: Proc. ICIP, vol. 1, pp. 353–357 (1998)
6. Smeaton, A.F., Lehane, B., O'Connor, N.E., Brady, C., Craig, G.: Automatically selecting shots for action movie trailers. In: Proceedings of the 8th ACM International Workshop on Multimedia Information Retrieval, pp. 231–238 (2006)
7. Giannakopoulos, T., Makris, A., Kosmopoulos, D., Perantonis, S., Theodoridis, S.: Audio-Visual Fusion for Detecting Violent Scenes in Videos. In: Konstantopoulos, S., Perantonis, S., Karkaletsis, V., Spyropoulos, C.D., Vouros, G. (eds.) SETN 2010. LNCS, vol. 6040, pp. 91–100. Springer, Heidelberg (2010)
8. ICTCLAS, http://ictclas.org/ictclas_download.aspx
9. Raghavan, V.V., Wong, S.K.M.: A critical analysis of vector space model information retrieval. JASIS 37(5), 279–287 (1986)
10. Zhang, H.J., Kankanhalli, A., Smoliar, S.W.: Automatic partitioning of full-motion video. Multimedia Systems 1(1), 10–28 (1993)
11. Sinha, D., Tewfik, A.H.: Low bit rate transparent audio compression using adapted wavelets. IEEE Trans. Signal Processing 41(12), 3463–3479 (1993)

Discriminant Analysis
Based on Nearest Feature Line

Lijun Yan[1], Cong Wang[1], Shu-Chuan Chu[2], and Jeng-Shyang Pan[1]

[1] Harbin Institute of Technology Shenzhen Graduate School
Xili University Town, NanShan, Shenzhen, China
yanlijun@126.com, jengshyangpan@gmail.com
[2] School of Computer Science, Engineering and Mathematics
Flinders University of South Australia
GPO Box 2100, Adelaide, South Australia 5001, Australia
scchu@csem.flinders.edu.au

Abstract. A novel feature extraction algorithm based on nearest feature line is proposed in this paper. The proposed algorithm can extract the local discriminant features of the samples. The performance of the proposed algorithm is directly associated with the parameter, so we use two discriminant power criterions to adaptively determine the parameter. Some experiments are implemented to evaluate the proposed algorithm and the experimental results demonstrate the efficiency of the proposed algorithm.

Keywords: Nearest Feature Line, Feature Extraction, Dimensionality Reduction.

1 Introduction

Image classification and related technologies [1] have a variety of potential applications in information security, smart card, access control, etc. However, there exists some difficulties in image classification. For example, in the face recognition task, the number of training images points per person is smaller than that of the dimensionality of face images. High-dimensional face images result in high computational complexity and overfitting. Dimensionality reduction technology is an effective way to alleviate it, and subspace-based dimensionality reduction algorithms have been widely used.

Principal Component Analysis (PCA)[2], Linear Discriminant Analysis (LDA) [3,4], and Margin Criterion (MMC)[5] are several most popular subspace-based dimensionality reduction algorithms. PCA projects the original data to a low dimensional subspace, which is spanned by the eigenvectors associated with the largest eigenvalues of the covariance matrix of all prototype samples. PCA is the optimal representation of the input samples in the sense of minimizing the mean squared error. However, PCA is an unsupervised algorithm, which may impair the recognition accuracy. LDA finds a transformation matrix U that linearly maps high-dimensional sample $x \in R^m$ to low-dimension data $y = U^T x \in R^n$, where $n \ll m$. LDA can calculate an optimal discriminant projection by

J. Yang, F. Fang, and C. Sun (Eds.): IScIDE 2012, LNCS 7751, pp. 356–363, 2013.

maximizing the ratio of the trace of the between-class scatter matrix to that of the within-class scatter matrix. LDA takes consideration of the labels of the prototype samples and improves the discriminant ability. However, LDA suffers from the famous small sample size (SSS) problem. Many effective approaches have been designed to solve the problem. Some nonlinear extensions using kernel trick of these algorithms are proposed recently[6]. Besides, there are some works on feature fusion presented[8] to improve the performance of feature extraction.

Nearest feature line (NFL)[9] is a novel classifier, proposed by Li et. al in 1999, firstly. In particular, it performs better when only few prototype samples are available. The basic idea of the NFL approach is to use all the possible hyperlines consisting of any pair of feature vectors within the same class in the prototype set to encode the feature space in terms of the ensemble characteristics and the geometric relationship. As a simple but effective algorithm, NFL has shown its good performance in face recognition, audio classification, image classification, and retrieval. NFL takes advantage of both the ensemble and the geometric features of samples for pattern classification. In contrast to the nearest neighbor (NN) classifier, NFL makes better use of the ensemble information for classification[10,11,12].

While NFL has achieved reasonable performance in the classification, most current NFL-based algorithms just use the NFL for classification and not in the training phase. While classification can be enhanced by NFL to a certain extent, the learning ability of existing subspace-based dimensionality reduction methods remains to be poor when the number of prototype samples is few. To address this issue, a number of enhanced subspace-based dimensionality reduction algorithms based on the NFL have been developed, recently. For instance, Zheng et al. proposed a nearest neighbour line nonparametric discriminant analysis (NNL-NDA)[13] algorithm, Pang et al. presented a nearest feature line-based space (NFLS)[14] method, and Lu et al. put forward an uncorrelated discriminant nearest feature line analysis (UDNFLA)[15]. Neighborhood discriminant nearest feature line analysis (NDNFLA)[16] is proposed to extract the local discriminant features of prototype samples. However, in NDNFLA approach, the difference of between class scatter and within class scatter are used to evaluate the scatter of the samples. A very large between class scatter or a very small within class scatter may lead to some misclassification. A novel feature extraction algorithm based on NFL is proposed in this paper to avoid this issue.

The rest of the paper is organized as follows. In section 2, some preliminaries are given. In section 3, we give an introduction of the proposed methods. In section 4, a number of experiments are implemented to justify the superiority of the proposed algorithms. Conclusions are made in section 5.

2   Outline of NFL and NDNFLA

2.1   Nearest Feature Line

Nearest feature line is a classifier. It is first presented by Stan Z. Li and Juwei Lu. Given a training samples set, $X = \{x_n \in R^M : n = 1, 2, \cdots, N\}$, denote the class label of x_i by $l(x_i)$, the training samples sharing the same class label with x_i by $P(i)$, and the training samples with different label with x_i by $R(i)$. NFL generalizes each pair of prototype feature points belonging to the same class: $\{x_m, x_n\}$ by a linear function $L_{m,n}$, which is called the feature line. The line $L_{m,n}$ is expressed by the span $L_{m,n} = sp(x_m, x_n)$. The query x_i is projected onto $L_{m,n}$ as a point $x^i_{m.n}$. This projection can be computed as

$$x^i_{m,n} = x_m + t(x_n - x_m) \tag{1}$$

where $t = [(x_i - x_n)(x_m - x_n)]/[(x_m - x_n)^T(x_m - x_n)]$.

The Euclidean distance of x_i and $x^i_{m,n}$ is termed as FL distance. The less the FL distance is, the more probability that x_i belongs to the same class as x_m and x_n. Fig. 1 shows a sample of FL distance. In Fig. 1, the distance between y_p and the feature line $L_{1,2}$ equals to the distance between y_q and y_p, where y_p is the projection point of y_q to the feature line $L_{1,2}$.

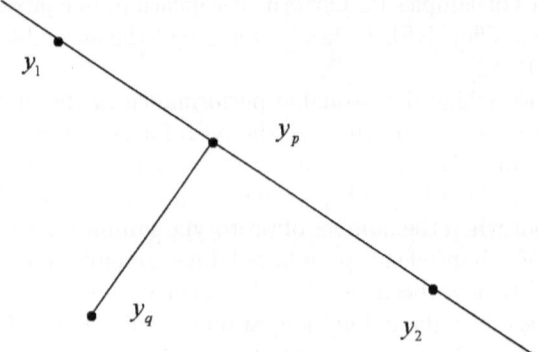

Fig. 1. NFL metric

2.2   NDNFLA

Let's introduce two definitions firstly.

Definition 1. *Homogeneous neighborhoods: For a sample x_i, its k nearest homogeneous neighborhood N^o_i is the set of k most similar data which are in the same class with x_i.*

Definition 2. *Heterogeneous neighborhoods: For a sample x_i, its k nearest Heterogeneous neighborhoods N^e_i is the set of k most similar data which are not in the same class with x_i.*

In NDNFLA approach, the optimization problem is as follows:

$$\max J(W) = (\sum_{i=1}^{N} \frac{1}{NC^2_{|N^e_i|}} \sum_{x_m, x_n \in N^e_i} \left\| W^T x_i - W^T x^i_{m,n} \right\|^2$$
$$- \sum_{i=1}^{N} \frac{1}{NC^2_{|N^o_i|}} \sum_{x_m, x_n \in N^o_i} \left\| W^T x_i - W^T x^i_{m,n} \right\|^2) \tag{2}$$

Using matrix computation,

$$
\sum_{i=1}^{N} \frac{1}{NC_{|N_i^e|}^2} \sum_{x_m, x_n \in N_i^e} \left\| W^T x_i - W^T x_{m,n}^i \right\|^2
$$

$$
= \sum_{i=1}^{N} \frac{1}{NC_{|N_i^e|}^2} \sum_{x_m, x_n \in N_i^e} \mathrm{tr}[W^T (x_i - x_{m,n}^i)(x_i - x_{m,n}^i)^T W] \tag{3}
$$

$$
= \mathrm{tr}\{W^T \sum_{i=1}^{N} \frac{1}{NC_{|N_i^e|}^2} \sum_{x_m, x_n \in N_i^e} [(x_i - x_{m,n}^i)(x_i - x_{m,n}^i)^T] W\}
$$

where tr denotes the trace of a matrix. Similar with the above,

$$
\sum_{i=1}^{N} \frac{1}{NC_{|N_i^o|}^2} \sum_{x_m, x_n \in N_i^o} \left\| W^T x_i - W^T x_{m,n}^i \right\|^2
$$

$$
= \mathrm{tr}\{W^T \sum_{i=1}^{N} \frac{1}{NC_{|N_i^o|}^2} \sum_{x_m, x_n \in N_i^o} [(x_i - x_{m,n}^i)(x_i - x_{m,n}^i)^T] W\}
\tag{4}
$$

Then the problem becomes

$$
\max J(W) = \mathrm{tr}[W^T(A - B)W] \tag{5}
$$

where

$$
A = \sum_{i=1}^{N} \frac{1}{NC_{|N_i^e|}^2} \sum_{x_m, x_n \in N_i^e} [(x_i - x_{m,n}^i)(x_i - x_{m,n}^i)^T] \tag{6}
$$

$$
B = \sum_{i=1}^{N} \frac{1}{NC_{|N_i^o|}^2} \sum_{x_m, x_n \in N_i^o} [(x_i - x_{m,n}^i)(x_i - x_{m,n}^i)^T] \tag{7}
$$

A length constraint $w^T w = 1$ is imposed on the proposed NDNFLA. Then, the optimal projection W of NDNFLA can be obtained by solving the following eigenvalue problem.

$$
(A - B)w = \lambda w \tag{8}
$$

Let $w_1, w_2, \cdots, w_q$ be the eigenvectors of formula(8) corresponding to the q largest eigenvalues ordered according to $\lambda_1 \geq \lambda_2 \geq \cdots \geq \lambda_q$. An $M \times q$ transformation matrix $W = [w_1, w_2, \cdots, w_q]$ can be obtained to project each sample $M \times 1$ x_i into a feature vector $q \times 1$ y_i as follows:

$$
y_i = W^T x_i, \qquad i = 1, 2, \cdots, N \tag{9}
$$

3   Proposed Algorithm

In NDNFLA approach, the optimal criterion is to maximize the difference between between class scatter and within class scatter. If the within class is small enough, the difference will be large even the between class scatter is not very large. To avoid this problem, In this section, a novel image feature extraction algorithm based on NFL is proposed.

Suppose there are c pattern classes. N is the total number of training samples, and N_i is the number of the samples in the ith class. X_i denotes the ith prototype sample. $l(i)$ denotes the label of X_i.

$$
\begin{aligned}
\max J(W) = \alpha(\sum_{i=1}^{N} \frac{1}{NC^2_{|N_i^e|}} \sum_{x_m, x_n \in N_i^e} \left\| W^T x_i - W^T x_{m,n}^i \right\|^2 \\
- (1-\alpha) \sum_{i=1}^{N} \frac{1}{NC^2_{|N_i^o|}} \sum_{x_m, x_n \in N_i^o} \left\| W^T x_i - W^T x_{m,n}^i \right\|^2)
\end{aligned} \tag{10}
$$

Calculating by the same way with NDNFLA, formula (10) can be simplified as follows.

$$
(\alpha A - (1-\alpha)B)w = \lambda w \tag{11}
$$

where A and B are defined as formula (6) and formula (7).

Let $w_1, w_2, \cdots, w_q$ be the eigenvectors of formula(11) corresponding to the q largest eigenvalues $\lambda_1 \geq \lambda_2 \geq \cdots \geq \lambda_q$. An $M \times q$ transformation matrix $W = [w_1, w_2, \cdots, w_q]$ can be obtained to project each sample $M \times 1$ x_i into a feature vector $q \times 1$ y_i as follows:

$$
y_i = W^T x_i, \qquad i = 1, 2, \cdots, N \tag{12}
$$

From the formula (10), we can find the parameter α controls the influence of within class scatter and between class scatter in the feature extraction. If choosing a reasonable α, the problem referred in the beginning of this section can be alleviated. Here, two discriminant power criterions are presented as follows.

(1) Discriminant power criterion based on k-nearest neighbor (k-NN): Let l_i denote the number of nearest neighbors in the same class with x_i among its k-NN. Then, let $L = \sum_{i=1}^{c} \sum_{j=1}^{n_i} l_i^j$. At last, let

$$
J_{kNN} = \frac{L}{k * N} \tag{13}
$$

According to the formula(13) , it is clear that the bigger J_{kNN} is, the more discriminant features are. The main idea of the proposed feature extraction algorithm, is to find a most discriminant α using J_{kNN}. The detailed procedure of proposed method is as follows:

Training stage:

Step 1, let α vary from 0 to 1, using the formula (11) to calculate the transformation matrix W;

Step 2, project all the prototype samples with the transformation matrix W;

Step 3, calculate J_{kNN} using formula(13) and find the optimal parameter α;

Step 4, extract the feature of prototype samples following formula(12).

Classification stage:

Step 1, extract the feature of query following formula(12);

Step 2, classify with NFL.

4   Experimental Results

In this section, a number of experiments on ORL face database[17] and AR face database[18] are implemented to evaluate the effectiveness of the proposed algorithm, which is also compared with some conventional subspace learning methods, including PCA, as well as the latest NFL-based subspace learning algorithms, such as NFLS, UDNFLA and NDNFLA. The experiments are implemented on a PC with 1.6-GHz CPU and 1G RAM. NFL classifier is used for classification on the features extracted by the NFL-based learning algorithms. To reduce the computation complexity, PCA is used before the NFL-based learning algorithms. All the energy is retained in the PCA phase.

4.1   Experiments on ORL Face Database

In ORL face database, there are 10 images for each of the 40 human subjects, which were taken at different times, varying the lighting, facial expressions (open/closed eyes, smiling/not smiling) and facial details (glasses/no glasses). The images were taken with a tolerance for some tilting and rotation of the face up to $20°$. These images have size of 112×92. In this experiment, 5 images per person from the ORL database are selected randomly for training and the rest are used for testing. The system runs 20 times. Table 1 tabulates the maximum average recognition rate (MARR) of these algorithms on ORL face database. Clearly, MARR of the proposed algorithm is higher than other popular approaches.

Table 1. MARR of different algorithms on ORL face database

Algorithms	MARR	Feature dimension
fisherface	0.9154	39
PCA+NN	0.9271	40
PCA+NFL	0.9449	80
UDNFLA	0.8870	180
NFLS	0.9284	190
NDNFLA	0.9690	150
Proposed algorithm using J_{kNN}	0.9789	140
Proposed algorithm using J_{NFL}	0.9793	155

4.2   Experiments on AR Face Database

AR face database was created by Aleix Martinez and Robert Benavente in the Computer Vision Center (CVC) at the U.A.B. It contains over 4,000 color images corresponding to 126 people's faces (70 men and 56 women). Images feature frontal view faces with different illumination conditions, facial expressions, and occlusions (sun glasses and scarf). The pictures were taken at the CVC under strictly controlled conditions. Each person participated in two sessions, separated by two weeks (14 days) time. The same pictures were taken in both sessions. In

the following experiments, only nonoccluded images of AR face database are
selected. Five images per person are randomly selected for training and the
other images are for testing. This system also runs 20 times. Table 2 tabulates
the maximum average recognition rate (MARR) of these algorithms on AR face
database. Clearly, MARR of the proposed algorithm is higher than other popular
approaches.

Table 2. MARR of different algorithms on AR face database

Algorithms	MARR	Feature dimension
fisherface	0.9481	120
PCA+NN	0.7604	120
PCA+NFL	0.8521	190
UDNFLA	0.9353	120
NFLS	0.9126	190
NDNFLA	0.9690	150
Proposed algorithm using J_{kNN}	0.9802	160
Proposed algorithm using J_{NFL}	0.9784	150

5 Conclusion

A novel feature extraction scheme based on NFL is proposed in this paper. The
proposed algorithm can extract some local discriminant feature of the samples.
The parameter in the algorithm can be defined adaptively. Compared with UD-
NFLA, NDNFLA and so on, the proposed algorithm has the higher recognition
accuracy. Experimental results confirm the efficiency of the proposed algorithm.

References

1. Zhou, X., Nie, Z., Li, Y.: Statistical Analysis of Human Facial Expressions. Journal of Information Hiding and Multimedia Signal Processing 1, 241–260 (2010)
2. Jolliffe, I.T.: Principal Component Analysis. Springer, New York (2002)
3. Belhumenur, P.N., Hepanha, J.P., Kriegman, D.J.: Eigenfaces vs Fisherfaces: Recognition Using Class Specific Linear Projection. IEEE Trans. Pattern Analysis Machine Intelligence 19, 711–720 (1997)
4. Xu, Y., Yang, J.Y., Jin, Z.: A Novel Method for Fisher Discriminant Analysis. Pattern Recognition 37, 381–384 (2004)
5. Li, H., Jiang, T., Zhang, K.: Efficient and Robust Feature Extraction by Maximum Margin Criterion. IEEE Trans. Neural Networks 17, 157–165 (2006)
6. Xu, Y., Zhang, D., Jin, Z., Li, M., Yang, J.Y.: A Fast Kernel-Based Nonlinear Discriminant Analysis for Multi-Class Problems. Pattern Recognition 39, 1026–1033 (2006)
7. Li, J.B., Pan, J.S., Lu, Z.M.: Face Recognition Using Gabor-Based Complete Kernel Fisher Discriminant Analysis with Fractional Power Polynomial Models. Neural Comput. and Applic. 18, 613–621 (2009)

8. Xu, Y., Zhang, D.: Represent and Fuse Bimodal Biometric Images at the Feature Fevel: Complex-Matrix-Based Fusion Scheme. Opt. Eng. 49, 037002 (2010)
9. Li, S.Z., Lu, J.: Face Recognition Using the Nearest Feature Line Method. IEEE Trans. Neural Networks 10, 439–443 (1999)
10. Chien, J.T., Wu, C.C.: Discriminant Waveletfaces and Nearest Feature Classifiers for Face Recognition. IEEE Trans. Pattern Analysis and Machine Intelligence 24, 1644–1649 (2002)
11. Chen, K., Wu, T.Y., Zhang, H.J.: On the Use of Nearest Feature Line for Speaker Identification. Pattern Recognition Letters 23, 1735–1746 (2002)
12. Gao, Q.B., Wang, Z.Z.: Using Nearest Feature Line and Tunable Nearest Neighbor Methods for Prediction of Protein Subcellular Locations. Computational Biology and Chemistry 29, 388–392 (2005)
13. Zheng, Y.-J., Yang, J.-Y., Yang, J., Wu, X.-J., Jin, Z.: Nearest Neighbour Line Non-parametric Discriminant Analysis for Feature Extraction. Electronics Letters 42, 679–680 (2006)
14. Yang, Y., Yuan, Y., Li, X.: Generalised Nearest Feature Line for Subspace Learning. Electronics Letters 43, 1079–1080 (2007)
15. Lu, J., Tan, Y.P.: Uncorrelated Discriminant Nearest Feature Line Analysis for Face Recognition. IEEE Signal Processing Letter 17, 185–188 (2010)
16. Yan, L., Pan, J.S., Chu, S.C., Roddick, J.F.: Neighborhood Discriminant Nearest Feature Line Analysis for Face Recognition. In: Second International Conference on Innovations in Bio-inspired Computing and Applications (IBICA), pp. 344–347. IEEE Press, Shenzhen (2011)
17. Olivetti and Oracle Research Laboratory Face Database of Faces, http://www.cam-orl.co.uk/facedatabase.html
18. Martinez, A.M., Benavente, R.: The AR Face Database. CVC Technical Report 24 (1998)

A Novel Matching Strategy for Finger Vein Recognition

Rongyang Xiao, Gongping Yang[*], Yilong Yin, and Lu Yang

School of Computer Science and Technology, Shandong University,
Jinan, 250101, China
ygpsdu@126.com

Abstract. Finger vein recognition is a promising biometric recognition technology, which verifies identities via the vein patterns in the fingers. The vein vessel network is a very important vein pattern for finger vein recognition. Based on this pattern, in the matching stage, the matched pixel ratio (MPR), Hamming distance (HD) and the mismatched ratio are commonly used as the matching algorithms to evaluate the similarity between two finger vein images. But these matching algorithms are calculated pixel by pixel, they are sensitive to the image translation and rotation. In this paper, a novel matching strategy region-based axis projection (RAP) is proposed for finger vein recognition. We first divide the vein pattern into small regions, then concatenate the projection of the vein distribution curves on the x-axis and y-axis of each region, and finally evaluate the similarity by calculating the projections of the whole vein pattern. Experimental results show that the proposed method can avoid image translation and rotation to some extent and achieve a better performance.

Keywords: Finger vein recognition, matching algorithm, region-based axis projection.

1 Introduction

Biometric recognition, or simply biometrics, refers to the use of distinctive anatomical and behavioral characteristics or identifiers (e.g., fingerprints, faces, iris, voices, and hand geometries) for automatically recognizing an individual [1,2]. Recently, finger vein recognition was proposed and has been well studied. In [3], the authors prove that each finger has a unique vein pattern that can be used for personal verification. The finger vein recognition has some advantages over other hand-based biometric authentication techniques [4,5]:(1) non-contact: finger vein patterns are not influenced by surface conditions. Non-invasive and contactless data capture ensures both convenience and cleanliness for the users, and it is more acceptable for the users; (2) live body identification: finger vein patterns can only be identified on a live body; (3) high security: finger vein patterns are internal features that are difficult to forge; (4) small device size: as compared to palm vein based verification devices, most finger vein recognition devices are smaller in size.

---

[*] Corresponding author.

J. Yang, F. Fang, and C. Sun (Eds.): IScIDE 2012, LNCS 7751, pp. 364–371, 2013.

According to the extracted patterns, we can divide the finger vein extraction methods into two categories: gray distribution-based and vein vessel network-based. About the first kind of methods, Principal component analysis (PCA)[6], linear discriminant analysis (LDA)[7] and Two dimensional principal component analysis ((2D)2PCA) [8] first are adopted to extract finger vein features and then use back-propagation(BP) network, ANFIS and support vector machine (SVM) for pattern classification. Local binary pattern (LBP) and local derivative pattern (LDP) are two similar format translators, and they were used as feature extraction method for finger vein in [9]. Based on LBP, A general framework personalized best bit map (PBBM) is proposed to use the best bits only for recognition [10]. In the matching stage, they all use the Hamming distance to evaluate the similarity between two patterns.

However, the aforementioned methods are applied based on the fixed schema, which are indifferent to the really content within the image. That is two images are seen as the same only if they have the similar model whether it is a dog or a cat. These methods neglect the intrinsic distribution information of the vein vessel network and meanwhile increase the interference of the background. In order to better utilize the features from the segmented blood vessel network for recognition, [11] extracts the finger vein pattern from the unclear image with line tracking, which starts from various positions, then gain a dissimilarity between two patterns by using the mismatched ratio. In [12], the minutiae features including bifurcation points and ending points are extracted from these vein patterns. These feature points are used for geometric representation of the vein patterns' shape. A modified Hausdorff distance algorithm is provided to evaluate the identification ability among all possible relative positions of the vein patterns' shape. In [13] the authors propose a mean curvature method, which regards the vein image as a geometric shape and finds the valley-like structures with negative mean curvatures; then use the matched pixel ratio (MPR) to evaluate the similarity between two images. In [14] the patterns of finger veins are extracted by combining two segmentation methods, including Morphological Operation and Maximum Curvature Points in Image Profiles, subsequently, the mismatch ratio is used to quantify the differences of two patterns.

In the matching stage, the aforementioned matching algorithms (e.g. the mismatch ratio, MRP) are sensitive to the image translation and rotation, meanwhile, the distribution information of the vein vessel network is neglected. In order to diminish the influence by these factors, we propose a new matching strategy: Region-based Axis Projection (RAP). Experimental results show that the RAP method can better utilize the vein vessel network distribution information and avoid image translation and rotation to some extent.

The rest of this paper is organized as follows: Section 2 presents two previous commonly used matching algorithms and introduces the definition of the RAP method. Section 3 presents two experiments to verify the proposed method. Finally, Section 4 concludes this paper.

2   The Proposed Method

A typical finger vein recognition system mainly includes image capturing, image preprocessing, feature extraction and matching. A binary vein pattern (vein vessel network) is gained after the feature extraction as shown by Fig. 1(a), in which the white parts denote the vein vessel network and the black parts are background. Fig. 1 (b) shows that in the store matrix the vein patterns and background are represented by '1' and '0', respectively.

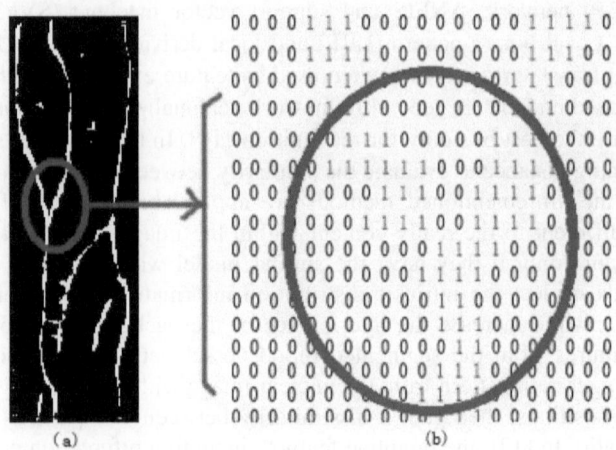

Fig. 1. A binary finger vein pattern and its store format in a matrix

2.1   Related Work

In the matching stage, as the patterns to be matched here are binary. In [13] the matched pixel ratio (MPR) is adopted to measure similarities between the two binary patterns. Let S denote the MPR between the input binary image i and the template t. It is represented as follows:

$$S = 2 \frac{\sum_{x,y} i(x, y) \cdot t(x, y)}{\sum_{x,y} i(x, y) + \sum_{x,y} t(x, y)} \qquad (1)$$

In [14], an enrolled pattern t is overlapped with the input vein image i. The values of overlapping pixels are compared pixel by pixel. The pairs of pixels, where one is a vein and the other a background pixel are counted. Such a pair is called a mismatch. The ratio of the number of mismatched pairs to the total number of vein pixels is defined as the mismatch ratio R_m, which can be expressed in Equation (2):

$$R_m = \frac{\sum_{x,y} t(x, y) \otimes i(x, y)}{\text{sizeof}(t)} \qquad (2)$$

In Equation (2), $\otimes$ is a Boolean exclusive-OR operator between two binary patterns. Because the S and R_m are calculated pixel by pixel, it is sensitive to the image translation and rotation extremely.

2.2   Region-Based Axis Projection

As mentioned above, the binary vein patterns in a store matrix are described with '0' and '1'. So the distribution information of '1' (the vein vessel network) can be a useful discriminator for two binary patterns. We first divide the binary vein pattern into small regions, the shape and the size of each region are determined according to the gathered sample which can be seen as two variable parameters. Here we select a rectangle which side length L is 20 pixels. Then in each region we describe the distribution information of '1' use Equation (3) and (4) as follows. In which B represents the selected region, B_h and B_v are the statistics of the frequency of '1' in the jth column and ith row, respectively.

$$B_h(j) = \sum_{i=1}^{L} B(i, j) \quad 1 \le j \le L \tag{3}$$

$$B_v(i) = \sum_{j=1}^{L} B(i, j) \quad 1 \le i \le L \tag{4}$$

In order to diminish the impact of image translation and rotation, we add another parameter T (statistic step length) into Equation (3) and (4), and then calculate the frequency of '1' not column by column (or row by row) but in a step length T. The modified Equation (3)' and (4)' are shown as follows.

$$B_h(k) = \sum_{j=(k-1)*T+1}^{j+T} \sum_{i=1}^{L} B(i, j) \quad 1 \le k \le \frac{L}{T} \tag{3'}$$

$$B_v(k) = \sum_{i=(k-1)*T+1}^{i+T} \sum_{j=1}^{L} B(i, j) \quad 1 \le k \le \frac{L}{T} \tag{4'}$$

In fact, the B_h is a L-length (or K-length) row vector, in which stores the calculated frequency information of '1'. It can be interpreted as a discrete distribution curve projection on the x-axis. In the same way, the B_v can be interpreted as a discrete distribution curve projection on the y-axis. If we regard the location of x-axis or y-axis as the independent variable and the corresponding statistic value as the dependent variable, we can get a histogram-like distribution graph. Subsequently we concatenate the x-axis projection and y-axis projection of each region together, and then evaluate the similarity between two binary patterns by histogram intersection or evaluate the dissimilarity by calculating the Euclidean distance. The procedure of this matching algorithm is illustrated in Fig. 2.

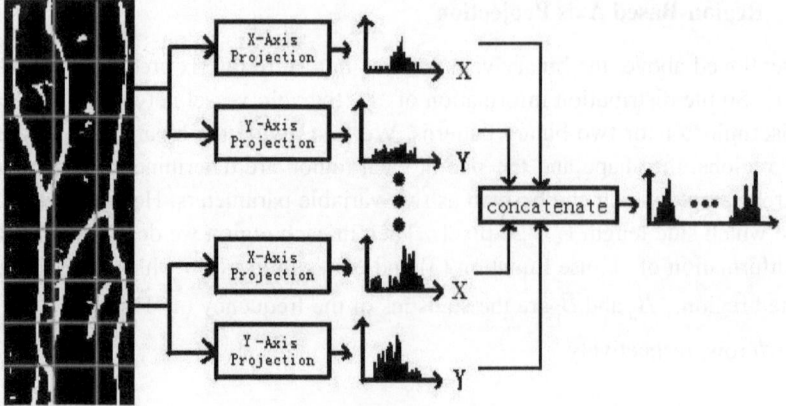

Fig. 2. The procedure of proposed Region-based Axis Projection

3   Experimental Results and Analysis

We conduct the experiments in the verification mode using our finger vein database which collects 106 fingers, where each finger contributes 14 finger vein images. In order to avoid class imbalances, we use all the 14 vein images in intra-class matching meanwhile the first 6 vein images in interclass matching. Consequently, there are 9,646 ($106\times13\times14/2$) intra-class matching and 200,340 ($6\times105\times6\times106/2$) interclass matching in total. In this paper, the performance of our proposed method is evaluated by the EER (equal error rate). EER is the error rate when the FRR (false rejection rate) equals the FAR (false acceptance rate) and which is suited for measuring the overall performance of biometrics systems because the FRR and FAR are treated equally.

3.1   Experiment 1

The original spatial resolution of the finger vein image is 320×240, After ROI (region of interesting) extraction and size normalization, the size of the region used for feature extraction is reduced to 240×80 which are shown in Fig. 3(a) and (b), respectively. Then we extract the finger vein patterns using three methods: local thresholding method, maximum curvature points (MCP) method and mean curvature method. Fig. 3(c)~(e) are the binary vein patterns extracted by various methods. Fig. 3(c) is by the local thresholding method; Fig. 3(d) is by the MCP method; Fig. 3(e) is by the mean curvature method.

In the matching stage, we use the matched pixel ratio (MPR) and Region-based Axis Projection (RAP) to evaluate the performance of aforementioned three vein extraction methods respectively. The ROC curves are shown in Fig. 4. From Fig. 4 we can see that based on the same extraction algorithm the RAP method achieves a much lower EER than the MPR method. This indicates that the RAP method can better gather vein vessel network statistic distribution information and illustrate the differences between the individuals.

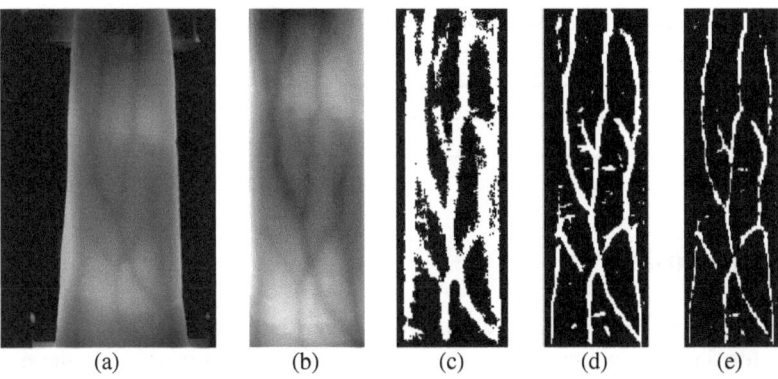

(a)   (b)   (c)   (d)   (e)

Fig. 3. Original image, normalization and vein patterns extracted by various methods

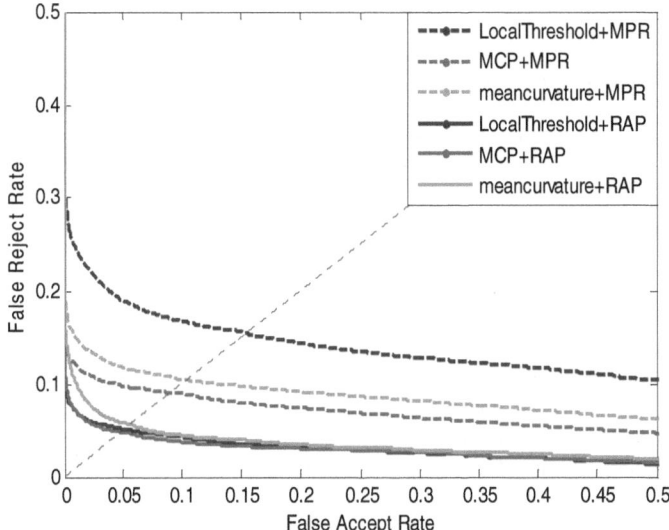

Fig. 4. ROC curves by different matching algorithms

3.2   Experiment 2

This experiment is conduct to show the advantages of the proposed method in rotation and translation compared with the MPR method. We first construct three simulated rotated databases based on the source database with rotating all the samples within different rotated degree ranges (e.g., 1° 3° 5°). For example, in the simulated 3°database, the rotated degree may range from 0° to 3°. Then we extract the vein pattern using the mean curvature and take the MPR and the RAP as the matching algorithm respectively. Table 1 shows the EER of the RAP method and the MPR method in different rotated databases. From Table 1 we can see the RAP method can overcome translation and rotation better compared with the MPR method. Besides, the EER keeps stable in small degree rotation database.

Table 1. Matching results (EER) with different rotated degree

Rotated degree	MPR	RAP
1°	0.1509	0.0561
3°	0.1679	0.0603
5°	0.1814	0.0810

4 Conclusion

In this paper we propose a new matching algorithm region-based axis projection (RAP) for finger vein recognition. Based on the binary vein patterns, in the matching stage, this method can better utilize the vein vessel network distribution information and the divided regions are independence with each other, so it is robustness to the image translation and rotation. Experimental results show that a better performance is obtained by this method.

Acknowledgments. This work is supported by National Natural Science Foundation of China under Grant No.61173069 and 61070097. The authors would like to thank the anonymous reviewers for their helpful suggestions.

References

1. Ross, A.A., Nandakumar, K., Jain, A.K.: Handbook of Multibiometrics, 1st edn. Springer, Berlin (2006)
2. Maltoni, D., Maio, D., Jain, A.K., Prabhakar, S.: Handbook of Fingerprint Recognition, 2nd edn. Springer, Berlin (2009)
3. Yanagawa, T., Aoki, S., Ohyama, T.: Human finger vein images are diverse and its patterns are useful for personal identification. MHF Prepr. Ser. 12, 1–7 (2007)
4. Liu, Z., Yin, Y.L., Wang, H.J., Song, S.L., Li, Q.L.: Finger vein recognition with manifold learning. J. Netw. Comput. Appl. 33, 275–282 (2010)
5. Wu, J.D., Ye, S.H.: Driver identification using finger-vein patterns with radon transform and neural network. Expert Syst. Appl. 36, 5793–5799 (2009)
6. Wu, J., Liu, C.-T.: Finger-vein pattern identification using principal component analysis and the neural network technique. Expert Systems with Applications 38, 5423–5427 (2011)
7. Wu, J., Liu, C.-T.: Finger-vein pattern identification using SVM and neural network technique. Expert Systems with Applications 38, 14284–14289 (2011)
8. Guan, F., Wang, K., Liu, J., et al.: Bi-direction weighted (2D)2PCA with eigenvalue normalization one for finger vein recognition. Pattern Recognition and Artificial Intelligence 24, 417–424 (2011)
9. Lee, E.C., Jung, H., Kim, D.: New finger biometric method using near infrared imaging. Sensors 11, 2319–2333 (2011)
10. Yang, G., Xi, X., Yin, Y.: Finger Vein Recognition Based on a Personalized Best Bit Map. Sensors 12, 1738–1757 (2012)

11. Miura, N., Nagasaka, A., Miyatake, T.: Feature extraction of finger-vein patterns based on repeated line tracking and its application to personal identification. Mach. Vis. Appl. 15, 194–203 (2004)
12. Yu, C.B., Qin, H.F., Zhang, L., Cui, Y.Z.: Finger-vein image recognition combining modified hausdorff distance with minutiae feature matching. J. Biomed. Sci. Eng. 2, 261–272 (2009)
13. Song, W., Kim, T., Kim, H.C., Choi, J.H., Kong, H.J., Lee, S.R.: A finger-vein verification system using mean curvature. Patt. Recogn. Lett. 32, 1541–1547 (2011)
14. Hoshyar, A.N., Sulaiman, R., Houshyar, A.N.: Smart access control with finger vein authentication and neural network. J. Am. Sci. 7, 192–200 (2011)

A Fast Vanishing Point Detection Method in Structured Road

Ting Zhao, Wankou Yang, and Changyin Sun

School of Automation, Southeast University, Nanjing 210096, China
zhao_ting1987@sina.com, {wkyang,cysun}@seu.edu.cn

Abstract. Vanishing point detection plays a very important role in road detection, since road boundaries pass the vanishing point and the line of boundaries can be extracted by the information of the vanishing point location. In this paper, a fast vanishing point detection method based on texture features is presented for structured road, which uses joint activities of only four Gabor filters to estimate the local dominant orientation at each pixel and votes to estimate the location of vanishing point. Furthermore, a step of edge detection is adopted before vanishing point voting which greatly reduce the computational. Experimental results show that the proposed algorithm has a good performance on the detection of the vanishing point of structured road.

Keywords: Gabor transform, edge detection, vanishing point detection.

1 Introduction

In the field of intelligent transportation research, computer vision-based lane detection is a key technology to implement the functions such as lane departure warning and lane keeping. So far, a variety of different vision-based lane detection algorithms have been proposed[1]and most of these methods can be classified into three categories, which are edge based methods, region based[2]methods and texture based methods.

Edge based method is more applicable for structured road and it is difficult for region based method to recognize the road due to variant illuminations and inclement weather[3]. To overcome the problems, a detection method based on texture features was presented. Texture based methods find the local oriented textures[4]-[7] first and vote for the location of the road's vanishing point. A location with the most votes is considered as the vanishing point of the road. The information of vanishing point can be used to verify the boundries of the road, thus it can improve lane detection. Here we mainly use this kind of method.

According to the salient features of the structured road edge information, first, the method of joint activity of four Gabor filters to estimate the local dominant orientation is used to detect the dominant orientation of grayscale images. Second, we use OTSU threshold segmentation and LoG(Laplassian Gaussian algorithm) edge detector to detect the edge of grayscale images. Finally, we vote the points in the edge with the weight strategy to estimate robust vanishing point.

J. Yang, F. Fang, and C. Sun (Eds.): IScIDE 2012, LNCS 7751, pp. 372–379, 2013.

2   Vanishing Point Detection Based on Gabor Transform[8]

2.1   Local Dominant Orientation Estimation

In this paper, we use Gabor filters to estimate local dominant orientation at each pixel. Gabor transform is a windowed Fourier, and a 2D Gabor kernel g for the orientation φ_n and radial frequency $\omega_0 = 2\pi / \lambda$ [9] can be written as

$$g_{\omega_0, \varphi_n}(x, y) = \frac{\omega_0}{\sqrt{2\pi}K} e^{-\frac{\omega_0^2}{8K^2}(4a^2 + b^2)} \cdot [e^{i\omega_0 a} - e^{-\frac{K^2}{2}}] \tag{1}$$

where $a = x\cos\varphi_n + y\sin\varphi_n$, $b = -x\sin\varphi_n + y\cos\varphi_n$, $K = \pi/2$, $\lambda = 4\sqrt{2}$. To estimate the local dominant orientation of each pixel, the grayscale image $I(p)$ is convolved with four Gabor filters with predefined orientations,

$$\hat{I}_{\varphi_n}(p) = I(p) \otimes g_{\varphi_n}(p)$$

$$\varphi_n = \frac{(n-1)\pi}{4} \quad \text{n=1, 2, 3, 4} \tag{2}$$

The Gabor energy is calculated as the magnitude of the complex filter responses as follows:

$$E_{\varphi_n}(p) = \sqrt{\text{Re}(I_{\varphi_n}(p)^2 + \text{Im}(\hat{I}_{\varphi_n}(p))^2} \tag{3}$$

As every vector in R^2 can be represented by a combination of two linearly independent vectors, every local texture orientation in the image represented by a vector can be obtained by two independent Gabor energy which strongly respond to the texture vector orientation. More specifically, a grayscale image is convolved with four oriented Gabor filters g_φ, $\varphi \in \{0°, 45°, 90°, 135°\}$, and Gabor energy responses are computed for each pixel. Each energy response is considered as a vector whose magnitude is Gabor energy response value and its direction corresponds to its preferred orientation φ. Then, these Gabor energy responses E_{φ_n} are sorted based on their magnitudes in descending order($E_\varphi^1 > E_\varphi^2 > E_\varphi^3 > E_\varphi^4$) for each pixel. In this way, we can use vector V of the two most dominant filter activation strengths to represent the local texture orientation as follows:

$$V(p) = V_x(p) + jV_y(p) = \sum_{i=1}^{2} E_\varphi^i(p)e^{j\varphi_i} \tag{4}$$

where φ_1 and φ_2 represent the corresponding angle of the two dominant Gabor energy responses $E_\varphi^1(p)$ and $E_\varphi^2(p)$ respectively. Then the estimated dominant orientation $\hat{\theta}$ at pixel p is determined by

$$\hat{\theta}(p) = \tan^{-1}\frac{V_y(p)}{V_x(p)} \tag{5}$$

However, if the pixel has no apparent dominant orientation such as background pixel, the Gabor energy response values maybe very similar for all four orientations. In such a case, relying on only two strongest filter responses may result in an estimation error. To solve this problem, we can use joint activities of all four Gabor energy responses by introducing two new vectors $S_1(p)$ and $S_2(p)$ defined as follows:

$$\|S_1(p)\| = E_\varphi^1(p) - E_\varphi^4(p) \qquad \varphi_{S1}(p) = \varphi_1(p) \tag{6}$$

$$\|S_2(p)\| = E_\varphi^2(p) - E_\varphi^3(p) \qquad \varphi_{S2}(p) = \varphi_2(p) \tag{7}$$

Then, V can be represented by these two new vectors as follows:

$$V(p) = \sum_{i=1}^{2} S_i(p)e^{j\varphi_{Si}} \tag{8}$$

In that way, the local dominant orientation at pixel p can be defined by (5).

2.2   Vanishing Point Voting

The dominant orientation $\hat{\theta}(p)$ of pixel p can be obtained by the above steps, then a ray whose vertex is p and direction is $\hat{\theta}(p)$ can be drawn in the accumulation space which is initialized to zero and whose size is the same as the original image. If the pixel falls on the ray, the corresponding accumulator increases one. After counting all rays, the pixel location with the most votes is the vanishing point. Since there is a relation between position of vanishing point and road, we provide some strategies for voting patterns as follows:

A. Weighting Method
Because the vanishing point is obtained by votes of most nearly vertical rays, the rays which is close to horizontal direction should be suppressed. To get this goal, a trigonometric function of its dominant orientation $\sin\hat{\theta}(p)$ is assigned to each ray. Thus, a ray which is more close to horizontal direction has a weight closer to zero, which can effectively suppress the ray which is close to horizontal and horizontal direction.

B. Voting scheme
Although the weighting method mentioned above can solve a part of problems, since it allow only the pixels to vote upward, the pixels in the upper part of the image receive more votes than lower image pixels[10]. To address this problem, we propose a new distanced-based voting scheme. The distance function is computed as

$$y_i(\hat{d}) = e^{-\frac{\hat{d}^2}{2\sigma^2}}, \quad \hat{d} = \frac{d}{D_p}, \quad d = \sqrt{(x-x_i)^2 + (y-y_i)^2} \tag{9}$$

where D_p is the distance between the ray vertex and the intersection of the ray and the image boundries, d is the distance between the ray vertex and any point (x_i, y_i) on its ray, $\hat{d}$ is the normalized distance, and σ^2 is the variance that is experimentally set as 0.25. Road in our testing images is in two-thirds part of the image below and the location of vanishing point is approximately in the middle third part, so point on the ray obtained by pixels of lower third part get more votes if it is farther from the ray vertex; point on the ray obtained by pixels of middle third part is just as the opposite.

The final weight can be defined as follows:

$$weight = \begin{cases} \sin \hat{\theta}(p) + y_i(\hat{d}), & y > height / 3 \\ \sin \hat{\theta}(p) + 1/ y_i(\hat{d}), & y < height / 3 \end{cases} \tag{10}$$

3   Edge Detection Based on OTSU Threshold Segmentation and LoG Edge Detector

3.1   OTSU Threshold Segmentation

OTSU[11]has a good segmentation performance when the criterion function of the image interclass variance is unimodal. Its theory is as follows: given that t is the threshold, the ratio of foreground to image is w_0, and the average gray value of foreground is u_0; the ratio of background to image is w_1, and the average gray value of background is u_1. So the mean gray value of the image is as follows: $u=w_0*u_0+w_1*u_1$, the variance of foreground and background of the image is as follows: $g=w_0*(u_0-u)^2+w_1*(u_1-u)^2$, make t take the value from minimum gray value to the maximum and the value which make g the biggest is the optimal threshold. This threshold segmentation method can distinguish foreground from background very well and can meet the requirements of subsequent edge detection. Fig.1 and Fig.2 show the original grayscale image and segmentation image respectively.

3.2   LoG Edge Detection

The edge detection algorithms based on zero-crossing point of the image second derivative (e.g. Laplace operator) are very sensitive to noise, to solve this problem Marr and Hildreth filter out the noise before edge enhancement and combined Gaussian filtering and Laplace edge detection which is known as LoG [12](Laplacian of Guassian).

2D Guassian filter response function is as follows:

$$G(x, y) = \frac{1}{2\pi\sigma^2} \exp(-\frac{x^2 + y^2}{2\sigma^2}) \tag{11}$$

where σ is space scale factor.

Given that $f(x, y)$ is a grayscale image, we interchange the convolution and derivation in linear system and the equation is as follows:

$$\nabla^2[G(x, y) * f(x, y)] = [\nabla^2 G(x, y)] * f(x, y) \tag{12}$$

Gaussian smoothing filter can be combined with Laplace differential operator into a convolution operator[13]as follows:

$$\nabla^2 G(x,y) = \frac{1}{2\pi\sigma^4}(\frac{x^2 + y^2}{\sigma^2} - 2)\exp(-\frac{x^2 + y^2}{2\sigma^2}) \tag{13}$$

In the practical application, we first select a spacial scale factor for an image to get the LoG template, and then calculate the convolution between the image and the template. The common LoG operator of 5×5 is as follows:

$$\begin{bmatrix} -2 & -4 & -4 & -4 & -2 \\ -4 & 0 & 8 & 0 & -4 \\ -4 & 8 & 24 & 8 & -4 \\ -4 & 0 & 8 & 0 & -4 \\ -2 & -4 & -4 & -4 & -2 \end{bmatrix}$$

We can get second-order gradient image by the convolution of convolution kernel and the image. In this paper, OTSU threshold segmentation method is used to preprocess the image first and as road in our testing images is in roughly two-thirds part of the image below, we detect edges in only two-thirds part of the image below. Then for pixel whose gradient value is negative, we set its grayscale value zero and for the rest, we set its grayscale value 255. Fig. 1 show a gray road image, Fig. 2 shows a threshold image and the corresponding edge image is shown in Fig. 3.

Fig. 1. Original image   **Fig. 2.** OTSU segmentation   **Fig. 3.** Edge image
image

4   Experiment

4.1   Our Algorithm

Step 1: Get dominant orientation of each pixel by Gabor convolution with the grayscale image of road;
Step 2: Detect edges of grayscale image by OTSU segmentation and LoG edge operator;
Step 3: Vote for dominant orientation of each edge points in edge image using some voting strategies, the point with the most votes is the vanishing point location.

4.2   Experimental Results

150 structured road of 352×288 that all photographed by camera in front of the car are used to evaluate our algorithm. To reduce the computing time, the original data set are first downsized to 88×72 by the Guassian pyramid, and then the proposed vanishing point detection method is implemented to find the vanishing point location. The first row images in Fig. 4 show the voting information, and the corresponding vanishing points are shown in the second row in Fig. 4.

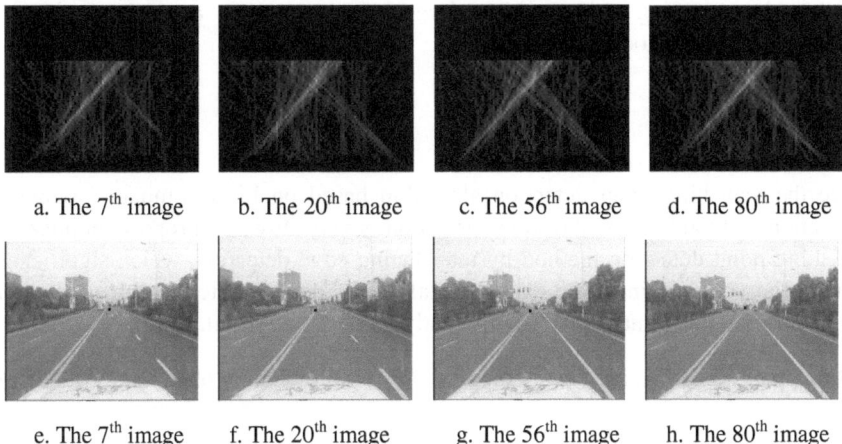

a. The 7th image   b. The 20th image   c. The 56th image   d. The 80th image

e. The 7th image   f. The 20th image   g. The 56th image   h. The 80th image

Fig. 4. (a)~(d) are accumulator spaces, (e)~(h) are the vanishing points that found by our algorithm

In the top of the image, the red dot is the ground truth vanishing points and the black dot is the estimated vanishing point. We can find that the error between the two points is very small. Experimental results show that our method can estimate the vanishing point locations successfully in about 80% structured road images and we evaluate the performance of the method by equation in (14). Also it has a faster speed than the method in Ref [8] and experimental results are shown in Table 1.

Table 1. Rate Comparison

	The 7th image	The 20th image	The 56th image	The 80th image
Ref [8]	59.563	70.000	60.219	57.234
Our method	4.656	4.641	5.157	5.234

To measure the vanishing point estimation error, we use the normalized Euclidean distance where the Euclidean distance is normalized by the size of the diagonal of the image resolution as follows[8]:

$$NormDist = \frac{\|P - P_0\|}{diag} \tag{14}$$

where P_0 is the estimated vanishing point location and P is the ground truth vanishing point location, while diag is the size of the diagonal of the image. We can see that the smaller the value is, the estimated vanishing point location is closer to the location of the ground truth. By calculating the Normlist, our method performs well in which 80% images have the Normlist of less than 0.1.

According to results shown above, our method is ten times faster than the method in Ref [8]. As we have added a step of edge detection before vanishing point voting step which reduces voting points with a large margin, we get a more considerable processing rate with good performance.

5   Conclusion

Since the vanishing point detection algorithm based on local dominant orientation estimation method using Gabor transformation is very slow, we propose an improved vanishing point detection method by introducing edge detection, which significantly reduces the voting points. The experimental results demonstrate that our method can achieve a good performance and significantly increase the speed.

Acknowledgment. This project is supported by NSF of China (61005008, 61273023).

References

1. Thorpe, C., Hebert, M., Kanade, T., Shafer, S.: Vision and navigation for the Carnegie-Mellon Navlab. IEEE Trans. Pattern Anal. Mach. Intell. 10(33), 362–373 (1988)
2. Alon, Y., Ferencs, A., Shashua, A.: Off-road path following using region classification and geometric projection constraints. In: Proc. IEEE Conf. Comput. Vis. Pattern Recog., pp. 689–696 (2006)
3. Thrun, S., Montemerlo, M., Dahlkamp, H., Stavens, D., Aron, A., Diebel, J., Fong, P., Gale, J., Halpenny, M., Hoffmann, G.: Stanley: The robot that won the DARPA grand challenge. J. Field Robot. 23(9), 661–692 (2006)

4. Nieto, M., Salgado, L.: Real-Time Vanishing Point Estimation in Road Sequences Using Adaptive Steerable Filter Banks. In: Blanc-Talon, J., Philips, W., Popescu, D., Scheunders, P. (eds.) ACIVS 2007. LNCS, vol. 4678, pp. 840–848. Springer, Heidelberg (2007)
5. McCall, J.C., Trivedi, M.M.: Video-based lane estimation and tracking for driver assistance:Survey, system, and evaluation. IEEE Trans. Intell. Transp. Syst. 7(1), 20–37 (2006)
6. Kong, H., Audibert, J., Ponce, J.: General road detection from a single image. IEEE Trans. Image Process. 19(8), 2211–2220 (2010)
7. Rasmussen, C.: RoadCompass: Following rural roads with vision+ladar using vanishing point tracking. Auton. Robots 25(3), 205–229 (2008)
8. Moghadam, P., Starzyk, J.A., Wijesoma, W.S.: Fast vanishing point detection in unstructured environments. IEEE Trans. Image Process. 21(1), 425–430 (2012)
9. Lee, T.: Image representation using 2D Gabor wavelets. IEEE Trans. Pattern Anal. Mach. Intell. 18(10), 959–971 (1996)
10. Rasmussen, C.: Grouping dominant orientations for ill-structured road following. In: Proc. IEEE Conf. Comput. Vis. Pattern Recog., pp. 470–477 (2004)
11. Otsu, N.A.: Threshold selection method from gray-level histogram. IEEE Transactions on Systems, Man and Cybernetics 9(1), 62–66 (1979)
12. Gonzalez Rafael, C., Woods Richard, E.: Digital Image Pracessing, 2nd edn., pp. 581–585. Publishing House of Electronics Industry, Beijing (2003)
13. Yang, D., Li, J., Bian, Z.: A Research on Edge Detection by Marr Algorithm. Journal of Image and Graphics 11(6), 823–826 (2006)

Tetrolet Regularization and Learning for Single Frame Image Super-Resolution

Liang Xiao, Heng Li, Huixia Wang, and Liqian Wang

School of Computer Science, Nanjing University of Science and Technology,
210094 Nanjing, China
xiaoliang@mail.njust.edu.cn

Abstract. A single frame image super-resolution reconstruction technique is proposed with two stages contains tetrolet regularization and tetrolet learning. In the first stage, the tetrolet regularization is used to estimate an initial high-resolution image. In the second stage, the tetrolet coefficients at finer scales of the estimated high-resolution image are learned locally from a set of high-resolution training images. Finally the fusion of tetrolet reconstruction produces the super-resolution image. Experimental results demonstrated that the proposed method outperforms state-of-the-art super-resolution methods in terms of PSNR index and visual quality.

Keywords: Image Super-resolution, Regularization, Tetrolet Learning, Example based Learning, Texture Synthesis.

1   Introduction

Super-resolution (SR) refers to the process of producing a high resolution (HR) image than what is afforded by the physical sensor through post-processing, making use of single or multiple frame low-resolution (LR) images. It includes up-sampling the image, thereby increasing the maximum spatial frequency, and removing degradations that arise during the image capture, namely, aliasing and blurring. Super-resolution (SR) has been one of the most active research areas. Generally, SR techniques can be divided into two broad categories: reconstruction-based methods and learning-based methods [1].

In reconstruction-based methods, single or several LR images are used to recovery a HR image. The basic idea of reconstruction-based method is to treat SR as an inverse problem, thus regularization method is often used to overcome the ill-posed [2]. Researchers have produced many extended algorithms, such as nonlocal-means (NLM) based approach [3], multidimensional kernel regression based approach [4], the joint formulation of reconstruction and registration [5], etc. The reconstruction-based method can deal with blurring and noise very well; however, their use has limitations in the texture and details recovery.

In learning-based method, the new information for predicting the HR image is obtained from a set of training images rather than from the subpixel shifts among LR observations. The idea behind the methods is that lots of self-similarity exists in an image. Joshi and Chaudhuri [6] have proposed a learning-based method for image

J. Yang, F. Fang, and C. Sun (Eds.): IScIDE 2012, LNCS 7751, pp. 380–389, 2013.

super-resolution from zoomed observations. They model the high-resolution image as a Markov random field (MRF), the parameters of which are learned from the most zoomed observation. In [7], they have proposed a single-frame super-resolution algorithm using a wavelet-based learning technique where the HR edge primitives are learned from the HR data set locally. Later, they extend this method from wavelet coefficients learning to contourlet coefficients learning. Their experiments shown that contourlet leaning is able to learn a high-resolution representation of oriented edge primitive and partial textures from training data. However, the limitations inherent in learning-based methods, when the blurring occurs in the LR image.

To take full advantage of reconstruction and learning based approaches, in this paper, we propose a two stage framework combined tetrolet learning and regularization for zoomed-out and linear space-invariant noisy blurred LR image frame which subsume several well-known SR models. In the first stage, we use a compound model to estimate a HR image, in which sparse prior and total variation are combined to formulate the SR image reconstruction as an optimization problem. In the second stage, the tetrolet [8] coefficients at finer scales of the estimated high-resolution image are learned locally from a set of high-resolution training images, thus the tetrolet reconstruction of which produces the super-resolution image.

The remainder of the paper is organized as follows. In Section 2 we present the image observation model of the SR problem. We discuss the proposed tetrolet regularization and learning approach In Section 3. We present experimental results on different types of images in Section 4, and the paper concludes with Section 5.

2   Problem Formulation

We start by describing the problem model. The image observation model is employed to relate the desired referenced HR image to all the observed LR images. The image observation model is employed to relate the desired referenced HR image to all the observed LR images. Consider the desired HR image of size $N \times N$ written in lexicographical notation as the vector $\mathbf{u} = [u_1, u_2 .... u_{N^2}]$. If $\mathbf{f}$ denotes the $M^2 \times 1$ lexicographical ordered vector containing pixels from the low-resolution observation, then it can be modeled as

$$\mathbf{f} = \mathbf{SHu} + \mathbf{n} \tag{1}$$

Where $\mathbf{H}$ and $\mathbf{S}$ are the blur and decimation operator. For a decimation factor of q, $N = qM$ and the decimation matrix $\mathbf{S}$ consists of q^2 nonzero elements of value $1/q^2$ along each row at appropriate locations and has the form [6-7]:

$$\mathbf{S} = \frac{1}{q^2} \begin{bmatrix} 11...1 & & 0 \\ & 11...1 & \\ & ... & \\ 0 & & 11...1 \end{bmatrix} \tag{2}$$

The recovery of $\mathbf{u}$ from $\mathbf{f}$ is thus an inverse problem, combining motion compensation, denoising, deblurring, scaling-up operation, all merged to one. The quality of the desired SR image depends on the assumption that $\mathbf{S}, \mathbf{H}$ are known, or the accuracy in estimating the degraded operators. The decimation $\mathbf{S}$ is dependent on the resolution scale-factor we aim to achieve, and as such, it is easily fixed. In most cases, the blur $\mathbf{H}$ refers to the camera point spread function (PSF), and therefore it is also accessible. Even if this is not the case, the blurring is typically dependent on few parameters, and those, in the worst case, can be manually set.

3   Tetrolet Regularization and Learning: A Two Stage Approach

3.1   Tetrolet Transform

We start with a brief review of the tetrolet transform [8], which is a geometric adaptive wavelet transform. Tetrolet is an extension of Haar-type wavelets whose supports are the shapes called tetrominoes. The tetrominoes are some geometric shapes in the famous computer game 'Tetris'. All the tetrominoes are made by connecting four equal-sized squares. For the tetrolets, it has been proved that there are 117 solutions for disjoint covering of a 4 × 4 board with any four tetrominoes. Fig.1 shows the 22 fundamental forms tiling a 4 × 4 board disregarding rotations and reflections.

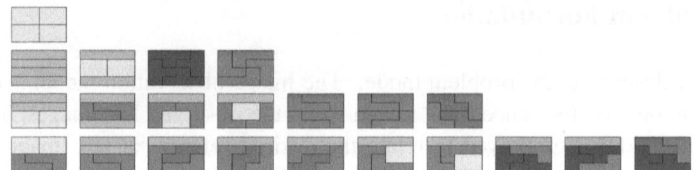

Fig. 1. The 22 fundamental forms tiling a 4 × 4 board disregarding rotations and reflections

When apply the tetrolet transform to an image $\mathbf{a} = [a(i,j)]_{i,j=1}^{n}$, the image will be divided into $4{\times}4$ blocks. Depending on the local structure in the block, each block will be covered by four of the five tetrominoes. These four tetrominoes, as the adaptive basis, are denoted by $\{I_0, I_1, I_2, I_3\}$ and their corresponding order can be set as $\{0,1,2,3\}$ by a bijective mapping L. According to these definitions, the tetrolets are defined as the following discrete basis functions

$$\phi_{I_v}[i,j] := \begin{cases} 1/2, & (i,j) \in I_v, \\ 0, & \text{else} \end{cases}, \tag{3}$$

$$\psi_{I_v}^l[i,j] := \begin{cases} \varepsilon[l, L(i,j)], & (i,j) \in I_v, \\ 0, & \text{else} \end{cases}, \tag{4}$$

where $l = 1, 2, 3$ and the function values $\varepsilon[l, L(i, j)]$ come from the Haar wavelet transform matrix

$$W := \left(\varepsilon[m,n]\right)_{m,n=0}^{3} = \frac{1}{2}\begin{pmatrix} 1 & 1 & 1 & 1 \\ 1 & 1 & -1 & -1 \\ 1 & -1 & 1 & -1 \\ 1 & -1 & -1 & 1 \end{pmatrix}. \tag{5}$$

The two-dimensional Haar wavelets fix four 2×2 squares for disjoint covering of a 4×4 board. The advantage of the tetrolets is that the tetrolets allow more partitions than the conventional Haar wavelets such that the tetrolet basis can adapt to local structures.

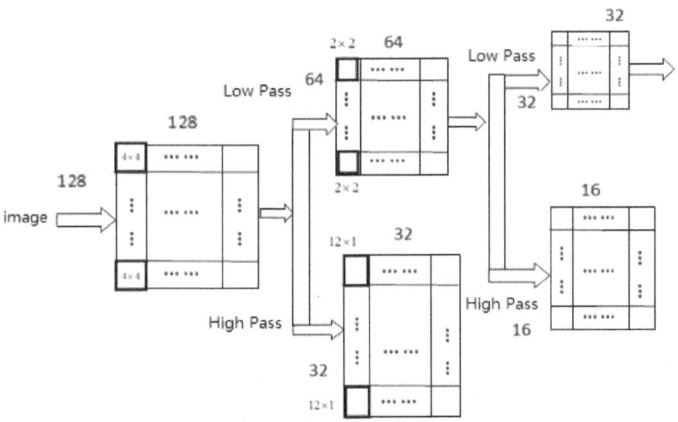

Fig. 2. The pipeline of the tetrolet transform

In order to decompose an image into a sparse representation, the sparse transform coefficients should be obtained. In the tetrolet filter bank, a two dimensional Haar wavelet transform will be applied to the four tetrominoes $\{I_0, I_1, I_2, I_3\}$ to obtain four lowpass and twelve highpass coefficients in each 4×4 block. Furthermore if the tetromino covering of the 4×4 block matches the corresponding local structure of an image, the obtained tetrolets coefficients become sparsest. The tetrolet transform finds the most appropriate covering in the 117 solutions such that the L^1 norm of the twelve highpass coefficients becomes minimal. Thus the tetrolet transform produces more sparse coefficients. Different coverings of these free tetrominoes characterize different local structures of an image. The algorithm of tetrolet transform contains 4 steps as illustrated in Fig.2: 1). divide the image into 4×4 blocks; 2) find the sparsest tetrolet representation in each block; 3) rearrange the low- and high-pass coefficients of each block into a 2×2 block; 4) store the tetrolet coefficients (high-pass part); 5) apply step 1 to 4 to the low-pass image for further decomposition. For more details, see [8].

3.2   A Two Stage Framework

As illustrated in Fig.3, the pipeline of the proposed approach has the following two stages:1) Initial estimate of the HR image by tetrolet regularization; 2) Refinement by tetrolet learning from a set of HR examples. For stage one, we use a compound regularization model inspired by our recent work in [9]. For stage two, the estimated HR image is further refined by the tetrolet learning. The advantage of this stage is the subtle texture will be recovered by the tetrolet learning from the HR images set.

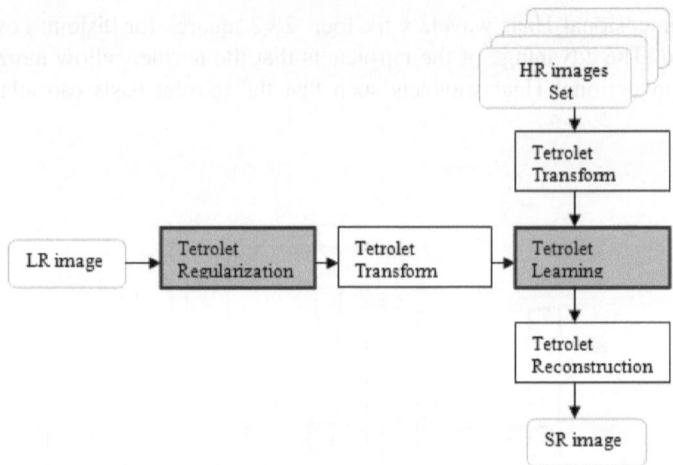

Fig. 3. The proposed two stage framework for image super-resolution

3.3   Tetrolet Regularization to Estimate the Initial HR Image

In general, smoothness and sparsity are two kinds of image prior for image modeling, and both of them have been applied broadly in image restoration, super-resolution and image segmentation, etc.. On one hand, the smoothness regularization can effectively remove noise and keep sharp edges, but they often eliminate the local details such as textures; on the other hand, the sparsity priori based regularization can protect the local details very well. Therefore, in order to achieve good performance in both noise removal and fine-scale details preservation, we combine these two regularizations to formulate a new compound regularization model for image restoration as follows:

$$\mathbf{u} = \arg\min_{u}\left\{ R_{TV}(\mathbf{u}) + \lambda_1 \left\| \Psi_{tetrolet}\mathbf{u} \right\|_1 + \frac{\lambda_2}{2}\left\| \mathbf{S}\mathbf{H}\mathbf{u} - \mathbf{f} \right\|_2^2 \right\} \tag{6}$$

Here, the first term $R_{TV}(\mathbf{u})$ is the TV regularization term. In this paper, the TV term is defined by $R_{TV}(\mathbf{u}) = \sum_i \sqrt{(D_h\mathbf{u})_i^2 + (D_v\mathbf{u})_i^2}$, where D_h and D_v are the horizontal and the vertical gradient operators respectively. The second term $\left\| \Psi_{tetrolet}\mathbf{u} \right\|_1$ is the sparsity priori based regularization. In this paper, we let $\Psi_{tetrolet}$ denotes the tetrolets basis. The third term is the data fidelity term to control the measurement error between the LR image and the unknown SR image. By using the variable splitting method [10], Eq. (6) can be rewritten as

$$u = \arg\min_{u}\left\{\|\mathbf{w}\|_1 + \lambda_1\|\Psi_{tetrolet}\mathbf{u}\|_1 + \frac{\lambda_2}{2}\|\mathbf{SHu}-\mathbf{f}\|_2^2\right\} \qquad s.t. \quad \mathbf{w} = D\mathbf{u} \qquad (7)$$

where $\mathbf{w}$ is an auxiliary variable which denotes the approximation of $D\mathbf{u}$, $R_{sparse}(u)=\|\Psi_{tetrolet}u\|_1$. Using the Lagrange multiplier method, (7) can be rewritten as

$$(\mathbf{u},\mathbf{w}) = \arg\min_{u,\mathbf{w}}\left\{\|\mathbf{w}\|_1 + \frac{\beta}{2}\|\mathbf{w}-D\mathbf{u}\|_2^2 + \lambda_1\|\Psi_{tetrolet}\mathbf{u}\|_1 + \frac{\lambda_2}{2}\|\mathbf{SHu}-\mathbf{f}\|_2^2\right\} \qquad (8)$$

By using the alternate iterative technique, Eq. (8) can be solved by the following two sub-problems with respect to $\mathbf{w}$ and u:

$$\begin{cases} \mathbf{w}^{k+1} = \arg\min_{\mathbf{w}}\left\{\|\mathbf{w}\|_1 + \frac{\beta}{2}\|\mathbf{w}-D\mathbf{u}^k\|_2^2\right\} \\ \mathbf{u}^{k+1} = \arg\min_{u}\left\{\frac{\beta}{2}\|\mathbf{w}^{k+1}-D\mathbf{u}\|_2^2 + \lambda_1\|\Psi_{tetrolet}\mathbf{u}\|_1 + \frac{\lambda_2}{2}\|\mathbf{SHu}-\mathbf{f}\|_2^2\right\} \end{cases} \qquad (9)$$

Furthermore, to solve two sub-problems in Eq.(9), the closed-form solution of the first sub-problem is equivalent to soft shrinkage on gradient, while the Douglas-Rachford splitting and frame shrinkage scheme can be applied to the second sub-problem. The above two sub-problems are iteratively solved and finally the estimated HR image $\hat{\mathbf{u}}_R$ can be obtained.

3.4   Tetrolet Learning to Recover Super-Resolution Image

Using the estimated HR image from the first stage as the input image, we want to learn a further SR image. The learning methods used here is similar to best matching in texture synthesis. The dependence of tetrolet subbands is exploited to predict the high frequency tetrolet coefficients. Let $\hat{\mathbf{u}}_R$ be the resulted image of size $N \times N$ obtained from tetrolet regularization as discussed in the previous sub-section and $\{\mathbf{u}_i\}_{i=1,2,...M}$ be the training images of size $N \times N$. In the learning process, the same three-level tetrolet decomposition is applied both on the estimated HR image and the HR image data sets. The index of tetrolet subbands are denoted by {0, I, II,III,IV, V, VI, VII, VIII,IX}, where the subbands indexed by {VII, VIII,IX} of $\hat{u}_R$ will conducted coefficients learning from the corresponding subbands of HR image data sets. The procedure of coefficients learning is demonstrated by Fig.4.

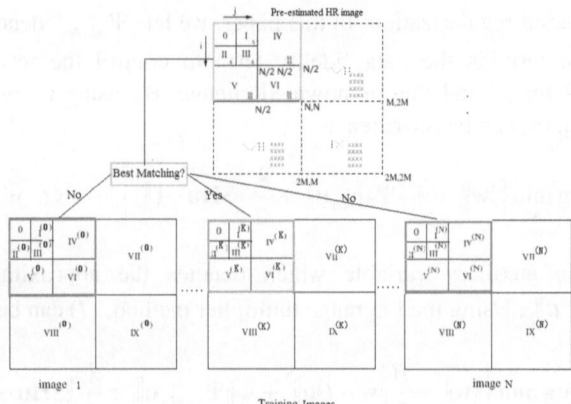

Fig. 4. The illustration of coefficients learning from HR image data sets

Step 1. Perform three-level tetrolet decomposition with four directional subbands on the resulted image $\mathbf{u}_R$ of size $N \times N$ obtained from tetrolet regularization and three-level decomposition on all training images $\{\mathbf{u}_i\}_{i=1,2,...M}$ each of size $N \times N$.

Step 2. Consider the tetrolet coefficients at locations (i, j), (i, j +N/4), (i+N/4, j), and (i+N/4, j +N/4) in subbands I, II, III and the corresponding 2×2 blocks in IV–VI of the $\hat{\mathbf{u}}_R$ as well as the high-resolution training set $\{\mathbf{u}_i\}_{i=1,2,...M}$.

Step 3. Obtain the sum of absolute difference (MAD) between the tetrolet coefficients in the image $\hat{\mathbf{u}}_R$ and all the coefficients for each of the training images.

Step 4. Best matching and coefficients fusion

1) If MAD < threshold, obtain the learning high resolution tetrolet coefficient (4×4 block) from the training image offering the best match locally in subbands VII–IX; then we apply tetrolet coefficient fusion using the corresponding learning coefficients and the tetrolet coefficient (4×4 block) from the image $\hat{\mathbf{u}}_R$ locally in subbands VII–IX by average operator. Finally, the fusion tetrolet coefficients are set as the unknown high resolution tetrolet coefficient.

2) Else, the tetrolet coefficients from the image $\hat{\mathbf{u}}_R$ locally in subbands VII–IX are set as the unknown high resolution tetrolet coefficient.

Step 5. Repeat steps 2–4 for every tetrolet coefficient in bands I–IV of the $\hat{\mathbf{u}}_R$.

Step 6. Perform inverse tetrolet transform to obtain the final high-resolution image of the given test image.

In the above algorithm, we actually search for the best matching training coefficients at given location (i, j) that matches to the tetrolet coefficients for the regularized image in the bands I, II, III, and IV,V,VI in MAD sense and fuse the corresponding high tetrolet bands for the final result image.

$$MAD = |d_I(i,j) - d_I^m(p,q)| + |d_{II}(i,j) - d_{II}^m(p,q)|$$
$$+ |d_{III}(i,j) - d_{III}^m(p,q)| + S_{IV} + S_V + S_{VI}]$$

(10)

Here S_{IV} is defined by

$$S_{IV} = |d_{IV}(2i-1, 2j-1) - d_{IV}^m(2p-1, 2q-1)|$$
$$+ |d_{IV}(2i-1, 2j) - d_{IV}^m(2p-1, 2q)|$$
$$+ |d_{IV}(2i, 2j-1) - d_{IV}^m(2p, 2q-1)|$$
$$+ |d_{IV}(2i, 2j) - d_{IV}^{index}(2p, 2q)|$$

and S_V and S_{VI} are the corresponding sums for subbands V, and VI, respectively, and m= 1, 2, . . . K. Here $d_J^m(\cdot)$ denotes the tetrolet coefficients for the mth training image at the J th subband.

4   Experiments

In this section, one hundred high-resolution images are used for tetrolet learning. In order to evaluate the performance of the proposed method, we design two experiments. The PSNR measure is used to quantify the reconstruction performance.

The first experiment is conducted to verified the efficiency of our method for the image super-resolution when the degraded images are generated by down-sample them by a factor using decimation matrix $\mathbf{S}$ in (1). Fig. 5(a) is one cropped ground truth image from 'Barbara' and Fig. 5(b) shows the LR image of size 64×64. Fig.5(c) shows the result image using the bi-cubic interpolation technique, while Fig.5(d) and Fig5.(e) show the super-resolved image obtained by wavelet learning[6] and contourlet learning[7], respectively. The super-resolved image obtained using our method is presented in Fig.5(f). From these figures we can see that bi-cubic interpolation only produce image multiplication with jaggy and artifacts. Pure leaning based method remove most of such artifacts through texture synthesis, however the smoothness along contours is not very good. Without any exceptions, the proposed approach reconstructs the most visually pleasant HR images.

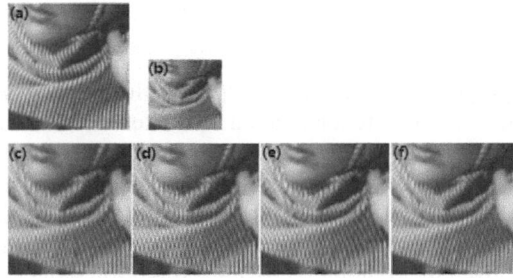

Fig. 5. Barbara ground truth, LR image and SR results according to different methods: (a) ground truth; (b) LR image; (c) Bicubic interpolation; (d) Wavelet learning in [6]; (e) Contourlet learing in [7]; (f) Our method

Table 1. Comparison of PSNRs for the zoom factor $q = 2$ expressed in dB

Ex.	Bicubic	Wavelet learning[6]	Contoulet learning[7]	Our method
Barbara	20.01dB	20.52dB	22.33dB	24.65dB

In the second experiment, the degraded LR images are generated by first applying a blur kernel and then down-sampling (Fig.6(b)). The blurring kernel in the simulations is a 5×5 Gaussian filter with standard deviation of 1.4. We magnify the LR images by a factor of 2. The HR images reconstructed by the TV (total variation) based method remove most of jaggy artifacts but they are over-smoothed and many image details are eliminated. The approach by Tetrolet and TV compound regularization is competitive in visual quality (Fig.6(d)). The edges and textures reconstructed by our approach (Fig.6(e)) are much sharper and cleaner than others. Also, more image details are recovered by our approach.

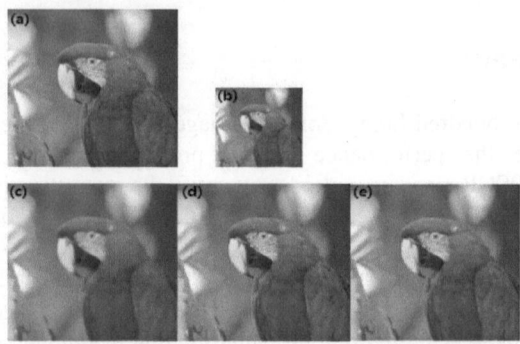

Fig. 6. Parrot Ground truth, LR image and SR results according to different methods:(a) ground truth; (b) LR image; (c)TV reconstruction in [10]; (d) Tetrolet +TV in [9]; (e) Our method

Table 2. Comparison of PSNRs for the zoom factor q = 2 and Gaussian Blur expressed in dB

Ex.	TV reconstruction [10]	Tetrolet+TV [9]	Our method
Parrot	32.83dB	33.2dB	34.34dB

5   Conclusion

In this paper, we propose a novel image super-resolution reconstruction approach, which exploits the tetrolet regularization and tetrolet learning. The tetrolet regularization can adaptively preserve the intensity discontinuities along four directions rather than only two directions. The tetrolet coefficients at finer scales, learned from a set of several high-resolution training images, are used to further improve the super-resolved image. The learning process ensures capturing partial textures and details from the training set given a low-resolution observation. Thus the local details and edges can be recovered by the collaborative regularization and learning mechanism. Experiments demonstrated that the proposed approach can reconstruct sharp edges and fine image details.

Acknowledgments. This work was supported in part by the Natural Science Foundation of China under Grant No. 61171165 and 60802039 and also sponsored by Natural Science Foundation of Jiangsu under Grant No.BK2010488 and Qing Lan Project of Jiangsu Province.

References

1. Tian, J., Ma, K.-K.: A survey on super-resolution imaging. Signal, Image and Video Processing 5(3), 329–342 (2011)
2. Chung, J., Haber, E., Nagy, J.: Numerical methods for coupled super-resolution. Inverse Problems 22(4), 1261–1272 (2006)
3. Protter, M., Elad, M., Takeda, H., Milanfar, P.: Generalizing the nonlocal- means to super-resolution reconstruction. IEEE Trans. Image Process. 16(1), 36–51 (2009)
4. Takeda, M.P., Protter, M., Elad, M.: Super-resolution without explicit subpixel motion estimation. IEEE Trans. Image Process. 18(9), 1958–1975 (2009)
5. Hardie, R.C., Barnard, K.J., Armstrong, E.E.: Joint MAP registration and high-resolution image estimation using a sequence of undersampled images. IEEE Transactions on Image Processing 6(12), 1621–1633 (1997)
6. Jiji, C.V., Joshi, M.V., Chaudhuri, S.: Single-frame image super-resolution using learned wavelet coefficients. International Journal of Imaging Systems and Technology 14(3), 105–112 (2004)
7. Jiji, C.V., Subhasis, C.: Single-Frame Image Super-resolution through Contourlet Learning. EURASIP Journal on Applied Signal Processing 2006, Article ID 73767, 1–11 (2006)
8. Krommweh, J.: Tetrolet Transform: A new Adaptive Haar Wavelet Algorithm for Sparse Image Representation. Journal of Vision Communication and Image Representation 21(4), 364–374 (2010)
9. Wang, L., Xiao, L., Wei, Z.: Compound tetrolet sparsity and total variation regularization for image restoration. Multispectral Image Acquisition, Processing, and Analysis
10. Wang, Y., Yang, J., Yin, W., Zhang, Y.: A new alternating minimization algorithm for total variation image reconstruction. SIAM Journal on Imaging Sciences 1(3), 248–272 (2008)
11. Afonso, M., Bioucas-Dias, J., Figueiredo, M.: Fast image recovery using variable splitting and constrained optimization. IEEE Trans. Image Process. 19(9), 2345–2356 (2010)
12. Setzer, S.: Split Bregman Algorithm, Douglas-Rachford Splitting and Frame Shrinkage. In: Tai, X.-C., Mørken, K., Lysaker, M., Lie, K.-A. (eds.) SSVM 2009. LNCS, vol. 5567, pp. 464–476. Springer, Heidelberg (2009)

Optimization and Fiber-Centered Prediction of Functional Network ROIs

Can Feng[1], Tianming Liu[2], Liang Xiao[1], and Zhihui Wei[1]

[1] School of Computer Scicence, Nanjing Unieversity of Science & Technology
210094 Nan Jing, China
[2] Department of Computer Science, The University of Georgia,
30605, Athens, GA, USA

Abstract. Study of functional and structural brain networks via fMRI and DTI data has received significant interest recently. A fundamental and challenging problem to identify a specific brain networks is how to localize the best possible regions of interests (ROIs). In this paper, we firstly propose a new approach to quantitatively describe fiber bundle and measure the similarity of two fiber bundles. Then we present a novel framework to optimize the shape of ROIs by maximizing fiber bundles similarity cross subjects and predict brain network ROIs in individual brain based only on DTI data. Our experimental results show that optimized ROIs have significantly improved consistency in structural profiles across subjects and demonstrated that fiber bundle description model derived from DTI data is a good predictor of functional ROIs. This capability of accurately predicting brain network ROIs would open up many applications in brain imaging that rely on identification of functional ROIs.

Keywords: fMRI, DTI, structural connectivity, shape optimization, ROI prediction.

1   Introduction

It is widely believed that the brain's function is integrated via structural and functional connectivities [1-3]. Construction of brain networks based on vivo brain imaging data offers an exciting and unique opportunity to understand cortical architecture. In brain networks, network nodes ROIs provide the structural substrates for connectivity measurement within individual brains and for pooling data across populations [2]. Therefore, a fundamental question in constructing structural and functional network is how to define the best possible regions of interests (ROIs). In our view, this task is challenging for several critical reasons. 1) The boundaries between cortical regions are unclear. 2) Individual variability of cortical anatomy, connection, and function is remarkable. Quantitative mapping of the regularity, while accounting for the variability, of cortical structure and function is a challenging task. 3) The properties of ROIs are highly nonlinear [4, 5].

Current approaches to identify ROIs can be broadly classified into four categories [6, 7]. The first is manual labeling by experts using their domain knowledge. The second is a data-driven clustering of ROIs from the brain image itself. The third is to

J. Yang, F. Fang, and C. Sun (Eds.): IScIDE 2012, LNCS 7751, pp. 390–397, 2013.

predefine ROIs in a template brain, and warp them back to the individual space using image registration. Lastly, ROIs can be defined from the activated regions observed from an activation map.

Identifying ROIs using an activation map is regarded as the standard framework for ROI identification [8]. The most common approach in this framework is to create small ROIs (usually spheres) at local maxima in the activation map. Our rationale is that the activation peaks are close to the true functional ROIs, but the accuracy of their sizes and shape is dependent on several factors such as the spatial normalization procedure and individual diversity. Therefore, in this paper, we present a novel framework to optimize the shape of ROIs based on maximizing fiber bundles similarity cross subjects. In particular, we focus on optimizing the shape of default mode network ROIs using rest state fMRI (rsfMRI) data.

Additionally, the human brain is composed of many functional networks, such as default model, working memory, vision, auditory and emotion systems. Extensive acquisition of fMRI data for all these networks is both time consuming and expensive, which makes it impractical for wide use. Instead, a typical DTI images scan needs less than 10 min, is much less demanding, and is widely available. Those reasons strongly encourage us to identify and predict functionally meaningful ROIs based only on DTI data. The close relationship between structural connectivity pattern and brain function has been reported in the literature [9, 10]. An interesting observation from our recent results in [7] is that white matter (WM) fiber connection patterns of the same functional cortical ROI are reasonably consistent across different subjects, suggesting that fiber connection pattern might be a good predictor of functional ROI. Hence, in this paper, as a sequel to ROIs optimization procedure, we use the locations, shape and fiber bundles of optimized functional ROIs as the prior knowledge, and propose a new model to predict functional ROIs based only on DTI data.

The arrangement for the rest of the paper is as follows. In section 2, we firstly detail the data acquisition and preprocessing of the multimodal data including fMRI and DTI data; then, we present the fiber bundles description model and similarity measurement method; lastly, we formulate the energy function for ROI optimization and prediction. Section 3 presents some experiment results and their interpretations. Discussions and conclusion are provided in section 4.

2 Materials and Methods

2.1 Overview of the Framework

The pipeline of our framework is composed of three stages. The first stage is rsfMRI and DTI data preprocessing including independent component analysis, default mode network activation map selection and DTI tractography. The second one is group-wise optimization: after pre-processing of DTI and rsfMRI dada, rsfMRI default mode network activation map and white matter fibers were used to optimize the shape of default mode network ROIs. The third stage is prediction: given DTI data of a new subject, we predict default mode network ROIs of this subject by proposed functional ROIs prediction mode.

2.2   Data Acquisition and Preprocessing Method

Seven university students were recruited to participate in this study. Subjects were instructed simply to keep their eyes closed and not to think of anything in particular while fMRI data was acquired. DTI scans were also acquired for each participant. FMRI and DTI scans were acquired on a 3T GE Signa HDx scanner. Acquisition parameters were as follows, fMRI: 128x128 matrix, 2mm slice thickness, 256mm FOV, 60 slices, TR=1.5s, TE=25ms, ASSET=2; DTI: 128x128 matrix, 2mm slice thickness, 256mm FOV, 60 slices, TR=15.1s, TE= variable, ASSET=2, 3 B0 images, 30 optimized gradient directions, b-value=1000.

The following steps were done on rsfMRI data we acquired.

Independent Component Analysis (ICA). For each subject, after pre-processing (including brain skull removal, motion correction, spatial smoothing, temporal pre-whitening, slice time correction, global drift removal, and band pass filtering (0.01Hz~0.1Hz)),the 4D rsfMRI data was then analyzed with FSL MELODIC ICA software (http://www.fmrib.ox.ac.uk/fsl/melodic/index.html). ICA is a statistical technique that separates a set of signals into independent uncorrelated and no-Gaussian spatiotemporal components [11]. In this paper we used the default settings of MELODIC to automatically estimate the number of components from the data. And in the experiment, the number of components ranged from 29 to 35.

Selection of the Best-Fit Component. We select the component in each subject that most closely matched the default mode network by experts using their domain knowledge.

DTI pre-processing consisted of skull removal, motion correction, and eddy current correction. After the pre-processing, fiber tracking was performed using MEDINRIA (http://www-sop.inria.fr/asclepios/software/MedINRIA/). Fibers were extended along their tangent directions to reach into the gray matter when necessary. Brain tissue segmentation was conducted on DTI data by the method in [12] and the cortical surface was reconstructed from the tissue maps using the marching cubes algorithm. The cortical surface was parcellated into anatomical regions using the HAMMER tool [13]. DTI space was used as the standard space from which to generate the GM (gray matter) segmentation and report the ROI locations on the cortical surface. Co-registration between DTI and fMRI data was performed using FSL FLIRT [14].

2.3   Bundle Description Based on Gradient-Spherical Surface Mapping Model

Many algorithms, such as the spectral clustering, normalized cut clustering and atlas-based clustering, have been developed to cluster white matter fibers into different bundles. However, an open problem remains: how can a fiber bundle be described quantitatively? In this paper, we proposed a novel method called Gradient-Spherical Surface Mapping (GSSM) model to describe fiber bundle quantitatively and measure similarity of two fiber bundles.

Consider one fiber bundle $F = \{f_i, i = 1, 2, \cdots, T\}$, T is the number of fibers in the bundle, f_i is the i-th fiber which is composed of a collection of space points, denoted by $f_i = (X_1, X_2, \cdots, X_N)$, X_j is the 3D coordinate of j-th point, N is the number of points. We define $N-1$ unit gradient directions of f_i as follows:

$$\Psi_i = \left\{ g_j = \frac{X_{j+1} - X_j}{\| X_{j+1} - X_j \|_2}, j = 1, 2, \cdots, N-1 \right\} \tag{1}$$

We can see that $\| g_j \|_2 = 1$ and g_j is a point in the unit spherical surface. We perform the same procedure on all fibers in the bundle F, and define **gradient-map** of fiber bundle F as:

$$G(F) = \{ \Psi_i, i = 1, 2, \cdots T \} \tag{2}$$

Two issues should be noted here. First one is that we must make sure all subjects' brains are aligned. In this paper, we align different brains by the principal direction which is calculated using PCA. The second issue is that we need to explicitly assign one of two ends of every fiber in each fiber bundles as the starting point. Since each fiber was extracted from a small region in the brain, we select the end that is closer to the center of the region as the starting point.

After representation of fiber bundles by GSSM model, two bundles can be compared by calculating the similarity between the distributions of respective gradient-map.

As we know, unit spherical surface can be described by the following equation:

$$\begin{cases} x = \sin \alpha \cos \beta \\ y = \sin \alpha \sin \beta \\ z = \cos \alpha \end{cases} \quad \alpha \in [0, \pi], \beta \in [0, 2\pi] \tag{3}$$

Given a positive integer M, unit spherical surface can be divided into $2M^2$ sub-surfaces by:

$$S_{ij} = \left\{ \begin{cases} x = \sin \alpha \cos \beta \\ y = \sin \alpha \sin \beta \\ z = \cos \alpha \end{cases} \quad \alpha \in \left[\frac{i-1}{M} \pi, \frac{i}{M} \pi \right], \beta \in \left[\frac{j-1}{M} \pi, \frac{j}{M} \pi \right] \right\} \tag{4}$$

$$i = 1, 2, \cdots, M, j = 1, 2, \cdots, 2M$$

Given gradient-map G, we calculate the point density of sub-surface S_{ij}, denoted by $\rho_G(S_{ij})$, as follows:

$$\rho_G(S_{ij}) = n_{ij} / |G| \tag{5}$$

where n_{ij} is the number of points located in S_{ij} and $|G|$ is total number of points in the gradient-map. Fig. 2 (e) show point density distribution of fiber bundles in Fig. 2(a), here $M = 12$, and each $\rho_G(S_{ij})$ is rearranged into a vector. The similarity of two gradient-maps G_1 and G_2 is defined as

$$D(G_1, G_2) = \sum_{i=1}^{M} \sum_{j=1}^{2M} | \rho_{G_1}(S_{ij}) - \rho_{G_2}(S_{ij}) | \tag{6}$$

Note that the point density $\rho_G(\cdot)$ is normalized so that we do not require that the numbers of points in different trace-maps are equal.

2.4   Optimization of ROIs across Subjects

In the data preprocessing stage, we selected default model network activation map manually in each subject. To construct default model network ROIs using those activation maps, one approach that has often been used is to threshold the activation map and construct network node ROI by activated voxels. However, this approach can be very sensitive to the specific threshold. Can we optimize the shape of ROIs by select optimal thresholds? In this section, we optimize the shape of ROIs by selecting group-wise optimal thresholds through maximize the similarity of the fiber bundles across subject and formulate the problem of optimization ROI shape as an energy minimization problem.

Taking ROI i for example, given a threshold λ_j and default mode network activity map, we define $R_i^j(\lambda_j)$ as the local activated region (activation value great than λ_j) centered at location of ROI i on subject j's surface, and fiber bundle penetrating $R_i^j(\lambda_j)$ was extracted and denoted as $f_i^j(\lambda_j)$. Then, mathematically, the energy function to minimize is defined as:

$$\left(\tilde{\lambda}_1,\tilde{\lambda}_2,\cdots,\tilde{\lambda}_p\right)=\arg\min_{\lambda_1,\lambda_2,\cdots,\lambda_p}\sum_{m=1}^{p}\sum_{n=m+1}^{p}D\Big(G\big(f_i^m(\lambda_m)\big),G\big(f_i^n(\lambda_n)\big)\Big) \qquad (7)$$

Where p is number of subjects, $G\big(f_i^m(\lambda_m)\big)$ is the gradient-map of fiber bundle $f_i^m(\lambda_m)$. By solving problem (7), we can find the optimal threshold $\left(\tilde{\lambda}_1,\tilde{\lambda}_2,\cdots,\tilde{\lambda}_p\right)$ and corresponding optimal ROIs $R_i^1(\tilde{\lambda}_1),R_i^2(\tilde{\lambda}_2),\cdots,R_i^p(\tilde{\lambda}_p)$.

2.5   ROIs Prediction

If the shape of fiber bundles of ROIs are descriptive enough and consistent across different brains, they can be used as good morphological signatures to predict functional ROIs in the absence of fMRI data. Therefore, an ROI prediction frame work is developed based on group-wise fiber bundles characteristics. After group-wise ROIs optimization in section 2.5, optimal ROIs and corresponding fiber bundles (we called it reference ROIs and reference fiber bundles) are used to prediction the localization and size of ROIs of a new subject (we called it target subject). Here we use a ball to describe a ROI, and average size (number of voxels) of reference ROIs is taken as the size of ball.

Suppose there are q reference subject, reference fiber bundles of ROI i denoted as $f_i^1, f_i^2, \cdots, f_i^q$, we formulated prediction progress as following energy function minimization problem:

$$\tilde{B}(x,y,z)=\arg\min_{x,y,z}\sum_{j=1}^{q}D\Big(G\big(f_i^j\big),G\big(f_{B(x,y,z)}\big)\Big) \qquad (8)$$

where $G\left(f_i^{\,j}\right)$ is the gradient-map of fiber bundle $f_i^{\,j}$, $f_{B(x,y,z)}$ is the fiber bundle of ball $B(x,y,z)$, (x,y,z) is the center of the ball, $G\left(f_{B(x,y,z)}\right)$ is the gradient-map of fiber bundle $f_{B(x,y,z)}$.

In our implementation, we first align all reference ROIs to DTI space of target subject; then calculate the center of each reference ROI to form a search space; at last, we find the solution of question (8) by whole space searching.

3   Experimental Results

In this section, we present some experimental results. Our results consist of 2 parts. First, we test our default mode network ROI optimization framework using dataset described in section 2.1, and show optimized ROIs and corresponding fiber bundles in section 3.1. Second, we performed leave-one-out prediction experiments and show the prediction results and some quantitative measurements in section 3.2.

3.1   Optimization Results of 7 Subjects and 8 ROIs

Fig 4 shows the optimized default mode network ROIs and corresponding fiber bundles of 7 subjects after optimization. From the figure it can be seen that optimized ROIs from different subjects have consistent structural connectivity profiles.

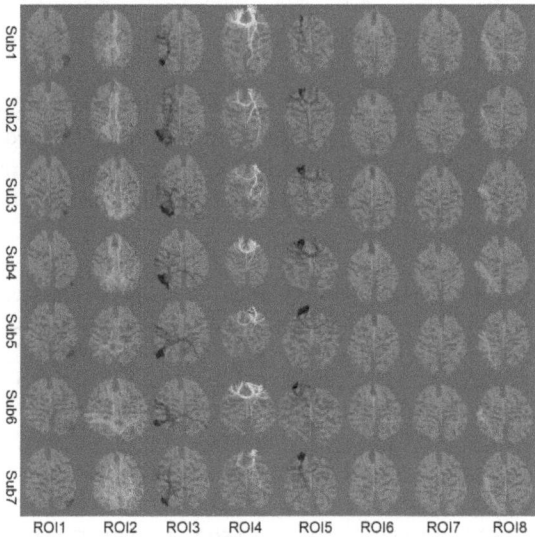

Fig. 1. Visualization of optimized ROIs and corresponding fiber bundles of 7 subjects

3.2   Leave-One-Out Prediction

We used the leave-one-out strategy to evaluate the ROI prediction framework on the dataset described in section 2.1. To visualize the consistency of fiber bundles of the predicted ROIs, we showed the fibers emanating from predicted ROIs in Fig 5.

It is evident that the fiber bundles of the predicted ROIs are quite similar to those of optimized ROIs which are shown in Fig 4.

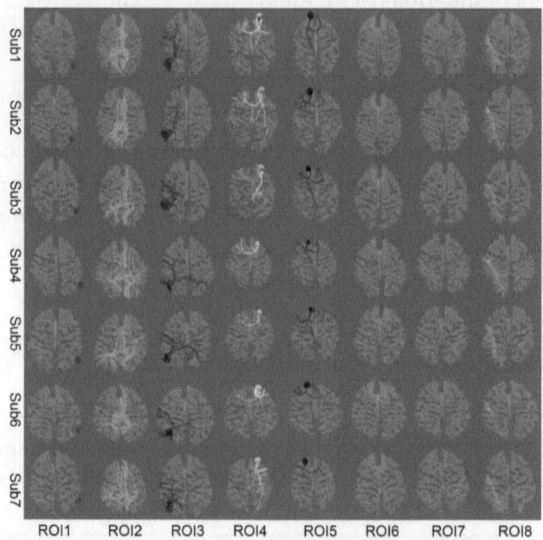

Fig. 2. Visualization of predicted ROIs and corresponding fiber bundles

In order to further evaluate the performance of our prediction framework, we show the Euclidean distances between centers of optimized and predicted ROIs in Table 1. We can see that most of the prediction errors are approximately 2--6 mm, which is 1--3 voxels in DTI volumes. On average, the average prediction error for ROIs is 5.5078, which is considered as very accurate.

Table 1. Euclidean distances between centers of optimized and predicted ROIs

(mm)	ROI 1	ROI 2	ROI 3	ROI 4	ROI 5	ROI 6	ROI 7	ROI 8	mean
sub 1	2.9986	4.6785	4.0883	5.8943	6.3434	4.2054	4.6815	1.8549	4.3431
sub 2	3.0651	3.9306	10.670	2.8391	3.6169	5.5907	5.4525	4.4219	4.9484
sub 3	5.5276	6.486	10.982	1.902	5.4555	5.7313	4.4985	6.1455	5.8411
sub 4	6.2541	4.8263	3.4572	4.3281	4.7349	7.8706	2.2865	6.5534	5.0389
sub 5	1.9175	2.4672	3.0958	13.309	4.1456	6.9007	9.1755	3.1937	5.5257
sub 6	9.0583	5.5533	14.688	2.6011	9.1592	14.040	7.1509	4.7839	8.3795
sub 7	5.0527	2.6292	4.3583	8.4163	0.5767	2.9615	7.6921	4.1348	4.4777
mean	4.8391	4.3673	7.3345	5.6129	4.8617	6.7573	5.8482	4.4412	5.5078

4   Conclusion

In this paper, we presented a novel framework for functional brain network ROI optimization and prediction using rsfMRI data and DTI data. This framework has been extensively evaluated on 8 ROIs across 7 subjects. Our optimization results indicated that the structural connectivity patterns of each individual's functional ROI are very consistent after optimization, and prediction results demonstrated that our fiber bundle description model of functional brain ROIs have remarkable prediction capability. In the future, we plan to apply and evaluate this ROI prediction framework in other brain networks, such as working memory, attention and semantics memory systems and validate this framework on clinical data sets such as the DTI data sets of Alzheimer's disease and Autism.

References

1. Bharat, B.B., et al.: Toward discovery science of human brain function. Proc. Natl. Acad. Sci. 107(10), 4734–4739 (2010)
2. Dijk, K.R.A.V., Hedden, T., Venkataraman, A., Evans, K.C., Lazar, S.W., Buckner, R.L.: Intrinsic functional connectivity as a tool for human connectomics: theory, properties, and optimization. J. Neurophysiol. 103(1), 297–321 (2010)
3. Hagmann, P., et al.: MR connectomics: Principles and challenges. J. Neurosci. Methods 194(1), 34–45 (2010)
4. Zhang, T., Guo, L., Li, K., Jing, C., Yin, Y., Zhu, D., Cui, Z., Li, L., Liu, T.: Predicting functional cortical ROIs via DTI-derived fiber shape models. Cereb. Cortex (2011)
5. Zhu, D., Li, K., Faraco, C., Deng, F., Zhang, D., Jiang, X., Chen, H., Guo, L., Miller, S., Liu, T.: Optimization of functional brain ROIs via maximization of consistency of structural connectivity profiles. NeuroImage 55(2), 1382–1393 (2012)
6. Liu, T.: A few thoughts on brain ROIs. Brain Imaging and Behavior 5(3), 189–202 (2011)
7. Li, K., Guo, L., Zhu, D., Hu, X., Han, J., Liu, T.: Individualized ROI Optimization via Maximization of Group-wise Consistency of Structural and Functional Profiles. Neuroinformatics 10(3), 225–242 (2012)
8. Poldrack, R.A.: Region of interest analysis for fMRI. Soc. Cogn. Affect. Neurosci. 2(1), 67–70 (2007)
9. Passingham, R.E., Stephan, K.E., Kötter, R.: The anatomical basis of functional localization in the cortex. Nat. Rev. Neurosci. 3, 606–616 (2002)
10. Honey, C.J., Sporns, O., Cammoun, L., Gigandet, X., Thiran, J.P., Meuli, R., Hagmann, P.: Predicting human resting-state functional connectivity from structural connectivity. Proc. Natl. Acad. Sci 106(6), 2035–2040 (2009)
11. Beckmann, C., Smith, S.: Probabilistic independent component analysis for functional magnetic resonance imaging. IEEE Trans. Med. Imag. 23(2), 137–152 (2004)
12. Liu, T., Li, H., Wong, K., Tarokh, A., Guo, L., Wong, S.T.C.: Brain Tissue Segmentation Based on DTI Data. NeuroImage 38(1), 114–123 (2007)
13. Shen, D., Davatzikos, C.: HAMMER: hierarchical attribute matching mechanism for elastic registration. IEEE Trans. Med. Imaging 21(11), 1421–1439 (2002)
14. Jenkinson, M., Bannister, P., Brady, M., Smith, S.: Improved optimization for the robust and accurate linear registration and motion correction of brain images. Neuroimage 17, 825–841 (2002)

A Projected Conjugate Gradient Method for Compressive Sensing

Yuanying Qiu[1], Wei Xue[2,*], and Gaohang Yu[2]

[1] School of Foreign Languages, Gannan Normal University, Ganzhou 341000,
People's Republic of China
[2] School of Mathematics and Computer Science, Gannan Normal University,
Ganzhou 341000, People's Republic of China
wxmaths@163.com

Abstract. Frequently, the most important information in a signal is much sparser than the signal itself. In this paper, we study a projected conjugate gradient method for finding sparse solutions to an undetermined linear system arising from compressive sensing. The construction of this method consists of two main phases: (1) reformulate a l_1 regularized least squares problem into an equivalent nonlinear system of monotone equations; (2) apply a conjugate gradient method with projection strategy to the resulting system. The derived method only needs matrix-vector products at each step and could be easily implemented. Global convergence result is established under some suitable conditions. Numerical results demonstrate that the proposed method can improve the computation time while obtaining similar reconstructed quality.

1   Introduction

Compressive sensing (CS) is an emerging field and is attracting considerable research interest in signal processing community. The fundamental principle of CS is that a sparse signal $\bar{x} \in R^n$ can be recovered from the undetermined linear system $y = \Phi\bar{x}$, where $\Phi \in R^{m \times n}$ (often $m \ll n$). By defining l_0 norm ($\|x\|_0$) of a vector as the number of nonzero elements in x, one natural way to recover $\bar{x}$ from the system is to solve the following problem

$$\min_{x \in R^n} \|x\|_0 \text{ s.t. } y = \Phi x. \tag{1}$$

However, the l_0 norm problem is computationally intractable. An alternative model is to replace l_0 norm by l_1 norm, which is defined as $\|x\|_1 = \sum_{i=1}^{n} |x(i)|$. The resulting adaptation of (1) is the Basis Pursuit (BP) problem [1]

$$\min_{x \in R^n} \|x\|_1 \text{ s.t. } y = \Phi x. \tag{2}$$

Optimization methods often find a solution of (1) by solving the following closely related l_1 regularized least squares problem

$$\min_{x \in R^n} \frac{1}{2}\|y - \Phi x\|_2^2 + \mu\|x\|_1. \tag{3}$$

---

* Corresponding author.

J. Yang, F. Fang, and C. Sun (Eds.): IScIDE 2012, LNCS 7751, pp. 398–406, 2013.

Here, $\mu > 0$ is related to the Lagrange multiplier of the constraint in (2).

It follows from some existing results that if a signal is sparse or approximately sparse in some orthogonal basis, then an accurate recovery can be obtained when Φ is a random matrix projections [3]. Various types of methods have been proposed to solve the l_1 regularized minimization problem. Recently, some first-order methods are popular for solving (3), such as the projection steepest descent method [2], the gradient projection algorithm (GPSR) proposed by Figueiredo et al. [5], and so on. In this paper, we mainly focus on developing an iterative method for solving l_1 regularized problem arising in CS. Among all the methods mentioned above, GPSR method firstly splits vector x into two vectors, reformulates (3) into a bound-constrained quadratic programming problem, and solves it by using the well-known BB stepsize. In [8], the authors notice that the quadratic programming problem is equivalent to a system of nonlinear equations. We use a projected conjugate gradient method to solve the resulting monotone equations in this paper. Our method has two main phases. In the first phase, a l_1 regularized least squares problem (3) is transformed into an equivalent nonlinear system of monotone equations. And then a projected conjugate gradient method is introduced to solve the equivalent system in the second phase.

The rest of this paper is organized as follows. We present the full description of the proposed algorithm in the next section. In Section 3, we establish its global convergence under some suitable conditions. We report some numerical experiments to illustrate the efficiency of the proposed method in Section 4. Some conclusions are drawn in Section 5.

2   Proposed Algorithm

We state our algorithm in this section. Firstly, we recall the approach of constructing a quadratic programming problem in [5]. Making a substitution, for any vector $x \in R^n$, it can be formulated as $x = u - v$, where $u \geq 0, u \in R^n$, $v \geq 0, v \in R^n$ and $u_i = \max\{0, x_i\}$, $v_i = \max\{0, -x_i\}$. Consequently, (3) can be formulated by the following bound-constrained quadratic programming

$$\min_{u,v \in R^n} \frac{1}{2}\|y - \Phi(u - v)\|_2^2 + \mu(\mathbf{I}_n^T u + \mathbf{I}_n^T v) \text{ s.t. } u \geq 0, \ v \geq 0, \qquad (4)$$

where $\mathbf{I}_n^T$ represents the transpose of $\mathbf{I}_n$, and $\mathbf{I}_n = [1, 1, \ldots, 1]^T$ is a vector consisting of n ones. Particularly, it follows from [5] that (4) can be rewritten as the following form

$$\min_{p \in R^{2n}} \frac{1}{2}p^T \Gamma p + q^T p \text{ s.t. } p \geq 0, \qquad (5)$$

where $p = [u \ v]^T$, $b = \Phi^T y$, $q = \mu \mathbf{I}_{2n} + [-b \ b]^T$ and $\Gamma = \begin{bmatrix} \Phi^T \Phi & -\Phi^T \Phi \\ -\Phi^T \Phi & \Phi^T \Phi \end{bmatrix}$.

Recently, Xiao et al. [8] pointed out that (5) can be transformed into the following form

$$F(p) = \min\{p, \Gamma p + q\} = 0, \qquad (6)$$

where function F is vector value, and the "min" is interpreted as componentwise minimum. Without specific statements, $\| \cdot \|$ denotes the Euclidean norm in the following paper.

The following lemma shows that $F(\cdot)$ is Lipschitz continuous [6].

Lemma 1. *There exists a positive constant L such that*

$$\|F(x) - F(y)\| \le L\|x - y\|, \ \forall \ x, y \in R^{2n}. \tag{7}$$

The following lemma shows that $F(\cdot)$ is monotone [8].

Lemma 2. *The mapping $F(\cdot)$ is monotone, i.e.,*

$$(F(x) - F(y))^T (x - y) \ge 0, \ \forall \ x, y \in R^{2n}. \tag{8}$$

The above two lemmas illustrate that the system of nonlinear equations has nice properties, and it can be solved efficiently by some derivative-free methods [4,9,10].

In this paper, we propose a projected conjugate gradient method for the minimization of l_1 regularized minimization problem with application to CS. Particularly, the search direction is generated by the following way

$$d_k = \begin{cases} -F(p_1) & \text{if} \ \ k = 1, \\ -F(p_k) + \alpha_k d_{k-1} - \beta_k y_{k-1} & \text{if} \ \ k \ge 2, \end{cases} \tag{9}$$

where $\alpha_k = \frac{F(x_k)^T y_{k-1}}{\|F_{k-1}\|^2}$, $\beta_k = \frac{F(x_k)^T d_{k-1}}{\|F_{k-1}\|^2}$ and $y_{k-1} = F(x_k) - F(x_{k-1})$.

The full description of our method, PCG Algorithm (short for " projected conjugate gradient algorithm"), can be formally presented as follows now.

Algorithm 1. *(PCG Algorithm)*

Date: Give initial point $p_1 \in R^{2n}$, set parameters $\sigma_1 > 0$, $\sigma_2 > 0$ and $\rho \in (0,1)$.

Convergence test: If $\|F(p_1)\| = 0$, then stop. Else set $d_1 = -F(p_1)$. Let $k := 1$.

Line search update: Determine the steplength λ_k and set $z_k = p_k + \lambda_k d_k$, where $\lambda_k = \sigma_1 \rho^{m_k}$ with m_k being the smallest nonnegative integer m satisfying

$$- F(z_k)^T d_k \ge \sigma_2 \sigma_1 \rho^m \|F(z_k)\| \|d_k\|^2. \tag{10}$$

Projection update: Compute

$$p_{k+1} = p_k - \frac{F(z_k)^T(p_k - z_k)}{\|F(z_k)\|^2} F(z_k). \tag{11}$$

If $\|F(p_{k+1})\| = 0$, then stop. Else let $k := k + 1$ and compute d_k defined by (9). Then go to the Convergence Test.

The following lemma states that PCG Algorithm is well-defined, which can be proved in a way similar to the proof of Lemma 1 in [10].

Lemma 3. *Suppose that $F(p_k) \ne 0$ for all k, then there exists a nonnegative integer m_k satisfying (10) for all k.*

3   Global Convergence of PCG Algorithm

We prepare to show our main global convergence result of PCG Algorithm. Throughout this section, we assume that the solution set of (6) is nonempty.

3.1   Some Properties

In this subsection, we derive some useful properties of PCG Algorithm.

Lemma 4. *Suppose that the sequence $\{p_k\}$ is generated by PCG Algorithm, then for any $\hat{p}$ such that $F(\hat{p}) = 0$, it holds that*

$$\lim_{k\to\infty} \lambda_k \|d_k\| = 0. \tag{12}$$

Proof. By the line search process (10), we have

$$\begin{aligned}
F(z_k)^T(p_k - z_k) &= -\lambda_k F(z_k)^T d_k \geq \sigma_2 \lambda_k^2 \|F(z_k)\| \|d_k\|^2 \\
&= \sigma_2 \|F(z_k)\| \|p_k - z_k\|^2 > 0.
\end{aligned} \tag{13}$$

By (11) and the monotonicity of F, it is easy to deduce that

$$\begin{aligned}
\|p_{k+1} - \hat{p}\|^2 &= \left\| p_k - \frac{F(z_k)^T(p_k - z_k)}{\|F(z_k)\|^2} F(z_k) - \hat{p} \right\|^2 \\
&= \|p_k - \hat{p}\|^2 - 2F(z_k)^T(p_k - \hat{p})\frac{F(z_k)^T(p_k - z_k)}{\|F(z_k)\|^2} + \frac{[F(z_k)^T(p_k - z_k)]^2}{\|F(z_k)\|^2} \\
&\leq \|p_k - \hat{p}\|^2 - 2F(z_k)^T(p_k - z_k)\frac{F(z_k)^T(p_k - z_k)}{\|F(z_k)\|^2} + \frac{[F(z_k)^T(p_k - z_k)]^2}{\|F(z_k)\|^2} \\
&= \|p_k - \hat{p}\|^2 - \frac{[F(z_k)^T(p_k - z_k)]^2}{\|F(z_k)\|^2} \\
&\leq \|p_k - \hat{p}\|^2 - \sigma_2^2 \|p_k - z_k\|^4.
\end{aligned} \tag{14}$$

Hence the sequence $\{\|p_k - \hat{p}\|\}$ is decreasing and convergent. Furthermore, the sequence $\{\|p_k\|\}$ is bounded. By the Cauchy-Schwarz inequality and the monotonicity of F, we have

$$\|F(p_k)\| \geq \frac{F(p_k)^T(p_k - z_k)}{\|p_k - z_k\|} \geq \frac{F(z_k)^T(p_k - z_k)}{\|p_k - z_k\|} \geq \sigma_2 \|F(z_k)\| \|p_k - z_k\|. \tag{15}$$

Moreover, we obtain that the sequence $\{z_k\}$ is bounded too. It follows that

$$\sum_{k=1}^{\infty} \|p_k - z_k\|^4 \leq \frac{1}{\sigma_2^2} \sum_{k=1}^{\infty} (\|p_k - \hat{p}\|^2 - \|p_{k+1} - \hat{p}\|^2) < \infty, \tag{16}$$

which implies

$$\lim_{k\to\infty} \|p_k - z_k\| = 0, \text{ namely, } \lim_{k\to\infty} \lambda_k \|d_k\| = 0. \quad \blacksquare \tag{17}$$

The following lemma can be proved in a way similar to the proof of Lemma 2.1 in [9].

Lemma 5. *Suppose that the sequences $\{p_k\}$ and $\{z_k\}$ are generated by PCG Algorithm, then it holds that*

$$\lambda_k \geq \min\{\sigma_1, \frac{\rho\|F(p_k)\|^2}{(L + \sigma_2\|F(z_k')\|)\|d_k\|^2}\}, \tag{18}$$

where $z_k' = p_k + \lambda_k' d_k$ and $\lambda_k' = \lambda_k \rho^{-1}$.

The following lemmas come from Lemma 2.4 in [9] and Lemma 3.1 in [11], respectively.

Lemma 6. *Suppose that sequence $\{p_k\}$ is generated by PCG Algorithm, $\hat{p}$ satisfies $F(\hat{p}) = 0$, $z_k' = p_k + \lambda_k' d_k$ and $\lambda_k' = \lambda_k \rho^{-1}$, then there exists a constant $M_1 > 0$ such that $\|F(p_k)\| \leq M_1$ and $\|F(z_k')\| \leq M_1$.*

Lemma 7. *If there exists a constant $\varepsilon > 0$ such that $\|F(p_k)\| \geq \varepsilon$ for all k, then there exists a constant $M_2 > 0$ such that $\|d_k\| \leq M_2$ for all k.*

3.2 Convergence Result

In this subsection, we establish the global convergence of the PCG Algorithm proposed in the previous section.

Theorem 1. *Suppose that the sequence $\{p_k\}$ is generated by PCG Algorithm, then it holds that*

$$\lim_{k\to\infty} \inf \|F(p_k)\| = 0. \tag{19}$$

Proof. Suppose that $\lim_{k\to\infty} \inf \|F(p_k)\| \neq 0$, then there exists a constant $\varepsilon > 0$ such that $\|F(p_k)\| > \varepsilon$, for $k \geq 1$. Notice that d_k defined by (9) satisfies $F(p_k)^T d_k = -\|F(p_k)\|^2$ and $\|F(p_k)\| \leq \|d_k\|$, which implies

$$\|d_k\| \geq \varepsilon, \text{ for } k \geq 2. \tag{20}$$

For all k sufficiently large, by Lemma 5, Lemma 6, Lemma 7, $\|F(p_k)\| \geq \varepsilon$ and (20), we deduce that

$$\begin{aligned}
\lambda_k\|d_k\| &> \min\{\sigma_1, \frac{\rho\|F(p_k)\|^2}{(L+\sigma_2\|F(z_k')\|)\|d_k\|^2}\}\|d_k\| \\
&= \min\{\sigma_1\|d_k\|, \frac{\rho\|F(p_k)\|^2}{(L+\sigma_2\|F(z_k')\|)\|d_k\|}\} \\
&\geq \min\{\sigma_1\varepsilon, \frac{\rho\varepsilon^2}{(L+\sigma_2 M_1)M_2}\} \\
&> 0.
\end{aligned} \tag{21}$$

Obviously, (21) contradicts with (12). Similarly, we can derive a contradiction when $k = 1$. Hence the proof is complete. ∎

4   Experimental Results

In this section, numerical experiments are presented to show the performance of the PCG Algorithm for reconstructing sparse signals. These experiments are all tested in Matlab R2012a. Mean squared error (MSE) is used to measure the quality of the reconstructive signals which is defined as MSE $= \|\hat{x} - \bar{x}\|^2/n$, where $\hat{x}$ denotes the reconstructive signal, $\bar{x}$ denotes the original signal and n is the length of the signal.

In our experiments, we consider a typical compressive sensing scenario, the goal is to reconstruct a n length sparse signal from m observations. Random Φ is the Gaussian matrix whose elements are generated from shape $i.i.d.$ normal distributions $\mathcal{N}(0,1)$ (randn(m,n) in Matlab). For y, we add some noises such as $y = \Phi x + \eta$, where η is the Gaussian noise distributed as $\mathcal{N}(0,\sigma^2 I)$.

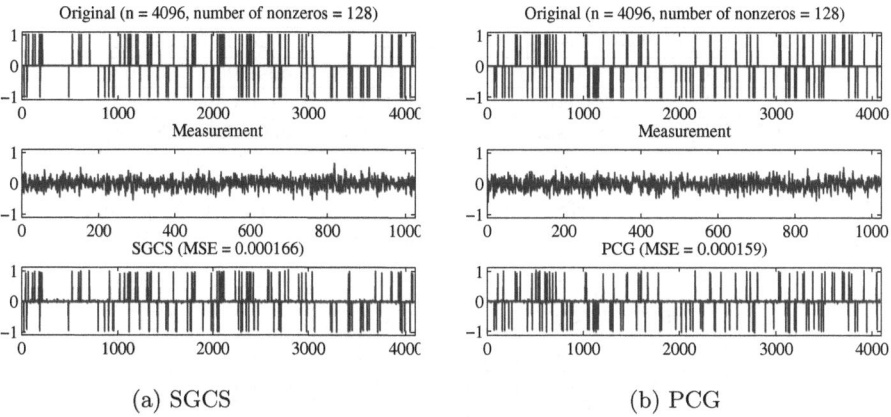

(a) SGCS   (b) PCG

Fig. 1. (a) Top: original signal. Middle: noisy measurement with length 1024. Bottom: recovered signal by SGCS when $\sigma^2 = 10^{-2}$. (b) Top: original signal. Middle: noisy measurement with length 1024. Bottom: recovered signal by PCG when $\sigma^2 = 10^{-2}$.

It should be emphasized that we are mainly concerned with the speed of reconstructing the true signal $\bar{x}$ from the noisy measurement y in this paper. We restrict our attention to the penalized least squares model (3), and use $f(x) = \frac{1}{2}\|y - \Phi x\|_2^2 + \mu\|x\|_1$ as the merit function. Additionally, μ is forced to decrease as in [5] in order to avoid the solution of the quadratic penalty function (3) going to the BP problem while $\mu \to 0$. We compare the performance of PCG method with SGCS method [8]. According to [8], we let $\beta = 1.0$, $\rho = 0.1$, $\gamma = 1.2$ and $\xi = 10^{-4}$ in SGCS Algorithm. However, in PCG Algorithm, we let $\rho = 0.1$, $\sigma_1 = 0.95$ and $\sigma_2 = 0.93$. The common stopping criterion of both methods is

$$\frac{\|f(x_k) - f(x_{k-1})\|}{\|f(x_{k-1})\|} < 10^{-4}. \tag{22}$$

We choose three different signals and four different values of σ^2 in our experiments. In order to test the speed of the algorithms more fairly, we list the average of the five results in the following tables, respectively. Numerical results are listed in Tables 1 2 3, in which we report the number of iterations (Iter), the CPU time in seconds (Time) required for the whole reconstructing process, the means of squared error to every original signal $\bar{x}$ (MSE) and the final objective function value (Obj). From Tables 1 2 3, we can see that the PCG method is faster than SGCS method, and the number of iteration of PCG method is less than that of the SGCS method. Moreover, we note that the MSE and Obj values attained by the PCG and SGCS method are very similar.

Table 1. SGCS v.s. PCG: performance of signal reconstruction. Original signal with length 1024 and 32 non-zero elements, noisy measurement with length 256.

	SGCS				PCG			
σ^2	Iter	Time	MSE	Obj	Iter	Time	MSE	Obj
10^{-4}	196	0.59	7.756e-06	7.365e-02	162	0.47	1.202e-05	6.725e-02
10^{-3}	245	0.62	1.671e-05	6.240e-02	166	0.49	1.304e-05	6.753e-02
10^{-2}	198	0.58	1.293e-04	7.738e-02	161	0.46	1.731e-04	6.829e-02
10^{-1}	255	0.66	1.357e-02	1.290e-01	193	0.61	1.666e-02	1.158e-01

Table 2. SGCS v.s. PCG: performance of signal reconstruction. Original signal with length 2048 and 64 non-zero elements, noisy measurement with length 512.

	SGCS				PCG			
σ^2	Iter	Time	MSE	Obj	Iter	Time	MSE	Obj
10^{-4}	213	2.53	8.252e-06	1.347e-01	160	1.97	1.181e-05	1.348e-01
10^{-3}	188	2.22	7.414e-05	1.333e-01	161	2.01	1.622e-05	1.391e-01
10^{-2}	201	2.43	1.327e-04	1.542e-01	169	2.15	1.647e-04	1.514e-01
10^{-1}	279	3.28	1.314e-02	2.524e-01	218	2.71	1.434e-02	2.415e-01

Table 3. SGCS v.s. PCG: performance of signal reconstruction. Original signal with length 4096 and 128 non-zero elements, noisy measurement with length 1024.

	SGCS				PCG			
σ^2	Iter	Time	MSE	Obj	Iter	Time	MSE	Obj
10^{-4}	191	7.95	7.061e-06	2.834e-01	161	6.82	1.169e-05	2.891e-01
10^{-3}	203	8.31	9.753e-05	2.685e-01	182	7.89	1.689e-05	2.961e-01
10^{-2}	191	8.03	1.459e-04	3.038e-01	175	7.63	1.838e-04	3.058e-01
10^{-1}	283	11.8	1.396e-02	6.235e-01	165	6.88	1.367e-02	4.952e-01

Fig. 1 shows simulation results of SGCS and PCG for a signal sparse reconstruction when $\sigma^2 = 10^{-2}$, respectively. As we can see from Figure 1 (b), all the original sparse signals are restored exactly by PCG method. These experiment results show that the PCG method can work well in an efficient manner.

5   Concluding Remarks

We have proposed a projected conjugate gradient method for solving a convex quadratic programming problem arising from compressed sensing. Our motivation for developing the method mainly comes from [8], where the authors point out that (5) can be transformed into an equivalent nonsmooth nonlinear system of monotone equations, namely, $F(p) = 0$. This system is monotone and Lipschitz continuous, and it can be solved efficiently with some derivative-free methods. In this paper, we adopt the recent conjugate gradient method of Zhang, Zhou and Li [11] with projection strategy of Solodov and Svaiter [7]. We name our method PCG (the abbreviation of " projected conjugate gradient") and establish its global convergence under some suitable conditions. Numerical results show that the PCG method can significantly improve the CPU time for solving the nonlinear system of monotone equations in sparse signals reconstruction while obtaining similar reconstructive quality.

Acknowledgments. This work was supported by the NSFC grants (No. 11001060 and No. 61262026), the NSF of Jiangxi Province (No. 2009GQS0007), the programm of JGZX (No. 20112BCB23027) and the Postgraduate Innovation Fund of Gannan Normal University (No. YCX10B006).

References

1. Chen, S., Donoho, D.L., Saunders, M.: Atomic decomposition by basis pursuit. SIAM Journal on Scientific Computing 20, 33–61 (1998)
2. Daubechies, I., Fornasier, M., Loris, I.: Accelerated projection gradient method for linear inverse problems with sparsity constraints. Journal of Fourier Analysis and Applications 14, 764–792 (2008)
3. Duarte, M.F., Eldar, Y.C.: Structured compressed sensing: from theory to applications. IEEE Transactions on Signal Processing 59, 4053–4085 (2011)
4. Grippo, L., Sciandrone, M.: Nonmonotone derivative-free methods for nonlinear equations. Computational Optimization and Applications 37, 297–328 (2007)
5. Figueiredo, M.A.T., Nowak, R.D., Wright, S.J.: Gradient projection for sparse reconstruction: Application to compressed sensing and other inverse problems. IEEE Journal of Selected Topics in Signal Processing 1, 586–598 (2007)
6. Pang, J.S.: Inexact Newton methods for the nonlinear complementary problem. Mathematical Programming 36, 54–71 (1986)
7. Solodov, M.V., Svaiter, B.F.: A globally convergent inexact Newton method for systems of monotone equations. In: Fukushima, M., Qi, L. (eds.) Reformulation: Nonsmooth, Piecewise Smooth, Semismooth and Smooth Methods, pp. 355–369. Kluwer Academic Publishers (1998)
8. Xiao, Y.H., Wang, Q.Y., Hu, Q.J.: Non-smooth equations based method for l_1-norm problems with applications to compressed sensing. Nonlinear Analysis: Theory, Methods & Applications 74, 3570–3577 (2011)
9. Yan, Q.R., Peng, X.Z., Li, D.H.: A globally convergent derivative-free method for solving large-scale nonlinear monotone equations. Journal of Computational and Applied Mathematics 234, 649–657 (2010)

10. Yu, G.H., Niu, S.Z.: Multivariate spectral gradient projection method for large-scale nonlinear systems of monotone equations,
    http://www.paper.edu.cn/index.php/default/
    en_releasepaper/downPaper/201201-778
11. Zhang, L., Zhou, W.J., Li, D.H.: A descent modified Polak-Ribiere-Polyak conjugate gradient method and its global convergence. IMA Journal of Numerical Analysis 26, 629–640 (2006)

High-Quality Synthetic Aperture Auto-imaging under Occlusion

Zhengxi Song, Yanning Zhang, Tao Yang, and Xiaoqiang Zhang

School of Computer Science, ShaanXi Provincial Key Laboratory of Speech and
Image Information Processing, Northwestern Polytechnical University, Xi'an, China
{lirpa,vantasy}@mail.nwpu.edu.cn, ynzhang@nwpu.edu.cn,
yangtaonwpu@163.com

Abstract. This paper proposes a novel method to see-through occlusion and automatically focuses on object with high-level imaging quality
using camera array. Even with the amazing perspective identity, synthetic aperture imaging still suffers from blurs and disturbance caused
by occlusion. The novelties of the approach include: (1) Rather than
using the direct observed images to achieve synthetic aperture image,
this paper raises the idea to synthesize edge image, which synthetic binary images after edge detection. (2) Based on the special data identity
of camera array, this paper proposes an "Auto-Cut" segmentation idea,
which could upgrade interactive cut method, such as GrowCut, GrabCut
and Graph Cut, to a totally automatic method. (3) This paper proposes
an automatically selecting the focal depth method which could yield a
convincing estimation even under serious occluded situation. The feasibility of our approach is experimentally demonstrated. A multi-view
images based improved synthetic aperture imaging system has been set
up, and experimental results with qualitative and quantitative analysis
demonstrate that the method can improve imaging quality and resist
occlusion in challenge scene.

Keywords: Synthetic aperture image, visibility analysis,
"Auto-GrowCut".

1 Introduction

Occluded object imaging is a challenging problem in the fields of computer vision
and pattern recognition, and it plays a significant role in the application of
camera array. By using synthetic aperture imaging, camera array could see-
through occlusions like dense crowd or foliage, when we synthetically "focus" this
virtual lens on objects behind occlusions. Artifacts often occur when synthesizing
occluded areas in each view, see Fig.1(b). Prevalent imaging system usually
tries to solve degraded imaging problem by building a denser and larger virtual
lens camera array which unnecessarily increases the hardware expense. Many
cut and alpha matting algorithms inspired the idea of labeling the occluded
areas and synthesizing only visible areas to upgrade imaging quality, either of
each asking for a manual trimap or interactive seeds. Under the SAI translation

J. Yang, F. Fang, and C. Sun (Eds.): IScIDE 2012, LNCS 7751, pp. 407–416, 2013.

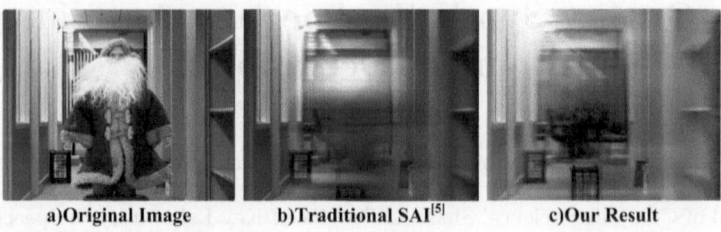

a)Original Image   b)Traditional SAI[5]   c)Our Result

Fig. 1. Traditional synthetic aperture imaging result and our result

condition, calibrated images from camera array could focus on any objects when the focusing depth is obtained. Previous method simplifies this problem within un-occluded object situation and hypothesis that low variance at a depth plane implies that a number of features are aligned and thus an object is present; However, visibility analysis masks the original images which make each pixel of synthetic image averaging from different number views, therefore make the variance evaluation incomparable.

This paper proposes a novel system that can make visibility analysis and see-through occluded object automatically with improved imaging quality, see Fig.1(c). Inspired by the work of Neel Joshi[1], which ingeniously take the uniqueness of camera array captured data and computed a trimap based on synthetic variance image to later matting process, we propose an idea of upgrading the regular manual cut method to an automatic algorithm on the synthetic image. Afterward with the known parallax we could project this segmentation result to each view and mask the occluded pixels. Rather than sweep a synthetic variance focal plane through the scene, we scan a synthetic edge image to automatic select focal plane. The result is an automatic algorithm that avoid troubles like computing the background depth and reconstructing the 3D scene at a more arbitrarily environment. Additionally, it could significantly improve the image quality.

The main contributions of our paper are as following: (1) the paper proposes a new synthetic aperture imaging model, synthetic edge images, which could provide special information for later process. (2) By observing the unique multi-view from camera array, we modify the interactive grow-cut to totally automatic algorithm, which could get rid of user input. (3) This paper presents a method through synthetic edge imaging to estimate the depth of occluded object in the scene, when traditional method could not give an accurate estimation. (4) We design a new multi-view synthesis method, where only interested pixels put into a synthetic process which efficiently promote the imaging quality and remove unnecessary blurs. (5) We design an improved synthetic aperture imaging framework which could handle occlusion situation with no approximation of background, which for the first time integrate color and edge information together.

In the next section, we will discuss some of the related work in this area. In section 3, we will present our occluded object perspective automatic imaging algorithm and propose two improved synthetic imaging model. Lastly, we present experimental results with UCSD multi-view database in section 4, followed by a discussion of our method and our conclusions.

2   Related Work

Followed by the pioneer work of Levoy et al, who proposed the concept of synthetic aperture imaging, quite a few camera array capture systems have been built to promote the development of computer vision and computer graphics: the Stanford multi-camera array [2], the SAIIP camera array [3] and the U of Alberta camera array system[4]. Concerning the new identity of data from camera array, researchers take efforts to improve the performance of synthetic aperture imaging itself meanwhile they also explore new approach to solve some existing challenges like image-based rendering, natural video matting, high speed videography and synthetic aperture tracking.

Vaish et al.[5] created a simple but robust procedure to calibrate camera arrays, known a plane + parallax framework, which totally replaces the full metric calibration. Despite the perspective view ability is fantastic, averaging method also lead to ghost image with ungraded quality. Wilburn et al are the first to set effort to solve the ghost problem [6]. Through computing the temporal variance of each pixel of every view in a time interval and creating a binary mask to identify the visibility, the method works well when static occlusion against moving objects after training. This resembles our work in spirit, in that it using a binary mask to instructive imaging process, but it differs from our method in that it need training process and require the static occlusion. Pei et al.[7] refine the synthetic aperture image method through background subtraction. By labeling the occluded areas in each camera, encouraging result is yield when it applied in scenes with moving occlusion.

Image cut is an old problem which has been research for almost half a century. Works like Graph Cut, GrabCut and GrowCut have been put into practical use. Graph Cut is a combinatorial optimization technique, which was applied by [8] to the task of image segmentation. By treating the image as a graph where each pixel as a graph node, the global optimization can be efficiently computed by maxflow/min-cut algorithms. However, it needs user to specify object and background seed pixels. The Graph Cut method then has been extended by GrabCut[9], which introduce the concept of iterative segmentation scheme. But it is also an interactive algorithm which calls for user draws rectangle around the object of interest. The GrowCut[10] method combines the advantages, distributed between the mentioned methods (multi-label segmentation, N-dimensional images processing, speed high enough for interactive segmentation), and offers more - algorithm extensibility by varying the automaton evolution rule, more interactivity and user control of the segmentation process. Therefore, we use GrowCut in our occluded cut system. Despite all the priorities of the GrowCut method, it's still need the aid of users. It seems like the bottleneck of all cut problems which are all under intuitive user interaction scheme even with outstanding performance.

Neel Joshi for the first time use the camera array to automatically matte in the natural scene which is a good example to take advantages of the unique data identity. The system uses high frequencies present in natural scenes to compute mattes by creating a synthetic aperture image that is focused on the foreground object, which reduces the variance of pixels reprojected from the foreground while increasing the variance of pixels reprojected from the background.

3   High-Quality Occluded Object Auto-imaging

Our work is similar with the method above in spirit, which we used data identity of camera to provide part of intuitive seed before cut. Where our work differs is as follow: Rather than only variance, we combine variance and synthetic edge image to generate the trimap which direct the following segmentation; after segmenting synthetic image which focused at the occlusion by Auto-GrowCut, we use the known parallax to label occlusion; different from scanning higher order statistics (i.e., variances) of image measurements and computing the depth of minimum variances, we synthesis images (i.e., means) on edge images. Specifically, our algorithm proceeds as the Fig.2 shows, which mainly include two parts: Visibility analysis by using auto-GrowCut and occluded focal plane estimation.

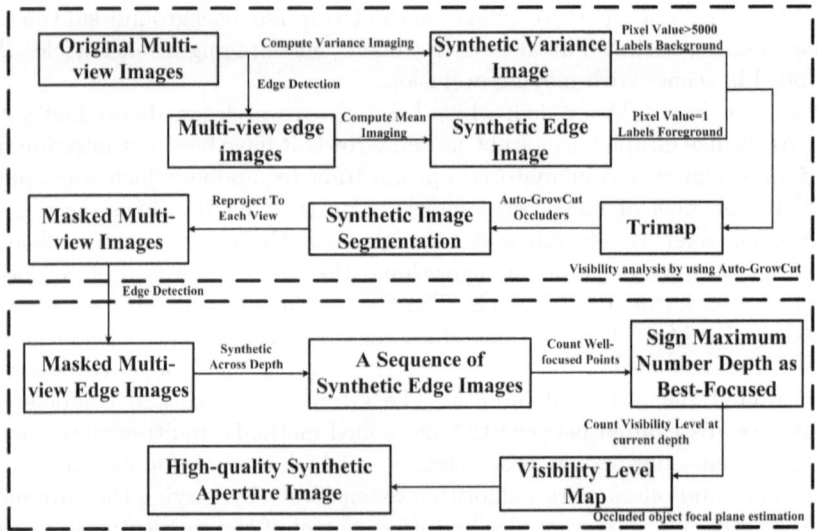

Fig. 2. Flow chart of the high-quality occluded object auto-imaging algorithm

In the first module, by extracting the edge information of original images for each view, we could combine the traditional Synthetic Aperture Imaging model and Synthetic Edge Aperture Imaging model which this paper proposed to construct a trimap. Then the trimap can be input to GrowCut Algorithm and eliminate the corresponding pixel on the reference view. Based on the known parallax, such segmentation can be projected to each view. In the second module, as the Fig.2 shows, by re-extracting the edge information of masked images for each view, we automatically find the object depth plane and apply the effective SAI to remove the effect of occluders on focal depth plane. The per-pixel running time of the algorithm (including the computation of the edge detection and the trimap) is linear in the number of cameras in the array. We will first describe the two new synthetic aperture imaging method, as they form the core of our contribution. Discussion of the whole steps of the algorithm implement and its result is deferred to section 4.

3.1 Synthetic Edge Aperture Imaging

Given n images of a scene, we consider the following traditional synthetic aperture imaging equation of a given scene point p:

$$I_{SA}(p) = \frac{1}{n} \sum_{i=1}^{n} H_i \circ I_i(p) \tag{1}$$

Where $I_i(p)$ corresponds to the intensity of point p recorded in image i and H_i denotes the homography required to project I_i on the desired focal plane. And then $I_{SA}(p)$ is the synthetic aperture image of a scene point p and $H_i \circ I_i$ reference the projection of image I_i on to the focal plane.

The intensity is changing smoothly in two dimensions original observed images and have large similarity area. Synthetic aperture image, usually refer to the first-order moments (i.e., mean) of image, is sharp and clear on the focal plane while blur and aliasing off this plane. Common method prefer to use higher order moments (i.e., variance) value to evaluate the clarity, which pixel on this plane is relatively small near to zero while off this plane is quite large. This method suffers from two problems, which make it invalid on some situation. At the segmentation process, when the scene have large similarity area in color, pixel belong to different object would have same color value which cause the variance is unexpectedly small. When upgrading the variance image to trimap, it would have error areas which have relative small value while they belong to background. The other problem is that when original images are masked in the visibility process variance would be incomparable in a synthetic image. Some pixels could be seen by all views while some pixels could be seen by several views. In this sense, when sweeping the scene to find focal plane, some relative small variance value pixel could not only mean it is well focused and it would be could be too barely in visible level to arouse high variance.

So we introduce the concept of synthetic aperture edge images. In order to maintain the continuity, we choose "canny" operator to extract the edge information of each image. Then we construct the synthetic edge image equation:

$$E_{SA}(p) = \frac{1}{n} \sum_{i=1}^{n} H_i \circ E_i(p) \tag{2}$$

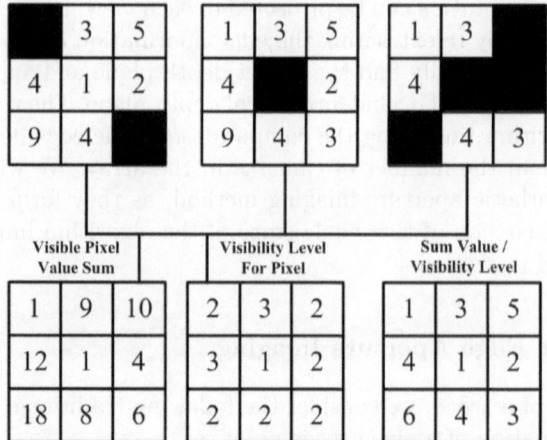

Fig. 3. Key diagram of effective imaging model

where E_i is the edge images of each view and E_{SA} is the synthetic edge image. Since E_i is binary image which only reflect the structure in each view and eliminate the disturbance of color. Under this circumstance, how could this synthetic model would be better than above method? By synthesize binary image, the well-focused points would be show extremely near to 1, see Fig.4(d), which could exactly meet the GrowCut input need. Such points can be label as foreground seeds. On the other hand, by synthesizing edge image in each depth, we could clearly find that focal plane would gather more focused points. So we could count the number of focal points to estimating the focal plane. Moreover, since we use the first-order moment measurement of the edge image, well-focused point would also be relatively high to 1 when areas of some views have been masked.

3.2 Effective Synthetic Aperture Imaging

After visibility analysis, we obtain images which eliminate pixels of occluders. We promote the synthetic model to be a more effective synthetic aperture imaging, where we synthetic after calculating a visibility level map at focal plane, so-called visibility analysis. Follow equation is the improved synthetic imaging model:

$$I_{SA}(p) = \frac{1}{visible(p,depth)} \sum_{i=1}^{visible(p,depth)} H_i \circ I_i(p)$$

$$I_i = \begin{cases} I_i \to mask(I_i) = 1 \\ 0 \to mask(I_i) = 0 \end{cases} \tag{3}$$

where $visible(p, depth)$ is based the visibility analyzing map, who is a function of point p of a given scene and targeting focal depth. We reproject each point p to each corresponding point in every view and judge whether it is belong to masked area or view area. In the Eq.3, $mask(I_i) = 1$ means it is a visible point and the intensity of this point could be used to synthesize. Fig.3 illustrates the

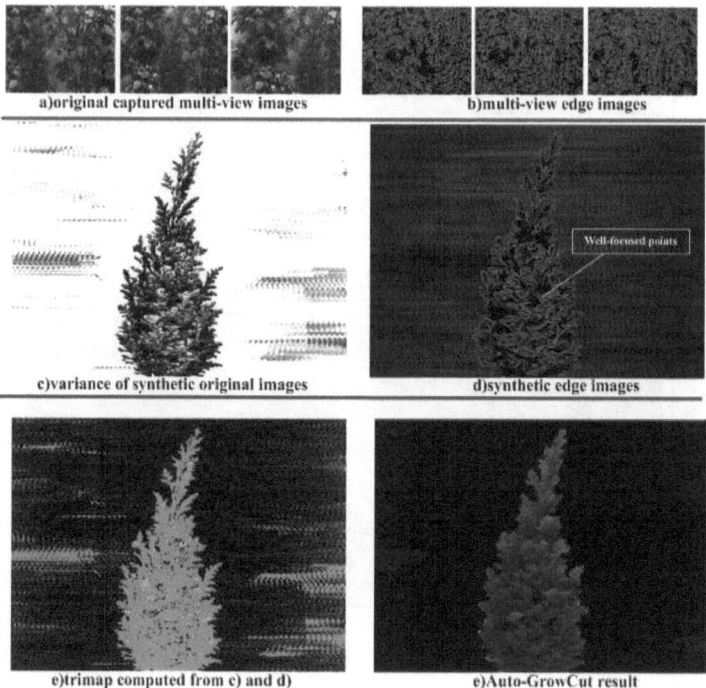

a)original captured multi-view images   b)multi-view edge images

c)variance of synthetic original images   d)synthetic edge images

e)trimap computed from c) and d)   e)Auto-GrowCut result

Fig. 4. "Auto-GrowCut" result of Tree scene in the database

key diagram of this imaging model, the upper three grids(3*3) represents three multi-view images from camera array after parallax compensation, where the dark squares represent the masked pixels and the digits mean the pixel value. After counting the visibility levels($1-3$) for each pixel and summing each valid pixel up, we could achieve a occlusion removal image by taking division for each part. Finally, we experiments demonstrate such synthetic model would highly improve the imaging quality.

4   Experimental Results

We test our results on the UCSD/Light Field data base: Trees scene and Santa-doll office scene, which contain 120 different views of a single scene. The central view #60 is defined as the reference camera and we use views from #40 to #80 to automatically remove the blurs of foreground objects and achieve high-quality imaging on the target object. We will demonstrate our results for each step and compare it with traditional synthetic result.

After extracting edge information from each view, we could find that the well focal point would be synthesized from all edge points(value 1) whose value should also be 1, see Fig.4(d).So we label these points as the foreground seeds which instead the user input. Moreover, we refer to the color variance image

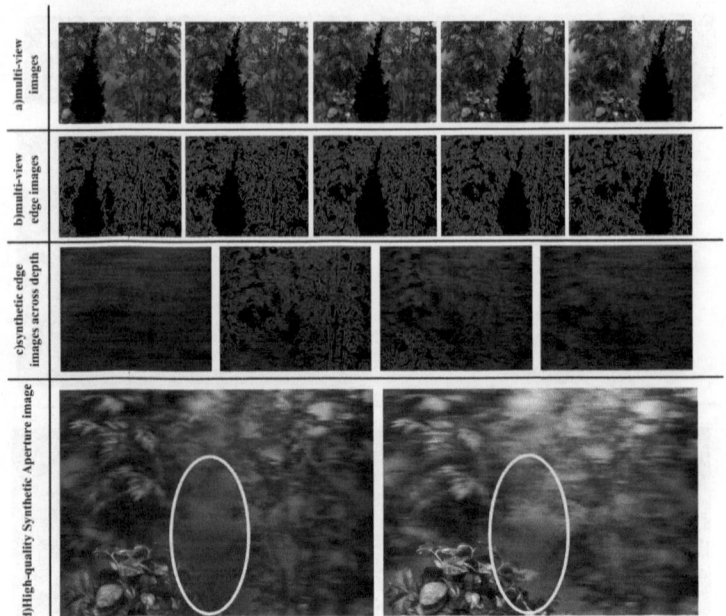

Fig. 5. Occluded object synthetic aperture imaging computing precess and result

to label points whose value higher than 5000. Therefore, we compute a trimap automatically to instruct following GrowCut, as Fig.4(e). By serving this trimap into standard GrowCut algorithm, we can get cut result as Fig.4(f). Such segmentation could be directly masked the corresponding pixel on reference view. Then use the parallax to reproject these pixels on each view, Fig.5(a). Again, we re-extract the edge on the image after visibility analysis, Fig.5(b). By scanning this space with tiny step near to far, we calculate the number of well-focused points for each depth and sign the first maximum number show at the best focal plane, see Fig.5(c)(d).

Under the occlusion, traditional estimation which prior define some square as the A, B, C, D Fig.6(a)right and sum the variance value of all square across depth to find the depth of minimum value would provide ridiculous result, as the Fig.6(a)left. Despite these squares are all belong to target, some show a maximum value while others show a minimum value because of the different visibility level. Therefore, their sum could not provide a convincing result. However, in Fig.6(b),we could see that estimation on synthetic edge image would be more accurate and stable result than estimation based on color variance value.

We also test our algorithm in the other multi-view data from this data, which captured a santa-doll in the office scene. Fig.7 shows the results of our whole procedure. From the comparison of (d) and (e), we could see that our High-quality imaging could effectively remove the blur caused by occlusion.

Fig. 6. The comparison of two different focal plane estimation method

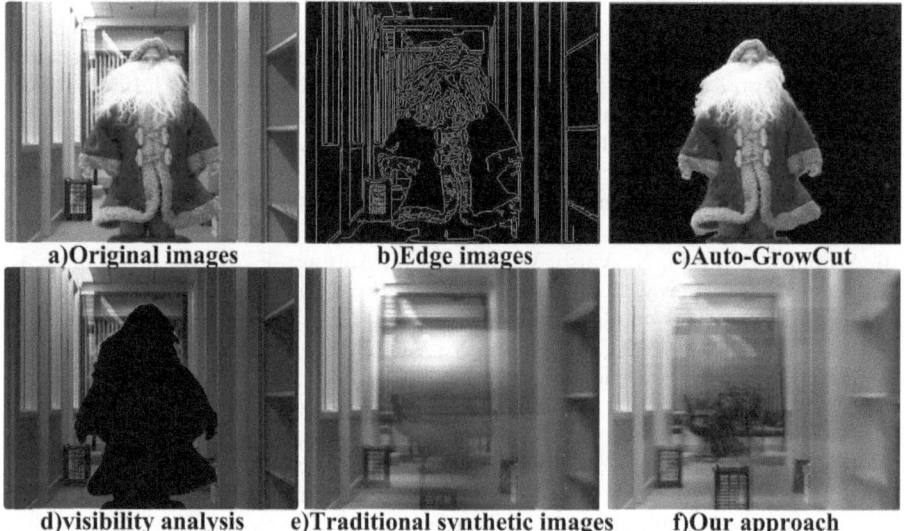

Fig. 7. Occluded object high-quality synthetic aperture imaging result of Santa-doll in office scene

5   Conclusion

We have presented an automatic system for high-quality occluded object imaging using camera array. In the visibility analysis process, we propose an "Auto-GrowCut" algorithm which works well on difficulty scenes without user input. By raising the synthetic edge model, robust focusing depth estimation could be made even under occlusion. The experimental results show that our method can synthesize the occluded object completely clear.

Acknowledgments. This work is supported by the National Natural Science Foundation of China (No.61272288,No.61231016,No.60903126), China Postdoctoral Special Science Foundation(No.201003685), China Postdoctoral Science Foundation(No.20090451397),the NPU Foundation for Fundamental Research (No.JC201120), and Plan of Soaring Star of Northwestern Polytechnical University(No.12GH0311).

References

1. Joshi, N., Matusik, W., Avidan, S.: Natural video matting using camera arrays. ACM Transactions on Graphics (TOG) 25, 779–786 (2006)
2. Vaish, V., Levoy, M., Szeliski, R., Zitnick, C., Kang, S.: Reconstructing occluded surfaces using synthetic apertures: Stereo, focus and robust measures. In: 2006 IEEE Computer Society Conference on Computer Vision and Pattern Recognition, vol. 2, pp. 2331–2338. IEEE (2006)
3. Yang, T., Zhang, Y., Tong, X., Zhang, X., Yu, R.: Continuously tracking and see-through occlusion based on a new hybrid synthetic aperture imaging model. In: 2011 IEEE Conference on Computer Vision and Pattern Recognition (CVPR), pp. 3409–3416. IEEE (2011)
4. Lei, C., Da Chen, X., Yang, Y.: A new multiview spacetime-consistent depth recovery framework for free viewpoint video rendering. In: 2009 IEEE 12th International Conference on Computer Vision, pp. 1570–1577. IEEE (2009)
5. Vaish, V., Wilburn, B., Joshi, N., Levoy, M.: Using plane+ parallax for calibrating dense camera arrays. In: Proceedings of the 2004 IEEE Computer Society Conference on Computer Vision and Pattern Recognition, CVPR 2004, vol. 1, pp. 1–2. IEEE (2004)
6. Wilburn, B., Joshi, N., Vaish, V., Talvala, E., Antunez, E., Barth, A., Adams, A., Horowitz, M., Levoy, M.: High performance imaging using large camera arrays. ACM Transactions on Graphics 24(3), 765–776 (2005)
7. Pei, Z., Zhang, Y., Yang, T., Zhang, X., Yang, Y.: A novel multi-object detection method in complex scene using synthetic aperture imaging. Pattern Recognition (2011)
8. Boykov, Y., Funka-Lea, G.: Graph cuts and efficient nd image segmentation. International Journal of Computer Vision 70(2), 109–131 (2006)
9. Rother, C., Kolmogorov, V., Blake, A.: Grabcut: Interactive foreground extraction using iterated graph cuts. ACM Transactions on Graphics (TOG) 23, 309–314 (2004)
10. Vezhnevets, V., Konouchine, V.: Growcut: Interactive multi-label nd image segmentation by cellular automata. In: Proc. of Graphicon, pp. 150–156 (2005)

Convergence Properties of Perceptron Learning with Noisy Teacher

Kazushi Ikeda[1], Hiroaki Hanzawa[1], and Seiji Miyoshi[2]

[1] Nara Institute of Science and Technology
Ikoma, Nara 630-0192 Japan
kazushi@is.naist.jp
[2] Kansai University
Suita, Osaka 564-8680 Japan
miyoshi@kansai-u.ac.jp

Abstract. This paper analyzed convergence properties of an online learning method when teacher's signal includes noise in the thermodynamic limit. The learning curve was analytically derived using a statistical mechanical method and its validity was confirmed by computer simulations. In this case, the learning curve shows an overshoot phenomenon. In order to elucidate why and how it occurs in this case, the asymptotic analysis of dynamical systems was applied to the differential equations that expresses the dynamics of the learning curve and showed that the phenomenon results from the properties of the system matrix of the equations.

Keywords: Learning curve, perceptron, online learning, statistical mechanics, asymptotic analysis.

1 Introduction

Statistic mechanical methods can apply to problems in information science such as neural networks [1] and communication theory [2]. One successful application is the analyses of the perceptron learning algorithm [3, 4] and its variations [6–11].

The perceptron learning is an online algorithm where the student updates its weight vector of a linear dichotomy according to the teacher's signal [3]. Biehl and Schwarze introduced the statistical mechanics to the analysis of the perceptron learning [4] and Inoue and Nishimori applied the method to the AdaTron learning in unlearnable cases [6]. Hara and Okada discussed the perceptron learning with a margin [7] and Miyoshi and his colleagues extended the analysis to the ensemble learning and/or noisy cases [8–11].

In this paper, we analyze the case where the teacher has noise in its output while the student does not. In this case, the learning curve (the average prediction error) is not monotonically decreasing, differently from other noisy cases, but has an overshoot. Because the overshoot depends on the learning coefficient, the analysis (and control) of the phenomenon enables the student to perform better.

J. Yang, F. Fang, and C. Sun (Eds.): IScIDE 2012, LNCS 7751, pp. 417–424, 2013.

Our analysis consists of two steps. In the first step, we apply the statistical mechanical method to our problem. In other words, we introduce three order parameters assuming the thermodynamic limit, and derive a system of differential equations. In the second step, we apply the asymptotic analysis of dynamical systems to the differential equations. More specifically, we linearize the equations around their convergence point and analyze their behaviors by the eigenvalues and eigenvectors of the system matrix (state-transition matrix). The two steps elucidate how and why the overshoot phenomenon occurs.

The rest of this paper is organized as follows. Section 2 formulates the problem we treat. Sections 3 and 4 are devoted to statistical mechanical analysis and asymptotic analysis of dynamical systems, respectively. In Section 5, we discuss the results and conclude in Section 6.

2   Problem Statement

Two linear perceptrons are treated: a teacher and a student, whose connection weights are $B = (B_1, \ldots, B_N) \in R^N$ and $J = (J_1, \ldots, J_N) \in R^N$, respectively. The initial value of each of the components are independently drawn from the normal distribution $N(0, 1)$, that is,

$$\langle B_i \rangle = 0, \qquad\qquad \langle (B_i)^2 \rangle = 1, \qquad\qquad (1)$$

$$\langle J_i \rangle = 0, \qquad\qquad \langle (J_i)^2 \rangle = 1, \qquad\qquad (2)$$

where $\langle \cdot \rangle$ denotes the mean of $\cdot$.

The mth input vector $x^m = (x_1^m, \ldots, x_N^m) \in R^N$ is independently drawn from the N-dimensional normal distribution $N(0, I/N)$ and the corresponding output y^m of the teacher is

$$y^m = \mathrm{sgn}(v_m), \qquad\qquad v_m = B^m \cdot x^m + n_B^m, \qquad\qquad (3)$$

where n_B^m is an observation noise obeying $N(0, \sigma_B^2)$.

Given the mth input vector x^m, the student updates its weight vector J^m when its output does not coincide with the teacher's. Thus,

$$J^{m+1} = J^m + f^m x^m, \qquad\qquad f^m = \eta y^m \Theta(-y^m J^m \cdot x^m), \qquad\qquad (4)$$

where η is a learning coefficient and $\Theta(\cdot)$ is the Heaviside function,

$$\Theta(t) = \begin{cases} 1 & t \geq 0, \\ 0 & t < 0. \end{cases} \qquad\qquad (5)$$

As the learning proceeds, the weight vector J^m of the student approaches the teacher's B. The problem of learning curves is to derive how fast the covariance R^m between J^m and B,

$$R^m = \frac{B \cdot J^m}{\|B\| \|J^m\|}, \qquad\qquad (6)$$

approaches unity.

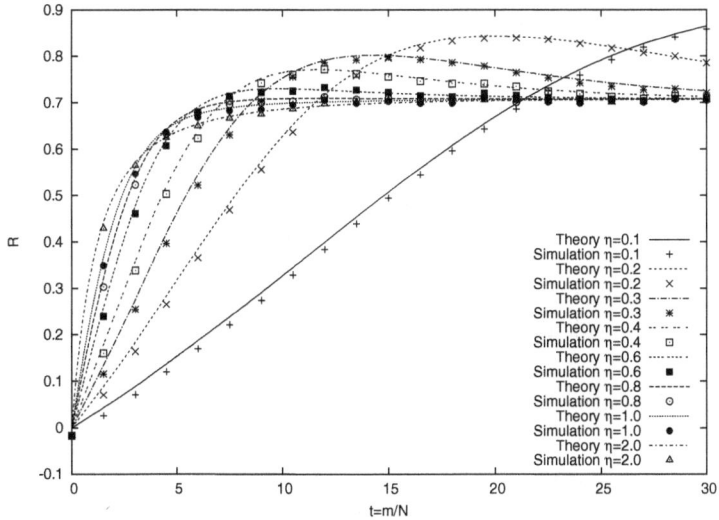

Fig. 1. Dynamics of R. $\sigma_B^2 = 1.0$, $\eta = 0.1, \ldots, 2.0$, plots: experiments, lines: theory

3   Statistical Mechanical Analysis

3.1   Theory

Introduce auxiliary order parameters, R^m and

$$l^m = \|J^m\|/\sqrt{N}, \tag{7}$$

and consider the thermodynamic limit, $N, m \to \infty$ and $m/N = t$. Then,

$$\|B\| = \sqrt{N}, \qquad\qquad \|J^0\| = \sqrt{N}, \qquad\qquad \|x^m\| = 1, \tag{8}$$

hold and the random vector of the inner products

$$v^m = B \cdot x^m, \qquad\qquad u^m l^m = J^m \cdot x^m \tag{9}$$

obeys the two-dimensional normal distribution $N(0, \Sigma)$ where

$$\Sigma = \begin{pmatrix} 1 & R^m \\ R^m & 1 \end{pmatrix}. \tag{10}$$

By self-averaging and omitting the step index m in (4), we get the simultaneous differential equations of the order parameters,

$$\dot{l} = \langle fu \rangle + \frac{\langle f^2 \rangle}{2l}, \tag{11}$$

$$\dot{R} = \frac{\langle fv \rangle - \langle fu \rangle R}{l} - \frac{R}{2l^2} \langle f^2 \rangle, \tag{12}$$

where $\langle \cdot \rangle$ expresses the average over (u, v) and $n_B \sim N(0, \sigma_B^2)$ [1]. Here, the ensemble average $\langle fv \rangle$ is calculated as

$$\langle fv \rangle = \langle \eta \Theta(-u(v + n_B)) \text{sgn}(v + n_B) v \rangle \tag{13}$$

$$= \eta \int_{-n_B}^{\infty} dv \int_{-\infty}^{0} du \int_{-\infty}^{\infty} dn_B P(u, v) P(n_B) v$$

$$- \eta \int_{-\infty}^{-n_B} dv \int_{0}^{\infty} du \int_{-\infty}^{\infty} dn_B P(u, v) P(n_B) v, \tag{14}$$

where $P(u, v)$ and $P(n_B)$ are Gaussian distributions [5]. In the same way, $\langle fu \rangle$ and $\langle f^2 \rangle$ are also calculated.

3.2   Experiments

To confirm the validity of the theory above, we conducted some computer simulations. The experimental values (plots) of R coincided well with the theoretical values (lines) for any learning coefficient η (Fig. 1).

The value of R converged to 0.70 for any η due to the noise on the teacher's output. One notable property was that R was not monotonically increasing but had an overshoot. This overshoot phenomenon does not occur in other cases (Table 1). In the next section, we quantitatively analyze this phenomenon using an asymptotic analysis of dynamical systems.

Table 1. Overshoot phenomena

Teacher \ Student	noisy	noise-free
noisy	×	○
noise-free	×	×

4   Asymptotic Analysis of Dynamical Systems

The ensemble averages in (11) and (12) cannot explicitly be calculated since it includes integrals such as (14). However, they are approximated well by the following linear functions of R (Fig. 2):

$$\langle fu \rangle \approx 0.28\eta R - 0.40\eta, \tag{15}$$

$$\langle fv \rangle \approx -0.40\eta R + 0.28\eta, \tag{16}$$

$$\langle f^2 \rangle \approx -0.24\eta^2 R + 0.50\eta^2. \tag{17}$$

These approximations reduce the differential equations (11) and (12) to

$$\dot{l} = 0.28\eta R - 0.4\eta + \frac{-0.24\eta^2 R + 0.5\eta^2}{2l}, \tag{18}$$

$$\dot{R} = \frac{0.28\eta(1 - R^2)}{l} - \frac{R}{2l^2}(-0.24\eta^2 R + 0.5\eta^2), \tag{19}$$

which still expresses the experimental results well (Fig. 3).

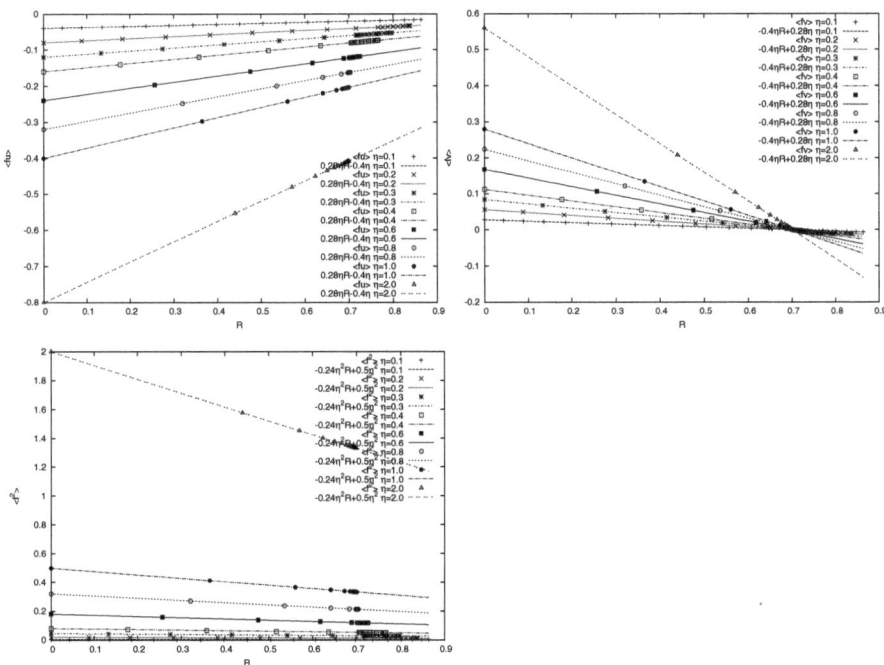

Fig. 2. Approximation of ensemble averages. The linear functions of R approximates the points well.

In order to analyze the behaviors of differential equations (18) and (19), we change their variables R and l to

$$e = (1 - R) - 0.30, \qquad\qquad d = 1/l - 1.23/\eta, \qquad (20)$$

so that $e, d \to 0$ as $t \to \infty$. Then, (18) and (19) are rewritten as

$$\dot{e} = 0.1162\eta^2 d^2 + 0.1428\eta d - 0.593542\eta ed - 0.605574e$$
$$- 0.082\eta^2 ed^2 - 0.01494\eta e^2 d + 0.163172e^2 - 0.12\eta^2 e^2 d^2 \qquad (21)$$
$$\dot{d} = -0.408\eta d^2 - 0.250699d - 0.16241\eta ed^2 + 0.144509ed$$
$$- 0.166\eta^2 d^3 + 0.200152\frac{e}{\eta} - 0.12\eta^2 ed^3. \qquad (22)$$

The linear approximation of the above around the origin is

$$\begin{pmatrix} \dot{e} \\ \dot{d} \end{pmatrix} = \begin{pmatrix} -0.606 & 0.143\eta \\ 0.200/\eta & -0.251 \end{pmatrix} \begin{pmatrix} e \\ d \end{pmatrix} \qquad (23)$$

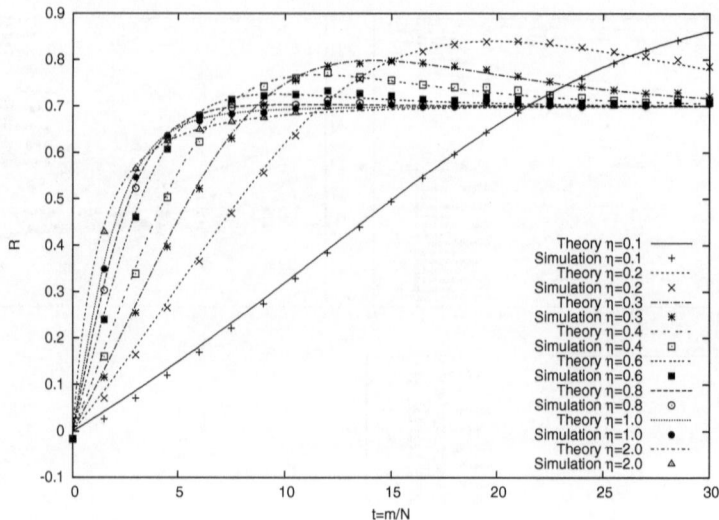

Fig. 3. Dynamics of R. $\sigma_B^2 = 1.0$, $\eta = 0.1, \ldots, 2.0$, plots: experiments, lines: theory with approximation.

in matrix form. The eigenvalues and corresponding eigenvectors of the system matrix are

$$\lambda_1 = -0.67, \qquad u_1 = \begin{pmatrix} \eta \\ -0.43 \end{pmatrix}, \tag{24}$$

$$\lambda_2 = -0.18, \qquad u_2 = \begin{pmatrix} \eta \\ 3.07 \end{pmatrix}. \tag{25}$$

Because $|\lambda_2| < |\lambda_1|$, the component along u_2 decreases more slowly than that along u_1. Moreover, u_2 is in the first/third quadrant of the e, d-plane while u_1 is in the second/fourth quadrant. Hence, the points in the second quadrant move to the third quadrant once and then go to the origin along u_2 (Fig. 4). Since $R = 0.70 - e$, the above explains why the overshoot of R occurs.

5   Discussion

The asymptotic analysis of dynamical systems (21) and (22) elucidated that the overshoot phenomenon originated from the difference of the decay speed along u_1 and u_2. Since the linear approximation in the previous section is experimental, we need to develop a new theory that justifies the approximation.

As for the dependency on the learning coefficient η, Fig. 3 suggests that we can achieve a faster convergence and a lower residual error (a larger R) if we adjust η so that the learning curve traces the envelop of the curves with fixed coefficients.

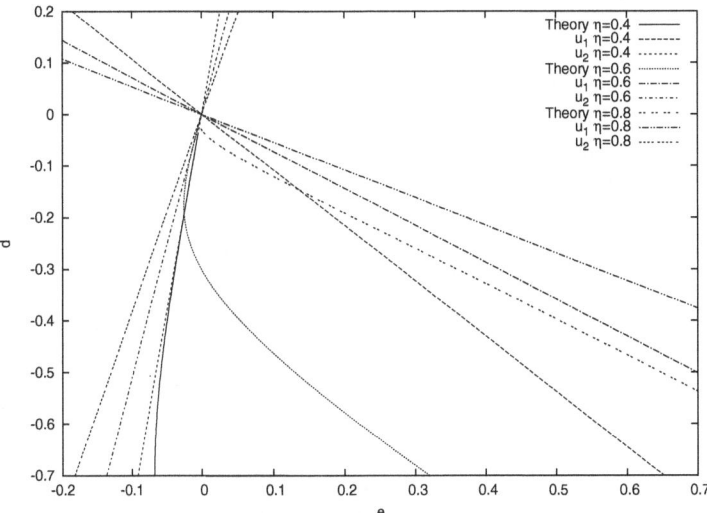

Fig. 4. Eigenvectors and traces of (e, d). $\sigma_B{}^2 = 1.0$. The difference of the eigenvalues and the direction of the corresponding eigenvectors induce the curves.

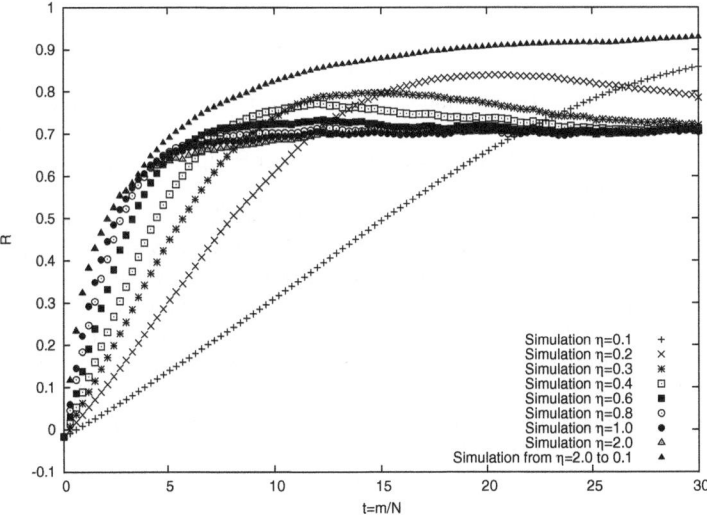

Fig. 5. Dynamics of R with variable η. $\sigma_B{}^2 = 1.0$. R with variable η (triangle) is larger than any other R with fixed η.

In fact, if we decrease the learning coefficient η from 2 to 0.1 so that R is maximized at each time step, then the value of R traces higher than any learning curves with fixed η (Fig. 5).

6   Conclusions

In this paper, we analyzed convergence properties of the perceptron learning with noisy teacher. The learning curve in this case was analytically derived using a statistical mechanical method and was consistent with the experimental results in our simulation. Differently from other cases, the learning curve has an overshoot, that is, the covariance R of the teacher and the student exceeds the convergence value once. We showed that this phenomenon results from the difference of the eigenvalues and eigenvectors of the system matrix using the asymptotic analysis of dynamical systems. This result may give a method for controlling the learning coefficient η to achieve a faster convergence speed and a lower residual error in the future.

References

1. Nishimori, H.: Statistical Physics of Spin Glasses and Information Processing: An Introduction. Oxford Univ. Press, Oxford (2001)
2. Tanaka, T.: A Statistical-Mechanics Approach to Large-System Analysis of CDMA Multiuser Detectors. IEEE Trans. Information Theory 48(11), 2888–2910 (2002)
3. Rosenblatt, F.: Principle of Neurodynamics. Spartan, Washington, D.C. (1961)
4. Biehl, M., Schwarze, H.: Online Learning of a Time-Dependent Rule. Europhysics Letters 2, 733–738 (1992)
5. Hanzawa, H.: An Asymptotic Analysis of Perceptron Learning with Noisy Teacher. Bachelor's thesis, Kansai University (2009)
6. Inoue, J., Nishimori, H.: On-line AdaTron Learning of Unlearnable Rules. Physical Review E 55(4), 4544–4551 (1997)
7. Hara, K., Okada, M.: On-line Learning through Simple Perceptron Learning with a Margin. Neural Networks 17, 215–223 (2004)
8. Miyoshi, S., Hara, K., Okada, M.: Analysis of Ensemble Learning Using Simple Perceptrons Based on Online Learning Theory. Physical Review E 71, 036116 (2005)
9. Miyoshi, S., Okada, M.: Analysis of On-line Learning When a Moving Teacher Goes around a True Teacher. J. Physical Society of Japan 75(2), 024003 (2006)
10. Miyoshi, S., Okada, M.: Statistical Mechanics of Online Learning for Ensemble Teachers. J. Physical Society of Japan 75(4), 044002 (2006)
11. Uezu, T., Miyoshi, S., Izuo, M., Okada, M.: Theory of Time Domain Ensemble On-line Learning of Perceptron under the Existence of External Noise. J. Physical Society of Japan 76(11), 114006 (2007)

An Improved Approximate K-Nearest Neighbors Nonlocal-Means Denoising Method with GPU Acceleration

Wenchao Jin and Jinqing Qi

School of Information and Communication Engineering,
Dalian University of Technology, Dalian, China
`jinwenchao@mail.dlut.edu.cn, jinqing@dlut.edu.cn`

Abstract. The nonlocal-means(NLM) is a denoising algorithm which takes advantage of the redundancy of similar patches in the image. While producing state-of-the-art denoising results, the NLM algorithm has high computational complexity. In [1] Ce Liu *et al.* introduced approximate K-nearest neighbors(AKNN) matching to classical NLM for reducing the complexity of the algorithm. In this paper an improved AKNN-NLM algorithm with NVIDIA GPU acceleration is proposed. The experiments show that the improved GPU based AKNN-NLM algorithm have excellent performance on both image and video denoising. The GPU based implementation is up to 40 times faster than the CPU counterparts.

Keywords: Denoising, Nonlocal-means, GPU, Approximate K-nearest neighbors.

1 Introduction

Image denoising is still a challenging problem despite large amount of research has devoted to them. Buades *et al* .introduced the nonlocal-means filter to image denoising [2]. It explicitly exploits self-similarities in natural images. Although it has very good performances on image denoising, it has high computational complexity which is due to the cost of weights calculation. Many methods have been proposed to accelerate the original NLM. The main focus is on the patch selection [3,4]. In [1] Ce Liu *et al.* proposed the approximate K-nearest neighbors match to the NLM algorithm. The AKNN-NLM method breaks the original NLM framework which averages out the pixels in one search window. Instead, it uses a randomized correspondence [1,5] algorithm to obtain K-nearest neighbors for every pixel, then uses these K pixels to average out the denoised pixel. The algorithm reduces the complexity of computation.

Although the AKNN-NLM algorithm proposed by *Liu* reduces the computational complexity, it is still quite slow. It has three steps including initialization, propagation and random search. The pixels are processed one by one in a scanline or reverse scanline order serially. So in this paper we consider to implement the AKNN-NLM algorithm by GPU. As the fast development of GPU, more and

J. Yang, F. Fang, and C. Sun (Eds.): IScIDE 2012, LNCS 7751, pp. 425–432, 2013.

more people tend to use it for more general purposes than its original graphic related work [6,7]. NVIDIA CUDA (Compute Unified Device Architecture) technology is a fundamentally new computing architecture that enables the GPU to solve very complex computational problems. It provides a CUDA C language programming interface to the NVIDIA GPUs. In this paper, we use NVIDIA Geforce GT 430 GPU for our experiments.

The rest of the paper is organized as follows. In Section 2 we will briefly introduce the GPU architecture and CUDA programming model. In Section 3 we review the AKNN-NLM denoising algorithm and GPU implementation details will be described. In Section 4 we show the experimental results on GPU using CUDA C language and on Matlab. In Section 5 we give some conclusions.

2   GPU and CUDA

Graphics Processing Unit(GPU) is a hardware unit which is used in computer to render graphic information for users. GPU is specialized for compute-intensive, highly parallel computation. In November 2006, NVIDIA introduced CUDA, a general purpose parallel computing architecture with a new parallel programming model and instruction set architecture that leverages the parallel compute engine in NVIDIA GPUs to solve many complex computational problems in a more efficient way than on a CPU.

The CUDA parallel programming model is simply exposed to the programmer as a minimal set of language extensions. GPU has a large amount of streaming multi-processors(SM). One SM contains several scalar processors, shared memory, and a multi-threaded instruction unit. When a kernel (GPU function) is launching on the GPU, a 2D grid is distributed to available SMs. The grid can divided into blocks logically, and each block can also divided into threads. The threads in the same block can communicate with each other by shared memory. The threads in different blocks usually use global memory to share data.

In our experiments, we use 4.0 version of CUDA and SDK. The hardware we choose NVIDIA graphics card Geforce GT 430 which has 96 streaming multi-processors.

3   GPU Based AKNN-NLM Implementation

In order to have a clear comparison between classic NLM and AKNN-NLM algorithm, we first briefly review these two methods then the GPU implementation details will be presented.

3.1   Nonlocal-Means Method

Nonlocal-means estimates the pixel i with a weighted average of all pixels in the image. We denote the noisy image by $v(i)$, where i is pixel index. NLM is performed according to the following formula.

$$NL(v)(i) = \sum_{j \in I} w(i, j) v(j). \tag{1}$$

$$w(i,j) = \frac{1}{W} e^{-\frac{||u(N^d(i))-u(N^d(j))||^2_{2,Ga}}{h^2}}. \tag{2}$$

Where $NL(\bullet)$ is the operator of NLM. $N^d(i)$ represents the square patch of size $(2d+1)\times(2d+1)$ centered at pixel i ,and W is a normalizing term, $W = \sum_j w(i,j)$, The parameter h will referred to as the filter parameter. I is a square search window centered around pixel i .Usually, the search window is 21×21.

3.2 AKNN-NLM Method

For reducing the complexity of the classical NLM method, In [1] Ce Liu *et al.* denoted approximate K-nearest neighbors(AKNN) to NLM video denoising. The follows are AKNN-NLM algorithm steps.

Step1 Initialization. Every pixel initializes K neighbors by randomization.

Step2 Propagation. This idea is based on the fact that neighboring pixels tend to have similar AKNN structure. This procedure intertwines between scanline order and reverse scanline order. In scanline order we use $v(x-1, y)$ and $v(x, y-1)$ to improve $v(x, y)$, while in reverse scanline order we use $v(x+1, y)$ and $v(x, y+1)$ to improve $v(x, y)$.We use $v(x - 1, y)$ to improve $v(x, y)$ as an example. There is no need to recalculate the patch distance as illustrated in Fig.1.

Step3 Random Search. After the propagation, we allow every patch to match other patches in the image for M times. Then iterating step2 and step3 for several times.At last K-nearest neighbors will be obtained.

Step4 Weighted Average. Obtain the K-nearest neighbors for every pixel and compute the weighted average to get the denoisied image.

3.3 Implementation Based on GPU

In CUDA image processing application, CPU and GPU always cooperate to fulfill a task. CPU usually acts the host role and GPU acts the server role. So we must extract the parallelism of the algorithm as much as possible to implement it by CUDA parallel programming.

Initially, one noisy image is read on the host (CPU) side and is transfer to the GPU for processing. Texture memory is a good option to save the raw image data for non-coalesced readings. Then the algorithm is executing by 4 steps mentioned above, we will describe each step for GPU implementation as follows.

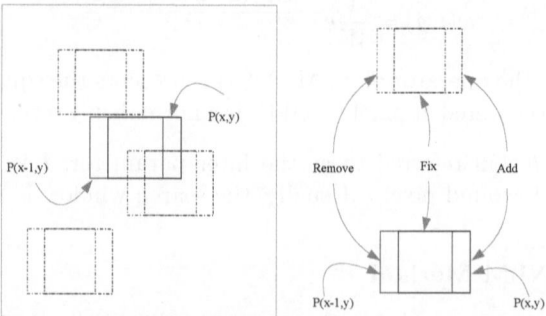

Fig. 1. The approximate K-nearest neighbors of P(x, y) (blue) can be improved by the AKNN of P(x-1, y) (red). The dotted line patches are the AKNN of solid line patch with the same color. Left: the K-nearest neighbors of P(x-1,y)(*dotted red pathch*) shift one pixel to the right and generate K neighbors of P(x, y) (*dotted blue patch*). Right: The distance of the patch P(x, y) and its new K neighbors can simply take the distance between P(x-1,y) and its neighbors, then remove the left column's distance and add the right column's distance.

Initialization. The AKNN-NLM algorithm relies heavily on random numbers. The CUDA 4.0 toolkit provides the CURAND library [8] that focus on the simple and efficient generation of high quality pseudorandom and quasirandom numbers. We use the device side API which can avoid generating random number in host side and transferring to GPU memory. We firstly generate the random seeds by GPU side and store them in the global memory. We use 16×16 block, so there is 256 threads in every block. Both the grids and blocks are 2 dimensions. GPU kernels always use this structure. Then every pixel is responsible for one thread, and generates random seed for its later use.

After that, every pixel should randomly generate K neighbors in a certain ranges. Three 2D arrays with size of $[imageW * imageH, K]$ are needed. Here the $imageW * imageH$ means the image width multiply the image height. The results coordinates of the neighbor and the distances are stored separately in the global memory. As it is said above non-coalesced access to global memory will decrease the performance. CUDA provides us cudaMallocPitch() function [9] to allocate the 2D array which meet the coalesced access requirement.

Propagation. Propagation plays the key role in the whole algorithm. In Matlab implementation this procedure starts with the pixel in left up, then serially propagates to the last pixel. Every pixel's K neighbors are improved by the updated pixel next to them. The propagation and the random search are performed in interleaving manner. This structure is obviously not suitable for GPU parallel implementation. We must optimize the algorithm's procedure to make it proper for GPU implementation. We make the propagation and random search to execute in a totally separate way. All pixels execute propagation in one kernel, after that we sort the 3K neighbors' distance for every pixel, at last random search is performed and sorting is done once again.

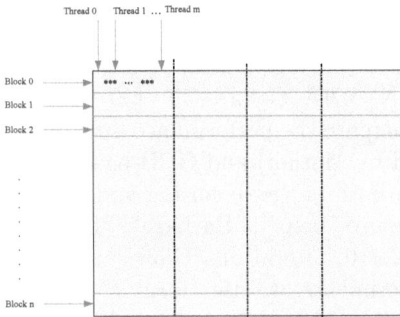

Fig. 2. We distribute n blocks which equals the image height. Every block is responsible for processing one row of the image. The row is divided into 4 segments. The first m pixels(*first segment of one row*)will be processed, and they run in parallel. After they finish, the next segment of the row will be processed. Until the whole row is finished, this GPU kernel will return.

So firstly, we describe the propagation kernel. In this step, every pixel is responsible for improving its own K neighbors by its two neighbor pixels'(left pixel and up pixel or right pixel and under pixel) 2K neighbors. We use 1 dimension block instead, every block process one row of the image, also the grid is 1 dimension too. Because every pixel's neighbor information will be frequently read during propagation. Shared memory is used for temporary storage of pixel's neighbors' information such as coordinates and distance. Fig.2 shows the blocks and threads structure in propagation kernel. We can see that the length of the grid equals to the image height, while the length of the block is shorter than the image width. For example in our experiment 256×256 image is used, and the length is determined to 64 which costs 15KB shared memory. Every row is divided into 4 segments, so it needs 4 times to finish the whole propagation task in one block.

Random Search. After the propagation, every thread does the ascending sort for every 3*K distances (K for own initialization, the other 2*K for the neighbors' propagation), and get the closest K neighbors. Then every neighbor can match the patches in the image randomly for M times. In our experiments we choose M=3. The grid structure can be the same as initialization.

Rendering with OpenGL. The propagation and the random search will iterate several times, In [1] it is said 4 times is enough to get the best result. Finally, we get the approximate K-nearest neighbors for every pixel, and use them to compute the weighted-average results. We choose OpenGL to rendering. It provides useful tools such as split window display.

4   Experimental Results

Our experiments are done on the machine with Dual-Core 3.3GHz, 2G RAM. The graphics card is NVIDIA Gefore GT 430 with 96 multi-processors and 512MB memory. We compare the performance and speed among classical NLM, AKNN-NLM(processed by Matlab) and GPU based AKNN-NLM methods. We did the experiments on both image denoising and video denoising. In image denoising experiments, we use "Lena", "Barbara", "Peppers", "Child" images with size of 256×256, and K=10, 4 iterations. Gaussian white noise($\sigma = 20$) is added. In video denoising experiments, we use "forman", "bus", "mobile" videos with size of 352×288. Video denoising requires reliable motion estimation[10,1]. We use 5 temporal frames in our experiments. Table 1 to Table 2 show the PSNR and Elapsed time comparison for these three methods.

In our experiments the PSNR is defined as follows:

$$PSNR = 10 \log(\frac{255^2}{MSE}).$$ (3)

From Table 1 and Table 2, we can see that AKNN-NLM algorithm does have a soft acceleration comparing with the original NLM method. While GPU based method will have a huge acceleration rate comparing with Matlab based implementation. The largest acceleration rate is up to 40 times.

Table 1. PSNR and elapsed time comparison for image denoising

Image	NLM(Matlab)		AKNN-NLM(Matlab)		AKNN-NLM(CUDA)	
	PSNR	Time(s)	PSNR	Time(s)	PSNR	Time(s)
Lena	33.01	130	32.76	90	32.68	10
Barbara	32.5	134	32.44	94	32.47	10
Peppers	32.4	130	32.46	90	32.48	9
Child	31.9	125	31.97	94	32.01	10

Table 2. PSNR and elapsed time comparison for video denoising

Video	NLM(Matlab)		AKNN-NLM(Matlab)		AKNN-NLM(CUDA)	
	PSNR	Time(s)	PSNR	Time(s)	PSNR	Time(s)
Forman	33.6	384	33.9	183	33.93	11
Bus	31.0	389	31.4	184	31.2	12
Mobile	30.9	400	31	190	31.2	11

From Fig.3, we can see that the performance of Matlab and CUDA implementation is almost the same, while GPU implements a 10 times acceleration rate.

From Fig.4, over smoothness is easily seen in classical NLM results. The tree in "bus" frame are becoming blurring. But by the AKNN-NLM method, edges are mostly prevented. The GPU based AKNN-NLM also performs well and gain a maximally 40 acceleration rate. If we use higher configuration GPU, the results would be even faster.

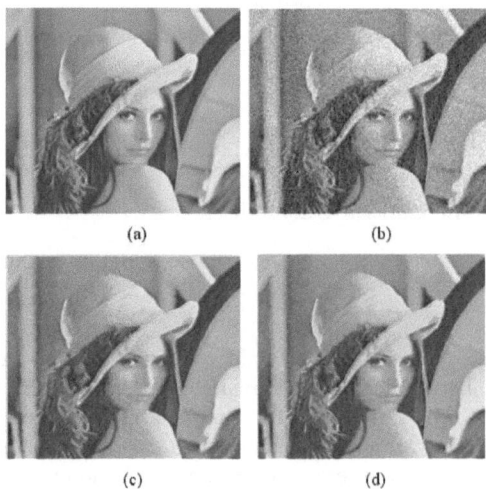

Fig. 3. (a) is the original "Lena" image, (b) is the noisy image($\sigma = 20$), (c) is the result with AKNN-NLM method on Matlab, (d) is the GPU based AKNN-NLM result.

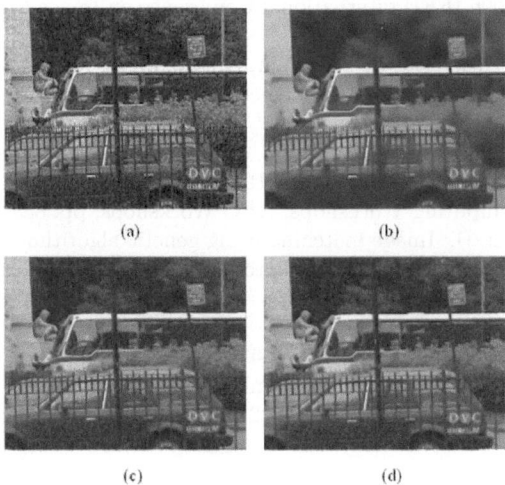

Fig. 4. (a) is the 4th frame of bus video and Gaussian White noise($\sigma = 20$) is added, (b) result with NLM algorithm, (c) result with AKNN-NLM method, (d) result with GPU based AKNN-NLM method.

5   Conclusions

The AKNN-NLM method largely reduces the complexity of the classical NLM. It gets a soft acceleration with excellent denoising performance. Throught NVIDIA GPUs, we totally extract the parallel characters of the AKNN-NLM algorithm and implement a maximal 40 acceleration rate on Geforce GT 430 graphics card. The experiments show the good denoising performance both on image and video denoising.

Acknowledgments. The authors would like to thank Professor Huchuan Lu for his valuable comments. This work was supported in part by "the Fundamental Research Funds for the Central Universities".

References

1. Liu, C., Freeman, W.T.: A High-Quality Video Denoising Algorithm Based on Reliable Motion Estimation. In: Daniilidis, K., Maragos, P., Paragios, N. (eds.) ECCV 2010, Part III. LNCS, vol. 6313, pp. 706–719. Springer, Heidelberg (2010)
2. Buades, A., Coll, B., Morel, J.: A non-local algorithm for image denoising. In: IEEE Computer Society Conference on Computer Vision and Pattern Recognition, CVPR 2005, vol. 2, pp. 60–65. IEEE (2005)
3. Mahmoudi, M., Sapiro, G.: Fast image and video denoising via nonlocal means of similar neighborhoods. IEEE Signal Processing Letters 12(12), 839–842 (2005)
4. Wang, J., Guo, Y., Ying, Y., Liu, Y., Peng, Q.: Fast non-local algorithm for image denoising. In: 2006 IEEE International Conference on Image Processing, pp. 1429–1432. IEEE (2006)
5. Barnes, C., Shechtman, E., Finkelstein, A., Goldman, D.: Patchmatch: a randomized correspondence algorithm for structural image editing. ACM Transactions on Graphics (TOG) 28, 24 (2009)
6. Li, J., Wang, X., He, R., Chi, Z.: An efficient fine-grained parallel genetic algorithm based on gpu-accelerated. In: IFIP International Conference on Network and Parallel Computing Workshops, NPC Workshops, pp. 855–862. IEEE (2007)
7. Ke, Y., Li, Y., Li, D.: Image matching using genetic algorithm on gpu. In: 2011 International Conference on Control, Automation and Systems Engineering (CASE), pp. 1–4. IEEE (2011)
8. Cuda toolkit 4.0 curand guide (2011)
9. Nvidia cuda c programming guide version 4.2 (2012)
10. Buades, A., Coll, B., Morel, J.M.: Nonlocal image and movie denoising. International Journal of Computer Vision 76(2), 123–139 (2008)

Frameworks for Multimodal Biometric
Using Sparse Coding

Zengxi Huang, Yiguang Liu[*], Ronggang Huang, and Menglong Yang

Vision and Image Processing Laboratory, College of Computer Science,
Sichuan University, Chengdu, P.R. China
lygpapers@yahoo.com.cn

Abstract. In this paper, we will introduce three frameworks for multimodal biometric using sparse representation based classification (SRC), which has been successfully used in many classification tasks recently. The first framework is multimodal SRC at match score level (MSRC_s), in which feature of each modality is sparsely coded independently, and then their representation fidelities are used as match scores for multimodal classification. The other two frameworks are multimodal SRC at feature level (MSRC_f1, MSRC_f2), where features of all modalities are first fused and then classified by using SRC. The difference between them is that MSRC_f1 fuses the features to form a unique multimodal feature vector, while MSRC_f2 implicitly combines the features in an iterative joint sparse coding process. As a typical application, the fusion of face and ear for human identification is investigated by using the three frameworks. In our experiments, Principal Component Analysis (PCA) based feature extraction is applied. Many results demonstrate that the proposed multimodal methods are significantly better than the multimodal recognition using common classifiers. Among the SRC based methods, MSRC_s gets the top recognition accuracy in almost all the test items, which might benefit from allowing sparse coding independence for different modalities.

Keywords: Multimodal biometric, Sparse Representation, Match score level, Feature level.

1 Introduction

The original goal of sparse representation (or coding, SR) was for representation and compression of signals, potentially using lower sampling rates than the Shannon-Nyquist bound [1]. Nevertheless, Wright et al. [2] reckoned that the sparse representation is naturally discriminative and then designed a novel classification scheme, namely sparse representation based classification (SRC), which was employed in face recognition (FR) and achieved impressive performance. SRC could be seen as a more general model than the previous nearest classifiers, like Nearest Neighbor (NN), Nearest Feature Line (NFL) [3] and Nearest Subspace (NS) [4, 5], and it uses the samples from all classes to collaboratively represent the query sample to overcome the small-sample-size problem in FR [6].

---

[*] Corresponding author.

J. Yang, F. Fang, and C. Sun (Eds.): IScIDE 2012, LNCS 7751, pp. 433–440, 2013.

SRC based techniques have been widely applied to various object classification tasks, such as FR [2, 6], flower classification [7]. In almost all of these applications, using sparsity as a prior leads to state-of-the-art results. However, to the best of our knowledge, there is still not any report of applying sparse coding to multimodal biometric. Of course, some multi-features or multi-samples based classifications, which share similar fusion mechanism with multimodal biometric, using sparsity constraint have been reported [7, 8]. In multimodal biometric systems, evidences can be fused at five different levels, i.e., sensor data, feature, match score, rank, and decision levels. Since fusion at the later level, the less information can be used for classification, thereby rank and decision levels' fusions have rarely been adopted in community. Compared with match score and sensor data levels, fusion at feature level can exploit the most discriminative information and eliminate the redundant/adverse information from the original biometric data, and hence is much more popular recently. Overall, for the multimodal biometric using sparse coding, fusions at match score and feature levels might be the most reasonable choices.

In this paper, we will introduce three frameworks for multimodal biometric using sparse coding. The first framework is multimodal SRC at match score level (MSRC_s), in which feature of each modality is sparsely coded independently, and then their representation fidelities are used as match scores. The other two frameworks are multimodal SRC at feature level (MSRC_f1, MSRC_f2) where all features are first fused and then classified by using SRC. The difference between them is that MSRC_f1 fuses the features to form a unique multimodal feature vector, while MSRC_f2 implicitly combines the features in an iterative joint sparse coding process. As a typical application, the fusion of face and ear for human identification is investigated by using the three frameworks. In our experiments, Principal Component Analysis (PCA) [9] based feature extraction is applied. Many results demonstrate that the proposed methods are significantly better than commonly used classifiers, like NN and NFL. Among the frameworks, MSRC_s gets the top performance in almost all the test items, which may benefit from allowing sparse coding independence for different modalities. Match score level fusion based method appears to be more robust. From the viewpoint of sparse coding, we also give many discussions about their working mechanisms and the experimental results, and deduce a conclusion that striking a good balance between the similarity and distinctness of the coding vectors may bring better robustness for multimodal biometric recognition.

The rest of this paper is organized as follows: Section 2 gives the details of the proposed multimodal biometric frameworks, including MSRC_f1, MSRC_f2 and MSRC_s. Section 3 conducts experiments on multimodal databases for evaluating the frameworks. Finally, our conclusions are summarized in Section 4.

2   The Proposed Multimodal Biometric Frameworks

Face is considered to be the most acceptable and promising biometric. However, the face by itself is not yet as accurate and flexible as desired due to makeup, eyeglasses, illumination and expressions. On the other hand, the ear has some appealing advantages over the face: a) ear has a rich and stable structure; b) ear has a uniform distribution of color; c) ear is small, and requires less computational time [10]. In this paper, for the simplicity and convenience of introducing our multimodal frameworks, the face and ear based multimodal biometric is investigated as a case study.

2.1   Feature Extraction

The pioneer SRC research in FR [2] had shown that with sparsity properly harnessed, the choice of features becomes less important than the number of features used. Besides, SRC methods are relatively time-consuming compared to the commonly used classification methods. Hence, a simple, time-saving and general feature extraction method is desired for our SR based multimodal biometric. In our frameworks, PCA is utilized for face and ear feature extractions.

Suppose that we have c classes of subjects in a multimodal database, $A^f = \begin{bmatrix} A_1^f, A_2^f, \cdots, A_c^f \end{bmatrix}$ and $A^e = \begin{bmatrix} A_1^e, A_2^e, \cdots, A_c^e \end{bmatrix}$ separately denote the face and ear training sample sets, where $A_i = \begin{bmatrix} a_{i,1}, a_{i,2}, \cdots; a_{i,m} \end{bmatrix}$ ($i = 1, 2, \cdots, c$, m samples per class) is the subset from class i. According to the PCA technique, the training feature sets can be computed by: $D^f = \left(P^f \right)^T A^f$ and $D^e = \left(P^e \right)^T A^e$, where P^f and P^e are PCA projection matrices that calculated from the face and ear training datasets, respectively. And the feature vectors of face and ear query images (y^f and y^e), can be calculated by: $z^f = \left(P^f \right)^T y^f$, $z^e = \left(P^e \right)^T y^e$.

2.2   Our Frameworks

MSRC_f1 and MSRC_f2. The general feature fusion techniques include serial concatenation, parallel fusion using a complex vector, and CCA-like methods that extracts correlation feature of two modalities. Compared with the other methods, serial concatenation that our multimodal frameworks adopt is simple, effective and easy to extend for combining more than two modalities. In MSRC_f1, the multimodal feature of query data is obtained by $z = \begin{bmatrix} z^f; z^e \end{bmatrix}$. Likewise, the multimodal feature dictionary is constructed by $D = \begin{bmatrix} D^f; D^e \end{bmatrix}$. MSRC_f1 seeks to find a sparse representation of z in terms of dictionary D, the l_1-norm minimization problem can be formulated as

$$\hat{a} = \arg \min \|a\|_1 \text{ s.t. } \|z - Da\|_2 < \varepsilon, \tag{1}$$

where $\hat{a} = [\hat{a}_1; \hat{a}_2; \cdots; \hat{a}_c]$, $\hat{a}_i$ is the coefficient vector associated with class i.

The classification of MSRC_f1 is based on the multimodal feature coding error that yielded by each class. The classification rule can be defined by

$$g(y) = \arg \min_i \left\{ \left\| z - D_i \, \hat{a}_i \right\|_2 \right\}, \tag{2}$$

where D_i is the subset of multimodal features from class i.

Unlike in MSRC_f1, the face and ear features, in MSRC_f2, are not directly combined to form a unique multimodal feature vector, but are jointly represented in an iterative sparse coding process instead, which is an implicit feature fusion way. The goal of joint sparsity of MSRC_f2 can be achieved by imposing $l_{1,2}$ mixed-norm on the coding coefficients. Suppose $\boldsymbol{a}^f = \left[\boldsymbol{a}_1^f ; \boldsymbol{a}_2^f ; \cdots ; \boldsymbol{a}_c^f \right]$, $\boldsymbol{a}^e = \left[\boldsymbol{a}_1^e ; \boldsymbol{a}_2^e ; \cdots ; \boldsymbol{a}_c^e \right]$, where $\boldsymbol{a}_i^f$ and $\boldsymbol{a}_i^e$ are the face and ear coding vectors associated with class i. Let $\boldsymbol{a}_i = \left[\boldsymbol{a}_i^f , \boldsymbol{a}_i^e \right]$ ($\boldsymbol{a}_i \in R^{m \times 2}$). MSRC_f2 can be defined by

$$\min_{\boldsymbol{a}^f , \boldsymbol{a}^e} \left\| \boldsymbol{z}^f - \boldsymbol{D}^f \boldsymbol{a}^f \right\|_2^2 + \left\| \boldsymbol{z}^e - \boldsymbol{D}^e \boldsymbol{a}^e \right\|_2^2 + \lambda \sum_i^c \left\| \boldsymbol{a}_i \right\|_2 , \tag{3}$$

which can be solved by using the $l_{1,2}$ mixed-norm APG algorithm introduced in [7]. And its classification rule is

$$g(\boldsymbol{y}) = \arg \min_i \left(\left\| \boldsymbol{z}^f - \boldsymbol{D}_i^f \hat{\boldsymbol{a}}_i^f \right\|_2^2 + \left\| \boldsymbol{z}^e - \boldsymbol{D}_i^e \hat{\boldsymbol{a}}_i^e \right\|_2^2 \right). \tag{4}$$

MSRC_s

In SRC, the representation fidelity is often measured by the l_2-norm of sparse coding error, which can also be used as a distance score to denote how "different" between the query sample and the training samples of a class. So, in MSRC_s, we first sparsely represent the face and ear features ($\boldsymbol{z}^f , \boldsymbol{z}^e$) in terms of their associated dictionaries ($\boldsymbol{D}^f , \boldsymbol{D}^e$), respectively. After that, we fuse the sparse coding errors of face and ear features by using Sum-rule. And the two l_1-minimization problems can be formulated as follows:

$$\hat{\boldsymbol{a}}^f = \arg \min \left\| \boldsymbol{a}^f \right\|_1 \quad \text{s.t.} \quad \left\| \boldsymbol{z}^f - \boldsymbol{D}^f \boldsymbol{a}^f \right\|_2 < \varepsilon , \tag{5}$$

$$\hat{\boldsymbol{a}}^e = \arg \min \left\| \boldsymbol{a}^e \right\|_1 \quad \text{s.t.} \quad \left\| \boldsymbol{z}^e - \boldsymbol{D}^e \boldsymbol{a}^e \right\|_2 < \varepsilon , \tag{6}$$

The match score fusion can be defined by

$$r_i = \left\| \boldsymbol{z}_i - \boldsymbol{D}_i^f \hat{\boldsymbol{a}}_i^f \right\|_2 + \left\| \boldsymbol{z}_i - \boldsymbol{D}_i^e \hat{\boldsymbol{a}}_i^e \right\|_2 , \quad i = 1, 2, \cdots, c \tag{7}$$

Since r is a distance score, the classification would be ruled in favor of the class which has the lowest r.

2.3   Theoretical Comparison

From the viewpoint of information fusion, MSRC_f1 and MSRC_f2 are the frameworks that combine face and ear at feature level, while MSRC_s integrates them at match score level. Because of its capability of utilizing more information for classification, feature level fusion has been considered to be the most promising avenue for multimodal biometric. Therefore, intuitively, the former two approaches might perform better than MSRC_s.

From the viewpoint of sparse coding, their differences mainly lie on the constraint imposed on the coding vectors of face and ear. In MSRC_f1, the modalities are enforced to share the same coding vector, that is to say, $\alpha^f = \alpha^e$ is required. This requirement may help to correctly find the coding vector especially when some query samples of modalities are of good quality while the others are not, so as to make the multimodal recognition more robust than the unimodal recognition. However, on the other hand, if these query samples encounter the degeneration severe to some certain extent, this requirement may deteriorate the estimation of sparse coding vector. Thus, such strong requirement in MSRC_f1 can be seen as a double-edged sword. In contrast to MSRC_f1, MSRC_s sparsely codes the features independently. Since each sparse coding process does not affect any others, MSRC_s may lead to the lowest overall reconstruction error of all modalities, which, though, does not necessarily result in the best classification performance. Compared with MSRC_f1 and MSRC_s, MSRC_f2 does not require the coding vectors to be the same or completely independent, but allows flexibility between the similarity and distinctness of the coding vectors, which is a moderate constraint.

3   Experiments and Discussions

For evaluating the performance of our multimodal biometric methods, we build up two virtual multimodal databases with 38 and 79 subjects, respectively, based on publicly available databases, including the Extended Yale B [11], AR [12] face databases and the USTB ear database III [13], of which sample images of one person are showed in Fig. 1, where the images in red rectangle are used as gallery images and the rest are used for testing. We name the databases as Multimodal Database I and II (MD I and II), respectively, whose constitutions are described in details in Table 1. For obtaining more instances for testing, each face test image is paired with every ear test image. All the face and ear images are normalized to have a size of 50×40.

In the experiments on each multimodal database, we use the same dimension of face feature vector and ear feature vector. The dimensionalities are selected empirically as 120 and 200 for MD I and II, respectively. The commonly used NN, NFL classifiers are used as references in comparison. We call their multimodal extensions as Multimodal NN (MNN) and Multimodal NFL (MNFL), respectively, which employ the same feature fusion scheme as that MSRC_f1 used.

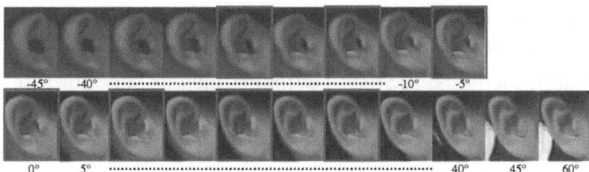

Fig. 1. Sample images of one subject of USTB ear database III

Table 1. The constitutions of MD I and II

Multimodal database		Instances		Face	Ear
MD I (38)	Gallery	7×38 = 266	Extended Yale B	Subset 1 (7 images per subject)	Gallery (38)
	Probe	38×13×38 = 18772		Subset 2, 3, 4 (38 images per subject)	Probe (38)
MD II (79)	Gallery	7×79 = 563	AR	Subset 1 (7 images per subject without occlusion from Session 1)	Gallery (79)
	Probe Subset 1	7×13×79 = 7189		Subset 2 (7 images per subject without occlusion from Session 2)	Probe (79)
	Probe Subset 2	6×13×79 = 6162		Subset 3 (6 images per subject with sunglasses)	
	Probe Subset 3	6×13×79 = 6162		Subset 4 (6 images per subject with scarf)	

Recognition without Occlusion

In this part, we will evaluate the performance of the proposed SR based methods with variations like illumination, pose and expression changes but without occlusion. Hence, on MD II, only the Subset 1 is used for testing. The experimental results are listed in Table 2. On MD I, a best recognition accuracy of 97.839% is yielded by MSRC_s, which is slightly better than the 97.552% of MSRC_f2. Compared with them, MSRC_f1 is obviously worse by about 4%. However, this disadvantage almost disappears on MD II, where MSRC_s achieves the top performance of 99.33%, only slightly better than MSRC_f1. Overall, MSRC_f2 and MSRC_s are comparable to each other, while they are better than MSRC_f1. Compared with the biometrics using common classifiers, i.e., MNN and MNFL, the superiority of our SR based methods can be seen clearly on all the multimodal databases.

Table 2. Multimodal recognition without occlusion

	MNN	MNFL	MSRC_f1	MSRC_f2	MSRC_s
MD I	72.171%	76.662%	93.663%	97.552%	**97.839%**
MD II	79.018%	82.25%	98.92%	99.027%	**99.33%**

Table 3. Multimodal recognition with real face disguise

	MNN	MNFL	MSRC_f1	MSRC_f2	MSRC_s
Subset 2	72.99%	78.469%	89.365%	92.365%	**95.344%**
Subset 3	41.344%	46.74%	90.375%	95.396%	**97.677%**

Recognition Despite Real Face Disguise

Here we use Subset 2 (face images with sunglasses) and Subset 3 (face images with scarf) of MD II to evaluate the robustness of the proposed methods against real face disguise. The detailed performance of all competing methods is listed in Table. 3. Clearly, the proposed SR based methods significantly outperform MNN and MNFL. Among the SR based methods, MSRC_s demonstrates its superior robustness again, it achieves the top performance of 95.344% and 97.677% on the two datasets, respectively. MSRC_f2 achieves the second best recognition accuracies on both datasets, which are 92.365% and 95.396%, respectively.

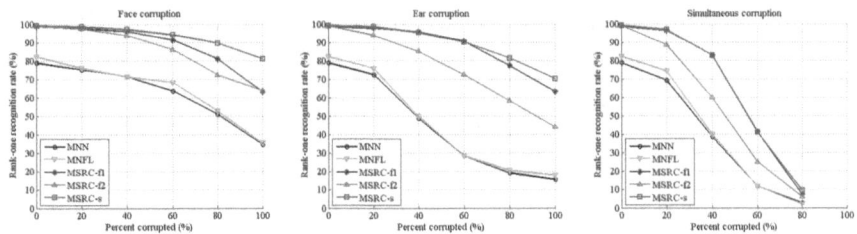

Fig. 2. Recognition on MD II versus different percentage of corrupted pixels. (a) Face corruption. (b) Ear corruption. (c) Simultaneous corruption.

Recognition Despite Random Pixel Corruption

Fig. 2 plots the recognition performance of all methods under the percentage of corrupted pixels of face or ear or both of them varies from 0% to 100% on MD II. With the increase of corrupted pixels, all the methods' recognition-rate curves fall, but their descend speeds are obviously different. In the case of face corruption, MSRC_s is significantly better than others. When the corruption is 100% of pixels, it still achieves recognition rate of 81.116%, which is much better than the 63.298% and the 63.83% of MSRC_f1 and MSRC_f2, respectively. When ear test images suffer from corruption, the superiority of MSRC_s becomes tight over MSR_f1, but their advantages over MSRC_f2 become more obvious. When test samples of both the two modalities suffer from corruption, it can be seen that MSRC_f1 and MSRC_s are comparable to each other, while they are still much better than MSRC_f2.

According to the above three series of experiments, the comparison of the proposed SR based methods appears to be very clear. The match score level fusion method, MSRC_s, is much more robust than the two feature level fusion based methods, especially MSRC_f1. Except in the random pixel corruption case, allowing flexibility of feature codings, such as in MSRC_f2 and MSRC_s, can lead to better biometric recognition performance. In the sparse coding frameworks, compared with the flexibility of feature codings of different modalities, it seems that the capability that utilizing more information of feature level fusion seems to become secondary. Although the complete independence of coding vectors leads to the best recognition performance, intuitively, such complete independence may bring deterioration to the multimodal recognition under some conditions, which just haven't been found in our experiments. It still deserves expectation that striking a good balance between the similarity and distinctness of the coding vectors may bring better robustness.

4   Conclusion

In this paper, we have introduced three frameworks for multimodal biometric using sparse coding. MSRC_f1 and MSRC_f2 are the frameworks that combine face and ear at feature level, while MSRC_s integrates them at match score level. In our experiments, face and ear based multimodal biometric is investigated by using the frameworks. Many results demonstrate that the proposed multimodal methods are significantly better than those using common classifiers, like NN and NFL classifiers.

Among the SR based multimodal methods, MSRC_s gets the top recognition accuracy in almost all the test items, which may benefit from allowing sparse coding independence for different modalities. However, striking a balance between the similarity and distinctness of the coding vectors should be expected to bring better robustness for multimodal biometric system.

Acknowledgment. This work is supported by the National Natural Science Foundation of China (NSFC) under Grants 61173182 and 61179071, and by Applied Basic Research Project (2011JY0124) and International Cooperation and Exchange Project (2012HH0004) of Sichuan Province.

References

1. Candes, E.: Compressive sampling. Proc. Int'l Congress of Mathematicians (2006)
2. Wright, J., Yang, A., Ganesh, A., Sastry, S., Ma, Y.: Robust face recognition via sparse representation. IEEE Trans. Pattern Anal. Mach. Intell. 31(2), 210–227 (2009)
3. Li, S., Lu, J.: Face recognition using the nearest feature line method. IEEE Trans. Neural Netw. 10(2), 439–443 (1999)
4. Watanabe, S., Lambert, P., Kulikowski, C., et al.: Evaluation and selection of variables in pattern recognition. In: Computer and Information Science II. Academic, New York (1967)
5. Liu, Y., Ge, S., Li, C., You, Z.: k-NS: a classifier by the distance to the nearest subspace. IEEE Trans. Neural Netw. 22(8), 1256–1268 (2011)
6. Yang, M., Zhang, L., Yang, J., Zhang, D.: Regularized robust coding for face recognition. arXiv: 1202.4207v2 [cs.CV] (2012)
7. Yuan, X., Yan, S.: Visual classification with multi-task joint sparse representation. In: IEEE Conf. on Computer Vision and Pattern Recognition, pp. 3493–3500 (2010)
8. Zhang, H., Nasrabadi, N., Zhang, Y., Huang, T.: Joint dynamic sparse representation for multi-view face recognition. Pattern Recognition 45(4), 1290–1298 (2012)
9. Turk, M., Pentland, A.: Eigenfaces for recognition. Journal of Cognitive Neuroscience 3(1), 71–86 (1991)
10. Abate, A., Nappi, M., Riccio, D.: RBS: a robust bimodal system for face recognition. International Journal of Software Engineering and Knowledge Engineering 17(4), 497–513 (2007)
11. Georghiades, A., Belhumeur, P., Kriegman, D.: From few to many: illumination cone models for face recognition under variable lighting and pose. IEEE Trans. Pattern Analysis and Machine Intelligence 23(6), 643–660 (2001)
12. Martinez, A., Benavente, R.: The AR face database, CVC Technical Report, No. 24 (1998)
13. University of Science & Technology Beijing (USTB),
    http://www1.ustb.edu.cn/resb/

Facial Expression Recognition
Using a New Image Representation
and Multiple Feature Fusion*

Zhiming Liu[1], Wei Wu[1,**], Qingchuan Tao[1], and Jian Yang[2]

[1] School of Electronics and Information Engineering, Sichuan University, China
zl9@njit.edu, {wuwei,taoqingchuan}@scu.edu.cn
[2] School of Computer Science and Technology,
Nanjing University of Science and Technology, China
csjyang@mail.njust.edu.cn

Abstract. This paper proposes a novel method for facial expression recognition using a new image representation and multiple feature fusion. First, the new image representation is derived from the normalized hybrid color space, by principal component analysis (PCA) followed by Fisher linear discriminant analysis (FLDA). Second, multi-scale local phase quantization (LPQ) features and patch-based Gabor features are applied to the new image representation and gray-level image, respectively, to extract multiple feature sets. Finally, due to the complementary characteristic between the new image representation and gray-level image, combining the classification results of multiple feature sets at the score level can improve recognition performance further. Experiments on Multi-PIE show that the proposed method achieves state-of-the-art performance for facial expression recognition.

1 Introduction

Facial expressions are a form of nonverbal communication, playing a primary role in conveying emotions and intentions among humans. Automatic recognition of facial expressions by computer vision technology is required by many applications such as human-computer interaction and computer graphics animation. Numerous methods have been proposed in the past two decades to address the problem of automatic facial expression recognition [1],[2]. Generally speaking, these methods fall into two major categories: geometric based methods and appearance based methods. As there exists commonness to a certain extent between facial expression recognition methods and facial identity recognition methods, some appearance based methods used widely in face recognition have been successfully applied to facial expression recognition. Some representative methods

---

* This work is supported by the National Natural Science Foundation of China under grant no. 61271330.
** Corresponding author.

J. Yang, F. Fang, and C. Sun (Eds.): IScIDE 2012, LNCS 7751, pp. 441–449, 2013.

include principal component analysis (PCA) and Fisher linear discriminant analysis (FLDA), which basically apply the global scatter information embedded in samples, and Gabor wavelets and local binary patterns (LBP) features, which primarily extract image information in a local region.

While most face recognition methods use gray-level images, a few research recently have advocated utilizing color information for boosting face recognition performance and have achieved promising results [3],[4],[5],[6]. In this paper, we investigate the effect of color information upon improving the performance of facial expression recognition. Specifically, we propose a new image representation that is derived from an uncorrelated color space, and then we propose an effective framework of multiple feature fusion for facial expression recognition.

The proposed method has been evaluated on the Multi-PIE database [7], a challenging facial expression database that contains real-world variabilities such as facial hair (mustaches and bears), glasses, and illumination. Experiments use 129 subjects who were recorded in all four sessions and experimental results show that the new image representation outperforms gray-level image for facial expression recognition and the proposed method can achieve state-of-the-art performance.

2   A New Image Representation for Facial Expression Recognition

Most, if not all, facial image analysis methods use gray-level images for face recognition and facial expression recognition. Gray-level images have been deduced from three primary colors, R, G, and B, i.e., the luminance image $Y = 0.299R + 0.58G + 0.114B$ and the intensity image $I = (R + G + B)/3$. However, no efforts have been dedicated to generating an image representation consisting of other color components. Furthermore, due to strong correlation among R, G, and B, a natural question is whether gray-level images can guarantee the best recognition accuracy for facial image analysis such as face recognition and facial expression recognition. Based on these motivations, this section derives a new image representation from the normalized hybrid color space $R\tilde{G}\tilde{Z}$ instead of RGB for facial expression recognition.

2.1   Normalized Hybrid Color Space: $R\tilde{G}\tilde{Z}$

Recently, Yang et al. [3] proposed the concept of color space normalization (CSN) to significantly reduce the correlation between color component images, thus enhancing their discriminating power for face recognition. When the across-color-component normalization (CSN-II) is applied to the hybrid color space RGZ, the normalized $R\tilde{G}\tilde{Z}$ can result in excellent face recognition accuracy [3]. Specifically, $R\tilde{G}\tilde{Z}$ is derived from RGB by the following transformation [3]:

$$\begin{bmatrix} R \\ \tilde{G} \\ \tilde{Z} \end{bmatrix} = \begin{bmatrix} 1 & 0 & 0 \\ -0.5774 & 0.7887 & -0.2113 \\ -0.4600 & -0.1986 & 0.6586 \end{bmatrix} \begin{bmatrix} R \\ G \\ B \end{bmatrix} \quad (1)$$

2.2   A New Image Representation

Given three color components R, $\tilde{G}$, and $\tilde{Z}$, the statistical learning scheme of PCA followed by FLDA is proposed to generate a new image representation. In the $R\tilde{G}\tilde{Z}$ color space, an image of resolution $m \times n$ consists of three color components R, $\tilde{G}$, and $\tilde{Z}$. Without loss of generality, let R, $\tilde{G}$, and $\tilde{Z}$ be column vectors: $R, \tilde{G}, \tilde{Z} \in \mathbb{R}^N$, where $N = m \times n$. Each vector is normalized to have zero mean and unit variance. A data matrix $\mathbf{X} \in \mathbb{R}^{3 \times Nl}$ can be formed using the training images:

$$\mathbf{X} = \begin{bmatrix} R_1, \tilde{G}_1, \tilde{Z}_1 \\ R_2, \tilde{G}_2, \tilde{Z}_2 \\ \vdots, \vdots, \vdots \\ R_l \ \tilde{G}_l, \tilde{Z}_l \end{bmatrix}^t \tag{2}$$

where l is the number of training images. In $\mathbf{X}$, each column is an observation and each row is a variable. The covariance matrix Σ_X may be formulated as $\Sigma_X = \frac{1}{Nl-1}\tilde{\mathbf{X}}\tilde{\mathbf{X}}^t \in \mathbb{R}^{3 \times 3}$, where $\tilde{\mathbf{X}}$ is the centered data matrix. PCA is used to factorize Σ_X into the following form: $\Sigma_X = \Phi \Lambda \Phi^t$, where $\Phi = [\Phi_1, \Phi_2, \Phi_3] \in \mathbb{R}^{3 \times 3}$ is an orthonormal eigenvector matrix and $\Lambda = diag\{\lambda_1, \lambda_2, \lambda_3\} \in \mathbb{R}^{3 \times 3}$ a diagonal eigenvalue matrix with diagonal elements in decreasing order $(\lambda_1 \geq \lambda_2 \geq \lambda_3)$.

By projecting the three color component images $R, \tilde{G}, \tilde{Z}$ of an image onto Φ_1, a new image representation $\mathbf{U} \in \mathbb{R}^N$ can be obtained:

$$\mathbf{U} = [R, \tilde{G}, \tilde{Z}]\Phi_1 \tag{3}$$

According to the properties of PCA, $\mathbf{U}$ is optimal for data representation but not for data classification. To obtain an image representation that enhances discrimination for facial expression recognition, one needs to handle the within- and between-class variations separately. FLDA is a well-known subspace learning method that yields an optimal subspace to separate the different classes as far as possible and compress the same classes as compactly as possible. Thus, a more powerful image representation can be obtained based on $\mathbf{U}$ via the FLDA framework.

Let $\bar{\mathbf{U}}_i$ be the mean vector of class ω_i and $\bar{\mathbf{U}}$ be the grand mean vector. Then the between- and within-class scatter matrices $\mathbf{S}_b$ and $\mathbf{S}_w$ are defined as follows:

$$\mathbf{S}_b = \sum_{i=1}^k P(\omega_i)(\bar{\mathbf{U}}_i - \bar{\mathbf{U}})(\bar{\mathbf{U}}_i - \bar{\mathbf{U}})^t \tag{4}$$

$$\mathbf{S}_w = \sum_{i=1}^k P(\omega_i)\mathcal{E}\left\{(\mathbf{U} - \bar{\mathbf{U}}_i)(\mathbf{U} - \bar{\mathbf{U}}_i)^t | \omega_i\right\} \tag{5}$$

where $P(\omega_i)$ is the prior probability of class ω_i, and k is the number of classes, and $\mathbf{S}_b, \mathbf{S}_w \in \mathbb{R}^{N \times N}$. The general Fisher criterion in the U image space can be defined as follows:

$$J(\mathbf{P}) = \frac{|\mathbf{P}^t \mathbf{S}_b \mathbf{P}|}{|\mathbf{P}^t \mathbf{S}_w \mathbf{P}|} \tag{6}$$

Maximizing this criterion can be solved by deriving the optimal transform matrix $\mathbf{P} = [\psi_1, \psi_2, \ldots, \psi_d] \in \mathbb{R}^{N \times d}$, where $\psi_1, \psi_2, \ldots, \psi_d$ are chosen from the generalized eigenvectors of $\mathbf{S}_b \Psi = \lambda \mathbf{S}_w \Psi$ corresponding to the d largest eigenvalues.

Let $\mathbf{C} = [R, \tilde{G}, \tilde{Z}]$ be the original color configuration with the normalization of zero mean and unit variance for each vector. By projecting $\mathbf{C}$ onto the matrix $\mathbf{P}$, one can define the general color-space between-class scatter matrix $\mathbf{L}_b$ and color-space within-class scatter matrix $\mathbf{L}_w$ as follows:

$$\mathbf{L}_b = \sum_{i=1}^{k} P(\omega_i)(\bar{\mathbf{C}}_i - \bar{\mathbf{C}})^t \mathbf{P}\mathbf{P}^t (\bar{\mathbf{C}}_i - \bar{\mathbf{C}}) \tag{7}$$

$$\mathbf{L}_w = \sum_{i=1}^{k} P(\omega_i)\mathcal{E}\left\{(\mathbf{C} - \bar{\mathbf{C}}_i)^t \mathbf{P}\mathbf{P}^t (\mathbf{C} - \bar{\mathbf{C}}_i)|\omega_i\right\} \tag{8}$$

where $\bar{\mathbf{C}}_i$ is the mean of class ω_i and $\bar{\mathbf{C}}$ is the grand mean, and $\mathbf{L}_b, \mathbf{L}_w \in \mathbb{R}^{3 \times 3}$. By obtaining the generalized eigenvectors ξ_1, ξ_2, ξ_3 of $\mathbf{L}_b \Xi = \lambda \mathbf{L}_w \Xi$, ξ_1 corresponding to the largest eigenvalue is selected as the optimal color transformation, which can generate the discriminating image representation suitable for facial expression recognition. Finally, a novel image representation $\mathbf{D} \in \mathbb{R}^N$ can be derived by projecting the three color component images $R, \tilde{G}, \tilde{Z}$ of an image onto ξ_1:

$$\mathbf{D} = [R, \tilde{G}, \tilde{Z}]\xi_1 \tag{9}$$

3   Multiple Feature Fusion for Facial Expression Recognition

Different features have different discriminating capabilities for pattern recognition. When such features are complementary to each other, the diversity of misclassifications in statistical pattern recognition could be enhanced. As a result, feature-level or decision-level fusion can be utilized to improve the performance of pattern recognition. This section thus designs a framework of multiple feature fusion for improving facial expression recognition accuracy, based on the new image representation and local image features.

3.1   Multi-scale Local Phase Quantization (LPQ)

Local binary patterns (LBP) describe well local image texture and have been demonstrated to be a successful image descriptor for face recognition. Fourier phase information has a good property that is insensitive to the spatial blurring of images. Similar to the coding mechanism of LBP operator, local phase quantization (LPQ) [8] encodes local phase information in frequency domain and outperforms LBP operator in texture classification and face recognition. Specifically, LPQ computes short-term Fourier transform (STFT) in a window and for each pixel local frequency coefficients are computed at four frequency points. Each coefficient consists of one real part and one imaginary part. After quantization, there is an eight-bit codeword from these four coefficients. As LBP does, codewords are converted into decimal numbers and histogram is used as a

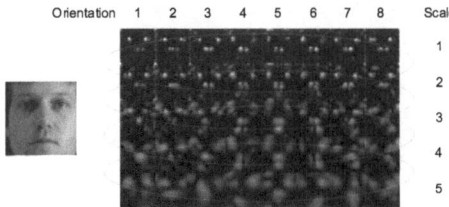

Fig. 1. The patch-based Gabor image representation

feature vector. Furthermore, a decorrelation step is applied in LPQ to reduce the dependency between adjacent pixels. One of the important parameters in LPQ is filter size that can help LPQ operator capture different local image information with its varying values. Thus, the strategy of multi-scale LPQ has been proposed to enhance the discriminative capacity of LPQ for face recognition [9]. In our implementation, two sets of LPQ features corresponding to the filter sizes 3 and 5 are extracted respectively. Then they are projected into the FLDA space separately for classification. Finally, two sets of similarity scores are fused at the score level by means of the sum rule with z-score normalization.

3.2   Patch-Based Gabor Image Representation (GIR)

Gabor image representation (GIR) of an image captures salient visual properties such as spatial location, orientation selectivity, and spatial frequency characteristics, displaying robust characteristics in dealing with image variabilities. Commonly used Gabor kernels contain five different scales and eight orientations that form the high-dimensional GIR. It is intractable to handle GIR directly. To provide a trade-off between computational efficiency and recognition accuracy, downsampling by a factor ρ is applied to GIR [1]. GIR actually contains rich information to be exploited for pattern classification. This paper presents a better method than downsampling to utilize GIR for facial expression recognition.

Fig. 1 shows the GIR of a face image, consisting of 40 filtered images (the magnitude representation). The whole GIR is decomposed into an ensemble of the patches along the horizontal direction, with the images in two adjacent scales and all the eight orientations forming one sub-GIR as illustrated in Fig. 1. The rationale of grouping two adjacent scales is due to the assumption that the redundancy in GIR is caused mainly by the similarities between all of the adjacent Gabor filters. Discrete cosine transform (DCT) can thus be used to reduce dimensionality and redundancy for improving computational efficiency and recognition performance. To facilitate the DCT feature extraction, each sub-GIR patch image is first reshaped to a square as illustrated in [4]. After transforming a reshaped sub-GIR patch image to the DCT domain, a frequency set located in the upper-left corner of the DCT domain is selected as features, which are used for facial expression recognition by FLDA. Finally, four recognition outputs corresponding to four sub-GIR patches are fused at the score level by means of the sum rule with z-score normalization.

Fig. 2. Example images from the Multi-PIE database. Expressions from the top row to the bottom row: neutral, smile, surprise, squint, disgust, and scream. Illumination labels from the left column to the right column: 0, 1, 4, 8, 10, 13, and 15.

The above presented multi-scale LPQ and patch-based GIR features are applied to the new image and gray-level image Y, respectively. Because the new image is complementary to Y image, as shown in experiments, such two classification outputs can be fused at the score level for enhancing performance.

4   Experiments

Experiments have been carried out on Multi-PIE [7] to assess the proposed method for facial expression recognition. Multi-PIE, a challenging facial database, contains six facial expressions: neutral, smile, surprise, squint, disgust, and scream. 129 subjects are selected in experiments so that all subjects are available in all four sessions for all expressions. 129 subjects are equally divided into three data sets: training, validation, and test, without overlapping subjects between data sets. Various illuminations are included in experiments to increase challenge for recognition. To the end, each facial expression of each subject in the training set contains illuminations 0, 2, 4, 6, 8, and 10, while different illuminations 1, 3, 5, 9, 13, and 15 are selected for the validation and test sets. Thus, each data set consists of 1548 frontal face images (43 subjects × 6 expressions × 6 illuminations). Face images in the experiments are geometrically rectified by first automatically detecting face and eye locations through the Viola-Jones face and eye detectors and then aligning the centers of the eyes to predefined locations for extracting 64 × 64 face region. Fig. 2 shows some example Multi-PIE images used in the experiments. Ten independent runs of experiments are performed. The validation set is used for selecting the optimum parameters during training and the test set for reporting recognition accuracy. For the FLDA classification, the nearest neighbor rule with a cosine measure is used.

Table 1. Recognition results of the Y and new images

	Y image	New image	Score level fusion
Accuracy	67.52%	70.31%	72.99%

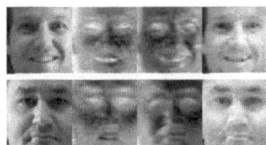

Fig. 3. The left column through the right column: R, $\tilde{G}$, $\tilde{Z}$, and new images

Table 2. Recognition results using different methods

	Patch-based GIR for Y image	Multiscale LPQ for new image	Score level fusion
Accuracy	76.59%	78.79%	**81.59%**
	LGBP for Y image [2]	Gabor with $\rho = 4$ for Y image	
Accuracy	76.57%	74.36%	

The essence of the new image is learning a transformation from the uncorrelated color space $R\tilde{G}\tilde{Z}$ to a space of most discriminative image via (9). One example of the learnt transformation coefficient sets is [0.6822, -0.6113, -0.3940] that can generate the new images as shown in Fig. 3. It should be noted that the R, $\tilde{G}$, and $\tilde{Z}$ color images should be normalized to zero mean and unit variance before applying the linear combination. Table 1 shows the facial expression recognition accuracy of the Y and new images using FLDA. The results clearly show that the new image representation outperforms Y image for facial expression recognition. Furthermore, the fusion of two recognition results using the sum rule at the score level improves facial recognition accuracy, indicating complementary information between two different image representations.

The next set of experiments is performed by applying multi-scale LPQ to the new image and patch-based GIR to Y image. For LPQ, an image is divided into 144 (12×12) overlapping windows of 9×9 pixels (3 pixels overlapping). For GIR, the DCT features of size 64×64 are selected from each sub-GIR patch. Their recognition results are tabulated in Table 2. For comparison we have also implemented two state-of-the-art methods: the Gabor features with the downsampling factor $\rho = 4$ using the FLDA method and local Gabor binary patterns (LGBP) with a SVM classifier [2]. The facial expression recognition results in Table 2 show that the best accuracy (81.59%) is achieved by the proposed multiple feature fusion, i.e., fusing patch-based GIR for Y image and multi-scale LPQ for the new image at the score level by means of the sum rule. It should be noted that the patch-based GIR method can achieve comparable recognition accuracy to the LGBP method [10]; nevertheless, the former is more computationally efficient, since LGBP is of very high dimensionality. Table 3 shows the confusion matrix for the proposed method. As shown in Fig. 2, scream is the most distinct

Table 3. Confusion matrix for the proposed method

	Neutral	Smile	Surprise	Squint	Disgust	Scream
Neutral	78.80%	9.65%	2.36%	4.81%	3.18%	1.20%
Smile	7.56%	82.48%	2.44%	1.94%	4.88%	0.70%
Surprise	0.35%	3.02%	89.07%	0	0.04%	7.52%
Squint	1.55%	3.18%	0.19%	68.37%	23.99%	2.71%
Disgust	0.97%	3.35%	0.27%	18.29%	72.68%	4.26%
Scream	0	0.97%	0.19%	0.19%	0.43%	98.22%

from other expressions, while squint is the most similar to disgust than other expressions. As a result, recognition rates in Table 3 indicate that scream is the easiest expression for recognition, while squint is the most difficult expression for recognition since it is most likely to be confused with disgust.

5   Conclusions

This paper proposes a novel method for facial expression recognition using a new image representation and multiple feature fusion. The new image is derived via a learning scheme from the normalized hybrid $R\tilde{G}\tilde{Z}$ color space instead of the traditional combination of RGB. The new image is not only more discriminative than gray-level image, it is also an excellent complement to gray-level image for facial expression recognition. The multiple feature fusion consisting of multi-scale LPQ and patch-based GIR is used to improve classification accuracy further. Experiments on Multi-PIE show the effectiveness of the proposed method for facial expression recognition under challenging conditions.

References

1. Donato, G., Bartlett, M., Hager, J., Ekman, P., Sejnowski, J.: Classifying facial actions. IEEE Trans. Pattern Analysis and Machine Intelligence 21(10), 974–989 (1999)
2. Moore, S., Bowden, R.: Local binary patterns for multi-view facial expression recognition. Computer Vision and Image Understanding 115, 541–558 (2011)
3. Yang, J., Liu, C., Zhang, L.: Color space normalization: Enhancing the discriminating power of color spaces for face recognition. Pattern Recognition 43, 1454–1466 (2010)
4. Liu, Z., Yang, J., Liu, C.: Extracting multiple features in the CID color space for face recognition. IEEE Trans. on Image Processing 19(9), 2502–2509 (2010)
5. Choi, J.Y., Ro, Y.M., Platanioits, K.N.: Boosting color feature selection for color face recognition. IEEE Trans. on Image Processing 20(5), 1425–1434 (2011)
6. Lee, S.H., Choi, J.Y., Ro, Y.M., Platanioits, K.N.: Local color vector binary patterns from multichannel face images for face recognition. IEEE Trans. on Image Processing 21(4), 2347–2353 (2012)
7. Gross, R., Matthews, I., Cohn, J., Kanade, T., Baker, S.: Multi-PIE. Image and Vision Computing 28, 807–813 (2010)

8. Ojansivu, V., Heikkilä, J.: Blur insensitive texture classification using local phase quantization. In: Proc. ICISP (2008)
9. Chan, C.H., Kittler, J., Poh, N., Ahonen, T., Pietikäinen, M.: Multiscale local phase quantisation histogram discriminant analysis with score normalisation for robust face recognition. In: Proc. ICCV Workshops, pp. 633–640 (2009)
10. Zhang, W., Shan, S., Chen, X., Gao, W.: Local Gabor binary pattern histogram sequence (LGBPHS): a non-statistical model for face representation and recognition. In: Proc. ICCV, pp. 786–791 (2005)

Face Recognition Using Fast Neighborhood Component Analysis with Spatially Smooth Regularizer

Faqiang Wang, Hongzhi Zhang, Wangmeng Zuo, and Kuanquan Wang

Biocomputing Research Centre, School of Computer Science and Technology,
Harbin Institute of Technology, Harbin 150001, China
tshfqw@163.com, {zhanghz0451,cswmzuo}@gmail.com,
wangkq@hit.edu.cn

Abstract. For the robust recognition of noisy face images, this paper proposed an improved fast neighborhood component analysis (FNCA) method by introducing a spatially smooth regularizer (FNCA-SSR). The SSR can penalize large differences between adjacent pixels by enforcing local spatially smoothness, and makes FNCA-SSR model robust to Gaussian and pepper-salt noises in face image. Experimental results on the ORL and FERET face data sets show that, for the recognition of noisy face images, FNCA-SSR is very robust and can achieve much higher recognition accuracy than FNCA and other subspace methods.

Keywords: Face recognition, Metric learning, Neighborhood components analysis, Robust recognition.

1   Introduction

Face recognition can be used for many real world applications, e.g., public security, human-computer interaction, and biometrics, and has received lot of research interests in the field of computer vision [1, 2]. In the past fourty years, varieties of face recognition approaches have been proposed, but face recognition is remained a challenging problem which deserves in-depth studies.

Among various approaches, subspace method is one of the most representative subclass of face recognition methods which has been very successful and widely studied. Generally, since the dimension of face image is very high, subspace method first learns a mapping to transform face image into a low dimensional feature subspace, and designs a classifier (e.g., nearest neighbor, NN) in the low dimensional subspace for face recognition. The studies of subspace methods for face recognition can be traced back to 1987, Sirovich and Kirby first used principal component analysis (PCA) for face representation [3, 4]. Turk and Pentland developed an Eigenfaces method for dimensionality reduction of face images [5]. Subsequently, linear discriminant analysis (LDA) has been proposed for the same purpose [6]. By far, various linear and nonlinear, unsupervised and supervised subspace methods have been developed for face recognition [7, 8, 9, 10]. Based on the graph embedding or least-squares framework, some researchers also intended to unify subspace methods [11, 12].

J. Yang, F. Fang, and C. Sun (Eds.): IScIDE 2012, LNCS 7751, pp. 450–457, 2013.
© Springer-Verlag Berlin Heidelberg 2013

After dimensionality reduction, NN or k-NN are usually adopted for face recognition. So it would be interesting to study subspace method as a problem of learning an effective Mahalanobis distance metric in the k-NN framework,

$$D_{\mathbf{M}}(\mathbf{x},\mathbf{y}) = \sqrt{(\mathbf{x}-\mathbf{y})^T \mathbf{M}(\mathbf{x}-\mathbf{y})} \overset{\mathbf{M}=\mathbf{L}^T\mathbf{L}}{=} \sqrt{(\mathbf{x}-\mathbf{y})^T \mathbf{L}^T\mathbf{L}(\mathbf{x}-\mathbf{y})} = \left\|\mathbf{L}(\mathbf{x}-\mathbf{y})\right\|_2, \quad (1)$$

where $\mathbf{M}$ is a $d \times d$ low rank positive semi-definite matrix. Since the rank of $\mathbf{M}$ is low, we can write $\mathbf{M} = \mathbf{L}^T\mathbf{L}$, where $\mathbf{L}$ is a $d_0 \times d$ matrix with $d_0 < d$.

By far, a number of metric learning methods have been proposed. Neighborhood components analysis (NCA) aims to learn a Mahalanobis distance measure to maximize the expected number of samples correctly classified [13]. Large margin component analysis (LMCA) has been proposed to learn a distance metric to separating points in different classes by a large margin [14]. Large margin nearest neighbor (LMNN) model learns a metric that penalizes both large distances between samples whose labels are the same, and small distances between samples with different labels [15]. To improve the computational efficiency of NCA in the training stage, we modified the NCA model and proposed a fast neighborhood component analysis (FNCA) model [16].

Previous metric learning methods for face recognition, however, usually do not take into account the spatially correlation characteristics of face images, and their performance would significantly degrade for the recognition of noisy face images. Because noise is unavoidable to be introduced during the acquisition and communication of face images, it would be valuable to develop robust method for the recognition of noisy face images.

In this paper, we study the robust noisy face recognition problem by proposed a FNCA with spatially smooth regularizer (FNCA-SSR) method. Based on our FNCA model, we further introduce a spatially smooth regularizer (SSR). SSR can penalize large differences between adjacent pixels for enforcing spatially smoothness of the projection vector, and thus FNCA-SSR is robust for the recognition of noisy face images. Actually, this regularizer has been successfully adopted for image denoising and restoration [17, 18]. Compared with the total variation (TV) regularizer [19, 20], SSR is smooth and differentiable, making that FNCA-SSR can be efficiently solved. Moreover, experimental results on several face data sets also verify the effectiveness of the proposed method for robust noisy face recognition.

The remainder of this paper is organized as follows. Section 2 describes the FNCA-SSR method. Section 3 provides the experimental results. Finally, Section 4 concludes the paper.

2   The FNCA-SSR Model for Face Recognition

In this section, we first introduce the FNCA method, then discuss its disadvantages for noisy face recognition, and finally propose a spatially smooth regularizer to improve the robustness for noisy face recognition.

2.1 The FNCA Model

Let $\mathbf{X}$ be the set of training samples $\{(\mathbf{x}_1, y_1), (\mathbf{x}_2, y_2), \cdots, (\mathbf{x}_n, y_n)\}$, where $\mathbf{x}_i$ is the vector representation of a $m \times n$ image to denote the ith training sample, and y_i is the corresponding class label of $\mathbf{x}_i$. In FNCA, we define $\mathbf{x}_j$ as a hit of $\mathbf{x}_i$ if $\mathbf{x}_j (j \neq i)$ has the same class label with $\mathbf{x}_i$. Otherwise, we call $\mathbf{x}_j (j \neq i)$ as a miss of $\mathbf{x}_i$. The probability of $\mathbf{x}_i$ selects $\mathbf{x}_j$ as its reference point is defined as [16],

$$p_{ij} = \begin{cases} \dfrac{\exp(-d_{ij}(\mathbf{L})/\sigma)}{\displaystyle\sum_{k \in M_i \cup H_i} \exp(-d_{ik}(\mathbf{L})/\sigma)} & \forall j \in M_i \cup H_i \\ 0 & \text{otherwise} \end{cases}, \qquad (2)$$

where $M_i = \{j | 1 \leq j \leq N, \mathbf{x}_j = \mathrm{NM}_k(\mathbf{x}_i), 1 \leq k \leq K\}$, $H_i = \{j | 1 \leq j \leq N, \mathbf{x}_j = \mathrm{NH}_k(\mathbf{x}_i), 1 \leq k \leq K\}$. $\mathrm{NM}_k(\mathbf{x}_i)$ and $\mathrm{NH}_k(\mathbf{x}_i)$ are the kth nearest miss and hit of $\mathbf{x}_i$, respectively. So the probability of $\mathbf{x}_i$ being misclassified is [16]:

$$p_i = \sum_{j \in M_i} p_{ij} \qquad (3)$$

So the loss function of FNCA is defined as the leave-one-out classification error. For face recognition, the dimension of face image usually is much larger than the number of training samples, and the simple FNCA model usually suffers from the overfitting problem for face recognition. In order to address this problem, we further introduce a Frobenius norm regularizer into the FNCA loss function [16]:

$$\xi_{FNCA}(\mathbf{L}) = \sum_{i=1}^{n} p_i + \frac{\lambda}{2} \|\mathbf{L}\|_F^2 = \sum_{i=1}^{n} \sum_{j \in M_i} p_{ij} + \frac{\lambda}{2} \|\mathbf{L}\|_F^2. \qquad (4)$$

where λ is a coefficient of the regularizer, $\|\cdot\|_F$ denotes the Frobenius norm. Here we choose $\lambda = 1$.

In order to minimize the loss function, the gradient descent method is used to derive the optimal transform matrix $\mathbf{L}$. The gradient of $\xi_{FNCA}(\mathbf{L})$ with respect to $\mathbf{L}$ can be computed by [16],

$$\frac{\partial \xi_{FNCA}(\mathbf{L})}{\partial \mathbf{L}} = \frac{1}{\sigma} \sum_i \left(p_i \sum_{j \in H_i} \frac{p_{ij}}{d_{ij}(\mathbf{L})} \mathbf{L} (\mathbf{x}_i - \mathbf{x}_j)(\mathbf{x}_i - \mathbf{x}_j)^T + (p_i - 1) \sum_{j \in M_i} \frac{p_{ij}}{d_{ij}(\mathbf{L})} \mathbf{L} (\mathbf{x}_i - \mathbf{x}_j)(\mathbf{x}_i - \mathbf{x}_j)^T \right) + \lambda \mathbf{L}. \quad (5)$$

Then we can use gradient descent to obtain the minimum value of $\xi_{FNCA}(\mathbf{L})$ together with the optimal transform matrix $\mathbf{L}$.

However, the Frobenius norm regularizer only imposes a loose constraint on the solution of FNCA to avoiding overfitting, and does not take the characteristics of image into account. Thus FNCA is not robust for the recognition of noisy images.

We conduct a toy experiment to reveal the disadvantage of FNCA with the Frobenius norm regularizer. Note that $\mathbf{L}$ is a $d_0 \times mn$ matrix. We call each row $\mathbf{l}_i$ of $\mathbf{L}$ a projection vector, and can transform $\mathbf{l}_i$ into an $m \times n$ image. In our experiment, we use half of images from the ORL face data set for training, add salt and pepper noise with standard density of 0.05 to each training samples, and use the noisy images to train FNCA. Then, we choose the first ten projection vectors and transform them to images, as shown in Fig. 1. From the figure, one can see that, the appearance of the projection vector is still noisy. So FNCA with the Frobenius norm regularizer would not achieve robust recognition accuracy for noisy face recognition.

Fig. 1. The appearance of the first 10 projection vectors obtained by FNCA with the Frobenius norm regularizer

2.2   FNCA with the Spatially Smooth Regularizer

In this section, we proposed a method of FNCA with the spatially smooth regularizer to improve the robustness of FNCA for noisy face recognition. We argue the poor robust performance of the Frobenius norm regularizer should be attributed to that it neglects the spatial correlation between adjacent pixels. Actually, local spatial correlation has been widely applied for image denoising. Local smoothing filtering is one of the major classes of image denoising methods. Many popular denoising algorithms, e.g., Gaussian smoothing, anisotropic diffusion, total variation (TV), and wavelet thresholding, belong to the category of local smoothing filtering.

Motivated by the success of TV regularizer in image denoising and restoration, we propose our spatially smooth regularizer for robust face recognition. The TV of the image $\mathbf{x}$ is:

$$\Phi(\mathbf{x}) = TV(\mathbf{x}) = \iint_{\Omega} \|\nabla \mathbf{x}\| \, d\Omega \tag{6}$$

where Ω is the region that image $\mathbf{x}$ occupies. For discrete image, the discrete version of $TV(\mathbf{x})$ is defined as,

$$TV_{iso}(\mathbf{x}) = \sum_{i=1}^{m} \sum_{j=1}^{n} \sqrt{(x_{i+1,j} - x_{i,j})^2 + (x_{i,j+1} - x_{i,j})^2}, \tag{7}$$

where $\mathbf{x} \in \mathbb{R}^{m \times n}$. Since TV regularizer is nonsmooth, it would be difficult to solve TV regularization optimization problem. For computational efficiency, by referring to [18, 21], we adopt a TV-like regularizer by modifying Eq. (7) as,

$$SSR(\mathbf{x}) = \sum_{i=1}^{m} \sum_{j=1}^{n} (x_{i+1,j} - x_{i,j})^2 + (x_{i,j+1} - x_{i,j})^2 = \|\mathbf{D}_v \mathbf{x}\|_F^2 + \|\mathbf{D}_h \mathbf{x}\|_F^2 \tag{8}$$

So the gradient of $SSR(\mathbf{x})$ respect to $\mathbf{x}$ is:

$$\frac{\partial SSR(\mathbf{x})}{\partial \mathbf{x}} = 2 \left(\mathbf{D}_v^T \mathbf{D}_v \mathbf{x} + \mathbf{D}_h^T \mathbf{D}_h \mathbf{x} \right). \tag{9}$$

In practice, we do not require to compute $\mathbf{D}_v\mathbf{x}$ and $\mathbf{D}_v^T\mathbf{x}$ explicitly, but compute it more efficiently by,

$$\left(\mathbf{D}_v\mathbf{x}\right)_{i,j} = x_{i+1,j} - x_{i,j}, \quad \left(\mathbf{D}_v^T\mathbf{x}\right)_{i,j} = x_{i,j} - x_{i-1,j}, \tag{10}$$

Analogously, we can compute $\mathbf{D}_h\mathbf{x}$ and $\mathbf{D}_h^T\mathbf{x}$ efficiently.

For simplicity, we call FNCA with spatially smooth regularizer as FNCA-SSR, whose objective function is,

$$F_{FNCA-SSR}\left(\mathbf{L}\right) = \xi_{FNCA}\left(\mathbf{L}\right) + \frac{\mu}{2}\sum_{i=1}^{d}\mathrm{SSR}\left(\mathbf{l}_i\right) = \sum_{i=1}^{n}\sum_{j\in M_i}p_{ij} + \frac{\lambda}{2}\left\|\mathbf{L}\right\|_F^2 + \frac{\mu}{2}\sum_{i=1}^{d_0}\mathrm{SSR}\left(\mathbf{l}_i\right) \tag{11}$$

where λ and μ are nonnegative regularization parameters. In order to minimize the objective function and to obtain the optimal transform matrix $\mathbf{L}$, we differentiate the objective function with respect to $\mathbf{L}$,

$$\frac{\partial F_{FNCA-SSR}\left(\mathbf{L}\right)}{\partial \mathbf{L}} = \frac{\partial \xi_{FNCA}\left(\mathbf{L}\right)}{\partial \mathbf{L}} + \frac{\mu}{2}\sum_{i=1}^{d}\frac{\partial \mathrm{SSR}\left(\mathbf{l}_i\right)}{\partial \mathbf{l}_i}. \tag{12}$$

Compared with FNCA, FNCA-SSR incorporates a spatially smooth regularizer to take into account the local spatial correlation of projection vector. Considering the effectiveness of spatially smooth regularization in image restoration and denoising, we believe that the introduction of SSR would suppress the adverse influence of noise on projection vectors of $\mathbf{L}$, and thus improve the robustness for noisy face recognition.

Fig. 2. The appearance of the first 10 projection vectors obtained by FNCA-SSR

Using FNCA-SSR, we conduct the same experiment in Section 2.1 to show the impact of SSR on projection vectors. Fig. 2 shows the appearance of the first ten projection vectors. One can see that, the appearances of the projection vectors of FNCA-SSR are much smoother than that of FNCA, and thus we expect FNCA-SSR would also achieve more robust performance for noisy face recognition.

3   Experimental Results

In this section, we evaluate the performance of FNCA-SSR for the recognition of noisy face image. Two face datasets, the ORL database, and a subset of the FERET database, are used in our experiments. We also compare FNCA-SSR with several subspace methods, e.g. FNCA, Fisherfaces and discriminant common vectors (DCV) In our experiments, we consider two kinds of image noise, additive Gaussian noise and salt and pepper noise. The FNCA-SSR model involves two regularization parameters, λ and μ. Here we choose $\lambda = 1$ and $\mu = 0.4$.

The ORL face data set contains 400 face images taken from 40 individuals, where each individual has 10 face images. To construct the training and test data sets, we randomly choose five images from each individual, and take these 200 face images as training set and the remained 200 images as test set. We construct 10 pairs of training and test sets using that process. For each pair, we use the training set to train a model, and then use the corresponding test set to obtain a recognition rate. After we get the recognition rates from those 10 pairs, we take the average of the 10 recognition rates as the final recognition rate.

We also use a subset of the FERET face database to evaluate FNCA-SSR. The subset contains 1400 face images. Those images are taken from 200 individuals and each individual has 7 images. In our experiments, we randomly select four images from each individual and construct a training set of 800 images, and use the remained 600 images as a test set. By this way, we construct 10 pairs of training and test sets to evaluate the recognition method, and take the average of the 10 recognition rates as the final recognition accuracy.

We evaluate the influence of feature dimension on recognition rates of FNCA and FNCA-SSR. We first normalize the gray level of images to [0,1], and then add Gaussian noise with mean zero and variance 0.002, or add salt and pepper noise with noise density 0.05. Figs. 3-4 show the plots of recognition rates against feature dimensions on the ORL and FERET datasets with Gaussian noise (variance of 0.002) and salt and pepper noise (density of 0.05), respectively. From Fig. 3 and Fig. 4, one can see that FNCA-SSR is consistently better than FNCA for noisy face recognition.

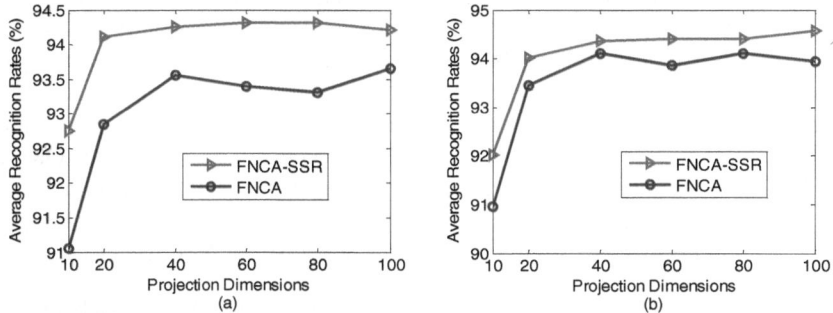

Fig. 3. Average recognition rates vs. projection dimensions on ORL dataset with (a) salt and pepper noise and (b) Gaussian noise

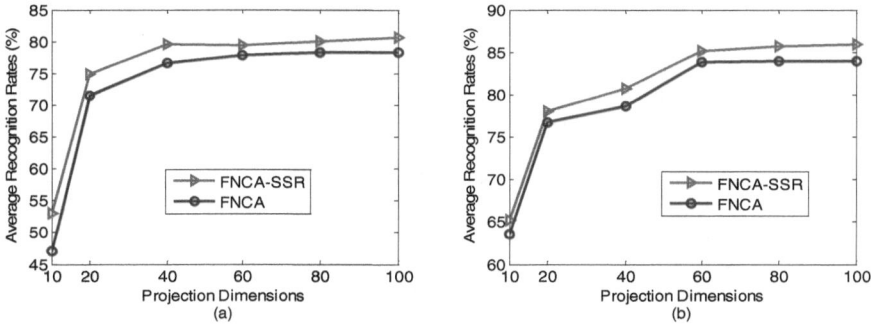

Fig. 4. Average recognition rates vs. projection dimensions on FERET dataset with (a) salt and pepper noise and (b) Gaussian noise

456   F. Wang et al.

Furthermore, we also compare the recognition accuracy of FNCA-SSR with other subspace methods, e.g. Fisherfaces and DCV. In FNCA-SSR, we choose the projection dimension as 100. In Fisherfaces, we choose the projection dimension as 40 for ORL database and 30 for FERET database. Table 1 lists the recognition rates of FNCA-SSR, Fisherfaces and DCV on the ORL and the FERET data sets. From Table 1, we can see that FNCA-SSR can achieve higher recognition rate than DCV and Fisherfaces in these data sets.

Table 1. Comparison of recognition rates (%) obtained by FNCA-SSR, DCV and Fisherfaces

Subspace Method	ORL with salt and pepper noise	FERET with salt and pepper noise	ORL with Gaussian noise	FERET with Gaussian noise
FNCA-SSR	**94.20**	**80.50**	**94.55**	**85.87**
DCV [9]	93.00	42.78	92.05	40.60
Fisherfaces [6]	86.05	75.45	91.40	76.50

4   Conclusion

In this paper, we propose a spatially smooth regularizer to improve the robustness of FNCA for noisy face recognition. With the introduction of the spatially smooth regularizer, FNCA-SSR can effectively penalize the large differences between adjacent pixels to enforce local spatially smoothness of the projection vectors, and thus can achieve robust performance for the recognition of noisy face image. To evaluate FNCA-SSR, a series of experiments are conducted on two face datasets, the ORL and the FERET databases. Experimental results indicate that, for the recognition of images with either Gaussian noise or salt and pepper noise, FNCA-SSR is much superior to and more robust than other subspace methods, e.g. FNCA, DCV and Fisherfaces.

Acknowledgements. The work is partially supported by the NSFC funds of China under Contract No.s   60902099, 61071179, 61001037, and Harbin Special Funds for Innovative Talents of Science and Technology Research Project.

References

1. Zhao, W., Chellappa, R., Phillips, P.J., Rosenfeld, A.: Face Recognition: A Literature Survey. ACM Computing Surveys 35, 399–458 (2003)
2. Abate, A.F., Nappi, M., Riccio, D., Sabatino, G.: 2D and 3D face recognition: A survey. Pattern Recognition Letters 28(14), 1885–1906 (2007)
3. Sirovich, L., Kirby, M.: Low-Dimensional Procedure for Characterization of Human Faces. J. Optical Soc. Am. 4(3), 519–524 (1987)
4. Kirby, M., Sirovich, L.: Application of the KL Procedure for the Characterization of Human Faces. IEEE Trans. Pattern Analysis and Machine Intelligence 12(1), 103–108 (1990)
5. Turk, M., Pentland, A.: Eigenfaces for recognition. J. Cogn. Neurosci. 3(1), 71–86 (1991)

6. Belhumeur, V., Hespanha, J., Kriegman, D.: Eigenfaces vs. Fisherfaces: Recognition using class specific linear projection. IEEE Trans. Pattern Analysis and Machine Intelligence 19(7), 711–720 (1997)
7. Liu, C., Wechsler, H.: Gabor feature based classification using the enhanced Fisher linear discriminant model for face recognition. IEEE Trans. Image Processing 11(4), 467–476 (2002)
8. Yang, J., Zhang, D., Frangi, A.F., Yang, J.-Y.: Two-Dimensional PCA: A New Approach to Appearance-Based Face Representation and Recognition. IEEE Trans. Pattern Analysis and Machine Intelligence 26(1), 131–137 (2004)
9. Cevikalp, H., Neamtu, M., Wilkes, M., Barkana, A.: Discriminative common vectors for face recognition. IEEE Transaction on Pattern Analysis and Machine Intelligence 27(1), 4–13 (2005)
10. He, X., Yan, S., Hu, Y., Niyogi, P., Zhang, H.-J.: Face recognition using Laplacianfaces. IEEE Trans. Pattern Analysis and Machine Intelligence 27(3), 328–340 (2005)
11. Yan, S., Xu, D., Zhang, B., Zhang, H., Yang, Q., Lin, S.: Graph Embedding and Extensions: A General Framework for Dimensionality Reduction. IEEE Trans. Pattern Analysis and Machine Intelligence 29(1), 40–51 (2007)
12. De la Torre, F.: A Least-Squares Framework for Component Analysis. IEEE Trans. Pattern Analysis and Machine Intelligence 34(6), 1041–1055 (2012)
13. Goldberger, J., Roweis, S., Hinton, G., Salakhutdinov, R.: Neighbourhood components analysis. Advances in Neural Information Processing Systems 17, 513–520 (2005)
14. Torresani, L., Lee, K.-C.: Large margin component analysis. Advances in Neural Information Processing Systems 19, 1385–1392 (2007)
15. Weinberger, K.Q., Blitzer, J., Saul, L.K.: Distance Metric Learning for Large Margin Nearest Neighbor Classification. Journal of Machine Learning Research 10, 207–244 (2009)
16. Yang, W., Wang, K., Zuo, W.: Fast neighbourhood component analysis. Neurocomputing 83, 31–37 (2012)
17. Gonzalez, R.C., Woods, R.C.: Digital Image Processing. Addison-Wesley Publishing Company, Inc. (1992)
18. Cho, S., Lee, S.: Fast Motion Deblurring. ACM Trans. Graphics 28(5), 1–8 (2009)
19. Rudin, L., Osher, S., Fatemi, E.: Nonlinear total variation based noise removal algorithms. Phys. D 60(1-4), 259–268 (1992)
20. Chambolle, A.: An Algorithm for Total Variation Minimization and Applications. J. Math. Imaging Vision 20(1-2), 89–97 (2004)
21. Hirsch, M., Schuler, C.J., Harmeling, S., Scholkopf, B.: Fast Removal of Non-Uniform Camera Shake. In: IEEE International Conference on Computer Vision, ICCV 2011 (2011)

Autoencoder for Polysemous Word

Cheng-Yuan Liou[1,*], Chen-Wei Cheng[1,2], Jiun-Wei Liou[1], and Daw-Ran Liou[3]

[1] Department of Computer Science and Information Engineering
National Taiwan University, Taiwan, Republic of China
[2] Institute of Statistical Science, Academia Sinica, Taiwan, Republic of China
[3] Sibley School of Mechanical and Aerospace Engineering, Cornell University
cyliou@csie.ntu.edu.tw

Abstract. Instead of training a single code vector for a word by using Elman network [1], this work presents a method to train multi-code for the polysemous word where each code represents a different meaning of the word. These multiple codes can accommodate different meanings of a word and facilitate the operation of word-sense disambiguation in semantic space.

Keywords: autoencoder, Elman network, stylish analysis, computational linguistics, semantic ranking, semantic indexing, semantic categorization.

1 Introduction

Elman network [2] has been redesigned for semantic encoding [1,3]. The trained codes are used in ranking, indexing, and categorizing literal works. To facilitate various operations of word sense disambiguation (WSD) [4], this work further presents a method to encode each polysemous word with multiple codes by using the redesigned network. This redesigned network is briefly reviewed in this section. The method and an experiment on Chinese novels are in the next two sections.

Elman network is a simple recurrent network that has a context layer, see Fig. 1. It was designed to find the hidden structure of sequential patterns [2]. Let L_o, L_h, L_c, and L_i be the number of neurons in the output layer, the hidden layer, the context layer, and the input layer, respectively. In the network, L_h is equal to L_c. During operation, at each training step t, the output of the hidden layer will be loaded to the context layer and together with the input layer to activate the hidden layer at time $t + 1$.

Let the two weight matrices between layers be W_{oh} and W_{hic}, where W_{oh} is an $L_h + 1$ by L_o matrix and W_{hic} is an $L_i + L_c + 1$ by L_h matrix. Consider a sequence of input patterns, $\{p(t), t = 1, 2, 3, \ldots\}$, the output vector of the hidden layer is denoted as $h(p(t))$ when $p(t)$ is fed to the input layer. $h(p(t))$ is an L_h by 1 column vector with L_h elements. Let $o(p(t))$ be the output vector of the output layer when $p(t)$ is fed to the input layer. $o(p(t))$ is an L_o by 1 column vector. The function of the hidden layer is $h(p(t)) = \varphi(W_{hic}^T[in(p(t))])$, where

---

* Corresponding author.

J. Yang, F. Fang, and C. Sun (Eds.): IScIDE 2012, LNCS 7751, pp. 458–465, 2013.

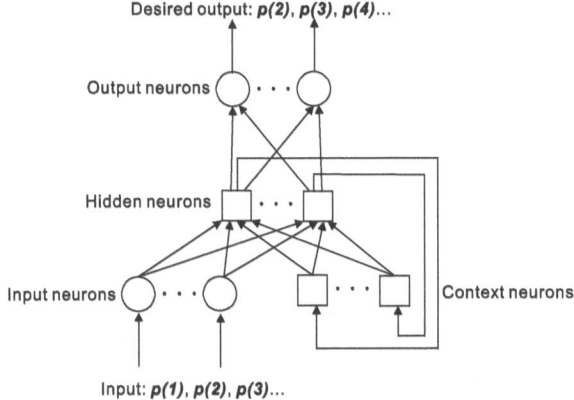

Desired output: **p(2), p(3), p(4)...**

Output neurons

Hidden neurons

Input neurons

Context neurons

Input: **p(1), p(2), p(3)...**

Fig. 1. Illustration of Elman network

$[in(p(t))]$ is an $L_i + L_c + 1$ by 1 column vector and φ is a sigmoid activation function that operates on each element of a vector [5]. The input vector $[in(p(t))]$ has the form $[in(p(t))] = [p^T(t), h^T(p(t-1)), 1]^T$. The sigmoid function used in this work is $\varphi(x) = [2/(1 + exp(-0.5x))] - 1$. The function of the output layer is $o(p(t)) = \varphi(W_{oh}^T [h^T(p(t)), 1]^T)$.

The back-propagation learning algorithm (BP) is commonly employed to train the weights, W_{oh} and W_{hic}, to reduce the difference between $o(p(t))$ and its desired output. The next input pattern $p(t+1)$ is served as the desired output [2]. For example, consider a sequence of word stream, $\{p(1), p(2), p(3), ...\}$, the input at time $t = 1$ is $p(1)$, and its desired output at time $t = 1$ is $p(2)$. The input at time $t = 2$ is $p(2)$, and its desired output at time $t = 2$ is $p(3)$. Therefore, all the attempts are aimed at minimizing the error between the network's outputs and their desired outputs, $\|o(p(t)) - p(t+1)\|^2$, to satisfy the prediction $o(p(t)) \approx p(t+1)$.

Consider a corpus with N different words $\{c_n; n = 1, 2, ..., N\}$, each word c_n initially has a random lexical code vector with R dimensions, $c_n = [c_{n1}, c_{n2}, ..., c_{nR}]^T$. R is equal to L_i in this paper. The input word stream is encoded accordingly, $p(t) = c_i$; $c_i \in \{c_n; n = 1, 2, ..., N\}$. This stream is the same as that used in Elman's training method. The redesigned network [1] minimizes the prediction error $\|o(p(t)) - p(t+1)\|^2$ between the network's output $o(p(t))$ and its desired output $p(t+1)$ to satisfy the prediction $o(p(t)) \approx p(t+1)$ by the BP algorithm. The weights W_{hic} and W_{oh} are updated after each word presented.

Set one epoch as when all words in the corpus were presented. We renew the codes every k epochs. We call a "pass" for each k epochs. Let g be the number of passes in the training process. The first pass is $g = 1$. After the training in each pass, a new raw code is calculated as follows:

$$c_n^{raw} = \frac{1}{freq_n} \sum_{\{t|p(t)=c_n^g\}} o(p(t-1)), n = 1 \sim N, \tag{1}$$

where $freq_n$ is the number of times that the word c_n appears in the current training pass. Note that Elman averaged all the hidden output vectors for each word c_n, but we averaged all the prediction vectors for it instead.

All renewed codes are normalized before using them in the next pass by the equation, $c_n^{g+1} = c_n^{nom} = \|c_n^{ave}\|^{-1} c_n^{ave}$, where the norm function is $\|c_n\| = (c_n^T c_n)^{0.5}$. This equation sets the norm of each code vector as 1 and is able to prevent a diminished solution, $\{\|c_n\| \sim 0, n = 1 \sim N\}$, derived by the BP algorithm. c_n^{ave} is the nth column of the matrix $C_{R\times N}^{ave}$,

$$C_{R\times N}^{ave} = C_{R\times N}^{raw} - \frac{1}{N} C_{R\times N}^{raw} \begin{bmatrix} 1 \cdots 1 \\ \vdots\ 1\ \vdots \\ 1 \cdots 1 \end{bmatrix}_{N\times N}. \tag{2}$$

In (2), $C_{R\times N}$ is a code matrix and has the form $[c_1, c_2, ..., c_N]$. The equation (2) makes each row of the code matrix $C_{R\times N}^{ave}$ become zero-mean.

The initial coding for $c_n^{g=1}$ in the first pass is randomly assigned under the restriction that different words have different codes and they are normalized. Then use the BP algorithm to reduce the prediction error $\|o(p(t)) - p(t+1)\|^2$ in each training step.

Note that the training step is operated after each word presented and the renewing step is operated after each pass. Iterations on these two steps constitute the training process. After the training process, we expect that each element of the code vector c_n contains a well isolated attribute of the word c_n. The meaning of each attribute can be calibrated by using the shard property of similar words.

2   Iterative Multi-code Re-encoding

Since a polysemous word has more than one meaning, such as Chinese characters, it should occupy more than one position in the semantic space. The averaged code vector (1) of a polysemous word can not represent it precisely. Although a code structure for accumulating the richness of semantic meanings of the polysemous word has been developed in [1], it can not be used in certain word space models, such as the word-based co-occurrence model [6] and the syntax-based semantic space model [7]. The popular cosine and Euclidean distance cannot be applied to this structure. This work proposes an automatic method to assign more than one code for a polysemous word. This may benefit the research of text mining, ranking, and indexing.

Consider a corpus with N different words, $\{c_n, n = 1 \sim N\}$. Instead of assigning one lexical code vector with R dimensions to each word c_n, we prepare a meaning pool matrix $(M_n)_{R\times B}$ for each of them. In this matrix, one word c_n can have at most B different codes. During the training process, the specific code vector used in the input word sequence $p(t+1) = c_j$ is the sth column of the pool M_j^g which receives minimum prediction error, see Fig 2. The sth meaning among all codes in the pool M_j^g is the best matched one of the prediction of its precedent word $p(t)$. Denote the sth column vector of M_j^g as $M_j^g(s)$.

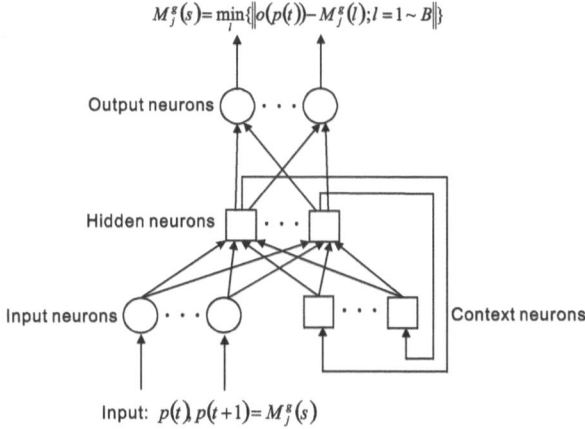

$$M_j^g(s) = \min_l \{\|o(p(t)) - M_j^g(l)\|; l = 1 \sim B\}$$

Output neurons

Hidden neurons

Input neurons

Context neurons

Input: $p(t), p(t+1) = M_j^g(s)$

Fig. 2. Illustration of Elman network for multi-code

Note that we start from $p(0) = c_n$, where $n =$ "end of sentence". Set the selected input as

$$p(t+1) = M_j^g(s), \qquad (3)$$

where the sth column vector minimizes $\{\|o(p(t)) - M_j^g(l)\|^2 \text{ for all } l = 1 \sim B\}$.

One word may have a set of different code vectors when it appears in different locations of the training sequence. All code vectors are fixed in each pass until we apply the re-encoding method to renew it at the end of the pass. The renewed vectors will be used in the next pass.

After every k training epochs (or a pass), a new raw code for the sth meaning, $c_n(s)$, of the word c_n is calculated as follows:

$$c_n^{raw}(s) = M_n^g(s) = \frac{1}{freq_{n,s}} \sum_{\{t | p(t) = M_n^g(s)\}} o(p(t-1)), \qquad (4)$$

where $freq_{n,s}$ is the number of times the code vector $c_n(s)$ appears in the training sequence of the current pass. $M_n^g(s)$ is the sth code vector in the word pool M_n^g in the current g pass. All raw codes are further normalized before using them in the next pass by the equation $c_n^{g+1}(s) = \|c_n^{ave}(s)\|^{-1} c_n^{ave}(s)$, where $\|c_n(s)\| = (c_n^T(s)c_n(s))^{0.5}$ and

$$C_{R \times S}^{ave} = C_{R \times S}^{raw} - \frac{1}{S} C_{R \times S}^{raw} \begin{bmatrix} 1 & \cdots & 1 \\ \vdots & 1 & \vdots \\ 1 & \cdots & 1 \end{bmatrix}_{S \times S}. \qquad (5)$$

In the normalization, S is the total number of different code vectors presented in the input sequence of the current pass and $S > N$ for the polysemous case.

Note that only the code vectors (or, all those matched vectors s) presented in the input sequence during the current pass will be updated and the normalization

Dream of the Red Chamber

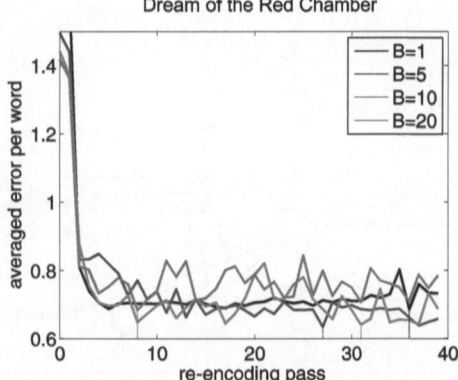

Fig. 3. Training errors using different pool sizes. Color vertical lines mark the minimum pass.

is only applied to them as well. In the next pass, all B vectors in each pool will be the candidates for the next input, $p(t+1)$, according to the minimization. The re-encoded vectors in the pool M_n^g after the minimum pass are the accomplished vectors for word meanings. They will be used to represent the meanings of the polysemous words. The minimum pass is the pass that reaches the minimum prediction error during the all training passes. The last training pass may not be the minimum training pass.

3   Experiment on Real Corpus

Each Chinese character is a word. Two greatest classical novels in Chinese literatures are analyzed. The first novel is *Dream of the Red Chamber*, which is composed by Cao Xueqin in the 18th century. The second is *Romance of the Three Kingdoms*, which is written by Luo Guanzhong in the 14th century. In our corpus, *Red Chamber* has more than 841 thousands of characters and uses 5069 different Chinese characters, including punctuation marks. *Three Kingdoms* has more than 570 thousands of characters and uses 5071 Chinese characters. Each kind of punctuation mark is treated as a single Chinese character. There are 27 kinds of such marks. In our training, each of the names, which consists of several Chinese characters, has a single matrix M_n^g.

One redesigned network is trained separately for one novel. All sentences and characters in each novel are concatenated together, sentence after sentence, into one single character sequence. This sequence is the training corpus and will be used as the network input. In each epoch, the network starts from the first character of the novel and reads the characters one after one until the end character of the novel.

From experiences, we set $R = 15$ for each code. The number of dimensions in the input layer is 15 as well as the output layer, $L_i = L_o = 15$. Both context layer and hidden layer have 15 neurons. The initial values of all synapse weights

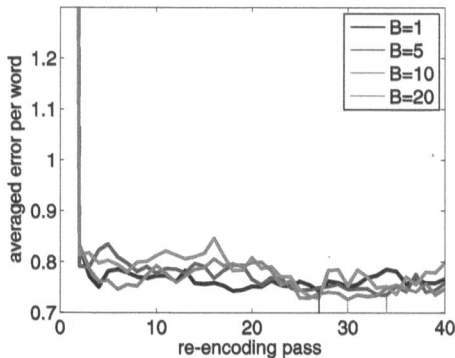

Fig. 4. Training errors using different pool sizes. Color vertical lines mark the minimum pass.

in W_{hic} and W_{oh} are randomly assigned in the range $[-1,1]$. The initial values of the neurons in the context layer are set to zero. The initial codes for all characters in the first pass $M_n^{g=1}$ are randomly assigned under the restriction that there are no repeated codes and they are normalized. Use the BP algorithm to reduce the prediction error, $\|o(p(t)) - p(t+1)\|^2$. The learning rate is fixed to 0.01. The W_{hic} and W_{oh} are updated after each input character presented. We renew the codes after every $k = 100$ epochs.

To ease the training, for the character with too high or too low frequency in the corpus, we set one single code vector c_n to represent it, that is $B = 1$. We prepare the code vector pools M_n for the rest characters whose appearing frequency within the range $\{ \geq 300$ and $\leq 1200\}$ in *Red Chamber*. There are 246 such characters and 246 pool matrices M_n with $B > 1$. As for the novel *Three Kingdoms*, there are 258 character pools with $B > 1$ whose appearing frequency within the range ≥ 225 and ≤ 525.

We compare the experiment results with different pool sizes, $B = 1, 5, 10,$ and 20. The results for *Red Chamber* are recorded in Fig. 3. This figure recorded the averaged prediction error for each word during the 40 training passes. The four color vertical lines mark the minimum passes for the four pool sizes. The training errors decrease roughly when we enlarge the pool size. The averaged prediction errors for *Three Kingdoms* are plotted in Fig. 4.

We study the accomplished codes in $B = 5$ after $g = 27$ re-encoding pass. The left part of Fig. 5 lists twelve of characters in *Red Chamber* that have multiple meanings. The red underlines mark the characters that have the highest number of meanings. The total number of meanings of a character is labeled next to its character. The right portion of this figure lists twelve characters in the novel *Three Kingdoms* that have multiple meanings obtained after the minimum pass, $g = 35$.

One multi-meaning character is shown in Fig. 6. This character has two meanings "walk" or "row" in different contexts. Six samples are listed in the figure. The sth meaning of the character in M, $1 \leq s \leq B = 5$, is labeled next to

紅樓夢	三國演義
張（2）、真（4）、輕（2）、 花（4）、分（2）、長（3）、 把（3）、思（2）、答（2）、 紅（2）、經（3）、方（5）、	車（2）、發（2）、差（2）、 合（2）、陣（2）、投（2）、 成（3）、禮（3）、飛（2）、 老（2）、騎（2）、勢（2）、

Fig. 5. Characters have multiple codes. The total number of meanings of a character is labeled next to its character.

紅樓夢

Sentences	Meaning
士隱接了看時，原來是塊鮮明美玉，上面字跡分明，鐫著「通靈寶玉」四字，後面還有幾行（4）小字。	排
說畢，在前導引，大家攀藤撫樹過去。只見水上落花愈多，其水愈清，溶溶蕩蕩，曲折縈迂。池邊兩行（4）垂柳，雜著桃杏，遮天蔽日，真無一些塵土。	排
說著，大家出來。行（1）不多遠，則見崇閣巍峨，層樓高起，面面琳宮合抱，迢迢複道縈紆；青松拂檐，玉欄繞砌，金輝獸面，彩煥螭頭。	走
回頭再走，又有窗紗明透，門徑可行（1）；及至門前，忽見迎面也進來了一群人，都與自己形相一樣，卻是一架玻璃大鏡相照。	走

Fig. 6. Sentences in *Red Chamber* which contain the same character with two different meanings, s=1 and s=4

三國演義

Sentences	Meaning
張寶遣副將高昇出馬搦戰。玄德使張飛擊之。飛縱馬挺矛，與昇交戰，不數合（4），刺昇落馬。玄德麾軍直衝過去。	一場
堅曰：「汝言正合（2）吾意。明日便當託疾辭歸。」商議已定，密諭軍士勿得洩漏。	相配
操見其人威風凜凜，心中暗喜，分付典韋，今日且詐敗。韋領命出戰；戰到三十合（4），敗走回陣。壯士趕到陣門中，弓弩射回。	一場
漢以火德王，而明公乃土命也。許都屬土，到彼必興。火能生土，土能旺木：正合（2）董昭、王立之言。他日必有興者。	相配

Fig. 7. Sentences in *Three Kingdoms* which contain the same character with two different meanings, s=2 and s=4

its character. A multi-meaning character in *Three Kingdoms* is shown in Fig. 7. This character has two meanings, "a round" and "match". Six samples are listed in the figure.

Red Chamber has 15 names that have multiple codes and *Three Kingdoms* has 6 names that have multiple codes. Some samples are listed in the Fig. 8. The different codes for the same name imply different roles or characteristics in different contexts.

A portion of error reduction for large size B in Fig. 3 and 4 is caused by the decrease of difference between code vectors. This is because there are too many vectors in the pool and it is easy to find a similar one in the pool. Similar vectors will give low error in the BP training process. Based on BP, when all the code vectors are similar, the training error must be low, and there is no need to assign another different code vector to a polysemous character.

紅樓夢	三國演義
周瑞家的不敢驚動，遂進裏間來，只見薛寶釵穿著家常衣服，頭上只散挽著鬐兒，坐在炕裏邊，伏在小炕兒上同丫鬟鶯兒正描花樣子呢。見她進來，寶釵（２）便放下筆，轉過身來，滿面堆笑讓：「周姐姐坐。」……寶釵（４）笑道：「那裏的話！只因我那種病又發了兩天，所以靜養兩日。」……周瑞家的聽了點頭兒，因又說：「這病發了時到底覺怎樣？」寶釵（４）道：「也不覺甚什麼，只不過喘嗽些，吃一丸也就罷了。」……寶玉亦湊了上去，從項上摘了下來，遞在寶釵手內。寶釵（２）托於掌上，只見大如雀卵，燦若明霞……	城中忽有一將，引數百人遶上城樓，大喝：「蔡瑁、張允，賣國之賊！劉使君乃仁德之人，今為救民而來投，何得相拒！」眾觀其人，身長八尺，面如重棗；乃義陽人也，姓魏，名延，字文長。當下魏延（２）輪刀砍死守門將士，開了城門，放下弔橋，大叫：「劉皇叔快領兵入城，共殺賣國之賊！」張飛便躍馬欲入。玄德急止之曰：「休驚百姓！」魏延（４）只管招呼玄德軍馬入城。只見城內一將飛馬引軍而出，大喝：「魏延（４）無名小卒，安敢造亂！認得我大將文聘麼！」……魏延（２）與文聘交戰，從巳至未，手下兵卒，皆已折盡。

Fig. 8. Samples of two names having multiple codes

Word sense disambiguation (WSD) impacts the accuracy of semantic search, machine translation, etc. and has been described as an AI-complete problem. This work provides the iterative multi-code re-encoding method to assign different meanings to one word automatically. To our knowledge, this method is the only one that automatically encodes multiple codes for the polysemous word. This method has been applied to analyze many English literal works and web databases.

Acknowledgement. This work was supported by National Science Council under project NSC 100-2221-E-002-234-MY3.

References

1. Liou, C.-Y., Huang, J.-C., Yang, W.-C.: Modeling word perception using the Elman network. Neurocomputing 71, 3150–3157 (2008)
2. Elman, J.L.: Finding structure in time. Cognitive Sci. 14, 179–211 (1990)
3. Huang, J.-C., Cheng, W.-C., Liou, C.-Y.: Distributed Representation of Word. In: Nguyen, N.T., Kim, C.-G., Janiak, A. (eds.) ACIIDS 2011, Part I. LNCS (LNAI), vol. 6591, pp. 169–176. Springer, Heidelberg (2011)
4. Navigli, R.: Word sense disambiguation:a survey. ACM Comput. Surv. 41(2), 1–69 (2009)
5. Rumelhart, D.E.: Parallel distributed processing: Explorations. In: McClelland, J.L. (ed.) The Microstructure of Cognition: Foundations. MIT Press, Cambridge (1986)
6. Lowe, W.: Towards a theory of semantic space. In: Proceedings of the 23rd Annual Conference of the Cognitive Science Society, Edinburgh, UK, pp. 576–581 (2001)
7. Grefenstette, G.: Explorations in Automatic Thesaurus Discovery. Kluwer Academic Publishers, Dordrecht (1994)

Pedestrian Detection on Moving Vehicle Using Stereovision and 2D Cue

Yang Yang[1,2], Jingyu Yang[1], and Dongyan Guo[1]

[1] School of Computer Science and Technology,
Nanjing University of Science and Technology
[2] School of Computer and Information Engineering, Henan University

Abstract. We present a novel approach for pedestrian detecting on moving vehicle which equipped with low-cost cameras. Our approach is working in a framework which combines two-dimensional human body characteristics and three-dimensional information such as parallax and distance. By constructing a SPM (surface parallax map), it calculates parallax of object which do not belong to the road plane such as human body and obstacles. After recording the scores of all road area, an occlusion image is created, in which high density area indicates people's most likely appearance. Then a SVM (support vector machine) classifier is trained to classify pedestrian and non-pedestrian windows in candidate area. We also propose an algorithm to maintain SPM in real time. We evaluate our approach on real data which are taken from crowded city areas, the efficient and accurate results are demonstrated.

Keywords: stereo vision, support vector machine, pedestrian detection, hog, surface parallax map.

1 Introduction

Nowadays, Along with the development of auto industry, car ownership in China is rising straight up. By the end of 2011, this number reached 104 million. But there was another fact behind this number, 6.2 million people were killed in traffic accidents in 2011. In this decade, a new study field called ADAS [1] (advanced driver assistance systems) attracted more and more attention which can reduce accidents by adding auxiliary equipment on vehicle. The ADAS is designed to prevent or at least reduce the harm from people in traffic accidents. Pedestrian detection system is a typical ADAS, and it is a very challenging task because irregular movement of human and the complex background where people appear have increased the difficulty of detecting.

Computer vision is widely used in the pedestrian detection systems, which make conclusion simulating human analysis. By processing the data acquired from video capture device, such systems perceive the traffic area to determine the location of the pedestrian. Because of the complexity of pedestrian characteristics and background in real traffic, only 2D information is insufficient to provide enough clues. Nowadays, more and more pedestrian detection system

J. Yang, F. Fang, and C. Sun (Eds.): IScIDE 2012, LNCS 7751, pp. 466–474, 2013.

introduced stereo vision. By extracting scene depth information, stereo vision makes a significant contribution to detecting pedestrian. After candidate areas which contain people have been determined, detection algorithms use classifiers to verify the windows containing pedestrian or not. Classifiers commonly use 2-D feature as human characteristics such as Harr [2], EOH [3] and HOG [4], etc. We present pedestrian detecting method combined 3-D information of scene with 2-D human features, which can detect pedestrians and obstacles accurately and efficiently.

2   Related Work

ROIs (region of interest) are hot zones with pedestrian appearing. ROI extraction is a common pre-operation in pedestrian detection system, which can avoid exhaustive searching from the whole image region and identify an accurate location of the pedestrian. Solid objects can be extracted as pedestrian candidates with disparity segmentation and fixed-size window is used to remove non-pedestrian [5]. Image is segmented into sub-image object candidates using disparities discontinuity, then a connected-component grouping operator is applied to find the pedestrian regions with smoothly varying range [6]. GAVRILA uses a feature-based, multi-resolution stereo algorithm to generate stereo-based ROI [7]. A multi-resolution approach is used to perform stereo analysis by finding correspondences on coarse level that can be recursively refined, and in order to reduce the computational cost Franke uses edge feature to match local correlation [8]. David presents a three module system in literature [1] which uses 3D information to estimate the road plane parameters and select ROIs, uses Real AdaBoost combined Haar wavelets and EOH to classify ROI windows. 3D information is extracted using stereo reconstruction algorithm which finds pairs of left-right correspondent edge points and maps them into the 3D world, then these 3D points are grouped into objects such as human body according to their distance to each other [9]. In order to reduce the search space, only edge points in left-right images are matched in stereo reconstruction algorithm. A 3D lane model is introduced to distinguish between road surface and target points by projecting those points in three specified planes [10]. In this paper, an efficient onboard pedestrian detection method is proposed. It equips with low-cost cameras, calculates occlusion image according to SPM, extracts ROIs from occlusion image by analysing the parallax pattern, adopts SVM training by using HOG feature to classify pedestrian or non-pedestrian.

3   SPM-Based ROI Generation

Within stereo vision the disparity map is generally used to calculate depth information of scene. We propose a variant of disparity map by restricting matching strategy on a specific plane–road surface to improve the efficiency of matching. This variant is so called Surface Parallax Map. Parallax is the object's different position in left-right images. Consider pedestrians and other non-road objects are located between cameras and road surface, some parts of the road surface must be occluded. By analysing occlusion pattern, locations of object on road can be calculated.

3.1   Surface Parallax Map

Traditional pedestrian detection methods filter feature points by considering their coordinates. Using our methodpedestrians within range of 50 meters from vehicle can be detected by using low-cost cameras. By introducing GCP (Ground control point) in matching algorithm, accuracy of matching can be improved. GCP was first proposed by AARON [11] in order to improve sensitivity of DSI (Disparity-space image). In this paper, GCPs are used to help synthesizing disparity of pixels on road surface into a SPM. The theory of SPM is described as follows, as shown in Fig. 1.

(a)   (b)

Fig. 1. Surface Parallax Map Initialization. (a) A frame in video 1 with manually calibrated GCP. (b) Sketch of GCPs and Control Lines.

Fig.1a shows an overlap image which has obvious parallax. The original images are captured by two fixed-angle cameras at same time. As shown in Fig.1b, O is a line which passes the midpoint of two optical centers and is perpendicular to the base line. Markers are GCPs on road surface, and control lines lying at right and left side pass their GCPs respectively. All the control lines are a set of parallel lines in road plane. These bold lines which called CameraL belong to left camera and CameraR belong to right camera, d is the parallax in horizontal direction in current position. When cameras are calibrating, two optical axes of cameras must be restricted in the same plane in order to avoid parallax in vertical direction. Considering driving area in front of vehicle within a certain distance can be approximated as a plane, parallax and distance from camera to target are linear correlation. That is, parallax and ordinate of target in image are linear correlation. Because the camera equipped can only provide resolution of 320X240, details provided by pixels are limited in far distance. In order to produce visible parallax at 50 meters, we set the two optical axes intersecting behind the camera.

SPM algorithm is described in formula 1, Parallax d_y is given by:

$$d_y = SPM(y) = \frac{1}{n} \sum_{i=1}^{n} \frac{(y - \Delta y)\frac{1}{m} \sum_{j=1}^{m} d_j}{y_i - \Delta y} \ . \tag{1}$$

Where y is the ordinate in image, the origin of image is on the upper left corner, Δy is the ordinate of the horizon, m is the number of control lines, the minimum value is 2, n is the number of GCPs on every control line, y_i is ordinate of ith GCP.

By giving d_y it adopts cost function to calculate occlusion image. Literature [8] introduces classical SSD (Sum-of-squared Differences) and SAD (Sum-of-absolute Differences). Literature [12] evaluates window-based matching technique called ZSAD (Zero-mean Sum of absolute Differences), NCC (Normalized Cross Correlation) and ZNNC (Zero-mean Normalized Cross Correlation). We compare the performance of these cost functions in experiment, the results are demonstrated in Section 5.

In general, SSD gives a better representation of difference between feature points than SAD. ZSAD subtracts the mean intensity of the window form each intensity inside the window before computing the sum of absolute differences. As described in formula 2, where I_L and I_R are pixel values at coordinate (x, y) in left and right images, Ω is the neighborhood of current pixel, using rectangle with longer horizontal edges, $\overline{I}(x, y)$ expresses mean intensity of neighborhood Ω.

$$OI_{ZSAD}(x,y) = \sum_{(x,y \in \Omega)} (I_L(x,y) - \overline{I}_L(x,y) - I_R(x - d_y, y) + \overline{I}_R(x - d_y, y)) \ . \tag{2}$$

$$\overline{I}(x,y) = \frac{1}{n} \sum_{(x,y \in \Omega)} I(x,y) \ . \tag{3}$$

NCC compensates for gain changes and is statistically the optimal method for dealing with Gaussian noise. ZNCC introduces zero mean in NCC and can compensate for differences in both gain and offset within the correlation window. Fig. 2 are the occlusion image using cost function mentioned above which have obvious occluded area on road surface. In Section 4, we train a SVM to classify pedestrian or non-pedestrian area from occlusion image.

3.2   Real-Time SPM Correction

In Section 3.1, we propose a hypothesis that the driving area in front of vehicle is always belong to the same plane. However, this assumption is not fully satisfied in actual road. For example, the angle between camera orientation and road surface changes when vehicle going uphill or downhill. Fig.3 shows the details, where Z orientation is the forward direction of car, as the road surface changes, driving direction will change around three axes correspondingly. Because cameras are fixed to the vehicle, camera rotating around Y and Z axis does not affect SPM,

   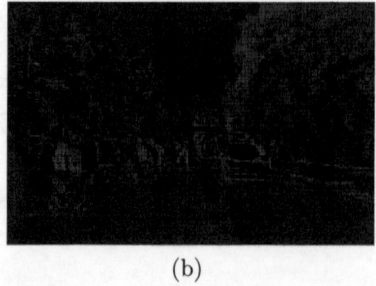

(a)   (b)

Fig. 2. Occlusion image generated by using SPM. (a) The 150 frame in video 2. (b) Occlusion image of this frame.

but rotating around X axis will influence results of formula 1 . In this paper, a real-time road surface fitting algorithm is proposed which using angle between road and horizontal to correct SPM output.

In the previous literature, stereovision-based road fitting technique is divided into three categories. The first category extracts road surface information by detecting the lane and roadside. A lane model is proposed in literature [9] [10] [13], surface parameters are calculated by lane model which derives from road geometry. The second category detects road feature points, calculates 3D coordinates of these points by using stereo vision technique, then, fits road surface. A RANSAC based fitting is applied over 2D barycenters intended for removing outlier cells, road surface parameters are computed by means of a least squares fitting over all 3D points contained in inlier cells[1]. A RANSAC based technique is used for finding the best fitting plane to those points which belong to the road [14]. The third category apply SFM (Structure from motion) based technique to fit road plane. By recording contact points between road and wheels when the car is moving, road plane is reconstructed between roadside limit lines[15]. The limitations of such methods is that only passed area can be rebuilt. The proposed method first matches edge feature points in images to obtain point cloud of road area, then projects those points to YOZ plane and analyses the angle between road and horizontal, uses formula 8 to correct SPM output. Since the upper part of the image is usually occupied by meaningless background such as trees and buildings, our matching algorithm only consider the lower half of images, which is the road region. Because of the perspective effect of road in images, texture of road surface is oblique. The results of matching horizontal or vertical edge feature separately are not satisfying. In this paper, we use Canny operator as edge detector. By finding the local maximum of gradient, Canny algorithm can generate continuous edges which are conducive to matching road feature points. In experiment,$\sigma = 1$, T_H is decided by the image itself, $T_L = 0.5T_H$.

Matching algorithm of feature points adopts Cost Function used to calculate occlusion image in Section 3.1. As shown in Fig.4, the 3D coordinates of feature points are projected to YOZ plane, β expresses the angle between road and horizontal, the line with minimum β and passing maximum points is selected as the road surface projection.

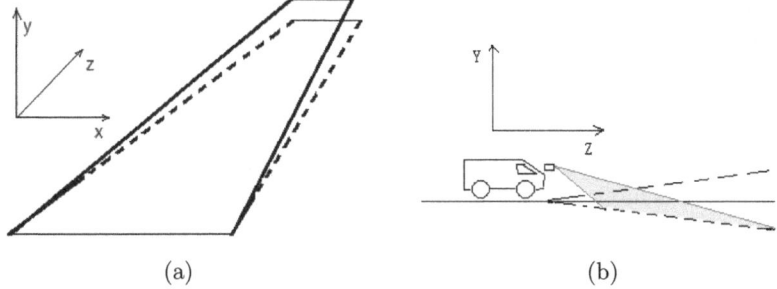

(a)   (b)

Fig. 3. SPM affected by Angle between road surface and horizontal. (a) Driving direction of vehicle. (b) Side view of uphill and downhill.

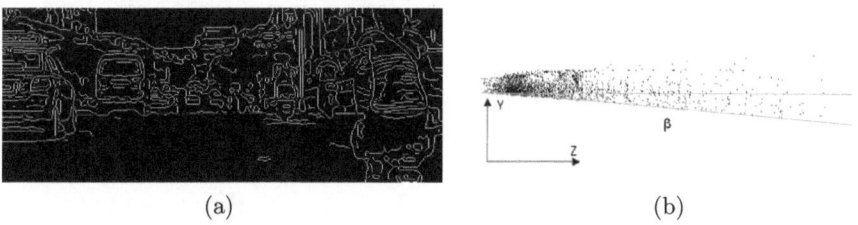

(a)   (b)

Fig. 4. Edge feature based road fitting. (a) Edge feature of this frame. (b) Feature points projected to YOZ plane, β is the angle between road and horizontal.

After correcting the formula of parallax d_y is modified as:

$$d_y = SPM(sy) = \frac{1}{n} \sum_{i=1}^{n} \frac{(sy - \Delta y)\frac{1}{m}\sum_{j=1}^{m} d_j}{y_i - \Delta y} . \qquad (4)$$

Where s is the correction factor, H is the height from camera to road surface, L_i is the distance from each GCP to the camera center, L is the road surface distance from current target to camera, β is the angle between road and horizontal which is positive uphill and negative downhill.

$$s = \frac{\tan \alpha}{\tan \alpha + \tan \beta} . \qquad \tan \alpha = \frac{H}{L} . \qquad L = \frac{1}{n} \sum_{i=1}^{n} \frac{y - \Delta y}{y_i - \Delta y} L_i . \qquad (5)$$

4   Pedestrian Verification in ROI

Many human features are proposed to classify detecting windows as pedestrian or non-pedestrian in previous literatures. Some detecting methods exhaustively search the entire image and cost most of the time. Under the premise of ROIs, detecting can be concentrated in the most likely area. GAVRILA proposes a shape-based detector and a texture-based classifier in literature [7]. David uses a Real

AdaBoost classifier combined with HW(Haar wavelets) and EOH (Edge orientation histograms) to locate pedestrians. In our approach, cameras are calibrated
using Zhangs technique [16], uses triangulation to calculate size of the detecting
window on ROIs position, a line SVM classifier [17] is trained using HOG as pedestrian feature. During the training step, the size of window is 128x64 pixels, cell is
8x8 pixels, histogram is calculated in every cell, a block is made up by 2x2 cells, L2-
Hys normalize is implemented in every block. Non-maximum suppression is used
on classification results to determine the accurate location of pedestrians.

Fig.5 shows the result images.

(a)   (b)

Fig. 5. Final results and candidate ROIs. (a) is a frame from video 3. (b) are the
candidate ROIs proposed by analysing Occlusion pattern.

5   Experiment

In experiments, We evaluate our approach on four challenging video sequences
which are captured from the urban traffic. The experiment platform is equipped
with Intel Core2 2.0GHz CPU and 2.0GB RAM. The main purpose of this approach is to detect pedestrian in front of the vehicle. Although our system is
equipped with low-cost cameras, it can detect targets at the range of 50 meters
with viewing angle of 50 degrees. The five cost functions mentioned in Section
3.1 are evaluated in experiments in both execution time and disparity between
ROI and Neighborhood. ZSAD is chosen for its best overall performance which
can process three frames in one second. Specialized commercial vision system
and 3D reconstruction software have been used to extra 3D information of the
region in front of the host vehicle [1]. On the contrary, our method generates
ROI by matching parallax differences of road surface. Under the premise of using
the same image capture device, the performance of SPM based method is significantly better than stereo matching based method [1] [5]. Experiment details
are shown in Table 1.   Since our method only detects pedestrian in ROI, it can
improve the detection speed than method adopting exhaustive strategy [4].Part
of experiment results are shown in Fig.5. Occlusion image based ROI extraction
provides candidate regions. At the same time, some non-pedestrian regions are
also proposed. After the candidate regions are verified by HOG feature based
SVM, the false detection is wiped out.

Table 1. Performance of ROI generation algorithm using SPM and stereo match

Algorithm	Max hit rate	Max effective distance	Processing time
SPM-based	96%	50 m	3 fps
Stereo-match-based	60%	15 m	0.4 fps

6 Conclusion

In this paper, we have presented a novel approach for pedestrian detection from moving vehicles. Our approach relies on the difference between parallax of pedestrian projecting to the road plane and the parallax of road itself. Because of SPM that we propose in section 3, the ROIs extraction algorithm based on parallax comparison is efficiently performed. A linear SVM classifier is trained with HOG features to label the selected ROIs as pedestrians or non-pedestrians. In addition, we have proposed an algorithm to maintain SPM by fitting road plane in real time.

We evaluate our approach on several video sequences taken in urban traffic. The experiments demonstrate that our approach is able to obtain reliable detecting results, even with low-cost cameras. In this paper, we have focused on detecting pedestrians. This can be extended to other object categories for which have different parallax from road plane, above or below.

References

1. Geronimo, D., Sappa, A.D., Ponsa, D., Lopez, A.M.: 2d-3d-based on-board pedestrian detection system. Computer Vision and Image Understanding 114(5), 583–595 (2010)
2. Papageorgiou, C., Poggio, T.: A trainable system for object detection. International Journal on Computer Vision 38(1), 15–33 (2000)
3. Levi, K., Weiss, Y.: Learning object detection from a small number of examples: the importance of goodfeatures. In: Proceedings of the IEEE Conference on Computer Vision and Pattern Recognition, Washington DC, CO, USA, pp. 53–60 (2004)
4. Dalal, N., Triggs, B.: Histograms of oriented gradients for human detection. In: Proc. IEEE Conf. Computer Vision and Pattern Recognition, vol. 1(38), pp. 886–893 (2005)
5. Soga, M., Kato, T., Ohta, M., Ninomiya, Y.: Pedestrian detection with stereo vision. In: Proceedings of the IEEE International Conference on Data Engineering, Tokyo, Japan (2005)
6. Zhao, L., Thorpe, C.E.: Stereo and neural network based pedestrian detection. IEEE Transactions on ITS 1(3), 148–154 (2000)
7. Gavrila, D., Munder, S.: Multi-cue pedestrian detection and tracking from a moving vehicle. International Journal on Computer Vision 73(1), 41–59 (2007)
8. Franke, U., Joos, A.: Real-time stereo vision for urban traffic scene understanding. In: Proc. of the IEEE Intelligent Vehicle Symposium, Detroit, USA, pp. 273–278 (2000)

9. Nedevschi, S., Schmidt, R., Graf, T., Danescu, R., Frentiu, D., Marita, T., Oniga, F., Pocol, C.: High accuracy stereo vision system for far distance obstacle detection. In: Proc. of IEEE Intelligent Vehicles Symposium, Parma, Italy, pp. 161–166 (2004)

10. Nedevschi, S., Danescu, R., Marita, T., Oniga, F., Pocol, C., Sobol, S., Graf, T., Schmidt, R.: Driving environment perception using stereovision. In: Proc of IEEE Intelligent Vehicles Symposium, Las Vegas, USA, pp. 331–336 (2005)

11. Bobick, A.F., Intille, S.S.: Large occlusion stereo. International Journal of Computer Vision 33(3), 181–200 (1999)

12. Hirschmuller, H., Scharstein, D.: Evaluation of stereo matching costs on images with radiometric differences. IEEE Trans. Pattern Analysis and Machine Intelligence 31(9), 1582–1599 (2009)

13. Danescu, R., Sobol, S., Nedevschi, S., Graf, T.: Stereovision-based side lane and guardrail detection. In: Proceedings of the IEEE International Conference on Intelligent Transportation Systems, Toronto, Canada, pp. 1156–1161 (2006)

14. Sappa, A., Geronimo, D., Dornaika, F., Lopez, A.: On-board camera extrinsic parameter estimation. Electronics Letters 42(13), 745–747 (2006)

15. Cornelis, N., Cornelis, K., Gool, L.V.: Fast compact city modeling for navigation pre-visualization. In: Proc. IEEE Conf. Computer Vision and Pattern Recognition (2006)

16. Zhang, Z.: A flexible new technique for camera calibration. IEEE Transactions on Pattern Analysis and Machine Intelligence 22(11), 1330–1334 (2000)

17. Yang, Y., Yang, J.Y.: Pedestrian detection based on compound feature. Journal of Image and Graphics 17(5), 671–675 (2012)

Multiscale Sample Entropy Analysis of Wrist Pulse Blood Flow Signal for Disease Diagnosis

Lei Liu[1], Naimin Li[1,2], Wangmeng Zuo[1], David Zhang[1,3], and Hongzhi Zhang[1]

[1] Biocomputing Research Centre, School of Computer Science and Technology,
Harbin Institute of Technology, Harbin, 150001, China
[2] Harbin Binghua Hospital,
Harbin, 150001, China
[3] Biometrics Research Centre, Department of Computing
The Hong Kong Polytechnic University, Kowloon, Hong Kong

Abstract. Recent study reported that wrist pulse blood flow signal is effective for disease diagnosis. The multiscale entropy, which was developed for quantifying the complexity of a time series of physiological signals over a range of scales, had been widely applied for feature extraction from medical signals. In this paper, using the multiscale sample entropy (Multi-SampEn) algorithm, we compute the value of SampEn of wrist pulse blood flow signal that includes 83 samples healthy persons, 45 samples of patients with liver diseases (LD), and 45 with sugar diabetes (SD). Then we use the values of SampEn as the feature input of the support vector machine classifier for disease diagnosis. Experimental results show that the proposed method could achieve the classification accuracy of 76.30% with the dimension $m = 2$ and the threshold $r = 0.6$, which is promising in diagnosing the healthy subjects, liver diseases, and sugar diabetes.

Keywords: Wrist pulse blood flow signal, multiscale sample entropy, feature extraction, pulse diagnosis.

1 Introduction

Various methods have been adopted for the analysis and diagnosis of the healthy condition of people in medical community, which could provide rich information of the health status and are great assistance to doctors in clinical disease diagnosis. But in many examinations the cost and the complexity are so high, and especially some examinations are harmful to the patients. Recently in virtue of its noninvasive and low cost diagnosis characteristics, the analysis of wrist pulse signal has received great attention for the diagnosis of many diseases (e.g. pancreatitis and Duodenal Bulb Ulcer, etc.) [1, 2].

The studies of wrist pulse signal have shown that it contains rich information and is great importance in the analysis of the health status and pathologic changes of a person. Several quantitative methods have been developed for the analysis of wrist pulse signal to standardize and objectify the disease diagnosis technique. For

J. Yang, F. Fang, and C. Sun (Eds.): IScIDE 2012, LNCS 7751, pp. 475–482, 2013.

example, Zhang et al. [3, 4] used the Hilbert-Huang transform and the wavelet methods to extract pulse features including wavelet powers, wavelet packet powers, and Doppler ultrasonic diagnostic parameters. Chen et al. [1, 2] developed several models to extract the features from wrist pulse blood flow signals. Using the extracted features as the inputs, some statistical classification methods, e.g., bayesian, support vector machine (SVM), fuzzy neural network (FNN), and artificial neural network (ANN) are adopted to identify the health status of a person.

Generally wrist pulse signals are analyzed by using traditional methods in time and frequency domains. But some nonlinear dynamics of wrist pulse signals are not disclosed. Therefore approximate entropy (ApEn) is adopted to quantify the dynamics of wrist pulse signals and has achieved some significant results which could not be disclosed by other conventional methods [5].

In this paper, sample entropy (SampEn) as a modification of ApEn is used for the analysis of wrist pulse blood flow signal which is proposed by Richman and Moorman et al. [6, 7]. Since sample entropy is a measure of the complexity and could be used to more accurately describe the nonlinear characteristics of time series, therefore it can be adopted to capture the important nonlinear features of wrist pulse blood flow signal. But in some cases a higher entropy value only reflects an increase in the degree of randomness and not an increase in the complexity of the time series, e.g., uncorrelated random signals (white noise) which may be highly unpredictable, whereas are not really "complex". The reason is the traditional entropy-based algorithms to quantify the complexity of a time series that these measures are single-scale based. Therefore, motivated by the multiscale entropy method which is proposed by Zhang et al. [8], we take into account multiple scales for more meaningful measure of complexity of wrist pulse blood flow signal. Compared with the traditional definition of sample entropy we could acquire more desirable complexity measures for disease diagnosis [9].

In this paper, we adopt the multiscale sample entropy (Multi-SampEn) method for the analysis of wrist pulse blood flow signals which include 83 samples of healthy persons, 45 samples of patients with liver diseases (LD) and 45 with sugar diabetes (SD). First the value of SampEn of wrist pulse blood flow signal is calculated at different scales. Then using the extracted features as the inputs, we construct an individual SVM classifier, resulting in Mutil-SampEn-SVM. Finally we use our constructed dataset to validate the effectiveness of the proposed method in the disease diagnosis.

The remainder of the paper is organized as follows. Section 2 describes the dataset of wrist pulse blood flow signal and the multiscale sample entropy (Multi-SampEn) method, respectively. Section 3 provides the experimental results. Finally, Section 4 ends this paper with concluding comments.

2   Material and Methods

In this section, we describe the dataset of wrist pulse blood flow signal and Multi-SampEn method for the analysis of wrist pulse blood flow signal in the disease diagnosis, respectively.

2.1   Subjects and Wrist Pulse Blood Flow Signal

In this paper we construct a wrist pulse blood flow signal dataset of 173 samples by using CBS 2000 that the sampling frequency is 110 Hz. Specifically, the samples in dataset are grouped into three categories, which include 83 samples of healthy persons, 45 samples of patients with liver diseases (LD) and 45 with sugar diabetes (SD). The healthy persons are chosen from the staffs and the students from Harbin Institute of Technology who have been diagnosed as healthy persons in the yearly physical examination. The patients are collected from Harbin Binghua Hospital where the diseases are diagnosed by the doctors according to the clinical data. For each subject, only wrist pulse blood flow signal of the left hand is acquired where the location of acquiring wrist pulse blood flow signal is depicted in Fig. 1, and we select a stable segment of 1200 points for subsequent feature extraction and classification. The age distribution and the number of subjects of each category are summarized in Table 1.

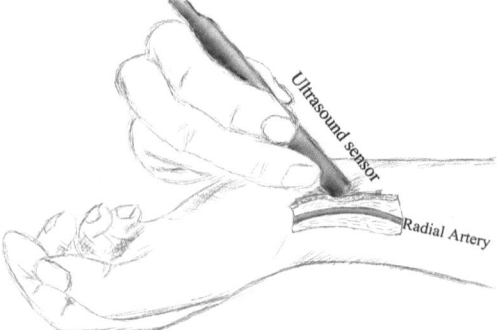

Fig. 1. The location of acquiring wrist pulse blood flow signal

Table 1. Sample Distribution of Our Dataset

Age / Diseases	0~20	21~40	41~60	61~	Total
Healthy	3	80	0	0	83
LD	0	10	21	14	45
SD	0	4	25	16	45

After constructing the database, a preprocessing step usually is adopted for denoising, baseline wander removal, and segmentation on wrist pulse blood flow signal. First wrist pulse blood flow signal is acquired by using the Doppler ultrasonic device, where an example of the raw Doppler spectrogram is shown in Fig. 2 (a). Then as shown in Fig. 2 (b) the maximum velocity envelope of Doppler spectrogram of wrist pulse blood flow signal is detected and extracted. Finally using the 7-level 'db6' wavelet transform, we remove the baseline wander by suppressing the 7th level 'db6' wavelet approximation coefficients, and reduce the 1st level wavelet detail coefficients for denoising. The final waveform for the analysis of disease is shown in Fig. 2 (c).

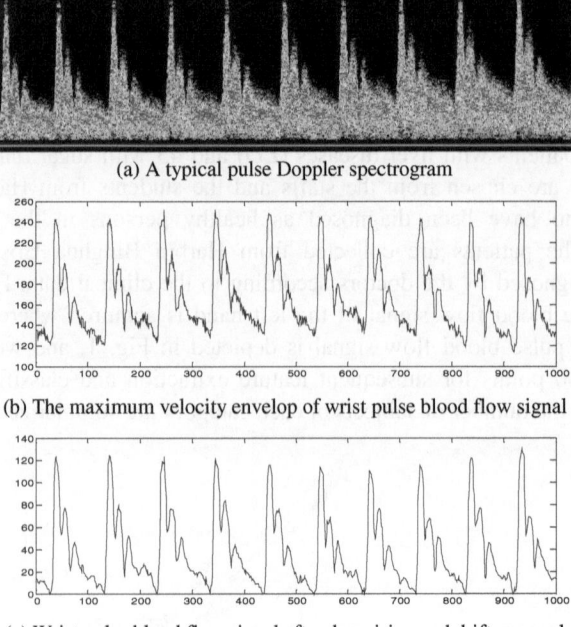

(a) A typical pulse Doppler spectrogram

(b) The maximum velocop of wrist pulse blood flow signal

(c) Wrist pulse blood flow signal after de-noising and drift removal

Fig. 2. The pre-processing of wrist pulse blood flow signal. (a) A typical pulse Doppler spectrogram, (b) the maximum velocity envelop of wrist pulse blood flow signal, and (c) wrist pulse blood flow signal after de-noising and drift removal.

2.2   Multiscale Sample Entropy (Multi-SampEn)

1) Sample entropy. In [6, 7], Richman and Moorman et al. proposed a modification of approximate entropy (ApEn) algorithm named sample entropy (SampEn). SampEn is a measure of the complexity and predictability of a time series and thus can be used to describe the nonlinear characteristics of wrist pulse blood flow signal. Compared with ApEn there are two properties of SampEn that are improved.

First, the value of SampEn is very robust against the values of the input parameters.

Second, the probability measure is calculated directly as the logarithm of conditional probability instead of the ratio of the logarithmic sums, which could increase the accuracy.

In the following, we describe the procedure of the SampEn in more detail.

First, given N data points of original signal $x(i)$ $i = 1, 2, \cdots N$, the new series $X_m(i)$ with the dimension m is then constructed by,

$$X_m(k) = [x(k), x(k+1), x(k+2), \cdots, x(k+m-1)], \tag{1}$$

where $k = 1, 2, \cdots N - m + 1$.

Second the quantity of constructed sequences could be calculated by,

$$B_i^m(r) = \frac{1}{N-m} num\{d[X(i), X(j)] < r\}, \ i = 1, 2, \cdots N - m + 1, \qquad (2)$$

where $d[X_m(i), X_m(j)]$ is the Euclidean Distance between the vectors with m dimensions and r is the threshold. Referred to [7], the input parameters of m and r could be fixed.

Finally the regularity parameter of sample entropy is defined as

$$SampEn(m, r) = \lim_{N \to \infty} \{-\ln[B^{m+1}(r) / B^m(r)]\}, \qquad (3)$$

where $B^m(r) = (N - m + 1)^{-1} \sum_{i=1}^{N-m+1} B_i^m(r)$.

2) *Multiscale approach.* In virtue of only supplying a single index concerning the general behavior of the time series by ApEn and SampEn, Costa et al. [10, 11] introduced the so-called multiscale entropy approach to reveal the underlying dynamics of the generating system and quantify the regularity of time series. Compared with the traditional definition of entropy it has the desirable property of yielding higher complexity and is a more meaningful measure of complexity by calculating entropy over multiple scales.

Then we briefly describe the multiscale approach. Based on N data points of original signal $x(i)$ ($i = 1, 2, \cdots N$) the consecutive coarse-grained time series could be constructed by,

$$y_j^{(\tau)} = \frac{1}{\tau} \sum_{i=(j-1)\tau+1}^{j\tau} x_i, 1 \le j \le \frac{N}{\tau}, \qquad (4)$$

where τ is the scale factor. For scale 1 the length of each coarse-grained time series is the original time series. Then we calculate the entropy for each one of the coarse-grained time series $\{y^{(\tau)}\}$. Then the new coarse-grained series $Y_m(i)$ with the dimension m is then constructed by

$$Y_m(k) = [y^{(\tau)}(k), y^{(\tau)}(k+1), y^{(\tau)}(k+2), \cdots, y^{(\tau)}(k+m-1)]. \qquad (5)$$

Finally by computing the sample entropy of the each new coarse-grained series we obtain the Multi-SampEn for the study of wrist pulse blood flow signal in the disease diagnosis.

3   Results

In this section Multi-SampEn was applied in the analysis of wrist pulse blood flow signal for the disease diagnosis of different categories, i.e., 83 samples of healthy persons, 45 samples of patients with liver diseases (LD) and 45 with sugar diabetes (SD). In our experiment, first the value of SampEn for wrist pulse blood flow signal was calculated at different scales, i.e., $\tau = 1, 5, 10, 15, 20, 25$ and 30 with m = 2 and r = 0.6 that a 7-dimensional feature vector of wrist pulse blood flow signal was derived. As depicted in Fig. 3 Multi-SampEn could achieve more meaningful measure of complexity of wrist pulse blood flow signal. Besides the complexity of diseases are more than healthy states on different scales. The pathological interpretation could be found in [10] that pathologic dynamics is associated with either increased regularity/decreased variability or with increased variability due to loss of correlation properties which are both characterized by a reduction in complexity. Therefore we could interpret that healthy states of wrist pulse blood flow signal may be more regular; as well as LD and SD states of wrist pulse blood flow signal may be more variable.

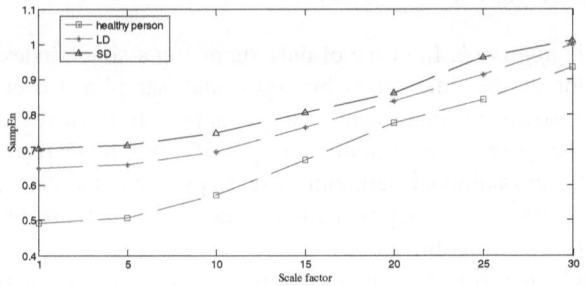

Fig. 3. The mean values of SampEn of healthy persons, LD and SD with $m = 2$ and $r = 0.6$ at different scales

Then in order to verify the effectiveness of Multi-SampEn in disease diagnosis, we constructed an individual SVM classifier, resulting in Mutil-SampEn-SVM which was trained by using the constructed data with the value of SampEn of each coarse-grained time series. To the end we tested the classification methods for classifying healthy persons and patients with two kinds of diseases. Specifically, we adopted the 10-fold cross-validation method to evaluate the diagnostic accuracy of the proposed method, where the proposed method achieved the classification accuracy of 76.30%.

In order to provide a comprehensive evaluation, we compared the performance of the proposed method with SVM with other individual feature extractor, i.e., ApEn, Cross-ApEn, SampEn [6, 7] and time warp edit distance (TWED) [12, 13], thus we constructed an individual SVM classifier, resulting in four SVM classifiers: ApEn-SVM, Cross-ApEn-SVM, SampEn-SVM and GTWED-SVM [14]. Then the 10-fold cross-validation method was adopted to assess the classification accuracy. From Table 2, one could see that, the proposed method could obtain much higher classification accuracy than any individual classifier, which verified that the Multi-SampEn was much more effective for the analysis of wrist pulse blood flow signal for the disease diagnosis. Besides, the McNemar test was adopted to evaluate the statistical significance of the difference between the proposed method and GTWED-SVM. The result showed that, the statistic value of the McNemar test was 4.22, which indicated that the performance difference was statistically significant at $\alpha = 0.05$.

Table 2. The Rusults of Different Classification Methods

Methods	Size/Class	Accuracy (%)
ApEn-SVM	173/3	65.32
Cross-ApEn-SVM	173/3	70.52
SampEn-SVM	173/3	69.36
GTWED-SVM	173/3	72.83
Mutil-SampEn-SVM	173/3	76.30

Table 3. The Confusion Matrix of the Proposed Method

		Predicted		
		H	**LD**	**SD**
Actual	**H**	73	7	3
	LD	9	31	5
	SD	10	7	28

We list the confusion matrices of the proposed method in Table 3. As show in Table 3, the proposed method could achieve comparable classification accuracy for each class.

4   Conclusions

In this paper, we adopt an effective nonlinear analysis method, i.e., Multi-SampEn, for the classification of wrist pulse blood flow signal from persons with different health status, i.e., healthy, liver diseases, and sugar diabetes. The application of the method could measure the different degree of complexity of wrist pulse blood flow signal on different scales, which could provide useful methods for the analysis of wrist pulse blood flow signal in disease diagnosis. Compared with other nonlinear analysis method, e.g. ApEn and Cross-ApEn, Multi-SampEn could provide more valuable information about signals' properties which remain hidden. According to the diagnostic accuracy the proposed method has been proven quite helpful in the understanding of deeper mechanisms of wrist pulse blood flow signal about healthy persons, liver diseases and sugar diabetes.

To evaluate the classification performance of Multi-SampEn, experiments are carried out to classify healthy people, patients with liver diseases and patients with sugar diabetes. Experimental results show that the proposed method achieves the classification accuracy of 76.30% with the parameters $m = 2$ and $r = 0.6$, and performs much better than the other methods, e.g., ApEn-SVM, Cross-ApEn-SVM, SampEn-SVM, and GTWED-SVM.

For future work, we will further investigate proper nonlinear analysis method for disease classification based on wrist pulse blood flow signal, and develop more effective classification methods for computerized pulse diagnosis.

Acknowledgment. The work is partially supported by the NSFC funds of China under Contract No.s 61001037, 61071179, and 60902099.

References

1. Chen, Y., Zhang, L., Zhang, D.: Wrist pulse signal diagnosis using modified Gaussian models and Fuzzy C-Means classification. Medical Engineering & Physics 31, 1283–1289 (2009)
2. Chen, Y., Zhang, L., Zhang, D.: Computerized Wrist Pulse Signal Diagnosis Using Modified Auto-Regressive Models. Journal of Medical Systems 35, 321–328 (2011)
3. Zhang, D., Zhang, L., Zhang, D., Zheng, Y.: Wavelet based analysis of doppler ultrasonic wrist-pulse signals. In: BioMedical Engineering and Informatics: New Development and the Future - 1st International Conference on BioMedical Engineering and Informatics, BMEI 2008, May 27-30, pp. 539–543. Inst. of Elec. and Elec. Eng. Computer Society (2008)
4. Zhang, D.Y., Zuo, W.M., Zhang, D., Zhang, H.Z., Li, N.M.: Wrist blood flow signal-based computerized pulse diagnosis using spatial and spectrum features. Journal of Biomedical Science and Engineering 3, 361–366 (2010)
5. Xu, L., Meng, M.Q.H., Qi, X., Wang, K.: Morphology Variability Analysis of Wrist Pulse Waveform for Assessment of Arteriosclerosis Status. Journal of Medical Systems 34, 331–339 (2010)
6. Lake, D.E., Richman, J.S., Griffin, M.P., Moorman, J.R.: Sample entropy analysis of neonatal heart rate variability. American Journal of Physiology-Regulatory, Integrative and Comparative Physiology 283, R789 (2002)
7. Richman, J.S., Moorman, J.R.: Physiological time-series analysis using approximate entropy and sample entropy. American Journal of Physiology-Heart and Circulatory Physiology 278, H2039 (2000)
8. Zhang, Y.C.: Complexity and 1/f noise. A phase space approach. Journal de Physique I 1, 971–977 (1991)
9. Valencia, J.F., Porta, A., Vallverdu, M., Claria, F., Baranowski, R., Orlowska-Baranowska, E., Caminal, P.: Refined Multiscale Entropy: Application to 24-h Holter Recordings of Heart Period Variability in Healthy and Aortic Stenosis Subjects. IEEE Transactions on Biomedical Engineering 56, 2202–2213 (2009)
10. Costa, M., Goldberger, A.L., Peng, C.K.: Multiscale entropy analysis of complex physiologic time series. Physical Review Letters 89, 68102 (2002)
11. Costa, M., Goldberger, A.L., Peng, C.K.: Multiscale entropy to distinguish physiologic and synthetic RR time series. In: Computers in Cardiology 2002, September 22-25, pp. 137–140. Institute of Electrical and Electronics Engineers Computer Society (2002)
12. Marteau, P.F.: Time warp edit distance with stiffness adjustment for time series matching. IEEE Transactions on Pattern Analysis and Machine Intelligence 31, 306–318 (2009)
13. Liu, L., Zuo, W., Zhang, D., Li, N., Zhang, H.: Classification of Wrist Pulse Blood Flow Signal Using Time Warp Edit Distance. Medical Biometrics, 137–144 (2010)
14. Zhang, D., Zuo, W., Zhang, D., Zhang, H.: Time series classification using support vector machine with Gaussian elastic metric kernel. In: 2010 20th International Conference on Pattern Recognition, ICPR 2010, August 23-26, 2010, pp. 29-32. Institute of Electrical and Electronics Engineers Inc. (2010)

Interactive Segmentation with Recommendation of Most Informative Regions

Canxiang Yan, Dan Wang, Shiguang Shan, and Xilin Chen

Key Laboratory of Intelligent Information
Processing of Chinese Academy of Sciences(CAS),
Institute of Computing Technology, Beijing 100190, China
{canxiang.yan,dan.wang,shiguang.shan,xilin.chen}@vipl.ict.ac.cn

Abstract. Compared to automatic segmentation, interactive segmentation is a flexible method to separate the interesting object from background. However, satisfactory results may not be achieved even with lots of interactions since user's operation may not provide enough information to decide the labels of ambiguous regions. To deal with this problem, we present an interactive segmentation approach based on active learning scheme, which can automatically recommend the most informative regions to guide the user interactions. Our method employs a two-step strategy. Firstly, based on initial user interactions, it adopts active learning to iteratively select the most crucial regions and query the oracle for their true labels. In the second step, we minimize an energy function, which combines low-level features extracted from total interactions, to segment the object. Experimental results demonstrate our method can achieve high segmentation accuracy within desirable interactions.

Keywords: active learning, interactive, segmentation, energy minimization.

1 Introduction

Interactive segmentation has gained great concerns in the field of computer vision. A lot of methods [3,5,6,10,12,13,14,15,17] have been proposed for interactive segmentation. They describe an interactive operation between an annotator and a machine to achieve segmentation results. More specifically, given an input image with single or multiple interesting objects, the oracle is queried for labeling limited pixels contained in foreground objects and backgrounds. Then this prior information is utilized to obtain the full segmentation results of desirable objects.

In this paper, we strive to address above-mentioned problems, via presenting a novel interactive segmentation algorithm based on active learning scheme [1,7,18]. The overview of the method is illustrated in Figure 1. Given an input image, a user is asked to give initial interactions by providing scribbles on objects (red lines) and backgrounds (blue lines). Then, the active learning method is employed to iteratively recommend a series of candidate regions (denoted as green regions) for guiding interactions. In each iteration, only one crucial region

J. Yang, F. Fang, and C. Sun (Eds.): IScIDE 2012, LNCS 7751, pp. 483–490, 2013.
© Springer-Verlag Berlin Heidelberg 2013

with greatest information is chosen for users to receive further annotation. we define region entropy(RE), distance ratio(DR) and neighbour similarity(NS) to describe regions, then use combined information map to find most crucial region in each iteration. After T iterations, we combine feature maps into our graphical CRF model, and minimize an energy function to obtain final segmentation result.

The rest of our paper is organized as follows. Sec. 2 reviews related work. Sec. 3 presents our method. Sec. 4 is our active learning algorithm. Sec. 5 discusses experimental results. Finally, Sec. 6 concludes the paper.

2   Related Work

2.1   Interactive Segmentation

The current literatures about interactive segmentations [10,12,14,17]are roughly classified into two categories: 1) The boundary-based methods, like active contour models [10] and intelligent scissors [14], use an adaptive curve to fit the object's boundary to pop out target object. 2) Graph-based methods, such as GrabCut [17] and Lazy Snapping [12], formulate interactive segmentation into an energy minimization problem. More recently, some methods extend the previous efforts to improve the performance of interactive segmentation. Delong et al. [5] introduce a two-level MRF to model object appearance. In comparison, our target is to make user interactions more effective and reduce user effort.

2.2   Active Learning

The process of active learning is defined as querying the oracle for the true label of input data. It has been widely used in image Retrieval [8], image classification [16], object categorization [9], object segmentation [1,7,18]. A. Fathi et al. [7] propose an incremental self-training approach by iteratively labeling the least uncertain frame for video segmentation. In [18], active learning is used to address the substantial gap between segmentation accuracy of fully and weakly supervised methods. Instead, our paper proposes an active learning method to iteratively select most crucial regions to guide the user interactions. The contributions of this paper are mainly summarized as follows: 1).we use active learning in our interactive segmentation framework to recommend most crucial regions. This method can achieve high segmentation accuracy within desirable interactions. 2).we define region entropy, distance ratio and neighbor similarity, which are helpful to recommend the most crucial regions.

3   Our Method

We formulate the interactive segmentation problem as a binary labeling problem. Denote $L=\{a_p\}$ is the final result; for each component a_p , if pixel p belongs to foreground objects, $a_p = 1$; otherwise, $a_p = 0$. Given an input observation image

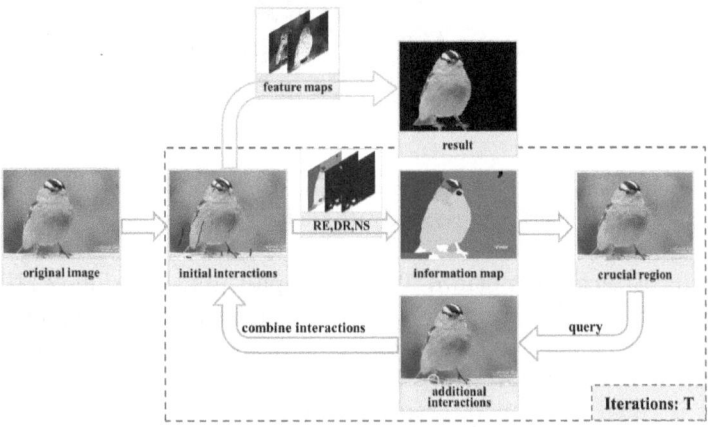

Fig. 1. An overview of our method

I, we build a graph $G = (V, E)$, whose vertex corresponds to pixel and edges are defined in an 8-connected neighborhood system. Our objective is to infer the optimal result L^* that leads to the minimization of the following Gibbs energy function:

$$E(L|I, A_0) = \sum_{p \in V} \sum_{m=1}^{M} \lambda_m F_m(a_p|I, A_0) + \sum_{(p',p) \in E} S(a_p, a_{p'}|I, A_0), \quad (1)$$

where A_0 is user interactions; F_m is the local-based energy function trained based on low-level appearance features and λ_m is the associated weights. We use a linear combination of M local-based energy function. $S(\cdot)$ denotes the pairwise function to smooth the label configuration of L. Immediately below, we elaborate these two energy functions, respectively.

Specifically, F_m is a local-based function, written as:

$$F_m(a_p) = \begin{cases} f_m(p) & \text{if } a_p = 1; \\ 1 - f_m(p) & \text{if } a_p = 0. \end{cases} \quad (2)$$

$S(a_p, a'_p)$ penalizes the disagreement between adjacent pixels. We define it as:

$$S(a_p, a_{p'}) = l(a_p \neq a_{p'}) \cdot exp(-\beta d_{pp'}^2), \quad (3)$$

3.1 Energy Minimization Based on Feature Maps

Suppose that user interactions A_0 are acquired, then we extract two low-level feature maps: 1) An object and background Gaussian Mixture Model, which are learned from object pixels and background pixels in A_0, respectively. 2) Color spatiality extracts spatiality from the color model and becomes a global feature to describe the object.

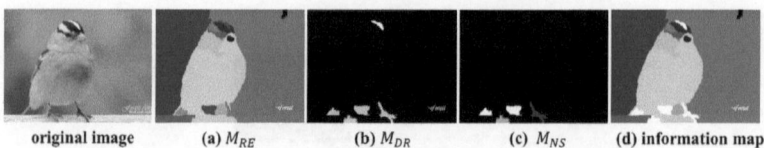

original image   (a) M_{RE} (b) M_{DR} (c) M_{NS} (d) information map

Fig. 2. Information map for crucial regions. (a), (b), (c) are region entropy, distance ratio, and neighbor similarity respectively. (d) is final information map.

Minimizing Energy Function. We use Graph-Cuts algorithm [2,11] to minimize the energy function in our graphical CRF model. In our graph $G = (V, E)$ over all pixels, each node $v \in V$, is connected to the source s and the sink t and their adjacent nodes, respectively. Based on graph theory, each configuration $L = \{a_p\}$ is equal to a cut of graph G, which makes each node only connect to one of the two terminal nodes. Thus, for obtaining the optimal segmentation result L^*, our target is to find a cut which has minimum cost. So far, our energy minimization problem is equal to minimum cut of the G.

4   Active Learning for Most Informative Regions

Active learning starts when the initial user interactions are finished. Firstly, we over-segment the image I into homogeneous regions with irregular size and shape using mean shift technique [4]. After initial user interactions, some object regions and background regions are initialized respectively. The image I then can be represented as a set of all irregular regions from over-segmentation: $I = \{(A_{r_1}, r_1), (A_{r_2}, r_2), ..., (A_{r_N}, r_N)\}$, where N is the number of regions; In each pair (A_{r_i}, r_i), the label A_{r_i} takes a value from $\{A_u, A_{obj}, A_{bg}\}$ which means that the initial annotation of region r_i is unknown region, object or background. Using homogeneous regions in our method can improve the computational efficiency and allow us to compute complex statistics (e.g. texture and color).

The active learning method is conducted iteratively. In t^{th} iteration, the following three steps are performed one by one: 1) it learns region-based features: region entropy, distance ratio and neighbor similarity based on previous interactions and combines them to form an information map. 2) Then most informative regions are recommended to user; additional interactions are made on those regions, and then are combined with previous interactions as the user inputs of the $(t+1)^{th}$ iteration. After T iterations, the total interactions A_0 are obtained. The whole process is presented below.

4.1   Information Map

We use three maps to describe how informative a region is: region entropy, distance ratio and neighbor similarity.

Region Entropy. We define initial foreground object pixels: $P_{obj} = \{p | A_{r_i} = A_{obj}, p \in r_i \text{ and } r_i \in I\}$, and use raw RGB color of P_{obj} to learn GMM_{obj}.

Similarly, we use initial background pixels to get GMM_{bg}. The components of GMM_{obj} and GMM_{bg} are 5 and 8 respectively. Then, for each pixel p_{ik} in region r_i, we normalize its object and background likelihoods from $GMMs$ to get a two-class distribution and calculate its Shannon entropy $H(p_{ik})$. The region entropy then can be written as:

$$M_{RE}(r_i) = \frac{1}{k} \sum_{k=1}^{K} H(p_{ik}),\tag{4}$$

where K is the number of pixels included in the region r_i, $M_{RE}(r_i)$ is the average entropy of region r_i. Region entropy is defined based on a simple idea that the label of a region may be ambiguous to decide if the appearance of the region is mixed by foreground object and background or its appearance has large differences to both. Shannon entropy can measure the uncertainty of a pixel label a_p and the average entropy $M_{RE}(r_i)$ of pixels in a region will be high if the total uncertainty is great. Thus, the more ambiguous a region is, the higher entropy it gets. We then get an entropy map M_{RE} by normalizing $M_{RE}(r_i)$ over all regions in image I to the range $[0, 1]$, shown in Figure 2(a). A region which has complex appearance at the bird body is bright, while another region which has consistent appearance at the bird head is darker than the former.

Distance Ratio. First we define the set of initial foreground regions $O = \{r_i | A_{r_i} = A_{obj}, r_i \in I\}$. For a region r' which has the initial user annotation A_u, we calculate Chi-square distance between the histogram of RGB color of r_i and r':

$$D(r_i, r') = \frac{1}{2} \sum_k \frac{(H_{r_i}^k - H_{r'}^k)^2}{H_{r_i}^k - H_{r'}^k}\tag{5}$$

where H^k is the kth bin of the color histogram; we quantify RGB colors to $5 \times 5 \times 5$, thus H has 125 bins. We then define the minimum distance from O as:

$$D^*(O, r') = min_{r_i \in O} D(r_i, r').\tag{6}$$

Similarly, the distance $D^*(B, r')$ from r' to background regions B is defined. The distance ratio of the region r' is written as:

$$M_{DR}(r') = \frac{1}{|D^*(O, r') - D^*(B, r')| + 1}.\tag{7}$$

We then get a ratio map M_{DR} which is also normalized to $[0, 1]$. As shown in Figure 2(b), the unknown region at the tail of the bird is crucial because it is both similar to the wood with background label, which the bird stands on, and some labelled parts of the bird body.

Neighbor Similarity. Appearance information extracted from initial object regions O is often not effective enough to decide the whole object. There are two key reasons: 1) O only contains some parts of an object; 2) each part of the desired object may have different appearance. With this knowledge, we use neighbor similarity to find crucial regions which are adjacent to object regions O

and have different appearance to O. Denote an unknown region r'' is adjacent to *object* regions O', where O' is a subset of O and consists of labelled regions. We calculate the Chi-square distance of RGB histogram $D(O', r'')$ Equ. (5). Then neighbor similarity can be written as:

$$M_{NS}(r'') = w \cdot D(O', r''), \tag{8}$$

where w is a distance weight with the range $[0.5, 2]$:

$$w = \begin{cases} 0.5 & \text{if } R(O', r'') \in (0, 0.5]; \\ R(O', r'') & \text{if } R(O', r'') \in (0.5, 2]; \\ 2 & \text{if } R(O', r'') \in (2, +\infty]. \end{cases} \tag{9}$$

where $R(O', r'')$ is the area ratio between O' and r''. In Figure 2(c), the small regions near the bird paw are more crucial.

4.2   Interactions

Then we combine those maps using same weights to get the final information map $\gamma : \gamma(r_i) = M_{RE}(r_i) + M_{DR}(r_i) + M_{NS}(r_i)$, for $i = 1, 2, , N$. The most crucial region is the one which has the highest score in γ; the brightness of regions in Figure 2(d) is proportional to their score.

After the above operations, the method queries the oracle for the true label of the most crucial region. Then this additional interaction on crucial region is combined with initial interactions. Denote the additional interaction in the tth iteration is A^t, thus the updated initial interactions in the $(t+1)^{th}$ iteration are $A_0 = A_0 \cup A^t$. After T iterations, the total interactions A_0 is the combined set of initial interactions and all additional user interactions.

5   Experiments

In our experiments, we present results of our method and compare them with the state of the art algorithm [17]. All the experiments use unified parameters: we set the associated weights λ_1 and λ_2 to $\{0.5, 0.5\}$ in energy function Eq.(1). The parameters of mean shift over-segmentation code are set to $\{sigmaS = 9, sigmaR = 9, minRegion = 100\}$. Only one crucial region is recommended in each iteration. Then we use min-cut algorithm [2] to get an optimal solution of our energy minimum problem.

We show several experimental results in Figure 3. The desired object has great appearance variances. Given an input image, a user marks on the foreground object(red scribbles) and background(blue scribbles) respectively; the second column are the result of initial interactions. All 5 recommended crucial regions are overlapped onto the original image and highlighted with green color. The additional interactions are made in yellow circles. The last column shows the optimal results by graph-cuts algorithm. Under the parameter setting of over-segmentation algorithm, there are about 50 regions per image with 260×200

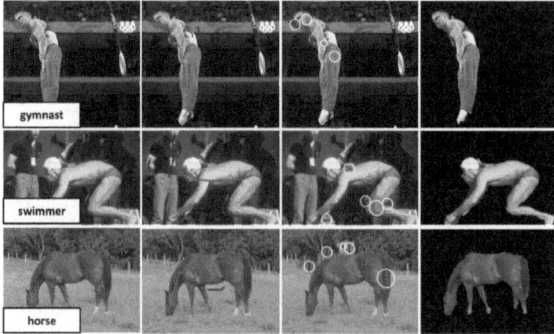

Fig. 3. Experimental results of our method. From the left to right: 1) input images; 2) images overlapped by initial interactions; 3) all recommended crucial regions with their true labels in yellow circles; 4) segmentation results. (Best viewed in color)

Fig. 4. Comparison of our segmentation results with GrabCut. Our method in the second row is initialized by object and background scribbles while GrabCut in the first row is initialized by a bounding box. Results are highlighted with black background.

resolution and the method only recommends about 10% regions to guide the user. The experiment on a variety of images shows that it can lead to significant improvements in segmentation accuracy within limited iterations.

Figure 4 is the comparison of our segmentation result with GrabCut. GrabCut starts with an initial bounding box. Then the iteratively scribbles are marked by the user until obtaining the desirable result. Instead, most crucial regions are recommended iteratively in our method; segmentation results of GrabCut and our method show that 1) the most crucial regions are effective to guide the user. 2) By recommending only 10% regions, we can get comparable segmentation results.

6   Conclusion

We present an active learning method for interactive segmentation. First, it iteratively selects the most crucial regions. In each iteration, region entropy, distance ratio and neighbor similarity are learned to form information map, and our method recommends the most crucial region from this map to guide user. Second, we combine color model and color spatiality feature maps acquired from interactions into our model, and minimize associate energy function to obtain the final segmentation result. Experimental results show that our method is effective to achieve high segmentation accuracy in limited interactions. Our future work will focus on the learning of associate weights to make them object-specific.

Acknowledgments. This paper is partially supported by National Basic Research Program of China (973 Program) under contract 2009CB320902; Natural Science Foundation of China (NSFC) under contracts Nos. 60833013 and No. 60832004.

References

1. Batra, D., Kowdle, A., Parikh, D., Luo, J., Chen, T.: icoseg: Interactive co-segmentation with intelligent scribble guidance. In: CVPR. pp. 3169–3176 (2010)
2. Boykov, Y., Kolmogorov, V.: An experimental comparison of min-cut/max-flow algorithms for energy minimization in vision. TPAMI 26(9), 1124–1137 (2004)
3. Brunne, G., Chittajallu, D., Kurkure, U., Kakadiaris, I.: Patch-cuts: A graph-based image segmentation method using patch features and spatial relations. In: BMVC (2010)
4. Comaniciu, D., Meer, P.: Mean shift: A robust approach toward feature space analysis. TPAMI 24(5), 603–619 (2002)
5. Delong, A., Gorelick, L., Schmidt, F., Veksler, O., Boykov, Y.: Interactive segmentation with super-labels. In: EMMCVPR. pp. 147–162 (2011)
6. Ding, L., Yilmaz, A.: Enhancing interactive image segmentation with automatic label set augmentation. In: ECCV. pp. 575–588 (2010)
7. Fathi, A., Balcan, M., Ren, X., Rehg, J.: Combining self training and active learning for video segmentation. In: BMVC. pp. 78–1 (2011)
8. Hoi, S., Jin, R., Zhu, J., Lyu, M.: Semi-supervised svm batch mode active learning for image retrieval. In: CVPR. pp. 1–7 (2008)
9. Kapoor, A., Grauman, K., Urtasun, R., Darrell, T.: Active learning with gaussian processes for object categorization. In: ICCV. pp. 1–8 (2007)
10. Kass, M., Witkin, A., Terzopoulos, D.: Snakes: Active contour models. IJCV 1(4), 321–331 (1988)
11. Kolmogorov, V., Zabin, R.: What energy functions can be minimized via graph cuts? TPAMI 26(2), 147–159 (2004)
12. Li, Y., Sun, J., Tang, C., Shum, H.: Lazy snapping. TOG 23(3), 303–308 (2004)
13. Mishra, A., Aloimonos, Y., Fah, C.: Active segmentation with fixation. In: ICCV. pp. 468–475 (2009)
14. Mortensen, E., Barrett, W.: Intelligent scissors for image composition. In: CGIT. pp. 191–198 (1995)
15. Ning, J., Zhang, L., Zhang, D., Wu, C.: Interactive image segmentation by maximal similarity based region merging. Pattern Recognition 43(2), 445–456 (2010)
16. Qi, G., Hua, X., Rui, Y., Tang, J., Zhang, H.: Two-dimensional active learning for image classification. In: CVPR. pp. 1–8 (2008)
17. Rother, C., Kolmogorov, V., Blake, A.: Grabcut: Interactive foreground extraction using iterated graph cuts. TOG 23(3), 309–314 (2004)
18. Vezhnevets, A., Buhmann, J., Ferrari, V.: Active learning for semantic segmentation with expected change. In: CVPR (2012)

Illumination Normalization for Face Recognition under Extreme Lighting Conditions

Yong Cheng[1], Zuoyong Li[2], and Liangbao Jiao[1]

[1] School of Communication Engineering, Nanjing Institute of Technology,
Nanjing 211167, China
[2] Department of Computer Science, Minjiang University, Fuzhou 350108, China
chyo_200908@hotmail.com

Abstract. An effective illumination normalization method based on human visual system is presented for extreme lighting face recognition. One contribution is that illumination normalization based on retinal modeling is mainly executed on low frequency band considering lighting conditions, the other is the introduction of discrete wavelet transform into human visual modeling for illumination normalization. The proposed method not only gets better illumination normalized result, but also preserves more image details. Both of them are very important for face recognition under complex lighting conditions. Experimental results on extended Yale B face databases demonstrate that our method is effective for dealing with variable lighting, especially for extreme lighting variation situation.

Keywords: illumination normalization, face recognition, extreme lighting conditions.

1   Introduction

Over the last many years, face recognition research has obtained significant progresses. However, there are still many challenges [1] for face recognition under uncontrolled environments. Illumination variation, which seriously degrades the performance of face recognition, is still an unresolved problem to a certain degree for an effective face recognition system. Many methods have been proposed to solve this problem, including multiscale retinex (MSR) [2], logarithmic total variation (LTV) [3] and Gradient faces [4], etc. Unfortunately, varying lighting effects cannot be completely eliminated by these methods. Recently, many methods inspired by human vision system have been exploited for image processing and feature extraction. Meylan et al. [5] developed a modified Naka-Rushton equation, and utilized two consecutive nonlinear operations based on the equation to model human vision system for avoiding halos and improving global appearance. Later, an illumination normalization method based on retinal modeling [6] was presented by using the aforementioned two consecutive nonlinear operations and difference of Gaussian filters. It achieves good performance on varying lighting face recognition. To further eliminate the effect of illumination variation on face recognition, we present a novel illumination normalization method for face recognition under extreme lighting conditions.

J. Yang, F. Fang, and C. Sun (Eds.): IScIDE 2012, LNCS 7751, pp. 491–497, 2013.

2   Illumination Normalization for Extreme Lighting Conditions

2.1   Retina Model

The primate's retinal model has three functional layers, namely, the photoreceptors layer, the Outer Plexiform Layer (OPL) and the Inner Plexiform Layer (IPL)[7,8].

The photoreceptors layer is composed of two cells: cones and rods. It is responsible for visual data acquisition and is related to a local image illumination compression with respect to the illumination of neighborhood, which is achieved by the cellular network: the Horizontal cells. This local compression has been modeled by the authors based on the Naka-Rushton equation, and the model has been used to illumination normalization.

The Outer Plexiform Layer includes Bipolar and Horizontal cells. The functions of the Bipolar cells are simply to transmit signal from the OPL to IPL. The Horizontal cells are responsible for obtaining the illumination of neighborhood (local illumination). It feeds back photoreceptors layer for local compression of image illumination. In [5] and [6], the local illumination is achieved by Gaussian filtering.

The Inner Plexiform Layer consists of the Ganglion cells and the Amacrine cells. This layer has two functions. The first one is to modify illumination like the photoreceptors layer does. In this stage, the Amacrine cells provide local illumination like the Horizontal cells do. The second is to perform contour enhancement. This can be modeled by a spatial band-pass filter.

As mentioned above, illumination normalization based on retinal model can be established. It includes two consecutive local image illumination compressions and a spatial band-pass filtering.

2.2   Naka-Rushton Equation

Naka-Rushton equation proposed by Naka and Rushton [9] is a nonlinear function for modeling retinal illumination normalization. It is defined as the following equation,

$$Y = X / (X + X_0). \tag{1}$$

Where X is input light intensity, adaption factor X_0 is a fixed value determined by average light intensity of entire visual area, and Y is the adapted signal. Recently, an improved Naka-Rushton equation [5] is proposed by adjusting the adaptation factor X_0 as a local variable, which is mainly computed for each pixel by performing Gaussian filtering on its neighborhood. It can be described as following:

$$X_0 = X(p) * G_H + \bar{X}/2. \tag{2}$$

where p is a pixel in the image, X_0 is the adaptation factor at pixel p, $X(p)$ is the intensity of the input image, $*$ denotes the convolution operation, G_H is a two-dimensional Gaussian filter, $\bar{X}$ corresponds to the mean value of the X image pixel intensities. Illumination normalization based on the improved equation can avoid halos phenomenon and improve global appearance. However, there still have some defects. For instance, Applying Gaussian filtering to original images for estimating local light intensity often produces inaccurate results on the edges of an image. The mean value of an image $\overline{X}$ as part of the adaptation factor to all pixels is not appropriate, especially in complex illumination conditions. These might cause distorted results in the later process of illumination normalization. Hence, in this paper, to gain accurate adaptation factors we mainly perform illumination normalization on low frequency band of an image, which is achieved by 2-D wavelet transform. Moreover, the adaptation factor X_0 is estimated considering lighting differences.

2.3 Estimating the Adaptation Factor

Let I be an original image. A, DH, DV and DD are its low frequency sub-band and high frequency sub-bands respectively. Firstly, illumination classification based on Otsu (maximum variance between various classifications) is performed on low frequency sub-band A. Supposing t is the threshold obtained by Otsu. Lighting conditions of A can be approximately divided into two categories:

$$M_1 = \{A(x,y)|A(x,y) \geq t\}$$
$$M_2 = \{A(x,y)|A(x,y) < t\}$$

where $A(x,y)$ is the intensity of A at point (x,y).

The adaptation factor $A_{0\rho}$ of A can be defined as follows:

$$A_{0\rho}(x,y) = A(x,y) * G_\rho + \overline{mean_A}/2. \qquad (3)$$

$$\overline{mean_A} = \begin{cases} \overline{M_A^1} & A(x,y) \in M_1 \\ \overline{M_A^2} & A(x,y) \in M_2 \end{cases}. \qquad (4)$$

where G_ρ is Gaussian filtering and ρ be its standard deviation. $\overline{M_A^1}$ and $\overline{M_A^2}$ are the average value of M_1 and M_2 respectively.

2.4 Illumination Normalization

A novel effective human visual modeling for illumination normalization of face recognition is established. Firstly, single-level discrete 2-D wavelet transform is implemented on a face image to extract its low and high frequency information. Then, low frequency information is processed by imitating retinal information processing mechanism considering illumination conditions and large high frequency coefficients is truncated by a threshold. In the end, illumination normalization result is obtained by inverse discrete 2-D wavelet transform on adapted low and high frequency information. The detailed process of low and high frequency information is as follows:

Low Frequency Processing
Step 1. Modeling the photoreceptors layer: low frequency sub-band A is processed by the following equation,

$$A^p(x,y) = \left(max_A + A_{0\rho_1}(x,y)\right)A(x,y)/\left(A(x,y) + A_{0\rho_1}(x,y)\right). \tag{5}$$

where max_A is the maximum of $A,A_{0\rho_1}(x,y)$ is the adaptation factor of A at point (x,y), ρ_1 is standard deviation of the Gaussian filtering and $A^p(x,y)$ is the corresponding adapted output.

Step 2. Modeling the Outer Plexiform: A^p is processed as following,

$$A^o(x,y) = \left(max_{A^p} + A^p_{0\rho_2}(x,y)\right)A^p(x,y)/\left(A^p(x,y) + A^p_{0\rho_2}(x,y)\right). \tag{6}$$

where $max_{A'}$ be the maximum of $A^p, A^p_{0\rho_2}(x,y)$ is the adaptation factor of A^p at point (x,y), ρ_2 is standard deviation of the Gaussian filtering and $A^o(x,y)$ is the corresponding adapted output.

Step 3. Edge detection: the spatial band-pass filtering is adopted to model Inner Plexiform information processing. It can be defined as

$$A^e(x,y) =$$
$$(exp(-(x^2+y^2)/2\rho_L^2)/2\pi\rho_L^2 - exp(-(x^2+y^2)/2\rho_H^2)/2\pi\rho_H^2) * A^o. \tag{7}$$

where ρ_L and ρ_H are corresponding standard deviations of two Gaussian functions, A^e denotes the edge of image.

High Frequency Treatment
Since large wavelet coefficients in high frequency sub-bands are not stable for a face image under different lighting conditions, high frequency information used for face recognition will produce negative influence without processing. So, to enhance illumination robustness of high frequency information, we perform a truncation operation on big high frequency coefficients.

Let H is DH, DV or DD. The truncation operation of high frequency sub-band is defined as following:

$$H'(x,y) = \begin{cases} t & H(x,y) > t \\ H(x,y) & |H(x,y)| \le t. \\ -t & H(x,y) < -t \end{cases} \tag{8}$$

$$t = median(|H(x,y)|). \tag{9}$$

where $median(\cdot)$ is to return the median value of array $|H|$. Height frequency sub-bands treated by the truncation are DH', DV' and DD' respectively.

In this paper, ρ_1, ρ_2, ρ_L and ρ_H are set to be 1, 3, 0.5 and 4, respectively. This coincides with parameter selection in [6]. Fig. 1 shows the results by using our method and Vu's approach to five images with different illumination conditions. It can be seen that the proposed method produces better natural images and keeps more details than Vu's. In addition, data normalization is implemented to ensure that its mean and variance are 0 and 1 respectively before image recognition.

3   Experimental Results

The extended Yale B [10] contains 38 human subjects under 9 poses and 64 lighting conditions. Its extreme lighting conditions still make it a challenging task for most face recognition methods. In this paper, only frontal images under varying lighting conditions are chosen as samples and all images are resized to 192×168. PCA (reserved 90% energy) is employed to reduce dimension, and the Nearest Neighbor Classifier based on Euclidean distance is used for classification. Wavelet function db1 is used for discrete 2-D wavelet transformation.

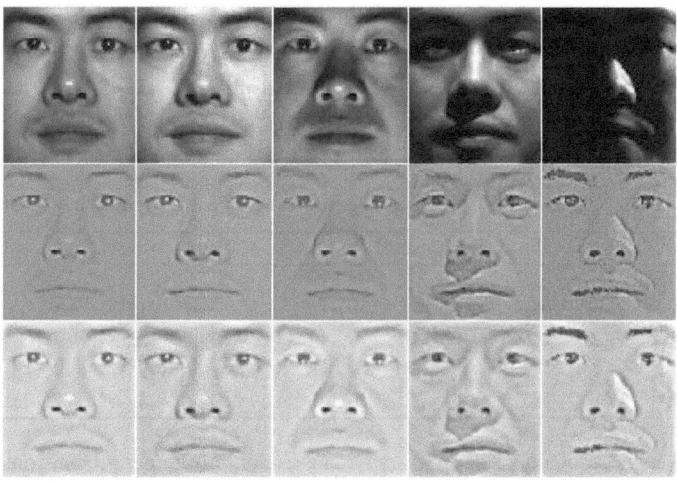

Fig. 1. Results achieved by our method and Vu's method: the top row is originals of one person from subset 1 to 5; the middle row is Vu's; the bottom row is ours

To evaluate the performance of the proposed method, we perform three groups of experiments. Firstly, subset 1 or 5 of five subsets is used for training, and remainder for testing. Corresponding results are listed in Tables 1 and 2. From the tables, it can be observed that our algorithm achieves higher recognition performance than other methods under two different training sets. Secondly, images with the most neutral light sources ('A+000E+00') are used as the gallery (38 images), and all other frontal images of the standard subsets 1-5 are used as probes (in all, 2 414 images of 38subjects).Table3 lists the corresponding results. From the table, one can conclude that our average recognition rate reaches 96.03%, implying better recognition performance. Thirdly, a random image per person from subset 1 is chosen as training set, and the rest of images as testing set. Since the above training set and testing set are random, we run the simulation 30 times and average the results over them. The average recognition rates of various methods are listed in Table 4. From the table, one can observe that recognition rates of the proposed method are higher than those of other methods, higher than others, which further demonstrates the effectiveness of our method. In addition, all Tables also show that our algorithm is little affected by selection of training set, and has good stability.

Table 1. Recognition rates (%) using subset 1 for training

Methods	Subset 2	Subset 3	Subset 4	Subset 5
Original	83.33	18.80	1.31	0.14
LTV	90.79	97.93	92.11	95.43
Gradient faces	100.00	95.87	82.68	94.60
Vu [3]	96.93	99.25	96.71	98.89
Ours	99.78	99.44	98.47	97.78

Table 2. Recognition rates (%) using subset 5 for training

Methods	Subset 1	Subset 2	Subset 3	Subset 4
Original	4.89	4.39	6.39	8.11
LTV	93.06	88.16	97.56	97.59
Gradient faces	94.36	89.47	84.59	99.34
Vu [3]	98.12	94.30	98.68	99.12
Ours	98.12	97.15	99.45	99.34

Table 3. Recognition rates (%) using one image with neutral light sourcesfor training

Methods	Subset 1	Subset 2	Subset 3	Subset 4	Subset 5	All subsets
Original	2.63	2.63	2.82	3.95	2.91	3.01
LTV	96.05	75.44	97.18	85.53	91.97	89.14
Gradient faces	99.56	98.03	92.11	79.61	90.58	91.10
Vu [3]	97.73	88.60	97.18	91.45	93.07	93.23
Ours	98.25	96.27	97.73	95.61	94.46	96.03

Table 4. Average recognition rates (%) using one random image per person from subset 1 for training

Methods	Subset 1	Subset 2	Subset 3	Subset 4	Subset 5	All subsets
Original	2.63	2.78	3.06	4.13	3.33	3.25
LTV	92.06	80.22	92.18	85.99	91.46	88.50
Gradient faces	99.47	98.36	89.15	77.71	90.25	90.04
Vu [3]	96.05	90.15	95.65	90.84	94.11	93.26
Ours	97.75	96.00	96.33	95.24	95.22	95.86

4   Conclusions

In this paper, we propose an effective illumination normalization method based on human visual system for variable lighting face recognition. This method can preserve more image details while getting better illumination normalized result. Experimental results on extended Yale B face databases show that the proposed method is robust to extreme lighting conditions.

Acknowledgments. This work is partially supported by Nanjing Institute of Technology Innovation Fund (Grant Nos.CKJ2010009), National Natural Science Foundation of China (Grant Nos. 61202318) and the Natural Science Foundation of Shandong Province (Grant Nos. ZR2011FM004).

References

1. Phillips, P.J., Scruggs, W.T., O'Toole, A.J., Flynn, P.J., Bowyer, K.W., Schott, C.L., Sharpe, M.: FRVT 2006 and ICE 2006 large-scale experimental results. IEEE Trans. Pattern Anal. Mach. Intel. 32, 831–846 (2010)
2. Jobson, D.J., Rahman, Z., Woodell, G.A.: A multiscaleRetinex forbridging the gap between color images and the human observation ofscenes. IEEE Trans. Image Process. 6, 965–976 (1997)
3. Chen, T., Yin, W., Zhou, X.S., Comaniciu, D., Huang, T.S.: Total variation models for variable illumination face recognition. IEEE Trans. Pattern Anal. Mach. Intel. 28, 1519–1524 (2006)
4. Zhang, T., Tang, Y.Y., Fang, B., Shang, Z., Liu, X.: Facerecognition under varying illumination using Gradientfaces. IEEE Trans. Image Process. 18, 2599–2606 (2009)
5. Meylan, L., Alleysson, D., Susstrunk, S.: Model of retinal local adaptation for the tone mapping of color filter array images. J. Opt. Soc. Amer. 24, 2807–2816 (2007)
6. Vu, N., Caplier, A.: Lighting robust face recognition busing retina modeling. In: Proc. IEEE Int'l Conf. Image Processing, pp. 3289–3292. IEEE Press, New York (2009)
7. Hérault, J., Durette, B.: Modeling Visual Perception for Image Processing. In: Sandoval, F., Prieto, A.G., Cabestany, J., Graña, M. (eds.) IWANN 2007. LNCS, vol. 4507, pp. 662–675. Springer, Heidelberg (2007)
8. Benoit, A., Caplier, A., Durette, B., Herault, J.: Using Human Visual System modeling for bio-inspired low level image processing. J. Comput. Vis. Image Understand 114, 758–773 (2010)
9. Naka, K.-I., Rushton, W.A.H.: S-potentials from luminosity units in the retina of fish (cyprinidae). Journal of Physiology 185, 587–599 (1966)
10. Georghiades, A.S., Belhumeur, P.N., Kriegman, D.J.: From few to many: Lighting cone models for face recognition under variable lighting and pose. IEEE Trans. Pattern Anal. Mach. Intel. 23, 643–660 (2001)

A Finger Vein Image Quality Assessment Method Using Object and Human Visual System Index

Hui Ma[1], Kejun Wang[2], Linlin Fan[1], and Fengpeng Cui[3]

[1] College of Electronic Engineering, Heilongjiang University, Harbin 150001
[2] College of Automation, Harbin Engineering University, Harbin 150001
[3] China Ship Heavy Industrial Group Company 703th Research Institute Harbin 150001

Abstract. A finger vein image quality assessment method which uses object and human visual system index is proposed. This is based on both the human visual characteristics and finger vein image characteristics captured by the use of contactless and near infrared rays. We present an *HSNR* (signal to Noise ratio based on human visual system) finger vein image quality evaluation index by simulating the human visual system, and integrating the *HSNR* with effective area score, clarity score, finger shifting score and contrast score to obtain the total image quality score of the finger vein image. Experimental results demonstrate that the proposed method is consistent with subjective assessment by humans, and thus can be used to describe the visual perception of the image effectively.

Keywords: finger vein image, image quality assessment, human visual system.

1   Introduction

Finger vein image recognition makes use of short near infrared transmission on finger to obtain vein features for personal identification. It is a very reliable biometric recognition method with merits of being rapid, precise and non-contact [1-2].There are various forms of distortion in finger vein image after the process of capture, coding and transmission. These distortions have different influence on the image quality and this leads to a degradation in the recognition system's performance. Thus, it is necessary to evaluate the quality of finger vein image evaluation in the recognition system.

At present, there are a few special image quality assessment methods on finger vein image [3]. Image quality assessments methods are mainly on fingerprint image, iris image, and single frame video image [4-6]. Reference [4] use fast Fourier transform on fingerprint image to obtain spectrum distribution map then use the moment of inertia and eccentricity ratio of spectrum distribution map to evaluate the quality of fingerprint image. With the aid of iris image quality assessments, Reference [5] used Wavelet-based contourlet transform to reflect visual characteristics of iris textures. Reference [6] use grayscale extreme value and discrete wavelet transform to evaluate the quality of infrared image.

J. Yang, F. Fang, and C. Sun (Eds.): IScIDE 2012, LNCS 7751, pp. 498–506, 2013.

Images have some quality issues including (but not limited to) overexposure, underexposure, blur edges as a result of change in light intensity, and sensor noise coming from the finger image capture process. Such image quality issues can lead to low correct identification ratios. The major function of the eyes is to extract structural information from the field of vision and the human visual system (HVS) is suitable for extracting such information [7]. On this basis, reference [8] provides an image quality assessment method based on structural similarity. Experimental results show that this method has a good correlation compared with the subjective assessment method. Based on the above, this paper proposes a finger vein image quality assessment method using object and human visual system index. This method is founded on the cognitive function of human visual system in combination with various factors that affect finger vein image quality. We fuse four objective appraisal indices including contrast, effective area, finger shifting and clarity with a index *HSNR* based on HVS to establish a finger vein image quality assessment system. Experimental results demonstrate that the proposed method is consistent with subjective assessment by humans.

2   Objective Finger Vein Image Quality Assessments

The performance of biometric identification is closely related to an image's quality. To determine the effect of image quality on recognition result and to ensure the accuracy of the recognition system, we need a systematic and objective evaluation of finger vein images. The finger vein image acquisition is done using near infrared wavelengths without the limitations of fingerprint image capture like scar, dirt, too dry and too wet conditions, which increase the difficulty in image evaluation. Thus in this method, we design finger vein image quality assessment according to various factors influencing image quality and its characteristics. Through analyzing the experimental results of a large number finger vein images, the main factors that have effect on image quality are as follows:

1)   Contrast: the more the image contrast, the better the finger vein image quality.
2)   Position offset: different finger placement produce calculation errors which result in high false reject rate.
3)   Effective area: the bigger the effective area of finger vein image, the more amount of information i.e. better finger vein image quality.
4)   Clarity: If an image is fuzzy, then the vein and background have no clear distinction. In such circumstances, subsequent enhancement and matching operations will be directly affected.
5)   Signal to noise ratio: the less noise, the higher the value of signal to noise; that is, a better finger vein image quality.

Building on previous research [9], this paper fuses weighted indices to obtain the final finger vein image quality score.

2.1   Contrast Index

Contrast of finger vein image means the grey difference that can be represented by mean square deviation:

$$M = \sqrt{\frac{\sum_{i=1}^{N}(p_i - \bar{p})^2}{N}} \tag{1}$$

Where N represents the total pixels of the image region. p_i represents grey value of a pixel and $\bar{p}$ is the average grey value of the whole image.

The contrast score S_1 is given by:

$$S_1 = \begin{cases} \dfrac{M}{M'} \times 100 & \dfrac{M}{M'} < 1 \\ 100 & otherwise \end{cases} \tag{2}$$

where M' is the standard contrast that is optimized to get the vein recognition system performance. We get the optimal recognition system performance when M' is equal to 40.

2.2   Finger Shifting Index

Finger shifting refers to the deviation of foreground areas' center of mass with respect to the geometric centre of the whole image and includes horizontal and vertical offset. A finger shifting score S_2 is estimated by horizontal offset score P_H and vertical offset score P_V. The detailed calculation of P_H and P_V is described as follows:

$$P_H = (1 - \frac{\left| \sum_{i \subset I} x_i / N - I_x \right|}{I_x}) \times 100 \tag{3}$$

$$P_V = (1 - \frac{| \sum_{j \subset I} y_j / N - I_y |}{I_y}) \times 100 \tag{4}$$

Where x_i is the horizontal coordinate of pixel i and y_i is the vertical coordinate of pixel j of the image region. N represents the total pixels of image. I_x is the geometric centre of the horizontal coordinate and I_y is the geometric centre of vertical coordinate of the vein image. The finger shifting score S_2 is given by:

$$S_2 = P_H \times P_V / 100 \tag{5}$$

2.3   Effective Area Index

The valid region of vein image is the area of finger region in finger vein image. So the area score S_3 is determined by the ratio of the foreground areas S' to the total image area S. The background region's average grey value is close to zero while the average grey value of finger region is relatively large. The original image is transformed into 4 × 4 blocks to determine from its average grey value, whether the block constitutes a foreground region. These finger regions are added to S'. S_3 has a maximum score if the foreground region is more than 1/5th of the whole image area. S_3 is calculated as follows:

$$S_3 = \begin{cases} 100 & , \quad (S' > \frac{1}{5}S) \\ \dfrac{S'}{S \times \frac{1}{5}} \times 100, & otherwise \end{cases}$$

(6)

2.4   Clarity Index

If an image is fuzzy, the average values of pixels along the vein ridge are closer to the values of pixels perpendicular to the vein ridge. The difference between these two average values is large in a clarity image. Thus, we divide the finger region into some 4 × 4 blocks .Then we calculate the above two average values to determine whether the block is a blur block or not. S_M is calculated by adding together the area of all blur blocks. So, the clarity area score S_4 used in this paper is determined as follows:

$$S_4 = (1 - \frac{\sum_{i=1}^{n} S_M}{S'}) \times 100$$

(7)

2.5   Signal to Noise Ratio Index

Signal to noise ratio means the ratio of useful signal to noise. If image is affected by noise seriously then the image has a low signal to noise value. In other words, image quality is poor. Traditional signal-to-noise ratio is described as follows:

$$SNR = 10 \times \lg(\frac{\sigma_f^2}{MSE})dB$$

(8)

Where M and N represents the length and the breadth of finger vein image respectively. σ_f^2 represents grayscale value's variance of original image. MSE (Mean square error) is calculated as follow:

$$MSE = \frac{1}{MN} \sum_{i=1}^{M} \sum_{j=1}^{N} (f'_{i,j} - f_{i,j})^2$$

(9)

Where (i,j) is the coordinate of pixel. $f_{i,j}$ represents pixel's gray value of reference image. $f'_{i,j}$ is pixel's gray value of evaluation image. The signal to noise ratio score S_5 is given by:

$$S_5 = SNR$$

(10)

3   Object and HVS Index Image Quality Assessment Method

Human visual system sensitivity is selective on images. That is to say, the HVS sensitivity varies for different image regions and different distortion signals. The above signal to noise index does not take the HVS sensitivity into consideration for different regions. Ultimately, perception and hence assessment of the image quality is performed by the human observer and many times, subjectivity plays a great role in this assessment. An effective image quality assessment method needs to be fit for the visual characteristic of HVS [10].

Traditional image quality assessment methods based on HVS evaluate image quality simply by subjective visual inspection. Its evaluation process takes a long time and its evaluation result is easily affected by the working experience and other mental factors of the person doing the evaluation. The above reasons make it hard to meet the actual specification of the recognition system. Because the commonly used SNR function doesn't include the HVS parameter, there will be different evaluation results based on subjective image quality assessment even when two images have the same signal to noise ratios.

Based on this analysis, we combine the four above mentioned objective appraisal indices namely: contrast, effective area, finger shifting and clarity with an index *HSNR* (signal to Noise ratio based on HVS) to establish a robust finger vein image quality assessment system.

The *HSNR* based on HVS is established by the contrast sensitivity weights matrix and the impulse noise distribution matrix of the finger vein image.

1) To obtain contrast sensitivity weights matrix:
   The contrast sensitivity function of HVS is described as follows:

$$S(f) = 2.6(0.0192 + 0.114f)\exp^{-(0.114f)^{1.1}} \tag{11}$$

Where *f* represents the spatial frequency of finger vein image, determined by

$$f = \sqrt{f_x^2 + f_y^2} \tag{12}$$

Where f_x and f_y represents the row spatial frequency and the column spatial frequency respectively. f_x and f_y is given by:

$$f_x = \sqrt{\frac{1}{MN}\sum_{i=0}^{M-1}\sum_{k=1}^{N-1}[F(i,k) - F(i,k-1)]^2} \tag{13}$$

$$f_y = \sqrt{\frac{1}{MN}\sum_{k=0}^{N-1}\sum_{j=1}^{M-1}[F(k,j) - F(k,j-1)]^2} \tag{14}$$

Where *M* is the number of row and *N* is the number of column.
   Normalized spatial frequency is determined by:

$$\overline{f} = \frac{f - f_{min}}{2 \times (f_{max} - f_{min})} \tag{15}$$

Where f_{max} and f_{min} is the maximum and minimum value of image's spatial frequency respectively. Bringing $\overline{f}$ into formula (11) gives the contrast sensitivity weights matrix $S(i,j)_\circ$

2) Impulse noise distribution matrix

We use a 5×5 template to detect noise point all over the image. $\overline{f}(x,y)$ represents the gray value mean of all pixels' in the 5×5template. It is calculated as:

$$\overline{f}(x,y) = \sum_{0\leq x,y\leq 4} f(x,y)/25 \tag{16}$$

To determine whether or not a pixel is a noise input, we compute the difference between pixel's gray value from the middle of the template and the mean gray value of all pixels in the 5×5 template. If there is a remarkable difference between it and the threshold we consider the pixel in the middle of the template is a noise point.

Determine the impulse noise distribution matrix $Q(x,y)$, the values of this matrix are all first initialized to zero. The number of zeros is equal to multiplication of M-1 by N-1; where M is the number of rows and N is the number of columns in the image. The value of matrix element equal to 1 corresponds to the position of a noise point; else is its value remains zero. Thus we can obtain the whole noise distribution matrix $Q(x,y)$.

3) HSNR

Noisy level *HMSE* based on HVS is obtained by $S(i,j)$and $Q(x,y)$ and is calculated as follows:

$$HMSE = \frac{\sum_{m=1}^{M-1}\sum_{n=1}^{N-1} S(m,n)Q(m,n)}{(M-2)\times(N-2)} \tag{17}$$

Where N is the number of row and M is the number of columns in image. (m,n) representing the image pixel.

We can get the *HSNR* index on the basis of *HMSE*:

$$HSNR = 10\lg\frac{G^2}{HMSE} \tag{18}$$

Where G is the gray level of image.

Thus the signal to noise ratio index is modified as follows:

$$S_5 = HSNR \tag{19}$$

After obtaining the five quality assessment parameters, the finger vein image has a comprehensive evaluation. The lower the quality scores S_i $(i=1, 2, 3, 4, 5)$,the higher the weight and the influence it has on total score. The weight of S_i is correlated with four other quality scores. Total quality score S is given by:

$$S = \sum_{i=1}^{5}[S_i\times(100-S_i)\times(100-S_i)/\sum_{i=1}^{5}(100-S_i)\times(100-S_i)]\times[(\sum_{i=1}^{5}S_i-S_i)/400]$$

$$\tag{20}$$

4   Experimental Results

To verify the effectiveness of the proposed method, we test the algorithm using images from the finger vein image database captured by our laboratory. The database includes five forefinger vein images each of 150 females. Each finger vein image size is 320 × 240. Partial images from our experimental database are shown in Figure 1.

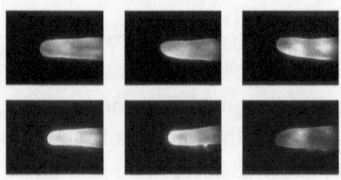

Fig. 1. Some original finger vein images of our experimental database

4.1   Effectiveness Testing

Three typical images are selected from our finger vein image database with good quality image, finger shifting image and overexposed image are shown in figure 2. The experimental results of our image quality assessment method are shown in Tab. 1.

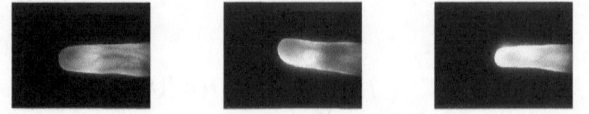

(a) high quality image   (b) finger shifting image   (c) exposure image

Fig. 2. Example finger vein image

Table 1. Quality evaluation results of above finger vein images

	Fig.(a)	Fig.(b)	Fig.(c)
area score	96	87	76
finger shifting score	92	74	89
contrast score	88	82	65
fuzzy score	93	90	68
signal-to-Noise score	79	76	79
image total score	91	79	67

Experimental results demonstrate that the proposed method has a better performance for finger vein image quality assessment and it is consistent with subjective assessment by humans.

4.2   Rationality Testing

To verify the effectiveness of the proposed method, we use the database finger vein images to build two kinds of distortion images: images with finger region becoming smaller and smaller and images with Gaussian noise. (The mean of Gaussian noise image is zero). As shown in figure 3, the quality assessment score decreases with increased image distortion.

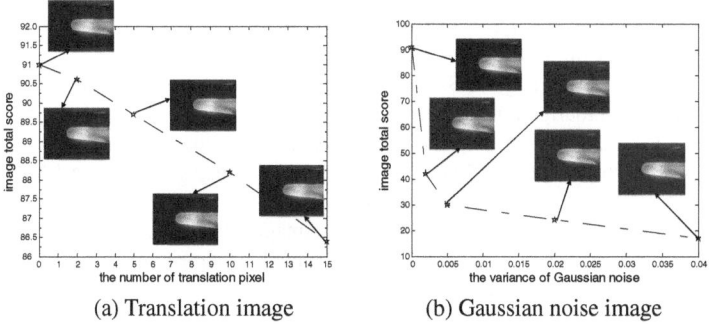

(a) Translation image   (b) Gaussian noise image

Fig. 3. The quality score curve of our method under different distortion cases

As shown in Figure 3(a), finger region translation mostly affects effective area score but has no significant effect on other quality score; thus the total quality score fell slower. As can be seen in figure 3(b), Gaussian noise mostly affects clarity index and signal to noise index. It makes these two quality scores quite low. In other words, the weights of above two quality score is more in the total quality score formula causing the total quality score to become low. This clearly demonstrates that the proposed method is reasonable.

5   Conclusion

The proposed method of finger vein image quality assessment is founded on the cognitive function of human visual system combined with various effect factors affecting finger vein image quality to evaluate image quality. Experimental results demonstrate that the proposed method is consistent with subjective assessment by humans, and thus can be used to describe the visual perception of the image effectively. Our quality assessment result can be used to guide the finger vein image capture process towards optimal positioning of fingers in order to obtain very good quality images. We can also use the result as a weighting component of the recognition result in order to make the recognition result more reliable. Further research is needed in the future, for a complete finger vein image assessment system and reducing the computation complexity of the method proposed in this work.

References

1. Kumar, A., Prathyusha, K.V.: Personal authentication using hand vein triangulation and knuckle shape. IEEE Transactions on Image Processing 9(18), 2127–2136 (2010)
2. Kang, W., Li, H., Deng, F.: Direct gray-scale extraction of topographic features for vein recognition. Scientia Sinica (Informationis) 41(3), 324–337 (2011)

506 H. Ma et al.

3. Cui, J.-J., Wang, L.-H., Chen, D.-L., Pan, F.: On the Vein Image Capturing System Based on Near-Infrared Image Quality Assessment. Journal of Northeastern University (Natural Science) 30(8), 1099–1102 (2009)
4. Zhao, X., Cai, A.: Fingerprint Image Quality Analysis. Journal of Computer-Aided Design & Computer Graphics 18(5), 644–650 (2006)
5. Chen, R., Lin, X., Ding, T.: Assessment of Iris Image Quality Based on Wavelet-based Contourlet Transform. Acta Automatica Sinica 35(5), 618–622 (2009)
6. Lu, W., Gao, X.-B., Wang, T.-S.: A natural image quality assessment metric based on wavelet-based contourlet transform. Acta Electronica Sinica 36(2), 303–308 (2008)
7. Wang, Z., Bovik, A.C., Sheikh, H.R., Simoncelli, E.P.: Image quality assessment: from error visibility to structural similarity. IEEE Trans Image Processing 13(4), 600–612 (2004)
8. Wang, Z., Simon Celli, E.P.: Reduced-reference image quality assessment using a wavelet-domain natural image statistic model. In: Annual Symposium on Electronic Image, pp. 149–159. SPIE Press, San Jose (2005)
9. Wang, K., Ma, H., Popoola, O.P.: Decision fusion for fingerprint and finger vein recognition based on image quality evaluation. International Journal of Biometrics 4(2), 189–201 (2012)
10. Mu, X., Yang, S.: Image quality assessment method based on human visual system. Radio Communications Technology 27(5), 32–33 (2011)

Synaesthetic Correspondence between Auditory Clips and Colors: An Empirical Study

Lihan Chen

Department of Psychology and Key Laboratory of Machine Perception
(Ministry of Education), Peking University, Beijing 100871, China
Clh20000@gmail.com

Abstract. This study investigated cross modal matching between some common auditory clips and colors, with two groups of participants: normal control group and the expertise group. Participants were required and encouraged to select a specific color to match a given auditory clip (normal onomatopoeia sounds and music clips) and give appraisal of their fondness to the clip, using a 7-Likert point scale. The results showed that two groups showed similar fondness of the auditory clips. However, the expertise group showed higher variance in the clip-color matching and were more sensitive to the luminance of the colors. The expertise group tended to match the low luminance color to the clips, although both groups did not differ in the average score of hue selection in this mapping.

Keywords: Auditory clip, color, crossmodal, matching.

1 Introduction

Synaesthetic correspondences- "joining of the senses" [1-2] refer to the perceived correspondence between basic features (e.g., pitch, lightness, brightness, size) in different modalities, in which synaesthetic congruency usually refers to correspondences between putatively non-redundant stimulus attributes or dimensions that happen to be shared by many people [3]. Synaesthetic correspondence is widely implemented between visual and auditory modalities. For example, people often match high-pitched sounds with small and/or bright objects. Both adults and children paired light grey color patches with louder sounds and darker grey color patches with quieter sounds [4-6]. Go beyond unidimensional sensory stimuli, people can also reliably match more complex stimuli, such as music with pictures [7-8]. However, it is important to note that no synaesthetic correspondence has so far been observed between pitch and hue [9] or between loudness and lightness [10] . A recent study also failed to demonstrate any crossmodal association between auditory pitch and visual contrast, using a speeded classification task [11].

Studies in this line has also indicated the individual differences in crossmodal synaesthetic correspondence. For example, Marks (1974) summarized that approximately half of the population tested matched loud sounds to darker grey surface while the rest

J. Yang, F. Fang, and C. Sun (Eds.): IScIDE 2012, LNCS 7751, pp. 507–513, 2013.

took the opposite mapping instead (i.e., matching louder sounds to lighter grey surfaces) [12]. By using simple pure tones, the studies mentioned above did not reveal certain associations between auditory pitch and visual hue (contrast). The materials used in the above studies might limit the generalization of the findings. In our daily life, we are confronted with more complex auditory scenes. Henceforth, the current study aims to use relatively complex sound material (normal onomatopoeia sounds and music clips) to examine synaesthetic correspondence between auditory clips and colors. Moreover, here I included two groups: normal control group and the expertise group (received systematic music training) for comparison, to investigate the group differences in the auditory clip-color matching task.

2   Method

2.1   Participants

A total of 102 undergraduate students attended the experiments. The participants were recruited from two local universities and separated into two groups: normal group (70 participants, 40 female, averaged 20.3 years old) and expertise group (musician group, 32 participants, 17 female, averaged 21.7 years old), who have received specialized music training above six years. They attended the experiments with informed consent and were paid for participation.

2.2   Materials

29 auditory clips in a sub-battery from *synesthete.org* were used, developed by Dr. David Eagleman from Department of Neuroscience and Psychiatry at Baylor College of Medicine. The labels (contents) for the auditory clips were as follows: TwenCent-Fox, Boxer, Seductive, Airport, Hitchcock, Applause, Bee, Bird, Bleep, Bluejay, Boing, Classictwo, Bottle, Bubble, Coke, Overture, Buzzer, Cello, Guitar, Haydn, Horse, Entertainer, Giggling, Ouch, Plop, Drums, Engine, Elephant, Chirps. A color palette appeared on the monitor screen after each clip was presented.

2.3   Design and Procedure

A between-subjects design was used. The stimuli of visual color palette and auditory clips were presented and controlled by Matlab 7.1 (MathWorks Inc., Natick, MA) and Psychophysics Toolbox [13-14]. Each participant was required to finish two tasks: *First*, they made a cross-modal matching task by picking a color from color palette in the monitor screen to pair with an auditory clip; *Second*, they evaluated the clips and rated their fondness of the given auditory clip (with score 1 to score 7, "1" represents that they do not like the clip at all and "7" shows that they like the clip very much). The experimental procedure was as follows: each clip was presented three times and the orders of presentation were randomized. After each clip, a color palette appeared on the monitor screen which prompts the participants to pick a color by mouse to optimally match the clip they just heard. The arrangements of the segment colors within the palette were also fully randomized across trials to prevent potential response bias. After that, participants were encouraged to rate the fondness of the auditory clip, by inputting the corresponding number (from 1 to 7) in the keyboard. The subjective report of fondness was also used to exclude the response bias due to general group difference of preference in the given auditory clips.

3   Results

The averaged mappings of colors with each auditory clip were plotted in Figure 1 (normal control group) and Figure 2 (expertise group). SPSS 16.0 and Matlab7.1 were used to analyze the data. Figure 3 depicts the averaged fondness ratings of 29 clips in the two groups. One-way analysis of variance (ANOVA) was used to compare the results under different conditions.

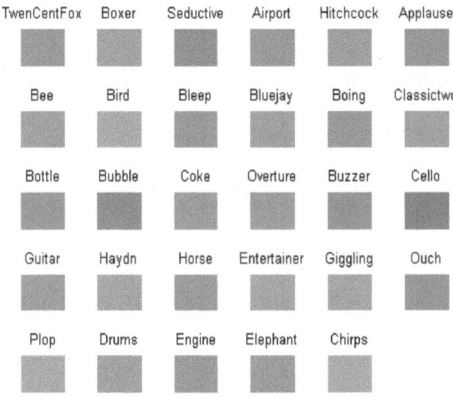

Fig. 1. The auditory clip- color mapping in normal control group

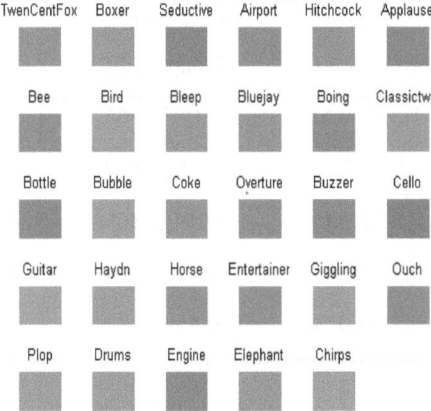

Fig. 2. The auditory clip- color mapping in expertise group

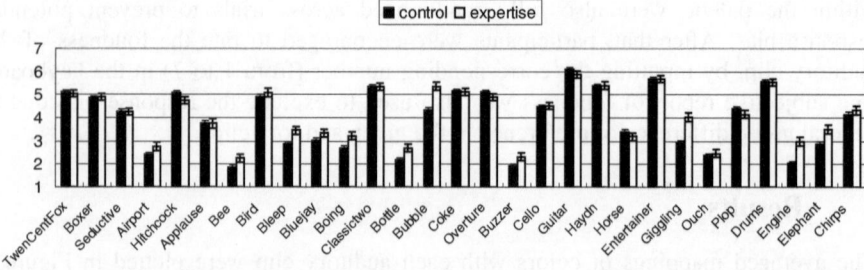

Fig. 3. The fondness rating scores (Y-axis) for 29 auditory clips (X-axis) across control group (black bars) and expertise group (white bars)

I calculated the variance scores of the values of the R (red), G (green), B(blue) vectors of each auditory clip-color mapping, summed and averaged over three trials per clip. The luminance of color was obtained by equation:

$$Luminance = 0.3*R+0.59*G+0.11*B. \tag{1}$$

The averaged hue was obtained by equation:

$$hue = \sqrt{R^2 + G^2 + B^2} \tag{2}$$

The (post hoc) saturation was given by:

$$saturation = 0.5 * R - 0.4187 * G - 0.0813 * B + 128 \tag{3}$$

Table 1 shows the descriptive results (The results in saturation is normalized) and the results of statistical analysis of ANOVA, the data format is mean value associated with one standard deviation.

Table 1. The descriptive results and one-way between subjects ANOVA results

	Control Group (n=70)	Expertise Group (n=32)	$F(1,101)$	p
Variance score	1.51±0.55	1.98±0.66	14.111	0.000
appraisal	3.94±0.57	4.15±0.77	2.303	0.132
luminance	0.51±0.03	0.49±0.05	5.761	0.018
hue	0.96±0.04	0.97±0.04	0.207	0.650
saturation	0.55±0.04	0.55±0.06	0.470	0.495

For the averaged collapsed scores over all 29 clips, one way ANOVA analysis showed that the two groups differed in the variance scores for the auditory clip-color matching. The expertise group showed higher variance in performing this task,

$F(1,101)=14.111$, $p<0.001$. The expertise group also showed lower luminance scores, that is, they tended to tag colors with relatively darker luminance to the given auditory clips. However, in the dimensions of averaged color (hue) mapping and saturation as well as the subjective fondness of the auditory materials, the two groups demonstrated the similar patterns.

On the other hand, across the 29 clips, we examined the group difference in each auditory clip with the four dimensions. For the *appraisal* scores, the main effect of clips was significant, $F(28, 2800)=100.19$, $p<0.001$; the main effect of group was not significant, $F(1,100)=2.303$, $p=0.132$. This indicates that to a large extent the pitch-color cross-modal matching is not based on personal preference (fondness) of a given auditory clip. The interaction between clip and group was significant, $F(28, 2800)=2.520$, $p<0.001$; Further comparisons showed that expertise group gave higher appraisals to "Bleep", "Boing", "Bottle", "Bubble", "Giggling", "Engine" than did in control group ($ps<0.05$). For the *variance* scores, the main effect of clips was significant, $F(28, 2800)=13.30$, $p<0.001$; The main effect of group was significant, $F(1,100)=14.111$, $p<0.001$. The interaction between clip and group was significant, $F(28, 2800)=2.060$, $p<0.001$; Further comparisons showed that expertise group showed higher variance scores in the following clips: "Boxer", "Seductive", "Airport", "Hitchcock", "Applause", "Bee" , "Bleep", "Bluejay", "Boing", "Bottle", "Guitar", "Plop", "Engine", "Elephant" and "Chirps" than did in control group ($ps<0.05$). For the *luminance* scores, the main effect of clips was significant, $F(28, 2800)=8.458$, $p<0.001$; The main effect of group was significant, $F(1,100)=5.757$, $p<0.05$. The interaction between clip and group was significant, $F(28, 2800)=1.773$, $p<0.01$; Comparisons showed that expertise group tagged darker color to the following clips: "Bee", "Bluejay", "Entertainer", "Engine" but lighter color to "Bubble" than did in control group ($ps<0.05$). For the *hue* scores, the main effect of clips was significant, $F(28, 2800)=13.674$, $p<0.001$; The main effect of group was not significant, $F(1,100)=0.208$, $p=0.650$. The interaction between clip and group was significant, $F(28, 2800)=1.943$, $p<0.01$; Further comparisons showed that expertise group tagged warmer color to the following clips: "Bubble", "Coke", "Horse", "Engine" and "Elephant" than did in control group ($ps<0.05$). For the *saturation* scores, the main effect of clips was significant, $F(28, 2800)=39.76$, $p<0.001$; the main effect of group was not significant, $F(1,100)=0.469$, $p=0.495$. The interaction between clip and group was significant, $F(28, 2800)=1.654$, $p<0.05$; Further comparisons showed that expertise group picked higher saturation to match "Bird" ($p<0.05$), "Entertainer" ($p<0.05$), "Drums" ($p<0.01$) than did in control group, but picked lower saturation to "Bubble" ($p<0.05$), "Ouch" ($p<0.05$) than did in control group.

In general, there are large variances for the crossmodal matching task between the groups. But there remains a certain signature(clip-"Engine") which tears apart the two groups in most dimensions (variance score, appraisal, luminance and hue) in investigation. However, the dimension of "saturation" exhibits a mixed picture and is not as consensus as the other dimensions.

4   Discussion

Synaesthetic correspondence refers to correspondences between basic features in different modalities and suggests an automatic perceptual binding effect, rather than higher level of cognitive categorization between visual and auditory objects. Evidence suggests existence of synaesthetic correspondence at very early in life. For example, 29-day-old infants are already able to match the shape of a pacifier seen visually to that explored orally beforehand [15], suggesting an automatic nature of synaesthetic correspondence.

The variance score reflects the stability as well as flexibility in performing the clip-color matching task. The expertise group showed a high variance, perhaps due to their subtle differentiations towards the auditory pieces, and henceforth a fine and careful mapping between clip and color. Note that although a certain clip was presented exactly three times (the order of clips was randomized), post-experiments debriefing with participants revealed that they sometimes perceived the differences in the "qualities" of the same clip across the three presentations (but absent in the control group), which might lead to high variance in the matching. Moreover, the expertise group seemed to be "optimistic" in evaluating the clips. For some clips, their rating score of fondness was higher, although the gross statistical significance has not been reached. Since the auditory materials are of complex timbre and elicits a kind of "dull" percept, and the expertise group were readily able to catch the intricate features and chose corresponding "darker" colors, this leads to the difference of luminance scores among the two groups. However, at the dimension of averaged color (hue) and saturation, the two groups were same in the obtained scores, it was consistent with findings in Bernstein et al. and Evans and Treisman [9, 11].

This study leaves much space for improvement, such as to examine further into synaesthetic correspondences between different categories of auditory clips (such as clips with different pitches and spectra/envelopes, pure tones and complex tones, tones with different semantic categories) and colors. Since for the dimension of "saturation", the results pattern for the signature clips showed a mixed picture, It would be worthwhile to directly manipulate the different levels of saturation in future studies. The current study is theoretically important in which it shows that for different consumers, subjective appraisals and correspondence between visual and auditory events, typified in auditory clip and its synaesthetically induced colors, could be diversified. It may help to design appropriate multimedia products to the target user, for example, to develop auditory image presentation system for blind people, using auditory clip-color correspondence, especially the signature mappings to facilitate communications [16, 17].

5   Conclusion

Normal control group without specific music training and expertise group with systematic music training showed similar fondness of the auditory clips. However, the expertise group showed larger variance in the crossmodal matching between auditory clips and colors. The expertise group tended to match the colors with darker luminance to the clips, although both groups did not differ in the average scores of hue and saturation in this matching.

Acknowledgements. This research was supported by grants from the Natural Science Foundation of China (31200760) and National High Technology Research and Development Program of China (863 Program) (2012AA011602).

References

1. Melara, R.D., O'Brien, T.P.: Interaction between synesthetically corresponding dimensions. Journal of Experimental Psychology: General 116, 323–336 (1987)
2. Wicker, F.W.: Mapping the intersensory regions of perceptual space. The American Journal of Psychology 81, 178–188 (1968)
3. Spence, C.: Crossmodal correspondences: A tutorial review. Attention, Perception & Psychophysics 73, 971–995 (2011)
4. Bond, B., Stevens, S.S.: Cross-modality matching of brightness to loudness by 5-year-olds. Perception & Psychophysics 6, 337–339 (1969)
5. Stevens, J.C., Marks, L.E.: Cross-modality matching of brightness and loudness. Proceedings of the National Academy of Sciences 54, 407–411 (1965)
6. Root, R.T., Ross, S.: Further validation of subjective scales for loudness and brightness by means of cross-modality matching. The American Journal of Psychology 78, 285–289 (1965)
7. Cowles, J.T.: An experimental study of the pairing of certain auditory and visual stimuli. Journal of Experimental Psychology 18, 461–469 (1935)
8. Karwoski, T.F., Odbert, H.S., Osgood, C.E.: Studies in synesthetic thinking: II. The rôle of form in visual responses to music. Journal of General Psychology 26, 199–222 (1942)
9. Bernstein, I.H., Eason, T.R., Schurman, D.L.: Hue–tone interaction: A negative result. Perceptual and Motor Skills 33, 1327–1330 (1971)
10. Marks, L.E.: On cross-modal similarity: Auditory–visual interactions in speeded discrimination. Journal of Experimental Psychology: Human Perception and Performance 13, 384–394 (1987)
11. Evans, K.K., Treisman, A.: Natural cross-modal mappings between visual and auditory features. Journal of Vision 10(1), 6:1–6:12 (2010)
12. Marks, L.E.: On associations of light and sound: The mediation of bright-ness, pitch, and loudness. The American Journal of Psychology 87, 173–188 (1974)
13. Brainard, D.H.: The Psychophysics Toolbox. Spatial Vision 10, 433–436 (1997)
14. Pelli, D.G.: The VideoToolbox software for visual psychophysics: transforming numbers into movies. Spatial Vision 10, 437–442 (1997)
15. Meltzoff, A.N., Borton, R.W.: Intermodal matching by human neonates. Nature 282, 403–404 (1979)
16. Meijer, P.B.: An experimental system for auditory image representations. IEEE Trans. Biomed. Eng. 39(2), 112–121 (1992)
17. Amedi, A., Stern, W.M., Camprodon, J.A., Bermpohl, F., Merabet, L., Rotman, S., Hemond, C., Meijer, P.B., Leone, A.P.: Shape conveyed by visual-to-auditory sensory substitution activates the lateral occipital complex. Nature Neurosci. 10, 687–689 (2007)

Learning Attribute Relation in Attribute-Based Zero-Shot Classification

Mingxia Liu[1,2], Songcan Chen[1], and Daoqiang Zhang[1]

[1] School of Computer Science and Engineering,
Nanjing University of Aeronautics & Astronautics, Nanjing, 210016, P.R. China
[2] School of Information Science and Technology,
Taishan University, Taian, 271021, P.R. China
{mingxialiu,s.chen,dqzhang}@nuaa.edu.cn

Abstract. Recently, zero-shot learning has attracted increasing attention in computer vision community. One way of realizing zero-shot learning is by resorting to knowledge about attributes and object categories. Most existing attribute-centric approaches focus on attribute-class relation artificially derived by linguistic knowledge base or mutual information. In this paper, we aim to learn the attribute-attribute relation automatically and explicitly. Specifically, we propose to incorporate the attribute relation learning into attribute classifier design in a unified framework. Furthermore, we develop a new scheme for attribute-based zero-shot object classification, such that the learned attribute relation can be reused to boost the traditional attribute classifiers. Extensive experimental results demonstrate that our proposed method can enhance the performance of attribute prediction and zero-shot learning.

Keywords: attribute relation, attribute-based classification, zero-shot learning.

1 Introduction

In recent years, the problem of object classification in zero-shot scenarios (i.e. no training samples are available for the target classes) has attracted increasing attention in computer vision community [1-6]. This problem is challenging and on the other hand very useful, especially for such a real-world setting where the number of object classes is large but samples are unable or costly to be acquired. In contrast, high-level semantic attributes for each class can be obtained more conveniently, such as color and texture for arbitrary objects.

As the training and the test classes are commonly disjoint in zero-shot learning, researchers have developed several alternative methods for knowledge transfer from training to unseen test classes. Among them, attribute-centric methods based on high-level attributes have been proven effective in leveraging knowledge between attribute and object category experimentally [1-3].Existing attribute-centric methods focus on exploitation of the semantic relation between *attribute* and *object class* [4-6]. Likewise, intuitively, the relation between attributes of an object can also convey supplementary information from another aspect. However, in traditional methods,

J. Yang, F. Fang, and C. Sun (Eds.): IScIDE 2012, LNCS 7751, pp. 514–521, 2013.

attribute classifiers are trained separately with respect to individual attributes, and seldom consider the *attribute-attribute relation*.

In fact, there are often some correlation relations between attributes. Thus taking such relation into account in the process of training attribute classifiers should be helpful. For example, researchers in [2,4,13] have considered the attribute relation to some degree. However, the relations they adopted are largely derived artificially or predefined. In contrast, in this paper, we propose to train attribute classifiers and learn the attribute-attribute relation simultaneously in a unified objective function.

In this paper, we first propose a joint learning method for attribute classifiers coupled with the attribute-attribute relation. And then we develop a new attribute-based zero-shot learning scheme by incorporating the attribute relation learned in previous step, and demonstrate that it enhance the performance of attribute prediction.

The rest of this paper is organized as follows. Section 2 introduces related works of zero-shot and attributes relation learning. The proposed attribute relation learning (ARL) method is described in Section 3. In Section 4, we propose our attribute-based zero-shot object classification scheme with attribute relation incorporated. Extensive experiments are carried out in Section 5. Finally, conclusion is given in Section 6.

2   Related Works

To the best of our knowledge, the concept of zero-shot learning can at least be traced back to one of the early works proposed by Larochelle et al.[7]. It attempts to solve the problem of predicting novel samples that were unseen in the training data set. In [8, 9], researchers proposed methods to obtain the intermediate class description such as semantic knowledge base to perform zero-shot learning. In computer vision community, Farhadi et al.[3] described objects by their semantic or discriminative attributes, and revealed the potential to predict novel classes in zero-shot scenarios. In order to tackle the problem of learning with disjoint training and testing classes, Lampert et al.[14] proposed attribute-based classification, as well as Direct Attitude Prediction (DAP) and Indirect Attitude Prediction (IAP) to perform zero-shot learning. More recent related works can be found in [4, 5, 11, 12].

On the other hand, as for attribute-centric object classification, some researchers have demonstrated considering attribute-class relation can result in improved generalization performance [4-6, 13]. However, the attribute relation needs to be pre-computed, which are then used in a latent discriminative model for classification. Rohrbach et al.[2] used external linguistic knowledge bases and proper semantic relatedness to capture attribute-class relation, which requires extra expert knowledge for natural language processing. Siddiquie et al.[5] demonstrated modeling pairwise correlations between attributes brings better results in image ranking and retrieval.

Different from the previous works, we model the attribute relation in a totally new perspective. First, we learn an attribute covariance matrix that models the relation between attributes in the form of matrix variant normal distribution, motivated by multi-task relation learning [14-16]. Second, the attribute relation we learned can be incorporated into traditional classifiers separately and has been proven effective to enhance the performance of zero-shot learning in our experiments.

3   Proposed Approach

Suppose we are given a mid-level representation in the form of inventories of attributes $\{A_m\}_{m=1}^M$ for N object classes. Given a set of training images$\{x_i\}_{i=1}^N$, $x_i \in \mathbb{R}^d$ and class labels $l_i^m = \{1, \cdots, N\}$ as well as attributes labels $y_i^m \in \{1, 0\}$. Binary attributes are considered in this paper. For each attribute classifier, we learn a linear function $f_m(\mathbf{x}) = w_m^T \mathbf{x} + b_m$, where $\mathbf{W}$ is the weight vector and $\mathbf{b}$ is the bias term.

3.1   Problem Formulation

Since data for each attribute classifier is in the same pool, we denote as $\mathbf{X} = (x_1^1, \ldots, x_N^1, \ldots, x_1^M, \ldots, x_N^M)^T$, the attribute labels $\mathbf{y} = (y_1^1, \ldots, y_N^1, \ldots, y_1^M, \ldots, y_N^M)^T$ and the bias term $\mathbf{b} = (b_1, \ldots, b_M)^T$. Then the posterior distribution for $\mathbf{W}$ can be obtained through the prior and the likelihood in the following function [22]:

Let $N(\mathbf{m}, \mathbf{\Sigma})$ denotes multivariate normal distribution with mean$\mathbf{m}$ and covariance matrix $\mathbf{\Sigma}$. Given x_i^m, w_m, b_m and ε_m, the likelihood of y_i^m is given in the following form:

$$y_i^m \mid x_i^m, w_m, b_m, \varepsilon_m \sim N(w_m^T x_i^m + b_m, \varepsilon_m^2) \tag{1}$$

Denote $\mathbf{I}_d$ as the $d \times d$ identity matrix, and the prior on $\mathbf{W} = (w_1, \ldots, w_M)$ is defined as

$$\mathbf{W}|\epsilon_m \sim (\prod_{m=1}^M N(w_m|\mathbf{0}_d, \epsilon_m^2 \mathbf{I}_d)) q(\mathbf{W}) \tag{2}$$

where the first term is employed to control the column complexity and second one is for structure modeling of $\mathbf{W}$. As $\mathbf{W}$ is a matrix variable, the matrix variant normal distribution [17] is used as $q(\mathbf{W})$. So $q(\mathbf{W})$ can be defined as

$$q(\mathbf{W}) = \mathrm{MN}_{d \times m}(\mathbf{W}|\mathbf{0}_{d \times m}, \mathbf{I}_d \otimes \mathbf{\Omega}) \tag{3}$$

Here, $\mathbf{I}_d$is a row covariance matrix modeling the features relation and $\mathbf{\Omega}$ is a column covariance matrix for the relation between w_m's.

Since data for each attribute classifier is in the same pool, we denote as $\mathbf{X} = (x_1^1, \ldots, x_N^1, \ldots, x_1^M, \ldots, x_N^M)^T$, the attribute labels $\mathbf{y} = (y_1^1, \ldots, y_N^1, \ldots, y_1^M, \ldots, y_N^M)^T$ and the bias term $\mathbf{b} = (b_1, \ldots, b_M)^T$. Then the posterior distribution for $\mathbf{W}$ can be obtained through the prior and the likelihood in the following function [22]:

$$p(\mathbf{W}|\mathbf{X}, \mathbf{y}, \mathbf{b}, \varepsilon, \epsilon, \mathbf{\Omega}) \propto p(\mathbf{y}|\mathbf{X}, \mathbf{W}, \mathbf{b}, \varepsilon) p(\mathbf{W}|\epsilon, \mathbf{\Omega}) \tag{4}$$

By taking the negative logarithm of Eq. (4) and combing it with Eqs. (1)-(3), the maximum likelihood estimation of $\mathbf{\Omega}$ and $\mathbf{b}$, as well as the maximum a posterior estimation of W, can be obtained through the following: .

$$\min_{\mathbf{W}, \mathbf{b}, \mathbf{\Omega}} \sum_{m=1}^M \frac{1}{\varepsilon_m^2} \sum_{i=1}^N \left(y_i^m - (w_m^T x_i^m + b_m) \right)^2 + \sum_{m=1}^M \frac{1}{\epsilon_m^2} w_m^T w_m$$
$$+ \mathrm{tr}(\mathbf{W}\mathbf{\Omega}^{-1}\mathbf{W}^T) + d \ln|\mathbf{\Omega}| \tag{5}$$

\

For convenience to optimize the problem Eq. (5), the last term $d \ln|\Omega|$ can be replaced by the constraint $tr(\Omega) = 1$ which is convex, with the same aim to restrict the complexity of Ω. Then the model can be rewritten as the following form:

$$\min_{W,b,\Omega} \sum_{m=1}^{M} \frac{1}{N} \sum_{i=1}^{N} \left(y_i^m - (w_m^T x_i^m + b_m)\right)^2 + \frac{\lambda_1}{2} tr(WW^T) + \frac{\lambda_2}{2} tr(W\Omega^{-1}W^T)$$

$$\text{s.t. } \Omega \geq 0, \ tr(\Omega) = 1 \tag{6}$$

where $\lambda_1 = \frac{2\varepsilon^2}{\epsilon^2}$, and $\lambda_2 = 2\varepsilon^2$. The constraint $\Omega \geq 0$ in Eq. (6) is used to restrict Ω as positive semi-definite because it is the attribute covariance matrix. The first term in Eq. (6) gives the empirical loss on the training data. And the second one is to penalize the complexity of W. It is worth noting that the last term $tr(W\Omega^{-1}W^T)$ is to model the relation between all attributes.

3.2   Alternating Optimization Algorithm

In Eq. (6), three variables to be optimized are jointly convex, which can be achieved through an alternating optimization method. The first step is to optimize W and b given a fixed Ω, and the second one is to optimize Ω when W and b are fixed.

Optimizing W and b when Ω is fixed

As is shown in [16], the dual problem defined in Eq. (6) can be written as

$$\min_{\alpha} \ \frac{1}{2}\alpha^T \widetilde{K}\alpha - \sum_{m=1}^{M} \sum_{i=1}^{N} \alpha_i^m y_i^m \ \text{s.t. } \sum_{i=1}^{N} \alpha_i^m = 0, \ \forall m, \ m = 1, \dots, M \tag{7}$$

where $\widetilde{K} = K + \frac{1}{2}\Lambda$, $\alpha = (\alpha_1^1, \dots \alpha_N^1, \dots, \alpha_1^M, \dots \alpha_N^M)^T$ and. Note that K is the kernel matrix on all data points for all attributes classifiers, whose element is $k(x_{i1}^{m1}, x_{i2}^{m2}) = e_{m1}^T \Omega(\lambda_1\Omega + \lambda_2 I_M)^{-1} e_{m2}(x_{i1}^{m1})^T x_{i2}^{m2}$, and Λ as a diagonal matrix with elements value N if the corresponding data point belong to the m-th attribute classifier.

Optimizing Ω when W and b are fixed

If W and b are fixed, the problem in Eq. (6) can be has an analytical solution

$$\Omega = \frac{(W^T W)^{\frac{1}{2}}}{tr((W^T W)^{\frac{1}{2}})} \tag{8}$$

The above two steps are performed alternatively, until the optimization procedure converges or the maximal iteration number is reached.

4   Incorporating Attribute Relation to Zero-Shot Learning

After we capture the attribute relation automatically in previous section, we want to use it explicitly for improving the zero-shot learning performance. Fig.1 illustrates the overall flowchart of our method. First of all, we use all the training data points and their attributes to train M attribute classifiers. And then, these classifiers are employed to predict the attribute values of unseen test images, and each test image will be given M attributes of real values. With specific inventory of attributes for each object class, predicted attributes will be mapped into class labels through DAP technique [10].

It's worth noting that the relation learned by our proposed method can be incorporated into the attribute-label prediction process easily. The underlying intuition is to join traditional dependent attribute classifiers together through our learned attribute relation. As is shown in Fig.1, Attribute value vector Y predicted by traditional classifiers, e.g. SVM and KRR, can be modified by attribute relation through $Y\_modi = Y * Cor$, where $Y \in \mathbb{R}^U$, U is the number of test images, and $Cor \in \mathbb{R}^{M \times M}$ is the attribute correlation matrix we learned from ARL method.

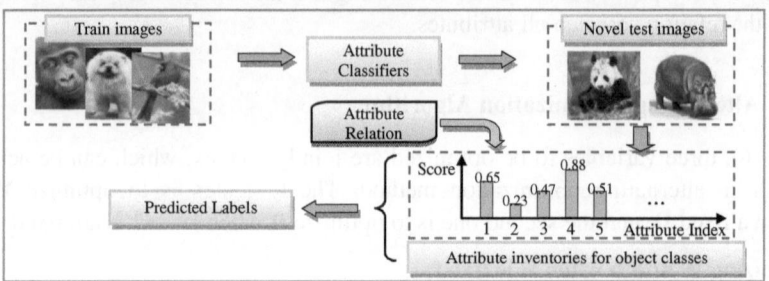

Fig. 1. Overview of our attribute-based zero-shot object classification

5   Experiments

5.1   Experimental Setup

We use a subset of the *Animals with Attributes* (*AWA*) dataset [14], which consists of 50 animal categories and 85 attributes. Following [10], only 59 attributes are employed in our experiments. For computational reasons, we down-sample this data set to 100 images per category in our experiments. Six feature types are used as image descriptors. And 10 classes are used as testing set and the other 40 classes as training set. We also use the *a-Pascal-train* data set [1] which consists of 20 classes and 64 attributes. As is given in [3], color and texture features are employed as image descriptors. Five classes (*bus, car, cat, dog and motorbike*) are used as test data while the other 15 ones as training data. We employ the χ^2-kernel [21] for image representation, i.e. kernel matrixes are based on χ^2-kernel of individual feature type.

Two baselines are employed, i.e. SVM and kernel ridge regression (KRR). Normalized mutual information (NormMI) used in [4,13] are employed as baseline for attribute relation learning. We adopt area under ROC curve (AUC) to evaluate the attribute prediction results. Average classification accuracies are used as performance measure for zero-shot learning methods. The regularization parameters λ_1 and λ_2 for our ARL method are both selected on from {0.001, 0.01, 0.05, 0.1, 0.5, 1}, and the parameter C for SVM and λ for KRR are chosen from {0.001, 0.01, 0.1, 1, 10, 100, 1000}. Optimal parameter values are confirmed on a validation set of training classes.

A sigmoid transform [22] maps the outputs of attribute classifiers into probabilistic scores for DAP model, where optimal values for parameter A and B are set on the same validation set.

5.2 Results and Discussion

5.2.1 How about Classifiers Learned from ARL in Attribute Relation?

Next, we examine how the proposed ARL method performs in attribute prediction, with comparison to SVM and KRR. The quality of the individual attribute predictors is shown in Fig.3. And Table1 reports the AUC values of each attribute.

Table 1. Average AUC of attribute prediction on two data sets (%)

Data Sets	SVM	KRR	ARL
aPascal	47.2	63.8	**68.5**
AWA	65.3	72.9	**74.4**

From Fig.3, it's obvious to find that our ARL method outperforms SVM and KRR on average. Due to the fact that testing classes are different from the training ones, the across category generalization of the attribute classifiers learned from our ARL method is quite reliable. From Table1, we can see a significant improvement in average accuracy is obtained by ARL model comparing to SVM and KRR.

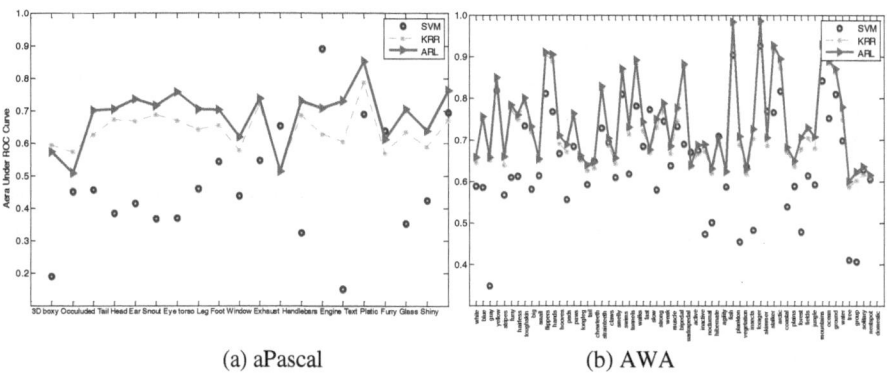

(a) aPascal (b) AWA

Fig. 2. Attribute prediction Results on *aPascal* and *AWA* data sets

5.2.2 How Does the Attribute Relation Influence Zero-Shot Learning Accuracy?

We now study the influence of attribute relation on the performance of zero-shot learning problem. Attribute relation learned from ARL method is used to modify the output of SVM and KRR. Table 2 reports the average accuracy of test classes in *AWA* data set. Note the term "*+Corr*" means all correlation coefficients (positive, negative and uncorrelated) are incorporated, while the "*+posCorr*" means only the positive ones are employed. From the results inTable 2, we draw the following conclusions.

First, our ARL model is superior to SVM and KRR in most cases, with the highest accuracy of 17.3% using Phog features. As is expected, ARL considers the attribute relation during the procedure of training attribute classifiers while SVM and KRR not.

Second, KRR and SVM with attribute relation are generally superior to those without correlation. It further validates our intuition that mining the associations between attributes helps improve performance of zero-shot object classification tasks.

Table 2. Zero-shot performance of different methods on *AWA* data set (%)

Method	Cq	Lss	Sift	rgbSift	Phog	Surf	All features
SVM	15.4	15.4	15.4	14.3	14.0	15.4	13.0
SVM +posCorr	**20.7**	13.9	14.5	**22.3**	**22.2**	6.7	13.8
SVM +Corr	12.9	9.2	**20.9**	7.6	16.1	9.1	9.2
KRR	12.0	10.0	5.6	13.9	9.7	10.5	12.8
KRR+posCorr	17.2	10.6	15.2	16.5	13.4	7.5	18.1
KRR+Corr	13.8	10.1	18.7	13.3	9.8	**17.8**	13.5
ARL	15.4	**17.0**	14.5	15.4	17.3	10.0	**15.7**

6   Conclusion

In this paper, we focus on attribute relation learning for attribute-based zero-shot object classification. We first propose an attribute relation learning (ARL) method to learn the correlation between attributes explicitly. Then, we present an attribute-based zero-shot object classification scheme with attribute relation incorporated. Finally, we further study how the attribute relation learning help improve the performance of zero-shot learning. Experimental results have validated the effectiveness of our proposed method.

Acknowledgement. This work is supported by the National Natural Science Foundation of China under grant Nos. 60875030, 60973097 and 61072148, and by the Specialized Research Fund for the Doctoral Program of Higher Education under grant No. 20123218110009.

References

1. Farhadi, A., et al.: Describing Objects by Their Attributes. In: Proceedings of the IEEE Computer Society Conference on Computer Vision and Pattern Recognition (2009)
2. Rohrbach, M., et al.: What Helps Where - and Why? Semantic Relatedness for Knowledge transfer. In: Proceedings of the IEEE Computer Society Conference on Computer Vision and Pattern Recognition (2010)
3. Lang, H., Ling, H.: Classifying Covert Photographs. In: Proceedings of the IEEE Computer Society Conference on Computer Vision and Pattern Recognition (2012)
4. Wang, Y., Mori, G.: A Discriminative Latent Model of Object Classes and Attributes. Perspectives in Neural Computing, 155–168 (2010)

5. Siddiquie, B., et al.: Image Ranking and Retrieval based on Multi-attribute Queries. In: Proceedings of the IEEE Computer Society Conference on Computer Vision and Pattern Recognition (2011)
6. Rohrbach, M., et al.: Evaluating Knowledge Transfer and Zero-Shot Learning in a Large-Scale Setting. In: Proceedings of the IEEE Computer Society Conference on Computer Vision and Pattern Recognition (2011)
7. Larochelle, H., et al.: Zero-data Learning of New Tasks. In: Proceedings of the 23rd AAAI Conference on Artificial Intelligence (2008)
8. Mitchell, T.M., et al.: Predicting Human Brain Activity Associated with the Meanings of Nouns. Science 320, 1191–1195 (2008)
9. Palatucci, M., et al.: Zero-Shot Learning with Semantic Output Codes. In: Advances in Neural Information Processing Systems (2009)
10. Lampert, C.H., et al.: Learning To Detect Unseen Object Classes by Between-Class Attribute Transfer. In: Proceedings of the IEEE Computer Society Conference on Computer Vision and Pattern Recognition (2009)
11. Farhadi, A., et al.: Attribute-centric Recognition for Cross-category Generalization. In: Proceedings of the IEEE Computer Society Conference on Computer Vision and Pattern Recognition (2010)
12. Parikh, D., Grauman, K.: Interactively Building a Discriminative Vocabulary of Nameable Attributes. In: Proceedings of the IEEE Computer Society Conference on Computer Vision and Pattern Recognition (2011)
13. Kovashka, A., et al.: Actively Selecting Annotations among Objects and Attributes. In: IEEE International Conference on Computer Vision (2011)
14. Argyriou, A., et al.: Convex Multi-task Feature Learning. Machine Learning 73, 243–272 (2008)
15. Bonilla, E., et al.: Multi-task Gaussian Process Prediction. In: Proceedings of NIPS (2007)
16. Zhang, Y., Yeung, D.-Y.: A Convex Formulation for Learning Task Relationships in Multi-Task Learning. In: Proceedings of UAI, pp. 733–742 (2010)
17. Rukhin, A.L.: Matrix Variate Distributions. Journal of the American Statistical Association 98, 462, 495–496 (2003)
18. Bishop, C.M.: Pattern Recognition and Machine Learning. Springer, New York (2006)
19. Van Gestel, T., et al.: Benchmarking Least Squares Support Vector Machine Classifiers. Machine Learning 54, 5–32 (2004)
20. Keerthi, S.S., Shevade, S.K.: SMO Algorithm for Least-squares SVM Formulation. Neural Computation 15, 487–507 (2003)
21. Teytaud, O., Jalam, R.: Kernel-based Text Categorization. In: International Joint Conference on Neural Networks, vol. 3, pp. 1891–1896 (2001)
22. Platt, J.C.: Probabilistic Outputs for Support Vector Machines and Comparison to Regularized Likelihood Methods. In: Smola, A.J., Bartlett, P., Schölkopf, B., Schuurmans, D. (eds.) Advances in Large Margin Classifiers, pp. 61–74 (1999)

Calculating Vanishing Points in Dual Space

Yong-Gang Zhao[1], Xing Wang[1], Liang-Bing Feng[1],
George Chen[1], Tai-Pang Wu[2], and Chi-Keung Tang[3]

[1] Shenzhen Institutes of Advanced Technology,
Chinese Academy of Sciences, Shenzhen 518055, China
{zhao.yg,lb.feng,qian.chen}@siat.ac.cn
[2] Enterprise and Consumer Electronics (ECE) Group,
Hong Kong Applied Science and Technology Research Institute
tpwu@astri.org
[3] Department of Computer Science and Engineering,
Hong Kong University of Science and Technology, Clear Water Bay, Hong Kong
cktang@cse.ust.hk

Abstract. Vanishing points can be used to exploit the parallel and orthogonal lines in 3D scenes thus the cameras' orientation parameters for vision processing. This paper proposed a vanishing point detection and estimation method in the dual image space. First, edge line segments are extracted. Second, based on the point-line duality theory, lines are transformed into points in the dual space where the transformed points belong to the same vanishing point form collinear clusters. Third, vanishing points are estimated by grouping and fitting straight lines across those clusters. The novel points of our method are: 1) automatically grouping the edge line segments that are the support of a vanishing point; 2) calculating the vanishing points by fitting straight lines in the dual space. Experiment results validated the proposed method.

Keywords: Vanishing Point, Dual Space, Line Detection, Homography.

1 Introduction

Parallel lines in the 3D space are projected into the 2D image space forming the so called pencil-of-lines. The intersection point of a group of convergent lines in the image plane is called a vanishing point which is actually the projection of a point-at-infinity corresponding to the aforementioned parallel 3D lines. The related convergent lines are called the support of the vanishing point. Vanishing points are widely used in parsing 3D scene structures. In 1, a 3D reconstruction approach from a single image was proposed using vanishing points. In 2, Foroosh *et al.* described how 3D Euclidean measurements could be made in a pair of un-calibrated images using vanishing points when only minimal geometric information is available. Vanishing points also play key roles in estimating the camera parameters for camera calibration. To quickly and accurately detect the vanishing points, three main problems need to be solved: i) detecting edge line segments in the image, ii) clustering of the lines belong to a pencil

J. Yang, F. Fang, and C. Sun (Eds.): IScIDE 2012, LNCS 7751, pp. 522–530, 2013.
© Springer-Verlag Berlin Heidelberg 2013

(or the convergent lines), and iii) estimating the intersection of the lines of a cluster. Sometimes ii) and iii) are intertwined.

Common clustering methods first create a partition of the image then tally the lines that pass through each region. Line clusters are found by identifying the regions with peak counts. Different methods varied in ways of dividing the images. For example, Almansa et al. 4 partitioned an image into "vanishing regions" such that the probability that a random line of the image meets a vanishing region is constant for all regions. In the work of Tuytelaars et al. 5 and Rother 6, a partition method was used such that the projection of each vanishing region on the Gaussian sphere has a quasi-constant area. The recent work by Li et al. 3 suggested replacing the 2D partition by two 1-D partitions thus greatly improved the detection speed. The partition based methods are appropriate for cases where the vanishing points are close to the image center but facing difficulties in dealing with vanishing points at infinity. Schmitt et al. 11 used the intersection point neighborhood instead of image plane partition for clustering. The issue of point at infinity is handled in an ad hoc way by substituting the intersection point of two parallel lines (at infinity) by one or two faraway points between the parallel lines.

Other methods perform line counting in some transformed space, often Hough space or Gaussian sphere. Barnard 7 first proposed a vanishing point detection method based on the Gaussian sphere in 1983. Antone et al. 14 simplified the idea and used three orthogonal planes in place of the Gaussian sphere. Based on Barnard' work, Lutton et al. 8 proposed a method using the Hough transform to calculate the vanishing point. Ebrahimpour et al. 10 used the start and end points of the lines to find the vanishing points though Hough transform and K-means clustering. Using the Gaussian mixture model (GMM) of vanishing points and vanishing lines, Akihiro Minagawa et al. were able to solve the line clustering problem and vanishing point estimation problem simultaneously 9. The joint point and line random process was modeled in GMM and the MAP solution was found using the standard EM algorithm which produced line clusters, vanishing points as well as vanishing lines all at once.

Most of the previous algorithms either rely on the camera calibration information (e.g. those based on Gaussian sphere) or encounter singularities when dealing with points at infinity (e.g. the partition or GMM methods). In this paper, we propose a new vanishing point detection method based on dual space theory. Our key observation is that pencil-of-lines appear in the dual space as collinear points. Thus vanishing point detection and calculation are solved jointly as robust line fitting in the dual space. Straight line segments belong to distinct vanishing points are automatically classified. There is no need of any form of line accumulation. Furthermore, points at infinity are naturally represented.

Below is the organization of the paper: Section 2 describes the vanishing point calculation theory in dual space. Section 3 illustrates the vanishing point detection algorithm. Section 4 shows the experimental results. Section 5 summarizes the paper and discusses some directions for future work.

2   Vanishing Point Calculation Theory in Dual Space

According to the duality principle, lines in the image space correspond to points in the dual space. Lines meet at one point in the image space are equivalent to collinear points in the dual space. For a quick explanation, we express the 2D line equation as:

$$(a \quad b \quad c) \begin{pmatrix} x \\ y \\ 1 \end{pmatrix} = 0 \tag{1}$$

which can be equivalently written as:

$$(x \quad y \quad 1) \begin{pmatrix} a/c \\ b/c \\ 1 \end{pmatrix} = 0 \ \text{ for } c \neq 0. \tag{2}$$

Thus the coefficients of the line equation $(a \ b \ c)$ in the image space are equivalent to the point $\left(\frac{a}{c}, \frac{b}{c}\right)^T$ in the dual space. A pencil-of-lines in the image space means that a, b, c are variable while x, y fixed. Therefore in the dual space, points $\left(\frac{a}{c}, \frac{b}{c}\right)^T$ are collinear with the line parameters $(x, y, 1)$. For this representation to hold, the parameter c cannot be zero which in practice can be achieved by a translation of the image and later compensated by applying the inverse translation to the resulted vanishing points.

This paper takes advantage of the principle that the intersecting lines in the image space form a set of collinear points in the dual space. Therefore the problem of calculating intersection of lines is converted to the problem of fitting a straight line across a set of collinear points in the dual space as shown in Figure 1.

Based on the above observation, a new kind of vanishing point detection algorithm is proposed in this paper. First, detect edge line segments in the image space. Second, calculate the line equations of those lines and transform them into the dual space. Third, group dual points and fit lines in the dual space. The coefficients of the resulting line equations are the wanted vanishing points. Using the homogeneous point representation, this method deals with the vanishing point at infinity nicely.

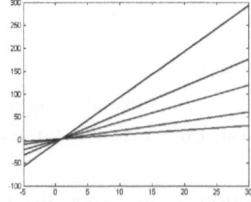 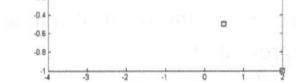

(a) Intersecting lines in image space   (b) Collinear points in dual space

Fig. 1. Pencil-of-lines in image space and the corresponding collinear points in dual space

3   The Vanishing Point Detection Algorithm

3.1   Line Segments Detection

In most of the vanishing point detection algorithms, the first step is line detection. Among the line detection approaches, Hough transform 12 and Line Segment Detector (LSD) 13 are widely used. In this paper we adopted the LSD method for line detection.
    Using polar form for lines:

$$\rho = x\cos\theta + y\sin\theta, \tag{3}$$

we represent a line in the dual space by

$$\left(-\frac{\cos\theta}{\rho}, \quad -\frac{\sin\theta}{\rho}\right)^{T} \tag{4}$$

To avoid the case of zero denominator, i.e. $\rho = 0$, the image is shifted initially so that none of the lines of our interest passes through the (shifted) image origin. The inverse shifting is later applied to the obtained vanishing points.

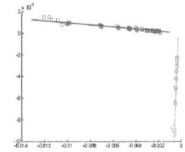

(a) Checkerboard image   (b) Lines in image space   (c) The collinear point clusters
                                                                           in dual space

Fig. 2. Checkerboard image and its collinear dual

    In Figure 2, line segments detected by LSD are shown in (b). The vertical and horizontal directions in (a) become the red and green collinear point clusters in (c). Each point cluster in the dual space corresponds to the almost collinear edge segments of the checkerboard. The spread of the cluster is due to the inaccuracy in the LSD process. So the points in dual space do not completely focus at one position but appear scattered.

3.2   Grouping the Dual Points and Getting Vanishing Points

Usually the parallel lines of the architectural buildings are in three axial directions, forming 2 to 3 vanishing points by the camera's perspective projection. In the dual space, the appearance is 2 or 3 collinear point clusters. In this paper, we propose RANSAC and Gustafson-Kessel (GK) 1617 methods for grouping the dual points and use RANAC to fit lines in each group.
    RANAC and GK each has its respective advantages and disadvantages. RANSAC is fast but hard to determine the threshold which may produce wrong groupings. GK is relatively more accurate but slow especially when using random initialization. Deciding the number of groups may also be problematic sometimes. In this paper, we combine these two methods as follows: we use RANSAC as an initialization step, and use GK to refine the grouping iteratively. Afterwards, we use RANSAC again within each group to find the straight lines. Below is the pseudo code:

> ■ Loop
>   ✧ Grouping the point clusters using RANSAC
>   ✧ Refine the previous grouping using GK method
>   ✧ Fit a straight line using RANSAC in the largest group
>   ✧ Output the straight line coefficients
>   ✧ Remove the inliers from the point clusters
> ■ Until no more meaningful groups can be found in the remaining points

Let the detected line equations be

$$a_{1,2,3}x + b_{1,2,3}y + c_{1,2,3} = 0 \tag{5}$$

Then, the three vanishing points are $(a_{1,2,3}, b_{1,2,3}, c_{1,2,3})^T$ respectively. Notice the homogeneous formulation here. It allows for dealing with the vanishing point at infinity.

3.3   Calculating the Rotation Matrix and Focal Length from Vanishing Points

The pinhole camera model is

$$\begin{pmatrix} u \\ v \\ 1 \end{pmatrix} \sim M[R \quad T]\begin{pmatrix} X \\ 1 \end{pmatrix} \tag{6}$$

Let r_1, r_2, r_3 be the column vectors of R, $(a_{1,2,3}, b_{1,2,3}, c_{1,2,3})^T$ the vanishing points of the three axial directions in homogeneous coordinates. We have:

$$\begin{pmatrix} a_1 \\ b_1 \\ c_1 \end{pmatrix} \sim [M \quad 0]\begin{bmatrix} r_1 & r_2 & r_3 & T \\ 0 & 0 & 0 & 1 \end{bmatrix}\begin{pmatrix} 1 \\ 0 \\ 0 \\ 0 \end{pmatrix} = Mr_1, \text{ Thus,}$$

$$r_{1,2,3} \sim M^{-1}\begin{pmatrix} a_{1,2,3} \\ b_{1,2,3} \\ c_{1,2,3} \end{pmatrix} \tag{7}$$

In case only two directions are observed, the third one can be obtained by $r_3 = r_1 \times r_2$. Since these vectors are recovered up to a scale, they should subsequently be normalized. In practice, it is often safe to assume that the camera's principal axes are orthogonal. Therefore we use the model

$$M = \begin{bmatrix} f & 0 & C_x \\ 0 & f & C_y \\ 0 & 0 & 1 \end{bmatrix}, \quad \text{Thus,}$$

$$M^{-1} = \begin{bmatrix} 1/f & 0 & -C_x/f \\ 0 & 1/f & -C_y/f \\ 0 & 0 & 1 \end{bmatrix} \tag{8}$$

Knowing $r_1 \cdot r_2 = 0$, we have,
$$\begin{pmatrix} \dfrac{a_1 - c_1 C_x}{f} & \dfrac{b_1 - c_1 C_y}{f} & c_1 \end{pmatrix} \begin{pmatrix} \dfrac{a_2 - c_2 C_x}{f} \\ \dfrac{b_2 - c_2 C_y}{f} \\ c_2 \end{pmatrix} = 0$$

which results in

$$f = \sqrt{-\left(\dfrac{a_1}{c_1} - C_x\right)\left(\dfrac{a_2}{c_2} - C_x\right) - \left(\dfrac{b_1}{c_1} - C_y\right)\left(\dfrac{b_2}{c_2} - C_y\right)} \qquad (9)$$

where c_1, $c_2 \neq 0$, i.e., the two vanishing points must not be at infinity which means that the camera must have certain tilt angle with respect to the object in the scene. We assume $C_x = W/2$, $C_y = H/2$ where W, H are the width and height of the image respectively. If all three axial directions are observed, then two additional estimations of the focal length are available from $r_3 \cdot r_2 = 0$, and $r_1 \cdot r_3 = 0$. We can use the average as the final estimation. We can use the homography matrix to rectify images as followed (10).

$$H = M R^T M^{-1} \qquad (10)$$

The validity of the detected vanishing points can be judged by observing the orthogonality of the rectangular features in the rectified image.

4   Experimental Results

4.1   Dual Points Grouping

We have tested the RANSAC method and the improved GK method to grouping dual vanishing points as follows.

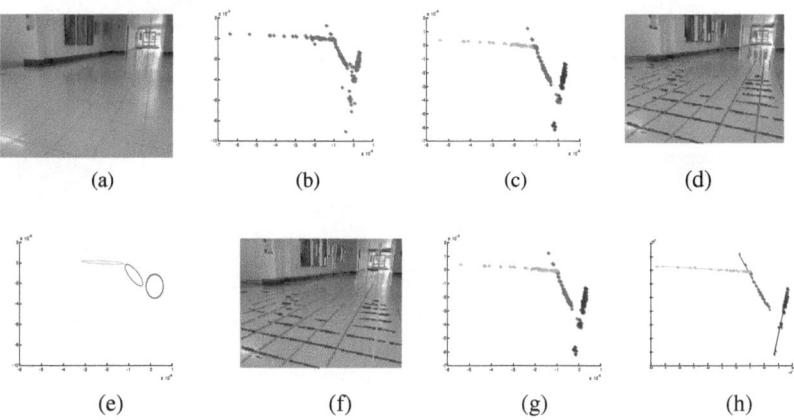

(a)   (b)   (c)   (d)

(e)   (f)   (g)   (h)

Fig. 3. Visualization of RANAC and GK grouping. (a) is the input image. (b) is the converted dual points. (c) and (d) show the result of grouping by RANSAC. (e) show the process of grouping by GK initiated by the previous RANSAC. It represent the cluster covariance matrices. (f) is Visualization of GK grouping result. (g) is the result of grouping by GK. (h) shows the final fitted lines. The vanishing points are actually the coefficients of the line equations.

We can see from Figure 3 that the two methods have different grouping effects. Those lines at the upper right of (d) were grouped wrongly (colored in red but should have been green) due to the fact that in the dual space (c), they are close to the intersection of the red and the green groups. Notice that in (c) the red group is atop of the green group near the intersection. This grouping error did not happen in the GK method as shown in (f) and (g). Notice differently from (c), in (g) the green group is atop the red one. However the GK method failed to group the long line towards the right side (green one which should have been grouped as red). But this error somehow "makes sense" because the line indeed appears parallel to those green ones above it.

4.2   Image Rectification

In order to validate the detected vanishing points, the experiments of image rectification were carried out using the homography matrix given in (10). We compare our results with the homography matrices obtained from two other methods: 1) M and R provided by camera calibration with the MATLAB calibration toolbox, 2) M and R provided by camera calibration with the method implemented in OPENCV 15. Figure 4 demonstrates that our method using a single image produces comparable results with those calibration methods using multiple images. Figure 5 provides two more rectification examples: one indoor image and one outdoor building image.

(a) Original images   (b) By OpenCV   (c) By Matlab toolbox   (d) By our method

Fig. 4. The image rectification experiment performed on checkerboard images

(a) Indoor image   (b) Rectified by our method (c) Outdoor image (d) Rectified by our method

Fig. 5. Image rectification experiment performed on indoor and outdoor images

5   Conclusions

In this paper, we proposed a vanishing point detection method by fitting lines in the dual space based on the point-line duality principle. This method can detect vanishing points in distinct directions and group them automatically. Experiment results verified the validity of the proposed method. Image rectification experiments further confirmed the feasibility of calculating camera intrinsic parameters and rotation matrix using the detected vanishing points.

One shortcoming of the proposed method is the shifting of the image in order to avoid zero denominators in the dual space transform. We plan to address this issue in the future. We also plan to device a more robust grouping algorithm in dealing with more complicated images.

Acknowledgment. This work is supported in part by the National Natural Science Fund of China under grant 61070147 and 61171146, Shenzhen Scientific & Research Development Fund under grant JC201005270331A and Shenzhen Internet Industry Development Fund under grant ZD201006110037A.

References

1. Guillou, E., Meneveaux, D., Maisel, E., Bouatouch, K.: Using Vanishing Points for Camera Calibration and Coarse 3D Reconstruction from a Single Image. The Visual Computer (S0178-2789) 16(7), 396–410 (2000)

2. Foroosh, H., Cao, X., Balci, M.: Metrology in Uncalibrated Images Given One Vanishing Point. In: IEEE International Conference on Image Processing, pp. III-361-4. IEEE, USA (2005)

3. Li, B., Peng, K., Ying, X., Zha, H.: Simultaneous Vanishing Point Detection and Camera Calibration from Single Images. Advances in Visual Computing (2010)

4. Almansa, A., Desolneux, A., Vamech, S.: Vanishing Point Detection without Any A Priori Information. IEEE Transactions on Pattern Analysis and Machine Intelligence 25(4), 502–507 (2003)

5. Tuytelaars, T., Proesmans, M., Van Gool, L.: The Cascaded Hough Transform. In: Proc. Int'l Conf. Image Processing (ICIP 1997), vol. 2, pp. 736–739 (1997)

6. Rother, C.: A New Approach for Vanishing Point Detection in Architectural Environments. In: Proc. British Machine Vision Conf. (2000)

7. Barnard, S.T.: Interpreting perspective images. Artificial Intelligence 21, 435–462 (1983)

8. Lutton, E., Maitre, H., Lopez-Krahe, J.: Contribution to be the Determination of Vanishing Points using Hough Transform. IEEE Transaction on Pattern Analysis and Machine Intelligence 16(4), 430–438 (1994)

9. Minagawa, A., Tagawa, N., Moriya, T., Gotoh, T.: Line Clustering with Vanishing Point and Vanishing Line. In: Proceedings of International Conference on Image Analysis and Processing, pp. 388–393 (1999)

10. Ebrahimpour, R., Rasoolinezhad, R., Hajiabolhasani, Z., Ebrahimi, M.: Vanishing point detection in corridors: using Hough transform and K-means clustering. IET Computer Vision, 40–51 (2012)

11. Schmitt, F., Priese, L.: Vanishing Point Detection with an Intersection Point Neighborhood. In: Brlek, S., Reutenauer, C., Provençal, X. (eds.) DGCI 2009. LNCS, vol. 5810, pp. 132–143. Springer, Heidelberg (2009)

12. Duda, R.O., Hart, P.E.: Use of the Hough Transformation to Detect Lines and Curves in Pictures. Communications of Association for Computing Machinery 15(1), 11–15 (1972)

13. Grompone von Gioi, R., Jakubowicz, J., Morel, J., Randall, G.: LSD: A Fast Line Segment Detector with a False Detection Control. IEEE Transactions on Pattern Analysis and Machine Intelligence 32(4), 722–732 (2010)

14. Antone, M.E., Teller, S.: Automatic Recovery of Relative Camera Rotations for Urban Scenes. In: Proc. CVPR 2000, pp. 282–289 (2000)

15. Bradski, G., Kaehler, A.: Learning OpenCV, pp. 432–437. O'Reilly Media, Inc. (2008)

16. Gustafson, D.E., Kessel, W.C.: Fuzzy clustering with a fuzzy covariance matrix. In: Proc. IEEE CDC, San Diego, CA, USA, pp. 761–766 (1979)

17. Babuska, R., van der Veen, P.J., Kaymak, U.: Improved covariance estimation for Gustafson–Kessel clustering. In: Proc. of the IEEE Internat. Conf. on Fuzzy Systems, pp. 1081–1085 (2002)

SAR Image Segmentation Based on Gabor Filter Bank and Active Contours

Weiping Ni[1], Xinbo Gao[1], and Weidong Yan[2]

[1] School of Electronic Engineering, Xidian University, Xi'an 710071, China
nihao_wpni@163.com, xbgao@mail.xidian.edu.cn
[2] Northwest Institute of Nuclear Technology, Xi'an 710024, China

Abstract. Image segmentation is a fundamental step and foundation for automatic SAR image interpretation. By combining the Gabor filter bank (GFB) and active contours, this paper proposes a new SAR image segmentation method. Firstly, GFB is used to efficiently suppress speckles in SAR image and modify the gray histogram into Gaussian mixture model (GMM). Then GMM-based pixel classification is employed to pre-segment the filtered image. Finally, the active contours, initialized with the pre-segmented regions, are applied to unfiltered image for the final segmentation with several iterations. Experiments are conducted to vivificate the efficiency and effectiveness of the proposed method.

Keywords: SAR image segmentation, Gabor filter bank, pre-segmentation, Gaussian mixture model, parametric active contours.

1 Introduction

Acquiring data at all-climate, day and night, Synthetic aperture radar (SAR) has been widely applied in surface surveillance, disaster monitoring and etc [1], which make it urgent to realize the automatic interpretation of the obtained SAR images. Image segmentation establishes the basis of automatic target recognition (ATR), proven by MSTAR program of DARPA [2,3]. It has always been a research hotspot to design simple and efficient methods, and various methods have been proposed.

Generally, SAR image segmentation approaches include three categories: clustering methods, thresholding methods, and model-based methods. The former classifies pixels into regions by textures with difficulty in discrimination determining [4]. The middle divides an image into several regions by some thresholds with noise restraining being a key step [5]. There are many de-noising methods for SAR images [6, 7]. The model based algorithms can be also divided: statistics model and geometric model. The former, such as Markov random field (MRF) models, provides links between local and global characteristics, but time-consumption still is a problem in optimization [8]. The latter, such as active contours, searches control points of a curve fitting the boundaries of regions [9], but it is hard to initialize.

Motivated by the aforementioned discussions, we propose a new segmentation method based on Gabor filter bank (GFB) and parametric active contours (PAC). Firstly,

J. Yang, F. Fang, and C. Sun (Eds.): IScIDE 2012, LNCS 7751, pp. 531–538, 2013.

GFB is used to significantly restrain the speckles and then the statistical characteristics of filtered images are analyzed with gray histograms, and approximated them with Gaussian mixture models (GMM). GMM-based algorithm is employed to pre-segment the filtered SAR image, which is used as initializations of PAC to the original SAR images to obtain the final segmentation.

The rest of this paper is organized as follows. Section 2 presents details of proposed method with the GFB based speckle restraining and active contours based segmentation. Section 3 reports the experimental results about the comparison between the proposed method and MRF-based method. Finally, Section 6 concludes this paper.

2   The Proposed SAR Image Segmentation Method

As mentioned above, active contours have good advantage in image segmentation. However, the random guess of the initial contours, as well as the presence of speckles, lead to false fitting or too slow convergence. So, we can restrains the speckles firstly, and then uses effective initialized active contours to fulfill the segmentation.

Fig.1 shows the conceptual flowchart of proposed method, which mainly consists two parts.. In the speckle restraining part, we adopt GFB to restrain the speckles. In the segmentation part, we use the GMM to approximate the histograms of filtered SAR images, and estimate model parameters with EM iterations. Then, pixel classification, based on the probability of every pixel belonging to each Gaussian distribution, is carried out to pre-segment the filtered SAR images, which is in turn used as the initialization of parametric active contours to make the final segmentation processing efficiently. In the following, each module will be described detailedly.

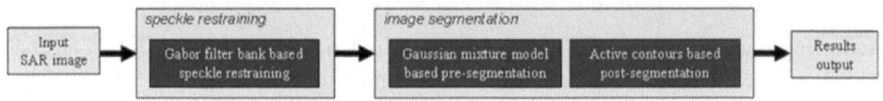

Fig. 1. The flowchart of the proposed method

2.1 Speckle Restraining Based on Gabor Filter Bank

The Gabor filters are band-pass filters with tunable orientation and radial frequency bandwidths [7]. A 2-D Gabor function is defined as

$$g(x, y) = \frac{1}{2\pi\sigma_x\sigma_y} \exp\left[-\frac{1}{2}\left(\frac{x^2}{\sigma_x^2} + \frac{x^2}{\sigma_x^2} \right) + 2\pi j\omega x \right] \tag{1}$$

The Gabor filters bank can be obtained by scale and rotation of $g(x, y)$, i.e.

$$g_{m,k}(x, y) = a^{-m} g(x', y') \tag{2}$$

Where, $x' = a^{-m}(x\cos\theta + y\sin\theta)$ and $y' = a^{-m}(-x\cos\theta + y\sin\theta)$ with $\theta = n\pi/k$. k is number of orientations. Given an image $I(x, y)$, filtering with GFB is given by:

$$W_{mk}(x, y) = |I(x, y) \otimes g_{mk}(x, y)| \tag{3}$$

Usually, the bigger the scales, the more the speckle suppressed and when orientation of local texture nearing to orientation of Gabor filters, the outputs are strong. So, we synthesize the filtered images together according to (4), which can further reduce speckles and get better contrast between different regions, as shown in Fig.2.

$$J(x, y) = \sum_m \sum_k |W_{mk}(x, y)| = \sum_m \sum_k |I(x, y) \otimes g_{mk}(x, y)| \tag{4}$$

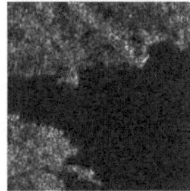 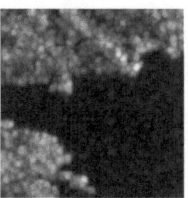

(a) Original SAR image   (b) Synthesizing result

Fig. 2. Illustration of filtered images synthesizing based on Gabor filter bank

2.2 Segmentation Based on Active Contours

2.2.1 Pre-segmentation with Gaussian Mixture Model

It can be seen by Fig.3 and Fig.4 that histograms of images unfiltered maybe different, but histograms of images filtered all look like mixture of some Gaussian functions, which can be well modeled by Gaussian mixture models (GMM) [10, 11]. The probability density function for a Gaussian distribution is given as:

$$p(\chi | \theta) = \frac{1}{(2\pi)^{d/2} \sqrt{|\Sigma|}} \exp\left(-\frac{(\chi - \mu)^T \Sigma^{-1}(\chi - \mu)}{2}\right) \tag{5}$$

Where, $\chi = (x_1, x_2 \cdots, x_n)$ is a n d-dimensional data and $\theta = (\mu, \Sigma)$. μ and Σ are the mean and covariance matrix. The density of each vector is:

$$p(x_t | \Theta) = \sum_{i=1}^k \alpha_i p_i(x_t | \theta_i) \tag{6}$$

Where, $\Theta = (\alpha_1, \alpha_2, \cdots, \alpha_k; \theta_1, \theta_2, \cdots, \theta_k)$ and $(\alpha_1, \alpha_2, \cdots, \alpha_k)$ are k mixed components coefficients such that $\sum_{i=1}^k \alpha_i = 1$; p_i is density function parameterized by θ_i .

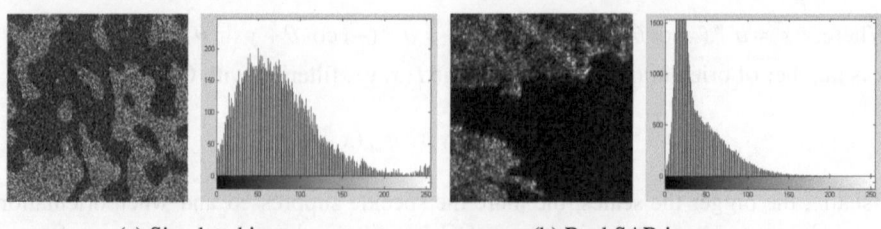

(a) Simulated image   (b) Real SAR image

Fig. 3. Unfiltered images and histograms

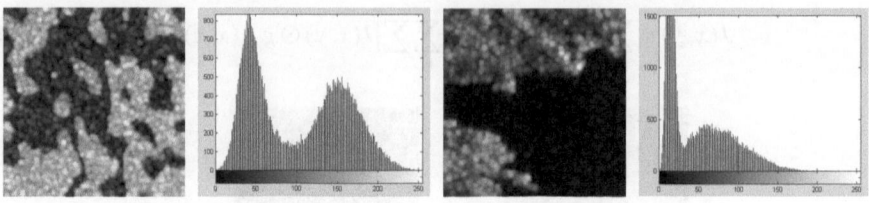

(a) Simulated image   (b) Real SAR image

Fig. 4. Filtered images and histograms

We use EM algorithm to estimate GMM parameters. Table 1 and Fig.5 gives the corresponding results, which well consistent with histogram-curves.

Table 1. GMM parameters estimation for histograms of filtered SAR image

Simulated Image	Real SAR image
α = 0.4549; 0.5451	α = 0.5712; 0.4288
μ = 45.8064; 150.3073	μ = 17.2318; 110.4847
σ = 15.7505; 32.4360	σ = 6.9007; 36.3182

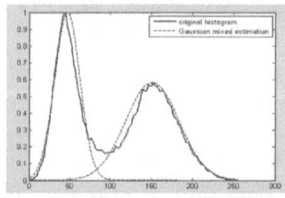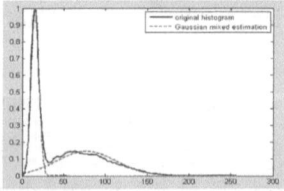

(a) Simulated image   (b) Real SAR image

Fig. 5. Histograms estimated based on GMM

Then, the pre-segmentation of filtered SAR images is as follows.

Step 1: Compute $p(I(x, y) | \theta_i)$ of each pixel of filtered image with GMM;

Step 2: Compare $p(I(x, y) | \theta_i)_{|i=1,2,...,k}$, then label pixel with maximum probability:

$$class\ label_{|I(x,y)} = \arg \max_i p(I(x, y) | \theta_i) \qquad (7)$$

Step 3: Represent different class of pixels with different values ranging in 0~255.

Fig.6 gives the segmented results for the filtered images. Obviously, most homogeneous regions have been segmented. But for inevitably blurring effect of GFB, some false classification also exist in the segmented results, , which need refinement.

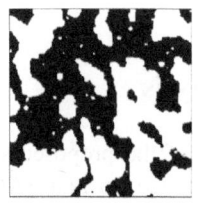

(a) Simulated image   (b) Real SAR image

Fig. 6. Pre-segmentation results with GMM based pixel classification

2.2.2   Post-segmentation Based on Active Contours

Generally, active contours include two types: parametric active contours (PAC) and level set active contours [12, 13]. Here, we use the former, defined internal energy as:

$$E_{\text{int}}(u(s)) = \alpha|u_s(s)|^2 + \beta|u_{ss}(s)|^2 \tag{8}$$

Where, $u(s) = (x(s), y(s))$, $s \in [0,1]$. $\alpha|u_s(s)|^2$ and $\beta|u_{ss}(s)|^2$ are membrane and plate energy. $u_s(s)$, $u_{ss}(s)$ are the 1st and 2nd derivatives. External energy is defined as.

$$E_{ext}(u(s)) = \int_0^1 P(u(s))ds \tag{9}$$

Summing the external and internal energy, we get the total energy along $u(s)$:

$$E_{snake} = \int [E_{\text{int}}(u(s)) + E_{\text{ext}}(u(s))]ds \tag{10}$$

The solution of PAC is to find the minimization of above energy function. Using Fig.6 to initialize PAC, we get final segmentation of original images shown in Fig.7.

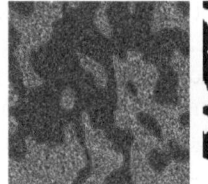

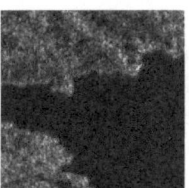
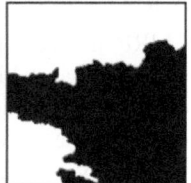

(a) Simulated image   (b) Real SAR image

Fig. 7. Post-segmented results based on PAC

It is clear that the regions with different textures have been segmented in high accuracy, with contours well corresponding to the boundaries between regions. In addition, by using the pre-segmentation based initialization, the calculation procedure has been greatly reduced to tens of iterations, much less than that of the traditional random guess based initialization. Thus, our algorithm has good potential both in segmentation accuracy and time consumption.

3   Experimental Results and Analysis

In this section, we examine the performance of the proposed segmentation method more carefully through the experiments conducted on real SAR images, and make comparisons with the traditional MRF method.

Fig.8 gives some SAR images and ground truth made manually for the subjective and objective evaluation. Figure 9 and show segmentation results with MRF method and the proposed method.

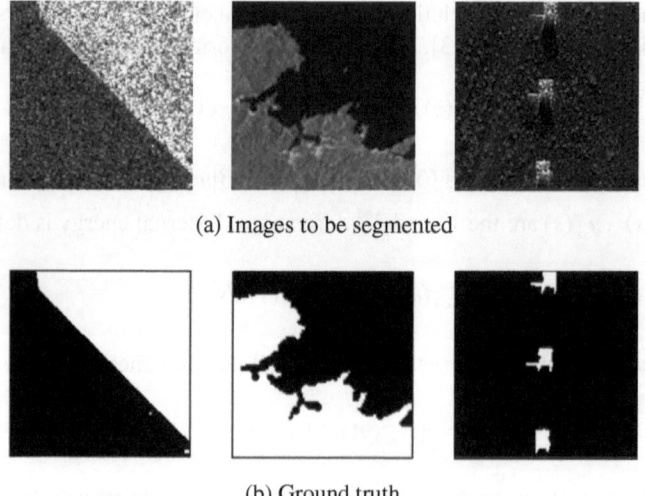

(a) Images to be segmented

(b) Ground truth

Fig. 8. SAR images to be segmented and ground truth

Visually, results of our method are much more consistent with ground truth than that of MRF method. The reason is that MRF method can't capture more characteristics of SAR image; while our method use GFB to reduce uncertainty of GMM based pixel classification, which in turn decreases randomicity during PAC initialization.

We evaluate the performance quantitatively using two global quality indicators, the overall accuracy and Kappa index [8], which, listed in Table 2 and Table 3, also prove the ascendant performance of the proposed method compared with MRF method. Table 4 presents comparison of time consuming. It is obviously that our algorithm is faster than that of MRF-based method by about one order.

  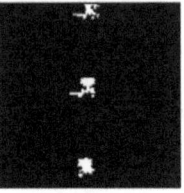

(a) Segmented with the MRF method

 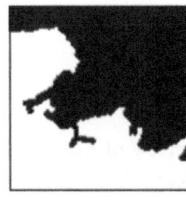 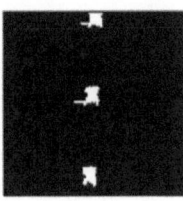

(b) Segmented with the proposed algorithm

Fig. 9. SAR images segmentation results comparison

Table 2. Overall accuracy (%)

	Image 1	Image 2	Image 3
our method	99.05	98.48	99.56
MRF	98.77	97.21	98.14

Table 3. Kappa Index (%)

	Image 1	Image 2	Image 3
our method	98.10	96.94	90.99
MRF	97.49	94.40	79.58

Table 4. Comparison on segmentation time (s)

	Image 1	Image 2	Image 3
Image size	185×183	115×111	173×154
our method	2.3358	1.3144	2.7195
MRF	30.7325	7.5159	40.6414

4   Conclusions

Segmentation is a key step in SAR image interpretation, and the challenges mainly arise from the presence of speckle. In this paper, we developed a combined method based on GFB and PACs. Before segmentation, SAR image is smoothed with GFB to effectively suppress the speckle, and modify the histograms into ones with multiple peaks, and then adopt GMMs to estimate the parameters with EM algorithm, based on which the filtered images is pre-segmented, which is subsequently used to initialize parametric active contours to make a final segmentation processing. The experimental results indicate the effectiveness of the proposed method, and compared with the common used MRF method, our method performs better in segmentation accuracy and time efficiency. More experiments will be done by using more real SAR data to evaluate the performance of our method with in further work.

Acknowledgments. This work was supported by the Fundamental Research Funds for the Central Universities, and the National Defense Foundation of China (No. 513310601).

References

1. Jiao, L., Wang, S., Hou, B.: A Review of SAR Images Understanding and Interpretation. Acta Electronica Sinica 33(12), 2423–2434 (2005)
2. El Zaart, A., Ziou, D., Wang, S., Jiang, Q.: Segmentation of SAR Images. Pattern Recognit. 35(2), 713–724 (2002)
3. Ross, T., Worrell, S., Velten, V., Mossing, J., Bryant, M.: Standard SAR ATR Evaluation Experiments Using the MSTAR Public Release Data Set. In: SPIE Conf. on Algorithms for Synthetic Aperture Radar Imagery, vol. 3370, pp. 566–573 (1998)
4. Chumsamrong, W., Thitimajshima, P., Rangsanseri, Y.: Synthetic Aperture Radar (SAR) Image Segmentation Using a New Modified Fuzzy C-means Algorithm. In: Proceedings of IEEE Symp. Geosci. Remote Sens., Honolulu, pp. 624–626 (2000)
5. Petrou, M., Matrucceli, A.: On the Stability of Thresholding SAR Images. Pattern Recognit. 31(11), 1791–1796 (1998)
6. Hua, X., Pierce, L.E., Ulaby, F.T.: SAR Speckle Reduction Using Wavelet Denoising and Markov Random Field Modeling. IEEE Trans. Geosci. Remote Sens. 40(10), 2196–2212 (2002)
7. Yan, X., Jiao, L., Xu, S.: SAR Image Segmentation Based on Gabor Filters of Adaptive Window in Overcomplete Brushlet Domain. In: Proceedings of Asia-Pacific Conference on Synthetic Aperture Radar, pp. 660–663 (2009)
8. Weisenseel, R.A., Karl, W.C., Castanon, D.A., Brower, R.C.: MRF-based Algorithms for Segmentation of SAR Images. In: Proceedings of the 1998 International Conference on Image Processing, vol. 3, pp. 770–774 (1998)
9. Horritt, M.S.: A Statistical Active Contour Model for SAR Image Segmentation. Image and Vision Computing 17, 213–224 (1999)
10. Kristan, M., Leonardis, A., Skocaj, D.: Multivariate Online Kernel Density Estimation with Gaussian Kernels. Pattern Recognit. 44(10), 2630–2642 (2011)
11. Ari, C., Aksoy, S., Arkan, O.: Maximum Likelihood Estimation of Gaussian Mixture Models Using Stochastic Search. Pattern Recognit. 45, 2804–2816 (2012)
12. Kass, M., Witkin, A., Terzopoulos, D.: Snakes: Active Contour Models. Int. J. Compution Vision 1(4), 321–331 (1987)
13. Peng, R., Wang, X., Lu, Y.: SAR Imagery Segmentation Based on Integrated Active Contour. In: Proceedings of the Int. Conf. on Advanced Computer Control, pp. 43–47 (2010)

Measuring the Attentional Effect of the Bottom-Up Saliency Map of Natural Images

Cheng Chen[1,3], Xilin Zhang[2,3], Yizhou Wang[1,3], and Fang Fang[2,3,4,5]

[1] National Engineering Lab for Video Technology and School of Electronics
Engineering and Computer Science
[2] Department of Psychology
[3] Key Laboratory of Machine Perception (Ministry of Education)
[4] Peking-Tsinghua Center for Life Sciences
[5] IDG/McGovern Institute for Brain Research
Peking University, Beijing 10081, P.R. China
chencheng880829@gmail.com,
{zhangxilin,Yizhou.Wang,ffang}@pku.edu.cn

Abstract. A saliency map is the bottom-up contribution to the deployment of exogenous attention. It, as well as its underlying neural mechanism, is hard to identify because of the existence of top-down signals. In order to exclude the contamination of top-down signals, invisible natural images were used as our stimuli to guide attention. The saliency map of natural images was calculated according to the model developed by Itti *et al.* [1]. We found a salient region in natural images could attract attention to improve subjects' orientation discrimination performance at the salient region. Furthermore, the attraction of attention increased with the degree of saliency. Our findings suggest that the bottom-up saliency map of a natural image could be generated at a very early stage of visual processing.

Keywords: Bottom-up saliency map, Natural image, Visual Attention, Unconsciousness.

1 Introduction

Because of the limited resources of the visual system, visual attention is essential for us to select the most valuable information from extremely complex natural scenes, and thus plays an important role in understanding the world. The information selection process can be achieved by directing visual attention to a target under a top-down goal, or be triggered by a salient stimulus. The former process is executed voluntarily, while the latter process is automatic and guided by the saliency map. Relative to extensive studies on the neural basis of top-down selection, the neural basis of bottom-up saliency map is controversial because of the possible contamination by top-down signals in higher brain areas.

In this paper, we measured the attentional effect of the bottom-up saliency map of natural images. Low luminance natural images were presented very

J. Yang, F. Fang, and C. Sun (Eds.): IScIDE 2012, LNCS 7751, pp. 539–548, 2013.

briefly, which rendered them invisible to subjects and also excluded the contamination by top-down signals. Natural images were used here instead of simple textures because of their rich naturalistic low-level features that the human visual system is tuned to. Although these natural images were invisible, the difference of visual saliency (calculated by a famous computational saliency model [1]) between inside and outside a local region could attract attention to improve the performance of an orientation discrimination performance at the salient region. We use the degree of saliency to refer to the saliency difference in the paper. Then we could measure the attentional effect generated by different degrees of saliency. Investigating this topic not only provide evidence for the bottom-up saliency map in our brain, but also is helpful to many important applications, such as object detection.

1.1 Related Work

The representation of the strength of the bottom-up attention attraction from our visual input [2] is a saliency map. It is constructed in our brain and can direct our attention along with top-down signals. Several studies had tried to measure the effect of visual saliency, and also found brain regions that realize the saliency map. For example, Geng and Mangun found that anterior intraparietal sulcus could realize the saliency map [3]. Mazer and Gallent found a goal-related activity in V4, which provided evidence that V4 could realize the saliency map [4]. However, these studies can not rule out the top-down attentional control, which makes it hard to identify the neural basis of the bottom-up saliency map. So it's important to probe the bottom-up attention attraction free from the top-down influence.

Several methods can be used to reduce the top-down signals influence, such as backward masking, binocular rivalry and continuous flash suppress (CFS). Zhang *et al.* had adopted the backward masking method to investigate the neural substrate of the bottom-up saliency map [5]. In their study, stimuli were presented so briefly and followed by a high contrast mask so that subjects could not perceive the stimuli. Similarly, we also used backward masking to make low-luminance stimuli invisible. Stimuli in our experiment were natural images collected from the Internet instead of simple pattern or texture, for natural images contain multiple and naturalistic low-level features [6]. Consider that some studies had used checkerboard to mask objects [7], random checkerboard was used as mask in our experiment.

In our experiment, we adopt a revised version of the cueing effect paradigm proposed by Posner *et al.* [8]. In this paradigm, a target appears in one of two locations randomly, and subjects need to finish a discrimination task about this target. Prior to this target, a cue indicates the location of the following target. Trials with a correct cue are called valid cue trials, while trials with an incorrect cue are called invalid cue trials. A classical result demonstrates that performance (response time or accuracy) in the valid cue trials is significantly higher than that in the invalid cue trials [9]. The salient region of a natural image was used as a cue in our experiment.

 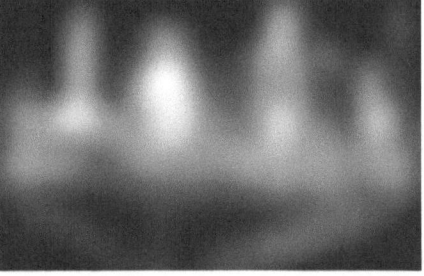

Fig. 1. An example of a color image (left) and its saliency map (right). White region in the right image indicate its salient region.

Many studies had also proposed a computational model to generate the saliency map of an image. An example can be seen in Fig. 1, the value of each pixel in the saliency map ranges from 0 to 1, higher value correlated with more saliency. Itti *et al.* proposed a biologically-plausible saliency model based on a center-surround mechanism, by combining information from three channels: color, intensity and orientation [1]. According to the spectrum of natural images, Hou *et al.* compute the spectral residual of an input image and transform the spectral residual to spatial domain to obtain its saliency map [10]. By simulating the information transmitting between neurons, Wang *et al.* proposed a saliency model based on information maximization [11]. These saliency models can provide a prediction about the attentional effect of a bottom-up saliency map.

Moreover, the underlying neural mechanism of the bottom-up saliency map has been subject to debate. A dominant view assumes that saliency results from pooling different visual features (e.g. [2], [12]), thus could be realized by higher cortical areas such as parietal cortex. However, Li proposed the V1 theory which claimed the saliency map was created by V1 (e.g. [13], [14]). It was completed via intra-cortical interactions that are manifest in contextual influences [15]. By combing psychophysical data and brain imaging results, Zhang *et al.* found that neural activities in V1 could create a bottom-up saliency map of simple texture [5], which supports the V1 theory. But evidence on natural images is still lack.

The rest of this paper is organized as follows. In Section 2 we introduce the details of our approach, including the information of subjects, the stimuli and the procedure of psychophysical experiment. The results of our experiment are given in Section 3. Finally, we conclude and discuss our work in Section 4.

2   Our Approach

2.1   Subjects

16 human subjects (7 male and 9 female) participated in the psychophysical experiments. All subjects were right-handed, reported normal vision or corrected-to-normal vision, and had no known neurological or visual disorders. Ages ranged from 19 to 26. All of them were naive to the purpose of our study except for

one subject who was one of the authors. They were given written, informed consent in accordance with the procedures and protocols approved by the human subjects review committee of Peking University.

2.2   Stimuli

We collected a large number of grayscale images about natural scenes from the Internet, resized them into the same size (384×1024 pixels), and decreased the luminance of these images to a low level (about 2.9 cd/m^2), Fig. 2 (a) shows a sample image. To quantitatively measure the attentional effect, we adopt a visual saliency model proposed by Itti $et\ al.$ [1] and calculate the saliency map of each image. After that we selected 50 images, and each of them had a round salient region centered at about 7.2° eccentricity in the lower left quadrant(called left-salient images). The diameter of the salient region was about 4°. By flipping each image across its vertical midline, we can generate 50 new images, each of them had a local salient region in the lower right quadrant(called right-salient images). Notice that the content between the two groups of images were totally the same, the only difference between the two groups was the location of the salient region. The average saliency map of the 50 left-salient images can be seen in Fig. 2 (b).

Based on the bottom-up saliency, we classified all images into two groups: high salient images and low salient images. We proposed a salient index to measure the degree of saliency based on the following formulation:

$$Index(n) = \frac{S_I(n) - S_O(n)}{S_O(n)} . \tag{1}$$

In the above formulation, n denoted the index of an image. For left-salient images, S_I denoted the averaged saliency value of the round region in Fig. 2 (b), and S_O denoted the averaged saliency value of the residual region. The higher $Index$ value indicated the higher saliency. We selected half of images with a higher $Index$ in left-salient images as the high salient images, and selected the other half as the low salient images. The same manipulation was adopted on right-salient images. Thus, stimuli used for psychophysical experiment had two groups: high salient and low salient groups. Each group contained 50 images, half of them were left-salient and the other were right-salient.

Mask stimuli were high contrast checkerboards that randomly arranged (see Fig. 3), the size of each checker was about 0.25° × 0.25°. The luminance of a black checker was 1.8 cd/m^2, while the luminance of a white checker was 79 cd/m^2.

2.3   Psychophysical Experiment

In the psychophysical experiment, all stimuli were displayed on a Gamma-corrected Iiyama HM204DT 22 inches monitor, with a spatial resolution of

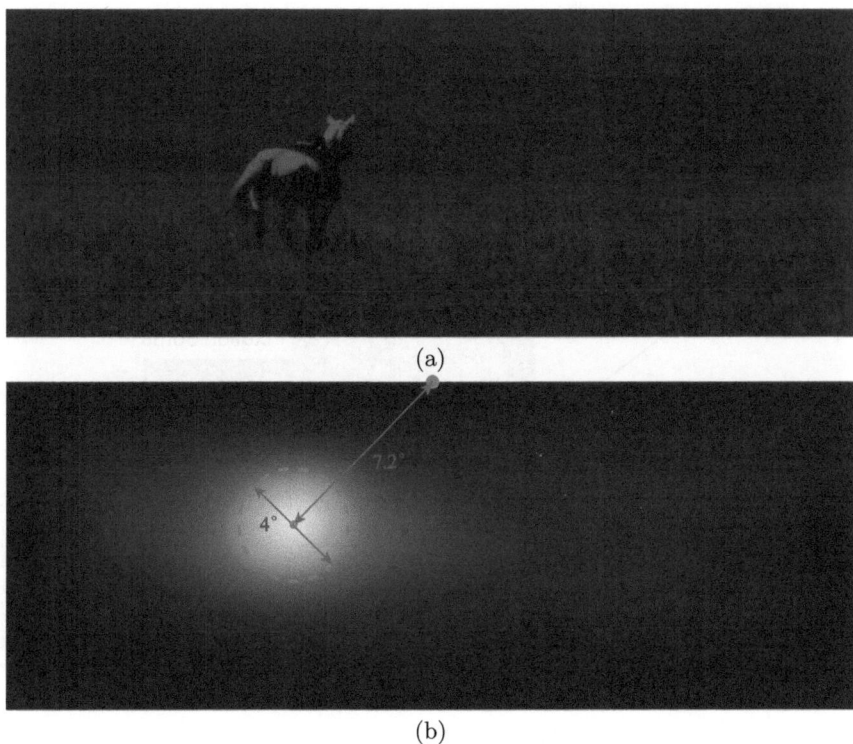

(a)

(b)

Fig. 2. (a) A sample of a low-luminance image used as our stimulus. (b) The averaged saliency map of left-salient images, a circle-like local salient region can be seen on this map.

1024×768 and a refresh rate of 60Hz. The viewing distance was 83 cm, and their head position was stabilized using a chin rest and a head rest. A white fixation cross was always present at the center of the monitor, and subjects were asked to fixate the cross throughout the experiment.

We adopt a modified version of the cueing effect paradigm proposed by Posner to measure the attentional effect of the visual saliency of invisible natural images. Each trial started with a fixation. A low-luminance (2.9 cd/m^2) image was presented on the lower half of the screen for 50 ms, followed by a 100ms mask at the same position, and another 50ms fixation interval. The bottom-up saliency map of the image served as a cue to attract spatial attention, and the mask could ensure that the image was invisible to subjects. Then a grating oriented at about ± 1.5? which centered at about $7.2°$ eccentricity from the fixation was presented randomly at either the lower left quadrant or lower right quadrant with equal probability for 50 ms. The location of the grating was either at or symmetric with the salient region of the previous image, thus indicated the valid cue condition or the invalid cue condition. The grating had a spatial frequency of

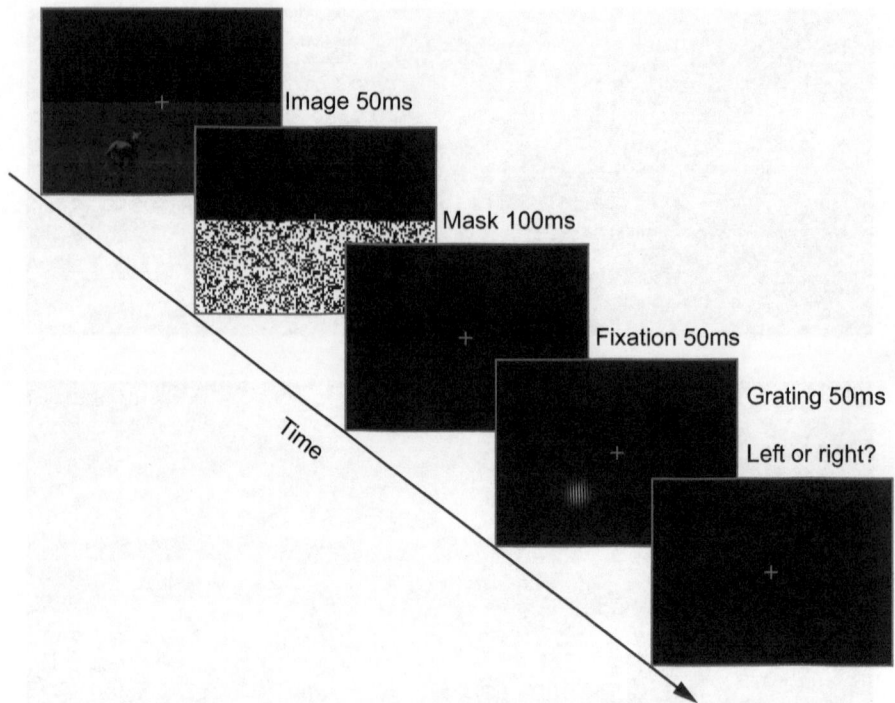

Image 50ms

Mask 100ms

Fixation 50ms

Grating 50ms

Left or right?

Time

Fig. 3. The procedure of our experiment

5.5 cpd (cycle per degree) and its diameter was 2.5 with full contrast. Subjects were asked to press one of the two keys to indicate the orientation of the grating. The duration of each trial was 2s, Fig. 3 shows the procedure of our experiment.

The experiment consisted of 10 blocks. Each block contained 100 trials with two conditions: high salient condition and low salient condition. Images for the first condition were selected randomly from the high salient group, and images for the second condition were selected randomly from the low salient group. The attentional effect of bottom-up saliency maps of invisible images for each condition was measured by the difference between the performance of the valid cue condition and invalid cue condition in the grating orientation discrimination task (see Section 3.2 for details).

Moreover, in order to determine whether the image was indeed invisible, subjects were asked to complete a two-alternative forced choice (2AFC) experiment in a criterion-free way before the attentional effect experiment. Each trial began with either a low-luminance image or a blank, followed by a mask. Subjects were asked to make a forced choice response to judge whether there was an image presented before the mask. The performance at chance level in this experiment could provide an objective confirmation that the masked images were indeed invisible.

3   Experimental Result

3.1   Images Invisibility

The purpose of the 2AFC experiment was to evaluate whether those natural images used as the cue in the attentional experiment were indeed invisible. High salient images and low salient images were counterbalanced in this task. Subjects had to report whether they can see an image before the mask (details can be found in Section 2.3.

We found that percentages of correct detection (mean ± std) were 48.6 ± 6.0% and 50.9 ± 5.7% for high salient and low salient images respectively. Paired t-test results showed that the percentages of correct detection were statistically indistinguishable from the chance level for both high salient and low salient images(paired t-test: high salient images: $t_{15} = -0.934$, $p = 0.365$; low salient images: $t_{15} = 0.6324$, $p = 0.537$; significant level $\alpha = 0.5$), indicated that natural images in both groups were indeed invisible for subjects in our experiment.

3.2   Attentional Effect

The attentional effect of bottom-up saliency maps of invisible images was measured by the difference between the accuracy of grating orientation discrimination performance in the valid cue condition, and that in the invalid cue condition. The grating appeared at randomly either the same location with the salient region of an image (valid cue condition) or its contralateral counterpart (invalid cue condition) with equal probability.

We found that the discrimination accuracy was higher in the valid cue condition than that in the invalid cue condition (see Fig. 4 (a)), for both high salient images (Valid: 81.31 ± 3.93%; Invalid 72.88 ± 3.92%) and low salient images (Valid: 77.86 ± 3.7%; Invalid 76.54 ± 3.52%). The results indicated that the bottom-up saliency map exhibited a positive cueing effect even when the image was invisible, which suggested that subjects' attention was attracted to the salient region of an invisible image, so that they performed better in the valid cue condition than in the invalid cue condition.

Moreover, we measured the attentional effect of bottom-up saliency maps for both high salient and low salient images(see Fig. 4 (b), the left two green bars), the results suggested that the attentional effect of high salient images (8.43 ± 1.32%) and that of low salient images (1.486 ± 1.89%) were both significantly higher than zero(high salient:$t_{15} = 18.126$, $p < 0.001$; low salient: $t_{15} = 2.782$, $p = 0.014$; significant level $\alpha = 0.05$). The attentional effect of high salient images was significantly higher than that of low salient images ($t_{15} = 9.665$, $p < 0.001$).

We also calculated the proposed index of the high salient and the low salient images(high salient: 10.67 ± 4.20%; low salient: 4.03 ± 1.02%), the index predicted the degree of the attention attraction of a bottom-up saliency map(see Fig. 4 (b), the right two yellow bars). Psychophysical data were consistent with the prediction from the computational model.

546 C. Chen et al.

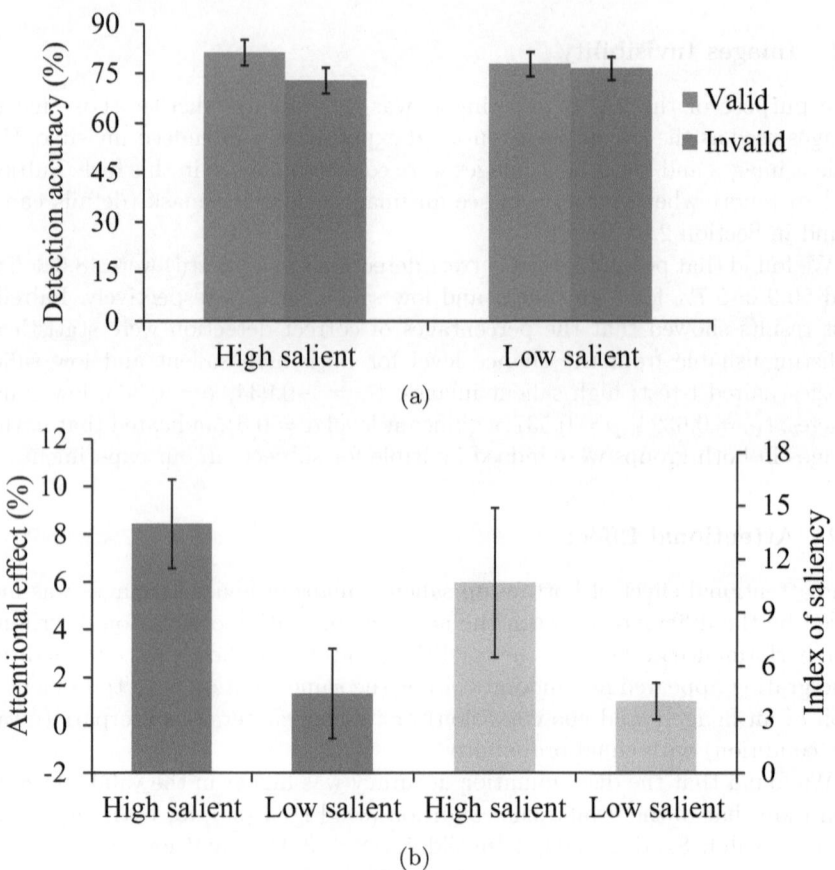

Fig. 4. Results of our experiment. (a) The performance of the grating orientation discrimination task for high salient images and low salient images. (b) The left two green bars indicate the attentional effect of bottom-up saliency maps in high salient and low salient groups. The right two yellow bars indicate the predication of the attentional effect in two groups.

4 Conclusion and Discussion

In this paper, we proposed a method to measure the attentional effect of bottom-up saliency maps. By using backward masking, we could eliminate the contamination of top-down signals. We selected natural images which had a local round salient region and found that even those natural images were invisible, the salient region could attract attention to improve the orientation discrimination performance on a grating in the cueing effect paradigm. Furthermore, we found that the attraction of attention increased with the degree of saliency.

In our experiment, we assume that the absence of awareness to the whole image could maximally reduced top-down signals, even if it did not completely abolish them [5]. These top-down signals may include feature and object perception, as well as subjects' intentions [16]. Compared to previous studies, such manipulation could help us observe the attentional effect based on a relatively pure bottom-up saliency signal. Our findings may suggest that the bottom-up saliency map of a natural image could be generated at a very early stage of visual processing.

In the future, we will extend our study to find the neural substrate of bottom-up saliency maps of natural images. Moreover, consider it's difficult to modulate the degree of saliency on the same content, we will also extend our work on synthesized textures so that we could quantitatively change the degree of saliency on one image.

Acknowledgement. This work was supported by the Ministry of Science and Technology of China (2011CBA00400 and 2009CB320904) and the National Natural Science Foundation of China (Project 30925014, 31230029 and 90920012).

References

1. Itti, L., Koch, C., Niebur, E.: A model of saliency based visual attention for rapid scene analysis. IEEE Trans. Patt. Anal. Mach. Intell. 20, 1254–1259 (1998)
2. Koch, C., Ullman, S.: Shifts in selective visual attention: towards the underlying neural circuitry. Hum. Neurobiol. 4, 219–227 (1985)
3. Geng, J.J., Mangun, G.R.: Anterior intraparietal sulcus is sensitive to bottom-up attention driven by stimulus salience. J. Cogn. Neurosci. 21, 1584–1601 (2008)
4. Mazer, J.A., Gallent, J.L.: Goal-related activity in V4 during free viewing visual search: evidence for a ventral stream visual salience map. Neuron 40, 1241–1250 (2003)
5. Zhang, X., Zhaoping, L., Zhou, T., Fang, F.: Neural activities in V1 create a bottom-up saliency map. Neuron 73, 183–192 (2012)
6. Bogler, C., Bode, S., Haynes, J.D.: Decoding Successive Computational Stages of Saliency Processing. Curr. Biol. 21, 1667–1671 (2011)
7. Fang, F., He, S.: Cortical responses to invisible objects in human dorsal and ventral pathways. Nature Neuroscience 8, 1380–1385 (2005)
8. Posner, M.I., Snyder, C.R.R., Davidson, B.J.: Attention and the detection of signals. J. Exp. Psychol. 109, 160–174 (1980)
9. Eckstein, M.P., Schimozaki, S.S., Abbey, C.K.: The footprints of visual attention in the Posner cueing paradigm revealed by classification images. J. Vis. 2, 25–45 (2002)
10. Hou, X., Zhang, L.: Saliency detection: A spectral residual approach. In: IEEE Computer Vision and Pattern Recognition (2007)
11. Wang, W., Wang, Y., Huang, Q., Gao, W.: Measuring visual saliency by site entropy rate. In: IEEE Computer Vision and Pattern Recognition (2010)
12. Itti, L., Koch, C., Niebur, E.: Computational Modelling of Visual Attention. Nat. Rev. Neurosci. 2, 194–203 (2001)
13. Li, Z.: Contextual influences in V1 as a basis for pop out and asymmetry in visual search. Proc. Natl. Acad. Sci. USA. 96, 10530–10535 (1999)

14. Li, Z.: A saliency map in primary visual cortex. Trends Cogn. Sci. 6, 9–16 (2002)
15. Allman, J., Miezin, F., McGuinness, E.: Stimulus specific responses from beyond the classical receptive field: neurophysiological mechanisms for local-global comparisons in visual neurons. Annu. Rev. Neurosci. 8, 407–430 (1985)
16. Jiang, Y., Costello, P., Fang, F., Huang, M., He, S.: A gender-and sexual orientation-dependent spatial attentional effect of invisible images. Proc. Natl. Acad. Sci. USA. 103, 17048–17052 (2006)

Balance between Diversity and Relevance for Image Search Results

Zhong Ji[1,2], Jing Li, Yuting Su, Yuqing He, and Yanwei Pang

[1] School of Electronic Information Engineering
Tianjin University, Tianjin, 300072, China
[2] State Key Laboratory for Novel Software Technology
Nanjing University, Nanjing, 210093, China
{jizhong,jingli,ytsu,heyuqing,pyw}@tju.edu.cn

Abstract. Image search reranking has received great attention since it overcomes the drawback of "only textual features utilization" in nowadays web-scale image search engines. Most of existing methods focus on relevance reranking, that is reordering the returned results according to their relevance with the query. However, in many cases, users cannot precisely and exhaustively describe their requirements by several query words. Therefore, relevant results with more diversity are more easily meet the users' ambiguous purpose. To address this problem, in this paper, we proposed a DIR (DDrank-based Image search Rerank) algorithm, which can enrich the topic coverage while keeping the relevance with minimal impact. DIR is based on vertex reinforced random walk and further curbing neighbor items' growing rate. Therefore, DIR can automatically balance the relevance and diversity of the top ranked vertices in a principle way. Extensive experiments are performed in a popular image dataset, and the results demonstrate the superiority against other existing methods in the criterion of both AP and ADP.

Keywords: Image search reranking, diversity, random walk.

1 Introduction

The rapid expansion of image data makes their search and ranking technologies more and more important than before. There are mainly two image retrieval schemes: text-based image retrieval and content-based image retrieval. The former searches images with the annotated keywords labeled manually or automatically. While the latter retrieves the images with their visual contents, such as color, texture, and shape. In these years, great progresses have been made from both academic and industry [1]-[3]. However, because of the hindrance of "semantic gap" between low-level image features and high-level semantic concepts, it is still a challenging work needed painstaking study.

Recently, image search reranking technique has emerged to apply visual information to refine the initial text-based search results to get search performance improvement. One kind of methods focuses on relevance reranking [4]-[6], which

J. Yang, F. Fang, and C. Sun (Eds.): IScIDE 2012, LNCS 7751, pp. 549–556, 2013.
© Springer-Verlag Berlin Heidelberg 2013

reorder the returned results according to their relevance with the query. The other focuses on diversity reranking [7]-[9], which deals with the situation that only an ambiguous query provided. For example, a query "fox" may refer to many topics, such as an animal, a car, an American superstar or a film corporation logo, and so on. Therefore, it is better to provide diverse results to cover multiple topics. Currently, most diversity reranking methods are based on clustering algorithms [7]-[8]. For example, Deselaers et al. [7] defined a criterion to measure the diversity of results and propose a greedy selection method and a dynamic programming method to jointly optimize the diversity and the relevance. Leuken et al. [8] proposed a visual lightweight clustering method to visually diversify image search results, which was applied to both ambiguous and non-ambiguous queries. The reranked images were selected from each cluster, which together formed a diverse results set. However, the clustering-based diversity reranking approaches suffer the problem of determining the appropriate cluster number and controlling the shape of clusters.

Recently, random walk-based approaches show its effectiveness in the diversity reranking domain [10]-[11]. For example, the random walk algorithm was applied separately for ever visual feature to avoid any normalization issue between visual similarity metrics in [10]. In [11], the authors adopted the algorithm of absorbing random walks to enhance the diversity as well as relevance of the initial search results. Inspired by the success of Decayed Divrank (DDrank) algorithm in text domain [12], we propose a novel image reranking algorithm, named as DIR (DDrank-based Image search Reranking), which enhances the reranking diversity as well as preserves the relevance characteristics. Our purpose is to achieve the balance of relevance and diversity in image search reranking.

The paper is organized as follows. Section 2 describes the proposed DIR algorithm in detail. Experimental results are presented in section 3 and section 4 concludes the paper.

2   Image Reranking with the Proposed DIR Algorithm

The flowchart of the proposed DIR method is illustrated in Fig. 2. Take the query term "Paris" as an example, it mainly contains 4 steps. 1) When "Paris" is submitted to the web image search engine, an initial text-based search result is returned to the user. The results are unsatisfactory because some irrelevant images are retrieved as top ones. Moreover, it covers only few sub-topics. 2) The images are used to build an ergodic graph, where each image is a node and the similarity between them is used as an edge. 3) The algorithm of DDRank is utilized in the graph to find the diversity results while keeping the relevance. 4) The images are reordered with the items' stationary probability. In the following, we first formulate the problem, and then present the proposed DIR method in detail.

2.1   Problem Formulation

The proposed DIR algorithm contains three inputs: a graph G, an initial ranking probability distribution r, and a weight $\lambda \in [0,1]$. In the task of image reranking, there is a list of search results for each query q by text-based image search engine, we take the top n images as the candidate set $C = \{c_1, c_2 ... c_n\}$. Generally, C covers a kind of sub-topics relevant to q, i.e. $T = \{t_1, t_2 ... t_m\}$. By using the visual similarity as the transition probability, we build a graph $G = (V, E, W)$, in which V represents the node set, E is a set including all the edges which are formed between every two nodes, and W is an adjacency matrix whose each element is a weight to the according edge in E. Since the similarity is for both sides, therefore the graph is undirected. w_{ij} stands for the similarity between the image i and j, and it is always greater or equal to zero. Specifically, the weights are non-negative, w_{ii} equals to 0.

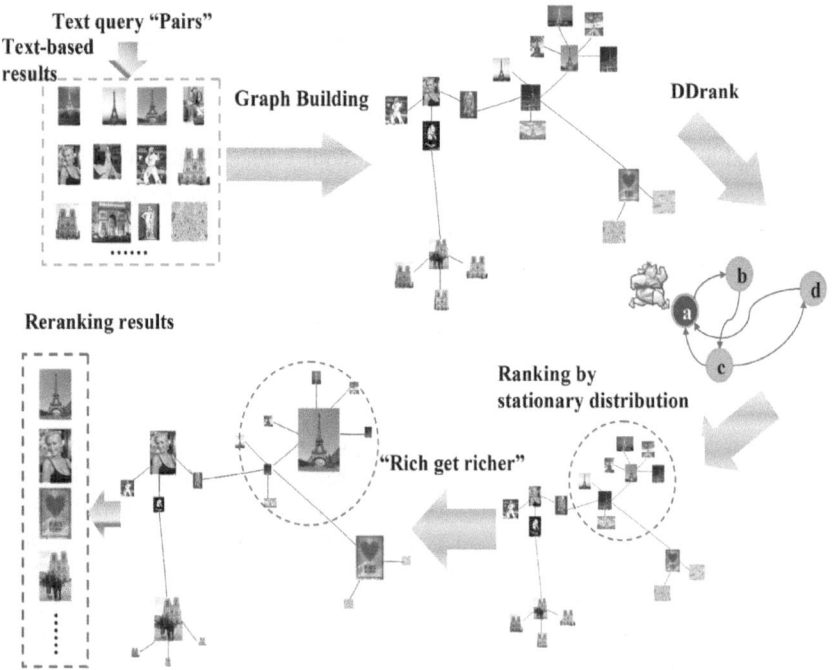

Fig. 1. The flowchart of the proposed DIR method

2.2   DDRank Modal

DDRank is an improvement to traditional random walks by highlighting the diversity and centrality simultaneously, and has been successfully applied in text summarization [12]. In DDrank model, the visiting times of each node is increasing as the time going but with different growing rate. Therefore, the visiting times of each item vary from growing rate, thus encouraging diversity. Since the DDRank model is developed from DivRank algorithm [13], we first make a brief introduction to it. DivRank algorithm applies the vertex-reinforced mechanism which is originally used in the Markov chain on the graph modal. It is believed that the transition probability would be increasing rather than remaining constant over time. Therefore, the process will be eventually form into a "rich get richer" phenomena and "absorb" the score of neighbors and present diversified results.

The general form of DivRank is described as below:

$$P_T(u,v) = (1-\lambda)p*(v) + \lambda \frac{p_0(u,v)N_T(v)}{D_T(u)} \tag{1}$$

$$\sum_{v \in V} p*(v) = 1, \quad p*(v) \geq 0 \tag{2}$$

$$D_T(u) = \sum_{v \in V} p_0(u,v)N_T(v) \tag{3}$$

where $P_T(u,v)$ is the transition probability from state u to any state v at time T. $p*(v)$ represents the initial ranking probability distribution of visiting vertex v. $p_0(u,v)$ is the "organic" transition probability prior to any reinforcement, which can be estimated as a regular time-homogenous random walk. Matrix $N_T(v)$ means the visiting times of each node, and it is the reinforcing factor during the random walk process. The reinforced random walk defined above by (1) converges to a stationary distribution $f(v)$:

$$f(v) = \sum_{v \in V} p_t(u,v)f(u), \quad \forall t \geq T \tag{4}$$

where $\sum_{v \in V} f(v) = 1$.

However, in DivRank, we usually obtain a diversified result at a price of changing the structure of the information network. So the relevance between the object and the query which fully depend on the original ranking results may not be well preserved. Therefore, it is reasonable to consider to re-weighting the reinforcement on each item's relevance score based on how much it relevant to the query. DDRank algorithm improves DivRank algorithm by preserving the local relationship between objects and query. Since the process includes both preserving and reinforcing operations, therefore relevance and diversity can be both obtained.

The task of DDRank algorithm is to preserve relevance and diversity simultaneously. There are usually two requirements to be met. On one hand, in order to keep the relevance, we have to assure that the higher the relevance score between the query and the object is, the more competitive the object is. On the other hand, since diversity is also an important role playing in current ranking field, the competition between the nodes which are near the query has to be weaker than those away from the query.

As a result, the general form of DDRank is as follows.

$$P_{t+1}^T = (1-\lambda)p*(v) + \lambda P_t^T (P_0 N_T^{1-p*(v)}) D_t^{-1} \tag{5}$$

$$D_T(u) = \sum_{v \in V} p_0(u,v)N_T(v)^{1-p*(v)} \tag{6}$$

$$P_0 = \beta P + (1-\beta)I , \quad 0 \leq \beta \leq 1 \tag{7}$$

where P_0 is the new transition matrix on which the vertex-reinforced random walk depend, and matrix D_T assures the convergence of the process. If the network is ergodic, the reinforced random walk defined by (4) also converges to a stationary distribution $f(v)$ after a sufficiently large time t. Then we used this distribution to rank the vertices in the image network.

It can be seen from the (6), the rate of the N_T between a couple of neighbors $\rho = [N_T(1)/N_T(2)]^{(1-r)}$ gets curbed based on the original relevance score r. When $0 < N_T(1)/N_T(2) < 1$, we can get $\partial \rho / \partial r > 0$, which means that if $N_T(1)$ is smaller than $N_T(2)$, the gap between $N_T(1)$ and $N_T(2)$ will be reduced with the increase of r. In this way, the relevance will be preserved to some extent on the algorithm of DivRank. And the reinforcement part is similar. We also consider the first derivative of the $l = N_T^{(1-r)}$ on r. Since N_T is always smaller than 1, the $l_r' \geq 0$ means the object would obtain more competitiveness if its original ranking score is higher, in other words more relevant.

3   Experiment and Results

We use MASA-MM_V1.0 dataset [14] to evaluate our approach. The images are all collected from the Microsoft Live search. The data set covers a variety of queries including angels, baseball, bees, chocolates, cowboys, dragons, earth, flowers etc. The images for each query are all collected together with their associated information including each image's relevance score, the downloading website link. And the more important side is several features of the images extracted are supplied. For each image, we use 255D block-wise color moment features generated from 5×5 fixed partition of the image, 128D wavelet texture features, 64DHSV color histogram, 256D RGB color histogram, 144D color correlogram, 75D edge distribution histogram and 7D face features. The ground truth of the relevance of each image is already supplied by the data set. We set the parameter β in equation (7) to 0.2 and weighting parameter λ in equation (5) is empirically set to 0.9 for all the queries.

We use corresponding feature vectors to calculate similarity between images. And the distance between two images is defined below:

$$d(i,j) = \exp(-\frac{1}{N}\sum_{k=1}^{N} d_k(i,j)/\beta) \tag{8}$$

$$d_k(i,j) = \frac{f_{ik} - \mu_k}{\sigma_k} - \frac{f_{jk} - \mu_k}{\sigma_k} \tag{9}$$

where k represents the k th dimensional feature, N is the feature dimension, β is a constant number, μ_k and σ_k are its mean and variance respectively. We adopt AP (Average Precision) and ADP (Average Diverse Precision) which proposed in [9] to measure the performance.

The AP and ADP are defined below:

$$AP(\tau, D) = \frac{1}{R} = \sum_{j=1}^{n} [y(\tau(j)) \frac{\sum_{k=1}^{j} y(\tau(k))}{j})] \qquad (10)$$

$$ADP(\tau, D) = \frac{1}{R} [\sum_{j=2}^{n} y(\tau(j))Div(\tau(j)) \times (\frac{1}{j}[\sum_{j=2}^{n} y(\tau(k))Div(\tau(k)))) + a(1)] \qquad (11)$$

$$Div(\tau(k)) = \min_{1 \le t \le k}(1 - d(\tau(t), \tau(k))) \qquad (12)$$

$$a(1) = \begin{cases} 1 & if \quad y(\tau(1)) = 1 \\ 0 & if \quad y(\tau(1)) = 0 \end{cases} \qquad (13)$$

where $D = \{x_1, x_2...x_n\}$ represents a set of images. $y(x_i)$ takes 1 if x_i is relevant to the query and 0 if not. τ is the images' ranking, R stands for how many real relevant images in the D. $d(\tau(t), \tau(k))$ measures the distance between images t and k. Since the ADP adds $Div(\tau(k))$ into the primitive expression, therefore, both relevance and diversity can be considered.

As for the probability distribution r, we make an extension to the relevance score function proposed in [9], and define r as follows:

$$r(t) = \frac{h(t)}{\sum_{t=1}^{n} h(t)} \qquad (14)$$

$$h(t) = \frac{2e^{-(t-1)/z}}{1 + e^{-(t-1)/z}} \qquad (15)$$

where t is the initial position number, z is a constant number. Here we set it to 50 and the sum of it is 1.

We compare DIR method with the DivRank algorithm [13] and the GrassHopper algorithm [11]. 1) DivRank ranking is based on vertex-reinforced mechanism so that the times of the vertex got visited would increase as the time going. Therefore, the graph would in a way present the diverse effect. 2) The ranked item in GrassHopper reduces the importance of the unranked nodes so as to "absorbing" the unranked nodes' importance. Specifically, the original ranking results are referred as the "Baseline". Fig. 2 illustrates the AP and ADP performances obtained by different methods at the depth of {10, 20, 30, 40, 50}. The concept of depth normally means the number of the top images. We can clearly see that our algorithm performs better both on diversity and relevance part than other methods. This indicates that the proposed DIR method can hold the relevance as well as diversity by suppressing the visiting times of each node. It also means the method can preserve the original data structure at its best from being affected by the side effect and behave diversity results within the allowed range.

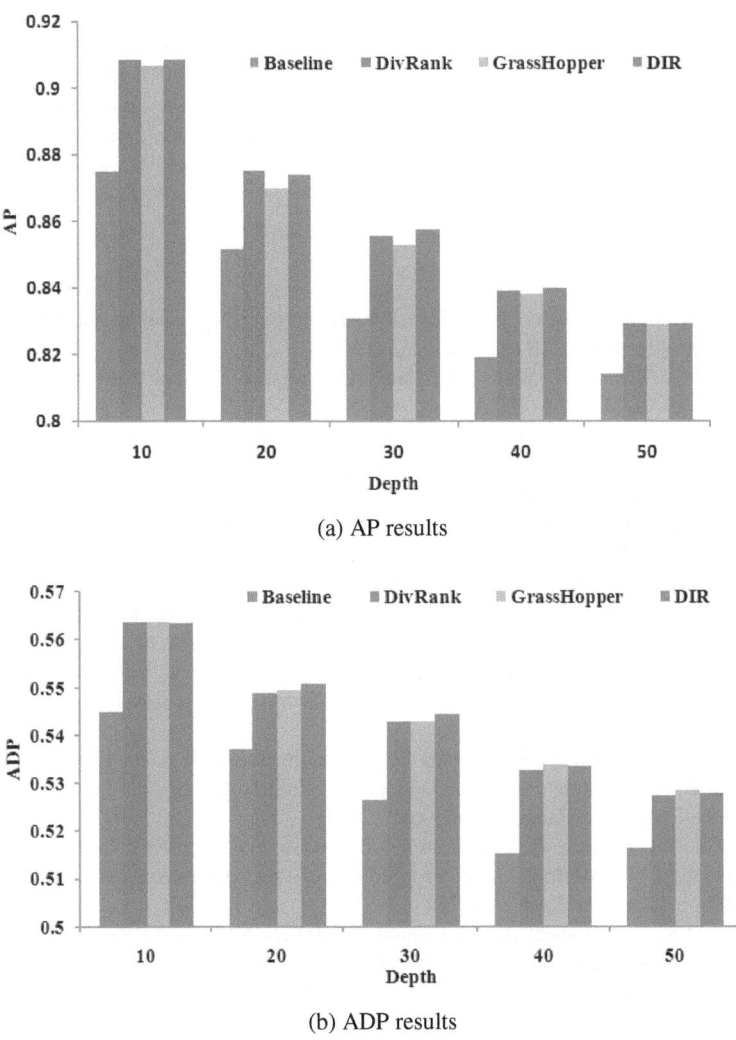

(a) AP results

(b) ADP results

Fig. 2. ADP and AP performance comparisons

4   Conclusions

In this paper, we present a novel image reranking approach considering both relevance and diversity. It extends DDrank to image domain, utilizing visual information to reorder the original image search results, which has been proved to be very effective. In the future, we will try to use the method to new applications, such as image and video summarization.

Acknowledgments. This work was supported by the National Natural Science Foundation of China (Nos. 61271325, 60975001, 61172121, 61222109, 61271412), the Tianjin Research Program of Application Foundation and Advanced Technology (No. 10JCYBJC07700), the Specialized Research Fund for the Doctoral Program of Higher Education (No. 200900321100028), the Program for New Century Excellent Talents in University (No. NCET-10-0620), and the Open Project Program of the State Key Laboratory for Novel Software Technology (No. KFKT2012B35), Nanjing University.

References

1. Jiao, L., Wang, S., Hou, B.: A Review of SAR Images Understanding and Interpretation. Acta Electronica Sinica 33(12), 2423–2434 (2005)
2. El Zaart, A., Ziou, D., Wang, S., Jiang, Q.: Segmentation of SAR Images. Pattern Recognit. 35(2), 713–724 (2002)
3. Ross, T., Worrell, S., Velten, V., Mossing, J., Bryant, M.: Standard SAR ATR Evaluation Experiments Using the MSTAR Public Release Data Set. In: SPIE Conf. on Algorithms for Synthetic Aperture Radar Imagery, vol. 3370, pp. 566–573 (1998)
4. Chumsamrong, W., Thitimajshima, P., Rangsanseri, Y.: Synthetic Aperture Radar (SAR) Image Segmentation Using a New Modified Fuzzy C-means Algorithm. In: Proceedings of IEEE Symp. Geosci. Remote Sens., Honolulu, pp. 624–626 (2000)
5. Petrou, M., Matrucceli, A.: On the Stability of Thresholding SAR Images. Pattern Recognit. 31(11), 1791–1796 (1998)
6. Hua, X., Pierce, L.E., Ulaby, F.T.: SAR Speckle Reduction Using Wavelet Denoising and Markov Random Field Modeling. IEEE Trans. Geosci. Remote Sens. 40(10), 2196–2212 (2002)
7. Yan, X., Jiao, L., Xu, S.: SAR Image Segmentation Based on Gabor Filters of Adaptive Window in Overcomplete Brushlet Domain. In: Proceedings of Asia-Pacific Conference on Synthetic Aperture Radar, pp. 660–663 (2009)
8. Weisenseel, R.A., Karl, W.C., Castanon, D.A., Brower, R.C.: MRF-based Algorithms for Segmentation of SAR Images. In: Proceedings of the 1998 International Conference on Image Processing, vol. 3, pp. 770–774 (1998)
9. Horritt, M.S.: A Statistical Active Contour Model for SAR Image Segmentation. Image and Vision Computing 17, 213–224 (1999)
10. Kristan, M., Leonardis, A., Skocaj, D.: Multivariate Online Kernel Density Estimation with Gaussian Kernels. Pattern Recognit. 44(10), 2630–2642 (2011)
11. Ari, C., Aksoy, S., Arkan, O.: Maximum Likelihood Estimation of Gaussian Mixture Models Using Stochastic Search. Pattern Recognit. 45, 2804–2816 (2012)
12. Kass, M., Witkin, A., Terzopoulos, D.: Snakes: Active Contour Models. Int. J. Compution Vision 1(4), 321–331 (1987)
13. Peng, R., Wang, X., Lu, Y.: SAR Imagery Segmentation Based on Integrated Active Contour. In: Proceedings of the Int. Conf. on Advanced Computer Control, pp. 43–47 (2010)

Learning with Weak Views Based on Dependence Maximization Dimensionality Reduction

Qing Zhang[1,2], De-Chuan Zhan[2,3,*], and Yilong Yin[1]

[1] School of Computer Science and Technology,
Shandong University, Jinan, 250101, China
[2] National Key Laboratory for Novel Software Technology,
Nanjing, 210046, China
[3] Shenzhen Key Laboratory of High Performance Data Mining,
Shenzhen, 518055, China
{zhangqing2008,ylyin}@sdu.edu.cn,
zhandc@nju.edu.cn

Abstract. Large number of applications involving multiple views of data are coming into use, e.g., reporting news on the Internet by both text and video, identifying a person by both fingerprints and face images, etc. Meanwhile, labeling these data needs expensive efforts and thus most data are left unlabeled in many applications. Co-training can exploit the information of unlabeled data in multi-view scenarios. However, the assumptions of co-training, i.e., sufficient and redundant are so strong to be held in most situations. It is notable that different views often have different discrimination ability, while views with strong discrimination ability are usually hard to be obtained. As a consequence, it is a promising way to exploit unlabeled multi-view training data to integrate the information of the strong view into the weak view so that the weak view's discrimination ability can get improved. Only classifiers trained on the weak view will be used to do the classification tasks afterwards. In this paper, based on dependence maximization, we propose a framework to inject the information of strong views into weak ones. Experiments show that the framework outperforms co-training in improving the performances of classifiers trained on the weak view.

Keywords: multi-view, dimensionality reduction, semi-supervised learning, co-training.

1 Introduction

Many real applications involve more than one modal of data, and abundant data with multiple views are at hand. A representative example is that the Internet news on web pages are always presented by text, audio, video, simultaneously. For real applications, unlabeled training data are readily available but labeled ones are fairly expensive to be obtained because labeling unlabeled data requires expensive human efforts and time. Semi-Supervised Learning (SSL) [1] methods were proposed to make use of the unlabeled data. With single view data, the classifier trained on one view tries to exploit the

---

* Corresponding author.

J. Yang, F. Fang, and C. Sun (Eds.): IScIDE 2012, LNCS 7751, pp. 557–564, 2013.

unlabeled data by itself. In semi-supervised scenario, when data are presented by multiple views, co-training [2] is proposed to exploit the disagreements between the multiviews so as to improve views' generalization abilities. However, Blum and Mitchell [2] pointed out that in co-training style methods, the data should be assumed to have two sufficient and redundant views, where each view is sufficient for training a strong classifier and the views are conditionally independent given the class label. Nevertheless, in real-world applications, it is rarely that the two views are conditionally independent given the class label.

There is a ubiquitous fact in multi-view data to which researchers have never paid much attention, that is, different views always have different discrimination ability. Views can be categorized into strong views which have strong discrimination ability and weak views which are with weak discrimination ability. Classifiers trained on strong views usually have better performance (generalization ability) than those trained on weak views. For example, fingerprints are accurate and robust in identifying a person, while face images show less discrimination ability and are vulnerable to variances. It seems that we can rely on strong views only, however, in real applications, strong view data are always harder to be acquired. E.g., to capture a person's fingerprints needs expert devices and more human efforts than to take the person's face photographs. As a consequence, it is expected that only weak view data are used for the classification task, e.g., face recognition is preferred in human identification. Many efforts have been made to raise the performance of the weak view, e.g. [5-6]. However, there has hardly been any researches so far concerning integrating information of the strong view into the weak view.

In this paper, we propose a framework to improve the weak view in semi-supervised scenario. This framework attempts to exploit unlabeled multi-view training data to integrate information of the strong view into the weak view. Dependence maximization is used to make information of strong and weak views mostly aligned. A dimensionality reduction method DMDR (Dependence Maximization Dimensionality Reduction method) is proposed to project the weak view data to a lower-dimensional space in which the dependence of information from both strong and weak views is maximized. We verify the effectiveness of the proposed framework by experiments on real and synthetic datasets. The proposed framework outperforms co-training in promoting the classifiers trained on the weak view.

This paper is organized as follows. Section 2 shows some related works. Section 3 describes the proposed framework in detail. Section 4 shows the experiments on real and synthetic datasets. Section 5 concludes the paper and points out the future work.

2   Related Works

Researches on multi-view data learning are mostly concentrated on how to exploit unlabeled information in semi-supervised scenarios. A representative multi-view semi-supervised learning approach is the co-training approach [2] which works with two views. Co-training initializes a classifier on each view, respectively, using the original labeled data. Then, each of the two classifiers selects and labels a certain number of highly-confident unlabeled instances to refine the other classifier. Blum and Mitchell

[2] proved that if the two views are sufficient and redundant, the predictive accuracy of an initial weak classifier can be boosted to arbitrarily high using unlabeled data by co-training. Zhou et al. [7] also showed that, with sufficient and redundant views, it is possible to execute an effective semi-supervised learning with a single labeled training example. In real-world tasks, however, the requirement of sufficient and redundant views is too luxury. Thus, researchers tried to find relaxed conditions for co-training style methods that work in real situations. Abney [3] showed that the two views are not needed to be conditionally independent, and a weak independence assumption is sufficient. Balcan et al. [4] proved that even the weak independence is not needed if PAC classifiers can be obtained on each view, and a weaker assumption of expansion of the underlying data distribution is sufficient. In spite of these findings, co-training can only get effective results under certain assumptions.

In this paper, we emphasize the fact that in multi-view data, different views always have different discrimination ability. Since strong view data are harder to be obtained, it is necessary and important to improve the weak view. Our work is related to dimensionality reduction as the core of the proposed framework to improve the weak view is a dimensionality reduction method for the weak view. Traditional dimensionality reduction methods can be classified into supervised or unsupervised, depending on whether the label information is used or not. A representative unsupervised method is PCA [8], and advances include ISOMAP [9], ICA [10], LPP [11], LLE [12], etc. Representative supervised dimensionality reduction methods are LDA [13], PLS [14], etc. All these methods work on a single view, while the proposed method in this paper takes two views into consideration. The dimensionality reduction method CCA [15] also concerns aligning information of two views. However, since our work concentrates on the promotion of the weak view, it seems more straightforward to only inject information of the strong view into the weak one. Actually, we testified in our experiments that injecting information of both views into each other somehow inhibits the promotion of the weak view. The proposed dimensionality reduction method can be categorized into unsupervised method since no label information is needed.

3 Proposed Framework

In this section, we firstly introduce the DMDR method followed by the full picture of the framework.

3.1 DMDR Method

Let $\mathcal{X}$ and $\mathcal{Y}$ denote the original feature space of the weak view and the strong view, respectively. Labeled data are presented by $\{(\boldsymbol{x}_1, \boldsymbol{y}_1, l_1), (\boldsymbol{x}_1, \boldsymbol{y}_2, l_2), \cdots, (\boldsymbol{x}_n, \boldsymbol{y}_n, l_n)\}$, where $\boldsymbol{x}_i$ denotes an instance of the weak view, and $\boldsymbol{y}_i$ denotes an instance of the strong view, l_i is the corresponding label, $i = 1, 2, \cdots, n$. Unlabeled data are presented by $\{(\boldsymbol{x}_{n+1}, \boldsymbol{y}_{n+1}), (\boldsymbol{x}_{n+2}, \boldsymbol{y}_{n+2}), \cdots, (\boldsymbol{x}_N, \boldsymbol{y}_N)\}$, where n is the number of labeled data, and N is the total number of labeled and unlabeled data. The total dataset is presented by $Dn = \{(\boldsymbol{x}_1, \boldsymbol{y}_1), (\boldsymbol{x}_2, \boldsymbol{y}_2), \cdots, (\boldsymbol{x}_N, \boldsymbol{y}_N)\}$. By assuming that the weak view contains discriminative information implicitly, we need to extract these discriminations with the supervision of the strong view. In detail, we attempt to find a lower-dimensional feature space for the weak view in which the dependence of information of both strong and weak views is maximized. So that, the lower-dimensional feature space can stress the discriminations more explicitly and classifiers built on the feature space can be with better performances. By denoting the projection vector of the weak view data as p, an instance x is projected into a new space $\mathcal{F}$ by $\phi(x) = p^\top x$ and the deduced kernel function is $\kappa(x_i, x_j) = \langle \phi(x_i), \phi(x_j) \rangle = \langle p^\top x_i, p^\top x_j \rangle$. For instances of the strong view, we define the kernel function $\ell(y_i, y_j) = \langle y_i, y_j \rangle$. Given the dataset Dn with joint distribution P_{xy}, we define the kernel matrix for the weak view and the strong view as $\mathbf{K} = [\kappa_{ij}]_{N \times N}, \kappa_{ij} = \kappa(x_i, x_j)$ and $\mathbf{L} = [\ell_{ij}]_{N \times N}, \ell_{ij} = \ell(y_i, y_j)$, respectively. Then, we try to maximize the dependence of the weak view data in the projected feature space $\mathcal{F}$ with the strong view data.

For facilitating the dependence maximization and make full use of the potential nonlinearities in both views, we define a dependence between the kernels of different views as

$$\mathfrak{D}(\mathbf{K}, \mathbf{L}) = tr(\mathbf{KL}) \tag{1}$$

by assuming both $\mathbf{K}$ and $\mathbf{L}$ are centralized and normalized. For general kernels of data, if we define $\mathbf{H} = \mathbf{I} - \frac{1}{N} \times \mathbf{ee}^\top$, where $\mathbf{I}$ is an identity vector and $\mathbf{e}$ is an all-one column vector, the equation above becomes

$$\mathfrak{D}(\mathbf{K}, \mathbf{L}) = tr(\mathbf{HKHL}) \tag{2}$$

where $\mathbf{H}$ can be regarded as a centralized operator. Our dependence criterion is closely related to a kind of independence criterion called Hilbert-Schmidt Independence Criterion [16] . The dependence criterion computes the square of the norm of the cross-covariance operator over the domain $\mathcal{X} \times \mathcal{Y}$ in Hilbert Space. Due to the neat theoretical properties, we maximize the dependence of the information of both views, i.e.,

$$\max \quad \mathfrak{D}(\mathbf{K}, \mathbf{L}) \quad = \quad \max \quad tr(\mathbf{HKHL}) \tag{3}$$

By representing the instances in $\mathcal{X}$ as $\phi(x)$, we can rewrite the target function in eq. 3 as

$$p^* = \arg\max_{p} \quad tr(\mathbf{HX}^\top pp^\top \mathbf{XHL}) \tag{4}$$

To avoid the scaling problem, we add the constraint that the $l_2 - norm$ of p should be bounded. Therefore, we reformulate the optimization problem as

$$p^* = \arg\max_{p} \quad tr(\mathbf{HX}^\top pp^\top \mathbf{XHL})$$

$$s.t. \qquad p^\top p = 1$$

Notice that

$$tr(\mathbf{HX}^\top pp^\top \mathbf{XHL}) = p^\top (\mathbf{XHLHX}^\top)p \tag{5}$$

Since $\mathbf{XHLHX}^\top$ is symmetric, the eigenvalues are all real. Without any loss of generality, we can assume that the eigenvalues of $\mathbf{XHLHX}^\top$ are sorted as $\lambda_1 \geq \lambda_2 \geq \cdots \lambda_D$. Thus, if d is the dimensionality of the new feature space, the optimal projection matrix P^* can be defined as $P^* = [p_1^*, p_2^*, \cdots, p_d^*]$, where p_i^* is the normalized eigen-vector corresponding to the i-th largest eigenvalue λ_i, $i = 1, \cdots, d$ $(d \ll D)$. Since the eigenvalues reflect the contribution of the corresponding dimensions, we can control d by setting a threshold $thr(0 \leq thr \leq 1)$ and then choose the first d eigenvectors such that

$$\sum_{i=1}^{d} \lambda_i \geq thr \times \sum_{i=1}^{D} \lambda_i \qquad (6)$$

3.2 Framework Summarization

To give a clear picture of the framework, we summarize the framework in Algorithm 1.

Algorithm 1. The proposed framework
---
1. Get a training dataset of two views $Dn = \{(x_1, y_1), (x_2, y_2), \cdots, (x_N, y_N)\}$ including labeled and unlabeled data.
2. Perform DMDR and project the weak view data to a lower-dimensional space.
3.1 Train a classifier $\mathcal{A}$ by labeled weak view data.
3.2 Train a classifier $\mathcal{B}$ by labeled strong view data.
4. Use $\mathcal{B}$ to predict labels for unlabeled data.
5. Re-train $\mathcal{A}$.

---

4 Experiments

In experiments, the framework is compared with co-training in the performances of the classifiers trained on the weak view. Five dimensionality reduction methods PCA, ICA, ISOMAP, LPP, and LLE are used in co-training for comparison. CCA is also adopted by the framework to compare with DMDR. Binary SVM is used as the basic classifier.

4.1 Datasets

We use three multi-view datasets in our experiments - two real datasets and a synthetic dataset constructed from a real single view dataset. For each dataset, two views are selected as the strong view and the weak view according to their discrimination ability. The two real datasets are Ads dataset [17] and WebKB dataset [18]. The Ads dataset has five views from which two views are selected. The WebKB dataset contains two views naturally. The synthetic two-view dataset is constructed from the Newsgroup dataset [19], in which the first view is selected by PCA, and the second view is constructed by selecting 700 features from all features randomly. Classification problems are confined to be two class problems. The WebKb dataset is processed to include two classes - one class is course page of 230 instances, and the other is non-course page of 821 instances. For Newsgroup dataset, we choose the first two classes among all the 20 classes.

4.2   Configuration

The average accuracies of SVMs trained on different views in each dataset evaluated by 10 times10-Crosses Validation (CV) are listed in Table 1. Results show that no matter

Table 1. Accuracies of SVMs trained on different views in each dataset

Datasets	Views	Accuracy of SVMs		
		Euclidean Space	PCA	LLE
Ads	View1	89.0%	93.8%	89.1%
	View2	86.8%	91.2%	86.3%
	View3	91.1%	95.9%	94.9%
	View4	84.2%	89.0%	89.3%
	View5	81.7%	86.7%	88.6%
WebKB	View1	75.4%	82.4%	86.9%
	View2	89.2%	95.1%	92.9%
Newsgroup	View selected by PCA	82.8%	88.7%	93.6%
	View selected randomly	74.8%	78.0%	72.1%

in the original Euclidean space or space selected by linear and nonlinear dimensionality reduction methods, classifiers trained on some views always have better performances than those trained on some other views. In the Ads dataset, View 3 presents anchor text attached to hyper-links pointing to a web page and shows the best discriminative ability; View 5 presents the caption of a web page and is the weakest view in all situations. The situation is the same for the WebKB dataset. It indicates that strong views are always harder to be obtained. Gathering information from other web pages which link to a web page is always harder than getting information from a web page itself. In Newsgroup dataset, the view selected by PCA is always stronger than the view selected randomly.

Table 2. Configurations in co-training

Datasets	Proportion	Labeled data		Unlabeled data labeled in each round	
		Positive Class	Negative Class	Positive Class	Negative Class
Ads	1:6	5	30	1	6
WebKB	1:5	5	25	1	5
Newsgroup	1:1	50	50	5	5

In our framework, 50 instances of each class in the training set are selected as the labeled data, the rest data are treated as unlabeled. In co-training, the number of labeled data and the number of unlabeled data labeled by each classifier in each round in the training set is decided by the proportion of data in two classes. The numbers are listed in Table 2. Co-training process goes until all unlabeled data are labeled.

4.3 Experimental Results

In the experiment, both the proposed framework and co-training are performed by 10 times 10-CV. Average accuracies of SVMs trained on the weak view of each dataset are listed in Table 3. The listed results in co-training are the results obtained in the last round. Performances obtained in the proposed framework with DMDR are higher than the best results in co-training. For the proposed framework, using CCA results in lower performance than using DMDR which implies injecting information of both views into each other will inhibit the promotion of the weak view.

Table 3. Accuracies of SVMs trained on weak view in each dataset

Datasets	Proposed framework		Co-training				
	DMDR	CCA	PCA	ICA	ISOMAP	LPP	LLE
Ads	**86.7%**	86.6%	86.2%	85.9%	86.1%	85.9%	86.1%
WebKB	**84.2%**	79.2%	79.0%	73.9%	78.1%	78.1%	81.0%
Newsgroup	**71.9%**	59.8%	50.0%	45.5%	45.1%	46.5%	47.5%

5   Conclusion and Future Work

Nowadays, abundant multi-view data are available. Meantime, labeling data needs expensive human efforts and time, and more data at hand are left unlabeled. In this paper, we emphasize a ubiquitous fact in multi-view data that different views always have different discrimination ability. In most situations, more efforts will be paid to collect strong view data, so it is often expected that only weak view data are used for classification. As a result, it is necessary and important to improve the discrimination ability of the weak view. A framework is proposed to improve the weak view through the help of the strong view in semi-supervised mode. We show the superiority of the framework by experiments. In the future work, we will employ the framework on real applications such as biometric recognition and try to extend it for data with more than two views.

Acknowledgements. This research is supported by Shenzhen Key Laboratory for High Performance Data Mining with Shenzhen New Industry Development Fund under grant No.CXB2010052 50021A, NSFC under grant No. 61105043, Baidu Open Research Fund and National Natural Science Foundation of China under Grant No.61070097, as well as Jiangsu Natural Science Foundation, Grant No. BK2011566.

References

1. Chapelle, O., Schölkopf, B., Zien, A.: Semi–Supervised Learning. MIT Press, Cambridge (2006)
2. Blum, A., Mitchell, T.: Combining labeled and unlabeled data with co–training. In: Proceedings of the 11th Annual Conference on Computational Learning Theory, Madison, WI, pp. 92–100 (1998)

3. Abney, S.: Bootstrapping. In: Proceedings of the 40th Annual Meeting of the Association for Computational Linguistics, Philadelphia, PA, pp. 360–367 (2002)
4. Balcan, M.F., Blum, A., Yang, K.: Co-training and expansion: Towards bridging theory and practice. In: Saul, L., Weiss, Y., Bottou, L. (eds.) Advances in Neural Information Processing Systems 17, pp. 88–96. MIT Press, Cambridge (2004)
5. Roli, F., Didaci, L., Marcialis, G.L.: Template Co-update in Multimodal Biometric Systems. In: Lee, S.-W., Li, S.Z. (eds.) ICB 2007. LNCS, vol. 4642, pp. 1194–1202. Springer, Heidelberg (2007)
6. Roli, F., Didaci, L., Marcialis, G.L.: Adaptive biometric systems that can improve with use. In: Ratha, N., Govindaraju, V. (eds.) Advances in Biometrics: Sensors, Systems and Algorithms, vol. 3, pp. 447–471. Springer, Heidelberg (2008)
7. Zhou, Z.-H., Zhan, D.-C., Yang, Q.: Semi-supervised learning with very few labeled training examples. In: Proceedings of the 22nd AAAI Conference on Artificial Intelligence, Vancouver, Canada, pp. 675–680 (2007)
8. Jolliffe, I.T.: Principal Component Analysis. Springer, New York (1986)
9. Tenenbaum, J.B., Desilva, V., Langford, J.C.: A global geometric framework for nonlinear dimensionality reduction. Science 290(5500), 2319–2323 (2000)
10. Comon, P.: Independent component analysis-A new concept? Signal Processing 36(3), 287–314 (1994)
11. He, X., Niyogi, P.: Locality Preserving Projections. In: Sebastian, T., Saul, L., Schölkopf, B. (eds.) Advances in Neural Information Processing Systems 16, Vancouver, Canada, pp. 153–160 (2003)
12. Roweis, S.T., Saul, L.: Nonlinear dimensionality reduction by locally linear embedding. Science 290(5500), 2323–2326 (2000)
13. Fisher, R.: The use of multiple measurements in taxonomic problems. Annals of Eugenics 7(2), 179–188 (1936)
14. Wold, H.: Partial least squares. Encyclopedia of the Statistical Sciences 6, 581–591 (1985)
15. Hardoon, D., Szedmak, S., Taylor, J.: Canonical correlation analysis: An overview with application to learning methods. Neural Computing 16(12), 2639–2664 (2004)
16. Gretton, A., Bousquet, O., Smola, A.J., Schölkopf, B.: Measuring Statistical Dependence with Hilbert-Schmidt Norms. In: Jain, S., Simon, H.U., Tomita, E. (eds.) ALT 2005. LNCS (LNAI), vol. 3734, pp. 63–77. Springer, Heidelberg (2005)
17. Kushmerick, N.: Learning to remove internet advertisements. In: Proceedings of the 3rd Annual Conference on Autonomous Agents, Seattle, WA, pp. 175–181 (1999)
18. Craven, M., DiPasquo, D., Freitag, D., McCallm, A., Mitchell, T., Nigam, K., Slattery, S.: Learning to extract symbolic knowledge from the world wide web. In: Proceedings of 15th National Conference on Artificial Intelligence, Madison, WI, USA, pp. 509–516 (1998)
19. Joachims, T.: A probabilistic analysis of the Rocchio algorithm with TFIDF for text categorization. Computer Science Technical Report CMU-CS-96-118. Carnegie Mellon University (1996)

Schizophrenia Candidate Genes Specific to Human Brain Region Are Restricted to Basal Ganglia

Xinguo Lu[1,2], Ping Liu[3], Ling-li Zeng[1], Renfa Li[2], and Dewen Hu[1]

[1] College of Mechatronics and Automation,
National University of Defense Technology, Changsha, 410073, China
hnluxinguo@126.com, dwhu@nudt.edu.cn
[2] School of Information Science and Engineering,
Hunan University, Changsha, 410082, China
[3] Hunan Want Want Hospital, Changsha, 410016, China

Abstract. Gene expression profiles produced by Microarray present a potential method for exploring the relationship between genes and brain. Here we analyzed the human gene expression levels among eight brain regions, including frontal lobe, Insula, limblic lobe, occipital lobe, parietal lobe, temporal lobe, basal ganglia and other cerebral nuclei. Top 45 schizophrenia candidate genes(179 probes) expression levels were compared between each two of these regions. Strikingly, 66.7% gene(83.2% probe) expression in basal ganglia distinguished from other regions. The great variation of the schizophrenia candidate gene expression in different brain regions indicated there is different neuropathologic mechanisms regulated by these candidate genes to this disease and basal ganglia would be the expression center of schizophrenia candidate genes.

1 Introduction

Schizophrenia is a disabling psychiatric disorder which is considered to be inherent disease affecting the brain. However, the mechanisms by which specific gene defects lead to cognitive impairment remain obscure[1]. In recent years, computational neurobiology analyzes neuroimaging data to explore brain variance between schizophrenic patients and healthy controls. Abnormalities of the prefrontal, superior temporal cortices and hippocampus have been frequently reported in patients with schizophrenia Some consistent findings such as cortical atrophy, particularly of the prefrontal cortex, with related ventricular enlargement, shrinkage of the hippocampus and of the superior temporal gyrus, and reduction of the cerebral asymmetry also have been presented.

Gene expression profiling is the measurement of the activity (the expression) of thousands of genes at once, to create a global picture of cellular function. The altered mRNA expression identified to underlie the pathophysiology of schizophrenia with RNA from postmortem tissues. Several studies have used gene arrays to profile differentially expressed genes between schizophrenic and healthy human post-mortem brain. Unique alterations in gene expression were reported in

J. Yang, F. Fang, and C. Sun (Eds.): IScIDE 2012, LNCS 7751, pp. 565–572, 2013.
© Springer-Verlag Berlin Heidelberg 2013

prefrontal cortex of schizophrenics compared with matched controls[2][3]. The expression of the genes found to be significantly altered in the superior temporal gyrus were compared to those previously reported to be altered in peripheral blood lymphocytes[4]. Some comprehensive studies comparing gene expression changes associated with schizophrenia from multiple brain regions[5].

Collectively, these findings provide evidence for unique brain region specific patterns of altered gene expression in human brain. However, no studies have directly compared gene expression profiles associated with schizophrenia across different brain regions in human. Recently, in animal models after chronic inter- mittent ethanol exposure subtractive hybridization experiments have suggested that such brain subregion-restricted genes do exist[6].

Here we use gene expression profiles produced by oligonucleotide microarrays and statistics to examine the differently expressed genes across different regions in human post-mortem brain. The top 45 schizophrenia candidate genes(179 probes) expression levels were compared across eight brain regions including frontal lobe, Insula, limblic lobe, occipital lobe, parietal lobe, temporal lobe, basal ganglia and other cerebral nuclei. The relationship between schizophrenia candidate genes and brain region was analyzed. 73.3% gene(88.3% probe) ex- pressed differently between one region to any other regions. These brain region specific changes in gene expression would play a significant role in schizophrenia. Strikingly, 66.7% gene(83.2% probe) expression in basal ganglia distinguished from any other regions. The great variation of the schizophrenia candidate gene expression between different brain regions indicates the different neuropathologic mechanisms regulated by these candidate genes to this disease.

2   Materials and Methods

2.1   Microarray Data Generation

An Agilent 8x60K array, custom-designed by Beckman Coulter Genomics in conjunction with the Allen Institute, was used to generate microarray data. The array design included the existing 4x44K Agilent Whole Human Genome probe set supplemented with an additional 16,000 probes. The normalized expression values can be download from allen institute for brain science[1].

2.2   Anatomical Parcellation

The whole brain was divided into 8 regions including frontal lobe, Insula, limblic lobe, occipital lobe, parietal lobe, temporal lobe, basal ganglia and other cerebral nuclei[7].

2.3   Top 45 Schizophrenia Candidate Genes

The SZGene database[2] contains information from 1727 studies, reporting data on 1008 genes, and 8788 polymorphisms; this database has 287 meta-analyses.

---

[1] http://human.brain-map.org
[2] http://www.szgene.org/

SZGene ranks its top results using the HuGENet interim guidelines published by Ioannidis and colleagues, which consider the amount of evidence[8].The top 45 schizophrenia candidate genes are listed in the website and selected as studying objects in the following work.

2.4   Data Analysis

After normalization, the top 45 schizophrenia candidate gene(179 probe) expression levels were compared across these brain regions using t test. For each probe the gene expression considered as specific to one region compared with other regions met the following criteria: (1) there were significantly different gene expression values between the two regions at the threshold(P) by using a two-sample two-tailed t test, (2) the gene expression values in one region were significantly different to any other regions at the threshold(P) by using a two-sample two-tailed t test.

3   Results

3.1   Variation between Regions in Human Brain

Difference of schizophrenia candidate gene expression values between the eight regions in the human brain was analyzed by determining if the gene was specific to the region at threshold P of 0.05,10E-2,10E-3,10E-5,10E-7(Table.1). In Table.1, the first column is the region name compared to other regions, in the followed 5 columns are numbers of specific genes to the corresponding region at the threshold P. For example, when threshold P=0.05 the number of genes specific to one region compared to other regions were: 1 gene (1 probe) in Frontal Lobe(Fro) 2 genes (2 probes) in Insula (Ins); 4 genes (4 probes) in Limbic Lobe(Lim); 9 genes (19 probes) in Occipital Lobe(Occ); 0 genes (0 probes) in Parietal Lobe (Par); 1 genes (1 probes) in Temporal Lobe(Tem); 30 genes (149 probes) in Basal Ganglia(BasG); 4 genes (11 probes) in other Cerebral nuclei(othC) .

Table 1. Number of brain region specific genes with two-sample t test at different threshold P

Region	P=0.05	P=10E-2	P=10E-3	P=10E-5	P=10E-7
Frontal Lobe	1(1)	0(0)	0(0)	0(0)	0(0)
Insula	2(2)	0(0)	0(0)	0(0)	0(0)
Limbic Lobe	4(4)	2(2)	0(0)	0(0)	0(0)
Occipital Lobe	9(19)	5(8)	0(0)	0(0)	0(0)
Parietal Lobe	0(0)	0(0)	0(0)	0(0)	0(0)
Temporal Lobe	1(1)	0(0)	0(0)	0(0)	0(0)
Basal Ganglia	30(149)	26(109)	21(80)	18(55)	11(29)
other Cerebral nuclei	4(11)	2(3)	0(0)	0(0)	0(0)

3.2   Significantly Different Gene Expression across Brain Regions

By setting threshold P=10E-7 significant difference of gene expression values between the eight regions in the human brain were analyzed. Compared to other regions in basal ganglia there are 11 specific genes(29 probes)(Table.2). But in any region from other seven regions there were not specific genes. So in Table.2, there list the significantly different expressed genes in basal ganglia in comparison with other regions. The first and fourth columns are significantly differently expressed gene names, the second and fifth columns are significantly differently expressed probe names, and the third and sixth columns are P value of t test distinguished basal ganglia from other regions. These different gene expression are shown in a heat map (Fig.1). From each region five tissues were selected and there are 40 tissues(5*8) in the expression. In Fig.1, the first five columns are gene expression from tissues in region of Basal Ganglia(BasG), and the followed are Frontal Lobe(Fro), Insula (Ins), Limbic Lobe(Lim), Occipital Lobe(Occ), other Cerebral nuclei(othC), Parietal Lobe (Par), Temporal Lobe(Tem). The last column is the significantly differently expressed probe name.

Table 2. Significantly different expressed genes across brain regions (P=10E-7)

geneName	probeName	P value	geneName	probeName	P value
HP	CUST_8877_PI416261804	5.54E-17	COMT	CUST_2057_PI416261804	8.57E-08
DRD2	CUST_240_PI416408490	3.75E-15	DE4B	CUST_506_PI416408490	7.42E-08
DTNBP1	A_24_P145316	3.25E-14	OPCML	A_23_P72663	1.19E-07
HP	A_23_P5831	1.78E-13	HP	A_32_P54442	1.36E-07
DRD2	CUST_1494_PI417557136	1.73E-12	HTR2A	CUST_1602_PI417557136	1.49E-07
PPP3CC	A_23_P157495	2.05E-12	HTR2A	CUST_1609_PI417557136	1.90E-07
OPCML	A_32_P160563	1.49E-11	PPP3CC	CUST_15969_PI416261804	2.83E-07
HP	CUST_7320_PI416261804	9.87E-11	HTR2A	CUST_1617_PI417557136	2.90E-07
IL10	A_23_P203173	5.91E-10	DRD2	A_24_P283834	3.05E-07
HTR2A	CUST_1610_PI417557136	6.68E-10	DTNBP1	A_23_P42435	3.37E-07
HTR2A	CUST_1606_PI417557136	7.02E-10	DRD2	CUST_1807_PI417557136	3.68E-07
HP	A_23_P11372	1.10E-09	HTR2A	CUST_1614_PI417557136	4.63E-07
RPP21	A_23_P315602	1.22E-08	HTR2A	CUST_512_PI417557136	5.82E-07
HTR2A	CUST_1607_PI417557136	1.43E-08	DISC1	A_23_P62967	6.07E-07
HTR2A	A_24_P355967	3.06E-08			

4   Discussion

In this study, we detected the expression level of schizophrenia candidate genes in different human brain regions including frontal lobe, Insula, limblic lobe, occipital lobe, parietal lobe, temporal lobe, basal ganglia and other cerebral nuclei, and expected to evaluate the role of schizophrenia candidate genes in these regions. Table.3 list the specific genes to these brain regions excluding basal ganglia. Here we analyzed the expression of specific genes to different brain region and the significantly different expression in basal ganglia compared to other regions.

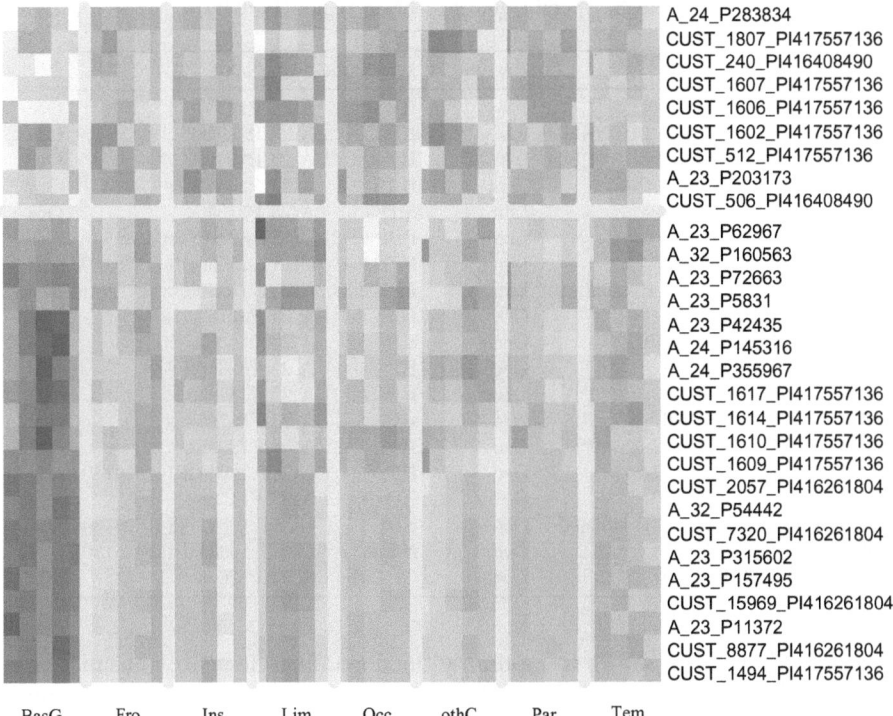

A_24_P283834
CUST_1807_PI417557136
CUST_240_PI416408490
CUST_1607_PI417557136
CUST_1606_PI417557136
CUST_1602_PI417557136
CUST_512_PI417557136
A_23_P203173
CUST_506_PI416408490

A_23_P62967
A_32_P160563
A_23_P72663
A_23_P5831
A_23_P42435
A_24_P145316
A_24_P355967
CUST_1617_PI417557136
CUST_1614_PI417557136
CUST_1610_PI417557136
CUST_1609_PI417557136
CUST_2057_PI416261804
A_32_P54442
CUST_7320_PI416261804
A_23_P315602
A_23_P157495
CUST_15969_PI416261804
A_23_P11372
CUST_8877_PI416261804
CUST_1494_PI417557136

BasG   Fro   Ins   Lim   Occ   othC   Par   Tem

Fig. 1. Significantly different gene expression in basal ganglia compared with other regions(P=10E-7)

4.1   Specific Genes to Different Brain Regions

In frontal lobe a significant genetic predisposition of the plexin A2 gene(PLXNA2) was observed. The PLXNA2 gene is one of the receptor genes for axonal guidance factors. PLXNA2 is located at chromosome 1q32, a region frequently reported as a susceptibility locus in linkage studies of schizophrenia. Mah et al. identified PLXNA2 as a candidate for schizophrenia in a whole-genome association study[9]. Plexin members have been implicated in the development of axonal projections and neural regeneration. Plexin A2 heterodimerizes with either neuropilin (NRP) 1 or 2 and forms a receptor for the class 3 semaphorins (sema3), which are secreted axonal chemorepellents. Four single nucleotide polymorphisms (SNPs), rs841865, rs752016, rs1327175 and rs2498028 from the PLXNA2 genomic interval have been reported to be associated with schizophrenia[10]. This finding indicated that the PLXNA2 gene would be the mainly expressed schizophrenia candidate gene in frontal lobe then affect the anatomical brain abnormalities which approved to associate with this disease.

In temporal lobe, the expression of interleukin-1beta (IL-1B) were significant different in comparison with other regions. Interleukin-1beta (IL-1B) is a pleiotropic cytokine contributing to inflammation, cell growth, and tissue repair.

Table 3. Specific genes to brain region(P=0.05)

geneName	probeName	Region	geneName	probeName	Region
PLXNA2	A_23_P46618	Fro	HP	A_24_P139665	Occ
COMT	CUST_2057_PI416261804	Ins	HP	A_23_P119464	Occ
HP	CUST_7320_PI416261804	Ins	HP	A_23_P256107	Occ
IL1B	A_23_P79518	Tem	IL10	A_23_P203173	Occ
HP	A_23_P109442	Lim	MDGA1	A_23_P310460	Occ
MDGA1	A_23_P310460	Lim	PPP3CC	CUST_522_PI416408490	Occ
RGS4	A_23_P433855	Lim	PPP3CC	CUST_15969_PI416261804	Occ
RPP21	CUST_16633_PI416261804	Lim	PPP3CC	A_23_P157495	Occ
AHI1	A_24_P38143	Occ	DISC1	A_24_P83787	othC
AHI1	A_24_P213710	Occ	HP	A_23_P80362	othC
AHI1	A_23_P70746	Occ	HP	A_23_P328652	othC
APOE	A_23_P164650	Occ	HP	A_23_P137630	othC
APOE	A_32_P93036	Occ	HP	A_23_P397055	othC
APOE	A_24_P788772	Occ	HP	A_23_P75299	othC
CMYA5	A_24_P925413	Occ	HP	CUST_16644_PI416261804	othC
CMYA5	A_23_P124946	Occ	HP	A_23_P115683	othC
DTNBP1	CUST_243_PI416408490	Occ	NRG1	CUST_16501_PI416261804	othC
DTNBP1	A_23_P42435	Occ	NRG1	CUST_16981_PI416261804	othC
GSTM1	A_23_P115407	Occ	OPCML	A_23_P72663	othC

Cytokines have been shown to modulate the dopaminergic neurotransmission in the central nervous system (CNS), which is thought to be aberrant in schizophrenia. In addition, some studies demonstrated that IL-1b produces dose-dependent decreases in the number of neurons in embryonic rat cortical slices, suggesting decreased cerebral cortical neuron survival. This finding suggests that IL-1b plays an important role in temporal lobe and then a heightened risk of schizophrenia.

The specific gene to limblic lobe and occipital lobe is MAM domain containing glycosylphosphatidylinositol anchor 1(MDGA1). MDGA1 which is a unique cell surface glycoprotein is expressed predominantly in the developing nervous system, both central and peripheral. The MDGA1 gene maps to 6p21.2, within the schizophrenia linkage region 6p22.3Cp21.1.In addition, in occipital lobe another three specific genes, including APOE (apolipoprotein E), CMYA5 (cardiomyopathy associated 5) and AHI1 (Abelson helper integration-1)were discovered. These genes are required for both cerebellar and cortical development[1] which indicates these genes regulate nervous system development in the occipital lobe and play a important role in cognitive (dys)function.

4.2 Significantly Different Expression in Basal Ganglia Compared to Other Regions

The main finding of the present study was that the expression of schizophrenia candidate genes was focus on basal ganglia. There was about 66.7% gene(83.2% probe) expression in basal ganglia distinguished from other regions(P=0.05). When threshold P was set to 10E-7, there were 11 significantly different expression genes(29 probes) in basal ganglia(Table.2). And these gene expression was shown in Fig.1. However, in other 7 brain regions there were not any significantly differently expressed gene.

The basal ganglia consist of several functionally related nuclei involved in corticalsubcortical circuits that modulate motor, cognitive and emotional behavior. The basal ganglia includes the following nuclei: caudate nucleus, putamen, nucleus accumbens, globus pallidus (GP), subthalamic nucleus, and the mesencephalic nuclei of the substantia nigra and ventral tegmental area.

Some authors found increased caudate, putamen and globus pallidus volumes, whereas others found no abnormalities in putamen or caudate volumes [11]. These findings imply that basal ganglia structural abnormalities observed in subjects with schizophrenia are at least in part an intrinsic feature of the illness. Anatomical brain abnormalities have been reported from magnetic resonance imaging (MRI) studies of schizophrenia patients.

The present study explored the relationship between the expression of schizophrenia candidate genes and human brain regions and found these differently expressed genes were mainly located in basal ganglia. Strikingly, when the threshold was set 10E-7, there sre significantly different expressed genes only in basal ganglia. These significantly different expression may imply the basal ganglia shape or volume abnormality and regulate dopamine transmission within the basal ganglia and therefore has been linked to schizophrenia.

5   Conclusion

In the present study, we analyzed the human gene expression levels among eight brain regions, including frontal lobe, Insula, limblic lobe, occipital lobe, parietal lobe, temporal lobe, basal ganglia and other cerebral nuclei and explored the relationship between the expression of schizophrenia candidate genes and human brain regions. Schizophrenia specific genes to different regions were observed. Strikingly, these significantly different expressed genes were mainly located in basal ganglia. We therefore conclude that basal ganglia would be the expression center of schizophrenia candidate genes and provided the intrinsic pathophysiology of schizophrenia.

Acknowledgment. This work is supported by the National Natural Science Foundation of China (61202288), the Ph.D. Programs Foundation of Ministry of Education of China (20100161120023), the National Science Foundation for Postdoctoral Scientists of China (20100471790), the Fundamental Research Funds for the Central Universities and the Young Teachers Program of Hunan University.

References

1. Barnes, M.R., Huxley-Jones, J., Maycox, P.R., et al.: Transcription and pathway analysis of the superior temporal cortex and anterior prefrontal cortex in schizophrenia. Journal of Neuroscience Research 89(8), 1218–1227 (2011)

2. Vawter, M.P., Crook, J.M., Hyde, T.M., et al.: Microarray analysis of gene expression in the prefrontal cortex in schizophrenia: a preliminary study. Schizophrenia Research 58(1), 11–20 (2002)
3. Potkin, S.G., Turner, J.A., Brown, G.G., et al.: Working memory and dlpfc inefficiency in schizophrenia: the fbirn study. Schizophr. Bull. 35(1), 19–31 (2009)
4. Bowden, N.A., Scott, R.J., Tooney, P.A.: Altered gene expression in the superior temporal gyrus in schizophrenia. Bmc Genomics 9, 199 (2008)
5. Lauriat, T.L., Shiue, L., Haroutunian, V., et al.: Developmental expression profile of quaking, a candidate gene for schizophrenia, and its target genes in human prefrontal cortex and hippocampus shows regional specificity. J. Neurosci. Res. 86(4), 785–796 (2008)
6. Melendez, R.I., McGinty, J.F., Kalivas, P.W., et al.: Brain region-specific gene expression changes after chronic intermittent ethanol exposure and early withdrawal in c57bl/6j mice. Addict. Biol. 17(2), 351–364 (2012)
7. Liang, M., Zhou, Y., Jiang, T., et al.: Widespread functional disconnectivity in schizophrenia with resting-state functional magnetic resonance imaging. Neuroreport 17(2), 209–213 (2006)
8. Allen, N.C., Bagade, S., McQueen, M.B., et al.: Systematic meta-analyses and field synopsis of genetic association studies in schizophrenia: the szgene database. Nat. Genet. 40(7), 827–834 (2008)
9. Mah, S., Nelson, M.R., Delisi, L.E., et al.: Identification of the semaphorin receptor plxna2 as a candidate for susceptibility to schizophrenia. Mol. Psychiatry 11(5), 471–478 (2006)
10. Takeshita, M., Yamada, K., Hattori, E., et al.: Genetic examination of the plxna2 gene in japanese and chinese people with schizophrenia. Schizophrenia Research 99(1-3), 359–364 (2008)
11. Oertel-Knochel, V., Knochel, C., Matura, S., et al.: Cortical-basal ganglia imbalance in schizophrenia patients and unaffected first-degree relatives. Schizophrenia Research 138(2-3), 120–127 (2012)

Logistic Tensor Regression for Classification

Xu Tan, Yin Zhang, Siliang Tang, Jian Shao, Fei Wu, and Yueting Zhuang

College of Computer Science and Technology, Zhejiang University,
Hangzhou, China
{tanxu,zhangyin98,siliang,jshao,wufei,yzhuang}@zju.edu.cn

Abstract. Logistic regression is one of the classical approaches for classification which has been widely used in computer vision, bioinformatics as well as multimedia understanding. However, when it is applied to high-dimensional data with structural information such as facial images or motion data, traditional *vector-based* logistic regression suffers from two main weaknesses: one is its negligence of structural information, and the other is its trend of overfitting. In this paper, we propose *Logistic Tensor Regression* (LTR) for classification of high-dimensional data with structural information. The proposed LTR not only reserves the underlying structural information embedded in data by tensorial representations, but also avoids overfitting by the introduction of a sparsity regularizer. Experiments on classification of facial images and motion data show that our proposed Logistic Tensor Regression approach outperforms the state-of-the-art algorithms.

Keywords: Logistic Regression, Tensor, Sparse.

1 Introduction

In the era of information explosion, multimedia data is overwhelmingly growing. How to effectively organize the immense amounts of data is an urgent need. Therefore classification has been becoming increasingly important to solve this issue. Logistic regression is one of the popular approaches for classification, in which a type of regression analysis is used for predicting the outcome of a categorical criterion variable based on one or more predictor variables. Logistic regression has been widely used in various applications such as computer vision, bioinformatics, natural language processing, and multimedia understanding.

In multimedia applications the features of objects can be naturally represented by multiple views. For example, we can extract high-dimensional heterogeneous features from one given image, such as global features (color, shape, and texture) or local features (scale-invariant feature transform, shape context, and gradient location and orientation histogram). How to utilize these structure of features has become a focal point. [6] explored structured visual feature selection to boost the performance of image annotation. [12] using NOVA, which introduced a non-convex penalty for group selection in high-dimensional heterogeneous features space. However, traditional vector-based logistic regression commonly neglects

J. Yang, F. Fang, and C. Sun (Eds.): IScIDE 2012, LNCS 7751, pp. 573–581, 2013.

these structural information among high-dimensional data, the classification performance of logistic regression will be deteriorated and lead to the problem of overfitting [5].

Genkin *et al.*[4] proposed a Bayesian approach to avoid overfitting. The Bayesian Logistic Regression uses a prior probability distribution that favors sparseness in the fitted model, along with an optimization algorithm and implementation tailored to that prior. Nonetheless, it hasn't solved the substantial problem and the structural information among the data is still disregarded.

Recently, several classic vector-based approaches have been extended to deal with tensorial [5]. For examples, Principle Component Analysis (PCA) was extended to Two Dimensional Principle Component Analysis (2DPCA) in [13]. Guo *et al.*[5] proposed a regression framework for linear tensor learning. In these tensor-based approaches, original objects are naturally represented as multi-dimensional arrays. In comparison with vector-based approaches, tensor-based methods take the underlying structural prior into account with lower dimensions. It has been shown that the tensor-based approaches outperform their vector-based counterparts.

Motivated by the recent success in tensor research, we extend logistic regression from vector-space to tensor-space and propose an approach named *Logistic Tensor Regression* (LTR) in this paper. LTR tends to keep the intrinsic structural information in data and overcome the problem of overfitting. In LTR, we use the CANDECOMP/PARAFAC Decomposition[2] on the parameter (coefficient) tensor, which acts as a scheme of feature selection or dimension reduction. This leads to fewer parameters to be estimated and takes the structure of the feature space into consideration. To avoid the problem of overfitting, we use the ℓ_1-norm regularizer to make the solution sparse and interpretable.

2   Notations and Preliminaries

In this section, we will briefly describe some notations and concepts of tensorial algebra [7].

A *tensor* is a multidimensional array. Tensors include vectors and matrices as first-order and second-order as special cases. The *order* of a tensor, also known as ways or modes, is the number of dimensions. Matrices (tensors of order two) will be denoted by boldface capital letters, e.g., $\mathbf{A}$, vectors (tensor of order one) by boldface lowercase letters, e.g., $\mathbf{a}$, scalars (tensors of order zero) by lowercase letters, e.g., a. Higher-order tensors (order three or higher) are denoted by boldface Euler script calligraphic letters, e.g., an M-order tensor $\mathcal{X} \in \mathbb{R}^{I_1 \times I_2 \cdots \times I_M}$.

The ith element of a vector $\mathbf{a}$ is denoted by a_i, the (i, j)th element of a matrix $\mathbf{A}$ is denoted by a_{ij}. while the jth column of a matrix $\mathbf{A}$ is denoted by $\mathbf{a}_j$. Similarly, the elements of an M-order tensor $\mathcal{X} \in \mathbb{R}^{I_1 \times I_2 \cdots \times I_M}$ will be denoted by $x_{i_1 i_2 \dots i_M}, i_l = 1, 2, \dots, I_l, l = 1, 2, \dots, M$.

Matricization, also known as unfolding or flattening, transforms a tensor into a matrix by reordering the elements of an N-way array into a matrix. For instance, a $2 \times 3 \times 4$ tensor can be arranged as a 12×2 matrix or $8 \times$

3 matrix, and so on. In this paper, we consider only the useful special case of **mode-n matricization**. The mode-n matricization of an M-order tensor $\mathcal{X} \in \mathbb{R}^{I_1 \times I_2 \cdots \times I_M}$, denoted by $\mathbf{X}_{(d)} \in \mathbb{R}^{I_d \times (I_1 \cdots I_{d-1} I_{d+1} \cdots I_M)}$ arranges the mode-n fibers to be the columns of the resulting matrix. Tensor element $(i_1, i_2, \ldots, i_N)$ maps to matrix element (i_n, j), where

$$j = i + \sum_{k=1, k \neq n}^{N} i_k - 1 J_k \quad \text{with} \quad J_k = \prod_{m=1, m \neq n}^{k-1} I_M$$

Similarly, the *vectorization* of a matrix $\mathbf{A}$ is to stack its elements into a vector, which is denoted by $\text{vec}(\mathbf{A})$.

The inner product of two tensors of the same size $\mathcal{X}, \mathcal{Y} \in \mathbb{R}^{I_1 \times I_2 \cdots \times I_M}$ is the sum of the products of their enyties, defined by

$$\langle \mathcal{X}, \mathcal{Y} \rangle = \sum_{i_1=1}^{I_1} \cdots \sum_{i_M=1}^{I_M} x_{i_1 \ldots i_M} y_{i_1 \ldots i_M}.$$

From the unfolded tensor equivalents, we have

$$\langle \mathcal{X}, \mathcal{Y} \rangle = \langle \mathbf{X}_{(j)}, \mathbf{Y}_{(j)} \rangle = \text{trace}(\mathbf{X}_{(j)} \mathbf{Y}_{(j)}^T). \tag{1}$$

The Khatri-Rao product of two matrices $\mathbf{A} \in \mathbb{R}^{J_1 \times L}$ and $\mathbf{B} \in \mathbb{R}^{J_2 \times L}$, denoted by $\mathbf{A} \odot \mathbf{B}$, is defined as

$$\mathbf{A} \odot \mathbf{B} = \begin{bmatrix} a_{11}\mathbf{b}_1 & a_{12}\mathbf{b}_2 & \ldots & a_{1L}\mathbf{b}_L \\ a_{21}\mathbf{b}_1 & a_{22}\mathbf{b}_2 & \ldots & a_{2L}\mathbf{b}_L \\ \vdots & \vdots & \ddots & \vdots \\ a_{J_1 1}\mathbf{b}_1 & a_{J_1 2}\mathbf{b}_2 & \ldots & a_{J_1 L}\mathbf{b}_L \end{bmatrix}.$$

The CANDECOMP/PARAFAC (CP) decomposition factorizes an M-order tensor $\mathcal{X} \in \mathbb{R}^{I_1 \times I_2 \cdots \times I_M}$ into a sum of R rank-one tensors, denoted as

$$\mathcal{X} \approx \sum_{r=1}^{R} \mathbf{u}_r^{(1)} \circ \mathbf{u}_r^{(2)} \cdots \circ \mathbf{u}_r^{(M)} \triangleq [\![\mathbf{U}^{(1)}, \mathbf{U}^{(2)}, \ldots, \mathbf{U}^{(M)}]\!]. \tag{2}$$

The operator "$\circ$" is the outer product of vectors and the factor matrices $\mathbf{U}^{(k)} = [\mathbf{u}_1^{(k)}, \ldots, \mathbf{u}_R^{(k)}] \in \mathbb{R}^{I_k \times R}, k = 1, 2, \ldots, M$.

In terms of unfolded tensors, the CP decomposition can be expressed as

$$\mathbf{X}_{(d)} = \mathbf{U}^{(d)} (\mathbf{U}^{(M)} \odot \cdots \odot \mathbf{U}^{(d+1)} \odot \mathbf{U}^{(d-1)} \odot \cdots \odot \mathbf{U}^{(1)})^T. \tag{3}$$

The rank of a tensor $\mathcal{X}$, denoted $R = rank(\mathcal{X})$, is defined as the smallest number of rank-one tensors whose sum is equal to $\mathcal{X}$.

3   The Proposed Approach

In this section, we first review the Logistic Regression in vector space. Then we propose the model of Logistic Tensor Regression (LTR) and give out the details of our proposed LTR algorithm.

3.1   Sparse Logistic Regression in Vector Space

Logistic regression is a well-known binary classification method. Given a set of training examples $\{(\mathbf{x}_i, y_i)\}_{i=1}^{N}$, $\mathbf{x}_i \in \mathbb{R}^p$ denotes samples and $y_i \in \{-1, +1\}$ is the corresponding binary class labels. The conditional probability model[4] is of the form

$$p(y_i = +1 | \boldsymbol{\beta}, \gamma, \mathbf{x}_i) = \Psi(\boldsymbol{\beta}^T \mathbf{x}_i + \gamma),$$

where $\boldsymbol{\beta} \in \mathbb{R}^p$ is the vector of regression coefficients in the model and $\gamma \in \mathbb{R}$ is the intercept. In what follows we use the logistic link function

$$\Psi(r) = \frac{\exp(r)}{1 + \exp(r)},$$

thereby the logistic regression model in vector space is based on the following expression for the conditional probabilities[4]:

$$p(y_i = +1 | \boldsymbol{\beta}, \gamma, \mathbf{x}_i) = \frac{1}{1 + \exp(-\boldsymbol{\beta}^T \mathbf{x}_i - \gamma)}.$$

An estimate of the parameters $\boldsymbol{\beta}$ and γ can be obtained by solving the corresponding maximum (log-)likelihood problem, which is equivalent to:

$$(\widehat{\boldsymbol{\beta}}, \widehat{\gamma}) = arg\ min_{\boldsymbol{\beta}, \gamma} -\mathcal{L}(\boldsymbol{\beta}, \gamma).$$

where

$$\mathcal{L}(\boldsymbol{\beta}, \gamma) := -\sum_{i=1}^{m} \log(1 + \exp(-y_i(x_i^T \boldsymbol{\beta} + \gamma)))$$

is the log-likelihood function.

However, when it is applied to high dimensional data, the performance would be deteriorated due to the lack of regularization in the original formulation [10]. Some regularization norms are often applied to cover this weakness. The ℓ_0-norm is a good regularized term to remedy overfitting and make regression model both sparse and interpretable but it's solution is NP-hard. ℓ_2-norm will lead to a stable and smooth(differentiable) unconstrained convex optimization problem, however it's hard to make the regression model interpretable.

As pointed in [11], the ℓ_1-norm (namely *lasso*, least absolution shrinkage and selection operator) regularizer can be introduced to deal with high-dimensional features (predictors) and obtain a robust classifier. The ℓ_1-norm regularized logistic regression has a direct Bayesian explanation[8], which could be interpreted as posterior mode estimates, and makes the regression model sparse. As a result, we get

$$(\widehat{\beta}(\lambda), \widehat{\gamma}(\lambda)) = arg\ min_{\beta, \gamma} -\mathcal{L}(\beta, \gamma) + \lambda(\sum_k |\beta_k| + |\gamma|),$$

where $\lambda > 0$ is a tuning parameter used to control the sparsity.

3.2   Logistic Tensor Regression

In this paper, we concern with the extension of the logistic regression model from the vector space to the tensor space. Given a set of training examples $\{(\mathcal{X}_i, y_i)\}_{i=1}^{N}$, $\mathcal{X}_i \in \mathbb{R}^{I_1 \times I_2 \cdots \times I_M}$ denote samples with tensorial representations and $y_i \in \{-1, +1\}$ are the associated binary class labels. The proposed Logistic Tensor Regression (LTR) model is based on the following expression for the conditional probabilities:

$$p(y_i = +1 | \mathcal{W}, \gamma, \mathcal{X}_i) = \frac{1}{1 + \exp(-\langle \mathcal{W}, \mathcal{X}_i \rangle - \gamma)},$$

where $\mathcal{W} \in \mathbb{R}^{I_1 \times I_2 \cdots \times I_M}$ is the parameter tensor of regression coefficients and the scalar $\gamma \in \mathbb{R}$ is an intercept.

Thus, the corresponding maximum (log-)likelihood problem can be expressed as

$$(\widehat{\mathcal{W}}, \widehat{\gamma}) = arg\, min_{\mathcal{W}, \gamma} -\mathcal{L}(\mathcal{W}, \gamma).$$

where

$$\mathcal{L}(\mathcal{W}, \gamma) := -\sum_{i=1}^{m} \log(1 + \exp(-y_i(\langle \mathcal{W}, \mathcal{X} \rangle + \gamma))) \qquad (4)$$

is the log-likelihood function.

According to the principle of CP decomposition, we constrain the regression coefficients $\mathcal{W} \in \mathbb{R}^{I_1 \times I_2 \cdots \times I_M}$ to be a sum of R rank-one tensors. By substituting Eq.(2) into Eq.(4), we get

$$\mathcal{L}(\{\mathbf{U}^{(1)}, \mathbf{U}^{(2)}, \ldots, \mathbf{U}^{(M)}\}, \gamma) := -\sum_{i=1}^{M} \log(1 + \exp(-y_i(\langle \mathcal{W}, \mathcal{X} \rangle + \gamma))). \qquad (5)$$

The optimization problem mentioned above can be solved by a coordinate-descent approach or alternative projections in an iteration manner. Concretely, we solve for the parameter $\mathbf{U}^{(j)}$ at each iteration, while keeping the parameters $\{\mathbf{U}^{(k)}\}|_{k=1, k \neq j}^{M}$ fixed. Substituting Eq.(1) and Eq.(3) into Eq.(5), we obtain that the $\widehat{\mathbf{U}}^{(j)}$ which maximizes Eq.(5) is the one which maximizes

$$\mathcal{L}_j(\mathbf{U}^{(j)}, \gamma) = -\sum_{i=1}^{M} \log(1 + \exp(-y_i(\text{trace}(\mathbf{U}^{(j)} \mathbf{U}^{(-j)T} \mathbf{X}_{i(j)}^{T}) + \gamma))),$$

where $\mathbf{U}^{(-j)} = \mathbf{U}^{(M)} \odot \cdots \odot \mathbf{U}^{(d+1)} \odot \mathbf{U}^{(d-1)} \odot \cdots \odot \mathbf{U}^{(1)}$.

Let $\widetilde{\mathbf{X}}_{i(j)} = \mathbf{X}_{i(j)} \mathbf{U}^{(-j)}$, and note that $\text{trace}(\mathbf{U}^{(j)} \widetilde{\mathbf{X}}_{i(j)}^{T}) = [\text{vec}(\mathbf{U}^{(j)})]^T [\text{vec}(\widetilde{\mathbf{X}}_{i(j)})]$, we can vectorize the problem as

$$\mathcal{L}_j(\text{vec}(\mathbf{U}^{(j)}), \gamma = -\sum_{i=1}^{M} \log(1 + \exp(-y_i([\text{vec}(\mathbf{U}^{(j)})]^T [\text{vec}(\widetilde{\mathbf{X}}_{i(j)})] + \gamma))).$$

As discussed in the previous subsection, we apply the ℓ_1-norm to make the solution sparse and interpretable. As a result, we get

$$(\text{vec}(\widehat{\mathbf{U}}^{(j)})(\lambda), \widehat{\gamma}(\lambda)) = arg\,min_{\text{vec}(\mathbf{U}^{(j)}), \gamma} - \mathcal{L}(\text{vec}(\mathbf{U}^{(j)}), \gamma) + \lambda(\sum_k |\text{vec}(\mathbf{U}^{(j)})_k| + |\gamma|),$$

where $\lambda > 0$ is a tuning parameter used to control the sparsity.

The above logistic regression with the ℓ_1-norm regularizer can be implemented by BBR software described in [4]. Our proposed LTR is summarized in Algorithm 1.

Algorithm 1: Logistic Tensor Regression

Input: The training data with a tensorial representation and their corresponding class labels, i.e., $\{(\mathcal{X}_i, y_i)\}|_{i=1}^N$, $\mathcal{X}_i \in \mathbb{R}^{I_1 \times I_2 \cdots \times I_M}$ and $y_i \in \{-1, +1\}$
Output: The regression coefficient tensor $\mathcal{W}$ and the intercept γ.

1   Initialize randomly $\{\mathbf{U}^{(1)}, \ldots, \mathbf{U}^{(M)}\}|_{(0)}$.
2   **repeat**
3   $t \leftarrow t + 1$
4   **for** $k = 1$ to M **do**
        Solve with respect to $\mathbf{U}^{(k)}|_{(t)}$:
        Use BBR with $(\text{vec}(\widetilde{\mathbf{X}}_{i(j)}), y_i)$ as input
      end
5   **until** $\|\mathcal{W}^{(t)} - \mathcal{W}^{(t-1)}\|/\|\mathcal{W}^{(t-1)}\| \leqslant \epsilon$ or $t \geqslant T_{\max}$

---

4   Experiments

In this section, we first analyze the performance of LTR on high dimensional data with structural information under different settings of the rank R, then compare our proposed LTR with other *state-of-the-art* algorithms.

4.1   Experimental Datasets and Evaluation Criteria

To evaluate the performance of the proposed LTR algorithm, we conduct experimental comparisons on two public benchmark datasets in the real world: facial images and human motion data.

1. *Facial Images:* We conduct experiments on facial images for the classification according to age. In this paper, we use one publicly available dataset, the FG-NET dataset[1]. The FG-NET dataset comprises of 1002 facial images of 82 people being from 0 to 69 years old. All the acquired images are resized to 40×30 pixels, thus each image forms a $40 \times 30 \times 3$ tensor. We randomly pick 544 of them as training set and the rest as testing set.

2. *Motion Data:* In the case of motion data, we conduct experiments on classifying the human pose, and carry experiments using publicly available dataset, the Carnegie Mellon University's Graphics Lab motion capture database[9]. We process the data to form each motion a 3×49 tensor. We randomly choose 48 classes for the experiment, including 30000 motions for training and another 39363 motions for testing.

We evaluate the performance on the task of classification in terms of two kinds of metrics in this paper. The first one is the area under the ROC curve, called AUC [3], we use the MacroAUC (average on AUC of all the classes) and the MicroAUC (the global calculation of AUC regardless of classes). The second one is the harmonic mean of precision and recall, called F1 score[3], the MacroF1 (average on F1 scores of all the classes) and the MicroF1 (the global calculation of F1 regardless of classes) are presented.

4.2 Rank Analysis

The rank of a tensor is the smallest number of components in an exact CP decomposition, where "exact" means that there is equality in Eq.(2). However, as pointed in [7], there is no straightforward algorithm to determine the rank of a specific given tensor; in fact, the problem is NP-hard.

We conduct experiments to evaluate the performance with respect to the influence of the "rank" R of the coefficient tensor that controls the number of simultaneous projections along each mode.

The experiments are repeated ten times to obtain the average MacroAUC, MicroAUC, MacroF1 and MicroF1. The results are depicted in Table 1. We can observe that for the case of Facial Images the highest performance is achieved when the "rank" is among $R = 8$ and $R = 10$; while for the Motion Data case, the "rank" $R = 2$ achieves the highest performance. We will therefore report the value of the optimal rank from now onwards.

4.3 Performance Comparison on Classification

We compare our proposed LTR algorithm with the following six methods, of which two are vector-based methods, i.e., Linear Support Vector Regression (SVR), Bayesian Logistic Regression (BBR), and the other four are tensor-based methods, i.e., hrTRR,hrSTR, orTRR, orSTR[5]. The parameters of all the methods are tuned using cross-validation:

Table 2 shows the performance in terms of the average MacroAUC, MicroAUC, MacroF1 and MicroF1 after repeating the experiments ten times. We can observe that all of the tensorial approach outperform their vector-based counterparts in terms of the two kinds of evaluation metrics. What's more, the proposed LTR outperforms the linear tensor regression methods in the task of classification.

Table 1. The performance of LTR with different "rank"s in terms of MacroAUC, MicroAUC, MacroF1 and MicroF1. The results shown in boldface are better than others.

A. Performance on Facial Images.

R	2	4	6	8	10	12
MacroAUC	0.6929	0.6624	0.7173	**0.7482**	0.7353	0.7056
MicroAUC	0.8249	0.8195	0.8491	0.8692	**0.8779**	0.8306
MacroF1	0.5129	0.5324	0.5273	**0.5697**	0.5453	0.5456
MicroF1	0.5092	0.5469	0.5457	**0.5717**	0.5491	0.5308

B. Performance on Motion Data.

R	2	4	6	8	10	12
MacroAUC	**0.6976**	0.6687	0.6723	0.6679	0.6680	0.6694
MicroAUC	**0.9080**	0.8697	0.8733	0.8689	0.8690	0.8703
MacroF1	**0.5218**	0.4691	0.3989	0.4984	0.4674	0.5063
MicroF1	**0.5031**	0.4623	0.4038	0.4687	0.4695	0.4709

Table 2. Performance comparison of different algorithms in terms of MacroAUC, MicroAUC, MacroF1 and MicroF1. The results shown in boldface are better than others.

A. Comparison on Facial Images

Method	MacroAUC	MicroAUC	MacroF1	MicroF1
LTR	**0.7482**	**0.8692**	**0.5697**	**0.5717**
SVR	0.4956	0.6731	0.4073	0.3631
BBR	0.5702	0.7984	0.4417	0.5231
hrTRR	0.6153	0.8152	0.4568	0.4321
hrSTR	0.6916	0.8250	0.5213	0.5037
orTRR	0.5780	0.7756	0.3513	0.3231
orSTR	0.5184	0.7555	0.3913	0.3114

B. Comparison on Motion Data

Method	MacroAUC	MicroAUC	MacroF1	MicroF1
LTR	**0.6976**	**0.9080**	**0.5218**	**0.5031**
SVR	0.5413	0.7231	0.4006	0.4268
BBR	0.5609	0.7527	0.4332	0.4367
hrTRR	0.6708	0.8344	0.4359	0.4135
hrSTR	0.6684	0.8686	0.4965	0.4631
orTRR	0.6594	0.8894	0.5063	0.4792
orSTR	0.5517	0.7932	0.3993	0.4010

5   Conclusions

This paper proposes a tensor-based logistic regression approach called Logistic Tensor Regression (LTR). LTR not only preserves the underlying structural information, but also overcomes the problem of overfitting. Experiments with six other state-of-the-art methods for classification on two kinds of high dimensional data with structural information, i.e. Facial Images and Motion Data, show the effectiveness of our algorithm.

Acknowledgments. This work is supported by 973 program (2010CB327904), National Natural Science Foundation of China (No. 60833006, No. 61103099), 863 program2012AA012505

References

1. Agarwal, A., Triggs, B., Rhone-Alpes, I., Montbonnot, F.: The fg-net aging database (2010), `http://www.fgnet.rsunit.com`
2. Carroll, J., Chang, J.: Analysis of individual differences in multidimensional scaling via an n-way generalization of "eckart-young" decomposition. Psychometrika 35(3), 283–319 (1970)
3. Fawcett, T.: An introduction to ROC analysis. Pattern Recognition Letters 27(8), 861–874 (2006)
4. Genkin, A., Lewis, D., Madigan, D.: Large-scale bayesian logistic regression for text categorization. Technometrics 49(3), 291–304 (2007)
5. Guo, W., Kotsia, I., Patras, I.: Tensor learning for regression. IEEE Transactions on Image Processing 21(2), 816–827 (2012)
6. Han, Y., Wu, F., Tian, Q., Zhuang, Y.: Image annotation by input-output structural grouping sparsity. IEEE Transactions on Image Processing (99), 1–1 (2010)
7. Kolda, T., Bader, B.: Tensor decompositions and applications. SIAM Review 51(3), 455–500 (2009)
8. Park, T., Casella, G.: The bayesian lasso. Journal of the American Statistical Association 103(482), 681–686 (2008)
9. Piazza, T., Lundström, J., Kunz, A., Fjeld, M.: Predicting Missing Markers in Real-Time Optical Motion Capture. In: Magnenat-Thalmann, N. (ed.) 3DPH 2009. LNCS, vol. 5903, pp. 125–136. Springer, Heidelberg (2009)
10. Schütze, H., Hull, D., Pedersen, J.: A comparison of classifiers and document representations for the routing problem. In: Proceedings of the 18th annual international ACM SIGIR Conference on Research and Development in Information Retrieval, pp. 229–237. ACM (1995)
11. Tibshirani, R.: Regression shrinkage and selection via the lasso. Journal of the Royal Statistical Society. Series B (Methodological), 267–288 (1996)
12. Wu, F., Yuan, Y., Rui, Y., Yan, S.: Annotating web images using nova: Nonconvex group sparsity. In: ACM International Conference on Multimedia (2012)
13. Yang, J., Zhang, D., Frangi, A.F., Yang, J.: Two-dimensional pca: a new approach to appearance-based face representation and recognition. IEEE Transactions on Pattern Analysis and Machine Intelligence 26(1), 131–137 (2004)

A Patch-Based Non-local Means Method for Image Denoising

Shaobo Dang, Yanning Zhang, and Dong Gong

Shaanxi Provincial Key Laboratoryof Speech & Image information Processing,
School of Computer, Northwest Polytechnical University, Xi'an, Shaanxi, P.R.C
dangshaobo@mail.nwpu.edu.cn, ynzhang@nwpu.edu.cn,
Edward.dong.gong@gmail.com

Abstract. In this paper, a revised version of non-local means denoising method is proposed. Different from the original non-local means method in which the algorithm is processed on a pixel-wise basis, the proposed method using image patches to implement non-local means denoising. Given that some details, texture and structure information will be smoothed out when performing weighted averaging, we carry out a pre-processing procedure to classify the image patches into several clusters according to their feature similarities. Later, the weights needed in non-local means algorithm is calculated between image patches in the same cluster. By this means, the above mentioned detail, texture and structure can be preserved due to the redundancy in similar image patches. We illustrate the overall algorithm's performance via several experiments. The results indicate the effectiveness of the proposed method.

Keywords: Image denoising, Non-Local Means, Patch-based, K-means Clustering.

1 Introduction

Image Denoising has long been a fundamental but important issue in image processing. Despite of the significant improvement in hardware, such as denser pixels per inch, noise always exists. In fact, a high resolution image is more prone to affected by noises since each pixel receives less photon. Hence, image denoising remains to be a problem requires further research.

Generally, the purpose of image denoising problem is to recovers the original image from an observed noisy version,

$$y(i) = x(i) + n(i) \tag{1}$$

Where, $y(i)$ is the observed value, $x(i)$ is the true value and $n(i)$ denotes the noise at each pixel. Often, the noise $n(i)$ is assumed to be an i.i.d standard Gaussian distribution with zero means and variance σ^2.

In the past few decades, lots of efforts have been made based on different interpretations of the problem. But the very basic remarks resemble, i.e. by averaging.

J. Yang, F. Fang, and C. Sun (Eds.): IScIDE 2012, LNCS 7751, pp. 582–589, 2013.

Pixel-wise processing has been adopted in the beginning. In [1], a Gaussian smoothing model is employed to estimate the processing pixel using its local neighbors. In [2], an anisotropic filtering is used instead. These methods are all implemented in space domain, while in frequency domain, there are also some classical methods proved to be effective in noise removal, such as the Wiener filter proposed by Yaroslavsky[3] and Wavelet thresholding methods by Coiffman-Donoho[8]. In these methods, the input image is decomposed into multiple scales, representing different time-frequency components of the original signals. Operations such as thresholding and statistical modeling are performed at each scale. Then the processed coefficients, filtering out the small values considered to be noise, are transformed back to spatial domain. Later, some improvements are made on the original wavelet methods including ridgelet and curvelet methods in order to preserve line structures [9]. But the wavelet method was handicapped by the very essence that it uses a set of fixed basis. However, in natural images, some important local texture information could not be represented by wavelet basis efficiently, thus loss of local continuity, such as artifacts, may occur in the final results. Non-Local Means method handles the denoising problem by weighted averaging similar pixels ranging around the whole image. The weight is inversely related to the distance between current processing pixel and its similarities. Further improvement could be found in bilateral filtering methods where the averaging is performed.

Block-wise or patch-wise method has been a new trend recently. The initiative is to take advantage of the redundancy of patches, often extracted from the original image as sub-blocks, to recover the image and meanwhile preserving the texture or geometry information. Recently proposed methods has proved the high efficiency of patch-wise method, such as in [12], the overlapping patches are processed using Principle Component Analysis in three different ways- global, local and hierarchical. The state of the art algorithm BM3D [4], is a successful implementation of image denoising in sparse domain. Patch-wise denoising has gain significant success in image denoising since it trickily preserve the texture information, which is always the major shortcoming of other method, by a redundant representation.

In this paper, we propose a revised Non-Local Means of global patch denoising algorithm. First, overlapping and redundant patches are extracted from the noise image, thus possibilities of loss of important but trivial texture information could be lessened. The final obtained version will be clearer with rich amount of details instead of being too smooth. Then with these patches, we perform a global classification operation on them so that each patch is classified to a set consisting of patches having similar features to each other. Finally, each patch is updated by weighted averaging of patches in classification set it belongs to. After all these procedures, each patch is reallocated back to its initial position with indexes recorded in extracting procedure.

The main contributions of this paper lie in that we propose a revised version of Non-local means method to make it applicable to image patch-base other than the traditional approach of pixel-base. In order to prove the effectiveness as well as maintain the efficiency of the non-local means algorithm, we perform a pre-clustering procedure to filter out the less connective relationship between image patches.

The rest of this paper is organized as follows. Section 2 mainly focuses on the globally extracted patches clustering procedure. Section 3 gives a brief description of the revised Non-local means method compatible to patch-based procedure. The experiment results of the proposed Non-Local Means are represented in Section 5 and finally, a conclusion and future works are given in Section 6.

2   K-means Patch Clustering

Patch-wise method was initially devised for image texture synthesis [6] and image inpainting [7], and has been proven highly efficient, especially in recent over-complete dictionary learning algorithms.

Fig. 1. Overlapping image patches are extracted, though maybe near in term of position (the red,green and yellow blocks), on an global scope, patches share some common features in the same cluster may located relatively faraway

However, calculating the weights between patches, as will be introduced later, on a global scope requires a large amount of calculating resources, usually $O(N^2)$, depending on the patch size, while for some less similar patches. We perform a pre-clustering step to identify the patches that containing similar structure, resulting in reallocating the patches into several classes to improve the efficiency and lessen the unnecessary weight calculating between less similar patches (see Figure 1).

To perform clustering, low level features such as gradient or gray value...etc, is often used. However, these feature maybe instable due to the presence of noise. In[5], the steering weights computing in a neighborhood that is robust to a significant amount of noise is proposed. These weights are representative of the underlying local data structure to a certain extent. Hence in this paper, we employ a method by vectorizing each patch and calculating their mutual weights to perform clustering. In detail, a certain patch $N(i)$ with size of $\sqrt{N} \times \sqrt{N}$ is reshaped to a $N \times 1$ vector, thus, the original patch is mapped to a high-dimension space where the clustering is performed. The final result of clustering procedure is expected to be that each patch is divided into a specified class containing other elements are patches having similar features to it. Thus, the original noisy image could be denoted as

$$x = \bigcup_{k=1}^{K} \{x_i, x_i \in \Psi_k\} \tag{2}$$

In which, x_i is a specific image patch x_i, with the position of its upperleft corner; k is meant to denote the number of desired classes; Ψ_k signifies the kth class that patch x_i belongs to. We denote $N(\Psi_k)$ the number of elements in kth cluster. Suppose we got an image sized $M \times M$ and the patch size is $m \times m$, the total number of overlapping patches extracted from the image is $PN(x) = (M - m + 1)^2$, then we could obtain that $\sum_{i=1}^{K} N(\Psi_i) = PN(x)$.

In this paper, we chose k – means algorithm to perform the clustering procedure. As the simplest unsupervised clustering algorithm, when applying it to image patches, it simply classifies the image patches to a predefined number of classes (K), so that in each class, the squared distance of any member to the class center is minimized. As for the distance metric, one can just calculate the L_1 or L_2 norm of the feature vector or their weighted version. Since we have chosen the Euclidean distance as the distance metric, the objective function for a clustering procedure would be:

$$L = \sum_{k=1}^{K} \sum_{x_i \in \Psi_k} d^2(x_i - \mu_k) = \sum_{k=1}^{K} \sum_{x_i \in \Psi_k} \|x_i - \mu_k\|^2 \tag{3}$$

where μ_i is the arithmetic center of kth cluster.

3   Patch-Based Non-local Means

In this section, a brief introduction to NL-means algorithm in [10] is presented as well as a revised version compatible to patch-wise denoising algorithm.

Let x_i and y_i be the same patch on a noisy image and its original version respectively, then under the context of image denoising, the estimated version of the ground truth x_i could be computed by a weighted averaging on those in the same cluster as it, which could be represented as

$$\hat{x}_i = \sum_{x_i, x_j \in \Psi_k} w(i, j) x_i \tag{4}$$

where $\hat{x}_i$ is the estimated value of patch i; the weights $w(i, j)$, expressing the similarities between patches, is determined by the similarities between patch i and j, both of which belong to the same cluster Ψ_k, and also the weights satisfy $0 \le w(i, j) \le 1$ and $\sum w(i, j) = 1$.

In order to determine the weight $w(i, j)$ representing the similarity level between two image patches x_i and x_j, we employed a Gaussian formed equation. Normally, this could be interpreted as those more similar patches get higher weights of similarity while on a Euclidian space, they demonstrate shorter distance between each other, which means the level of similarity is inversely proportional to the distance. The weights is then can be calculated by

$$w(i, j) = \frac{1}{Z_i} e^{-\frac{d_{i,j}^2}{h^2}} \tag{5}$$

Where $d_{i,j}^2$ is the Euclidian distance between two patch x_i and x_j, i.e. $d_{i,j}^2 = \left\| x_i - x_j \right\|_2^2$, Z_i is a normalizing factor $Z_i = \sum_{N(\Psi_k)} w(i, j)$ and parameter h controls the decay rate of the weight function.

A complete description of the proposed patch-clustering based Non-local Means denoising algorithm is shown in Table 1 below.

Table 1. The main flow of the proposed method

---

1. **Input:** add white Gaussian noise $n \sim (0, \sigma^2)$ to a clear noise-free image x to generate a noisy image y

2. **Initializing:** set up patch size $\sqrt{N} \times \sqrt{N}$, decay factor h and cluster number K

3. **Denoising processing:**
   - ➤ **Patch Extraction:** Extracting overlapping image patches of predefined size $\sqrt{N} \times \sqrt{N}$ from the obtained noisy image, each indexed by the upperleft position in the image (i, j)
   - ➤ **Patches pre-clustering:** Performing an unsupervised clustering procedure to classify the image patches, which in this step is vectorized to a $N \times 1$ row vector, into a predefined number (K) clusters, each containing the most similar image patches in terms of feature.
   - ➤ **Weights calculation:** Calculating the similarity weights of each patch using the distance between this patch and others in the same cluster.
   - ➤ **Non-local Mean Restoration:** Restore each image patches using Non-local means method.

4. **output:** the estimated image $\hat{x}$

---

4   Experiment Results

In this section, we conduct several experiments on gray value images to validate the proposed method, both qualitatively and quantitatively. The noisy image used for experiment is generated by adding zero mean white Gaussian noise of different standard to the original clear image. The experiments include implementation of several classical as well as state-of-the-art works on the same noisy image. Finally, the method is evaluated quantitatively. The Peak Signal to Noise Ratio (PSNR) is selected as the evaluation metric which is defined by

$$PSNR(x,\hat{x}) = 10\log_{10} \frac{255^2}{\frac{1}{m}\sum_{i=0}^{m-1}(x_i - \hat{x}_j)} \tag{6}$$

Where $\hat{x}$ denotes the estimated denoising image and x denotes the original clear image. m is the total number of pixels in x. The noisy image used for experiment is generated by adding zero mean white Gaussian noise of different standard to the original clear image.

To further verify the effectiveness of the proposed method, we compare the performance of the proposed method to several other algorithms: BM3D [4], patch-based PCA [12], and Non-local means [13]. In the experiment, we use eight pieces of standard test images, and conduct parallel experiments with different level of noise, $\sigma = 5, \sigma = 10$. The parameters in the proposed method have been chosen for each level of noise as those maximizing the PSNR, i.e. windowsize $= 7 \times 7$, and cluster number $k = 10$ for results presented. For the other three methods, we have used the default parameters and the implementations provided by the authors. The results in terms of PSNR are summarized in Table 2.

Table 2. Denoising Results and Comparison

Method		Patch-based PCA			BM3D	NLM	Proposed method
		PGPCA	PHPCA	PLPCA			
Barbara	$\sigma=5$	37.6	38.3	38.5	38.2	37.0	38.1
	$\sigma=10$	33.6	34.5	34.8	34.9	33.0	34.7
Cameraman	$\sigma=5$	37.8	37.8	38.0	38.2	37.7	37.8
	$\sigma=10$	33.3	34.5	33.5	34.0	33.2	33.4
Couple	$\sigma=5$	37.2	37.3	37.4	37.4	36.8	37.1
	$\sigma=10$	33.5	33.5	33.6	34.0	32.7	33.3
House	$\sigma=5$	39.1	39.3	39.5	39.6	38.5	39.0
	$\sigma=10$	35.4	35.7	35.8	36.6	34.8	35.7
Hill	$\sigma=5$	35.8	35.9	36.0	36.0	35.7	35.8
	$\sigma=10$	35.8	31.7	31.8	31.8	31.5	31.4
Peppers	$\sigma=5$	37.7	37.7	37.9	38.0	37.4	27.9
	$\sigma=10$	33.8	33.9	34.1	34.6	33.3	29.9
Lena	$\sigma=5$	38.4	38.7	38.8	38.6	37.9	38.4
	$\sigma=10$	35.3	35.4	35.6	35.9	34.2	35.4
Man	$\sigma=5$	37.4	37.6	37.7	37.7	37.1	37.3
	$\sigma=10$	33.5	33.6	33.7	33.9	32.9	33.4
Fingerprints	$\sigma=5$	36.5	36.6	36.7	36.4	34.8	36.5
	$\sigma=10$	32.2	32.3	32.4	32.4	30.6	32.1

As can been seen from Table 2, the denoising performance of the proposed method outperform the original Non-local Means algorithm on almost all settings showed clearly by comparing the last two columns of table 2, while only some results performs better comparing to patch-based PCA algorithm and still gets a long way to go compared to *the-state-of-the-art* BM3D. The results show that the proposed method does improve the effectiveness compared to the original Non-local Means.

For qualitative results，we select three typical test images, one abundant of edges (*house*), and the other of structure (*boat*), to demonstrate the denoising performance of the proposed method (Figure 2, 3). As can be seen from the results, features on selected subareas are well reconstructed in the results compared to its counterpart in the original images.

| (a) orginal image | (b) denoising with σ=5 | (c) denoising with σ=10 |

Fig. 2. Denoising result on *boat*

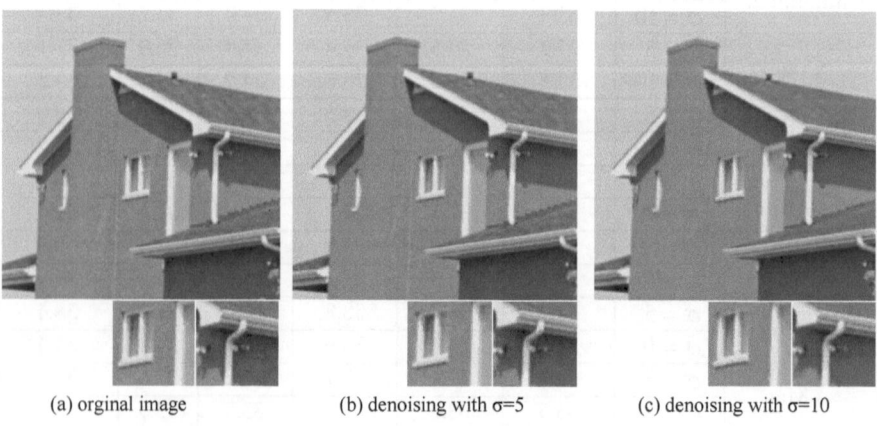

| (a) orginal image | (b) denoising with σ=5 | (c) denoising with σ=10 |

Fig. 3. Denoising results on house

5   Conclusion and Future Works

In the paper, we proposed a revised non-local means algorithm applicable to image denoising problem based on image patches other than pixel in the original non-local means algorithm. The proposed method is evaluated experimentally and compared to some of the state of the art methods for denoising. The results prove our approach to be effective and practical.

The work presented here provides an appealing and promising framework of image denoising. Some important issue should be addressed in our later work would be that:

1. The cluster number may be decided adaptively so that a satisfactory balance would be strike between the calculation efficiency and final performance.
2. Some more informative feature of image patch should be investigated to support a more effective clustering.
3. A more flexible and adaptive patch extraction method should be discovered so that the local feature would be more informative.

References

1. Lindenbaum, M., Fischer, M., Bruckstein, A.M.: On Gabor Contribution To Image-Enhancement. Pattern Recognition 27, 1–8 (1994)
2. Perona, P., Malik, J.: Scale space and edge detection using anisotropic diffusion. IEEE Trans. Pattern. Analysis. Mach. Intell 12, 629–639 (1990)
3. Yaroslavsky, L.P.: Digital Picture Processing - An Introduction. Springer (1985)
4. Dabov, K., Foi, A., Katkovnik, V., Egiazarian, K.O.: Image denoising by sparse 3-D transform domain collaborative filtering. IEEE Trans. Image Process. 16(8), 2080–2095 (2007)
5. Takeda, H., Farsiu, S., Milanfar, P.: Kernel regression for image processing and reconstruction. IEEE Trans. Image Process. 16(2), 349–366 (2007)
6. Efros, A.A., Leung, T.K.: Texture synthesis by non-parametric sampling. In: ICCV, pp. 1033–1038 (1999)
7. Criminisi, A., Pérez, P., Toyama, K.: Region filling and object removal by exemplar-based image inpainting. IEEE Trans. Image Processing 13(9), 1200–1212 (2004)
8. Coifman, R.R., Donoho, D.: Translation-invariant de-noising, Wavelets and Statistics, pp. 125–150. Springer (1995)
9. Zhang, L., Dong, W., Zhang, D., Shi, G.: Two-stage image denoising by principal component analysis with local pixel grouping. Pattern Recognition 43(4), 1531–1549 (2010)
10. Buades, A., Coll, B., Morel, J.M.: On image denoising methods, CMLA Preprint (2004), http://www.cmla.ens-cachan.fr/Cmla/ (to appear in SIAM Multiscale Modeling and Simulation)
11. Bradley, P., Fayyad, U.: Refining initial points for K-means clustering. In: Proc. 15th Int. Conf. Machine Learning, pp. 91–98 (July 1998)
12. Deledalle, C.-A., Salmon, J., Dalalyan, A.: Image denoising with patch based PCA: local versus global. In: BMVC, Britain (2011)
13. Buades, A., Coll, B., Morel, J.-M.: A review of image denoising algorithms, with a new one. Multiscale Model. Simul. 4(2), 490–530 (2005)

Contributions to the Horn-Schunck Optical Flow Equations-Part II : Decoupling via Linear Transformations

Guohua Dong, Xiangjing An, and Dewen Hu

College of Mechatronics and Automation, National University of Defense Technology
Deya Road, Changsha, 410073, China
{ghdong,anxiangjing,dwhu}@nudt.edu.cn

Abstract. The Horn-Schunck optical flow equations, as a coupled system of partial differential equations, are proved to *not* be decouplable under linear transformations. This negative result, on one side, forbids any further attempt in this direction, on the other side, motivates our alternating algorithms which are in decoupled form at each single iteration, as will be carried out in the third part. Both the consequence and limitations are discussed.

Keywords: Optical Flow, Horn-Schunck Equations, Decoupling.

1 Introduction

Recent results [1] we have shown that the Horn-Schunck system is well-posed, mainly due to its variational structure and uniformly elliptic nature, thus it has a unique solution with proper regularity. Furthermore, the approximate solution sequence of the classical algorithm exponentially converges to the true solutions under mild conditions [1]. The system, together with the classical iterative algorithm, are both in coupled forms, i.e., each of the two components of optical flow can not be determined *per se* and depends on the other one. A natural question arises: can we decouple the system via some linear transformation? More precisely, can we rewrite the system in a different form, where the two new unknowns are linear combinations of the old ones, such that the each equation is autonomous?

The motivation of this question comes from both computational and performance analysis considerations. Given positive answer, practical computation and theoretical analysis would become much easier, a common experience shared by practical and theoretical workers. The answer is *no*, as will be shown in this paper.

This question reminds us of several familiar situations: diagonalization in linear algebra, decoupling in linear ordinary differential equations, and decoupling problem in control theory, to mention just three. The first two are contained in almost every textbook on linear algebra; here we only give a brief description of the last one.

The question of decoupling is an extensively studied topic in the control literature [3-5], where decoupling roughly means as follows: given a multi-input and multi-output control system, each output usually dependents on several, sometimes all,

J. Yang, F. Fang, and C. Sun (Eds.): IScIDE 2012, LNCS 7751, pp. 590–596, 2013.
© Springer-Verlag Berlin Heidelberg 2013

inputs, however, after some possible state and input transformation, each output may then depend on a single corresponding input but not on other inputs anymore. If this is the case, then the control system is said to be decouplable and is decoupled by the corresponding transformation. Necessary and sufficient conditions for a given control system being decoupled, and algorithms make a given system to be transformed into decoupled form, are known. While the meaning of decoupling described here is rather restricted and the above description is somewhat vague and brief, readers interested in control theory should refer to the above mentioned books for a complete grasp of this issue.

It is immediately noted that the difference between our situation and the other three: we are dealing with neither linear algebraic nor ordinary differential equations (with or without control inputs), but with system of partial differential equations. Another difference lies in the fact that the order of the Horn-Schunck system is two. These differences indicate that different mathematical method must be adopted: indeed, the method used here is analytic while those used in the mentioned areas are algebraic or geometric in nature.

The organization of this paper is as follows: In Section 2, the Horn-Schunck equations is briefly recalled with notions fixed. In Section 3, the Decoupling Problem is formulated and solved in Section 4. Section 5 concludes the paper with discussions on the limitations of this research and relations with the algorithms proposed in the third part [2] of this series of papers.

2   The Horn-Schunck Optical Flow Equations

For the sake of convenience, we recall some notions used in [1, 2, 6, 7].

At time t, the gray value of the image sequence at location (x, y) is $I(x, y, t)$ (abbreviated as I once there is no risk of confusion). The two components of the optical flow are

$$u(x, y, t) = \tfrac{dx}{dt}, v(x, y, t) = \tfrac{dy}{dt} \tag{1}$$

where (x, y) corresponds to some point on a moving object.

The Horn-Schunck optical flow equations [6, 7] are

$$\begin{aligned} \Delta u &= \lambda[I_x u + I_x v + I_t]I_x \\ \Delta v &= \lambda[I_x u + I_x v + I_t]I_y \end{aligned} \tag{2}$$

where $\Delta = \nabla^2 = \frac{\partial^2}{\partial x^2} + \frac{\partial^2}{\partial y^2}$ is the Laplacian. The system is the Euler-Lagrange system of the Dirichilet functional

$$\iint [|\nabla u|^2 + |\nabla v|^2] dxdy + \lambda \iint [I_x u + I_y v + I_t]^2 dxdy \tag{3}$$

where $\lambda > 0$ is the regularization parameter which tends to balance the two requirements, i.e., smoothness of the optical flow and the basic constraint on (u, v).

3   Formulation of the Linear Decoupling Problem

Decoupling Problem for the Horn-Schunck equations asks whether there exists, if any, transformation of the following form

$$u_1 = \alpha_{11}u + \beta_{12}v$$
$$v_1 = \alpha_{21}u + \beta_{22}v$$

(4)

such that the *coupled* system (2) can be written into the following *decoupled* form?

$$u_{1,xx} + c_{12}u_{1,xy} + c_{22}u_{1,yy} + c_{10}u_{1,x} + c_{01}u_{1,y} + c_0 u_1 + c = 0$$
$$v_{1,xx} + d_{12}v_{1,xy} + d_{22}v_{1,yy} + d_{10}v_{1,x} + d_{01}v_{1,y} + d_0 v_1 + d = 0$$

(5)

Naturally, the undetermined $\alpha_{11}, \alpha_{21}, \beta_{12}$ and β_{22} are C^2 functions of (x, y). The regularity requirements guarantee that all partial derivatives up to order two exist, and are such that the transformation is invertible. Note that we have not make any assumption on the unknown functions c's and d's. Now the difficulty lies in the facts that the original equations are coupled partial differential equations of order two, and that a lot of unknown coefficient function are involved: indeed, there are sixteen altogether!

4   Solution of the Linear Decoupling Problem

If the linear decoupling problem for the Horn-Schunck equations is solvable, some constraints on the equations, i.e. on coefficients, especially on the images, are expected to arise. This is indeed the case, as will be shown now.

Suppose a linear combination of u and v, say, $w = \alpha u + \beta v$ (where α and β being some functions of (x, y) to be determined), satisfied the flowing general form of *autonomous* partial differential equation of order two

$$w_{xx} + c_{12}w_{xy} + c_{22}w_{yy} + c_{10}w_x + c_{01}w_y + c_0 w + c = 0$$

(6)

Obviously, if this is true, w can be solved independently and then u and v can be deduced from each other. Without loss of generality, suppose $\alpha \neq 0$, when $u = {(w-\beta v)}/{\alpha}$ is inserted back into each of the original two equations, *two* equations of v will be obtained. These two equations of v must thus be consistent and the redundancy in the consistency conditions will give the expected constraints. The details are as follows.

Note first that eight coefficients, $(\alpha, \beta, c_{12}, c_{22}, c_{10}, c_{01}, c_0, c)$, are involved. All these unknown must be eventually canceled in order to obtain the desired constraints.

Insert $w = \alpha u + \beta v$ into the above assumed equation of w, after some easy but tedious computation, we obtain

$$u[\alpha_{xx} + c_{12}\alpha_{xy} + c_{22}\alpha_{yy} + c_{10}\alpha_x + c_{01}\alpha_y + c_0\alpha]$$
$$+v[\beta_{xx} + c_{12}\beta_{xy} + c_{22}\beta_{yy} + c_{10}\beta_x + c_{01}\beta_y + c_0\beta]$$
$$+u_x[2\alpha_x + c_{12}\alpha_y + c_{10}\alpha] + v_x[2\beta_x + c_{12}\beta_y + c_{10}\beta] \tag{7}$$
$$+u_y[c_{12}\alpha_x + 2c_{22}\alpha_y + c_{01}\alpha] + v_y[c_{12}\beta_x + 2c_{22}\beta_y + c_{01}\beta]$$
$$+[u_{xx} + c_{12}u_{xy} + c_{22}u_{yy}]\alpha + [v_{xx} + c_{12}v_{xy} + c_{22}v_{yy}]\beta + c = 0.$$

From the Horn-Schunck equations (2) we have

$$\alpha(u_{xx} + u_{yy}) + \beta(v_{xx} + v_{yy}) = \lambda(I_xu + I_yv + I_t)(\alpha I_x + \beta I_y) \tag{8}$$

The last two equations must be the *same*, and we therefore obtain *nine* constraints among the *eight* unknowns, with one redundant, as expected,

$$c_{12} = 0, c_{22} = 1, c = -\lambda I_t(\alpha I_x + \beta I_y)$$
$$2\alpha_x + c_{10}\alpha = 0, 2\beta_x + c_{10}\beta = 0$$
$$2\alpha_y + c_{01}\alpha = 0, 2\beta_y + c_{01}\beta = 0 \tag{9}$$
$$-\lambda I_x(\alpha I_x + \beta I_y) = \Delta\alpha + c_{10}\alpha_x + c_{01}\alpha_y + c_0\alpha$$
$$-\lambda I_y(\alpha I_x + \beta I_y) = \Delta\beta + c_{10}\beta_x + c_{01}\beta_y + c_0\beta$$

Since three of the unknowns, i.e. c_{12}, c_{22} and c, have already been determined, the rest *five*, i.e. $\alpha, \beta, c_{10}, c_{01}$ and c_0 must satisfy other *six* constraints.

From the fourth to the seventh constraints in (9) we can deduce that

$$\frac{\partial c_{10}}{\partial x} = \frac{\partial c_{01}}{\partial y} \quad \text{(from the equality between mixed partials of order two)} \tag{10}$$

and

$$\nabla \ln \alpha = -\frac{1}{2}\begin{bmatrix} c_{10} \\ c_{01} \end{bmatrix} = \nabla \ln \beta, \quad \text{thus } \frac{\alpha}{\beta} = S \quad (S \text{ being some constant}) \tag{11}$$

Solve all the partials of order one in the fourth, fifth, sixth, and seventh constraint in (9) and insert them into the last two constraints, we find that all partials vanish and obtain

$$\alpha[-\frac{1}{2}\frac{\partial c_{10}}{\partial x} - \frac{1}{2}\frac{\partial c_{01}}{\partial y} + (c_0 - \frac{c_{10}^2 + c_{01}^2}{4}) + \lambda I_x^2] + \beta\lambda I_xI_y = 0$$
$$\alpha\lambda I_xI_y + \beta[-\frac{1}{2}\frac{\partial c_{10}}{\partial x} - \frac{1}{2}\frac{\partial c_{01}}{\partial y} + (c_0 - \frac{c_{10}^2 + c_{01}^2}{4}) + \lambda I_y^2] = 0 \tag{12}$$

These are *linear algebraic equations* in (α, β). Necessary and sufficient condition for the system to have a nontrivial solution is that the determinant is zero, i.e.

$$(\lambda I_xI_y)^2 = [\lambda I_x^2 + (c_0 - \frac{c_{10}^2 + c_{01}^2}{4}) - \frac{1}{2}\frac{\partial c_{10}}{\partial x} - \frac{1}{2}\frac{\partial c_{01}}{\partial y}][\lambda I_y^2 + (c_0 - \frac{c_{10}^2 + c_{01}^2}{4}) - \frac{1}{2}\frac{\partial c_{10}}{\partial x} - \frac{1}{2}\frac{\partial c_{01}}{\partial y}]$$

Denote $\gamma = (c_0 - \frac{c_{10}^2 + c_{01}^2}{4}) - \frac{1}{2}\frac{\partial c_{10}}{\partial x} - \frac{1}{2}\frac{\partial c_{01}}{\partial y}$, we obtain immediately that

$$\gamma = 0 \quad \text{or} \quad \gamma = -\lambda |\nabla I|^2 \tag{13}$$

Let us summarize the results obtained up to now: besides three functions those have been determined, i.e. c_{12}, c_{22} and c, the remaining five $\alpha, \beta, c_{10}, c_{01}$ and c_0 satisfy the following six constraints (where S is some constant):

$$\frac{\partial c_{10}}{\partial x} = \frac{\partial c_{01}}{\partial y}$$

$$\gamma = (c_0 - \frac{c_{10}^2 + c_{01}^2}{4}) - \frac{1}{2}\frac{\partial c_{10}}{\partial x} - \frac{1}{2}\frac{\partial c_{01}}{\partial y} = 0 \quad \text{or} \quad -\lambda |\nabla I|^2$$

$$\frac{\alpha}{\beta} = S, \nabla \ln \alpha = \nabla \ln \beta = -\frac{1}{2}\begin{pmatrix} c_{10} \\ c_{01} \end{pmatrix} \tag{14}$$

$$\alpha(\gamma + \lambda I_x^2) + \beta \lambda I_x I_y = 0$$

$$\alpha \lambda I_x I_y + \beta(\gamma + \lambda I_y^2) = 0$$

Insert $\frac{\alpha}{\beta} = S$ into the last two equations, we find

$$\frac{\alpha}{\beta} = S = -\frac{\lambda I_x I_y}{\gamma + \lambda I_x^2} = -\frac{\gamma + \lambda I_y^2}{\lambda I_x I_y} \tag{15}$$

Recall that $\gamma = 0$ or $\gamma = -\lambda |\nabla I|^2$, we have finally

$$S = -\frac{I_x}{I_y} \quad \text{or} \quad -\frac{I_y}{I_x} \quad \text{(some constant)} \tag{16}$$

This is an obvious contradiction since both of these equalities are impossible for *generic* images; therefore we arrive at the conclusion that there does *not* exist any nontrivial linear combination $w = \alpha u + \beta v$ such that w satisfies an autonomous partial differential equation of order two. This certainly implies that the Horn-Schunck system of equations can not be decoupled via linear transformation.

Remark: In the above derivation, we use implicitly that both α and β are not zero when viewed as functions, since otherwise the linear combination would be trivial, however, does there exist any chance that they take values of zero simultaneously at some (x, y)? This is also impossible since this will lead to the conclusion that the transformation from the pair u, v to the pair w, w', say, is not invertible, another contradiction.

We summarize the conclusion obtained hitherto in the following theorem.

Theorem. For *generic* images the Horn-Schunck equations can *not* be decoupled via linear transformation.

5   Conclusions and Discussions

We have shown that linear transformation of u, v definitely will not lead to decoupling of the Horn-Schunck equations for generic images. Our method can easily adapt to other system of PDEs in the optical flow literature.

We have not exhausted all possibilities such as nonlinear transformations and/or changes in the independent variables, so our conclusion has a restricted meaning. We conjecture that the same conclusion still holds but we have not proven at current time.

Return back to our original interest, i.e. the algorithmic aspect of the Horn-Schunck equations. Given the negative conclusion, does it mean that we can not devise an algorithm in some decoupled form? Our answer is *no* even if the above general conjecture holds. Indeed, in the next part of this series of papers, we proposed two alternating algorithms and both take the *decoupled* forms at each step of iterations. These new algorithms take the following forms

Alternating Algorithm 1:
$$\Delta u^{(n)} - \lambda I_x^2 u^{(n)} = \lambda (I_y v^{(n-1)} + I_t) I_x$$
$$\Delta v^{(n)} - \lambda I_y^2 v^{(n)} = \lambda (I_x u^{(n-1)} + I_t) I_y$$
(17)

Alternating Algorithm 2:
$$\Delta u^{(n)} = \lambda [I_x u^{(n-1)} + I_y v^{(n-1)} + I_t] I_x$$
$$\Delta v^{(n)} = \lambda [I_x u^{(n-1)} + I_y v^{(n-1)} + I_t] I_y$$
(18)

At each step, the two components of approximate optical flow can be solved independently in explicit and analytical form. The existence of these decoupled algorithms does not contradict the main Theorem of this paper since "decoupling" now has different meaning: the solution sequence of the each component surely depends on the initial value of the other component, so when we look the iteration process as a whole, they are in fact coupled though they are in decoupled forms at each single iteration step. The "freezing" of the previous approximate solutions leads to "temporary decoupling". Convergence, exponential stability and other properties of these algorithms are also established in that paper [2]. These results, together with the well-posedness result obtained in [1], imply that limits of both sequences of the approximate solutions uniquely exist and do not depend on special choices of initial optical flow.

In the previous and the current part, we freely take derivatives of the images but do not explicitly make assumptions on the proper regularity of natural images. This is currently a hot and controversial topic in the computer vision filed; a brief and relevant surrey has been given at the Section 5 of the previous part [1].

Acknowledgments. The author is grateful to the three anonymous referees for careful checking of the details and for helpful comments that greatly improved this paper. One of referee suggested that "it would be better if some experiments are presented to illustrate the method". Experiments have in fact been carried out for the classical Horn-Schunck algorithm in the first part of this series where assertions on both convergence and exponential stability were validated under nine different choices of algorithmic parameters, and the approximate sequence of solutions did approach to the ground truth.

This research is supported by National Science Fund under Grant No. NSF60835005 and NSF90820302, and supported by 973 Plan Fund under Grant No. 2007CB311001.

References

1. Dong, G.H., An, X.J., Hu, D.W.: Horn-Schunck optical flow equations. Part I: Stability and Rate of Convergence of the Classical Algorithm (accepted by Journal of Central South University, under minor modification)
2. Dong, G.H., An, X.J., Hu, D.W.: Horn-Schunck optical flow equations. Part III: Alternating Iteration Algorithms (submitted)
3. Wonham, W.M.: Linear Multivariable Control: A Geometric Approach. Springer, Berlin (1979)
4. Nijmeijer, H., van der Schaft, A.: Nonlinear Dynamical Control Systems. Springer, New York (1990)
5. Isidori, A.: Nonlinear Control Systems, 3rd edn. Springer, London (1995)
6. Horn, B.K.P., Schunck, B.G.: Determining Optical Flow. AI 17, 185–203 (1981)
7. Wu, L.D.: Compuetr Vision. Fudan University Press, Shanghai (1993) (in Chinese)
8. Chen, Y.Z., Wu, L.C.: Elliptic Partial Differential Equations and Systems of Order Two. China Academic Press, Beijing (1997) (in Chinese)
9. Courant, R., Hilbert, D.: Methods of Mathematical Physics, vol. 2. Interscience, New York (1962)

Fast Restoration of Nonuniform Blurred Images

Hong Deng, Wangmeng Zuo, and Hongzhi Zhang

Biocomputing Research Centre, School of Computer Science and Technology,
Harbin Institute of Technology, Harbin, China
denghong_hit@163.com, {cswmzuo,zhanghz0451}@gmail.com

Abstract. Nonuniform blurring is general for image degradation. Either defocus, camera shaking, or motion would result in nonuniform blurring. However, most current image restoration algorithms were developed for restoration from image blurred with one single space-invariant convolution kernel. The computational inefficiency would be significant if we directly extend these algorithms for restoration of nonuniform blurred image. In this paper, we propose a novel fast restoration algorithm for restoration of nonuniform blurred images. In our method, we first model nonuniform blurring as a space-variant weighted summation of images blurred by a group of basis filters, and use principal component analysis (PCA) to obtain the basis filters in advance. Then, based on the total variation (TV) based model, we adapt the generalized accelerated proximal gradient (GAPG) algorithm for image restoration. Experimental results indicate that the proposed method can dramatically improve the computational efficiency while achieving satisfactory restoration performance.

Keywords: nonuniform blurring, restoration, generalized accelerated proximal gradient, total variation.

1 Introduction

Image restoration aims to recover the original image from a corrupted and degraded input, and has been a long-standing problem in a broad field of image processing. Even if the blur kernel is known, image restoration task, such like deblurring, denoising, super-resolution and defocus, is still a challenging ill-posed problem, and has been intensively studied over the last four decades.

In many image restoration tasks, e.g., motion deblurring, and shape from defocus, it is more proper to model the image degradation process as non-uniform blurring rather than space-invariant convolution. In blind deconvolution, Levin et al. had shown that real camera shake cannot be treated as one single spatially uniform filter [1]. In shape from defocus, the defocused image can be regarded as the non-uniform blurring of the intrinsic image [2]. As a result, nonuniform deblurring methods are highly desired in image processing applications.

Image deconvolution is a kind of image restoration problem where the known blur kernel is space-invariant. Among various image deconvolution algorithms, augmented Lagrange (AL) method and fast gradient-based method had received

J. Yang, F. Fang, and C. Sun (Eds.): IScIDE 2012, LNCS 7751, pp. 597–604, 2013.

considerable recent research interests. With the help of operator splitting, a number of state-of-the-art AL methods have been developed for image deconvolution [3]. These methods usually should solve a matrix inverse problem. Unfortunately, for nonuniform blurred image deblurring, since the image size is huge and the blurring is nonuniform, it would be impossible to directly utilize FFT and be very time consuming for solving the matrix inverse problem. Thus, we do not consider to use AL methods for restoration of nonuniform blurred image in this paper. Fast gradient-based method is another kind of method which has been intensively studied recently. In early work, Chambolle et al. proposed a gradient-based method, the iterative shrinkage thresholding algorithm (IST) [4], which has been frequently adopted for deconvolution due to its simplicity and efficiency. Beck and Teboulle suggested an accelerated algorithm, the fast iterative shrinkage thresholding algorithm (FISTA) [5], which belongs to the family of accelerated proximal gradient (APG) algorithms. Most recently, Zuo and Lin proposed a GAPG algorithm to accelerate the convergence rate of APG algorithm [6]. For gradient based algorithm, we do not required to solve the matrix inverse problem, and only need to compute the gradient term. Thus, to extend fast gradient method to cope with the restoration problem from nonuniform blurred images, we need to develop appropriate models and algorithms for computing the blur operator and its adjoint operator.

In computer vision and image processing, several models had been suggested for modeling nonunifiorm blurring. In computer vision, Tai et al. proposed a projective motion path blur model, where camera ego motion is modeled as the average of multiple planar projective transformed images [7]. Whyte et al. proposed a parameterized geometric blurring model for camera motion [8]. These methods, however, are focused on the effectiveness for modeling camera ego motion, but consider little on the computational efficiency of the blur model. In image processing, several methods has been proposed for modeling nonuniform blurring. The filter banks - based method [9] models nonuniform blurring by discretizing the blur levels, and thus is not suitable for smooth nonuniform blurring. Recently, based on PCA, Popkin et al. [10] proposed a fast algorithm to approximate space-variant Gaussian blurring with high-precision.

In this paper, we propose a novel efficient method for TV-based restoration of nonuniform blurred images. First, we adapt the method proposed by Popkin et al. [10] for modeling general nonuniform blurring. Then we extend the GAPG algorithm for the restoration of nonuniform blurred images. Moreover, experimental results show the effectiveness and efficiency of the proposed method.

The remainder of this paper is organized as follows. Section 2 describes the fast restoration algorithm based on GAPG method for the restoration of nonuniform blurred images. Section 3 presents our model for the nonuniform blur operator and the adjoint operator. Section 4 provides the experimental results of deblurring the camera motion blurred image and Gaussian blurred image using the proposed method. Finally, Section 5 concludes the paper.

2   Restoration of Nonuniform Blurred Images

2.1   Nonuniform Blurring

Given an $m \times n$ gray-level image $\mathbf{x}$ with the blur kernel $k_{i,j}$ for the (i,j)th pixel, the gray level of the blurred image $\mathbf{b}$ can be defined as,

$$b_{i,j} = (\mathbf{x} * \mathbf{k}_{i,j})_{i,j} \tag{1}$$

where $\mathbf{k}_{i,j}$ is the point spread function (PSF) of the (i,j)th pixel of the image. By introducing an $m \times n$ 0-1 weighting matrix $\mathbf{w}_{i,j}$ for the (i,j)th pixel,

$$\mathbf{w}_{i,j}(k,l) = \begin{cases} 1, & \text{if } k = i \text{ and } l = j \\ 0, & \text{else} \end{cases} \tag{2}$$

nonuniform blurring can be modeled as,

$$\mathbf{b} = F(\mathbf{x}, \mathbf{m}) = \sum_{i=1}^{m} \sum_{j=1}^{n} \mathbf{w}_{i,j} \circ (\mathbf{x} * \mathbf{k}_{i,j}) \tag{3}$$

where $\circ$ denote the element-wise multiplication operator. Nonuniform blurring is also a linear operator. Generally, we can rewrite the process as,

$$\mathcal{A}_{\mathbf{m}}(\mathbf{x}) = \sum_{i=1}^{m} \sum_{j=1}^{n} \mathbf{w}_{i,j} \circ (\mathbf{x} * \mathbf{k}_{i,j}) = \mathbf{A}_{\mathbf{m}} \mathbf{x} \tag{4}$$

Lemma 1. $\mathcal{A}_{\mathbf{m}}(\mathbf{x})$ is a linear operator on $(\mathbf{x})$.

Proof. For $\forall \mathbf{x}_1, \mathbf{x}_2 \in \mathbb{R}^{mn}$, and $\forall \alpha_1, \alpha_2 \in \mathbb{R}$,

$$\mathcal{A}_{\mathbf{m}}(\alpha_2 \mathbf{x}_1 + \alpha_1 \mathbf{x}_2) = \sum_{i=1}^{m} \sum_{j=1}^{n} \mathbf{w}_{i,j} \circ [(\alpha_2 \mathbf{x}_1 + \alpha_1 \mathbf{x}_2) * \mathbf{k}_{i,j}]$$

$$= \sum_{i=1}^{m} \sum_{j=1}^{n} \mathbf{w}_{i,j} \circ \{[\alpha_1 (\mathbf{x}_1 * \mathbf{k}_{i,j})] + [\alpha_2 (\mathbf{x}_2 * \mathbf{k}_{i,j})]\} \tag{5}$$

$$= \alpha_1 \mathcal{A}_{\mathbf{m}}(\mathbf{x}_1) + \alpha_2 \mathcal{A}_{\mathbf{m}}(\mathbf{x}_2)$$

The second quality follows from the fact that convolution is a linear operator, and the result proves that $\mathcal{A}_{\mathbf{m}}$ is a linear operator.   □

Based on the definition of $\mathbf{A}_{\mathbf{m}}$, we can obtain the following inequality:

Proposition 1. $\|\mathbf{A}_{\mathbf{m}}\|_2 \leq 1$

Proof. In matrix analysis [11], we have:

$$\|\mathbf{A}_{\mathbf{m}}\|_2 \leq \sqrt{\|\mathbf{A}_{\mathbf{m}}\|_1 \|\mathbf{A}_{\mathbf{m}}\|_\infty} \tag{6}$$

where $\|\cdot\|_1$ and $\|\cdot\|_\infty$ denote the 1-norm and ∞-norm of the matrix. First, $\|\cdot\|_\infty$ is defined as the largest l_1-norm of the rows of the matrix. According to Eq.(1),

$$b_{i,j} = \sum_{k,l} x_{k,l} \mathbf{k}_{i,j}(i-k, j-l) \tag{7}$$

So the l_1-norm of the row of the matrix $\mathbf{A_m}$ that corresponds to the (i,j) th entry of $\mathbf{b}$ is,

$$\sum_{k,l} |\mathbf{k}_{i,j}(i-k, j-l)| = \|\mathbf{k}_{i,j}\|_{l_1} = 1 \tag{8}$$

Hence, $\|\mathbf{A_m}\|_\infty = 1$. On the other side, $\|\cdot\|_1$ is defined as the largest l_1-norm of the rows of the matrix.

The adjoint operator of $\mathbf{A_m}$ is defined as: $x_{i,j} = \sum_{k,l} b_{k,l} \mathbf{k}_{i,j}(k-i, l-j)$. So the l_1-norm of the column of the matrix $\mathbf{A_m}$ that corresponds to the (i,j) th entry of $\mathbf{x}$ is,

$$\sum_{k,l} |\mathbf{k}_{i,j}(k-i, l-j)| = \|\mathbf{k}_{i,j}\|_{l_1} = 1 \tag{9}$$

Hence, $\|\mathbf{A_m}\|_1 = 1$. Thus, the inequality is proved.   □

2.2   Image Restoration Model and the GAPG Solution

The real process of nonuniform blur can be modeled as,

$$\mathbf{b} = \mathbf{A_m}\mathbf{x} + \varepsilon \tag{10}$$

where ε is an additive normally distributed noise. Because the linear operator $\mathcal{A}_\mathbf{m}$ is ill-conditioned, typical image restoration model usually includes a confidence term and a regularization term. Here we choose the isotropic TV regularizer, and image restoration is formulated as the following convex optimization problem,

$$\min_{\mathbf{x}} \tfrac{1}{2} \|\mathbf{A_m}\mathbf{x} - \mathbf{b}\|_2^2 + \lambda \|(\mathbf{D}_v\mathbf{x}, \mathbf{D}_h\mathbf{x})\|_{iT} \tag{11}$$

where the isotropic TV is defined the same as in [6].

Rather than directly solving the problem defined in Eq.(11), we introduce two variables $\mathbf{d}_v$ and $\mathbf{d}_h$ to approximate $\mathbf{D}_v\mathbf{x}$ and $\mathbf{D}_h\mathbf{x}$, respectively, and define $\hat{\mathbf{x}}^T = [\mathbf{x}^T, \mathbf{d}_v^T, \mathbf{d}_h^T]^T$. According to [6], the diagonal matrix $\mathbf{L}_f$ in GAPG is defined as,

$$\mathbf{L}_f = \mathrm{diag}\,(\lambda_{\max}\mathbf{I}, \eta\mathbf{I}, \eta\mathbf{I}) \tag{12}$$

where $\lambda_{\max} \leq \left(\sqrt{\mu}\|\mathbf{A_m}\|_2 + \sqrt{\eta}\|\mathbf{D}_h\|_2 + \sqrt{\eta}\|\mathbf{D}_v\|_2\right)^2$ with $\eta \leq 2$, $\|\mathbf{D}_h\|_2 \leq 2$, and $\|\mathbf{D}_v\|_2 \leq 2$. According to Proposition 1, $\|\mathbf{A_m}\|_2 \leq 1$.

Given the diagonal Lipschitz matrix $\mathbf{L}_f$ defined in Eq.(12), our GAPG-based algorithm for the restoration of nonuniform Gaussian blurred images is exactly the same with that in [6] except the updating of $\mathbf{x}$, which is updated by solving the following sub-problem:

$$\min_{\mathbf{x}} \frac{\lambda_{\max}}{2} \|\mathbf{x} - \{\mathbf{y}_\mathbf{x}^k - \lambda_{\max}^{-1} [\mathbf{A}_\mathbf{m}^T (\mathbf{A_m}\mathbf{y}_\mathbf{x}^k - \mathbf{b}) $$
$$+ \mathbf{D}_v^T (\mathbf{D}_v\mathbf{y}_\mathbf{x}^k - \mathbf{y}_{\mathbf{d}_v}^k) + \mathbf{D}_h^T (\mathbf{D}_h\mathbf{y}_\mathbf{x}^k - \mathbf{y}_{\mathbf{d}_h}^k)]\}\|_F^2 \tag{13}$$

where the definition of $\mathbf{y}_\mathbf{x}^k, \mathbf{y}_{\mathbf{d}_v}^k, \mathbf{y}_{\mathbf{d}_h}^k$ are the same as in [6].

3   Efficient Computation of the Nonuniform Blurring Model

Direct computation of $\mathbf{A}_m\mathbf{x}$ and $\mathbf{A}_m^T\mathbf{b}$ is computationally expensive. In this section, we modify the method in [10] for fast nonuniform blurring, and then extend it for the fast computation of $\mathbf{A}_m^T\mathbf{b}$. Finally, we discuss the computational complexity of the fast scheme.

3.1   Fast Nonuniform Blurring

In this subsection, we use principal component analysis (PCA) to model the PSFs in a compact subspace.

Given the blur kernel parameters,we first calculate the mean of PSFs as,

$$\mathbf{k}_0 = \frac{1}{m * n} \sum_{i,j} \mathbf{k}_{i,j} \tag{14}$$

Then, we define the covariance matrix $\mathbf{S}_t$ as,

$$\mathbf{S}_t = \sum_{i,j} (\mathbf{k}_{i,j} - \mathbf{k}_0)(\mathbf{k}_{i,j} - \mathbf{k}_0)^T \tag{15}$$

Based on eigenvalue decomposition theory, we have,

$$\mathbf{S}_t = \mathbf{U}\mathbf{\Lambda}\mathbf{U}^T \tag{16}$$

where $\mathbf{U} = [\mathbf{k}_1, \mathbf{k}_2, \ldots, \mathbf{k}_d]$, $\mathbf{k}_p$ is the p th eigenvector, and $\mathbf{\Lambda} = diag(\lambda_1, \lambda_2, \ldots, \lambda_d)$, $(\lambda_d > 0)$, λ_p is the p th positive eigenvalue.

Given $[\mathbf{k}_0, \mathbf{k}_1, \mathbf{k}_2, \ldots, \mathbf{k}_d]$, each PSF can be represented as,

$$\mathbf{k}_{i,j} = \mathbf{k}_0 + \mathbf{U}\mathbf{w}^{(i,j)} \tag{17}$$

where $\mathbf{w}^{(i,j)}$ is a $d \times 1$ vector defined by $\mathbf{w}^{(i,j)} = \mathbf{U}^T(\mathbf{k}_{i,j} - \mathbf{k}_0)$.

Then we can reorganize all $\mathbf{w}^{(i,j)}$ with respect to (i,j) to d $m \times n$ matrixes, $\mathbf{w}_1, \mathbf{w}_2, \ldots, \mathbf{w}_d$, with

$$\mathbf{w}_q(i,j) = \mathbf{w}^{(i,j)}(q), 1 \leq i \leq m, 1 \leq j \leq n, 1 \leq q \leq d \tag{18}$$

We further introduce a $m \times n$ matrix $\mathbf{w}_0$:

$$\mathbf{w}_0(i,j) = 1 \tag{19}$$

The nonuniform blurring is equivalent to,

$$\mathbf{b} = \mathbf{A}_m\mathbf{x} = \sum_{q=0}^{d} \mathbf{w}_q \circ (\mathbf{x} * \mathbf{k}_q) \tag{20}$$

For practical application, we need not to keep all the positive eigenvalues, but rather to choose first t components that corresponds to the normalized energy larger then 99%. The normalized energy of the first i eigenvalue is defined as

$$\hat{E}(j) = \left(mn + \sum_{p=1}^{j} \lambda_p\right) \Big/ \left(mn + \sum_{p=1}^{d} \lambda_p\right) \tag{21}$$

In our experiment, we use $t = 4$ for Gaussian nonuniform blurring and y axis rotation, $t = 12$ for z axis rotation according to Eq.(21).

3.2   Fast Computation of $\mathbf{A_m^T b}$

One can easily derive the adjoint operator of $\mathbf{A_m}(\cdot)$

$$\mathbf{A_m^T x} = \sum_{q=0}^{d} (\mathbf{w}_q \circ \mathbf{x}) * \mathbf{k}_q^* \qquad (22)$$

where $\mathbf{k}_q^*$ is the adjoint operator of $\mathbf{k}_q$.

3.3   Computational Complexity

Based on the nonuniform Gaussian blurring, we discuss the computational complexity of the proposed model. The computational complexity of the direct model in Eq.(3) is $O((2w+1)^2 mn)$, while the proposed model defined in Eq.(20) is $O(\max(\log m, \log n)tmn)$. In practical applications, the filter size is large, $(2w+1)^2$ usually is much higher than $\max(\log m, \log n)t$, and thus the proposed model can significantly enhance the computational efficiency.

4   Experimental Results

In this section, we conduct experiments on Lena image with nonuniform blurring to verify the efficiency of the proposed method. First, we apply the proposed method for the restoration of nonuniform Gaussian blurred image, and compare it with the direct model. In nonuniform Gaussian blurring, the variance parameter σ is space-variant, and thus can be represented by an blur map $\mathbf{m}$. So we model nonuniform Gaussian blurring using 15×15 Gaussian PSFs constructed according to the parameters in the three maps shown in Fig. 1. Second, we apply the proposed method for the restoration of nonuniform blurred images caused by camera shake, and compare the proposed method with the project homography based method proposed by Whyte et al. [8]. The blurred images are contaminated by normally distributed noise with mean zero and standard deviation 10^{-3}. In the main algorithm, the parameters were set the same as in [6].

Fig. 1 shows the synthetic degraded Lena images and the restoration results obtained by the proposed method. One can see that the proposed method can obtain satisfactory restoration quality. Fig. 2 shows the restoration results by the proposed method and Whyte et al.'s method. For the blurring caused by either z axis or y axis rotation, the proposed method can achieve satisfactory restoration results, and be much more efficient than Whyte et al.'s method. Table 1 lists the peak signal-noise ratio (PSNR) and the CPU time of two restoration methods. In

Table 1. PSNR values and CPU time

Image-maps	Direct Model		Proposed	
	PSNR	Time	PSNR	Time
Lena-map1	29.51	379.07s	29.56	15.11s
Lena-map2	33.43	369.83s	33.51	15.62s
Lena-map3	31.60	367.49s	31.66	15.59s

Fig. 1. First line left to right: image blurred with map 1-foveation map with parameter range:$\sigma \in [1/3, 5]$; image blurred with map 2-blur increasing steadily from left to right; image blurred with map 3-based on the visual fields of a glaucoma patient; Second line: corresponding restoration result. The corresponding blur maps are located in the bottom-right corner of the blurred images.

Fig. 2. Comparation on the restoration results of camera motion blurred images. Left to right: The restoration results of the proposed method for images blurred by z axis rotation and y axis rotation; the corresponding restoration results using Whyte et al.'s method.

the direct model, we use the simple pixel-by-pixel strategy to calculate $\mathbf{A_m x}$ and $\mathbf{A_m^T b}$. In the proposed model, we use the model defined in Eq. (21) in calculating $\mathbf{A_m x}$ and $\mathbf{A_m^T b}$. As shown in Table 1, although these two methods can obtain similar restoration quality, the proposed method is much more efficient than the direct model. Table 2 lists the CPU time and the PSNR values of Whyte et al.'s and the proposed methods.

Table 2. PSNR values and CPU time

Rotation	Whyte et al.		Proposed	
	PSNR	Time	PSNR	Time
z-axis rotation deblur	24.01	237.43s	30.46	28.86s
y-axis rotation deblur	20.99	238.94s	24.37	14.28s

5   Conclusions

In this paper, we proposed a fast method for the TV-based restoration of nonuniform blurred images. First, since the process of nonuniform blur cannot be modeled by a circular matrix, conventional ALM methods cannot be adopted for image restoration. Thus, we use a fast gradient-based method, GAPG. Second, since the direct computation of nonuniform blurring is computationally expensive, we proposed a PCA-based method where the nonuniform blurring process is modeled by the weighted summation of the images filtered by a set of atom filters. Moreover, the proposed method can be directly extended to other gradient-based image deblurring algorithms and can cooperate with various kinds of regularization terms. Experimental results show that the proposed method can be used for the restoration of many types of nonuniform blurred images, e.g., nonuniform Gaussian blurring and camera shake, and obtain satisfactory restoration results with high computational efficiency.

Acknowledgment. The work is partially supported by the NSFC funds of China under Contract No.s 60902099, 61001037, and 61071179, and Harbin special funds for innovative talents of science and technology research project.

References

1. Levin, A., Weiss, Y., Durand, F., et al.: Understanding and evaluating blind deconvolution algorithms. In: Computer Vision and Pattern Recognition, pp. 1964–1971. IEEE Press, Florida (2009)
2. Favaro, P., Soatto, S.: A geometric approach to shape from defocus. IEEE Transactions on Pattern Analysis and Machine Intelligence 27, 406–417 (2005)
3. Afonso, M.V., Bioucas-Dias, J.M., Figueiredo, M.A.T.: An Augmented Lagrangian Approach to the Constrained Optimization Formulation of Imaging Inverse Problems. IEEE Transactions on Image Processing 20, 681–695 (2011)
4. Chambolle, A., De Vore, R.A., Lee, N.Y., Lucier, B.J.: Nonlinear wavelet image processing: Variational problems, compression, and noise removal through wavelet shrinkage. IEEE Transactions on Image Processing 7, 319–335 (1998)
5. Beck, A., Teboulle, M.: A fast iterative shrinkage-thresholding algorithm for linear inverse problems. SIAM Journal on Imaging Sciences 2, 183–202 (2009)
6. Zuo, W., Lin, Z.: A Generalized Accelerated Proximal Gradient Approach for Total-Variation-Based Image Restoration. IEEE Transactions on Image Processing 20, 2748–2759 (2011)
7. Tai, Y., Tan, P., Brown, M.: Richardson-lucy deblurring for scenes under projective motion path. IEEE Transactions on Pattern Analysis and Machine Intelligence, 1603–1618 (2010)
8. Whyte, O., Sivic, J., Zisserman, A., Ponce, J.: Non-uniform deblurring for shaken images. In: Computer Vision and Pattern Recognition, pp. 491–498. IEEE Press, San Francisco (2010)
9. Vetterli, M., Herley, C.: Wavelets and filter banks: theory and design. IEEE Transactions on Signal Processing 40, 2207–2232 (1992)
10. Popkin, T., Cavallaro, A., Hands, D.: Accurate and efficient method for smoothly space-variant Gaussian blurring. IEEE Transactions on Image Processing 19, 1362–1370 (2010)
11. Horn, R.A., Johnson, C.R.: Matrix analysis. Cambridge Univ. Pr. (1990)

Chaotic Random Projection for Cancelable Biometric Key Generation

Xi Chen, Xiangwei Bai, Xie Tao, and Xiaolu Pan

The School of Information Engineering and Automation,
Kunming University of Science and Technology, Kunming 650051
xcbiometrics@126.com, {23276387,77183118,xlvc}@qq.com

Abstract. In this paper, we propose a novel finger vein based cancelable biometric key generation scheme. Three main steps are involved in the proposed scheme: (i) Finger vein feature is extracted by maximum margin locality preserving projection (MMLPP); (ii) Chaotic random projection was used for cancelable biometric template generation; (iii) Construct a fuzzy commitment based key generation system. Experiment and analysis show our proposed scheme is a secure and efficient key generation scheme.

Keywords: Finger Vein, Maximum Margin Locality Preserving Projection, Chaotic Random Projection, Fuzzy Commitment, Cancelable Biometric key.

1   Introduction

Although biometric keys have been studied for several years, there are still some shortcomings in classical biometric key generation system. Face and voice are not very stable and not suitable for key generation. Iris recognition is known for low error rates of authentication, but direct application of light into eyes will result in psychological burden to users. Palmprint is exposed to skin surface and may be forged by attackers. Fingerprints have been known that it can be easily wearied out and forged.

Although Finger vein in biometric recognition have been studied for several years [1] [2], Finger vein in biometric key generation is a completely new field. Best of our knowledge, there is no previous work contributed to key generation based on finger vein. To overcome the within-class variation, we extract discriminative feature of finger vein feature using maximum margin locality preserving projection algorithm (MMLPP) and user specific chaotic random projection (CRP). Furthermore we design BCH to correct the within-class difference. Unlike some typical systems which store biometric templates persistently, our system stores recovery information on a smart card carried by the user. The key is generated from a subject's finger vein with the aid of recovery information, which does not reveal the key. The reproduction of the key depends on two factors: the finger vein and the token. The attacker has to produce both of them to compromise the key.

The rest of this paper is organized as follows: Section 2 introduces maximum margin locality preserving projection and chaotic random projection for cancelable finger vein feature extraction. Experimental results and security analysis are given in Section 3. Section 4 highlights the conclusions.

J. Yang, F. Fang, and C. Sun (Eds.): IScIDE 2012, LNCS 7751, pp. 605–612, 2013.

2   Finger Vein Feature Extraction

2.1   Maximum Margin Locality Preserving Projection

To enhance the discriminating power, modified maximizing margin criterion [3] was integrated into the objective of LPP and constructs the objective of MMLPP. MMLPP solves the following constrained optimized problem:

$$\min \left\{ a^T (XLX^T - (S_b - \mu S_w) a \right\} \quad s.t. \quad a^T XDX^T a = 1 \tag{1}$$

where $S_b = \sum_{i=1}^{C} n_i (m_i - m)(m_i - m)^T$ and $S_w = \sum_{i=1}^{C} \sum_{x_k \in M_i} (x_k - m_i)(x_k - m_i)^T$ are between-class scatter matrix and within-class scatter matrix, respectively. m_i is the average vector of the ith class, $m_i = (1/n_i) \sum_{i=1}^{n_i} x_i$. m is the average vector of all samples, $m = (1/N) \sum_{i=1}^{N} x_i$. n_i is the number of samples in the ith class. M_i represents the set of the ith class, μ is a positive constant.

The above constrained minimization problem can be done using the method of Lagrange multipliers:

$$L(a_i) = a_i^T (XLX^T) a_i - a_i^T (S_b - \mu S_w) a_i + \gamma \times (1 - a^T XDX^T a) \tag{2}$$

Compute the gradients with respect to a_i and set the gradients to zero, we have the following minimization general eigenvalue problem:

$$(XLX^T - (S_b - \mu S_w)) a_i = \lambda_i XDX^T a_i \tag{3}$$

The vectors a_i $(i = 1, 2, \cdots, l)$ that minimize the above objective function correspond to the eigenvectors associated with the l smallest eigenvalues. Thus, the embedding is described as the following:

$$x_j \to y_j = A^T x_j, \quad A = (a_1, a_2, \cdots, a_l) \ , \ (j = 1, 2, \cdots, M) \tag{4}$$

2.2   Random Projection Based Cancelable Biometric Template Generation

In this section, we provide the approach of how to construct generating cancelable biometric templates utilizing random projection matrix and chaotic mapping.

2.2.1   Cancelable Biometric Template Generation
To construct cancelable template for cancelable biometric key generation, we project the biometric templates extracted by MMLPP to the user specific random matrix which is constructed by logistic chaotic mapping. Eq. (5) shows the logistic chaotic mapping:

$$y_{n+1} = g(y_n) = \mu y_n (1 - y_n) \tag{5}$$

where $n=0,1,2,3,....$ is the iteration number, μ is the system parameter and y_0 is the initial value. For $3.57 < \mu \le 4.0$, the sequence is non-periodic, non-convergent, and very sensitive to the initial value. In our paper, we choose $\mu = 4.0$. The random vector generated by Eq. (5) is consisted of real number entries distributed in the interval $[0,1]$, which was shown as $\{\beta_1,\beta_2,\beta_3,...,\beta_d\}$. After generating the pseudorandom vector, we transform the basis $\{\beta_1,\beta_2,\beta_3,...,\beta_d\}$ into an orthonormal vector set $\{u_1,u_2,u_3,...,u_d\}$ by the Gram-Schmidt algorithm. The chaotic Random Projection matrix was defined as $R \overset{def}{=} [u_1,u_2,u_3,...,u_d]$.

Our cancelable biometric templates are constructed by MMLPP and chaotic random projection (CRP), which we called MMLPP & CRP algorithm. Suppose the projection subspace obtained from MMLPP and CRP are R_{MMLPP} and R_{CRP}, respectively. If two finger vein vectors are v_1 and v_2, the projection of v_1 and v_2 on subspace R_{MMLPP} and R_{CRP} are $u_1 = (v_1 \times R_{MMLPP}) \times R_{CRP}$ and $u_2 = (v_2 \times R_{MMLPP}) \times R_{CRP}$, respectively.

3   Experimental Evaluation

In this section, we experimentally evaluate results and discussions on the proposed scheme for finger vein based key generation on a subset of PKU finger vein databases. The number of nearest neighbors used in MMLPP and LPP are both the number of training samples of each subject. The neighborhood of each training sample in those algorithms consisted of its same class samples, which boosts the classification performance. In MMLPP and LPP, PCA was used as a preprocessing means to reduce the length of biometric data to $M - C$, where M are the number of total training samples and C are the number of class. All experiments are executed on a computer system of PIV 3.0 GHz and 512MB RAM with Matlab 7.1.

3.1   Database

The PKU Finger Vein Database (V2) contains 4574 grayscale images corresponding to 431 different fingers, which were collected on the first semester of 08-09 school years [4]. In this paper, only a subset which contains 400 images of 40 subjects was used in our experiments with 10 images per user. Figure 1 shows the preprocessing process of finger vein images. The size of each cropped image in all experiments is 96×64 pixels, with 256 gray levels per pixel. Multi-scale retinex algorithm (MSR) [5] is used to enhance the cropped image (Fig.1 (d). Some of samples of three subjects are shown in Figure.2.

3.2   Biometric Key Based Encrypting/Decrypting System

The framework of biometric key based encrypting/decrypting system is constructed by two phases. In encrypting phase, finger vein feature was extracted by MM

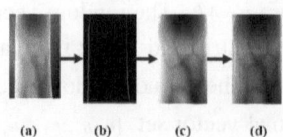

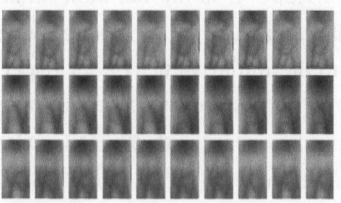

Fig. 1. Feature extraction from finger vein
(a) a original   finger vein image;
(b) the edges of finger;
(c) the cropped finger vein image;
(d) enhanced finger vein image

Fig. 2. the preprocessed sample images of three subjects from a subset of PKU finger vein database

LPP&CRP, and then quantizing by threshold value and obtaining the binary vector. The binary vector was further processed by hash function to generate the biometric key which can be used for encrypting plaintext. Besides, the binary vectors are encoded by BCH algorithm and generated the Error Correction Code. The biometric key and Error Correction Code all both stored in the smart card for decrypting. In decrypting phase, after feature extraction by MMLPP&CRP, and quantizing by threshold value, a binary vector can be generated from user's finger vein image, then utilizing the error correction code stored in smart card; the binary vector generated in the encrypting phase can be refreshed. The refreshed binary vector was encoded by hash function to generated biometric key for decrypting.

3.2.1   Evaluation Feature Extraction by MMLPP & CRP

In our biometric key generation scheme, biometric feature extraction is a necessary step. Generally speaking, more discriminative the biometric templates are, more efficient the generated biometric key will be. So we should extract the most discriminative biometric templates for key generation. For evaluation feature extraction of LPP and MMLPP&CRP algorithm, we random choose 5 samples of each subject for training and the rest for testing. We repeat our experiments 10 times and calculate the mean error rate under different feature dimensions. The mean error rate versus feature dimensions is plotted in Fig.3. From the experimental results, we find the best result of LPP and MMLPP are 6% and 2%, respectively. To generate long enough biometric key, the length of biometric key should be as long as possible, so the feature dimension of MMLPP was kept as 150, with which the mean error recognition rate is 4%. In CRP algorithm, there are two cases, one is each user possesses a different token (Genuine token) which control the chaotic sequences. The other is all users have a common token (Stolen token). MMLPP&CRP (Stolen token) and MMLPP&CRP (Genuine token) can be obtained with the feature dimension of 80 and 134, respectively. The minimum mean error recognition rate of MMLPP&CRP (Stolen token) and MMLPP&CRP (Genuine token) are 2.5% and 1%, respectively. To get long enough feature vectors for key generation, the maximum feature dimension of MMLPP&CRP (Stolen token) and MMLPP&CRP (Genuine token) are both kept as 150, with which the mean error recognition rate is 4% and 3%, respectively.

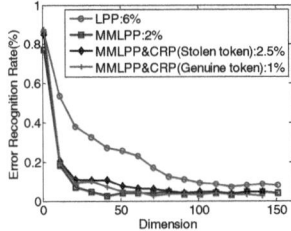

Fig. 3. Performance evaluations of LPP and MMLPP&CRP

3.2.2   Evaluation MMLPP&CRP for Key Generation

After obtaining the biometric feature vectors from biometric data by MMLPP&CRP, we quantize the obtained biometric template to generate binary vector as follows with threshold value 0. The maximum feature dimension of MMLPP is 150. As our proposed algorithm is associated with an external input which is used as chaos initial value in our paper, a unique random matrix is derived from the chaos initial value. Thus, two scenarios may be taken place [6]. One class of scenarios is compromise biometric (Scenario 1) in which an imposter possesses intercepted biometric data of sufficiently high quality to be considered authentic. Another class is Compromised external input (Scenario 2); imposter has access to the external input and can hence reproduce the user-specific random matrix. Here, the same chaos initial value is used to generate a common random matrix for all 200 images in testing set. The Receiver Operating Curve (ROC) of our proposed MMLPP, MMLPP&CRP, MMLPP&CRP under scenario 1 and scenario 2 are plotted in Fig 4. From Fig.4, we can see the genuine accept rate of our method is very high. As for the compromised external input scenario, the performance does not seriously be deteriorated. The genuine accept rate of Scenario 1 is close to that of MMLPP.

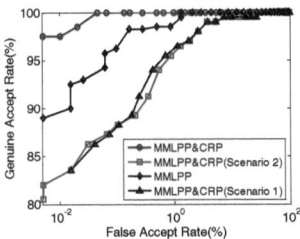

Fig. 4. ROC curves for MMLPP, MMLPP&CRP and its two compromised scenarios

As there are still some within-class variation between generated binary biometric vectors, we can use error-correction code to rectify within-class variations, such as $BCH(n,k,t)$, where n is the length of code, k is the length of key K and t is the number of bits which BCH can be corrected. For our system, t can be computed using its distance threshold T as following:

$$t = 150 \times T \qquad (6)$$

Therefore, a suitable distance threshold is needed in our system to obtain high performance. The FAR and FRR at different Hamming Distance thresholds and the numbers of the error bits are listed on Table 1.

Table 1. FAR and FRR under different length of key

Threshold	Error-correction	FRR	FAR
T	capability t (bits)	(%)	(%)
0.26	39	37.25	0
0.28	42	47.25	0
0.3	45	25.75	0
0.313	47	21.5	0
0.353	53	8.0	0.5

3.3   Security Analysis about Key Generation System

3.3.1   Key Sensitivity

A good encryption system must be sensitive to the keys and key space should be large enough to make brute force attacks infeasible. As an example, we used $y_0 = 0.002$, and $y_0 = 0.002 + 10^{-15}$ to generate two random matrixes and a feature vectors generated by MMLPP is projected to these two random matrixes, respectively. The diffusion of projections at two different keys is shown in Fig.5. From Fig.5, we can see that our proposed scheme is very sensitive to a small change in the key and without knowing the key; it is very difficult to generate the same template.

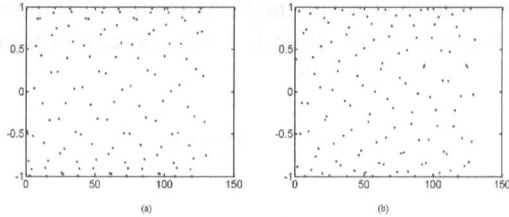

Fig. 5. Diffusion of template (normalized) at two different keys (a) Diffusion when key y_0 =0.002 (b) Diffusion when key y_0 =0.002+10^{-15}

A secure biometric template should be robust against statistical attacks. To evaluate our approach, we perform statistical analysis to demonstrate its superior confusion and diffusion properties, statistical analysis is evaluated by the following equations [7]:

$$B_{\min} = \min(\{B_i\}_1^N) \text{, denotes the minimum of } B_i$$

$$B_{\max} = \max(\{B_i\}_1^N) \text{, denotes the maximum of } B_i$$

$$\bar{B} = \frac{1}{N}\sum_{i=1}^{N} B_i \text{, denotes the mean of } B_i$$

$$\Delta B = \sqrt{\frac{1}{N-1}\sum_{i=1}^{N}(B_i - \overline{B})^2} \text{ , denotes the standard variance of } B_i$$

$$P = (\overline{B}/130) \times 100\% \text{ , denotes the probability of } \overline{B}$$

$$\Delta P = \sqrt{\frac{1}{N}\sum_{i=1}^{N}(B_i /130 - P)^2} \text{ , denotes the standard variance of } P.$$

The initial key is $y_0 = 0.2317853426$, we randomly selected a bit and replace it with another random digital, and then a new template will be generated. These two templates are compared and the number of changed bit is counted as B_i. If the size of template is 150 bits, then one bit changed in the key concentrates around the ideal changed bit number-75bit, it shows that the template has very strong capability for diffusion and confusion. Through the tests with N = 512, 1024, 2048, respectively, experimental results are listed on Table 2.

Table 2. Number of changed bits B_i

	$N = 512$	$N = 1024$	$N = 2048$
B_{min}	52	53	54
B_{max}	96	94	95
$\overline{B}$	74.0	73.5	74.5
ΔB	5.52	5.61	5.74
$P(\%)$	49.66	49.60	49.36
$\Delta P(\%)$	4.28	4.34	4.53

4   Conclusions

Eliminating within-class difference and enhance between-class difference is the primary process for biometric key, in this paper, maximum margin locality preserving projection was used to extract discriminative feature of finger vein. Chaotic random projection was used to generate cancelable biometric vector which results new key can be released when key compromised.   Experiment and analysis show our proposed scheme is a secure and efficient key generation approach.

Acknowledgments. This work was supported by National Science Foundation of P.R.China(Grant: 61262040), the applied basic research projects of Yunnan Province(Grant:KKSY201203062) and the Scientific Research Fund of Hunan Provincial Education Department (Grant:11C1067).

References

[1] Miura, N., Nagasaka, A., Miyatake, T.: Feature extraction of finger-vein patterns based on repeated line tracking and its application to personal identification, pp. 194–203 (2004)
[2] Zhang, Z., Ma, S., Han, X.: Multiscale feature extraction of finger-vein patterns based on curvelets and local interconnection structure neural network. In: The 18th International Conference on Pattern Recognition (ICPR 2006), pp. 145–148 (2006)

[3] Li, B., Zheng, C.-H., Huang, D.-S.: Locally linear discriminant embedding: An efficient method for face recognition. Pattern Recogn. 41, 3813–3821 (2008)

[4] PKU Finger Vein Database (V2), http://ai.pku.edu.cn/

[5] Jobson, D.J., Rahman, Z., Woodell, G.A.: A multi-scale retinex for bridging the gap between color images and the human observation of scenes. IEEE Trans. Image Processing 6(7), 965–976 (1997)

[6] Jin, A., Goh, A., Ling, D.: Random multispace quantization as an analytic mechanism for biohashing of biometric and random identity inputs. IEEE Trans. Pattern Anal. Mach. Intell. 28(12), 1892–1901 (2006)

[7] Xiao, D., Xiaofeng, L., Shaojiang, D.: One-way Hash function construction based on the chaotic map with changeable-parameter. Chaos, Solitons and Fractals 24, 65–71 (2005)

An Efficient Video Copy Detection Method Combining Vocabulary Tree and Inverted File

Xuan Li[1], Bing Li[2], Weiming Hu[2], and Jinfeng Yang[1]

[1] College of Aviation Automation, Civil Aviation University of China, Tianjin, China
{xli_yjs10,jfyang}@cauc.edu.cn
[2] Institute of Automation, Chinese Academy of Sciences, Beijing, China
{bli,wmhu}@nlpr.ia.ac.cn

Abstract. In this paper, we present an efficient content-based video copy detection method based on vocabulary tree and inverted files. The copy detection system exploits complementary local features and video sequence matching. Using two different local features, vocabulary trees and inverted files are built respectively to get keyframes matching result. Histogram-based and diagonal-based sequence matching approaches are applied to detect the copy video sequences. The experimental results on the TRECVID 2011 video copy detection dataset show that the proposed system is effective and efficient.

Keywords: Content-based copy detection, Bag-of-Words, Vocabulary tree, Inverted file.

1   Introduction

Along with the development of computer, web, and other multimedia technologies, the amount of videos increases exponentially nowadays. The spread of the videos leads to many copies on the Internet or TV. Therefore, a video copy detection system is necessary for the digital copyright protection, information (illegal content) monitoring, video retrieval and so on.

Given a query video, video copy detection aims at finding the copy clips in the database of reference videos [1]. A copy is a transformed video sequence derived from an original video. The transformation contains scaling, compression, picture-in-picture (PIP), pattern insertion and cropping, etc [1].

The video copy detection is challenging due to the following factors. (1) effective and robust feature extraction. Many kinds of video transformations change the videos seriously, such as PIP, spatial shifts, blur and so on. (2) high dimension of features. For instance, Scale-invariant feature transform (SIFT) [2] that is robust, is a 128-d descriptor. The high dimension inevitably make the computational cost of matching processing be high. (3) context cues in video. The temporal information should be considered effectively and reasonably for video copy detection.

In this paper, we present a video copy detection system which can effectively address these problems. An overview of our system is shown in Fig. 1. We extract two kinds of local features for a subsample of frames from the video. One is the 128-d

J. Yang, F. Fang, and C. Sun (Eds.): IScIDE 2012, LNCS 7751, pp. 613–621, 2013.
© Springer-Verlag Berlin Heidelberg 2013

SIFT descriptor, which is employed to cope with spatial content-altering transformations. The other is the 32-d Oriented Features from Accelerated Segment Test (FAST) and Rotated BRIEF (ORB) descriptor [3]. It is a very fast binary descriptor based on Binary Robust Independent Elementary Features (BRIEF), which is rotation invariant and resistant to noise. These two descriptors are combined together, achieving total robustness to various transformations. In order to reduce dimensionality and index efficiently, the local descriptors are hierarchically quantized in a vocabulary tree. This vocabulary tree allows a larger and more discriminative vocabulary to be used efficiently. Each keyframe is represented by a vocabulary histogram. The entire set of the representations of the video dataset is stored in a structure similar into the inverted file, which accelerates feature matching process. We proposed three sequence matching methods for video copy detection, two of them are based on the histogram, and   the other is based on diagonal.

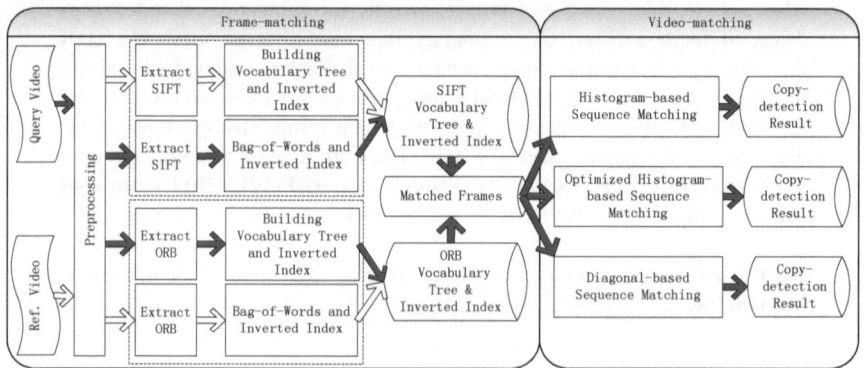

Fig. 1. An overview of our CBCD system

2   Related Work

Due to the definition of copy video, a kind of robust and discriminative feature should be extracted to detect copy videos. The feature is widely used for CBCD before it can be classified into two broad categories, global features and local features. Global features include the global statistics of low-level features. Many research studies have used them to deal with some simple video transformations. For example, Coskun et al. [4] use the Discrete Cosine Transform (DCT) to identify copies. Hampapur and Bolle [5] compare global features based on color, motion and spatio-temporal distribution of intensities. This ordinal measure has been proved to be robust to different resolutions, illumination shifts and display formats, but it does not robust to pattern insertion, shifting, or cropping. Overall, although those features are more compact and can be extracted more efficiently, they are only robust to some simple transformations. Considering this situation, local features have demonstrated their effectiveness for detecting copy videos. Many approaches extract local patches with the Hessian-Affine region detector and describe them with the SIFT or Speed Up Robust Features (SURF) descriptors [6]. The high dimension leads to heavy computational cost. The most popular method for reducing the dimension is the |

bag-of-words (BoW) approach [7] [8]. It first clusters a large amount of training features to generate a visual word vocabulary. And every cluster center is a visual word. Then every features extracted from the database are mapped to those word. So every frame is represented by a visual word histogram. Moreover, combining with inverted file, the BoW representation can improve the effectiveness and efficiency of the approaches using local features.

The video matching methods are paid more attention. Wei et al. [9] proposes a frame fusion scheme which is based on a Viterbi-like algorithm, comprising an online back-tracking strategy with three relaxed constraints. Tian et al. [8] propose a sequential pyramid matching method which is adopted to aggregate frame level results into video level results.

3 Overview of Our Propose Approach

The CBCD generally involves two parts: frame matching and video matching. For frame matching, there are three steps: (1) extracting features of the keyframes, (2) utilizing the Bag-of-Words model to compact the features, (3) indexing the similar frames using the inverted files. For video matching, we proposed three methods of matching sequences. Two are based on the histogram, the rest one is based on diagonal.

4 Frame-Matching

Frame-matching aims at getting the similar frames of the query frame. We use local features and their Bag-of-Words representations, and then index the similar frames with inverted files.

4.1 Preprocessing

Since many adjacent frames are quite similar, it is redundant to use every frame in a video for copy detection. Two approaches are widely used to extract keyframe. One is based on shot boundary, the other is to sample keyframe uniformly. The shot boundary can not be detected accurately enough, and it is proved that the result of uniform sampling is better than the shot boundary based method. Consequently, the keyframes in our video copy detection system are obtained by uniformly sampling at a rate of 2 frames per second.

In addition, we detect the black frames by computing the average luminance values and remove them to avoid affecting the feature extraction. For PIP transformation, it is difficult to find this kind of copies directly. We preprocess them, and match the foreground, background and the original frames respectively.

4.2 Feature Extraction

We use two kinds of local features for our CBCD system, SIFT and ORB. For SIFT, we firstly detect Hessian-Laplace regions using the software of [10] with default parameters. This region detector is invariant to many image transformations, such as scale changes, image rotation and noise. SIFT are invariant to image scale and rotation. They are also robust to changes in illumination, noise, and minor changes in viewpoint. The dimension of SIFT is 128.

Another local feature is a very fast binary descriptor based on BRIEF, called ORB, which is rotation invariant and resistant to noise. First, FAST is used to detect the interest points. Then we use BRIEF as the feature descriptor, which uses simple binary tests between pixels in a smoothed image patch. Its performance includes robustness to lighting, blur, and perspective distortion. The dimension of ORB is 32, and the speed of using ORB is almost 100 times faster than using SIFT.

4.3   Building Vocabulary Tree

Directly matching the frames using the Euclidian distance between two feature vectors with 128-d or 32-d is not scalable at all. In this work, we utilize Bag-of-Words to generate compact feature representation for efficient feature matching.

The first step is to convert the feature vectors to visual words, which also produces a visual vocabulary. We perform hierarchical k-means clustering on a training set of frames' descriptors to generate a visual vocabulary. Visual words are then defined as the centers of the learned clusters. After generating a vocabulary, each descriptor is mapped to a visual word by searching the nearest centroid in the visual vocabulary.

In order to obtaining a larger and more discriminatory vocabulary efficiently, we build a vocabulary tree [11], which is shown to result in a significant improvement of retrieval quality. This vocabulary tree defines a hierarchical quantization by hierarchical k-means clustering. K defines the branch factor of the tree, and L defines the level of the tree. In our experience, the number of branch factor is 10, and the number of level is 6. 1,000,000 leaf nodes are used for our method. Every descriptor in the database is mapped to the vocabulary tree. So the frames of reference videos are represented by a set of visual words in the vocabulary tree. Moreover, the representation of database is saved in a file.

For two kinds of features, we build two different vocabulary trees respectively. Each tree has 1,000,000 leaf nodes. The representations using different features are also generated respectively. We get two vocabulary trees, and two representations for every frame in the database.

4.4   Indexing by Inverted File

We use a score function [11] to measure the relevance between the query frame and the reference frames based on how similar the paths down the vocabulary tree. The score function between q and d is defined as

$$s(q,d) = \left\| \frac{q}{\|q\|} - \frac{d}{\|d\|} \right\|$$ (1)

Moreover, we use inverted files to index the similar frames efficiently with the large database. The inverted files store the id-numbers of the images in which a particular node occurs, as well as for each image the term frequency.

Given a query video, for each frame, we use the top A similar reference frames by ranking the score. The output of a query video is a collection M_f which contains a series of frame level matches $m_f^{(a)}$:

$$m_f^{(a)} = \left\langle q, t_q, r_f^{(a)}, t_r, s_f^{(a)} \right\rangle, \tag{2}$$

where q and r identify the query and reference video, t_q and t_r are the timestamps of matched query and database frames, s_f is the score of the frame match, computed by(1), and a defines the top A matching frames and $a=1,2,\cdots,A$.

5   Video Matching

We proposed two approaches for video matching based on frame-matching results. One is based on histogram, the other is based on diagonals.

5.1   Histogram-Based Sequence Matching

After Frame-matching, we get the collection M_f which contains a series of frame level matches $m_f^{(a)}$.

We respectively sum the numbers of matched frames which belong to the same reference video, denoted h_r :

$$h_r = \sum_{a,f} Sig_f^{(a)} \tag{3}$$

$$Sig_f^{(a)} = \begin{cases} 1, r_f^{(a)} = r \\ 0, others \end{cases} \tag{4}$$

where r is the name of the reference video in the database.

After getting a set of h_r , we get a histogram of them, and the video matching result is the maximum of h_r that must be larger than a threshold τ_h .

$$h_{max} = \max\left\{ h_r, r \in Database \right\} \tag{5}$$

$$N_r = \left\{ r \mid h_r = h_{max} \right\} \tag{6}$$

$$R_h = \begin{cases} N_r, h_{max} > \tau_h \\ none, others \end{cases} \tag{7}$$

where N_r is the name of reference video which get the largest h_r, R_h defines the result of video matching. If the maximum of h_r is larger than the threshold τ_h, the result R_h returns the name of copy video, otherwise, the result R_h returns *none* which means there is no copy of the query video.

5.2 Optimized Histogram-Based Sequence Matching Using Scores

In the subsection 5.1, the information of scores, which may be an important factor detecting the copy video, is not considered. We optimize our model of histogram-based sequence matching by combining the scores.

We respectively compute the sum of scores for each reference video which may be a copy of query video, denoted s_r:

$$s_r = \sum_{a,f} \tilde{s}_f^{(a)} \tag{8}$$

We use $s_f^{(a)}$ to replace $Sig_f^{(a)}$ in Section 5.1. The following steps are same. Then we get the video matching result R_s for the query video.

5.3 Diagonal-Based Sequence Matching

In the section 5.1 and section 5.2, the histogram doesn't consider the context information of videos. The copy sequence is consistent in spatio. So we proposed a sequence matching method based on diagonal scheme.

Considering the frame-matching result M_f, we build a matrix $M = \left[m_{i,j} \right]$ ($i=1,\cdots,l_q$, $j=1,\cdots,l_r$) of each reference video for a query video,

$$m_{i,j} = \begin{cases} s_f^{(a)}, t_q=i \& t_r=j \\ 0, \text{others} \end{cases} \tag{9}$$

where l_q (l_r)defines the number of query (reference) video keyframes.

The diagonal we define in our method is the k-*th* diagonal of matrix M, which contains continuous non-zero numbers. We first line all the diagonals in the matrix. Then the diagonals whose length is less than a threshold are ignored. That is, those single-frame matches and short sequence matches are filtered out. A score $s_d^r = \sum s_f^{(a)}$ is computed for each diagonal left. A threshold is used for filtering the better diagonals whose s_d^r is larger than it. For the remaining diagonals, we connect the neighbor diagonals to form a longer one by operations of insertions and deletion. Then we compute the score of these new diagonals, and return the result that the copy video is the reference video with the largest score which is larger than a threshold τ_d.

$$s_{d\text{-max}} = \max\left\{s_d^r, r \in Database\right\}, \tag{10}$$

$$N_d = \left\{r \,\middle|\, s_d^r = s_{d\text{-max}}\right\}, \tag{11}$$

$$R_d = \begin{cases} N_d, s_{d\text{-max}} > \tau_d \\ none, others \end{cases} . \tag{12}$$

6   Experiment

In this section, we evaluate our CBCD system on the TREC Video Retrieval Evaluation (TRECVID) 2011 dataset [1]. We first introduce the dataset, and give the details of experimental settings. Finally, the experiment result is given and analyzed.

6.1   Dataset and Experimental Setting

TRECVID is sponsored by the National Institute of Standards and Technology (NIST), devoting to research in automatic segmentation, indexing, and content-based retrieval of digital video. Since 2008, TRECIVD has organized the task CBCD, and the dataset we use in this paper is TRECVID 2011 CBCD.

Table 1. Video transformations

Transformation	Description
T1	Simulated camcording
T2	Picture in picture
T3	Insertions of pattern
T4	Compression
T5	Change of gamma
T6	Decrease in quality (a mixture of 3 transformations among blur, gamma, frame dropping, contrast, compression, ratio, white noise)
T8	Post production (a mixture of 3 transformations among crop, shift, contrast, text insertion, vertical mirroring, insertion of pattern, picture in picture)
T10	Change to randomly choose 1 transformation from each of the 3 main categories

In TRECVID 2011 CBCD task, the reference dataset contains more than 12000 videos. More than 10000 query videos are created. For each query, the tools will take a segment from the reference dataset, optionally transform it, embed it in some video segment which does not occur in the reference dataset, and then finally apply a video transformation on the entire query segment. There are ten kinds of video transformations listed in Table 1.

For video copy detection, we first extract keyframes uniformly from the video at a rate of 2 frames per second. Then the SIFT and ORB features are extracted from the keyframes. We build the vocabulary tree with 6 levels and 1,000,000 leaf nodes, and index the similar frames by ranking the score using inverted files. We store the top 20 similar frames and their scores for each frame. After matching the frames, we match the video using the three methods respectively. Our experiment is done on the subset of TRECVID dataset, which are sampled randomly. The distributions of type of queries and transformations are same with the whole dataset.

6.2   Experiment Result and Analysis

The evaluations we use are precision and recall. In our paper, the **true positives** (TP) are the copies that are correct and the **false positives** (FP) are unexpected copies. The **false negatives** (FN) are the missing copies, and **true negatives** (TN) are correct non-copies.

We use precision, recall, true negative rate, and accuracy to evaluate our CBCD system. The results we get using three different video matching methods are shown in the Table 2.

Table 2. Precision, Recall, TNR, and Accuracy of Matching Method

	Method 5.1	Method 5.2	Method 5.3
Precision	0.444	0.600	0.765
Recall	0.727	0.692	0.867
TNR	0.500	0.667	0.750
Accuracy	0.581	0.677	0.806

From the result we can find that the matching method based on diagonal is proved better than the method based on histogram. The result using scores is proved better than the result without scores. The reason is that the score information and spatio information are significant for sequence matching.

Table 3. Precisions for different transformations

Transformation	T1	T2	T3	T4	T5	T6	T7	T8	T9	T10
Precision	0.78	0.77	0.83	0.83	0.88	0.81	0.72	0.71	0.67	0.64

As shown in Table 3, it is obvious that the difficulty of copy detection for different transformations varies a lot. T5 is the easiest one. Most of T3, T4 and T6 can be matched accurately as well. The more challenging transformations are T2, T8 and T10. Without preprocessing on PIP transformation, our system does not perform well on T2 and T8 which may contain PIP transformation. It is proved that the preprocessing for PIP is required to detect the foreground and background. For this reason, we improve our system by detect the PIP transformation during the preprocessing. It turns out that the result is better than before.

The processing time of our system is appropriate. The usage of vocabulary tree and inverted files decreases the processing time.

7   Conclusions and Future Work

We have proposed a content-based copy detection system based on vocabulary tree and inverted file. By introducing the SIFT and ORB features, the proposed system utilizes the Bag-of-Words model to compact the features, and indexes with the inverted files. Then it uses the methods based on histogram and diagonal to matching the video sequences respectively. The evaluation result on TRECVID 2011 CBCD dataset shows that the proposed system is effective and efficient. Our future work will be devoted on the usage of audio information and accurate copy localization. However, there are a few thoughts on further reducing the computational complexity as following. (1) Decreasing the number of SIFT features and using the most representational interest points. (2) Increasing the size of vocabulary tree to adopt a more efficient way to read in the large inverted files. Moreover, the video sequences matching method should be improved using more context information.

Acknowledgement. This work is partly supported by the National Nature Science Foundation of China (No.61005030, 60935002, and 60825204) and Chinese National Programs for High Technology Research and Development (863 Program) (No. 2012AA012503 and No. 2012AA012504).

References

1. TRECVID, TREC Video Retrieval Evaluation (2011),
   http://www-nlpir.nist.gov/projects/trecvid
2. Lowe, D.G.: Distinctive Image Features from Scale-Invariant Keypoints. IJCV 60(2), 91–110 (2004)
3. Rublee, E., Rabaud, V.: ORB: an efficient alternative to SIFT or SURF. In: International Conference on Computer Vision, pp. 2564–2571 (2011)
4. Coskun, B., Sankur, B., Memon, N.: Spatio-temporal transform-based video hashing. IEEE Transactions on Multimedia 8(6), 1190–1208 (2006)
5. Hampapur, A., Bolle, R.: Comparison of sequence matching techniques for video copy detection. In: Conference on Storage and Retrieval for Media Databases, pp. 194–201 (2002)
6. Bay, H., Tuytelaars, T., Van Gool, L.: SURF: Speeded Up Robust Features. In: Leonardis, A., Bischof, H., Pinz, A. (eds.) ECCV 2006. LNCS, vol. 3951, pp. 404–417. Springer, Heidelberg (2006)
7. Douze, M., Jegou, H., Schmid, C.: An image-based approach to video copy detection with spatio-temporal post-filtering. IEEE Transactions on Multimedia 12(4), 257–266 (2010)
8. Tian, Y., Jiang, M., Mou, L., Fang, X., Huang, T.: A multimodal video copy detection approach with sequential pyramid matching. In: IEEE International Conference on Image Processing, Brussels, Belgium, September 11-14 (2011)
9. Wei, S., Zhao, Y., Zhu, C., Xu, C., Zhu, Z.: Frame fusion for video copy detection. IEEE Transactions on Circuits and Systems for Video Technology 21(1) (2011)
10. Vedaldi, A., Fulkerson, B.: VLFeat: An Open and Portable Library of Computer Vision Algorithms, http://www.vlfeat.org
11. Nister, D., Stewenius, H.: Scalable recognition with a vocabulary tree. In: Proc. IEEE Conf. Comput. Vision Pattern Recognit., pp. 2161–2168 (October 2006)

Estimation Based on RBM from Label Proportions in Large Group Case

Kai Fan, Hongyi Zhang, Yu Zang, and Liwei Wang

Key Laboratory of Machine Perception, MOE,
School of Electronics Engineering and Computer Science, Peking University
{kaifan,hongyi.zhang,zangyu}@pku.edu.cn, wanglw@cis.pku.edu.cn

Abstract. Learning a classifier when only knowing about the features and *marginal distribution of class labels* in each of the data groups is both theoretically interesting and practically useful. Specifically, we are interested in the case where the ratio of the number of data instances to the number of classes is large. For this problem, we show that the performance of a previously proposed discriminative classifier will deteriorate quickly as the ratio grows. In contrast, we formulate a density estimation framework to learn a generative classifier by RBM in this scenario with guaranteed performance under mild assumption.

Keywords: Proportion Learning, Density Estimation, RBM.

1 Introduction

Sometimes we encounter a dataset with a permanent key attribute (label, class, etc.) and some variable features changing with environments. So the proportion for key attribute will be strictly controlled or known previously. As noted in [1], a useful case of steel factory where steel sticks are processed sequentially at several production stations is presented with the purpose of discarding the unqualified sticks as early as possible, but the quality information for certain stick is neither available nor necessary. Likewise, all dataset details are required regardless of what machine learning algorithm is. However, large expense for obtaining all information of dataset has been considered recently. In many data selling deals, partial information is more probably to be sold and commercial corporations prefer not to provide all consumers information. Also in [6], useful cases such as e-commerce, politics vote and spam filter in real world are included in the domain of this problem, which we would like to mining useful information here.

For this problem, we present a Bayesian probabilistic framework for reducing it to estimating posterior probability $P(X|Y)$. Since the contrastive divergence learning introduced by [7,9] are probabilistic generative models that can be learned and approximated inference with efficient greedy algorithms. In this paper, we design a combined deep networks estimator for the generally acknowledged difficult task: computing the probability $P(X|Y)$, which is enlightened by the Quantitative Analysis of Deep Belief Networks in [10].

J. Yang, F. Fang, and C. Sun (Eds.): IScIDE 2012, LNCS 7751, pp. 622–629, 2013.
© Springer-Verlag Berlin Heidelberg 2013

Definition 1 (Proportions Learning). *Assume that $\mathcal{X}$ is an instance space (features product space) and Y is the set of labels. Let $P(X, Y)$ be a fixed but unknown probability distribution and Y be some discrete values, i.e. $Y = \{y_1, ..., y_l\}$. Given a set of unlabeled observations $\{x_1, ..x_N\}$, which is drawn i.i.d. from P and divided into n disjunct subsets $X_1, ...X_n$, where we denote the size of $|X_i| = m_i, \forall i \in [n]$[1]. In addition, we do not know the label of each observation but know the label proportions π_{ij} for each subset, where π_{ij} means the label y_j proportion in subset X_i. Based on this information, our goal is to design a learning algorithm which is able to predicted y for certain observation $x \in \mathcal{X}$, i.e. constructing a good estimator to acquire $P(Y|X)$. For convenience, we combine all label proportions π_{ij} as a $n \times l$ matrix $\boldsymbol{\Pi}$, which is denoted as **Proportion Matrix**. An obvious result is that the elements in each row is sum up to one.*

Related Works. Here, we propose the most competitive algorithm Inverse SVM for binary case of this domain in [5], which will also be executed in experiments as comparison with our methods. Denote the π_i as the positive label proportion without loss of generality. So the proportion matrix is transformed to proportion vector $\boldsymbol{\pi}$. The algorithm uses the scaling function $\forall i, \pi_i = \sigma(y) = \frac{1}{1+\exp{(-y)}}$ and its its inverse $y = \sigma^{-1}(\pi_i) = -\log(\frac{1}{\pi_i} - 1)$. And to circumvent the unawareness of labels which would be used in SVM, a term is used to approximate the label $\frac{1}{|X_i|}\sum_{x \in X_i}(wx + b) \approx y_i$, where $f(x) = wx + b$ is the hyperplane for classification. And the objective is to solve the primal problem formulated as $\frac{1}{2}\|w\|^2 + C\sum_{i=1}^{n}(\xi + \xi^*) \to min$, subject to $\forall i \in [n] : \xi, \xi^* \geq 0, y_i - \epsilon_i - \xi \leq \frac{1}{|X_i|}\sum_{x \in X_i}(wx + b) \leq y_i + \epsilon_i + \xi^*$.

2   Sample Complexity Analysis

A crucial problem for Inverse SVM is that, when each label proportion is almost $P(Y = 1)$ or $1 - P(Y = 1)$, the SVM method does not work obviously due to the approximation term is almost the same for each subset when no assumption is added. Therefore, we could see intuitively that the bias of label proportions is a significant measurement for generalizing the supervised learning algorithms to this problem. From this intuition, we define a sufficient condition of learnable case, **learnable proportion vector** π_l, if it is combined with some elements of $\boldsymbol{\pi}$ with the property that $\delta_i = \min\{1 - \pi_i, \pi_i\}$ and the new vector keeps the index that satisfying $\delta_i \leq \delta$, where δ given previously is a constant close to 0 but is not necessarily very close. But we could see an intuition that it is obviously as learnable as supervised learning if $\delta = 0$. Then we have the following result.

Theorem 1 (Sample Complexity). *In binary case, we assume that $p = P(Y = 1)$ and $\delta = \max_{i \in [n]} \min\{\pi_i, 1 - \pi_i\}$, $\delta \neq \min\{p, 1 - p\}$, and let the subset size be the same and denote as m. The probability of obtain at least one subset satisfying the learnable proportion vector condition is $O(m^{-\frac{1}{2}}e^{-\frac{m}{2}})$, and the number of samples we need to draw from the X is $O(m^{\frac{3}{2}}e^{\frac{m}{2}})$.*

---

[1] [n] is equal to 1,...,n.

The proof can be seen in a journal version later. Here, we will first see in practical case, when m is large and chosen as 32, 64 or 128, the distribution could be finely approximated by normal distribution. And we can see that when the group size m increases, the sample complexity for learning a small-error model as supervised case is a scale of exponential increment. The reason is that no assumption is undertaken in SVM calibration algorithm and **random sampling** without any technique will lead to positive label proportion close to p with larger probability. So we need more observations to get a learnable proportion vector. If data size is not sufficient, the performance will convergence to random guess quickly when m is large as shown in Fig 1.

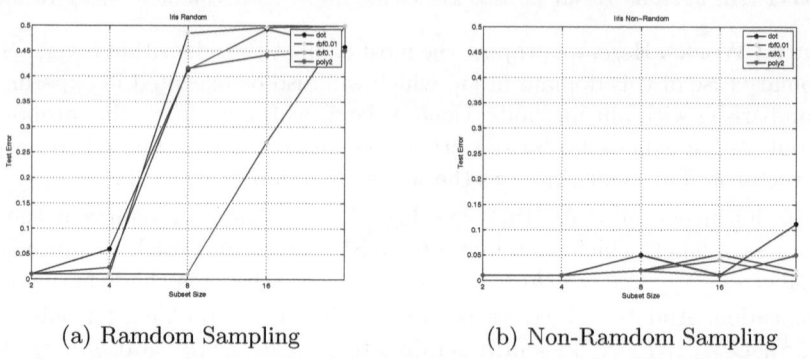

(a) Ramdom Sampling   (b) Non-Ramdom Sampling

Fig. 1. The horizontal axis represents different group sizes divided for learning and the vertical axis represents classification error. In this experiment, Three types of kernel functions with different parameters are attempted for Inverse SVM learning. The left figure shows that the dataset is divided randomly, while the right figure shows that the dataset is divided with $\delta_i \leq \frac{1}{4}, \forall i \in [n]$.

Without losing generality, we can assume $p \leq \frac{1}{2}$. In addition, we can find that the relation between δ and p is of great significance. On the circumstance that m is large, random sampling leads to $\pi_i \approx p$, which means $\delta_i \approx p$. If p is large, the learnable proportion vector condition is difficult to hold. If p is small, which means the dataset is inherently with large bias related in label attribute, we could acquire high predict accuracy by guessing negative label.

3   Learning Framework

In this section, we give a Bayesian model by acquiring the conditional probability $P(Y|X)$. Here, we put forward a estimator from deep network methods enlightened by the idea of [10]. First of all, according to assumption necessity, we have to **assume** that the conditional distribution of x is independent of the subset index i as long as we know the label y. And the assumption will be an important sampling technique, rather than random sampling in experiments.

Bayesian Framework. As we noted in definition 1, we want to acquire $P(Y|X) = \frac{P(X|Y)P(Y)}{P(X)}$. We can see that this is the perfect theoretical framework without any approximation. What we have to do is to compute the variables in the left side of equation. Note that $P(Y)$ (i.e. $p_i, ...p_n$) is trivially estimated through the term $\hat{p}_j = \frac{1}{N}\sum_{i=1}^{n} m_i \pi_{ij}$, where $p_j = P(Y = y_j)$. Also, $P(X)$ can be estimated by $\sum_{j=1}^{l} p_j P(X|y_j)$, if $P(X|Y)$ is known. Once $P(X|Y)$ can be approximately estimated, this classification problem can be solved by attaching the final class label of certain instance x, $\arg\max_{j\in[l]} \frac{P(X=x|Y=y_j)p_j}{\sum_{j=1}^{l} p_j P(X=x|Y=y_j)}$.

Therefore, estimating $P(X|Y)$ is the key point for this problem. For convenience, we denote that $p(x|y_j) = P(X = x|Y = y_j)$. Then for some certain observation x, what we have to compute is to compare l numerators $p_j p(x|y_j)$ due to the same denominator $\sum_{j=1}^{l} p_j p(x|p_j)$.

Proportion Learning Framework. As we have assumed in the beginning of this section, we paraphrase the assumption to mathematical express $(x \perp i|y)$. Before we compute term $p(x|y)$, a new probability $p(x|i)$ $(i = 1, ..., n)$ is introduced naturally, which means the probability density in the ith subset. According to our assumption, $p(x|y, i) = p(x|y)$, we have the important formula and its matrix form by denoting $\boldsymbol{p}_{set} = (p(x|1), ..., p(x|n))'$ and $\boldsymbol{p}_{lab} = (p(x|y_1), ..., p(x|y_l))'$

$$p(x|i) = \sum_{j=1}^{l} p(x|y_j, i)p(y_j|i) = \sum_{j=1}^{l} p(x|y_j)\pi_{ij}, \quad \boldsymbol{\Pi}\boldsymbol{p}_{lab} = \boldsymbol{p}_{set} . \tag{1}$$

According to Equation (1), we will acquire exactly what we are seeking for $\boldsymbol{p}_{lab} = \boldsymbol{\Pi}^+ \boldsymbol{p}_{set}$, where $\boldsymbol{\Pi}^+$ is the Moore-Penrose Inverse of $\boldsymbol{\Pi}$. So if we could give a good estimation of $\boldsymbol{p}_{set}$, then the posterior probability $\boldsymbol{p}_{lab}$ will be estimated well.

Proportion Matrix Analysis. Now we assume that we have the estimation (the detailed algorithm is to be presented in the next section) of $\boldsymbol{p}_{set}$, which is denoted as $\hat{\boldsymbol{p}_{set}}$. And we will see that the estimation quality of $\boldsymbol{p}_{set}$ immediately contributes to the final Bayesian model. Considering the $\boldsymbol{p}_{lab}$ as unknown variables, then we attempt to solve the linear equations $\boldsymbol{\Pi}\boldsymbol{p}_{lab} = \hat{\boldsymbol{p}_{set}}$.

First, we analysis the stability bounds in the situation that a tiny disturbance Δ appears in $\boldsymbol{\Pi}$, when proportion matrix is full column rank. We will see that the influence can be constrained by the lemma of Stability of Pseudo-Iverse [4] and our second main theorem.

Theorem 2 (Stability Bound of under Permutation). *For linear equations's solution* $\hat{\boldsymbol{p}_{lab}}$, *the ineqation hold.*

$$\|\tilde{\boldsymbol{p}_{lab}} - \hat{\boldsymbol{p}_{lab}}\|_{\sigma\infty} \leq \mu \|\boldsymbol{\Pi}\|_{\sigma\infty} \|(\boldsymbol{\Pi} + \Delta)^+\|_{\sigma\infty} \|\Delta\|_{\sigma\infty} \left[\sum_{i,j} < \hat{p}(x|i), \hat{p}(x|i) >\right]^{\frac{1}{2}},$$

where $\tilde{\boldsymbol{p}_{lab}}$ *is the solution with permutation matirx* $(\boldsymbol{\Pi} + \Delta)\boldsymbol{p}_{lab} = \hat{\boldsymbol{p}_{set}}$.

The proof is similar to the lemma in [4]. However, we can show that it does not differ much by a significant amount by bounding the errors made in computing an (pseudo-) inverse of a matrix. Normally, we assume that the matrix has the full column rank and more details about whether the condition of full column rank could be loosen is in supplemental material.

Estimation in Boltzmann Machines. Kernel density estimation has been implemented and shown worse performance than MAP and Inverse SVM in [6]. So we present a kind of deep belief networks approach for estimating the conditional density of $p(x|i), \forall i \in [n]$, which in fact estimating the log-probability of testing data according to the generative model. Without loss of generality, denote any observation as vector form $\boldsymbol{x} = (x_1, ... x_t)' \in R^t$.

A Boltzmann Machine is a kind of network of binary stochastic units which contains a bottom visible layer with $\boldsymbol{x} \in \{0,1\}^t$, and a hidden layer with $\boldsymbol{h} \in \{0,1\}^s$. The energy of the state $\{\boldsymbol{x}, \boldsymbol{h}\}$ is: $E(\boldsymbol{x}, \boldsymbol{h}) = -\frac{1}{2}(\boldsymbol{x}^T, \boldsymbol{h}^T)\boldsymbol{W}\left(\begin{smallmatrix}\boldsymbol{x}\\\boldsymbol{h}\end{smallmatrix}\right)$, where $\boldsymbol{W}$ contains parameters as the weights connected any couple of units. According to Gibbs distribution, we can define probability of the networks (assume temperature T is a equilibrium) at state $\{\boldsymbol{x}, \boldsymbol{h}\}$ is $p(\boldsymbol{x}, \boldsymbol{h}) = \frac{1}{Z}e^{-\frac{E(\boldsymbol{x},\boldsymbol{h})}{T}}$ where Z is a normalizing factor. What we concern about here is the margin distribution $p(\boldsymbol{x}) = \frac{1}{Z}\sum_{\boldsymbol{h}} e^{-\frac{E(\boldsymbol{x},\boldsymbol{h})}{T}}$, where $Z = \sum_{\boldsymbol{x},\boldsymbol{h}} e^{-E(\boldsymbol{x},\boldsymbol{h})}$ is normalizing factor. So our **goal** is to compute this margin probability in every subset.

Usually, a special BM, Restricted Boltzmann Machines (RBM) with energy function $E(\boldsymbol{x}, \boldsymbol{h}; \theta) = -\boldsymbol{x}^T \boldsymbol{W}\boldsymbol{h} - \boldsymbol{b}^T\boldsymbol{x} - \boldsymbol{a}^T\boldsymbol{h}$ is focused. Contrastive divergence learning proposed by Hinton makes a contribution to architecture of RBM. Here we will give a less consuming version of margin computation method than CD learning.

Margin Computation in RBM. The key of estimation $p(\boldsymbol{x})$ is to computing the normalizing factor Z, while the $p^*(\boldsymbol{x}) = \sum_{\boldsymbol{h}} \exp\left(-E(\boldsymbol{x}, \boldsymbol{h})\right)$ can be tractably computed in the case of RBM due to its conditional independence.

The idea to compute the normalizing factor Z is to construct a distribution sequence close to $p(\boldsymbol{x})$ step by step. If another distribution function $p_0(\boldsymbol{x}) = \frac{p^*(\boldsymbol{x})}{Z_0}$ and $p_0(\boldsymbol{x}) \neq 0$ whenever $p(\boldsymbol{x}) \neq 0$, we have $\frac{Z}{Z_0} = \frac{\int p^*(\boldsymbol{x})d\boldsymbol{x}}{Z_0} = \int \frac{p^*(\boldsymbol{x})}{p_0^*(\boldsymbol{x})}p_0(\boldsymbol{x})d\boldsymbol{x} = E_{P_0}\frac{p^*(\boldsymbol{x})}{p_0^*(\boldsymbol{x})}$. A natural thought for acquiring the ratio of normalizing factors is to draw independent samples from p_0 by using Monte Carlo method. However, if p_0 and p are not sufficiently close, the approximation $\frac{1}{M}\sum_{i=1}^{M} \frac{p^*(\boldsymbol{x}^i)}{p_0^*(\boldsymbol{x}^i)}$ will be very poor, where $\boldsymbol{x}^i$ is draw i.i.d from p_0. It is been proved that the variance of this term is very large, but annealing important sampling can be applied to circumvent this problem. By defining a distribution sequence $p_k(\boldsymbol{x}) \propto p_0^*(\boldsymbol{x})^{1-\beta_k} p^*(\boldsymbol{x})^{\beta_k}$ with $0 = \beta_0 < ... < \beta_K = 1$, we will get the AIS Algorithm in [10].

According to AIS, we can approximate $\frac{Z}{Z_0} \approx \frac{1}{M}\sum_{i=1}^{M} Mw^i = \hat{r}_{AIS}$. And the variance of $\hat{r}_{AIS}$ will be proportion to $\frac{1}{MK}$. And by setting the RBM of first distribution p_0 with parameters $\theta = \{0, b, a\}$, we can trivially obtain $Z_0 = 2^{M_0} \prod_i (1 + e^{b_i})$ and $p_0(x) = \prod_i 1/(1 + e-b_i)$. In addition, due to the special conditional independence in RBM, we can compute the normalizing distribution density $p_k(x)$ and obtain the estimate

$$\hat{p(x)} = \frac{\exp\left((1 - \beta_K)x^T b\right)}{\hat{r}_{AIS} Z_A} \sum_h \exp\left(-\beta_K\right) E(x, h) . \tag{2}$$

Algorithm 1. Deep Networks for Proportion Learning (DNPL)

Input: Proportion Matrix Π, subsets $\{X_1, ... X_n\}$, labels $Y = \{y_1, ..., y_l\}$ and test data x.

Output: label for x.

1. Compute $p(y_j)$, $\forall j \in [l]$, according to $\boldsymbol{Pi}$ and $\sum_i |X_i|$.
2. Compute $\boldsymbol{p}_{set}$ in each subset for x.
   (a) Construct a RBM according AIS;
   (b) For $i = 1$ to n;
       i. In ith subset, compute log-probability $\log(p(x))$ using Equation (2);
       ii. Scaling the $p(x)$;
3. Solving linear equations $\Pi \boldsymbol{p}_{label} = \boldsymbol{p}_{set}$ according to proportion matrix analysis.

Return $\arg \max_{j \in [l]} p(x|y_j) p_j$.

DNPL Algorithm. With all discussion above, we will put forward Algorithm 1 to solve the label proportions learning problem. **Note** that we call it DNPL because a refined version for density estimation can be obtained by Deep Networks combined by some RBMs. We also refine the primal algorithm in AIS for computing normalizing factor, because its version is designed for binary units, which mostly cannot be satisfied. One trick is to scale continuous-valued inputs to the $(0, 1)$ interval and consider each input them as the probability for a binary random variable to take the value 1. This has worked well for pixel gray levels, but it may be inappropriate for other kinds of input variables. A more advanced approach is to apply an adjust of the Gaussian units for fitting the continuous-valued inputs according to [2].

4   Experimental Results

Datasets Selection. We split dataset with high ratio between subset size and the entity according to the discussion in Section 1. This way we can distinguish whether the result of performance is caused by some learnable subsets of random sampling. For the sake of comparing performance with Inverse SVM [5] which has been shown statistically better than another famous algorithm Mean Map in [6], we use modified five UCI datasets for 15 experiments. For the training set, we partition the samples of each class into groups according to a specified class-conditional distribution based on our assumption, then calculate each group distribution of samples in each class. We then discard all the labels in the training set, and feed the unlabeled training data with features alone and label proportions matrix to the learning algorithm. Finally, we test the classification accuracy using the unlabeled data in the test set. Previous algorithms are mostly designed for subset size 2 to 64. Due to our focus on large ratio, the partition strategy of naturally depends on the size of dataset, for $m = 32, 64, 128$ and 256 respectively, which is shown in Table 1.

Table 1. Dataset Partition Strategy and DNPL Performance

Exp. No.	Dataset	Training Size	Subset Size	Test Size	Test Error (%)
01,02	Breast	512	32,64	171	2.92+-0.00
03			128		3.51+-0.00
04			256		5.26+-0.00
05	Re	196	32	58	18.97+-0.00
06			64		23.62+-0.84
07	Wdbc	512	32	57	3.51+-0.00
08,09			64,128		5.26+-0.00
10			256		13.64+-0.00
11,12	Wpbc	128	32,64	66	28.79+-0.00
13	Iris	128	16	22	18.18+-0.00
14			32		13.64+-0.00
15			64		36.82+-9.45

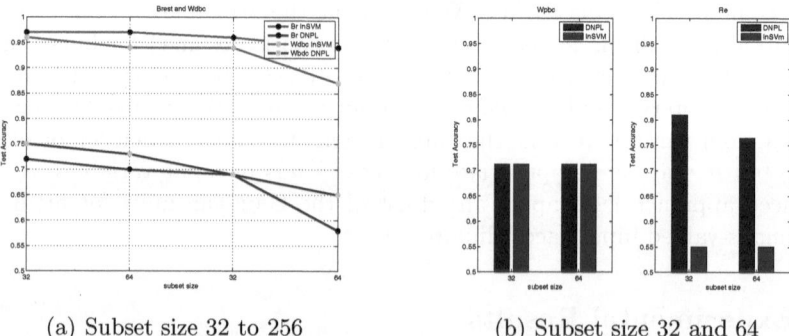

(a) Subset size 32 to 256   (b) Subset size 32 and 64

Fig. 2. We test binary-label case in four dataset with 10 experiments in large size ratio partition strategy. And the classification accuracy comparison shows the performance of our method is not worse than Inverse SVM. The dataset Iris is a multi class case, so we cannot compare the result with Inverse SVM.

For the density estimation approach to proportions learning, we use the same datasets as used in the SVM approach. Our model uses the same classic RBM model as the one used in [9]. In the following experiments, we fix the number of hidden units in RBM to be 2, and use CD-25 (i.e. doing blocked Gibbs sampling for 25 rounds before one batch update of RBM parameters) to approximate maximum likelihood learning. After training the RBMs with data from one group, we calculate the log-likelihood of each test sample generated from the trained model. Finally we combine these subset-conditional data likelihood with subset-conditional label proportions to get the class-conditional likelihood of each test sample, and do classification by MAP inference. The compared result with Inverse SVM is shown in Fig 2.

5 Conclusion and Future Work

In this paper we presents a theoretical analysis between label proportions learning and supervised learning by giving a sufficient condition of learnable case on binary classification. And an exponential relationship between entire dataset size and subset size has been depicted, though it is negative on the circumstance of random sampling. However, we have known that assumptions in this domain are as necessary as they are in that of semi-supervised learning. Meanwhile, a new approach in straightforward way has been put forward to solve this problem and performed well in experiments.

Acknowledgments. This work was supported by NSFC(61222307, 61075003) and a grant from Microsoft Research Asia.

References

1. Stolpe, M., Morik, K.: Learning from Label Proportions by Optimizing Cluster Model Selection. In: Gunopulos, D., Hofmann, T., Malerba, D., Vazirgiannis, M. (eds.) ECML PKDD 2011, Part III. LNCS (LNAI), vol. 6913, pp. 349–364. Springer, Heidelberg (2011)
2. Bengio, Y., Lamblin, P., Popovici, D., Larochelle, H.: Greedy Layer-Wise Training of Deep Networks. In: NIPS, pp. 153–160 (2007)
3. Kueck, H., de Freitas, N.: Learning about individuals from group statistics. In: Uncertainty in Artif (UAI), pp. 332–339. AUAI Press, Arlington (2005)
4. Wedin, P.A.: Perturbation theory for pseudo-inverses. BIT Numerical Mathematics 13(2) (1973)
5. Rueping, S.: SVM Classifier Estimation from Group Probabilities. In: Proceedings of the 27th Annual International Conference on Machine Learning, ICML 2010 (2010)
6. Quadrianto, N., Smola, A., Caetano, T., Le, Q.: Estimating labels from label proportions. JMLR 10, 2349–2374 (2009)
7. Carreira-Perpinan, M., Hinton, G.: On contrastive divergence learning. In: 10th Int. Workshop on Artificial Intelligence and Statistics, AISTATS 2005 (2005)
8. Hinton, G.: A Practical Guide to Training Restricted Boltzmann Machines. UTML TR, 2010-003 (August 2010)
9. Hinton, G., Osindero, S., Teh, Y.: A fast learning algorithm for deep belief nets. Nerual Computation 18, 1527–1554 (2006)
10. Salakhutdinov, R., Murray, I.: On the Quantitative Analysis of Deep Belief Networks. In: ICML (2008)

Neighborhood-Preserving Estimation Algorithm for Facial Landmark Points

Yin Liu, Ying Cui, and Zhong Jin

School of Computer Science and Engineering
Nanjing University of Science and Technology, Nanjing, China
seraphever@126.com,
{cuiying,jinzhong}@patternrecognition.cn

Abstract. In this paper, we propose a neighborhood-preserving estimation (NPE) algorithm for facial landmark points at arbitrary poses. The proposed NPE algorithm is based on the following assumption: the neighboring structure of the face shapes in non-frontal view is consistent with that in frontal view. A face shape both in frontal and non-frontal view is represented as a linear combination of its neighbors. It is assumed the neighbors both in frontal and non-frontal view are from same persons and with same weights in the combinations. Extensive experiments are conducted on IMM and BU_3DFE to validate the performance of the proposed algorithm.

Keywords: neighborhood-preserving estimation, regression model, facial landmark points.

1 Introduction

Automatic face annotation plays an essential role in many applications. Accurate face annotation is a premise for virtual generation of face images and face recognition. It is an interesting and challenging problem to develop an automatic annotation method, which can effectively predict the spatial locations of the landmark points for face images under varying poses and expressions.

In [1], modeling images and related visual objects as bags of pixels or sets of vectors was proposed, which made the computation of correspondences for all images simultaneously possible. In [2], the automatic face annotation problem was viewed as an energy-minimizing image coding problem which can be solved by an efficient gradient descent algorithm. [3] and [4] made full use of image sequences to annotate face images automatically. Recently, an approach for automatic annotation of the faces with any arbitrary pose and expression from annotated frontal faces only was put forward by Asthana [5]. The relevant work was extended in [6-8], in which a regression based algorithm to predict the landmarks of unseen images was presented and demonstrated on the examples of automatically annotating face images. The approach based on regression learning drastically simplifies the process of deformable modeling building and decreases the computation.

Manifold learning algorithms are based on the idea that the data points are actually samples from a low-dimensional manifold that is embedded in a high-dimensional

J. Yang, F. Fang, and C. Sun (Eds.): IScIDE 2012, LNCS 7751, pp. 630–638, 2013.

space. Locally Linear Embedding (LLE) [9] is one of the manifold learning algorithms, which attempts to reduce the dimension of the data while preserving the relationships between neighboring data points. Motivated by LLE, we propose the neighborhood-preserving estimation algorithm for facial landmark points.

The rest of the paper is organized as follows. Section 2 introduces regression based learning methods briefly. Section 3 describes the proposed method in detail. In section 4, we exhibit the experimental results and section 5 provides the discussion and conclusion of the paper.

2   Regression Based Learning

The core idea of the regression based learning approach is to learn the correspondences of the frontal and non-frontal face images in a regression framework.

Given a set of training set $\{(s_1, s_1'), ...(s_m, s_m')\} \subset R^{2n} \times R^{2n}$, where m is the number of the frontal and the corresponding non-frontal images and $s_i, s_i' (i = 1, ..., m)$ is the shapes of the i-th person for frontal and non-frontal images respectively:

$$s = [x_1, y_1, ..., x_n, y_n]^T \tag{1}$$

where $x_j, y_j (j=1, \cdots, n)$ is the coordinate of the j-th landmark point of the face image. What we desire to do is to minimize the objective function:

$$\sum_{i=1}^{m} \left| s_i' - P(s_i) \right| \tag{2}$$

where $P(s_i)$ means the shape we estimate. The error between the manual and estimating annotation is expected to be as small as possible.

Viewed the shape vector as an integral whole, the correspondences can be posed in a linear framework and the location and local movements of the landmark points are completely dependent, then we'll get the dense linear regression (DLR) approach, in which the regression function can be represented as

$$s' = P(s) = Ws + v \tag{3}$$

where W is the regression matrix, s' is the shape vector estimated in non-frontal view and v is the noise. The problem becomes a ridge regression problem,

$$\min\{\|P - WG\|_2^2 + \lambda \|W\|_2^2\} \tag{4}$$

where $G = [s_1, s_2, ..., s_m]$ and $P = [s_1', s_2', ..., s_m']$, $\lambda > 0$ is a regularization factor and $\|a\|_2^2 = (\sum_i a_i^2)^{1/2}$ represents the L_2 norm of a. Eq. (4) can be solved by

$$W = PG^T (GG^T + \lambda I)^{-1} \tag{5}$$

Similar to DLR, if the correspondences are put in the fully sparse representation framework, we'll get the fully sparse linear regression (FSLR) approach. What's more, if the regression function is non-linear, the dense non-linear regression (DNLR) and the fully sparse non-linear regression (FSNLR) are generated. Reference [8] expands on all these methods, so we skip here.

3   Neighborhood-Preserving Estimation

In this section, we take full account of the neighboring information of the testing face shapes, and based on the hypotheses that the the neighborhood structure in the frontal and non-frontal view are consistent, we propose the neighborhood-preserving estimation algorithm to predict the facial shape vectors.

3.1   Motivation

As a new unsupervised learning method, manifold learning is capturing increasing interests of researchers in the field of machine learning. Manifold learning is a popular recent approach to non-linear dimensionality reduction, which maps the data into a lower-dimensional space and preserves the information in the original space simultaneously.

As one of the manifold learning methods, LLE is based on simple geometric intuition [9]: each data point and its neighbors are lie on or close to a locally linear patch of the manifold, so the data point can be represented as a linear combination of its neighbors. What's more, when the original data point is mapped into a certain high-dimensional space, the corresponding data point can also be reconstructed with its corresponding neighbors in the high-dimensional space and the reconstruction weights are invariant to the mapping.

LLE preserves the neighborhood structure of the data sets between the observed and the mapping space. Regression based learning methods consider the shape of the face as a whole and the goal is to find the regression function between the face shapes of the frontal and non-frontal view. Here we look the face shape as a data point in 2n space. Since regression methods look for some relationship to map the data point in frontal view into the data point in non-frontal view, then inspired by LLE, we have a similar assumption: the neighborhood structures of the face shapes are invariant to the head pose changing. That is to say, given a testing face shape in frontal view and its neighbors, the corresponding counterpart in non-frontal view preserve the same neighboring information. The regression weights computed from the frontal view are the same to those of the non-frontal view. Besides, the neighboring face shapes in frontal and non-frontal view are from the same subject.

3.2   NPE Algorithm

In order to illustrate our idea clearly and intuitively, we conduct a small experiment. We collect a number of annotated face images, and then choose a frontal shape as the testing sample and the other shapes in frontal and non-frontal views as the training samples. Firstly, we compute the distance between the testing sample and the training samples in frontal view and find ten nearest neighbors of the testing sample. Then we reconstruct the testing sample from its neighbors with linear coefficients. Finally, we use the linear coefficients and the shapes of chosen neighbors in non-frontal view to estimate the testing shapes in non-frontal view. Fig.1 shows the scattered dots distribution of the testing and the training shapes, and the estimated shape as well. From Fig.1 (a) and (b), we can find that the shape of the testing sample is close to the shapes of its neighbors in both frontal and non-frontal view. The estimated shape for the testing sample in non-frontal view is shown in Fig.1 (c). It can be observed that the estimated shape is similar to the manually annotated shape of the testing sample. So, the proposed method can estimate the facial landmark points in non-frontal view.

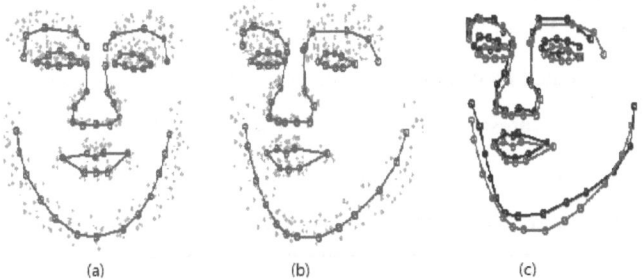

(a)   (b)   (c)

Fig. 1. A small experiment conducted with neighborhood-preserving estimation method: red and green dots represent the landmark points in testing and training shapes, blue dots represent the estimating landmarks, (a) testing shape and the shapes of the ten nearest samples for testing sample in frontal view; (b) the shapes of the testing sample and its ten nearest neighbors in non-frontal view; (c) the estimated shape with our method

Given an annotated frontal face image, we estimate its face shape t' in non-frontal view. Firstly, we calculate the distance between the testing face shape and the training face shapes to find K nearest neighbors. Thus, we can use the K nearest neighbors to represent the testing face shape as

$$t = w_1 s_1 + ... + w_K s_K = SW \tag{6}$$

where $s_i (i=1,2, \cdots, K)$ are the *ith* nearest neighbor in frontal view and we can rewrite them into $S=[s_1, \cdots, s_K]$, and all the linear coefficients w_i $(i=1,2, \cdots, K)$ can be constructed as the reconstruction weight vector $W=[w_1, \cdots, w_K]^T$. And W can be acquired by using

$$W = (S^T S + \alpha I)^{-1} S^T t \tag{7}$$

where α is a non-negative constant and I is an unit. The reconstruction weight vector implies the K nearest neighbors' contribution to representing the testing shape in frontal and non-frontal view. Lastly, we predict the face shape with W and the K nearest neighbors in non-frontal view. The consequence is

$$t^{'} = S^{'}W = \sum_{i=1}^{K} w_i s_i^{'}$$ (8)

where $s_i^{'} (i = 1,..., K)$ is the face shape of the ith person in non-frontal view and $t^{'}$ is the shape vector predicted by NPE approach. Since the shapes of the subjects in non-frontal and frontal views keep the same neighboring structure, $s_i^{'}$ and s_i are the face shapes corresponding to the same person.

The complete framework of our algorithm is summarized in Algorithm I.

Algorithm I (Neighborhood-Preserving Estimation Algorithm)

Input: An annotated frontal image for testing, the fixed pose rotation

Output: An estimated face shape with the determined pose rotation

  Step 1: Find K nearest neighbors for the testing sample in frontal view;

  Step 2: Use Eq. (7) to compute the reconstruction weight vector;

  Step 3: Use Eq. (8) to estimate the face shape t'.

---

4   Experiments

4.1   Experiments on IMM

We conduct experiments on the IMM database [10]. IMM has a total of 240 images across 40 persons and each of them is manually annotated with 58 landmark points. Fig.2 (a) represents one example with the annotated landmark points in IMM.

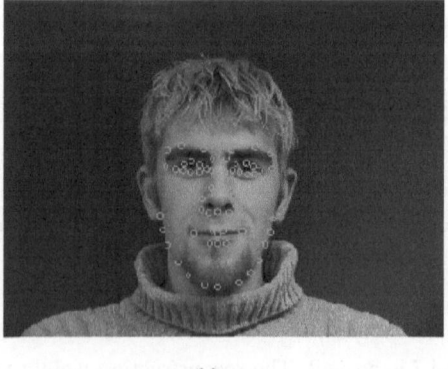 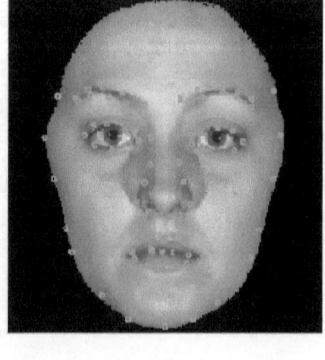

(a)   (b)

Fig. 2. Example faces with manually landmark points in IMM and BU_3DFE: (a) one sample with 58 landmark points in IMM; (b) one sample with 83 landmark points in BU_3DFE

To demonstrate the performance of the proposed method, we estimate the shape vectors for the face images, and compute the error between the manually annotated and the estimating landmark points.

We adopt the leave-one-out cross validation scheme throughout the experiment. Each time we choose one sample in the database as the testing sample and the others are used for training. In order to evaluate the performance of the proposed approach, we calculate the mean point-to-point error between the manually annotated and the estimated landmark points for every image.

Fig.3 (a) shows the error distributions over IMM obtained from NPE. It is clear that NPE approach can be used to estimate facial landmark points under varying poses. The error between the automatic and the manually annotated landmark points are less than 2 pixels.

For the sake of showing a visual result, we give an example of the estimated facial landmark points on IMM, which are shown in Fig.4 (a). It is noted that the estimated landmark points are accurate under changing head poses. And the result of the fourth image is good as well, which means NPE can predict accurate shapes under changing illuminations. But the location of the landmark points in the fifth image in Fig.4 (a) is not as precise as that under the other conditions.

4.2   Experiments on BU_3DFE

The aforementioned NPE approach provides a satisfactory result on IMM, so NPE is able to estimate the facial landmark points under varying poses. If we regard the neutral expression as the frontal view and the expressions, such as happy, as the non-frontal view, we'll have the similar thought under changing expressions. The neighboring information of the shapes with the neutral and the other expressions keeps the same. To validate the thought, we make an experiment on BU_3DFE database [11], which contains 2500 images at seven expressions in frontal view and each of them is annotated with 83 landmark points. Fig.2 (b) represents an example with the annotated landmarks in BU_3DFE.

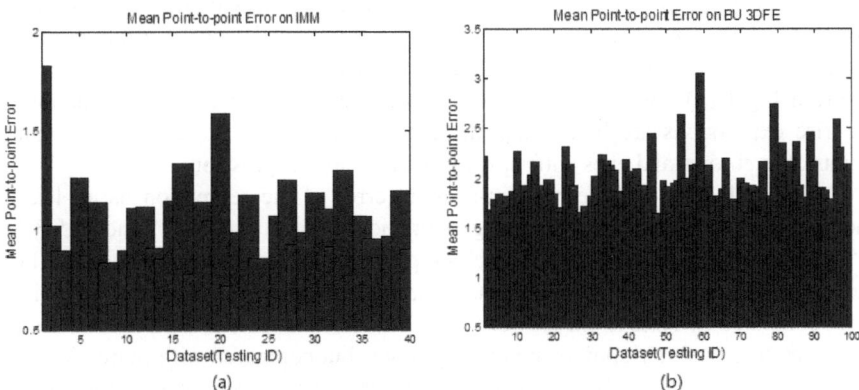

Fig. 3. Estimated annotation results for IMM and BU_3DFE with NPE: (a) mean point-to-point errors on IMM database; (b) mean point-to-point errors on BU_3DFE database

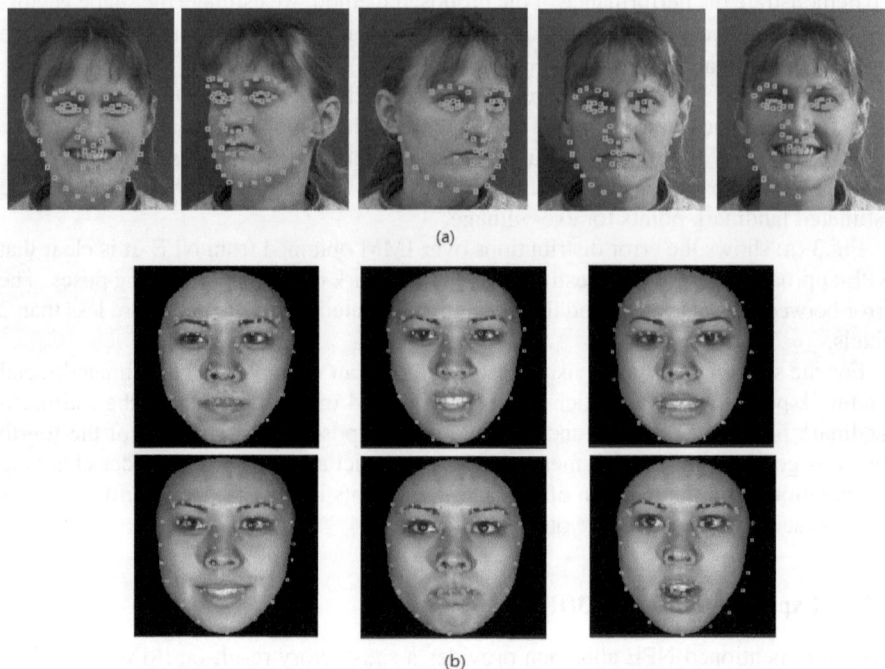

Fig. 4. Sample estimated results via NPE from IMM and BU_3DFE: (a) the shapes of the sample estimated in IMM, every image represent full frontal face with happy expression, face rotated approx 30 degrees to the person's right and left, full frontal face spot light added at the person's left side and full frontal face with arbitrary expression respectively; (b) the shapes of the sample estimated in BU_3DFE, each image represents the result with a certain expression (angry, disgust, fear, happy, sad and surprise)

Fig.3 (b) shows the the error distribution over BU_3DFE. It is noted that the errors between the estimated and the manually annotated landmark points are less than 3.5 pixels. So, NPE can estimate shapes at varying poses well, but the errors are a litter bigger when estimating the shapes with arbitrary expressions.

From Fig.4 (b) , we can find that the estimated results with disgust, happy and surprise expressions are better than the other expressions. The estimated landmark points of eyebrows and eyes shift upward with the fear expression.

Table 1 tabulates the mean point-to-point error of the regression based learning methods and neighborhood-preserving estimation method over IMM and BU_3DFE. We can see that the proposed approach outperforms the regression methods in IMM. However the results on BU_3DFE dataset are not as good as the results on IMM. The error computed by NPE is only lower than DLR. To some extent, NPE is more suitable for estimating landmark points at arbitrary poses, but compared with regression based learning methods, its ability of estimating the landmark points under varying expressions is not good enough.

Table 1. Mean estimation errors obtained for the IMM and BU_3DFE

Mean Error	DLR	FSLR	DNLR	FSNLR	NPE
IMM	1.9895	1.4519	1.7362	1.5889	1.0195
BU_3DFE	2.3685	1.5883	1.7098	1.5731	1.9381

5   Discussion and Conclusion

An approach based on neighborhood-preserving has been proposed for simplifying the Active Appearance Model (AAM) fitting process. NPE makes use of the neighboring information of the testing face shape, which uses local linear regression to reflect the global non-linear information. The experiments mentioned above indicate that NPE can estimate the face shapes with varying poses well, while the results on BU_3DFE show NPE is not very suitable to estimate the face shapes with changing expressions. NPE views the face shapes as data points, which implies NPE does not take the local information of the face landmark points into account. The face shapes with varying expressions may be influenced greatly with the local landmark points, so we get poor relatively performance on BU_3DFE.

Future work will focus on exploring an algorithm that can model the subtle changes of the landmark points at arbitrary expressions and studying an approach which not only use the neighboring structure of the testing face shape, but also take the local information of the face shapes to get more accuracy results.

Acknowledgments. This work is partially supported by the Natural Science Foundation of China under grant No. 60973098, 61005005.

References

1. Jebara, T.: Images as Bags of Pixels. In: Proc. IEEE International Conference on Computer Vision, New York, vol. 1, pp. 265–272 (2003)
2. Baker, S., Matthews, I., Schneider, J.: Automatic Construction of Active Appearance Models as an Image Coding Problem. IEEE Transactions on Pattern Analysis and Machine Intelligence 26(10), 1380–1384 (2004)
3. Walker, K., Cootes, T., Taylor, C.: Automatically Building Appearance Models from Image Sequences Using Salient Features. Image and Visual Computing 20(5-6), 435–440 (2002)
4. Saragih, J., Goecke, R.: Learning Active Appearance Models from Image Sequences. In: HCSNet Workshop on the Use of Vision in HCI (VisHCI 2006), Conferences in Research and Practice in Information Technology (CRPIT), Canberra, pp. 51–60 (2006)

5. Asthana, A., Khwaja, A., Goecke, R.: Automatic Frontal Face Annotation and AAM Building for Arbitrary Expressions from a Single Frontal Image Only. In: IEEE International Conference on Image Processing, Cairo, pp. 2445–2448 (2009)
6. Asthana, A., Goecke, R., Quadrianto, N., Gedeon, T.: Learning Based Automatic Face Annotation for Arbitrary Poses and Expressions from Frontal Images Only. In: IEEE Conference on Computer Vision and Pattern Recognition, pp. 1635–1642 (2009)
7. Asthana, A., Sanderson, C., Gedeon, T., Goecke, R.: Learning-based Face Synthesis for Pose-Robust Recognition from Single Image. In: Proceedings of the British Machine Vision Conference, Dublin, pp. 2–4 (2009)
8. Asthana, A., Lucey, S., Goecke, R.: Regression Based Automatic Face Annotation for Deformable Model Building. Pattern Recognition 44(10-11), 2598–2613 (2011)
9. Roweis, S.T., Saul, L.K.: Nonlinear Dimensionality Reduction by Locally Linear Embedding. Science 290(5500), 2323–2326 (2000)
10. Nordstrom, M.M., Larsen, M., Sierakowski, J., Stegmann, M.B.: The IMM Face Database-An annotated Dataset of 240 Face Images. Informatics and Modeling, Technical University of Denmark, DTU (2004)
11. Yin, L., Wei, X., Sun, Y., Wang, J., Rosato, M.J.: A 3D Facial Expression Database for Facial Behavior research. In: The 7th International Conference on Automatic Face and Gesture Recognition (FG 2006), IEEE Computer Society TC PAMI, Southampton, pp. 211–216 (2006)

An Efficient Algorithm for Feature Selection with Feature Correlation

Li-li Huang[1,2], Jin Tang[1,2], Si-bao Chen[1,2], Chris Ding[1], and Bin Luo[1,2]

[1] School of Computer Science and Technology, Anhui University, 230601, China
[2] Key Lab of Industrial Image Processing & Analysis of Anhui Province,
Hefei 230039, Anhui, China
huaiyangok@126.com, ahhftang@gmail.com,
{sbchen,luobin}@ahu.edu.cn, chqding@uta.edu

Abstract. Feature selection is an important component of many machine learning applications. In this paper, we propose a new robust feature selection method for multi-class multi-label learning. In particular, feature correlation is added into the sparse learning of feature selection so that we can learn the feature correlation and do feature selection simultaneously. An efficient algorithm is introduced with rapid convergence. Our regression based objective makes the feature selection process more efficient. Experiments on benchmark data sets illustrate that the proposed method outperforms many state-of-the-art feature selection methods.

Keywords: feature selection, feature correlation, multi-label classification, sparse learning, machine learning.

1 Introduction

In machine learning, more and more high dimensional data need to be processed. However, high dimensional data usually are hard for classification due to the curse of dimensionality. A common way to resolve this problem is feature selection, that is to select a subset of the most representative or discriminative features from the input feature set (feature dimensions). The main challenge is to select a set of features, as small as possible, that help the classifier to accurately classify the learning examples.

Depending on how the classifiers are involved in the procedure of feature selection, feature selection methods can be divided into three general categories. The first category is *filter methods* where the selection is independent of classifiers[1]. Widely used filter-type feature selection methods include *F*-statistic, ReliefF, mRMR[2], and information gain [3]. The second category is *wrapper* methods where a classifier is used as a black box to evaluate subsets of features [4]. There are "forward selection", "backward selection", and "floating selection". The advantage of wrapper methods is that the feature subset is the most effective when combined with the classifier used in the wrapper. This may also indicate that the selected feature subset could overfit, i.e., the performance degrades when using different test data. Another drawback is that the selected features are less optimal for other classifiers. There are other feature selection methods that do not fall into above categories. The embedded method [5] is one type where the procedure of feature selection is embedded directly in the training process. Decision tree is one such example.

J. Yang, F. Fang, and C. Sun (Eds.): IScIDE 2012, LNCS 7751, pp. 639–646, 2013.

Recently, sparsity regularization receives increasing attention in feature selection studies. Tibshirani[6] introduced LASSO method where the L_1 regularization is used to shrink/suppress variables to achieve the goal of variable selection. Group LASSO [7] was proposed where the covariates are assumed to be clustered in groups, and the sum of Euclidean norms of the loadings in each group is utilized. Obozinsky et. al. [8] and Argyriou et. al. [9] proposed a similar model for $L_{2,1}$-norm regularization.

In this paper, motivated by the previous sparse learning based research, we propose to add *feature correlation* into the sparse learning feature selection approach. We note that in previous LASSO type feature selection, feature correlations are not taken into account, while in most real-life data, features dimensions are often correlated. If three feature dimensions p, q, r of data X are highly correlated, then the shrink of dimension p will very likely lead the shrink of dimensions q and r. On the other hand, the purpose of feature selection is to select one feature dimension out of several high correlated feature dimensions, i.e., selecting one dimension out of p, q, r dimensions. Therefore we need to take into account the feature correlation in feature selection. To our knowledge, existing LASSO type of feature selection methods have not considered feature correlation.

In the following, we present our formulation of LASSO type feature selection. We first discuss the two-class case using standard LASSO. We then present our method for multi-class with structured sparsity using $L_{2,1}$ norm. We then propose an efficient algorithm to solve the proposed sparse learning problem and present vigorous convergence analysis of the algorithm. This method is tested on several multi-label multi-class data sets. Experimental results suggest that our method outperforms other commonly used methods.

2   Robust Feature Selection Based on Feature Correlation

2.1   Two-Class Case

We start with two-class classification. Given training data $X = [x_1, x_2, \cdots, x_n] \in \Re^{p \times n}$ and the associated class labels $y = [y_1, y_2, \cdots, y_n]^T \in \Re^n$, where $y_i = \pm 1$, $i = 1, \cdots, n$ are class labels. The classical LASSO problem for two-class learning is:

$$\min_{\beta} \left\| y^T - \beta^T X \right\|^2 + \lambda \|\beta\|_1 . \tag{1}$$

The key point of LASSO is that when λ becomes large, some elements of β become zero. If the k-th element of β is zero, then the k-th row of X are always multiplied by a zero and therefore do not contribute to $\beta^T X$. In other words, the k-th feature is eliminated and the remaining features are *selected*. However, LASSO can not deal with the condition that the features selected may highly correlate with each other. That is to say, there are redundant features among the selected features.

In this paper, we propose to minimize the following term

$$\min_{\beta} \beta^T C \beta \tag{2}$$

where the (k, l)-th element of C, C_{kl}, is the absolution value of the correlation coefficient between feature dimensions k and l,

$$C_{kl} = \left| \frac{\sum_i (x_i^k - \bar{x}^k)(x_i^l - \bar{x}^l)}{\sqrt{\sum_i (x_i^k - \bar{x}^k)^2} \sqrt{\sum_i (x_i^l - \bar{x}^l)^2}} \right| . \tag{3}$$

This term enforces that when C_{kl} is large, β_k, β_l can not be large simultaneously, thus achieving the goal of minimizing the un-wanted correlation between elements (feature dimensions) of β.

2.2   Generalization to Multi-class

Flat Sparsity Formulation. Now we generate this to multi-class multi-label learning with class labels $Y = [y_1, y_2, \cdots, y_n] \in \mathfrak{R}^{c \times n}$ where $y_i = \{0,1\}^{c \times 1}$ describes the class labels of x_i. The LASSO formulation can be extended to

$$\min_B \sum_{k=1}^c [\sum_{i=1}^n (y_i^k - \beta_k^T x_i)^2 + \lambda \|\beta_k\|_1] = \|Y - B^T X\|_F^2 + \lambda \|B\|_1 \tag{4}$$

where $B = [\beta_1, \cdots, \beta_c]$ and $\|B\|_1 = \sum_{ij} |B_{ij}|$. However, in this form (often called *flat sparsity*), it often happens that the features selected for different classes are not consistent --- no unique feature set fits all c classes. What we need is to select the same feature set for all c classes.

Structured Sparsity Formulation. We can achieve this goal of selecting the same feature set for all c classes by the following method: suppose there is a method that enforces all elements in one row of B to be zero; thus all c classes eliminate this feature dimension of X. Suppose the method can enforce several rows of B to be zero, then these feature dimensions of X are eliminated by all c classes.

The goal to enforce entire rows of B to be zero can be achieved by using the group LASSO or the $L_{2,1}$ norm. $L_{2,1}$ norm of matrix $M \in \mathfrak{R}^{p \times q}$ is defined as $\|M\|_{2,1} = \sum_{j=1}^q \sqrt{\sum_{i=1}^p m_{ij}^2} = \sum_{j=1}^q \|m_j\|_2$. This will enforce several $\|m_j\|_2$ to be zero, which is identical to setting several columns of M to be all zero. Then the structural LASSO is formed:

$$\min_B \|Y - B^T X\|_F^2 + \lambda \|B^T\|_{2,1} \tag{5}$$

where $L_{2,1}$ norm enforces some rows of B to be all zero, and the associated rows of X are not used (feature eliminated).

Following the above discussion of minimizing feature correlation in Eqs.(1,2), noting that $\beta_1^T C \beta_1 + \cdots + \beta_c^T C \beta_c = Tr(B^T CB)$, we thus propose the feature selection model with feature de-correlation focus:

$$\min_B J(B) = \|Y - B^T X\|_F^2 + \lambda \|B^T\|_{2,1} + \gamma Tr(B^T CB) \tag{6}$$

In the next section, we propose an efficient algorithm to solve this problem.

2.3   An Efficient Algorithm

Since all three terms in Eq.(6) are convex, and addition of convex functions is also convex. Therefore the optimization problem of Eq.(6) is a convex optimization, and B has a unique global optimum solution. We propose an iterative algorithm in this paper to obtain the solution B. The algorithm is described in Algorithm 1. In each iteration step, D is calculated with the current B, and then B is updated based on the current calculated D. The iteration procedure is repeated until the algorithm converges.

Algorithm 1. An efficient iterative algorithm to solve the optimization problem in Eq.(6)

Data (Input): $X \in \Re^{p \times n}$, $Y \in \Re^{c \times n}$

Result (Output): $B \in \Re^{p \times c}$

Initialize: $B = (XX^T + \gamma C)^{-1} XY^T$

Repeat

 Calculate the diagonal matrix D with $D_{jj} = \left\| B^j \right\|_2$.

 Calculate $B_{t+1} \leftarrow D^{\frac{1}{2}} \left[D^{\frac{1}{2}} (XX^T + \gamma C) D^{\frac{1}{2}} + \lambda I \right]^{-1} D^{\frac{1}{2}} XY^T$.   (7)

Until Converges

We have done analysis of the proposed algorithm with the following results:

Theorem 1. The objective function in Eq.(6) decreases monotonically, $J(B^{t+1}) \le J(B^t)$ with each update of Eq.(7) in Algorithm 1.

Due to the length limit of the paper, the proof of Theorem 1 is omitted here. In Fig. 1, we show the behavior of objective function value of Eq.(6) during iterations of our proposed algorithm on four data sets (details are given in Sec 3). In all data sets, the objective decreases monotonically and rapidly, confirming the result of Theorem 1.

3   Experiments

3.1   Data Sets and Evaluation Metrics

In order to validate the performance of the proposed method, we apply our method into multi-label classification. Four publicly available data sets are used: MSRC, Mediamill, Scene, and Yeast. Table 1 summarizes these data sets. The K Nearest Neighbor (KNN) classifier [10] is employed to these data sets, using 5-fold cross-validation.

 MSRC data set includes 591 images with 23 classes (i.e., "building", "car", "road", "grass") [11]. It is provided by the computer vision group at Microsoft Research Cambridge. We extract the 384-dimensional color moment features, where each image is divided into 64 blocks by an 8×8 grid and the first and second moments (mean and variance) of each color band are calculated.

 Mediamill data set contains 43907 sub-shots with 101 classes. We follow [12] to characterize each image by a 120-dimensional vector.

Scene data set comprises 2407 images with 6 classes. We extract 294-dimensional features of Luv space as recommended by the researches in [13].

Yeast data set is formed by micro-array expression data and phylogenetic profiles [14]. The input feature dimension is 103, while label dimension is14.

Multi-label classification requires different evaluation metrics than conventional single-label classification problem. We adopt three measures: classification average accuracy, Micro F1 and Macro F1 to evaluate the performance of different algorithms for multi-label learning. For multi-label classification average accuracy, we first compute the classification accuracy for each class, and then compute the averaged classification accuracy over all the classes. The higher the three measures are, the better the performance is.

Table 1. Description of the data sets

Datasets	MSRC	Mediamill	Scene	Yeast
Domain	images	video	images	biology
#training	591	43907	2407	2417
#classes	22	101	6	14
#features	384	120	294	103

3.2   Classification Results

All data sets are standardized to be zero-mean and normalized by standard deviation. The proposed feature selection method (called as FCFS) is compared with several popularly used feature selection methods, such as Information Gain (IG) [3], mRMR [2], and RFS [15]. We first select a subset of the most representative or discriminative features from the original feature set, and then make use of the subset to classify.

Fig. 2 shows the classification accuracy comparisons of all four feature selection methods on different feature number. Table 2 shows the detailed experimental results using the top 20 and top 80 features for all feature selection approaches. As we can see from Fig. 2, the line denoting FCFS method is much smoother than other lines, which demonstrates the proposed FCFS method is not only efficient, but also robust. Furthermore, with the same amount of features, the classification accuracy of FCFS is always higher than that of other three feature selection methods. Obviously our approach outperforms other methods significantly due to its consideration of feature correlation.

F1-score considers both the precision and the recall and can be viewed as a harmonic mean of precision and recall. Therefore, Micro F1 and Macro F1 are good measures of the classification accuracy.

The Micro F1 and Macro F1 results are given in Table 3. We firstly compute the Micro F1 using the top 20 and top 80 features respectively, and then take their average as the final results shown in Table 3. The way we compute Macro F1 is the same to Micro F1. From Table 3, we can draw similar conclusions as that of table 2. The Micro F1 and Macro F1 results of FCFS are both higher than other methods.

Table 2. The detailed classification average accuracy on the four data sets

Data set	Average accuracy of top 20 features				Average accuracy of top 80 features			
	IG	mRMR	RFS	FCFS	IG	mRMR	RFS	FCFS
MSRC	0.8657	0.8276	0.8914	0.9474	0.8925	0.8907	0.9261	0.9561
Mediamill	0.6962	0.6902	0.7072	0.7523	0.7179	0.7090	0.7008	0.7440
Scene	0.6963	0.6276	0.6572	0.8024	0.7916	0.7650	0.7996	0.8101
Yeast	0.8881	0.8584	0.8764	0.9317	0.9228	0.8976	0.9147	0.9261

Table 3. The Micro F1 and Macro F1 results (average) on the four data sets

Data sets	Measures	Compared methods			
		IG	mRMR	RFS	FCFS
MSRC	Macro F1	0.8739	0.8318	0.8912	0.9436
	Micro F1	0.8762	0.8261	0.9023	0.9383
Mediamill	Macro F1	0.7084	0.6901	0.6964	0.7328
	Micro F1	0. 6914	0.6722	0.6845	0.7293
Scene	Macro F1	0.6958	0.6539	0.6451	0.7841
	Micro F1	0.7193	0.6281	0.6872	0.7973
Yeast	Macro F1	0.8962	0.8465	0.8730	0.9219
	Micro F1	0.8794	0.8618	0.8836	0.9137

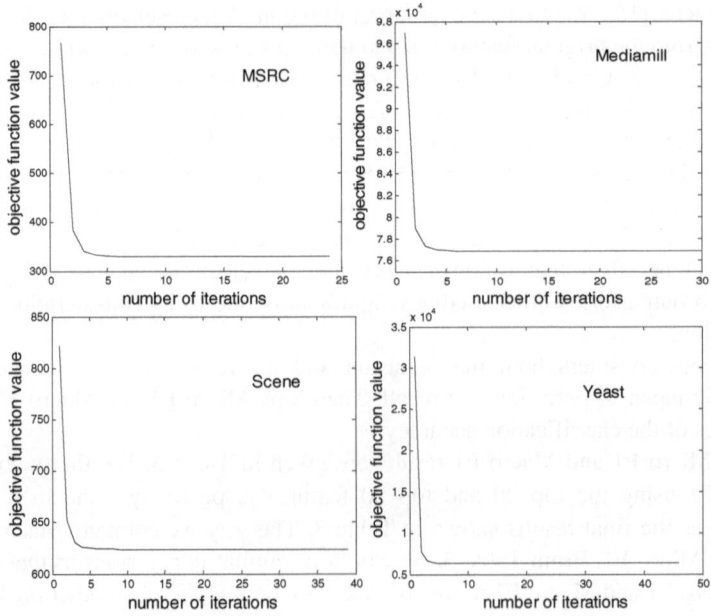

Fig. 1. The behavior of objective function value of Eq.(6) during iterations of our proposed algorithm on four data sets

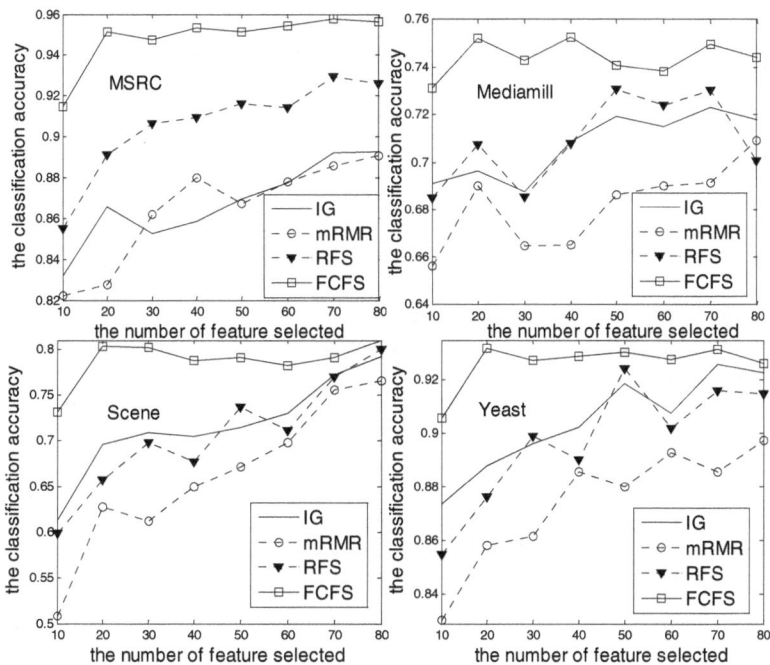

Fig. 2. Classification accuracy comparisons of four feature selection algorithms on different feature number

Acknowledgements. This research is supported in part by National Nature Science Foundation of China (NO.61073116) and the Key Natural Science Project of Anhui Provincial Education Department (KJ2010A006 & KJ2012A004).

References

1. Langley, P.: Selection of relevant features in machine learning. In: AAAI Fall Symposium on Relevance, pp. 140–144. AAAI Press, New Orleans (1994)
2. Peng, H., Long, F., Ding, C.: Feature selection based on mutual information: Criteria of max-dependency, max-relevance, and min-redundancy. IEEE Trans. Pattern Analysis and Machine Intelligence 27, 1226–1238 (2005)
3. Raileanu, L.E., Stoffel, K.: Theoretical comparison between the gini index and information gain criteria. Annals of Mathematics and Artificial Intelligence 41, 77–93 (2000)
4. Kohavi, R., John, G.H.: Wrappers for feature subset selection. Artificial Intelligence 97, 273–324 (1997)
5. Guyon, I., Elisseeff, A.: An introduction to variable and feature selection. J. Machine Learning Research 3, 1157–1182 (2003)
6. Tibshirani, R.: Regression shrinkage and selection via the LASSO. J. Royal Statist. Soc. B. 58, 267–288 (1996)
7. Yuan, M., Lin, Y.: Model selection and estimation in regression with grouped variables. J. Royal Statist. Soc. B 68, 49–67 (2006)

8. Obozinski, G., Taskar, B., Jordan, M.: Multi-task feature selection. Technical report, Department of Statistics, University of California, Berkeley (2006)
9. Argyriou, A., Evgeniou, T., Pontil, M.: Multi-task feature learning. In: 19th Ann. Conf. Neural Information Processing Systems, pp. 41–48. MIT Press, Cambridge (2007)
10. Zhang, M.L., Zhou, Z.H.: ML-KNN: A lazy learning approach to multi-label learning. Pattern Recognition 40, 2038–2048 (2007)
    Criminis, A.: Microsoft research cambridge object recognition image dataset. version 1.0 (2004), http://researchMicrosoft.com/en-us/projects/objectclassrecognition/default.htm
11. Snoek, C., Worring, M., Gemert, J.V., Geusebroek, J.M., Smeulders, A.W.M.: The Challenge Problem for Automated Detection of 101 Semantic Concepts in Multimedia. In: 14th Annual ACM International Conference on Multimedia, pp. 421–430. ACM Press, New York (2006)
12. Boutell, M.R., Luo, J., Shen, X., Brown, C.M.: Learning multi-label scene classification. Pattern Recognition 37, 1757–1771 (2004)
13. Elisseeff, A., Weston, J.: A kernel method for multi-labelled classification. In: 14th Ann. Conf. Neural Information Processing Systems, pp. 681–687. MIT Press, Cambridge (2002)
14. Nie, F.P., Huang, H., Ding, C.: Efficient and Robust Feature Selection via Joint $l_{2,1}$-Norms Minimization. In: 22th Ann. Conf. Neural Information Processing Systems, pp. 1813–1821. MIT Press, Cambridge (2010)

Classification of Three Wine Varieties
Based on ELM and PCA

Yuhui Zhao[*], Suxia Yu, Bingbing Chu, Nan Zhang, and Xin Hu

Northeastern University at QinHuangdao, Qinhuangdao City, China
yuhuizhao@mail.neuq.edu.cn

Abstract. Grape varieties have a decisive impact on the quality of the wine, but it still is a difficult problem to differentiate wine quality which is made from the variety of grapes, the paper suggests a grape varieties detection method using principal component analysis (PCA) and Extreme Learning Machine (ELM). Firstly, in order to get a more stable model, the k-fold cross-validation method is used to determine the training data sets and test data sets. Then, the PCA algorithm is adopted to process chemical components of publicly available data sets wine, and lastly classification predict train is performed using the ELM. The train and test experimental results show that the proposed model is better than the separately ELM in the three wine varieties classification ability.

1   Introduction

The history of wine is more than seven thousand years. Its quality is decided by the grape species and origin and the brewing technique. Recently, people have been using chemometrics to evaluate the wine quality. CortezIxl[1] have build classification models on the quality of wineWith SVM. Based on the content of miniral substance, Moreno[2] have successfully used Probabilistic neural network to divide the 54 samples into 2 types. Daniel Cranato[3] have identified the wine quality with antioxidative activity and sensory quality which results a good outcome, but it's more complex than Artificial Intelligence Method when taking into enforcement. Paulo Cortez[4] use data mining on wine quality assessment. Zheng[5] and others came up the way, Hopfield memory ,which was based on SLFN, to identify wine quality.

Aiming at this, we suggest a modified ELM model, which is a Wine variety identification method based on PCA and ELM, and offer the following solutions: firstly, adopting the variance measurement for processing each of the components to eliminate the influence of the measurement units on the experimental results; secondly, using PCA analysis constituents and reduce the dimension; thirdly, a K fold cross validation method is used to generate training set and test set; and lastly, building an effective prediction model of the wine species with ELM training, and the experiment result proved the effectiveness of the prediction model.

---

[*] Corresponding author.

J. Yang, F. Fang, and C. Sun (Eds.): IScIDE 2012, LNCS 7751, pp. 647–654, 2013.

2   Wine Data

This experiment used public data set wine for experiments[6]. Wine data is a group of chemical analysis data sets, a total of 173 samples, including 13 attributes. Chooses is Italy the same area 3 different kinds of wine data, use category 1, 2, 3 respectively represent different grape variety, all kinds of samples were as follows: class 1,59; Class 2 use; Class 3, 48 (see chart 1).

3   Structure of PCA-ELM Model

PCA-ELM classification model is showed in Figure 1, including three parts of extracting the principal component with PCA, K-fold cross-validation and ELM classification model.

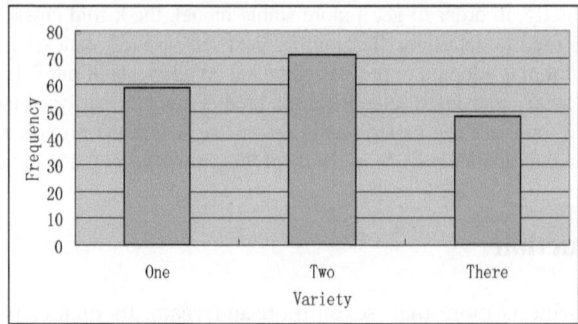

Fig. 1. Distribution of grape varieties

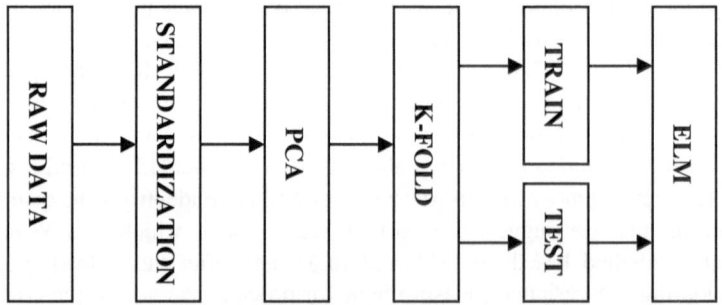

Fig. 2. PCA-ELM classification model structure

3.1   PCA

Given a raw data Matrix: $X_{N \times K} = \left[X_1^T, X_2^T \cdots, X_N^T \right] = (x_{ij})_{N \times K}$, which contains N samples of data, X_1^T is a row vector indicating the first sample, which has

K attributes. X_{ij} indicates the value of the i-th sample of the j-th attribute. PCA steps are as follows[7]:

Standardizing the Data
Different units of measurement variance due to the inconsistency of the unit of measurement of various properties of the grape, such as grape alcohol concentration units can g/ml and mg/ml, obviously the same concentration has a great variance difference. In PCA analysis, in order to eliminate the bad effect on variance due to the unit of measurement, each attribute can divided by the variance of the corresponding property. Then the matrix displayed with column vector, the variance of the *j-th* column:

$$D(X_j) = E(X_j - E(X_j)) \tag{1}$$

$$X_{N \times K} = \left[\frac{X_1}{D(X_1)}, \frac{X_2}{D(X_2)}, \cdots \frac{X_K}{D(X_K)} \right] \tag{2}$$

Formula (1) $E(x_j)$ is the average of x_j. Formula (2) denotes each column of the raw data matrix divided by the variance of its property.

Establish a Standardized Data Covariance Matrix
Compute the covariance of the attribute and the *j*-th attribute:

$$\sigma_{ij} = COV(X_i^T, X_j^T) = \left. \frac{E(X_i - E(X_i)) E(X_j - E(X_j))}{\sqrt{D(X_i)D(X_j)}} \right. \tag{3}$$

And the covariance matrix of X is,

$$\Sigma_X = \begin{bmatrix} D_1 & \sigma_{12} & \cdots & \sigma_{1K} \\ \sigma_{21} & D_2 & \cdots & \sigma_{2K} \\ \vdots & \vdots & \ddots & \vdots \\ \sigma_{K1} & \sigma_{K2} & \cdots & D_K \end{bmatrix} \tag{4}$$

Calculate Eigenvalues of the Covariance Matrix
As Σ_X is a non-negative definite matrix, according to the knowledge of linear algebra, there must be an orthogonal matrix U, can realize

$$U^T \Sigma_X U = \begin{pmatrix} \lambda_1 & \cdots & 0 \\ \vdots & \ddots & \vdots \\ 0 & \cdots & \lambda_K \end{pmatrix} \tag{5}$$

$\lambda_1, \lambda_2 \cdots \lambda_K$ are the eigenvalues of covariance matrix Σ_X. Assume $\lambda_1 \geq \lambda_2 \geq \cdots \geq \lambda_K$, and A is the orthogonal matrix consist of eigenvector which is corresponding to eigenvalue. There are AA'=A'A=I.

$$A = (a_1, a_2, \cdots, a_K) = \begin{bmatrix} a_{11} & a_{12} & \cdots & a_{1K} \\ a_{21} & a_{22} & \cdots & a_{2K} \\ \vdots & \vdots & \ddots & \vdots \\ a_{K1} & a_{K2} & \cdots & a_{KK} \end{bmatrix} \tag{6}$$

$$a_i = \begin{pmatrix} a_{1i} & a_{2i} & \cdots & a_{Ki} \end{pmatrix}^T, \quad i=1, 2, \ldots, K$$

$$F_1 = a_1^T X \quad , \qquad F_i = a_i^T X \tag{7}$$

Formula (7) calculate the principal component score.

Cumulative Contribution Rate

$$V(F_1) = a_1^T \sum a_1 = \lambda_1 \quad , \quad V(F_i) = a_i^T \sum a_i = \lambda_i \tag{8}$$

Formula (8) calculate the variance of the first principal component. The proportion of the i-th principal component variance of the total variance is the contribution rate of the i-th principal component, which reflect containing how much the information, of the original K attributes. How much the information the first P principal components containing was represent by the proportion of the first P principal component variances of the total variances.

Contribution Rate:

$$\lambda_i \Big/ \sum_{i=1}^{k} \lambda_i$$

Cumulative Contribution Rate :

$$\eta_P = \sum_{i=1}^{P} \lambda_i \Big/ \sum_{i=1}^{K} \lambda_i \tag{9}$$

Determine the number of principal components: when $\eta_P \geq 85\%$, and the eigenvalue are $\lambda_1, \lambda_2, \cdots, \lambda_P$,we called the corresponding eigenvector, $A' = [a_1, a_2, \cdots, a_P]$, projection matrix.

Calculate the Principal Component Scores
You need to center the raw data before the calculation of the principal component score, as for the i-th sample:

$$X_i^* = X_i^T - \overline{X} = \left(x_{i1} - \overline{x}_1, x_{i2} - \overline{x}_2, \cdots, x_{ik} - \overline{x}_k \right) \tag{10}$$

X_i^* is the centered data of the i-th sample and X_i^T is the

ra $X_i^* = X_i^T - \overline{X} = \left(x_{i1} - \overline{x_1}, x_{i2} - \overline{x_2}, \cdots, x_{ik} - \overline{x_k} \right)$ w data of the i-th

sample. $\overline{x_1}, \overline{x_2}, \cdots, \overline{x_k}$ is the average of each column. $X_1^*, X_2^*, \cdots, X_N^*$ make up the

matrix we call X*. Then the principal component score matrix $S = X^* A'$.

3.2   K-Fold Cross-Validation

In the choice of training and test set, in order to get a more stable model, K-fold cross-validation method [8] was used for this experiment. K-fold cross-validation: the initial sample is divided into K sub-samples, a separate sub-sample was retained as the validation model data, and other K-1 samples used for training. Cross-validation is repeated K times, Each sub-sample validated once, the results of the average K times, or other combination, and ultimately gets a single estimate. The advantage of this method is the repeated use of a random sub-sample training and validation, each of result verified once.

3.3   ELM Modeling Method

After 3.1 Principal Component Analysis, we can gain the training of N data samples $\left(X_i^T, t_i \right) \in R_N \times R_P$, ELM mathematical model [9-10].

$$\sum_{i=1}^{L} \beta_i G(a_i, b_i, x_j) = t_j, N = 1, 2 \cdots n \tag{11}$$

The (a_i, b_i) is the hidden node parameters, a_i are the weights which connect the i-th hidden layer node and the input layer nodes, b_i is the i-th hidden layer node thresholds, they are not only independent of the samples, but also independent of each other[11]. Assume that a training data sample properties are 14-dimensional (13-dimensional input attributes and one-dimensional varieties category attribute), then the ELM input nodes are 13 and the i-th hidden node parameter a_i are form a_{i01} to a_{i13}. β_i is the i-th hidden layer node and output layer connection weight.

Given h(x_i)=G((a_i,b_i, x_i),which is the output of the i-th hidden note, and the function G can be sigmod and so on. Then we can calculate the H matrix.

$$H = \begin{bmatrix} h(x_1) \\ \vdots \\ h(x_N) \end{bmatrix} = \begin{bmatrix} G(a_1, b_1, x_1) \cdots G(a_L, b_L, x_1) \\ \vdots \\ G(a_1, b_1, x_N) \cdots G(a_L, b_L, x_N) \end{bmatrix}_{N \times L} \tag{12}$$

L means the hidden layer having L nodes. Combination of formula (11), $H\beta=T$ can be gained.

$$\beta = \begin{bmatrix} \beta_1^T \\ \vdots \\ \beta_L^T \end{bmatrix}_{L\times M} \qquad T = \begin{bmatrix} t_1^T \\ \vdots \\ t_L^T \end{bmatrix}_{L\times M} \qquad (13)$$

According to the formula (12), the hidden layer nodes and output into the connection weights β will be obtained.

When HH^T is nonsingular,

$$f_L(x) = h(x)\beta = h(x)H^T\ (HH^T)^{-1}T \Rightarrow h(x)H^T(\frac{I}{C} + HH^T)^{-1}T \qquad (14)$$

When H^TH is nonsingular,

$$f_L(x) = h(x)\beta = h(x)\ (H^TH)^{-1}H^TT \Rightarrow h(x)(\frac{I}{C} + H^TH)^{-1}H^TT \qquad (15)$$

A positive value I/C can be added to the diagonal of H^TH or HH^T of the Moore-Penrose generalized inverse H the resultant solution is stabler and tends to have better generalization performance. This article don't use I/C .

EML modeling steps can be summarized as follows:

1. Randomly generated hidden node parameters (a_i, b_i), i = 1 ... n, where a_i is the i-th input layer node and hidden layer connection weights, bi is the threshold.
2. Choose the activation function: such as sigmod function, and calculate the hidden layer node output matrix H.
3. Calculate the hidden nodes and output connection weights β: β = H⁺T, H⁺ is generalized inverse matrix.

4   Experimental Results

We made K 10 and hidden notes 1000, then use K-fold cross-validation to select the training set and test set of samples for training, establishing a classification model based on PCA-ELM of wine variety. The compare of PCA-ELM experimental results and ELM experimental results show in Figure 2.

The prediction accuracy of testing data set with PCA-ELM algorithm classification model of wine variety is nearly 100%, which is significantly better than the ELM in Figure 3. You can also visually gain that the prediction model based on PCA-ELM has better classification ability than ELM in Figure 4 and Figure 5.

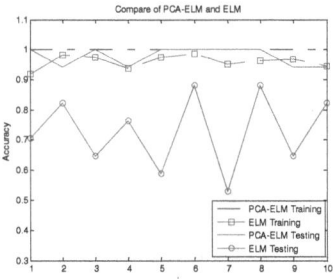

Fig. 3. Compare of PCA-ELM and ELM

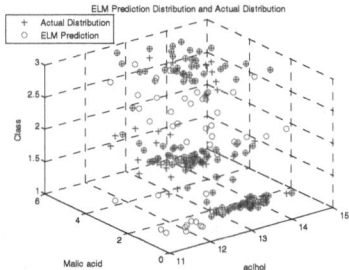

Fig. 4. ELM Prediction Distribution and Actual Distribution

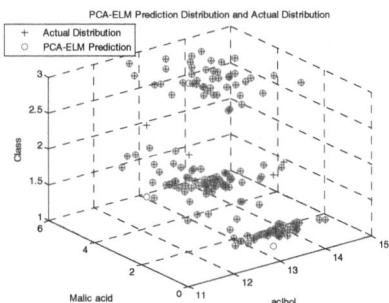

Fig. 5. PCA-ELM Prediction Distribution and Actual Distribution

5   Conclusion

In this paper, we discuss the classification methods of the three wine varieties and put forward a forecasting method based on the PCA-ELM classification modeling. We use the PCA to extract the principal components in the chemical constituents of the wine to eliminate the correlation between the multi-features, which removal the interference information, reduce the ELM input dimensions, and improve the prediction accuracy.

Acknowledgments. This work is based on the Chinese National Natural Science Foundation 61,100,021 project.

References

1. Cortez, P., Cerdeira, A., Almeida, F.: Modeling wine preferences by data mining from physicechemieal properties. Decision Support Systems 47(4), 547–553 (2009)
2. Moreno, I., Gonzadez-Weller, D., Gutierrez, V.: Differentiation of two Canary DO red wines according to their metal content from inductively coupled plasma optical emission spectrometry and graphite furnace atomic absorption spectrometry by using pmbabilistic neural networks, Talanta (2007)
3. Daniel, G., Flávia, C.U.K., Inar, A.: C.: Phenolic composition of South American red wines classified according to their antioxidant activity, retail price and sensory quality. In: Food Chemistry. SCI (2011)
4. Cortez, P., Cerdeira, A., Almeida, F.: Using Data Mining for Wine Quality Assessment. Springer, Heidelberg (2009)
5. Zheng, P.S., Zhang, J.X., Tang, W.S.: Learning associative memories by error backpmpagation. IEEE Transactions on Neural Networks 22(3), 347–355 (2011)
6. Wine Data Set, http://archive.ics.uci.edu/ml/datasets/Wine
7. Jolliffe, I.T.: Principal Component Analysis, 2nd edn. Springer (2002)
8. Dietterich, T.: Approximate Statistical Tests for Comparing Supervised Classification Learning Algorithms. Neural Computation 10(7) (1985)
9. Guangbin, H., Dianhui, W., Yuan, L.: Extreme learning machines: a survey. Int. J. Mach. Learn. & Cyber. 2, 107–122 (2011)
10. Huang, G.-B.: Extreme Learning Machine: Learning Without Iterative Tuning Nanyang Technological University. In: WCCI 2012, Singapore (2012)
11. Huang, G.: Extreme Learning Machines-Learning Without Iterative Tuning. School of Electrical and Electronic Engineering, Nanyang Technological University, Singapore

On Non-Euclidean Metrics Based Clustering

Hong Cao[1], Ping Wang[1], Runing Ma[1], and Jundi Ding[2]

[1] School of Science, Nanjing University of Aeronautics and Astronautics,
Nanjing 211100, China
[2] School of Computer Science and Technology,
Nanjing University of Science and Technology, Nanjing 210094, China

Abstract. In this paper, non-Euclidean metrics, such as kernel metric, Mahalanobis distance and the metric based on the shortest weighted path, are introduced into PAM and CURE clustering algorithms. The purpose is to have a detailed research on non-Euclidean metrics based clustering. Firstly, modified algorithms are established by replacing Euclidean metric with non-Euclidean metrics. Then these modified algorithms are applied on various data sets including UCI data sets as well as artificial data sets. Detailed evaluations and analysis have been made about the performances of different metrics. Experimental results demonstrate that the application scope of these clustering algorithms has been extended by adopting non-Euclidean metrics. As a result, we can conclude that the application of non-Euclidean metrics is of great importance.

Keywords: Clustering, Kernel Metric, Mahalanobis Distance, Shortest Weighted Path.

1   Introduction

Clustering has extensive applications in many aspects [1–5]. The clustering algorithm design usually accompanies with the definition of the proximity metric, and the clustering results are affected directly by the selected metric. Most conventional clustering algorithms adopt Euclidean distance [6], so that they are just appropriate for convex or sphere data set. However, non-Euclidean metrics based clustering algorithms have the benefit of not confining the space shape of data set.

Non-Euclidean metrics such as kernel metrics have been introduced into some clustering algorithms, such as k-means [7]. Kernel-based method transforms the data patterns from input space to a feature space by a kernel function [8, 9]. In this way, nonlinearly patterns in the input space can be transformed into linearly patterns in the feature space [10, 11], without the curse of dimensionality [1, 12]. Mahalanobis distance [13] is relative with correlations between variables, whereas, Euclidean distance treats them equally. In this paper, a proximity metric based on the the shortest weighted path is taken into account. In graph theory, the shortest path problem is to find a shortest path from a source node to a destination. Dijkstra's algorithm [14] solves the single-source shortest-path

J. Yang, F. Fang, and C. Sun (Eds.): IScIDE 2012, LNCS 7751, pp. 655–663, 2013.

problem in weighted graphs. Based on it, the shorted weighted path between any two data points can be obtained. PAM [15] is effective in clustering linear separable data patterns. It presents a cluster by one centriod. CURE [16] represents each cluster by a fixed number of well scattered points and then shrinking them toward the center of the cluster. Having more than one representative point per cluster, CURE has a good performance on non-spherical shapes.

In our work, by introducing non-Euclidean metrics into PAM and CURE, we have a good research on these non-Euclidean metrics based clustering.

2   Non-Euclidean Metrics

2.1   Kernel Metric

Given a data set $X = \{x_1, \cdots, x_n\}$, where x_i belongs to the D-dimensional space R^D, and let Φ be a non-linear mapping function which maps $\mathbf{x_i}$ from the input space R^D to feature space Q [9]:

$$\Phi : R^D \longmapsto Q \qquad x \longmapsto \Phi(x)$$

The inner product $(x_i \cdot x_j)$ is expressed by the dot product(x_i, x_j). By employing Φ, the $(x_i \cdot x_j)$ is mapped to $\Phi(x_i) \cdot \Phi(x_j)$, and the kernel function is expressed by the dot product in the new space Q:

$$K(x_i, x_j) = \Phi(x_i) \cdot \Phi(x_j). \tag{1}$$

Let $u_i = \phi(x_i)$ denote the transformation from input space to feature space, the distance in feature space is represented as

$$\begin{aligned} d^2(u_i, u_j) &= \| u_i - u_j \|^2 \\ &= \| \phi(x_i) - \phi(x_j) \|^2 \\ &= K(x_i, x_i) - 2K(x_i, x_j) + K(x_j, x_j). \end{aligned} \tag{2}$$

Therefore, the dot product in feature space can be expressed by the dot product in input space without the concrete form of the mapping function. Two commonly applied kernel functions are the Gaussian kernel function and the polynomial kernel function. They are respectively listed below:

$$K(x_i, x_j) = exp(-\frac{\| x_i - x_j \|^2}{\sigma^2}), \tag{3}$$

where $\sigma > 0$;

$$K(x_i, x_j) = (x_i \cdot x_j + 1)^d, \tag{4}$$

where $c \geq 0$, $d \in N$.

2.2   Mahalanobis Distance

The Mahalanobis distance matrix of a multivariate vector $x = \{x_1, x_2, \cdots, x_n\}^T$ from a group of values with mean $\mu = (\mu_1, \mu_2, \cdots, \mu_n)^T$ and covariance matrix S is defined as:

$$D\_M(x) = \sqrt{(x - \mu)^T * S^{-1} * (x - \mu)}. \tag{5}$$

Given $X = \{x_1, \cdots, x_n\}$, the distance between any two points can be expressed as

$$d(x_i, x_j) = D\_M(x)(i, j). \tag{6}$$

2.3   Newly Proposed Metric Based on the Shortest Weighted Path

The weights of two direct-connected and indirect-connected points are respectively defined as:

$$D(x_i, x_j) = exp(-\frac{\| x_i - x_j \|^2}{\sigma^2}). \tag{7}$$

$$D(x_i, x_j) = D(x_i, x_{k1}) + D(x_{k1}, x_{k2}) + \cdots + D(x_{kn}, x_j), \tag{8}$$

where $x_{k1}, x_{k2}, \cdots, x_{kn}$ are the nodes on this path. Considering a data set X with n samples, the distance between x_i and x_j can be expressed as:

$$d(x_i, x_j) = min\{D(x_i, x_j)|D(x_i, x_j) : \text{ weight of any path from } x_i \text{ to } x_j\} \tag{9}$$

Note that any path from x_i to x_k combines any path from x_k to x_j constitute a path from x_i to x_j. So

$$d(x_i, x_j) \leq d(x_i, x_k) + d(x_k, x_j). \tag{10}$$

From Eq.(10), this distance metric conforms to the Triangle equality condition in the definition of a distance, and obviously conforms to the other two conditions: Symmetry and Positivity. Based on this method, the distance of any two data points can be obtained for clustering.

3   PAM and CURE Clustering Algorithms

By replacing Euclidean metric with non-Euclidean metrics, the non-Euclidean metrics based clustering algorithms can be easily obtained.

3.1   PAM Clustering Algorithm

PAM [15] utilizes real samples as medoids, replaces the medoids with nonmedoids one by one until reaches the best results. To express the effect of a swap, the cost is calculated for all nonmedoids, and it is defined differently in different cases. Given a data set with n samples, PAM aims to find the best set of k samples as medoids, then nonmedoids are grouped with the most similar medoids. More explicitly, we use $\mathbf{m}$ to denote a current medoid to be replaced, $\mathbf{p}$ to denote a nonmedoid which is treated as a new medoid to replace $\mathbf{m}$, $\mathbf{j}$ to denote other nonmedoids which may or may not need to be moved, $\mathbf{m}'$ to denote a current medoid that is nearest to $\mathbf{j}$ without $\mathbf{m}$ and $\mathbf{p}$, and $d(\mathbf{x_i}, \mathbf{x_j})$ to denote the distance of two samples. The PAM clustering algorithm works as follows:

Step 1. Select k initial medoids arbitrarily.

Step 2. Compute the total cost of replacing $\mathbf{m}$ with $\mathbf{p}$ for all pairs of $\mathbf{m}$ and $\mathbf{p}$. For each $\mathbf{j}$, different cases should be taken into account to compute the cost C_{jmp}.

Case 1. If $\mathbf{j}$ currently belongs to the cluster represented by $\mathbf{m}$ and $\mathbf{j}$ is more similar to $\mathbf{m}'$ than to $\mathbf{p}$, i.e., $d(\mathbf{m}, \mathbf{p}) < d(\mathbf{m}', \mathbf{p})$ and $d(\mathbf{j}, \mathbf{p}) \geq d(\mathbf{j}, \mathbf{m}')$, then the cost is given by

$$C_{jmp} = d(\mathbf{j}, \mathbf{m}') - d(\mathbf{j}, \mathbf{m}). \tag{11}$$

Case 2. If $\mathbf{j}$ currently belongs to the cluster represented by $\mathbf{m}$ and $\mathbf{j}$ is less similar to $\mathbf{m}'$ than to $\mathbf{p}$, i.e., $d(\mathbf{m}, \mathbf{p}) < d(\mathbf{m}', \mathbf{p})$ and $d(\mathbf{j}, \mathbf{p}) < d(\mathbf{j}, \mathbf{m}')$,

$$C_{jmp} = d(\mathbf{j}, \mathbf{p}) - d(\mathbf{j}, \mathbf{m}). \tag{12}$$

Case 3. If $\mathbf{j}$ currently belongs to the cluster represented by $\mathbf{m}'$ and $\mathbf{j}$ is more similar to $\mathbf{m}'$ than to $\mathbf{p}$, i.e., $d(\mathbf{m}, \mathbf{p}) \geq d(\mathbf{m}', \mathbf{p})$ and $d(\mathbf{j}, \mathbf{p}) \geq d(\mathbf{j}, \mathbf{m}')$,

$$C_{jmp} = 0. \tag{13}$$

Case 4. If $\mathbf{j}$ currently belongs to the cluster represented by $\mathbf{m}'$ and $\mathbf{j}$ is less similar to $\mathbf{m}'$ than to $\mathbf{p}$, i.e., $d(\mathbf{m}, \mathbf{p}) \geq d(\mathbf{m}', \mathbf{p})$ and $d(\mathbf{j}, \mathbf{p}) < d(\mathbf{j}, \mathbf{m}')$,

$$C_{jmp} = d(\mathbf{j}, \mathbf{p}) - d(\mathbf{j}, \mathbf{m}'). \tag{14}$$

Combine the four cases above, the total cost of the swap between $\mathbf{m}$ and $\mathbf{p}$ is

$$TC_{mp} = \sum_{j} C_{jmp}. \tag{15}$$

Step 3. Select the pair of $\mathbf{m}$ and $\mathbf{p}$ corresponding to the minimum value of TC_{mp}. If the minimum TC_{mp} is negative, replace $\mathbf{m}$ with $\mathbf{p}$ and go back to Step 2. Otherwise, go to Step 4.

Step 4. For each nonmedoid, find the most similar medoids. The referenced nonmedoid belongs to the cluster represented by the medoid which is corresponding to the minimum $d(\mathbf{x_i}, \mathbf{m_k})$.

In traditional PAM clustering algorithm, Euclidean distance is employed as metric for clustering, and the distance between two samples is expressed as

$$d^2(\mathbf{x_i}, \mathbf{x_j}) = \| \mathbf{x_i} - \mathbf{x_j} \|^2 . \tag{16}$$

3.2   CURE Clustering Algorithms

CURE [16] employs a novel hierarchical clustering algorithm. In CURE, the number c of well scattered points in a cluster and the cluster number k have to be determined at first. Each cluster has a centroid and a number of representative points determined from the merging procedure. Let C_i denote each cluster, C_{im} denote the centroid and C_{ip} denote the representative points. The clusters with the closest pair of representative points are the clusters that should be merged at each step of CURE's algorithm. The centroid of the new cluster C_k merged with C_i and C_j is calculated as

$$C_{km} = \frac{|C_i| * C_{im} + |C_j| * C_{jm}}{|C_i| + |C_j|}, \tag{17}$$

where $|C_i|$ and $|C_j|$ are the numbers of data points of the cluster C_i and C_j, respectively. The representative points of C_k can be calculated as

$$C_{kp} = C_p + \alpha * (C_{km} - C_p), \tag{18}$$

where C_p stands for all of the representative points of C_i and C_j, α is a fraction used in the shrinking. Clusters with the closest pair of representative points are merged step by step until k clusters are left.

4   Experimental Results

In this section, we concentrate on evaluations and analysis for clustering results of algorithms based on different metrics. UCI data sets iris, wine, machine and artificial data sets normrand3, block, semicircle, circled are selected. Moreover, due to the sensitivity to parameters, the involved parameters are made adjustment constantly for best clustering results.

The parameter σ controls the radial scope of the function, we have:

$$\sigma = \sigma_0 \cdot max\|x_i - x_j\|^2, \tag{19}$$

where x_i and x_j denote the data points. By making adjustment to σ_0, σ can reaches a value for a better clustering. Other parameters are also taken into account in the experiments.

Firstly, we compare kernel functions based PAM with its traditional version as classifier. σ_0 and d are set to be different values for better results. Table 1 presents their test results on iris, wine and machine. As is shown in the table, the highest classification accuracy of iris, wine and machine achieve 90.67%, 72.48% and 82.30%, respectively. Compared with its traditional version, the Gaussian kernel function works much better. Clearly, for iris, wine and machine, as the data points are described in a higher dimensional space, they are more clearly separable. Therefore, for the data patterns which are not easy separable in the original space, the kernel based metric can be applied. And, the adjustment to parameter makes this method more applicable.

Table 1. Classification accuracy of the kernel metrics based PAM and its traditional version on UCI data set iris, wine and machine

Metric	σ_0	d	Classification accuracy		
			iris	wine	machine
Euclidean	-	-	89.33%	70.79%	75.12%
Gaussian-Kernel	0.07	-	86.67%	61.24%	82.30%
	0.1	-	89.33%	72.47%	81.82 %
	0.4	-	90.67%	71.91%	79.43%
	0.7	-	89.33%	71.91%	75.12%
Poly-Kernel	-	1	89.33%	70.79%	79.43%
	-	2	85.33%	69.10%	79.43%
	-	3	85.33%	67.42%	75.60%
	-	4	84.00%	65.73%	75.60%

Firstly, we apply the traditional and the new proposed schemes on norm-rand3. Fig.1 shows that the patterns of normrand3 are all well classified, and the centroids of the entire data set are clearly marked out by black five-pointed star. This experiment illustrates that these non-Euclidean metrics based PAM keeps the original advantages of being effective for convex data sets. Or rather, these non-Euclidean metrics are also effective for clustering convex or sphere data sets.

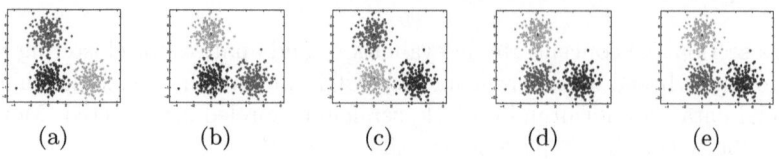

| (a) | (b) | (c) | (d) | (e) |

Fig. 1. (a)-(e) are the clustering results of the Euclidean distance, Mahalanobis distance, Gaussian kernel, polynomial kernel and weighted shortest path based PAM on normrand3, respectively

The following two experiments are to demonstrate the superiority of the Mahalanobis distance and the proposed metric based PAM algorithm. Firstly, the Euclidean distance and non-Euclidean metrics based PAM are employed into data set block. The clustering performances are shown in Fig.2, it can be observed that the Mahalanobis distance based PAM algorithm is effective for block. As for the distance measure to data, Mahalanobis distance, the correlations of the data points are taken into account. In other words, Mahalanobis distance is invariant to any nonsingular linear transformation, if the data points are not correlated, the Mahalanobis distance is equivalent to the Euclidean distance. As a consequence, it tend to form hyperellipsoidal clusters, that's why the Mahalanobis distance based PAM algorithm is suitable for clustering block-shaped data set instead of just for convex data set.

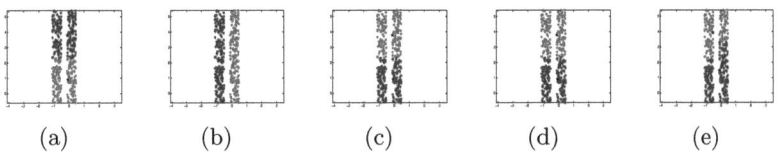

Fig. 2. (a)-(e) are the clustering performances of the Euclidean distance, Mahalanobis distance, Gaussian kernel, polynomial kernel and the new metric based PAM on block, respectively

Secondly, Fig.3 and Fig.4 show clustering performances of non-Euclidean metrics based PAM on circled and semicircle, respectively. Although the kernel metrics and Mahalanobis distance are not effective for nonlinear data set circled and semicircle, the metric based on the shortest weighted path is absolutely effective for them. Fig.3(a)(e) and Fig.4(a)(e) show that Euclidean distance leads to local misclassification. Take semicircle as an example, when the distances are described by Euclidean distance, the centriod in the up arc are closer to the rightmost part of down arc rather than leftmost part of the up arc. Whereas, the shortest weighted path overcomes the limitation of the theorem that, in a triangle, the length of any two sides is greater than the third one. So we can obtain the minimum cumulative weights of the centriod in the up arc to any point in leftmost part of the up arc is much smaller than to the point in the rightmost part of the down arc. This explains why the leftmost part of the up arc and the rightmost part of the down arc are misclassified and the proposed metric based on the shortest weighted path is effective for semicircle. From the experiment, the proposed metric based PAM is effective in clustering nonlinear separable data set, such as circled and semicircle.

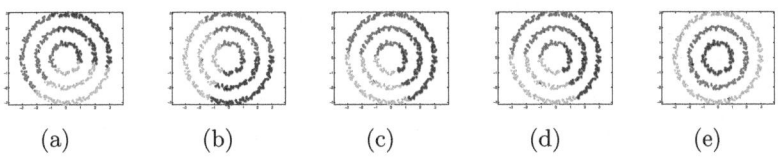

Fig. 3. (a)-(e) are the clustering performances of the Euclidean distance, Mahalanobis distance, Gaussian kernel, polynomial kernel, and the shortest weighted path metric based PAM on circles, respectively

As is shown above, by introducing non-Euclidean metrics into PAM, the application scope has been largely extended. In the same way, we apply Euclidean distance based CURE on artificial data sets. Furthermore, for best results, $c = 2$ and $\alpha = 0.8$ are chosen for normrand3, $c = 2$ and $\alpha = 0.2$ for circled and semicircle. For the block, c varies from 2 to 10, α varies from 0.2 to 0.8. Fig.5 the Euclidean distance based CURE is not effective for the block-shaped data set,

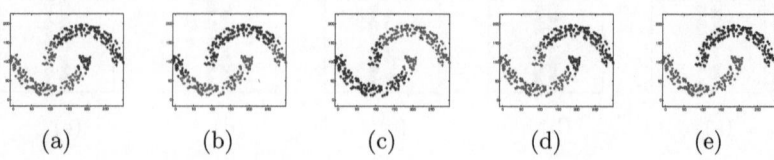

Fig. 4. (a)-(e) are the clustering performances of the Euclidean distance, Mahalanobis distance, Gaussian kernel, polynomial kernel, and the shortest weighted path metric based PAM on semicircle, respectively

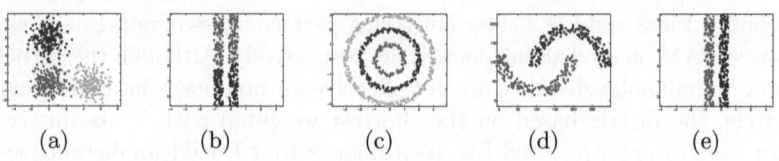

Fig. 5. (a)-(d) are the results of the Euclidean distance based on CURE on normrand3, block, circled and semicircle respectively. (e) shows the result of Mahalanobis distance based on CURE on block.

but block can be well classified by the Mahalanobis distance based CURE. This test demonstrates the superiority of the Mahalanobis distance again.

PAM represents a cluster by one centriod, which leads not be effective for the non-spherical shapes, such as circled and semicircle. For discussing the superiority of this metric and extending the application scope of PAM, we introduce this metric into PAM. However, Fig.5 shows that the CURE has had a good performance on non-spherical shapes, so finally, we only discuss the clustering results of the kernel metrics based CURE on UCI data set iris. Table 2 shows the best classification accuracy of non-Euclidean metrics based CURE on iris. Due to the sensitivity for non-Euclidean metrics based CURE to parameters, the best classification accuracy is obtained by making adjustment to parameters constantly. Obviously, compared with the traditional version, the polynomial kernel based CURE works much better on iris, the Gaussian kernel based CURE is not appropriate on iris.

Table 2. Classification accuracy of the kernel metrics based CURE and its traditional version on iris

Metric	c	α	σ_0	d	Classification accuracy
Euclidean distance	2-15	0.2-0.8	-	-	67.3%
Gaussian kernel	2-15	0.2-0.8	0.4	-	34.7%
Polynomial kernel	2-15	0.2-0.8	-	2	74.7%

5   Conclusions

In our work, some non-Euclidean metrics are introduced into PAM and CURE. To discuss the role of these metrics in clustering, various data sets including UCI data sets as well as artificial data sets are tested on these metrics based clustering algorithms. In this paper, the advantages of non-Euclidean metrics are presented, and give the detailed interpretations about the situations under which these metrics should be used. So based on all the test results above, it has been found that the application scope of PAM and CURE has been extended, and we can draw the conclusion that, although a new algorithm is of importance, the applications of non-Euclidean metrics can not be ignored.

References

1. Burges, C.J.C.: A Tutorial on Support Vector Machines for Pattern Recognition. Data Min. Knowl. Disc. 2, 121–167 (1998)
2. Kaufman, L., Rousseeuw, P.J.: Finding Groups in Data: An Introduction to Cluster Analysis. Wiley, New York (1990)
3. Jain, A.K., Murty, M.N., Flynn, P.J.: Data clustering: a review. ACM Comput. Surv. 31, 264–323 (1999)
4. Shi, J., Malik, J.: Normalized cuts and image segmentation. IEEE T. Pattern Anal. 22, 888–905 (2000)
5. Pal, N.R., Pal, S.K.: A review on image segmentation techniques. Pattern Recogn. 26, 1277–1294 (1993)
6. Bezdek, J.C., Keller, J.M., Krishnapuram, R., Pal, N.R.: Fuzzy Models and Algorithms for Pattern Recognition and Image Processing. Kluwer Academy Publishers, Boston (1999)
7. Dhillon, I.S., Guan, Y.Q., Kulis, B.: Kernel k-means: spectral clustering and normalized cuts. In: Proceedings of the Tenth ACM SIGKDD International Conference on Knowledge Discovery and Data Mining, pp. 551–556. ACM, New York (2004)
8. Scholkopf, B., Smola, A.J.: Learning with Kernels. The MIT Press, Cambridge (2002)
9. Cristianini, N., Shawe-Taylor, J.: Support Vector Machines and Other Kernel Based Learning Methods. Cambridge University Press, New York (2000)
10. Girolami, M.: Mercer kernel-based clustering in feature space. IEEE T. Neural Networ. 13, 780–784 (2002)
11. Muller, K.R., Mika, S., Ratsch, G., Tsuda, K., Scholkopf, B.: An Introduction to Kernel-Based Learning Algorithms. IEEE T. Neural Networ. 12, 181–201 (2001)
12. Hearst, M.A., Dumais, S.T., Osman, E., Platt, J., Scholkopf, B.: Support vector machines. IEEE Intell. Syst. App. 13, 18–28 (1998)
13. De Maesschalck, R., Jouan-Rimbaud, D., Massart, D.L.: The Mahalanobis distance. Chemom. Intell. Lab. Syst. 50, 1–18 (2000)
14. Dijkstra, E.W.: A note on two problems in connexion with graphs. Numer. Math. 1, 269–271 (1959)
15. Ng, R.T., Han, J.W.: CLARANS: A Method for Clustering Objects for Spatial Data Mining. IEEE T. Knowl. Data En. 14, 1003–1016 (2002)
16. Guha, S., Rastogi, R., Shim, K.: Cure: an efficient clustering algorithm for large databases. Inform. Syst. 26, 35–58 (2001)

Learning Compact Representation
for Image with Tensor Manifold Perspective

Songsong Wu[1,2], Zhisen Wei[1], Xiaoyuan Jing[2], Jian Yang[1], and Jingyu Yang[1]

[1] School of Computer Science and Technology,
Nanjing University of Science and Technology, Nanjing, P.R. China
sswu@yahoo.cn, weizsnjust@163.com, {csjyang,yangjy}@mail.njust.edu
[2] School of Automation,
Nanjing University of Posts and Telecommunications,
Nanjing, P.R. China
cnjingxy@njupt.edu.cn

Abstract. In this paper, Local Tensor Subspace Alignment algorithm (LTESA) is proposed to explore the substantial geometry of image manifold by regarding images as tensor objects. LTESA characterizes local geometry of tensor in local tensor subspace with rank-one tensor approximation, then align the local tensor subspaces to achieve a global low-dimensional representation for images. LTESA obtains the intrinsic latent variables of image through nonlinear dimensionality and achieves compactness of image representation in vector form. Moreover, Landmark-LTESA is proposed to reduce computational complexity of LTESA and a generalization version of LTESA is proposed to solve the out-of-sample problem for image feature extraction. LTESA is evaluated in applications of data visualization for face images and face recognition. Experimental results suggest that the proposed approaches provide a strong capability of detecting complex image manifold and is effective on unsupervised nonlinear feature extraction of image.

Keywords: Manifold Learning, Image Manifold, Multilinear Subspace, Feature Extraction.

1 Introduction

Exploratory analysis of image is crucial to many appearance-based approaches in pattern recognition and computer vision. As there often exists redundancy among pixels, a fundamental problem of image analysis is to discover fewer latent variables that actually control the pixel intensity. Image manifold [7][13] presumes images are sampled from a low-dimensional manifold that can be parameterized by independent variables. And this assumption is supported by evidences in human vision perception research [9]. Manifold learning techniques are beneficial to image analysis task and have been wildly applied in many correlative domains, such as image coding [10], image retrieval [3] and image classification [4] [15] [5].

Generally, the traditional methods of image manifold learning treat an image as a point in observation space. The space consists of "image vectors", which

J. Yang, F. Fang, and C. Sun (Eds.): IScIDE 2012, LNCS 7751, pp. 664–671, 2013.
© Springer-Verlag Berlin Heidelberg 2013

are obtained by rearrangement of the pixel identities. In this case, the objective of image manifold learning is to find a meaningful low-dimensional manifold which is embedded in the high-dimensional image-vector space. Linear subspace methods [12] [1] aim to find linear representation for images with the assumption that image manifold owns linear structure. While recent research of vision perception indicates image manifold may be intrinsically curved [9]. So the called manifold learning methods emerge for nonlinear manifold detection, such as the well known Isomap [11], LLE [8], Laplacian Eigenmap [2], and LTSA [16] algorithms. These nonlinear approaches attain a compact nonlinear representation of image through spectral analysis to a matrix specially constructed from observed images. Nonlinear manifold learning methods perform in a sophisticated way of global nonlinear with local linear, so they can uncover a more faithful structure of image manifold compared with linear subspace methods.

Instead of converting an image into a vector, some recent works have paid attention to taking the original matrix-format of image as input. 2DPCA [14] calculate linear subspaces in the both directions of image based on their matrix-form. MPCA and ROTA view images as high-order tensor objects and use multilinear algebra to reveal linear image manifold. The difference between them lies in that MPCA [6] represents an image by a low-dimensional tensor without change its order, while ROTA [10] represents an image by a low-dimensional coordinate vector in a certain rank-one tensor space. Besides, tensor-based graph-embedding method [3] has been developed for nonlinear image manifold and achieved impressive results.

Our main contribution in this paper is a novel unsupervised learning algorithm for detecting image manifold via local tensor subspace alignment(LTESA). Our method owns three advantages: (1) LTESA overcomes the problem of previous vector-based image manifold learning methods by holding internal structure of image. (2) LTESA provides a strong capability to reveal the nonlinear structure of image manifold. (3) LTESA provides a potential approach for image recognition as its ability of nonlinear feature extraction.

The organization of this paper is as follows. In Section 2, we present the details of computing local tensor subspaces and the alignment trick for global coordinates. In Section 3, we show how to extend LTESA on computational issue and generalization. In Section 4, we evaluate the proposed method in application of face image visualization and face recognition. Finally, we provide some concluding remarks in section 5.

2   LTESA

2.1   Local Tensor Subspace

For tensor set $\{\mathcal{X}_i\}_{i=1}^N$, where $\mathcal{X}_i \in R^{I_1 \times \cdots \times I_M}$, a weighted graph $\mathcal{G}$ is constructed. The vertexes of $\mathcal{G}$ are $\mathcal{X}_i$ and the weight of edge connecting $\mathcal{X}_i$ and $\mathcal{X}_j$ is their tensor distance $dist(\mathcal{X}_i, \mathcal{X}_j) = \left(\sum_{i_1=1,\ldots,i_M=1}^{I_1,\ldots,I_M} (\mathcal{X}_i - \mathcal{X}_j)_{i_1,\ldots,i_M}^2 \right)^{\frac{1}{2}}$. Based on the adjacent matrix of $\mathcal{G}$, k tensors $\mathcal{X}_{i1} \ldots \mathcal{X}_{ik}$ can be selected to form a local neighborhood to any certain tensor $\mathcal{X}_i$. In this local neighborhood, we want to find a series of rank-one tensors $\mathcal{P}_{i,1}, \ldots, \mathcal{P}_{i,D}$ that behave as D basis of a low-dimensional tensor subspace $\mathcal{T}_i$. Consequently, the local low-dimensional representation of each neighboring tensor of $\mathcal{X}_i$ is provided by the following tensor-to-vector projection $\lambda_{ij} = \mathcal{X}_{ij} \prod_{d=1}^{D} \times (\mathcal{P}_{i,d})^T$, $j = 1, \ldots, k$.

In order to ensure $\mathcal{T}_i$ retain as much of local geometrical information as possible, the local-reconstruction error should be minimized. Recall the definition of rank-one tensor, we get the following model that seeks the optimal local tensor subspace with respect to projection vectors

$$\mathbf{u}_{i,d}^l \ ^* = \arg \min_{\mathbf{u}_{i,d}^l} \sum_{j=1}^{k} \left\| \mathcal{X}_{i,j} - \sum_{d=1}^{D} \lambda_{i,j}^d \prod_{l=1}^{M} \otimes \mathbf{u}_{i,d}^l \right\|_F^2 \tag{1}$$
$$\text{s.t.} \ \ \left\| \mathbf{u}_{i,d}^l \right\|^2 = 1, \ l = 1, \ldots, M \text{ and } d = 1, \ldots, D$$

The model(1) does not have closed-form solution. Here, we takes the strategy of "iteration plus alternating projection "similar to [10] to solve the optimal problem. We firstly initialize the M-mode basis vectors someway, and then seek one optimal projection vector with fixing the others. Specifically, once we have obtained $\{\mathbf{u}_{i,p}^l, 1 \leq l \leq M\}_{p=1}^{d-1}$ as at the end of the $d-1$th iteration, the dth rank-one tensor basis can be obtained by separately computing $\mathbf{u}_{i,d}^l$ from $l = 1$ to $l = M$ one after another. The lth basis vector $\mathbf{u}_{i,d}^l$ is given as the eigenvector according to the largest eigenvalue of the matrix

$$\mathbf{S}_{i,d}^l = \tfrac{1}{k} \sum_{j=1}^{k} \left(\mathcal{X}_{i,j}^d \prod_{p \neq l} \times (\mathbf{u}_{i,d}^p)^T \right) \left(\mathcal{X}_{i,j}^d \prod_{p \neq l} \times (\mathbf{u}_{i,d}^p)^T \right)^T .$$

2.2   Global Alignment of Local Tensor Subspace

Because the local coordinate systems in different tensor subspaces are not comparable with each other, we need to align the obtained local tensor subspaces to construct a global coordinate. First of all, we denote the global representation as $\{y_i\}_{i=1}^N$ and denote the local affine transformation matrix associated with ($\mathcal{X}_i$'s neighborhood as L_i. A reasonable criterion that can ensure $\{y_i\}_{i=1}^N$ reflect the intrinsic structure of image manifold is to keep local geometry in the global coordinate system. So, we get the following alignment model

$$\mathbf{Y}^* = \arg \min_{\mathbf{Y}} \sum_{i=1}^{N} \sum_{j=1}^{k} \| \mathbf{y}_{ij} - (\bar{\mathbf{y}}_i + \mathbf{L}_i \lambda_{ij}) \|^2 \tag{2}$$
$$\text{s.t.} \ \ \mathbf{Y}\mathbf{Y}^T = I_D$$

where $\bar{\mathbf{y}}_i = \tfrac{1}{k} \sum_{j=1}^k \mathbf{y}_{ij}$. By simple algebra formulation, we can solve $\mathbf{L}_i$ by $\mathbf{L}_i = \mathbf{Y}\mathbf{S}_i(I - \tfrac{1}{k}ee^T)\Lambda_i^+ \Lambda_i$ where $\mathbf{S}_i$ is a selection matrix that satisfies $\mathbf{Y}\mathbf{S}_i = (\mathbf{y}_{i1}, \ldots, \mathbf{y}_{ik}), \Lambda_i^+$ is the Moore-Penrose pseudoinverse of Λ_i.

Using Lagrange Multiplier, the solution of (2) is translated to the following eigenvalue problem

$$\Phi\eta = \theta\eta \tag{3}$$

Where $\Phi = [\mathbf{SWW}^T\mathbf{S}^T]$, $\mathbf{S} = [\mathbf{S}_1, \ldots, \mathbf{S}_N]$, $\mathbf{W} = diag(\mathbf{W}_1, \ldots, \mathbf{W}_N)$ and $\mathbf{W}_i = (I - \frac{1}{k}ee^T)\Lambda_i^{\dagger}\Lambda_i$. Let $\eta_1, \eta_2, \ldots, \eta_D$ be the eigenvectors of (3) according to the smallest D nonzero eigenvalues, then the optimal global representation is given by $\mathbf{Y} = [\eta_1, \eta_2, \ldots, \eta_D]^T$.

3   Landmark and Generalization

Landmark LTESA. Consider tensor set $\mathbf{X} = \{\mathcal{X}_i\}_{i=1}^N$, we want to divide it into several overlapping subsets $\mathbf{X}_j$. The subsects must satisfy two demands: (1) each $\mathbf{X}_j$ contains k nearest neighbors and (2) the union of $\mathbf{X}_j$ covers the whole tensor set, i.e. $\bigcup_j \mathbf{X}_j = \mathbf{X}$. The division is implemented by a iterative procedure. At each iteration, two landmark points are selected, then their local k−nearest neighborhoods are constructed. Particularly, a point $\mathcal{X}^0$ is randomly selected from $\mathbf{X}$ under uniform distribution. The first landmark point $\mathcal{L}_p$ is determined to be the farthest point in $\mathbf{X}$ apart from $\mathcal{X}^0$, and the second landmark point $\mathcal{L}_q$ is determined to be the farthest point in $\mathbf{X}$ apart from $\mathcal{L}_p$. Then, k nearest neighbors are found to form a landmark section for each one of the two landmarks, and the tensor set $\mathbf{X}$ is subtracted by eliminating $r(r < k)$ nearest neighbors of each landmark point, where r is a preset threshold parameter. Finally, if $\mathbf{X} = \phi$(e.t. $\mathbf{X}$ is an empty set), the division procedure is over. Otherwise, another iteration is carried out for two more landmark points.

Once the landmark sections are obtained, the following process is similar to the original LTESA. The difference lies that we calculates local tensor subspace according to each landmark points rather than in the local neighborhood of each tensor objects. Concretely, for a landmark-section $\mathbf{X}_j = \{\mathcal{X}_{j\kappa}\}_{\kappa=1}^k$, its corresponding D-dimensional local tensor subspace $span\left(\prod_{l=1}^M \otimes u_{j,d}^l, d = 1, \ldots, D\right)$ and the local coordinates $\Omega_j = \{\lambda_{j\kappa}\}_{\kappa=1}^k$ are obtained by performing ROTA to $\mathbf{X}_j$. The global alignment matrix Φ is constructed based on the local coordinates Ω_j, then we can obtain a global coordinate system for the entire tensor objects.

Generalization of LTESA. In order to make LTESA be a feature extraction method which is suitable to classification task, it is valuable to explore the generalization extension of LTESA. To this end, the obtained local tensor subspace information need to be stored. It includes: landmark points $\{\mathcal{L}_j\}$, rank-one tensor projection to local tensor subspace $\{\mathcal{P}_{j,d} = \prod_{l=1}^M \otimes u_{j,d}^l, d = 1, \ldots, D\}$, the local coordinate systems $\{\Omega_j\}$. For a new coming tensor object $\mathcal{X}'$, we assign it to the local neighborhood of its nearest landmark point $\mathcal{L}_{j'}$. By projection $\mathcal{X}'$ onto the local tensor subspace, its local coordinate is given by

$$\lambda' = \mathcal{X}' \prod_{d=1}^D \times (\mathcal{P}_{j',d})^T \tag{4}$$

Then by performing alignment on λ' with local affine transformation matrix, we get the global representation of $\mathcal{X}'$ as

$$\mathbf{y}' = \bar{\mathbf{y}}_{j'} + \mathbf{L}_{j'} * \lambda' \tag{5}$$

where $\mathbf{L}_{j'} = \mathbf{YS}_{j'}(I - \frac{1}{k}ee^T)(I - \Lambda_{j'}^+)$ is the local affine transformation matrix.

4   Evaluation

4.1   Visualization of Face Images

Yale faces data set[1] is employed for visualization in this experiment. We compare LTESA(and its landmark version) to four well known manifold learning methods: Isomap [11], LLE [8], LTSA [2], and MVU [13]. We show the resulting two-dimensional embedding as a scatterplot in Fig.1.

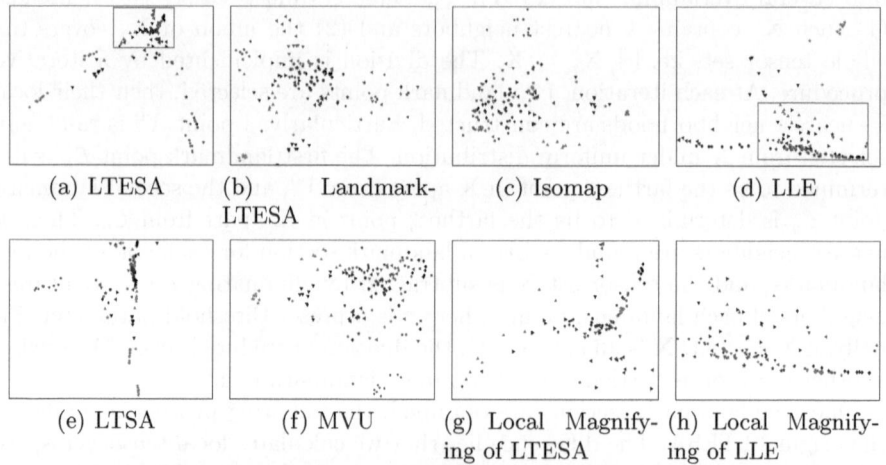

(a) LTESA	(b) Landmark-LTESA	(c) Isomap	(d) LLE
(e) LTSA	(f) MVU	(g) Local Magnifying of LTESA	(h) Local Magnifying of LLE

Fig. 1. Visualization of 165 face images from the Yale faces data set

Fig.1 reveals the competence of LTESA on discovering face image manifold, in comparison with the other manifold learning methods. Specifically, the four comparative methods could discover a certain amount of submanifolds corresponding to different face class, whereas they mix up the other classes. We can hardly find obvious structure about the face image manifold from their results. In contrast, LTESA does much better on discovering the subtaintial structure of face image manifold. The leading manifold structure according to face images are captured by LTESA in its scatterplot. We notice that there exist overlaps in the results of LTESA similar to LLE, but the degree and scope of overlap in LTESA are notably ameliorated. Through magnifying overlapping part in the scatterplots of LTESA and LLE, we find LTESA still holds a obviously separation between different classes, while LLE remains weak separation. As to Landmark-LTESA, it achieves a compromise of clustering and separation.

4.2   Face Recognition

ORL Database. We conducted experiment of face image classification over the ORL faces image database[2]. For each individual, $p(2,3,4)$ face images are

---

randomly selected to form training samples and the rest are used for testing. We repeated this process 20 times and calculated the average classification accuracy as the final result. The maximal recognition rate of each method and the corresponding dimensionality are given in Tab. 1. The comparative results in the Tab. 1 show the superiority of LTESA over NPE, LPP, LLTSA and MPCA on the best face recognition performance. Note that LTESA achieves considerable recognition accuracy when feature dimension is less than 30. It indicates that LTESA is able to extract compact feature from face image by grasping the key character of face relative to identification with a few features.

Table 1. The Maximal Recognition Rates (Percent) on the ORL Data set

Method	2Train	3Train	4Train
NPE	73.6(22)	80.0(23)	85.4(29)
LPP	77.5(44)	78.6(67)	79.6(112)
LLTSA	74.7(31)	72.9(51)	77.1(53)
MPCA	80.3(50)	82.5(57)	88.3(91)
LTESA	83.4(18)	85.0(21)	89.6(26)

AR Database. The AR face database[3] is used in our experiment for algorithm evaluation in large-scale database. We selected $C = 120$ individuals(65 males and 55 females) with their non-occluded images for the experiment. The training sample set comprises the 1th image and the 7th image of each person, which represent the neutral expression from two sessions, and the remaining $14 - 2$ images form the testing sample set. Thus, the training sample set contains 240 images and the testing sample set consists of 1440 images.

Table 2. The Maximal Recognition Rates (Percent) on a subset of AR Database

	NPE	LPP	LLTSA	MPCA	LTESA
MRR(%)	79.86	77.78	74.31	75.21	82.08
Dim	81	119	42	116	28

Tab.2 gives the maximal recognition rates of each method and the corresponding dimension. As can be seen, LTESA enhances the maximal recognition rate for at least 2 percents compared with other four methods. This result demonstrates that LTESA can extract effective feature from face images even in small sample size case(only two samples of each person for training). Besides, we should notice that the NPE, LPP and MPCA achieve their best recognition performances with pretty high dimensional feature, while LLTSA and LTESA achieve the best recognition using much lower feature. This phenomenon can be comprehended as local neighboring path alignment strategy makes LTESA and LLTSA are both

---

[3] http://rvl1.ecn.purdue.edu/ARdatabase/ARdatabase.html

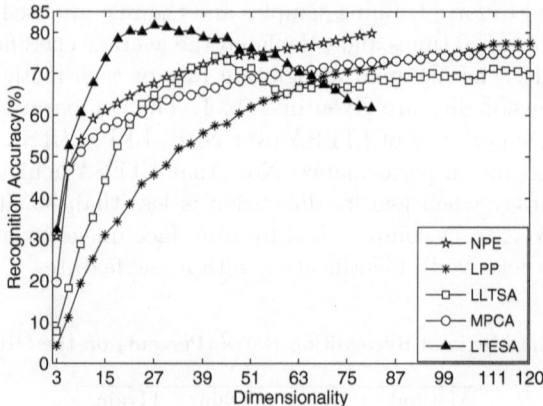

Fig. 2. The recognition rates versus the dimensions on a subset of AR database

based on, and LTESA treats face image as tensor so that the internal structure of image can be used to boost face image manifold learning.

Fig. 2 shows the recognition curve versus the variation of dimensions. From Figure 2, we can see that LTESA achieves decent recognition accuracy using feature within low-dimension, indicating that LTESA captures the key structure of face image manifold and LTESA are suitable for memory-sensitive applications. Besides, we observe that the recognition accuracy descends along with the increasing feature dimensionality. The reason may be the oversize neighborhood holds no longer exact local geometry information.

5   Conclusion

This paper presents a novel nonlinear dimensionality reduction algorithm, called Local Tensor Subspace Alignment(LTESA), with its two extensions, to discover the intrinsic structure of image manifold. LTESA owns several good characteristics: (1)LTESA take images as tensor objects, therefore avoid information loss of internal structure of an image; (2)LTESA obtain a compact representation of image via a low-dimensional vector; (3)LTEDA is sensitive nonlinear structure of image manifold. LTESA is applied into face image data visualization and face recognition, and the experimental results demonstrate the strong capability of LTESA on dimensionality reduction for discovering the substantial structure of image manifold in an unsupervised way.

Acknowledgments. This work was partially supported by National Natural Science Foundation of China under Grant No.61125305, No.60973098, No.61073113.

References

1. Bartlett, M., Movellan, J., Sejnowski, T.: Face recognition by independent component analysis. IEEE Transactions on Neural Networks 13(6), 1450–1464 (2002)
2. Belkin, M., Niyogi, P.: Laplacian eigenmaps for dimensionality reduction and data representation. Neural Computation 15(6), 1373–1396 (2003)
3. He, X., Cai, D., Niyogi, P.: Tensor subspace analysis. In: Advances in Neural Information Processing Systems 19, vol. 18, pp. 499–505. MIT Press (2006)
4. He, X., Cai, D., Yan, S., Zhang, H.: Neighborhood preserving embedding. In: IEEE International Conference on Computer Vision (ICCV), vol. 2, pp. 1208–1213 (2005)
5. He, X.H., Niyogi, P.: Locality preserving projections. In: Advances in Neural Information Processing Systems 16 (2003)
6. Lu, H., Plataniotis, K., Venetsanopoulos, A.: Mpca: Multilinear principal component analysis of tensor objects. IEEE Transactions on Neural Networks 19(1), 18–39 (2008)
7. Lu, H.M., Fainman, Y., Hecht-Nielsen, R.: Image manifolds. In: Applications of Artificial Neural Networks in Image Processing III. Proceedings of SPIE, vol. 3307, pp. 52–63 (1998)
8. Roweis, S., Saul, L.: Nonlinear dimensionality reduction by locally linear embedding. Science 290(5500), 2323 (2000)
9. Seung, H., Lee, D.: The manifold ways of perception. Science 290(5500), 2268 (2000)
10. Shashua, A., Levin, A.: Linear image coding for regression and classification using the tensor-rank principle. In: IEEE Computer Society Conference on Computer Vision and Pattern Recognition (CVPR), vol. 1, pp. 42–49 (2001)
11. Tenenbaum, J., Silva, V., Langford, J.: A global geometric framework for nonlinear dimensionality reduction. Science 290(5500), 2319 (2000)
12. Turk, M., Pentland, A.: Eigenfaces for recognition. Journal of Cognitive Neuroscience 3(1), 71–86 (1991)
13. Weinberger, K., Saul, L.: Unsupervised learning of image manifolds by semidefinite programming. In: IEEE Computer Society Conference on Computer Vision and Pattern Recognition (CVPR), vol. 2, pp. I-988–II-995 (2004)
14. Yang, J., Zhang, D., Frangi, A., Yang, J.: Two-dimensional pca: a new approach to appearance-based face representation and recognition. IEEE Transactions on Pattern Analysis and Machine Intelligence 26(1), 131–137 (2004)
15. Zhang, T., Yang, J., Zhao, D., Ge, X.: Linear local tangent space alignment and application to face recognition. Neurocomputing 70(7-9), 1547–1553 (2007)
16. Zhang, Z., Zha, H.: Principal manifolds and nonlinear dimension reduction via local tangent space alignment. SIAM Journal of Scientific Computing 26(1), 313–338 (2002)

Efficient ESL-Event-to-SQL Translation

Narat Phucharoen and Juggapong Natwichai

Computer Engineering Department, Faculty of Engineering
Chiang Mai University, Chiang Mai, Thailand
narat.phu@gmail.com, juggapong@eng.cmu.ac.th

Abstract. Expressive Stream Language-Event (ESL-Event), which is based on the traditional SQL, is a language developed for streaming data management. It can handle the data streams and temporal event queries effectively. However, it has yet to be implemented commercially. In this paper, we propose an efficient ESL-Event-to-SQL translation. Since the SQL language can be used widely on the traditional DBMSs. Thus, our proposed work allows the users to leverage the features of the ESL-Event on the current systems with no effort. Our approach firstly parses an ESL-Event statement into the intermediate representation, the parse tree. Subsequently, the tree is converted to the SQL-syntax-complied parse tree, in which the semantic of all the ESL-Event features is well preserved and implemented efficiently. Once the tree is traversed, the SQL statement is generated. From the experiment results, our work is highly efficient.

1 Introduction

In order to manage a data stream, which is a very large amount of ordered data and flows in a very high speed, an effective and efficient data management system is required. Not only the data are difficult to be stored and monitored in the traditional approach, but also the temporal events in the stream should be able to be queried effectively. For example, the arrival time of the items read by the RFID readers in a factory at any checked-stations should be able to be monitored along with its non-temporal data. In addition, the correctness of the order of the item-checking should be able to be queried effectively. Obviously, the query language for the data management systems should be effective and expressive enough for such tasks.

There are several attempts proposed to address the mentioned issues [8]. Among the proposed work, Expressive Stream Language-Event (ESL-Event), which is one of the most prominent data stream query languages, was proposed in [4]. The language can handle temporal event querying effectively with its various features e.g. sequencing or tuple pairing enforcement. It is also user-friendly, since it is based on the standard SQL. In addition, it can handle multiple data sources both in the traditional tabular data and streaming data formats. However, the ESL-Event is yet to be implemented commercially.

In this paper, we propose an efficient ESL-Event-to-SQL translation. Since the SQL language can be used widely on the traditional DBMSs. Thus, our proposed

J. Yang, F. Fang, and C. Sun (Eds.): IScIDE 2012, LNCS 7751, pp. 672–682, 2013.

work allows the users to leverage the features of the ESL-Event on the current systems. Our approach firstly parses an ESL-Event statement into the parse tree [6]. Subsequently, the tree is converted to the SQL-syntax-complied parse tree, in which the semantic of all the ESL-Event features is well preserved and implemented efficiently by the sub-query approach [5]. Last, the tree is traversed, and the SQL statement is generated. The experiment results are presented to evaluate our work.

2   Related Work

Data stream is vast amount of data that flows into the system in very high speed. Generally, it is difficult to process the data stream in database management systems (DBMSs). Because the DBMSs are developed initially for processing persistent data. In order to cope with the data stream, a new type of data management system is proposed, so called data stream management systems (DSMSs) [9,10].

In order to utilize DSMSs effectively, there are many issues to be addressed. For example, not like the data in DBMSs, the data in the data stream have their order to be considered. Therefore, the querying can be more complex. Additionally, the continuous nature of the streaming data may degrade the efficiency of some data operators. Furthermore, Structured Query Language (SQL), which is designed for the persistent data, might not be appropriated for the data stream.

There are several work addressed the data management for data stream as follows. In [11], the authors proposed the StreamSQL, a query language for streaming data. This language is extended from the standard SQL. By the way, a few operators to process data stream are extended, e.g. selecting from a data stream, stream-relation join, union merge, window aggregation, and window joining. In [2], a continuous query language was proposed for manipulating data stream. It is also based on the expressive SQL. In [7], the authors developed a system to work with large volume of RFID event streams efficiently. The system can query on the events using the proposed operators in a limited time. However, the system does not support the temporal event querying, i.e. an event that has a pattern of sequences and time constraints.

Among the proposed work, in [1], the authors proposed a prominent query language to support data stream query, so called Expressive Stream Language (ESL). The ESL supports complex querying on data stream as well as traditional table-like data. It also allows the users to improve the performance of the windows aggregation. It can manipulate several data streams as the order sequences of tables [3]. Furthermore, it is user-friendly since it is based on the traditional SQL. Nevertheless, the ESL can be complex when deals with temporal event.

In [4], an extension to the ESL, i.e. ESL-Event, is proposed. It provides several features, such as sequence function, sliding window, and tuple pairing modes. The sequence function provides a mechanism to enforce the order constraints for the data in data stream. The sliding window is a function that can be used to control the finished time of the event. And, the tuple pairing mode is a method to both maintains less data history, and reduces unwanted data subjected to the time constraints. However, ESL-Event has not been implemented commercially.

3   ESL-Event to SQL

3.1   Background: Expressive Stream Language Event

Expressive Stream Language Event (ESL-Event) [4] is a language developed from the Expressive Stream Language (ESL) [3]. Such ESL has the capability to support data streams, i.e. it can filter, search, and eliminate duplicate data from the data stream. For the ESL-Event, the event or sequence manipulating functionalities are added. Its event-based functions can be utilized widely where the temporal order is needed to be enforced. The syntax of the ESL-Event is similar to the traditional SQL as shown in Figure 1. Note here that the data in a "table" mentioned in this paper can refer to both traditional persistent data as well as the streaming data.

SELECT $< attribute\_list >$
     FROM $< table\_list >$
     WHERE SEQ $(< object\_id >,< C_1, C_2, \ldots, C_n >)$
          MODE $[< time\_stamp >,< mode >]$
          OVER $[< time\_window > PRECEDING < last\_table >]$
          $< condition >;$

Fig. 1. ESL-Event Syntax

From Figure 1, the important details of syntax are as follows.

- $< object\_id >$ is an attribute for multiple-tables joining.
- $< C_1, C_2, \ldots, C_n >$ is a ordered sequence of the tables.
- $< time\_stamp >$ is a time stamp attribute for multiple-tables joining.
- $< mode >$ is the tuple pairing mode.
- $< time\_window >$ is a sliding-window time constraints for data filtering.
- $< last\_table >$ is the last table for data filtering.
- $< condition >$ is a conditional expression that identifies the records to be retrieved by the query.
- *SEQ Operator* is an operator for enforcing the order in multiple data streams. For example, suppose that there are ten dependent procedures for quality checking in a factory before the good can be shipped, such constraints can be implemented by the SEQ operator. The operator, which is to be used in the WHERE clause of the statement, has the following syntax: SEQ $(C_1, C_2, \ldots, C_n)$. If the timestamp from table C_j is higher than the timestamp from table C_i, for each $i < j$, the operator will return true, and show the result of the query condition.
- *Tuple Paring Mode (MODE)* is an event operator modifier which can reduce the generation of unwanted combination of data e.g. the laboratory staffs need to know the recent time in the overall laboratories composed of each laboratory event. In [4], the authors proposed four modes as follows.

- *Recent mode*: The SEQ operator will return true on only the sequences that are the most recent qualifying tables.
- *Chronicle mode*: The operator will return true on only the sequences that are the earliest qualifying tables.
- *Consecutive mode*: The SEQ operator will return true on all possible sequences that are in consecutive order.
- *Unrestricted mode*: The operator will return true on all possible orders of the sequences.

− *Sliding Window on SEQ (OVER)* is an operator for filtering out the undesirable objects defined by the specified window. This operator can help reducing unwanted data, for example; in a scout camp, the scouts must finish the activities depending on the sequence within a certain time window in order to pass the test. Let Time-Window is a time for the filtering, C_n is the last table for the filtering, and Time-Type is a unit of time, e.g. minute and hour. The operator can be used by the keyword OVER. It has the following syntax: Over[Time-Window Time-Type PRECEDING C_n]. If the first and the last events finish within a defined Time-Window, the operator will return true value and show the result of the data which are not filtered out.

3.2   ESL-Event-to-Parse Tree

In order to translate ESL-Event statements into the SQL statements, we propose to apply the parse tree [6] as the ESL-Event intermediate representation. The parse trees can be used to parse the languages complying with Backus-Naur Form (BNF) into the representation of tree. The algorithm to translate ESL-Event statements into the tree is shown in Figure 2.

From Figure 2, there are two sets of the keywords. The SQL-Keywords set includes the basic keywords of SQL, i.e. SELECT, FROM, WHERE, GROUP BY, and HAVING. And, the ESL-Event-Keywords set includes additional keywords for ESL-Event, i.e. SEQ, MODE, and OVER. The given input statements are to be parsed by white spaces and symbols. The algorithm begins with creating the *root node* of the parse tree. If the keywords is founded, the *Child node* linked to the *root node* is created. Then, the *current node* pointer is set at such node. If the keyword is an element of ESL, it will be created as a child of the *current node*. Then, the ESL-Event boolean flag is set as true value, such flag maintains the *in-processing* literal of ESL-Event keywords. If a non-keyword literal is founded, a child is created and set as the leaf node of the tree.

3.3   ESL-Event Parse Tree to SQL Statement

After the ESL-Event parse tree is generated for the input statement, the SQL statement can be generated by our approach which composes of two procedures. First, the ESL-Event parse tree is converted into SQL parse tree. The structure is to be adjusted according to the syntax of the SQL. Subsequently, the SQL parse tree is traversed, and the SQL statement can be obtained from the traversed result. The conversion can be categorized into three types as follows.

Algorithm 1. Translate ESL-Event to Parse Tree
Input: ESL-Event Statement
Notation:
   SQL-Keywords:{"SELECT", "FROM", "WHERE", "GROUP BY", "HAVING"}
   ESL-Event-Keywords:{"SQL","MODE","OVER"}
Output: Parse Tree
Create *root node* of the parse tree.
For each *literal* in ESL-Event Statement **do**
   If *literal* ∈ **SQL-Keyword then**
      Create a new node, *t*, as a child node of *root node*.
      Set *t.value* as the *literal*.
      Set *t* as *current node*.
   Else if *literal* ∈ **ESL-Event-Keywords then**
      Create a new node, *t*, as a child node of *current node*.
      isESL-Event:= true
      Set *t.value* as the *literal*.
      Set the *current node* as node *t*.
   Else
      Set the *current node* as node *t*.
      Set *t.value* as the *literal*.
      If (isESL-Event == true)and(*literal* == ")" or *literal* == "]") **then**
         isESL-Event:= false
         Set *current node.parent* as *current node*.
      End if
   End if
End for

Fig. 2. ESL-Event to Parse Tree Algorithm

– *General Parse Tree Conversion*: Our proposed approach firstly identifies the SEQ operator in the ESL-Event parses tree since it indicates the general structure of the SQL parse trees. Once the operator is identified, the attribute for multiple-tables joining, *object_id*, is then examined. It is considered along with the corresponding tables, $C_1, C_2, \ldots, C_n$. The SEQ operator is represented by a natural join in SQL parse tree. The structure depends on the number of corresponding tables, i.e. the joining conditions are generated $N-1$ times. Such that the order of the joins follows the SEQ operator. Then, the SEQ node is deleted from the input tree. After the SQL parse tree is generated, the pre-oder traversal creates the SQL statement. Figure 3 shows the overview of the conversion using the proposed approach.

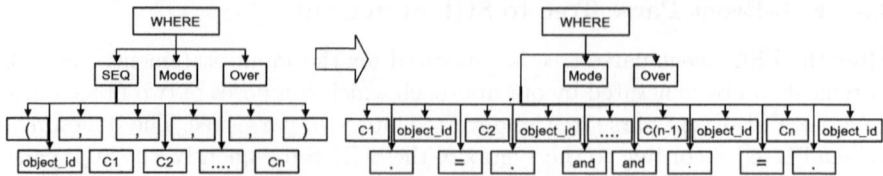

Fig. 3. General Parse Tree Conversion

- *Tuple Pairing Mode Conversion*: After the general tree has been converted, the tuple pairing mode is considered, if it exists. The conversion approach for each mode is as follows.
  - *Unrestricted mode*: Since there is no further restriction on the orders, any two tuples satisfying the SEQ operator can be paired. Thus, the conversion is straightforward as the general conversion. In Figure 4, the approach of the conversion for this mode is shown.
  - *Consecutive mode*: In this mode, the same approach as the unrestricted mode can be applied. Though, the consecutive condition is to be enforced by the "less than" operator in SQL. Such condition is enforced by the pairs of joining condition, however, the *time_stamp* attributes are considered. The approach for this mode conversion is shown in Figure 5.
  - *Recent mode*: The same approach as in the consecutive mode is applied for the recent mode. However, the MAX function is used to provide the recent characteristic. Also, the GROUP BY parse tree is generated for grouping the result according to the *object_id*, such tree is to be traversed after the SEQ parse tree. The conversion approach is shown in Figure 6

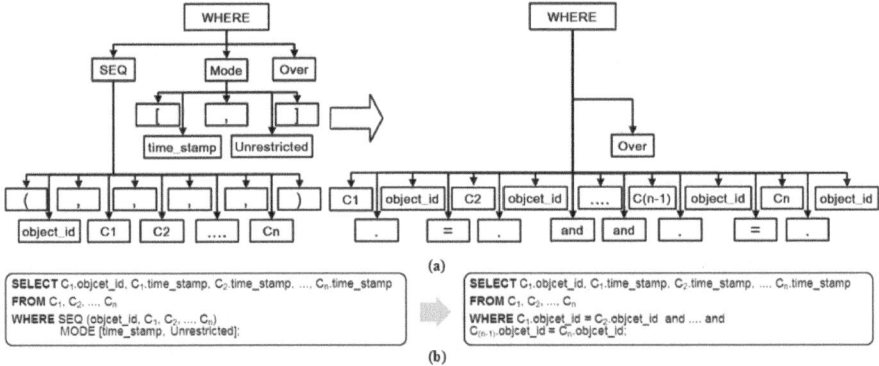

Fig. 4. Unrestricted Mode Conversion

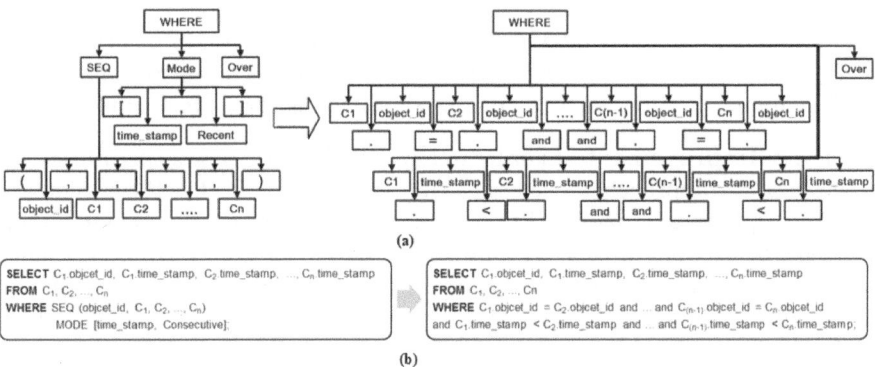

Fig. 5. Consecutive Mode Conversion

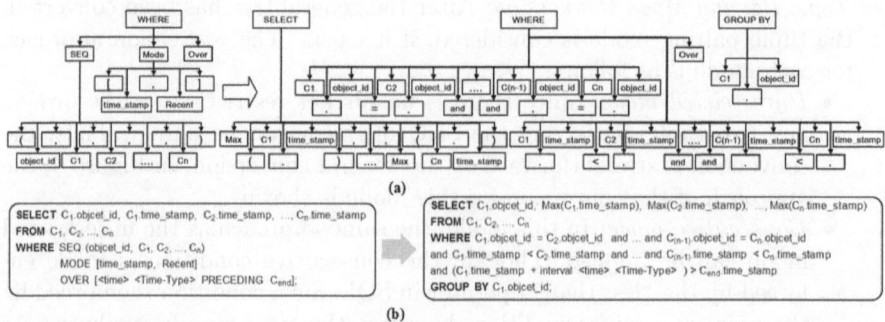

Fig. 6. Recent Mode Conversion

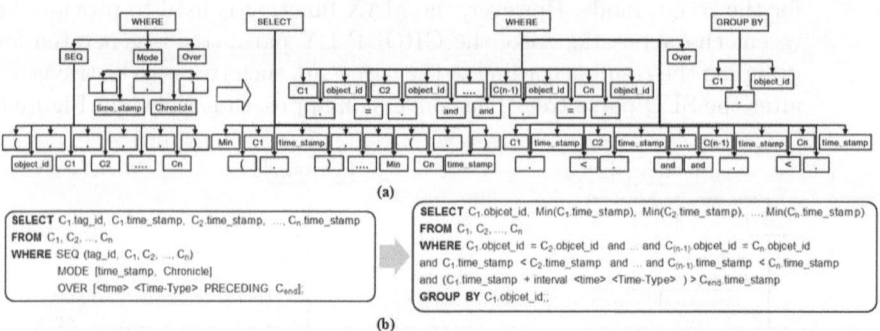

Fig. 7. Chronicle Mode Conversion

- *Chronicle mode*: The same approach as in the recent mode is applied, however the MIN is used instead. In Figure 7, an approach to illustrate the conversion in this mode is shown.
- *Sliding Window Conversion*: In case the sliding window operator exists in a parse tree, a condition to determine whether the results are in the window is generated. Specifically, the time stamp of the records from the first table is added to the *time_period*, then compared with the time stamp of the records from the last table. So, a WHERE condition is added to the SQL parse tree. The approach of the sliding window conversion is shown in Figure 8.

Efficiency Improvement. In order to improve the efficiency of the translation, we propose an approach to generate the SQL parse tree for the recent and chronicle tuple pairing modes. Since both of them relies on the maximum and minimum values from the pairing tuples respectively. Thus, retrieving such values before the natural joins can be more efficient. This can be implemented by the sub-query approach [5]. In order to apply such approach, a sub tree of the SELECT sub-query is created as a child of the FROM node. The approaches to convert the parse trees by the sub-query approach for recent and chronicle modes are shown in Figure 9 and 10 respectively.

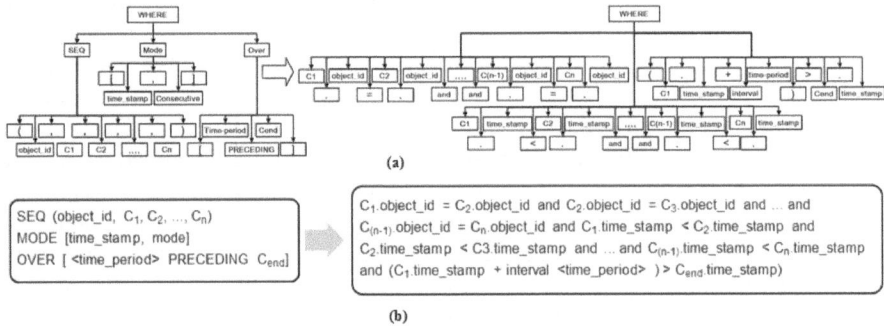

Fig. 8. Sliding Window Conversion

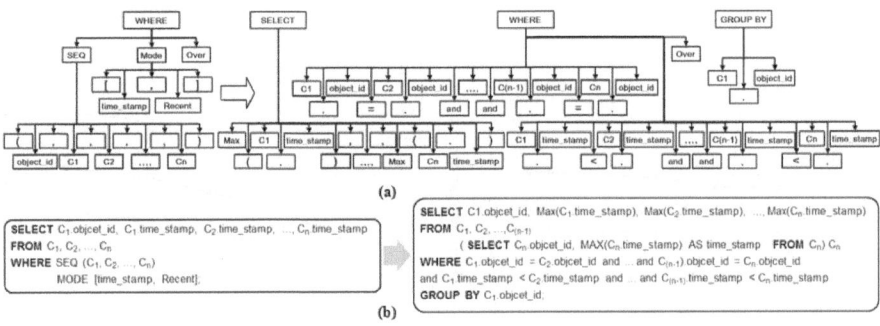

Fig. 9. Sub-Query - Recent Mode Conversion

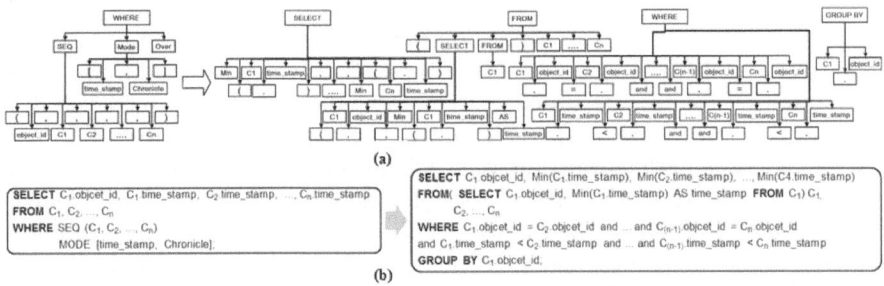

Fig. 10. Sub-Query - Chronicle Mode Conversion

4   Experiment Results

After our approach to translate the ESL-Event to SQL statements is proposed, in this section, we present the experiment results to evaluate the efficiency of the translated queries.

The experiments were conducted on a 3.0 GHz Intel Core 2 PC running Windows 7 operating system with 4 GB memory. The translator was implemented by Visual C♯ 2010 Express with Oracle 11g Express as the DBMS. IBM Quest Market-Basket Synthetic Data Generator was used as a data generator. In the experiments, there are five data streams for consideration. For each data stream, a record contains 2 attributes. The first attribute is "object_id" ranged from 10 to 50, this value will determine the number of objects in the stream. The second attribute is "time_stamp" ranged from 1 to 120. The number of records depends on the number of objects, which ranges from 500 to 2,500 records. The temporal event queries are randomly generated for the evaluation. The efficiency is evaluated in terms of the execution time when the two parameters change, i.e. the number of objects in the data streams for the scalability evaluation, and the redundancy of the retrieved data in each stream. The redundancy here is the maximum number that a record can be duplicate. The resulting numbers reported are three-time average.

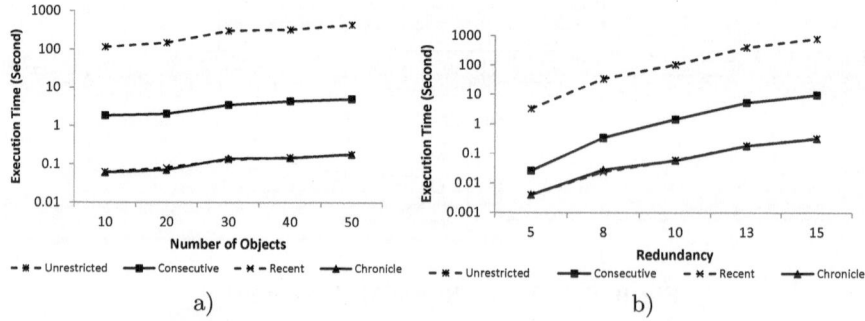

Fig. 11. Execution Time of Each Tuple Pairing Mode

In the first experiment, we evaluate the efficiency of the translation when the number of objects in the data stream changes. Such value is varied from 10, 20, 30, 40, and 50 to evaluate its effect. The redundancy of of the retrieved data is fixed at 10. In Figure 11a), the result of the experiment is presented. Note that the y-axis is in the logarithm scale. Obviously, when the number of objects is increased, the execution time of the queries is also increased. The rationale behind this is that when the number of the objects is higher, the amount of data satisfying the sequencing enforcement is also increased. More data are to be joined to the other data sources. Also, it can be seen that the results of the unrestricted tuple pairing mode have the highest execution time. Since such mode generates results without any constraints. Meanwhile, it can be seen that the recent and chronicle tuple pairing modes uses less execution time since they have the maximum and minimum constraints on the results.

In Figure 11b), the redundancy of the retrieved data is varied from 5 to 15 to evaluate the effect. Obviously, the redundancy degrades the efficiency of the translation. The higher level of redundancy causes the higher execution time. The reason is that the redundancy can increase the size of the result, as well as require more joining data. Overall, it can be seen that the SQL statements generated from our approach are efficient. Though, the statements with the unrestricted mode might not be as efficient as the other modes, the execution times are still acceptable considering its high-computational expense nature.

The next experiment evaluates the proposed sub-query for the recent and chronicle tuple pairing modes. The execution time of such two modes with regard to the number of objects is reported in Figure 12. From the figure, it can be seen that when the number of objects is increased, the execution time of the sub-query approach is much less than the non sub-query approach. The reason behind this is that the maximum and minimum bounding values are retrieved before the natural joins are performed. It can eliminate the table scan for the efficiency. Also, Figure 13 shows the execution time of the sub-query approach when the redundancy is varied. Obviously, both the sub-query and non sub-query use more execution time when the redundancy is increased. However, the proposed approach is much more efficient for both modes by the mentioned reason.

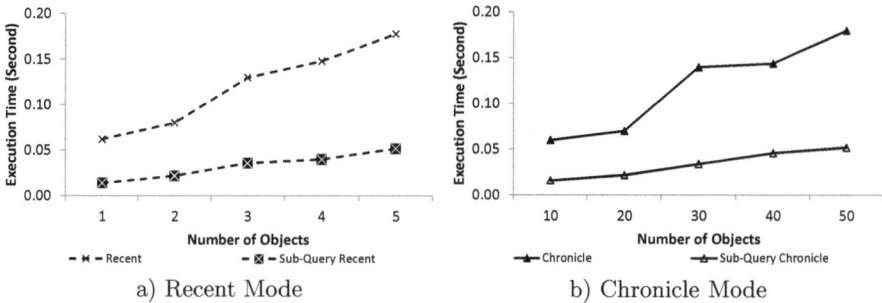

a) Recent Mode   b) Chronicle Mode

Fig. 12. Effects of Object Number on Sub-Query Approach Improvement

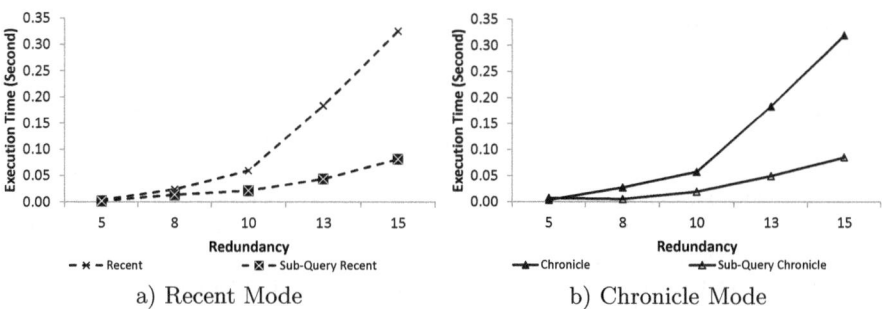

a) Recent Mode   b) Chronicle Mode

Fig. 13. Effects of Redundancy on Sub-Query Approach Improvement

5   Conclusion

In this paper, we have proposed an approach to realize the ESL-Event on the traditional DBMSs. Specifically, our work can translate ESL-Event statements into SQL statements. The sequence enforcing operators, as well as all the tuple pairing mode are implemented. The proposed approach firstly generates a parse tree for the given ESL-Event statement. Such intermediate representation can provide a mean for the translation. After the tree is converted into the corresponding SQL parse tree, the SQL version of the input statement is generated. In addition to the proposed approach, we also apply the sub-query for two tuple pairing modes for efficiency improvement. From the experiment results, the translated statements can achieve a high efficiency, particularly, the statements with tuple pairing modes implemented by sub-query approach.

References

1. Abadi, D., Carney, D., Çetintemel, U., Cherniack, M., Convey, C., Erwin, C., Galvez, E., Hatoun, M., Maskey, A., Rasin, A., Singer, A., Stonebraker, M., Tatbul, N., Xing, Y., Yan, R., Zdonik, S.: Aurora: a data stream management system. In: Proceedings of the 2003 ACM SIGMOD International Conference on Management of Data, SIGMOD 2003, pp. 666–666. ACM, New York (2003)
2. Arasu, A., Babu, S., Widom, J.: The cql continuous query language: semantic foundations and query execution. The VLDB Journal 15(2), 121–142 (2006)
3. Bai, Y., Thakkar, H., Wang, H., Luo, C., Zaniolo, C.: A data stream language and system designed for power and extensibility. In: Proceedings of the 15th ACM International Conference on Information and Knowledge Management, CIKM 2006, pp. 337–346. ACM, New York (2006)
4. Bai, Y., Wang, F., Liu, P., Zaniolo, C., Liu, S.: Rfid data processing with a data stream query language. In: 2007 IEEE 23rd International Conference on Data Engineering, pp. 1184–1193. IEEE (2007)
5. Bellamkonda, S., Ahmed, R., Witkowski, A., Amor, A., Zait, M., Lin, C.-C.: Enhanced subquery optimizations in oracle. Proc. VLDB Endow. 2(2), 1366–1377 (2009)
6. Buehrer, G.T., Weide, B.W., Sivilotti, P.A.G.: Using parse tree validation to prevent sql injection attacks. In: Proceedings of the International Workshop on Software Engineering and Middleware (SEM), pp. 106–113 (2005)
7. Choi, S.Y., Jung, H.M., Bang, K.S., Lee, W.Y., Ko, Y.W.: Real-time data stream management system for large volume of rfid events. Design Issues, 515–521 (2008)
8. Golab, L., Özsu, M.T.: Issues in data stream management. SIGMOD Rec. 32(2), 5–14 (2003)
9. Plagemann, T., Goebel, V., Bergamini, A., Tolu, G., Urvoy-keller, G., Biersack, E.W.: Using data stream management systems for traffic analysis - a case study. In: Passive and Active Measurements
10. Surdu, S.: Data stream management systems: a response to large scale scientific data requirements. Annals of the University of Craiova, Mathematics and Computer Science Series 38, 66–75 (2011)
11. Zeitler, E., Risch, T.: Scalable Splitting of Massive Data Streams. In: Kitagawa, H., Ishikawa, Y., Li, Q., Watanabe, C. (eds.) DASFAA 2010. LNCS, vol. 5982, pp. 184–198. Springer, Heidelberg (2010)

An Efficient Active Learning Method Based on Random Sampling and Backward Deletion

Hoyoung Woo and Cheong Hee Park

Dept. of Computer Science and Engineering Chungnam National University
220 Gung-dong, Yuseong-gu, 305-764, Korea
{Whyo,cheonghee}@cnu.ac.kr

Abstract. Active learning aims to select data samples which would be the most informative to improve classification performance so that their class labels are obtained from an expert. Recently, an active learning method based on locally linear reconstruction(LLR) has been proposed and the performance of LLR was demonstrated well in the experiments comparing with other active learning methods. However, the time complexity of LLR is very high due to matrix operations required repeatedly for data selection. In this paper, we propose an efficient active learning method based on random sampling and backward deletion. We select a small subset of data samples by random sampling from the total data set, and a process of deleting the most redundant points in the subset is performed iteratively by searching for a pair of data samples having the smallest distance. The distance measure using a graph-based shortest path distance is utilized in order to consider the underlying data distribution. Experimental results demonstrate that the proposed method has very low time complexity, but the prediction power of data samples selected by our method outperforms that by LLR.

Keywords: Active Learning, Optimal Experimental Design, Random Sampling.

1 Introduction

In many applications of pattern recognition and data mining, collecting a huge volume of unlabeled data is getting easier due to the rapid technical advance, but the manual labeling process is very costly and time-consuming. When a classifier should be trained on a small number of labeled data, the construction of a reliable classifier is difficult. Hence, it is important to select a subset of data samples for labeling which can effectively represent the underlying data distribution. Active learning is the method to aggressively select data samples which would be the most informative to train a classifier[1]. By acquiring the exact class label of a data sample from an expert, active learning enlarges the size of a labeled data set on which a new classifier is trained. The most important factor in active learning is to select the data sample that would be the most helpful if its class label is known. Active learning methods select either the most difficult data sample for the current classifier[2,3] or the most informative data sample that maximizes some expected gain[4,5].

J. Yang, F. Fang, and C. Sun (Eds.): IScIDE 2012, LNCS 7751, pp. 683–691, 2013.

On the other hand, when only unlabeled data is given and any labeled data samples are not available for training an initial classifier, most of the existing active learning methods are difficult to apply. In statistics, a data sample and its class label are referred to as *an experiment* and *measurement*, respectively. *Optimal experimental design* is concerned to select the most informative experiments to measure[6]. In optimal experimental design, selection of experiments is performed by the criterion of minimizing the variance of model parameters or the variance of prediction value[7,8]

Recently, an active learning method based on locally linear reconstruction (LLR) has been proposed, which can be applied when no labeled data samples are given. LLR computes the coefficients of every samples by taking into account the local structure of the data space and selects the most representative points whose coefficients can be used to best reconstruct the whole data set[9]. The performance of LLR was demonstrated well in the experiments comparing with other optimal experimental design and active learning methods. However, the time complexity of LLR is very high due to matrix operations required repeatedly for data selection.

A fundamental and simple way to select a data subset for manual labeling is to sample randomly a small number of data samples from the pool of unlabeled data. Random sampling is very easy and fast to implement, but the points selected by random sampling are not guaranteed to be optimal. Therefore, for the performance of a classifier to reach a certain level, a significant number of data samples should be sampled. In this paper, we propose an efficient active learning method whose time complexity is low as much as random sampling, but the prediction power of selected samples by our method outperforms that by LLR. We select a small subset of data samples by random sampling from the total data set, and the most redundant points in the subset are deleted. The deletion process is performed repeatedly by searching for a pair of data samples having the smallest distance and removing one of them from the subset. Euclidean distance can be used as a distance measure. However, a small subset by random sampling is deficient to represent total data distribution. In order to overcome this weakness, we use the shortest path distance based on a graph structure of the total data set. The proposed method reduces the time complexity by applying random sampling in the first stage and achieves the predictive performance by removing redundant data samples using the distance measure which reflects the total data distribution.

Our work is related with optimal experimental design in that the goal is to select a subset of samples on which an optimal classification model is trained. However, data samples are not initially accompanied with class labels. Instead, class labels should be obtained from an outside expert. In this respect, our work also resembles active learning. However, the approach is a little different, since we are not given any labeled set initially, differently from that very small labeled set is usually given and active learning is used to increase the size of labeled set by moving data samples from a large volume of unlabeled data.

In Section 2, related works are reviewed. In Section 3, an active learning method based on random sampling and deletion process is presented. Experimental results demonstrate the performance of the proposed method in Section 4, and discussions follow in Section 5.

2   Related Works

In statistics, sampling is to select a subset of individuals from a population to estimate characteristics of the entire population[10]. In a simple random sampling, each sample of the data has an equal probability of selection and a subset of samples are selected by using a table of random numbers. In many studies, simple random sampling is commonly used in various forms. Random sampling is not only very simple but its performance also does not fall below the average.

Optimal experimental design is to select a subset of the pairs of experiments and measurements (corresponding to data samples and class labels in a classification problem) to learn a prediction function so that the expected prediction error can be minimized[6]. In a simple linear regression, the parameter w in $f(x) = w^T x$ needs to be designed and a subset of experiments which optimize some objective criterion such as the minimization of the variance of model parameters or the variance of prediction value are found.

Active learning aims to minimize expert's efforts for labeling required for a supervised classifier[1]. When the number of labeled data samples is very small, the most informative data sample is selected from the pool of unlabeled data samples and it is labeled by an expert. A new classifier is trained on the increased labeled set and this process is repeated until achieving a satisfactory performance. Selection criteria in active learning are usually divided to two categories. One of them is to select a data sample which is the most difficult to predict by the current classifier or whose predicted label is most uncertain[2,3]. The other category of selection criteria is to select the most informative data sample that maximizes some expected gain[4,5]. For each unlabeled data sample, the expected gain to obtain if its class label is known is computed, and the sample with the maximum expected gain is selected.

3   The Proposed Method

We are given an unlabeled data set X. The goal is to choose an optimal subset S of X for manual labeling. Since the size of S is usually small, the elements of S should be representative to reflect the data distribution of X so that a classifier can be effectively trained on S. Selection of an informative data sample from X is usually made sequentially by evaluating an objective criterion. However, the computational complexity of selection process is very high, especially when the size of X is large, since search is performed on the total data set X. We propose a method which selects a subset of data set by random sampling, usually much smaller than the total data set, and then reduces the size of the selected set by removing the most redundant data samples.

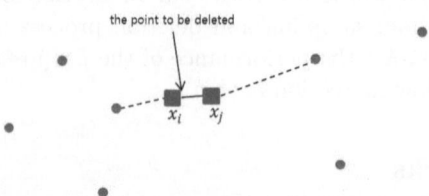

Fig. 1. Determining the point for deletion using 1-NN distance

When we want to select t data samples for manual labeling, we compose a subset $S = \{x_1, x_2, \cdots, x_n\}$ from the unlabeled data set X by random sampling. While n can be any number which is larger than t, we set $n = 2t$ in our experiments. Let $d(x_i, x_j)$ be a distance measure between x_i and x_j. The smaller $d(x_i, x_j)$ is, the nearer two points x_i and x_j are located. Then the information obtained from x_i and x_j may be overlapped. We remove one of them from S. Let (x_i, x_j) be a pair of data samples in S whose distance measure is the smallest among all the pairs of points in S. We can delete any point among two points. However, we can make better choice by taking into account the neighborhood of two points. Let $N_k(x_i)$ be the set of the k-nearest neighbors of x_i in S, excluding x_j. $N_k(x_j)$ is defined similarly. The average distance $f(x_i, k)$ from x_i to its k-nearest neighbors in $N_k(x_i)$ is defined by

$$f(x_i, k) = \frac{1}{k} \sum_{x \in N_k(x_i)} d(x_i, x). \tag{1}$$

The one which has the smallest neighborhood distance $f(x_i, k)$ is chosen for deletion from S. This deletion process is repeated until the size of S becomes t. Figure 1 illustrates the situation to choose a point for deletion using 1-nearest neighbor distance. Among all the points in S, two points marked by red rectangles have the smallest distance. By deleting x_i from S which has shorter 1-nearest neighbor distance, the scatterness in remaining points can get bigger. In the next subsections, we present two distance measures which are used in the proposed method, Euclidean distance and graph-based shortest path distance.

3.1 Using Euclidean Distance

Distance measure $d(x_i, x_j)$ which computes the distance between x_i and x_j should satisfy the following properties[11].

[Positivity] $d(x, y) \geq 0$ for all x and y, $d(x, y) = 0$ only if $x = y$.
[Symmetry] $d(x, y) = d(y, x)$ for all x and y.
[Triangle inequality] $d(x, z) \leq d(x, y) + d(y, z)$ for all x, y, and z.

Euclidean distance is usually used in many applications. The proposed algorithm using Euclidean distance $d_1(x, y)$ as a distance measure is summarized in Table 1.

Table 1. [Algorithm 1] Active learning based on random sampling and backward deletion using Euclidean distance

Input X: the set of unlabeled data
      t: the number of samples to be selected
      k: the number of neighbors considered in the deletion process
Output S: a set of data samples for manual labeling

Select a subset S from X at random such as $\lvert S \rvert = 2t$
repeat
      Find a pair of data samples x_i and x_j in S where $d_1(x_i, x_j)$ is the smallest
      For x_i and x_j, compute $f(x_i, k)$ and $f(x_j, k)$ which is defined in Eq. (1)
      From S, delete the point which has smaller value among $f(x_i, k)$ and $f(x_j, k)$
until $\lvert S \rvert = t$

3.2   Using Graph-Based Shortest Path Distance

In Algorithm 1 of Table 1, when t is very small, the set S selected by random sampling is insufficient to represent the data distribution of X. Hence it is difficult to obtain the best result only by the removal of redundant data samples from S. In order to solve this problem, we adapt a distance measure to reflect the data distribution in X. Based on graph structure on X, the distance between points in S is computed via the shortest path distance.

We first build a weighted graph on X, representing data points as vertices and placing weights on an edge between points. For each vertex, only the edges connected to its l-nearest neighbors are preserved with positive weights and the other edges are deleted by putting zero weight. The weight $w(x_i, x_j)$ can be defined with any similarity measures, we simply put weight 1 on the edge connected by l-nearest neighborhood relationship.

$$w(x_i, x_j) = \begin{cases} 1 & \text{if } x_i \text{ (or } x_j) \text{ is in } l\text{-nearest neighborhood of } x_j (\text{or } x_i). \\ 0 & \text{otherwise.} \end{cases} \quad (2)$$

The distance $d_2(x_i, x_j)$ between any points in S is defined by the sum of weights of the edges on the shortest path between them. It is easy to check that $d_2(x_i, x_j)$ satisfies the distance properties given in Section 3.1. The algorithm using $d_2(x_i, x_j)$ as a distance measure is summarized in Table 2.

Distance measure $d_2(x_i, x_j)$ defined by the shortest path distance contains more information than Euclidean distance $d_1(x_i, x_j)$, since it reflects neighborhood structure in the total data set. Figure 2 illustrates a case that five points marked by red rectangles are contained in S. If Euclidean distance measure is used, x_1 and x_2 have the shortest distance and x_1 with smaller $f(x_1, 1)$ is deleted. The deletion of x_1 makes no points left in the cluster marked with small blue rectangles. On the other hand, by using graph-based distance $d_2(x_i, x_j)$, the distance between x_1 and other points in S becomes ∞ and therefore the point x_4 is deleted from S.

It takes more time to construct a graph from X and compute the shortest path distances between points in S than to simply use Euclidean distance on S.

Table 2. [Algorithm 2] Active learning based on random sampling and backward deletion using graph-based distance

Input X: the set of unlabeled data
   t: the number of samples to be selected
   k: the number of neighbors considered in the deletion process
   l: the number of neighbors considered in graph construction on X
Output S: a set of data samples for manual labeling

Construct the weighted graph on X which is defined in (2).
Select a subset S from X at random such as $\|S\| = 2t$
Compute the distance $d_2(x_i, x_j)$ between any two points in S
repeat
  Find a pair of data samples x_i and x_j in S where $d_2(x_i, x_j)$ is the smallest
  For x_i and x_j, compute $f(x_i, k)$ and $f(x_j, k)$ which is defined in Eq. (1)
  From S, delete the point which has smaller value among $f(x_i, k)$ and $f(x_j, k)$
until $\|S\| = t$

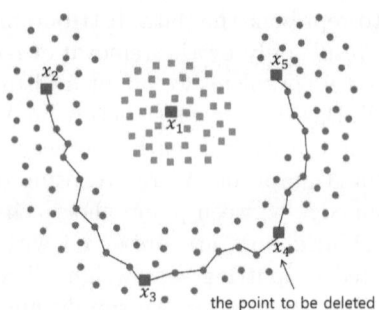

Fig. 2. Graph-based Distance

But the increased computational complexity can be compensated by the improved predictive power which will be shown in the experiments of Section 4. Also we show that the computational complexity of Algorithm 2 is still much lower than LLR.

4   Experimental Results

In order to test the performance of the proposed algorithms, we evaluate and compare the proposed method with random sampling and LLR using real data sets. The detailed information for data sets is described in Table 3. Segment, optical digits and pen digits were downloaded from UCI[12], and USPS is the data set used in [9] whose data dimension is reduced from 256 to 64 by averaging pixel values in 2×2 blocks. We randomly split the data set into training and test sets. For segment and optical digits, training set was set as 60% of the data set and test set was 40%. In pen digits and USPS, for manageable training size, the split ratio of training and test sets was 40%:60%.

Table 3. Data Sets

data sets	samples	cases	features
segment	2,310	7	19
optical digits	5620	10	64
pen digits	10,992	10	16
USPS	11,000	10	64

Table 4. Perforamnce comparison

			SVM						MLP			
			Algo.1		Algo.2				Algo.1		Algo.2	
t	Random	LLR	1NN	3NN	1NN	3NN	Random	LLR	1NN	3NN	1NN	3NN
					Segment data set							
10	51.89	54.63	59.69	58.79	63.40	62.59	50.22	53.93	55.41	55.25	57.62	58.80
30	75.37	73.60	79.59	81.41	80.56	81.57	70.68	72.42	80.17	80.75	80.38	81.33
50	82.99	81.35	85.59	85.93	85.72	86.11	81.26	80.83	86.35	85.97	85.67	86.05
70	86.40	86.63	88.86	88.57	87.58	88.04	85.69	85.83	88.88	88.73	88.00	88.21
					Optical digit data set							
10	51.08	55.24	53.91	54.78	58.68	59.40	46.63	51.61	51.17	51.68	54.86	55.19
30	74.41	69.90	80.37	81.48	82.68	81.64	71.52	68.45	76.17	77.98	79.28	77.62
50	82.41	71.96	86.95	87.66	87.98	88.10	79.57	70.87	84.75	85.36	85.93	86.17
70	86.74	72.72	90.25	90.78	90.46	90.87	84.43	71.75	89.04	88.93	89.30	89.82
					Pen digit data set							
10	42.48	64.46	51.80	51.88	53.54	53.88	40.00	61.62	49.12	49.70	49.40	50.50
30	71.48	80.71	78.40	78.75	77.53	78.45	68.59	78.03	74.77	75.83	74.13	74.57
50	80.35	85.54	84.25	84.49	84.82	84.67	77.95	83.86	80.91	80.22	81.73	80.95
70	83.27	87.87	88.13	88.87	89.31	89.18	80.64	84.18	83.74	83.58	85.07	85.71
					USPS data set							
10	33.42	34.34	35.08	34.90	37.10	36.72	32.81	35.99	34.37	33.86	35.93	36.41
30	53.30	55.80	57.08	56.62	59.38	60.37	61.82	59.57	64.27	65.19	68.04	66.65
50	65.29	60.42	66.53	67.29	69.82	70.35	70.09	62.97	72.19	72.03	73.01	73.29
70	73.48	63.27	75.01	72.61	73.87	75.64	73.96	63.55	73.63	74.25	74.84	76.07

After selecting data samples from training set for manual labeling by the compared methods, a classifier is trained on the labeled set and prediction accuracies are measured on test set. Prediction accuracies are averaged over 10 runs with different splitting of training and testing data. To test the effectiveness of our method across different classifier methodologies, we used SVM by libSVM[13] and Multi-layer Perceptron(MLP) implemented in WEKA[14]. The parameters were set as default values. In SVM, linear kernel was used. For MLP, learning rate 0.3, the number of epoch 500, and momentum rate 0.2 were set.

For our proposed algorithms using Euclidean and graph-based distance, Algo. 1 and Algo.2, the number k for k-NN in the deletion process were set as 1 and 3. In Algo.2, a weighted graph was constructed with the number of edges $l = 10$. For LLR, the number of neighbors and the parameter μ were set as 10 and 0.1, respectively. Random sampling, LLR, and the proposed algorithms are applied to select $t(=10,30,50,70)$ data samples from training samples. In Table 4, prediction accuracies(%) using SVM and MLP classifiers are reported. The best accuracies are emphasized in a bold face. In most of cases, Algorithm 2 using graph-based distance outperformed or comparable to other methods. However, for pen digits and optical digits, when selecting data samples less than 50, LLR shows the better performance. The performance of Algorithm 1 using Euclidean distance is not bad compared with that of other methods, especially if we consider very low computational complexity of Algorithm 1, which will be shown in the next experiment to compare computational complexities of all the methods.

Table 5. comparison of time complexity using USPS data set

method size of training set	Algorithm1	Algorithm 2	LLR
1000	0.0873606	28.46394246	601.065613
2000	0.0842405	136.7473566	4738.88606
4000	0.0811205	790.9094699	35017.8409

In order to compare computational complexity, we used USPS data set and measured cpu time of LLR and the proposed algorithms with 1NN. Since random sampling only requires index manipulation for sampling, it was excluded in this test. The experiments were performed on Intel Core I7-2600k CPU 3.40GHz with RAM 8G using Matlab version 11. Varying the size of training set as 1000, 2000 and 4000, the cpu time which takes to select 30 data samples was measured. The test was repeated 10 times and average time in seconds is shown in Table 5. The computational complexity of LLR is very high compared with the proposed methods, and Algorithm 1 using Euclidean distance is much faster than other methods as expected.

5   Discussions

We have proposed an active learning method based on random sampling and deletion of the least informative data samples. Two distance measures to compute distances between data samples of the subset selected by random sampling were presented, Euclidean distance and graph-based shortest path distance. Using graph-based distance requires more computation than using Euclidean distance, but the selection method using graph-based distance produces more representative points, since graph-based distance gives the effects of taking entire data into consideration.

Acknowledgments. This research was supported by Basic Science Research Program through the National Research Foundation of Korea(NRF) funded by the Ministry of Education, Science and Technology(2012-0004189).

References

1. Settles, B.: Active learning literature survey: Computer sciences technical report 1648, University of Wisconsin-Madison (2009)
2. Freund, Y., Seung, H.S., Shamir, E., Tishby, N.: Selective sampling using the query by committee algorithm. Machine Learning 28(2-3), 138–168 (1997)
3. Tong, S., Keller, D.: Support vector machine active learning with applications to text classifications. Journal of Machine Learning Research 2, 45–66 (2002)
4. Cohn, D.A., Ghahramani, Z., Jordan, M.I.: Active learning with statistical models. Journal of Artificial Intelligence Research 4, 129–145 (1996)
5. Zhu, X., Ghahramani, Z., Lafferty, J.: Semi-supervised learning using Gaussian Fields and Harmonic functions. In: Proceedings of ICML, pp. 912–919 (2003)
6. Atkinson, A., Donev, A., Tobias, R.: Optimum Experimental Designs, with SAS. Oxford Univ. Press (2007)
7. Asprey, S.P., Macchietto, S.: Designing robust optimal dynamic experimetns. J. Process Control 12(4), 545–556 (2002)
8. Hardin, R.H., Sloane, N.J.A.: A new approach to the construction of optimal designs: J. Statistical Planning and Inference 37(3), 339–369 (1993)
9. Zhang, L., Chen, C., Bu, J., Cai, D., He, X., Huang, T.S.: Active Learning Based on Locally Linear Reconstruction. IEEE Trans. Pattern Anal. Machine Int. 33(10), 2026–2038 (2011)
10. Groves, R.M., Fowler Jr., F.J.: Survey methodology. Wiley (2010)
11. Tan, P., Steinbach, M., Kumar, V.: Introduction to Data Mining. Addison Wesley, Boston (2006)
12. UCI Machine Learning Repository, http://archive.ics.uci.edu/ml/
13. A Library for SVM, http://www.csie.ntu.edu.tw/~cjlin/libsvm/
14. Machine Learning Group at University of Waikato, http://www.cs.waikato.ac.nz/ml/weka/

Fuzziness-Preserving Attribute Reduction from Hybrid Data

Wei Wei, Jiye Liang, Yuhua Qian, and Junhong Wang

Key Laboratory of Computational Intelligence and Chinese Information Processing
of Ministry of Education, School of Computer and Information Technology,
Shanxi University, Taiyuan, 030006, Shanxi, China
{weiwei,ljy,jinchengqyh,wjhwjh}@sxu.edu.cn

Abstract. In this paper, we devote to present a fuzziness-preserving
attribute reduction in fuzzy rough set framework. Through constructing
the membership function of an object, we first introduce a fuzzy measure
to assess the fuzziness of a fuzzy rough set and a fuzzy rough decision,
which underlies a foundation for attribute reduction algorithm. Then,
we derive an attribute significance measure based on the proposed fuzzy
measure and design a forward greedy algorithm (ARBF) for attribute
reduction from hybrid. Numerical experiments show the validity of the
proposed algorithm from search strategy and heuristic function in the
meaning of fuzziness-preserving.

Keywords: Fuzzy rough set, fuzzy decision tables, attribute reduction,
fuzzy measure, hybrid data.

1 Introduction

In recent years, we encounter databases in which both the number of objects becomes higher and their dimensionality (number of attributes) gets larger as well.
Tens, hundreds, and even thousands of attributes are stored in many real-world
application databases, which brought great challenge for data mining and knowledge discovery from large scale data [2]. As was pointed out in several literature,
attributes that are irrelevant to knowledge discovery tasks may deteriorate the
performance of learning algorithms [6]. That is to say, storing and processing
all attributes (both relevant and irrelevant) could be computationally very expensive and impractical. To deal with this issue, we can omit some attributes,
which will not seriously impact the target of knowledge discovery. Therefore, the
omission of some attributes could not only be tolerable but even desirable relatively to the costs involved in such cases [4]. This task is often called attribute
reduction or feature selection.

Rough set theory, proposed by Pawlak in [7] has been conceived as a tool to
conceptualize and analyze various types of data. It can be used in attribute value
representation models to describe the dependencies among attributes, evaluate
the significance of attributes and derive decision rules. The theory shows important applications to intelligent decision-making and cognitive sciences, as a

J. Yang, F. Fang, and C. Sun (Eds.): IScIDE 2012, LNCS 7751, pp. 692–699, 2013.

tool for dealing with vagueness and uncertainty of information [9]. In [3], Hu et al. established a new rough set framework called fuzzy probabilistic approximation spaces, which introduces probability into fuzzy rough set model and leads to a tool for dealing with randomness, roughness and fuzziness in real-world applications. In the fuzzy rough set model, for a fuzzy approximation space, if condition attributes and decision attributes are distinguished, then it is called a fuzzy decision table (or a fuzzy decision information system).

Attribute reduction is an outstanding contribution made by rough set research to data analysis, which does not attempt to maximize the class separability but rather attempts to retain the discernible ability of original features for the objects from the universe [8]. In this paper, from the viewpoint of preserving the fuzziness of a rough decision, we will construct a feature selection method from fuzzy decision tables in the framework of fuzzy rough sets, which is called fuzzy measure-based attribute reduction. We will also perform a series of experimental analyses for illustrating the validity of the proposed approach from search strategy and heuristic function.

In the next section, some preliminary concepts are briefly reviewed. In Section 3, we will introduce a so-called fuzzy measure to calculate the fuzziness of a fuzzy rough set and that of a fuzzy rough decision in a general fuzzy decision table. In Section 4, based on the proposed fuzzy measure, we develop a fuzziness-preserving attribute reduction approach in fuzzy rough set framework. In Section 5, we present a series of experimental studies that focus on the quantification of fuzziness of the selected attributes from search strategy and heuristic function on ten hybrid data sets. Finally, Section 6 concludes this paper.

2   Preliminaries

In this section, we review some basic concepts and give some notations in fuzzy-rough set model.

Given a nonempty finite set U, $\widetilde{R}$ is a fuzzy binary relation over U, denoted by a matrix

$$M(\widetilde{R}) = \begin{pmatrix} r_{11} & r_{12} & \cdots & r_{1n} \\ r_{21} & r_{22} & \cdots & r_{2n} \\ \cdots & \cdots & \cdots & \cdots \\ r_{n1} & r_{n2} & \cdots & r_{nn} \end{pmatrix},$$

where $r_{ij} \in [0,1]$ is the relation value between x_i and x_j.

We say $\widetilde{R}$ is a fuzzy equivalence relation if $\forall x, y, z \in U$, $\widetilde{R}$ satisfies

1) Reflexivity: $\widetilde{R}(x,x) = 1$;
2) Symmetry: $\widetilde{R}(x,y) = \widetilde{R}(y,x)$;
3) Transitivity: $\widetilde{R}(x,z) \geq min_y\{\widetilde{R}(x,y), \widetilde{R}(y,z)\}$.

A fuzzy equivalence relation generates a fuzzy partition of the universe and a series of fuzzy equivalence classes, which are also called fuzzy knowledge granules.

The fuzzy partition of the universe generated by a fuzzy equivalence relation $\widetilde{R}$ is defined as $U/\widetilde{R} = \{[x_i]_{\widetilde{R}}\}_{i=1}^n$, where $[x_i]_{\widetilde{R}} = \{(r_{i1}/x_1) + (r_{i2}/x_2) + \cdots + (r_{in}/x_n)\}$. $[x_i]_{\widetilde{R}}$ is the fuzzy equivalence class containing x_i and r_{ij} is the degree of x_i equivalent to x_j. Here, "+" means the union of elements.

The cardinality of the fuzzy equivalence class $[x_i]_{\widetilde{R}}$ can be calculated with

$$|[x_i]_{\widetilde{R}}| = \sum_{j=1}^n r_{ij},$$

which appears to be a natural generalization of the cardinality of a crisp set.

In this case, $[x_i]_{\widetilde{R}}$ is a fuzzy set and the family of $[x_i]_{\widetilde{R}}$ forms a fuzzy concept system of the universe. This system will be used to approximate the object subset of the universe.

Definition 1. *A two-tuple $\langle U, \widetilde{R} \rangle$ is a fuzzy approximation space or a fuzzy information system, where U is a nonempty and finite set of objects, called the universe, and $\widetilde{R}$ is a family of fuzzy equivalence relations defined on U.*

Let $\widetilde{X}$ be a fuzzy set. Then, it is represented as $\widetilde{X} = \frac{\mu_{\widetilde{X}}(x_1)}{x_1} + \frac{\mu_{\widetilde{X}}(x_2)}{x_2} + \cdots + \frac{\mu_{\widetilde{X}}(x_n)}{x_n}$, where $\mu_{\widetilde{X}}(x_j)$ denotes the membership degree of the object x_j in X.

For convenience, we give another denotation of a fuzzy information system. From now, denoted by $S = (U, \widetilde{A})$ be a fuzzy information system and $\widetilde{A}$ a fuzzy attribute set in U. It can generate a fuzzy equivalence relation $\widetilde{R}_A$ on U. The fuzzy relation matrix $M(\widetilde{R}_A)$ is denoted by

$$M(\widetilde{R}_A) = \begin{pmatrix} r_{11} & r_{12} & \cdots & r_{1n} \\ r_{21} & r_{22} & \cdots & r_{2n} \\ \cdots & \cdots & \cdots & \cdots \\ r_{n1} & r_{n2} & \cdots & r_{nn} \end{pmatrix},$$

where $r_{ij} \in [0, 1]$ is the relation value between x_i and x_j.

A fuzzy decision table is a fuzzy information system $S = (U, \widetilde{C} \cup \widetilde{d})$, where $\widetilde{C}$ is called a fuzzy condition attribute set and $\widetilde{d}$ is called a fuzzy decision attribute. In practical decision-making issues, in general, the decision attribute $\widetilde{d}$ can induce an equivalence partition, i.e., a crisp classification. In this paper, we only focus on this kind of fuzzy decision tables.

In the following, we review the definition of a fuzzy rough set in fuzzy information systems, which is shown as follows.

Definition 2. *[1] Let $\langle U, \widetilde{R} \rangle$ be a fuzzy approximation space and $\widetilde{X}$ a fuzzy subset of U. The lower approximation and upper approximation are denoted by $\underline{\widetilde{R}}\widetilde{X}$ and $\overline{\widetilde{R}}\widetilde{X}$, respectively. Then, the membership degree of x to $\widetilde{X}$ are defined as*

$$\begin{cases} \mu_{\underline{\widetilde{R}}\widetilde{X}}(x) = \wedge\{\mu_{\widetilde{X}}(y) \vee (1 - \widetilde{R}(x, y)) : y \in U\}, x \in U \\ \mu_{\overline{\widetilde{R}}\widetilde{X}}(x) = \vee\{(\mu_{\widetilde{X}}(y) \wedge \widetilde{R}(x, y)) : y \in U\}, x \in U \end{cases},$$

where $\wedge$ and $\vee$ mean min and max operators, respectively, and $\mu_{\widetilde{X}}(y)$ means the membership of y to $\widetilde{X}$. The order pair $\langle \underline{\widetilde{R}}\widetilde{X}, \overline{\widetilde{R}}\widetilde{X} \rangle$ is called a fuzzy rough set.

3   Fuzzy Measures in Fuzzy Decision Tables

In this section, we deal with the fuzzy measure of a fuzzy rough set and that of a fuzzy rough decision in the context of fuzzy decision tables.

Given a fuzzy approximation space $\langle U, \widetilde{A} \rangle$, a fuzzy subset $\widetilde{X}$ of U and a fuzzy equivalence relation $\widetilde{R}$ denoted by a relation matrix $M(\widetilde{R})$ on U, a membership function of $x_i \in U$ in $\widetilde{X}$ is defined as $\delta_{\widetilde{X}}^{\widetilde{R}}(x_i) = \frac{\sum_{j=1}^{n}(\mu_{\widetilde{X}}(x_j) \wedge r_{ij})}{\sum_{j=1}^{n} r_{ij}}$, where $\mu_{\widetilde{X}}(x_j) \wedge r_{ij}$ is equivalent to $min\{\mu_{\widetilde{X}}(x_j), r_{ij}\}$. In fact, this formula can be expressed equivalently as $\delta_{\widetilde{X}}^{\widetilde{R}}(x_i) = \frac{|\widetilde{X} \cap [x_i]_{\widetilde{R}}|}{|[x_i]_{\widetilde{R}}|}$.

Let $\langle U, \widetilde{A} \rangle$ be a fuzzy approximation space and $\widetilde{X}$ a fuzzy subset of U. From the above notations, we know that $\delta_{\widetilde{X}}^{\widetilde{A}}(x)$ is also a fuzzy concept and it can induce a fuzzy set $F_{\widetilde{X}}^{\widetilde{A}} = \{(x, \delta_{\widetilde{X}}^{\widetilde{A}}(x)) \mid x \in U\}$ on the universe U. Naturally, one can induce the following theorem.

Theorem 1. $\widetilde{X} \subseteq \widetilde{Y} \Rightarrow F_{\widetilde{X}}^{\widetilde{A}} \subseteq F_{\widetilde{Y}}^{\widetilde{A}}$.

As follows, we will give a definition of fuzzy measure of a fuzzy rough set in the context of fuzzy approximation spaces.

Definition 3. Let $\langle U, \widetilde{A} \rangle$ be a fuzzy approximation space, $\widetilde{X}$ be a fuzzy set and $M(A)$ be the fuzzy equivalence relation matrix induced by $\widetilde{A}$. Then, a fuzzy measure of the fuzzy rough set $\langle \underline{\widetilde{A}}\widetilde{X}, \overline{\widetilde{A}}\widetilde{X} \rangle$ is defined as $F(F_{\widetilde{X}}^{\widetilde{A}}) = \frac{1}{n} \sum_{i=1}^{n} \delta_{\widetilde{X}}^{\widetilde{A}}(x_i)(1 - \delta_{\widetilde{X}}^{\widetilde{A}}(x_i))$.

From Definition 3, one can obtain the following theorem.

Theorem 2. If X is a crisp and definable set on a fuzzy approximation space, then the fuzzy measure of $\langle \underline{\widetilde{A}}X, \overline{\widetilde{A}}X \rangle$ is equal to zero.

In the following, we investigate a fuzzy rough classification and its fuzziness on the fuzzy approximation space. Let $\langle U, \widetilde{A} \rangle$ be a fuzzy approximation space and $D = \{D_1, D_2, \cdots, D_r\}$ an equivalence classification (i.e., an equivalence partition). We call

$$\begin{cases} \underline{\widetilde{A}}D = \{\underline{\widetilde{A}}D_1, \underline{\widetilde{A}}D_2, \cdots, \underline{\widetilde{A}}D_r\} \\ \overline{\widetilde{A}}D = \{\overline{\widetilde{A}}D_1, \overline{\widetilde{A}}D_2, \cdots, \overline{\widetilde{A}}D_r\} \end{cases}$$

a lower approximation and an upper approximation of D with respect to $\widetilde{A}$, respectively. The order pair $\langle \underline{\widetilde{A}}D, \overline{\widetilde{A}}D \rangle$ is said to be a fuzzy rough classification of U with respect to $\widetilde{A}$. For any $x_i \in U$, a membership function of x_i in D is defined as $\mu_D(x_i) = \frac{|D_k \cap [x]_{\widetilde{A}}|}{|[x]_{\widetilde{A}}|}$, $x_i \in D_k$, $D_k \in D$.

Through using a relation matrix, this membership function can be equivalently written as $\delta_D^{\widetilde{A}}(x_i) = \frac{\sum_{j=1}^{n}(\mu_{D_k}(x_j) \wedge r_{ij})}{\sum_{j=1}^{n} r_{ij}}$, $x_i \in D_k$, $D_k \in D$.

Theorem 3. *Let $\langle U, \widetilde{A} \rangle$ be a fuzzy approximation space and C, D with $C \subseteq D$ two equivalence partitions. Then, $F_C^{\widetilde{A}} \subseteq F_D^{\widetilde{A}}$.*

Definition 4. *Let $\langle U, \widetilde{A} \rangle$ be a fuzzy approximation space and $D = \{D_1, D_2, \cdots, D_r\}$ an equivalence partition. Then a fuzzy measure of the fuzzy rough classification $\langle \underline{\widetilde{A}}D, \overline{\widetilde{A}}D \rangle$ is defined as $F(F_D^{\widetilde{A}}) = \frac{1}{n} \sum_{i=1}^{n} \delta_D^{\widetilde{A}}(x_i)(1 - \delta_D^{\widetilde{A}}(x_i))$.*

From Definition 4, one can draw the following conclusion.

Theorem 4. *The fuzzy measure of a crisp classification on fuzzy approximation spaces is equal to zero.*

From Definition 4 and the above discussions, one can know that the fuzzy measure is the degree of fuzziness of the rough classification approximated by the fuzzy condition attributes in a fuzzy decision table. When only one fuzzy condition attribute is considered, then the fuzzy measure of the given rough classification induced by this attribute can be calculated.

4 Fuzziness-Preserving Attribute Reduction

Given an attribute set, in rough set theory, the task of attribute reduction can be seen as a search for an "optimal" attribute subset through the competing candidate subsets. The definition of what an optimal subset is may vary depending on the problem to be solved. Although an exhaustive method may be used for this purpose in theory, this is quite impractical for most data sets. Usually attribute reduction algorithms involve heuristic or random search strategies in an attempt to avoid this prohibitive complexity.

As the above discussions, we always need to acquire a rough decision with much smaller fuzziness from a given fuzzy decision table in practical decision applications. Therefore, the fuzzy measure can be used to heuristically obtain the rough decision of a target decision in a fuzzy decision table. It is deserved to point out that the fuzzy measure also can be used to evaluate the decision performance of a fuzzy decision table.

If we take the smaller fuzziness as the target of feature selection, then it is very important to keep the fuzziness of a given data set for efficient knowledge discovery. The key problem of this task is how to get a subset of attributes that keeps the fuzziness of an original data set or satisfies user's requirement (it is always given by a threshold).

Let $S = (U, \widetilde{C} \cup \widetilde{d})$ be a fuzzy decision table, where $\widetilde{C}$ and $\widetilde{d}$ be the condition attribute set and decision attribute. $\widetilde{B} \subseteq \widetilde{C}$, $\forall \widetilde{a} \in \widetilde{C} - \widetilde{B}$, we define a coefficient $Sig(\widetilde{a}, \widetilde{B}, \widetilde{d}) = F(F_{\widetilde{d}}^{\widetilde{B}}) - F(F_{\widetilde{d}}^{\widetilde{B} \cup \widetilde{a}})$ as the significance of attribute $\widetilde{a}$ in $\widetilde{C} - \widetilde{B}$ relative to decision $\widetilde{d}$, which measures the reduction of the fuzziness if attribute $\widetilde{a}$ is introduced in $\widetilde{B}$. This measure can be used in a forward feature selection algorithm, while $Sig(\widetilde{a}, \widetilde{B}, \widetilde{d})$ is applicable to determine the significance of every attribute in the context of fuzziness.

Starting with the attribute with minimal fuzziness, we take the attribute with the maximal significance into the attribute subset in each loop until the entire fuzziness of this feature subset satisfies the target requirement (or is less than a given threshold), and then we can get a feature subset. This forward greedy feature selection algorithm can be designed as follows.

Algorithm 1. Attribute reduction based on the fuzzy measure (ARBF)
Input: $S = (U, \widetilde{C} \cup \widetilde{d})$, a threshold of fuzzy measure δ and a range λ.
Output: A feature subset $\widetilde{B}$ of $\widetilde{C}$.
Step 1.   Compute the fuzzy measure of each attribute with $F(F_{\widetilde{d}}^{\widetilde{a}})$, $\forall \widetilde{a} \in \widetilde{C}$
Step 2. $\widetilde{B} \leftarrow \widetilde{a}_0$, $F(F_{\widetilde{d}}^{\widetilde{a_0}}) = \min_i\{F(F_{\widetilde{d}}^{\widetilde{a_i}}), \forall \widetilde{a}_i \in \widetilde{C}\}$
Step 3.   Do
$\quad\quad\quad \{\forall \widetilde{a} \in \widetilde{C} - \widetilde{B}$, compute $Sig(\widetilde{a}, \widetilde{B}, \widetilde{d})$
$\quad\quad\quad\quad$ If $Sig(\widetilde{a}_j, \widetilde{B}, \widetilde{d}) = \max_i\{Sig(\widetilde{a}_i, \widetilde{B}, \widetilde{d})\}$
$\quad\quad\quad\quad\quad \widetilde{B} = \widetilde{B} \cup \{\widetilde{a}_j\}\}$
$\quad\quad\quad$ Until $|F(F_{\widetilde{d}}^{\widetilde{B}}) - \delta| \leq \lambda$
Step 4.   Return $\widetilde{B}$

Table 1. Data description

Data sets	Samples	Numerical features	Nominal features	Classes
1 Pima-indians-diabetes (Pima)	768	8	0	2
2 Glass (Glass)	214	9	0	7
3 Yeast (Yeast)	1484	8	0	10
4 E.coli (Ecoli)	336	7	0	8
5 Credit (Credit)	690	6	9	2
6 Heart (heart)	270	6	7	2
7 Cmc (Cmc)	1473	2	7	3
8 Auto-mpg (Auto)	392	5	2	3
9 Zoo (Zoo)	101	0	16	7

5   Experimental Analysis

Nine data sets from the University of California at Irvine (UCI) Machine Learning Repository are used in the empirical study. The information about these data sets is shown in Table 1. The objective of these experiments is to show the power of the proposed method to select attribute subsets from hybrid data.

For numeric data, firstly, we normalize the numerical attribute x into the interval $[0, 1]$ with $a' = \frac{a - a_{min}}{a_{max} - a_{min}}$.

The value of the fuzzy similarity degree r_{ij} between objects x_i and x_j with respect to numerical attribute a is computed as

$$r_{ij} = \begin{cases} 1 - 4 \times |x_i - x_j|, & |x_i - x_j| \leq 0.25, \\ 0, & otherwise. \end{cases}$$

As $r_{ij} = r_{ji}$ and $r_{ii} = 1$, $0 \leq r_{ij} \leq 1$, the matrix $M = (r_{ij})_{n \times n}$ is a fuzzy similarity relation. We can get a fuzzy equivalence relation from M with max-min transitivity operation. In practice the operation cannot be conducted and we directly search a feature subset with a similarity relation.

For nominal data, given a nominal decision table $S = (U, C \cup d)$, we have that

$$r_{ij} = \begin{cases} 1, & f(x_i) = f(x_j), \ \forall a \in C \\ 0, & otherwise. \end{cases}$$

The matrix $M = (r_{ij})_{n \times n}$ is clearly an equivalence relation.

In order to emphasize the advantage of ARBF algorithm, we will use heuristic functions (the proposed fuzzy measure, fuzzy information entropy and positive region) on nine data sets. Table 2 presents the results. In this experiment, we employ the framework of ARBF algorithm to investigate the performance of each of the fuzzy measure, fuzzy information entropy and positive region for feature selection. In this table, the value from each lattice is the mean value of sum of fuzzy measures on various features, and the last row presents the mean value of fuzzy measures induced by each heuristic function on nine data sets.

Table 2. Variation of fuzziness with three heuristic functions on nine data sets

	Data sets	Heuristic functions		
		Fuzzy measure	Fuzzy information entropy	Positive region
1	Pima	0.1731	0.1804	0.1829
2	Glass	0.1848	0.1931	0.2011
3	Yeast	0.1773	0.1944	0.1817
4	Ecoli	0.1389	0.1399	0.1663
5	Credit	0.0492	0.0641	0.1864
6	Heart	0.0553	0.0717	0.1413
7	Cmc	0.1916	0.1923	0.2065
8	Auto	0.1229	0.1346	0.1524
9	Zoo	0.0213	0.0170	0.0718
*	Average	0.1122	0.1202	0.1517

From Table 2, it can be seen that the ARBF algorithm based on the fuzzy measure is almost better than that based on the fuzzy information entropy and that based on positive region as to the nine data sets in the meaning of fuzziness. Furthermore, the attribute reduction based on the fuzzy information entropy is better than that based on the positive region, which illustrates the heuristic function also greatly influences the result of feature selection. Therefore, from the viewpoint of preserving the fuzziness of a rough decision, the proposed algorithm based on the fuzzy measure outperforms the other heuristic functions.

6 Conclusions and Further Work

In this paper, we have introduced a fuzzy measure to assess the fuzziness of a fuzzy rough set and a rough decision, which underlies a foundation for attribute reduction algorithm. Using the proposed fuzzy measure, we have defined the significance of an attribute which is used to select candidate attributes in attribute reduction process. Based on these strategies, we have presented a fuzziness-preserving attribute reduction in fuzzy rough set framework. This approach does not require discretizing the numerical data in fuzzy rough set framework, called ARBF, and can obtain an attribute reduct from a given hybrid data set.

Acknowledgements. This work was supported by the national natural science foundation of China (Nos. 61202018,60903110), the national key basic research and development program of China (973)(No. 2007CB311002), and the natural science foundation of Shanxi province (Nos. 2009021017-1, 2010021017-3).

References

1. Dübois, D., Prade, H.: Rough fuzzy sets and fuzzy rough sets. International Journal of General Systems 17, 191–209 (1990)
2. Düntsch, I., Gediga, G.: Uncertainty measures of rough set prediction. Artificial Intelligence 106, 109–137 (1998)
3. Hu, Q.H., Yu, D.R., Xie, Z.X., Liu, J.F.: Fuzzy probabilistic approximation spaces and their information measures. IEEE Transactions on Fuzzy Systems 14(2), 191–201 (2006)
4. Hu, Q.H., Xie, Z.X., Yu, D.R.: Hybrid attribute reduction based on a novel fuzzy-rough model and information granulation. Pattern Recognition 40, 3509–3521 (2007)
5. Kryszkiewicz, M.: Rough set approach to incomplete information systems. Information Sciences 112, 39–49 (1998)
6. Kwak, N., Choi, C.H.: Input feature selection for classification problems. IEEE Transactions on Neural Networks 13, 143–159 (2002)
7. Pawlak, Z.: Rough Sets: Theoretical Aspects of Reasoning about Data. Kluwer Academic Publishers, Dordrecht (1991)
8. Qian, Y.H., Liang, J.Y., Li, D.Y., Wang, F., Ma, N.N.: Approximation reduction in inconsistent incomplete decision tables. Knowledge-Based Systems 23(5), 427–433 (2010)
9. Qian, Y.H., Liang, J.Y., Li, D.Y., Zhang, H.Y., Dang, C.Y.: Measures for evaluating the decision performance of a decision table in rough set theory. Information Sciences 178, 181–202 (2008)
10. Slezak, D., Ziarko, W.: The investigation of the Bayesian rough set model. International Journal of Approximate Reasoning 40, 81–91 (2005)

An Improved Fisher Discriminant Dictionary Learning for Video Object Tracking[*]

Ji Zhang[1,2], Hong-yuan Wang[1,2,**], and Fu-hua Chen[3]

[1] School of Information Science and Engineering, ChangZhou University,
ChangZhou 213164, China
[2] ChangZhou Key Laboratory for Process Perception and Interconnected Technology,
ChangZhou 213164, China
[3] Department of Natural Science and Mathematics, West Liberty University,
West Virginia 26074, United States
{zhangji,hywang}@cczu.edu.cn, fuhua@ufl.edu

Abstract. Video object tracking plays an important role in modern computer vision, and many algorithms have been proposed in recent years. $\ell1$-tracker, a novel generative tracking method based on sparse coding, has demonstrated very promising performance in numerous challenging sequences. But the high computational cost, which is caused by the large size of dictionary, influences its application in tracking severely. In this paper, based on original Fisher discriminant dictionary learning(FDDL) and our improved version, we present a novel tracking algorithm, called FD^{2}LT. In our framework, tracking is considered as a problem consisting of three components, including object location, training samples selection, and dictionary updating. Experimental results demonstrate the effectiveness of the proposed tracking algorithm.

Keywords: Fisher Discrimination Dictionary Learning, Sparse Coding, Video Object Tracking.

1 Introduction

Visual object tracking, in essence, deals with video streams which change frame by frame. In video sequence, target changes dynamically and uncertainly, which is caused by occlusion, noisy and varying illumination, viewpoints, and so on[1]. Recently, many tracking algorithms have been proposed, which can be divided into two categories: generative tracking and discriminant tracking[1].

Based on sparse coding, a new signal reconstruction method[2][3], Mei proposed $\ell1$-tracker for generative tracking[4][5]. Occlusion,corruption and other challenging issues in tracking are addressed seamlessly. However, computational cost of $\ell1$-tracker is too expensive to achieve efficient tracking, because of the large size of dictionary(see Section 2.2 for detail). An alternative way is to construct the dictionary not only having rich representation ability, but also having few atoms, which is called dictionary learning(DL)[6][9][10][11].

---

[*] The National Natural Foundation of China under Grant No. 61070121,60973094.
[**] Corresponding author.

J. Yang, F. Fang, and C. Sun (Eds.): IScIDE 2012, LNCS 7751, pp. 700–710, 2013.
© Springer-Verlag Berlin Heidelberg 2013

As a dictionary learning algorithm used in face recognition, fisher discriminant dictionary learning(FDDL) aims to learn a structured dictionary whose sub-dictionaries have specific class labels[6]. Our work is motivated by FDDL, and a novel tracking framework based on FDDL is proposed. The rest of this paper is organized as follows: ℓ1-tracker and FDDL are introduced in section 2; In section 3, FDDL is modified and introduced into the framework of discriminant tracking, called FD^{2}LT; Experimental results with FD^{2}LT and some competitive algorithms are reported in section 4; Section 5 is the conclusion.

2   Brief Introduction of ℓ1-Tracker and FDDL

2.1   Sparse Coding and Sparse Coding Based Tracking (ℓ1-Tracker)

The main task of sparse coding is to reconstruct a signal $y \in \mathbb{R}^{d \times 1}$ over dictionary $D \in \mathbb{R}^{d \times n}$ (which contains n d-dimension atoms) with sparse coefficient vector $x \in \mathbb{R}^{n \times 1}$[2][3]. The formulation of sparse coding, which is NP-hard, can be written by the optimization problem $\min \|x\|_0$ under the constraint $y = Dx$, where $\|x\|_0$ counts the number of non-zero elements of vector x. Candès has proven that[2][3], $\|x\|_1$ is the tightest upper bound of $\|x\|_0$. Simultaneously, consider the existence of observation noise, sparse coding problem can be written as:

$$min_x \|x\|_1 \quad , \quad s.t. \quad \|y - Dx\| \leq \varepsilon \tag{1}$$

Sparse coding based ℓ1-tracker is one of the robust tracking algorithms proposed by Mei[4][5](see Fig.1). Suppose that, target is located in #205(not shown here). Tracking in #206 is initializing with the location of object in #205, and N candidate regions are generated with Bayesian inference around it, as shown in Fig.1(a). With n templates and $2d$ trivial templates in Fig.1(b)(including d positive ones and d negative ones, and d is the dimension of $1D$ stretched image), Eq.(1) is solved as Fig.1(c). Reconstruction errors of all candidates can be used to determine the weights for each candidate, and object in #206 can be located with the sum of weighted candidates.

2.2   Fisher Discriminant Dictionary Learning

Computational cost of ℓ1-tracker heavily depends on the number of candidate regions N and the size of dictionary D. In order to achieve robust tracking, N must be large, while the dimension of dictionary D is fixed to $d \times (2d+n)$ in Eq.(1). It is quite nature that, how to reduce the number of candidate regions and the size of dictionary without loss of tracking accuracy, are two important issues in ℓ1 tracking. The former depends on the improvement of particle filtering, which is not mentioned in this paper; and the latter, specifically, how to construct dictionary which not only contains few atoms, but also has good ability of representation, is exactly the main task of dictionary learning[6][8].

Yang's FDDL learns a structured dictionary $D = [D_1, D_2, ..., D_c]$ instead of the whole dictionary, where D_i is class-i sub-dictionary, and c is the class number. Let $Y = [Y_1, Y_2, ..., Y_c]$ and $X = [X_1, X_2, ..., X_c]$ denote the set of training

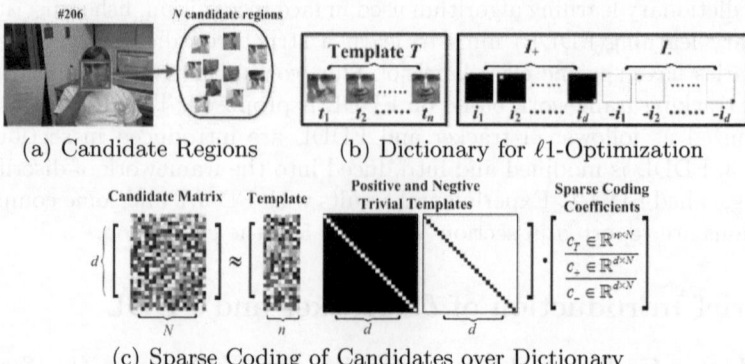

(a) Candidate Regions   (b) Dictionary for ℓ1-Optimization

(c) Sparse Coding of Candidates over Dictionary

Fig. 1. Original ℓ1-tracker Algorithm

samples and the coding coefficient matrix of Y over D respectively, and $Y \approx DX$. Yang's FDDL can be formulated as following[6]:

$$J_{(D,X)} = \arg \min_{(D,X)} \{r(Y, D, X) + \lambda_1 \|X\|_1 + \lambda_2 f(X)\} \tag{2}$$

where $\|X\|_1$ measures the sparsity; $f(X) = tr(S_W(X)) - tr(S_B(X)) - \eta\|X\|_F^2$ is a discriminative constraint imposed on X, which makes D discriminative for the samples in Y; $S_W(X)$ and $S_B(X)$ are within-class and between-class scatters of X, respectively; λ_1, λ_2 are scalar parameters for tuning the influences of three terms; $r(Y, D, X) = \sum_{i=1}^{c} r(Y_i, D, X_i)$ is the discriminative fidelity term, and

$$r(Y_i, D, X_i) = \|Y_i - DX_i\|_F^2 + \|Y_i - D_i X_i^i\|_F^2 + \sum_{j=1, j \neq i}^{c} \|D_j X_i^j\|_F^2 \tag{3}$$

where $\|.\|_F$ is the Frobenius norm. The first two terms in Eq.(3) guarantee that Y_i can be represented by D and D_i approximately with X_i and X_i^i, respectively; the third one ensures that the representation of Y_i over $D_j(i \neq j)$ is small. Consider a query signal $y \in \mathbb{R}^d$, $\tilde{y} = Dx$ and $\hat{y} = D_i x_i$ are reconstruction results of y over D and D_i with the optimization of sparse coding as shown in Eq.1, respectively. Reconstruction errors $\tilde{e} = y - \tilde{y}$ and $\hat{e} = y - \hat{y}$ are shown in Fig.2.

3   FDDL Based Tracking

3.1   Improved FDDL

Based on the above analysis, Eq.(3) minimizes $\tilde{e}, \hat{e}$ and $\sum_j y_j(i \neq j, j = 1, \ldots, c)$ in Fig.2. But we find that, it is not sufficient for the reconstruction of query signal y based on the minimization of Eq.(3). Denote by $y' = D_i x'$ the approximation of $\tilde{y}$ over D_i, and $e' = \hat{y} - y'$, $e = y - y'$, $e^* = \tilde{y} - y'$. Here, we use AR face database, which contains 700 face images uniformly from 100 individuals, to validate the insufficiency of $r(Y_i, D, X_i)$ in Eq.(3). We select 100 images randomly as query images, and the rest 600 ones as dictionary atoms. For each query image, $\|e\|_F$, $\|e^*\|_F$, $\|e'\|_F$, $\|\tilde{e}\|_F$ and $\|\hat{e}\|_F$ are calculated and plotted in Fig.3. It is clear to see that: (1)Minimization of $\|\tilde{e}\|_F$ and $\|\hat{e}\|_F$ cannot guarantee minimization of $\|e\|_F$, $\|e^*\|_F$ and $\|e'\|_F$; (2)Because $\|e'\|_F > 0$(unless the total dictionary D is consisting of i-th class dictionary D_i merely, which is not practical), $\|\hat{e}\|_F < \|e\|_F$; (3)For each image, $\|e\|_F$ is maximum among five error terms, and its minimization can be considered as the upper bound of them and Eq.(3) can be rewritten by:

$$r'(Y, D, X) = \sum_{i=1}^{c} r'(Y_i, D, X_i) = \sum_{i=1}^{c} (\|Y_i - D_i X'^i_i\|_F^2 + \sum_{j=1, j\neq i}^{c} \|D_j X'^j_i\|_F^2) \quad (4)$$

where, Y_i is the class-i subset of the training samples, X'^j_i is the coding coefficient matrix of $\tilde{Y}_i$ (reconstruction of Y_i over D_j). Note that, we use X'^j_i in the second term of Eq.(4) instead of X^j_i in Eq.(3). Thus Eq.(2) can be re-written as,

$$J_{(D,X')} = \arg \min_{(D,X')} \{r'(Y, D, X') + \lambda_1 \|X'\|_1 + \lambda_2 f(X')\} \quad (5)$$

where, $f(X') = tr(S_W(X')) - tr(S_B(X')) - \eta \|X'\|_F^2$.

3.2 FDDL Based Tracking

Target Location and Training Samples Selection. According to target location in last frame(red box in Fig.4(a)), we can extract a number of candidate regions with Bayesian inference(dotted boxes in Fig.4(b)), then distinguish object from background regions with sparse coding classification[7][8][6]. But dictionary learning with large number of samples(Y in Eq.(2) (5)) will inevitably lead to the tremendous computional cost in online tracking.

In order to achieve robustness and efficiency, we seek the most likely candidate region as target in the current frame, then generate object/positive samples and background/negative samples, as shown in Tab.1.

We select 10 regions extremely nearby the object (small red rectangle in Fig.4(c)) as positive samples Y_O, and 10 regions(including up-left, up, up-right, left, right, left-down, down, left-right, 1/3 bigger and 1/3 smaller than the small red box) as negative samples Y_B. Notice that, as shown in Fig.4(d), in order to remain information of target during tracking, we always fix the 10th samples in Y_O with target selected artificially in #1; and the last two background regions are used to assist our algorithm to deal with the scale changing of target.

Dictionary Learning. Most of discriminant tracking methods are based on following assumption: appearance of object and background(near the object) do not change seriously frame by frame during tracking. According to this, we can represent candidate regions selected in current frame using the dictionary D_{old} learned in last frame, and locate target as shown in last section; afterwards, update D_{old} with the 20 selected samples to help tracking in the successive frame. So, how to update the dictionary is also a critical problem in our FD^2LT.

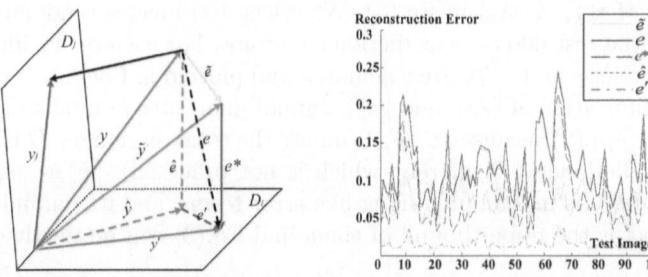

Fig. 2. Some Terms in Eq.(3)   **Fig. 3.** Error Comparison

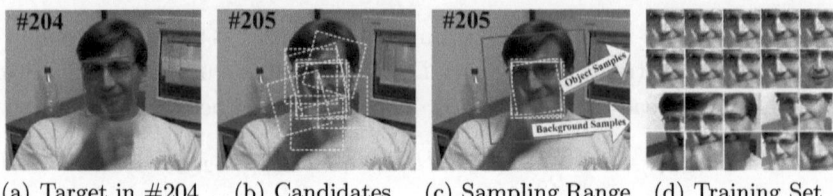

(a) Target in #204   (b) Candidates   (c) Sampling Range   (d) Training Set

Fig. 4. Object Location and Training Samples Selection

Suppose that $D_{old}=[D_{old\_O}, D_{old\_B}]$ is the last learned dictionary, and labeled training samples $Y=[Y_O, Y_B]$ are selected as Fig.4(d). The problem of DL is how to update D_{old}, such that Y can be represented by new dictionary D_{new} with error as less as possible. According to improved FDDL proposed in Section3.1, minimization function of improved FDDL for tracking is:

$$J_{(D',X')} = \arg\min_{(D',X')}\{\|Y_O - D'_O X'^O_O\|^2_F + \|Y_B - D'_B X'^B_B\|^2_F + \|D'_B X'^B_O\|^2_F$$
$$+\|D'_O X'^O_B\|^2_F + \lambda_1\|X'\|_1 + \lambda_2(S_W(X') - S_B(X') - \eta\|X'\|^2_F)\} \tag{6}$$

where $\tilde{Y}=[\tilde{Y}_O, \tilde{Y}_B]$ are approximation of $Y=[Y_O, Y_B]$ over D_{old}, $X'_O=[X'^O_O, X'^B_O]$ and $X'_B=[X'^O_B; X'^B_B]$ are sparse coding coefficients of $\tilde{Y}_O$ and $\tilde{Y}_B$ over updated D', respectively. And $X'=[X'_O, X'_B]$.

Solving Modified FDDL. $J_{(D',X')}$ in Eq.(6) can be divided into two subproblems: updating X' by fixing D'; and updating D' by fixing X'. This procedures are iteratively implemented like that in FDDL[6]. But these are also some differences, as shown below.

Update X'. Suppose that D' is fixed, and the objective function $J_{(D',X')}$ in Eq.(6) is reduced to a sparse coding problem to update $X' = [X'_O, X'_B]$. Like that in original FDDL[6], updating of X' can be divided into two components:

$$J_{(X'_O)} = \arg\min_{(X'_O)}\{\|Y_O - D'_O X'^O_O\|^2_F + \|D'_B X'^B_O\|^2_F + \lambda_1\|X'_O\|_1$$
$$+\lambda_2(S_W(X'_O) - S_B(X'_O) - \eta\|X'_O\|^2_F)\} \tag{7}$$

Table 1. Object Location and Training Samples Selection

Input: Learned dictionary $D_{old} = [D_{old\_O}, D_{old\_B}]$ in last frame, location of object c_{old} in last frame, pre-definited factor α
Step.1 Generate N candidate regions c_i around c_{old} using Bayesian inference, and extract the candidate features F_i of all regions, $i=1,\ldots,N$; **Step.2** For each F_i, solve $\ell 1$-problem over D_{old}: $\min_{X_i} \|F_i - D_{old}X_i\|_F^2 + \|X_i\|_1$ **Step.3** For each F_i, compute reconstruct errors $e_i = \|F_i - D_{old}X_i\|_F^2$ **Step.4** For each F_i, compute weights $w_i = \exp(-e_i/\alpha)$ and normalize $w_i = w_i/\Sigma_i w_i$. **Step.5** Compute object location in current frame, $c_{new} = \Sigma_i w_i c_i$; **Step.6** Select 10 regions that extremely nearby the object as positive samples Y_O; Select 10 regions that far away from the object as negative samples Y_B.
Output: Object location c_{new} and samples $Y=[Y_O, Y_B]$ for FDDL in current frame.

Table 2. Algorithm of Updating X' by Fixing D'

Input: Dictionary $D'=[D'_O, D'_B]$, training samples $Y=[Y_O, Y_B]$, pre-definited factors $\sigma, \tau > 0$
Step.1 Initialize $X'^{(1)}_i = 0$ and $h = 1$ **Step.2** while convergence and the maximal iteration number are not reached, do $\quad h \leftarrow h+1 \quad , \quad X'^{(h)}_i = S_{\tau/\sigma}(X'^{(h-1)}_i - \nabla Q(X'^{(h-1)}_i/2\sigma)$ $\quad\quad$ where $\nabla Q(X'^{(h-1)}_i)$ is the derivative of $Q(X_i)$, w.r.t. $X'^{(h-1)}_i$ and $S_{\tau/\sigma}$ is a soft thresholding function defined in [9].
Output: $X'_i \leftarrow X'^{(h)}_i$.

Table 3. Our Improved FDDL

Input: Original dictionary $D_{old}=[D_{old\_O}, D_{old\_B}]$, training samples $Y=[Y_O, Y_B]$, and coding coefficients $X_{old}=[X_{old\_O}, X_{old\_B}]$
Step.1 Initialize all the atoms of D' as random vectors; **Step.2** Code $\tilde{Y}(\tilde{Y}$ is sparse approximation of Y over D' with X') with D'_O and D'_B, denote the coding coefficients with X'_O and X'_B, respectively. $X'=[X'_O, X'_B]$; **Step.3** while convergence and the maximal iteration number are not reached, do $\quad\quad$ **Update** the sparse coding coefficients X'. Fix D', then update X' by solving Eq.(7) and Eq.(8) with algorithm in Tab.2 $\quad\quad$ **Update** dictionary D'. Fix X', then update D' by solving Eq.(9) and Eq.(10) with the algorithm in[10].
Output: Updated X' and D'.

Table 4. FDDL based tracking(FD^{2}LT)

Input: Video stream, location of object c_1 in #1
Step.1 According to the object template in #1 of video sequence, extract 10 object regions and 10 background regions as D_O^{init} and D_B^{init} in section3.2; denote $D' = [D_O^{init}, D_B^{init}]$; **Step.2** While the maximal number of video frames are reached, $\quad\quad$ **Input** a new video frame; $\quad\quad$ **Get** location of object in current frame, and generate training samples $Y = [Y_O, Y_B]$ with algorithm shown in Tab.1; $\quad\quad$ **Update** X' and D' iteratively with improved FDDL in Tab.3 until convergence.
Output: Tracking result of each frame.

$$J_{(X'_B)} = \arg\min_{(X'_B)}\{\|Y_B - D'_B X'^B_B\|_F^2 + \|D'_O X'^O_B\|_F^2 + \lambda_3\|X'_B\|_1$$
$$+ \lambda_4(S_W(X'_B) - S_B(X'_B) - \eta\|X'_B\|_F^2)\}$$
(8)

Let n_O, n_B denote the number of object and background samples respectively, Yang has proved that, if $\eta > 1 - n_O/(n_O + n_B)$, the last terms in Eq.(7) and (8) are strictly convex[6]; while other two terms, except $\|X'_O\|_1$ and $\|X'_B\|_1$, are differentiable. So we can see that, Eq.(7) and (8) are convex, and iterative project method (IPM)[9] can be used to solve them. Here, we take Eq.(7) for example, to show how to update X' when fixing D'.

First of all, Eq.(7) can be rewritten as $J_{(X'_O)} = \arg\min_{(X'_O)}\{Q(X'_O) + \lambda_1\|X'_O\|_1\}$, where, $Q(X'_O)$ is the sum of five differentiable terms except $\|X'_O\|_1$, and:

$\nabla_{X'_O}(\|Y_O - D'_O X'^O_O)\|F^2 = 2(D'_O)^T D'_O X'^O_O - 2(D'_O)^T Y_O$

$\nabla_{X'_O}(\|X'_O\|_F^2) = 2(X'_O)^T$

$\nabla_{X'_O}(\|D'_B X'^B_O\|F^2) = 2(D'_B)^T D'_B X'^B_O$

$\nabla_{X'_O}(S_B(X'_O)) = -2P_{n_O}P_{n_O}{}^T + 2P_{n_O}G - 2C_{n_O,n_B}C_{n_O,n_B}{}^T(X'_O)^T + 2C_{n_O,n_B}Z$

$\nabla_{X'_O}(S_W(X'_O)) = 2N_{n_O}N_{n_O}{}^T(X'_O)^T$

where, $N_{n_O} = I_{n_O \times n_O} - E_{n_O \times n_O}/n_O$; $P_{n_O} = E_{n_O \times n_O}/n_O - E_{n_O \times n_O}/(n_O + n_B)$; $C_{n_O,n_B} = E_{n_O \times n_B}/(n_O + n_B)$; $Z = X'_O E_{(n_O + n_B) \times (n_O + n_B)}/(n_O + n_B) - X'^B_O C_{n_B,n_O}$; $C_{n_B,n_O} = E_{n_B \times n_O}/(n_O + n_B)$; $G = X'^B_O C_{n_B,n_O}$; $I_{n \times n}$ is $n \times n$ unit matrix, $E_{m \times n}$ is $m \times n$ all-ones matrix. $n_O = n_B = 10$. Thus, the algorithm of updating X' is summarized in Tab.2.

Update D'. When X' is fixed, update $D' = [D'_O, D'_B]$ as follows:

$$J_{(D'_O)} = \arg\min_{(D'_O)}\{\|Y_O - D'_O X'^O_O\|_F^2 + \|D'_O X'^O_B\|_F^2\}$$
(9)

$$J_{(D'_B)} = \arg\min_{(D'_B)}\{\|Y_B - D'_B X'^B_B\|_F^2 + \|D'_B X'^B_B\|_F^2\}$$
(10)

In general, we require that each column of D' is a unit vector. Eq.(9)(10) are quadratic programming problems and can be solved by the algorithm in [10][11], which update D' atom by atom. Our modified FDDL is summarized in Tab.3.

FDDL Based Tracking(FD²LT). We give our FD²LT in Tab.4. Fig.5 shows some tracking results with FD²LT and our improved FD²LT in #205(Fig.5(b)(b)) and #355(Fig.5(c)) of Dudek video sequence. Tracking results of two method are similar in #205. But it is clear that, the former remain object information well, while the latter destroys almost all object information, which can be seen in 10 object atoms(updated dictionary in current frame) in the lower-left corner of Fig.5(b)(b), respectively; so, we can assert that, coding coefficients used to represent target in #205 with improved FD²LT is sparser than that with original FD²LT. The button row of Fig.5(b)(b) shows the coding results and coding coefficients of these two methods. After a lot of test experiments, we have similar

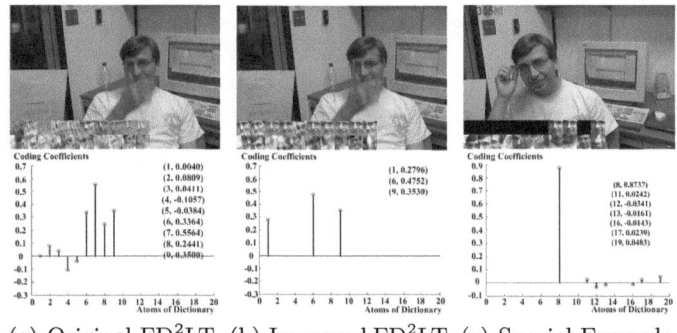

(a) Original FD²LT   (b) Improved FD²LT   (c) Special Example

Fig. 5. Some Results of FD²LT on Dudek

conclude and list as following: (1)Our improved FDDL is much sparser than original FDDL, when coding same signal; (2)Dictionary learned using Our improved FDDL has stronger representation ability than original FDDL.

The second conclusion can be expressed in Fig.5(c). With our method, after a few successive frame, some atoms in dictionary are almost close to zeros. As shown in Fig.5(c), only three object atoms and ten background atoms are used to represent the object with our improved FDDL, while it appears rarely when we use original FDDL in tracking. But it does not always benefit for tracking, especially when object changes heavily, as if the number of dictionary atoms is too small, they cannot remain the object information and adapt the change of object, simultaneously. In order to maintain rich ability of representation, we set those atoms with all zero elements as mean of all non-zero atoms.

4 Experiment

4.1 Experiment Setting

In order to evaluate our original and improved FD²LT, we conduct experiments on six challenging video sequences, including Surfer(375 frames), Dudek(1145 frames), faceocc2(819 frames), animal(71 frames), girl(500 frames) and car11(393 frames). These sequences cover many challenges in tracking: occlusion, motion blur, rotation and scale variation, etc. For comparison, we use two state-of-the-art algorithms, incremental learning based tracking(IVT, one of the discriminant tracking methods[12]) and sparse coding based original ℓ1-tracker (one of the generative tracking methods[4][5]), with the same initial target position and 200 particles in bayesian inference. Programs are coded with Matlab7.0, and experiments are running on computer with 2.67GHz CPU and 2GB memory.

4.2 Experimental Results

We evaluate the above-mentioned algorithms using the center location error, and the results are shown in Fig.6 and Fig.7. Overall, our original and improved FD²LT trackers performs well against the other state-of-the-art algorithms.

708   J. Zhang, H.-y. Wang, and F.-h. Chen

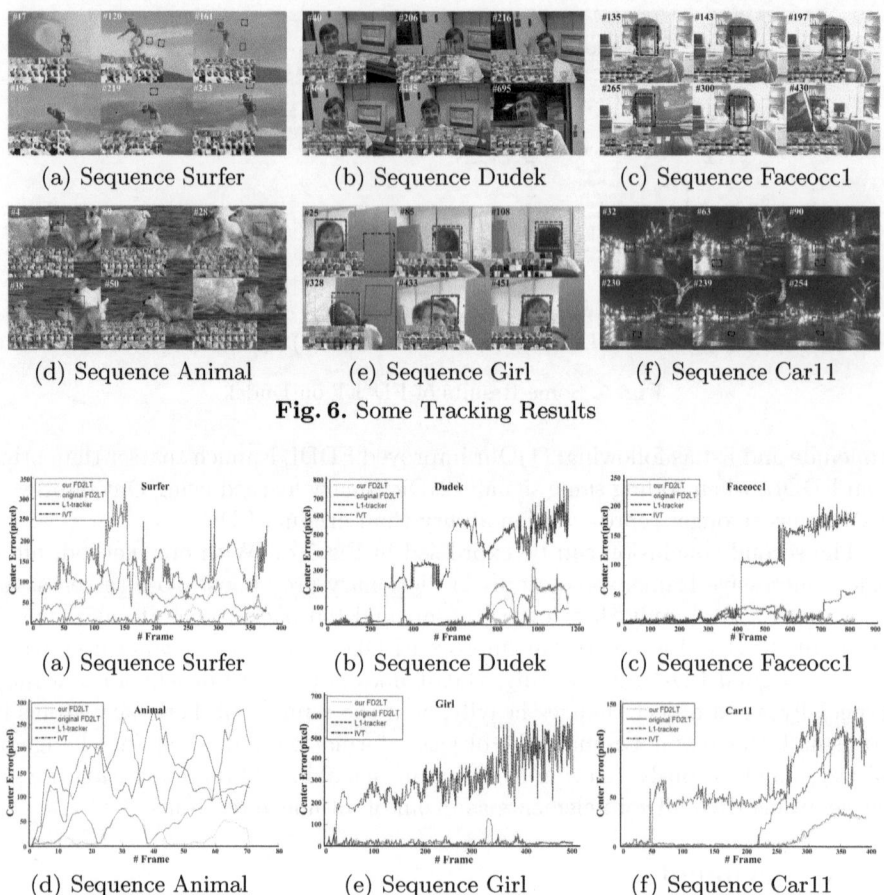

(a) Sequence Surfer   (b) Sequence Dudek   (c) Sequence Faceocc1

(d) Sequence Animal   (e) Sequence Girl   (f) Sequence Car11

Fig. 6. Some Tracking Results

(a) Sequence Surfer   (b) Sequence Dudek   (c) Sequence Faceocc1

(d) Sequence Animal   (e) Sequence Girl   (f) Sequence Car11

Fig. 7. Quantitative Evaluation in terms of Center Location Error (in pixel)

For occlusion, three algorithms except IVT work steadily roughly, especially at #206,#366 of faceocc2 sequence in Fig.6(b)(head for tracking is covered by hand and glasses) and at #85, #108, #433 of girl sequence in Fig.6(e)(head for tracking turns right, turns back and blocks by someone else). IVT works poorly, even lost the target from #10 in girl sequence(see Fig.7(e)), because of the number of positive and negative samples are limited(in consideration of the learning efficiency), and incremental updating of classifier in IVT is less effective; while original ℓ1-tracker and the derivative two FD^{2}LT trackers, which have strong representative abilities based on the large size of dictionary and good performances of dictionary learning, respectively, performance excellently.

For motion blur, our trackers work better than other two trackers, see #4, #9, #38, and #50 in Fig.6(d). The animal runs and jumps fast(motion blur) with a lot of water splashing(occlusion). IVT and ℓ1-tracker fail both from #4, and never recover again. Original and improved FD^{2}LT lost target at #28 and recover at #38 (Fig.6(d)). And at #12-#21 and #43-#71, improved FD^{2}LT works better than original FD^{2}LT.

For rotation and scale variation, our trackers also work better than other ones, see Fig.6(a), 7(a), and Fig.6(e), 7(e). And for complex background, as shown in Fig.6(f), the car is driven in the dark with bright lamplight and car light affecting the tracking, ℓ1-tracker and our two trackers work well before #220 but IVT losts the target from #50, see Fig.7(f).

In general from above analysis, our original and improved FD^{2}LT trackers work nearly the same, and the latter is slightly better, especially in Dudek and Animal sequences, see Fig.7(b),7(d). Original ℓ1-tracker performances good in most frames, but also fail sometimes. IVT is sensitive, when the occlusion, rotation, motion blur of target is appeared in tracking.

We also find that, improved FD^{2}LT remains object information in dictionary atoms, see the button-two rows of lower-left of each frame in Fig.6(a), 6(b), while original FD^{2}LT destroys almost all object information, which are shown in top-two rows of lower-left of each frames. The same conclusions can be obtained when investigating the rest sequences in Fig.6. Moreover, the fixed 10th object training samples also prevent the procedure of DL from degeneration. See #28 in Fig.6(d), all four trackers fail, and most of the dictionary atoms after updating are confused. With our selection of training samples for DL in section3.1, our two trackers retrieves the animal's head in #38, but the other two methods fail.

Table.5 is the time comparison of four algorithms used in our experiments. Our two FD^{2}LT frameworks work much faster than ℓ1-tracker, and the improved FD^{2}LT is slightly fast than original one, because of the simpler optimal problem in Eq.5 than original problem in Eq.2.

Table 5. Comparison of Average Tracking Time(frame per second)

Method Video	IVT	ℓ1-tracker	Original FD^{2}LT	Improved FD^{2}LT
Surfer	2.8694	0.0668	2.0886	2.3469
Dudek	3.3211	0.0270	1.8748	2.3154
Faceocc2	2.7886	0.0327	1.7103	1.8979
Animal	1.8979	0.0417	1.3596	1.4620
Girl	1.6548	0.0389	1.6909	1.7391
Car11	2.7901	0.0377	1.7473	2.3624

5   Conclusion and Future Works

In this paper, we present a modified version of FDDL based on Yang's FDDL, then introduce original and modified FDDL into video object tracking, called original and modified FD^{2}LT, respectively. In our FD^{2}LT framework, three important components, including object location, training samples selection, and dictionary learning, are introduced and discussed detaily in Section3. Experiments demonstrate the effectiveness and efficiency of these two trackers.

But there are also some aspects required to be studied in future, including:(1) some conclusions proposed in this paper are short of strict proving, e.g. two conclusions at the end of section3.2 and the convergence of FD^2LT; (2) our FD^2LT framework is much faster than $\ell 1$-tracker, but it is still far away from real-time tracking(more than 20 fps in common video sequence). How to accelerate is one of our and even all computer vision researchers' further goals.

References

1. Yilmaz, A., Javed, O., Shah, M.: Object Tracking: A Survey. Acm Computing Surveys (CSUR) 38(4), 13 (2006)
2. Candès, E.J., Wakin, M.B.: An Introduction to Compressive Sampling. IEEE Signal Processing Magazine 25(2), 21–30 (2008)
3. Candès, E.J., Romberg, J., Tao, T.: Robust Uncertainty Principles: Exact Signal Reconstruction from Highly Incomplete Frequency Information. IEEE Transactions on Information Theory 52(2), 489–509 (2006)
4. Mei, X., Ling, H.B., Wu, Y., et al.: Minimum Error Bounded Efficient l1 Tracker with Occlusion Detection. In: 2011 IEEE Conference on Computer Vision and Pattern Recognition, CVPR (2011)
5. Mei, X., Ling, H.B.: Robust Visual Tracking and Vehicle Classification via Sparse Representation. IEEE Transactions on Pattern Analysis and Machine Intelligence 33(11), 2259–2272 (2011)
6. Yang, M., Zhang, L., Feng, X.C., et al.: Fisher Discrimination Dictionary Learning for Sparse Representation. In: 2011 IEEE 13th International Conference on Computer Vision (2011)
7. Wright, J., Yang, A.Y., Ganesh, A., et al.: Robust Face Recognition via Sparse Representation. IEEE Transactions on Pattern Analysis and Machine Intelligence 31(2), 210–227 (2009)
8. Jiang, Z., Lin, Z., Davis, L.S.: Learning a Discriminative Dictionary for Sparse Coding via Label Consistent K-SVD. In: 2011 IEEE Conference on Computer Vision and Pattern Recognition, CVPR 2011 (2011)
9. Rosasco, L., Mosci, S., Santoro, M., et al.: Iterative Projection Methods for Structured Sparsity Regularization. Technical Report MIT-CSAIL-TR-2009-050, MIT (2009)
10. Yang, M., Zhang, L., Yang, J., Zhang, D.: Metaface learning for sparse representation based face recognition. In: 2010 17th IEEE International Conference on Image Processing (ICIP 2010), pp. 1601–1604. IEEE, Hong Kong (2010)
11. Yang, M., Zhang, L., Yang, J., et al.: Robust Sparse Coding for Face Recognition. In: 2011 IEEE Conference on Computer Vision and Pattern Recognition (CVPR 2011), pp. 625–632 (2011)
12. Ross, D.A., Lim, J., Lin, R.S.: Incremental learning for robust visual tracking. International Journal of Computer Vision 77, 125–141 (2008)

Performance Analysis of Matrix Gait Recognition under Linear Interpolation Framework

Xianye Ben[1], Wankou Yang[2], Hongyuan Wang[1], and Shengdi Wang[1]

[1] School of Information Science and Engineering,
Shandong University, Jinan 250100, China
benxianyeye@163.com
[2] School of Automation, Southeast University, Nanjing 210096, China
wankou_yang@yahoo.com.cn

Abstract. A novel gait period detection algorithms based on pseudo-Zernike moments was proposed in this paper. Pseudo-Zernike moments can directly detect gait periodicity because of its characteristic of describing movement images. As features in one frame were only relevant to those in prior and subsequent frames during walking, a framework for matrix gait recognition based on linear interpolation was proposed. Then, Trace transform and Fan-Beam projection were used as instantiation in CASIA(B) gait database to prove the validity of gait recognition framework, which has brought new ideas to solve gait recognition problem.

Keywords: Gait recognition, linear interpolation framework, gait period detection, pseudo-Zernike moment.

1 Introduction

Biometrics include physiological and behavior characteristics [1]. Physiological characteristics includes face, fingerprint, vein (finger vein, dorsal hand vein), iris, human ear, hand shape, palm print, retina, lip, DNA, body odour, etc. Behavior characteristic includes gait, voice, keystroke, signatures, etc. All these characteristics have the properties of collectability, universality, uniqueness and stability. Therefore, they can be used for identification and certification in different situations. Gait has dynamic or static information which can be caught without any perception.

Many researchers pay attention to gait recognition. Boulgouris et al. [2] compared gait recognition with the other biometrics, and indicated that gait recognition can be a part of biometrics recognition technology. They also presented the constitution of gait recognition system, and discussed the gait period detection, the method based on model, appearance, the frequency domain transform of time sequence, template matching and the hidden markov model. The methods of gait recognition generally fall into two categorical types: computer vision and sensor. Gafurov [3] summarized the factors of influencing the performance of gait recognition, and assessed the security and being aggressive of gait. The changes of pedestrian's walking speed will affect the performance of gait recognition, which is based on computer vision. A few parts of sampling rates can be selected; consequently, the matching amongst samples

J. Yang, F. Fang, and C. Sun (Eds.): IScIDE 2012, LNCS 7751, pp. 711–718, 2013.

presents dynamic features. Thus there are two ways of time normalization proposed by Boulgouris et al., namely Dynamic time warping (DTW) [4] and Linear time normalization (LTN) [5]. Yu et al. [6] used Fourier descriptor of the key frame to express gait features. This method doesn't need period alignment. Because of the complicated adjustment process of time normalization, and insufficient information included in the key frame, this paper addressed gait period detection algorithm based on pseudo Zernike moment, proposed a matrix gait recognition under linear interpolation framework within an entire gait period, and finally analyzed and summarized the recognition performance.

2   Gait Period Detection

We extracted single frame images from a video and performed gray-scale transformation, and then we selected the image which doesn't include human body to be the original background image of the entire video and used the updating background subtraction to extract the silhouette. The method of Kapur Entropy threshold was used for binary image sequence processing. We filled the cavity of binary images by mathematical morphology and extracted silhouette by single connected analysis. And then we made the silhouette centered, made the size of the images into 64×64 pixels. Finally, we removed the redundancy frames which include the imperfect human image. Fig.1 shows the results of the above-mentioned gait image preprocessing. We calculated persudo Zernike moments frame by frame which can be used as the foundation of detecting the gait period. It means that the starting frame is the first appearance of local extremum, and the ending frame is the previous one of the third time when the local extremum appears.

Fig. 1. Results of gait image preprocessing

n-order and m-fold pseudo-Zernike moment is defined as

$$pZ_{nm} = \frac{n+1}{\pi} \int_0^{2\pi} \int_0^1 R_{nm}(\rho) \exp(-jm\theta) f(\rho,\theta) \rho d\rho d\theta \qquad (1)$$

where $f(\rho,\theta)$ is the polar coordinate of a gray-scale image, and $R_{nm}(\rho)$ is a pseudo radial polynomial, defined as

$$R_{nm}(\rho) = \sum_{s=0}^{n-|m|} (-1)^s \frac{(2n+1-s)!}{s!(n-|m|-s)!(n+|m|+1-s)!} \rho^{n-s} \frac{-b \pm \sqrt{b^2 - 4ac}}{2a} \qquad (2)$$

where n is a natural number; m can be a positive or negative number, and satisfying $|m| \le n$.

Obviously, the pseudo-Zernike moments have rotation invariance and orthogonality. Because the pseudo-Zernike moment is expressed by the polar coordinate (ρ, θ), and $\|\rho\| \leq 1$, the original image needs to be mapped into the unit circle when calculated. The discrete formula of the pseudo-Zernike moment is

$$pZ_{nm} = \frac{n+1}{\pi} \sum_{x=-1}^{1} \sum_{y=-1}^{1} R_{nm}(\rho_{xy}) \exp(-jm\theta_{xy}) f(x, y) \tag{3}$$

where $\rho_{xy} = \sqrt{x^2 + y^2}$ is the polar distance; θ_{xy} is the polar angle, and $f(x, y)(-1 \leq x \leq 1, -1 \leq y \leq 1)$ represents the gray-scale image of a unit circle whose center is regarded as the coordinate origin.

The advantages of the pseudo-Zernike moments: (1) rotation invariance; (2) robustness, which means it is robust to the minimal changes of the shape and the noise. Compared with the Zernike moments in the same order, the pseudo-Zernike moments have more low-order moments. For instance, when the order is five, the pseudo-Zernike moments have 21 moments, but the Zernike moments only have 12 ones. And the more low-order moments are, the stronger the noise resisting ability is; (3) small redundancy of the information expression; (4) high effectiveness of the information expression. An image can be expressed well by a small number of the pseudo-Zernike moments sets, which have a very small mean square error. (5) multiple levels expression. The relevant group of the pseudo-Zernike moments sets can express all the shapes of a pattern efficiently, where the low-order moment describes the whole shape, while the high-order moment describes the details; (6) it can describe movement images, and can be used in movement image sequence analysis.

We tested the gait sequence shown in Fig.1, using the preprocessing of the image sequences as the foregoing. And then we calculated the modulus of the pseudo Zernike moment pZ_{10} frame by frame. The result of the test is shown in Fig.2(1), the first subfigure of which corresponds the result without body mediacy and size normalization and the second subfigure of which corresponds the result with body mediacy and size normalization. The advantage (6) of the pseudo Zernike moments was used to calculate the modulus of the pseudo-Zernike moment pZ_{53} frame by frame before standard centralization of the preprocessing as the foregoing, and the result of the test is shown in Fig. 2(2).

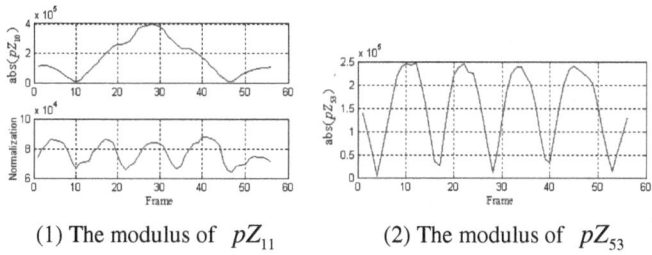

(1) The modulus of pZ_{11} (2) The modulus of pZ_{53}

Fig. 2. Detection results using Pseudo-Zernike moments

3   The Conformation of the Framework of the Matrix Gait Recognition Based on Linear Interpolation

Considering that the features between the two frames during walking only have relationship with the features of its former frame and its latter frame, we can employ the Newton linear interpolation to approximately simulate the discrete gait sequence as the continuous gait. Through the method of the feature templates conformation, a period of the gait sequence image is expressed as a form of matrix in order to match quickly during the stage of recognition. Newton linear interpolation can be expressed as a two-point model, so

$$g(x) = f(x_0)\frac{x-x_1}{x_0-x_1} + f(x_1)\frac{x-x_0}{x_1-x_0} \tag{4}$$

where $(x_0, f(x_0))$ and $(x_1, f(x_1))$ denote two points on the line.

We use f_n to express the nth signal frame gait image and F_n is the corresponding feature of this image. That is when the time is t, it just corresponds the nth signal frame gait image.

$$F(t) = F_n, t = n \tag{5}$$

The feature between frame n and frame $n+1$ constructed by Newton liner Interpolation is

$$F(t) = (n+1-t)F_n + (t-n)F_{n+1}, n < t < n+1 \tag{6}$$

where F_{n+1} is the gait image feature of the frame $n+1$.

Then, the feature of one gait period of sequence images was obtained by weighted integral in one gait circle:

$$C_T = \frac{1}{T}\int_0^T F(t)Q(w,t)dt \tag{7}$$

where T is the period of this sequence; $Q(w,t)$ is weighting function; w is frequency; t is time.

$$Q(w,t) = 1 + j + \cos(wt) + j\sin(wt) \tag{8}$$

$Q(w,t)$ includes not only the static structure information which corresponds to the first two terms, but also the dynamic ones. We will talk about the suitable w value through experiments.

From the continuous gait discretization of previous three equations, we can obtain

$$C_T = \frac{1}{N}\sum_{n=1}^{N-1}\int_{t=n}^{t=n+1} [(n+1-t)F_n + (t-n)F_{n+1}]\times(1+\cos(wt)+j(1+\sin(wt))dt \tag{9}$$

Eq.(9) is the matrix gait recognition algorithm framework of linear interpolation. It provides a new solution to gait recognition and it is easy to achieve. It also makes the feature F_n (F_{n+1}) instantiate to achieve a new gait recognition algorithm.

4   Instantiation

In this section, $F(t)$ was instantiated into Trace transform feature and Fan-Beam mapping feature. We used Two-Dimensional Principal Component Analysis (2DPCA) for reducing the dimensionality. We carried out the experiments on PC configured with Pentium(R) Dual-Core CPU E5200 @2.50GHz, 2.00GB RAM, choosing the lateral view (the target movement direction is perpendicular to the optical axis of the camera) from the CASIA(B) database for study. We collected 6 samples for one person in a normal walking condition. The first three are used for training and the others for test. To assess the effectiveness of every feature extraction, we tested the recognition rate by Nearest Neighbor (NN) classifier.

4.1   Trace Transform Feature

The features of Trace transform are abundant and flexible. T is the functional chosen in Trace transform. Set of trace t in all directions constitute the trace domain Υ. F is the value along the chosen trace t. So Trace transform is defined on Υ. It can be described as

$$\Gamma(\varphi, \rho) = T(F(\varphi, \rho, t)) \tag{10}$$

Under the effect of functional T, the parameter t has been cancelled by Trace transform.

The use of Trace transform is to calculate the image projection in the appointed direction and line integral of that image function in a certain direction. The first column pixels of Trace transform image correspond to the line integral of original image in perpendicular direction. The middle column pixels of Trace transform image correspond to the line integral of original image in horizon direction.

We talk about one form of functional here.

$$F_1 : T(f(t)) = \int e^{-jt} f(t) dt$$

F_1 is equal to mapping the image to the complex field, and then real parts and imaginary parts are obtained. F_1 can be treated as the inner products of the image multiplied by the real template $\cos t$ and the imaginary template $\sin t$ respectively. As shown in Fig.3, angle is the phase angle of every element in the complex matrix, and it can be found that the phase difference of the real template and the imaginary template is $\pi / 2$.

The optimum recognition rates are respectively 83.60%、 83.87% and 83.87%, using the real part, the imaginary part and the phase angle as features in Fig. 3. The final dimensions are all 14×95, using 2DPCA to reduce the dimensions of the three features.

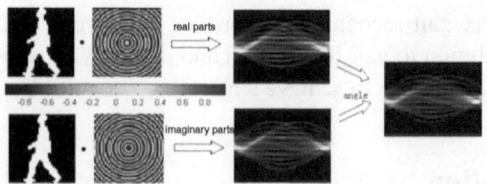

Fig. 3. Trace transform using functional F_1

4.2   Fan-Beam Mapping Feature

Fan-Beam mapping which is very close to Trace transformation is also to calculate the integral along a group of lines with specific directions, but Fan-Beam mapping is to calculate the line integral of the sector lines. Each column of the pixels in the Fan-Beam mapping corresponds to the mapping feature on specific angles, and each row of it corresponds to the mapping feature on specific distances.

The adjustable parameters of the Fan-Beam mapping are the distance d, which is from the point light source s to the origin o, and the density e of rays. How to get effective Fan-Beam mapping features? Here, we can observe through the effect of the reconstructed image by inverse Fan-Beam transform, because the more sufficient the Fan-Beam mapping features are, the closer the effect of corresponding reconstructed is to the original gait image.

Through the experiments, the size of the original gait image is of 64×64 pixels, and the minimum value of d is 49 pixels. When $d=49$, we altered e into $e = \{0.5, 1, 2, 4, 10\}$ and tested the Fan-Beam features under every group of parameters, the time of feature extraction, the dimensions of the original Fan-Beam features and the relevant reconstructed results, as shown in Table 1.

Then we fixed e to be 2, altered d as $d = \{49, 55, 60, 65\}$ and tested the Fan-Beam feature under every group of parameters, the time-consumed of feature extraction, the dimensions of the original Fan-Beam feature and the relevant reconstructed results, as shown in Table 2. We can find that, when e remains to be constant and d is increased gradually, Fan-Beam feature began to present mosaic appearance, and corresponding reconstruction effect was worse and worse. To trade off the time-consumed of feature extraction, the dimensions of the original Fan-Beam feature and the relevant reconstructed results, it is better to fix d to be 49 and e to be 2. With parameters $d = 49$ $e = 2$ $w = 5\pi / 16T$, the best recognition rate 88.71% has been obtained.

Even though we have considered about the static gait information, we give more attention to the dynamic one. And finally it makes different weights among frames in one gait period. Therefore, the feature constructed by this kind of framework is another sort of energy form with different weights. Compared with Trace transform, Trace transform is with the addition of Euclidean distance measuring. And it has a slightly higher recognition rate than Radon transform, while the feature dimension is almost the same. Fan-Beam mapping has yielded the highest recognition, since there are the largest feature dimensions even after data reduction. So the matching time is longest. At last, Table 3 shows the comparison of this proposed framework and other algorithms. The classifier is NN.

Table 1. Feature, time-consumed and dimension from Fan-Beam projection and corresponding reconstruction results under parameters $d = 49$, $e = \{0.5, 1, 2, 4, 10\}$

Parameter	Fan-Beam feature	Time-consumed (/s)	Feature dimension	Reconstruction results
$e = 0.5$		1.391	313×360	
$e = 1$		0.907	157×360	
$e = 2$		0.656	79×360	
$e = 4$		0.531	39×360	
$e = 10$		0.453	15×360	

Table 2. Feature, time-consumed and dimension from Fan-Beam projection and corresponding reconstruction results under parameters $e = 2$, $d = \{49, 55, 60, 65\}$

Parameters	Fan-Beam feature	Time-consumed (/s)	Feature dimension	Reconstruction results
$d = 49$		0.656	79×360	
$d = 55$		0.593	61×360	
$d = 60$		0.587	53×360	
$d = 65$		0.547	47×360	

Table 3. Comparison of this proposed framework and other algorithms

Methods	Recognition rate
Reference[7] : Baseline	0.7984
Reference [6] : KFD	0.7500
Reference [4]: DTW	0.8091
Fan-Beam under the proposed framework	0.8871

The calculation amount of Baseline [7] and DTW [4] are larger than the proposed framework, while the recognition results are even poor. Ignoring the feature of other frames, Ref.[6] uses only KFD to extract gait features. So it is not strange that it has got the poorest recognition performance.

5   Conclusion

In this paper, for the gait period detection method based on moments, the pseudo Zernike moment has the feature that can describe movement images, and it can avoid the process of body mediacy and size normalization to directly test the periodicity of gait. Because the features between the two frames during walking only have relationship with the features of its former frame and its latter frame, this paper also proposed a set of construction of the matrix gait recognition algorithm framework based on the linear interpolation, and verified the effectiveness of this construction by instantiating into Trace transform features and Fan-Beam mapping features. The recognition rate of Fan-Beam mapping is higher, but it is on the cost of higher feature dimension, and the dimension after data reduction is still higher, so the time for matching is also longer.

Acknowledgments. We sincerely thank the Institute of Automation Chinese Academy of Sciences for granting us permission to use the CASIA(B) gait database. This project is supported by the Natural Science Foundation of China (Grant No. 61201370, 61100103), the Independent Innovation Foundation of Shandong University (Grant No.2012GN043, 2012DX007) and the Specialized Research Fund for the Doctoral Program of Higher Education of China (Grant No. 20120131120030).

References

1. Ben, X.Y., Xu, S., Wang, K.J.: Review on Pedestrian Gait Feature Expression and Recognition. Pattern Recognition and Artificial Intelligence 25(1), 71–81 (2012)
2. Boulgouris, N.V., Hatzinakos, D., Plataniotis, K.N.: Gait Recognition: A challenging signal processing technology for biometric identification. IEEE Signal Processing Magazine, 78–90 (2005)
3. Gafurov, D.: A survey of biometric gait recognition: Approaches, security and challenges. In: Proc. Annual Norwegian Computer Science Conference, Oslo, Norway, pp. 19–31 (2007)
4. Boulgouris, N.V., Plataniotis, K.N., Hatzinakos, D.: Gait recognition using dynamic time warping. In: Proc. of IEEE Int. Symp. Multimedia Signal Processing, pp. 263–266 (2004)
5. Boulgouris, N.V., Plataniotis, K.N., Hatzinakos, D.: Gait recognition using linear time normalization. Pattern Recognition 39(5), 969–979 (2006)
6. Yu, S.Q., Wang, L., Huang, K.Q., et al.: Gait Analysis for Human Identification in Frequency Domain. In: Proc. of the 3rd International Conference on Image and Graphics, pp. 282–285 (2004)
7. Sarkar, S.: The Human ID gait challenge problem: data sets, performance and analysis. IEEE Trans on Pattern Analysis and Machine Intelligence 27(2), 162–177 (2005)

Discriminative GMM-HMM Acoustic Model Selection Using Two-Level Bayesian Ying-Yang Harmony Learning

Zaihu Pang[1], Shikui Tu[2], Xihong Wu[1,3,*], and Lei Xu[1,2,*]

[1] Speech and Hearing Research Center, Key Laboratory of Machine Perception
(Ministry of Education), Peking University, Beijing, China
[2] Department of Computer Science and Engineering,
The Chinese University of Hong Kong, Hong Kong, China
[3] College of Computer Science and Technology, Jilin University, Changchun, China
{pangzh,wxh}@cis.pku.edu.cn, {sktu,lxu}@cse.cuhk.edu.hk

Abstract. This paper proposes a two-level Bayesian Ying-Yang (BYY) harmony learning based acoustic model discriminative training method. In this method, a rival penalized competitive learning (RPCL) simplified BYY harmony learning based discriminative training is conducted at the HMM state level to optimizing the state boundaries, while a BYY based model selection is conducted at the Gaussian mixture components level to determine the Gaussian mixture components within the same HMM state. Two levels of learning work coordinately and have good convergence. Experiments show that the trained model is more discriminative with better recognition performance, and also more compact with smaller number of Gaussian components.

Keywords: Discriminative training, hidden Markov model, Bayesian Ying-Yang harmony learning, rival penalized competitive learning, large vocabulary continuous speech recognition.

1 Introduction

Parameters of acoustic model in HMM-based speech recognition systems are usually estimated using maximum likelihood estimation (MLE). The weakness of MLE lies in that it cannot directly optimize recognition error rates due to its strong assumptions on sufficient training data and model-correctness [1]. To solve this problem, a variety of discriminative training methods have been extensively investigated to improve automatic speech recognition (ASR) system for decades. Typical ones include maximum mutual information (MMI) estimation [2], minimum classification error (MCE) [3], and minimum word/phone error (MWE/MPE) [4].

Generally, these discriminative methods achieve good performances when acoustic conditions in a testing set match well with those in the training set.

---

* Corresponding authors.

J. Yang, F. Fang, and C. Sun (Eds.): IScIDE 2012, LNCS 7751, pp. 719–726, 2013.
© Springer-Verlag Berlin Heidelberg 2013

However, in most practical conditions, the test speech does not match the training set well. Smoothing techniques have been proposed to improve generalization ability of these methods, such as smoothing sigmoid function in MCE [3], acoustic scaling and weaken language modeling in [2], and I-smoothing in MPE [4]. The Baum-Welch (BW) algorithm is extended to update HMM parameters for implementing these techniques.

In our previous study [8], the BYY harmony learning has been also introduced into estimating GMM components by a two-stage procedure. The transfer probability matrix of hidden Markov chain is still trained by the Baum-Welch algorithm, while the GMM components are trained by a batch mode iterative algorithm that takes the place of the classic expectation maximization (EM) algorithm, with the number of GMM components determined automatically during learning. Experiments on the Hub4 broadcast news database have shown better performances than ones by not only the standard MLE training but also plus selecting GMM components in help of the classic Bayesian information criterion (BIC) and Akaike information criterion (AIC) criterions.

In [8], the BYY harmony learning acts on GMM components within the same hidden Markov state, without targeting at a discriminative training. To enhance discriminative ability, we can make the BYY harmony learning at the levels of states, phones, and words for training the GMM components across different states and also transfer probabilities across states, for which we may adopt the Ying-Yang alternative learning algorithm for a typical HMM model given in Sect.5.3 and especially Fig.14 of [10]. Towards this purpose, we proceed step by step in order to examine the effects as the MLE training is replaced by the BYY harmony learning part by part.

In implementation, each speech frame x_t is regarded as coming from a hierarchical mixture of Gaussian components as shown in Fig.1 [10]. That is, each speech frame x_t probabilistically comes from each of states and each state consists of a mixture of Gaussian components. This paper considers not only automatic model selection to determine Gaussian components within the same state, but also discriminative training at the level of hidden Markov states, with the help of the following two-stage procedure. First, we train the entire model by the Baum-Welch algorithm, with the resulted model as an initialization. Second, we jointly make the RPCL simplified BYY harmony learning [6] across a number of different states coordinately, and automatic model selection on Gaussian components as in [8]. It should be noted that the bottom level (Gaussian component level) is implemented in an unsupervised way while the top level (HMM state level) is implemented in a supervised way, with the teaching information obtained from the Viterbi force alignment for every speech frame x_t. Taking the advantage of BYY as a unified framework, the BYY harmony learning accommodates unsupervised learning, supervised learning, and semi-supervised learning in a same formulation (see Sect.4.4 in [10], Sect.5.3.1 in [11]).

Experiments on TIMIT phonetic recognition task show that implementing, not only the RPCL simplified BYY harmony learning on the hidden Markov state level, but also the BYY based automatic model selection on the Gaussian

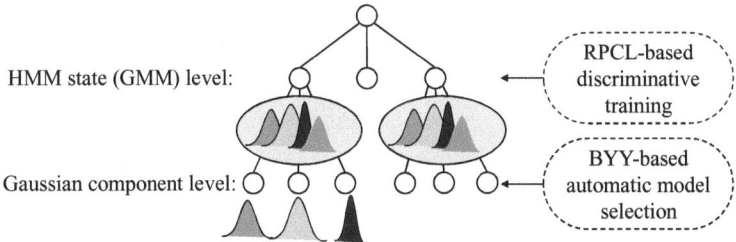

Fig. 1. Two level hierarchical mixture of Gaussians. Specifically, the above two level hierarchy can be regarded as degenerated either from the three level hierarchy in Fig.12 of Ref.[10] by discarding one level or from the HMM model in Fig.13 of Ref.[10] by ignoring the transient links between states.

component level, leads to a more compact model with fewer Gaussian components under each hidden Markov state and better recognition performance than the previous methods in [8] and [6], as well as the model trained by the classical MLE criterion.

The rest of the paper is organized as follows. Section 2 tackles the learning tasks on the two level hierarchical GMM, by the RPCL discriminative training on the HMM state level and the BYY automatic model selection on the Gaussian component level. Section 3 conducts experiments to justify the performance of the training algorithm. Finally, concluding remarks are given in Section 4.

2 Learning the Two Level Hierarchical Mixtures of Gaussians

In [8], for speech recognition via Gaussian mixture models (GMM) based hidden Markov models (HMMs), a two-stage procedure was proposed on the two level hierarchical GMM in Fig.1, with the hidden Markov state level trained still under the maximum likelihood principle by the Baum-Welch algorithm but with the GMM level trained under the Bayesian Ying-Yang (BYY) best harmony [7,11]. For each state j, we have $p(x_t|\theta_j) = \sum_{k=1}^{K} \alpha_{jk} \mathcal{N}(x_t|\mu_{jk}, \Sigma_{jk})$, where $\mathcal{N}(x_t|\mu_{jk}, \Sigma_{jk})$ is the k-th Gaussian distribution of the j-th GMM with mean μ_{jk} and covariance matrix Σ_{jk}, and K is the Gaussian components number of every GMM.

Specifically, an unsupervised BYY learning algorithm was implemented in [8] on GMM under one hidden Markov state with Gaussian components of GMM determined by automatic model selection, due to a favorable feature of least complexity of BYY [10]. The algorithm iteratively alternates Yang-step and Ying-step. Given the weight of the k-th component of the j-th GMM $p_{jk,t}$ fixed, the Ying-step has a same format as the M-step of the well-known EM algorithm on Gaussian mixture, i.e., renewing the parameters of the k-th component $\theta_{jk}^{old} = \{\alpha_{jk}^{old}, \mu_{jk}^{old}, \Sigma_{jk}^{old}\}$ into θ_{jk}^{new} as follows:

$$\alpha_{jk}^{new} = \frac{\sum_t p_{jk,t}}{\sum_t \sum_k p_{jk,t}}, \mu_{jk}^{new} = \frac{\sum_t p_{jk,t} x_t}{\sum_t p_{jk,t}},$$

$$\Sigma_{jk}^{new} = \frac{\sum_t p_{jk,t} (x_t - \mu_{jk}^{old})(x_t - \mu_{jk}^{old})^T}{\sum_t p_{jk,t}}. \tag{1}$$

The key difference comes from how $p_{j,t}$ is computed. If we simply let

$$p_{jk,t} = p(k|x_t, \theta_j^{old}) = \frac{\alpha_{jk} \mathcal{N}(x_t|\mu_{jk}, \Sigma_{jk})}{\sum\limits_{k=1}^{K} \alpha_{jk} \mathcal{N}(x_t|\mu_{jk}, \Sigma_{jk})}, \tag{2}$$

we get the E-step of the EM algorithm. That is, alternatively iterating Eqs.(1) and (2) actually implements the standard EM algorithm for the MLE on Gaussian mixture.

In contrast, the E-step of the EM algorithm is replaced by the following Yang-step in BYY harmony learning (see e.g., Eq.(3) in [8]),

$$p_{\ell,t} = p(\ell|x_t) + \Delta_{\ell,t}, \quad \Delta_{\ell,t} = p(\ell|x_t)\Delta\pi_{\ell,t}, \quad \pi_t(\theta_\ell) = \ln[\alpha_\ell G(x_t|\mu_\ell, \Sigma_\ell)], \quad (3)$$

$$\Delta\pi_{\ell,t} = \pi_t(\theta_\ell) - \sum\nolimits_k p(k|x_t)\pi_t(\theta_\ell) = \ln p(\ell|x_t) - \sum\nolimits_k p(k|x_t)\ln p(k|x_t), \quad (4)$$

$$p(\ell|x_t) = p(\ell|x_t, \theta^{old})\delta(\ell, J_t), \quad \delta(\ell, J_t) = \begin{cases} 1, \text{ (a) if } \ell \in J_t, \\ 0, \text{ (b) if } \ell \notin J_t, \end{cases} \tag{5}$$

where J_t is a set which consists of all the labels. For a GMM, $J_t = \{1, \cdots, K\}$ consists of all Gaussian components and ℓ denotes the identity of the ℓ-th component.

The above Ying-Yang iteration only acts on bottom level of Fig.1 in [8]. In general, it can also be implemented on the top level of Fig.1, named RPCL simplified BYY harmony learning, where the learning algorithm is implemented by a hierarchical harmony flow from the bottom-up by the Yang-step and the top-down by the Ying-step. Then, the $p_{jk,t}$ in Eq.(1) is replaced by

$$p_{jk,t} = p_{j,t} p(k|j, x_t) + p(j|x_t)\Delta_{k|j,t}, \tag{6}$$

where $\Delta_{k|j,t}$ is the updating step size generated from the competition of all Gaussian components from j-th state as defined in Eq.(3).

For an incremental study step by step, we implement in the top level a RPCL simplified BYY harmony learning which was shown to have better performance than the classical MLE, minimum phone error (MPE), maximum mutual information estimation (MMIE) in [6], and minimum classification error (MCE) in [5]. For every frame speech x_t, the state that corresponds to the identity of this input by the Viterbi force alignment is regarded as the correct state c_t, while the rival state is given by $r_t = \arg\min_{j \neq c_t}\{-\ln q(x_t|j, \theta)\}$. Particularly, we consider $p_{j,t}$ in Eq.(6) by

$$p_{j,t} = \begin{cases} 1 + p(r_t|x_t), & \text{if } j = c_t \\ -p(r_t|x_t)\gamma, & \text{if } j = r_t \\ 0, & \text{otherwise} \end{cases} \tag{7}$$

where γ, playing a similar role as in standard RPCL (see e.g., Eq.(4) of [9]), denotes the de-learning rate. The bigger the γ is, the more strengthen the de-learning is. Noticing that, different from the standard RPCL[9], the Eq.(7) not only de-learned the components associated with the rival state, but also enhance the description of the components on the correct state. The above Eq.(7) is obtained by approximately simplifying a general rival penalizing mechanism of the BYY harmony learning by Eq.(3). The details are referred to [6].

One other issue needs to be addressed is that an acoustic model chosen for a practical speech recognition purpose involves several thousands of tied states. Considering all the states in the top level of Fig.1 is not only computationally unfeasible but also unnecessary. Instead, we only consider a subset C_s of states, called the candidate rival set consisting of top-N nearest states that compete with the teaching state j_t^*. Also, the states mapped to the same monophone with the state j_t^* are excluded in the candidate set. The candidate set can be considered in two levels too. A large candidate set is prepared before the iteratively learning in Fig.1, while a much smaller such set is selected from this big set to be used as C_s, consisting of a few rival candidates, which will significantly reduce computational complexity.

There could be difference ways for judging whether one state competes with the teaching state j_t^*. We use the following approximation of a KL divergency measure:

$$D(f\|g) \approx \sum_{i=1}^{K} \alpha_i \min_{j=1..k} KL(f_i\|g_j) \tag{8}$$

which is based on a matching function between each component of the GMM under the states f and g.

In a summary, we get the working flow chart in Fig.2.

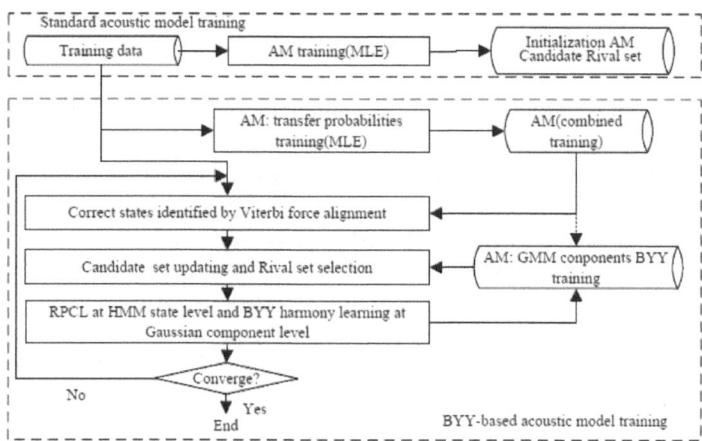

Fig. 2. The flow chart of discriminative training on GMM-HMM acoustic model by BYY harmony learning

Table 1. Recognition PER(%) of GMMs trained by MLE criterion with different values for the number of Gaussian components within each hidden Markov state

#mix	1	2	4	8	16	32
PER(%)	40.01	35.72	33.35	32.42	32.79	34.05

Table 2. A brief summary of the algorithms

Abbreviations	implementation scenarios on top/bottom levels of Fig. 1
alg(m,m)	(top&bottom) MLE, maximum likelihood estimation
alg(r,m) [6]	(top) RPCL simplified BYY learning by Eq.(7); (bottom) MLE, i.e., set $\Delta_{\ell\mid j,t} = 0$;
alg(m,b) [8]	(top) MLE, i.e., set $\Delta_{j,t} = 0$; (bottom) BYY learning by Eq.(3);
alg(r,b)	(top) RPCL simplified BYY learning by Eq.(7); (bottom) BYY learning by Eq.(3);

3 Experiments and Results

3.1 Experiment Setup and Implementation Details

The speech corpus employed in this paper is the TIMIT database. Excluding the "sa" utterances in TIMIT, 3696 utterances and 192 core-test utterances were used as training set and test set respectively. The input speech data is made up of Mel-frequency cepstral coefficients (MFCCs), with 13 cepstral coefficients including the logarithmic energy and their first and second-order differentials. The baseline system is built using HTK toolkit. The acoustic models chosen for speech recognition were triphones models built using decision-tree state clustering. After clustering, the resulted HMMs had 995 tied states. All experiment results were obtained through single pass recognition on test speech using a phone-loop network (without language model). The performance evaluation metric used in phonetic recognition experiments is the phonetic error rate (PER). In the phonetic recognizer's evaluation, 48 monophones are merged into 39 monophones and confusion among the merged phones is not considered as errors.

In MLE training, the number of mixture components per tied state is uniformly set to a constant empirically. After decision-tree based state clustering, Gaussian splitting strategy is used to increase model size from single Gaussian distribution to mixture of Gaussian distributions. We examine recognition performance of models trained using MLE criterion in different model size and results are given in Tab.1. As can be seen from this table, the performance seems saturate when the mixture number increase to 8.

To investigate the effects of replacing the MLE parts in the top and/or bottom levels by BYY harmony learning, we conduct incremental implementations as listed in Tab.2. To save computing cost, the candidate set for every state is calculated using Eq.(8) with a size $N = 50$, which is updated before every iteration.

Table 3. Recognition PER(%) results of different algorithms

alg(m,m)		alg(r,m)		alg(m,b)		alg(r,b)	
#*mix*	PER(%)	avg #*mix*	PER(%)	avg #*mix*	PER(%)	avg #*mix*	PER(%)
8	32.42	8	32.04	6.89	31.98	6.79	31.71
16	32.79	16	32.56	13.05	32.54	11.50	32.06
32	34.05	32	33.86	25.48	33.15	15.32	32.27

3.2   Experiment Results

Figure 3 shows the BYY harmony function value in the training procedure of
alg(r,b). It can be observed that alg(r,b) have good convergence in the training
procedure.

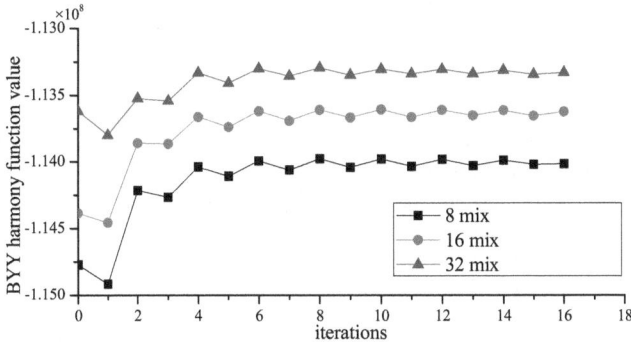

Fig. 3. The value of BYY harmony function for alg(r,b) with different model size

Table 3 shows the recognition performance and average number of Gaussian
components (#*mix*) trained by the algorithms in Tab.2, which are initialized
from MLE-based models with Gaussian component number being 8, 16 and 32,
respectively. All algorithms improve the MLE, and get best results when initializing #*mix* = 8. The alg(m,b) is comparable to alg(r,m) in terms of PER,
but the alg(m,b) obtains more compact models with a smaller average number
of Gaussian components for the hidden Markov states. Moreover, the alg(r,b)
further improves either alg(r,m) or alg(m,b) with the smallest average numbers
of Gaussian components. This implies that BYY (or its simplified RPCL approximation) is effective in discriminative training on either of or both levels of
the two level hierarchical mixture of Gaussians in Fig.1, and automatic model
selection also plays a helpful role in speech recognition system.

4   Conclusions

We focus on a two-stage BYY harmony learning based discriminative acoustic
model training method for speech recognition system. Experiments on phonetic

recognition tasks showed that the RPCL simplified BYY discriminative training on the hidden Markov state level and the BYY automatic model selection on the Gaussian components level within the same hidden Markov state, work better in a combined way, leading to a more compact model and a better recognition performance, than their individual implementation as well as MLE implementation.

In the future, we will adopt the whole BYY harmony learning on the levels of hidden Markov states, phones, words, etc., and further improvements may be expected.

Acknowledgments. The work was supported in part by the National Natural Science Foundation of China (No.91120001 and No.90920302), a HGJ Grant of China (No. 2011ZX01042-001-001), a research program from Microsoft China and by a GRF grant from the Research Grant Council of Hong Kong SAR (Project CUHK 4180/10E). Lei Xu is a Chang Jiang Chair Professor in Peking University.

References

1. Brown, P.: The acoustic-modeling problem in automatic speech recognition. Ph.D. dissertation, Carnegie Mellon University (1987)
2. Bahl, L., Brown, P., De Souza, P., Mercer, R.: Maximum mutual information estimation of hidden Markov model parameters for speech recognition. In: Proc. of 1986 IEEE Intl. Conf. on ICASSP, pp. 49–52 (1986)
3. Juang, B.H., Katagiri, S.: Discriminative learning for minimum error classification. IEEE Trans. on Signal Processing 40(12), 3043–3054 (1992)
4. Povey, D., Woodland, P.C.: Minimum phone error and I-smoothing for improved discriminative training. In: Proc. of 2002 IEEE Intl. Conf. on ICASSP, pp. 105–108 (2002)
5. Pang, Z.H., Wu, X.H., Xu, L.: A Comparative Study of RPCL and MCE Based Discriminative Training Methods for LVCSR. In: Zhang, Y., Zhou, Z.-H., Zhang, C., Li, Y. (eds.) IScIDE 2011. LNCS, vol. 7202, pp. 27–34. Springer, Heidelberg (2012)
6. Pang, Z.H., Tu, S.K., Su, D., Wu, X.H., Xu, L.: Discriminative training of GMM-HMM acoustic model by RPCL learning. Frontiers of Electrical and Electronic Engineering in China 6(2), 283–290 (2011)
7. Xu, L.: Bayesian-kullback coupled ying-yang machines: unified learning and new results on vector quantization. In: Proc. Intl. Conf. on Neural Information Processing, pp. 977–988 (1995)
8. Su, D., Wu, X.H., Xu, L.: GMM-HMM acoustic model training by a two level procedure with Gaussian components determined by automatic model selection. In: Proc. of 2010 IEEE Intl. Conf. on ICASSP, pp. 4890–4893 (2010)
9. Xu, L.: Rival penalized competitive learning. Scholarpedia 2(8), 1810 (2007)
10. Xu, L.: Bayesian Ying-Yang system, best harmony learning, and five action circling. Frontiers of Electrical and Electronic Engineering in China 5(3), 281–328 (2010)
11. Xu, L.: On essential topics of BYY harmony learning: Current status, challenging issues, and gene analysis applications. Frontiers of Electrical and Electronic Engineering in China 7(1), 147–196 (2012)

Lightly Supervised Acoustic Model Training for Mandarin Continuous Speech Recognition

Xiangang Li[1,2], Zaihu Pang[1,2], and Xihong Wu[1,2,3,*]

[1] Speech and Hearing Research Center,
[2] Key Laboratory of Machine Perception (Ministry of Education),
Peking University, Beijing, 100871, China
[3] College of Computer Science and Technology,
Jilin University, Changchun, 130012, China
{lixg,pangzh,wxh}@cis.pku.edu.cn

Abstract. This paper investigates a kind of lightly supervised acoustic model training method for Mandarin continues speech recognition system. The speech materials with rough transcription, which provide some light supervision for acoustic model training, are available in various forms these days. In this work, the quality problem of this kind of data is classified into two types: the first is non-speech and low-quality speech in the corpora, while the second is the transcription errors. A framework is proposed to tackle these two types separately: the speech recognition with transcription-relevant language model is adopted to remove the first type, while with general language model to provide candidate transcription errors which are checked by the final automatic verification process. The performance of proposed framework was evaluated from two aspects: the data quality has significantly improved, and the speech recognition results show that a 21.88% relative CER reduction was obtained.

Keywords: Lightly supervised acoustic model training, automatic speech recognition, Mandarin broadcast news, Mandarin broadcast conversation.

1 Introduction

The last decades have witnessed significant progress in automatic speech recognition (ASR),with many ASR systems having been successfully deployed in everyday-life applications. In these systems, a large amount of accurate manually transcribed training speech is always desired to achieve good performance. Obtaining this necessary accurate transcribed data is an expensive process of both manpower and time. There are certain audio sources, such as radio and television broadcasts, can provide an unlimited supply of acoustic model training materials with manually derived transcription, which is just rough, but not exact transcription of what is being spoken [1]. Thus, how to build reliable acoustic models on this kind of data becomes an important topic.

---

* Corresponding author.

J. Yang, F. Fang, and C. Sun (Eds.): IScIDE 2012, LNCS 7751, pp. 727–734, 2013.
© Springer-Verlag Berlin Heidelberg 2013

In recent years, many technologies have been studied to reduce the supervision required for acoustic model training. These technologies, studied for the tasks where the transcription is not available, is called unsupervised training[2–4], while for the cases that rough transcriptions can provide some supervision, is called lightly supervised training[1]. In [1, 5, 6], the speech recognizer is used to automatically transcribe raw audio data, and then compare with the rough transcriptions to filter words that are potentially incorrect. In [7–9], verification-based methods are introduced: after the ASR or Viterbi alignment results were first carried out with a low resolution acoustic model, some criteria or information is applied to measure the transcription quality.

In this paper, a lightly supervised framework is proposed to automatically cleansing these broadcast speech materials and obtaining a high quality training data. The key idea of this framework is to conduct a two-pass speech recognition and an additional verification, where the first pass recognition is based on a transcription-relevant language model while a general language model is employed in the second pass recognition. We describe these methods that the transcription-relevant language model is a "biased" expert, while the general language model is an "unbiased" expert. These two experts here "discuses" the data, and the disagreements need to be carefully checked. Thus, some other experts in particular fields, such as acoustic likelihood, are introduced to conduct the verifications. In this framework, multi-source information can be used to clean the broadcast speech materials. In addition, in order to investigate the performance of the proposed framework, a 10-hour data quality evaluation set and a 3-hour ASR evaluation set were built. A quality assessment criterion was proposed and conducted on the 10-hour set. The statistics show that data quality was significantly improved. With ML estimation, we trained the acoustic models and then compared the ASR performance. A 21.88% relative CER reduction was obtained with the proposed method.

The rest of the paper is organized as follows. The next section presents the corpora used in this work and the detail of the 10-hour data quality evaluation set. Section 3 details the proposed framework. Section 4 describes the experiments, followed by the results in section 5. The discussions and conclusions are drawn in the last section.

2   Speech Corpora

In the corpora used in this paper, the mandarin speech transcriptions are firstly recognized automatically, and then corrected manually. This corpora consist of BN and BC speech coming from 49 television broadcasts, and add up to 1345-hour speech. A 10-hour data quality evaluation set was built to evaluate the data quality of the corpora. This database was random selected from the corpora, in which 5-hour speech was selected from BN programs in 10 televisions, and the other 5-hour speech from BC programs in 5 televisions. The quality assessment was conducted on the 10-hour set, which can reflect the quality the corpora from the side. Moreover, another 3-hour speech data set was built to conduct the ASR evaluation, which were also random selected from the corpora.

In order to evaluate the quality of the corpora, a quality assessment method is proposed in two aspects. First, reference to the focus condition of 1996 CSR Hub4 evaluation set, the speech in our corpora is classified into 4 types: clean speech, speech in low signal-to-noise ratio (SNR) environment, non-mandarin speech, and non-speech. Only the clean speech is suitable for acoustic model training. Thus, the speech length percentage (SLP) of clean speech can reflect the speech quality of the corpora. On the other hand, in order to evaluate the transcription quality of the corpora, just like the evaluation in ASR, the label character error rate (LCER) is introduced. The speech in the 10-hour data quality assessment set was manually segmented, and carefully label out its type. The quality assessment was conducted on BN speech and BC speech separately, while the LCER evaluation was conducted on clean speech and speech in low SNR environment. All the results are listed in Table 1.

Table 1. SLP and LCER for 4 types of speech in the quality assessment dataset

	BC		BN	
	SLP(%)	LCER(%)	SLP(%)	LCER(%)
Clean Speech	88.29	2.41	93.57	9.93
Speech in low SNR environment	4.09	15.03	2.97	32.76
Non-Mandarin	3.48	-	1.06	-
Non-speech	4.14	-	2.39	-

It can be figured out that, more than 11% speech is not suitable for acoustic model training in the BN speech, while about 6% speech is not clean speech in the BC speech. BN speech contains many outdoor recordings, making environments and speakers very complex. BC speech is always recoded in the recording studios, thus, the SLP of clean speech in BN is less than that in BC. However, BN speech is mainly announcer speech that is standard Mandarin, while BC speech contains is mainly conversation dialog speech, which makes it difficult to transcribe BN speech than BC speech, which has been reflected in LCER clearly.

3 The Proposed Framework

In this paper, our interest is focused on how to train a reliable acoustic model from the speech materials with rough transcription. In this section, the lightly supervised training is first revisited. Followed an analysis of the problem, a framework is proposed, and is described in details in this section.

3.1 Revisiting Lightly Supervised Training

In [1], an initial acoustic model was used to transcribe the training data, where the related language models were employed. The inconsistent parts between the recognition results and original transcription are considered as the possible transcription errors. Due to the ML based acoustic model training and the hypothesis that most of the transcriptions are correct, the initial acoustic model can be used to detect the transcription errors. However, while using the ASR results to detect the transcription errors, the misrecognizing problems in the ASR technology need to be carefully considered. The "biased" language used in [6], where the rough transcriptions are added with a heavy weight, helps in selecting words the recognizer could have misrecognized if using a fair language model. In addition, there are some methods focused on introducing the acoustic or prosody information to do verification. These methods are mainly based on confidence measurement technology [10] and the results of ASR or forced Viterbi alignment. The units in the results are checked and measured to determine whether they are correct or not. From another point, these verifications are used to determine the undetermined parts.

Through the above analysis, the ASR results based lightly supervised method can efficiently locate the possible error parts, and the verification based methods can determine whether this part is an error. In addition, the "biased" language method may ignore some transcription errors, but can effectively locate the large tracts of continuous ones. The fair language model would lead to a bad ASR results, making too many correct transcriptions are mistakenly seen as errors, but can provide candidates for verification.

3.2   Problem Analysis

Our purpose is to build high quality corpus with correct transcriptions for acoustic model training. More specifically, the main task is to locate non-speech, low-quality speech and transcription errors.

These quality problems are already reflected in Table 1, which can be classified into two types: the first is the non-speech and low-quality speech, while the second is the transcription errors. The low-quality speech contains non-Mandarin, speech in low SNR environment. In most cases, the non-speech materials are music, and that do not have any transcriptions or labels. In the non-Mandarin materials, they may speak dialect or foreign language, which is acoustically mismatch Mandarin. The speech in low SNR environment is somehow contaminated by the background noise. Besides, there is no clear border between the clean speech and low SNR speech. One possible way is to locate the units that are acoustically mismatching. Even if non-speech and low-quality speech is removed, there are still many transcription errors. The transcription quality problem seems to be serious for the BC speech, where about 10% transcriptions are errors. The speech quality problem and transcription quality problem can not be confused: they should be tackled respectively.

According to the two problems, the task is divided into two parts: the first is to remove non-speech and low-quality speech, while the second is to detect the transcription errors in the high quality speech.

3.3   Framework Description

This section details the proposed framework. The framework is divided into two procedure, the first procedure is introduced to locate non-speech and low-quality speech, while the second is to detect the transcription errors.

The low-quality speech would lead to a large tracts of misrecognizing problem. Inspired from this "biased" language model method, in the proposed framework, a transcription-relevant language model was trained for every record's transcription, and used for speech recognition. The speech recognition with transcription-relevant language model utilizes global acoustic model information and the local internal constraints of transcriptions. Specifically, if the speech is non-Mandarin or non-speech, there would be a long segment of speech that acoustically mismatches standard Mandarin. If the speech is in low SNR environments, there are still some disagreements between the recognition results and transcriptions while employing these biased language models. On the other hand, in the clean speech, the transcription errors mainly come from annotator's careless, which makes the errors have some regular patterns, and hardly occur successively. A threshold is introduced to ignore the local recognition errors and to make the remaining speech not too fragmentary. Thus, the first procedure can be summarized as follows:

1. Train acoustic model on the whole corpora
2. Train the transcription-relevant language model for every record
3. Conducting the speech recognition with transcription-relevant language model on all the train set
4. Align the closed-captions with ASR results (using a dynamic programming algorithm) and remove speech segments where these two transcriptions disagree

The second procedure is to detect the transcription errors. In [1], the inconsistent parts between ASR results and rough transcriptions are regarded as transcription errors. However, the inconsistent parts can only provide the labels need to be confirmed. Therefore, an automatic verification process is necessary. From another perspective, the verification process is used to recall some speech that is deleted mistakenly by the second pass ASR. The readers may argue that the second pass ASR is not necessary and the verification can conduct on the ASR results based on transcription-relevant language model. This argument makes some sense, but the ASR with fair language model can give more candidate errors. Besides, the acoustic model trained in the last procedure, can be employed to replace the initial acoustic model in the first procedure. Through this iteratively way, the performance can get further improvement. These steps summarized as follows:

1. Conducting speech recognition with fair language model on the remaining speech, the inconsistent parts between the two ASR results are identified as candidate errors
2. A verification process is followed to determine whether candidates are really transcription errors
3. The remaining speech is used to training acoustic model.

In the proposed method, these speech quality problem and transcription errors are tackled respectively. The readers may argue that the first procedure is not necessary because the ASR with fair language model can handle both speech quality and transcription error problems. However, in the ASR with fair language model, non-speech and low-quality speech will impact the recognition performance on high quality speech, due to the dynamic programming technology used in the recognizer. Based on this argument, if the first pass speech recognition with transcription-relevant language model is not conducted, the recognizer may misrecognize too many, making the verification too difficult.

4   Experiments and Results

Our experiments were all conducted on the corpora mentioned in section 2. The initial acoustic model was trained with all the train set by ML estimation. For each record in the train set, the transcription-relevant 3-gram language model was trained with the corresponding transcription document. In the 1st procedure, every record was recognized with its corresponding language model. Then the recognition results were aligned with the transcriptions. The inconsistent parts partition the original transcriptions into fragments. The fragments, that have less than 10 characters, were removed. The general language model training materials come from the People's Daily of China (from 1998 to 2007). The recognition with general language model was conducted on the remaining speech. In this way, the influence of low-quality speech on ASR can be avoided, and can also accelerate recognition. The inconsistent parts between these two recognition results were identified as the candidate errors. Finally, the acoustic model scores were used to check the candidate errors. Those segments, where the acoustic model score in the first procedure is less than that in the second procedure, are considered as including some transcription errors.

Some statistics, such as the remaining speech length, the speech length percentage (SLP) of clean speech and the label character error rate (LCER), were obtained from the 10-hour data quality assessment set to investigate the performance of proposed framework. Besides, after the every procedure was finished, an acoustic model was trained and tested. In the ASR evaluation, the general language model was employed.

After the 1st procedure, about 185-hour speech is considered as low-quality speech, and then 365-hour speech is viewed as including some transcription errors. The quality assessment results are listed in Table 2. From these results, we can easily find out that the qualities of the database, both the speech length percentage (SLP) of clean speech and the LCER of the clean speech, have improved. The SLP have relative increased 9.05% in the BN speech and 5.20% in the BC speech. Besides, about 58.92% and 56.80% LCER relative reduction can be obtained in the BN and BC speech. Although the proposed framework was designed to tackle the two problems in the corpora respectively, the speech and transcription quality can improve at every procedure. After the 1st procedure, an acoustic model was trained with the remaining 1160-hour speech materials, and

Table 2. Speech length percentages of clean speech and label character error rate in the speech quality assessment set

Corpora	BN		BC	
	SLP (%)	LCER (%)	SLP (%)	LCER (%)
The original corpora	88.29	2.41	93.57	9.93
The corpora after the 1st procedure	95.01	1.03	97.95	4.97
The corpora after the 2nd procedure	98.60	0.76	99.38	2.55

Table 3. ASR results

Acoustic models	CER (%)	Rel.CER.Reduction(%)
The initial acoustic model	15.72	-
The acoustic model trained after the 1st procedure	13.60	13.48
The acoustic model trained after the 2nd procedure	12.28	21.88

Table 3 shows the recognition performance can get 13.88 relative CER reduction. Moreover, although the final acoustic model was trained with only 795-hour speech materials, the system finally yielded a 21.88% relative CER reduction.

5   Discussions and Conclusions

This paper is focused on how to obtain clean speech with correct transcription for acoustic model training. The problems of the corpora coming from an enterprise standardization process are analyzed and summarized into two types: the speech quality problem and transcription quality problem. Though investigating the conventional lightly supervised training methods, a framework, which tackles these two problems respectively, is proposed. In this framework, the speech recognition with transcription-relevant language model is used to remove the low-quality' speech, while the speech recognition with general language model and the verification process are used to detect the transcription errors.

In the multi-pass ASR technology, a general language model is always adopted in the first-pass recognition to provide a candidate lattice. The lattice can help determine the domain or topic of the test speech, thus in the second-pass recognition, the domain-relevant or topic-specific language model can be employed. However, quite the opposite, the proposed framework adopts the biased language model first, and the general language model second. In this way, the impact on ASR performance that the low-quality speech has can be avoided to some extent. The experiments were conducted on 1345-hour corpora. From the quality assessment results, we can clearly find out that, the proposed method can provide a cleansing data set that contains clean speech with correct transcription. Besides, the ASR results show that the proposed method can achieve a significant improvement on the Mandarin BC and BN corpora.

Though comparing the quality assessment results in section 2 and the experiment results, we can find out that, our method remove too many speech materials. Although the unlimited supply of BN and BC speech materials makes it not a big problem, our further work focus on how to keep more and remove less high quality speech materials. In our further work, confidence measurements on many acoustic and prosody information, such as phone duration and word-level posterior probability, can be adopted in the verification process. In such case, more correct transcriptions that are mistakenly regarded as errors can be recalled.

Acknowledgments. The work was supported in part by the National Natural Science Foundation of China (No.91120001,No.90920302), a HGJ Grant of China (No.2011ZX01042-001-001) and a research program from Microsoft China.

References

1. Lamel, L., Gauvain, J., Adda, G.: Lightly Supervised and Unsupervised Acoustic Model Training. Computer Speech and Language 16, 115–129 (2002)
2. Wessel, F., Ney, H.: Unsupervised Training of Acoustic Models for Large Vocabulary Continuous Speech Recognition. IEEE Transactions on Speech and Audio Processing 13(1), 23–31 (2005)
3. Wang, L., Gales, M.J.F., Woodland, P.C.: Unsupervised training for Mandarin Broadcast News and Conversation Transcription. In: Proc. ICASSP, vol. 4, pp. 353–356 (2007)
4. Fraga-Silva, T., Gauvain, J., Lamel, L.: Lattice-based unsupervised acoustic model training. In: Proc. ICASSP, pp. 4656–4659 (2011)
5. Kawahara, T., Mimura, M., Akita, Y.: Language model transformation applied to lightly supervised training of acoustic model for congress meetings. In: Proc. ICASSP, pp. 3853–3856 (2009)
6. Nguyen, L., Xiang, B.: Light Supervision in Acoustic Model Training. In: Proc. of ICASSP, vol. 1, pp. 185–188 (2004)
7. Chen, B., Kuo, J.W., Tsai, W.H.: Lightly Supervised and Data-Driven Approaches to Mandarin Broadcast News Transcription. In: Proc. of ICASSP, vol. 1, pp. 770–780 (2004)
8. Pitz, M., Molau, S., Schluter, R., Ney, H.: Automatic Transcription Verification of Broadcast News and Similar Speech Corpora. In: Proc. DARPA Broadcast News Workshop, pp. 157–159 (1999)
9. Kurata, G., Itoh, N., Nishimura, M.: Acoustic Model Training with Detecting Transcription Errors in the Training data. In: Proc. of INTERSPEECH, pp. 1689–1692 (2011)
10. Jiang, H.: Confidence measures for speech recognition: A survey. Speech Communication 45(4), 355–470 (2005)

A Hierarchical Representation Policy Iteration Algorithm for Reinforcement Learning

Jian Wang[1], Lei Zuo[1], Jian Wang[2], Xin Xu[1], and Chun Li[1]

[1] College of Mechatronics and Automation, National University of Defense Tech.,
Changsha, 410073, P.R. China
{jian_wang,xinxu}@nudt.edu.cn,
zuo1986lei@163.com
[2] Xi'an Air Force Military Representative Office
wangjian0313025@163.com

Abstract. This paper presents a hierarchical representation policy iteration (HRPI) algorithm. It is based on the method of state space decomposition implemented by introducing a binary tree. Combining the RPI algorithm with the state space decomposition method, the HRPI algorithm is proposed. In HRPI, the state space is decomposed into multiple sub-spaces according to an approximate value function, then the local policies are estimated on each sub-space and finally the global near-optimal policy is obtained by combining these local policies. The simulation results indicate that the proposed method has better performance compared to the conventional RPI algorithm.

Keywords: Reinforcement learning, HRPI, State space decomposition.

1 Introduction

In recent years, reinforcement learning (RL) has been attracted in the community of artificial intelligence and machine learning because of its self-learning and self-adaptive characteristics. However, the time-consuming problem and the curse of dimensionality will occur when the classic table methods, such as Q-learning [1] and Sarsa learning [1], are used in dealing with the large scale problems with continuous spaces. Presently, there are two main approaches to address the continuous space problems. One technique is called value function approximation (VFA) [2–7] and the other is hierarchical reinforcement learning (HRL). The former is divided into linear [2–6] and nonlinear [7], while the latter is implemented in the ways of state space decomposition [8, 9], temporal abstraction [10] and state abstraction [11].

In order to address the curse of dimensionality and achieve good generalization performance in continuous spaces, VFA-based RL has been widely studied, especially those based on the linear architecture [4–6]. Combining the least-square technique with the policy iteration, Lagoudakis [4] proposed least-square policy iteration. Subsequently, Xu [5] and Mahadevan [6] proposed kernel-based least-square policy iteration (KLSPI) and representation policy iteration (RPI),

J. Yang, F. Fang, and C. Sun (Eds.): IScIDE 2012, LNCS 7751, pp. 735–742, 2013.
© Springer-Verlag Berlin Heidelberg 2013

respectively. Both the methods are based on the LSPI framework, but have different techniques to construct the basis functions. KLSPI uses kernel methods, while RPI employs a manifold approach (the *graph Laplacian*). However, final policies of VFA-based methods are greatly influenced by the approximation precision of value functions and the constructed basis functions.

HRL is another major technique to deal with the curse of dimensionality while solving Markov decision process (MDP) problems with large scale or continuous spaces. It has become a key issue in RL in recent years [12]. The core of HRL is to introduce the abstract and hierarchical mechanisms to decompose a complex MDP into different simple sub-MDPs and then solve each sub-MDP by using RL methods on their own spaces. By this means, a complex MDP problem is simplified and the curse of dimensionality can be alleviated to some extent. Based on a reasonable abstract mechanism, HRL can simplify definitions of the state, action and reword function and reduce the requirement of data storage and the computational complexity. Accordingly, it helps to accelerate the learning speed or obtain a near-optimal policy with a higher precision.

The main idea of RPI is to automatically construct basis functions for solving MDPs by diagonalizing symmetric diffusion operators [6]. The diffusion operators are obtained from sample data and used to build the *graph Laplacian* on an undirected graph. The final policy obtained by RPI is a near-optimal policy, which means there are errors between the near-optimal policy and the optimal. Thus, in order to obtain a more precise policy, a HRL approach based on the state space decomposition and RPI is developed. Motivated by the binary tree used in [9], this paper proposes a hierarchical representation policy iteration (HRPI) algorithm in which the state space decomposition is performed by using the binary tree.

Xu [9] proposed the hierarchical approximate policy iteration (HAPI) framework through hierarchical kernel-based least-square policy iteration (HKLSPI), while this paper is an extension for both HAPI and RPI. On the one hand, the basis functions of HKLSPI are obtained through kernel functions which are need to be selected manually and empirically, while HRPI uses the manifold approach to automatically construct the basis functions. On the other hand, compared to RPI, HRPI combines the state space decomposition with the original RPI method, which will help to obtain a more precise policy.

The contribution of this work is to integrate the RPI method into the HAPI framework. The proposed HRPI algorithm firstly uses a k-means clustering method to obtain a subset of samples on which the basis functions are constructed. Secondly, it employs the binary tree to decompose the original MDP into multiple sub-MDPs with smaller state spaces. After that, the RPI algorithm is implemented on each subspace to obtain better near-optimal local policies, respectively. Finally, the final global policy is derived by combining the obtained local policies. In the spaces of the sub-MDPs, RPI can estimate local near-optimal policies with higher precision, and the global near-optimal policy can be evaluated as a combination of the local near-optimal policies.

2   The HRPI Algorithm

Details about the binary tree have been presented in [9], please see it for more information. The flow chart of the HRPI algorithm is shown in Fig. 1. The algorithm includes five main steps: sample collection procedure, basis function construction, state space decomposition, local policies combination and testing the combined policy.

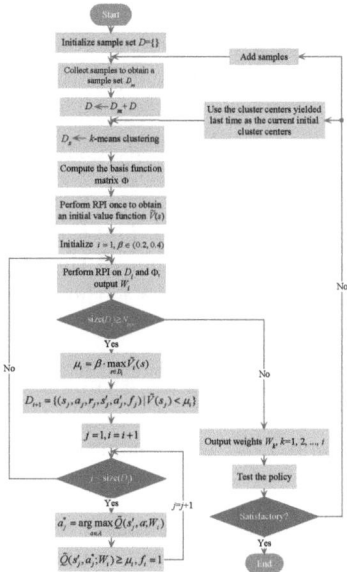

Fig. 1. The flow chart of the HRPI algorithm

2.1   The Procedure of Sample Collection

In order to obtain better performance, the procedure of sample collection needs to explore the state-action space of an absorbing MDP uniformly. In the simplest case, samples are collected by using a random walk and the simplest random policy.

The initial sample set D can be very large if the MDP has a large scale or continuous state space. However, a much smaller set of samples is usually sufficient to construct the graph and derive the basis functions. Thus, a k-means clustering method is employed to select the subset D_s of the initial sample set D.

In this paper, the selection of D_s is based on the Euclidean distance which is employed to evaluate the distance between two samples in the clustering method. From the initial set $D = \{d_1, d_2, \ldots, d_t, \ldots, d_N\} \subset \mathbf{R}^n$, the k-means algorithm allocates each sample to one of k clusters to minimize the within-cluster sum of squares J:

$$J = \sum_{j=1}^{k} \sum_{i \in A_k} ||d_i - c_j||_2 \tag{1}$$

where $d_t = \{(s_t, a_t, r_t, s'_t, a'_t, f_t)\}$, meaning that at time step t the MDP is in state s_t, action a_t is taken, a reward r_t is received, the resulting next state is s'_t and the next action to be selected is a'_t, f_t is the flag for terminal states, N is the size of the initial sample set D, k is the number of clusters, A_k is the set of objects (samples) around the kth cluster, and c_k is the average value of those samples in A_k. The calculation of c_k is by using the following formula:

$$c_k = \frac{1}{n_k} \sum_{j=1}^{n_k} d_j, d_j \in A_k, \tag{2}$$

where n_k is the number of samples in A_k. The k-means clustering algorithm is presented as follows, where k is the number of the clusters, q is the maximum iteration number, and $D = \{d_1, d_2, \ldots, d_N\}$ is the initial collected sample set.

Algorithm 1. The k-means clustering algorithm

---

1. Input: k, q, D and $i = 1$;
2. Select k samples randomly from D as the initial centers;
3. Repeat

   1) Assign each samples to the closest cluster center according to equation 1;
   2) For $i = 1, 2, \ldots, k$, use equation 2 to compute the new cluster center c_i.

4. Until the centers do not change or $i > k$;
5. For $i = 1, 2, \ldots, k$, select the sample s_i closest to the c_i from the original sample set D to construct the subset D_s;
6. Return D_s.

The resulting subset D_s consists of the collected samples closest to the corresponding cluster centers and is far smaller than the initial sample set D.

2.2   Basis Function Construction

The basis functions construction is based on the *graph Laplacian*. During this process, different weighted graphs are associated to the subset D_s. Given D_s, edges are inserted between pairs of states s_i and s_j if s_j are among the k nearest neighbors of s_i, where $k > 0$ is a scalar and denotes the number of samples nearest to each cluster center. Weight $W(i, j)$ are assigned to the edges according to equation 3:

$$W(i, j) = \exp(-||s_i - s_j||^2 / \sigma) \tag{3}$$

where $\sigma > 0$ is also a scalar.

The combinatorial Laplacian L is defined as $L = D_{dm} - W$, where D_{dm} is a diagonal matrix called the valency matrix whose entries are row sums of the weight matrix W composed of $W(i, j)$. Compute the eigenvectors corresponding to the k' smallest eigenvalues of L, and collect them as the columns of the basis function matrix Φ, a $|D_s| \times k'$ matrix.

To extend eigenfunctions computed on the subset D_s to new unexplored samples, the Nyström method is used:

$$\phi_i(x) = \frac{1}{1 - \lambda_i} \sum_{y \sim x} \frac{w(x, y)}{\sqrt{d(x)d(y)}} \phi_i(y) \tag{4}$$

where x is a new sample, y belongs to the k nearest neighbors of x in D_s and $w(x, y)$ is the weight of the edge between x and y. $d(x)$ is computed as the following formula:

$$d(x) = \sum_{y \sim x} w(x, y) = \sum_{y \sim x} \exp(-\frac{||x - y||^2}{\sigma}) \tag{5}$$

The state action bases $\phi(s, a)$ can be generated by duplicating the state base $\phi(s)$ $|\mathcal{A}|$ times and setting all the elements of this vector to zero except for the ones corresponding to the number of the chosen action a, where $|\mathcal{A}|$ is the size of the action set $\mathcal{A}$.

2.3   State Space Decomposition

Let $D_0 = \{(s_j, a_j, r_j, s'_j, a'_j, f_j)\}(j = 1, 2, \ldots, N)$ be the set of data samples collected in the initial state space S_0 and initialize the iteration number $i = 1$. Let D_i denote the sample subset for sub-state space S_i. The first step of each iteration is to obtain a near-optimal policy in S_i by using RPI based on D_i. After the convergence of RPI, a weight vector W_i determines the value function of the near-optimal policy π_i. To decompose S_i into sub-state spaces iteratively, the estimated value functions $\tilde{V}_i(s)$ in S_i divides the sample set D_i into two subsets, where one subset D_{i+1} is used for state space decomposition in the next iteration. The threshold μ_i can be automatically determined by the following equations:

$$\mu_i = \beta \cdot \max_{s \in D_i} \tilde{V}_i(s) \tag{6}$$

$$\tilde{V}_i(s) = \max_a \tilde{Q}_i(s, a; W_i) = \max_a \phi^T(s, a)W_i \tag{7}$$

where, $\tilde{Q}_i(s, a; W_i)$ is the estimated action-value function of π_i with respect to W_i. The subset D_{i+1} for M_i are determined as follows:

$$D_{i+1} = \{(s_j, a_j, s'_j, a'_j, r_j, f_j) \in D_i | \tilde{V}_i(s_j) < \mu_i\} \tag{8}$$

For every sample $(s_j, a_j, r_j, s'_j, a'_j, f_j)$ in D_{i+1}, a state is identified as a boundary state if the following inequation is hold:

$$\tilde{Q}_i(s'_j, a^*_j; W_i) \geq \mu_i \tag{9}$$

where $a^*_j = \arg\max_a \tilde{Q}_i(s'_j, a; W_i)$.

A boundary state is a terminal state or absorbing state in M_i, where the absorbing flag is set as $f_j = 1$. The prior procedure is iterated until the number of remaining samples is smaller than the given number N_{min} or the iteration number i reaches the maximum iteration number. When the HRPI algorithm is completed, a set of policies $\{\pi_1, \pi_2, \ldots, \pi_{n+1}\}$ is obtained, where π_i is the local near-optimal policy of $M_i(1 \leq i \leq n)$ obtained by RPI and π_{n+1} is the local near-optimal policy of S_{n+1}. The final policy to be tested is the combination of the local policies $\{\pi_1, \pi_2, \ldots, \pi_{n+1}\}$.

2.4   Local Policies Combination

The algorithm of local policies combination of HRPI is the same as that of HAPI, please see [9] for more details. According to [9], under certain assumptions, the final policy obtained by combination of the local policies can be guaranteed to be hierarchically optimal. Thus, in the sense of the hierarchical optimality, the performance of the combined policy in HRPI can be improved.

2.5   Test the Combined Policy

Apply the final combined policy in some application. If the result of the control performance is satisfactory, terminate the algorithm. Otherwise, re-conduct the procedure of the sample collection, add the new samples to the previous sample set, select the previous cluster centers as the current initial centers, implement the k-means clustering algorithm to obtain a new subset and then perform the HRPI algorithm again.

Although the used time for the learning of HRPI is longer than that of RPI, the performance errors of the final policy obtained by HRPI can be significantly reduced by decomposing the original MDP into sub-MDPs through the binary-tree structure and combining the obtained local policies together.

3   Simulation Results

In this section, the HRPI and RPI algorithms were tested in the mountain-car task. The mountain-car task [1] is a minimum time problem and its goal is to get a simulated car to the top of a hill as quickly as possible, as shown in Fig. 2.

The car does not have enough power to get there immediately, and so must oscillate on the hill to build up the necessary momentum. The state space includes the position x_t and velocity v_t of the car, where t is the time step. There are three actions: full throttle forward (+1), full throttle reverse (-1), and zero throttle (0). Its position, x_t and velocity v_t, are updated by

$$\begin{cases} x_{t+1} = bound[x_t + v_t] \\ v_{t+1} = bound[v_t + 0.001\dot{v}_t - 0.0025 \cdot \cos(3x_t)] \end{cases} \tag{10}$$

where the *bound* operation enforces $-1.2 \leq x_{t+1} \leq 0.5$ and $-0.07 \leq v_{t+1} \leq 0.07$. The car's velocity v_{t+1} is reset to zero when $x_{t+1} = -1.2$. Each episode starts from the lowest point O: $x_0 = -0.5$ with a velocity of 0, and ends when the car successfully reaches the goal position, defined as $x_{t+1} \geq 0.5$. The reward r_t is -1 if $x_{t+1} < 0.5$, while 100 if $x_{t+1} \geq 0.5$.

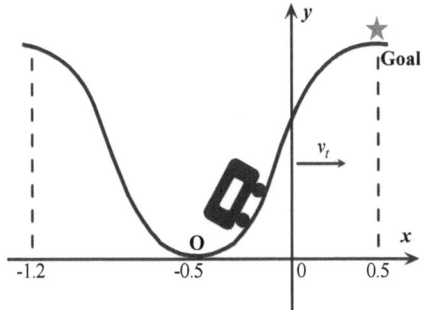

Fig. 2. The mountain-car task

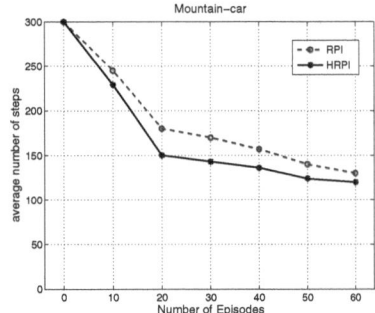

Fig. 3. The comparison of the performance between the HRPI and RPI algorithms

During the course of the sample collection, training samples were collected uniformly and randomly. The initial state of each episode was distributed at random over the entire state space. Each sampling episode terminated when the car arrived at the goal or the number of steps was maximum.

The experiment was repeated 20 times. Each time, the HRPI and RPI algorithms were respectively used to build the basis functions and estimate the near-optimal policy at every 10 episodes. And then, the quality of the obtained near-optimal policy was evaluated in terms of the average number of steps to reach the goal state. The samples were accumulated after each episode.

The common parameters of HRPI and RPI are the total episodes, maximum number of steps, number of nearest neighbors, number of basis functions, number of clusters, scaling and maximum number of iteration, whose value were 60, 200, 10, 30, 600, [1,3] and 10, respectively. In HRPI, another two parameters β and N_{min} were used for decomposing the state space, where β was the threshold value for decomposition and N_{min} was the minimal number of remaining samples. In this experiment, β was 0.3 and N_{min} was 100.

The simulation results are shown in Fig. 3. It can be seen that with the same sample set, the HRPI approach has less steps to arrive at the goal state, meaning that HRPI can obtain a near-optimal policy with a higher precision on the same set of the experience data. Therefore, the proposed HRPI approach can improve the quality of the final policy for the conventional RPI algorithm.

4   Conclusions and Future Work

This paper presents a HRPI algorithm which is based on the state space decomposition. The original state space is decomposed into multiple sub-spaces by using a binary tree and an initial near-optimal value function. By implementing the RPI algorithm in each sub-space and then combining the obtained local near-optimal policies, the global optimal policy can be approximated well. The simulation results show the effectiveness of the proposed HRPI algorithm. Our future work is to design RL algorithms with the ability of automatic constructing the abstract structures. Furthermore, the approximation and generalization of the HRL algorithms still need to be further studied.

Acknowledgments. This work is supported by National Natural Science Foundation of China under Grant 61075072, 90820302 and the Program for New Century Excellent Talents in University under Grant NCET-10-0901.

References

1. Sutton, R.S., Barto, A.G.: Reinforcement Learning: An Introduction. The MIT Press, Cambridge (1998)
2. Sutton, R.S.: Generalization in reinforcement learning: Successful examples using sparse coarse coding. In: Neural Information Processing Systems 8. The MIT Press, Cambridge (1996)
3. Xu, X., He, H.G., Hu, D.: Efficient reinforcement learning using recursive least-squares methods. Journal of Artificial Intelligence Research 16(1), 259–292 (2002)
4. Lagoudakis, M.G., Parr, R.: Least-squares policy iteration. Journal of Machine Learning Research 4, 1107–1149 (2003)
5. Xu, X., Hu, D., Lu, X.: Kernel-based least squares policy iteration for reinforcement learning. IEEE Transactions on Neural Networks 18(4), 973–992 (2007)
6. Mahadevan, S., Maggioni, M.: Proto-value functions: A laplacian framework for learning representation and control in markov decision processes. Journal of Machine Learning Research 8, 2169–2231 (2007)
7. Xu, X., He, H.G.: Residual-gradient-based neural reinforcement learning for the optimal control of an acrobat. In: Proceedings of the 2002 IEEE International Symposium on Intelligent Control, Vancouver, Canada, pp. 758–763 (2002)
8. Dietterich, T.G.: Hierarchical reinforcement learning with the max-q value function decomposition. Journal of Artificial Intelligence Research 13, 227–303 (2000)
9. Xu, X., Liu, C., Yang, S.X., Hu, D.: Hierarchical approximate policy iteration with binary-tree state space decomposition. IEEE Transactions on Neural Networks 22(12), 1863–1877 (2011)
10. Sutton, R.S., Precup, D., Singh, S.: Between mdps and semi-mdps: A framework for temporal abstraction in reinforcement learning. Artificial Intelligence 112, 181–211 (1999)
11. Andre, D., Russell, S.J.: State abstraction for programmable reinforcement learning agents. In: Proceedings of the 18th National Conference on Artificial Intelligence, CA, pp. 119–125 (2002)
12. Barto, A.G., Mahadevan, S.: Recent advances in hierarchical reinforcement learning. Discrete Event Dynamic Systems: Theory and Applications 13, 41–47 (2003)

Shape Retrieval Based on Parabolically Fitted Curvature Scale-Space Maps

Yinan Cui and Baojiang Zhong

School of Computer Science and Technology,
Soochow University, Suzhou 215006, China
{20114527002,bjzhong}@suda.edu.cn

Abstract. Shape retrieval is to find shapes in a database similar to the query one. For this purpose two components are usually needed, shape representation and shape matching. In this paper, a new shape retrieval algorithm based on the curvature scale-space (CSS) representation and the parabolic fitting technique is proposed. First, by a parabolic fitting of the arc-shaped contours in CSS maps, several sets of parameters are obtained for shape representation, which are the coefficients of the fitting polynomials. Then, shape matching is achieved by a comparison of the parameter sets of the query shape and the database shapes. The distance between every pair of compared shapes is measured and the database shapes are ranked according to their distance to the query shape. The proposed algorithm is tested and evaluated on two shape databases.

Keywords: Shape retrieval, scale-space, shape representation, shape matching, polynomial fitting.

1 Introduction

The number of images stored in computers becomes much bigger than before with a considerable improvement in memory technologies and CPU processing speed. The problem of organizing images has therefore been brought to popular attention. In many fields (for example, biology) users wish to get similar images to an known one. Shape retrieval can be used to satisfy this requirement, which usually includes two aspects 1) robust shape representation, 2) rapid and effective shape matching.

The CSS map of a shape usually consists of a set of arc-shaped contours, which are the scale-space trajectories of curvature zero-crossing points on the shape in a curve evolution (curve smoothing) process. In this paper, a CSS-based shape retrieval algorithm is proposed. It first represents the CSS map with two global parameters and several sets of parameters, which are got by a polynomial fitting of the arc-shaped CSS contours. Following, shape matching is then achieved by estimating the similarity between two CSS maps according to the parameters got in the first stage. The new algorithm is evaluated on two shape databases and its performance is compared with the algorithm in [1] which we called the *reference algorithm* in this paper.

J. Yang, F. Fang, and C. Sun (Eds.): IScIDE 2012, LNCS 7751, pp. 743–750, 2013.

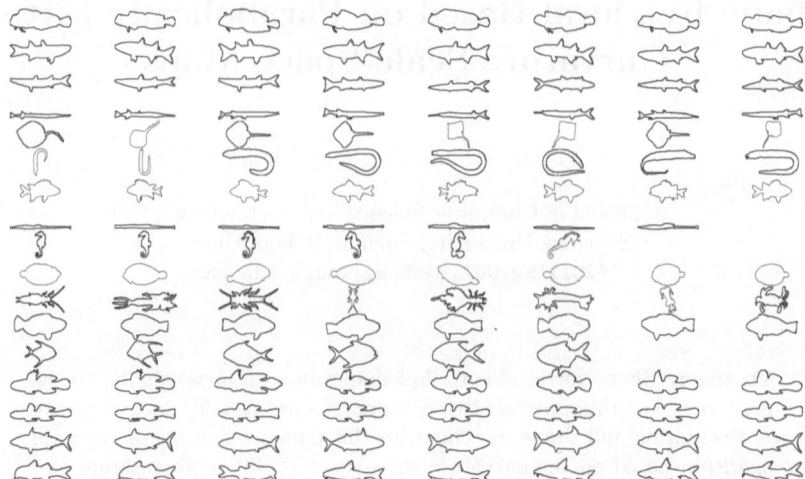

Fig. 1. A database with 131 objects in 17 classes, each row represents a class of fish. The first class is numbered as '00', and the following classes are numbered in turn. The database is called SQUID database, which was developed to test the reference CSS algorithm (see, http://www.ee.surrey.ac.uk/CVSSP/demos/css/squid.html).

The rest of the paper is organized as follows. In Section 2, we introduce the CSS technique and explain how to track arc-shaped counters of the CSS map. In Section 3, the tracked point sets are parabolically fitted and the coefficients of the polynomial fitting are used for shape representation. In Section 4, the shape matching method utilizing the idea of circular correlation are described. Finally, the new algorithm is evaluated and compared with the reference algorithm in Section 5, and the paper is concluded in Section 6.

2   CSS Maps for Shape Representation

2.1   CSS Representation

The CSS technique was developed for shape retrieval [3]. Let $C(s) = \{x(s), y(s)\}$ represents the boundary curve of a shape, where s is the arc-length parameter. Fig. 1 shows the boundary curves of shapes in a database. The scale-space of the shape is constructed by a curve evolution operation. The evolved curve is written as $C(s, \sigma) = \{x(s, \sigma), y(s, \sigma)\}$, where σ is the scale parameter of the curve evolution operation. Owing to the nonuniform shrinkage of the evolved curves [6], a standard curvature expression [7] is thus used to compute the CSS maps in this article.

In the discrete case, a curve is represented by a set of equi-distant samples $\{p_i = (x_i, y_i) | i = 1, 2, ..., N\}$. The curve evolution can be implemented by exploiting an iterative scheme as follows:

$$p_{i,m+1} = 0.25 p_{i-1,m} + 0.5 p_{i,m} + 0.25 p_{i+1,m}, \tag{1}$$

where $m = 0, 1, 2, ...$ is the iteration number, and $p_{i,0} = p_i$, for $i = 1, 2, ...N$.

The CSS representation is invariant under translation, rotation, and scaling transformations of the considered shape. It has also proven robust in the presence of noise [4]. Each contour on the CSS map is associated to a convexity or concavity of the shape. The deeper the convexity/ concavity, the taller the contour. Also, the longer the convexity/concavity, the wider the counter [5]. Fig. 2 shows two similar shapes, and their CSS maps as well. We can see that the similarity of the two shapes has been well expressed by the similarity of their CSS maps.

2.2 Tracking the Arc-shaped Contours on the CSS Map

Every arc-shaped contour on the CSS map consists of a set of points, which are not ready being grouped. We need to propose a tracking step to group them. First, the peak point of each CSS contour is detected. Then we track its two branches from the peak to the bottom, respectively. The tracked points are grouped and associated to the considered CSS contour. Note that sometimes the contour could be cut into two segments by the horizontal edge of the CSS map, and a small horizontal shifting of the CSS map is needed before tracking.

To suppress noise, in the tracking step those CSS contours whose heights are less than 1/6 of the highest peak are not processed. As a result, only the major concavities and convexities of the shape are included in the new representation described in the following section.

3 Parabolical Fitting of CSS Maps

The CSS peaks is used to simplify the arc-shaped CSS contours in the reference algorithm. To exploit more information, we introduce a new method to utilize the CSS map contours. The basic idea is to describe every arc-shaped contour with several parameters, which are the coefficients of a fitting polynomial of the contour. *Polynomial fitting* is the process of constructing a curve, that has the best fit to a series of data points, possibly subject to constraints.

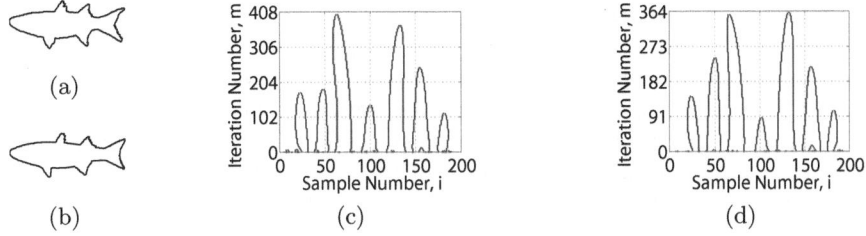

Fig. 2. (a)(b) test shapes; (c)(d) the corresponding CSS maps

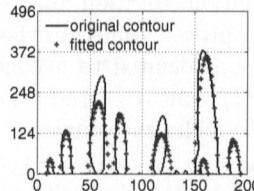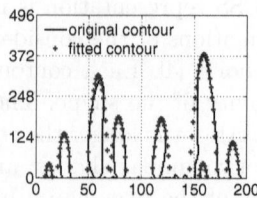

Fig. 3. The effect of the fitting different point sets

3.1 Polynomial Fitting

Polynomial fitting determines the coefficients of a polynomial $p(x)$ of degree n that fits the data y best in a least-squares sense [2], where $p(x)$ is defined as

$$p(x) = \alpha_n x^n + \alpha_{n-1} x^{n-1} + ... + \alpha_1 x + \alpha_0. \tag{2}$$

From Fig. 2, it can be seen that the CSS contour looks like a parabola. We therefore choose parabolic fitting. In other words, every CSS contour is fitted by a binomial (i.e. second order polynomial). Denote the binomial by $p(x) = \alpha_2 x^2 + \alpha_1 x + \alpha_0$. The outcome of the parabolic fitting is the parameter set $\{\alpha_i\}_{i=0,1,2}$. Thus, a shape can be described with several parameter sets finally.

3.2 Choosing Important Points for Fitting

Although the parameter sets can be obtained by a direct polynomial fitting of all points on the original CSS contour, however as it shown in Fig. 3(left) the deviation between the fitted curve and the CSS contour could be serious. To reduce the deviation as much as possible, only points in important location such as the peak of the contour are used to determine the coefficients of the binomial as it shown in Fig. 3(right). It is easily to observe that in Fig. 3(right) the fitted curves can cover more points on the original contours. That is to say the second strategy works much better, and it is selected to be used in our algorithm.

4 Shape Matching

Shape matching is achieved through a comparison of the CSS representation between the query and database shapes. As mentioned in Section 3, each shape in the database is represented by the coefficient sets of the quadratic polynomials. With $\{\alpha_i\}_{i=0,1,2}$, the height and width of the corresponding CSS contour can both be measured. The matching step can be eciently conducted based on them.

4.1 Basic Steps of Shape Similarity Computation

For convenience, in this paper we call the query shape *input* and the database shapes *target*. The matching algorithm is described as follows.

1) Compute the height h and width w of each useful CSS contour on the CSS maps of both the input and the target: $h = 4\alpha_2\alpha_0 - \alpha_1^2/4\alpha_2, w = \sqrt{\alpha_1^2 - 4\alpha_2\alpha_0}/\alpha_2$. where $\{\alpha_i\}_{i=0,1,2}$ is the corresponding parameter set.

2) Seek the corresponding pair. Begin with the highest CSS contour on the input CSS map, we seek the nearest contour to it on the target. The nearest means the straight-line distance between the two peaks of the counters is the lowest. If the horizontal distance between the two peaks is less than 0.4 of the maximum possible distance, then the contour find its corresponding pair, i.e, it is matched, otherwise, it is not.

3) Repeat step 2) from higher contours to lower ones to make sure each contour on the input has found its corresponding pair no matter the contour is matched or not finally.

4) Compute the similarity value, which consists of two parts, the matched and the unmatched. The matched similarity value is the summation of the matching cost divide the number of the matched pairs. The matching cost of each matched pair is defined as

$$cost = \frac{\min(h_i, h_t)}{\max(h_i, h_t)} \cdot \frac{\min(w_i, w_t)}{\max(w_i, w_t)}, \qquad (3)$$

where h_i, h_t, w_i, w_t refer to the height and the weight of the input and the target, respectively. The unmatched part is the summation of the matching cost of the unmatched contours of the input and the target. The matching cost of the unmatched contours is computed by $-h/h_{\max}$, where h, $h_{\max}$ is the height of unmatched contour and the highest contour on the input CSS map, respectively.

The above are the basic steps for shape similarity computation. The final similarity value is compute by a *circular correlation operation* as described below.

4.2 Circular Correlation Operation for Final Shape Similarity Computation

Different start point or a rotation of the shape can cause a circular shift to its CSS map. In the reference algorithm, a shifting of the CSS map before matching is used to compensate this effect. However, the best shifting value is unknown and cannot estimated easily. In our algorithm, a *circular correlation* operation is used to solve the problem.

In view of the basic idea of the circular correlation operation, we define a similar operation to solve the circular shift problem when computing the shape similarity. Let $x(n)$ and $y(n)$ represent parameter series of two CSS maps, $y(n)_N$ is the new series after periodic continuation by periodic N of $y(n)$. Then we left shift $y(n)_N$ by m, and get $y(n+m)_N$. The similarity value of every shift parameter (i.e., m) can be denoted as

$$r_{xy}(m) = x(n) \odot y(n+m)_N, \qquad (4)$$

where '$\odot$' is to compute the shape similarity, as explained in step 2) to step 4) in Section 4.1. We take $m \in [0, 200]$ due to the width of the CSS map, and

Table 1. Effects of using circular correlation operation. The first row is the class number, and the second and third rows show the retrieval efficiency. 'T' represent the average of the retrieval efficiency. 'Y' and 'N' mean with and without using the operation, respectively.

	00	01	02	03	04	05	06	07	08	09	10	11	12	13	14	15	16	T
Y	86	86	52	95	100	88	78	77	50	94	69	86	75	94	100	67	41	79
N	81	70	53	94	100	88	72	77	50	94	69	78	75	89	100	69	42	76
Diff	5	16	-1	1	0	0	6	0	0	0	0	8	0	5	0	-2	-1	3

therefore r_{xy} is a vector that consists of 200 similarity values. The component of r_{xy} most close to 1 is taken to be the final similarity value between the input shape and the target. The effect of the circular correlation operation is shown in Table 1. It is clear that the shape retrieval efficiency can be improved.

4.3   Global Parameters

The size of shape database is usually very big. If we match the input with every database shape, the run-time of the retrieval could be too long. To make the matching more effective, some global parameters need to be employed to reject dissimilar shapes to the input firstly, and then the number of candidates can significantly reduce. In this paper, two parameters e and c are used, which refer to eccentricity and circularity of the shape, respectively [3]. To reject the dissimilar shapes, r_e, r_c are calculated first:

$$r_e = |e_i - e_t| / \max(e_i, e_t), \quad r_c = |c_i - c_t| / \max(c_i, c_t).$$

Note that r_e and r_c range from zero to one. The smaller the values of r_e and r_c, the higher similarity of the shapes would be. We set both the thresholds of r_e and r_c as 0.3 in our experiments. If any one of them is above the relevant threshold, the corresponding shape will be rejected. Note that if the value of threshold is too low, some similar shapes may be missed, otherwise the number of the candidates may not reduce efficiently. By using global parameters, the average of the retrieval efficiency is increased by 17%.

5   Experiments

The performance of our algorithm is evaluated and compared with the reference algorithm in this section. The algorithms are applied on two database. The first database consist of a set of 131 fish divided into 17 classes of 6-8 fish, called the SQUID database, see Fig. 1. The second database is the MPEG-7 shape database (Set B), which contains 70 classes of 20 shapes each, i.e, the size of the database is 1400.

5.1   Performance of the Proposed Algorithm

Fig. 4 is an example of the proposed algorithm without and with using the global parameters. In the example, the test shape is taken from the SQUID

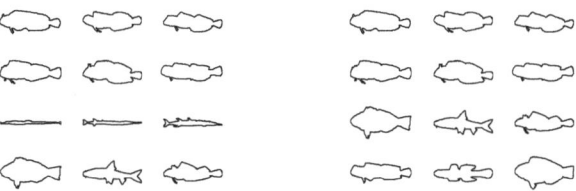

Fig. 4. An example of the proposed algorithm: without (left) and with (right) using the global parameters

database, and the return number k is taken to be 12. With the effect of the global parameters, the dissimilar shapes are rejected effectively. The shapes dissimilar to the input in global appearance which is displayed in Fig. 4(left) (such as the seventh and the eighth) is not in the outcome in Fig. 4(right).

Recall of the retrieval is used as the evaluation standard of the performance, which is defined as the ratio of the number of the output relevant shapes r to the total number m of shapes in the same class. In Fig. 4(left) $Recall = 7/8 = 87.5\%$ and in Fig. 4(right) $Recall = 8/8 = 100\%$. It is obvious the performance in Fig. 4(left) is better than Fig. 4(right) which benefit from the use of the global parameters. *Recall* measures the robustness of the retrieval performance.

5.2 Comparison with Reference Algorithm

Table 2 shows a comparison between the proposed and reference algorithms with statistical results. Every shape in the two databases are input as the query in turn. It should be pointed out that evaluation standard *Recall* of the the reference algorithm without global parameters mentioned in [1] is 76%, which is better than 65% as shown in Table 2. This difference is caused by whether the employed parameters have all been elaborately designed for the used database or not. In this paper we are not attempt to find the best parameters which might

Table 2. A comparison of reference algorithm [3] and the proposed algorithm on the SQUID and MPEG-7 shape databases.

Algorithm	Return Number	Global Parameters	Average%
SQUID Database: 131 shapes, 17 classes			
Reference algorithm	15 shapes	No	65%
	15 shapes	Yes	75%
Proposed algorithm	15 shapes	No	68%
	15 shapes	Yes	79%
MPEG Database: 1400 shapes, 70 classes			
Reference algorithm	40 shapes	No	32%
	40 shapes	Yes	41%
Proposed algorithm	40 shapes	No	47%
	40 shapes	Yes	57%

take a lot of time and are also impractical for common users and common shape databases. From the statistical results in Table 2, with the same parameters the *Recall* of the proposed algorithm is better than the reference one.

6   Conclusion

A shape retrieval algorithm based on the CSS representation and the parabolic fitting technique is proposed. The basic idea is to apply a parabolic fitting to the arc-shaped contours on the CSS map, and represent the corresponding shape by the coefficients of the fitting polynomials. Shape matching is then achieved by comparing the parameter sets of the query shape with those of the database shapes. The retrieval accuracy also benefit from a circular correlation operation. Experimental results show that the proposed algorithm has a better performance than the reference algorithm which is currently being used in MPEG-7.

Acknowledgements. This research was supported in part by the National Natural Science Foundation of China under grants 61075040 and 61033013 (Key Project), the Key Project of Natural Science Foundation of Higher Education of Jiangsu Province under grant 10KJA520047, the Natural Science Foundation of Jiangsu Province under grant BK2012645, and the Soochow Scholars Programme of Soochow University.

References

1. Abbasi, S., Mokhtarian, F., Kittler, J.: Curvature scale space image in shape similarity retrieval. Multimedia Systems 7(6), 467–476 (1999)
2. Fan, J.Q., Gijbels, I.: Data-driven bandwidth selection in local polynomial fitting: variable bandwidth and spatial adaptation. Journal of the Royal Statistical Society. Series B (Methodological) 57(2), 371–394 (1995)
3. Mokhtarian, F., Bober, M.: Curvature scale space representation: theory, applications, and MPEG-7 standardization. Kluwer Academic Publishers, Dordrecht (2003)
4. Mokhtarian, F., Mackworth, A.: Scale-based description and recognition of planar curves and two-dimensional shapes. IEEE Transactions on Pattern Analysis and Machine Intelligence 8(1), 34–43 (1986)
5. Zhang, D.S., Luo, G.J.: A comparative study of curvature scale space and fourier descriptors for shape-based image retrieval. J. Visual Communication and Image Representation 14(1), 39–57 (2003)
6. Zhong, B.J., Ma, K.K.: On the convergence of planar curves under smoothing. IEEE Trans. Image Processing 19(8), 2171–2189 (2010)
7. Zhong, B.J., Ma, K.K., Liao, W.H.: Scale-space behavior of planar-curve corners. IEEE Transactions on Pattern Analysis and Machine Intelligence 31(8), 1517–1524 (2009)

Markerless Tracking Algorithm Based on 3D Model for Augmented Reality System

Peng Yao[1,2], Can Chen[1], and Dongdong Weng[1]

[1] School of Optoelectronics, Beijing Institute of Technology, Beijing, China
[2] Science and Technology on Complex Land Systems Simulation Laboratory, Beijing, China
yp_718@139.com

Abstract. We present a markerless augmented reality(AR) system based on 3D model. First, the feature of environment was extracted using SIFT operator, then the method of stratified reconstruction was used to reconstruct the 3D scene, after that we constructed the database of prior knowledge using the KD-Tree. Finally, we tracked the 3D model based on these prior knowledge via feature matching and pose estimation in real time. Experimental results demonstrated that this method is sufficient for markerless tracking registration. With the prior knowledge, key frame selection problem can be avoided and the running speed is also increased.

1   Introduction

With the rapid development of science and technology, augmented reality, as a branch of virtual reality, is widely used in military, civil, commercial and other fields. One of the key techniques of AR system is camera location technique, by which the 6 Degree of Freedom (DOF) parameters of camera can be obtained. Nowadays, most registration methods are supposed to be used in static environment. The registration technology can be classified into two classes so far: one is based on hardware, while the other based on vision.

According to the differences of scene features extracting method, vision-based method is divided into marker and markerless approaches. For vision-based method, the problem is how to find the correct and stable features to be tracked. Since having enough pre-known information, artificial markers can be tracked stably and also the algorithm is relatively simple [3]. However, for most real applications, it is unrealistic to use markers in the scene, which makes it necessary to study robust markerless algorithm. For markerless tracking, the problem can be simplified if there is special structure such as planar structure in the scene. Literature [5] adopted cumulative calculation. First, it initializes a frame by selecting a planar structure in the scene, then to calculate the position of subsequent frames through the method of planar homography. But it results in cumulative error inevitably. Literature [6] proposed a tracking algorithm based-on key frames. Instead of initializing the first frame, the method initializes many key frames, thus cumulative error can be eliminated effectively. However, planar structure doesn't always exist in reality. So for complicated scene, it is necessary to obtain some structure information as prior, so

J. Yang, F. Fang, and C. Sun (Eds.): IScIDE 2012, LNCS 7751, pp. 751–758, 2013.

that to provide a stable tracking system. Besides, many researchers have proposed tracking approaches based on 3D model. Those approaches can not only eliminate cumulative error, but also result in relatively high precision registration. Unlike corresponding algorithm, which needs to know 3D model of the scene and a few key frames beforehand, the system can overcome the defects of key frame-based algorithm effectively with these prior knowledge.

In the model-based tracking theory, the technical difficulty lies in how to find "markers" constructed using natural features and then to recognize and track them stably. For the human-vision, the more prior information and the preciser data, the better for recognizing and tracking of the scene. However, model itself is the prior information for model-based tracking theory. The model not only contains the geometric information of the scene, but also the organizational form of these information. The model data precision and its organizational form greatly influence complexity, tracking precision and efficiency of the system. Thus it is necessary to study the method to obtain these prior information.

Considering the problems above, our paper proposed the framework of markerless AR registration method based on model. SIFT operator[11] is used first to extract scene features and large amounts of data is used to mark the scene. Then theory of hierarchical reconstruction is adopted to calculate 3D information of scene features. Finally, we use KD-Tree to construct the database of prior information. The adoption of SIFT operator here ensured the stability of features. Compared to [9], this paper used the triangulation method to estimate scene parameters in hierarchical reconstruction, and thus guaranteed the stability of the initial value of the system and the precision of the results. The prior information in this paper only includes descriptors of features and three-dimensional coordinates of points comparing to [7][8][9], which can improve operation speed of the system and reduce the complexity of system calculation, and at the same time also avoid the on-line selection of key frame.

2   Framework Overview

The prior knowledge defined in this paper consists of two parts. One is descriptors of feature points, and the other is 2D and 3D information corresponding to the scene. The basic idea of the reconstruction model-based tracking algorithm is as follows: first to obtain the reconstructed 3D model of object through three-dimensional reconstruction technique so as to build the 3D-2D relationship; then in real-time tracking stage, we get the relationship between current frame feature points and database feature points through matching, thus build 3D structure relationship between current frame and database; finally, the current frame camera position is calculated using optimization algorithm to track the camera. Figure 1 describes the framework of the system.

2.1   Selection of Feature

For AR system, stable features are very important. So in terms of image feature, first it should be easy to extract, and also to distinguish, not to be sensitive to illumination, rotation and scale changes. However, the traditional Harris point do not has all such characteristics. David G. Lowe first proposed the scale space-based local feature descriptor – Scale Invariant Feature Transform (SIFT) in 1999, and it has properties of scale invariance, rotation invariance and even affine transformation invariance. Later on, he summarized and improved the theory in 2004[11][12]. This kind of feature point can maintain a certain invariance in terms of zoom, rotate, scale and affine transformation, and also perspective changes, illumination changes, while maintain a good match in situations of objects' moving, occlusion, noise and other factors. With these properties, feature matching between two images with relatively large differences can be realized.

This paper adopts KD-Tree based BBF algorithm[13] in terms of matching, and meantime uses Ransac robust algorithm[14] in order to reduce false matching rate.

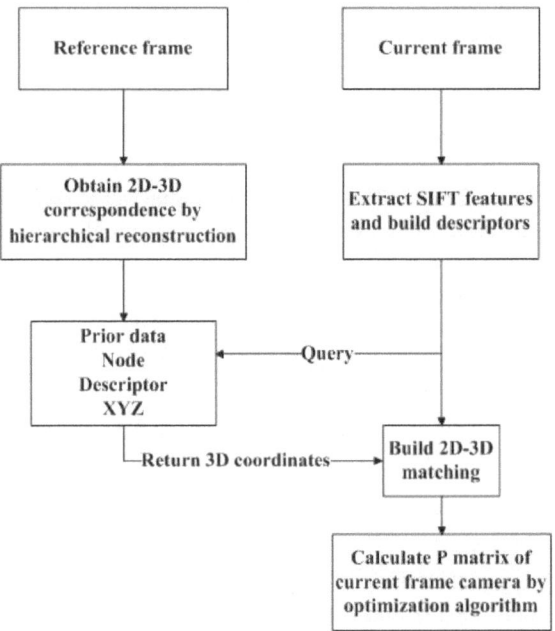

Fig. 1. Framework overview

2.2 Building of 3D Structure Information

By the aforementioned method, the scene features descriptor, as well as part of the scene structure information have already been obtained. In the following, the correspondence between the descriptors and three-dimensional scene will be built through movement structure reconstruction method. The idea of stratified reconstruction[14][16] is used here, what is different is that the camera used in this paper is calibrated, thus the reconstruction process is simplified. 7 reference frames is used here, and the origin of world coordinate system is defined in the first frame camera. The second frame camera position can be obtained using normalized 7 points algorithm[14]. There are measure points $x = PX$, $x' = P'X$ on each image, and these equations can be combined into the $AX = 0$ form. It is linear equations about X. First eliminate homogeneous scalar factor by cross product, so that each image point gives two linearly independent equations. These equations are linear about the components of X, and can build equations with a form of $AX = 0$. Then 3D coordinates of points corresponding to the first two frames are calculated through SVD decomposition.

After the above process, we have already got a preliminary 3D point cloud. However, all possible scene feature points should be covered for prior information. So it is necessary to continue to calculate the 3D points corresponding to following-up frames. For the third frame, for example, 2D-3D correspondence of model and first two frames already exists, so the 2D-3D correspondence between the third frame and natural scene can be obtained through 2D points matching between the third and second frames. Thus the correspondence between feature point m_t of the third frame and its three-dimensional space coordinates M_t is built. Then we can get the rotation and translation matrix of camera using the PNP algorithm or nonlinear optimization algorithm, so as to get the 3D point cloud of the third frame. And so on for later coming frames.

When all frames are calculated, we get the initial value of 3D point cloud structure corresponding to all frames and camera position initial value for each frame. The above process would introduce cumulative error inevitably, to overcome this problem, we use bundle adjustment nonlinear optimization method to optimize the estimation. The bundle adjustment is actually a high-dimensional parameter optimization Levenberg-Marquardt method. Its objective function is:

$$\min_{a_{ij}} \sum_i \sum_j \left\| \omega_j (x_{ij} - \tilde{x}_{ij}) \right\|^2 \tag{1}$$

in which x_{ij} is the feature point extracted from each frame, $\tilde{x}_{ij}$ is its corresponding match point's projection, ω_j is the weight value of each feature point, and a_{ij} is the vector that contains all of the structure from motion variables, including both intrinsic parameters and pose parameters of camera.

2.3   Definition of Prior Information Structure

After the above-mentioned 3D reconstruction process, the scene's prior information has already been obtained actually. For real-time tracking requirements, an appropriate form should be selected to organize and use these priori information. For the purpose of easy query and match, also fast speed, the basic principle is to use as little description as possible to cover more information.

In this paper, we adopts the KD-Tree method to organize prior information. KD-Tree algorithm is an extension of the binary search tree, its each layer divides the space into two, and nodes of the tree are classified by the corresponding dimension. KD-Tree method is relatively effective in low-dimensional situations, while very slow in face of high-dimensional vectors. So BBF(best bin first) algorithm is used here to improve KD-Tree algorithm in two aspects. First, the search nodes number is limited to 200; second, the distances between nodes searched and target points are sorted to generate a priority queue, which makes the nearest neighbor searching easier. The data structure of KD-Tree node is as follow:

Table 1. Node structure

Node
Descriptor (128 Dimension)
X Y Z

The composition method is first to compare all node descriptors, and to find the most-changing dimension by calculating the variance on each dimension of the descriptor. Then feature points are divided into two parts according to this most-changing dimension. Repeat this process until all of the nodes are assigned to the leaf nodes of binary tree. Then we'll have a binary tree of depth $d = log_2 N$.

In real-time tracking stage, first the SIFT feature of the current frame is extracted, and then based on the descriptors, use the BBF algorithm to search the match point and obtain its corresponding X, Y, Z coordinates, thus establishing the 2D-3D correspondence. Since 3D points number is much less than that of 2D points in key frame, the method in this paper reduces the amount of data storage, and avoids a lot of data query work.

2.4   Real-Time Tracking

After the real-time image is captured, the scene marking information is queried first, then feature extraction of the current frame is guided with the returning marking information. After this, we'll get 2D-3D matching through data searching in KD-Tree. Then a Ransac will be done to further optimize the match to get the initial value of P matrix. Lastly, the final solution is obtained using LM algorithm to solve the following objective function:

$$\min_{P_t} \sum_j \left\| w_{tj}(x_{tj} - \mathcal{K}_{tj}) \right\|^2 \qquad (2)$$

which depicts the sum of 2D-3D re-projection error of each current frame at time t.

3   Experiment

3.1   Feature Extraction

As we already know, in markerless tracking based on model, the system mismatch has a great influence on the pose calculation. Especially for complex structure outdoor scenes, robust methods such as RANSAC and M estimate[15] can only remove most of the mismatch. For some very similar points, the above robust methods can not help either, as illustrated in figure 2. After using Ransac to remove some of the outliers, we also artificially remove the unstable match points in Dashuifa scene of Yuanming Yuan, so as to ensure the reconstruction accuracy in next step.

Fig. 2. Mismatch of the similarities

3.2   Model Calculation

In off-line stage, the selected input image and video should cover as large as possible a scene so that the tracking area is large enough. Meantime, the image baseline should not be too broad and the image should be clear, which helps to obtain a relatively precise 3D structure. 14 image frames covering the vision of the Yuanming Yuan Dashuifa from left to right were taken as reference images, as shown in figure 3. Then the model of Yuanming Yuan was built via the hierarchical reconstruction algorithm described in this paper, and its result is shown in figure 4, in which the black point cloud with 892 points represents 3D point position corresponding to recovered scene features, and the blue cones represent recovered camera pose and position. The re-projection error of each frame is less than 0.6 pixel, with an average value of 0.438 pixel.

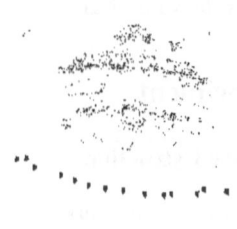

Fig. 3. Reference images   **Fig. 4.** Recovered scene structure

Fig. 5. First frame camera to seventh frame camera respectively, from right to left

Fig. 6. Augmented Yuanming Yuan

3.3   Tracking Result

With the scene 3D model obtained, prior model is built first using the method in 2.3, then real-time tracking is conducted using the method in 2.4. Figure 5 plots real-time camera position and orientation (red) and ground-truth (blue) for comparison. Figure 6 is operation result of the system.

4   Conclusion

In this paper, we developed a markerless tracking system based on model. First, the feature of environment was extracted using SIFT operator, then the method of stratified reconstruction was used to reconstruct the 3D scene, after that we constructed the database of prior knowledge using the KD-Tree. Finally, we tracked the 3D model based on these prior knowledge via feature matching and pose estimation in real time. Experimental results demonstrated that this method is sufficient for markerless tracking registration, and with the prior knowledge, key frame selection problem can be avoided effectively.

Due to the complexly-structured objects, the system has to handle large scale data, which results in lower frame rate. In future work, we'll use GPU for the extraction of SIFT feature, which can greatly reduce the computational time. At the same time, SIFT extraction can be implemented in multi-core machines with parallel computing.

Acknowledgement. The research work was supported by National Natural Science Foundation of China (No. 60903069).

References

1. Azuma, R., et al.: Recent Advances in Augmented Reality. IEEE Computer Graphics and Applications 12(11), 34–47 (2001)
2. Azuma, R.: A Survey of Augmented Reality. Teleoperators and Virtual Environments 4(6), 355–385 (1997)
3. ARToolKit, http://www.hitl.washington.edu/research/ sharedspace/download/
4. Lepetit, V., Fua, P.: Monocular Model-Based 3D Tracking of Rigid Objects: A Survey. Foundations and Trends in Computer Graphics and Vision 1(1), 1–89 (2005)
5. Simon, G., Berger, M.-O.: Pose estimation from planar structures. Computer Graphics and Applications 22, 46–53 (2002)
6. Chia, K., Cheok, A., Prince, S.: Online 6DOF Augmented Reality Registration from Natural Features. In: Proc. ISMAR 2002 (2002)
7. Klein, G., Drummond, T.: Robust Visual Tracking for Non-Instrumented Augmented Reality. In: Proc. Second IEEE and ACM International Symposium on Mixed and Augmented Reality (ISMAR 2003), Tokyo (2003)
8. Chen, J., Wang, Y., Li, Y., Hu, W., Zang, X.: Real-time Augmented Reality Registration Algorithm Based on Natural Feature Points. Journal of System Simulation 19 (November 2007)
9. Gordon, I., Lowe, D.G.: What and Where: 3D Object Recognition with Accurate Pose. In: Ponce, J., Hebert, M., Schmid, C., Zisserman, A. (eds.) Toward Category-Level Object Recognition. LNCS, vol. 4170, pp. 67–82. Springer, Heidelberg (2006)
10. Vacchetti, L., Lepetit, V., Fua, P.: Stable Real-time 3D Tracking Using Online and Offline Information. IEEE Transactions on Pattern Analysis and Machine Intelligence 26(10), 1385–1391 (2004)
11. Lowe, D.G.: Object recognition from local scale-invariant features. In: Proc.of the International Conference on Computer Vision ICCV, Corfu, pp. 1150–1157 (1999)
12. Lowe, D.G.: Distinctive image features from scale-invariant keypoints. International Journal of Computer Vision 2(60), 91–110 (2004)
13. Beis, J.S., Lowe, D.G.: Shape indexing using approximate nearest-neighbour search in high-dimensional spaces. In: Proceedings CVPR 1997, San Juan, Puerto Rico, pp. 1000–1006 (June 1997)
14. Hartley, R., Zisserman, A.: Multiple View Geometry in Computer Vision. Cambridge University Press (2003)
15. Hampel, F., et al.: Robust Statistics: The Approach Based on Influence Functions. John Wiley, s.l (1986)
16. Pollefeys, M.: Self-calibration and metric 3D reconstruction from uncalibrated image sequences. Ph.D thesis, Katholieke Universiteit Leuven (1999)

An Effective Lane Detection Algorithm for Structured Road in Urban

Weirong Liu[1,2] and Shutao Li[1]

[1] College of Electrical and Information Engineering,
Hunan University, 410082 Changsha, P.R. China
[2] Lanzhou University of Technology, 730050 Lanzhou, P.R. China
liu_weirong@163.com, shutao_li@yahoo.com.cn

Abstract. An effective and robust algorithm for structured road in urban is proposed in this paper. Firstly, the adaptive segmentation method is used to determine reasonable Region of Interest (ROI) that covers all candidate lane markings. Secondly, line segments are extracted by Line Segment Detector (LSD) and be used as low level feature to extract the structural information of road scene. Thirdly, non-lane markings are eliminated by clustering on orientation information and vanishing point. Finally, the lanes are extracted from the remaining candidate lane markings. Experimental results on Caltech lane datasets show that the algorithm can extract lanes in complex structured road scenarios.

Keywords: Lane detection, Line segment detector, K-means clustering.

1   Introduction

Lane detection is used to extract lane markings from clutter road scenes and filter them to produce a credible estimation of the lanes position, which plays a significant role in driver assistance systems [1].

Many approaches have been proposed for detection of road lane markings over the last decades [2]. Accompanied by the development of computer vision technology, vision-based lane detection and recognition algorithms have been an area of active research. However, it is still a challenge problem due to variations of lane markings appearance, illumination conditions, presence of shadows and other image artifacts.

A three-step process for lane detection based on visual cameras was summarized in [3]. The key step is a robust extraction of road features to initialize candidate lane markings and post-processing operation on the initialize candidate lane markings to remove outliers.

In most cases, lane markings on structured roads appear as well-defined straight lines or as curves which can be approximated by some short straight lines [4]. Therefore the commonly used methods for lane markings extraction are the Hough Transform (HT) [1, 2] and its variations, such as the randomized HT [5], hierarchical additive HT [4]. However, there are two major shortcomings of HT for lane markings extraction. Firstly, high computational cost is caused by the multiplications and trigonometric operations applied to each pixel in image. Secondly, high false detection rate result from ignoring orientation information and fixed thresholds.

J. Yang, F. Fang, and C. Sun (Eds.): IScIDE 2012, LNCS 7751, pp. 759–767, 2013.
© Springer-Verlag Berlin Heidelberg 2013

In order to improve validity of lane markings extraction and reduce the computation complexity, we propose an effective and robust algorithm for lane detection in this paper. Different from previous methods of image intensity or gradient, we use line segments as low-level features to analyze and detect lane markings on structured roads in urban. An efficient Line Segment Detector (LSD) presented in [6] can control false detection rate and require no parameter tuning. Therefore, lane markings on structured roads can be approximated by some line segments from LSD. Due to line segments as low-level features, we do not use any lane model in lane detection. As a result, prior knowledge of the scene constrains are not necessary except for orientation information.

The rest of this paper is organized as follows. Section 2 presents the effective lane detection algorithm. Experimental results with real world data sets are provided in Section 3 and Section 4 concludes this paper.

2   Lane Detection Algorithm

2.1   Extraction of ROI

As common scene, all candidate lane markings lie in the middle of the image captured from camera mounted on vehicle. Based on some priors, such as the parameters of camera mounted on vehicle, it is easy to roughly locate the position of Region of Interest (ROI) which covers all candidate lane markings. A rough ROI marked by the red rectangle box is shown in Fig. 1(a). The upper boundary of ROI is determined by fixed ratio and the lower boundary of ROI is determined by position of the front of vehicle.

Because scene dynamic varies with vehicle moving, a fixed ratio or position segmentation method can't precisely capture upper boundary. We analysis the mean of each row in various scenes as shown in Fig. 1(b) and proposed an adaptive ROI location algorithm as follow.

Let $\mathbf{F}_c \in \mathbb{R}^{m \times n}$ be an image captured from camera, where m is the number of rows in $\mathbf{F}_c$ and n is the number of columns in $\mathbf{F}_c$.

Firstly, the rough ROI $\mathbf{F} \in \mathbb{R}^{m_r \times n}$ is determined by some priors, where m_r is the number of rows of $\mathbf{F}$.

Secondly, the mean value $\overline{\mathbf{F}}_i$ of each row in $\mathbf{F}$ can be represented as (1)

$$\overline{\mathbf{F}}_i = \sum_{j=1}^{n} \mathbf{F}_{ij} / n , (i = 1, \cdots, m_r) . \tag{1}$$

and then row value d with minimal mean value can be calculated by (2)

$$d = \mathrm{E}(\min(\overline{\mathbf{F}}_i)) , (i = 1, \cdots, m_r) . \tag{2}$$

where the operator E is used to extract the row value with minimal mean value.

Finally, the reasonable ROI $\mathbf{F}_d \in \mathbb{R}^{(m_r - d + 1) \times n}$ extracted by (3)

$$\mathbf{F}_d = \mathbf{F}_{ij} , (i = d, \cdots, m_r , j = 1, \cdots, n) . \tag{3}$$

and the reasonable ROI $\mathbf{F}_d$ is shown by rectangle area within red box in Fig. 1(c).

As shown in Fig. 1, we can conclude that the ROI in Fig. 1(c) extracted by adaptive ROI segmentation algorithm is more accurate than the rough ROI in Fig. 1(a) extracted by some priors. It is obvious that a more accurate ROI can save computation times and reduce interference from sky, buildings and trees etc.

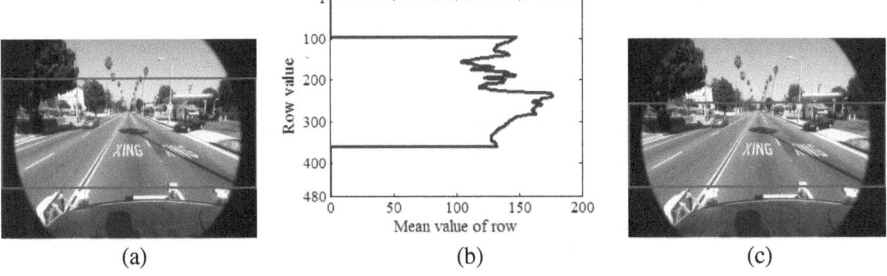

(a)   (b)   (c)

Fig. 1. ROI extraction. (a) the rough ROI extracted by some priors, (b) the mean value of each row in rough ROI, (c) the ROI extracted by adaptive ROI segmentation algorithm

2.2   Candidate Lane Markings Extraction Using LSD

Line segments can be used as low level feature to extract the structural information of images, and can help in several problems in image analysis and computer vision [6]. Due to high computational complexity and false detection, those HT methods and its variations are not adequate in many scenarios with real time requirement.

Aiming to overcome the disadvantages of traditional line segment detection methods, a good linear-time line segment detector was proposed in [6]. The LSD algorithm can gives accurate results and requires no parameter tuning. The key steps of the LSD are illustrated in Fig. 2.

The candidate lane markings extraction using LSD algorithm in ROI F_d are described as follow.

Let S define a set of line segments extracted from ROI F_d by (4),

$$S = LSD(F_d) \ . \tag{4}$$

where the operator LSD is the lines extraction algorithm as shown in Fig. 2, and $S=\{s_1, s_2, \cdots, s_k\}$ is the line segments extracted from ROI F_d.

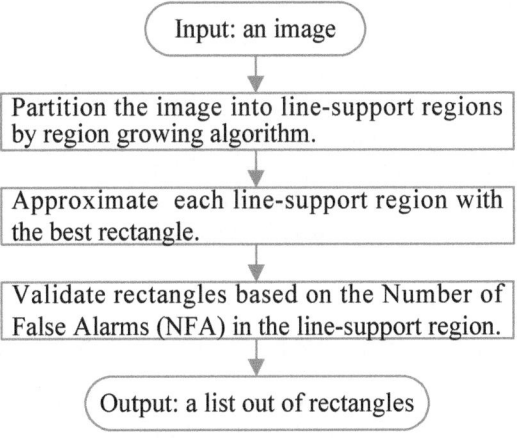

Fig. 2. The key steps of the LSD

Each line segment s_i, $(i = 1, \cdots, k)$ is defined as

$$s_i = \{x_i^1, y_i^1, x_i^2, y_i^2, \theta_i\}, \quad (i = 1, \cdots, k). \tag{5}$$

where (x_i^1, y_i^1) and (x_i^2, y_i^2) are the starting point coordinate and ending point coordinate of line segment s_i respectively. $\theta_i \in [-90°, 90°]$ is the angle of line segment s_i and calculate by (6).

$$\theta_i = \frac{180}{\pi} \times \arctan(\frac{y_i^2 - y_i^1}{x_i^2 - x_i^1}), \quad (i = 1, \cdots, k). \tag{6}$$

We use the images on different scenario in the first column of Fig. 3 to test the LSD algorithms. The images in first column show the ROI extracted by adaptive ROI segmentation algorithm. There are different types of lanes in ROI such as, straight lanes, curve lanes and dash lanes under the circumstance with shadows, writing, road marks and patches etc. The images in the second and third column show the line segments extracted by HT and LSD, respectively.

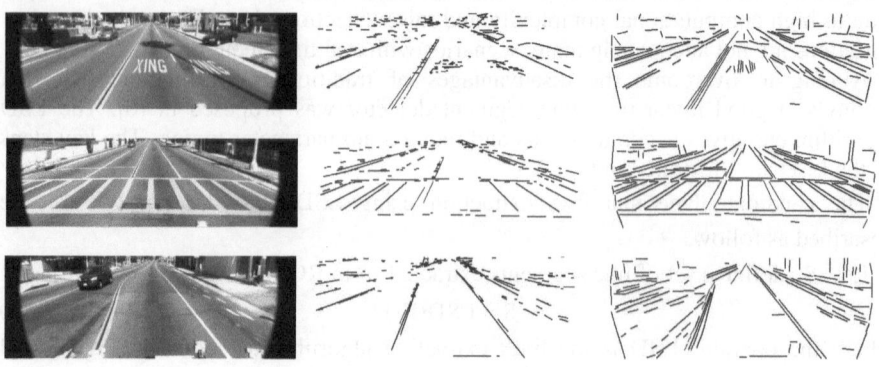

Fig. 3. A comparison of HT and LSD method on line segments detection. The images in first column are the ROI, and the images in the second and third column show the results extracted by HT and LSD, respectively.

We can see that firstly the almost full boundaries of different lanes in the ROI are detected by LSD while boundary detected by HT is poor. Secondly, more obvious traces on the road are captured by LSD while missed by HT. Hence the line segments are found by LSD algorithm are greater and more accurate than line segments extracted by HT method, and the detected line segments represent well the geometric structure of ROI F_d, such as candidate lane markings, other road marks and cluttered shadows.

2.3   Non-lane Markings Removed by Orientation Information

Because vehicles usually move on road in relative fixed lanes following the traffic rules, lane markings lie on limited direction in captured image. We find that a range of 45° is enough to cover almost lanes in captured image as shown in Fig. 3. In order to remove irrelevant line segments that represent non-lane markings, we retain the lines that have an orientation angel between $-25° \sim -70°$ in the left of ROI and between $25° \sim 70°$ in the right of ROI respectively, i.e.

$$S^L = \{s_i^L \mid x_i^1 < \frac{n}{2}, \ x_i^2 < \frac{n}{2}, \ \theta_i \in [-25°, -70°]\}, \quad (i = 1, \cdots, p) \ . \tag{7}$$

$$S^R = \{s_i^R \mid x_i^1 \geq \frac{n}{2}, \ x_i^2 \geq \frac{n}{2}, \ \theta_i \in [25°, 70°]\}, \quad (i = 1, \cdots, q) \ . \tag{8}$$

The retained line segments after non-lane markings removed by (7) and (8) are shown in Fig. 4 (a). By comparing the first row third column image in Fig. 3 with image in Fig. 4 (a), we can see that lots of non-lane markings are removed. The candidate lane marks can be more effective cover by the green lines S^L and the red lines S^R.

Furthermore, we find that the retained green lines S^L on left can be roughly divided into two categories based on direction. Therefore, we adopt the k-means algorithm to partition the retained lines S^L observations into two clusters. The retained red lines S^R on right can be separated by the same method.

Given a set of line segments $S^L = \{s_1^L, s_2^L, \cdots, s_p^L\}$ (or $S^R = \{s_1^R, s_2^R, \cdots, s_q^R\}$), k-means clustering aims to partition the p (or q) line segments into two sets $S^L = \{S_1^L, S_2^L\}$ (or $S^R = \{S_1^R, S_2^R\}$) so as to minimize the within-cluster sum of squares, i.e.

$$\arg\min_{S^L} \sum_{i=1}^{2} \sum_{s_j^L \in S_i^L} \| s_j^L - \mu_i^L \|^2, \quad (j = 1, \cdots, p) . \tag{9}$$

$$\arg\min_{S^R} \sum_{i=1}^{2} \sum_{s_j^R \in S_i^R} \| s_j^R - \mu_i^R \|^2, \quad (j = 1, \cdots, q) . \tag{10}$$

where μ_i^L, $(i = 1, 2)$ is the mean of points in S_i^L, $(i = 1, 2)$, and μ_i^R, $(i = 1, 2)$ is the mean of points in S_i^R, $(i = 1, 2)$.

Since the lanes close to vehicle are more useful than the lanes relative far from vehicle, the results from (9) and (10) which have larger angel are reserved. As shown in Fig. 4(b), the blue line segments reserved by k-means clustering algorithm can more accurately represent the lanes.

In some special cases, there are still some line segments from shadow, irregular patches etc. in reserved blue line segments set. These line segments may lie in any direction, while the lane marks always lie in relative fixed orientation and intersect at vanishing point in bird's eye view. For this reason, we can use vanishing point to eliminate the line segments indicating those interference as follows.

Let S_K define the reserved line segments through k-means algorithm and $V = (x_v^1, y_v^1, x_v^2, y_v^2)$ define the vanishing point calculated by

$$V = G(S_K) \ . \tag{11}$$

where the operator G extrapolate S_C across the whole ROI F_d, and select the point with maximum intersection number.

The final line segments S_V^L (or S_V^R) in the left (or right) of F_d are retained by

$$S_V^L = \{s_{V_i}^L \mid \parallel s_{V_i}^L - V \parallel < \varepsilon\}, \quad (i = 1, \cdots, u) \quad . \tag{12}$$

$$S_V^R = \{s_{V_i}^R \mid \parallel s_{V_i}^R - V \parallel < \varepsilon\}, \quad (i = 1, \cdots, w) \quad . \tag{13}$$

where ε is the tolerance between line segment and vanishing point, and it is set to 10 in experiments.

In Fig. 4 (c), the selected vanishing point V is marked with the red '+' and the reserved lines S_V^L and S_V^R are plotted in purple. It can be conclude that the remaining purple line segments are almost the part of lane markings.

(a)   (b)   (c)

Fig. 4. Non-lane markings removed by orientation information, (a) non-lane markings removed by (7) and (8), (b) non-lane markings removed by k-means clustering, (c) non-lane markings removed by vanish point.

2.4   Lane Extraction

After eliminating the non-lane markings by angles and position, the final step is extraction the left L_L (or right L_R) from candidate line segments remaining S_V^L (or S_V^L) following

$$L_L = \{L_L \mid \min(x_{V_i}^{L_1}), \max(y_{V_i}^{L_1}), \max(x_{V_i}^{L_2}), \min(y_{V_i}^{L_2}), \text{mean}(\theta_{V_i}^L)\}, \quad (i = 1, \cdots, u) \quad . \tag{14}$$

$$L_R = \{L_R \mid \max(x_{V_i}^{R_1}), \max(y_{V_i}^{R_1}), \min(x_{V_i}^{R_2}), \min(y_{V_i}^{R_2}), \text{mean}(\theta_{V_i}^R)\}, \quad (i = 1, \cdots, w) \quad . \tag{15}$$

where $\min(x_\bullet^\bullet)$ (or $\max(x_\bullet^\bullet)$) is the minimum (or maximum) of horizontal coordinate in all candidate lines $S_V^{(\bullet)}$, $\max(y_{(\bullet)}^{(\bullet)})$ (or $\max(y_{(\bullet)}^{(\bullet)})$) is the minimum (or maximum) of vertical coordinate in all candidate lines $S_V^{(\bullet)}$ and $\text{mean}(\theta_{(\bullet)}^{(\bullet)})$ is the mean of angle in all candidate lines $S_V^{(\bullet)}$.

The final detected left lanes L_L and right lanes L_R are painted in green and red respectively. For the convenience of validation results, the final detected lanes are superimposed on the original image as shown in Fig. 5.

3   Experiments

3.1   Data Sets

In order to evaluate the performance of the proposed method, experiments were performed on sequences with different types of urban streets from Caltech lane datasets in [7]. These data sets include four clips with size of 640×480, totaling 1225 color frames of road, and these clips captured in different scenarios are challenging, such as a lot of curvatures, some writings, different pavement types, lots of shadows and passing cars.

3.2   Results

We have completed the proposed algorithm in MATLAB R2010b and tested it on Caltech lane datasets. The experiments are implemented on Intel Core 2 Duo CPU T9400 at 2.53GHz with 3GB memory. The detected lanes superimposed on the original image with different scenarios are shown in Fig. 5. Because there is no ground truth of detection on Caltech lane datasets, we qualitatively evaluate performance of the proposed algorithm.

From left to right in Fig. 5, four columns of images come from washington1 clip, washington2 clip, cordova1 clip and cordova2 clip in Caltech lane datasets respectively. The results in first row show that the proposed method can accurately and completely detect left (highlighted in green) and right (highlighted in red) straight lane under the circumstance with no or a little interference. It can be seen from the images in the second row that the characters painted on road and irregular patches do not affect the performance of lane detection. Even under scenarios of the messy shadows and moving vehicles as shown in third row, the lanes are exactly extracted by the proposed method. Finally, the results in last row prove that the method is robust to the common road marks such as zebra stripes, stop lines and arrows.

Fig. 5. Examples of the detected lanes

The results for all clips in Caltech lane datasets are shown in Table 1. The first row shows the numbers of each clips, and the second row to fourth row show the correct detection rate, false detection rate and missing detection rate, respectively. It is should be noted that the average computation time on testing platform mentioned above is under 0.4s/frame.

Hence, the results in all cases shown in Fig. 5 and Table 1 clearly demonstrate the potential and practical utility of the proposed method for lane detection under different scenarios.

Table 1. The results for all clips in Caltech lane datasets

Clips	Washington1	Washington2	Cordova1	Cordova2
Numbers of frames	337	232	250	406
Correct detection rate	90.80%	93.50%	94.00%	94.30%
False detection rate	5.60%	4.80%	4.48%	3.80%
Missing detection rate	3.60%	1.70%	1.52%	1.90%

Because the proposed method adopts line segments as low-level and does not use any lane model, the detection results of curving lane are incomplete or false as shown in Fig. 6(a) and Fig. 6(b). For the scenarios with few lanes as shown in Fig.6 (c) or the sun facing the vehicle as shown in Fig.6 (d), the proposed method misses detection of all lanes.

| (a) | (b) | (c) | (d) |

Fig. 6. False detections and missing detections of lanes. (a) incomplete detection of curving lane, (b) false detection of curving lane, (c) missing detection of very short lanes, (d) missed detection of lanes with very high illumination.

4   Conclusion

In this paper, we have introduced an effective lane detection algorithm for structured road in urban. The algorithm has been tested on Caltech lane datasets and the experimental results show that the algorithm can extract lanes in complex traffic scenarios. The major advantages of the proposed algorithm are that lane markings on structured roads can be precisely represented by some line segments and prior knowledge of the scene constrains are not necessary except for orientation information.

Acknowledgement. This paper is supported by the National Natural Science Foundation of China (No. 61172161).

References

1. Borkar, A., Hayes, M., Smith, M.T.: A Novel Lane Detection System with Efficient Ground Truth Generation. J. IEEE T. Intell. Transp. 13, 365–374 (2012)
2. Gopalan, R., Hong, T., Shneier, M., et al.: A Learning Approach Towards Detection and Tracking of Lane Markings. J. IEEE T. Intell. Transp. (to appear, 2012)
3. McCall, J.C., Trivedi, M.M.: Video-Based Lane Estimation and Tracking for Driver Assistance: Survey, System, and Evaluation. J. IEEE T. Intell. Transp. 7, 20–37 (2006)
4. Satzoda, R.K., Sathyanarayana, S., Srikanthan, T.: Hierarchical Additive Hough Transform for Lane Detection. J. IEEE Embedded Systems Lett. 2, 23–26 (2010)
5. Xu, L., Oja, E., Kultanen, P.: A New Curve Detection Method: Randomized Hough Transform (RHT). J. Pattern Recognit. Lett. 11, 33–338 (1990)
6. Von Gioi, R.G., Jakubowicz, J., Morel, J.M., et al.: LSD: A Fast Line Segment Detector with a False Detection Control. J. IEEE T. Pattern. Anal. 32, 722–732 (2010)
7. Aly, M.: Real Time Detection of Lane Markers in Urban Streets. In: IEEE Intelligent Vehicles Symposium, pp. 7–12. IEEE Press, New York (2008)

A Scale-Space-Based System
for Time Series Prediction

Xiaoyong Hou and Baojiang Zhong*

School of Computer Science and Technology,
Soochow University, Suzhou215006, China
{20104227023,bjzhong}@suda.edu.cn

Abstract. Time series prediction is significant for various economical and financial applications. In this research the scale-space concept in computer vision is employed to construct a novel prediction system. The main idea is to process the time series data by wavelet and obtain data under different scales. Then the corresponding data sequence is transformed into a series of windows at every scale. Windows the most similar to the reference one are searched, with which the prediction value is computed. Finally, simulation results are presented.

Keywords: Time series prediction, Scale-space concept, Wavelet, Data sequence, Windows.

1 Introduction

Now time series prediction for various economical and financial applications (such as stock price prediction) has been considered one of the most challenging tasks. Due to the nonlinearity and volatility of the time series data in practice, traditional prediction method has high error rate. However, AI methods suffer a major deficiency in which they are slow, although they produce highly accurate forecasting results for volatile time series data [2].

The general character for economical and financial time series is that there are certain recurring patterns in its history data [7]. Based on this, In this paper a scale-space-based hybrid system integrating wavelet and curvature representing window appearance (CRWA) technology is proposed. The scale-space concept in computer vision has been widely used to represent images and describe shapes, and it is employed in the system in the present paper. We can not only obtain the time series data under different scales, but also remove the noise of the data by wavelet. The CRWA technology will be applied to retrieve windows possessing similar shapes with reference one under each scale.

The rest of this paper is organized as follows. Section 2 describes scale space theory, wavelet and their differences at multi-scale. Section 3 gives a detail account of the proposed method. Section 4 applies the proposed method to forecast stock price. Finally, conclusions are presented in Section 5.

---

* Corresponding author.

J. Yang, F. Fang, and C. Sun (Eds.): IScIDE 2012, LNCS 7751, pp. 768–775, 2013.

2   Theoretical Basis

2.1   Scale Space Theory

The scale-space concept in computer vision indicates that we can not foresee the structure of an object at only one scale. It is necessary to describe the structure under multiple scales in order to analyze the essential characteristic of the object. In view of this point, in the time series prediction a few typical situations are selected to represent the data. As mentioned by Tony Lindeberg [6], if someone aims at describing the deep structure of an object, a multi-resolution representation is most important. In this sense, a prediction system which can systematically removes high-frequency information from time series data and describe its deep information would be helpful and interesting.

2.2   Wavelet for Preprocessing Data

The Gaussian function is the only linear kernel to realize scale transform at scale-space [4]. In the research, wavelet is applied as kernel function instead of Gaussian kernel. In other word, data at different levels of scale are obtained by wavelet threshold de-noising [3]. The reason why we make such a choice can be explained by a comparison between Fig.1(b) and Fig.1(c). Fig.1(b) shows a section stock data processed by Gaussian filtering. Fig.1(c) shows the same data processed by wavelet de-noising. It can be found that the data series by wavelet filtering can retain main extremes with only a slight change of their positions. However, the data series processed by Gaussian shrink and some main extremes are missed. As a result, shape matching will be hard to achieve in the range. Therefore, in the paper the wavelet technique is used to obtain the data at different scales, while the scale-space concept is employed for shape matching.

   In this research, Haar wavelet is used as the major wavelet transform tool. According to [5], the sparse of signals we have in mind for applications suggests the use of Haar wavelet to represent underlying information. In general, Haar wavelet provides quickly processing time without losing much in performance than other wavelet bases [8].

   According to normalization rules, there are two types of wavelets, father wavelet and mother wavelet. Father wavelets are used for the lowest frequency. However, mother wavelets are used for the higher frequency. In practice, wavelet scaling factor and translation parameter are usually discretized by $a = 2^i$ and $b = ja$, where $i = \{1, 2, 3, ...\}$ and $j = \{0, 1, 2, ...\}$. When $\int \phi(t)dt = 1$ and $\int \psi(t)dt = 0$, they are written respectively by

$$\phi_{i,j} = 2^{-1/2}\phi(\frac{t - 2^i j}{2^i}), \tag{1}$$

$$\psi_{i,j} = 2^{-1/2}\psi(\frac{t - 2^i j}{2^i}). \tag{2}$$

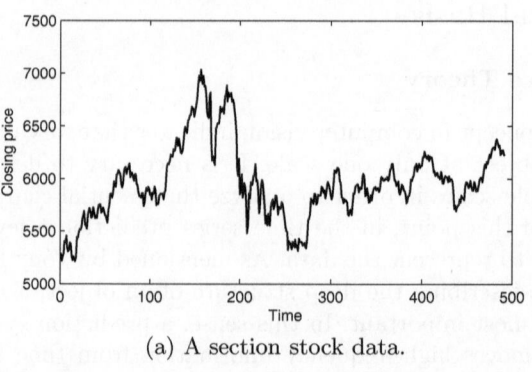

(a) A section stock data.

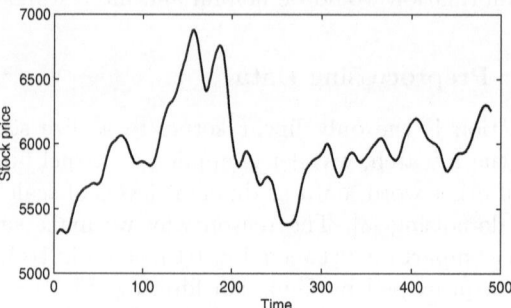

(b) The section stock data by Gaussian filtering, where Iterations are 200.

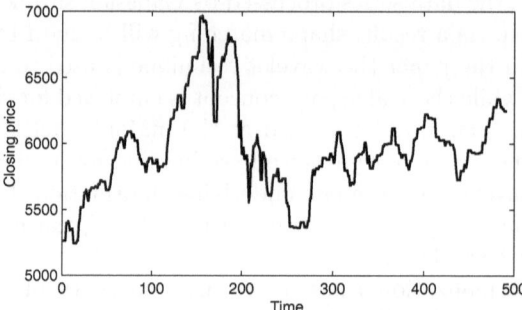

(c) the same stock data by wavelet de-noising, where the level values are taken to be 2.

Fig. 1. The difference between Gaussian filtering and wavelet de-noising

For convenience, a dyadic expansion can be used to represent the $f(t)$. The coefficients in the expansion are given as follows.

$$S_{A,j} = \int f(t)\phi_{A,j}(t)dt,$$

$$D_{i,j} = \int f(t)\psi_{i,j}(t)dt, \tag{3}$$

where A is the maximum of i, representing the maximum scale sustainable by the number of data points. The representation of the signal $f(t)$ can now be given by

$$f(t) = \sum_j S_{A,j}\phi_{A,j}(t) + \sum_j D_{A,j}\psi_{A,j}(t)$$

$$+ ... + \sum_j D_{1,j}\psi_{1,j}(t). \tag{4}$$

The corresponding multi-resolution decomposition of $f(t)$ is given by

$$\{S_A, D_A, D_{A-1}, ..., D_1\},$$

when j is constant. In this way, differentiated data for different scales could be obtained by regulating decomposition levels and how much high-frequency information is removed (namely wavelet threshold de-noising).

2.3 Performance Measure

To better valuate the prediction results, MAPE is used as a performance measure. It can be calculated as follows.

$$MAPE = \frac{1}{n}\sum_{t=1}^{n}\frac{|X_t - F_t|}{X_t} * 100, \tag{5}$$

where X_t is the true value and F_t is the predicted value at time t. The $MAPE$ is an average over n test sets.

3 CRWA Method

The idea of more flexible CRWA technology mainly derives from the k nearest neighbor algorithm (KNN) [1]. For the simpleness of KNN, we no longer describe it here.

To introduce the proposed prediction system, a few terms have to be defined firstly. Assume L is the sliding window size, which means there are L data points in each window and H is designated as the number of the most similar windows. Assign that the final L data points in the history data series are reference window. Rest of the data series generate numerous windows by sliding window one by one. And these windows are defined as candidate window whose size equals with reference window. Curvatures at every data points of the reference windows and candidate windows produce curvature reference and curvature candidate windows. The step-by-step procedure of the CRWA method is explained in details in the following.

Step 1) Use data at given scale as analysis data. Suppose we need to predict number i data from index number $t + 1$, i.e., number $\hat{Y}_{t+1}^i$ value.

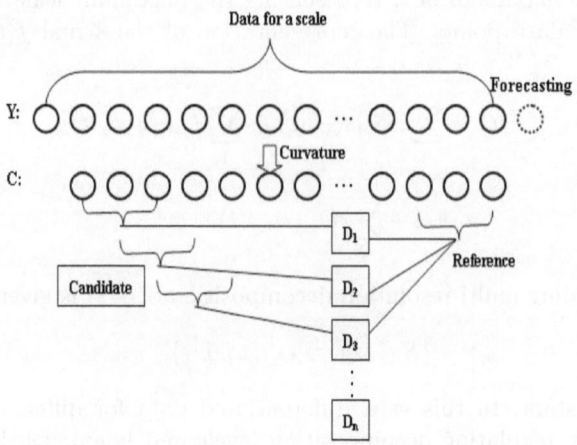

Fig. 2. CRWA forecasting with window size $L = 5$

Step 2) Use number $t - L + 1$ to t data for reference window. By sliding the window, the curvature C_j^i is computed one by one at every point. The curvature at window endpoints is ignored. C_j^i is defined by

$$C_j^i = \frac{\Delta X_j^i \Delta^2 Y_j^i - \Delta^2 X_j^i \Delta Y_j^i}{[(\Delta X_j^i)^2 + (\Delta Y_j^i)^2]^{1.5}}, \tag{6}$$

where $\quad j = 2, 3, ..., t - L - 1, \quad X^i = 1, 2, ..., t.$

$$\Delta X_j^i = (X_{j+1}^i - X_{j-1}^i)/2 = 1,$$

$$\Delta Y_j^i = (Y_{j+1}^i - Y_{j-1}^i)/2,$$

$$\Delta^2 X_j^i = (\Delta X_{j+1}^i - \Delta X_{j-1}^i)/2 = 0,$$

$$\Delta^2 Y_j^i = (\Delta Y_{j+1}^i - \Delta Y_{j-1}^i)/2,$$

thus

$$C_j^i = \frac{\Delta^2 Y_j^i}{[(1 + (\Delta Y_j^i)^2]^{1.5}}. \tag{7}$$

Step 3) Similarly, by sliding the window, the curvature candidate from 1 to $t - L$ is compared one by one with the curvature reference, and the Euclidean distance is computed from every interval D_j^i. The significant value U_j^i and F_j^i are then evaluated for prediction. D_j^i, U_j^i and F_j^i is defined by

$$D_j^i = \sqrt{\sum_{l=2}^{L-1} (C_{t-L+l}^i - C_{l+j-1}^i)^2}, \tag{8}$$

$$U_j^i = Y_{j+L-1}^i, \tag{9}$$

and

$$F_j^i = Y_{j+L}^i. \tag{10}$$

Step 4) Find the Hth smallest number in the set D_j^i. It is H option value of the method.

Step 5) Based on the latest data point, the weighted voting model presented in the following is used to compute the prediction value $\hat{Y}_{t+1}^i$. The model is given by

$$\hat{Y}_{t+1}^i = Y_t^i + \frac{\sum_{k=1}^{H}(F_k^i - U_k^i).e^{-W_k^i}}{\sum_{k=1}^{H} e^{-W_k^i}}, \tag{11}$$

where W_k^i represents the kth smallest D_j^i value, U_k^i and F_k^i respectively represent the U_j^i and F_j^i value corresponding to the kth smallest D_j^i value, $H = k$.

A simple example for the method with window size $L = 5$ is shown in FIG.2. If $H = 3$ and D_1, D_5 and D_{11} are the three windows whose distance values with current reference window are the smallest, the forecasted value at the given scale can be obtained with formula (11).

4 Simulation Results

The experiment is coded with MATLAB R2010a software. The time series data applied to evaluate the prediction system is obtained from Taiwan Stock Exchange (TSE). It is the closing price of TSE, including 614 records in all from 2003 7/18 to 2005 12/30. The data has been divided into a training data set and a test data set. The first 492 records are training set and the rest of the data, i.e., 122 records will be test data.

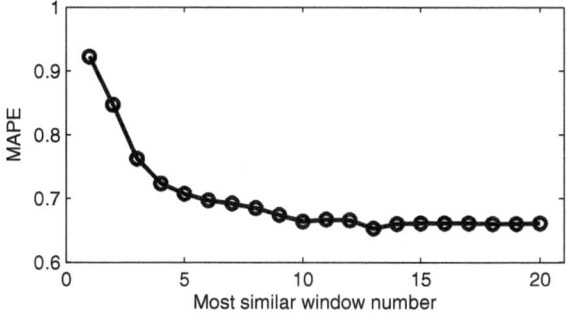

Fig. 3. $MAPE$ for different number of most similar windows when window size is fixed

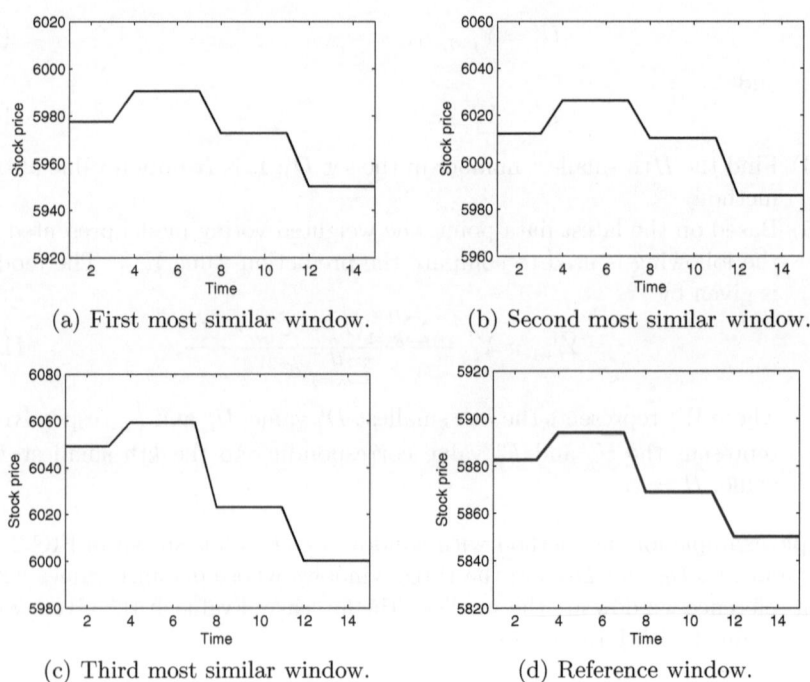

(a) First most similar window. (b) Second most similar window.

(c) Third most similar window. (d) Reference window.

Fig. 4. A example for three most similar windows with one reference window where the number of the most similar windows is 3 and the size of window is 15

Table 1. The results in 3 kinds of single scales and multi-scale

	Trial					
	1	2	3	4	5	AVE MAPE
Scale						
Scale 1	0.5145	0.5898	0.7769	0.8000	0.7203	0.669
Scale 2	0.6026	0.6514	0.6939	0.6956	0.7539	0.668
Scale 3	0.5607	0.6150	0.7170	0.7466	0.7214	0.661
Multiscale	0.5399	0.6108	0.7194	0.7405	0.7214	0.655

The experiment is conducted under three different scales. Firstly, wavelet threshold de-noising for one level, two levels and three levels is applied separately to the original data signal. When only the parameter H retain changed, a law can be discovered which indicates that with H increasing. The $MAPE$ firstly decreases and then stabilizes as shown in Fig3 where L is 18 and H varies from 1 to 20.

Then at every scale, we search for the most similar windows and obtain the prediction values by the descriptive method above. The Fig.4 gives an example at once searching that shows the reference window (Fig.4(d)) and searched three most similar candidate windows (Fig.4(a-c)) with it. TABLE 1 shows the results under single scale and mutil-scale. A decision can be reached that better results can be gained under mutil-scale.

5   Conclusion

This paper proposes a multiscale-based method integrating wavelet and CRWA technology for stock price prediction. To avail the prediction, wavelet threshold de-noising is applied to obtain the stock data at different scales, as well as remove noise. Then CRWA technique is used to search the most similar windows. And a specific model is used to gain prediction values. Finally, a series of experiment results are given, which show the feasibility of the idea of this paper.

Acknowledgements. This research was supported in part by the National Natural Science Foundation of China under grants 61075040 and 61033013 (Key Project), the Key Project of Natural Science Foundation of Higher Education of Jiangsu Province under grant 10KJA520047, the Natural Science Foundation of Jiangsu Province under grant BK2012645, and the Soochow Scholars Programme of Soochow University.

References

1. Chang, P.C., Fan, C.Y.: A hybrid system integrating a wavelet and tsk fuzzy rules for stock price forecasting. IEEE Transactions on Systems, Man, and Cybernetics, Part C 38, 802–815 (2008)
2. Choi, T.M., Hui, C.L., Yu, Y.: Intelligent time sereis fast forecasting for fashion sales: a research agenda. Machine Learning and Cybernetics 3, 1010–1014 (2011)
3. Donoho, D.: De-noising by soft-thresholding. IEEE Trans. Information Theory 41, 613–626 (1995)
4. Iijima, T.: Basic theory of pattern normalization (in case of typical one-dimensional pattern). Bulletin of the Electrotechnical Laboratory 26, 368–388 (1962)
5. Kolaczyk, E.D.: Methods for analyzing certain signals and images in astronomy using haar wavelets. Signals, Systems and Computers 1, 80–84 (1997)
6. Lindeberg, T.: Scale-space for discrete signals. IEEE Trans. Pattern Anal. Mach. Intell. 12, 234–254 (1990)
7. Mager, J., Paasche, U., Sick, B.: Forecasting financial time series with support vector machines based on dynamic kernels. Soft Computing in Industrial Applications 2, 252–257 (2008)
8. Renaud, O., Starck, J.L.: Prediction based on a multiscale decomposition. Int. J. Wavelets, Multiresolution Inf. Process. 1, 217–232 (2003)

An Affine Invariant Shape Retrieval Algorithm

Changzhong Li, Baojiang Zhong, and Yinan Cui

School of Computer Science and Technology,
Soochow University, Suzhou 215006, China
{20104227010,bjzhong,20114527002}@suda.edu.cn

Abstract. The Curvature Scale-Space (CSS) technique has been selected in MPEG-7 for shape similarity retrieval. While the technique
is invariant to shape transformations with respect to four parameters
namely zoom, rotation and translation (which needs two parameters
to represent), our algorithm proposed in the paper, called *Affine CSS*
(ACSS), treats the two left over parameters. Against any prognosis, simulating all views depend on these two parameters is feasible. The enriched
algorithm is used to find similar shapes from a very large prototype
database, and also a small classified database of marine creatures, which
consists of original as well as affine transformed shapes. An improvement
is observed over the conventional CSS algorithm.

Keywords: Shape retrieval, affine invariant, simulate, scale-space.

1 Introduction

A shape retrieval algorithm includes a shape representation step and a shape
matching step. Many approaches have been proposed in the literature for shape
representation and analysis, and they focus either on shape boundaries which use
curvature [1][2][3][4] or on interior region of shapes which use moments [5][6] and
Fourier descriptors [7][8]. The CSS representation [1] finds its roots in curvature
deformation and heat equation. This is carried out by convolving each coordinate
of a closed planar curve with a Gaussian function at different levels of scale and
extracting the location of inflection points of the resulting curves in order to
construct the CSS map (scale-space trajectories of the curvature zero-crossing
points of the shape).

Yu and Morel [9] [10] proposed an affine camera model in order to solve difficulty about image matching under affine transformations and the algorithm
exhibited excellent performance. The key to the success of the algorithm is that
the affine camera model simulated all the different viewpoints with limited number of sampling beautifully. Inspired by the idea, we propose an robust algorithm
for shape retrieval under affine transformations by introducing the affine camera
model into the original CSS-based shape retrieval method.

The following is the organization of the remainder of this paper. In Section
2, the construction of the CSS map and the CSS matching step are briefly
explained. The affine camera model is introduced in Section 3. The proposed

J. Yang, F. Fang, and C. Sun (Eds.): IScIDE 2012, LNCS 7751, pp. 776–783, 2013.

algorithm is described in Section 4, and an experimental proof that our algorithm is robustly affine invariant is given in Section 5. The concluding remarks are presented in Section 6.

2   The CSS Representation and Matching

Consider a parametric vector equation for a curve: $\mathcal{T}(v) = (x(v), y(v))$, where v is an arbitrary parameter. The formula for computing the curvature function can be expressed as

$$\kappa(v) = \frac{\dot{x}(v)\ddot{y}(v) - \ddot{x}(v)\dot{y}(v)}{(\dot{x}(v)^2 + \dot{y}(v)^2)^{3/2}}. \tag{1}$$

If we convolve each component of $\mathcal{T}$ with $g(v, \sigma)$, a one-dimensional (1D) Gaussian kernel of width σ, then $\mathcal{X}(v, \sigma)$ and $\mathcal{Y}(v, \sigma)$ represent the components of the resulting curve, $\mathcal{T}_\sigma$:

$$\mathcal{X}(v, \sigma) = x(v) * g(v, \sigma), \mathcal{Y}(v, \sigma) = y(v) * g(v, \sigma)$$

where * denotes a convolution operation. By the properties of convolution, the derivatives of every component can be calculated easily as following.

$$\dot{\mathcal{X}}(v, \sigma) = x(v) * \dot{g}(v, \sigma), \ddot{\mathcal{X}}(v, \sigma) = x(v) * \ddot{g}(v, \sigma)$$

and we have similar formula for $\dot{\mathcal{Y}}(v, \sigma)$ and $\ddot{\mathcal{Y}}(v, \sigma)$. Since the forms of $\dot{g}(v, \sigma)$ and $\ddot{g}(v, \sigma)$ are known, the curvature of the convolved curve is computed as

$$\kappa(v, \sigma) = \frac{\dot{\mathcal{X}}(v, \sigma)\ddot{\mathcal{Y}}(v, \sigma) - \ddot{\mathcal{X}}(v, \sigma)\dot{\mathcal{Y}}(v, \sigma)}{(\dot{\mathcal{X}}(v, \sigma)^2 + \dot{\mathcal{Y}}(v, \sigma)^2)^{3/2}}. \tag{2}$$

2.1   CSS Shape Representation

The curve of the shape is smoothed by a Gaussian function. The smoothed curve is called $\mathcal{T}_\sigma$, where σ denotes the width of the Gaussian function, $g(v, \sigma)$. The locations of curvature zero-crossings are determined at different levels of scale with (2). The process starts with $\sigma=1$, and at each level, σ is increased with $\delta\sigma$, chosen as 0.05 in our experiments. As σ increases, $\mathcal{T}_\sigma$ shrinks and becomes smoother, and the number of curvature zero-crossing points on it decreases. Finally, when σ is sufficiently high, $\mathcal{T}_\sigma$ will be a convex curve with no curvature zero-crossings (see Figure 1).

If we determine the locations of curvature zero-crossings of every $\mathcal{T}_\sigma$ during evolution, we can display the resulting points in (u, σ) plane, where u is the normalized arc length and σ is the width of the Gaussian kernel. The result of this process can be represented as a binary image called the *CSS map* of the curve (see Figure 2).

Every shape in the database is represented with the coordinates of its CSS contour peaks. For example, in Figure 2 there are seven peaks, and therefor the shape will be represented by seven pairs of integer numbers.

Fig. 1. Evolution of a shape (curvature zero-crossings are marked with dots)

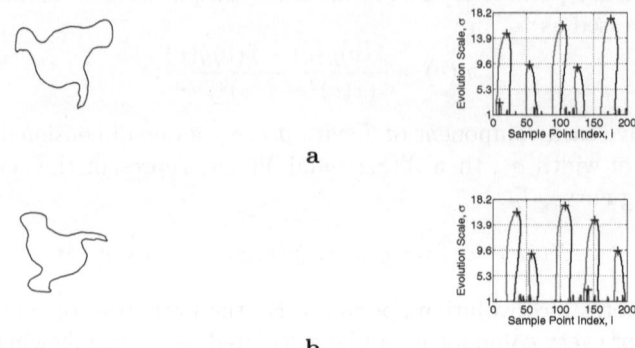

a

b

Fig. 2. (a) A shape and its CSS map. (b) Change in orientation causes a circular shift in CSS map.

2.2 CSS Shape Matching

Following the CSS peak exaction, the CSS matching algorithm compares two sets of peaks and assigns a matching value to them. The matching value represents the similarity between the two compared shapes. The algorithm has been described detailedly in Mokhtarian *et al.* [1].

3   The Affine Camera Model

Shape distortions arising from viewpoint changes can be locally modeled by affine planar transforms. Thus, the shape deformation model under a camera motion is $u(x,y) \rightarrow u(ax+by+e, cx+dy+f)$, Let A be the linear planar map with positive determinant. Any such map has a decomposition

$$A = \lambda \begin{pmatrix} \cos\psi & -\sin\psi \\ \sin\psi & \cos\psi \end{pmatrix} \begin{pmatrix} t & 0 \\ 0 & 1 \end{pmatrix} \begin{pmatrix} \cos\phi & -\sin\phi \\ \sin\phi & \cos\phi \end{pmatrix}, \qquad (3)$$

where we note $A = \lambda R_1(\psi)T_t R_2(\phi)$, with $\phi \in [0, 180°)$, $t = 1/\cos\theta(\theta \in [0, 90°))$, $\lambda > 0$. λt is the determinant of A, $R(\psi)$ denotes the planar rotation with angle ψ, and $T_t(t \geq 1)$ is called the tilt. Figure 3 shows a camera motion interpretation of (3): ϕ and $\theta = \arccos(1/t)$ are the camera viewpoint angles and ψ parameterizes the camera spin.

Fig. 3. Geometric interpretation of (3)

In this affine model the camera stands far away from a planar object. Starting from a frontal position, a camera motion parallel to the object's plane induces an image translation. The plane containing the normal and the optical axis makes an angle ϕ with a fixed vertical plane. For more detailed information, please refer to [9].

4 Affine Invariant CSS Algorithm

The idea of combining simulation and normalization is the main successful ingredient of the CSS technique. From the whole procedure of CSS map construction and CSS matching, we know that the technique normalizes translations and scale changes, and partly simulates rotation in the CSS matching. This actually provided the inspiration for our algorithm as described below.

4.1 The Main Procedure of ACSS

Now we describe our method for matching two shapes. ACSS proceeds by the following steps.

1. The two considered shapes are transformed by simulating all possible linear distortions caused by the change of orientation of the camera axis. These distortions depend upon two parameters: the longitude ϕ and the latitude θ. The shapes undergo ϕ-rotations followed by tilts with parameter $t = 1/\cos(\theta)$.

2. These rotations and tilts are performed for a finite and small number of latitudes and longitudes. The sampling steps of these parameters ensure that the simulated shapes keep close to any other possible view generated by other values of ϕ and θ.

3. All simulated shapes of one shape are compared with that of another by the CSS technique, which is invariant to scale, rotation and translations. The minimum matching value is regarded as a measure of their similarity.

4.2 Latitude and Longitude Sampling

The sampling of the latitude and longitude angles is specified below.

1. The latitudes θ are sampled so that the associated tilts follow a geometric series $1, a, a^2, ..., a^n$, with $a > 1$. The choice $a = \sqrt{2}$ is a good compromise between accuracy and sparsity. The value n can go up to 5 or more and the parameter is set to be 5 in our experiment.

2. The longitudes ϕ are for each tilt an arithmetic series $0, b/t, ..., kb/t$, where $b \approx 72°$ seems again a good compromise, and k is the last integer such that $kb/t < 180°$.

3. For visual description, the resulting sampling of the parameters $\theta = \arccos 1/t$ and ϕ on the observation hemisphere is illustrated in Figure 4.

For detailed explanation and proof and further theoretical analysis of the choice of parameters(a, b, t), please refer to [9]. More experimental analysis that proving the superiority of these parameters should be part of our future work.

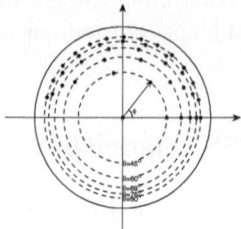

Fig. 4. Sampling of the parameters θ and ϕ. The samples are the black dots. Left: perspective illustration of the observation hemisphere. Right: zenith view of the observation hemisphere.

5   Experiments and Results

In this section, the performance of the proposed algorithm is evaluated and compared with the original CSS technique. The experiments are based on two different shape database involving general affine transformations.

The first database is the *MPEG-7 CE Shape-1 (Part B)* shape database which includes 1400 shapes classified in 70 different groups.

According to (3), we chose $t = 1, 2, 3, 4, 5, 6$, and $\phi = 0°, 20°, 40°, ..., 160°$ as the affine transformation parameters. Every original shape was transformed and selected as the input query to retrieval the most similar shapes to it, and the first n outputs of the system were observed to check whether the original versions in the shape group are retrieved. For example, we choose one shape (Heart-11.gif, see Figure 5(a)) and transform it acutely to a new shape (see in Figure 5(b)) as an input query. The results are illustrated in Figure 6.

The *success rate* denoted by η is employed as the performance measure which is defined as $\eta = m/m_{max}$, where m_{max} is the maximum possible value of m. Note that m_{max} is equal to n if n is less than or equal to the number of group members; otherwise, m_{max} is equal to the number of the group members. The *success rate* for the whole experiment is the average of the success rate for each input query. Note that, there are 20 shapes in a single shape group and the parameter n is set 40 in the experiment. Therefore, the parameter m_{max} is equal to 20 in the experiment.

Fig. 5. (**a**) The original shape (Heart-11.gif). (**b**) The transformed shape as input query shape ($\phi = 30°, t = 6$).

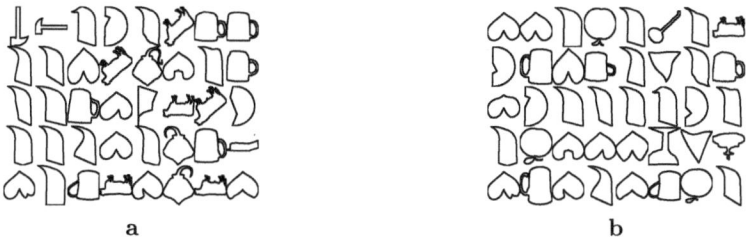

Fig. 6. (**a**) The retrieved shapes with the proposed CSS technique (seven correctly retrieved shapes). (**b**) The retrieved shapes with the proposed ACSS technique (ten correctly retrieved shapes)

Table 1. The result of the proposed ACSS technique

ϕ	t=1; η	t=2; η	t=3; η	t=4; η	t=5; η	t=6; η
0°	59.2	58.3	58.2	56.3	52.5	49.8
20°	58.6	57.6	57.6	56.7	53.7	50.2
40°	59.0	58.0	58.0	56.0	53.0	51.0
60°	58.4	58.6	58.2	56.8	53.8	50.8
80°	58.5	58.1	58.4	56.0	54.0	51.5
100°	58.3	57.6	57.5	57.3	53.3	50.3
120°	58.7	58.3	58.3	56.3	53.5	49.6
140°	58.6	57.9	56.5	56.5	52.6	51.6
160°	58.4	57.4	58.4	57.4	52.8	50.2
ave	58.6	58.0	57.9	56.6	53.2	50.6

The results of objective evaluation of the ACSS and original CSS techniques applied to the database are presented in the Table 1 and Table 2, respectively. The first column in each table presents the longitude ϕ and the first row presents the absolute tilt which has a relation with the latitude θ. The last row calculates the average value of the *success rate* under the same latitude. From the experiment results, we can see a good robustness of the proposed technique under affine transformation. In addition, the results also indicate that the CSS and ACSS techniques are invariant to shape rotation. The comparison is illustrated in Figure 8(a).

Table 2. The result of the original CSS technique

ϕ	t=1; η	t=2; η	t=3; η	t=4; η	t=5; η	t=6; η
0°	59.1	55.1	48.7	43.7	39.3	36.2
20°	58.5	54.5	49.5	45.6	40.8	38.0
40°	58.4	55.7	51.6	47.4	43.3	40.1
60°	58.2	55.4	50.7	45.4	41.5	38.9
80°	58.4	54.2	48.6	43.1	39.0	35.6
100°	58.5	55.1	48.9	44.0	39.1	35.7
120°	58.6	55.5	50.6	46.0	41.7	38.1
140°	58.4	55.3	51.5	46.9	43.5	39.8
160°	58.1	54.0	49.7	44.1	40.1	36.2
ave	58.5	55.0	50.0	45.1	40.9	37.6

Fig. 7. Test shapes for the second experiment

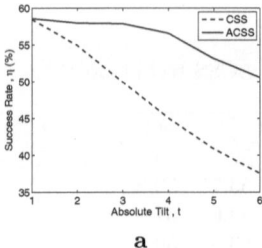 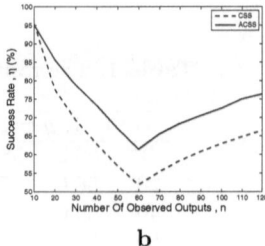

a   b

Fig. 8. (a) The result of the first experiments. (b) The result of the second experiment.

In the second experiment, the database is a classified subset of the 1100 images of marine creatures to evaluate the performance. We chose 90 shapes which belong to 15 different groups as presented in Figure 7. Every original shape is used to produce nine affine transformed shapes. As a result, a group with 6 members becomes 60 members after adding the transformed shapes. In the procedure of transformation of the original shapes, a random value between 1 and 6 for t and a random value between 0° and 360° for ϕ are chosen as the affine transformation parameters.

To evaluate the performance of the proposed technique, every shape in the database (including just produced ones) is treated as the input query. The best n similar shapes to the input one are then retrieved. The experimental result is illustrated in of Figure 8(b).

6   Concluding Remarks

The CSS technique has proven to be effective and robust for shape retrieval. However, in severely affine transformed environment, it cannot work well. In this paper, a technique to overcome the difficulty, called ACSS, is proposed through simulating all possible linear distortions caused by the change of orientation of camera axis.

Note that Sapiro and Tannenbaum [3] also investigated the affine invariant scale-space topic, while Mokhtarian and Abbasi [4] introduced and used affine length as alternative of the arc length parameter in affine transformed environments. The basic ideas of these authors' work and our present work are quite different. In brief, the affine invariance is realized via a simulation of the camera movement in our work, while it is realized via a mathematical improvement in [3] and [4]. A full comparison between ACSS and the existing approaches will be interesting. However, some key problems should be solved at first.

Acknowledgements. This research was supported in part by the National Natural Science Foundation of China under grants 61075040 and 61033013 (Key Project), the Key Project of Natural Science Foundation of Higher Education of Jiangsu Province under grant 10KJA520047, the Natural Science Foundation of Jiangsu Province under grant BK2012645, and the Soochow Scholars Programme of Soochow University.

References

1. Abbasi, S., Mokhtarian, F., Kittler, J.: Curvature Scale Space Image in Shape Similarity Retrieval. Multimedia Syst. 7(6), 467–476 (1999)
2. Mokhtarian, F., Bober, M.: Curvature Scale Space Representation: Theory, Applicatios, and MPEG-7 Standardization. Kluwer Academic Publishers, Dordrecht (2003)
3. Sapiro, G., Tannenbaum, A.: Affine Invariant Scale-Space. International Journal of Computer Vision 11(1), 25–44 (1993)
4. Mokhtarian, F., Abbasi, S.: Affine Curvature Scale Space with Affine Length Parametrisation. Pattern Anal. Appl. 4(1), 1–8 (2001)
5. Teague, M.R.: Image Analysis via the General Theory of Moments. J. Optical Soc. Am. 70, 920–930 (1980)
6. Zhao, D., Chen, J.: Affine Curve Moment Invariants for Shape Recognition. Pattern Recognition 30(6), 895–901 (1997)
7. Persoon, E., Fu, K.-S.: Shape Discrimination Using Fourier Descriptors. IEEE Trans. Pattern Anal. Mach. Intell. 8(3), 388–397 (1986)
8. Zhang, D., Lu, G.: Shape-Based Image Retrieval Using Generic Fourier Descriptor. Sig. Proc.: Image Comm. 17(10), 825–848 (2002)
9. Morel, J.-M., Yu, G.: ASIFT: A New Framework for Fully Affine Invariant Image Comparison. SIAM J. Imaging Sciences 2(2), 438–469 (2009)
10. Yu, G., Morel, J.-M.: A Fully Affine Invariant Image Comparison Method. In: ICASSP, pp. 1597–1600 (2009)

The Maximized Discriminative Subspace
for Manifold Learning Problem

Yuting Tao and Jian Yang

Nanjing University of Science and Technology,
Xiao Ling Wei 200, Nanjing, China, 210094
tao.yuting@yahoo.com.cn,
csjianyang@gmail.com

Abstract. Manifold learning problem, aims to seek some directions, which can keep the local structure and neighborhood of each sample as much as possible. In the conventional manifold learning approaches, feature extraction is performed in the original data space. In this paper, a new method called "the maximized discriminant subspace algorithm" (MDS) is implemented before feature extraction and classification. Extensive experiments show the better classification results than the conventional manifold approaches, due to projecting the original data onto the maximized discriminant subspace in a preliminary phase before feature extraction and classification in the transformed data space.

Keywords: manifold learning, original data space, transformed data space, the maximized discriminant subspace, preliminary phase.

1 Introduction

Pattern classification can be categorized into two general types—supervised and unsupervised. The only difference is that the former is offered with class label, such as linear discriminant analysis (LDA) [5, 6], local discriminant embedding (LDE) [8], marginal Fisher analysis (MFA) [7], etc.; while the latter without class label, such as principle component analysis (PCA) [10, 11, 12], locality preserving projection (LPP) [3, 4], locally linear embedding (LLE) [9], etc.

PCA is an optimal way for representation [3, 7]. Due to the centralization of the original data X, the projection axis (principle components) are supposed to be those directions after projection onto which the covariance of the whole data is maximized. Meanwhile, LDA seeks the discriminant vectors after projection onto which the ratio between inter-class covariance and the intra-class covariance are supposed to be as large as possible [3, 7].

In manifold learning problem, however, the projection axis we should seek play the role of distance minimizer for K nearest neighbors of each sample. E.g., if x_b is among the K nearest neighbors of x_a in the original space, then after projection onto the axis w, i.e. $\tilde{x}_a = w^T x_a$ and $\tilde{x}_b = w^T x_b$, the Euclidean distance $\| \tilde{x}_a - \tilde{x}_b \|^2$ should be minimized and $\tilde{x}_b$ is still one of the K nearest neighbors of $\tilde{x}_a$.

J. Yang, F. Fang, and C. Sun (Eds.): IScIDE 2012, LNCS 7751, pp. 784–792, 2013.
© Springer-Verlag Berlin Heidelberg 2013

Regarding the conventional manifold learning problems, no matter unsupervised (e.g. LPP, LLE) or supervised (e.g. LDE, MFA), the feature extraction should be done in the first step, after which is the classification task. For the large sample size and low dimensional case, such as digital character recognition, feature extraction can be done directly; while for the small sample size and high dimensional case, such as face recognition, PCA should be done to avoid the singularity before feature extraction.

In this paper, we focus on the four manifold learning problems (LPP, LLE, LDE and MFA), and propose a modified version, i.e. project the whole data set onto the maximized discriminant subspace (MDS) in the preliminary phase, then do feature extraction and classification. Extensive experimental results show that all these four manifold learning approaches under our proposed version outperform their conventional counterparts.

The following part of this paper is arranged as follows: Section.2 gives the previous work that underlies our proposed work; Section.3 illustrates the details of the proposed maximized discriminant subspaces; Section.4 shows the experiments and analysis and Section.5 draws the final conclusion.

2   Brief Outline of the Previous Work

Suppose the original data set $X \in R^{p \times n}$, where $X = [x_1, ...x_n]$, with the dimension p and the total sample number n. If there are C classes in total, and n_i is the number of samples in Class i, (i.e. C_i),then $n = \sum_{i=1}^{C} n_i$, and the original data set from the C_i is $X_i = [x_1, ...x_{n_i}]$, thus $X = [X_1, ..., X_C]$.

2.1   Orthogonal Projection from One Space to another Space

Let $X_{\sim i}$ denote the set consisting of all the data but X_i, i.e. $X = [X_i, X_{\sim i}]$. Suppose $rank(X_i) = s$, then X_i can be totally represented by the largest s principle components $u_1, ..., u_s$. So, the following SVD decomposition holds:

$$X_i = \hat{U} \hat{\Sigma} \hat{V}^T \qquad (1)$$

where $\hat{U} \in R^{p \times s}$, $\hat{\Sigma} \in R^{s \times s}$ and $\hat{V} \in R^{n \times s}$.
Since $X_i(X_i^T X_i)^{-1} X_i^T = \hat{U} \hat{U}^T$, the orthogonal projection of $X_{\sim i}$ onto X_i is:

$$X_{\sim i}^T (X_i(X_i^T X_i)^{-1} X_i^T) X_{\sim i} = X_{\sim i}^T \hat{U} \hat{U}^T X_{\sim i} \qquad (2)$$

2.2   The Supervised and Unsupervised Manifold Learning Problem

In this subsection, we only give a brief outline of the unsupervised manifold leaning approaches (LPP, LLE) and supervised manifold leaning approaches (LDE, MFA). For more details, please refer to these related references.

1. Locality preserving projection (LPP) It is to seek the transformed space z, after projection onto which the samples x_i and x_j that are in close proximity in the original space remain so in the new space [3, 4].
2. Linear local embedding (LLE) It performed in the following 3 steps [9]:
1).Find the k nearest neighbor of each sample x_i.

2). Construct the affinity matrix W which minimizes the residual sum of squares for reconstructing each x_i from its neighbors.

3). Find the coordinates z which minimize the reconstruction error.

3. Local discriminant embedding (LDE) It is a supervised manifold learning approach [8], and takes both intra-class and inter-class into account.

4. Marginal Fisher analysis (MFA) It aimed to construct the intra-class compactness and inter-class separability graphs [7].

3   The Maximized Discriminant Subspace Algorithm

We provide the details of the proposed method in Section.3.1, and then give this algorithm's framework in Section.3.2.

3.1   The Mechanism of the Maximized Discriminant Subspace

Seeking the Discriminant Subspace Class by Class

For the samples in C_i, i.e. X_i $(i = 1, ..., C)$, our goal is to seek the discriminant subspace $\Phi^i = [\varphi_{i1}, ..., \varphi_{ir}]$, $(r \leq s)$ i.e. the subspace spanned within X_i, after projection onto which the margin after homo-class samples X_i deducting the hetero-class samples $X_{\sim i}$ is maximized. In another word, such kind of subspace should be the linear combination of $[u_1, ..., u_s]$. Therefore, the original problem converts into seeking the optimal coefficient set $a_{ij} = [\alpha_{j1}, ..., \alpha_{js}]^T$, $(j = 1, ..., r)$ which satisfies $\varphi_{ij} = \hat{U} a_{ij}$.

The homo-class matrix is defined as:

$$H_1 = \frac{1}{n_i} X_i X_i^T$$

The hetero-class matrix is defined as:

$$H_2 = \frac{1}{n - n_i} X_{\sim i} X_{\sim i}^T$$

Motivated by the formula of Difference-LDA in [1, 2], which aimed to seek the discriminant vectors to achieve the maximized margin of $S_b - \Delta S_w$ (where Δ is a tunable positive parameter), the corresponding objective function is:

$$f_i = \max_{A_i} \left[tr(A_i^T \hat{U}^T (H_1 - H_2) \hat{U} A_i) \right] \quad s.t. \quad A_i^T A_i = I \tag{3}$$

where $A_i = [a_{i1}, ... a_{ir}]$ and $A_i \in R^{s \times r}$.

If we denote $E_i = \hat{U}^T (H_1 - H_2) \hat{U}$, then Eq.(3) can convert into:

$$a_{ij}^* = \underset{a_{ij}}{argmax} \left[a_{ij}^T E_i a_{ij} + \tau(a_{ij}^T a_{ij} - 1) \right] \quad j = 1, ..., r \tag{4}$$

where τ is a Lagrange multiplier.

Eq.(4) is equivalent to seeking r eigenvectors of E_i corresponding to the largest r eigenvalues [13].

Theorem 1. If every sample in X is of the same length, i.e. $\| x_1 \| = ... = \| x_n \|$, then there must exist at least one positive eigenvalue in E_i.

Proof. Since each sample in X is of the same length, $tr(H_1) = tr(H_2)$. Meanwhile, $\hat{U}$ is the range space of X_i, so X_i must be totally represented by the linear combination of $[u_1, ..., u_s]$, but $X_{\sim i}$ maybe cannot. Based on Eq.(1), $tr(\hat{U}^T H_1 \hat{U}) = \frac{1}{n_i} tr(\hat{\Sigma}^2) = tr(H_1)$. Therefore, it means: $tr(E_i) = tr(H_2) - tr(\hat{U}^T H_2 \hat{U}) = \frac{1}{n-n_i} tr(X_{\sim i}^T (I - \hat{U}\hat{U}^T) X_{\sim i}) \geq 0$. We can always find the coefficient vector a, to make $a^T \hat{U}^T (H_1 - H_2) \hat{U} a = \varphi^T (H_1 - H_2) \varphi > 0$. □

Nevertheless, it cannot theoretically guarantee that E_i is nonnegative definite. Regarding the negative eigenvalue λ_{iq} of E_i, the corresponding eigenvector a_{iq} can put side effect w.r.t. maximizing the margin between homo-class and hetero-class, since

$$a_{iq}^T \hat{U}^T (H_1 - H_2) \hat{U} a_{iq} = \varphi_{iq}^T (H_1 - H_2) \varphi_{iq} = \lambda_{iq} < 0$$

Therefore, such eigenvectors of E_i corresponding to negative eigenvalues should be removed before constructing the discriminant subspace Φ^i.

Maximization of All These Discriminant Subspaces as an Integral
In the statement above, we only provide the details of the discriminant subspace Φ^i in terms of the ith class, i.e.X_i. In fact, the discriminant subspaces of all the C classes should be worked out, before feature extraction and classification. And it is easy to verify that every discriminant vector φ_{ij} in Φ^i is of unit length (i.e.$\| \varphi_{ij} \| = 1$) and orthogonal to each other (i.e. $\varphi_{ip}^T \varphi_{iq} = 0$ for $p \neq q$). But $\varphi_{mp}^T \varphi_{nq} \neq 0$ for different Classes C_m and C_n.

After such a_{ij}^* (where $i = 1, ..., C$ and $\lambda_{ij} > 0$) in Eq.(4) solved, and based on Eq.(3), the following equation holds:

$$F = \sum_{i=1}^{C} f_i = \sum_{i=1}^{C} \sum_{j} \lambda_{ij} \qquad s.t. \lambda_{ij} > 0 \qquad (5)$$

In PCA, the eigenvalue λ denotes the covariance of the transformed data after projection onto the principle components. Here, however, if $\| \varphi_{ij} \| = \| x \| = 1$, the similarity is $\cos(x, \varphi_{ij}) = x^T \varphi_{ij}$, then $\varphi_{ij}^T H_1 \varphi_{ij} = \frac{1}{n_i} \sum_{x \in C_i} \cos^2(x, \varphi_{ij})$ and $\varphi_{ij}^T H_2 \varphi_{ij} = \frac{1}{n-n_i} \sum_{x \notin C_i} \cos^2(x, \varphi_{ij})$ therefore λ_{ij} stands for the average margin between the "homo-class to φ_{ij} similarity" and "hetero-class to φ_{ij} similarity".

In light of Eq.(5), F is the summation of these similarity margins. For the sake of optimizing the summation of the whole margins, we can additionally add a non-negative weight β_{ij} to each of these λ_{ij}, to maximize these margins. i.e.

$$\max_{\beta_{ij}} \sum_{i=1}^{C} \sum_{j} \beta_{ij} \lambda_{ij} \qquad s.t. \sum_{i} \sum_{j} \beta_{ij} = 1 \qquad \forall \beta_{ij} \geq 0 \qquad (6)$$

Eq.(6) is a linear optimization problem w.r.t. β_{ij}, and the optimal solution is:

$$\beta ij^* = \frac{\lambda_{ij}}{\sum_i \sum_j \lambda_{ij}} = \frac{\lambda_{ij}}{F}$$

For each $\lambda_{ij} > 0$, since $\lambda_{ij} = \varphi_{ij}^T(H_1 - H_2)\varphi_{ij}$, then $\beta_{ij}^*\lambda_{ij} = \eta_{ij}^T(H_1 - H_2)\eta_{ij}$. And here $\eta_{ij} = \sqrt{\beta_{ij}^*}\varphi_{ij}$ is the corresponding maximized discriminant vector that we desire for!

3.2   The Maximized Discriminant Subspace Algorithm (MDS)

Input: Original data X, each sample the unit length, i.e. $\| x_1 \| = ... = \| x_n \| = 1$
For $i = 1 : C$.
1). Do SVD of X_i, where $X_i = \hat{U}\hat{\Sigma}\hat{V}^T$, and extract the range space $\hat{U}$ of X_i.
2). Construct the matrix $E_i = \hat{U}^T(H_1 - H_2)\hat{U}$ based on Section 3.1.
3). Select eigenvectors $[a_{i1}, ..., a_{ir}]$ corresponding to largest r eigenvalues of E_i.
4). Remove the eigenvectors corresponding to negative eigenvalues of E_i in the selected set, to construct Φ^i, where $\Phi^i = \{\varphi_{ij} | \varphi_{ij}^T(H_1 - H_2)\varphi_{ij} > 0\}$.
End
5). Based on Eq.(5), work out F which is the summation of these $\lambda_{ij} > 0$.
6). Set $\beta_{ij} = \lambda_{ij}/F$, and $\eta_{ij} = \sqrt{\beta_{ij}}\varphi_{ij}$.
7). Construct the maximized discriminant set Θ, where $\Theta = \{\eta_{ij} | \lambda_{ij} > 0\}$.
8). Do SVD, i.e.$\Theta = P\Lambda Q^T$, and get the projected data Y, where $Y = P^T X$.
Output: the projected data Y

Consequently, P is the range space of Θ and in light of Eq.(2), $Y = [y_1, ..., y_n]$ is the corresponding data set after X orthogonally projected onto Θ.

4   Experiments and Analysis

In all of these following experiments, we compare the classification results achieved by the four manifold learning approaches (i.e. LLP, LLE, LDE, and MFA) that are performed on the original data space X and on the transformed space Y. All these performances adopt the nearest neighbor (NN) classifier.

4.1   The Experiments on the ORL Face Database

ORL face database (http://www.cam-orl.co.uk) contains 10 different images in each of the 40 subjects. The first five images for each subject were used for training while the rest five for testing. In order to save computational cost, we resize each original 112×92 image to 56×46 using the down-sampling method proposed in [4, 14]. To avoid the singularity, PCA should be done at beginning (we keep 100% energy in the experiments).

In all these 4 manifold approaches, we unify that the parameters $t = 20$ for LPP and LDE, and $k = 4$ for nearest neighbor in LPP and LLE, $k_1 = k_2 = 4$ for intra-class neighbor and inter-class neighbor respectively in LDE and MFA. Regarding the proposed version, let $r = 1, 2, 3, 4$ be the number of discriminant vectors for each class, as in Eq.(4).

Fig.1 showed the comparison of classification accuracies between the conventional manifold learning approaches and their proposed counterparts, with the axis number varying from 10 to the end at the interval of 2. In Table.1, the best classification results for all the cases are listed, and what inside the corresponding parenthesis is the dimension at which the best classification accuracy for each case is achieved. The bold font denotes the highest recognition result within each of the four manifold approaches.

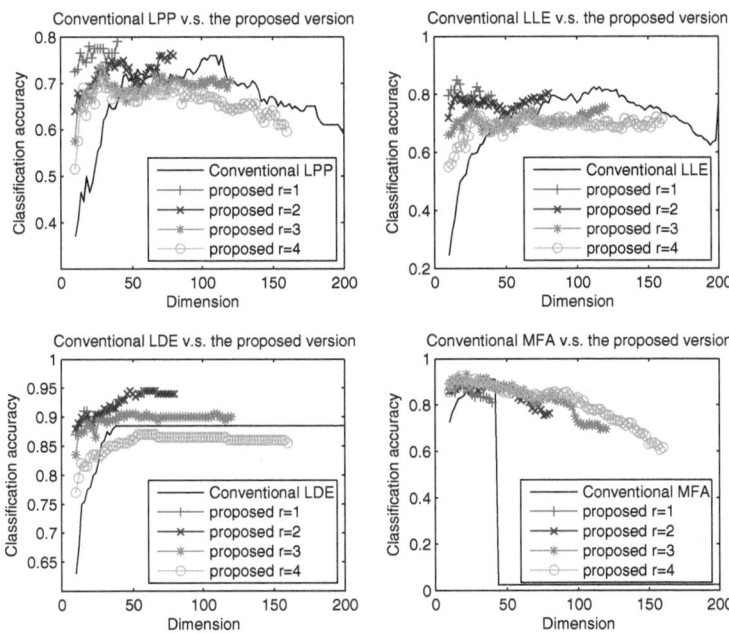

Fig. 1. The comparison between the conventional 4 manifold learning approaches (LPP, LLE, LDE and MFA) and their proposed versions (r=1, 2, 3 and 4) on the ORL face database

Table 1. The comparison of the best recognition results on the ORL face database

manifold approach	LPP	LLE	LDE	MFA
conventional version	0.760(106)	0.825(112)	0.885(38)	0.910(40)
proposed $r = 1$	**0.790(40)**	**0.850(16)**	0.930(40)	0.880(16)
proposed $r = 2$	0.765(78)	0.805(80)	**0.945(48)**	0.910(16)
proposed $r = 3$	0.735(28)	0.755(26)	0.905(42)	**0.930(22)**
proposed $r = 4$	0.725(28)	0.730(66)	0.870(54)	0.925(18)

4.2   Experiments on USPS Handwriting Digital Database

The USPS handwriting digital data (http://www.cs.toronto.edu/ roweis/data.
html) include 10 classes from 0 to 9. Each class has 1100 samples,with each

sample 256 dimensions. For saving the computational cost, we only choose the first 400 samples per each class in our experiment, i.e. the first 200 per class for training, while the rest 200 samples per each class for testing.

Similar to Section 4.1, $t = 1$ for LPP and LDE, and $k = 10$ for LPP and LLE, $k_1 = k_2 = 10$ for both LDE and MFA. Besides, let $r = 4, 6, 8, 10$ be the number of discriminant vectors for each class.

From Fig.2, we can see the comparison of classification accuracies between the conventional manifold learning approaches and their proposed counterparts, with the axis number varying from 10 to the end at the interval of 5. Likewise, Table.2 gave the best classification accuracy for each case.

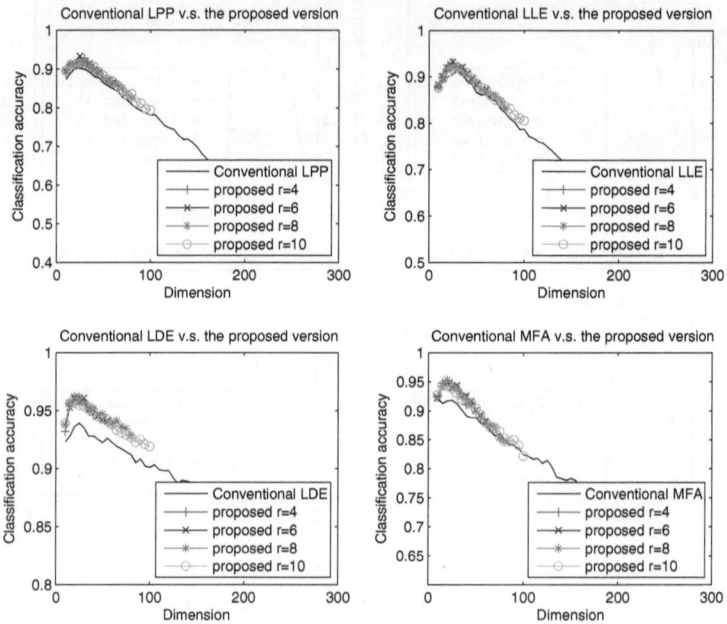

Fig. 2. The comparison between the conventional 4 manifold learning approaches (LPP, LLE, LDE and MFA) and their proposed versions (r=4, 6, 8 and 10) on USPS handwriting digital database

Table 2. The comparison of the best recognition results on the USPS handwriting digital database

manifold approach	LPP	LLE	LDE	MFA
conventional version	0.903(20)	0.914(30)	0.939(25)	0.919(10)
proposed $r = 4$	0.928(30)	0.931(25)	0.960(20)	0.950(20)
proposed $r = 6$	**0.934(25)**	**0.932(25)**	0.961(20)	0.948(20)
proposed $r = 8$	0.919(30)	0.921(25)	**0.962(20)**	**0.953(20)**
proposed $r = 10$	0.916(25)	0.921(30)	0.957(20)	0.944(15)

4.3   Analysis of These Experiments

Due to Theorem.1 in Section.3.1, the number of columns in Θ is no more than rC meanwhile no less than C. Since $Y = P^T X$, then $C \leq rank(Y) \leq rC$. Consequently, MDS transformation can be viewed as the dimensionality reduction, too. From the above experiments, we know that after the MDS transformation, the classification results both on the ORL database and on the USPS database improve to a large extent. Because in both of the two databases above, $p > n_i$, some columns of $\hat{U}$ are likely to be the null space of $X_{\sim i}$, which is favor of maximizing the optimization problem in Eq.(3), therefore facilitates the classification results. By contrast, if $p < n_i$, X_i is of full rank in terms of its feature space $\hat{U}$, and by which $X_{\sim i}$ can be totally linearly represented, thus it is likely to impair the margin between homo-class samples and hetero-class samples. In addition, from what inside the corresponding parenthesis of Table.1 and Table.2, it tells the fact that no matter in the conventional manifold learning approach or the proposed counterparts, small number of discriminant vectors can produce better classification results than the large ones.

5   Conclusion

In the conventional manifold learning problem, the unsupervised approaches such as LPP and LLE only take the nearest neighbors into account; meanwhile, the supervised ones such as LDE and MFA not only consider the neighbors but also the class labels. Nevertheless, in our proposed modified version, since in the preliminary phase the MDS transformation makes use of class label, all of the 4 approaches (LPP, LLE, LDE and MFA) are practically supervised. Anyway, the MDS transformation can shrink the scatter of the intra-class samples while enlarge the separability of inter-class samples, hence in such transformed space the better classification results can achieve.

Acknowledgments. The authors would like to thank the editor and the anonymous reviewers for their critical and constructive comments and suggestions. This work was partially supported by the Program for New Century Excellent Talents in University of China, the National Science Foundation of China under Grant No. 60973098 and the National Science Fund for Distinguished Young Scholars under Grant No. 61125305.

References

1. Song, F.X., Zhang, D., Mei, D.Y., Guo, Z.W.: A multiple maximum scatter difference discriminant criterion for facial feature extraction. IEEE Transactions on Systems, Man and Cybernetics, PartB: Cybernetics 37(6), 1599–1606 (2007)
2. Song, F.X., Zhang, D., Yang, J.Y., Gao, X.M.: Adaptive classification algorithm based on maximum scatter difference discriminant criterion. Acta Automatica Sinica 32(4), 541–549 (2006)

3. He, X.F., Yan, S.C., Hu, Y.X., Niyogi, P., Zhang, H.J.: Face recognition using lapla-cianfaces. IEEE Transactions on Pattern Analysis and Machine Learning 27(3), 328–340 (2005)
4. Xu, Y., Zhong, A., Yang, J., Zhang, D.: LPP solution schemes for use with face recognition. Pattern Recognition 43, 4165–4176 (2010)
5. Fisher, R.A.: The use of multiple measurements in taxonomic problem. J. Annals Eugenics 7(2), 179–188 (1936)
6. Welling, M.: Fisher Linear Discriminant Analysis, http://www.ics.uci.edu/~welling/classnotes/papers_class/Fisher-LDA.pdf
7. Yan, S., Xu, D., Zhang, B., Yang, Q., Lin, S.: Graph embedding and extension: a general framework for dimensionality reduction. IEEE Transactions on Pattern Analysis and Machine Learning 29(1), 40–51 (2007)
8. Chen, H.T., Chang, H.W., Liu, T.L.: Local discriminant embedding and its vari-ants. In: IEEE Computer Society Conference on Computer Vision and Pattern Recognition, CVPR (2005)
9. Roweis, S., Saul, L.: Nolinear dimensionality reduction by locally linear embedding. Science 290(22), 2323–2326 (2000)
10. Joliffe, I.: Principle component analysis. Springer (1986)
11. Martinez, A.M., Kak, A.C.: PCA versus LDA. IEEE Transactions on Pattern Anal-ysis and Machine Learning 23(2), 228–233 (2001)
12. Turk, M., Pentland, A.: Face recognition using eigenfaces. In: IEEE Computer Society Conference on Computer Vision and Pattern Recognition (CVPR), pp. 586–591 (1991)
13. Boyd, S., Vandenberghe, L.: Convex optimization. Cambridge University Press (2004)
14. Xu, Y., Jin, Z.: Down-sampling face images and low-resulotion face recognition. In: The 3th International Conference on Innonative Computing, Information and Control, pp. 392–395 (2008)

Ridge-Based Automatic Vascular Centerline Tracking in X-ray Angiographic Images

Ruoxiu Xiao, Jian Yang[*], Tong Li, and Yue Liu

Key Laboratory of Photoelectronic Imaging Technology and System,
Ministry of Education of China, School of Optics and Electronics,
Beijing Institute of Technology, Beijing, 100081
jyang@bit.edu.cn

Abstract. The extraction of vascular trees is very important for quantitative analysis of vascular structures. As angiographic image is the integration of X-ray through the whole body anatomy on the image plane, vascular structure loses most 3-D topological information. Hence, accurate vascular structure detection is of great help for clinical diagnosis. In this paper, a fully automatic vascular centerline extraction method is proposed. A self-adaptive morphological operator is combined with a multi-scale enhancement filter to enhance tubular-like structures. Then, points with local maximum intensity are extracted as seed points, while the initial track directions are determined by detecting prominent ridge points in the predefined range. By iteratively searching the connected ridge points, the centerlines are gradually extracted by connecting the ridge points. By statistically counting of connected components, fake connections are efficiently removed. And bifurcation points are discriminated from centerline skeletons by determining the connections of each centerline point. Our approach is automatic completely. Experimental results show that the proposed algorithm is very effective for the extraction of centerlines from angiographic images.

Keywords: Angiographic Image, Coronary artery, Centerline, Automatic extraction.

1 Introduction

In the latest World Health Organization survey for the "Top 10 Causes of Death," it has been reported that cardiovascular and cerebrovascular diseases are the biggest killer of human health in recent years. And the cardiovascular and cerebrovascular related deaths have accounted for 23.6% of all the deaths around the world [1]. In order to quantitatively identify the cardiac vascular diseases from the 2-D angiographic images, topology structures including the centerline, diameter and bifurcation should be extracted and measured. Among the quantitative analysis techniques, centerline extraction is the most important procedure, which directly decides the following analysis including motion estimation and 3-D reconstruction

---

[*] Corresponding author.

J. Yang, F. Fang, and C. Sun (Eds.): IScIDE 2012, LNCS 7751, pp. 793–800, 2013.
© Springer-Verlag Berlin Heidelberg 2013

[2]. However, due to the complexity of vascular structures and non-uniform gray-scale distribution of the contrast agent, centerline extraction is one of the most difficult techniques for the angiogram based analysis.

As vascular structure is perfused with contrast medium, vessels in the image show lower intensity than other tissues such as bones and myocardium. Hence, if the 2-D angiographic image is mapped to a higher 3-D gray-scale space, the vascular centerline corresponds to the ridge line of a hyper-surface. As demonstrated in [3] and [4], the ridge points are calculated by the gradient and Hessian matrix of the images. Due to the noise and un-even distribution of contrast medium, the ridge lines may scattered from the real centerlines. So, pre-processing techniques should be performed to rectify the ridge points. In this paper, the multi-scale vessel enhancement filter proposed by Frangi et al. [6] is used to enhance the vessels. Researchers [7-9] have proved that this multi-scale enhancement can effectively enhance the tubular-like structures, but more or less, requires some human intervention to be involved in.

Generally, the Hessian matrix of a point on the ridge has the property that two eigenvectors correspond to tangential and vertical directions of the vascular flow respectively. However, for the existence of noise and disturbance, the error introduced by the second-order derivative of the image is enlarged twice. Hence, the eigenvectors of Hessian matrix of the obtained ridge point cannot be used directly as the flow direction of vessels.

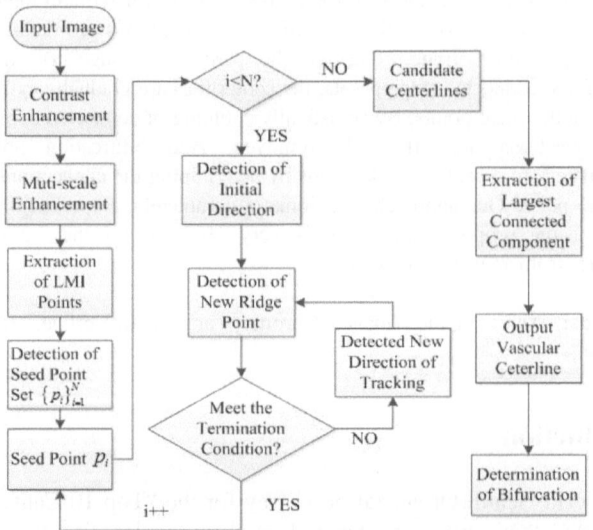

Fig. 1. Flow chart of the proposed algorithm

In this paper, we proposed a fully automatic vascular centerline extraction method. The angiographic image is adaptively enhanced by a multi-scale filter. A number of seed points and initial tracking directions are detected and determined by optimizing ridges of local maximum. Then, zero-order information of local ridges on every tracking step is used to calculate the newly added centerline points. Based on statistically counting of connected components, false connections are removed from the centerlines. Finally, bifurcations are discriminated from the centerline points by calculating connections of each centerline point. Fig.1 shows the calculation flow chart of this paper.

2   Methods

2.1   Vascular Enhancement

Generally, vascular centerline is the topological representation of vascular structure. And based on the X-ray imaging principle [2, 4], the vascular centerline should have the brightest or the darkest intensity distribution in the image in an ideal situation. However, as the existence of noise, the intensity distribution of centerline becomes very complex. Enhancement method is usually needed to strengthen the centerline features.

Suppose $TH(I)$ is a Top-Hat operator with a 3×3 structure element on image I, while $BH(I)$ is a Bottom-Hat operator. Then, the following operation is calculated to enhance the contrast of the image:

$$M(I) = I + TH(I) - BH(I) \qquad (1)$$

Based on the output of the above morphological operation, a multi-scale enhancement method proposed in [6] is utilized to enhance the tubular structures. Suppose V is a multi-scale vesselness of the Hessian matrix, and V_σ is the single-scale vesselness of the Hessian matrix at a specific scale σ, we have:

$$V = \max_{\sigma \in \{\sigma_1, \dots \sigma_N\}} \left(\sigma^\gamma \cdot V_\sigma \right) \qquad (2)$$

From the multi-scale calculation, the tubular structures can be effectively enhanced, and intensity of ridge points is more prominent compared to the neighboring points in the direction perpendicular to the flow direction.

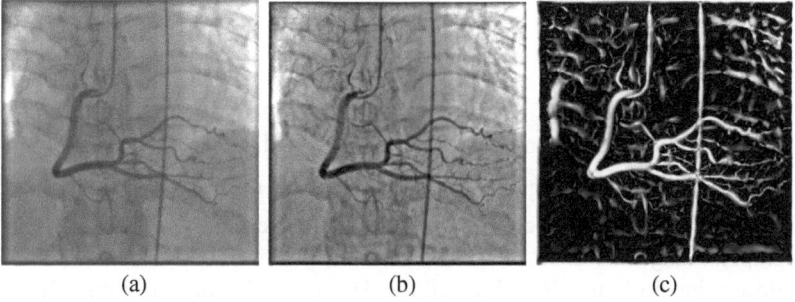

(a)   (b)   (c)

Fig. 2. (a) A routine coronary angiographic image. (b) Result of the morphology enhancement. (c) Result of the multi-scale enhancement.

Figure 2(a) shows a routine coronary angiographic image, while figure 2(b) and 2(c) show the enhanced results by the morphological operator and the multi-scale enhancement method respectively.

2.2   Automatic Seeding

After the enhancement calculation, pixels along the centerline become brighter than their neighboring points in the direction perpendicular to the vascular flow. If the pixel (x, y) is the point of local maximum intensity (LMI), its gradient is equal to zero ($\nabla(x, y) = 0$) and its Hessian matrix is negative [10]. However, the points satisfying these two conditions may not have integer coordinates. Hence, in this paper, a method to determine the LMI points is designed.

If the pixel (x, y) satisfies the equation (3), there must exist one point (ξ, η), which satisfies the equation (4). And the point (ξ, η) is a LMI point of the image.

$$\begin{cases} \nabla(x, y)\nabla(x+1, y+1) < 0 \quad or \quad \nabla(x+1, y)\nabla(x, y+1) < 0 \\ \lambda_i(x+m, y+n) < 0, (i = 1, 2, m = 0, 1, n = 0, 1) \end{cases} \tag{3}$$

$$\begin{cases} \nabla(\xi, \eta) = 0 \\ \lambda_1(\xi, \eta) < 0, \lambda_2(\xi, \eta) < 0 \end{cases} \tag{4}$$

As (ξ, η) needs to be interpolated by its neighbors and may introduce a great amount of unnecessary calculation, (x, y) is used to represent (ξ, η) in this paper.

Suppose (x_0, y_0) is one of the LMI points obtained by the previous processing, and g_1, g_2 are the two components of the normalized gradient $\nabla(x_0, y_0)$. Then, the adjacent LMI point can be found in the direction of its gradient in range d_1. Thus the detected point (x_1, y_1) can be obtained by:

$$I(x_1, y_1) = \sup_{0 \le d \le d_1} (I(x_0 + g_1 d_1, y_0 + g_2 d_1)) \tag{5}$$

Furthermore, the LMI point can be refined in the direction of the eigenvector in range d_2, so we have:

$$I(x_2, y_2) = \sup_{0 \le d \le d_2} (I(x_1 + v_{11} d_2, y_1 + v_{12} d_2)) \tag{6}$$

here, the eigenvalues of Hessian matrix at (x_1, y_1) are denoted as λ_1 and λ_2 ($|\lambda_1| \ge |\lambda_2|$). The corresponding eigenvectors of λ_1 and λ_2 are v_1 and v_2 respectively, and v_{11}, v_{12} are the two components of v_1. Then the refined point (x_2, y_2) can be regarded as a seed point which is located on the centerline of a blood vessel in the angiographic image.

Yellow points in figure 3(a) show all the detected LMI points in figure 2(c). It can be seen that most LMI points are located on blood vessels, while some are detected on the background. The reason is that the enhancement processes not only enhanced vessels but also enhanced the tubular-like noises. Figure 3(b) shows the refined LMI points. Figure 3(c) shows the locally enlarged result of figure 3(b). It can be seen that, all detected seed points are located on the centerlines of blood vessels, and they can be used as the initial points of the following ridge tracking.

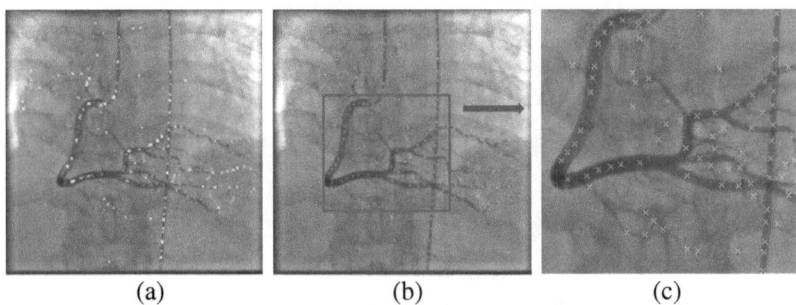

Fig. 3. (a) LMI points. (b) Extracted seed points. (c) Seed points on the locally enlarged view.

2.3   Initial Tracking Direction

Suppose p_1 and p_2 are two adjacent ridge points on the centerline, and l is the tangent of p_1. According to the definition of tangent, if p_2 is close enough to p_1, we have $l = \lim_{p_2 \to p_1} \overrightarrow{p_1 p_2}$. Then, $\overrightarrow{p_1 p_2}$ can be taken as the representation of l. As vascular centerline can be described as a relatively smooth and continuous curve, the direction from one ridge point to another can be approximated by the tangential direction.

According to the above analysis, for each seed point, two adjacent ridge points are calculated. These three points will be used to calculate both the forward and backward tracking direction. For any initial point $p(\theta_0)$ with coordinate $(x_{\theta_0}, y_{\theta_0})$, a circle is drawn with a radius of r. Obviously, the circle must have two intersections $p(\theta_{-1})$ and $p(\theta_1)$ with a centerline. If $p(\theta_1)$ is denoted as the point with the maximum point on the circle, we have:

$$I(p(\theta_1)) \geq I(p(\theta)), \forall \theta \in [0, 2\pi) \tag{7}$$

Then, in its opposite direction $2\pi - \theta_1$, the local maximum point $p(\theta_{-1})$ on the local arc $l(2\pi - \theta_1 - \Delta\theta, 2\pi - \theta_1 + \Delta\theta)$ is detected as following:

$$I(p(\theta_{-1})) \geq I(p(\theta)), \forall p(\theta) \in l(2\pi - \theta_1 - \Delta\theta, 2\pi - \theta_1 + \Delta\theta) \tag{8}$$

Figure 4(a) and 4(b) show the tracked forward and backward seed points (the forward is shown in red and the backward is shown in green). It can be seen that the detected initial tracking directions generally present the flow directions of the vascular structure.

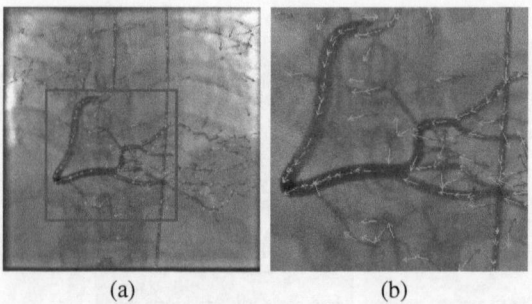

(a)   (b)

Fig. 4. (a) Detected initial tracking direction. (b) Locally enlarged view of (a).

2.4   Ridge Tracking

Based on the obtained initial seed points and tracking points, centerline points can be obtained by iteratively adding new ridge points onto the skeleton. If the current ridge point is p_1 with tracking direction θ_1, and the search angle encircle of p_1 is set as $2\Delta\theta$. The new LMI point p_2 can be detected on the local arc $l(\theta_1 - \Delta\theta, \theta_1 + \Delta\theta)$:

$$I(p_2) \geq I(p), \forall p \in l(\theta_1 - \Delta\theta, \theta_1 + \Delta\theta) \tag{9}$$

where p_2 is a ridge point, and θ_2 is the corresponding tracking direction. By iteratively adding new ridge points onto the centerlines in both forward and backward directions, the centerlines can be ultimately obtained.

The tracking process will be terminated by any of the following three conditions: (1) when the detected ridge point goes beyond the scope of the image; (2) when the intensity of the detected ridge point is smaller than a predefined threshold; and (3) when the distance between current point and the detected ridge point is less than a specified threshold. Under these termination conditions, the proposed algorithm can be carried out in the region of blood vessels without repetition.

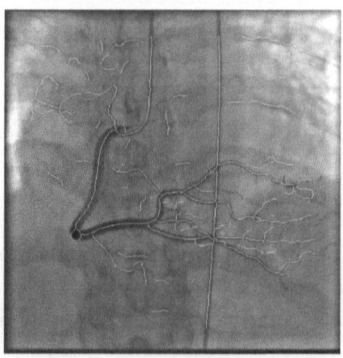

Fig. 5. Extracted all the tubular like structures

Figure 5 illustrates the result of the tracking algorithm. It can be seen that most of the vascular centerlines are extracted, while some background noise are also detected.

According to all above experiments, it can be seen that not all of the tracked objects are true centerlines, the pseudo vascular centerlines need to be removed.

Suppose $\{p_i\}_{i=1}^{N}$ represent the extracted ridge points, which can be divided into N connected ridge point sets $\left\{\{p_i^j\}_{i=1}^{M(i)}\right\}_{j=1}^{N}$. The largest connected ridge points set $\{p_i^j\}_{i=1}^{M_{max}}$ can be taken as a true centerline, and we have:

$$\left\|\{p_i^j\}_{i=1}^{M_{max}}\right\| = \sup_{1 \leq j \leq N}\left\|\{p_i^j\}_{i=1}^{M(i)}\right\| \tag{10}$$

Based on the refined vascular centerlines, bifurcations can be determined by counting connections of each skeleton point. We define that if the cardinality of the neighborhood ridges of a skeleton point p_0 is large than 3, this point can be taken as a bifurcation on the centerline. Then, we have:

$$\left|\{p_i | dist(p_i, p_0) < \frac{3}{2}r, p_i \in P\}\right| \geq 3 \tag{11}$$

Where r denotes the step length of the ridge tracking.

3   Experiments and Conclusions

To validate the effectiveness of the proposed algorithm, a series of experiments are designed and tested. Figure 6(a) shows the tracked centerlines superposed onto the original angiographic image, and figure 6(b) appends the detected bifurcations onto figure 6(a), while figure 6(c) shows the centerline extracting result of a commonly available algorithm obtained from the Insight application Toolkit from Kitware [9]. It can be seen that our method is very effective for the extraction of centerline in this angiographic image, which performs better than the algorithm from the Insight Application.

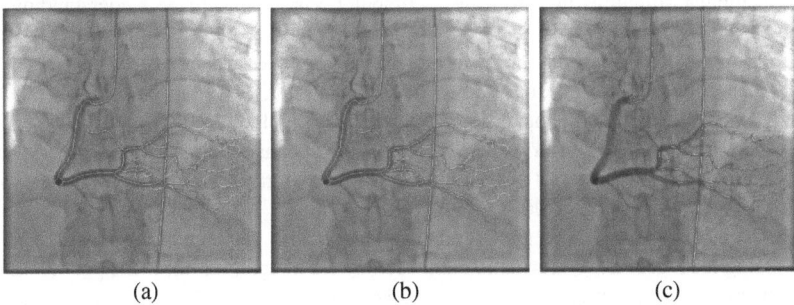

(a)   (b)   (c)

Fig.6. (a) Vascular centerlines extracted by the proposed method. (b) Detected bifurcations and extracted centerlines by the proposed method. (c) Centerline extraction results of the toolkit [9].

The extraction of vascular structure is important for computer-aided diagnosis and treatment of vascular diseases. Currently, manual operation is still the most commonly used method for vascular structure extraction in clinical practice. In this paper, an automatic vascular centerline extraction method for X-ray angiographic images is proposed. Our contributions in this paper are five-folds: First, compared to the traditional manual seed selection method, current method can automatically obtain a considerable amount of points distributed on the vascular structure. Second, the proposed method utilizes zero order information of ridge points to determine the tracking direction, which can effectively restrain the disturbance of hair-like errors. Third, compared with the multi-scale Hessian space, current method can greatly improve the calculation efficiency. Based on the seed points and the initial tracking direction, points with largest ridges are extracted and topologically connected. Fourth, the Hessian matrix and gradient distribution on the local tracking areas are combined to guarantee accurate centerline determination. By statistically counting continuity of the centerlines, false connections can be effectively removed. Fifth, bifurcations are determined by connection statistics of each skeleton point. Since there is not any intervention involved in the accurate vasculature extraction, the developed method can be utilized for computer-aided diagnosis of cardiovascular related diseases.

Acknowledgments. This work was supported by the National Basic Research Program of China (2010CB732505), National Science Foundation Program of China (60902103), New Century Excellent Talents in University of. Ministry of Education of China (NCET-10-0049) and the Plan of Excellent Talent in Beijing (2010D009011000004).

References

1. Organization, W.H.: The top ten causes of death-Fact sheet N310 (2008)
2. Kirbas, C., Quek, F.: A review of vessel extraction techniques and algorithms. ACM Computing Surveys 36(2), 81–121 (2004)
3. Damon, J.: Properties of ridges and cores for two-dimensional images. Journal of Mathematical Imaging and Vision 10, 163–174 (1999)
4. Eberly, D.: Ridges in image and data analysis. Kluwer Academic Pub. (1996)
5. Krissian, K., Malandain, G., Ayache, N., Vaillant, R., Trousset, Y.: Model-based detection of tubular structures in 3D images. Computer Vision and Image Understanding 80, 130–171 (2000)
6. Frangi, A.F., Niessen, W.J., Vincken, K.L., Viergever, M.A.: Multiscale Vessel Enhancement Filtering. In: Wells, W.M., Colchester, A.C.F., Delp, S.L. (eds.) MICCAI 1998. LNCS, vol. 1496, pp. 130–137. Springer, Heidelberg (1998)
7. Liao, R., Luc, D., Sun, Y., Kirchberg, K.: 3-D reconstruction of the coronary artery tree from multiple views of a rotational X-ray angiography. The International Journal of Cardiovascular Imaging 26, 733–749 (2010)
8. Schneider, M., Sundar, H.: Automatic global vessel segmentation and catheter removal using local geometry information and vector field integration. In: ISBI, pp. 45–48 (2010)
9. http://www.itk.org/ITK/applications/Curve2DExtraction.html
10. Velho, L., Carvalho, P.C.P.: Mathematical optimization in computer graphics and vision. Morgan Kaufmann (2008)

A Variational Model for Multiphase Image Segmentation on an Implicit Open Surface and Its Fast Algorithms

Jinming Duan, Zhenkuan Pan, Weibo Wei, Cunliang Liu, and Guodong Wang

College of Information Engineering, Qingdao University, Qingdao, P.R. China, 266071

Abstract. Based on the expression of a open surface on which images are defined as intersection of zero level set of a signed distance function and a binary label function and by making use of concepts of intrinsic gradient and divergence, the partitioning strategy of regions on a surface via m binary label functions for 2^m regions, a general varaitional model for multiphase image segmentation on an implicit open surface is proposed. Based on techniques of convex relaxation and thresholding, the gradient descent method, dual method, Split Bregman method, augmented Lagrange method are designed, where, the last three methods are fast ones. In order to improve its efficiency and make it implement easily, we propose another new method based on dual method without convex relaxation and thresholding of binary label functions, which is referred as direct dual method. Finally, numerical examples validate the model and its fast algorithms proposed in this paper.

Keywords: Multiphase segmentation, variational method, gradient descent method, dual method, Split Bregman method, augmented Lagrange method.

1 Introduction

Many tasks in computer vision and image processing can be regarded as a class of multiphase image segmentation problems on a surface. Contrast to traditional ones in rectangular domain, it depends not only on image features but also on local geometric surface on which the image is defined and is a typical complex problem coupling image and geometry.

Making use of the accurate calculation of gradient and curvature on zero level set, and the concepts of intrinsic gradient and intrinsic divergence of an image defined on a closed surface, [1] originally proposed a variational image diffusion model on an implicit surface using the variational level set method. Using the relevant concepts in [1], [2] realized the Chan-Vese[3] model on an implicit surface for two phase image segmentation using the intersection of a static zero level set and another dynamic level set to express the contour for region division on a surface.

The above mentioned models are defined on a closed surface which can be expressed using one level set function(LSF). But in some cases such as the images are defined on an open surface, and a LSF can not denote an open surface[4], so the surface expression technique mentioned above can not be used directly to image

J. Yang, F. Fang, and C. Sun (Eds.): IScIDE 2012, LNCS 7751, pp. 801–809, 2013.

segmentation on it. In order to deal with this problem, we denote the open surface using intersection of a static zero level set and a static binary label function for the multiphase image segmentation on it, and the main task of image partition in variational image segmentation is to design a characteristic function for every region. In this paper, we use m binary functions for 2^m regions proposed by [5] with less label functions and high computation efficiency.

The variational image segmentation model always leads to a series of partial differential equations(PDEs) which must be solved via numerical methods with low computation efficiency. For the variational image segmentation on an implicit surface, the time consuming is more severe, so we devote many efforts to design fast algorithms for the proposed model in this paper. Firstly, we extend the dual method[6], Split Bregman algorithm[7], augmented Lagrangian method(ALM)[8] proposed for image restoration to it, then we design another more simple and fast direct dual method(DDM) for the model in this paper.

2   The Variational Model of Two Phase Image Segmentation on Implicit Open Surface and Its Gradient Descent Equation

Let $\psi(x)$ be a static LSF as a continuous signed distance function, and its zero level set $\{x:\psi(x)=0\}$ denotes a close surface. Let $\Phi(x)\in\{0,1\}$ be a static binary function. The intersection $\{x:\Phi(x)=1,\psi(x)=0\}$ can be used to denote an implicit open surface s as in Fig.1.

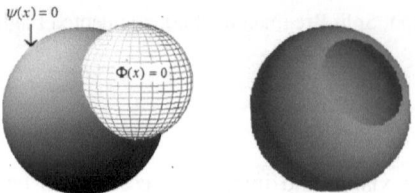

Fig. 1. An implicit open surface

If f is the image defined on this surface to be segmented, we introduce another dynamic binary function $\phi(x)\in\{0,1\}$ for image segmentation using its intersection with the open surface s. If it is divided into two regions s_1 and s_2, $s=s_1\cup s_2$, $s_1\cap s_2=\varnothing$, the characteristic functions of these two regions can be defined as $\chi_1(x)=\phi(x)=\begin{cases}1 & x\in S_1 \\ 0 & otherwise\end{cases}$

and $\chi_2(x)=1-\phi(x)=\begin{cases}1 & x\in S_2 \\ 0 & otherwise\end{cases}$ .The length of their interface is $\int_\Omega\left|P_{\nabla\psi}\nabla\phi\right||\nabla H(\psi)|\Phi dx$,

where, $P_{\nabla\psi}\nabla\phi$ is the intrinsic gradient of ϕ on s, and $P_{\nabla\psi}=I-\dfrac{\nabla\psi(x)\otimes\nabla\psi(x)}{|\nabla\psi(x)|}$ is a

projection operator[1] to project a vector onto the local tangent plane of s. Based on these definitions, the Chan-Vese model[3] for planar two phase piecewise constant image segmentation can be reformulated on the implicit open surface as

$$\min_{\phi \in \{0,1\}} \left\{ \begin{array}{l} E(\phi, u_1, u_2) = \alpha_1 \int_\Omega (f - u_1)^2 \phi \mid \nabla H(\psi) \mid \Phi dx + \alpha_2 \int_\Omega (f - u_2)^2 (1 - \phi \mid \nabla H(\psi) \mid \Phi dx \\ \quad + \gamma \int_\Omega \mid P_{\nabla \psi} \nabla \phi \mid\mid \nabla H(\psi) \mid \Phi dx \end{array} \right\}, \tag{1}$$

where, $\alpha_1, \alpha_2, \gamma$ are penalty parameters related with fidelity term and length term respectively, $H(\phi)$ is a typical Heaviside function. (1) is a minimization problem with multiple variables, it can be solved using alternating optimization procedure, i. e., fix ϕ to solve the sub-problem of minimization on u_1 and u_2, then fix u_1 and u_2 to solve the sub-problem of minimization on ϕ until the energy functional achieves the minimum. When u_1 and u_2 are obtained, (1) can be transformed into the optimization problem on ϕ as

$$\underset{\phi \in \{0,1\}}{Min} \left\{ E(\phi) = \int_\Omega Q_{12}(u_1, u_2) \phi \mid \nabla H(\psi) \mid \Phi dx + \gamma \int_\Omega \mid P_{\nabla \psi} \nabla \phi \mid\mid \nabla H(\psi) \mid \Phi dx \right\}, \tag{2}$$

where, $Q_{12}(u_1, u_2) = \alpha_1 (f - u_1)^2 - \alpha_2 (f - u_2)^2$. According to [9], we replace $\phi \in \{0,1\}$ with convex version $\phi \in [0,1]$, then apply variational method and gradient descent equation method to obtain the evolution of ϕ with artificial time t

$$\frac{\partial \phi}{\partial t} = \frac{\gamma}{|\nabla \psi|} \nabla \cdot \left(\frac{P_{\nabla \psi} \nabla \phi}{|P_{\nabla \psi} \nabla \phi|} |\nabla \psi| \Phi \right) - Q_{12}(u_1, u_2) \Phi . \tag{3}$$

Because ϕ must satisfy constraint $\phi \in [0,1]$, we can use the following projection formula to realize it

$$\phi = Max(0, Min(\phi, 1)) , \tag{4}$$

let the static solution of (3) be ϕ, the segmentation result can be obtained through the following thresholding technique

$$\phi = \begin{cases} 1 & \phi \geq \sigma \\ 0 & otherwise \end{cases}, \sigma \in (0,1) . \tag{5}$$

3   A Variational Multiphase Image Segmentation Model on Implicit Open Surface and Its Fast Algorithms

In this section, we partition an image f defined on the same implicit open surface as in section 2 into $n = 2^m$ regions using m binary label functions. The task of multiphase image segmentation is to divide an image on S into n regions $S_i (i = 1, 2, ... n)$ such that $S = \bigcup_{i=1}^n S_i$, $S_i \cap S_j = \varnothing, (i \neq j)$. Let $\chi_i(x) = \begin{cases} 1 & x \in S_i \\ 0 & otherwise \end{cases}$ be the characteristic function of region S_i, then $\chi_i(x)$ must fulfill $\sum_{i=1}^n \chi_i(x) = 1$ to avoid the overlapping and vacuum problem. The scheme of Vese and Chan[5] satisfies this condition naturally, thus gets rid of this constraint. To use it systematically, we define the vectorial binary label function $\phi=\{\phi_1,\phi_2...\phi_m\}$ and $\phi_i \in \{0,1\}(i=1,2,...,m)$. In order to express the characteristic functions in general form. For the ith region, we rewrite a natural number in binary form: $i-1=b_{i-1}^1 b_{i-1}^2 ... b_{i-1}^m$, where, $b_{i-1}^j =0 \vee 1$, then the characteristic function of the ith region can be designed according to the idea of [5] as $\chi_i(x)=\prod_{j=1}^{m}\left(b_{i-1}^j + (-1)^{b_{i-1}^j}\phi_j\right)$. Thus, the energy minimization problem of multiphase image segmentation on an implicit open surface can be written as

$$\underset{u,\phi\in[0,1]^m}{Min}\left\{E(\phi,u)=\sum_{i=1}^{n}\alpha_i\int_\Omega(f-u_i)^2\chi_i|\nabla H(\psi)|\Phi dx + \sum_{j=1}^{m}\gamma_j\int_\Omega|P_{\nabla\psi}\nabla\phi_j|\,|\nabla H(\psi)|\Phi dx\right\},\qquad(6)$$

where, u_i is the mean value of intensity of the ith region.

When $u_i(i=1,2,...,n)$ is known, we turn to the sub-problem of minimization on $\phi_j,(j=1,2,...m)$ based on the alternating optimization strategy. Also, $\phi_j \in \{0,1\}$ is replaced with its convex version $\phi_j \in [0,1]$ firstly, then we get the evolution equations of $\phi_j(j=1,2,...,n)$ by solving the following minimization problem using variational method (7) and gradient descent equation method (8)

$$\underset{\phi_j\in[0,1]}{Min}\left\{E(\phi_j)=\sum_{i=1}^{n}\alpha_i\int_\Omega(f-u_i)^2\chi_i|\nabla H(\psi)|\Phi dx + \gamma_j\int_\Omega|P_{\nabla\psi}\nabla\phi_j|\,|\nabla H(\psi)|\Phi dx\right\},\qquad(7)$$

$$\frac{\partial\phi_j}{\partial t}=\frac{\gamma_j}{|\nabla\psi|}\nabla\cdot\left(\frac{P_{\nabla\psi}\nabla\phi_j}{|P_{\nabla\psi}\nabla\phi_j|}|\nabla\psi|\Phi\right)-\sum_{i=1}^{n}\alpha_i\frac{\partial\chi_i}{\partial\phi_j}(f-u_i)^2\Phi.\qquad(8)$$

We can also use the projection formula (4) to realize the constraint $\phi_j \in [0,1]$. From the static solution ϕ_j of (8), we can use tresholding to get the final segmentation result.

In order to design the Dual Method(DM) [6] of (7), we introduce dual variables $\vec{p}_j(j=1,2,...,m)$ to transform it into the following Max-Min problem

$$\left(\phi_j^{k+1},\vec{p}_j^{k+1}\right)=\underset{\phi_j\in[0,1]}{ArgMin}\underset{\vec{p}_j,|\vec{p}_j|\le 1}{Sup}\left\{\sum_{i=1}^{n}\alpha_i\int_\Omega(f-u_i)^2\chi_i|\nabla H(\psi)|\Phi dx + \gamma_j\int_\Omega\phi_j P_{\nabla\psi}\nabla\cdot\vec{p}_j|\nabla H(\psi)|\Phi dx\right\},\qquad(9)$$

where, $P_{\nabla\psi}\nabla\cdot\vec{p}=\sum_{i=1}^{3}\sum_{j=1}^{3}\left(\frac{\partial\vec{p}_i}{\partial x_j}-\left(\nabla\psi_j\cdot\frac{\partial\vec{p}_i}{\partial x_j}\right)/|\nabla\psi|^2\nabla\psi_i\right)$. When $\vec{p}_j^k$ is fixed, the gradient descent equation of ϕ_j can be obtained as

$$\frac{\partial\phi_j}{\partial t}=-\sum_{i=1}^{n}\alpha_i(f-u_i)^2\frac{\partial\chi_i}{\partial\phi_j}-\gamma_j P_{\nabla\psi}\nabla\cdot\vec{p}_j^k,\qquad(10)$$

and its explicit iterative formula is

$$\phi_j^{k+1}=\phi_j^k-\tau_\phi\left(\sum_{i=1}^{n}\alpha_i(f-u_i)^2\frac{\partial\chi_i}{\partial\phi_j}+\gamma_j P_{\nabla\psi}\nabla\cdot\vec{p}_j^k\right),\qquad(11)$$

where τ_ϕ is the artificial time step. Due to the constraint $\phi_j^{k+1}\in[0,1]$, the result of (10) should be calculated using (4). When ϕ_j^{k+1} is fixed, the gradient ascent equation of $\vec{p}_j$

by solving the constrained optimization problem (9) using KKT condition and Chambolle's observation [6] is

$$\frac{\partial \vec{p}_j}{\partial t} = -A_j^k - \left|A_j^k\right| \vec{p}_j \Phi ,$$

(12)

where, $A_j^k = P_{\nabla \psi} \nabla \left(\phi_j^k \left|\nabla \Psi\right| \Phi\right)$. Using semi-implicit iterative scheme, the discrete version of (12) is

$$\vec{p}_j^{k+1} = \frac{\vec{p}_j^k - \tau_p A_j^k}{1 + \tau_p \left|A_j^k\right| \Phi} , \quad \vec{p}_j^0 = \vec{0} .$$

(13)

In order to ensure the convergence, two time steps must fulfill the condition $\tau_\phi \tau_p \le \frac{1}{8}$.

After the total energy achieve minimum, it is necessary to threshold the results of (11) using (5).

In order to design the Split Bregman Method(SBM) [7] of (7) , we introduce m auxiliary variables $\vec{\omega}_j (j=1,2,...m)$ and m Bregman iterative parameters $\vec{b}_j (j=1,2,...m)$, such that $\vec{\omega}_j \approx P_{\nabla \psi} \nabla \phi_j$ when the following energy functional achieves minimum

$$\left(\phi_j^{k+1}, \vec{\omega}_j^{k+1}\right) = \underset{\phi \in [0,1], \vec{\omega}_j}{ArgMin} \left\{ \begin{array}{l} \sum_{i=1}^{n} \alpha_i \int_\Omega (f-u_i)^2 \chi_i \left|\nabla H(\psi)\right| \Phi dx + \gamma_j \int_\Omega \left|\vec{\omega}_j\right| \left|\nabla H(\psi)\right| \Phi dx \\ + \frac{\theta_j}{2} \int_\Omega \left(\vec{\omega}_j - P_{\nabla \psi} \nabla \phi_j - \vec{b}_j^{k+1}\right)^2 \left|\nabla H(\Psi)\right| \Phi dx \end{array} \right\} ,$$

(14)

where, $\vec{b}_j^{k+1} = \vec{b}_j^k + P_{\nabla \psi} \nabla \phi_j^k - \vec{\omega}_j^k$, $\left(\vec{b}_j^0 = \vec{\omega}_j^0 = 0\right)$. Using alternating minimization strategy, we can get the following Euler-Lagrange equations on $\phi_j (j=1,2,...,m)$

$$\sum_{i=1}^{n} \alpha_i (f-u_i)^2 \frac{\partial \chi_i}{\partial \phi_j} \Phi + \frac{\theta_j}{\left|\nabla \psi\right|} \nabla \cdot \left(\left(\vec{\omega}_j - P_{\nabla \psi} \nabla \phi_j - \vec{b}_j^{k+1}\right) \left|\nabla \psi\right| \Phi\right) = 0 .$$

(15)

The results of (15) should be projected on $[0,1]$ as (4). The sub-optimization on $\vec{\omega}_j$ leads to the following generalized soft thresholding formula in analytical form

$$\vec{\omega}_j^{k+1} = Max\left(\left|P_{\nabla \psi} \nabla \phi_j^{k+1} + \vec{b}_j^{k+1}\right| - \frac{\gamma_j}{\theta_j}, 0\right) \frac{P_{\nabla \psi} \nabla \phi_j^{k+1} + \vec{b}_j^{k+1}}{\left|P_{\nabla \psi} \nabla \phi_j^{k+1} + \vec{b}_j^{k+1}\right|}, 0\frac{0}{0} = 0 .$$

(16)

The final segmentation result is obtained via thresholding formula (5).

In order to design the Augmented Lagrangian Method(ALM) [8] of (7), we introduce auxiliary variables $\vec{\omega}_j (j=1,2,...m)$, Lagrangian multipliers $\vec{\lambda}_j (j=1,2...m)$ and penalty parameter $\mu_j (j=1,2,...,m)$ to transform (7) into the following iterative minimization formulation

$$\left(\phi_j^{k+1}, \vec{\omega}_j^{k+1}, \vec{\lambda}^{k+1}\right) = \underset{\phi_j \in [0,1], \vec{\omega}_j, \vec{\lambda}_j}{ArgMin} \left(\begin{array}{l} \sum_{i=1}^{n} \alpha_i \int_\Omega (f-u_i)^2 \chi_i \left|\nabla H(\psi)\right| \Phi dx + \gamma_j \int_\Omega \left|\vec{\omega}_j\right| \left|\nabla H(\psi)\right| \Phi dx \\ + \frac{\mu_j}{2} \int_\Omega \left(\vec{\omega}_j - P_{\nabla \psi} \nabla \phi_j\right)^2 \left|\nabla H(\psi)\right| \Phi dx + \int_\Omega \vec{\lambda}_j \left(\vec{\omega}_j - P_{\nabla \psi} \nabla \phi_j\right) \left|\nabla H(\psi)\right| \Phi dx \end{array} \right) .$$

(17)

Here, $\bar{\lambda}_j^0 = 0$, $\bar{\omega}_j^0 = 0$. When $\bar{\omega}_j^k$, $\bar{\lambda}_j^k$ are fixed, ϕ_j^{k+1} can be obtained via solving the following Euler-Lagrange

$$\sum_{i=1}^n \alpha_i (f - u_i)^2 \frac{\partial \chi_i}{\partial \phi_j} \Phi + \frac{1}{|\nabla \psi|} P_{\nabla_\psi} \nabla \cdot (\bar{\lambda}_j^k |\nabla \psi| \Phi) + \frac{\mu_j}{|\nabla \psi|} \nabla \cdot ((\bar{\omega}_j^k - P_{\nabla_\psi} \nabla \phi_j) |\nabla \psi| \Phi) = 0 \tag{18}$$

with $\phi_j^{k+1} = Max(Min(\phi_j^{k+1}, 1), 0)$ as constraint projection. When ϕ_j^{k+1} and $\bar{\lambda}_j^k$ are fixed, $\bar{\omega}_j^{k+1}$ can be expressed as the following generalized soft thresholding formula

$$\bar{\omega}_j^{k+1} = Max\left(\left| P_{\nabla_\psi} \nabla \phi_j^{k+1} - \frac{\bar{\lambda}_j^k}{\mu_j} \right| - \frac{\gamma_j}{\mu_j}, 0 \right) \frac{P_{\nabla_\psi} \nabla \phi_j^{k+1} - \frac{\bar{\lambda}_j^k}{\mu_j}}{\left| P_{\nabla_\psi} \nabla \phi_j^{k+1} - \frac{\bar{\lambda}_j^k}{\mu_j} \right|} \quad 0 \cdot \frac{0}{|0|} = 0 \quad . \tag{19}$$

At the end of each loop, the Lagrange multipliers $\bar{\lambda}_j^k$ can be updated using

$$\bar{\lambda}_j^{k+1} = \bar{\lambda}_j^k + \mu_j \left(\bar{\omega}_j^{k+1} - P_{\nabla_\psi} \nabla \phi_j^{k+1} \right) . \tag{20}$$

The final segmentation result is obtained via thresholding formula (5) also.

4   The Direct Dual Method-DDM

Contrast to the methods in section 3 that employ convex relaxation and thresholding techniques, the new method in this section uses discrete binary label functions directly in the computation process, leading to a fast and simple implementation of (6). In order to do so, we reformulate (6) as

$$\underset{\phi_j \in \{0,1\}}{Min} \left\{ E(\phi_j) = \sum_{i=1}^n \alpha_i \int_\Omega (f - u_i)^2 \chi_i |\nabla H(\psi)| \Phi dx + \gamma_j \int_\Omega |P_{\nabla_\psi} \nabla \phi_j| |\nabla H(\psi)| \Phi dx \right\} . \tag{21}$$

Circumventing convex relaxation and thresholding processes, we introduce dual variables $\bar{p}_j (j = 1, 2, ..., m)$ to transform (21) into the following equivalent form

$$\left(\phi_j^{k+1}, \bar{p}_j^{k+1} \right) = \underset{\phi_j \in \{0,1\}}{ArgMin} \underset{\bar{p}_j |\bar{p}_j| \le 1}{Sup} \left\{ \gamma_j \int_\Omega \phi_j P_{\nabla_\psi} \nabla \cdot \bar{p}_j |\nabla H(\psi)| \Phi dx + \sum_{i=1}^n \alpha_i \int_\Omega (f - u_i)^2 \chi_i |\nabla H(\psi)| \Phi dx \right\} . \tag{22}$$

For every constraint $\phi_j (\phi_j - 1) = 0$, we add it to (22) via Lagrangian multiplier method, which results in the following minimization problem

$$\left(\phi_j^{k+1}, \bar{p}_j^{k+1} \right) = \underset{\bar{p}_j |\bar{p}_j| \le 1}{Arg \, Min \, Sup} \left\{ \begin{array}{l} \gamma_j \int_\Omega \phi_j P_{\nabla_\psi} \nabla \cdot \bar{p}_j |\nabla H(\psi)| \Phi dx + \sum_{i=1}^n \alpha_i \int_\Omega (f - u_i)^2 \chi_i |\nabla H(\psi)| \Phi dx \\ + \int_\Omega \lambda_j \phi_j (\phi_j - 1) |\nabla H(\psi)| \Phi dx \end{array} \right\} . \tag{23}$$

When $\bar{p}_j^k$ is fixed, according to the KKT condition of a constrained optimization problem[6], we know $\lambda_j \ge 0$ and

$$\gamma_j P_{\nabla_\psi} \nabla \cdot \bar{p}_j + \sum_{i=1}^n \alpha_i (f - u_i)^2 \frac{\partial \chi_i}{\partial \phi_j} + \lambda_j (2\phi_j - 1) = 0 . \tag{24}$$

Because $\phi_j \in \{0,1\}$ and $|2\phi_j - 1| = 1$, we can derive $\lambda_j = \left| \gamma_j P_{\nabla_\psi} \nabla \cdot \vec{p}_j + \sum_{i=1}^{n} \alpha_i (f - u_i)^2 \frac{\partial \chi_i}{\partial \phi_j} \right|$ through moving the third term to right hand side of (24) and then take absolute value from both sides. Thus

$$\phi_j^{k+1} = \frac{1}{2} \left[1 - sign \left(\sum_{i=1}^{n} \alpha_i (f - u_i)^2 \frac{\partial \chi_i}{\partial \phi_j} + \gamma_j P_{\nabla_\psi} \nabla \cdot \vec{p}_j^k \right) \right]. \tag{25}$$

is derived after λ_j is substituted into (24). The computation of $\vec{p}_j^{k+1}$ is the same as (13). (13) and (25) are very simple and fast to implement.

5   Experiments and Numerical Analysis

We set up 3 experiments to validate the model on an implicit open surface and to compare the efficiency of the proposed algorithms with the GDM. All experiments are performed using Matlab 2010b on a Windows 7 platform with an Intel Core 2 Duo CPU at 2.33GHz and 2GB memory.

Scar area measurement on human organs is a worthy work due to the inaccuracy using 2D plane image instead of 3D surface. Firstly, we will show this difference in Fig.2, where, the first row gives the result of scar segmentation on a real 3D feet, the second row gives the result of the same scar on 2D plane image. The scar areas in Fig.2(c) and Fig.2(f) are calculated using $S_{3D} = \int_\Omega \phi |\nabla H(\psi)| \Phi dx$ and $S_{2D} = \int_\Omega \phi dx$ respectively, where, ϕ is a label function. The results are $S_{3D} = 7.103 \times 10^3$, $S_{2D} = 6.596 \times 10^3$ which shows the significance of this work.

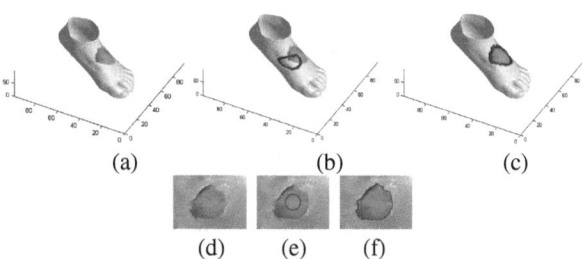

(a)   (b)   (c)

(d)   (e)   (f)

Fig. 2. Scar area measurement on a 3D open surface and on 2D image plane. (a) Scar on the 3D open surface; (b) Initial contour on the 3D open surface; (c) Final contour; (d) same scar on 2D plane; (e) Initial contour on the 2D image plane; (f) Final contour.

As the second example, we present a result of three phase image segmentation on an open vase surface in Fig.3, where, Fig .3(a) is an image defined on a open surface, Fig .3(d) shows the result of segmentation on an implicit open surface. Here, we use two binary label functions ϕ_1 and ϕ_2 to denote characteristic functions as

$\chi_1(x) = \phi_1(1-\phi_2)$, $\chi_2(x) = (1-\phi_1)\phi_2$, $\chi_3(x) = \phi_1\phi_2$, and $\chi_4(x) = (1-\phi_1)(1-\phi_2)$ to get 3 different regions with blue, green and red colors respectively.

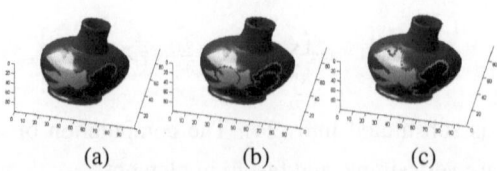

(a)   (b)   (c)

Fig. 3. Three phase image segmentation on an open vase. (a) An image to be segmented on the open surface; (b) Initial contours; (c) Final contours.

In the last experiment shown in Fig.4, we give a further illustration of the use of open surfaces. A interesting issue is when the objects to be segmented are inside of a closed surface like Fig.4 (a), we cannot see them without opening it. Unlike in Fig.3, we can get just right four phase segmentation by using two binary label functions ϕ_1 and ϕ_2 without making an empty phase. In Fig.4 (d), we get three correct contours labeled by $\chi_1(x)$ (blue), $\chi_2(x)$ (green), $\chi_3(x)$ (red) and a phase of background by $\chi_4(x)$.

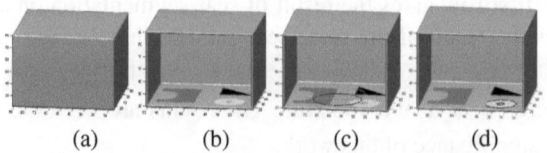

(a)   (b)   (c)   (d)

Fig. 4. Four phase image segmentation on an open cube. (a) A closed surface; (b) An image to be segmented inside the closed surface (a); (c) Initial contours; (d) Final contours..

We list the computing time using different algorithms for the three experiments in Tab.1. The last 4 methods use the convex relaxation and thresholding techniques, DDM does not use them. It is the main reason that the DDM is the most fast. Due to the computation of $\nabla \cdot \left((P_{\nabla_\psi}\nabla\phi)/|P_{\nabla_\psi}\nabla\phi|\right)$ with complex finite difference scheme and regularization of denominator, the GDM is the slowest. The SBM is equivalent with ALM theoretically [8], so their computational efficiencies are almost same.

Table 1. Comparison of iterations and computation time using different algorithms

Methods	Each iteration of CPU time/s			Total CPU time/s			Iteration numbers		
	Fig. 2	Fig. 3	Fig. 4	Fig. 2	Fig. 3	Fig. 4	Fig. 2	Fig. 3	Fig. 4
GDM	6.291	17.035	18.848	19.181	68.922	87.401	3	4	5
DDM	0.863	6.443	6.818	2.816	26.494	33.848	3	4	5
DM	1.528	7.537	7.515	4.601	29.741	34.721	3	4	5
SBM	4.076	14.333	14.165	16.609	54.387	70.886	3	4	5
ALM	4.112	14.397	15.956	16.825	54.513	72.778	3	4	5

6   Concluding Remarks

Taking advantage of level set method, vecrorial binary label function method and the concepts of intrinsic gradient, intrinsic divergence, a general varaitional model for multiphase image segmentation on an implicit surface is proposed. Based on techniques of convex relaxation and thresholding, its GDM, DM, SBM, ALM are designed, where, the last three methods are fast ones. To improve its efficiency and implement it easily, we propose another new method DDM without convex relaxation and thresholding of binary label functions, which is the most fast one.

Acknowledgments. This work is supported by the Natural Science Foundation of China(61170106).

References

1. Bertalmio, M., Cheng, L., Osher, S., Sapiro, G.: Variational problems and partial differential equations on implicit surface. J. of Com. Phy. 174(2), 759–780 (2001)
2. Li, J.-G., Pan, Z.-K., Wei, W.-B., Li, C.-F.: A variational level set method for image segmentation on implicit surfaces. Journal of Image and Graphics 15(5), 808–813 (2010)
3. Chan, T.F., Vese, L.A.: Active contours without edges. IEEE Tran. on Image Processing 10(2), 266–277 (2001)
4. Osher, S., Fedkiw, R.: The level set method and dynamic implicit surfaces. Springer (2002)
5. Vese, L.A., Chan, T.F.: A multiphase level set framework for image segmentation using the mumford and shah model. Int. J. of Com. Vis. 50(3), 271–293 (2002)
6. Chambolle, A.: An algorithm for total variation minimization and applications. Journal of Mathematical Imaging and Vision 20(1), 89–97 (2004)
7. Goldstein, T., Osher, S.: The Split Bregman algorithm for L1 regularized problems. SIAM Journal on Imaging Sciences 2(2), 323–343 (2009)
8. Wu, C., Tai, X.-C.: Augmented Lagrangian method, dual methods, and Split Bregman iteration for ROF, vectorial TV, and high order models. SIAM J. Ima. Sci. 3(3), 300–339
9. Bresson, X., Esedoglu, S., Vandergheynst, P., Thiran, J.-P., Osher, S.: Fast global minimization of the active contour/snake model. Journal of Mathematical Imaging and Vision 28(2), 151–167 (2007)

Fusing Discrete Cosine Transform and Multi-level Center-Symmetric Local Binary Pattern Features for Periocular Recognition

Zhe Chen[1], Yan Zhang[2], Changyin Sun[1], and Wankou Yang[1]

[1] School of Automation, Southeast University, Nanjing 210096, China
[2] The Third Research Institute of Ministry of Public Security, Shanghai 200031, China

Abstract. A novel method by fusing the features of discrete cosine transform (DCT) and multi-level center-symmetric local binary pattern (CS-LBP) is proposed for periocular recognition in this paper. Because that CS-LBP is used to extract features from the images only once, by which the extracted texture features are not adequate to represent the periocular images, we employ multi-level CS-LBP to extract more abundant and informative texture features for more times to get the spatial features. The primary information of the periocular image was centralized in a small number of DCT coefficients which were used as the frequency features of the image. The periocular image was divided regularly into small regions from which histograms were computed and concatenated into a spatial global histogram used as descriptor vector of the periocular image. Then the DCT features and the CS-LBP features were fused posterior to the normalization. Experimental results on ORL face database, AR face database and Morph periocular database demonstrate that the proposed method outperforms DCT or LBP features for periocular recognition.

Keywords: discrete cosine transform (DCT), multi-level center-symmetric local binary pattern (CS-LBP), features fusion, periocular recognition.

1 Introduction

The periocular biometric is gaining attention lately as a means of improving robustness of face or iris biometric modalities [1], [2]. It has been shown that periocular region can be independently used for recognition and can aid face or iris recognition when the inherent biometric content in the source images is poor [3]. Some researches based on testing the behavior of the eye to judge whether the driver is fatigue or not. As a biometircs, periocular analysis has also been applied in racial and gender recognition and achieves good results [4].

LBP has received considerable attentions in texture analysis field for its efficiency, simple theory and low computation [5, 6, 7, 8]. However, the dimensionality of the feature vector extracted by LBP is usually very high. On the contrary, CS-LBP can largely reduce the extracted feature dimensions. Therefore, in this paper, CS-LBP was employed to extract spatial features from periocular images. Because that DCT can describe the frequency characteristics of the image and then a new periocular

J. Yang, F. Fang, and C. Sun (Eds.): IScIDE 2012, LNCS 7751, pp. 810–816, 2013.

recognition algorithm by fusing multi-level CS-LBP and DCT was brought forward. Experimental results on ORL and Morph database demonstrate that compared with LBP, CS-LBP can achieve the comparable performance in terms of recognition rate with lower feature dimensions and higher identification accuracy.

2   Fusing Discrete Cosine Transform (DCT) and Multi-Level Center-Symmetric Local Binary Pattern (CS-LBP)

In this paper ,the nearest-neighbour classifier (NNC) is chosen for classification and identification. Finally, the proposed algorithm is shown in figure 1:

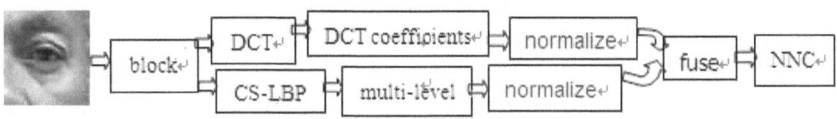

Fig. 1. The framework of the proposed algorithm

2.1   DCT

DCT which is a method of transformation has been widely applied to the filed of image and other data compression. Z.M.Haled etc [9] first used DCT for face recognition. As for a digital image X (x, y) with the size M×N, two dimensions (2D) DCT is defined as follows:

$$Y(u,v) = a(u)a(v)\sum_{x=0}^{M-1}\sum_{y=0}^{N-1} X(x,y)\cos[\frac{(2x+1)u\pi}{2M}]\cos[\frac{(2y+1)v\pi}{2N}] \quad (1)$$

Here u=0, 1, 2……M−1; v=0, 1, 2……N−1. Y(u,v) is the result of the DCT, which is namely DCT coefficients, $a(w)$ is defined as follows:

$$a(w) = \begin{cases} \sqrt{{1}/{M}} & w = 0 \\ \sqrt{{2}/{M}} & w = 1,2,3...M\text{-}1 \end{cases} \quad (2)$$

The figure shows the DCT coefficients of an original periocular image.

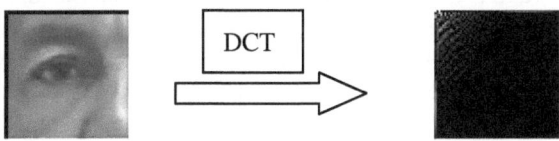

(a)   the original periocular image   (b) the DCT image

Fig. 2. Discrete cosine transform (DCT) coefficients

2.2   Multi-level Central-Semmetry Local Binary Pattern (CS-LBP)

An description of the traditional LBP operator is shown in Fig.3.

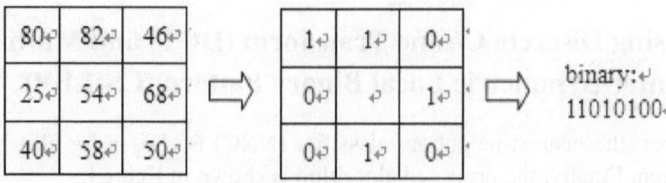

Fig. 3. The basic LBP operator

CS-LBP[10] introduces the thought of central symmetry to the theory of encoding image based on the traditional LBP. In the neighborhood as it defined, CS-LBP compares the gray-level difference of the pairs of opposite pixels in a region. If the gray-level of the pixel is greater than the symmetrical pixel, then replace the gray-level of the pixel with one, otherwise zero. Then convert it to a decimal number as the encoding value for the central pixel.

$$LBP = s(g_0 \text{-} g_c)2^0 + s(g_1 \text{-} g_c)2^1 + s(g_2 \text{-} g_c)2^2 + s(g_3 \text{-} g_c)2^3$$
$$+s(g_4 \text{-} g_c)2^4 + s(g_5 \text{-} g_c)2^5 + s(g_6 \text{-} g_c)2^6 + s(g_7 \text{-} g_c)2^7 \tag{3}$$

$$CS\text{-}LBP = s(g_0 \text{-} g_4)2^0 + s(g_1 \text{-} g_5)2^1 + s(g_2 \text{-} g_6)2^2 + s(g_3 \text{-} g_7)2^3 \tag{4}$$

$$LBP_{(P,R)}(x, y) = \sum_{i=0}^{N-1} s(g_i - g_c)2^i \tag{5}$$

$$s(x) = \begin{cases} 1 & x > 0 \\ 0 & else \end{cases} \tag{6}$$

$$CS\text{-}LBP_{(P,R)}(X,Y) = \sum_{i=0}^{(N/2)-1} s(g_i \text{-} g_{i+(N/2)})2^i \tag{7}$$

In the equation (5) and (7), (P, R) represents a round neighborhood, R is the radius of the round, P is the number of the pixels on the circle, N = P. In the (P, R) neighbourhood. CS-LBP gets a binary encoding string whose length is P/2, also converts the string into a decimal number. From the above analysis, we can conclude that the principle of CS-LBP and LBP are similar, but the length of CS-LBP histogram is much shorter than the LBP histogram.

CS-LBP is low complex, effective and fast, producing comparable results to LBP. Because the extraction can reflect the characteristics of gray image's local area [11], so we use it to extract features of periocular image. A gray image after an operation of CS-LBP can still be seen as a gray image, so we extract features by the same way of coding images for more times. Thus we can get multi-level CS-LBP features.

2.3   Feature Fusion

After the normalization of the image, first put the periocular image regularly into several pieces. Blocked images can enhance the local characteristics of space. Chen [12] has proved that DCT is orthogonal transformation. In this paper, we can choose a small square sub-block of the top left corner of the DCT image as the features.

The novel method mainly includes the following steps: (1) Block every source image and get the DCT coefficients for every blocked image and select parts of the coefficients as the feature vector of frequency domain; (2) Encode every original periocular image to get the multi-level coding feature images with the algorithm namely multi-level CS-LBP; (3) Divide each level coding feature image regularly into several regions which were not overlap each other, and statistics the histogram for each blocked image; (4) Concatenated the region histograms into a spatial global histogram. Assuming that the feature vectors of DCT and CS-LBP are respectively $v_1 \in R_{m1}$ and $v_2 \in R_{m2}$, the combination of the normalized feature vectors is $z \in R_{m1+m2}$, which is expressed as $z = [v_1/\sigma_1, v_2/\sigma_2]^T$. Here the σ_1 and σ_2 are respectively the standard deviation of v_1 and v_2 , which can be computed through the square root of the variances of feature vectors. At last a nearest-neighbour classifier is used to classify. Studies have shown that χ^2 measures are widely used and have some advantages.

Chi square statistic (χ^2) is defined as equation (8):

$$\chi^2(S,M) = \sum \frac{(S_i - M_i)^2}{S_i + M_i} \tag{8}$$

3   Experiments

To verify the effectiveness of the proposed method, we choose the standard ORL database, the AE database and a periocular database from Morph database for tests respectively with different kinds of methods .The standard ORL database contains 40 people, each have 10 images, a total of 400 face images. Among them, everyone's facial image includes the face expression, slight deflection, some degree of cover and other changes. We will adjust the size of the periocular region images to 64*64.

The first experiment on ORL database: we selected randomly N1 (here is :3,4... 9) images of each person as the training samples, the remaining as testing.

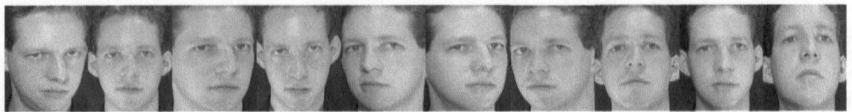

Fig. 4. 10 images of one person in ORL database

Table 1. The different results when the face images were divided by 4*4

samples / algorithm		recognition rate /%							
		3-7	4-6	5-5	6-4	7-3	8-2	9-1	average
LBP		89.2	91.2	93.0	93.7	95.0	96.2	95.0	93.3
CS-LBP		87.1	88.3	93.5	92.5	95.0	95.0	97.5	92.7
DCT+CS-LBP		87.5	90.4	94.5	93.5	95.8	95.0	97.5	93.4
DCT	5*5	80.3	82.9	83.5	87.5	90.0	88.7	90.0	86.1
	6*6	81.4	82.0	84.0	86.8	90.0	88.7	90.0	86.1
	7*7	81.7	82.0	84.0	86.8	89.1	88.7	90.0	86.0
DCT+multi-level CS-LBP	5*5	88.9	90.8	96.0	93.7	97.5	98.7	100.0	95.0
	6*6	88.5	90.4	96.0	94.3	96.6	100.0	100.0	95.1
	7*7	89.2	90.8	96.0	94.3	96.6	100.0	100.0	95.2

The second experiment on AR database: the images are showed as below:

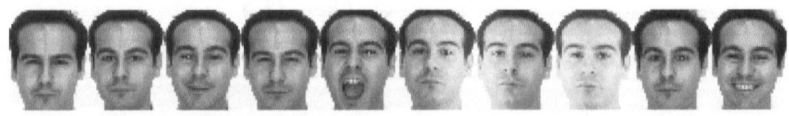

Fig. 5. 10 images of one person in the AR database

From table 1 and table 2, with the increasing of the training samples, recognition rate has been increasing. Compared with LBP, CS-LBP can get the close results with fewer dimensions. When the training samples increased, recognition rate has improved to a certain extent. From the results we can see that the method of fusing DCT and CS-LBP is better than it used CS-LBP or DCT alone.

Table 2. The experimental results on AR database when the images were divided by 5*4

samples algorithm	recognition rate /%‰							
	3-7	4-6	5-5	6-4	7-3	8-2	9-1	average
LBP	90.7	91.6	93.5	92.5	93.3	94.2	95.0	92.9
CS-LBP	90.0	92.0	92.5	91.5	89.1	93.8	92.5	91.6
DCT	78.9	82.3	85.0	87.0	87.3	88.2	90.0	85.5
DCT-CS-LBP	90.9	92.0	93.0	92.5	90.9	94.2	93.3	92.4
DCT+multi-level CS-LBP	**92.5**	**93.3**	**93.0**	**92.5**	**93.0**	**95.0**	**97.0**	**93.7**

The third experiment on Morph periocluar database showed as below:

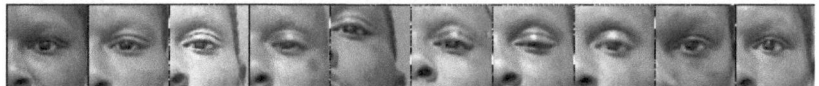

Fig. 6. 10 periocular images of one person in the Morph database

Table 3. The different results when the periocular images were divided by 4*4

samples algorithm	recognition rate /%‰							
	3-7	4-6	5-5	6-4	7-3	8-2	9-1	average
LBP	71.4	74.5	77.5	83.7	86.6	87.7	82.5	80.5
CS-LBP	68.5	69.5	75.0	83.1	85.8	90.0	87.5	79.9
DCT	55.0	58.7	64.0	70.0	70.8	76.2	67.5	66.0
DCT-CS-LBP	70.2	72.6	78.5	84.0	86.9	88.0	89.0	81.3
DCT+multi-level CS-LBP	**72.1**	**74.7**	**81.0**	**85.9**	**88.3**	**88.7**	**90.0**	**82.9**

From table 3, when the training samples are few, recognition rate is small, which is mainly because that the information of periocular images is not too much for distinguishing relative to face images. When choosing each person eight images for training, the rest two of each peraon for testing, the recognition rate achieved maximum, can reach 90.0%, indicating that this algorithm is feasible and successful.

4   Conclusion

We put forward a novel algorithm of fusing DCT and multi-level CS-LBP for periocular region recognition. Experiments on ORL face database, AR face database and Morph periocular database demonstrate that the proposed method has significant improvements compared with single DCT or LBP.

Acknowledgment. This project is supported by NSF of China (61005008, 61273023).

References

1. Park, U., Ross, A., Jain, A.K.: Periocular biometrics in the visible spectrum: a feasibility study. In: Biometrics: Theory, Applications and Systems (2009)
2. Woodard, D., Pundlik, S., Lyle, J., Miller, P.: Periocular region appearance cues for biometric identification. In: CVPR Workshop on Biometrics (2010)
3. Woodard, D., Pundlik, S., Miller, P., Jillela, R., Ross, A.: On the fusion of periocular and iris biometrics in non-ideal imagery. In: Proceedings of the IAPR International Conference on Pattern Recognition (2010)
4. Park, U., Ross, A., Jain, A.K.: Periocular biometrics in the visible spectrum: a feasibility study. In: Proceedings of the 3rd IEEE International Conference on Biometrics: Theory, Applications and Systems, pp. 153–158 (2009)
5. Park, Y.K., Kim, J.K.: Fast adaptive smoothing based on LBP for robust face recognition. IEE Electronics Letters 43(24), 1350–1351 (2007)
6. Heikkilä, M., Pietikäinen, M.: A texture-based method for modeling the background and detecting moving objects. IEEE Transactions on Pattern Analysis and Machine Intelligence 28(4), 657–662 (2006)
7. Ojala, T., Pietikäinen, M., Hardwood, D.: A comparative study of texture measures with classification based on feature distribution. Pattern Recognition 29, 51–59 (1996)
8. Ojala, T., Pietikainen, M., Maenpaa, T.: Multiresolution gray-scale and rotation invariant texture classification with Local binary patterns. IEEE Transactions on Pattern Analysis and Machine Intelligence 24(7), 971–987 (2002)
9. Hafed, Z.M., Levine, M.D.: Face recognition using the discrete cosine transform. International Journal of Compurer Vision 43(3), 167–188 (2001)
10. Heikkilä, M., Pietikäinen, M., Schmid, C.: Description of interest regions with local binary patterns. Pattern Recognition 42(3), 425–436 (2009)
11. Heikkilä, M., Pietikäinen, M., Schmid, C.: Description of Interest Regions with Center-Symmetric Local Binary Patterns. In: Kalra, P.K., Peleg, S. (eds.) ICVGIP 2006. LNCS, vol. 4338, pp. 58–69. Springer, Heidelberg (2006)
12. Chen, W., Er, M.J., Wu, S.: PCA and LDA in DCT domain. Pattern Recognition Letters 26(15), 2474–2482 (2005)

Artificial Neural Network Based Control Strategy Research and Simulation on Robot Uncalibrated Visual Servoing System

Jie Dong[1,2], Xuemei Jia[1,2], and Xu Zhang[1,2]

[1] School of Automation & Electrical Engineering, University of Science and Technology
Beijing, Beijing 100083, China
dongjie@ies.ustb.edu.cn
[2] Key Laboratory of Advanced Control of Iron and Steel Process (Ministry of Education),
Beijing 100083, China
jiaxuemeiok@139.com, zhangxu0415@126.com

Abstract. This paper presents a control strategy and new simulation model for the visual servoing control system with an eye-in-hand configuration from a 6-DOF robot of puma560. Because of the complexity of the calibration method on calculating image Jacobian matrix, an uncalibration method based on neural network is proposed. Firstly, simulation model of image-based robot visual servoing control system in Matlab7.8.0(R2009a) is established. Then, the concept of uncalibration is introduced and BP neural network controller is used as the visual controller instead of the calibration method to calculate image Jacobian matrix. Furthermore a convenient and wide range simulation model of robot uncalibrated visual servoing control system based on BP neural network is designed. The simulation results show that the simulation model is feasible, and can be achieved uncalibrated visual positioning.

Keywords: puma560, image Jacobian, BP neural network, uncalibrated visual servoing, visual positioning.

1 Introduction

Robot Visual Servo is firstly proposed by Hill and Park [1], which deals with the use of visual information into the control system. Traditional robot visual servo is based on calibrated techniques, including robotic calibration, the calibration of intrinsic and extrinsic camera parameters, and 3D geometry of the feature points [2]. However, accurate calibration is hardly realizable in real applications.

Therefore, uncalibration method has gradually become a research focus for many scholars[3]. There have been many solutions to this problem, such as, the online identification of the Jacobian matrix using Kalman filter, estimation of Jacobian matrix based on neural network, the ADRC dynamic compensation method and so on[4].

We choose BP neural network fitting image Jacobian by building a global nonlinear visual mapping model to realize the complex nonlinear mapping between the image coordinate system and the robot coordinate system. A model is also proposed to

J. Yang, F. Fang, and C. Sun (Eds.): IScIDE 2012, LNCS 7751, pp. 817–824, 2013.

simulate the control system and to verify the control strategy. Finally compared with calibrated visual servo, simulation result shows that uncalibrated visual servo control based on BP neural network could respond more quickly with the same steady-state error.

2   Image-Based Robot Visual Servo

In this paper, image-based robot visual servo (IBRVS) controls the movement of the manipulator according to the image characteristics, which fulfills the closed loop of image information [5]. So the structure diagram of the visual servo [6] is shown in Fig. 1.

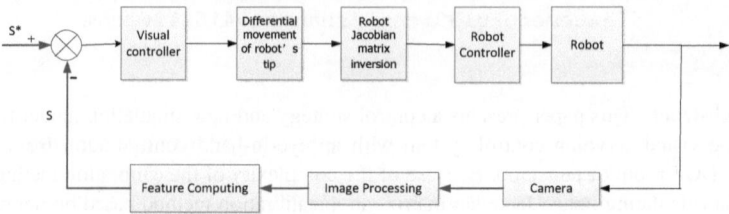

Fig. 1. Image-based robot visual servo structure diagram

Based on the IBRVS structure diagram (Fig.1) and robotics toolbox for MATLAB [7], we established a calibration simulation model using Matlab/Simulink shown in Fig.2.

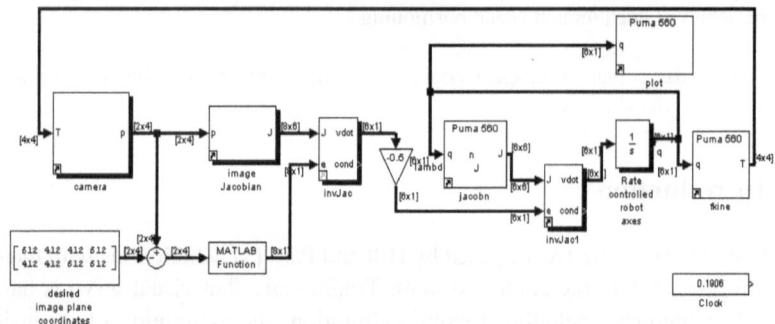

Fig. 2. Image-based robot visual servo simulation model

3   Design of Visual Controller Based on BP Network

3.1   Visual Mapping Relationship

The section presents a nonlinear visual mapping relationship for visual positioning with eye-in-hand robot [8].

In the camera coordinate system, the image features vector is $f = [f_1, f_2, \cdots, f_n]^T$, and the instantaneous velocity vector of the image features is $\Delta f = [\Delta f_1, \Delta f_2, \cdots, \Delta f_n]^T$; target movement velocity vector is $\Delta f_o = [\Delta f_{o1}, \Delta f_{o2}, \cdots, \Delta f_{on}]^T$; camera movement velocity vector is $\Delta f_c = [\Delta f_{c1}, \Delta f_{c2}, \cdots, \Delta f_{cn}]^T$.

In the base coordinates system, the location of the camera coordinate system is $X_c = [P_{cX} \quad P_{cY} \quad P_{cZ}]^T$, and the instantaneous velocity of the camera at that location is $\Delta X_c = [\Delta P_{cX} \quad \Delta P_{cY} \quad \Delta P_{cZ}]^T$; $X = [P_X \quad P_Y \quad P_Z]^T$ indicates the position of the robot's tip, and $\Delta X = [\Delta P_X \quad \Delta P_Y \quad \Delta P_Z]^T$ indicates the instantaneous velocity.

At time t, the velocity of image features is as follows:

$$\Delta f(t) = \Delta f_o(t) + \Delta f_c(t). \tag{1}$$

For visual positioning, target has the same position in the base coordinate system, which means $\Delta f_o(t) = 0$, therefore:

$$\Delta f(t) = \Delta f_c(t). \tag{2}$$

When the sampling frequency is high enough, we can obtain:

$$\Delta f(t) = \frac{f(t) - f(t-1)}{T}. \tag{3}$$

Where T is sampling period, to facilitate the solution, define $T = 1$.

Image features velocity components generated by the camera motion could be written as:

$$\Delta f_c(t) = F'_1\big(f(t), \Delta X_c(t)\big). \tag{4}$$

Since the camera is at the tip of the robot, ΔX_c and ΔX have one-to-one relationship, thus:

$$\Delta f_c(t) = F'(f(t), \Delta X(t)). \tag{5}$$

Based on (2) and (5), we can get:

$$\Delta f(t) = F'(f(t), \Delta X(t)). \tag{6}$$

Supposing $F(*,*)$ is a form of the unknown nonlinear function, we can obtain visual mapping relationship as follows [9]:

$$\Delta X(t) = F(f(t), \Delta f(t)). \tag{7}$$

3.2   Real-Time Motion Planning

The motion state of the target points could be planned in real-time basing on sequence image features, such as the target location on the image plane f , the velocity Δf and the desired location f^* [10].

At time k , the target location is $f(k) = (x(k), y(k))^T$, and the instantaneous velocity is $\Delta f(k) = (\Delta x(k), \Delta y(k))^T$. The velocity of the target after motion planning at time $k+1$ is $\Delta f'(k+1) = (\Delta x'(k+1), \Delta y'(k+1))^T$, desired location is $f^* = (x^*, y^*)^T$.

The target's velocity at time $k+1$ can be calculated by PI control law as:

$$\begin{bmatrix} \Delta x'(k+1) \\ \Delta y'(k+1) \end{bmatrix} = \begin{bmatrix} \Delta x(k) \\ \Delta y(k) \end{bmatrix} + c_1 \left[\begin{bmatrix} x(k) - x^* \\ y(k) - y^* \end{bmatrix} - c_2 \begin{bmatrix} x(k-1) - x^* \\ y(k-1) - y^* \end{bmatrix} \right]. \tag{8}$$

Where c_1 is proportional coefficient of PI Controller, and c_2 is integral coefficient.

3.3   Visual Controller

Input and Output for the BP Network

From equation (7), the inputs of the BP network are, image features position $f(t)$ and instantaneous velocity $\Delta f(t)$, and the outputs are the velocity of robot manipulator $\Delta X(t)$.

Number of the Hidden Layer Nodes

The primary of hidden nodes is determined the approximate range according to the empirical formula [11], as follows:

$$n_1 = \sqrt{n + m} + a . \tag{9}$$

Where n 、 m is the number of input and output layer nodes, and a is the constant between 0 and 10.

BP Network Offline Training and Simulation Model

(1) " training sample acquisition" model

BP network weight values are set through offline training methods, which use a standard gradient descent algorithm. Training samples could be get as follows: firstly, we build acquisition training sample model based on robotics toolbox for MATLAB shown in Fig.3, then add random number to the robot's tip, while ensuring that the target features are always located on the camera image plane. The robot's tip moves randomly and has a random speed, we record training samples as, the image features of target on the image plane f , the instantaneous velocity Δf , and the velocity of the robot's tip ΔX .

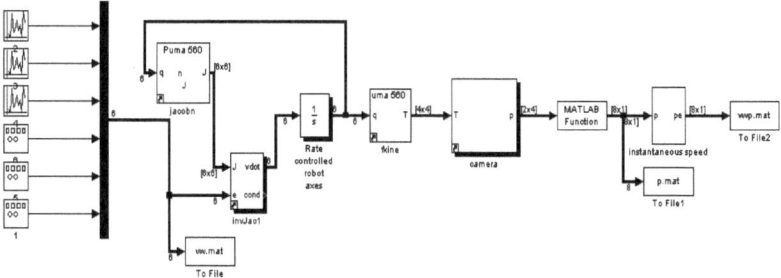

Fig. 3. Acquisition training sample model

(2) offline training of BP network

Basing on the training datas and neural network toolbox for MATLAB, we train network and take advantage of the "gensim" function to generate the neural network module "neural network1". We design BP neural network as the visual controller, and build robot uncalibrated visual servo simulation model based on BP network, which is shown in Fig.4.

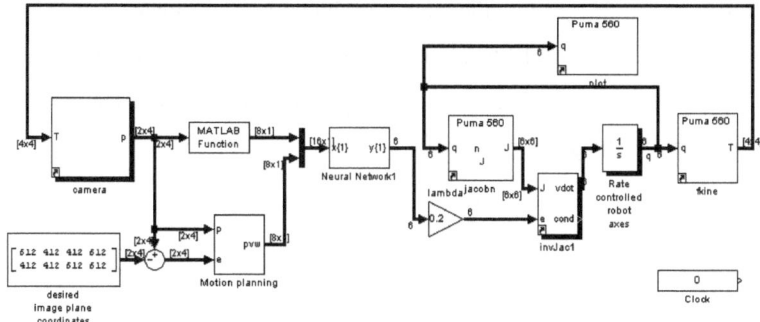

Fig. 4. Simulation model of robot uncalibrated visual servo based on BP network

"Motion planning" is a real-time motion planning module, which aims at planning the motion state of the target features. "Motion planning" module expansion plan is as shown in Fig.5.

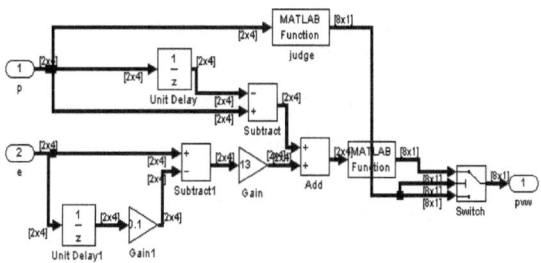

Fig. 5. "Motion planning" module expansion plan

In Fig.5, "judge" is a judgment module, which can be described as follows:

$$u = \begin{cases} 0 & |s^*-s| \le d \\ 1 & |s^*-s| > d \end{cases}.$$ (10)

Where u is the output of the module, d is threshold and can be set freely.

"Switch" is a switch control module, by the limiting condition ($u \ne 0$) concluded which module can be executed. If the condition is $u \ne 0$, real-time motion module continue to be executed, otherwise the output is 0.

This can be seen that the visual controller is composed by the real-time motion planning module, the neural network module, and proportion module.

4   Simulation and Results Analysis

In simulation, we select the four corners points of the target as feature points, then set the actual image location coordinates of the target as: $(463,403)$, $(314,552)$, $(463,700)$, $(611,552)$, and the expected location coordinates are: $(612,412)$, $(412,412)$, $(412,612)$, $(612,612)$. Initial six joint angles of robot are respectively: $(0 \quad 0.7856 \quad 3.1416 \quad 0 \quad 0.7854 \quad -0.7854)\,rad$ .The actual image features are shown in Fig.6, which the image coordinates (u,v) is the pixel values.

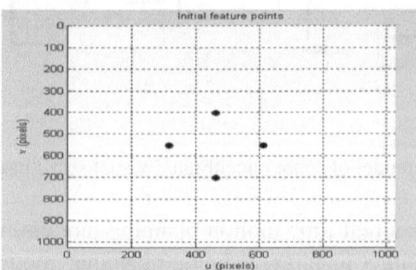

Fig. 6. The actual image feature points

According to the four feature points, we determine the dimension of the BP network inputs and outputs. The inputs of BP network are $[16\times1]$ column vector, and the network outputs are $[6\times1]$ column vector. BP network weights are get by 100 000 iterations from 543 training samples, and hidden layer nodes are 30 by the large number of experimental tests. After a number of simulation, real-time motion planning module: $k_p = 13$, $k_i = 0.1$. The proportion of modules: $k_p = 0.2$.

Target points trajectory is shown in Fig.7, and the initial coordinates are marked with "▷", the desired coordinates is "*". Therefore, it can be seen that the initial feature points are very close to the expected ones.

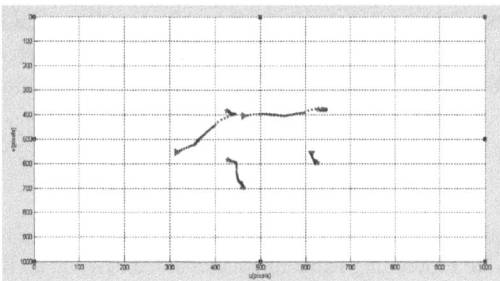

Fig. 7. Target points trajectory

In Fig.8, eight curves are respectively expressed horizontal axis and vertical axis error. The horizontal axis is time t, the vertical axis is the features error. From the figure, we know that the error tends to 0 about 0.85 s, which means the robot has reached the anticipated pose.

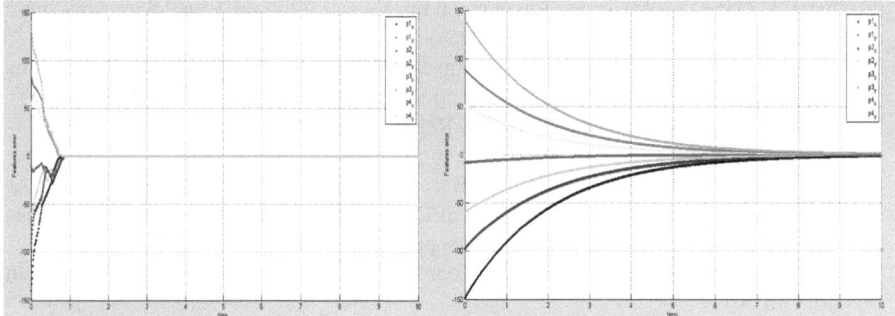

Fig. 8. Image features error trajectory   **Fig. 9.** Image feature error trajectory

When in phase synchronization and the same settings, we simulate IBRVS system, then get image features error trajectory as shown in Fig.9. Now it can be seen that the error tends to 0 about 9 s.

In short, from the simulation results, it can be seen the main advantage is that the small amount of calculation, fast, and the wide range of applications compared to the calibration method and calibration model. Meanwhile it also verifys the feasibility of uncalibration model, BP network visual controller can control the robot visual positioning.

5   Conclusion

This paper presents a convenient and wide range simulation model of robot uncalibrated visual servoing control system based on BP neural network, which eliminates complicated calibration processes and a large amount of computation. The simulation results show that the model can guarantee the accuracy and speed and realize the robot visual positioning in less time. However, many parameters need to be set in the simulation model. In the following studies, it needs to try to improve drawbacks.

Acknowledgements. This work is supported by the Fundamental Research Funds for the Central Universities (No. FRF-AS-09-006B). Beijing Key Discipline Development Program (No. XK100080537). National Natural Science Foundation (61074085).

References

1. Shademan, A., Farahmand, A., Jagersand, M.: Robust Jacobian Estimation for Uncalibrated Visual Servoing. In: Proceeding IEEE International Conference on Robotics and Automation, pp. 5564–5569 (2010)
2. Fu, Q., Zhang, Z., Shi, J.: Uncalibrated Visual Servoing Using More Precise Model. In: Proceeding IEEE Conference on Robotics, Automation and Mechatronics, pp. 916–921 (2008)
3. Xin, J.: Research on Robot Uncalibrated Visual Servoing Control System. Xi'an University of Technology, Xi'an (2007)
4. Li, M.: Research on Robot Uncalibrated Visual Servoing Key Technique. Journal of Harbin Institute of Technology, 13–15 (2008)
5. Zhong, J., Gang, X., Zhang, H.: Research on Image-based Robotic Visual Servoing Control System. Industrial Intelligent, 30–341 (2004)
6. Huang, G., Tung, C., Lin, H., Hsiao, S.: Inverse Kinematics Analysis Trajectory Planning for a Robot Arm. In: Proceeding IEEE Control Conference in Asian, pp. 965–970 (2011)
7. Corke, P.I.: Robotics Toolbox for Matlab, Release7.1 (2004),
   http://www.cat.csiro.au/cmst/staff/pic/robot
8. Pei, J., Yang, S.X., Mittal, G.S.: Vision Based Robot Control Using Position Specific Artificial Neural Network. In: Proceeding IEEE International Conference on Computational Intelligence and Communication Networks, pp. 110–115 (2010)
9. Su, J.: Uncalibrated Robotic Hand-eye Coordination. Publishing House of Electronics Industry (2010)
10. Pan, Q., Su, J., Xi, Y.: Uncalibrated 2D Robot Visual Tracking Based on Artificial Neural Network. Acta Automatica Sinica 27(2), 194–199 (2001)
11. Ramachandram, D., Rajeswari, M.: Neural Network-based Robot Visual Positioning for Intelligent Assembly. Journal of Intelligent Manufacturing 15(2), 219–231 (2004)

A Computational Personality-Based and Event-Driven Emotions Model in PAD Space

Shiqin Yang[1], Runhe Huang[2], and Jianhua Ma[2]

[1] Graduate School of Computer and Information Sciences, Hosei University, Tokyo, Japan
shiqin.yang.jk@stu.hosei.ac.jp
[2] Faculty of Computer and Information Sciences, Hosei University, Tokyo, Japan
{rhuang,jianhua}@hosei.ac.jp

Abstract. In this paper, we propose a new computational model of personality-based and event-driven mixed emotions called the PAEM (Personality Associated Emotion Model). The model delineates mixed emotions based on personality traits and updates dynamic emotional states by calculating the relative intensity of mixed emotions triggered by given events. The PAEM produces associated emotions by taking into account the personality traits of subjects, which have a significant influence on decision making processes.

Keywords: PAD space, mixed emotions, personality-associated emotion model.

1 Introduction

In the past decade, a remarkable change has occurred in the influence of affective neuroscience, combined with the psychological study of personality, emotion, and mood. As most of the computational models of emotion and personality are based on a general emotion model of influence on agent behavior, the relationship between personality and emotions in existing models is left open to debate. Since personality traits play an important role in emotion appraisal, a new generic emotion and personality model called the Personality-Associated Emotion Model (PAEM) is proposed in this paper. The PAEM, calculating the relative intensity of mixed emotions, is designed to implement the deduction of mixed emotions and provide a coherent dynamic emotional state.

2 Computational Models of Emotion and Personality

2.1 OCC Model and Five Factor Model

The OCC model, proposed by Ortony, Clore & Collins [1] is a psychological model based cognitive approach. It describes a hierarchy that contains three branches, namely consequences of events, actions of agents, and aspects of objects. The consequences of events are the focus in this paper.

J. Yang, F. Fang, and C. Sun (Eds.): IScIDE 2012, LNCS 7751, pp. 825–831, 2013.
© Springer-Verlag Berlin Heidelberg 2013

The Five Factor Model [2], the most widely accepted model of personality, is a purely descriptive model of personality. The five dimensional factors are Openness, Conscientiousness, Extraversion, Agreeableness and Neuroticism. A quantitative measurable personality can be expressed as a vector in which there are five personality traits (OCEAN), each of them has a value between -1 and 1. Let us denote P_i as a vector representing i^{th} person's personality traits, we have

$$P_i=[p_o\ p_c\ p_e\ p_a\ p_n],\quad -1\leqslant p_o,\ p_c,\ p_e,\ p_a,\ p_n\leqslant 1$$

2.2   The Pleasure-Arousal-Dominance Space

The PAD space is a bounded 3D space containing values ranged from -1 to 1 in each axis. The axes are Mehrabian's pleasure (P), arousal (A) and dominance (D) traits [3].

Emotions in PAD Space. Let us denote E_i as i^{th} emotion type. Each E_i of 22 emotion types in the OCC model can be projected on the PAD space.

$$E_i(\ e_p,\ e_a,\ e_d\ ),\ i=1,\ 2,\ ...,\ 22,\ -1\leqslant e_p,\ e_a,\ e_d\leqslant 1$$

Personality in PAD Space. PAD space is a mood space which projects emotions on different mood states. In fact, it is possible to find out the relationship between the personality trait and the PAD space by translating $P_i(p_o,\ p_c,\ p_e,\ p_a,\ p_n)$ into a corresponding PAD space mood state point $M_i(e_p,\ e_a,\ e_d)$ via the following mapping [3].

$$e_p=0.21*p_e+0.59*p_a+0.19*p_n \tag{1}$$

$$e_a=0.15*p_o+0.30*p_a-0.57*p_n \tag{2}$$

$$e_d=0.25*p_o+0.17*p_c+0.60*p_e-0.32*p_a \tag{3}$$

Fig. 1 shows the 22 types of emotions as circular red points and a personality point as a square blue point projected on the PAD space.

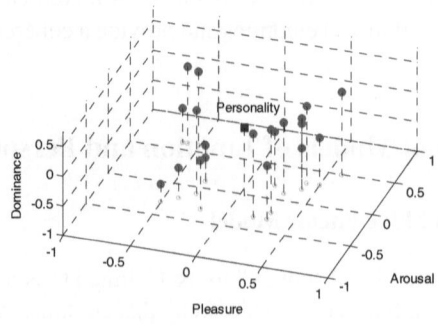

Fig. 1. Scattered emotions and personality point in 3-dimension space

3   Personality-Associated Emotion Model(PAEM)

Although many emotion models have been proposed and used in various applications, few have identified a specific link between emotions and individual personality traits. There are some models which resemble our work like Silas [4] for believable agents that model life-like synthetic characters including emotions and personality. Patrick Gebhard [5] proposed a layered model of affect that integrates three major affective characteristics: emotions, moods and personality. However, our work differs from all other work by trying to provide a dynamic emotion model for processing mixed emotions which are formed based on a subject's personality and events that have occurred.

3.1   Personality-Based Mixed Emotions

The proposed personality-based mixed emotions model is inspired by the concept of the "ecological niche". In general, emotions are generated together in a common environment, one emotion is triggered and other emotions may accompany it [6], while some other emotions may be triggered in sequence or simultaneously in the same niche environment. We take account of the concurrently triggered emotions and define $E_{mixed,}$ as a set of mixed emotions E_i.

$$E_{mixed} = \bigcup_{i=1}^{22} \alpha_i E_i \tag{4}$$

Where, α_i is correlative intensity of emotion E_i. E_{mixed} may vary between individual subjects even to the same trigger event.

3.2   Dynamical Mixed Emotions in PAEM

This model, the PAEM, showing the relationship of emotions and personality traits in PAD space, is based on two basic common sense assumptions.

Assumption 1. The intensity of the effect that personality traits have on an emotion varies according to their distance in the PAD space. The nearer they are, the more influence the personality traits have on the emotion. That is to say, the subject's emotions are more susceptible to his/her personality traits and more easily triggered by outside events.

Assumption 2. An outside event could trigger a main emotion. The neighboring emotions to the emotion triggered are likely jointly triggered. The probability of the emotion being triggered and its intensity along the time axis can be expressed based on the subject's personality trait point and the distances between the neighboring emotion and the main emotion in the PAD space.

Based on the assumptions above, the mixed emotions in the PAEM are composed of a triggered emotion and emotions associated with the triggered emotion at the same time. The intensity of each emotion decays along the time; however, it is hard to accurately define the intensity of an emotion in a single formula due to diverse personal cognition and experiences. So, the mixed emotions distribution and probability of being triggered and the intensity decay function are incorporated into the PAEM to reveal the dynamic mixed emotions.

The Emotions Space. An arbitrary point $u(x, y, z)$ in the PAD space is defined as an emotion point. Once an emotion is triggered by an event, the intensity of the emotion will decay as time passes. The intensity decay function is defined in this paper as:

$$y(t) = c \cdot e^{\lambda t} \tag{5}$$

Where c is constant parameter and λ is adjust parameter for different emotion. When emotion is triggered, it will be released as energy in an active field and the energy [7] will have an impact on other emotions through energy transfer. Both the triggered event and the personality trait will have an effect on the transferred energy.

The Subject's Personality Effect on His/Her Emotions. As stated in Assumption 1, a subject's personality point denoted as P_j, a blue point in the PAD space in Fig. 1 has its relatively steady effect on his/her emotions E_i. The effect of the personality traits on emotions can be expressed in formula (6).

$$I\left(P_j, E_i\right) = \frac{\varpi_1}{\left|p_j - u_i\right|^2} \tag{6}$$

where, ϖ_1 is the influence factor of P_j on E_i.

The Triggered Emotion's Effect on Other Emotions. Let us denote E_0 as the triggered emotion, E_i as one of other emotions in the PAD space which may be jointly triggered, $u_0(x, y, z)$ as the point of the emotion E_0 and $u_i(x, y, z)$ as the point of the emotion E_i in the PAD space, ϖ_2 as the influence factor of E_0 on E_i and $y(t)$ as the emotion decay function, the effect of the triggered emotion E_0 on E_i can be expressed in formula (7). Where, $y_e(t)$ is an event related decay function. It is suggested that there are a number of decay functions corresponding to a number of different types of events.

$$I\left(E_0, E_i\right) = \frac{\varpi_2 \cdot y_e(t)}{\left|u_0 - u_i\right|^2} \tag{7}$$

To sum up, when an event occurs, the total relative intensity of emotion E_i affected can be calculated in formula (8).

$$\eta\left(u_i, t\right) = I\left(P_j, E_i\right) + I\left(E_0, E_i\right) \tag{8}$$

As the mixed emotions can be explained as an emotions set, with a set of relative intensity values ranging from 0 to 1, they can be expressed by using $\eta(\mu_i)$ to substitute α_i in formula (4) and taking time into consideration, then we have

$$E_{mixed}(t) = \bigcup_{i=1}^{22} \alpha_i E_i = \bigcup_i^{22} \eta(u_i, t) E_i \qquad (9)$$

If we set a threshold for the intensity of an emotion and collect only a number of emotions, then we have

$$E_{mixed}(t) = \bigcup_i^{k(t)} \eta(u_i, t) E_i \qquad (10)$$

where, $k(t)$, the number of emotion types collected varies along time t.

4   Implementation of the PAEM

4.1   The Mechanism of the PAEM

The PAEM is the proposed human dynamic emotion model when an event or events occur. As stated above, a subject's personality is a critical factor in the manifestation of his/her emotions. It defines a subject's basic emotional state, which is described by a set of 22 weighted emotions with each weight value ranging from 0 to 1 to indicate the intensity of its associated emotion. At the same time, an outside event or events will trigger a main emotion intensively and other associated secondary concurrently emotions around the main triggered emotion. Therefore, a mixture emotion is the union of non-zero weighted emotions. Both the subject's personality traits and the emotion being triggered concurrently trigger other associated emotions.

4.2   Implementation of Event-Driven Emotions in the PAEM

To evaluate the PAEM, implementation was conducted on two groups of children at age 7. Their personality traits were evaluated by giving them well-designed questions and making long-term observations of their activities. With the result that their personality traits projected in PAD space is P_1 (-0.347,-0.482,-0.381), P_2 (0.21, 0.677, -0.63), P_3 (-0.121, 0.17, -0.331), P_4 (-0.366,-0.491,-0.362), P_5 (0.209, 0.662, -0.621), P_6 (-0.101, 0.181, -0.332).

Three children as a group were put in an experiment environment in which a teacher gives each of them an apple. At that moment, all of them were quite happy and joyous at having an apple.

In the PAEM, the effect of each child's personality traits P_i on 22 emotions using formulae (6) and the effect of the main triggered emotion, *"Joy"*, using formulae (7), can be calculated. In this experiment, the mixed emotions are calculated using formulae (10) and formed by selecting emotions with higher intensity values exceeds the specified threshold among 22 emotions.

According to the PAEM, when children are given an apple, all of their event-driven main emotions are the same emotion, which is *"Joy"*. The selected accompanying emotions that concurrently appear are *"Hope"*, *"Love"*, *"Happy"*, *"Gloating"* for child-1; *"Admiration"*, *"Hope"*, *"Gratitude"* for child-2; *and* *"Hope"*, *"Happy"*, *"Love"* for child-3 as shown in the left of Fig. 2.

The other experiments are conducted on another group of three children to test the triggered emotion "*Fear*" for negative personality traits. Similarly, the mixed emotions are calculated using formulae (10) for child-4, child-5 and child-6 are shown in the right of Fig 2.

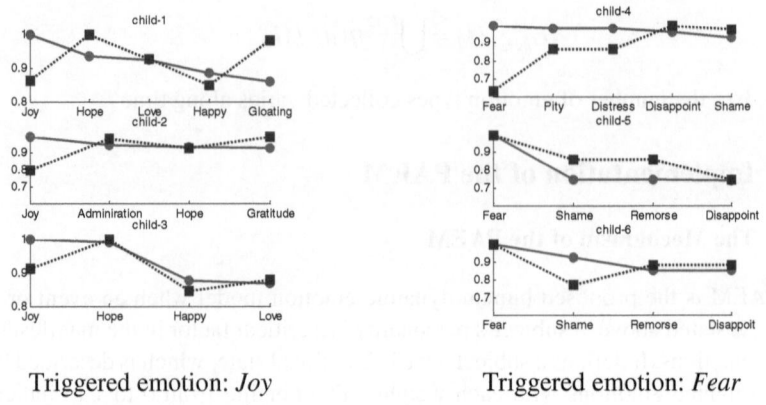

Triggered emotion: *Joy*   Triggered emotion: *Fear*

Fig. 2. Comparing first and updated mixed emotions of each child

The mixed emotions vary as time passes due to the emotion decay function. Fig. 2 shows the mixed emotions at the very first moment as a solid red line segment and the updated mixed emotions after a certain interval time as a dotted blue line segment. Taking child-4's "*Fear*" mixed emotions as an example, at the very first moment, the mixture emotions were *Fear→Pity→Distress→Disappointment→Shame* (indicated by red circles), and were updated to *Disappointment→Shame →Pity→Distress→Fear* (indicated by blue squares) after the interval. Which shows the mixed emotions and its distribution and dominance vary as time passes.

Table 1 shows the different triggered emotions by using the OCC model and the PAEM. As can be seen there is only triggered emotion in the OCC model. In contrast, there are accompanying emotions apart from the main triggered emotion in the PAEM.

Table 1. Comparisons of the resulting-Emotion in OCC Model and PAEM

Children	OCC model	PAEM
child-1	Joy	Joy+Hope +Love+Happy-for
child-2	Joy	Joy+Admiration +Hope +Gratitude
child-3	Joy	Joy+Hope + Happy-for + Love
child-4	Fear	Fear+Pity+Distress+Disappointment
child-5	Fear	Fear+Shame+Remorse+Disappointment
child-6	Fear	Fear+Shame+Remore+Disappointment

5   Conclusion and Future Work

In this paper, we have described a computational emotions model, PAEM, which is based on personality traits and driven by events occurring in an environment. We conclude that PAEM can model more complex emotions that resemble dynamic human emotions more reliably. Furthermore, PAEM can be applied in intelligent agents with personality, robots with emotions, and cyber individuals [8].

There is still much work to be done. One area of work is a better understanding of the effect of pre-existing memories on generating event-driven emotion of different emotion decay functions, and how these other factors can be incorporated into the computational emotion model. More elaborate experimentation is needed in order to justify our assumptions, and to verify and revise our proposed model.

References

1. Ortony, A., Clore, G.L., Collins, A.: The Cognitive Structure of Emotions. Cambridge University Press, Cambridge (1988)
2. McCrae, R.R., John, O.P.: An introduction to the five-factor model and its applications. Special Issue: The five-factor model: Issues and applications. Journal of Personality 60, 175–215 (1992)
3. Mehrabian, A.: Analysis of the Big-five Personality Factors in Terms of the PAD Temperament Model. Australian Journal of Psychology 48(2), 86–92 (1996)
4. Blumberg, B.M.: Old tricks, new dogs: ethnology and interactive creatures. MIT Media Lab, Cambridge, MA, Ph.D. Thesis (1996)
5. Gebhard, P.: ALMA-A layered model of affect. In: Autonomous agents & multiagent systems, pp. 29–36. ACM, New York (2005)
6. El-Nasr, M.S., Yen, J., Ioerger, T.: FLAME – A fuzzy logic adaptive model of emotions. US J. Autonomous Agents and Multi-agent Systems 3, 219–257 (2000)
7. Hu, B.-C., Chen, H.-S.: Research of Emotion Model Based on Particle System and Active Field. J. Mind and Computation 3(1), 36–44 (2009)
8. Ma, J., Wen, J., Huang, R., Huang, B.: Cyber Individual Meets Brain Informatics. IEEE Intelligent Systems 26, 30–37 (2011)

A Framework for Constructing Benchmark Databases and Protocols for Retinopathy in Medical Image Analysis

Tomi Kauppi[1], Joni-Kristian Kamarainen[1,2], Lasse Lensu[1], Valentina Kalesnykiene[3], Iiris Sorri[3], Hannu Uusitalo[4], and Heikki Kälviäinen[1]

[1] Machine Vision and Pattern Recognition Laboratory (MVPR), Lappeenranta University of Technology (LUT), P.O. Box 20, FI-53851, Lappeenranta, Finland
[2] MVPR Computational Vision Group, Kouvola Unit, LUT, Finland
[3] Department of Ophthalmology, University of Eastern Finland, Finland
[4] Department of Ophthalmology, University of Tampere, Finland

Abstract. We address performance evaluation practices for developing medical image analysis methods, and contribute to the practice to establish and to share databases of medical images with verified ground truth and solid evaluation protocols. This helps to develop better algorithms, to perform profound method comparisons, including the state-of-the-art methods, and consequently, supports technology transfer from research laboratories to clinical practice. For this purpose, we propose a framework consisting of reusable methods and tools for the laborious task of constructing a benchmark database. We provide a medical image annotation software tool which helps to collect and store ground truth for retinopathy lesions from experts, including the fusion of spatial annotations from several experts. The tool and all necessary functionality for method evaluation are provided as a public software package. For demonstration purposes, we utilise the framework and tools to establish the DiaRetDB1 V2.1 database for benchmarking diabetic retinopathy detection algorithms. The database contains a set of retinal images, ground truth from several experts, and a strawman algorithm for the detection of retinopathy lesions.

Keywords: Diabetic retinopathy detection, eye fundus imaging, benchmarking image database, eye fundus image processing, eye fundus image analysis, medical image processing, medical image analysis.

1 Introduction

Image databases and expert ground truth are in a regular use in medical image processing. However, it is common that the data is not public, and therefore, reliable comparisons and state-of-the-art surveys are difficult to conduct. In contrast to, for example, biometrics including face, iris, and fingerprint recognition, the research has been driven by public databases and solid evaluation protocols. These databases have been extended and revised resulting to continuous

J. Yang, F. Fang, and C. Sun (Eds.): IScIDE 2012, LNCS 7751, pp. 832–843, 2013.

pressure for the development of better methods. This could be adopted more also in medical image processing and analysis. During our research on diabetic retinopathy [1], we have experienced that developing databases from scratch is demanding, laborious and time consuming. However, certain tasks occur repeatedly and are reusable as such. Here, we discuss related practical issues, point out and solve repeatably occurring sub-tasks, and provide the solutions as open-source tools on our web site. In the experimental part, we utilise the proposed framework and devise a revised version of the diabetic retinopathy database DiaRetDB1 published originally in [2,3].

2   Benchmarking and Related Work

Recently Thacker et al. [4] studied the performance characterisation of computer vision methods, also transferable to medical image processing. The eight general considerations are adopted from [4], referred as the key questions:

C1: "How is testing currently performed?": If a commonly used database and protocol are available, their validity for development and evaluation needs to be examined. In the worst case, a new database needs to be constructed.

C2: "Is there a data set for which the correct answers are known?": Such a data set can be used to report the results enabling comparisons.

C3: "Are there data sets in common use?": See C1 and C2. Common data sets facilitate fair comparisons.

C4: "Are there experiments which show that algorithms are stable and work as expected?": This can be realised, if the expert ground truth is available.

C5: "Are there any strawman algorithms?": A strawman algorithm sets the baseline for method performance, also called as the baseline method. In this paper we call these kinds of baseline methods as strawman algorithms.

C6: "What code and data are available?": By publishing the code of a method, other research groups can avoid repeating the same work.

C7: "Is there a quantitative methodology for the design of algorithms?": This depends on the medical problem, but the methodology can be typically devised by following corresponding clinical work and practices.

C8: "What should we be measuring to quantify performance? Which metrics are used?": At least in image-wise (patient-wise) experiments, the receiver operating characteristic (ROC) curve together with several measurement points on the curve provide the means for the design and comparisons as in the medical practice where the sensitivity and specificity values (e.g., correctly classified normal images vs. correctly classified abnormal images) are in common use [5], [6].

There are three essential components for benchmarking medical image analysis algorithms: 1) true patient images, 2) ground truth from experts, and 3) an evaluation protocol. The key questions $C1 - C8$ are utilised here to acknowledge the current benchmarking practices in medical image analysis. In this paper, we focus on eye fundus images, containing a long-term research tradition. The most important public eye fundus databases are as follows: STARE (Structured analysis of the retina) [7], DRIVE (Digital retinal images for vessel extraction) [8],

MESSIDOR (Methods to evaluate segmentation and indexing techniques in the field of retinal ophthalmology) [9], CMIF (Collection of multispectral images of the fundus) [10], ROC (Retinopathy on-line challenge) [11], and REVIEW (Retinal vessel image set for estimation of width) [12]. A summary of the main properties (highlighted by the key questions) is given in Table 1. To compare the reference databases with the proposed framework and the DiaRetDB1 database in terms of the key questions, a corresponding summary is given in Table 2.

Table 1. Summary of the surveyed fundus image databases

Key questions	STARE (vessel)	STARE (disc)	DRIVE	MESSI DOR	CMIF	ROC	REV IEW
C2: *"Is there a data set for which the correct answers are known?"*	x		x	x		x	x
C3: *"Are there data sets in common use?"*	x	x	x	x	x	x	x
C4: *"Are there experiments which show algorithms are stable and work as expected?"*	x		x			x	
C5: *"Are there any strawman algorithms?"*	x	x	x				
C6.1: *"What code is available?"*						x	
C6.2: *"What data is available?"*	x	x	x	x	x	x	x
C7: *"Is there a quantitative methodology for the design of algorithms?"*							
C8.1: *"What should we be measuring to quantify performance?"*	x	x	x			x	x
C8.2: *"What metrics are used?"*		x	x			x	x
Σ	6	5	7	3	2	7	5

Table 2. Summary of the DiaRetDB1 V2.1 database

Key questions	DIARETDB1 V2.1
C2: *"Is there a data set for which the correct answers are known?"*	Yes
C3: *"Are there data sets in common use?"*	Yes (publicly available at [13]).
C4: *"Are there experiments which show algorithms are stable and work as expected?"*	Experimental results reported in Section 4.4 strawman algorithm.
C5: *"Are there any strawman algorithms?"*	Yes (description in Section 4)).
C6.1: *"What code is available?"*	Functionality for reading/writing images and ground truth, strawman algorithm, and annotation software (publicly available at [13,14])
C6.2: *"What data is available?"*	Images and ground truth (xml) (publicly available at [13]).
C7: *"Is there a quantitative methodology for the design of algorithms?"*	Medical practice used as a guideline at each development step.
C8.1: *"What should we be measuring to quantify performance?"*	Image- and pixel-based ROC-analysis (description in Section 4).
C8.2: *"What metrics are used?"*	Equal error rate (EER) defined in Section 4.

3   Patient Images and Ground Truth

3.1   Collecting Patient Images

Patient images are captured and selected by medical doctors or other trained persons. For a selected image set, two issues should be justified: 1) distribution correspondence with the desired population and 2) privacy protection of patient data. In DiaRetDB1, the ophthalmologists wanted to investigate the accuracy of automatic methods analysing fundus images of patients who are in a serious risk of having diabetic retinopathy, providing clear findings. This studied sub-population is routinely screened by the Finnish primary health-care. Privacy protection of patient data considers the ethics of clinical practice, medical research, including permissions from a national ethics committee and patients, and also to data security, meaning that all patient information, including hidden metadata, must be explicitly removed from images in a public database. In DiaRetDB1, the fundus images were acquired using a standard fundus camera and were converted to raw bitmaps and then saved to portable network graphics (PNG) format using lossless compression and hidden metadata removed.

3.2   Image Annotations as the Ground Truth

There is a need for computer assisted annotation as originally discussed in [3] and [2]. In general, image ground truth markings are essential for training supervised algorithms as well as for their evaluation and comparison. The information is typically constructed by manually annotating a set of images, and commonly, simple tailored tools are used to collect the data. Annotating medical images, the two essential considerations apply: 1) annotations must be performed by clinically qualified persons (specialised or specialising medical doctors, or other trained professionals for specific tasks), denoted as "experts", and 2) the ground truth should include annotations from several experts.

To avoid biasing the results, the experts should be given minimal guidance for their annotation work. Moreover, basic image manipulation for viewing the images is needed, and a set of geometric primitives is provided for making the spatial markings. During the process of collecting the DiaRetDB1 ground truth, ophthalmologists found the following polygon derived primitives useful: *small circle*, which can be quickly put on a small lesion, and *circle area* and *ellipse area* which are described by their centroid, radius/radii, and orientation (ellipse). Our system also requires at least one *representative point* for each lesion (the most salient cue, such as its colour or texture, describing the specific lesion). Furthermore, *confidence* from a set of three discrete values, low, moderate or high, is required for every marking. It is wise to define beforehand the types of markings, i.e., the class labels for the lesions (e.g., in DiaRetDB1: Hard exudates, Soft exudates, Microaneurysms, Haemorrhages).

Our tool is available at [14] as Matlab M-files and as a Windows executable. Matlab is not the optimal environment for developing GUI based applications, but it is widely used in scientific computing. The default GUI is shown in Fig. 1.

3.3   Medical Markings Data Format

To store the annotated markings and to be able to restore their graphical layout, we need to define a data format. The data is naturally structured, and therefore, structural data description languages are preferred. Several protocols for describing medical data exist, such as HL7 based on the extensible markup language (XML) [15], but these are complex protocols designed for patient information exchange between organisations and information systems. Since our requirements are considerably less comprehensive, we adopt our own light-weight data format based on the XML data description language, the Document Type Definition (DTD) description. The proposed data format was not used in [3,2], but the original data was converted to the XML format without loss of information.

3.4   Fusion of Annotations from Multiple Experts

Fusing multiple expert annotations was originally studied in [17], and is revised here. An important question for training, evaluating, and benchmarking is how the annotations from multiple experts should be combined: 1) How to resolve inconsistencies in the annotations from a single expert? 2) How to fuse equally trustworthy information from multiple sources (multiple expert co-fusion)?

In our data format, the available expert information is the following (Fig. 1): 1) spatial coverage (polygon area), 2) representative point(s) (small circle areas), and 3) the subjective confidence level. The representative points are distinctive "cue locations" that attracted the expert's attention to the lesion. The confidence level describes the expert's subjective confidence for the lesion to represent a specific class as shown in Fig. 2. Three intuitive solutions exist for the fusion problem: i) fixed size neighbourhoods around the representative points (Fig. 3(b)), ii) union of spatial coverage and thresholded by a fixed confidence (Fig. 3(a)), and iii) interest points masked with confidence regions (Fig. 3(c)).

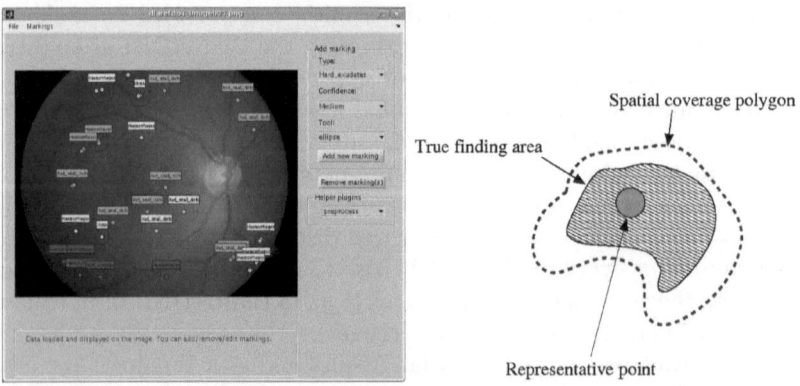

Fig. 1. GUI of the image annotation tool and parts of a single expert annotation

The area intersection provided the best fusion results in all experimental setups [17], and is computed in a straightforward manner as the sum of expert-annotated confidence images divided by the number of experts (see Fig. 3(d)).

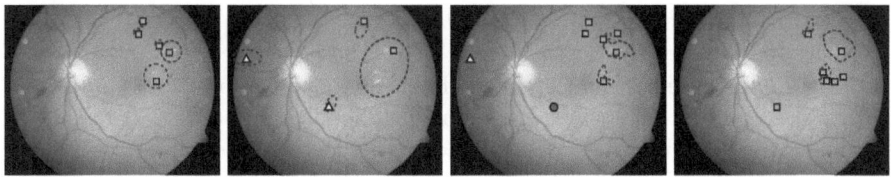

Fig. 2. Four independent sets of spatial markings (contours and representative points) for the same lesion type (Hard exudates). The representative point markers denote the confidence level (*square* = 100%, *triangle* > 50% and *circle* < 50%).

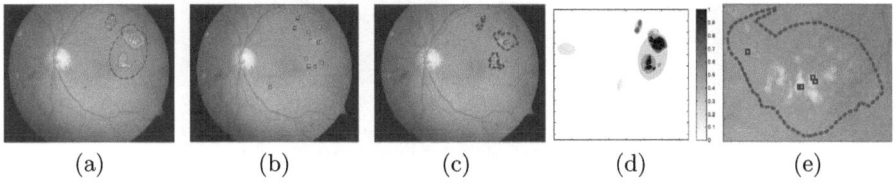

(a)   (b)   (c)   (d)   (e)

Fig. 3. Illustration of the strategies for fusing the annotations in Fig. 2: (a) Area intersection (blue denotes areas for the confidence level 0.25, red for 0.75, and green for 1.00); (b) Representative point neighbourhoods (5 × 5); (c) Representative point neighbourhoods masked with the area corresponding 0.75 confidence; (d) Summed area confidences; (f) Close-up of masked representative points

For DiaRetDB1, the fused confidence of 0.75 yielded to the best results [17], resolving the inconsistencies of annotations either from a single expert or multiple expert co-fusion problems.

4   Algorithm Evaluation

4.1   Evaluation Methodology

The ROC based analysis perfectly suits to medical decision making, being the acknowledged methodology in medical research [16]. An evaluation protocol based on the ROC analysis was proposed in [3] for image-based (patient-wise) evaluation and benchmarking, and the protocol was further studied in [17]. In clinical medicine, the terms *sensitivity* and *specificity* defined in the range [0%, 100%] or [0, 1], are used to compare methods and laboratory assessments. The *sensitivity* $= \frac{TP}{TP+FN}$ depends on the diseased population whereas the *specificity* $= \frac{TN}{TN+FP}$ on the healthy population, defined by true positive (TP), true negative (TN), false positive (FP), and false negative (FN). The x-axis of a ROC curve is 1-specificity, whereas the y-axis represents directly the sensitivity [6].

It is useful to form a ROC-based quality measure. The equal error rate (EER) [18] or weighted error rate (WER) [19] are preferred. The main difference between the two measures is that EER assumes equal penalties for both false positives and negatives, whereas in the WER, the penalties are adjustable.

In the image-based evaluation, a single likelihood value for each lesion should be produced for all test images. Using the likelihood values, a ROC curve can be automatically computed [17]. If a method provides multiple values for a single image, such as the full image likelihood map in Fig. 4(b), then the values must be fused to produce a single score.

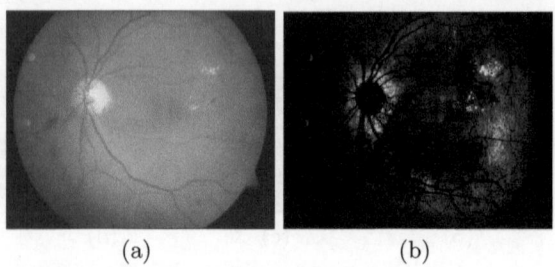

(a)   (b)

Fig. 4. Pixel-wise likelihoods for Hard exudates produced by the strawman algorithm (explained later): (a) Original image (hard exudates are the small yellow spots in the right part of the image); (b) "Likelihood map" for hard exudates

4.2   Image-Based Evaluation

We follow the medical practice where the decisions are "patient-wise" [20]. The image analysis system is treated as a black-box which takes an image as the input. The system produces a score that corresponds to the probability of the image being abnormal, and a high score corresponds to high probability. The objective of the image-based evaluation protocol is to generate a ROC curve by manipulating the score values of the test images. The image score based evaluation method is presented in Algorithm 1.

The image-based evaluation method is general since it requires only the scores for each test image. If we need to evaluate the performance in case of a method producing multiple values, e.g., a spatial likelihood map illustrated in Fig. 4, an additional procedure is needed to fuse the multiple values into a single score.

4.3   Pixel-Based Evaluation

To validate a design choice in method development, we also measure the spatial accuracy, i.e., whether the detected lesions are found in correct locations. Thefore, we propose also a pixel-based evaluation protocol (Algorithm 2) which is analogous to the image-based evaluation.

With a global pixel-wise score (curr_pix_score), the pixels in all test images are classified to either normal or abnormal. Now, the sensitivity and specificity

---

Algorithm 1. Image-wise evaluation based on image scores

```
 1: for each test images do
 2:     curr_score ← image score
 3:     for each test image do
 4:         if curr_score ≥ test image score then
 5:             assign "normal"
 6:         else
 7:             assign "abnormal"
 8:         end if
 9:     end for
10:     Compare test image assignments to the ground truth assignments and compute
        (sensitivity, specificity)-pair
11:     Add new ROC point (x, y) = (1-specificity, sensitivity)
12: end for
13: Return the final ROC curve (all points)
```
---

can be computed for each image by comparing the classified pixels to the disambiguated ground truth. Note that the sensitivity values are computed only for the abnormal test images whereas the specificity values are computed for all test images. This does not allow to determine the ROC curves for each test image, but if the procedure is repeated with a varying score, then the mean ROC curve can be computed.

Algorithm 2. Pixel-wise evaluation based on pixel scores

```
 1: Form a list of tested pixel scores
 2: for each tested pixel score (curr_pix_score) do
 3:     for each test image do
 4:         for each test image pixel score do
 5:             if curr_pix_score ≥ pixel score then
 6:                 assign pixel is "normal"
 7:             else
 8:                 assign pixel is "abnormal"
 9:             end if
10:         end for
11:     end for
12:     Compare test image assignments to the spatial ground truth assignments and
        compute (sensitivity, specificity)-pair (over all pixels in all images)
13:     Add new ROC point (x, y) = (1-specificity, sensitivity)
14: end for
15: Return the final ROC curve (all points)
```

4.4 Strawman Algorithm

We provide a baseline method in the form of a strawman algorithm (Algorithm 3) [17] which users of the database may find it easier to start to use the data and to self-evaluate the maturity and applicability of their methods. The strawman algorithm is based on the use of photometric cue. The strawman results for DiaRetDB1 are show in Fig. 5 (ROC curves) and in Table 3 (EER values).

The score fusion in the strawman algorithm is based on the following reasoning: If we consider M medical evidences (features) extracted from the image, $x_1, \ldots, x_M$, where each evidence is a vector, then we can denote the score value of the image as $p(x_1, \ldots, x_M | abnormal)$. The joint probability is approximated from the classification results (likelihoods) in terms of decision rules using the combined classifier theory (classifier ensembles) [23]. The decision rules for deriving the score were compared in the recent study [17] where the rules were devised based on Kittler et al. [23] and a new intuitive rank-order based rule "summax" which defines the image score $p(x_1 \ldots x_M | abnormal)$ using the compared decision rules when the prior values of the population characteristics are equal ($P(normal) = P(abnormal)$) as follows:

$$\text{SCORE}_{summax} = \sum_{m \in N_{Y\%}} p(x_m | abnormal) \tag{1}$$

where $N_{Y\%}$ are the indices of $Y\%$ top scoring pixel scores. Experimenting also with the max, mean, and product rules, strong empirical evidence supports the rank-order based sum of maxima (summax; portion fixed to 0.01). [17]

Algorithm 3. Strawman algorithm

1: Extract colour information (r, g, b) of the lesion from the train set images (Sec. 3.4).
2: Estimate $p(r, g, b | lesion)$ from the extracted colour information using a Gaussian mixture model determined by using the Figueiredo-Jain method. [21,22]
3: Compute $p(r, g, b | lesion)$ for every pixel in the test image (repeat step for every test image in the test set).
4: Evaluate the performance (Sec. 4).

---

5   Example – DiaRetDB1 Diabetic Retinopathy Database and Protocol V2.1

The authors have published two medical image databases with the accompanied ground truth: DiaRetDB0 and DiaRetDB1. The work on DiaRetDB0 provided us essential information how diabetic retinopathy data should be collected, stored, annotated and distributed. DiaRetDB1 was a continuation to establish a better database for algorithm evaluation. DiaRetDB1 contains eye fundus images selected by experienced ophthalmologists. The lesion types of interest were selected by medical doctors (see Fig. 6): microaneurysms (distensions in the capillary), haemorrhages (caused by ruptured or permeable capillaries), hard

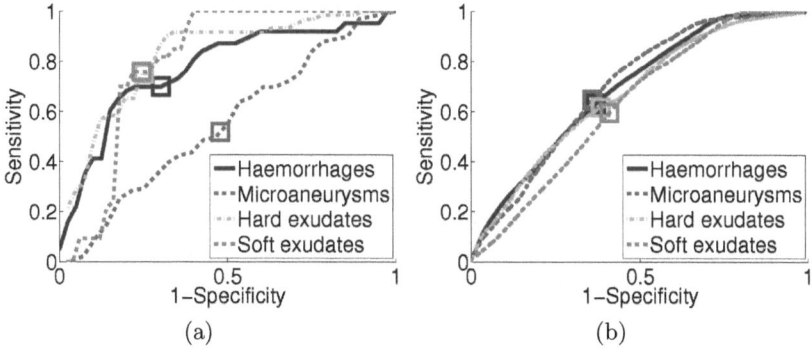

Fig. 5. ROC curves for the DiaRetDB1 strawman algorithm (*square = EER*): (a) image-based; (b) pixel-based

Table 3. EER results for the DiaRetDB1 strawman algorithm

	Ha			*Ma*			*He*			*Se*			*All lesions avg.*		
	min	max	mean	min	max	mean	min	max	mean	min	max	mean	min	max	mean
Image-based	0.25	0.33	0.30	0.43	0.50	0.48	0.22	0.27	0.24	0.18	0.28	0.24	0.27	0.35	0.31
Pixel-based	0.37	0.37	0.37	0.35	0.36	0.36	0.37	0.44	0.39	0.40	0.41	0.40	0.37	0.40	0.38

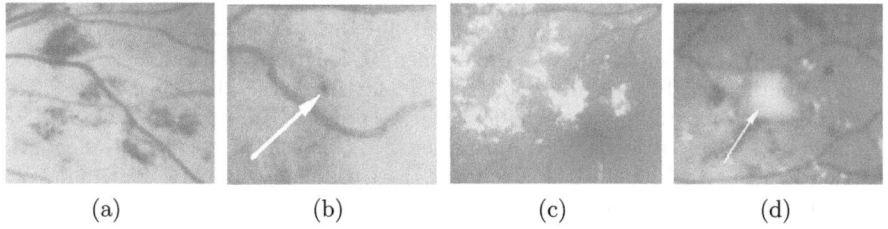

(a)   (b)   (c)   (d)

Fig. 6. Abnormal eye fundus findings caused by the diabetes (best viewed in colour): (a) haemorrhages; (b) microaneurysms (marked with an arrow); (c) hard exudates; (d) soft exudate (marked with an arrow)

exudates (leaking lipid formations), soft exudates (microinfarcts), and neovascularisation (new fragile blood vessels). These lesions are signs of mild, moderate, and severe diabetic retinopathy, and they provide evidence for the early diagnosis. The images were annotated by four independent and experienced medical doctors inspecting similar images in their regular work.

The images and ground truth can be downloaded from [13]. The images are in PNG format, and the ground truth annotations follow the XML format. Moreover, we provide a DiaRetDB1 kit containing full Matlab functionality (M-files) for reading and writing images and ground truth, fusing expert annotations, and generating image based evaluation scores. The whole pipeline from images to evaluation results (including the strawman algorithm) can be tested using the provided functionality on the weg page. The annotation software (Matlab files and executables) is available at [14].

6   Conclusions

We have discussed the problem of establishing benchmark databases for the development of medical image analysis. We have pointed out the importance of commonly accepted and used databases. We have proposed reusable tools needed to solve the important sub-tasks, put our implementations publicly available, and established the diabetic retinopathy database DiaRetDB1 to promote and help other researchers to collect and publish their data. We believe that public databases and common evaluation procedures significantly support the development and enable the best methods to be adopted in clinical practice.

Acknowledgements. The authors thank the Finnish Funding Agency for Technology and Innovation (TEKES Project Nos. 40430/05 and 40039/07), and the partners of the ImageRet project (http://www2.it.lut.fi/project/imageret/) for their support.

References

1. Kauppi, T.: Eye Fundus Image Analysing for Automatic Detection of Diabetic Retinopathy. PhD Thesis, Lappeenranta University of Technology (2010)
2. Kauppi, T., Kalesnykiene, V., Kamarainen, J.-K., Lensu, L., Sorri, I., Raninen, A., Voutilainen, R., Uusitalo, H., Kälviäinen, H., Pietilä, J.: The DIARETDB1 diabetic retinopathy database and evaluation protocol. In: Proc. BMVC (2007)
3. Kauppi, T., Kalesnykiene, V., Kamarainen, J.-K., Lensu, L., Sorri, I., Raninen, A., Voutilainen, R., Pietilä, J., Kälviäinen, H., Uusitalo, H.: DIARETDB1 diabetic retinopathy database and evaluation protocol. In: Proc. MIUA (2007)
4. Thacker, N.A., Clark, A.F., Barron, J.L., Beveridge, J.R., Courtney, P., Crum, W.R., Ramesh, V., Clark, C.: Performance characterization in computer vision: A guide to best practices. CVIU 109, 305–334 (2008)
5. Zou, K.H.: Receiver operating characteristic (ROC) literature research, http://splweb.bwh.harvard.edu:8000/pages/ppl/zou/roc.html
6. Fawcett, T.: An introduction to ROC analysis. Pattern Recognition Letters 27, 861–874 (2006)
7. STructured Analysis of the Retina (STARE), http://www.ces.clemson.edu/~ahoover/stare/
8. Digital Retinal Images for Vessel Extraction (DRIVE), http://www.isi.uu.nl/Research/Databases/DRIVE/
9. Methods to evaluate segmentation and indexing techniques in the field of retinal ophthalmology (MESSIDOR), http://messidor.crihan.fr
10. Collection of multispectral images of the fundus (CMIF), http://www.cs.bham.ac.uk/research/projects/fundus-multispectral/
11. Retinopathy Online Challenge (ROC), http://roc.healthcare.uiowa.edu/
12. REVIEW: Retinal Vessel Image set for Estimation of Widths (REVIEW), http://reviewdb.lincoln.ac.uk/

13. Diabetic retinopathy database and evaluation protocol (DIARETDB1),
    http://www2.it.lut.fi/project/imageret/files/diaretdb_v02_01/
14. Image annotation tool (IMGANNOTOOL),
    http://www.it.lut.fi/project/imgannotool
15. Application Protocol for Electronic Data Exchange in Healthcare Environments
    Version 2.5.1, ANSI Standard, http://www.hl7.org
16. Lasko, T.A., Bhagwat, J.G., Zou, K.H., Ohno-Machado, L.: The use of receiver
    operating characteristic curves in biomedical informatics. J. of Biomedical Infor-
    matics 38, 404–415 (2005)
17. Kauppi, T., Kamarainen, J.-K., Lensu, L., Kalesnykiene, V., Sorri, I., Kälviäinen,
    H., Uusitalo, H., Pietilä, J.: Fusion of Multiple Expert Annotations and Overall
    Score Selection for Medical Image Diagnosis. In: Salberg, A.-B., Hardeberg, J.Y.,
    Jenssen, R. (eds.) SCIA 2009. LNCS, vol. 5575, pp. 760–769. Springer, Heidelberg
    (2009)
18. Phillips, P.J., Moon, H., Rizvi, S.A., Rauss, P.J.: The FERET Evaluation Method-
    ology for Face-Recognition Algorithms. IEEE TPAMI 22, 1090–1104 (2000)
19. Bailliere, E., Bengio, S., Bimbot, F., Hamouz, M., Kittler, J., Mariethoz, J., Matas,
    J., Messer, K., Popovic, V., Poree, F., Ruiz, B., Thiran, J.P.: The BANCA Database
    and Evaluation Protocol. In: Kittler, J., Nixon, M.S. (eds.) AVBPA 2003. LNCS,
    vol. 2688, pp. 625–638. Springer, Heidelberg (2003)
20. Everingham, M, Zisserman, A.: The Pascal Visual Object Classes Challenge VOC
    2006 Results. In: Proc. ECCV Workshop of VOC (2006)
21. Figueiredo, M.A.T., Jain, A.K.: Unsupervised learning of finite mixture models.
    IEEE TPAMI 24, 381–396 (2002)
22. Paalanen, P., Kamarainen, J.-K., Ilonen, J., Kälviäinen, H.: Feature Representa-
    tion and Discrimination Based on Gaussian Mixture Model Probability Densities
    - Practices and Algorithms. Pattern Recognition 39, 1346–1358 (2006)
23. Kittler, J., Hatef, M., Duin, R.P.W., Matas, J.: On Combining Classfiers. IEEE
    TPAMI 20, 226–239 (1998)

Artificial Emotion Model of Child Playmate Robot Based on a Combined Method

Xuefei Shi, Qing Li, and Tao Feng

University of Science and Technology Beijing, Beijing 100083, China
{sxf1245,liqing,fengtao}@ies.ustb.edu.cn

Abstract. The bottom emotional model is designed by using fuzzy math and matrix style of state machine which is a combined method to solve indefinite and definite information. The model constructed in this way can reduce the memory spaces used only for storing the corresponding relations about stimulus, emotion and behaviour. The experiment is carried out based on child playmate robot. It verifies the effectiveness of bottom model.

Keywords: fuzzy relation, matrix style of state machine, artificial emotion, child playmate robot.

1 Introduction

Nowadays the functional role of emotions has been recently fully recognized as essential for intelligent systems. Indeed recent emergence of affective computing combined with artificial intelligence has made it possible to design intelligent agents that have emotions in order to interact with human being more nature.

In this article, we propose an emotion and behavior model for child playmate robotic. Combining fuzzy math and matrix style of state machine, we construct an efficient and simple first-level emotional model. Experimental result and conclusion are presented in the end.

2 The Bottom Emotional Model of Child Playmate Robot

2.1 Structure of the Model

For the application of child playmate robot the above preset emotion is the main intelligence. We construct a bottom-level emotion and behavior model structure shown in Fig.1. According to the emotional complexity mentioned above. An algorithm combined fuzzy mathematics and matrix style of state machine is presented in the following part. This method could generate a reasonable solution which considers those contradictions among emotion itself.

J. Yang, F. Fang, and C. Sun (Eds.): IScIDE 2012, LNCS 7751, pp. 844–849, 2013.

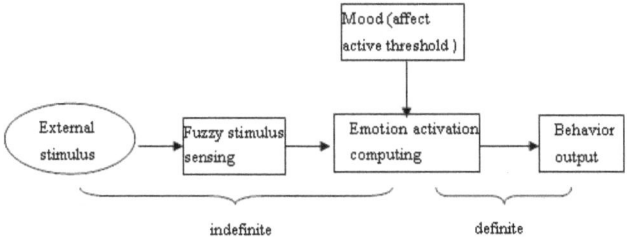

Fig. 1. Bottom-level emotion and behavior model for child playmate robot

For this level intelligence emotional activation is automatically completed in unintentional state. It undergoes non-cognition process. It is congenital temperament not acquired character that affects emotion generation. External stimulus in the first block of model is objective and fuzzy stimulus sensing matrix could be computed through parameter α which is related to individual child personality. The second part of model is indefinite and fuzzy. Current mood affects active threshold. Emotion activation computing decides whether some relevant emotion is generated. Behavior under active emotion is definite relatively.

2.2   Construction of Fuzzy Stimulus Sensing Matrix

Human sensing of external stimulus is fuzzy and uncertain. It is difficult to quantify whether stimulus events provoke a corresponding emotion definitely and how about its intensity of activation. Clearly the relationship between external events and active emotion should be a type of fuzzy relation.

Definition 1: $X = \{x_1, x_2, \cdots x_m\}$ is a definite external event set, $Q = \{Q_1, Q_2, \cdots Q_n\}$ is a basic emotion set. Then we get a fuzzy subset $R \in F(X \times Q)$. Its degree of membership $\mu_R(x, Q)$ represents the correlation degree of events and emotions. It could be expressed by the following equation:

$$\mu_R(x_i, Q_j) = r_{ij} \in [0,1] \ i = 1, \dots m; j = 1, \dots n \tag{1}$$

This fuzzy relation could be expressed by the fuzzy matrix $R = (r_{ij})_{m \times n}$. The bottom emotional model for child playmate robot is aimed at age-stage child. Personality in the model is dependent on nature temperament not acquired character. Currently A.Thomas and S.Chess's behavior pattern theory has the largest impact in the field of child temperament. We choose four dimensions which have close relationship with emotion from their research finding called New York longitudinal study (NYLS) [5]. A relationship between personality parameter α, temperament and emotion feature is designed as Table 1 shown.

Table 1. Relationship between individual parameter, temperament and emotion

Temperament dimensions	Easy type $\alpha \in [0,0.4]$	Delayed type $\alpha \in [0.4,0.7]$	Difficult type $\alpha \in [0.7,1]$
Avoidance of new situation	Active approach	Escape at first and approach later	escape
Sensing threshold	positive lower negative higher	medium	positive higher negative lower
Response strength	Medium or weak	Very weak	strong
Positive or negative mood	be positive mood	be quiet mood	be negative mood

2.3   Emotional Activation Computing

The model considers that mood might affect the threshold of emotional activation. We compute λ-cutting matrix of fuzzy stimulus sensing matrix R' and get the last matrix of active emotion. Whether emotions are provoked is dependent on this varying λ cutting matrix. Model computes as the following equation. The model changes an uncertain fuzzy relation matrix to a definite active matrix.

$$r_{ij}(\lambda) = \begin{cases} 1, r_{ij} > \lambda(t) \\ 0, r_{ij} \leq \lambda(t) \end{cases} \tag{2}$$

Emotion Q_j is active when $r_{ij}(\lambda)$ equals 1 and not active when $r_{ij}(\lambda)$ equals 0. Behaviour output equation of child playmate robot is computed by matrix form of state machine which will be introduced in detail in the following part.

2.4   Behaviour Output Equation Based on Matrix form of State Machine

State machine is a common-used mathematics model for discrete information data. Since the relationship between external event, emotional activation and behavior output is discrete, our model algorithm determines the output behavior through matrix style of state machine[6]. The following is the computing process for the special application of child playmate robot. Assume a given state transition is shown in the following table 2. Here $I = \{i_1, i_2, \ldots \ldots i_n\}$ is input event set and every element of this set is represented in Boolean vector $i = (x_1, x_2, \cdots x_n)^T$. $Q = \{s_1, s_2, \ldots \ldots s_l\}$ is internal state and every emotion state is also represented by vector $s = (Q_0, Q_1, \cdots Q_l)^T$. $Z = \{o_1, o_2, \ldots .. o_m\}$ is output behaviour set and every behaviour is expressed by vector $o = (z_0, z_1, \cdots z_m)^T$.

Table 2. Assumed state transition between event, emotion and behaviour

initial state	$i_1=(0,0)$	$i_2=(0,1)$	$i_3=(1,0)$	$i_4=(1,1)$
$Q_0\ s_0$	(s_1,o_0)	(s_2,o_1)	(s_4,o_0)	(s_1,o_1)
$Q_1\ s_1$	(s_2,o_1)	(s_0,o_0)	(s_3,o_0)	(s_3,o_2)
$Q_2\ s_2$	(s_3,o_2)	(s_2,o_2)	(s_2,o_0)	(s_4,o_0)
$Q_3\ s_3$	(s_3,o_2)	(s_2,o_3)	(s_3,o_3)	(s_0,o_1)
$Q_4\ s_4$	(s_3,o_1)	(s_1,o_4)	(s_3,o_4)	(s_3,o_2)

This transition could be expressed by the following matrix based on the method in reference 10:

$$
\begin{bmatrix} Z_0 \\ Z_1 \\ Z_2 \\ Z_3 \\ Z_4 \end{bmatrix} =
\begin{bmatrix}
\overline{x_2} & \overline{x_1}x_2 + x_1\overline{x_2} & x_1 & 0 & 0 \\
x_2 & \overline{x_1 x_2} & 0 & x_1 x_2 & \overline{x_1 x_2} \\
0 & x_1 x_2 & \overline{x_1} & \overline{x_1 x_2} & x_1 x_2 \\
0 & 0 & 0 & \overline{x_1}x_2 + x_1\overline{x_2} & 0 \\
0 & 0 & 0 & 0 & \overline{x_1}x_2 + x_1\overline{x_2}
\end{bmatrix}
\times
\begin{bmatrix} Q_0 \\ Q_1 \\ Q_2 \\ Q_3 \\ Q_4 \end{bmatrix}
\qquad (3)
$$

The behaviour outputs of child playmate robot are certain under some emotion activating. And for one emotion state playmate robot could generate multiple behaviors. The corresponding behaviors are computed on the basis of above method. The algorithm of model could save large memory space and speed up the computing time because outputs are produced directly by matrix computation not by stored in a form in advance.

3   Experimental Results

The bottom emotional intelligence studied in this paper is preset in neural system. Our implementation is focused on the first layer emotion generation. That means robotic agent show emotions like a child. Its input event set is X (30 events), emotion state set is Q (6 emotions) and output behaviour set is Z (20 behaviours). Child playmate robot shows its emotion by eyes and sound shown in Fig.2.

The fuzzy relation R between stimulus and emotional activation will be decided by survey data. Personality stimulus sensing matrix could be constructed according to different parameter α shown in the Fig. 3. The thresholds of active emotion are modified by current mood. Once special emotion is active the fitting output behaviors

Fig. 2. Child playmate robot shows its emotion by eyes and sound

could be computed according to matrix form of state machine. Trait behavior matrix of playmate robot is shown in the Fig.4. It is determined by the principle of Eq (3). The algorithm of this bottom model implements precise match of input event, active emotion and output behavior. So emotion intelligence is computed through this integrated method.

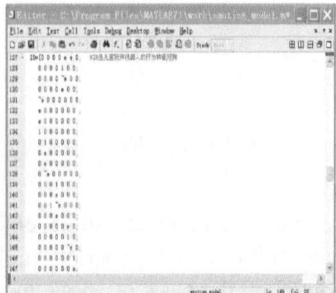
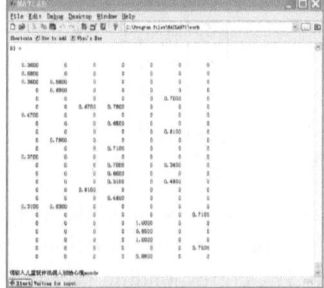

Fig. 3. Personality fuzzy sensing matrix   **Fig.4.** Trait behaviour matrix of playmate robot

4   Conclusion

The bottom emotional model is constructed to implement emotional intelligence of child playmate robot. The model algorithm proposed in this paper defines a fuzzy relation between external event and emotion state firstly. Next personality sensing matrix is designed based on NYLS temperament research finding. Mood effect on emotional activation is taken into account in our model. Fuzzy relation matrix is changed into definite active emotion matrix through λ -cutting set. The output behaviors are produced by definite and discrete model—state machine matrix style. An emotional intelligence computation method is put forward in this paper based on fuzzy mathematics and state machine matrix form. It could solve the conflict factors of uncertain and definite emerged in emotion itself. The algorithm proposed in this paper has the advantage of high real-time operability and short storage space. The original study of this paper is usage of simple and practice mathematics method to describe the emotional complexity. The model makes emotion of child playmate robot computational.

Acknowledgements. The research in this paper is partially co-supported by the Nation High-Tech Research and Development Plan of China (No.2010AA04Z218) and the Nature Science Foundation of Nation (No.60573059).

References

1. Liu, Z.: A personality based emotion model for intelligent virtual agents. In: Proceedings of the 4th International Conference on Natural Computation (ICNC 2008), October 18-20, pp. 13–16. IEEE Computer Society, Los Alamitos (2008)
2. Song, Y., Jia, P.: Humanlike Robot Control System Structure Based on artificial affection. Robot 26(6) (2004)
3. Yang, H., Yang, Z.: A Synthetic Computational Emotion Modeling Method. Journal of Computer Research and Development 45(4), 579–587 (2008)
4. Liu, X., Liang, B.: Complex System Modeling Theory, Method and Technology, 1st edn. Science Press, Beijing (2008)
5. Zhang, L.: Child Developmental Psychology, 1st edn. Central China Normal University Press, Wuhan (2006)
6. Zhu, Z., Zhu, Q.: Matrix Model Description of State Machine. Computer Science 28(4), 46–47 (2007)
7. Ruan, X.: Neural computing science—simulate brain function from cell level. National Defense Industry Press, Beijing (2006)
8. Marpaung, A.H., Lisetti, C.L.: Multilevel emotion modeling for autonomous agents. In: Proceedings of the Fall Symposium on the Intersection of Cognitive Science and Robotics: From Interfaces to Intelligence (AAAI FSS 2004), October 22-24. American Association for Artificial Intelligence, USA (2008)

Implementation of Data Acquisition System Using 89C2051

Sun Chen

School of Automation Engineering,
University of Electronic Science and Technology of China, Chengdu, 611731

Abstract. A type of MCUs, integrated with a precision analog comparator, provides the condition with which inexpensive DAQs can be made without using a special ADC. A scheme of inexpensive DAQs using a 89c2051 is presented. Two formulas of the time constant, parameters of the R-C circuit are deduced, using percentage expression and n-bits expression of resolution of ADC. The advantage of using assembler language to program a stand-alone DAQ is explained. And also, a distributed DAQ scheme is included. Both stand-alone and distributed projects were tested. Theory and experiment results in this paper have important reference value.

Keywords: Analog Comparator, Analog-to-Digital Convertor, Data Acquisition System, Resister and Capacitor circuit.

1   Introduction

The key component of data acquisition system is A/D converter, while the comparator is the core of the A/D converter. And now, many kinds of MCU have integrated comparator on chip[1]-[8], which makes it possible that low-cost data acquisition system can be made only by using a MCU without A/D convertor.

The AT89C2051 is a low voltage, high performance CMOS 8bit microcomputer with 2K bytes of Flash programmable and erasable read-only memory (PEROM). The device is manufactured using Atmel's high-density nonvolatile memory technology and is compatible with the industry-standard MCS-51 instruction set. By combining a versatile 8-bit CPU with Flash on a monolithic chip, the Atmel AT89C2051 is a powerful microcomputer which provides a highly-flexible and cost-effective solution to many embedded control applications.

The AT89C2051 provides the following standard features: 2K bytes of Flash, 128 bytes of RAM, 15 I/O lines, two 16-bit timer/counters, a five vector two-level interrupt architecture, a full duplex serial port, a precision analog comparator, on-chip oscillator and clock circuitry[2], which greatly facilitates the design of a data acquisition system.

A complete data acquisition system typically consists of the following parts:

1) Sensor/s – by which the physical parameter is converted to an electrical signal.
2) Signal conditioning circuits – by which signals output from the sensor/s are conditioned to accommodate for samples in the form of the digital signal.
3) Analog-digital converter unit (ADC), to convert the conditioned signal into digital form.

J. Yang, F. Fang, and C. Sun (Eds.): IScIDE 2012, LNCS 7751, pp. 850–857, 2013.

4) Computer hardware interface for the data processing.
5) Software for data acquisition system time coordination, data processing, displaying, and transferring, including the microcontroller program (firmware) and the host computer graphical user interface (GUI) program.

Having a precision comparator integrated, data acquisition system can be made easily by using resistor-capacitor circuit with the AT89C2051 microcontroller without special A/D converter. And also, only cooperate with an external D/A converter, the SAR structure A/D converter can be constructed using AT89C2051 [1]. With other resources on the MCU itself, such as the timer/counter, UART serial interface, a low-cost data acquisition system can be easily constructed.

2   Analysis of the Resistor-Capacitor Type A/D Converter Circuit

In Fig.1, K, the SPDT switch is under program control, the capacitor C is charged or discharged through the resistor R, according to whether the power supply VCC or "ground" is connected. When charging, assuming the voltage across the capacitor 0V, the capacitor voltage signal is called the zero state response [11], the response equation is

$$Vc = Vcc \times (1 - e^{-\frac{t}{RC}}) \tag{1}$$

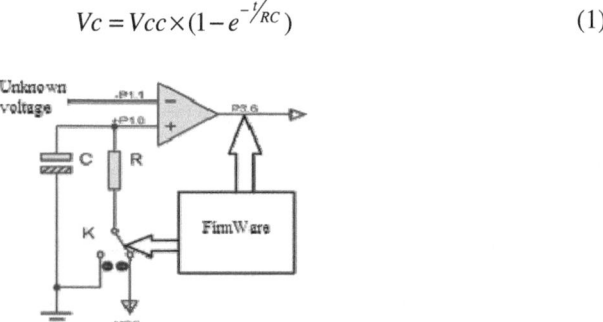

Fig. 1. A kind of cheap and practical R-C A/D converter

In (1), the product of the resistance and capacitance, RC, is called time constant of the circuit. When $t \to \infty$, $Vc = Vcc$ is called steady-state response; Fully charged, and then discharging, changing in voltage across the capacitor is called zero-input response [11], the response equation is

$$Vc = Vcc \times e^{-\frac{t}{RC}} \tag{2}$$

It can be seen from (1) and (2) that both zero-state and zero- input response of the voltage across the capacitor is a nonlinear function of time.

Fig.2 shows a charge-discharge curve which is the response of RC circuit activated by pulse train of which period is greater than 5RC.

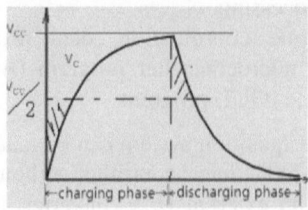
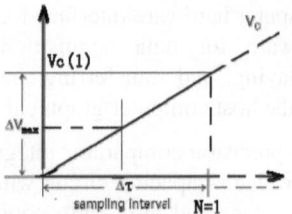

Fig. 2. Charging and discharging curve of RC circuit

Fig. 3. Enlarge part of the first sample

Assume that the unknown voltage, range from 0 to Vcc, is applied to the inverting input of the comparator, and the voltage across the capacitor is applied to the non-inverting input of the comparator, so during charging or discharging process the comparator output state will change, i.e. during the charging process the comparator output state will go from low to high and during the discharging process the comparator output state will go from high to low. Through constantly checking P3.6 state to see when the state change occur and make use of equations (1) and (2), the corresponding voltage across the capacitor can be calculated, but the look-up table method is common used in MCU projects [1]. Assume that the interval between two queries of P3.6 state is $\Delta\tau$, N times query corresponding to $t = N\Delta\tau$.

From (1) and (2) the corresponding capacitor voltage is calculated as,

$$V_c(N) = Vcc \times (1 - e^{-N\Delta\tau/RC}) \tag{3}$$

$$V_{disc}(N) = Vcc \times e^{-N\Delta\tau/RC} \tag{4}$$

Where $V_c(N)$ corresponding to charging phase and $V_{disc}(N)$ to discharging phase. $N = 0, 1, 2, \cdots, N_{max}$.

The shadow parts in Fig.2. show that the beginning part of the charge-discharge curve of the RC circuit is steepest, so, the first sample, the largest change of the capacitor voltage, determines the A/D conversion accuracy.

The enlarge part of the first sample is shown in Fig.3

Let N=1, in (3), there is

$$\Delta V_{max} = V_c(1) = Vcc \times (1 - e^{-\Delta\tau/RC}) \tag{5}$$

In Fig.3, take the intermediate point of the sampling interval to be calculated and the result is

$$\frac{\Delta V_{max}}{2} = Vcc \times (1 - e^{-\frac{\Delta\tau}{2RC}}) \tag{6}$$

So,

$$RC = \frac{\Delta\tau}{2[\ln 2y - \ln(2y-1)]} \tag{7}$$

Where, RC is the time constant of the RC circuit[12]; $\Delta\tau$ is the sampling interval; VCC is the supply voltage, in this case is the full-scale voltage range (FSR); ΔV_{max} is the maximum value of the minimum voltage that can be distinguished, corresponding to the ADC's increment [12]. $y = Vcc/\Delta V_{max}$ is defined, in this paper, as quantization number of the A/D converter. Obviously, it is the reciprocal of the resolution.

The equation (7) shows that the time constant, RC, of the circuit is determined by the sampling time interval $\Delta \tau$ and the quantization number y of the full-scale voltage.

Generally, the resolution of A/D converter can also be expressed with the number of bits of the A/D converter. Resolution of an n-bit ADC is the reciprocal of ($2^n - 1$). In (7), let y= $2^n - 1$, then

$$RC = \frac{\Delta \tau}{2\left[\ln\left(2^n - 1\right) - \ln(2^n - 1.5)\right]} \tag{8}$$

The equation (8) shows that the time constant, RC, of the circuit is determined by the sampling time interval $\Delta \tau$ and the resolution of the ADC.

3   Calculation for the Parameters of the RC Analog to Digital Convertor

Generally, the interval to query P3.6 is determined by the program instruction execution time. In the charging period, the assembly code segment to check the output state of the comparator is as follows:

Table 1. Instructions to determine the sample interval

No.	Instructions	Time needed(f = 12Mhz)
1st	Ad1:jb p3.6,ad4	2us
2ed	inc a	1us
3rd	cjne a, #SNUM, ad1	2us

1st instruction checks whether P3.6 is high, as soon as it is high, goes to instruction labeled ad4; if P3.6 is low at the moment it is checked a=a+1, where a is counter in 2ed instruction; 3rd instruction shows that the maximum sample number is #SNUM.

From Table 1, the sample interval is calculated is 5us when the frequency of the crystal is 12MHz.

In (7), let $\Delta \tau = 5\mu s$, $C = 2nF$, $Vcc = 5V$, $\Delta V_{max} = 0.05V$, it can be calculated that $y = 100$ and $R = 249.4\ K$ hom · Practically, choosing R=250 Khom, so RC=500 uS.

In (3), let $Vc(N) = \frac{Vcc}{2}$, i.e.

$$\frac{Vcc}{2} = Vcc \times (1 - e^{-0.01N}) \tag{9}$$

854   C. Sun

From (9), it can be calculated that N=70. Practically, choosing N=80.

In (3), let $Vc(N_m) = Vcc - \dfrac{\Delta V_{max}}{2}$, i.e. $Vc(N_m) = Vcc(1 - \dfrac{1}{200})$

$$Vcc \times (1 - \dfrac{1}{200}) = Vcc \times (1 - e^{-0.01N_m}) \qquad (10)$$

So, $N_m = 530$ implies that at least 530 samples have to be taken to complete the 0 to full scale scanning coverage. Practically, it can be chosen that $N_m = 600$.

4   Sample Strategy

The symmetry characteristic of charge-discharge curve can be used to reduce sample number needed to cover the full scale range (FSR) [1].

Fig.4 shows the sample strategy. The initial segments of the capacitor charging and discharging curve (shadowed area) together completely cover the entire full-scale range, so, if the measured voltage is constant, and is in the range of (0~Vcc), in the charging phase only the (0~Vcc/2) part the output state of comparator is monitored, and in the discharging phase only the (Vcc/2~Vcc) part the output state of the comparator is monitored. As soon as the change of the output state of the comparator is sensed the LUT is checked, and the value corresponding to the moment presents the value of the unknown voltage.

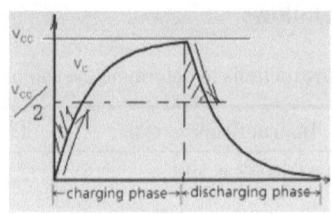

Fig. 4. Sampling strategy

From (3) and (4), the discrete values of samples of Vc can be calculated as follows:

$V_c()$=[0.0, 0.0, 0.1, 0.1, 0.2, 0.2, 0.3, 0.3, 0.4, 0.4, 0.5, 0.5, 0.6, 0.6, 0.7, 0.7, 0.7, 0.8, 0.8, 0.9, 0.9, 0.9, 1.0, 1.0, 1.1, 1.1, 1.1, 1.2, 1.2, 1.3, 1.3, 1.3, 1.4, 1.4, 1.4, 1.5, 1.5, 1.5, 1.6, 1.6, 1.6, 1.7, 1.7, 1.7, 1.8, 1.8, 1.8, 1.9, 1.9, 1.9, 2.0, 2.0, 2.0, 2.1, 2.1, 2.1, 2.1, 2.2, 2.2, 2.2, 2.3, 2.3, 2.3, 2.3, 2.4, 2.4, 2.4, 2.4, 2.5, 2.5, 2.5, 2.5, 2.6, 2.6, 2.6, 2.6, 2.7, 2.7, 2.7, 2.7, 2.8].

And also, the discrete values of samples of V_{cdis} can also be calculated correspondingly. The discrete values calculated are stored in BCD code in the FLASH ROM, form a BCD table, called a look-up-table, LUT.

In the LUT, 2.9 is stored as 29h, the data in the LUT is ordered by index N. The first 80 data represent charging phase, and the second 80 data represent the discharging phase. When the program of the MCU executed, the main loop checks the output state of the comparator, P3.6, the value of the counter is used as an index to look up the table.

5   A Low-Cost Data Acquisition System Constructed by Using Resistor and Capacitor

Fig.5 shows a resistor and capacitor-type low-cost data acquisition system (data acquisition part).

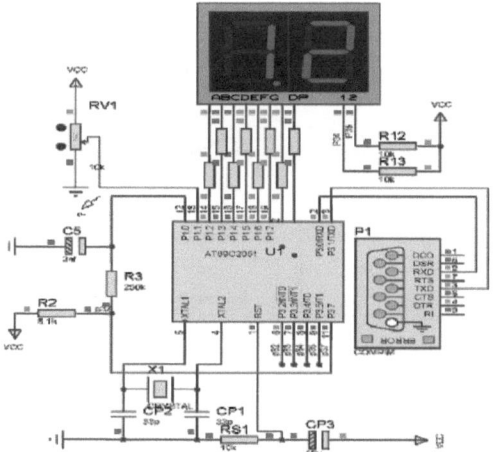

Fig. 5. A resistor and capacitor-type low-cost data acquisition system

The hardware is mainly constituted by the microcontroller 89C2051, resistor-capacitor circuit. Fig.6 shows the flow chart of microcontroller program, so called firmware.

A serial communication interface shown in Fig.5 is for real-time data transmission to the host computer, in order to record data changes or as the measured data of the control of the process involved in the control algorithm computing.

6   Distributed Data Acquisition System Detection Scheme and the C Program

As data acquisition front-end, or lower machine of a distributed data acquisition system, the microcontroller will collect real-time data and then transmitted to the host computer for further processing or as control parameter directly involved in the calculation.

Distributed data acquisition system can take full advantage of the powerful computing capabilities of the host computer. Timer0 as a timer of the charging or discharging processes. Once the status of P3.6 change is detected, the timer will stop and TH0 and TL0 of Timer 0 will be sent to the host computer via the serial port.

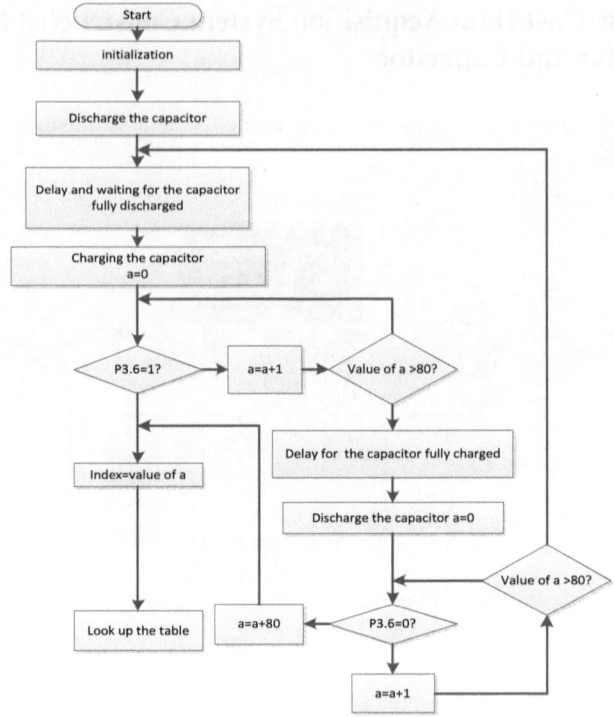

Fig. 6. Flow chart of the microcontroller program

The data communication frame structure is shown in table 2. PC program can be coded by using Visual Basic6 language. Communication program interface can use the MSComm control [10]. The GUI is different depending on application. Due to space constraints, the PC program, will not repeat here.

Table 2. Data communication frame structure

Header data	Data
55h AAh	TH0 TL0

7   Conclusions

As the state of the art, there are many kinds of MCUs in which precision comparator are integrated. Cooperating the comparator with a small number of components, such as resistors, capacitors, a low-cost data acquisition system can be constituted without using special ADC. The theoretical analysis shows that the resolution of the RC type data acquisition system depends on the time constant of the RC circuit. In this paper, two formulas by which circuit parameters are chosen according to the resolution expressions are deduced and included.

In the case of stand-alone data acquisition system, such as voltmeter, thermometer, assembly language is used.

As to distributed data acquisition system C language is more convenient. This paper gives a detailed flow chart of the C program, and C Program is included.

Assembly program and C program are tested, as well as the theoretical formulas are verified by experiments. In the experiments, PROTEUS7.10, Visual Basic6.0 and virtual serial port are used in the simulation process. The results of this paper have important reference value.

References

1. ATMEL Application Note, PC-Interfaced Data Acquisition System with the Atmel AT89C2051 Microcontroller. Atmel Corporation (2003)
2. AT89C2051 datasheet
3. ATMEL Application Note, Analog-to-Digital Conversion Utilizing AT89LP Microcontrollers without an ADC, Atmel Corporation (2011)
4. Liao, X.: Application of Analog Comparator Embedded in Microcomputer. International Electronic Elements (11), 59–61 (2001)
5. TI Application Report, Economic Measurement Techniques With the Comparator A Module, SLAA071 (October 1999)
6. Chen, S., Wu, L., Ding, G.: Design of Digital Capacitance Table Based on AT89C2051SCM. Modern Electronics Technique 18(305), 158–160 (2009)
7. Zhang, J., Yang, Y., Ma, L., Zhang, L.: The Use of Inside Comparing Device in PIC12F629. Journal of Inner Mongolia Agricultural University 28(4) (December 2007)
8. Microchip Technology Inc. A Comparator Based Slope ADC (2003)
9. ATMEL Application Note, Timers Program Examples
10. Fan, Y., Chen, L.: Visual Basic and RS-232 serial communication control
11. Li, H.: Fundamental of Circuit Analysis, 3rd edn. Higher Education Press (2000) ISBN: 7040037211
12. Luecke, J.: Analog and Digital Circuits for Electronic Control System Applications. Linacre House, Jordan Hill, UK, ISBN 0-7506-7810-0

A Modified BA Model with Tunable Clustering and Better Robustness

Guangming He, Haiyan Chen, and Jingyu Yang

Department of Computer Science, Nanjing University of Science and Technology,
Nanjing 210096, China
gmhe2006@gmail.com

Abstract. A modified BA modeling algorithm with "randomly chosen edges" and "triad formation step" is proposed in this paper. The basic characteristics and robustness against attack of the modified BA model and the standard BA model are analyzed by both analytically and numerical simulations, which shows our model possesses the same characteristics as the standard BA model, but with the high clustering coefficient and better robustness at the same time.

Keywords: BA Model, Modified BA Model, Tunable Clustering, Robustness.

1 Motivation

We know preferential attachment is the one of the most important characters of Barabási and Albert (BA) [1] model. In order to perform preferential attachment, the BA modeling algorithm should know the whole information of the network. That is to say, when a new node is going to be connected to the network, first, the degrees of all the existing nodes in the current whole network should be known, and then, basing on this information to carry out preferential attachment, finally, the new node is able to be connected to the network. Unfortunately, in most real-world networks, including such typical ones as the Internet, WWW, social networks, etc., a new node, to be added to the network, seldom can get the whole information of the existing network. Therefore, it's a very meaningful research on constructing the evolving models of scale-free networks under the part information of the network. There have been several promising attempts, such as X. Li and G. Chen proposed a local-world evolving network model [2], Z. Pan etc. generalized local-world models for weighted networks [3].

It should be noted that even though the BA model successfully explains the scale-free nature of many networks, it has the clustering coefficient goes to zero and thus fails to describe correctly networks with the high clustering, such as social networks. In the physics literature, networks with high clustering are commonly modeled by the small-world network model of Watts and Strogatz's model(WS) [4], while networks with the power-law degree distribution by the scale-free network model of BA. Yet each model lacks the property of the other model: the WS model shows a high clustering but without the power-law degree distribution, while the BA model with the scale-free nature does not posses the high clustering. Some attempts have been made to construct a specific model with both the power-law distribution and the high clustering [5]-[8].

J. Yang, F. Fang, and C. Sun (Eds.): IScIDE 2012, LNCS 7751, pp. 858–865, 2013.
© Springer-Verlag Berlin Heidelberg 2013

Motivated by the above discussion, we try to modify the standard BA model to incorporate both the power-law distribution and the high clustering by only using the part information of the network.

2   Models

The standard BA model is briefly reviewed first and then our modified BA modeling algorithm is proposed.

2.1   Standard BA Model

The BA model is defined as follows [1]:

(1) Initial condition: To start with, the network consists of $m_0 \geq 1$ nodes and no edges.
(2) Growth: Introduce one new node to the existing network each time, and this new node is connected to m existing nodes in the network, where $1 \leq m \leq m_0$.
(3) Preferential attachment (PA): The probability of the above new node is being connected to an existing node i of degree k_i is given by

$$\Pi_i = \frac{k_i}{\Sigma_{j=1}^{N} k_j} \tag{1}$$

where $N = m_0 + t - 1$ is the total number of existing nodes at the (t-1) step of the node-adding process.

Clearly, after t steps, the network will have a total of $N = m_0 + t$ nodes and mt edges. With continuum theory, Barabási and Albert calculated the degree distribution $p(k)$ analytically. The probability $p(k)$ that a randomly chosen node has k edges follows a power law $p(k) = \dfrac{2m^2}{k^3}$.

2.2   Modified BA Model

Our modified BA model is generated by the algorithm below.

(1) Initial condition: The network starts with a small number $m_0 \geq 1$ of nodes and small number $e_0 \geq 1$ of edges.
(2) Growth: Introduce one new node to the existing network each time step, and this new node is connected to some existing nodes in the network based on the following rules.
(3) Random attachment (RA): An edge is randomly chosen in the existing network first, and then the new node attaches to any one of the two nodes related to this chosen edge.

(4) Triad formation (TF) [5-6]: If the new node v is connected to an existing node u in the previous RA step, then add a new edge with probability p from v to a randomly chosen neighbor of u. If there remains no pair to connect, i.e., if all neighbors of u were already connected to v, stop TF step in this case.

(5) Repeat RA step and TF step $m\,(1 \le m \le m_0)$ times to every new node in each time step.

3   Basic Characteristics of the Models

In this section, basic characteristics of the models including degree distribution, clustering coefficient and average path length will be discussed. Numerical simulations will be also performed on the standard BA model and modified BA model.

3.1   Degree Distribution

The modified BA model gives the same degree distribution $p(k) \sim k^{-3}$ as the standard BA model, at least if every TF step follows a RA step. To see this, we first give some theoretic analysis, and then provide numerical simulation.

3.1.1   Analytical Results

The continuum theory is used to calculate the degree distribution.

First observe that in a RA step an arbitrary node i increases its degree k_i with the rate

$$\frac{\partial k_i}{\partial t} = \frac{1}{e_0 + m(1+p)t} \cdot \frac{1}{2} \cdot k_i \quad \text{for a RA step,} \quad (2)$$

where $e_0 + m(1+p)t$ is the numbers of edges in the network before the current new node is attached, $1/(e_0 + m(1+p)t)$ is the probability that each edge is being picked up when randomly choosing an edge from the network.

For a TF step the average increases of k_i is proportional to the probability that a node in the neighborhood j is linked in the RA step before, times the inverse of that node's degree (the probability that k is linked form j):

$$\frac{\partial k_i}{\partial t} = \sum_{j \in \Gamma_i} \left(\frac{1}{e_0 + m(1+p)t} \cdot \frac{1}{2} \cdot k_j \right) \cdot \frac{1}{k_j} \quad (3)$$

$$= \frac{1}{e_0 + m(1+p)t} \cdot \frac{1}{2} \cdot k_i \quad \text{for a TF step} \quad (4)$$

where Γ_i is the neighborhood of i, and the numbers of nodes in Γ_i is k_i.

From Eqs. (2) and (4) the total rate for one time step, composed of m RA steps and mp TF steps, is expressed as

$$\frac{\partial k_i}{\partial t} = m\left(\frac{1}{e_0 + m(1+p)t} \cdot \frac{1}{2} \cdot k_i\right) + mp\left(\frac{1}{e_0 + m(1+p)t} \cdot \frac{1}{2} \cdot k_i\right)$$

$$= m(1+p)\left(\frac{1}{e_0 + m(1+p)t} \cdot \frac{1}{2} \cdot k_i\right) \approx \frac{k_i}{2t} \quad (\text{as } t \to \infty). \tag{5}$$

The initial condition of Eq. (5) is

$$k_i(t_i) = m(1+p), \tag{6}$$

where node i is added to the network at instant t_i.

From Eqs. (5) and (6), we can obtain

$$k_i(t) = m(1+p)\left(\frac{t}{t_i}\right)^{1/2}, t \geq t_i \tag{7}$$

The above Eqs. lead to the degree distribution

$$P(k) = \frac{\partial P[k_i(t) < k]}{\partial k} = \frac{2m^2(1+p)^2 t}{m(1+p)+t} \cdot \frac{1}{k^3}$$

$$\sim 2m^2(1+p)^2 k^{-3} \quad (\text{as } t \to \infty). \tag{8}$$

Remark 1. It can be seen that the degree distribution of our model follows the same power law $p(k) \sim k^{-3}$ as the standard BA model. Moreover, if we let $p = 0$ in our model, the degree distribution of our model is equal to that of the standard BA model $P(k) \sim 2mk^{-3}$. In this sense, the standard BA model is a special case of our model.

Remark 2. In the above discussion, we have assumed that a RA step always follows a TF step, if not, the factor $1/k_j$ in Eq.(3) would be replaced by $1/(k_j - 1)$, that is a small correction when k_j is large. So the resulting degree distribution would not differ much from a power law.

3.1.2 Numerical Simulation
In the simulation for degree distribution, we start with a triangle whose three nodes connect one another, i.e., $n_0 = 3, e_0 = 3$, and add $N = 10000$ nodes to the network. Simulation results are shown in Fig. 1.

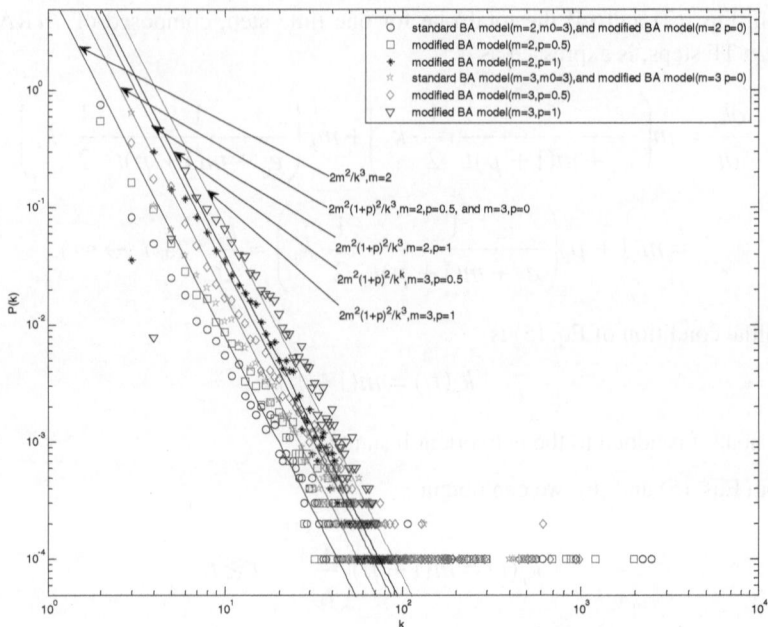

Fig. 1. Comparison of the degree distribution for the modified BA model and standard BA model

Fig. 1 shows comparison of the degree distribution for the modified BA model and standard BA model, where discrete points represent numerical simulations and solid lines are analytical expressions. Fig. 4 demonstrates all the degree distributions follow a power-law with the exponent $\gamma = 3$ in $P(k) \sim k^{-\gamma}$, which also verify the correction of the theoretical results.

3.2   Clustering Coefficient

In the simulation for clustering coefficient, we start with a triangle whose three nodes connect one another, i.e., $n_0 = 3, e_0 = 3$, and totally add $N = 9000$ nodes to the network with each time step repeating $m = 2$ times.

Simulation results are shown in Fig. 2 and Fig. 3.

Fig. 2 shows clustering coefficient C vs the network size N at various parameter values of p. For $p \neq 0$, C approaches a nonzero value as N is increased. And for $p = 0$, which is the standard BA model corresponds to the limiting case of $p = 0$ in our model, C goes to zero as N becomes larger.

Fig. 3 shows Clustering coefficient C vs parameter values of probability p at the network size $N = 9000$. We also observe that the relation between C and p is almost linear increasing, as depicted in Fig. 3.

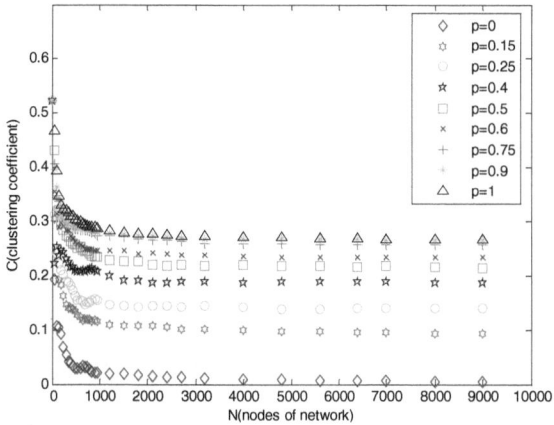

Fig. 2. Clustering coefficient C vs the network size N at various parameter values of p

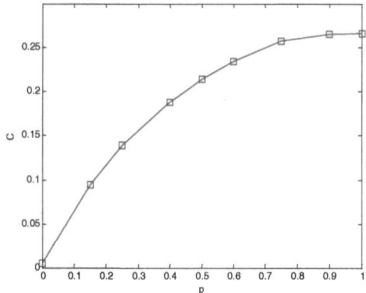

Fig. 3. Clustering coefficient C vs parameter values of probability p at the network size $N = 9000$

The parameter p in our model introduces the clustering effect into the system by allowing the formation of triads. As expected, we find that clustering coefficient C approaches to a finite nonzero value as N is increased at nonzero p, whereas the standard BA model, which corresponds to the limiting case of $p = 0$ in our model, is confirmed to have $C = 0$. Furthermore, the clustering coefficient can be varied systematically by changing p.

4   Attack Robustness and Fragility

In this section, we do some computer simulations on random attack and intentional attack for the standard BA model and modified BA model.

As usual, random attack or intentional attack in a network means random or specific removal of a small fraction of nodes from the network. We start the simulation with a triangle whose three nodes connect one another, i.e., $n_0 = 3, e_0 = 3$, and totally add $N = 100$ nodes to the network with each time step repeating $m = 2$ times.

Simulation results are shown in Fig. 4.

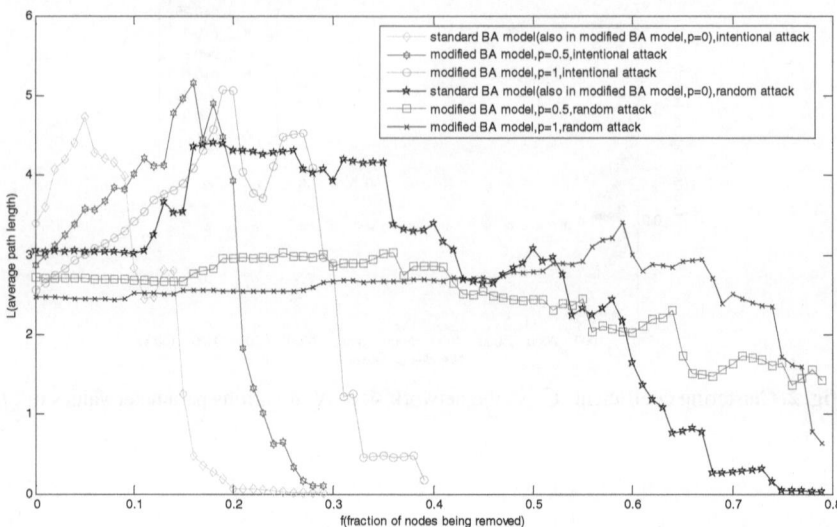

Fig. 4. Comparison of average path length vs fraction of nodes being removed for the modified BA model and standard BA

Fig. 4 illustrates the comparison of average path length vs fraction of nodes being removed for the modified BA model and standard BA model. From the observation of Fig. 4, we can conclude two things: (i) The scale-free network has the "robust yet fragile" property, which have been studied in [11-16]. This feature in mainly due to its significant heterogeneity, where a small number of "big" nodes play a critical role in maintaining the whole network's robustness. (ii) The parameter p takes an effect in maintaining the whole network's robustness: the network becomes more robustly as the parameter p is increased.

5   Conclusions and Discussion

In this paper, we propose A modified BA modeling algorithm with "randomly chosen edges" and "triad formation step". We analyze the basic characteristics and robustness against attack of the modified BA model and the standard BA model by both analytically and numerical simulations, and find that our model possesses the same characteristics as the standard BA model such as the power-law degree distribution, but with the high clustering coefficient and better robustness at the same time. In our model, the clustering coefficient is also shown to be tunable simply by changing the parameter p. Furthermore, we can improve the network robustness against attack by increasing the value of p.

The idea of "triad formation" in our model is enlightened by Refs. [5-6], which are both based on the same observation of triad formation and have been shown to possess similar network properties, i.e., the high clustering, small average path length, and a scale-free degree distribution. We believe, however, that our model has some advantage in describing network that adds new node by part information of the network—"randomly chosen edges", whereas the network model in Ref. [5] adds new nodes by the whole information of the network—"preferential attachment", and the network model in Ref. [6] has fixed network size. Our model provides an alternative possibility to achieve both the scale-free nature and the high clustering coefficient at the same time.

Recently, many generalized or improved versions of the standard BA model were developed to better capture the real-world network. In our model, there are still many limitations need to improve: give the theoretical formulas of clustering coefficient, average path length; think about the edge weight in the model; discuss synchronization in the network; consider improvements to allow for $2 < \gamma < 3$ in the power-law $P(k) \sim k^{\gamma}$ of the model, and et al.

References

1. Barabási, A.-L., Albert, R.: Energence of scaling in random networks. Science 286(5439), 509–512 (1999)
2. Li, X., Chen, G.: A local-world evolving network model. Physica A 328, 274–286 (2003)
3. Pan, Z., Li, X., Wang, X.: Generalized local-world models for weighted networks. Physical Review E 73, 056109 (2006)
4. Watts, D.J., Strogatz, S.H.: Collective dynamics of 'small-world' networks. Nature 393, 440–442 (1998)
5. Holme, P., Kim, B.J.: Growing scale-free networks with tunable clustering. Physical Review E 65, 026107 (2002)
6. Davidsen, J., Ebel, H., Bornholdt, S.: cond-mat/0108302
7. Klemm, K., Equiluz, V.M.: e-print cond-mat/0107606;e-print cond-mat/0107607
8. Puniyani, R., Lukose, R.M., Huberman, B.A.: e-print cond-mat/0107212
9. Bollobás, B., Riordan, O.: Mathematical results on scale-free random graphs. In: Bornholdt, S., Schuster, H.G. (eds.) Handbook of Graphs and Networks: From the Genome to the Internet, pp. 1–34. Wiley-VCH, Berlin (2003)
10. Cohen, R., Havlin, S.: Scale-free networks are ultrasmall. Phys. Rev. Lett. 90, 3682–3685 (2003)
11. Albert, R., Jeong, H., Barabási, A.-L.: Attack and error tolerance in complex networks. Nature 406, 387–482 (2000)
12. Barabási, A.-L., Bonabeau, E.: Scale-free networks. Scientific American, 50–59 (May 2003)
13. Carlson, J., Doyle, J.: Highly optimized tolerance: Robustness and power laws in complex systems. Phys. Rev. Lett. 84(11), 2529–2532 (2000)
14. Carlson, J., Doyle, J.: Complexity and Robustness. PNAS 99 (suppl.1), 2539–2545 (2002)
15. Broder, A., Kumar, R., Maghoul, F., et al.: Graph structure in the web. Comput. Networks 33, 309–320 (2000)
16. Doyle, J., Alderson, D., Li, L., et al.: The 'robust yet fragile' nature of the Internet. Proceedings of the National Academy of Sciences USA 102(41), 14497–14502 (2005)

Matrix-Variate Discriminative Analysis, Integrative Hypothesis Testing, and Geno-Pheno A5 Analyzer

Lei Xu

Dept. of Computer Science and Engineering, Chinese Univ. of Hong Kong
Chang Jiang Chair Professor Program, School of EE&CS, Peking Univ., Beijing
lxu@cse.cuhk.edu.hk

Abstract. A general perspective is provided on both on hypothesis testing and discriminative analyses, by which matrix-variate discriminative analyses are proposed based on the matrix normal distribution, featured by a bi-linear extension of Fisher linear discriminant analysis and a further extension to binary variables. Moreover, a general formulation is proposed for integrative hypothesis testing and five typical categories are summarized. Furthermore, major techniques for variable selection are briefly elaborated. Finally, taking analyses of gene expression and exome sequencing as examples, we further propose a general procedure called *Geno-Pheno A5 Analyzer* for integrative discriminant analysis.

Keywords: Matrix-variate discriminative analysis, Bi-linear Fisher mapping, Matrix-variate logistic regression, Confusion table testing, Geno-Pheno A5 analyzer, Gene expression analysis, Exome sequencing analysis.

1   Introduction

Fisher discriminative analysis works in a multidimensional space with samples presented as vectors. However, samples are usually in matrix format for tasks such as image classification, object recognition, and various gene analyses. Considering samples in vectors actually suffers some approximation to get an easy implementation. Improvements are expected if we make discriminative analysis directly on samples in matrix format.

Working on samples from two populations or classes, discriminative analysis is featured by finding a discriminating rule that classifies each sample into one appropriate class. In a complementary aspect, two sample hypothesis test examines whether two populations are significantly different via a statistics based on samples from the populations.

This paper provides a general perspective on both the aspects, with a road not only to generalize discriminative analyses and hypothesis testing to matrix-variate samples but also to a general formulation for testing hypotheses on a set of variables organized in structures. Also, five categories are summarized according to the characteristics of testing hypotheses.

Also, major techniques are briefly elaborated for identifying which subsets of variables are responsible to testing significance and discriminative ability. Taking gene expression analysis and exome sequencing analysis as examples, we propose a Geno-Pheno A5 analyzer for integrative genotype-phonotype discriminant analysis.

J. Yang, F. Fang, and C. Sun (Eds.): IScIDE 2012, LNCS 7751, pp. 866–875, 2013.

2   KL Perspective on Hotelling Statistics and Fisher Discriminant

We start from considering two multivariate Gaussian populations:

$$X^{(\ell)} = [x_1^{(\ell)}, \ldots, x_{N_\ell}^{(\ell)}], \quad x_t^{(\ell)} \text{ from } G(x \mid \mu^{(\ell)}, \Sigma^{(\ell)}), \ell = 0, 1, \tag{1}$$

with the proportional priori $\alpha^{(l)}$. The parameters are obtained by the maximum likelihood (ML) estimation with

$$\alpha^{(\ell)} = \frac{N_\ell}{N}, \ N = \sum_\ell N_\ell, \ \mu^{(\ell)} = \frac{1}{N_\ell} \sum_{t=1}^{N_\ell} x_t^{(\ell)}, \ \Sigma^{(\ell)} = \frac{1}{N_\ell - 1} \sum_{t=1}^{N_\ell} (x_t - \mu^{(\ell)})(x_t - \mu^{(\ell)})^T.$$

Under the null hypothesis

$$H_0: G(x_t \mid \mu^{(1)}, \Sigma^{(1)}) = G(x_t \mid \mu^{(0)}, \Sigma^{(0)}), \tag{2}$$

we have $\mu = \mu^{(l)}$, $\Sigma = \Sigma^{(l)}$. Putting $X^{(0)}$, $X^{(1)}$ together, the ML estimation becomes

$$\mu = \alpha^{(0)} \mu^{(0)} + \alpha^{(1)} \mu^{(1)}, \ \Sigma = \alpha^{(0)} \Sigma^{(0)} + \alpha^{(1)} \Sigma^{(1)}.$$

We further use the following Kullback–Leibler (KL) divergence

$$\mathrm{KL}(p \parallel q) = \int p(x) \ln[p(x) / q(x)] dx \tag{3}$$

to measure the deviation contributed by the difference of $\mu^{(l)}$, resulting in

$$s_{KL} = KL(G(x \mid \mu^{(1)}, \Sigma) \parallel G(x \mid \mu, \Sigma)) = \mathrm{Tr}[(\mu^{(0)} - \mu^{(1)})(\mu^{(0)} - \mu^{(1)})^T \Sigma^{-1}], \tag{4}$$

from which we observe the following Hotelling two-sample T-squared statistics [1]

$$T^2 = \alpha_0 \alpha_1 s_{KL} \tag{5}$$

Moreover, considering the following linear projection

$$y_t = w^T x_t, \ \mu_y^{(\ell)} = w^T \mu^{(\ell)}, \ \Sigma_y = w^T \Sigma w, \tag{6}$$

we use Eq.(4) to measure two resulted scalar Gaussian populations and get

$$s_{KL}(w) = w^T (\mu^{(0)} - \mu^{(1)})(\mu^{(0)} - \mu^{(1)})^T w (w^T \Sigma w)^{-1}, \tag{7}$$

which could be further maximized to obtain an optimal w^*. That is, we reach the popular Fisher linear discriminant analysis (LDA).

3   Matrix-Variate Discriminative Analysis

Following Sect. 3 in [2], we consider a $d \times m$ matrix variate X from the following matrix normal distribution [3]:

$$N(X \mid M, \Omega, \Sigma) = \frac{\exp\{-0.5 Tr[\Omega^{-1}(X - M)^T \Sigma^{-1}(X - M)]\}}{(2\pi)^{0.5md} |\Sigma|^{0.5d} |\Omega|^{0.5m}}, \ M = EX, \tag{8}$$

where an $m \times m$ matrix Ω describes the cross-column dependence of X and a $d \times d$ matrix Σ describes the cross-row dependence of X. This matrix distribution links to a multivariate Gaussian distribution $G(vec(X) \mid vec(M), \Sigma \otimes \Omega)$, where $\otimes$ denotes the Kronecker product and $vec[A]$ is the vectorization of a matrix A.

Given samples of the following populations:

$$\mathbf{X}^{(\ell)} = [X_1^{(\ell)}, \ldots, X_{N_\ell}^{(\ell)}], \quad X_t^{(\ell)} \text{ from } N(X \mid M^{(\ell)}, \Omega^{(\ell)}, \Sigma^{(\ell)}), \quad \ell = 0, 1, \tag{9}$$

the parameters are estimated as follows

$$\alpha^{(\ell)} = \frac{N_\ell}{\sum_\ell N_\ell}, \quad M^{(\ell)} = \frac{1}{N_\ell} \sum_{t=1}^{N_\ell} X_t^{(\ell)}, \Sigma^{(\ell)} = \frac{1}{N_\ell - 1} \sum_{t=1}^{N_\ell} (X_t^{(\ell)} - M^{(\ell)})(X_t^{(\ell)} - M^{(\ell)})^T, \tag{10}$$

where $\Omega^{(l)}$ depends on $\Sigma^{(l)}$ and one estimate is given by

$$\Omega^{(\ell)} = \frac{1}{N_\ell - 1} \sum_{t=1}^{N_\ell} (X_t - M^{(\ell)})^T \Sigma^{(\ell)-1} (X_t - M^{(\ell)}). \tag{11}$$

With the normal distribution, we get the following observations. First, we can get the following matrix variate Bayes discriminative rule:

$$\ell^* = \operatorname{argmax}_\ell g^{(\ell)}(X_t), \quad g^{(\ell)}(X_t) = \ln N(X^{(\ell)} \mid M^{(\ell)}, \Omega^{(\ell)}, \Sigma^{(\ell)}) + \ln \alpha^{(\ell)}. \tag{12}$$

Second, it follows from Eq.(4) that we similarly obtain

$$s_{KL} = KL(N(X \mid M^{(1)}, \Omega, \Sigma) \| N(X \mid M, \Omega, \Sigma)) = \operatorname{Tr}[\Omega^{-1}(M^{(1)} - M^{(0)})^T \Sigma^{-1}(M^{(1)} - M^{(0)})], \tag{13}$$

with the following estimates:

$$M = \frac{1}{N} \sum_{\ell=0,1} \sum_{t=1}^{N_\ell} X_t^{(\ell)}, \quad \Sigma = \frac{1}{N-1} \sum_{\ell=0,1} \sum_{t=1}^{N_\ell} (X_t^{(\ell)} - M)(X_t^{(\ell)} - M)^T,$$

$$\Omega = \frac{1}{N-1} \sum_{\ell=0,1} \sum_{t=1}^{N_\ell} (X_t^{(\ell)} - M')^T \Sigma^{-1}(X_t^{(\ell)} - M).$$

Third, we get a bi-linear projection by a $d_s \times d$ matrix W and a $m \times m_s$ matrix U

$$Y_t = W^T X_t U, \quad M_y^{(\ell)} = W^T M^{(\ell)} U. \tag{14}$$

It follows from Eq.(10) and Eq.(11) that we have

$$\Sigma_y = \frac{1}{N-1} W^T \sum_{\ell=0,1} \sum_{t=1}^{N_\ell} (X_t^{(\ell)} - M^{(\ell)}) U U^T (X_t^{(\ell)} - M^{(\ell)})^T W,$$

$$\Omega_y = \frac{1}{N-1} U^T \sum_{\ell=0,1} \sum_{t=1}^{N_\ell} (X_t^{(\ell)} - M^{(\ell)})^T W \Sigma_y^{-1} W^T (X_t^{(\ell)} - M^{(\ell)}) U. \tag{15}$$

Then, we use s_{KL} in Eq.(13) to measure the two matrix normal populations

$$s_{KL}(W, U) = \operatorname{Tr}[\Omega_y^{-1} U^T (M^{(1)} - M^{(0)})^T W \Sigma_y^{-1} W^T (M^{(1)} - M^{(0)}) U], \tag{16}$$

which is further maximized to obtain an optimal W^*, U^*. That is, we have

$$\{W^*, U^*\} = \operatorname{arg\,max}_{W,U} s_{KL}(W, U), \tag{17}$$

which is a bi-linear extension of Fisher linear discriminant analysis.

Particularly, when $d_s = m_s = 1$, we get the following Fisher criterion

$$s_{KL}(w,u) = \frac{(\mu_y^{(1)} - \mu_y^{(0)})^2}{\sigma_y^2}, \quad y_t = w^T X_t u, \quad \mu_y^{(\ell)} = w^T M^{(\ell)} u,$$

$$\sigma_y^2 = \Omega_y \Sigma_y = \frac{1}{N-1} \sum_{\ell=0,1} \sum_{t=1}^{N_t} [w^T (X_t^{(\ell)} - M^{(\ell)})u]^2. \tag{18}$$

Moreover, $y_t = w^T X_t u$ may also be turned into a regression to a label $I_t=1$ as X_t comes from $X_t^{(1)}$ or label $I_t=0$ as X_t comes from $X_t^{(0)}$, i.e.,

$$p(I_t = 1 \mid y_t) = 1/[1 + \exp(-\beta y_t)].$$

We may estimate w^*, u^* by maximizing the following likelihood

$$L(w,u) = -\sum_{t=1}^{N_1} \ln(1 + \exp[-\beta w^T X_t^{(1)} u]) - \sum_{t=1}^{N_0} \ln(1 + \exp[\beta w^T X_t^{(0)} u]). \tag{19}$$

Then, we further develop an algorithm for maximizing the measure either by Eq.(16) or Eq.(19). Simply, we may implement an alternative gradient descent updating to maximize $J(W,U)$, that is, we alternatively update

$$W^{new} = W^{old} + \eta \nabla_W J(W,U), \quad U^{new} = U^{old} + \eta \nabla_U J(W,U). \tag{20}$$

Moreover, a term of L1 norm or lasso penalty or Laplace priori may be added to $J(W,U)$ to regularize the coefficients of W, U such that extra parameters will be eliminated, e.g., adding to w by Eq.(7) leads to sparse LDA [4].

Last but not least, we may further consider two even general cases.

One is the following general matrix normal populations

$$s_{KL} = KL(N(X|M^{(1)}, \Omega^{(1)}, \Sigma^{(1)}) \| N(X|M, \Omega, \Sigma)), \tag{21}$$

$$\Sigma = \sum_{j=1}^{k} \lambda_j \phi_j \phi_j^T + \sigma^2 I, \quad k \le m,$$

where $\lambda_j, \phi_j, j = 1, \cdots, k$ are the first k largest eigenvalues of Σ and the corresponding eigenvectors. Typically, we consider $k < d$ for a small size of samples, which also provides modification to s_{KL} by Eq. (6) and s_{KL} by Eq.(13) at the special case $\Omega = I$.

The other considers that all the elements in X are binary valued, with help of the following Gibbs measure:

$$p(X|\beta, \Phi, \Psi) \propto e^{-\beta E(X|\Phi, \Psi)}, \quad E(X|\Phi, \Psi) = \sum_{i,j,k,\ell} \phi_{ij} \psi_{k\ell} x_{ik} x_{j\ell}, \tag{22}$$

which is used to replace $N(X|\cdot,\cdot,\cdot)$ in Eqs.(12) (21).

4   Integrative Hypothesis Testing and Variable Selection

4.1   One General Formulation and Typical Categories

Originally, studies on hypothesis testing consider samples from populations of one random variable, e.g., a SNP takes one of genotypes in the popular PLINK study [5].

For practical problems, a hypothesis is typically made on multi-variables and studies have also proceeded to multivariate hypothesis test [6,7], which recently gets ever-interested in gene analyses [8,9]. However, these studies work on vector samples while we actually encounter hypotheses on a set of variables organized in structures beyond vector. Thus, we need to perform testing by integrating information associated with these structured variables. Informally, we refer efforts towards this direction to a term called integrative hypothesis testing (IHT).

Similar to Eqs.(4), (13) & (21), we propose a general IHT formulation with the help of the KL divergence by Eq.(3). Given samples of the following populations

$$\mathbf{X}^{(\ell)} = [X_1^{(\ell)},...,X_{N_\ell}^{(\ell)}], \ X_t^{(\ell)} \text{ from } p(X \mid \Theta^{(\ell)}), \ \ell = 0,1,$$

with variables of $X_t^{(l)}$ in certain specific structure, we estimate $\Theta^{(l)}$ from $\mathbf{X}^{(l)}$ by a learning principle, e.g., by the following maximum likelihood estimation

$$\Theta^{(1)} = \arg\max_\Theta \ln p(\mathbf{X}^{(1)} \mid \Theta),$$

$$\Theta^{(0)} = \arg\max_\Theta \begin{cases} \ln p(\mathbf{X}^{(0)} \mid \Theta), & \text{Choice (a),} \\ [\ln p(\mathbf{X}^{(1)} \mid \Theta) + \ln p(\mathbf{X}^{(0)} \mid \Theta)], & \text{Choice (b).} \end{cases}$$

Then, we test the following null hypothesis

$$H_0: \ p(X \mid \Theta^{(1)}) \text{ is not different from } p(X \mid \Theta^{(0)}), \tag{23}$$

by the following statistics

$$s_{KL} = KL(p(X \mid \Theta^{(1)}) \| p(X \mid \Theta^{(0)})). \tag{24}$$

Also, there could be many other statistics for testing H_0 by Eq.(23). Even further, there could be various formulations for this testing, which are roughly classified into the following five categories.

(1) Decomposing H_0 by Eq.(23) with $X=\{\xi_1,...,\xi_d\}$ into the sub-hypotheses

$$H_0^{(j)}: \ p(\xi_j \mid \theta_j^{(1)}) \text{ is not different from } p(\xi_j \mid \theta_j^{(0)}), \ j = 1,...,d, \tag{25}$$

such that H_0 by Eq.(23) is equivalent to a composition of the sub-hypotheses. Testing H_0 by Eq.(23) is made by combining the tests of sub-hypotheses, e.g., combining the resulted p-values by the *Fisher's combined probability test* [10], the Kost's method [11], and others [12,13].

(2) Targeting at computing s_{KL} by effectively exploring the structure of X.

When X consists of independent variables $\xi_1, ..., \xi_d$, we have

$$s_{KL} = \sum_j s_j, \ s_j = KL(p(\xi_j \mid \theta_j^{(1)}) \| p(\xi_j \mid \theta_j^{(0)})), \tag{26}$$

that is, we make a test by a statistics that is simply a sum of individual ones, E.g., if each $s_j \sim \chi(k)$, we simply consider $s_{KL} \sim \chi(kd)$.

When $\xi_1, ..., \xi_d$ are joint variables in vector or matrix formats, not only the Hotelling T^2 statistics [1] can be computed from s_{KL} by Eq.(5) but also H_0 by Eq.(23) can be tested by an extension of the Hotelling statistics by Eq.(13).

For other dependence structures, e.g., $p(\xi_1, \xi_2, \xi_3) = p(\xi_1)p(\xi_2|\xi_1)p(\xi_3|\xi_2)$, we have $s_{KL}=s_1+s_{2|1}+s_{3|2}$ with $s_1=KL(p(\xi_1)\|q(\xi_1))$ and $s_{i|j}=KL(p(\xi_i|\xi_j)\|q(\xi_i|\xi_j))$.

Moreover, we may consider a statistics other than s_{KL}, e.g., Dempster's statistics [6], and others [7].

(3) Mapping H_0 by Eq.(23) into one or a set of hypotheses on certain inner representations of $X=\{\xi_1,...,\xi_d\}$. A rejection of these mapped hypotheses means that H_0 by Eq.(23) should be rejected since the truth of H_0 by Eq.(23) implies the truth of these mapped hypotheses. Being different from those tests directly made on the visible data domain of X (called Yang domain and thus named as Yang-test or shortly A-test), these mapped hypotheses are tested in an inner representation domain (called Ying domain), which forms another type of hypothesis testing that is named as Ying test or shortly I-test. In Sect. 6 of [2], two I-test approaches have been proposed which are elaborated below:

(a) FDA based one variable test   Considering the projection y_t from either Eq.(6) or Eq.(18), we test the following null hypothesis

$$H_0: G(y_t | \mu_y^{(1)}, \sigma_y^{(1)2}) \text{ is not same as } G(y_t | \mu_y^{(0)}, \sigma_y^{(0)2}), \qquad (27)$$

which can be performed by the Welch's t test. For Eq.(6), a best direction w^* is obtained by *Fisher discriminant analysis* (FDA) by Eq.(7).

(b) Confusion table test (CTT)   Instead of making a linear projection, we use a classifier to map the samples of $X^{(0)}, X^{(1)}$ into binary labels C_0 and C_1, resulting in a confusion table shown in Fig.1(a) on which we further test

$$H_0: \text{ two rows have no difference.} \qquad (28)$$

Again, its rejection means that H_0 by Eq.(23) should be rejected. This classifier is trained for a minimum classification error towards $N_{10}=0$ and $N_{01}=0$, i.e., two rows become as different as possible. Then, the hypothesis H_0 by Eq.(28) can be tested by the following Pearson chi-square statistics

$$t = \sum_{i=0,1}[\frac{(N_{i1}-pN_i)^2}{pN_i}+\frac{(N_{i0}-(1-p)N_i)^2}{(1-p)N_i}] \sim \chi^2(1), \ p=\frac{N_{01}+N_{11}}{N_0+N_1}. \qquad (29)$$

(4) Integrating tests on a number of the null hypotheses by Eq.(28) on the confusion tables obtained from a number of classifiers that either come from Eq.(6) or Eq.(18) for different values of w,u or come from different implementations (e.g., different learning methods, different initializations, or by A-3 given in the next section). The integration is made by one of the methods in the above Category (1).

(5)   Making integrative hypothesis testing by combining a number of classifiers with help of either of majority voting, Bayes voting, product rule, Dempster-Shafer rule (see Tab.2 in [14]), resulting in a final confusion table for testing the null hypotheses by Eq.(28). Moreover, we may learn a tree classifier to handle a tree-structured variables $\xi_1, ..., \xi_d$, resulting in a final confusion table for testing a general null hypotheses by Eq.(23).

4.2   Feature Selection: Bottom-Up Search versus Top-Down Search

Once H_0 is rejected on the joint variables of X, we further identify which one or subset of variables are responsible to this rejection, which is called feature selection and implemented by one of the following typical techniques.

Bottom-up search is featured by searching subset of variables with the size of subsets increasing from 1 to a desired one. Typical efforts may be summarized into the following two classes:

(1) *Examining all the combinations* by the J-value (i.e., the value of J). First, each variable of X is examined. Second, each pair of variables are examined. At each m, each of C_d^m subsets of X is examined, until m reaches to a desired one.

(2) *Incremental stepwise search* First, the best variable ξ_1^* (i.e., a biggest J value) is selected from X, resulting in X_{d-1}. Second, each pairing of ξ_1^* with each of $d-1$ variables of X_{d-1} is examined by the J-value, with the best $\xi_1^*\xi_2^*$ selected and X_{d-1} reduces to $X_{d-2}, ...$, so forth, until reaching a desired size.

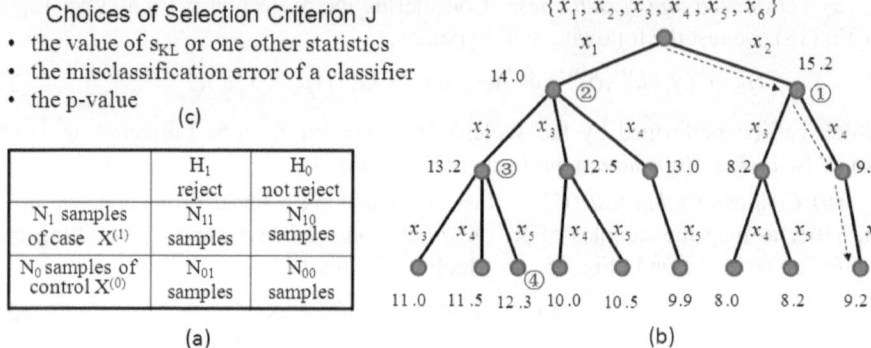

Choices of Selection Criterion J
- the value of s_{KL} or one other statistics
- the misclassification error of a classifier
- the p-value

(c)

	H_1 reject	H_0 not reject
N_1 samples of case $X^{(1)}$	N_{11} samples	N_{10} samples
N_0 samples of control $X^{(0)}$	N_{01} samples	N_{00} samples

(a)

Fig. 1. (a) A confusion table obtained by classification. (b) Illustration of the depth-first search (dashed line) versus the best-first search for feature selection. (c) Selection criterion J.

Top-down search is featured by searching subsets of variables with the size of subsets decreasing from d to a desired one. Typical efforts may be summarized into the following four classes:

(a) *Testing individual variable* Testing the significance of the coefficient associated with each variable, e.g., by the likelihood ratio test or the Wald statistics test [15].

(b) *Backward elimination* The simple one is the depth-first search. It gradually eliminates a variable that is the least important one, as d reduces to a desired one. At each m, we examine each of m subsets X_{m-1} resulted from eliminating x from X_m and keep the best subset by the J-value, as shown by the dashed line in Fig.1(b). However, this search is easy to fall in a local optimum. Instead, as shown by the circled numbers in Fig.1(b), a better alternative is the best-first search proposed in [16], which will find the optimum without enumerating all the possibilities when the J-value is monotonic to the cardinality of variable sets.

(c) *Sparse learning* that eliminates the coefficients associated with extra variables during learning, as previously discussed after Eq.(19).

Mixed search is featured by alternatively using bottom-up and top-down search in a specific procedure, e.g., forward-backward stepwise regression [17]. Moreover, we may make sparse learning with different initializations that lead different subsets of variables, and then use the union of these subsets to start a top-down search.

5   Integrative Geno-Pheno A5 Analyzer

The proposed methods in Sect.3 and Sect.4 are applicable to tasks such as image recognition, object detection, fault and disease diagnosis, with improvements that come from considering samples in matrix format. Particularly, the methods are more appealing to those real problems for which samples are actually in matrix format but used approximately in vectors merely for an easy implementation.

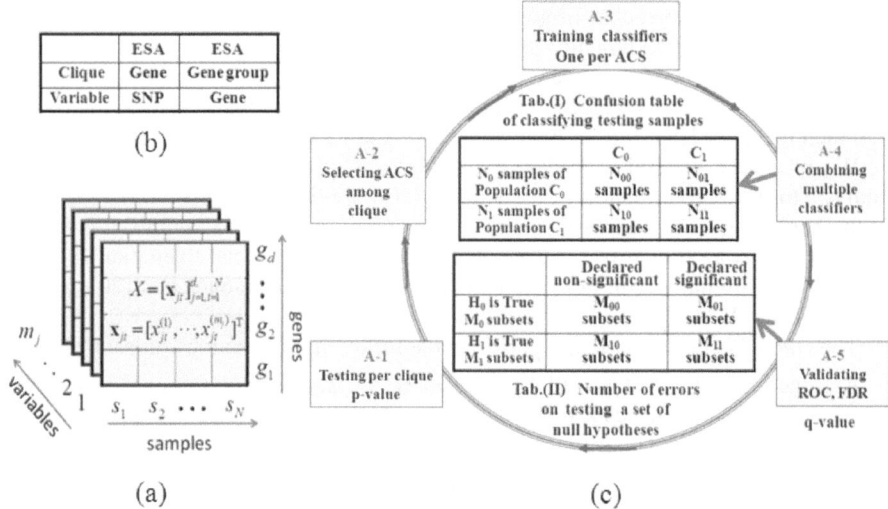

Fig. 2. (a) general data type widely encountered in gene analyses. (b) two levels of variable search. (c) integrative *Geno-Pheno A5 Analyzer*.

Many problems in gene analyses handle data as shown in Fig.2(a), e.g. eight types of data shown in Fig.5 of [18]. Here, we focus on two typical examples. One is *gene/RNA expression analysis (GEA)*. One simplest case consists of an array with each column representing a sample and each row corresponding gene. Moreover, each x_{jt} corresponds to the expression of one gene g_j. Generally, we may also consider the expressions of one gene under an additional control, represented by several variables for describing different times, conditions, ..., etc, where the number m_j may vary for different genes. In such cases, X is a data cubic with each sample being a $m_j \times d$ matrix.

The other is *exome sequencing analysis (ESA)*, e.g., see Sect.6.3 in [19]. For each gene, x_{jt} is a vector that consists of several SNPs with each in a genotype. Specifically, for the lth SNP of the gene j we take one of the following two choices for coding genotypes:

(a) $x_{jt}^{(l)}=1$ for AA, $x_{jt}^{(l)}=2$ for Aa, $x_{jt}^{(l)}=3$ for aa, and $x_{jt}^{(l)}=0$ if missing genotype;
(b) $x_{jt}^{(l)}=[1,0,0]$ for AA, $x_{jt}^{(l)}=[0,1,0]$ for Aa, $x_{jt}^{(l)}=[0,0,1]$ for aa, and $x_{jt}^{(l)}=[1/3,1/3, 1/3]$ when its genotype is missing, i.e., each x_{jt} is an $m_j \times 3$ matrix.

Among the problems with the data type shown in Fig.2(a), one widely studied family consists of association analyses on the relations of diseases or external phenomenon to genes or inner causes, e.g., whether a disease relates to some genes or SNPs within some genes. Such a study is featured by its scope of focusing. E.g., the scope of GEA considers all genes in *X*, which usually incurs for a huge computing cost; while the scope of GWAS covers the SNPs of all the genes and incurs an even huge computing cost such that enumerating the joint-effects of multiple SNPs is computationally not feasible.

To tackle the issue, we propose to divide the scope of focusing into two levels, as illustrated in Fig.2(b). A study starts from considering samples per clique, e.g., a clique for ESA is a gene that contains its SNPs while a clique for GEA is a group of genes in a same pathway. Each clique may also be formed according to some domain knowledge, e.g., molecular biology knowledge for ESA and GEA. Without this knowledge, we may technically divide data in Fig.2(a) along the vertical direction, e.g., clustering analysis on gene expression data to group genes into cliques or even randomly picking genes to form a clique. This level investigates cliques one by one, and then the other level further examines subsets of each clique.

We propose a general procedure that integrates hypothesis test and discriminative analysis for genotype-phonotype association (also applicable to those tasks mentioned at the beginning of this section), shortly called *Geno-Pheno A5 Analyzer*. It is featured by five actions as shown in Fig.2(c), which is actually a special exemplar of the A5 problem solving paradigm. For details, readers are referred to Sect.4 in [20], Sect.3.1 and Appendix B in [21].

The *Geno-Pheno A5 Analyzer* performs the following five actions on each clique:

A-1 (acquisition): test a null hypothesis H_0 on samples of this clique to check whether there is a significant difference between two populations C_0, C_1 (e.g, by one method of Category (3) introduced in Sect.4.1). Return to A-1 if the test is not significant, and put the clique into a set named REST.

A-2 (assumption): evaluate subsets of the clique for ones responsible to the difference between C_0, C_1. Each subset that passes a significance level is identified as one assumed candidate subset (ACS), resulting in a family of M_0 such ACSs.

A-3 (amalgamation): build up a classifier per ACS by training samples, resulting in a set of classifiers for the set of ACSs.

A-4 (apex-seeking) :   use each ACS classifier to assign testing samples into one of C_0, C_1 and combine the results by different ACS classifiers to get a final   assignment, together with the confusion table Tab.(1) in Fig.2(b), e.g, by one method of Category (5) introduced in Sect.4.1

A-5 (affirmation) :   validating two complementary hypotheses as follows:

(1)   *Vertical direction* Each of M_0 ACSs is tested by testing samples, resulting in the first row of Tab.(2) in Fig.2(b). Also, M_1 subsets from the set REST are tested as H_1 hypotheses, resulting in the second row of Tab.(2) in Fig.2(b). Then, a validation is made by the concept of false discovery rate (FDR), e.g., by the Benjamini–Hochberg procedure based on the resulted *p*-values [22, 23]. Also, we may return to A-4 for combining those ACS classifiers that passed this validation.

(2)   *Horizontal direction.* We validate the confusion table Tab.(1) by the rate of misclassification and the curve of Receiver Operating characteristics (ROC).

We may perform the above *Geno-Pheno A5 Analyzer* at the level of cliques one by one and then combine classifiers obtained from different cliques for a global test on whether two populations are significantly different.

Acknowledgments. This work was supported by the National Basic Research Program of China (973 Program) (No. 2009CB825404).

References

1. Hotelling, H.: The generalization of Student's ratio. Annals of Mathematical Statistics 2(3), 360–378 (1931)
2. Xu, L.: Semi-Blind Bilinear Matrix System, BYY Harmony Learning, and Gene Analysis Applications. In: Proc. of 6th International Conf. on New Trends in Information Science, Service Science and Data Mining, Taipei, October 23-25, pp. 661–666 (2012)
3. Dawid, A.P.: Some matrix-variate distribution theory: Notational considerations and a Bayesian application. Biometrika 68(1), 265–274 (1981)
4. Clemmensen, L., Hastie, T., Witten, D., Ersbøll, B.: Sparse discriminant analysis. Technometrics 53, 406–413 (2011)
5. Purcell, S., Neale, B., Todd-Brown, K., Thomas, L., Ferreira, M., Bender, D., Maller, J., Sklar, P., de Bakker, P., Daly, M.J., Sham, P.C.: PLINK: a toolset for whole-genome association and population-based linkage analysis. American Journal of Human Genetics 81 (2007), http://pngu.mgh.harvard.edu/purcell/plink/
6. Dempster, A.P.: A high dimensional two sample significance test. Ann. Math. Statist. 29, 995–1010 (1958)
7. Baringhaus, L., Franz, C.: On a new multivariate two-sample test. J. Multivariate Anal. 88, 190–206 (2004)
8. Glezko, G.V., Emmert-Streib, F.: Unite and conquer: univariate and multivariate approaches for finding differentially expressed gene sets. Bioinformatics 25, 2348–2354 (2009)
9. Hummel, M., Meister, R., Mansmann, U.: GlobalANCOVA: exploration and assessment of gene group effects. Bioinformatics 24, 78–85 (2008)
10. Fisher, R.A.: The Statistical Utilization of Multiple Measurements. Annals of Eugenics 8, 376–386 (1938)
11. Kost, J., McDermott, M.: Combining dependent P-values. Statistics & Probability Letters 60(2), 183–190 (2002)
12. Whitlock, M.C.: Combining probability from independent tests: the weighted Z-method is superior to Fisher's approach. J. Evol. Biol. 5(18), 1368–1373 (2005)
13. Chen, Z.: Is the weighted z-test the best method for combining probabilities from independent tests? J. Evol. Biol. 24(4), 926–930 (2011)
14. Xu, L., Amari, S.: Combining classifiers and learning mixture-of-experts. In: Ramón, J., et al. (eds.) Encyclopedia of Artificial Intelligence, pp. 318–326. IGI Global Pub. (2008)
15. Hilbe, J.M.: Logistic Regression Models. Chapman & Hall/CRC Press (2009)
16. Xu, L., Yan, P., Chang, T.: Best first strategy for feature selection. In: Proc. of 9th Intl Conf on Pattern Recognition, Rome, November 14-17, vol. 2, pp. 706–708 (1988)
17. Stepwise regression - Wikipedia, http://en.wikipedia.org/wiki/Stepwise_regression
18. Xu, L.: Codimensional matrix pairing perspective of BYY harmony learning: hierarchy of bilinear systems, joint decomposition. Front. Electr. Electron. Eng. China 6(1), 86–119 (2011)
19. Xu, L.: On essential topics of BYY harmony learning: Current status, challenging issues, and gene analysis applications. Front. Electr. Electron. Eng. 7(1), 147–196 (2012)
20. Xu, L.: A unified perspective and new results on RHT computing, mixture based learning, and multi-learner based problem solving. Pattern Recognition 40(8), 2129–2153 (2007)
21. Xu, L.: Bayesian Ying-Yang system, best harmony learning, and five action circling. Front. Electr. Electron. Eng. China 5(3), 281–328 (2010)
22. Benjamini, Y., Hochberg, Y.: Controlling the false discovery rate: a practical and powerful approach to multiple testing. J. of Royal Statistical Society B 57(1), 289–300 (1995)
23. False discovery rate - Wikipedia, http://en.wikipedia.org/wiki/False_discovery_rate

Author Index